THE LICHEN FLORA OF GREAT BRITAIN AND IRELAND

This volume is dedicated to the memory of

Ursula K. Duncan (1910–1985)

who did so much to keep British and Irish lichenology alive in the lean years

THE LICHEN FLORA OF GREAT BRITAIN AND IRELAND

Edited by

O. W. Purvis, B. J. Coppins, D. L. Hawksworth, P. W. James & D. M. Moore

Natural History Museum Publications
in association with
The British Lichen Society

Published by The British Lichen Society
c/o Department of Botany, The Natural History Museum,
Cromwell Road, London SW7 5BD

First published by The Natural History Museum on behalf of The British Lichen Society, 1992

Reprinted by The Natural History Museum for The British Lichen Society, 1994

British Cataloguing-in-Publication Data
A catalogue record for this book is available from the British Library

ISBN 0 9523049 0 2

Typeset by Spottiswoode Ballantyne Printers Ltd.
Printed by St Edmundsbury Press

Foreword

The Lichen Flora of Great Britain and Ireland is a work of very great scholarship, and represents a supreme achievement of modern British lichenology. It is a fitting and lasting tribute to those who have participated in the remarkable surge of interest and research in lichens which has occurred in the second half of the twentieth century. One of the seminal and guiding influences in this has been the British Lichen Society, founded at a meeting at the British Museum (Natural History) on 1 February 1958. The Society has subsequently provided an excellent focus for the varied types of interest in lichens as they have developed, it has encouraged the spread of knowledge about lichens among botanists in general, and its journal, *The Lichenologist*, is a major international forum for publications on lichens. This present volume would not have been possible without the Society.

The last comprehensive account of the lichens of the British Isles, the second edition of the *Monograph of British Lichens* by A. L. Smith, was published over half a century ago, between 1918 and 1926. This monograph, valuable though it was in its day, has been very much overtaken by the substantial wealth of new knowledge that has accumulated since it was published. The need for a new lichen flora has become even more acute as the relevance of lichenology to a range of biological disciplines has become realised. The importance of lichens in survey work is now generally accepted in descriptive ecology; lichens have proved uniquely interesting in aspects of physiological plant ecology; the marked sensitivity of lichens to atmospheric pollution has made them of central importance in environmental monitoring; and they have become classical experimental models for investigations into the nature of symbiosis. *The Lichen Flora of Great Britain and Ireland* is a distinguished successor to Smith's monograph, and it will last well into the twenty-first century. Its preparation has been a triumph of meticulously co-ordinated teamwork by the editors, assisted by the generous support and help of many colleagues around the world.

A major debt of gratitude is owed to all those whose dedicated work and support has enabled the publication of this book. Above all, our thanks are due to the editors, whose reward must be the knowledge that this Flora will bring much pleasure and great satisfaction to those interested in lichens for generations to come.

Professor Sir D. C. Smith, FRS, FRSE
University of Edinburgh, June 1992

British Lichen Society Flora Advisory Committee

Professor D. M. Moore (Chairman),
Department of Botany, Plant Science Laboratories, University of Reading, Whiteknights, P.O. Box 221, Reading, Berks, RG6 2AS

Mr. F. H. Brightman,
South London Botanical Institute, 323 Norwood Road, London, SE24 9AQ

Dr B. J. Coppins,
Royal Botanic Garden, Edinburgh, EH3 5LR

Mr F. S. Dobson,
The Richmond Publishishing Co. Ltd, P. O. Box 963, Slough, SL2 3RS

Professor B. W. Fox,
Tryfan, Longlands Road, New Mills, Stockport, Cheshire, SK12 3BL

Dr O. L. Gilbert,
Department of Landscape Architecture, University of Sheffield, Sheffield, S10 2TN

Professor D. L. Hawksworth,
International Mycological Institute, Bakeham Lane, Egham, Surrey, TW20 9TY

Mr. P. W. James,
Department of Botany, The Natural History Museum, Cromwell Road, London, SW7 5BD; 19 Edith Road, London, W14 0SU

Mr. J. R. Laundon,
Department of Botany, The Natural History Museum, Cromwell Road, London, SW7 5BD; 14 Victory Avenue, Morden, Surrey, SM4 6DL

Dr. O. W. Purvis,
Department of Botany, The Natural History Museum, Cromwell Road, London, SW7 5BD

Dr. F. Rose,
Rotherhurst, 36 St Mary's Road, Liss, nr Petersfield, Hants, GU33 7AH

Professor M. R. D. Seaward,
Department of Environmental Science, University of Bradford, Bradford, Yorkshire, BD7 1DP

Mr. R. G. Woods,
Countryside Council for Wales, 3rd Floor, The Gwalia, Ithon Road, Llandrindod Wells, Powys, LD1 6AA

Acknowledgements

The production of this *Flora* would never have been possible without the input provided by the British Lichen Society. Nor would it have been practicable without the financial support of the Natural Environment Research Council (NERC) and British Petroleum plc (BP). NERC enabled Dr O. W. Purvis to be employed on the project from 1986–89, and Mrs M. Hart as a part-time keyboarder from 1986–90 through Research Grant no. GR3/6133. BP generously provided half of the cost of a word-processing system. The Systematics Association contributed to a flora workshop at Ilfracombe in April 1988 for the testing of keys and discussions amongst contributors. Reading University provided facilities to Professor D. M. Moore, who administered the funding and Secretariat. The Natural History Museum generously provided bench space for Dr O. W. Purvis and access to its rich collections and library. The International Mycological Institute, National Museum of Wales and the Royal Botanic Garden Edinburgh were prominent among the institutes which afforded further support for this project.

We have been fortunate in receiving assistance from numerous lichenologists in Great Britain, Ireland and overseas in the preparation and revision of generic accounts. In particular, Professor P. M. Jørgensen generously made certain drafts from the Norwegian crustose lichen flora, now in preparation, available. Mr A. Orange kindly prepared all the line drawings included. The onerous keyboarding of drafts and revisions was undertaken almost entirely by Mrs M. Hart, to whom a special debt is owed.

Above all, however, the Editors are grateful to those 49 colleagues listed on the following page who prepared accounts, revised drafts and were a constant support throughout this project. It is their labours which we hope this volume will celebrate, and give a further impetus to much lichenological research in the coming years.

The Editors are indebted to the British Lichen Society for a generous subsidy towards the publication of the *Flora* mainly through a bequest to the Society from Mrs A. G. Side.

We are grateful to Mr J. R. Laundon for his valuable assistance with the compilation of the index.

Contributors

The following prepared draft accounts, assisted in their revision, or otherwise contributed to the production of this *Flora*. The primary authors(s) of each generic account is given on pp. 651–655. However, it must be stressed that the final versions presented are the responsibility of the Editors.

Dr B. Aguirre-Hudson (The Natural History Museum, U.K.), Professor T. Ahti (University of Helsinki, Finland), Professor G. Baron (London, U.K.), Dr O. Breuss (Vienna, Austria), Mr F. H. Brightman (South London Botanical Institute, U.K.), Dr M. Skytte Christiansen (Botanisk Centralbibliotek, Copenhagen, Denmark), Dr S. Clayden (New Brunswick Museum, Canada), Dr P. Clerc (Conservatoire et Jardin Botanique, Genève, Switzerland), Dr B. J. Coppins (Royal Botanic Garden, Edinburgh), Mr J. C. David (International Mycological Institute, U.K.), Dr D. H. Dalby (Imperial College, U.K.), Mr F. S. Dobson (Richmond Publishing Co. Ltd, U.K.), Mr T. Duke (Stourbridge, U.K.), Dr A. Fletcher (Leicestershire Museums, U.K.), Professor B. W. Fox (New Mills, U.K.), Dr D. J. Galloway (The Natural History Museum, U.K.), Mr V. Giavarini (Bournemouth, U.K.), Dr O. L. Gilbert (Universty of Sheffield, U.K.), the late Dr M. E. Hale (Smithsonian Institution, Washington D.C., U.S.A.), Dr R. C. Harris (New York Botanic Garden, U.S.A.), Professor D. L. Hawksworth (International Mycological Institute, U.K.), Professor Dr H. Hertel (Botanische Staatssammlung, München, Germany), Dr C. J. B. Hitch (Saxmundham, U.K.), Mr J. Holtan-Hartwig (University of Oslo, Norway), Mr P. W. James (The Natural History Museum, U.K.), Professor P. M. Jørgensen (University of Bergen, Norway), Dr J. G. Knoph (Freie Universität Berlin, Germany), Mr J. R. Laundon (The Natural History Museum, U.K.), Dr H. T. Lumbsch (Universität Essen, Germany), Dr P. M. McCarthy (National Botanic Garden, Melbourne, Australia), Dr H. Mayrhofer (Karl-Franzens Universität Graz, Austria), Professor D. M. Moore (University of Reading, U.K.), Mr A. Orange (National Museum of Wales, U.K.), Dr A. Pentecost (University of London, U.K.), Dr O. W. Purvis (The Natural History Museum, U.K.), Dr F. Rose (Petersfield, U.K.), Mr G. Salisbury (U.K.), Dr C. Scheidegger (Swiss Federal Institute of Forests Research, Switzerland), Professor M. R. D. Seaward (University of Bradford, U.K.), Dr J. W. Sheard (University of Saskatchewan, Canada), Mrs L. H. Skjolddal (University of Bergen, Norway), Dr L. Tibell (University of Uppsala, Sweden), Dr E. Timdal (University of Oslo, Norway), Dr T. Tønsberg (University of Bergen, Norway), Mr O. Vitikainen (University of Helsinki, Finland), Dr R. Watling (Royal Botanic Garden, Edinburgh), Dr M. F. Watson (University of Reading, U.K.), Mrs J. F. White (The Natural History Museum, U.K.) and Mr R. G. Woods (Countryside Council for Wales, U.K.).

Contents

Introduction

Contents

The Lichen Flora of Great Britain and Ireland is the first critical account of the lichen-forming fungi known to occur in the British Isles to have been prepared since the second edition of the *Monograph of British Lichens* by A. L. Smith published in 1918–26.

As "lichen" is a biological rather than a systematic category, the decision as to which taxa should be included in the present volume has, in certain cases, been rather arbitrary. The biological status of some species traditionally studied by lichenologists is in doubt, and others are now known not to be truly or consistently lichenized. This problem is discussed further by Hawksworth (1988).

All genera regularly studied by lichenologists have therefore been included, regardless of their actual biological status, and in the case of those genera all species reported from the region are covered, even if some are certainly not lichenized or are lichenicolous (lichen-inhabiting) fungi. Exclusively lichenicolous fungal genera have been omitted, although some which can cause confusion have been mentioned in the generic keys and notes; a key to the lichenicolous species known in the British Isles is provided by Hawksworth (1983). Lichenicolous lichens, i.e. those which have an independent lichenized thallus and grow obligately on other lichens are, however, treated appropriately in this *Flora*.

Arrangement

Generic names are placed in alphabetical order for ease of reference; the orders and families to which these are currently referred are shown on pp. 11–13. The ordinal and(or) familial position of many lichen-forming genera has still not been satisfactorily resolved and as a result any system now adopted would quickly require amendment and cause unnecessary difficulties for future users of this volume. Periodically revised recommended classifications of the ascomycetes, including lichen-forming genera, are published in *Systema Ascomycetum* (1982 on).

Species names within each genus are also arranged alphabetically, except in a few instances where extremely similar species have been placed together.

Nomenclature

The nomenclature adopted here is based on the most recent checklist, *The British Ascomycotina: an annotated checklist* (Cannon *et al.*, 1985), with additions and corrections since that time. All accepted names of lichen-forming fungi in that work, and those adopted in the previous checklist (Hawksworth *et al.*, 1980), are considered in this *Flora*. Earlier synonyms are excluded, but some later synonyms, especially for recent generic segregates used by some contemporary lichenologists which it has not been considered prudent to adopt at this stage, are also noted. The treatments of *Cetraria* s.lat. and *Parmelia* s.lat., where generic limits are actively undergoing revision, are particularly conservative.

Species which are considered to have been erroneously reported from Great Britain and Ireland, names which are of uncertain application, or taxa which are of dubious taxonomic validity, are mentioned at the ends of the appropriate generic accounts or below allied accepted taxa, as are a few species which were first recorded as the preparation of the *Flora* was nearing completion.

The places in which all names accepted in the 1980 and 1985 checklists appear in this new *Flora* can be located through the epithet index at the end of the volume.

The names of authors follow Laundon (1979), or are abbreviated according to the principles he recommends. In addition, the dates of publication of generic and specific names are included as an aid to the location of the precise references to their introduction using the catalogues of Zahlbruckner (1921–40), Lamb (1963), Hawksworth (1972a), and the twice-yearly *Index of Fungi* (1970 on).

The names used for lichens are controlled by the *International Code of Botanical Nomenclature* (Greuter *et al.*, 1988). The names strictly refer to the fungal partner (mycobiont), so that the photosynthetic partner (photobiont) can bear an independent name. The *Code* places major emphasis on historical priority in determining the name to be used for a particular species, the application of names being fixed by linking them to individual specimens, or "types". Name changes under the *Code* occur mainly because either older ones are uncovered in the literature, or they are discovered to have been used for species not conforming to their "type". The rule of priority has failed to deliver the lasting stability it was designed to achieve. Proposals to alter this provision of the *Code* are to be placed before the XIV International Botanical Congress in 1993, which will involve the granting of "protected" status to "Lists of Names in Current Use" (Hawksworth, 1991). In the light of this move, we have deliberately refrained in this *Flora* from taking up several old and unfamiliar names for common species that have come to our attention during its preparation. We believe it would have been irresponsible to take the alternative stance at this point in the history of biological nomenclature.

The *Code* does not legislate taxonomy, i.e. the way in which species are arranged into families and genera, or species are separated or united. These are a matter for scientific judgement based on new data from original research. Such taxonomic work leads to more changes than the *Code* in the current phase of lichen systematics, where generic concepts particularly are being refined according to modern mycological criteria.

No new species are described or new combinations made to transfer species between genera are made formally in this *Flora*. These are presented as a separate contribution in *The Lichenologist* **24**(4) (1992).

Descriptions

The generic descriptions are primarily based on the representatives known in Great Britain and Ireland, important differences relating to other species being indicated.

Species descriptions emphasize characters of value for species identification and do not repeat features from the generic accounts common to all species of the particular genus. The generic and species descriptions consequently need to be used in conjunction with one another.

Species descriptions have also been based on material from Great Britain and Ireland, significant differences seen in material from other countries being identified.

Where measurement ranges are given in the form (8-)10–12(-14) × (3-)4–5(-6) μm. The length or longest axis is that given first, the figures in parentheses indicate extremes occurring at a low frequency, and the other figures the ranges into which most of the structures fall.

Ecology

The account of each species in the *Flora* gives information on the substrata and habitats in which the species occurs in Great Britain and Ireland. Major differences may occur in the preferred habitats of some species elsewhere. In the case of species on trees the mention of generic names of trees indicates the genera and locally occurring species in the British Isles.

The term "old forest indicator species" refers to species recognized in the British Isles as largely restricted to ancient woodland sites (Rose, 1976; Harding & Rose, 1986). Where a significant number of such species occur together in one woodland, research has shown that there is a strong probability that such a woodland is ancient and has a long ecological continuity of habitat in terms of the continued presences of both mature and ancient trees, and often the persistence of enough shelter within the between trunk spaces to maintain a high degree of humidity. Old woodland in this sense includes uneven and high forests, pasture-woodlands, and parks of at least medieval origin. Old woodland lichens are not particularly characteristic of woodlands with a long history of coppice management. Woods with a dense "evergreen" understory of *Ilex* or *Rhododendron* are also usually poor in epiphytic lichens, especially low down on trees, as light is important.

Many lichens on trees are characteristic of, or even confined to, the relics of the Caledonian *Pinus/Betula* forests of the Scottish Highlands where they occur on either bark or wood. Some of these species (but not many) have spread into southern Britain over the last century, either on to worked timber rails and fences or on to bark with increased acidification due to pollution. Such species are indicated in the text where appropriate.

The pH of bark of trees varies not only with the species and genera but also genetically within populations of an individual tree species within a single wood (Gauslaa, 1985). Hence considerable natural variation can occur in the species on a host tree. Additionally, nutrient enrichment due to the presence of grazing animals or perching birds can effect changes in the bark, such that species of lichens (e.g. *Xanthoria*, *Physcia*, *Physconia*) found normally on base-rich bark of *Ulmus*, *Acer* or *Fraxinus*, may sometimes occur on *Quercus*, *Fagus*, etc.

Air pollution can also affect communities on trees. In addition to the wholesale destruction of normal communities by heavy "dry" sulphur dioxide pollution concentrations, and their replacement by such lichens as *Lecanora conizaeoides* and certain algae, "wet" pollution by sulphurous, sulphuric or nitric acids derived from the oxidation of pollutant gases may result in "acid rain" effects where lichen cover remains high on trees that normally have relatively high bark pH. Hence "calcifuge" communities of *Usnea* species, *Ochrolechia androgyna*, *Platismatia glauca*, *Parmelia saxatilis*, *P. laevigata*, etc. may replace *Lobarion* communities under conditions of acid rain and high rainfall on *Quercus*, *Fraxinus* or *Acer*. Pollution by agricultural chemicals has also impoverished or destroyed lichen communities on trees in whole areas of the agricultural lowland.

Communities on the ground in calcareous grasslands will normally only be found where the grassland is kept short and open by grazing or disturbance of some kind, such as trampling. Such communities cannot persist in dense, tall grasslands. Communities on acid heathland types (*Cladonia* heaths) are best developed in old, short, but rather open *Calluna* in inland heathland, on decalcified fixed dunes, or on windswept heaths on exposed cliffs. Where such communities have been severely burned within the last ten years, few species will be found as recolonization is slow.

Communities on rocks away from the coast are differentiated very much according to the chemical nature of the rocks present. Chalk or limestone areas tend to have completely different communities from siliceous rocks or granite. The presence of heavy metals in rocks, some of which are toxic to many species, encourages the development of specialized species.

Soft or readily weathered rocks in general support relatively poor saxicolous communities: most lichens require stable surfaces, and being slow colonists, need relatively hard rocks for effective establishment. Similarly, heavily shaded rocks support relatively few species as most lichens require plenty of light.

On coastal cliffs, there is a clear vertical zonation of species on rocks, related at the lower levels to the length of time of tidal immersion, and at the upper levels to incidence of sea spray, and exposure to wind. Examination of coasts at low tide is therefore advantageous. A parallel zonation is seen on lake margins and by upland unpolluted rivers. Dry underhangs, never directly wetted by rain, also have specialized lichen communities as do dry crevices and underhangs on tree bark. Consolidated shingle beaches, where sufficiently old and stable, may have rich and characteristic lichen floras with some species that do not occur elsewhere.

In Lowland Britain, extremely rich floras are often developed on artificial stonework, above all in the varied habitat niches of churchyards (walls, vertical memorials, roofs, chest-tombs, etc.).

Similar assemblages of species in many cases tend to develop on identical substrata. Familiarity with the major lichen community types present can then be also of great assistance in identification. A preliminary survey of the lichen communities present in the British Isles is provided by James *et al.* (1977).

Distribution

A scheme to map the distribution of lichens in Great Britain and Ireland, on a 10 × 10 km grid basis, was initiated by the British Lichen Society in 1964 (Seaward, 1988). This has been a great success, with records from 3298 grid squares, representing a coverage of 86% being achieved by September 1991 (Seaward, 1992). The first volume of 176 maps with explanatory legends was published by Seaward & Hitch (1982), and drafts of two further volumes have been circulated to specialists for comments prior to finalization. The published and unpublished data in the Society's Distribution Maps Scheme, held on computer at the University of Bradford, have been extensively utilized by the contributors to this volume. Draft maps provided from the database are the basis of the distributional remarks in this *Flora*, supplemented by herbarium and other sources as appropriate.

A checklist and list of the counties from which species have been recorded in Ireland is provided by Seaward (1984).

In the distributional remarks, the terms Central, North, South, East and West refer either to Great Britain and Ireland as a whole or to the component countries, and examples of species showing these distributional types are presented in Fig. 1. Lowland Britain approximates to a line from the mouth of the River Exe to the mouth of the River Tees. A valuable survey of the distributional types shown by corticolous lichens in the region is given by Coppins (1976).

Distributional patterns are substantially modified by the effects of pollutants, the erection of man-made substrata, woodland management practices, and other activities of man (Hawksworth *et al.*, 1974). The effects of sulphur dioxide have been particularly devastating, but during the last 10–15 years reduced emissions in urban areas have led to major improvements in the species present in many major cities, including Bristol, Leeds, London, Manchester and Sheffield.

Illustrations

Due to limitations of space, illustrations in this volume have been restricted to line drawings in those cases where character differences for species separations cannot be clearly expressed in print. Where appropriate, the application of terms has been illustrated in the Glossary.

However, several well-illustrated texts on lichens are already available which can usefully be used to complement this *Flora*: those of Dobson (1992), Foucard (1990), Jahns (1983), Moberg (1982), Ozenda & Clauzade (1970), Wirth (1980) and Wirth (1987); this list is extremely selective and some of the titles listed under "Further Reading" (p. 9) are also well-illustrated. Critical line drawings and photographs are also included in many of the papers cited under "Literature" below each generic description.

Microscopic Examination

While, with experience, the majority of the lichens found in Great Britain and Ireland can be recognized in the field with the aid of a × 10 hand-lens, the characters which are the basis for

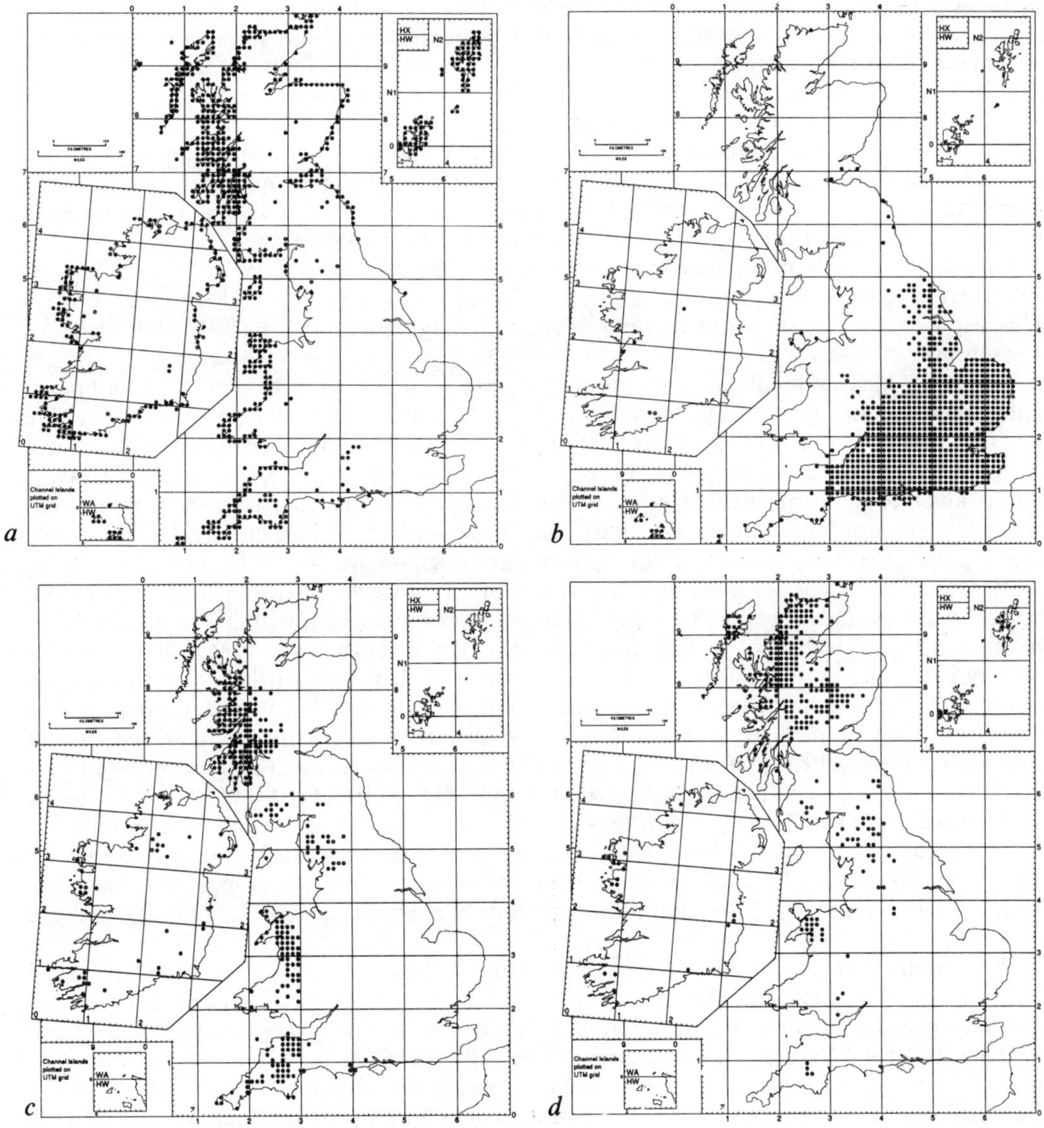

Fig. 1. Examples of distribution patterns. (a) *Anaptychia runcinata*; (b) *Caloplaca teicholyta* (south-eastern); (c) *Parmelia laevigata* (western); (d) *Umbilicaria torrefacta* (upland). Based on data from the British Lichen Society's Distribution Maps Scheme.

their placement in particular genera are, in most cases, based primarily on microscopic features. Knowledge of these is therefore a prerequisite for a thorough understanding of the genera and their positions, and also for the identification of unfamiliar taxa.

Material should first be studied with a binocular dissecting microscope capable of magnifications up to × 40–50 in order to observe the types of fruit bodies and also asexual vegetative reproductive organs. Samples to be examined further microscopically need to be removed while being observed through a dissecting microscope.

Dried material is best first moistened with water and left for 30 seconds to 1 minute. Thin vertical slices can then be cut by hand with the aid of a safety-razor blade or a surgical scalpel blade. New blades invariably give the best results, and should be replaced regularly. Sections should be cut as thinly as possible and placed in a drop of water, potassium hydroxide, lactophenol cotton-blue, etc., as required, on a microscope slide. Not more than 2–3 sections should be placed in each drop. After a coverslip has been applied, the arrangements and dimensions of tissues can be noted. In order to display the details of asci, paraphyses, and to encourage the release of ascospores, pressure should be applied by repeated tapping with the rounded end of a pencil or wooden needle-holder. The secret of a good squash preparation is to start with a very thin slice of material and apply pressure evenly. If a freezing-microtome is available, sections cut at a thickness of 10–15 μm are recommended.

For the examination of microscopic preparations, a microscope with × 10, × 40 and × 100 objectives is necessary. For the critical examination and measurement of ascospores, and also of ascospore wall and ascus tip structures, an overall magnification of × 1000 (oil immersion) is essential.

Ascospore measurements will be most accurate if made in water or in lactophenol cotton-blue (following warming). However, 10% potassium hydroxide is recommended here for routine work as it aids the separation of the tissues and causes little change in dimensions, except in the case of particularly thin-walled ascospores. Ascospores should be measured when mature and free of any pressures exerted by the ascus contents, i.e. when released from the asci in squash preparations. For general purposes, 10 sets of spore measurements are usually sufficient to adequately determine the main range of spore size in a collection.

Microscopic reagent tests are best made on the microscope stage itself so that colour changes can be noted. This is achieved by taking up excess mountant with tissue or filter paper and then applying a drop of reagent from one side of the cover slip so that it flows into the preparation. Note that changes with C can be very transient and, further, that when using iodine after pretreatment with K, the original reaction is sometimes reversed. Colours should be observed with a daylight filter on the microscope. It is often convenient in the case of crustose lecideoid lichens to mount sections on three slides, two in water and one in 10% KOH (K). The colour in K can then be compared directly, and C or 10% HNO_3 (N) be bled into the water-based sections. The K preparation can then be blotted and Lugol's iodine added to observe ascus reactions (see below). Medullary iodine reactions are also best studied under the microscope by squashing small amounts in K, blotting up as much as possible, and then adding Lugol's iodine (I). Note that iodine medullary reactions can be very patchy in some species.

The examination of the reaction to Lugol's iodine of different parts of the ascus apex (K/I reaction) provides important characters for identification, especially of crustose lichens with disc-like or lirellate ascomata. Examination should be carried out on healthy, preferably fresh material, but good results can be obtained from well preserved herbarium collections. The procedure below may well have to be varied according to the nature of the material being examined e.g. heavily gelatinized or carbonized ascomata may prove difficult, or according to the adeptness and experience of the examiner!

1. Moisten a healthy looking ascoma with water for a few minutes.
2. Remove a small portion of the ascoma including part of the hymenium (preferably a thin section) and place on a glass slide.
3. Add a drop of 10% KOH, and leave for 1–5 minutes (hymenia that have a dense gelatinous matrix require a longer time).
4. Carefully soak up most of the KOH with the edge of a tissue.
5. Add a drop of Lugol's iodine and soak up, being careful to leave the preparation *in situ* on the slide.
6. Repeat stage 5.
7. Repeat stage 5 again, but do not soak up; instead, carefully lower a cover slip on to the preparation and observe under the microscope at × 100 or × 400. If there is a bluing of the asci or hymenial gel, the slide is ready to examine more closely. N.B. in the case of species in

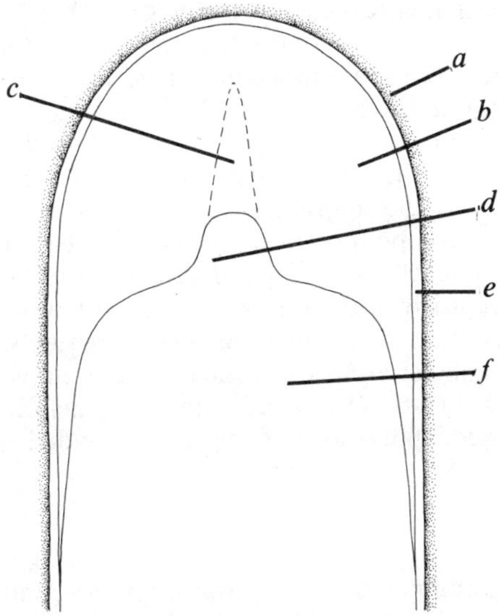

Fig. 2. Ascus structure (Lecanoralean, '*Bacidia*-type') (a) fuzzy coat (outer layer); (b) apical dome (tholus); (c) apical cushion (masse axiale); (d) ocular chamber; (e) ascus wall; (f) ascoplast.

which there is no bluing, the asci or gel will take on a pale yellowish or reddish coloration; all tissues will then be considered K/I−.

If there is no bluing, yellowing or reddening, add another drop of Lugol's iodine to the edge of the coverslip and allow the solution to flow over the preparation; this process can be assisted by drawing through the solution by applying a piece of tissue to the opposite edge of the coverslip, or by carefully raising the edge of the coverslip with a razor blade. Repeat until a coloration is obtained.

8. Examine preparation at ×400 or preferably × 1000 (oil-immersion), make notes and sketches. The preparation may need to be lightly tapped to spread out the tissues, but take care as asci are often easily disrupted. If the blue coloration gradually fades away, add another drop of Lugol's iodine to the preparation. Compare results with illustrations (Figs 2 and 33–35).

Chemical Examination

While chemical characters are rarely the sole basis on which species are separated, they are extremely valuable in identification since they can give unequivocal results and can be used particularly as an aid to the determination of sterile or fragmentary specimens. Reagent tests have been used in lichenology since the 1860s and can be performed directly on the thallus or soralia etc., but in some cases the compounds giving the reactions are localized and reactions from the medulla or exciple are specified; these are performed by first removing the cortical and algal layers by scraping with a razor or scalpel blade so that the white or yellowish medulla is exposed.

The results are best studied using a hand-lens or dissecting microscope; in some cases colours change in a characteristic manner and this also must be carefully noted (e.g. yellow to red). In the text, positive reactions are indicated by "+", negative by "−" and variable reactions by "±". Variable reactions may be due to concentration effects and(or) different chemical races ("chemotypes") being present within a species. If it is uncertain when testing dry material if a reaction is positive, it should be compared with a spot of water, or the colour taken up on to tissue or filter paper (especially useful for dark-coloured specimens); KC tests can be performed by taking up the result of the K reaction on to filter paper and then running C into the spot from the side, the same procedure in reverse is followed for the CK test. C reactions are often fleeting and are best studied by watching while the reagent is applied; when using Steiner's Pd solution, 2–3 minutes should be allowed for colours to develop fully. Where crystal formation is noted in the text (e.g. red needles of norstictic acid in K) this can also usually be seen only under the microscope.

Where the identity of the compounds present in a species are known, these have been indicated in parenthesis. Some of these respond vividly to ultra-violet light (UV) and the sensitivity and colour are recorded where important. In most cases these reports are based on thin-layer chromatography (t.l.c.), a method now routinely employed in specialist lichenological laboratories and essential for critical systematic work. In some cases t.l.c. is necessary for positive identification or sterile and fragmentary specimens. A valuable introduction to this technique, spot and other tests is provided by White & James (1985).

Stains and Reagents

This section summarizes the composition of the stains and reagents recommended above and used in this volume.

K = A solution of 10–35% potassium hydroxide in water. Valuable for making squash preparations as it causes gelatin binding ascomatal tissues to separate; 10% needed as a pre-treatment before making tests with I (see below). Gives yellow, yellow changing to red, red, or brownish reactions; in squash preparations the formation of reddish needle-like crystals occurs in some cases (e.g. norstictic acid).

C = Undiluted commercial bleach (e.g. Parazone) used as a colour-test reagent and giving red, rose, orange or green as positive reactions.

KC or CK = K or C is applied followed by the other reagent. The order of application is crucial; KC and CK are tests for quite different compounds. KC positive tests are pinks and reds, CK deep yellows.

Pd = Paraphenylenediamine. Most safely used as Steiner's Stable Pd solution: 1 g paraphenylenediamine, 10 g sodium sulphite, 5 ml detergent, 100 ml water. Needs to be used with care as will discolour paper, herbarium packets, clothing, etc. and is a possible carcinogen that can be absorbed through skin, remake at 3–4 month intervals. Gives reds, oranges and yellows. Discard fragments tested.

I = Iodine solutions are very useful in studying ascus structures as well as performing reagent colour tests. When studying asci, pre-treatment with 10% K is necessary (K/I) to ensure consistent results. Lugol's iodine solution is recommended for studying ascus reactions and those of hymenial tissues: 0.5 g iodine, 1.5 g potassium iodine, 100 ml distilled water.

N = 50% nitric acid. Gives red-purple colour with certain pigments in apothecia.

LCB = Lactophenol-cotton blue. Used for the preparation of semi-permanent slides; 20 g phenol (crystals), 20 g lactic acid (SG 1.21), 40 g glycerine, 20 ml water, + drops of cotton blue to give desired intensity. The cotton blue stains fungal hyphae blue and also algal cells deep blue or blue-green. After adding the coverslip and slight pressure, the slides need warming to almost boiling point to facilitate take-up of cotton blue, and counteract the tendency of the spores to shrink slightly in this reagent, and eliminate air bubbles. When all air is expelled and the slides cool they can be sealed with nail varnish (two coats at intervals) or a proprietary agent (e.g. Glyceel). As phenol is implicated as a carcinogen, this reagent should be made up in a fume cupboard and care taken *not* to boil slides.

Terminology

The editors have paid particular attention to the elimination of unnecessary jargon, particularly in the descriptions of genera and species. Otherwise, current mycological terminology has been employed and generally follows the most recent edition of *Ainsworth & Bisby's Dictionary of the Fungi* (Hawksworth *et al.*, 1983).

Collecting and Conservation

Lichens mostly grow extremely slowly, generally 1–5 mm in radius or length each year, and are notably sensitive to subtle changes in environmental factors. If lost from a habitat with a long history of ecological continuity, reestablishment may at worst never occur and at best take several human generations. Collecting, therefore, should be kept to the minimum necessary for species identification, and the building up of substantial personal collections is discouraged, except where the material comprises vouchers for records which will be published (in which case it should eventually be deposited in an appropriate local or national herbarium).

Specimens on bark or wood can be removed with a non-folding stout knife, care being taken not to cut into the living layers of tissue. Pruning secateurs and(or) a small hacksaw are useful for twig-inhabiting species.

The collection of specimens on rocks and other hard materials necessitates the use of a geological hammer and a 1 cm masonry chisel. Care must be taken not to damage walls, churchyard memorials and other man-made structures. When working in churchyards, for example, reagent tests and squash-slide preparations can be made in the field and, supplemented by notes of hand-lens features, identifications can then be made back in the laboratory.

Specimens on soil or sand easily become powdered so should be collected in tins and after collection stabilized by the use of a soil-binding agent. For example, Primal AC-55 brushed on to the underside of the specimens and left to dry overnight soil-side up.

Permission from land-owners should also be obtained. If it is not, courtesy aside, collectors could in some circumstances be liable to prosecution under the Conservation of Wild Creatures and Wild Plants Act 1975.

Habitat Keys

The following keys dealing with species in particular habitats can usefully be used as adjuncts to this *Flora*:

Lichenicolous fungi and lichens: Hawksworth (1983).
Limestone: Coppins (1984), McCarthy & Mitchell (1988).
Marine and maritime: Fletcher (1975).
Shaded acid rock crevices: James (1970).
Sterile crustose species with *Trentepohlia*: James & Coppins (1979).

Further Reading

For general introductions to the biology, ecology, chemistry, physiology and uses of lichens see Hale (1983) and Hawksworth & Hill (1984). More specialized texts include Ahmadjian & Hale (1974), Brown *et al.* (1976), Galun (1988), Henssen & Jahns (1973), Lawrey (1984), Kershaw (1983), Richardson (1975) and Seaward (1977).

References to the literature on lichens in different parts of Great Britain and Ireland are indexed by vice-county in Hawksworth & Seaward (1977) and Mitchell (1971). More recent major titles are listed in the *Bulletin of the British Lichen Society* (twice-yearly). The world systematic literature on lichen-forming fungi is covered by the *Bibliography of Systematic Mycology* (twice-yearly), which since 1987 has been provided with an index down to genus.

In addition to the works cited as sources of illustrations (p. 4) and under particular genera, the following titles are also of value for identification: Clauzade & Roux (1985), Dahl & Krog (1973), Foucard (1990), Gams (1987), Krog *et al.* (1980), Poelt (1969), Poelt & Vězda (1977, 1981), Thomson (1984), and Vainio (1921–34). A fuller list of titles relevant to European and world lichen identification is provided in Hawksworth & Ahti (1990).

Systematic Arrangement

The names of orders and families used in the Flora are compiled here in alphabetical order within the main subdivisions of the fungi, and the accepted genera are listed by the family to which they are assigned. The system of orders and families used here is essentially that of the *Dictionary of the Fungi* (Hawksworth *et al.*, 1983), with modifications for the Ascomycotina to take note of later work, mainly as compiled in Eriksson & Hawksworth (1991).

A "?" indicates that the placing of that family or genus is uncertain, and not doubt that the taxon should be accepted.

For places of publication of ordinal names see Hawksworth & Eriksson (1986), and for family names Hawksworth & David (1989).

ASCOMYCOTINA

ARTHONIALES Henssen ex D. Hawksw. & O. Eriksson (1986)

Arthoniaceae Reichenb. ex Reichenb. (1841): *Arthonia, Arthothelium*
Chrysotrichaceae Zahlbr. (1905): *Chrysothrix*

CALICIALES C. Bessey (1907)

Caliciaceae Chev. (1826): *Calicium, Cyphelium, Thelomma*
Coniocybaceae Reichenb. (1857): *Chaenotheca, Sclerophora*
Microcaliciaceae Tibell (1984): *Microcalicium*
Mycocaliciaceae A. Schmidt (1970): *Chaenothecopsis, Mycocalicium, Phaeocalicium, Stenocybe*
Sphaerophoraceae Fr. (1831): *Sphaerophorus*
Sphinctrinaceae M. Choisy (1950): *Sphinctrina*

DOTHIDEALES Lindau (1897)

Arthopyreniaceae W. Watson (1929): *Arthopyrenia, Leptorhaphis, Mycomicrothelia, Tomasellia*
Lichenotheliaceae Henssen (1986): *Lichenothelia*
Mycoporaceae Zahlbr. (1903): *Mycoporum*
No family: *Mycoglaena*

GRAPHIDALES C. Bessey (1907)

?Asterothyriaceae W. Watson ex R. Sant. (1952): *Gyalidea*
?Gomphillaceae W. Watson ex Hafellner (1984): *Gomphillus, Gyalideopsis, ?Sagiolechia*
Graphidaceae Dumort. (1822): *Graphina, Graphis, Phaeographis*
Thelotremataceae (Nyl.) Stizenb. (1862): *Diploschistes, Thelotrema*

GYALECTALES Henssen ex D. Hawksw. & O. Eriksson (1986)

Gyalectaceae (Massal.) Stizenb. (1862): *?Belonia, ?Bryophagus, Cryptolechia, Dimerella, Gyalecta, Pachyphiale, Ramonia*

LECANORALES Nannf. (1932)

Acarosporaceae Zahlbr. (1906): *Acarospora, Biatorella, Polysporina, Sarcogyne, Sarcosagium, Sporastatia, Strangospora, Thelocarpon*
Agyriaceae Corda (1838): *Agyrium, ?Ptychographa, Xylographa*
Alectoriaceae (Hue) Tomas. (1949): *Alectoria, Bryoria*
Arctomiaceae Th.Fr. (1860): *Arctomia*
Bacidiaceae W. Watson (1929): *Adelolecia, Bacidia, Biatora, Cliostomum, Herteliana, Lecania, Phyllopsora, Schadonia, ?Solenopsora, ?Squamarina, Toninia, Tylothallia*
Brigantiaceae Hafellner & Bellem. (1982): *Brigantiaea*

Candelariaceae Hakulinen (1954): *Candelaria, Candelariella, ?Claurouxia,*
Catillariaceae Hafellner (1984): *Catillaria, Halecania, Wadeana*
Cladoniaceae Zenker (1827): *Cladonia, Pycnothelia*
Coccocarpiaceae (Mont. ex K. Müller) Henssen (1986): *Spilonema*
Collemataceae Zenker (1827): *Collema, Leptogium*
Ectolechiaceae Zahlbr. (1905): *?Lopadium*
Eigleraceae Hafellner (1984): *Eiglera*
Haematommataceae Hafellner (1984): *Haematomma*
Hymeneliaceae Körber (1855): *Aspicilia, Hymenelia, Ionaspis, Megaspora, Tremolecia*
Lecanoraceae Körber (1854): *Bryonora, Carbonea, Clauzadeana, Lecanora, Lecidella, Megalaria, Miriquidica, Protoparmelia, Pyrrhospora, Scoliciosporum, Tephromela*
Lecideaceae Chev. (1826): *Hypocenomyce, Lecidea, ?Steinia*
Megalosporaceae Vězda ex Hafellner & Bellem. (1982): *Megalospora*
Micareaceae Vězda ex Hafellner (1984): *Micarea, Psilolechia*
Mycoblastaceae Hafellner (1984): *Mycoblastus*
Ophioparmaceae Rogers & Hafellner (1988): *Ophioparma*
Pannariaceae Tuck. (1872): *Degelia, Leproloma, Moelleropsis, Pannaria, Parmeliella, Psoroma*
Parmeliaceae Zenker (1827): *Allantoparmelia, Cavernularia, Cetraria, Cetrelia, Coelocaulon, Cornicularia, Evernia, Foraminella, Hypogymnia, Menegazzia, Parmelia, Parmeliopsis, Platismatia, Pseudephebe, Pseudevernia, Usnea*
Physciaceae Zahlbr. (1898): *Anaptychia, Buellia, Diploicia, Diplotomma, Heterodermia, Hyperphyscia, Phaeophyscia, Physcia, Physconia, Rinodina, Tornabea*
Pilocarpaceae Zahlbr. (1905): *Byssoloma, Fellhanera*
Porpidiaceae Hertel & Hafellner (1984): *Amygdalaria, Bellemerea, Clauzadea, Farnoldia, Koerberiella, Porpidia*
Psoraceae Zahlbr. (1898): *?Lecidoma, Protoblastenia, Psora*
Ramalinaceae Ag. (1821): *Ramalina*
?Rhizocarpaceae M. Choisy ex Hafellner (1984): *?Catolechia, ?Epilichen, Poeltinula, Rhizocarpon*
Rimulariaceae Hafellner (1984): *Lithographa, Mosigia, Rimularia*
Stereocaulaceae Chev. (1826): *Pilophorus, Stereocaulon*
Trapeliaceae M. Choisy ex Hertel (1970): *Placopsis, Placynthiella, Trapelia, Trapeliopsis*
Umbilicariaceae Chev. (1826): *Lasallia, Umbilicaria*
Vezdaeaceae Poelt & Vězda ex J.C. David & D. Hawksw. (1991): *Vezdaea*
No family: *Aphanopsis, Lepraria, Leprocaulon, Melaspilea, Siphula, Thamnolia*

LEOTIALES Carpenter (1988)

Baeomycetaceae Dumort. (1829): *Baeomyces, Icmadophila*
Leotiaceae Corda (1842): *Mniacea*

LICHINALES Henssen & Büdel (1986)

Lichinaceae Nyl. (1854): *Ephebe, Euopsis, Lemmopsis, Lempholemma, Lichina, Porocyphus, Psorotichia, Pterygiopsis, Pyrenopsis, Synalissa, Thermutis*
Peltulaceae Büdel (1986): *Peltula*

OPEGRAPHALES M. Choisy ex D. Hawksw. & Eriksson (1986)

Opegraphaceae Stizenb. (1862): *Bactrospora, Chiodecton, Enterographa, Lecanactis, Opegrapha, Schismatomma, Sclerophyton*
Roccellaceae Chev. (1826): *Dirina, Roccella*

OSTROPALES Nannf. (1932)

Odontotremataceae D. Hawksw. & Sherw. (1982): *Xerotrema*
Stictidaceae Fr. (1849): *Absconditella, ?Petractis, ?Thelopsis*

PATELLARIALES D. Hawksw. & O. Eriksson (1986)

Arthrorhaphidaceae Poelt & Hafellner (1976): *Arthrorhaphis*

PELTIGERALES W. Watson (1929)

Lobariaceae Chev. (1826): *Lobaria, Pseudocyphellaria, Sticta*
Nephromataceae Wetm. ex J.C. David & D. Hawksw. (1991): *Nephroma*
Peltigeraceae Dumort. (1822): ?*Massalongia, Peltigera, Solorina*
?Placynthiaceae Dahl (1950): *Placynthium,* ?*Polychidium, Vestergrenopsis*

PERTUSARIALES M. Choisy ex D. Hawksw. & O. Eriksson (1986)

Coccotremataceae Henssen ex J.C. David & D. Hawksw. (1991): *Coccotrema*
Pertusariaceae Körber ex Körber (1855): *Ochrolechia, Pertusaria*

PEZIZALES C. Bessey (1907)

?Schaereriaceae M. Choisy ex Hafellner (1984): *Schaereria*

PYRENULALES Fink ex D. Hawksw. & O. Eriksson (1986)

Pyrenulaceae Rabenh. (1870): *Acrocordia,* ?*Celothelium,* ?*Eopyrenula,* ?*Pyrenocollema, Pyrenula*
Trichotheliaceae (Müll. Arg.) Bitter & Schill. (1927): *Porina,* ?*Rhaphidicyrtis, Zamenhofia*
Trypetheliaceae Zenker (1927): *Parmentaria*

TELOSCHISTALES D. Hawksw. & O. Eriksson (1986)

?Fuscideaceae Hafellner (1984): *Fuscidea,* ?*Orphniospora*
Teloschistaceae Zahlbr. (1898): *Caloplaca, Fulgensia, Leproplaca, Teloschistes, Xanthoria*

VERRUCARIALES Mattick ex D. Hawksw. & O. Eriksson (1986)

Verrucariaceae Zenker (1827): *Agonimia, Catapyrenium, Dermatocarpon, Endocarpon, Leucocarpia, Leucocarpopsis, Macentina, Placidiopsis, Polyblastia,* ?*Sarcopyrenia, Staurothele, Thelidium, Trimmatothele, Verrucaria*

ORDER UNCERTAIN

Epigloeaceae Zahlbr. (1903): *Epigloea*
Phlyctidaceae Poelt ex J.C. David & D. Hawksw. (1991): *Phlyctis*
Protothelenellaceae Vězda, Mayhr. & Poelt (1985): *Protothelenella*
Strigulaceae Zahlbr. (1898): ?*Anisomeridium, Strigula*
Thelenellaceae Mayhr. (1986): ?*Chromatochlamys, Julella, Thelenella*
Thrombiaceae Poelt & Vězda ex J.C. David & D. Hawksw. (1991): *Thrombium*
No Family: *Normandina*

BASIDIOMYCOTINA

AGARICALES Clem. (1909)

Tricholomataceae Heim ex Pouzar (1983): *Omphalina*

STEREALES Jülich (1981)

Meruliaceae P. Karsten (1881): *Dictyonema*

DEUTEROMYCOTINA

AGONOMYCETES

Cystocoleus, Racodium

HYPHOMYCETES

Blarneya

Generic Key

General Synopsis

1 Thallus crustose, resembling either a crust which is continuous, rimose-cracked, areolate, evanescent, immersed, or powdery 2
Thallus not as above 11

2(1) Thallus fertile (fruits globose, disc-like, elongate, stalked, bracket, or mushroom-like) 3
Thallus sterile (often with isidia or ± delimited soredia, or entirely leprose) 5

3(2) Fruits basidiomata, producing basidiospores 4
Fruits ascomata, producing ascospores 8

4(3) Basidiomata stalked, mushroom-shaped .... ***Omphalina***
Basidiomata crust- to bracket-shaped .... ***Dictyonema interruptum***

5(2) Thallus leprose, entirely granular-powdery without any cortex, even when young, occasionally becoming indistinctly lobed at margin; mostly on vertical surfaces or underhangs 6
Thallus not entirely leprose, ± corticate, at least towards the margin and when young, lobes present or absent; on bark or wood .... **KEY 7** (p. 43)

[Sterile crustose lichens]

6(5) Thallus colour varied, never uniformly yellow and K+ red; ecology varied 7
Thallus dull brownish or bright yellow, K+ dark violet-red; on ± calcareous, vertical rocks .... ***Leproplaca***

7(6) Thallus of powdery, convex granules, often indistinctly lobed at margin of thallus; dibenzofurans present (t.l.c.) .... ***Leproloma***
Thallus of powdery or non-powdery granules, without lobes; dibenzofurans absent (t.l.c.) .... ***Lepraria***

8(3) Fruits sessile or immersed 9
Fruits distinctly stalked .... **KEY 4** (p. 28)
[Crustose lichens with stalked apothecia]

9(8) Fruits not perithecia 10
Fruits perithecia or perithecia-like, globose, opening by a minute pore at apex (occasionally several perithecia may occur within a receptacle with 3–15 minute surface pores) .... **KEY 3** (p. 25)
[Crustose lichens with perithecia]

10(9) Fruits elongate (lirellae), the disc exposed or a slit, sometimes short, straight, curved or serpentine, sometimes star-like or arranged in ± parallel lines .... **KEY 5** (p. 28)
[Crustose lichens with lirellate apothecia]
Fruits disc-shaped, the disc ± exposed, at least at maturity, concave to convex or globose (sometimes deeply immersed in the thallus or enclosed within warts) .... **KEY 6** (p. 29)
[Crustose lichens with rounded apothecia]

11(1) Thallus fruticose, shrub- or beard-like, rarely worm- or thread-like with erect to decumbent or hanging branches, terete or angular in section, if flattened then with photobiont cells above and below arranged round a central strand or hollow, fastened only at the base or attached by a few scattered hapters .... **KEY 1** (p. 15)
[Fruticose lichens]
Thallus foliose, large to minute, leaf-like, compacted or mostly horizontally spreading, often attached to substratum by rhizines, hapters or by a central disc, dorsiventral with photobiont cells arranged in a ± horizontal layer below the upper cortex, upper surface differing in colour from the lower; or squamulose and of small flattened, often compacted squamules or lobes; not dimorphic and without podetia and pseudopodetia .... **KEY 2** (p. 18)
[Foliose and squamulose lichens]

Generic Key 1

Fruticose lichens

Synopsis

1 Thallus yellow to orange-red, K+ purple-red 2
Thallus variously coloured, if yellow to orange-red then not K+ purple-red 3

3(1) Thallus dimorphic 4
Thallus not dimorphic 9

9(3) Thallus flattened, dorsiventral 10
Thallus not flattened, terete or, if flattened, algae distributed on both lower and upper sides 13

10(9) Marginal cilia present 11
Marginal cilia absent 12

13(9) Thallus gelatinous, dark green- to brown-black or black 14
Thallus not gelatinous, colour various 17

17(13) Thallus to *c.* 1 cm tall, consisting of very slender, minutely tufted, spreading filaments .. 18
Thallus *c.* 1–20 cm tall, filamentous or otherwise 23

Key 1

1 Thallus yellow to orange-red, K+ purple-red 2
Thallus variously coloured, if yellow to orange-red then not K+ purple-red 3

2(1) Thallus of small, flattened, compacted lobes, to 1 mm wide and 2 mm tall, without marginal fibrils; marginal soredia present .... ***Xanthoria candelaria***
Thallus either entirely terete, angular, richly branched and with small oval soralia, or with ± flattened lobes, to 2.5 mm wide with numerous marginal fibrils; without soralia, often fertile .... ***Teloschistes***

3(1) Thallus dimorphic, with a crustose, leprose or squamulose basal thallus from which arises an erect, secondary thallus with or without apothecia 4
Thallus not dimorphic, without a basal and erect secondary thallus 9

4(3) Basal thallus crustose, leprose or granular 5
Basal thallus squamulose or of rounded, peltate or flattened phyllocladia (sometimes disappearing early) 8

5(4) Basal thallus not leprose, erect thallus without soredia; apothecia often present 6
Basal thallus leprose, erect thallus of short, white podetia within a ± continuous, sorediate crust; apothecia unknown .... ***Leprocaulon microscopicum***

6(5) Podetia solid, mostly simple 7
Podetia hollow, simple to almost coralloid-branched .... ***Pycnothelia papillaria***

7(6) Apothecia brown-pink; cephalodia absent; podetia mostly decorticate, white, smooth or ± furrowed .... ***Baeomyces***
Apothecia black; cephalodia present; podetia corticate, granular ...... ***Pilophorus strumaticus***

8(4) Erect thallus hollow, often with squamules (at least at the base) and cups; phyllocladia and cephalodia absent; ascospores simple .... ***Cladonia***
Erect thallus solid, without cups and squamules; granular, peltate or finger-like phyllocladia and often pale brown to blackish cephalodia present; ascospores septate .... ***Stereocaulon***

9(3) Thallus flattened, strap-shaped, dorsiventral (with a distinct, differently coloured upper and lower surface); algae concentrated near the upper surface only 10
Thallus not flattened, terete or, if flattened, algae distributed on both lower and upper sides 13

10(9) Marginal cilia present 11
Marginal cilia absent 12

11(10) Upper surface pale grey-brown with pale marginal cilia .... ***Anaptychia ciliaris***
Upper surface white with black marginal cilia .... ***Heterodermia leucomelos***

12(10) Lower surface white, never black (see also *Ramalina lacera*) .... ***Evernia prunastri***
Lower surface at least partly black .... ***Pseudevernia furfuracea***

13(9) Thallus gelatinous, dark green- to brown-black or black 14
Thallus not gelatinous, colour various 17

14(13) Thallus on a variety of substrata, not forming dense swards on maritime rocks; apothecia various 15
Thallus forming dense swards on maritime rocks at or below high water mark; apothecia ± globose, sunken in lobe apices .... ***Lichina***

15(14) Thallus large, >4 mm tall; photobiont *Nostoc*, apothecia varied; ascospores septate or muriform; on various substrata 16
Thallus very small, 1–4 mm tall; photobiont *Gloeocapsa*; apothecia very rare, poriform, with swollen thalline exciple; ascospores simple; on damp limestone ***Synalissa symphorea***

16(15) Thallus without an upper cortex (microscope); thallus usually brown- or green-black ***Collema***
Thallus with an upper cortex consisting of a single layer of cells (microscope); thallus usually blue-grey, brown or grey-black .... ***Leptogium***

17(13) Thallus to *c.* 1 cm tall, consisting of very slender, minutely tufted, spreading filaments .. 18
Thallus *c.* 1–20 cm tall, filamentous or otherwise 23

18(17) Thallus with blue-green photobiont; filaments >0.05 mm diam.; on rocks in open, exposed sites 19
Thallus with *Trentepohlia*; filaments <0.02 mm diam.; on vertical or recessed rock faces 22

19(18) Thallus with *Stigonema*, or *Hyphomorpha*, olive-green to black, matt, ± decumbent; apothecia infrequent, in swellings in thallus; in seepage tracks on rocks 20
Thallus with *Nostoc*, shining brown, ± erect; apothecia frequent; disc red-brown; amongst mosses on trees and rocks .... ***Polychidium muscicola***

20(19) Thalline exciple absent, hymenium exposed; photobiont *Stigonema* or not; thallus small, fluffy, if spreading, then not combed in appearance 21
Thalline exciple well developed, ± enclosing hymenium in a swelling; photobiont *Stigonema*; thallus coarser, often matted and spreading with a combed appearance ***Ephebe***

21(20) Filaments to 50 μm wide; rhizine-like, N+ purple-red hyphae attaching thalli to substratum; photobiont *Stigonema* or *Hyphomorpha* .... ***Spilonema***
Filaments to 15 μm wide; rhizine-like hyphae absent; photobiont *Scytonema* ...... ***Thermutis***

22(18) Central *Trentepohlia* chain surrounded by elongate, straight, ± parallel-orientated septate hyphae; filaments smooth .... ***Racodium***
Central *Trentepohlia* chain surrounded by twisted, irregularly septate hyphae; filaments nodulose .... ***Cystocoleus***

23(17) Thallus and soralia (if present) C−; ecology various 24
Thallus and / or soralia C+ red; exclusively maritime, firmly attached to rock (if alpine, c.f. *Alectoria nigricans*) .... ***Roccella***

24(23) Thallus filamentous or otherwise, without a tough, axial strand (breaking cleanly when pulled apart) 25
Thallus filamentous, with a tough axial strand, not easily broken when branches are pulled; ascospores simple .... ***Usnea***

25(24) Thallus thread-like, with thin, often pointed, apices 26
Thallus not thread-like, apices mostly blunt 29

26(25) Terminal branchlets without hook-shaped apices; thallus pendent, erect or decumbent, colour various 27
Terminal branchlets with hook-shaped apices; thallus pendent or erect, pale grey to yellow-white .... ***Ramalina chondrina***

27(26) Thallus with or without indistinct pseudocyphellae which are never convex; colour varied, not yellow 28
Thallus with distinct, white, ± elongate, often convex pseudocyphellae; yellow-grey ***Alectoria***

28(27) Thallus densely divergently branched; soralia never present; dark brown-black; forming a spreading, decumbent mat attached by spreading hapters .... ***Pseudephebe pubescens***
Thallus not densely branched, or if so then ± erect; soralia present or absent; pale to brown-black; mostly only attached at one end .... ***Bryoria***

29(25) Thallus chalk white, yellow-grey or yellow-green 30
Thallus brown to brown-black, or white-, brown- or green-grey 34

30(29) Thallus chalk white 31
Thallus white- or yellow-grey 32

31(30) Thallus hollow, ± decumbent and worm-like, apices pointed .... ***Thamnolia vermicularis***
Thallus, solid, ± erect and compacted, apices blunt .... ***Siphula ceratites***

32(30) Thallus erect, tufted 33
Thallus decumbent, straggling over low vegetation .... ***Alectoria sarmentosa*** subsp. ***vexillifera***

33(32) Thallus solid, wrinkled and lacunose with wavy margins .... ***Cetraria nivalis***
Thallus hollow, without wrinkles or lacunose with wavy margins .... ***Cladonia***

34(29) Thallus green-grey to grey 35
Thallus brown to brown-black 41

35(34) Thallus not tomentose, smooth; thallus terete or flattened 36
Thallus distinctly pubescent-tomentose; thallus terete, often partly angular and flattened .... ***Tornabea scutellifera***

36(35) At least main branches distinctly flattened or channelled 37
Thallus branches ± terete 38

37(36) Branches entirely flattened; thallus erect to pendent; apothecial discs pale green-brown to fawn; mazaedium absent .... ***Ramalina***
Main branches flattened, lateral branches terete; apothecia globose, with a distinct black, powdery mazaedium, branching irregular .... ***Sphaerophorus melanocarpus***

38(36) Thallus hollow or solid; apothecia green-brown to black 39
Thallus of hollow, simple to richly branched podetia either with pointed apices or bearing cups; apothecia, if present, red or brown .... ***Cladonia***

39(38) Thallus without phyllocladia; cephalodia absent 40
Thallus covered with rounded, peltate or finger-like phyllocladia; cephalodia and/or soralia often present .... ***Stereocaulon***

40(39) Thallus erect, coralloid, sometimes brown-tinged, apothecia globose, with a distinct black, powdery mazaedium, branching irregular .... ***Sphaerophorus***
Thallus erect or pendent, tufted, green- or yellow-grey, apothecia disc-shaped, pale green-brown .... ***Ramalina***

41(34) Branches 1–3 mm wide, slightly flattened or ± terete 42
Branches >3mm wide with spinulose margins, distinctly flattened; stems erect, often reddish below .... ***Cetraria***

42(41) Thallus forming compact tufts to 2 cm tall, firmly attached to rock; apothecia terminal, frequent with brown-black discs often surrounded by short projections .... ***Cornicularia normoerica***
Thallus forming loosely interwoven, to 5 cm tall tufts amongst mosses on the ground; apothecia rare .... ***Coelocaulon***

Generic Key 2

Foliose and squamulose lichens

1 Thallus bright yellow to orange-red 2
Thallus not as above 7

2(1) Thallus K+ purple-red 3
Thallus K− or K+ yellow, not purple-red 5

3(2) On bark, wood or stone; schizidia absent; ascospores polarilocular 4
On calcareous soil, often spreading to mosses; schizidia often present; ascospores simple or 1-septate, not polarilocular .... ***Fulgensia***

4(3) Thallus closely attached to substratum by entire lower surface, often placodioid, rhizines and other attachment organs absent .... ***Caloplaca***
Thallus ± loosely attached to substratum by a basal sheath, attachment discs or rhizines .... ***Xanthoria***

5(2) Thallus lobes to *c.* 1 cm wide, degree of incision and presence of marginal soralia variable 6
Thallus lobes very small, 0.3–0.5(-1.5) mm wide, markedly incised with ± granular, marginal soralia .... ***Candelaria concolor***

6(5) Thallus irregularly granular-isidiate towards centre with white-pruinose marginal lobes, placodioid .... ***Candelariella medians***
Thallus with marginal soralia, lobes not pruinose .... ***Cetraria pinastri***

7(1) Thallus distinctly foliose, small to large, often rosette-forming or strap-shaped, attached to substratum by rhizines, hapters, tomentum, folds or by an umbilicus .... **Key 2a** (p. 18)
Thallus squamulose, sometimes minutely so, squamules mostly closely attached to substratum or elevated towards margins, scattered, contiguous or overlapping, or thallus placodioid and ± crustose at the centre with contiguous lobed margins .... **Key 2b** (p. 21)

Generic Key 2a

Foliose (not yellow to orange-red)

Synopsis

1 Thallus umbilicate 2
Thallus not umbilicate 4

4(1) Thallus not zoned in section (homoiomerous) 5
Thallus zoned in section (heteromerous) 6

6(4) Thallus on a distinct blue/black tomentose hypothallus 7
Thallus without, or with an indistinct blue/black hypothallus 8

8(6) Thallus of strap-shaped lobes 9
Thallus otherwise 10

10(8) Thallus consisting of medium to large spreading lobes, 0.5–4 cm wide 11
Thallus consisting of small lobes, <0.5 cm wide 28

11(10) Photobiont green 12
Photobiont blue-green 23

28(10) Photobiont blue-green 29
Photobiont green 30

30(28) Rhizines and tomentum absent on lower surface; marginal cilia absent 31
Rhizines or tomentum present on lower or upper surface; marginal cilia occasionally present 35

Key 2a

1 Thallus attached to substratum at a central point by an umbilicus 2
Thallus attached to substratum by other means, without an umbilicus (if perithecia present cf. *Dermatocarpon*) 4

2(1) Thallus with apothecia or sterile 3
Thallus dotted with abundant black perithecia on upper surface .... ***Dermatocarpon***

3(2) Thallus surface with radially arranged, coarse, oval pustules; isidia black, coralloid, very rarely fertile .... ***Lasallia pustulata***
Thallus surface without coarse pustules, surface smooth, wrinkled, ridged; soredia, phyllidia or thalloconidia sometimes present, or often fertile .... ***Umbilicaria***

4(1) Thallus not zoned in section (homoiomerous), often distinctly swollen when wet; blue-grey, brown-black or green-black 5
Thallus zoned in section (heteromerous), not distinctly swollen when wet; colour various 6

5(4) Thallus brown- or green-black; without a cellular upper cortex (microscope) .... ***Collema***
Thallus blue-grey, brown or grey-black; upper cortex of a single layer of well-defined cells (microscope) .... ***Leptogium***

6(4) Thallus on a distinct blue/black, tomentose hypothallus; photobiont blue-green 7
Thallus without, or with an indistinct blue-black hypothallus; photobiont green or blue-green 8

7(6) Thalline exciple absent; thallus Pd−; thallus plate-like .... ***Degelia***
Thalline exciple present; thallus Pd+ red; thallus with distinct lobes ... ***Pannaria rubiginosa***

8(6) Thallus of strap-shaped lobes 9
Thallus otherwise 10

9(8) Lower surface white, never black (see also *Ramalina lacera*), not isidiate .. ***Evernia prunastri***
Lower surface at least partly black, isidiate .... ***Pseudevernia furfuracea***

10(8) Thallus consisting of medium to large lobes, 0.5–4 cm wide 11
Thallus consisting of small lobes, <0.5 cm wide 28

11(10) Photobiont green 12
Photobiont blue-green 23

12(11) Thallus not vivid orange below, colour above various 13
Thallus vivid orange below, dark olive-green above when wet, dull grey-brown when dry .... ***Solorina crocea***

13(12) Cyphellae or pseudocyphellae present on lower surface 14
Cyphellae and pseudocyphellae absent on lower surface 15

14(13) Cyphellae present; medulla white; soredia absent .... ***Sticta canariensis*** (green algal morph)
Pseudocyphellae present; medulla yellow; soredia present .... ***Pseudocyphellaria aurata***

15(13) Lower surface with a ± well developed coarse network of white or brown veins 16
Lower surface without veins 17

16(15) Upper surface markedly ridged with distinct depressions, the ridges becoming sorediate or isidiate .... ***Lobaria pulmonaria***
Upper surface ± smooth, or with folioles or external cephalodia .... ***Peltigera***

17(15) Apothecia absent, emergent or sessile, or if immersed, then not in depressions; ascospores simple or septate, smooth, cephalodia internal or absent 18
Apothecia sunk in depressions in thallus, immarginate; ascospores 1-septate, ± finely warted; cephalodia internal or coralloid excrescences .... ***Solorina***

18(17) Lower surface of thallus not tomentose, with or without rhizines (see also *Umbilicaria*) .. 19
Lower surface of thallus finely tomentose, without rhizines (if montane and thallus yellow-green see *Nephroma arcticum*) .... ***Lobaria***

19(18) Lobes flat, or if convex, then solid; lower surface pale brown or black, usually smooth; rhizines present 20
Lobes convex, to ± tubular, hollow; lower surface (in part) black, wrinkled; rhizines absent .... ***Hypogymnia***

20(19) Lower surface ± naked or with a few, scattered rhizines; pseudocyphellae, if present, never sorediate 21
Lower surface ± densely rhizinose, at least towards centre; pseudocyphellae, if present, becoming sorediate .... ***Parmelia***

21(20) Upper surface without pseudocyphellae 22
Upper surface with numerous, minute, punctiform pseudocyphellae (×10); soralia marginal .... ***Cetrelia olivetorum***

22(21) Thallus uniformly brown when dry, greenish brown when wet, surface smooth .... ***Cetraria chlorophylla***
Thallus grey sometimes partly tinged brownish, ± unchanged in colour when wet, surface ± uneven to markedly ridged .... ***Platismatia***

23(11) Lower surface without veins 24
Lower surface with a ± well-developed network of veins .... ***Peltigera***

24(23) Cyphellae or pseudocyphellae present on lower surface 25
Cyphellae and pseudocyphellae absent 26

25(24) Cyphellae present .... ***Sticta***
Pseudocyphellae present .... ***Pseudocyphellaria***

26(24) Upper surface not as below, lower surface tomentose or not 27
Upper surface irregularly pitted, often with soredia along ridges, lower surface tomentose .... ***Lobaria scrobiculata***

27(26) Lower surface markedly tomentose; apothecia (if present) on upper surface of short marginal projections .... ***Peltigera***
Lower surface naked or very sparingly tomentose; apothecia (if present) on lower surface of recurved lobe ends .... ***Nephroma***

28(10) Photobiont blue-green 29
Photobiont green 30

29(28) Thalline exciple absent; thallus Pd−; thallus plate-like .... ***Degelia***
Thalline exciple present; thallus Pd+ red; thallus with distinct lobes ... ***Pannaria rubiginosa***

30(28) Rhizines and tomentum absent on lower surface; thallus attached by folds on lower cortex, or rarely attached to substratum by scattered, peg-like hapters; marginal cilia absent 31
Rhizines or tomentum present on lower or upper surface; marginal cilia occasionally present 35

31(30) Thallus ± hollow, or lobes ± terete, appearing ± inflated; medulla hollow or of ± loosely woven hyphae 32
Thallus solid, lobes flattened, not appearing terete, medulla compact, dense 34

32(31) Thallus larger, lower surface without indentations; soralia various 33
Thallus <3 cm diam.; lower surface with numerous, scattered indentations; soralia terminal .... ***Cavernularia hultenii***

33(32) Upper surface with scattered perforations .... ***Menegazzia terebrata***
Upper surface without perforations .... ***Hypogymnia***

34(31) Medulla KC+ red .... ***Allantoparmelia alpicola***
Medulla KC− .... ***Hypogymnia intestiniformis***

35(30) Lower surface not veined 36
Lower surface distinctly brown-veined .... ***Peltigera venosa***

36(35) Thallus shiny brown or shiny brown-black 37
Thallus not shiny brown or brown-black 38

37(36) Lower surface with numerous rhizines; lobe marginal projections absent .... ***Parmelia***
Lower surface with few rhizines; lobe marginal projections abundant .... ***Cetraria***

38(36) Upper cortex with hyphae running parallel to upper surface (microscope) 39
Upper cortex with a cellular structure (microscope) 40

39(38) Thallus K+ yellow; lobes grey-green to white .... ***Heterodermia***
Thallus K−; lobes pale to dark brown .... ***Anaptychia***

40(38) Thallus grey-white or brown-black, matt; apothecia black, matt, often pruinose; ascospores brown, 1-septate 41
Thallus yellow, grey or brown-black, shiny; apothecia brown to brown-black, often shiny; ascospores colourless, simple 44

41(40) Upper surface rarely pruinose; cortex K+ yellow or K− 42
Upper surface strongly pruinose; cortex at least in part, K− .... ***Physconia***

42(41) Thallus brown or grey-green, mostly greenish when wet; cortex K− 43
Thallus grey, ± unchanged when wet; cortex K+ yellow .... ***Physcia***

43(42) Lower cortex prosoplectenchymatous (lumina c. 2 μm wide), pale; lobes 0.2–0.5 mm wide, very adpressed and difficult to detach from substratum .... ***Hyperphyscia***
Lower cortex pseudoparenchymatous (lumina 3–7 μm wide), black, or if pale then thallus loosely attached and ± shrubby; lobes (0.05-)0.2–1.5 mm wide, mostly readily detached from substratum .... ***Phaeophyscia***

44(40) Lobes partly erect; thallus with a few, scattered rhizines towards centre 45
Lobes ± adpressed; thallus with rhizines almost to lobe margins 47

45(44) With marginal soredia and/or isidia; apothecia very rare 46
Without soredia and isidia; apothecia abundant .... ***Cetraria sepincola***

46(45) Upper and lower surface brown .... ***Cetraria***
Upper surface grey or brown with greyish parts; lower surface black, partly flecked brown or white .... ***Platismatia***

47(44) On acid bark and decorticated wood; lobes 0.5–1 mm wide 48
Saxicolous; lobes mostly >1 mm wide .... ***Parmelia***

48(47) Thallus sorediate; Pd−, UV+ white (divaricatic acid) .... ***Foraminella***
Thallus isidiate; Pd+ orange, UV− (thamnolic acid) .... ***Parmeliopsis aleurites***

Generic Key 2b

Squamulose/placodioid lichens (not yellow to orange-red)

Synopsis

1 Thallus and fruiting body various, never a mushroom 2
Thallus of pale green-grey, ± rounded squamules, becoming indented; fruiting body a yellow-white mushroom (basidioma) [see also *Baeomyces*] .... ***Omphalina hudsoniana***

2(1) Apothecia or perithecia present, or if absent then not as below; squamules variously shaped and coloured, not sorediate 3
Apothecia and perithecia absent (N.B. perithecia of the lichenicolous fungus *Lauderlindsaya borreri* may be present); squamules shell- or ear-like with raised, sorediate margins, glaucous to green-grey .... ***Normandina pulchella***

3(2) Perithecia present 4
Apothecia present, or fruits absent 7

7(3) Photobiont green 8
Photobiont blue-green 42

8(7) Thallus of adpressed or ascending, individual squamules, overlapping or ± dispersed, not forming radiating and elongated marginal lobes 9
Thallus ± forming rosettes or placodioid when ± crustose in centre with distinctly elongated radiating marginal lobes 29

9(8) Asci to 8-spored 10
Asci 50- to 300-spored .... ***Acarospora***

10(9) Thallus sorediate or with clusters of coarse, irregular isidia-like protuberances 11
Thallus without soredia and isidia 17

42(7) Thallus layered in section (heteromerous), not swelling when wet, blue-grey, brown or black 43
Thallus ± unlayered in section (homoiomerous), swelling when wet, black-green to black 49

Key 2b

1 Thallus and fruiting body various, never a mushroom 2
Thallus of pale green-grey, ± rounded squamules, becoming indented; fruiting body a yellow-white mushroom (basidioma) [see also *Baeomyces*] .... ***Omphalina hudsoniana***

2(1) Apothecia or perithecia present, or if absent then not as below; squamules variously shaped and coloured, not sorediate 3
Apothecia and perithecia absent (N.B. perithecia of the lichenicolous fungus *Lauderlindsaya borreri* may be present); squamules shell- or ear-like with raised, sorediate margins, glaucous to green-grey .... ***Normandina pulchella***

3(2) Perithecia present, mostly ± immersed; photobiont green 4
Apothecia present, emergent or sessile, or fruits absent; photobiont green or blue-green .. 7

4(3) Hymenial photobiont cells absent 5
Hymenial photobiont cells present .... ***Endocarpon***

5(4) Squamules small to medium, 0.5–4 (-7) mm wide; ascospores simple or septate, asci 8-spored 6
Squamules minute, 0.05–0.3 (-0.5) mm wide; ascospores muriform, asci 2- or 8-spored ***Agonimia***

6(5) Ascospores simple .... ***Catapyrenium***
Ascospores 1-septate .... ***Placidiopsis***

7(3) Photobiont green; thallus white, grey, green, yellow-green or brown, not gelatinous or swelling when wet 8
Photobiont blue-green; thallus blue-grey, brown, black-green, blue-black or black, sometimes gelatinous and swelling when wet 42

8(7) Thallus of adpressed or ascending, individual squamules, overlapping or ± dispersed, not forming radiating and elongated marginal lobes 9
Thallus ± forming rosettes or placodioid when ± crustose in centre with distinctly elongated radiating marginal lobes 29

9(8) Asci to 8-spored; ascospore size various; or thallus sterile 10
Asci 50- to 300-spored; ascospores minute, 3–6(-13) × 1–3(-6) μm .... ***Acarospora***

10(9) Thallus sorediate or with clusters of coarse, irregular, isidia-like protuberances 11
Thallus without soredia and isidia 17

11(10) Squamules ± closely adpressed to substratum, mostly rounded, with ± entire margin; podetia absent (see also *Baeomyces rufus*) 12

Squamules ± ascending, at least at edges, often ± elongate, margins frequently incised; small podetia frequently present .... ***Cladonia***

12(11) Squamules C+ red 13
Squamules C− 14

13(12) Squamules orientated in one direction, overlapping, the margin and lower surface sorediate .... ***Hypocenomyce scalaris***
Squamules irregular, clusters of coarse, irregular, isidia-like protuberances present on upper surface .... ***Trapeliopsis wallrothii***

14(12) Squamules not peltate, without a darker central area; ascospores simple or 1-septate 15
Squamules peltate, with a conspicuously darker central area; ascospores (1-)3- to 5-septate .... ***Stereocaulon***

15(14) Soredia lip-shaped, ± restricted to marginal soralia; ascospores simple; on peat or wood 16
Thallus ± uniformly covered in coarsely granular soredia; ascospores 1-septate; on coastal soil .... ***Solenopsora vulturiensis***

16(15) Cortex Pd+ red; on coniferous wood .... ***Hypocenomyce anthracophila***
Cortex Pd−; on peat or rotting wood .... ***Trapeliopsis***

17(10) Squamules ± closely adpressed to substratum, mostly rounded, with ± entire margin; podetia absent 18
Squamules ± ascending, at least at edges, often ± elongate, margins frequently incised; small podetia frequently present .... ***Cladonia***

18(17) Thallus on wood or bark 19
Thallus on rocks or soil 21

19(18) Squamules brown, green-brown or yellow-brown 20
Squamules greenish white, grey-green .... ***Cladonia***

20(19) Squamules very small, to 0.5 mm, granular, shiny; apothecia dark brown, ascospores simple .... ***Protoparmelia***
Squamules mostly >0.5 mm, bullate or flattened, matt or shiny; apothecia black; ascospores simple to 1(-3)-septate .... ***Hypocenomyce***

21(18) Squamules grey-green-brown; apothecia varied, ascospores not acicular and multiseptate 22
Squamules bright yellow-green; apothecia infrequent, black; ascospores acicular, multiseptate .... ***Arthrorhaphis alpina***

22(21) Apothecia emergent, or sessile, never stalked; squamules varied 23
Apothecia on short, pale stalks; squamules minute, forming a ± continuous crust ***Baeomyces***

23(22) Apothecia without a thalline margin 24
Apothecia with a thalline margin 27

24(23) Ascospores simple 25
Ascospores 1- to 5-septate or muriform 26

25(24) Apothecia brown-black; on soil amongst limestone rocks .... ***Psora***
Apothecia dark red-brown; on acid soil, montane .... ***Lecidoma demissum***

26(24) Ascospores pale brown, muriform .... ***Lopadium***
Ascospores colourless, 1- to 5-septate .... ***Toninia***

27(23) Thalline margin smooth; thallus with adpressed, ± overlapping squamules 28
Thalline margin ± granular-crenulate, thallus minutely granular-squamulose; cephalodia present, at times inconspicuous .... ***Psoroma hypnorum***

28(27) Squamules yellow-grey or grey-green; ascospores simple .... ***Squamarina***
Squamules grey-brown; ascospores 1-septate .... ***Solenopsora holophaea***

29(8) Squamules bright yellow-green 30
Squamules grey-white, grey-green or grey-brown 31

30(29) Marginal lobes distinctly elongate; surface smooth, ± shiny; ascospores brown, 1-septate .... ***Catolechia wahlenbergii***
Marginal lobes indistinct, surface ± matt; ascospores colourless, multiseptate .... ***Arthrorhaphis alpina***

31(29) Thallus sorediate or with irregular, isidia-like protuberances 32
Thallus not sorediate, without irregular, isidia-like protuberances 37

32(31) Cephalodia absent 33
Pink-grey cephalodia present on upper surface .... ***Placopsis gelida***

33(32) Thallus C+ red 34
Thallus C− 35

34(33) Soralia developing from the sides of areoles or cracks in thallus .... ***Trapelia placodioides***
Thallus with clusters of coarse, irregular, isidia-like protuberances .... ***Trapeliopsis wallrothii***

35(33) Soralia laminal; on rocks or bark 36
Soralia lip-shaped on lobe ends; on peaty turf or wood .... ***Trapeliopsis***

36(35) Thallus K+ yellow, thallus pruinose, with distinct lobes closely contiguous for most of their length; apothecia without a thalline exciple .... ***Diploicia canescens***
Thallus K−; thallus non-pruinose, with minute, ± discrete lobes; apothecia with a thalline exciple .... ***Hyperphyscia adglutinata***

37(31) On soil 38
On rocks 40

38(37) Thallus ± white, or white-pruinose; without papillae; apothecia frequent, sessile; on calcareous soil 39
Thallus green when wet, pale brown when dry; flattened papillae present towards centre; apothecia rare, ± stalked; on acid soil .... ***Baeomyces placophyllus***

39(38) Apothecia without a thalline exciple; ascospores brown, 1-septate .... ***Buellia epigaea***
Apothecia with a thalline exciple; ascospores colourless, simple .... ***Squamarina lentigera***

40(37) Thallus squamulose, apothecia sessile; asci <8-spored; thallus C− 41
Thallus ± crustose with elongated marginal lobes; apothecia immersed; asci >200-spored; thallus C+ red .... ***Sporastatia***

41(40) Ascospores 1-septate; on limestone and coastal serpentine .... ***Solenopsora***
Ascospores simple; on range of substrata, especially nutrient enriched rocks and building materials .... ***Lecanora muralis***

42(7) Thallus layered in section (heteromerous), not swelling when wet, blue-grey, brown or black 43
Thallus ± unlayered in section (homoiomerous), swelling when wet, black-green to black 49

43(42) Thallus thick, not papery, margins never lacerate, a blue-black hypothallus sometimes present; apothecia with or without a thalline margin; ascospores simple or septate 44
Thallus thin, papery, often ± wrinkled, margins sometimes lacerate; hypothallus absent, smooth or finely white-tomentose below; apothecia with a thalline exciple; ascospores muriform .... ***Leptogium***

44(43) Apothecia with a thalline margin or else thallus sterile and soredia, isidia or nodules present 45
Apothecia, when present, always without a thalline margin; soredia and isidia present or absent 46

45(44) Photobiont *Nostoc*; asci to 8-spored; without remaining combination of characters below (when sterile see also *Parmeliella*) .... ***Pannaria***

Photobiont *Scytonema*; asci (8-)12- to 16-spored; thallus lobes stellate-radiating with coarse, ± flattened nodules; lower surface pale (if lower surface black see *Placynthium pannariellum*) .... ***Vestergrenopsis elaeina***

46(44) Thallus squamulose; without remaining combination of characters below 47
Thallus placodioid, blue-grey with concentric, shell-like markings on upper surface; lower surface with felt-like tomentum; ascospores simple .... ***Degelia***

47(46) Thallus rosette-forming, small-lobulate at the centre with radiating marginal lobes or ± entirely small squamulose, blue-black when moist, rhizines absent but often with a well-developed dark prothallus; ascospores simple or 1- to 3-septate 48
Thallus of red-brown, ± overlapping squamules, dull dark green when moist; underside pale with a few concolorous rhizines; apothecia frequent, ascospores 1(-2)-septate .... ***Massalongia carnosa***

48(47) On siliceous rocks in streams, lake sides or in persistent seepage tracks; photobiont *Scytonema* or *Rivularia*; isidia coralloid or ± flattened; apothecia infrequent; ascospores septate .... ***Placynthium***
On tree trunks, sheltered and often mossy rocks or soil; photobiont *Nostoc*; often with isidia or granular soralia, apothecia rare; ascospores simple (when sterile see also *Pannaria*) .... ***Parmeliella***

49(42) Ascospores simple; apothecia ± deeply immersed, inconspicuous .... ***Lempholemma***
Ascospores septate or muriform; apothecia usually conspicuous .... ***Leptogium***

Generic Key 3

Crustose lichens with perithecia

Synopsis

1 Asci 30- to 300-spored 2
Asci to 8-spored 5

5(1) Perithecia combined in a receptacle, with a common cover (involucrellum), in which there are 3–15 minute pores; ascospores ellipsoid, ovoid-oblong or sole-shaped, simple to 5-septate 6
Perithecia usually single, or if united ascospores thread-like, 7- to 9-septate 8

8(5) Ascospores muriform 9
Ascospores simple or transversely septate (rarely with 1–2 longitudinal septa) 18

18(8) Ascospores simple 19
Ascospores 1-septate to transversely septate 22

Key 3

1 Asci 30- to 300-spored 2
Asci to 8-spored 5

2(1) Ascospores simple 3
Ascospores 1- to 3-septate 4

3(2) Perithecia mostly bright yellow-green pruinose, or dull yellowish to pale brown; asci 50- to 300-spored .... ***Thelocarpon***
Perithecia black, not yellow-green pruinose; ascospores 6–9 μm long; asci *c.* 65-spored .... ***Trimmatothele perquisita***

4(2) Photobiont chlorococcoid; ascospores 1-septate .... ***Epigloea grummannii***
Photobiont *Trentepolia*; ascospores 1-(3-) septate .... ***Thelopsis***

5(1) Perithecia combined in a receptacle, with a common cover (involucrellum), in which there are 3–15 minute pores; ascospores ellipsoid, ovoid-oblong or sole-shaped, simple to 5-septate 6
Perithecia usually single, or if united ascospores thread-like and 7- to 9-septate 8

6(5) Ascospores septate 7
Ascospores simple, >30 μm long .... ***Pertusaria***

7(6) Ascospores without longitudinal divisons when mature, 1- or 3-septate .... ***Tomasellia***
Ascospores with a few longitudinal divisions, 3- to 5-septate .... ***Mycoporum***

8(5) Ascospores muriform 9
Ascospores simple or transversely septate (rarely with 1–2 longitudinal septa) 18

9(8) Paraphyses persistent 10
Paraphyses soon becoming gelatinized and indistinct 16

10(9) Perithecia not inclined, with joined ostioles; photobiont various 11
Perithecia inclined, with ± joined ostioles; photobiont *Trentepohlia* .... ***Parmentaria chilensis***

11(10) Ascospores ellipsoid with ± rounded ends; thallus never wide-spreading, often effuse ... 12
Ascospores elongate with attenuated ends; thallus forming extensive pink-orange crust on vertical surfaces .... ***Belonia nidarosiensis***

12(11) Involucrellum present 13
Involucrellum absent 14

13(12) Ascomatal wall greenish brown pigments N−; ascospores with gelatinous perispore .... ***Julella sericea***
Ascomatal wall bright green pigments N+ red; ascospores without gelatinous perispore .... ***Mycoglaena acuminans***

14(12) Ascus apex I−; photobiont chlorococcoid 15
Ascus apex I+ blue; photobiont *Elliptochloris* .... ***Protothelenella***

15(14) Periphysoids present; ascus with an indentation in apex .... ***Chromatochlamys***
Periphysoids absent; ascus apex without an indentation .... ***Thelenella modesta***

16(9) Hymenial algal cells absent 17
Hymenial algal cells present, small, globose or cuboid .... ***Staurothele***

17(16) Perithecia colourless to pale .... ***Leucocarpia biatorella***
Perithecia dark brown to black .... ***Polyblastia***

18(8) Ascospores simple 19
Ascospores 1-septate to transversely septate 22

19(18) Paraphyses soon becoming gelatinized and indistinct 20
Paraphyses persistent (if soredia reacting bright golden yellow present see *Coccotrema citrinescens*) .... ***Thrombium epigaeum***

20(19) Ascospores ellipsoid to globose, ... 21
Ascospores thread-like with clavate apices, often twisted in the middle . ***Sarcopyrenia gibba***

21(20) Ascomatal wall dark brown to black, often dense and brittle .... ***Verrucaria***
Ascomatal wall colourless to pale orange, never dense and brittle ... ***Leucocarpopsis devensis***

22(18) Paraphyses soon becoming gelatinized and indistinct 23
Paraphyses persistent 26

23(22) Ascomatal wall dark brown, at least in upper part 24
Ascomatal wall pale .... ***Macentina***

24(23) Ascospores 1- to 3-septate 25
Ascospores (5-) 7-septate .... ***Lauderlindsaya***
(see page 397)

25(24) On smooth bark .... ***Arthopyrenia salicis***
On rock or soil .... ***Thelidium***

26(22) Photobiont not blue-green 27
Photobiont blue-green .... ***Pyrenocollema***

27(26) Ascospores dark brown 28
Ascospores colourless or pale brown 30

28(27) Ascospores 3- or more septate; perithecial wall K−; photobiont *Trentepohlia* or absent . 29
Ascospores 1-septate; perithecial wall K+ green; photobiont absent or else loosely associated .... ***Mycomicrothelia***

29(28) Ascospore walls markedly thickened with ± lens-shaped inner compartments; brown macroconidia absent .... ***Pyrenula***
Ascospore walls slightly thickened with inner compartments not lens-shaped; brown macroconidia often present .... ***Eopyrenula***

30(27) Ascospores >3-septate 31
Ascospores 1- to 3-septate 37

31(30) Hamathecium of branched paraphysoids 32
Hamathecium of simple paraphyses 33

32(31) Ascomata arising singly, hymenial gel I+ deep blue .... ***Rhaphidicyrtis***
Ascomata arising singly or commonly in small groups; hymenial gel I− .... ***Celothelium ischnobelum***

33(31) Photobiont *Trentepohlia*; substratum varied, if on soil ascospores <80 μm long 34
Photobiont chlorococcoid; on soil, mosses or algal mats; ascospores >120 μm long .... ***Belonia incarnata***

34(33) Thallus without isidia; ascospores varied 35
Thallus isidiate; ascospores (3-)5- to 17-septate .... ***Zamenhofia***

35(34) Ascus apex without apical dome 36
Ascus apex with a conspicuous apical dome .... ***Strigula stigmatella***

36(35) Ascospores 15- to 20-septate .... ***Belonia***
Ascospores to 14- septate .... ***Porina***

37(30) Hamathecium of branched and anastomosing pseudoparaphyses or paraphysoids; photobiont *Trentepohlia* or doubtfully lichenized 38
Hamathecium of simple paraphyses; photobiont chlorococcoid or *Trentepohlia* 42

38(37) Ascospores not uniseriate; if 1-septate, one cell smaller, shape varied; without warted epispore 39
Ascospores uniseriate; 1-septate, both ends of equal size; ellipsoid, with warted epispore .... ***Acrocordia***

39(38) Perithecial wall brown-black (water mounts); pigments N− 40
Perithecial wall blue-green to dark green (water mounts); pigments N+ red .... ***Mycoglaena myricae***

40(39) Ascospores ovoid, oblong, clavate-fusiform or fusiform 41
Ascospores acicular-fusiform, arcuate or sigmoid .... ***Leptorhaphis***

41(40) Perithecial wall cellular, not containing bark cells; asci cylindrical-clavate with broad ocular chamber; ascospores without perispore; paraphysoids long-celled, slender .... ***Anisomeridium***
Perithecial wall hyphal, containing bark cells; asci obpyriform to obclavate, **or** if cylindrical-clavate, then ascospores with distinct perispore; pseudoparaphyses short- or long-celled .... ***Arthopyrenia***

42(37) Photobiont *Trentepohlia* 43
Photobiont chlorococcoid .... ***Epigloea***

43(42) Thallus without isidia 44
Thallus with isidia .... ***Zamenhofia rosei***

44(43) Ascus apex with narrow apex without apical dome .... ***Porina***
Ascus apex with a conspicuous apical dome penetrated by a narrow ocular chamber ***Strigula***

Generic Key 4

Crustose lichens with stalked apothecia

1 Apothecia with a dry spore mass (asci disintegrating early) 2
Apothecia without a dry spore mass (asci persistent) 6

2(1) Ascospore mass black; ascospore walls dark brown 3
Ascospore mass greenish or brown; ascospore walls colourless, greenish or brown 4

3(2) Ascospores 1-septate .... ***Calicium***
Ascospores simple .... ***Sphinctrina***

4(2) Ascospore mass brown or pinkish; ascospore walls colourless or pale brown; ornamentation varied 5
Ascospore mass greenish; ascospore walls greenish with spirally arranged ridges ***Microcalicium***

5(4) Ascospores globose with colourless walls and warted, ± cog-like ornamentation; thallus immersed with *Trentepohlia* photobiont .... ***Sclerophora***
Ascospores globose, ellipsoid or oblong with pale brown walls, ornamentation varied; thallus immersed or usually superficial; photobiont chlorococcoid, *Stichococcus* or *Trentepohlia*. .... ***Chaenotheca***

6(1) Ascospores brown; mainly on twigs, bark, wood, or on lichens 7
Ascospores colourless; on rock, soil, peat or rotting wood or debris 11

7(6) Ascospores simple to 1-septate at maturity 8
Ascospores 3-septate at maturity .... ***Stenocybe***

8(7) Substratum varied, not on living twigs or small branches 9
On living twigs and small branches of deciduous trees (especially *Populus*) ... ***Phaeocalicium***

9(8) Ascospores simple to 1-septate, oblong to ellipsoid, < 4.5 μm wide 10
Ascospores simple, ± globose, or ellipsoid with pointed apices and > 4.5 μm wide ***Sphinctrina***

10(9) Ascus apex unevenly thickened; a canal penetrating the ascus apex visible at least in semi-mature stages; ascospores simple or 1-septate .... ***Chaenothecopsis***
Ascus apex unevenly thickened; ascospores simple .... ***Mycocalicium subtile***

11(6) Apothecia borne on tips of rarely branched, solid stalks 12
Apothecia borne on tips of simple or branched, hollow, turgid stalks . ***Pycnothelia papillaria***

12(11) Apothecia pink or brown .... ***Baeomyces***
Apothecia black .... ***Pilophorus strumaticus***

Generic Key 5

Crustose lichens with lirellate apothecia

1 Apothecia lacking true exciple or rarely rudimentary and developed only laterally; photobiont *Trentepohlia*, less often chlorococcoid 2
True exciple present; photobiont various 4

2(1) Apothecia single, at least not combined in a carbonaceous stromata 3
Apothecia crowded, immersed together in round-irregular, black, often pruinose stromata .... ***Chiodecton***

3(2) Ascospores 1- to 7-septate, ovoid to oblong-ovoid .... ***Arthonia***
Ascospores submuriform to strongly muriform, ovoid-ellipsoid .... ***Arthothelium***

4(1) Ascospores simple 5
Ascospores septate or muriform 8

5(4) Asci 8-spored; photobiont chlorococcoid; periphysoid filaments absent 6
Asci 100- to 200-spored; photobiont *Trentepohlia*; exciple with periphysoid filaments **Wadeana**

6(5) Outer surface of apothecium black and brittle; hamathecium various 7
Outer surface of apothecium pale or brownish; paraphyses simple or sparingly branched ..... **Xylographa**

7(6) On wood, rarely bark; thallus K− or K+ yellow; paraphyses simple or sparingly branched, persistent ..... **Ptychographa**
On rock; thallus K+ yellow to red; paraphyses branched, soon disappearing ..... **Lithographa tesserata**

8(4) Ascospores 1-septate 9
Ascospores multiseptate or muriform 10

9(8) Asci with K/I− apical dome; hypothecium colourless to brown; ascospores colourless becoming brown, N−; photobiont *Trentepohlia*, chlorococcoid or absent ..... **Melaspilea**
Asci with K/I+ blue apical dome (*Rhizocarpon*-type); hypothecium dark reddish brown; ascospores colourless becoming grey-green to violet-black, N+ red; photobiont chlorococcoid ..... **Poeltinula cerebrina**

10(8) Ascospores I+ violet, thick-walled, their cells with lens-shaped, round or oval lumina; hamathecium of simple paraphyses 11
Ascospores I−, with cylindrical or cuboid lumina; hamathecium of branched and anastomosing paraphysoids 13

11(10) Ascospores multiseptate, not muriform, colourless or not 12
Ascospores muriform, colourless ..... **Graphina**

12(11) Ascospores distinctly grey-brown when mature, but often remaining colourless for some time; apothecia flat with inconspicuous, scarcely raised margins and widely expanded discs ..... **Phaeographis**
Ascospores not coloured when mature, occasionally becoming brownish when over mature; apothecia with surfaces becoming longitudinally grooved or ridged and narrow discs ..... **Graphis**

13(10) True exciple black and friable 14
True exciple thin, brown or colourless 15

14(13) Apothecia surrounded by white thalline exciple; true exciple not prominent; ascospores 3-septate ..... **Schismatomma graphidioides**
Apothecia not surrounded by white thalline margin; true exciple usually prominent from the beginning; ascospores to 15-septate ..... **Opegrapha**

15(13) Ascospores colourless ..... **Enterographa**
Ascospores black-grey to dark brown ..... **Sclerophyton circumscriptum**

Generic Key 6

Crustose lichens with rounded apothecia

Synopsis

1 Photobiont green 2
Photobiont blue-green ..... **Key 6f** (p. 41)

2(1) Photobiont chlorococcoid 3
Photobiont *Trentepohlia* ..... **Key 6e** (p. 40)

3(2) Ascospores colourless 4
Ascospores brown, grey or black when mature ..... **Key 6d** (p. 38)

4(3) Ascospores simple 5
Ascospores septate or muriform ..... **Key 6c** (p. 36)

5(4) Asci >8-spored ..... **Key 6b** (p. 35)
Asci to 8-spored ..... **Key 6a** (p. 30)

Generic Key 6a

Crustose, apothecia rounded/photobiont chlorococcoid/ascospores colourless, simple/asci to 8-spored

1 Apothecia yellow to deep red and K+ purple (sometimes only apparent in sections) 2
Apothecia not yellow to deep red and K+ purple 5

2(1) Thalline exciple of apothecia absent 3
Thalline exciple of apothecia present 4

3(2) Thallus entirely finely granular-sorediate, C+ orange; apothecia dark purplish red ..... ***Pyrrhospora quernea***
Thallus smooth and continuous or ± evanescent, never sorediate, C−; apothecia yellow to orange-red ..... ***Protoblastenia***

4(2) Thallus yellow-orange, rather thick, granular-squamulose ..... ***Fulgensia***
Thallus grey, thin, minutely granular to film-like ..... ***Caloplaca nivalis***

5(1) Apothecia and thallus pale to deep yellow-orange, sometimes greenish 6
Apothecia and thallus not as above 7

6(5) Thallus entirely finely granular sorediate; apothecia convex without a thalline exciple, ascospores to 6 μm, tear-shaped ..... ***Psilolechia lucida***
Thallus granular; apothecia ± flat with a thalline exciple; ascospores >10 μm, ellipsoid ..... ***Candelariella***

7(5) Ascospores very large, >(30-)40 μm; asci often <8-spored 8
Ascospores small to medium, <30 μm; asci mostly 8-spored 13

8(7) Apothecia immersed in thallus, or sessile with well-developed true or thalline exciple; if sessile and immarginate then internally with greenish, violet, or purplish N+ red pigments; substrata various 9
Apothecia sessile, immarginate; thalline exciple absent; true exciple poorly developed, appearing ± pseudoparenchymatous at base of apothecium; internal pigments brownish, N−; on the ground ..... ***Aphanopsis coenosa***

9(8) Apothecia with a thalline exciple or immersed in thallus and bordered by areoles 10
Apothecia without a thalline exciple, sessile ..... ***Mycoblastus***

10(9) Apothecia without combination of characters below 11
Apothecia sessile, always single with a widely expanded disc and thick thalline exciple; ascospores thin-walled ..... ***Ochrolechia***

11(10) Cephalodia absent 12
Cephalodia present ..... ***Coccotrema citrinescens***

12(11) Thallus on mosses and decaying plants in calcareous environments, K−, KC−, Pd−; disc grey-black; epithecium K+ brownish; ascospores 30–60 μm long ***Megaspora verrucosa***
Thallus on bark, wood or rock, or if on mosses and decaying plant remains then epithecium K+ violet and thallus with a positive reaction to K, KC or Pd, ***or*** ascospores >65 μm long ..... ***Pertusaria***

13(7) Mature apothecia sessile with raised thalline exciple (in sections containing algae) 14
Mature apothecia immersed or sessile without a raised thalline exciple 20

14(13) Ascospores without a warted epispore 15
Ascospores with a distinctly warted epispore (×400) ..... ***Psoroma***

15(14) Hymenium colourless or variously pigmented in uppermost part (epithecium) 16
Hymenium purple-red or purple-violet throughout .... ***Tephromela***

16(15) Epithecium colourless or differently coloured, if greenish then not K+ violet 17
Epithecium dull greenish, K+ violet .... ***Pertusaria***

17(16) Ascus in K/I without dark blue axial tube (*Bacidia*-, *Lecanora*- or *Trapelia*-type) 18
Ascus in K/I with dark blue axial tube (*Porpidia*-type) .... ***Koerberiella wimmeriana***

18(17) Thallus C−, or if C+ red then cephalodia absent and ascus in K/I with dark blue apical dome 19
Thallus C+ red; reddish brown, wart-like cephalodia often present; asci in K/I with colourless or weakly blued apical dome .... ***Placopsis***

19(18) Cortex of thallus, thalline exciple and epithecium brown, K−; apothecia not pruinose; paraphyses mostly simple, with gelatinous apical hood that usually contains a dark brown cap; ascospores fusiform or narrowly ellipsoid, 7–13 × 2–5 μm; conidia bacilliform to short-acicular; atranorin absent .... ***Protoparmelia***
Without above combination of characters; atranorin often present .... ***Lecanora***

20(13) Apothecia urceolate or aspicilioid; disc concave to flat, at or below the level of the thallus surface; exciple thin or inconspicuous; on rocks (rarely worked timber) 21
Apothecia emergent to sessile; disc flat to convex at maturity, raised above level of thallus surface; exciple distinct or not; substratum various 40

21(20) Medulla I−, if I+ blue asci not of *Porpidia*-type, or exciple pigmented in section; ascospores without amyloid perispore 22
Medulla I+ blue; asci *Porpidia*-type; exciple thin or inconspicuous, colourless in section; ascospores ± with amyloid perispore .... ***Bellemerea***

22(21) Apothecial disc whitish to pale red-brown or pinkish, not pruinose 23
Apothecial disc black-brown, pruinose or not 24

23(22) Apothecia 0.1–0.4(-0.6) mm diam.; ascospores broadly ellipsoid to globose; paraphyses ± swollen at apices; on siliceous rocks or hard limestones .... ***Hymenelia***
Apothecia 0.1–0.2(-0.35) mm diam.; ascospores elongate-ellipsoid; paraphyses not swollen at apices; on chalk stones .... ***Lecidea lichenicola***

24(22) Apothecia > 0.3 mm diam.; if smaller then without following combination of characters 25
Apothecia 0.15–0.3 mm diam.; exciple distinct only in upper part, dark green; hypothecium colourless; ascus in K/I with uniformly blue apical dome; thallus scurfy to rimose, never distinctly areolate; on calcareous rocks or stones .... ***Eiglera flavida***

25(24) Epithecium N+ red, or if N− then without following combination of characters 26
Epithecium brownish to green, N− or N+ green; asci in K/I with colourless apical dome; exciple very thin, colourless or pale; ascospores broadly ellipsoid, 12–30 × 6–27 μm .... ***Aspicilia***

26(25) Asci not *Teloschistes*-type; apothecial pigments green or purplish in water, K or N, or paraphyses branched and coherent in K 27
Asci *Teloschistes*-type; apothecial and cortical pigments brown, K− or K+ yellowish brown, N−; paraphyses simple, separating in K .... ***Fuscidea***

27(26) Thallus and apothecia not on limestone, or if on limestone then thallus superficial or epithecium greenish (at least in K), N+ red 28
Thallus and apothecia immersed in limestone; epithecium brownish, N−; hypothecium colourless to pale reddish- or orange-brown .... ***Clauzadea***

28(27) Paraphyses and asci coherent and only slightly separating in K; asci with an apical dome 29
Paraphyses and asci lax and readily separating in K; asci without an apical dome ***Schaereria***

29(28) Asci *Porpidia*-type; apothecia mostly >0.5 mm diam. at maturity; ascospores mostly >6 μm wide 30
Asci not *Porpidia*-type; apothecia often <0.5 mm diam.; ascospores <6 μm wide 32

30(29) Medulla I+ blue (in section) 31
Medulla I− (in section) .... ***Porpidia***

31(30) Ascospores mostly >12 μm wide; cephalodia present or not; cortex C+ red or C− .... ***Amygdalaria***
Ascospores 6–10(-12) μm wide; cephalodia absent; cortex C− .... ***Porpidia***

32(29) Apothecial disc flat at maturity; asci in K/I apical dome (at least in uppermost part) blued; thallus rust-coloured or not, C+ red or C−; medulla reactions with Pd and K various 33
Apothecial disc mostly persistently concave; asci in K/I with colourless apical dome; thallus usually rust-coloured (very rarely dark grey), C−; medulla K−, Pd− .... ***Tremolecia atrata***

33(32) Thallus C+ red 34
Thallus C− 35

34(33) Apothecia 0.5–2(-3) mm diam.; disc flat to slightly convex; paraphyses mostly simple below but sparingly branched above .... ***Lecidea***
Apothecia 0.2–0.4(-0.5) mm diam.; disc concave or umbonate; paraphyses richly branched throughout .... ***Mosigia***

35(33) Ascospores mostly >9 μm long, or if shorter then without remaining combination of characters 36
Ascospores (4-)5–8(-9) × 3–5 μm; apothecia 0.2–0.4 mm diam.; exciple and hypothecium blackish, K+ red or purple; paraphyses simple, moniliform above; lichen products absent (t.l.c.) .... ***Claurouxia chalybeioides***

36(35) Hypothecium colourless 37
Hypothecium brown .... ***Lecidea***

37(36) Prothallus distinct, black, visible between areoles 38
Prothallus absent or visible only as a narrow line bordering the thallus 39

38(37) Areoles dark red-brown; paraphyses richly branched; ascospores mostly 8–10 × 3.5–5 μm; argopsin present .... ***Clauzadeana macula***
Areoles pale- to brown-grey; paraphyses sparingly branched; ascospores mostly 10–16 × 5–6 μm; miriquidic acid present .... ***Miriquidica***

39(37) Asci *Lecanora*-type; exciple containing algal cells .... ***Lecanora***
Asci *Lecidea*-type; exciple lacking algal cells .... ***Lecidea***

40(20) Apothecia dark coloured and ± marginate, or if pale and immarginate then either tissues bound by a gelatinous matrix, or asci with a distinct apical dome 41
Apothecia grey-white to pale brownish, convex, immarginate; tissues not bound by a gelatinous matrix, devoid of pigment; exciple absent; asci with thick, K/I+ blue walls with an apical pore (apical dome absent) .... ***Vezdaea***

41(40) Apothecia not on leafy liverworts, or if so then lichenized thallus conspicuous; ascus in K/I with or without colourless apical dome 42
Apothecia growing on leafy liverworts; thallus inconspicuous; ascus in K/I with colourless apical dome .... ***Mniacea***

42(41) Apothecia convex, pink, orange-red or red-brown; thallus Pd+ deep yellow, orange or red 43
Apothecia darker coloured, or if similar then thallus Pd− 45

43(42) Thallus warted or with smooth granules; Pd+ yellow or orange-red; on the ground, on rocks or on rotting wood or bark 44
Thallus of pubescent granules, Pd+ red; on trunks of old trees .... ***Phyllopsora rosei***

44(43) Apothecia constricted below, ± stalked; asci in K/I colourless or with a thin blue cap in the apex .... **Baeomyces**
Apothecia not or only slightly constricted below, adnate; asci in K/I with blue outer coat and amyloid apical dome (*Biatora*-type) .... **Biatora**

45(42) Asci with a distinct apical dome; or paraphyses coherent and only slightly separating in K 46
Asci without an apical dome; paraphyses and asci lax and readily separating in K **Schaereria**

46(45) Asci in K/I with colourless or weakly blued apical dome (± *Trapelia*-type) 47
Asci in K/I with a blued apical dome (i.e. *Bacidia*-, *Biatora*-, *Candelariella*-, *Catillaria*-, *Lecidea*-, *Porpidia*- and *Teloschistes*-types) 52

47(46) Not on other lichens 48
On other lichens, independent thallus inconspicuous .... **Lecidea**

48(47) Paraphyses not markedly swollen at apices, without dark brown caps; exciple of colourless or only weakly pigmented, short-celled, interwoven hyphae 49
Paraphyses with swollen apices, each with a dark brown cap; exciple of brown-walled pseudoparenchymatous hyphae .... **Placynthiella**

49(48) Ascospores 7–14 μm long 50
Ascospores 15–25 μm long .... **Trapelia**

50(49) Thallus lichenized; substrata various 51
Thallus not lichenized; on wood or bark .... **Agyrium rufum**

51(50) Apothecia 0.07–0.2 mm diam. .... **Trapelia**
Apothecia >0.2 mm diam. .... **Trapeliopsis**

52(46) Apothecial disc not umbonate or gyrose, or if so then ascospores <15 μm long 53
Apothecial disc umbonate or gyrose, black; ascospores *c.* 18–30 × 10–18 μm ..... **Rimularia**

53(52) Asci not *Teloschistes*-type; greenish or purplish apothecial pigments evident in water, K or N, and/or paraphyses branched and coherent in K 54
Asci *Teloschistes*-type; cortical and apothecial pigments brown, K− or K+ yellow-brown, N−; paraphyses mostly simple and lax in K .... **Fuscidea**

54(53) Exciple absent or poorly developed and then of paraphysis-like hyphae that ± separate in K; apothecia immarginate and convex from the beginning; epithecium without K+ dissolving granules 55
Exciple usually well-developed, and of hyphae that are coherent in K; apothecium usually marginate when young; epithecium sometimes with K+ dissolving granules .. 56

55(54) Ascospores tear-shaped, 3–7 × 1–2 μm; apothecia usually with whitish rim of protruding excipular hyphae at base; pycnidia absent, anamorph if present hyphomycetous **Psilolechia**
Ascospores ellipsoid or ovoid, if tear-shaped then mostly exceeding 7 × 2 μm; apothecial base without white rim of protruding hyphae; pycnidia often present [Note: a few species of *Lecidea* s. lat. may key out here] .... **Micarea**

56(54) Asci in K/I with pale blue apical dome and contrasting dark blue axial tube (*Porpidia*-type) 57
Asci in K/I without a distinct dark blue axial tube 61

57(56) On non-calcareous rocks, or if on calcareous substrata then without the combination of the characters below 58
On calcareous rock or stonework; thallus immersed or thin; apothecial pigment brown, K−, N−; hymenium and hypothecium without scattered violaceous (K+ green) granules; medulla I− .... **Clauzadea**

58(57) Epithecium brownish to green, N± reddish; paraphyses branched and anastomosing; hymenium and hypothecium without violaceous pigment granules; thallus often with a well-defined medulla; lichen products usually detected by t.l.c. 59
Epithecium colourless or brown, K−, N−; paraphyses mostly simple, occasionally sparingly branched above; hymenium and hypothecium often with scattered,

violaceous (K+ green) granules; thallus usually without a well-defined medulla; lichen products not detected by t.l.c. .... ***Lecidea*** (*L. hypnorum* group)

59(58) Hymenium >60 μm tall; apothecia 0.3–1 mm or more diam. 60
Hymenium <55 μm tall; apothecia mostly <0.5 mm diam. .... ***Lecidea*** s. lat.

60(59) On limestone, thallus immersed or thinly superficial; lichen products not detected by t.l.c. .... ***Farnoldia***
On acid substrata, or if on limestone then thallus superficial and lichen products detected by t.l.c. .... ***Porpidia***

61(56) Ascospores <17 μm long, or if longer then thallus not on rocks 62
Ascospores (16-)18–22(-25) μm long; on rocks .... ***Herteliana***

62(61) Apothecia dark coloured, or if pale then asci *Lecanora*-type and apothecia often with K+ dissolving granules 63
Apothecia whitish to pale red-brown; asci *Biatora*-type; epithecium without K+ dissolving granules .... ***Biatora***

63(62) True exciple without yellow-brown, K+ magenta pigment [dark brownish, K+ purple pigments sometimes present] 64
True exciple dark brown at outer edge, internally mottled yellow-brown, K+ magenta into solution .... ***Adelolecia pilati***

64(63) Epithecium differently coloured, if greenish then K− or K+ green intensifying; substrata various 65
Epithecium greenish, K+ violet; on wood .... ***Hypocenomyce xanthococca***

65(64) Asci in K/I *Lecanora-* or *Candelariella*-type [i.e. non-amyloid apical cushion reaching the upper ascus wall] 66
Asci in K/I *Bacidia-*, or *Biatora*-types [apical cushion conical, not reaching upper ascus wall], *Catillaria*-type [apical cushion absent], or *Lecidea*-type [I+ blue zone restricted to a cap-like structure at upper part of apical dome] 72

66(65) True exciple dark brown-black ('carbonaceous') throughout 67
True exciple pale, at least in inner part 69

67(66) Paraphyses simple or branched, scarcely swollen above or with a solitary swollen apical cell 68
Paraphyses simple, moniliform above with a series of swollen, short cells .... ***Claurouxia***

68(67) Ascospores 7–8 μm wide, ellipsoid .... ***Lecidella***
Ascospores 3.5–7 μm wide, oblong-ellipsoid .... ***Carbonea***

69(66) Paraphyses not swollen at apices, or if so then without combination of characters below 70
Paraphyses with swollen apices covered by a dark pigmented hood; ascospores oblong-ellipsoid, 4–7 μm wide; epithecium without K+ dissolving granules; apothecia dark brown to black; miriquidic acid often present; on rocks (rarely on worked timber) .... ***Miriquidica***

70(69) Paraphyses apices not swollen with a dark apical cap, or without combination of characters below 71
Paraphyses apices swollen, with a dark apical cap; hypothecium colourless; apothecial pigments brown, K−, N−; ascospores broadly ellipsoid to globose, thick-walled; on bark or over mosses on rocks .... ***Lecidea*** [*Japewia*]

71(70) Apothecia black, rarely dark blue-grey or red-brown; ascospores ellipsoid; epithecium without K+ dissolving crystals; hypothecium colourless or orange- to red-brown; usnic acid absent .... ***Lecidella***
Apothecia yellow-green to blue-grey or grey-black; ascospores narrowly ellipsoid to ellipsoid; epithecium often with K+ dissolving crystals; hypothecium colourless; usnic acid sometimes present .... ***Lecanora***

72(65) Asci *Biatora-*, *Catillaria-* or *Lecidea*-types, or if *Bacidia*-type then without combination of characters below 73

Asci *Bacidia*-type; apothecia 0.5–2 mm diam., immarginate; true exciple reduced; paraphyses simple or sparingly branched, each with a gelatinous coat strongly swollen in K; on rocks .... ***Tephromela***

73(72) Asci *Lecidea*-type .... ***Lecidea*** [s. str.]
Asci *Bacidia*-, *Biatora*- or *Catillaria*-type .... ***Lecidea*** [s. lat.]

Generic Key 6b

Crustose, apothecia rounded/photobiont chlorococcoid/ascospores colourless, simple/asci >8-spored

1 Asci (12-)16- to 32-spored 2
Asci >50-spored 3

2(1) Apothecia and thallus yellow, orange- to brown-orange; asci 12- to 32-spored . ***Candelariella***
Apothecia dark brown-black; asci 16-spored .... ***Steinia geophana***

3(1) Apothecia mostly with widely exposed discs, not perithecia-like, rarely punctiform and then thallus not yellow, variously shaped 4
Apothecia or thalline warts with apothecia with punctiform discs, globose to egg-shaped, perithecia-like, very small, pale yellow to greenish yellow pruinose, rarely black and non-pruinose .... ***Thelocarpon***

4(3) Apothecia lacking a thalline exciple in section (microscope); apical dome K/I−, + blue or absent 5
Apothecia with a thalline exciple in section (microscope) or immersed and bordered by thallus; apical dome well developed, K/I− except *A. heppii* .... ***Acarospora***

5(4) Thallus poorly developed, rarely areolate, C−; prothallus absent; apothecia smooth, or if wrinkled, sessile; apical dome K/I−, + blue or absent 6
Thallus well developed, areolate, C+ red; prothallus distinct; apothecia rough or wrinkled, immersed; apical dome K/I+ blue .... ***Sporastatia***

6(5) Apothecia convex, brownish, reddish to matt black; true exciple ± excluded (except in *Sarcosagium campestre*); species mostly ephemeral, on organic substrata 7
Apothecia convex to concave, black-brown to black, true exciple persistent; on rocky substrata 10

7(6) On soil, mosses, decaying plants, very rarely on decaying wood; ascospores ellipsoid or bacilliform 8
On bark or wood; ascospores globose 9

8(7) Apothecia pink-red to red-black, ± barrel-shaped, marginate; true exciple well developed, concolorous, but often grey-pruinose; asci thin-walled, apical dome absent .... ***Sarcosagium campestre***
Apothecia brown-yellow, yellow-red, convex, immarginate; true exciple very thin or absent; asci with a K/I− apical dome; paraphyses simple, richly branched at apices .... ***Biatorella***

9(7) Thallus well developed, minutely areolate, pale grey-green; paraphyses simple or a few forked above, with clavate apices 4–5 µm wide .... ***Biatorella monasteriensis***
Thallus indistinct, ± grey; paraphyses branched and anastomosing, apices scarcely swollen <3µm wide .... ***Strangospora***

10(6) Apothecial disc flat, smooth, not warty, folded or carbonaceous; paraphyses simple, distinctly septate, swollen at apices .... ***Sarcogyne***
Apothecial disc umbonate, often becoming gyrose; paraphyses richly branched and anastomosing .... ***Polysporina***

Generic Key 6c

Crustose, apothecia rounded/photobiont chlorococcoid/ascospores colourless, septate or muriform

1 Ascospores not polarilocular; the septum thin, continuous; epithecium K±, rarely K+ purplish ... 2
Ascospores polarilocular; a narrow to wide septum perforated by a narrow central channel; apothecia yellow, orange, red-brown or black, with or without a thalline exciple; epithecium usually (sometimes also the thallus) K+ violet, red or purple ***Caloplaca***

2(1) Thallus and apothecia yellow-orange or yellow-green, K− ... 3
Thallus and apothecia not yellow-orange or yellow-green, K− or K+ yellow to red ... 5

3(2) Apothecia yellow and marginate or black; ascospores 1-septate or 7- or more septate ... 4
Apothecia bright yellow to yellowish green, immarginate; ascospores (1-)3-septate ***Chrysothrix***

4(3) Apothecia black; ascospores 7- to 11-septate ... ***Arthrorhaphis citrinella***
Apothecia yellow; ascospores 1-septate ... ***Candelariella***

5(2) Ascospores distinctly muriform when mature ... 6
Ascospores septate, never muriform ... 12

6(5) Thallus not K+ red and/or apothecia not soralia-like or with remaining characters below ... 7
Thallus K+ red; apothecia soralia-like or else thallus sorediate; paraphyses simple or sparingly branched at apices; hypothecium colourless or pale ... ***Phlyctis***

7(6) Ascospores with a gelatinous outer wall (perispore); hymenium I−; apothecia not urceolate on wood ... 8
Ascospores with a compact outer wall or perispore present when young; hymenium I+ blue, or if I−, then apothecia urceolate on wood ... 9

8(7) Paraphyses and hyphae of exciple, branched-anastomosing in a gelatinous matrix; apothecia brown-black, ± translucent when wet; hyphophores often present ***Gyalideopsis***
Paraphyses simple; apothecia pale pink- to olive- or black-brown; hyphophores absent ... ***Gyalidea***

9(7) Paraphyses simple or forked, or when anastomosing the thallus on mosses, decaying plant materials or wood ... 10
Paraphyses distinctly branched and anastomosing; on rocks, rarely on wood, never on mosses or decaying plants ... ***Rhizocarpon***

10(9) Ascus I+ blue in some parts, thickened at apex or not; on moss and decaying plant materials ... 11
Ascus I− throughout, wall even in thickness; on wood ... ***Xerotrema megalospora***

11(10) Ascus apex thickened, with an I+ blue tholus; true exciple of branched, radiating hyphae; paraphyses branched and anastomosed, scarcely swollen and unpigmented at apex; asci 2- to 8-spored ... ***Schadonia fecunda***
Ascus apex not thickened, without an I+ blue tholus; true exciple paraplectenchymatous; paraphyses simple or forked, the end cells swollen and dark pigmented; asci 1-spored ... ***Lopadium***

12(5) Ascospores variously shaped, mostly <30 μm, if more, then acicular-fusiform; ecology varied ... 13
Ascospores ellipsoid and >70 μm; on mosses ... ***Megalospora tuberculosa***

13(12) True exciple not byssoid; ascus structure otherwise; ascospores and hypothecium varied ... 14
True exciple formed by loosely intricate, thick-walled hyphae, at times tomentose-arachnoid (byssoid); ascus *Byssoloma*-type; ascospores 3-septate; hypothecium dark brown ... ***Byssoloma***

14(13) Thalline exciple present, at least when young 15
Thalline exciple absent 20

15(14) Ascospores (1-) 3- to 7-septate, spindle- to needle-shaped or elongate-ellipsoid and curved 16
Ascospores 1- to 3-septate, mostly shortly ellipsoid, rarely ellipsoid to spindle-shaped 18

16(15) Apothecia red or brown; paraphyses branched and anastomosing or simple; on rocks or bark, rarely wood (see also *Lecania coerulescens*) 17
Apothecia dark brown to black-brown; paraphyses simple; on mosses in alpine situations .... ***Bryonora curvescens***

17(16) Thallus ± sorediate; asci *Lecanora*-type; paraphyses branched and anastomosed ***Haematomma***
Thallus not sorediate; asci with a uniformly I+ blue apical dome; paraphyses mainly simple .... ***Ophioparma***

18(15) Apothecia 0.4–0.6(-1) mm diam., brown to blackish brown, K−; on bark and stone 19
Apothecia 1–2(-3) mm diam., pink or pale orange-pink, K+ yellow to red-brown; on rotten stumps, peat, soil and mosses .... ***Icmadophila ericetorum***

19(18) Ascospores with distinct perispore; asci *Catillaria*-type .... ***Halecania***
Ascospores without thickened perispore; asci *Bacidia*-type .... ***Lecania***

20(14) Apothecia from beginning immarginate; true exciple strongly reduced 21
Apothecia usually at least marginate at beginning; true exciple strongly developed 23

21(20) Asci cylindrical or cylindrical-clavate; wall not noticeably thickened above 22
Asci ellipsoid, subglobose or clavate with a large apical dome and usually distinct ocular chamber; paraphyses richly branched and anastomosing .... ***Arthonia***

22(21) Apothecial tissues bound by a gel matrix; ascus apex with blue outer layer and apical dome and unstained wall, with an apical cushion at times surrounded by a dark cylindrical 'ring structure' .... ***Micarea***
Apothecial tissues lacking gel matrix; ascus wall I+ blue except for an apical pore ***Vezdaea***

23(20) Apothecia ± discoid, flat to convex; thallus areolate, granular or disappearing; ascospores varied 24
Apothecia top-shaped, black; thallus thin, translucent, gelatinous when wet; ascospores thread-like, to 100-septate .... ***Gomphillus calycioides***

24(23) Not on *Baeomyces*, without remaining combination of characters 25
On *Baeomyces*, apothecia black; thallus grey; ascospores 7- to 15-septate .... ***Arthrorhaphis***

25(24) Ascospores with a gelatinous outer wall (perispore); at least when young 26
Ascospores with a compact outer wall 28

26(25) Thallus mostly film-like or poorly developed, never areolate, prothallus absent; apothecia variously coloured; hypothecium pale; paraphyses branched or not 27
Thallus well developed, areolate, a black prothallus usually present; apothecia black; hypothecium dark brown; paraphyses richly branched and anastomosing .... ***Rhizocarpon***

27(26) Ascospores 15- to 20-septate; black scale-like hyphophores present; apothecia brown-black or red-brown; paraphyses richly branched and anastomosing in a gelatinous matrix .... ***Gyalideopsis muscicola***
Ascospores 1- to 3-septate; hyphophores absent; apothecia orange-brown or brown-black; paraphyses mostly simple .... ***Gyalidea***

28(25) Ascospores >1-septate; without remaining combination of characters below 29
Ascospores 6- to 7-septate; narrowly clavate and often attenuated at one end (tadpole-shaped); thallus thick, finely cracked-areolate .... ***Fuscidea lugubris***

29(28) Thallus granular scurfy, powdery or disappearing; apothecia variously coloured 30
Thallus of convex granules, areolate or squamulose, rarely absent and in other lichens; apothecia black .... ***Toninia***

30(29) Ascospores 1-septate ... 31
Ascospores 3- or more septate ... 38

31(30) Ascospores never urceolate, variously coloured; paraphyses varied ... 32
Apothecia deeply concave-urceolate, whitish or yellowish pink; paraphyses mostly simple ... ***Absconditella***

32(31) Ascospores <20 μm long; asci not *Lecanora*-type ... 33
Ascospores 20–30 × 10–15 μm; asci *Lecanora*-type ... ***Megalaria grossa***

33(32) Paraphyses mostly unbranched, or else branched above, not anastomosed; true exciple Pd− ... 34
Paraphyses branched and anastomosed; true exciple Pd+ yellow .... ***Tylothallia biformigera***

34(33) Paraphyses tips not strongly capitate and swollen; ascus apices not I+ blue throughout 35
Paraphyses tips abruptly swollen, at times with a dark brown apical cap; ascus apex with a uniformly blue apical dome ... ***Catillaria*** s. str.

35(34) Asci *Biatora*-type or otherwise, not *Bacidia-type* ... 36
Asci *Bacidia*-type ... ***Catillaria*** s.lat.

36(35) Asci *Biatora*-type or otherwise; pycnidia inconspicuous, <0.1 mm diam., whitish or if black then <0.1 mm diam. and immersed ... 37
Asci *Biatora*-type; pycnidia conspicuous, 0.1–0.6 mm diam., black ... ***Cliostomum***

37(36) Apothecia whitish; asci with a K/I+ blue tube in apical dome; on evergreen leaves, small twigs and siliceous rocks or stones ... ***Fellhanera bouteillei***
Apothecia blackish; asci *Biatora*-type; on various substrata, or apothecia pink, red-brown to reddish orange and on shaded limestones ... ***Catillaria*** s.lat.

38(30) Apothecia never entirely covered by true exciple, without remaining combination of characters below ... 39
Apothecia at first almost entirely covered by true exciple, later becoming emergent and concave; paraphyses simple; thallus membrane-like, gelatinous when moist ... ***Bryophagus gloeocapsa***

39(38) Ascospores mostly straight; paraphyses simple or sparingly branched; true exciple not paraphysis-like; asci mostly *Bacidia*-type ... 40
Ascospores acicular, sickle-shaped or bent spirally; paraphyses branched and anastomosing; true exciple of branched and anastomosing, paraphysis-like hyphae; asci *Lecanora*-type ... ***Scoliciosporum***

40(39) Apothecia dark-coloured, if pale then without remaining combination of characters below ... 41
Apothecia whitish; ascospores 11–16 × 2.5–4.5 μm, oblong-fusiform; pycnidia whitish; conidia pyriform, *c.* 3–4 × 1.3–1.7 μm ... ***Fellhanera subtilis***

41(40) Thallus of minute, pubescent granules or squamules, Pd+ red ... ***Phyllopsora rosei***
Thallus granular or sorediate, Pd− ... ***Bacidia***

Generic Key 6d

Crustose, apothecia rounded/photobiont chlorococcoid/ascospores coloured

1 Ascospores simple ... 2
Ascospores septate or muriform ... 4

2(1) Apothecia not deep red and K+ purple (in section); thallus C− or C+ red ... 3
Apothecia dark reddish brown; epithecium K+ purple (in section); thallus C+ orange ... ***Pyrrhospora quernea***

3(2) Hypothecium dark brown, K+ purplish; epithecium olive-green, K+ green intensifying, sometimes with additional brownish, K+ purplish pigment .... ***Orphniospora atrata***
Hypothecium colourless or pale straw; epithecium brown, never with K+ greenish or purplish pigments .... ***Fuscidea***

4(1) Ascospores soon extruded from asci, forming a dry, black, powdery spore mass, 1-septate, rarely submuriform; on bark or wood 5
Ascospores not quickly extruded from asci, nor leaving a black stain on finger when rubbed, 1- or more septate to muriform; ecology varied 6

5(4) Apothecia sessile to immersed, not in warts; true exciple raised; ascospore walls often distinctly ornamented .... ***Cyphelium***
Apothecia arising in swollen, raised warts; true exciple inapparent; ascospore walls smooth .... ***Thelomma***

6(4) Ascospores 1-septate or polarilocular 7
Ascospores 2- to 8-septate or muriform 12

7(6) Ascospores with a distinct perispore at least present before maturity, or the inner coloured wall layer coated by a colourless outer layer 8
Ascospores without a perispore or an outer colourless wall layer 9

8(7) Both ascospore cells each surrounded by a ring-like, dark wall with a thick, colourless outer layer; apothecia often somewhat crooked, ± umbonate .... ***Poeltinula cerebrina***
Dark ascospore wall uniform, entire spores, surrounded by a distinct halo when young; apothecia round, flat to weakly umbonate to wavy .... ***Rhizocarpon***

9(7) Paraphyses densely branched and anastomosing; asci with a small, thin, amyloid structure in the top of the only weakly amyloid tholus; thalli pale greenish yellow (rhizocarpic acid) 10
Paraphyses simple or branched above; asci with a thick, distinct, amyloid tholus; thalli mostly whitish, grey or brown, if yellow then containing usnic acid or xanthones 11

10(9) Thallus thick, wrinkled, mostly with elongate marginal lobes; lower surface black, attached by rhizines; apothecia to 2 mm diam.; montane .... ***Catolechia wahlenbergii***
Thallus thin, parasitic on *Baeomyces*; lower surface not black, not attached by rhizines; apothecia 0.2–0.8 mm diam. .... ***Epilichen scabrosus***

11(9) Thalline exciple absent; ascospores septate with thin walls .... ***Buellia***
Thalline exciple present; ascospores ± polarilocular or with thickened walls .... ***Rinodina***

12(6) Ascospores submuriform to muriform 13
Ascospores 2- to 8-septate 16

13(12) Paraphyses mostly simple, rarely forked; ascospore wall compact, perispore absent 14
Paraphyses branched and anastomosing; ascospores surrounded by perispore, at least when young .... ***Rhizocarpon***

14(13) Ascospores strongly muriform; ascus apex otherwise 15
Ascospores submuriform, with 1–2 longitudinal septa and internally thickened walls; asci *Lecanora*-type .... ***Diplotomma***

15(14) Apothecia at first perithecia-like, becoming urceolate; thalline exciple present, often scarcely indistiguishable from the verrucose thallus; ascospores 20–40(-54) μm long, dark brown .... ***Diploschistes***
Apothecia not perithecia-like, flat or concave; thalline exciple absent; ascospores 70–115 μm long, colourless or pale yellow-brown .... ***Lopadium***

16(12) Ascospores with a distinct swollen perispore; paraphyses branched and anastomosing or absent 17
Ascospores without a distinct swollen perispore; paraphyses mostly simple or else branched above 18

17(16) Thallus with an internal algal layer; paraphyses thin, branched and anastomosing (paraphysoids); apothecia disc-like from the first; ascus apex I+ blue in upper part .... ***Rhizocarpon***
Thallus without an internal algal layer, algae scattered on surface if present; paraphyses absent or cellular, broad; apothecia a cushion of tissue, becoming disc-like with age; ascus apex I− .... ***Lichenothelia convexa***

18(16) Distinct thalline exciple present 19
Thalline exciple absent .... ***Buellia***

19(18) Thallus thin and disappearing; apothecia sessile; hypothecium colourless; ascospores *Conradii*-type .... ***Rinodina conradii***
Thallus rimose-cracked or granular; apothecia immersed; hypothecium dark brown; ascospores with rounded lumina, often curved .... ***Diplotomma epipolium***

Generic Key 6e

Crustose, apothecia rounded/photobiont *Trentepohlia*

1 Hypothecium and/or true exciple dark, ± black and friable, if pale then true exciple absent; apothecia varied 2
Hypothecium and true exciple pale; apothecia mostly urceolate 10

2(1) Apothecia with an indistinct thalline exciple, or thalline exciple absent; without remaining combination of characters 3
Apothecia with a well-developed thalline exciple; paraphyses branched and anastomosing (paraphysoids); ascospores 3-septate, fusiform; asci *Opegrapha*-type; thallus C+ red .... ***Dirina massiliensis***

3(2) Hamathecium of simple or weakly forked paraphyses 4
Hamathecium of richly branched and anastomosing paraphysoids 6

4(3) Ascospores 1-septate, perispore absent; true exciple thick or not, not becoming fissured and crenate 5
Ascospores 3-septate at maturity, often with a thin perispore; true exciple very thick becoming strongly radially fissured and crenate in older apothecia .... ***Sagiolechia***

5(4) Ascospores brownish when mature; apothecia immersed to superficial .... ***Melaspilea***
Ascospores colourless; apothecia sessile .... ***Megalaria grossa***

6(3) True exciple absent, rarely rudimentary; thalline exciple absent; apothecia often irregular, elongate and divided 7
True exciple present, or if poorly developed then thalline exciple present 8

7(6) Ascospores 1- to 7-septate, the cells often unequal in size .... ***Arthonia***
Ascospores weakly or strongly muriform .... ***Arthothelium***

8(6) True exciple well developed; thalline exciple absent 9
True exciple poorly developed; thalline exciple present .... ***Schismatomma***

9(8) Ascospores not fragmenting into part-spores; asci 8-spored .... ***Lecanactis***
Ascospores fragmenting into part-spores; asci appearing multispored .... ***Bactrospora***

10(1) Ascospores with ± cylindrical cell lumina or ascospores simple 11
Ascospores with ± lens-shaped or globose cell lumina, mostly attentuate, transversely septate to muriform .... ***Thelotrema***

11(10) Ascospores 1- to multiseptate or muriform 12
Ascospores simple; apothecia aspicilioid .... ***Ionaspis***

12(11) Apothecial disc at first deeply immersed in thallus, after breaking through cortex, remnants of cortex remaining attached to disc margins; periphyses present or absent ... 13
Apothecial disc without attached cortical remnants after breaking through upper layer; periphyses absent ... 14

13(12) Apothecia when dry not separated from surrounding thallus through a crack; periphyses present; on bark or wood ... ***Ramonia***
Apothecia when dry separated from surrounding thallus through a crack; periphyses absent; on calcareous rocks ... ***Petractis clausa***

14(12) Ascus wall not thickened at apex; exciple not black and friable ... 15
Ascus wall distinctly thickened at apex; exciple at least partly black and friable, mostly deeply crenate ... ***Sagiolechia***

15(14) Ascospores 3- to 15-septate or muriform ... 16
Ascospores 1-septate ... ***Dimerella***

16(15) Asci >8-spored; ascospores 3- to 15-septate ... 17
Asci 8-spored; ascospores muriform, rarely 3- or 5- to 13-septate ... ***Gyalecta***

17(16) Asci to (8-)12(-16)-spored; apothecial disc pale yellow-brown ... ***Cryptolechia carneolutea***
Asci (8-)16- to 48-spored; apothecial disc red-brown to orange ... ***Pachyphiale***

Generic Key 6f

Crustose, apothecia rounded/photobiont blue-green

1 Thallus superficial, substratum various; apothecia various but never with radiating fissures ... 2
Thallus immersed in a calcareous substratum; apothecia ± conical, pale, with radiating fissures around a central pore-like disc ... ***Petractis clausa***

2(1) Photobiont bead-like, *Nostoc*, in lax chains suspended in a conspicuous, colourless, gelatinous matrix ... 3
Photobiont not in lax, bead-like chains, if *Nostoc*, then chains folded in packets, compacted, indistinct, gelatinous matrix colourless or yellow-brown to pink-purple ... 5

3(2) Ascospores septate, submuriform or muriform, ovoid to ellipsoid ... 4
Ascospores simple, subglobose to globose ... ***Lempholemma***

4(3) Upper cortex indistinct, never a single layer of clearly defined cells; thallus dark green-black, rarely with a red-brown tinge, often swelling when moist ... ***Collema***
Upper cortex a single layer of well-defined, rounded to cuboid cells; thallus red-brown-black, rarely blue-grey, not noticeably swelling when moist ... ***Leptogium***

5(2) Photobiont in chains of 2–3 cells or more, often folded, rarely single-celled or forming packets surrounded by individual pale or coloured gelatinous sheath ... 6
Photobiont single-celled or in packets surrounded by an individual, often coloured gelatinous sheath ... 13

6(5) Ascospores simple or 1- to 3-septate, rarely to 7-septate; globose to ellipsoid, rarely fusiform; thalline exciple absent or present or apothecia absent ... 7
Ascospores 7- to 14-septate, narrowly fusiform; thalline exciple absent .. ***Arctomia delicatula***

7(6) Thallus ± uniformly minutely squamulose or rosette-forming, with, at least in part, a distinctly radiating margin of ± flattened or convex lobes ... 8
Thallus crustose, continuous or ± areolate, sometimes with spreading, terete, filament-like lobes at the margin (best seen when moist) or entirely minutely fruticose 11

8(7) Thallus of scattered or overlapping squamules or, if forming rosettes, lobes never grooved and striate and pale below; isidia and soralia, if present, predominantly marginal 9

Thallus stellate-radiate, lobes longitudinally grooved and striate and pale below; coarse nodules present, laminal .... ***Vestergrenopsis elaeina***

9(8) Thallus rosette-forming, small-lobulate at the centre with radiating marginal lobes or ± entirely small-squamulose, blue-black when moist, rhizines absent but often with a well-developed dark prothallus; thalline exciple present or absent; ascospores simple or 1- to 3-septate 10

Thallus of red-brown, ± overlapping squamules, dull dark green when moist; underside pale with a few concolorous rhizines; apothecia frequent, thalline exciple absent; ascospores 11–23(-32) × 4.5–7(-8.5) μm, 1(-2)-septate .... ***Massalongia carnosa***

10(9) On siliceous rocks in streams, lake-sides or in persistent seepage tracks; photobiont *Scytonema* or *Rivularia*; isidia coralloid or ± flattened; apothecia very rare; ascospores septate .... ***Placynthium***

On tree trunks, sheltered and often mossy rocks or soil; photobiont *Nostoc*; often with isidia, granular soralia, or richly fertile; ascospores simple .... ***Pannaria* & *Parmeliella***

[If fertile, thalline exciple present at least in juvenile apothecia in *Pannaria*, but is absent in *Parmeliella*]

11(7) Thallus dark, red-brown to black or blue-black, amorphous or in part minutely filament-like (at least at the margin); photobiont *Calothrix*, *Scytonema*, or *Rivularia*; apothecia dark, with or without a thalline exciple; ascospores simple or 1- to 3-(-7)-septate; on rock, rarely basic soils 12

Thallus blue-grey or pale grey, entirely of small granules; photobiont *Nostoc*; apothecia with a widely expanded red-brown to dark brown disc with a granular thalline exciple; ascospores 0(-1)-septate .... ***Moelleropsis nebulosa***

12(11) Thallus red-brown-black, not pruinose, amorphous or indistinctly placodioid, subgelatinous when moist; photobiont *Calothrix*; apothecial disk poriform at first, later ± expanded, thalline exciple present; ascospores simple, globose to ellipsoid .... ***Porocyphus***

Thallus blue-black or black, sometimes grey- or blue-grey-pruinose, not subgelatinous when moist; photobiont *Rivularia* or *Scytonema*; apothecial disc fully expanded from the first; thalline exciple absent; ascospores (0-)1–3(-7)-septate, ellipsoid, shortly fusiform, rarely acicular .... ***Placynthium***

13(5) Photobiont *Gloeocapsa*, gelatinous sheaths pink-red-brown, especially towards the upper surface, K ± purplish 14

Photobiont not *Gloeocapsa*, gelatinous sheaths colourless, yellow-brown or red-brown towards the upper surface, K− 15

14(13) Apothecial disc fully expanded from an early stage; inner wall of ascus I+ blue, apical dome I−; paraphyses slender, not swollen and moniliform above .... ***Euopsis***

Apothecial disc scarcely expanded, apothecia ± perithecia-like or urceolate; apical dome of ascus I+ blue; paraphyses ± swollen and moniliform above .... ***Pyrenopsis***

15(13) Asci with distinctly thickened apices, either I+ blue (in part) or entirely brown; apothecia with an indistinct true exciple and a well-developed thalline exciple, often perithecia-like; mainly confined to wet, acid rocks 16

Asci uniformly thin-walled, entirely I−; apothecia often with a well-developed true exciple, not or partly obscured by an evanescent thalline exciple; perithecia-like or with a broad disc; on acid or calcareous rocks and soil 17

16(15) Thallus black, tinged red-brown, at least when moist; asci I−, but with an I+ blue apical dome; photobiont *Chroococcidiopsis* (*Xanthocapsa*) .... ***Pyrenopsis***

Thallus black, tinged greenish, at least when moist; asci uniformly I+ red-brown, apical structure I−; photobiont chroococcoid .... ***Pterygiopsis***

17(15) Apothecial disc expanding; on limestone, mortar and calcareous soils 18
Apothecial disc remaining pore-like; on acid to neutral rocks, often in seepage tracks or by streams and lake-sides; photobiont *Calothrix* .... ***Porocyphus***

18(17) Thallus ± areolate, surface and edges of areoles with ± globose isidia; ascomata open below; photobiont *Chroococcidiopsis* .... ***Psorotichia***
Thallus ± continuous or like a small *Leptogium*; isidia absent; ascomata closed below; photobiont *Nostoc*-like .... ***Lemmopsis***

Generic Key 7

Sterile crustose lichens

Sterile crustose lichens are difficult to identify, often even for the expert. In order to identify material correctly it is particularly important to pay attention to its ecology and specimens may require chemical analysis by t.l.c. for confirmation. No attempt is made here to key out sterile crustose species on rocks or walls, and consistently sterile genera *Lepraria, Leproloma* and *Leproplaca* should be checked. The key presented here is only a preliminary one and is in no way exhaustive. Nevertheless, it is included as a guide to identifying such awkward specimens on bark or wood.

Emphasis in this key is placed on the use of the standard spot test reactions Pd, K, C and KC. Lichen algae are also important in identification. The presence of *Trentepohlia* can usually be detected in fresh material by the pale to deep orange mark obtained when the thallus is scratched and the cortex or soralia abraded to expose the underlying algae. Unfortunately, the carotenoid pigment which imparts the orange colour to the contents of the algal cells is soon degraded, becoming colourless in the herbarium; similarly species with a pinkish tinge to the thallus when fresh tend to become uniformly pale grey a few weeks after collection, as, for example, in the case of *Schismatomma niveum*. As a rule, under the microscope, the cells of *Trentepohlia* tend to be larger and to have thicker walls than those of *Trebouxia* and other unicellular algae; sometimes the filamentous habit of the former is also discernible.

Note: the recently described *Micarea coppinsii, Schaereria corticola* and *Buellia violaceofusca* are not properly included (see Sterile key 2, couplet 19, and key 3 couplet 16).

Synopsis of species on bark or wood

1 Thallus or soredia not golden- or orange-yellow 2
Thallus and/or soredia golden- or orange-yellow (not pale yellow) when fresh .... **Sterile Key 1** (p. 44)

2(1) Thallus sorediate or leprose 3
Thallus without soredia, although occasionally secondarily derived from isidia 4

3(2) Thallus with at least some positive spot test reactions .... **Sterile Key 2** (p. 44)
Thallus Pd−, K−, C−, KC− .... **Sterile Key 3** (p. 48)

4(2) Thallus without isidia, hyphophores or papillae, often with pycnidia 5
Thallus with isidia, hyphophores, blastidia or papillae, in some species becoming sorediate, or appearing minutely squamulose, usually without pycnidia .... **Sterile Key 4** (p. 51)

5(4) Thallus with abundant pycnidia or sporodochia .... **Sterile Key 5** (p. 53)
Thallus without pycnidia or sporodochia .... **Sterile Key 6** (p. 56)

Sterile Key 1

Thallus and/or soredia golden- or orange-yellow

1 Thallus and/or soredia K+ bright purple (parietin) 2
Thallus and/or soredia K− or faint red (parietin absent) 4

2(1) Thallus immersed, thin and continuous, whitish grey with scattered to crowded, delimited, erose soralia 3
Thallus superficial, golden- to orange-yellow, of small sorediate or isidiate areoles or granules ..... ***Caloplaca citrina***

3(2) Soralia bright orange-yellow; on nutrient-enriched bark ..... ***Caloplaca chrysophthalma***
Soralia pale yellow to dirty yellowish brown; on dry, well-lit, non-nutrient-enriched oaks in parkland ..... ***Caloplaca lucifuga***

4(1) Photobiont chlorococcoid 5
Photobiont *Trentepohlia*; soralia orange when fresh, soon fading in the herbarium ..... **see key 2 or key 3**

5(4) Thallus ± immersed, brownish or dull orange-yellow, granular or areolate-squamulose, never entirely leprose 6
Thallus bright yellow, entirely leprose; on dry, rough-barked trees; calycin and/or pinastric acid present ..... ***Chrysothrix candelaris***

6(5) Thallus superficial, orange-yellow, granular or areolate-squamulose 7
Thallus ± immersed or weakly areolate, greenish brown to brown; soralia ± discrete, <0.4 mm diam., at first brown, later abrading ochraceous-yellow; on bark of *Alnus, Betula, Sorbus* and *Larix*; Scottish Highlands ..... ***Japewia subaurifera*** (see p. 326)

7(6) Granules minutely subsquamulose 8
Granules of numerous spherical, corticate granules to *c.* 0.1 mm diam. ..... ***Candelariella xanthostigma***

8(7) Squamules to 1 mm diam., usually becoming nodulose-crenate and breaking down into clusters of fine granular soredia ..... ***Candelariella reflexa***
Thallus of well-developed convex subsquamulose areoles or of scattered, warted granules, not becoming sorediate ..... ***Candelariella vitellina***

Sterile Key 2

Thallus sorediate with at least some positive spot test reactions

Synopsis

1 Thallus C+ bright orange (persisting) 2
Thallus C− or + red, orange-pink, carmine red or yellow (not persisting) 7

2(1) Thallus completely sorediate 3
Thallus with delimited or irregularly delimited soralia; soredia not completely covering thallus 4

7(1) Thallus C+ red, orange-pink or carmine-red 8
Thallus C− or C+ yellow 22

22(7) Thallus Pd+ red, Pd+ orange or Pd+ distinctly yellow 23
Thallus Pd− or Pd+ faintly yellowish 45

23(22) Thallus K+ yellow to bright red [crystals] (norstictic acid) 24
Thallus K−, K+ yellow or K+ brownish red [crystals absent] (norstictic acid absent) . 26

Sterile Key 2

1 Thallus C+ bright orange (persisting) 2
Thallus C− or + red, orange-pink, carmine red or yellow (not persisting) 7

2(1) Thallus completely sorediate 3
Thallus with delimited or irregularly delimited soralia; soredia not completely covering thallus 4

3(2) Soredia pale yellow-green, farinose; thallus thin, evanescent; thiophanic, usnic acid and zeorin .... ***Lecanora expallens***
Soredia brownish green, finely granular; thallus rather thick, continuous; isoarthothelin and thiophanic acid .... ***Pyrrhospora quernea***

4(2) Thallus distinct; soralia and soredia varied 5
Thallus evanescent; soralia becoming confluent; soredia farinose, pale yellow .... ***Lecanora expallens***

5(4) On rough bark, rotting bark or wood; thallus minutely granular-verrucose; soralia pale yellowish or green; various lichen products present 6
On smooth bark of trees and shrubs; thallus thin and even; soralia pale yellowish grey, efflorescent; arthothelin, granulosin and lichexanthone ***Lecidella elaeochroma*** f. ***soralifera***

6(5) Soralia round, convex, ± discrete or becoming confluent, pale yellow; soredia farinose; in woodland on acid bark; arthothelin, thiophaninic acid and atranorin ***Ochrolechia inversa***
Soralia irregular, yellow-green, often inconspicuous and best revealed by rubbing thallus with finger, coarsely granular; in open habitats on wood or enriched bark; atranorin, arthothelin and ± thuringione .... ***Lecidella scabra***

7(1) Thallus C+ red, orange-pink or carmine-red 8
Thallus C− or C+ yellow 22

8(7) Thallus of ± ascending, convex, often imbricate squamules; soredia on lower surface of squamules in lip-shaped soralia 9
Squamules absent; soredia on upper surface of thallus 10

9(8) Squamules mostly scattered, whitish, Pd+ yellow-orange .... ***Hypocenomyce leucococca***
Squamules mostly imbricate, grey-green to dark brown, Pd- .... ***Hypocenomyce scalaris***

10(8) Thallus Pd+ yellow-orange (alectorialic acid); on acidic substrata 11
Thallus Pd− (alectorialic acid absent); substratum varied 12

11(10) Thallus dull grey-green to dark brown-grey, with irregular patches of soredia ***Buellia pulverea***
Thallus whitish of ± immersed areoles and ± delimited to confluent, whitish grey to yellowish brown soralia flecked with dark grey .... ***Hypocenomyce sorophora***

12(10) Thallus without a conspicuous white prothallus; soralia yellowish, greenish or greyish white or brownish, often flat or irregular; soralia C+ orange-red (gyrophoric acid) 13
Thallus with a conspicuous white prothallus; soralia blue-grey-white, ± hemispherical, C+ carmine red (lecanoric acid) .... ***Pertusaria hemisphaerica***

13(12) Photobiont *Trentepohlia*; thallus pale to dark brown; soralia yellow-orange; on smooth barked trees in humid sites 14
Photobiont *Trebouxia* or chlorococcoid; thallus whitish, pale to medium grey, or grey-green 16

14(13) Soralia with yellow or orange tinge at least when fresh 15
Soralia pale fawn, mottled brown, never orange even when fresh .... ***Opegrapha fumosa***

15(14) Soralia yellow-grey often confluent on a dark brown thallus; dark prothallus often present; on hard, smooth bark; gyrophoric and schizopeltic acids ... ***Opegrapha gyrocarpa***
Soralia pale yellow to buff, small and ± discrete, on an olive to grey-brown thallus; dark prothallus sometimes present, on smooth or rough bark; gyrophoric acid .... ***Opegrapha sorediifera***

16(13) Soredia at least at first delimited in convex soralia; thallus thick, areolate or granular and irregular 17
Soredia scattered or aggregated in irregular, farinose-granular soralia; soralia not delimited; thallus thin; relict woodlands in E. Britain .... ***Biatora epixanthoidiza*** auct.

17(16) Soredia farinose, yellowish green or greenish grey; thallus rather thin, whitish pale grey, grey-green, olivaceous or brownish, inconspicuously areolate or verrucose-granular; photobiont chlorococcoid 18
Soredia granular, yellowish white; thallus thick, whitish grey, often granular or irregularly warted; photobiont chlorococcoid .... ***Ochrolechia androgyna***

18(17) Thallus without orange-red patches, all parts K− 19
Thallus with irregular patches of orange-red pigment, K+ purple .... ***Trapeliopsis pseudogranulosa***

19(18) Thallus a distinct, warted, granular or rimose-cracked crust, whitish pale grey or grey-green, never olivaceous or brown; soralia mostly 0.2–0.6 mm diam., or confluent and ± effuse; mostly on wood or rotting substrata 20
Thallus of inconspicuous, ± scattered, dark grey, olivaceous or brownish areoles; soralia 0.15–0.25 mm diam., punctiform, often totally obscuring areoles; mostly on bark .... ***Trapelia corticola***
[*Note:* see also *Micarea coppinsii* (p. 379) and *Schaereria corticola* (p. 558)].

20(19) Thallus granular to verrucose-areolate, granules >0.1 mm diam., often effigurate at edges; soralia usually ± discrete, if confluent rarely giving thallus leprose appearance 21
Thallus minutely granular, granules <0.1 mm diam., never effigurate; soralia soon confluent giving thallus a leprose appearance .... ***Trapeliopsis viridescens***

21(20) Thallus of pale grey to grey-green areoles; marginal areoles ± flattened and effigurate; soralia at first discrete, farinose to finely granular, grey-green to dark green; mainly on firm wood .... ***Trapeliopsis flexuosa***
Thallus of whitish to pale grey, sometimes ± pinkish granules or warted areoles; marginal areoles usually neither flattened nor effigurate; soralia ± irregular, granular, whitish to cream-yellow, rarely tinged grey-green; mostly on soft, decaying wood or bark .... ***Trapeliopsis granulosa***

22(7) Thallus Pd+ red, Pd+ orange or Pd+ distinctly yellow 23
Thallus Pd− or Pd+ faintly yellowish 45

23(22) Thallus K+ yellow to bright red [crystals] (norstictic acid) 24
Thallus K−, K+ yellow or K+ brownish red [crystals absent] (norstictic acid absent) . 26

24(23) Soralia genuine, not part of ascoma 25
'Soralia' in effect being the sorediate surface of delimited verrucae containing apothecia .... ***Phlyctis agelaea***

25(24) Soralia diffuse and erose; thallus ± thick .... ***Phlyctis argena***
Soralia delimited and excavate; thallus thin or immersed .... ***Buellia griseovirens***

26(23) Thallus Pd + yellow-orange (psoromic, stictic or thamnolic acids) 27
Thallus Pd+ red or dull orange (without psoromic, stictic or thamnolic acids) 30

27(26) Thallus K+ yellow; photobiont trebouxioid or chlorococcoid 28
Thallus K− (psoromic acid); photobiont *Trentepohlia* .... ***Schismatomma niveum***

28(27) Thallus with discrete soralia, K+ yellow or bright yellow (stictic or thamnolic acids) 29
Thallus entirely leprose-sorediate, K+ yellow (stictic acid) .... ***Lepraria lobificans***

29(28) Thallus grey-green, superficial; soralia convex to globular; K+ bright yellow (thamnolic acid); mainly on bark .... ***Haematomma elatinum***
Thallus whitish to pale grey, immersed; soralia ± flat, K+ pale yellow (stictic acid); on wood .... ***Xylographa vitiligo***

30(26) Thallus KC− (picrolichenic acid absent); not tasting bitter 31
Thallus KC+ violet (picrolichenic and protocetraric acid); tasting bitter .. ***Pertusaria amara***

31(30) Thallus a powdery crust without discrete soralia; ± rosette-forming with powdery marginal lobes 32
Thallus with discrete soralia, or if ± continuously sorediate then without powdery marginal lobes 33

32(31) Thallus with distinct lobes (pannaric and roccellic acids present) ***Leproloma membranaceum***
Thallus indistinctly lobed (methyl pannaric acid present) .... ***Leproloma vouauxii***

33(31) Soralia discrete, not part of ascoma 34
Soralia in effect being the sorediate surface of delimited verrucae containing apothecia .... ***Pertusaria multipuncta***

34(33) Photobiont *Trentepohlia* 35
Photobiont *Trebouxia* 36

35(34) Thallus superficial, cracked-areolate; soralia discrete, or as lines along cracks in the thallus; soredia farinose, *c.* 15–30 μm diam.; protocetraric acid ..... ***Enterographa sorediata***
Thallus mainly immersed, not distinctly cracked; soralia discrete, rounded; soredia finely granular, *c.* 30–70 μm diam.; fumarprotocetraric acid ...... ***Schismatomma quercicola***

36(34) Soredia at least partly forming discrete, delimited soralia, occasionally in lines 37
Soredia diffuse or forming irregular soralia 42

37(36) Thallus immersed to ± superficial, very thin 38
Thallus superficial, minutely warted-granular or areolate 40

38(37) Thallus K− (atranorin absent); mainly on bark 39
Thallus K+ yellow (atranorin present); on wood of fallen trunks and old fencing in sheltered, boggy areas .... ***Mycoblastus fucatus***

39(38) Soralia white, well-delimited, often retaining a thin rim .... ***Pertusaria pupillaris***
Soralia pale green, irregularly rounded, soon confluent; argopsin ***Biatora epixanthoidiza*** s.str.

40(37) Soralia 0.2–0.5 mm diam., areoles 0.1–0.5 mm diam., not fragile; fumarprotocetraric acid or pannarin 41
Soralia 0.1–0.2(-0.3) mm diam., areoles 0.05–0.1mm diam., fragile; argopsin .... ***Halecania viridescens***

41(40) Soralia often confluent and coalescing towards centre; thallus greyish to brownish green, when young often delimited by grey-brown hypothallus; fumarprotocetraric acid; on bark and wood of *Pinus*, rarely *Betula* and *Populus*, Scottish Highlands ***Fuscidea arboricola***
Soralia yellowish brown, scattered, not coalescing towards centre; thallus dark greenish grey, hypothallus absent; pannarin; on young trees or branches in moist woodlands and *Salix* carrs .... ***Rinodina efflorescens***

42(36) Soralia grey-white or bluish grey, not yellowish or greenish grey, K+ yellow 43
Soralia yellowish or greenish grey, K− or K+ brownish 44

43(42) Thallus a warty crust, often delimited by a black hypothallus; soredia at first delimited, ± concave, often becoming confluent; zeorin absent .... ***Mycoblastus sterilis***
Thallus disappearing, not delimited by a black hypothallus, becoming obscured by scattered or contiguous, coarsely granular efflorescent soralia; zeorin present .... ***Catillaria pulverea***

44(42) Thallus completely coarsely granular sorediate, cracked-areolate; ubiquitous in areas receiving SO_2 air pollution .... ***Lecanora conizaeoides***
Soralia often coalescing towards thallus centre, at periphery delimited and ulcerose, 0.15–0.4 mm diam.; bark and wood of *Pinus*, rarely *Betula*, and *Populus*, Scottish Highlands .... ***Fuscidea arboricola***

45(22) Thallus C− (variolaric and protolichesterinic acids absent); soralia varied, rarely in lines 46
Thallus C+ yellow (variolaric and protolichesterinic acids); soredia granular, soralia excavate, often in lines .... ***Ochrolechia turneri***

46(45) Thallus and soralia KC−, not tasting bitter; picrolichenic acid absent 47
Thallus and soralia KC+ violet, tasting bitter; picrolichenic acid .... ***Pertusaria amara***

47(46) Thallus without orange-brown patches which react K+ purple 48
Thallus grey mottled orange-brown, orange parts reacting K+ purple ***Chaenotheca ferruginea***

48(47) Soralia pale coloured 49
Soralia dark brown .... ***Rimularia fuscosora***

49(48) Soredia white, grey-white or pale fawn; thallus ± completely sorediate 50
Soredia yellowish green; thallus rarely completely sorediate 52

50(49) Photobiont trebouxioid 51
Photobiont *Trentepohlia*; thallus and soredia pure white;? lepraric acid; on dry bark of wayside trees .... ***Schismatomma cretaceum***

51(50) Soredia whitish to greenish grey, granular, 40–100 μm diam.; atranorin and zeorin .... ***Catillaria pulverea***
Soredia creamy white to pale fawn, farinose to finely granular, 20–50 μm diam., atranorin and roccellic acid .... ***Lecanora farinaria***

52(49) Soredia diffuse or forming irregular soralia, at times virtually completely covering thallus; ecology varied; chemistry otherwise 53
Soredia forming scattered, delimited, round soralia to 1 mm diam., decorticate wood in boggy places; gangaleodin and atranorin .... ***Lecanora jamesii***

53(52) Thallus pale grey, verrucose; on bark and wood of old *Pinus*, Scottish Highlands; atranorin, usnic and planaic acids .... ***Mycoblastus alpinus***
Thallus powdery-granular with a smooth, pale, continuous prothallus; on mossy bark of deciduous trees in old woodlands; usnic acid, zeorin and terpene(s) or sterol(s) .... ***Megalospora tuberculosa***

Sterile Key 3

Thallus sorediate with negative spot test reactions

Synopsis

1 Photobiont *Trentepohlia* 2
Photobiont chlorococcoid 6

6(1) Soredia white or pale grey 7
Soredia coloured 12

7(6) Soredia forming delimited, round or elongate soralia 8
Soredia diffuse or forming irregular soralia 9

12(6) Thallus with delimited or irregularly delimited soralia; soredia not completely covering thallus 13
Soredia completely covering thallus or else thallus entirely leprose or composed of goniocysts 28

Sterile Key 3

1 Photobiont *Trentepohlia* 2
Photobiont chlorococcoid 6

2(1) Thallus inconspicuous, grey to purple-grey, mauve or dark brown, individual thalli often small, sometimes mosaic-forming; soralia and ecology varied 3
Thallus conspicuous, pale grey to chalky-white, often wide-spreading, not distinctly mosaic-forming; 'soralia' (? ascomatal initials) pale brown, fleck-like, to 0.5 mm diam., ± scattered; on dry trunks of *Quercus* in old woodlands and parklands .... ***Lecanactis amylacea***

3(2) Soredia grey, pink, mauve- or lilac-grey 4
Soredia with a pronounced yellow, yellow-ochre or orange tinge 5

4(3) Thallus lilac- to pink-grey; soralia at first punctiform, becoming confluent; dry sides of old trees .... ***Schismatomma decolorans***
Thallus dark brown, often tinged mauve and delimited by a dark prothallus; soralia punctiform; on smooth bark, usually near streams; confluentic acid .... ***Opegrapha zonata***

5(3) Soralia pale fawn to ochraceous, at first 0.2–0.7 mm diam., becoming confluent and to 2–3 mm wide; old trees in open and woodland situations .... ***Opegrapha corticola***
Soralia bright orange (quickly fading in herbarium) 0.1–0.3 mm diam., sometimes becoming confluent; on exposed twigs and branches of trees and shrubs in boggy places, sometimes on shaded tree trunks .... ***Opegrapha multipuncta***

6(1) Soredia white or pale grey 7
Soredia coloured (yellowish grey, yellow, green, bluish grey, brown, ochraceous, orange or blackish) 12

7(6) Soredia forming delimited round or elongate soralia 8
Soredia diffuse or forming irregular soralia 9

8(7) Soralia genuine, not part of ascoma, round, >1 mm diam., white; thallus well developed with a zoned thalline margin; on bark .... ***Pertusaria albescens***
'Soralia' in effect being the sorediate surface of verrucae, containing apothecia; on acid bark in humid sites, N.W. Scotland .... ***Pertusaria ophthalmiza***

9(7) Thallus immersed, or if superficial, then with efflorescent soralia; fibrous prothallus absent; thallus or soralia UV+ white (divaricatic or perlatolic acids) 10
Thallus composed of pale grey, powdery granules surrounded by a white, fibrous prothallus; thallus UV− (atranorin, zeorin, porphyrilic acid) .... ***Haematomma ochroleucum*** var. ***porphyrium***

10(9) Thallus ± immersed or pale green-grey; soralia 0.2–0.6 mm diam., usually confluent and coalescing; rare; divaricatic or perlatolic acids 11
Thallus superficial, dark blue-grey; areoles becoming verrucose and bursting to form discrete or confluent, efflorescent, pale grey, irregular soralia; perlatolic acid .... ***Haematomma caesium***

11(10) Divaricatic acid present; on bark of *Betula, Juniperus* and *Pinus* in C. & E. Scottish Highlands .... ***Lecidea nylanderi***
Perlatolic acid present; mainly on wood, especially of *Pinus* in Scottish pine forests, less often on bark in C. & E. Scottish Highlands .... ***Lecidea botryosa***

12(6) Thallus with delimited or irregularly delimited soralia, soredia not completely covering thallus 13
Soredia completely covering thallus or else thallus entirely leprose or composed of goniocysts 28

13(12) Squamules absent; soralia not lip-shaped; ecology varied 14
Thallus of tiny squamules *c.* 0.3–0.8 mm wide; soralia lip-shaped at apices of squamules; on moist, rotting wood .... ***Trapeliopsis percrenata***

14(13) Soralia medium to dark brown or black, sometimes yellow-ochraceous internally (when abraded) 15
Soralia pale coloured 17

15(14) Thallus ± immersed; medulla I−; on bark 16
Thallus superficial, irregularly warted; medulla I+ blue; on wooden fencing .... ***Thelomma ocellatum***

16(15) Soralia medium to dark brown, internally ochraceous yellow .... ***Japewia subaurifera***
Soralia dark brown to black, internally concolorous .... ***Rimularia fuscosora***
[*Note:* see also *Buellia violaceofusca* (p. 130).]

17(14) Soralia developing from thallus pustules that split open so that a resultant concave soralium is delimited by a reflexed ring of thallus 18
Soralia not developing from thallus pustules 19

18(17) Thallus white, without bluish tinge; soralia 0.12–0.3 mm diam., pale yellowish green or greenish white, concave, discrete or confluent; on trunks of mature parkland and wayside trees .... ***Caloplaca ulcerosa***
Thallus pale grey often with bluish tinge; soralia pale green but often tinged blue-grey, often persisting as delimited concavities after becoming emptied of soredia; on nutrient-enriched bark of tree trunks and branches .... ***Caloplaca obscurella***

19(17) Soredia forming delimited, round or elongate soralia 20
Soredia diffuse or forming irregular soralia 22

20(19) Soralia greenish, or pale to blue-grey, UV− or UV+ white; thallus thin or disappearing, grey or blue-grey; on bark and wood in drier sites 21
Soralia pale yellow or greenish yellow, scattered, UV−; thallus thin, pale grey ***Lecanora jamesii***

21(20) Soralia 0.2–1 mm diam., often confluent, UV+ white (perlatolic acid); on acid bark and wood .... ***Haematomma caesium***
Soralia 0.14–0.3(-0.4) mm diam., mainly discrete UV− (atranorin and zeorin); on ± basic bark .... ***Rinodina griseosoralifera***

22(19) Thallus UV+ white or orange 23
Thallus UV− 26

23(22) Thallus and soralia UV+ white (lichexanthone absent); on moderately exposed bark of trunks and branches 24
Thallus and soralia UV+ orange (lichexanthone); on dry, shaded bark in woodlands .... ***Bacidia viridifarinosa***

24(23) Thallus indistinct, ± immersed or of convex, scattered to confluent verrucae, 0.08–0.16 mm diam., soredia efflorescent and irregular, rarely well developed and ulcerose; perlatolic or divaricatic acid 25
Thallus distinct, of smooth, corticate granules, 0.1–0.25 mm diam., thick; soredia often well developed and ulcerose, especially towards edge; divaricatic acid ***Fuscidea lightfootii***

25(24) Perlatolic acid present .... ***Fuscidea viridis***
Divaricatic acid present .... ***Fuscidea pusilla***

26(22) Thallus thin or powdery granular; soredia pale green, yellow-green or pale buff; overgrowing mosses in old woodlands 27
Thallus of convex verrucae often obscured by extensive sorediate patches; soredia blue-grey; at base of trunks of old parkland trees, especially *Ulmus* ... ***Caloplaca virescens***

27(26) Thallus thick, powdery granular with a smooth, pale, continuous prothallus; thalline granules aggregated, pale green, tinged yellow-green, often partly eroded and then paler; atranorin, usnic acid, zeorin, ?diterpene(s), sterol(s) .... ***Megalospora tuberculosa***
Thallus very thin, prothallus absent; soralia ± efflorescent, contiguous, granular, yellowish to pale buff; lichen products not detected by t.l.c. .... ***Bacidia epixanthoides***

28(12) Thallus without pseudopodetia 29
Thallus consisting of a primary leprose thallus from which simple or sparingly branched, granular pseudopodetia develop .... ***Leprocaulon microscopicum***

29(28) Thallus a blue-grey, servediate crust 30
Thallus white to grey with yellow tinge or yellow-green 31

30(29) Thallus granular sorediate; granules not loose and powdery; UV− ...... ***Caloplaca virescens***
Thallus leprose, of loose, powdery granules; UV+ white .... ***Lepraria incana***

31(29) Thallus yellow-white to yellow-grey 32
Thallus bright yellow-green 33

32(31) Thallus of yellow-grey, powdery granules surrounded by a white, fibrous prothallus; usnic acid and zeorin present .... ***Haematomma ochroleucum*** var. ***ochroleucum***
Thallus whitish to yellowish grey, forming irregular rosettes, ± with powdery, marginal lobes; methylpannaric acid present, usnic acid and zeorin absent .... ***Leproloma vouauxii***

33(31) On tree roots and wood or tree bases; rhizocarpic acid and/or vulpinic acid present 34
On bark or wood of tree trunks or on the undersurface of horizontal branches; rhizocarpic acid and an unidentified compound .... ***Chrysothrix chrysophthalma***

34(33) Rhizocarpic acid and unidentified compounds present; photobiont *Trebouxia*, rarely with *Stichococcus* but then <4 algal cells per chain .... ***Psilolechia lucida***
Vulpinic acid and ± rhizocarpic acid present; photobiont *Stichococcus*, >6 algal cells per chain .... ***Chaenotheca furfuracea***

Sterile Key 4

Thallus with isidia, hyphophores, blastidia or papillae, in some species becoming sorediate or appearing minutely squamulose

1 Thallus C+ bright orange or C+ orange-red 2
Thallus C− 4

2(1) Thallus grey-white to pale fawn, blastidia or coralloid isidia present; ecology varied; C+ persistent orange or fleetingly orange-red 3
Thallus yellowish green, upper surface coarsely warted with numerous, ± globose isidia; sunny sides of wayside trees; C+ persistent orange (xanthones) .... ***Pertusaria flavida***

3(2) Thallus grey-white; isidia small, cylindrical-coralloid, usually branched, forming a continuous crust in centre; on nutrient-enriched bark; C+ orange-red, not persistent (gyrophoric acid) .... ***Ochrolechia subviridis***
Thallus pale fawn brown; of small granules (blastidia), forming a thick, granular-areolate crust; rare on exposed, decorticated maritime wood; C+ persistent orange (xanthones) .... ***Lecidella prasinula***

4(1) Thallus KC− or KC+ yellow 5
Thallus KC+ violet; tasting bitter; upper surface warted-papillate, the apices and ridges breaking down into small punctiform soralia .... ***Pertusaria amara***

5(4) Thallus K+ yellow or red (stictic or norstictic acids) 6
Thallus K−, at times with orange, K+ purple pigments (anthraquinones) 7

6(5) Thallus K+ yellow to red [crystals], UV− (norstictic acid) .... ***Pertusaria coccodes***
Thallus K+ yellow [crystals absent], UV+ orange (coronaton, stictic and constictic acids). .... ***Pertusaria coronata***

7(5) Photobiont *Trentepohlia*; on *Quercus* in old woodlands and parklands 8
Photobiont trebouxioid; ecology varied 10

8(7) Thallus sandy brown or dull orange to pale green-brown or yellow-green; surface thickly isidiate and obscured by dispersed or a generally confluent crust of simple or frequently coralloid-branched isidia 9
Thallus dull grey, often with a bluish or purplish tinge when fresh; surface thinly to thickly minutely isidiate .... ***Zamenhofia coralloidea***

9(8) Thallus sandy brown to pale green-brown, ± continuously isidiate; isidia lacking cortex or when present very narrow and ill-defined; *Trentepohlia* cells 7–12 μm diam. .... ***Zamenhofia hibernica***
Thallus pale green-yellow; isidia forming a ± distinct, minutely branched, shrub-like aggregations; isidia with a wide well-defined algal free cortex; *Trentepohlia* cells 9–17 μm diam. .... ***Zamenhofia rosei***

10(7) Thallus squashes C− (gyrophoric acid absent); thallus paler; isidia morphology and ecology varied 11
Thallus squashes C+ red (gyrophoric acid); thallus black to dark brown, rarely pale green-brown (shade forms) of isidiose to minutely coralloid granules 25–150(-200) × 25–5- μm; on acid bark and wood .... ***Placynthiella icmalea***

11(10) Thallus appearing minutely squamulose, of ± cylindrical or nodular, branched, at times digitiform, isidia or filaments 12
Thallus not appearing squamulose or filamentous 15

12(11) Thallus pale grey to brown-green, bright green when wet, K− (atranorin absent); filament or squamule morphology varied; external cells of upper cortex with conical papillae; ecology varied 13
Thallus whitish grey, of scattered, branched, finger-like, cylindrical or compressed isidia, K+ yellow (atranorin); external cells of upper cortex without conical papillae; on trunks of mature *Quercus* in old woodlands .... ***Rinodina isidioides***

13(12) Thallus of cylindrical to flattened squamules >50 μm wide 14
Thallus of densely branched, pale green filaments, 12–35 μm wide; branches sometimes breaking down into soredia-like granules; on water-retentive bark, especially *Sambucus* .... ***Macentina stigonemoides***

14(13) Squamules to 0.5 mm wide, ± contiguous, often ascending, nodulose and subgranular, rarely dispersed and elongate, digitiform; on basic or nutrient-enriched bark .... ***Agonimia tristicula***
Squamules 0.05–0.1 × 0.05–0.25 mm, often dispersed, rarely contiguous and compacted into elongate, adpressed, simple, branched, or trilobed-digitiform clusters, amongst mosses on trunks of *Quercus* in relict woodlands .... ***Agonimia octospora***

15(11) Apices of isidia and pycnidia (if present) not orange, K−; ecology varied 16
Apices of isidia and pycnidia orange, K+ purple; on acid bark in well-lit woodlands and parklands .... ***Caloplaca herbidella***

16(15) Thallus whitish grey, grey-green to yellow-green; film-like or ± continuously isidiate; ecology varied 17
Thallus slate blue-grey to bluish black, totally scurfy-isidiate, cracked-areolate; on nutrient-enriched trees .... ***Caloplaca isidiigera***

17(16) Isidia scattered or forming a ± continuous crust; medulla I− 18
Isidia in black, soralia-like clusters; medulla I+ blue .... ***Thelomma ocellatum***

18(17) Thallus thin, wax- or skin-like, with rather inconspicuous, blackish or greenish 'isidia' (hyphophores) 19
Thallus conspicuously isidiate-granular 20

19(18) Hyphophores greenish grey, ± translucent, very slender, *c*. 0.02 mm diam., with paler, pointed apices; on horizontal boughs of *Salix* and smooth-barked, young trees in damp sites, even in polluted habitats .... ***Gyalideopsis anastomosans***
Hyphophores dark brown, bent over above with a flattened, expanded apex with a spiked margin; on pleurocarpous mosses, W. British Is. .... ***Gyalideopsis muscicola***

20(18) Thallus thin, grey-green to yellow-green, without a prothalline margin, granular-isidiate; granules 60–120 μm, rounded, or ± elongate, never becoming sorediate 21
Thallus thick, whitish grey, usually with conspicuous prothalline margin, coarsely verrucose or papillose to granular-isidiate, at times breaking down to form sorediate areas .... ***Pertusaria albescens*** var. ***corallina*** (cf. *Ochrolechia turneri*)

21(20) On trunks of mature, basic-barked trees in parklands, wayside situations and woodlands; granules never dark-coloured on upper surface .... ***Bacidia rubella***
On trunks of mature, acid-barked trees, especially *Quercus* in ancient parklands and woodlands; granules often brownish on upper surface .... ***Bacidia biatorina***

Sterile Key 5

Thallus without isidia, hyphophores, or papillae, with pycnidia or sporodochia

Only the most characteristic species occurring in this condition are included. Many other normally fertile (i.e. with ascomata) species occasionally occur sterile, but with pycnidia. Also not included are 'pycnidial morphs' of weakly lichenized species occurring on smooth bark e.g. species in *Arthonia, Arthopyrenia, Eopyrenula* and *Mycomicrothelia*.

Synopsis

1 Photobiont *Trentepohlia* 2
Photobiont micareoid or chlorococcoid 14

3(2) Pycnidia whitish to dull yellowish or white-pruinose 4
Pycnidia dark brown to black, not pruinose 8

14(1) Photobiont chlorococcoid, cells mostly >8 μm diam. 15
Photobiont 'micareoid', cells 4–7 μm, ± globose, thin-walled 19

15(14) Pycnidia white or pink-brown to rarely brown-black; conidia curved............................ 16
Pycnidia black or bluish black; conidia straight 17

17(15) Pycnidia sessile or immersed, often contorted, black; wall purple-violet in K; thallus smooth and cracked to granular-warted 18
Pycnidia stalked, bluish black, to 0.2 mm diam., often with a globose white spore mass extruded from ostiole; wall brown or green in K; conidia *c.* 3.7–4 × 1.5–1.7 μm, simple; thallus grey, leprose-granular; on acid bark and wood of old trees ***Lecidea doliiformis***

Sterile Key 5

1 Photobiont *Trentepohlia* 2
Photobiont micareoid or chlorococcoid 14

2(1) Thallus not lichenicolous, pale to violet-grey; conidiomata pycnidial; conidia simple; pycnidial pruina when present C− or C+ red 3

Thallus lichenicolous, on *Enterographa* and *Lecanactis*, pale pink with scattered, ± regular hemispherical warts (sporodochia), 0.3–1 mm diam., which develop over effete apothecia or pycnidia of the host; sporodochia composed of ± vertically orientated hyphae, forming 1-septate, ellipsoid conidia in chains; thallus and sporodochia C+ red (lecanoric and schizopeltic acids) .... ***Blarneya hibernica***

3(2) Pycnidia whitish to dull yellowish, or white-pruinose 4
Pycnidia dark brown to black, not pruinose 8

4(3) Pycnidia not pruinose, mostly dull yellowish; pycnidial wall ± colourless 5
Pycnidia white-pruinose; pycnidial wall dark-coloured 6

5(4) Conidia ellipsoid, 3–4(-5) × 1.3–1.8 μm ... ***Dimerella lutea***
Conidia oblong, ± with a median constriction, 6–7(-8) × 1.8–2.6 μm ... ***Dimerella pineti***

6(4) Apices of pycnidia and thallus C− (lecanoric and schizopeltic acids absent); conidia <10 μm long ... 7
Apices of pycnidia C+ red (lecanoric acid), thallus C− (schizopeltic acid); conidia 12–17 × 2–3 μm ... ***Lecanactis abietina***

7(6) Conidia 3.7–5 × 1.2–1.7 μm; pycnidia with a dark brown-black wall, K+ olive-black; pruina of pycnidia K+ yellow, KC−, C−, Pd−; thallus contains confluentic acid and accessory and lepraric acid; on predominantly acid substrata in old woodlands ... ***Lecanactis subabietina***
Conidia 4.5–6.7 × 1–1.4 μm; pycnidia with a brown wall, K−; pruina of pycnidia K−, KC−, C−, Pd−; thallus contains two unidentified pigments; on trees with a basic or nutrient-rich bark in parklands and woodlands ... ***Opegrapha vermicellifera***

8(3) Conidia globose, ellipsoid or oblong, not curved ... 9
Conidia curved ... 12

9(8) Conidia globose, ellipsoid or oblong-ellipsoid, simple, mostly <11 μm long; thallus whitish, grey-green or pale buff ... 10
Conidia oblong, 1-septate, 11–16.5 × 2.5–3 μm; thallus brown or dark olivaceous ... ***Strigula taylorii***

10(9) Conidia globose to ellipsoid, without a median constriction, <5 μm long ... 11
Conidia oblong-ellipsoid, ± with a slight median constriction, 7.6–12 μm long ... ***Arthopyrenia ranunculospora***

11(10) Pycnidia conical; conidia released in chains of cylindrical packets; conidia ellipsoid, 3.5–4.5 × 1.8–2 μm ... ***Anisomeridium nyssaegenum***
Pycnidia ± globose; conidia free, either subglobose to ellipsoid and 2.3–4 × 1.8–2.7 μm, or globose and 1–1.5 μm diam. ... ***Anisomeridium biforme***

12(8) Conidia short and curved <9 μm; colour varied ... 13
Conidia sickle-shaped, of two types: 9–15 × 1–1.3 μm or 9–14(-17) × 0.5–0.7 μm; thallus pale grey to dull olive or brownish ... ***Opegrapha vulgata***

13(12) Conidia 5–8 × 1–2 μm; thallus dull olive to reddish brown ... ***Opegrapha rufescens***
Conidia of two types: 4–7 × 1–1.5 μm or 7–9 × 0.7 μm; thallus dull grey-olive ... ***Opegrapha niveoatra***

14(1) Photobiont chlorococcoid, cells mostly >8 μm diam. ... 15
Photobiont 'micareoid', cells 4–7 μm, ± globose, thin-walled ... 19

15(14) Pycnidia white or pink-brown to rarely brown-black; conidia curved; on shaded bark ... 16
Pycnidia black or bluish black; conidia straight; usually on ± well-lit bark or wood ... 17

16(15) Pycnidia pink-brown, rarely brown-black, sessile ± globose or elongated vertically, 180–200 μm diam., to 300 μm tall; conidia (20-)30–43 × 0.8–1 μm; thallus pale grey-green, thin, scurfy; deciduous trees in woodlands including coppices and plantations ***Bacidia vezdae***
Pycnidia white, ± immersed, *c.* 100–200 μm diam.; conidia (21-)26–40(-47) × 1(-1.5) μm; thallus yellow-green to fawn, finely granular; on basic bark, especially *Sambucus* ... ***Bacidia arnoldiana/delicata***

17(15) Pycnidia sessile or immersed, often contorted, black; wall purple-violet in K; thallus smooth and cracked to granular-warted ... 18
Pycnidia stalked, bluish black, to 0.2 mm diam., often with a globose white spore mass extruded from ostiole; wall brown or green in K; conidia *c.* 3.7–4 × 1.5–1.7 μm, simple; thallus grey, leprose-granular; on acid bark and wood of old trees ***Lecidea doliiformis***

18(17) Pycnidia 0.2–0.6 mm diam.; thallus of pale brown or yellowish brown, warted granules, or disappearing; conidia 2–4 × 1–1.7 μm; on wood, especially near the sea, mainly E. England .... ***Cliostomum corrugatum***
Pycnidia 0.1–0.3 mm diam.; thallus whitish grey, usually irregular with ± smooth, areolate areas, sometimes becoming warted; conidia 3–5 × 1.5–2 μm; on ± acid bark and wood, widespread .... ***Cliostomum griffithii***

19(14) With stalked or sessile pycnidia containing mesoconidia (smaller ± immersed pycnidia containing microconidia sometimes present); never with curved or flexuose macroconidia 20
Pycnidia innate sometimes becoming emergent with gaping ostioles, *or* with small, sessile, apothecia-like sporodochia; pycnidia sometimes containing curved or flexuose macroconidia 25

20(19) Pycnidia black 21
Pycnidia whitish to reddish brown, never black 23

21(20) Pycnidial wall K− or K+ olivaceous 22
Pycnidial wall olivaceous brown, K+ violet; mesoconidia 3.5–5 × 1–1.5(-1.7) μm .... ***Micarea misella***

22(21) Pycnidial wall and stalk dark purple-brown, K+ dark green; mesoconidia 3.4–4.3 × 1.2–1.6 μm .... ***Micarea nigella***
Pycnidial wall dark olivaceous, K−; stalk fuscous or reddish-brown, K−; mesoconidia 3.5–4.8 × 1–1.5 μm .... ***Micarea botryoides***

23(20) Thallus indistinctly areolate or rimose, or endoxylic, without goniocysts, K−; pycnidia whitish without tomentum 24
Thallus of pale to dark green goniocysts. 12–40 μm diam.; goniocysts containing purple, oily substance in K; pycnidia brown with white tomentum; mesoconidia (4-)4.5(-6) × 1.3–1.7 μm .... ***Micarea hedlundii***

24(23) Pycnidia C+ red (gyrophoric acid), simple; mesoconidia 4–6 x 1–1.5 μm ***Micarea pycnidiophora***
Pycnidia C−, often branched; mesoconidia 6–8 × 1–1.8 μm .... ***Micarea stipitata***

25(19) Sporodochia absent; thallus and/or pycnidia C+ red (gyrophoric acid) or C− (t.l.c.: 'prasina unknowns') 26
With whitish, cushion-like, apothecia-like sporodochia *c.* 0.1–0.25 mm diam. bearing oblong-ellipsoid macroconidia 6–10 × 2–3 μm (immersed, inconspicuous pycnidia containing mesoconidia 4–5.3 × 1.2–1.5 μm may also be present; all parts C− (t.l.c.: nil) .... ***Micarea adnata***

26(25) Thallus areolate, or scurfy (invaded by dematiaceous hyphae and foreign algae), or ± endoxylic; sometimes with curved or flexuose macroconidia; thallus and/or pycnidia usually C+ red (gyrophoric acid) 27
Thallus of pale to dark green (sometimes K+ violet) goniocysts 12–60 μm; pycnidia, if present, 30–80 μm diam. containing either (5-)5.5–8 × 0.7–1 μm microconidia, or (3.5-)4–6 × 1.2–1.7 μm mesoconidia; all parts C− (t.l.c.: 'prasina unknowns') .... ***Micarea prasina***

27(26) Pycnidia usually with longer, curved or flexuose macroconidia; smaller pycnidia with microconidia sometimes present; mesoconidia unknown; pycnidial walls colourless or greenish, K−; on various substrata, rarely on worked timber 28
With numerous pycnidia containing mesoconidia, 3.5–4.5 × 1.3–2 μm, often extruded as conspicuous white blobs; in addition, pycnidia containing microconidia (4-)4.5–6(-6.5) × 0.7–1 μm, or curved macroconidia 12–21 × 1 μm, sometimes present: pycnidial walls with pale olivaceous pigment, K+ violet; usually on wood, commonly on worked timber .... ***Micarea denigrata***

28(27) Macroconidia curved or hamate, 21–40 × 1–1.5 μm; microconidia (5-)6–7(-7.7) × 0.4–0.7 μm; on bark of old trees (especially *Quercus*) in old woodlands ..... ***Micarea peliocarpa***
Macroconidia flexuose, 50–110 × 1 μm; microconidia (3.8-)4.5 × 0.5–0.7 μm; on shaded wood in woodlands .... ***Micarea cinerea***

Sterile Key 6

Thallus without pycnidia, sporodochia, soredia or isidia

Only the most characteristic species occurring in this condition are included. Sterile thalli (without ascomata, conidiomata, or distinctive vegetative dispersal structures) of many other species may be encountered - these being either juvenile or stunted or arrested in their development by environmental stress.

1 Thallus whitish grey to pale grey- or glaucous green 2
Thallus bright yellowish green, pale grey mottled orange-brown or fawn to deep brown ... 5

2(1) Photobiont chlorococcoid 3
Photobiont *Trentepohlia*; thallus pale grey to chalky white with pale brown flecks (? ascomatal initials) .... ***Lecanactis amylacea***

3(2) Thallus Pd+ orange-red; minutely granular-squamulose or ± smooth, continuous 4
Thallus Pd− ; forming superficial, granular verrucae; granules *c.* 80–120 μm diam., but granules never loose and soralia-like; forming large patches on mossy trunks in old woodlands, Lake District, N.E. England, Scottish Highlands .... ***Biatora vernalis***

4(3) Thallus smooth, ± continuous, very thin, often with erose patches; prothallus absent; K+ red [crystals] (norstictic acid); widespread .... ***Phlyctis argena***
Thallus of minute, flattened to granular, pale bluish grey, faintly white pubescent, squamules, ± delimited by a ± arachnoid prothallus; K− or faint yellow (argopsin); overgrowing mosses on mature tree trunks in old woodlands, W. & S. British Is. .... ***Phyllopsora rosei***

5(1) Thallus pale grey mottled orange-brown, fawn to deep brown, granular or bullate-areolate-squamulose 6
Thallus bright yellowish green, of convex granules to 0.5 mm diam.; wood and coniferous bark, mainly in E. Britain .... ***Chaenotheca chrysocephala***

6(5) Thallus fawn to deep brown, K− 7
Thallus grey mottled orange-brown, the orange patches reacting K+ purple; on dry, acid bark .... ***Chaenotheca ferruginea***

7(6) Thallus granular-squamulose; granules *c.* 0.1 mm diam. 8
Thallus squamulose; squamules 1(-1.5) mm diam., round or irregular, often crowded, bullate or irregularly ascending, margin ± crenulate; upper surface greyish green to dark brown, dull; on bark and wood tolerating atmospheric pollution .... ***Hypocenomyce caradocensis***

8(7) Granules matt, brown with paler, often white margins; on mossy, acid-barked trees, especially *Quercus*, often over bryophytes .... ***Lopadium disciforme***
Granules shiny brown, uniform; mainly on *Betula* and conifer trunks, Scottish Highlands .... ***Protoparmelia ochrococca***

ABSCONDITELLA Vězda (1965)

Thallus crustose, effuse, greenish or grey-green, usually intermixed with algal films, ± gelatinous when moist. **Photobiont** chlorococcoid. **Ascomata** apothecia; disc urceolate-concave, pale-coloured, immersed to emergent. **True exciple** entire, of ± pseudoparenchymatous or conglutinated parallel hyphae. **Hymenium** I− or I+ yellowish, K/I−. **Hypothecium** shallow, often indistinct. **Hamathecium** of simple, or rarely a few apically-branched, indistinctly septate, slender paraphyses, 0.5–1 μm wide, the apices slightly to distinctly swollen. **Asci** cylindrical to cylindrical-clavate, thin-walled, with a distinct apical dome that sometimes has a narrow ocular chamber, K/I− or rarely bluish, the apical dome K/I−, 8-spored. **Ascospores** ellipsoid, fusiform or fusiform-acicular, 1- to 3(-5)- septate, colourless, thin-walled, smooth; perispore absent. **Conidiomata** not known. **Chemistry**: all parts Pd−, K−, KC−, C−, UV−. **Ecology**: mostly on acidic, moist substrata which are usually covered by a ± gelatinous algal film.

All species are inconspicuous with minute apothecia and are rarely collected. Distinguished from *Dimerella* and *Gyalecta* by the absence of *Trentepohlia* as photobiont, and from *Bryophagus* by a thicker ascus apex and K/I− hymenium.

LITERATURE: Döbbeler & Poelt (1977), Kilias (1981), Vězda (1965a), Vězda & Pišút (1984), Vězda & Vivant (1975).

1 Ascospores 1-septate 2
Ascospores 3- or more septate 3

2(1) Apothecia 0.2–0.4 mm diam. .... **4. *sphagnorum***
Apothecia 0.07–0.2(-0.26) mm diam. .... **2. *delutula***

3(1) Ascospores fusiform to fusiform-ellipsoid, 4–6 μm wide 4
Ascospores acicular-fusiform, 1.5–2 μm wide .... **3. *pauxilla***

4(3) Apothecia red-brownish .... **1. *celata***
Apothecia whitish or pale yellowish .... **5. *trivialis***

1. A. celata Döbb. & Poelt (1977)

Apothecia 0.05–0.12 mm diam., red-brownish; true exciple 7–16 μm wide below, increasing to *c.* 30 μm wide above; hymenium 85–100 μm tall. Ascospores 13–19 × 4–5 μm, fusiform, 3-septate.

On rotting wood of a fallen *Fagus* branch and on compacted soil. S.E. Scotland (Selkirk), N.W. England (Cumbria). Swedish Lapland.

Differs from *A. trivialis* **(5)** in having darker coloured apothecia and shorter ascospores.

2. A. delutula (Nyl.) Coppins & Kilias (1981)

Apothecia 0.07–0.2(-0.26) mm diam., whitish; true exciple 15–20 μm wide below, increasing to *c.* 30 μm wide above; hymenium 70–90 μm tall. Ascus wall sometimes K/I+ bluish. Ascospores 10–15(-17) × 3–5(-6.5) μm, ellipsoid to oblong-ellipsoid, 1-septate.

On shaded stones, compacted soil and turf. S. England (Norfolk, N. Hampshire), N.W. England, W. Scotland, W. Ireland.

A. modesta (Stizenb.) Vězda (1965), from the Alps, possibly belongs in this species.

3. A. pauxilla Vězda & Vivant (1975)

Apothecia 0.1–0.25 mm diam., yellowish white; true exciple 10–12 μm wide below, increasing to 25–30 μm wide above; hymenium 45–55 μm tall. Ascospores 25–28 × 1.5–2 μm, fusiform-acicular, 3(-5)-septate.

On conifer wood of fallen trunks and stumps. N.E. Scotland (Aberdeen, Mar Forest) and S. England (Hampshire, New Forest). W. Pyrénées, amongst mosses on rocks.

Most likely to be confused with *Bryophagus gloeocapsa* which has larger apothecia, a thinner ascus apex and a K/I+ blue hymenium.

4. A. sphagnorum Vězda & Poelt (1965)

Apothecia 0.2–0.4 mm diam., entirely whitish or with a yellowish, flesh-coloured or pinkish disc; true exciple *c.* 20 μm wide below, increasing to *c.* 50 μm wide above; hymenium 65–85 μm tall. Ascospores 9.5–14.3 × 2.5–4(-5) μm, ellipsoid to oblong-ellipsoid, 1-septate.

On *Sphagnum capillaceum* at 685 m. C.E. Scotland (Perth). Sweden, Germany (Bavaria), U.S.A. (Ohio).

5. A. trivialis (Willey ex Tuck.) Vězda (1965)

Apothecia 0.18–0.4 mm diam., whitish to pale yellowish; true exciple 18–24 μm tall below, increasing to 45–50 μm wide above; hymenium 90–100 μm tall. Ascospores (17-)19–24(-28) × 4.7–5.5(-6) μm, fusiform, 3-septate.

On mosses on heavy metal-contaminated river gravels, wet rock faces and clay soil. C. Wales and S. England (N. Hampshire). France (Bretagne), Germany, N.E. U.S.A.

Two collections on moribund thalli of *Peltigera britannica* from S.E. Scotland (Selkirk) and on moribund moss on *Quercus* trunk from S. England (S. Devon), with smaller apothecia (0.1–0.15 mm diam.) and ascospores (14–16.5 × 4 μm), may represent a different species.

ACAROSPORA Massal. (1852)

Thallus crustose, often distinctly areolate or subsquamulose, rarely squamulose, areoles scattered to contiguous, rarely overlapping, angular to rounded, rarely lobate. **Upper surface** smooth or wrinkled, usually pale brown, red- to black-brown, occasionally greenish or rust-coloured, rarely ± white-pruinose. **Photobiont** chlorococcoid, algal layer continuous or ± discontinuous, lower part often very uneven. **Upper cortex** paraplectenchymatous. **Lower cortex** absent. **Ascomata** apothecia, one or several per areole, usually deeply immersed, rarely ± sessile. **Thalline exciple** ill-defined to ± prominent, ± elevated and darkened. **Disc** red-brown to brown-black, smooth or roughened, rarely including 1 or several thickened, sterile umbos. **Hamathecium** of paraphyses, numerous, thin, mostly simple, conglutinate. **Asci** more than 100-spored, clavate, apical dome distinct, I−, but I+ blue in *A. heppii* **(4)**. **Ascospores** rounded to ellipsoid, simple, colourless. **Conidiomata** pycnidia, immersed. **Conidia** ellipsoid-globose. **Chemistry**: lichen products often not detected by t.l.c., rarely with gyrophoric or norstictic acids. **Ecology**: usually on siliceous, more rarely calcareous, rocks, often in nutrient-rich or enriched habitats, very rarely on wood and compacted soil.

Some species are very variable and difficult to identify and more detailed field studies are needed to establish affinities between seemingly very different morphotypes and the response to different habitats and substrata.

The bright greenish yellow *Acarospora chlorophana* (Wahlenb.) Massal. (1852) with distinct, elongate, smooth marginal lobes and convex apothecia which soon become immarginate and tend to be clustered towards thallus centre, has recently been found on a sandstone gravestone in N. England (N.E. Yorks). Arctic-alpine, N. Europe.

LITERATURE: Clauzade & Roux (1981), Magnusson (1929, 1935, 1956).

1 Thallus C+ red 2
  Thallus C− 3

2(1) Thallus without pruina or if pruinose then areoles flat; apothecia persistently on the surfaces of the areoles .... **2. *fuscata***
  Thallus ± densely white-pruinose, areoles markedly convex; apothecia mostly between the areoles .... **13. *umbilicata***

3(1) Thallus K+ red 4
  Thallus K− 5

4(3) Areoles ± rounded, sub-conical, almost entirely occupied by apothecia, deep red-brown; thalline exciple distinctly finely pale striate; paraphyses 1.5–2 μm thick at base **12. *subrufula***
Areoles rounded to ± angular, flattened, not entirely occupied by apothecia, cream-coloured, yellow-brown, brown or greenish yellow; thalline exciple not striate; paraphyses 0.5–1 μm thick at base .... **11. *smaragdula***

5(3) Thallus pale grey to dark brown 6
Thallus orange to rust-red or ochraceous .... **10. *sinopica***

6(5) On soil 7
On rock 8

7(6) Thallus of scattered to imbricate, rounded to lobate subsquamules, with white, rooting rhizines, edges white; apothecia absent .... **8. *rhizobola***
Thallus ± contiguous, coherent, of interlocking areoles, without white rhizines, edges concolorous with upper surface; apothecia usually present .... **11. *smaragdula***

8(6) Asci 100- to 200-spored; ascospores 3–6 × 1–2.5 μm 9
Asci 30- to 100-spored; ascospores 6–13 × 3–6 μm .... **6. *macrospora***

9(8) Thallus ± well developed; apothecia usually >0.4 mm diam. 10
Thallus mostly thin, poorly developed, often visible only as a thin rim surrounding apothecia; apothecia 0.2–0.4 mm diam. .... **4. *heppii***

10(9) On siliceous rocks; apothecial disc and thallus not pruinose; algal layer regular and continuous, lower part ± even 11
On calcareous rocks; apothecial disc and/or thallus ± grey-white pruinose; algal layer irregular and discontinuous, lower part ± serrated .... **3. *glaucocarpa***

11(10) Apothecia not or little elevated, disc pale to dark; thalline exciple not well-defined or thickened; hymenium 60–120 μm tall 12
Apothecia prominent, elevated, dark; thalline exciple distinct, thick, black-brown, crater-like; hymenium 100–180 μm tall .... **15. *verruciformis***

12(11) Apothecia ± immersed; disc concave to flat; paraphyses 0.5–2 μm wide at the base; hymenium >70 μm tall 13
Apothecia often ± sessile; disc ± convex; paraphyses 2–3 μm wide at the base; hymenium 60–75 μm tall .... **1. *badiofusca***

13(12) Apothecia mostly uniform, ± rounded or only elongate due to mutual pressure; thallus of scattered or ± even and interlocking areoles, pale to dark brown, often with a red-brown tinge 14
Apothecia very variable on a single thallus, angular, rounded, elongate or deformed; thallus rather regular with small, interlocking areoles, black-brown .... **5. *impressula***

14(13) Apothecial disc ± concave, mostly smooth, even, not wrinkled or with a sterile umbo .. 15
Apothecial disc flat or ± elevated, uneven, neatly wrinkled or with a sterile umbo **7. *nitrophila***

15(14) Thallus dispersed, often of ± scattered, thick areoles; apothecia punctiform, flat to concave 16
Thallus closely contiguous, of regular, rarely scattered, thin areoles; apothecia rounded, punctiform, slightly elevated .... **9. *rufescens***

16(15) Areoles ± convex, pale cream to dark brown, often with several (1–12) darker, punctiform apothecia; thalline exciple indistinct, not swollen .... **11. *smaragdula***
Areoles flat, red-brown to dark brown, 1–3 apothecia per areole, with a concave disc; thalline exciple well-defined, somewhat swollen .... **14. *veronensis***

1. A. badiofusca (Nyl.) Th. Fr. (1861)

Thallus wide-spreading, often rather thin, areolate; areoles rather small, 0.5–1.5 mm wide, ± rounded, numerous, contiguous but not coherent, at times wavy, irregular, rarely imbricate, flat or ± convex, reddish brown. Algal layer continuous. Apothecia 0.4–2 mm diam., mostly single, rarely 2- to 4-contiguous per areole, rounded or gyrose-contorted, sessile; thalline exciple usually distinct, ±

elevated, entire, concolorous with disc; hymenium 60–75(-90) μm tall; disc flat-convex, often roughened, usually red-brown to brown-black, always darker than thallus. Paraphyses 2.5–3 μm wide at base, 4–5 μm at tips. Asci *c.* 200-spored. Ascospores 3–6 × 1.5–2.5 μm, ellipsoid. Thallus Pd−, K−, KC−, C− (lichen products not detected by t.l.c.).

On mica-schist boulders; very rare. Scotland (Perth, Breadalbane Mtns and Angus, Caenlochan). C. Europe, Italy, Poland, Scandinavia.

Characterized by the large, sessile apothecia which are often darker than the thallus. The British specimens correspond to subsp. *badiofusca* in which the apothecia are very dark.

2. A. fuscata (Schrader) Th.Fr. (1871)

Thallus areolate-squamulose, wide-spreading and often forming extensive patches, rarely scattered; areoles 0.5–3 mm wide, very variable, mostly contiguous, irregular in shape, ± flat, at times sublobulate or weakly imbricate, surface uneven, irregular, pale to dark reddish or yellowish brown, often with black margins and undersides. Apothecia 0.2–1 mm diam., 1(-5) per areole, usually numerous, at first punctiform, later concave-flat, usually angular or irregular, immersed, never elevated or emergent, reddish brown to blackish, usually ± darker than the thallus; disc smooth or slightly rough, level with thallus or ± depressed; hymenium (70-)80–120 μm tall. Paraphyses 1.5 μm thick at base, 2–3 μm thick at tips. Ascospores 4–6 × 1–1.5 μm, narrowly ellipsoid or ± cylindrical. Thallus Pd−, K−, KC+ red, C+ red (gyrophoric acid).

On nutrient-enriched siliceous rocks, farm walls, memorials, roofing tiles, gravestones, and birds' perching rocks; abundant. Throughout British Is. Europe, North America.

The commonest species of the genus; very variable but well-characterized by the continuous, widely spreading, areolate- to subsquamulose thallus, C+ red on the surface. *A. peliscypha* Th. Fr. (1860), also with a C+ red thallus, has ± imbricate subsquamules with prominent red-brown apothecia, the discs with one or several sterile umbos and with a ± elevated thalline exciple. It occurs on nutrient-rich, montane-alpine, ± basic rocks. The single record of this species in the British Is. is based on an exuberant morph of *A. fuscata* **(3)**.

3. A. glaucocarpa (Ach.) Körber (1859)

incl. *A. cervina* Massal. (1852), *A. glaucocarpa* var. *depauperata* (Körber) A. L. Sm. (1918)

Thallus polymorphic, areoles very variable, isolated, few, rounded or angular to elaborately lobate, contiguous or ± overlapping, 0.5–3(-4) mm wide, flat or convex, rather thick, pale brown-grey, surface smooth or slightly roughened, matt, partly to entirely blue-grey-pruinose, margin often eroded and white, underside white. Algal layer irregular, discontinuous, tooth-like below. Apothecia 0.7–1.5(-4) mm, few to abundant, sometimes absent, small and rounded or becoming lobulate, at times almost entirely occupying individual squamules; thalline exciple raised, swollen; disc flat or concave, ± concolorous or slightly darker than thallus, naked or often ± densely blue-grey-pruinose; hymenium 60–80(-100) μm tall. Paraphyses 3 μm thick at base. Ascospores 4–8 × 1.5–3 μm. Thallus Pd−, K−, KC−, C− (lichen products not detected by t.l.c.).

On hard limestones, epidiorite and other calcareous rocks, particularly in upland sites; local. Throughout British Is. Scandinavia.

Morphologically variable, often recognized as two distinct species: *A. cervina* has much reduced, scattered, rounded squamules which are not imbricate, sometimes pruinose, at least at their edges, and have ± naked apothecial discs; *A. glaucocarpa* has irregular to sublobulate, often imbricate squamules with densely chalky blue-grey-pruinose apothecia.

4. A. heppii (Naeg. ex Hepp) Naeg. ex Körber (1859)

Thallus effuse, very thin, mostly scarcely visible, indistinctly areolate, rarely minutely warted, pale brown to whitish grey. Apothecia 0.2–0.4(-0.6) mm diam., single, rarely 2- to 3-contiguous, regularly rounded, pale reddish brown, rarely grey-white, disc concave, surrounded by a ± concolorous or ± white- or green-pruinose true exciple; hymenium 80–120 μm. Paraphyses very thin, 0.5–1 μm at base, 1.5 μm at the tips, branched and anastomosing. Ascus with K/I+ blue apical dome. Ascospores 3–6 × 1.5–3 μm, ± shortly ellipsoid. Thallus Pd−, K−, KC−, C− (lichen products not detected by t.l.c.).

On calcareous rocks and stranded sea shells, occasionally on concrete or mortar in urban sites; widely distributed but very local. Throughout British Is. Europe, North America.

This diminutive species is distinguished by the tiny, dispersed apothecia connected by a thin, thallus network, *c.* 1.5 μm thick; its affinities with *Acarospora* need investigation.

5. A. impressula Th. Fr. (1871)

A. atrata Hue (1909)

Thallus to 2 cm diam., often in small patches, areolate; areoles rather small, (0.05-)0.5–1.5(-2.0) mm wide, contiguous, variously angular, rounded or elongate, dark reddish brown to black-brown, not all areoles fertile. Algal layer irregular, ± continuous, not tooth-like below. Apothecia 0.2–0.3(-0.5) mm diam., 1(-2) per areole, mostly punctiform, immersed and concave; disc round or variously angular, concolorous with thallus, without or with a thin, ± darker swollen margin; hymenium 60–80 μm tall. Paraphyses 1.5–2 μm at base, widening to 2–3 μm at apex. Asci *c.* 100-spored. Ascospores 3–4 × 2–2.5 μm, shortly ellipsoid. Thallus Pd−, K−, KC−, C− (lichen products not detected by t.l.c.).

On sunny, exposed, siliceous, at times heavy metal-rich, rocks and roofing slates, often on or near the coast; local. Throughout British Is. Europe.

Characterized by the aggregated, thin, small, sharply angled, flat, even, dark, interlocking areoles and small, deeply impressed apothecia with punctiform or ± oval-elongate discs.

6. A. macrospora (Hepp.) Massal. ex Bagl. (1857)

A. murorum Massal. (1853), ?*A. macrospora* var. *incusa* (Körber) Magnusson (1924)

Thallus forming extensive patches of interlocking areoles, rarely of ± isolated areoles; areoles 0.5–5 mm wide, flat, irregularly angular, mostly of a similar size, red-brown, pale brown or whitish brown, often sterile towards the edge of the thallus. Algal layer ± continuous. Apothecia 0.5–1 mm diam., 1–3(-8) per areole, ± immersed, at times confluent, red-brown or dark brown, almost always darker than thallus; thalline exciple inconspicuous or absent; hymenium 100–160 μm tall; disc naked, smooth or somewhat wrinkled. Paraphyses 2–3 μm thick at base, 3–5 μm at tips. Asci 30-to 100-spored. Ascospores 6–13 × 3–6 μm, ellipsoid. Thallus Pd−, K−, KC−, C− (lichen products not detected by t.l.c.).

On sunny, calcareous rocks, rarely mortar of old walls; very local. Mainly upland British Is. W. Europe.

In *subsp.* ***murorum*** (Massal.) Clauz. & Roux (1981) the squamules are greenish brown, green when wet, and the apothecia are usually solitary, prominent and have a thick thalline exciple. The species is easily distinguished by the large ascospores and habitat.

7. A. nitrophila Magnusson (1924)

A. muddii Magnusson (1926), *A. normanii* Magnusson (1924), *A. praeruptorum* Magnusson (1924)

Thallus of squamules 0.4–5 mm wide, scattered or in small groups; squamules flat or moderately convex, reddish brown to black-brown, usually with a darker margin or paler beneath, surface irregular, at times roughened with irregular, lobulate-crenulate edges. Apothecia 0.2–1.4 mm diam., 1–6 per areole; thalline exciple concolorous or pale, often prominent; disc finely roughened; hymenium 70–110 μm tall. Paraphyses to 3–6 μm wide at the tips. Ascospores 3–6 × 1.5–2 μm.

On siliceous rocks in upland areas; rare. Upland British Is. Europe.

8. A. rhizobola (Nyl.) Alstrup (1986)

Thallus of squamules to 4 mm diam.; squamules peltate and scattered or ± adpressed-imbricate, deep reddish to dull brown, bright green when wet, surface smooth or finely cracked-areolate, edges and underside white; distinctive whitish rhizines present below which firmly attach the squamules to the substratum. Apothecia unknown in British specimens, described as 0.3–0.4 mm diam., completely immersed. Asci 200–250 × 15–20 μm. Ascospores 4–5 × 2–2.2 μm. Pycnidia 0.08–0.15 mm diam., frequent; conidia 4–5 × 1 μm.

On mica-schist detritus; very rare. Scotland (Perth, Ben Lawers). Greenland.

The thallus, with rounded or ± extended squamules, and pale-edged has the aspect of a *Psora*.

9. A. rufescens (Ach.) Krempelh. (1861)

Thallus forming extensive, smooth, very even, dull or dark brown patches of ± closely contiguous areoles, gradually becoming thinner towards the edge; areoles 0.4–1 mm wide, rounded or angular, ± flat, thin. Apothecia 0.1–0.5 mm, 1(-5) per areole, punctiform, immersed; thalline exciple entire, slightly raised, ± prominent; disc concave or flat, concolorous or darker than thallus; hymenium 60–80 μm tall. Paraphyses 1–1.5 μm thick. Ascospores 3–4 × 1–1.5 μm. Thallus Pd−, K−, KC−, C− (lichen products not detected by t.l.c.).

On sandstone and other siliceous substrata, chiefly on monuments, gravestones, and wall tops; local. S. & S.E. England. France, Belgium, Sweden.

A. fuscata **(3)** with which it often grows, differs in the larger, more uneven, thicker, often paler, red-brown, squamules which are C+ red. *A. rufescens* has been considered a variety of *A. smaragdula* **(11)**, but the continuous, spreading thallus, K− reaction, and lower hymenium separate it from that species.

10. A. sinopica (Wahlenb.) Körber (1859)

Thallus in small patches, lines, or wide-spreading, rarely as isolated areoles between other lichens; areoles mostly contiguous, sometimes coherent, even, rounded to angular, separated by thin cracks, often sterile towards the margin, pale to dark rust-red, smooth, lower surface pale. Algal layer ± continuous. Apothecia usually abundant, (1-)2–5(-8) per areole, slightly raised to ± elevated, wart-like, punctiform; thalline exciple indistinct; disc 0.1–0.2(-0.4) mm diam., ± dark brown; hymenium (110-)120–150(-170) μm tall. Paraphyses 1–1.5 μm thick, not capitate. Ascospores 3–3.5 × 1–1.5 μm. Thallus Pd−, K− KC−, C− (lichen products not detected by t.l.c.).

On iron-rich rocks, mine-spoil heaps, often in ± exposed situations; local. W. Britain. Europe.

A distinctive species with a uniformly rust-red thallus and punctiform apothecia, often occurring with *Tremolecia atrata*, *Rhizocarpon oederi* and, sometimes *Lecidea silacea*, which also have rust-coloured thalli but with black apothecia and very different ascospores.

11. A. smaragdula (Wahlenb.) Massal. (1852)

A. amphibola Wedd. (1875), *A. benedarensis* Knowles (1913), *A. fusca* B. de Lesd. (1914), *A. scyphulifera* Vainio (1909), *A. smaragdula* var. *lesdainii* (Harm. ex A. L. Sm.) Magnusson (1929), *A. smaragdula* var. *murina* (Sandst.) Magnusson (1929), *A. scabrida* auct. brit. p.p., non Hedl. ex Magnusson (1924)

Thallus very polymorphic, typically pale fawn or whitish brown, rarely yellowish green, often of a waxy appearance, either of ± widely scattered, rounded, peltate, low convex, slightly swollen areoles, to 1 mm wide, or areoles ± contiguous forming a well developed, ± continuous crust of convex areoles; areoles smooth, warted or concentrically ridged. Algal layer ± continuous. Apothecia (1-)3–7(-12) per areole; thalline exciple absent to well developed, black and friable; disc dark or blackish brown, typically minute, 0.2–0.3 mm diam., punctiform, smooth, and immersed, occasionally widening to 0.7 mm diam., rarely ± sessile and becoming umbonate; hymenium 60–280 μm tall. Paraphyses 1–1.5 μm thick. Ascospores 2–5 × 1–2 μm. Thallus Pd± yellow, K+ red or rarely K−, KC+ red or rarely KC−, C− (norstictic acid in the medulla and algal layer, sometimes in small amount).

On siliceous or slightly basic, often heavy metal-rich rocks, often in somewhat sheltered situations, also on worked stone below metal grills protecting stained glass windows, rarely on soil and fence posts treated with wood preservatives; common. Throughout British Is. ?Cosmopolitan.

The thallus varies in colour according to the rock mineralogy; yellow-green morphotypes occur on copper-rich rocks. An extremely variable and confusing species in the form and disposition of the areoles and size of the apothecia. Individual thalli may combine, according to aspect, dark, minute, punctiform discs and elsewhere larger, wide, ± emergent apothecia with a dark, distinct thalline exciple; the K+ red (crystals) reaction of the medulla, best observed in squash preparations, serves to unite the many environmental modifications. See also the K+ red *A. subrufula* **(12)** and *A. verruciformis* **(15)**.

12. A. subrufula (Nyl.) H. Olivier (1900)

Thallus areolate; areoles 0.3–2 mm wide, markedly contiguous, often ± in lines, edges rounded or ± irregular by compression, swollen, convex, faintly pale striate, pale to deep reddish brown. Apothecia numerous 1(-3) per areole, dark grey-red-brown, darker than the thallus; thalline exciple somewhat prominent, ± swollen, striate; disc 0.1–0.3 mm diam., rounded, oblong, kidney-shaped, rarely ± contorted; hymenium 80–100 μm tall. Paraphyses 1.5–2 μm wide at the base, 2–4 μm at apex. Ascospores 4–6 × 1–2.5 μm. Thallus Pd+ yellow-orange, K+ red, KC+ red, C− (norstictic acid).

On sunny, exposed, granitic coastal rocks; rare. Channel Is., Is. of Scilly. France, Portugal, Spain.

A rare species characterized by the convex areoles mostly containing a single fruit, surrounded by a ± swollen, striate thalline exciple, the disc darker than the surrounding areole, the usually intense K+ red reaction, and the southern, coastal distribution. The thallus often forms lines of areoles in minute fissures of the substratum.

13. A. umbilicata Bagl. (1857)

Thallus continuous, areolate; areoles 0.5–1.5 mm diam., scattered to contiguous, markedly convex, densely white-pruinose. Apothecia less than 0.5 mm diam., 1 or 2–5 per areole, often coalescing, appearing as situated between areoles but at first immersed on sides of areoles; thalline exciple inconspicuous; disc concave becoming irregular, brown-black; hymenium 85–110(-135) μm tall. Paraphyses *c.* 1.5 μm wide. Ascospores 4–5 × 1.5–2 μm. Thallus Pd−, K−, KC+ red, C+ red (gyrophoric acid).

On siliceous rocks subjected to seepage from adjacent calcareous rocks or mortar, particularly on old walls; very local. England (Kent, Suffolk, Gloucester, Hereford, Yorkshire), Scotland (Perth). S. Europe.

The white-pruinose surface is due to a dense deposit of calcium oxalate crystals. The British specimens correspond to **f. *congredians*** Magnusson (1929) or possibly a distinct entity. The status of this species requires further study, many British specimens referred to it are pruinose morphs of *A. fuscata* **(3)**.

14. A. veronensis Massal. (1852)

A. magnussonii G. Samp. (1924)

Thallus of scattered or ± contiguous groups of areoles; areoles 0.2–1.4 mm wide, rounded or angular by compression, occasionally slightly lobulate, flat or ± convex, matt or shiny, dark red-brown. Apothecia numerous, 0.1–0.5 mm, (1-)2–3 per areole, sometimes confluent, concave, ± crater-like; thalline exciple inconspicuous; disc concolorous or blackish brown, rounded or oval or elongate by compression, smooth or wrinkled; hymenium 60–100 μm tall. Ascospores 3–5 × 1.5 μm. Thallus Pd−, K−, KC−, C− (lichen products not detected by t.l.c.).

On nutrient-rich siliceous rocks, pebbles in dunes and roofing tiles; rare. Mainly N. & W. British Is. W. Europe.

Superficially similar to *A. impressula* **(5)** which has more punctiform, less concave apothecia and generally darker areoles. *A. complanata* Magnusson (1924), to be expected in the British Is., has a squamulose-areolate rather than dispersed thallus and has angular rather than rounded and more squamule-like areoles.

15. A. verruciformis Magnusson (1924)

Thallus of scattered or ± contiguous areoles; areoles 0.4–3 mm wide, semi-globose or irregularly subsquamulose or warted, rather easily detached, grey-brown or greenish, especially when wet, pale below. Algal layer ± interrupted. Apothecia 0.2–0.4 mm diam., 1(-2) per areole, crater-like, immersed; thalline exciple thick, obtuse, dark brown to blackish, rising above the thallus; disc blackish brown, flat to concave, roughened; hymenium 100–130(-200) μm tall. Paraphyses *c.* 1.5 μm wide, apices not thickened. Ascospores 3–4 × 1.5 μm. Thallus Pd−, K−, KC−, C− (lichen products not detected by t.l.c.).

On siliceous, often heavy metal-rich rocks; rare. England (Yorkshire, Cornwall), N. Wales (Anglesey). Scandinavia.

Characterized by the ± scattered squamules, often distinctly greenish when wet, with large, dark, crater-like apothecia and the K− reaction. Specimens from sunny, serpentine rocks (Cornwall, Lizard; Shetland, Unst; Sutherland, Ben Hiant), previously identified as *A. verruciformis*, appear to represent an undescribed species with often ± poorly formed, concave to flat, dark red-brown squamules with raised, dark-edged margins and well-developed, conspicuous apothecia with a thick, shining, black, true exciple which may become laterally fissured; disc ± uneven-wrinkled, often with a prominent, central, sterile umbo.

ACROCORDIA Massal. (1854)

Thallus crustose, effuse, whitish or pale, thin or immersed in the substratum, cortex absent. **Photobiont** *Trentepohlia*. **Ascomata** perithecia, almost entirely immersed to ± sessile, black (rarely pink or whitish), compound, with a hemispherical to globose, brown-black involucrellum surrounding a ± globose, colourless or pale brownish exciple. **Hamathecium** of persistent, slender, sparingly branched or anastomosing, long-celled paraphysoids; periphyses absent. **Asci** (4-)8-spored, cylindrical, K/I−, fissitunicate; apical dome with a broad ocular chamber surmounted by a hemispherical, meniscus-like structure (Fig. 35a). **Ascospores** uniseriate, colourless, ellipsoid to oblong-ellipsoid, the ends usually rounded, 1(-3)-septate, the median septum thick and cells ± equal, not or slightly constricted at the septum, additional septa (if present) thin; epispore ornamented with minute warts that disappear in K. **Conidiomata** pycnidia; conidiogenous cells cylindrical, elongate; conidia acrogenous, ellipsoid to narrowly ellipsoid, simple, colourless. **Chemistry**: lichen products not detected by t.l.c. **Ecology**: ± basiphilous, on bark of broad-leaved trees or on mostly vertical surfaces of calcareous or basic rocks in humid situations.

Characterized by the cylindrical asci, with an apparently unique apical structure, and uniseriate, 1-septate, ellipsoid ascospores, with a warted epispore that disperses in K.

LITERATURE: Coppins & James (1978), Harris (1973), Swinscow (1970).

1 On rocks, rarely on soil 2
On bark, rarely on wood 4

2(1) Involucrellum spreading laterally away from base of true exciple, never continuous below 3
Involucrellum incurved under true exciple, often ± continuous below .... **5. *salweyi***

3(2) Ascospores 12–19 × 6–9 μm; perithecia often with distinct papillose ostioles; thallus usually immersed .... **2. *conoidea***
Ascospores 19–26(-28) × 9.5–12 μm; perithecia without distinct papillose ostioles; thallus not immersed .... **4. *macrospora***

4(1) Perithecia 0.5–1 mm diam.; ascospores 15–27(-30) × 9–12 μm .... **3. *gemmata***
Perithecia 0.3–0.6 mm diam.; ascospores 11–16.5 × 5.5–9.5 μm .... **1. *cavata***

1. A. cavata (Ach.) R. C. Harris (1974)

Thallus immersed, grey-white. Perithecia 0.3–0.6 mm diam., half-immersed; involucrellum hemispherical. Ascospores 11–16.5 × 5.5–9.5 μm. Pycnidia not known.

On wood of *Ilex* and on bark of *Corylus*. C.N. Scotland (Argyll, Perth). Europe, North America.

Similar to *A. gemmata* **(3)** but with smaller perithecia and ascospores. The ascus structure and uniseriate, somewhat larger, ornamented spores distinguish *A. cavata* from *Anisomeridium biforme*.

2. A. conoidea (Fr.) Körber (1855)

Thallus mainly immersed, smooth or finely granular, effuse, pale grey or brownish grey, often with a pinkish tinge when fresh. Perithecia 0.5–1 mm diam., ¼- to ½- immersed, conical-hemispherical, often flattened and with distinct papillose ostioles; involucrellum spreading outwards away from the exciple. Asci (Fig. 35a). Ascospores 12–19 × 6–9 μm, the ends mostly rounded but sometimes pointed, occasionally with a thin, secondary septum dividing each cell. Pycnidia 0.14–0.2 mm diam., frequent, often numerous.

On hard limestones and calcareous walls in ± shaded and moist situations. Abundant in areas with naturally occurring hard limestones (particularly Devonian and Carboniferous). C., N., & W. Britain, N.W. Ireland. Europe, North America.

When well developed, easily recognized by the flattened, ± conical appearance of the rather evenly dispersed perithecia. Difficult specimens can be distinguished from *A. macrospora* **(4)** and *A. salweyi* **(5)** in section by the more outwardly spreading base of the involucrellum and the smaller ascospores.

3. A. gemmata (Ach.) Massal. (1854)

Thallus immersed, white or pale grey. Perithecia 0.5–1 mm diam., black (rarely pink), $\frac{1}{4}$- to entirely immersed; ostiole often eccentric and sometimes papillate. Ascospores 15–30 × 8–12 μm. Pycnidia 0.1–0.25 mm diam., frequent but rarely numerous; conidia 3–5 × 0·8–1 μm.

In woodlands or open situations on rough bark of mature trees, especially *Fraxinus*, *Quercus* and *Ulmus*, more rarely on smooth bark of *Corylus*; widespread and locally common, but absent from central areas due to air pollution. Throughout British Is. but absent from much of C. Britain. Europe, ?North America.

Distinguished from the rare *A. cavata* **(1)** by the larger perithecia and ascospores. The distinctive ascus and ascospore morphology of the genus distinguish *A. gemmata* from other superficially similar, bark-inhabiting taxa, e.g. *Anisomeridium*, *Pyrenula* and the non-lichenized *Lophiostoma* species. Occasionally found with pycnidia only.

4. A. macrospora Massal. (1855)

Thallus not immersed, grey to grey-brown, smooth or ± rimose. Perithecia 0.8–1 mm diam., rather prominent, ostiole not papillose; involucrellum ± spreading laterally, never continuous below the exciple. Ascospores 19–26(-28) × 9.5–12 μm. Pycnidia 0.2–0.3 mm diam., usually few.

On sheltered, often shaded, siliceous or weakly calcareous rocks in coastal areas. N. & W. British Is. Italy, Norway.

Resembles *A. conoidea* **(2)** but the perithecia are not flattened, lack papillate ostioles and the ascospores are smaller. *A. salweyi* **(5)** is distinguished in section by the tightly incurved involucrellum which is often continuous below the true exciple; it occurs on strongly calcareous substrata.

5. A. salweyi (Leighton ex Nyl.) A. L. Sm. (1911)

Thallus ± immersed, whitish to pale brownish grey, thin, granular or smooth to rimose. Perithecia 0.7–1.5 mm diam., $\frac{1}{4}$- to $\frac{1}{2}$- immersed; involucrellum tightly incurved around the exciple and often continuous below. Ascospores 20–35 × 10–15 μm. Pycnidia 0.2–0.3 mm diam., usually few.

On soft, ± sheltered, often damp, highly calcareous rocks and especially old mortar, rarely terricolous. Throughout British Is., common in S. and W. England. Widespread in Europe.

ADELOLECIA Hertel & Hafellner (1984)

Thallus crustose, superficial and areolate or ± immersed. **Medulla** I−. **Photobiont** chlorococcoid. **Ascomata** apothecia, sessile, constricted at the base, black, the disc concave to flat or somewhat convex; margin prominent, persistent. **True exciple** dark brown at the surface but paler internally, of radially orientated, wide hyphae, almost pseudoparenchymatous in parts, K+ purplish red. **Thalline exciple** absent. **Epithecium** green in section, K−, N+ purple. **Hymenium** ± colourless to bluish tinged, I+ deep blue. **Hypothecium** pale to dark brown. **Hamathecium** of simple or sparingly branched and septate paraphyses, not markedly expanded at the tips, which are immersed in a greenish epithecial gel. **Asci** cylindrical-clavate, *Bacidia*-type, 8-spored. **Ascospores** narrowly ellipsoid, simple, colourless, smooth, lacking a distinct perispore. **Conidiomata** not observed. **Chemistry**: atranorin in the cortex and the pigment 2-chloroemodin in the true exciple and sometimes hypothecium; "*Lecidea* green" in the epithecium; unidentified products. **Ecology**: on somewhat basic, siliceous rocks in montane situations.

This monotypic genus is especially characterized by its *Bacidia*-type asci, and the intense K+ purple-red reaction of the true exciple when sections are flooded with K. It differs also from *Bacidia* in the simple ascospores.

LITERATURE: Hafellner (1984), Hertel (1975a).

1. A. pilati (Hepp) Hertel & Hafellner (1984)

Lecidea pilati (Hepp) Körber (1861)

Thallus of dark grey to whitish areoles, or ± immersed and scarcely apparent. Apothecia 0.5–1.5(-2) mm diam., sometimes aggregated together and then angular and mosaic-like, sometimes tuberculate; hymenium 35–50 μm tall. Ascospores (6-)8–11(-13) × (3-)3.5–5 μm. Thallus Pd−, K−, KC+ pink or KC−, C− (unidentified compound in some specimens); exciple K+ purple-red (2-chloroemodin); epithecium N+ purple-red ("*Lecidea* green").

On siliceous rocks in arctic-alpine situations, rare. N. Wales (Gwynedd), Scotland (Highlands). N. Europe, Alps.

A distinctive species easily identified by the colour reactions of the true exciple and epithecium, and further characterized by the narrowly ellipsoid spores.

Sometimes confused with the coastal species *Lecidea sarcogynoides* which, however, has *Lecidea*-type asci, narrower ascospores, and a weaker K+ purplish (not magenta) reaction to the hymenium and exciple.

AGONIMIA Zahlbr. (1909)

Thallus minutely squamulose to coarsely granular, pale green- or blue-grey to brown; squamules ± round or elongate, entire or divided, without soredia and isidia. **Cortex** pseudoparenchymatous; cells on upper surface with short papillae 1–2 × 1–1.5 μm. **Photobiont** chlorococcoid; cells 4.5–7(-10) μm diam. **Ascomata** perithecia, globose to barrel-shaped, black, often overgrown by the squamules; involucrellum absent. **True exciple** 3-layered, the outer dark brown, the inner two pale or colourless. **Hamathecium:** upper part of cavity lined with numerous simple or sparingly branched, septate periphyses, up to 60 × 2–4 μm; paraphyses absent. **Asci** 1- to 2- or 8-spored, clavate, thin-walled; wall 5–7 μm wide laterally, not or only slightly thickened at apex; without an ocular chamber; entirely I−; dehiscence fissitunicate. **Ascospores** colourless to pale yellow-brown, ± ellipsoid, strongly muriform. **Conidiomata** unknown. **Chemistry**: lichen products not detected by t.l.c. **Ecology**: on old, broad-leaved trees and ± basic soils, dunes, mortar and associated mosses.

Similar to *Polyblastia* but differs in having a squamulose thallus and thicker, 3-layered perithecium wall. The thallus granules of *Macentina* species also have cortical papillae, but the perithecia in that genus are smaller and pale-coloured and their asci and ascospores are very different. *Polyblastia allobata* (Stizenb.) Zsch. has recently been transferred to this genus as ***A. allobata*** (Stizenb.) P. James (1992).

LITERATURE: Coppins & Bennell (1979), Coppins & James (1978).

1 Perithecia smooth; asci 8-spored; squamules 0.05–0.1 mm wide, pale grey or grey-green, never brownish .... **1. *octospora***
Perithecia rough and plicate in upper part; asci (1-)2-spored; squamules 0.1–0.3(-0.5) mm wide, dull green, usually brown-tinged .... **2. *tristicula***

1. A. octospora Coppins & P. James (1978)

Thallus minutely squamulose; squamules 0.05–0.1 × 0.05–0.25 mm, ± elongate, sometimes branched, the margin entire or notched, scattered and dispersed amongst mosses or ± in contiguous groups, and becoming imbricate, closely adpressed, simple or ± irregularly divided; upper surface even, matt or slightly shiny, pale grey-green, blue-grey when wet. Perithecia 0.5–0.8 mm diam., ½- to ¾-emergent, scattered amongst squamules, often absent, globose or ± pyriform, black, matt or ± shiny; wall 120–160 μm thick, 3-layered; ostiole somewhat papillate, pale pink-grey (immature) to black (mature), ± eccentric. Asci 8-spored. Ascospores 60–75(-85) × 20–26(-30) μm, ellipsoid.

On the mossy bark of trunks of aged *Quercus* or *Fagus* in somewhat sheltered situations in old woodlands. S. Britain, S.W. Ireland. France (Brittany), Italy (Tuscany), Macaronesia.

The perithecia are sometimes isolated from the squamules. When sterile can be confused with corticolous morphs of *A. tristicula* **(2)**, *Macentina stigonemoides*, which has smaller, darker green, isidia-like squamules, or *Rinodina isidioides*, whose 'squamules' are K+ yellow, have larger photobiont cells and non-papillate cortical cells. *Polyblastia allobata* occurs in similar habitats, but has smaller perithecia and ascospores, and, at most, a minutely granular, rarely isidiate thallus.

2. A. tristicula (Nyl.) Zahlbr. (1909)

Thallus minutely squamulose; squamules to 0.1–1 × 0.1–0.3(-0.5) mm, crowded, ± contiguous and erect, sometimes nodulose and subgranular or rarely dispersed and elongate-digitiform, dull pale green-fawn to green-brown when dry, bright green when wet; upper cortex with irregular, low, abraded projections, rarely with small papillae (shade). Perithecia 0.4–0.5 mm diam., rather frequent, often in between and overgrown by squamules, barrel-shaped with plicate-rugose outer surface; wall thick, black, opaque and irregular externally, paler or brownish internally. Asci (1-)2-spored. Ascospores (60-)80–120(-150) × 26–50 μm, elongate-ellipsoid.

On soil, mosses and lichens (especially *Collema* species) in clefts of ± base-rich rocks, including limestones, mortar of walls, basalts, serpentines and soapstone, also on the ground on ± basic dunes, more rarely on bark of base-rich trees, especially *Acer*, *Ulmus* and *Fraxinus*; frequent. Throughout British Is. Europe, Azores.

Sterile, corticolous morphs are distinguished from *A. octospora* **(1)** by their larger, usually more distinctly flattened squamules which are darker olivaceous or brownish when dry and bright green when wet.

AGYRIUM Fr. (1822)

Thallus immersed within substratum but surface often paler. **Photobiont** absent. **Ascomata** apothecia. **Thalline exciple** absent. **True exciple** poorly developed. **Epithecium** red-brown, non-granular, K+ dissolving. **Hymenium** I+ blue. **Hypothecium** colourless to pale straw. **Hamathecium** of paraphyses, branched and entangled, especially towards the apices, slender, slightly swollen at apices. **Asci** 8-spored, *Trapelia*-type. **Ascospores** colourless, non-septate. **Chemistry**: lichen products not detected by t.l.c. **Ecology**: on bark and wood.

LITERATURE: Dennis (1981).

1. A. rufum (Pers.) Fr. (1822).

Lecidea grumosa Leighton (1876)

Apothecia to 1 mm diam., discrete, rounded to broadly ellipsoid, sessile, ± gelatinous when wet; disc convex, immarginate, faintly rugose, reddish- or orange-brown; hymenium 85–90 μm tall. Asci 70–80 × 10–12 μm, clavate, the apex broadly rounded, rather thick-walled, 8-spored. Ascospores 10–15 × 6–8 μm, ellipsoid, simple, biseriate, colourless, often with a large central oil drop.

On old wood of mainly conifers, rarely on bark and *Calluna* stems; ?frequent. Throughout Britain, especially the north. Europe.

Evidently not lichenized, but clusters of chlorococcoid algae ± obligately surrounded by hyphae of the fungus are occasionally present. *Agyrium rufum* forms spreading pale stains with scattered apothecia on smooth surfaces of rotting wood.

ALECTORIA Ach. (1810)

Thallus shrubby, erect, decumbent or pendent, greenish yellow or brown-black; branches generally terete and smooth, occasionally becoming compressed and striately ridged towards base and axils, in one subspecies markedly expanded and dorsiventrally compressed. **Isidia** and **soredia** absent. **Pseudocyphellae** always present, conspicuous, fusiform, white, flat to markedly raised. **Cortex** composed of periclinal, arachnoid, conglutinated hyphae; hyphae immersed in ± large amount of matrix. **Photobiont** trebouxioid. **Medullary** hyphae usually ornamented. **Ascomata** apothecia, lateral, rare. **Thalline exciple** persistent, concolorous with thallus. **Asci** clavate, thick-walled, 2- to 4-spored, *Lecanora*-type. **Ascospores** exceeding 20 μm in length, simple, ellipsoid, with distinct, colourless perispore, brown. **Conidiomata** pycnidia. **Conidia** 7–8 × 0.8 μm. **Chemistry:** Usnic acid present in all but one species **(1)**, orcinol and α-orcinol depsides frequent. **Ecology:** essentially arctic-alpine.

Readily separable from other genera, including *Bryoria* and *Pseudephebe*, even in the absence of apothecia, by the characteristic chemistry and cortical structure, as well as the abundance and structure of the pseudocyphellae.

LITERATURE: Brodo & Hawksworth (1977), Hawksworth (1972b).

1 Thallus ± uniformly pale yellow or yellowish green (usnic acid), sometimes blackened at apices; medulla C− 2
Thallus pink-grey to pale grey-brown at base, dark grey-brown to grey-black towards apices (without usnic acid); medulla C+ red .... **1. *nigricans***

2(1) Thallus ascending or prostrate; scrambling on low vegetation or over the ground 3
Thallus pendent; on bark or rocks .... **3. *sarmentosa***

3(2) Thallus ascending; main stems ± terete, usually <2.0 mm diam.; medulla KC− **2. *ochroleuca***
Thallus decumbent; main stems becoming dorsiventrally flattened and striately ridged, often >2.0 mm in width; medulla KC+ red or KC− .... **3. *sarmentosa***

1. A. nigricans (Ach.) Nyl. (1861)

Thallus to 5 (-8) cm tall, erect, sometimes decumbent; branching moderate, ± anisotomic-dichotomous; branches to 1.5 mm diam., mostly terete, becoming compressed basally, pink-grey to pale grey-brown towards the base, dark grey-brown to sooty grey-black towards apices, ± matt, smooth, or ± pitted at base. Pseudocyphellae to 0.8 mm long, white, ± fusiform, flat or slightly convex, usually conspicuous on thicker branches. Soralia and apothecia unknown in Britain. Cortex and medulla Pd+ yellow, K+ yellow, KC+ red, C+ red (alectorialic and barbatolic acids).

Locally abundant in arctic-alpine moss-lichen heaths, above 760 m. N. England, N. Wales, Scotland (Highlands). European mtns, Arctic, Japan, Tierra del Fuego, Australia, New Zealand, Antarctica.

Specimens stain paper pink-brown in the herbarium.

2. A. ochroleuca (Hoffm.) Massal. (1855)

Thallus to 8 (-12) cm, erect or becoming decumbent; branching moderate, ± anisotomic-dichotomous; branches 0.5–2.0 mm diam., the main stems terete, never strongly compressed, yellowish green to bright yellow, the apices often grey-black, ± matt. Pseudocyphellae 1 × 0.1 mm, abundant, ± fusiform, white. Apothecia unknown in Britain. Pycnidia frequent, to 0.15 mm diam., ± apical, black. Medulla Pd−, K−, KC−, C−, CK+ yellow, UV− (diffractaic acid).

In *Rhacomitrium - Empetrum* communites, above *c.* 800 m. C. Scotland (Cairngorm Mtns). European mtns, North America, India, Nepal, China, Japan, New Zealand.

Sometimes confused with *A. sarmentosa* subsp. *vexillifera* **(3)**, which is prostrate, has compressed and expanded main stems, and a UV+ ice-blue medulla reacting usually KC+ red but never CK+ yellow.

3. A. sarmentosa (Ach.) Ach. (1810)

Thallus to 20(-80) cm; branching moderate, isotomic-dichotomous, sparser towards the base, which may become pitted, greenish grey, yellow to bright golden yellow, the apices usually concolorous or ± blackened, matt. Pseudocyphellae 1 × 0.2 mm, elongate, ± fusiform, flat or convex, often abundant and conspicuous. Soralia unknown in Britain. Apothecia to 5 mm diam., lateral; thalline exciple concolorous with thallus; disc orange-yellow to dark brown or black, matt or ± shiny. Medulla Pd−, K−, KC− or KC+ red, C−, CK−, UV+ ice-blue (± alectoronic, squamatic, and perhaps α-collatolic acids).

Distinguished by the pendent or prostrate habit. Thin-branched morphs may resemble *Ramalina chondrina*, but that species has abundant, conspicuous, hook-shaped apices, is exclusively coastal in S.W. British Is., and lichen products have not been detected by t.l.c.

Subsp. *sarmentosa*: Thallus pendent; branches to 2.5 mm diam., terete, occasionally flattened at axils and towards base.

On *Pinus* bark in relict Caledonian woodlands, rarely on rock. N.E. England (Northumberland), Scotland (Highlands). Scandinavia, C. Europe, North America, South America (Patagonia).

Subsp. *vexillifera* (Nyl.) D. Hawksw. (1970): Thallus prostrate; branches to 8 mm wide, often markedly flattened, expanded, striately ridged and pitted; base dying, supported by hapters attached to surrounding vegetation.

Decumbent over low vegetation in montane areas, to 1100 m. N.E. England (Northumberland), Scotland. Scandinavia, S.W. Greenland, E. Canada, mtns. in Tanzania.

Subsp. *vexillifera* varies considerably in the degree of flattening and width of the main branches and is at times difficult to separate from subsp. *sarmentosa*, with which it intergrades in some parts of Europe but not in North America (and is therefore retained as distinct). May be confused with *A. ochroleuca* **(2)** which is, however, erect in habit, lacks compressed and expanded main stems and its UV− medulla is always KC− and CK+ yellow.

ALLANTOPARMELIA (Vainio) Essl. (1978)

Thallus foliose, closely adpressed, sometimes appearing ± crustose, attached to substratum by irregular, peg-like cortical outgrowths; rhizines absent; lobes convex, dorsiventrally flattened, corticate on both sides. **Upper surface** without pseudocyphellae, soredia or isidia. **Cortex** prosoplectenchymatous, arising from hyphae perpendicular to surface, K−, N−; upper and lower cortex similar in structure. **Photobiont** chlorococcoid. **Ascomata** apothecia, laminal, sessile; disc brown-black. **Thalline exciple** present. **Hymenium** colourless. **Asci** 8-spored. **Ascospores** colourless, ellipsoid to ± globose, simple. **Conidiomata** pycnidia, ± immersed in thallus. **Conidia** colourless, bifusiform to bacilliform or slightly clavate. **Chemistry:** 2 benzyl ester depsidones unknown elsewhere in the foliose Parmeliaceae.

Separated from *Parmelia* by the lower surface which lacks rhizines. The genus is chemically distinct from *Hypogymnia*.

LITERATURE: Esslinger (1977, 1978).

1. A. alpicola (Th. Fr.) Essl. (1978)

Thallus to 6 cm diam., forming rosettes or irregularly spreading, closely adpressed; lobes 0.2–1 mm wide, convex, elongate, somewhat gnarled, contiguous, often overlapping and entangled, particularly in older parts, at times so tightly compacted as to appear crustose towards the centre; upper surface dark brown to black, grey-black, rarely dark olive. Soralia and isidia absent; lower surface black, ± pleated; rhizines absent but occasionally peg-like outgrowths present. Apothecia frequent; disc to 5 mm diam.; margin entire, rarely crenate. Ascospores 7.5–10 × 5–7 μm. Conidia 4.5–6 × 1 μm. Medulla Pd+ bright yellow, K+ yellow, KC+ red, C+ rose-red, UV− (alectorialic and barbatolic acids).

On exposed siliceous boulders in montane situations, usually above 700 m. Locally abundant in Scotland, less frequent in N. England (Cumbria, Northumberland). Europe, North America.

Distinguished by the dark colour, absence of rhizines and Pd reaction; specimens stain paper red-brown in the herbarium. *Parmelia stygia* has rhizines and pseudocyphellae and a Pd− or Pd+ rust-red medulla.

AMYGDALARIA Norman (1853)

Thallus crustose, superficial, areolate, often coarsely so, often with a pinkish tinge, ± immersed to tuberculate cephalodia (Fig. 30b) between the areoles. **Photobiont** *Trebouxia*-like, with *Stigonema* in the cephalodia. **Ascomata** apothecia, deeply immersed in the areoles; discs black or brown-black, flat to concave. **True exciple** often poorly developed, of conglutinated, radially orientated hyphae. **Thalline exciple** brown. **Hymenium** often exceeding 140 μm tall, I+

blue. **Hypothecium** ± colourless to pale brown, or dark brown to black and carbonaceous, then well developed and cupulate, K+ reddish. **Hamathecium** a net of thin, branched and anastomosing paraphysoid-like hyphae, septate, especially above where they are almost moniliform, not or slightly swollen at apex. **Asci** elongate-clavate, *Porpidia*-type, with a thin outer amyloid layer and a thickened tholus penetrated by a pore, the sides of which stain deep blue in iodine, 8-spored. **Ascospores** ellipsoid, simple, colourless, smooth, with a compact gelatinous perispore (halonate), often exceeding 20 μm in length. **Conidiomata** pycnidia, immersed; conidia bacilliform. **Chemistry:** gyrophoric acid often present, sometimes with lecanoric acid (other orcinol depsides or β-depsidones in non-British representatives). **Ecology:** on somewhat base-rich, siliceous rocks, especially basalts, in upland and mountainous areas.

This genus is essentially an "aspicilioid" counterpart of *Porpidia*, with the apothecium sunk into the thallus areoles. However, some non-British taxa placed here do have elevated apothecia and well-developed true exciples. They are also distinguished by other quantitative features, such as the generally taller hymenium, large ascospores, less developed exciple, the tendency of the thalli to be pinkish or brownish, and the presence of cephalodia and(or) gyrophoric acid in some species. It will perhaps prove necessary to combine the genus with *Porpidia*, in which case *Amygdalaria* would be the earliest generic name.

LITERATURE: Brodo & Hertel (1987), Esnault & Roux (1987), Inoue (1984).

1 Cortex C−; thallus cracked-areolate; areoles ± plane; cephalodia, when present, usually immersed, brown .... **1. *consentiens***
Cortex C+ pink; thallus of distinct and sometimes dispersed and strongly convex areoles; cephalodia usually present, immersed to conspicuously tuberculate, red-brown to grey .... **2. *pelobotryon***

1. A. consentiens (Nyl.) Hertel, Brodo & Inoue (1984)

Huilia consentiens (Nyl.) Hertel (1980)

Thallus grey to ± rust-coloured, more rarely yellowish or pinkish, cracked-areolate, the areoles ± plane, the margin usually well delimited; medulla I−; cephalodia, when present, usually immersed, rarely convex, brown. Apothecia (0.3-)0.5–0.8(-1.5) mm diam., usually one per areole; disc flat, often with a central umbo; hymenium (100-)150–180(-200) μm tall. Ascospores (21-)23–30(-40) × (9.5-)12–18(-22) μm. Thallus Pd−, K−, C− (no substances detected by t.l.c.).

On exposed crystalline schists and other siliceous rocks in mountainous areas; local. N.W. England (Lake District), N. Wales, Scotland (Highlands), Ireland. Europe, Canada, Greenland, Japan.

Distinguished from *A. pelobotryon* **(2)** by the C− reaction of the thallus, the more continuous and generally greyer thallus, the frequent presence of an umbo in the apothecia, and, when present, the immersed cephalodia. A chemotype Pd+ orange, K+ yellow, (stictic and constictic acids), var. *japonica* Inoue (1984), is known from Japan and Canada.

2. A. pelobotryon (Wahlenb.) Norman (1953)

Thallus pale yellowish grey, fawn, to yellowish brown, often with a pinkish tinge, cracked-areolate, the areoles continuous to somewhat dispersed, strongly convex, the margin indistinct; medulla I−; cephalodia immersed to tuberculate, dark red-brown to grey (Fig. 30b). Apothecia (0.3-)0.5–1 mm diam., usually one per areole; disc concave to flat, exceptionally umbonate; hymenium (135-)140–180(-210) μm tall. Ascospores (17-)20–34(-40) × (8-)12–16(-19) μm. Thallus Pd−, K−, C+ pink (gyrophoric, ± lecanoric acids).

On ± damp siliceous rocks, especially basalts, in upland and mountainous areas. S.W. (Shropshire) & N. England, C. & N. Wales, Scotland, W. Ireland (Antrim). Europe, Canada, Greenland, Madeira, Azores, China, Japan.

A morphologically variable species, dispersed areolate morphs of which are doubtfully distinct from the non-European *A. subdissentiens* (Nyl.) Inoue & Brodo (1984), which contains the stictic acid complex (and ± gyrophoric acid).

ANAPTYCHIA Körber (1848)

Thallus foliose or ± shrubby, dorsiventral, closely adpressed throughout or ascending at apices; lobes branched, elongate. **Upper surface** matt, at times tomentose and with marginal cilia. **Lower surface** with rhizines when lower cortex present. **Soredia** and **isidia** absent. **Upper cortex** composed of conglutinated, thick-walled hyphae, periclinal, orientated mostly parallel to the surface, overlain by another, colourless, ± transparent layer also composed of hyphae orientated mostly parallel to the surface. **Photobiont** green. **Ascomata** apothecia, ± stalked, frequent. **Thalline exciple** persistent. **Epithecium** brown. **Hymenium** colourless. **Hypothecium** colourless or yellowish brown, **Hamathecium** of paraphyses, branched towards apices, septate, little or not conglutinate, ± swollen at apices, concolorous with the thallus. **Asci** 8-spored, *Lecanora*-type . **Ascospores** dark brown, 1-septate, ellipsoid, with ± rounded apices, medianly constricted, thin-walled, uniformly thickened, sculptured with minute spines or ridges. **Conidiomata** pycnidia, forming small warts. **Conidia** 3–5 × 1.5 μm, bacilliform, straight, colourless. **Chemistry:** atranorin and zeorin (trace).

Anaptychia is clearly separated from *Physconia* and other related genera by the cortex being composed of longitudinally arranged hyphae, which can be readily recognized in both transverse and longitudinal sections. Distinguished from *Heterodermia* by its relatively uniformly thin-walled ascospores, the walls of which are sculptured with minute spines or ridges.

LITERATURE: Kurokawa (1962a, 1973), Maas Geesteranus (1952).

1 Lobes closely adpressed, attached by rhizines below; marginal cilia absent .... **2. *runcinata***
Lobes ± ascending, partly unattached, not attached by rhizines below, with pale grey-white marginal cilia .... **1. *ciliaris***

1. A. ciliaris Körber ex Massal. (1853)

Thallus 3–5 (-10) cm diam., ± strap-shaped, loosely attached; lobes elongate, to 3–5 cm long, pale grey to grey-brown above, irregularly branched, ± horizontally spreading with ascending apices and numerous, pale, entangled cilia along their margins. Lower surface pale brownish white, channelled, without rhizines or lower cortex. Apothecia 2–5 mm diam., infrequent; disc deeply cup-shaped to discoid, brown-black, often blue-grey-pruinose; thalline exciple smooth or conspicuously deeply crenulate. Ascospores 18–24 × 40–45 (-54) μm.

This decreasing species is easily recognized by its pale grey, strap-shaped lobes with numerous lateral cilia and large apothecia with crenate margins. Sensitive to SO_2 air pollution >50 μg m^{-3}.

Subsp. *ciliaris*: Thallus lobes more than 2 mm wide, pale whitish grey to brown, unchanged when wet, upper surface finely tomentose.

Predominantly on bark of nutrient-rich, well-lit, mature, broad-leaved trees, rarely on ± calcareous rocks and tombstones, widespread but decreasing. E. England & Scotland extending to S.W. England (Devon) and S.W. Wales (Cardigan, Pembroke), S.E. Ireland. S. Scandinavia, S. & E. Europe, N. Africa, Canary Islands.

Subsp. *mamillata* (Taylor) D. Hawksw. & P. James (1980): Thallus lobes *c.* 2 mm wide, dark grey to brown, dark olive-green when wet, not tomentose.

Predominantly on maritime rocks; widespread but local. W. Great Britain, E. Scotland & Ireland. Scandinavia.

In Britain the 2 subspecies appear distinct in habit and habitat, but in Scandinavia intermediates are reported. Narrow-lobed morphs may be confused with *A. runcinata* **(2)**, but have marginal cilia.

2. A. runcinata (With.) Laundon (1984)

A. fusca (Huds.) Vainio (1899)

Thallus to 10 cm diam., foliose, irregularly spreading, ± forming rosettes, or as scattered fragments amongst other lichens, closely adpressed throughout, rather thick; lobes 0.3–2.5 mm wide, often slightly wider at the apices, flat to convex, contiguous, densely overlapping towards the centre of the thallus; upper surface dull olive-green when moist, pale (in shade) to dark brown when dry, not pruinose; lower surface pale to dark brown-black, with simple, brown to black, scattered rhizines. Apothecia 1–3 mm diam., frequent; disc black-brown, rough, with a coarsely crenulate thalline exciple. Ascospores 15–18 × 30–36 μm.

Abundant on hard, coastal rocks, also on turf and soil, decayed *Armeria* tufts and occasionally on wayside trees; in a few inland sites on rock outcrops, standing stones, church walls and trees. Throughout British Is. W. & Mediterranean Europe, Canary Is.

Distinguished by the adpressed, plate-like, light to dark brown thalli, olive-green when moist.

ANISOMERIDIUM (Müll. Arg.) M. Choisy (1928)

Thallus crustose, immersed, whitish or pale grey. **Photobiont** *Trentepohlia*. **Ascomata** perithecia, hemispherical to globose, simple to compound; wall composed of ± cellular hyphae without bark cells, thicker in upper part, often with a well-differentiated involucrellum, brown-black above, pale brown to ± colourless below; pigment K+ greenish. **Hamathecium** of slender, branched and anastomosed, long-celled pseudoparaphyses, *c.* 1 μm thick; periphyses absent; hymenial gel I−. **Asci** cylindric-clavate, K/I−, fissitunicate, the apical dome with indistinct or short and broad ocular chamber (Fig. 34c), 8-spored (sometimes a few spores aborting), the spores uni- to biseriate. **Ascospores** 1- to 3-septate, ovoid to clavate-fusiform, the first formed septum often towards lower end of ascospore, colourless, smooth, without a distinct perispore. **Conidiomata** pycnidia, immersed to sessile, ± globose or conical, black. **Conidiogenous cells** cylindrical, enteroblastic, with collarettes, often percurrently proliferating. **Macro-** and **microconidia** (produced in separate pycnidia) subglobose, ellipsoid, ovoid or rod-shaped. **Chemistry**: no lichen products detected in European species. **Ecology**: on bark.

A widely occurring subtropical to temperate genus of at least 15 species. Differs from *Arthopyrenia* in a combination of several features, especially the structure of the involucrellum, which is entirely fungal in composition and not clypeate. *Strigula* differs in having a narrow ocular chamber in the ascus apex and septate macroconidia, which are obliquely attached to the conidiogenous cells. *Acrocordia* has a different ascus structure, and ellipsoid to oblong-ellipsoid ascospores with a median septum and a warted epispore.

LITERATURE: Harris (1973).

1 Perithecia 0.3–0.4(-0.5) mm diam., the wall 50–100 μm thick in upper part; ascospores (9-)12–16 × 4.5–6(-7) μm, 1-septate .... **1. *biforme***
Perithecia 0.15–0.25 mm diam., the wall 30–50 μm thick in upper part; ascospores (12-)14–20(-23) × (3-)4.5–5(-6) μm, 1- to 3-septate .... **2. *nyssaegenum***

1. A. biforme (Borrer) R. C. Harris (1978)

Thallus usually conspicuous, whitish or pale greyish, effuse or delimited by a thin, blackish hypothallus. Perithecia 0.3–0.4(-0.5) mm diam., ¼ to almost totally immersed, usually numerous; upper wall with usually well-differentiated involucrellum, 50–100 μm thick; lower wall pale or colourless. Asci (Fig. 34c) 65–90 × 9–12 μm, cylindrical, the spores ± uniseriate. Ascospores (9-)12–16 × 4.5–6(-7) μm, 1-septate, ovoid- to fusiform-ellipsoid, the septum ½–⅔ from upper end. Pycnidia of two types: (a) 100–200 μm diam., with subglobose to ellipsoid macroconidia, 2.3–4 × 1.8–2.7 μm, extruded as a white blob; or (b) 40–100 μm diam., ± immersed, hemispherical to globose, with globose microconidia, 1–1.5 μm diam.

On smooth or rough bark of broad-leaved trees in woodland or sheltered open situations; common. Throughout Great Britain, especially in the west and north. Europe, North America, West Indies, E. Asia, New Zealand.

Often confused with *Acrocordia gemmata*, which has much larger perithecia and ascospores.

2. A. nyssaegenum (Ellis & Everh.) R. C. Harris (1985)

A. juistense (Erichsen) R. C. Harris (1980), *Sarcinulella banksiae* B. Sutton & Alcorn (1983) [anamorph].

Thallus often inconspicuous, effuse, whitish grey or pale grey-green. Perithecia 0.15–0.25 mm diam., rarely numerous, subconical to ± globose, initially immersed, becoming ± superficial; upper wall with a scarcely differentiated involucrellum, 30–50 µm thick; lower wall thinner and paler. Asci 55–75(–90) × 12–15 µm, cylindric-clavate. Ascospores (12-)14–20(-23) × (3-)4.5–5(-6) µm, 1- to 3-septate, clavate-fusiform, the part above the primary septum wider and about twice as long as the lower. Pycnidia black, of two types: (a) 100–150 µm diam., ± sessile, conical with an ostiolar neck comprising vertically arranged, brown-walled hyphae, which at the apex of the neck have free, colourless, pointed ends, with ellipsoid or ovoid macroconidia, 3.5–4.5 × 1.8–2 µm, extruded as a white cirrus, 10–13 µm wide, in which the conidia are bound by a gelatinous matrix; or (b) 50–100 µm diam., ± immersed, globose, with rod-shaped to narrowly ellipsoid microconidia, 2–3 × 1–1.3 µm.

On rough bark of broad-leaved trees, especially *Sambucus* and *Ulmus*, rarely on shaded rock; in humid woodland and tall scrub, tolerant of deep shade; common even in moderately polluted areas. Throughout British Is. Europe, North America, Australia.

Most often found with only abundant, conical macropycnidia. When fertile it differs from *A. biforme* **(1)** in its less conspicuous thallus, smaller perithecia with a thinner upper wall, and longer and narrower ascospores which are often 2- or 3-septate.

APHANOPSIS Nyl. ex Sydow (1887)

Thallus crustose, consisting of goniocysts. Goniocysts of ± pseudoparenchymatous hyphae, surrounding algal cells; outer hyphae with brown walls. **Photobiont** chlorococcoid. **Ascomata** apothecia, discoid, convex. **True exciple** poorly developed, not raised, colourless or in part brownish, very thin laterally, of vertically aligned, thin-walled hyphae, appearing ± pseudoparenchymatous at base of apothecium. **Hamathecium** of paraphyses, mostly simple, occasionally forked; apices not swollen. **Asci** 8-spored, cylindrical-clavate; wall I+ blue, apical dome sometimes with a broad K/I+ blue plug-like structure. **Ascospores** simple, colourless. **Conidiomata** unknown. **Chemistry**: lichen products not reported. **Ecology**: on soil.

Differs from *Placynthiella* in the apothecial and ascus structures and in the larger ascospores. The asci, with their broad I+ blue apical plug are not *Trapelia*-type, as previously thought.

LITERATURE: Coppins & James (1984).

1. A. coenosa (Ach.) Coppins & P. James (1984)

Thallus minutely granular, effuse, dull green or greenish brown, ± gelatinous when wet; photobiont cells either 7–12 µm diam. and ± globose or 7–14 × 5–8 µm and ellipsoid. Apothecia 0.2–0.6 mm diam., sessile, hemispherical, brown or brown-black, matt; hymenium 150–200 µm tall, colourless with irregular, reddish brown vertical streaks, I− but appearing blue due to amyloid ascal walls; hymenial gel absent or weak; hypothecium shallow, pale brown. Paraphyses 2–4 µm wide. Ascospores 25–38 × 13–18 µm, broadly ellipsoid but often one, or both, ends pointed or bluntly apiculate, thin-walled, smooth, usually with numerous oil droplets. No lichen products detected by t.l.c.

On humid, bare, clayey or fine-grained sandy soil on track sides or ditch margins in woodlands; very rare, E. Scotland (Angus, Forfar) and S.W. Ireland (Kerry). C. Europe.

ARCTOMIA Th. Fr. (1860)

Thallus crustose-granular or minutely foliose-squamulose. **Cortex** distinctly delimited, ± one cell thick, cells brown. **Photobiont** *Nostoc*, arranged in clusters between hyphal strands, the whole in a gelatinous matrix. **Ascomata** apothecia. **Thalline exciple** absent. **True exciple** of radiating hyphae, forming a pseudoparenchyma. **Hamathecium** of paraphyses, lax, simple, with clavate apices, 1- to 3-celled, brown. **Asci** broadly cylindrical, 8-spored. **Ascospores** colourless, multiseptate. **Conidiomata** pycnidia, ± immersed. **Conidia** bacilliform. **Chemistry**: lichen products not detected by t.l.c. **Ecology**: on mosses and liverworts.

LITERATURE: Henssen (1969), James (1971).

1. A. delicatula Th. Fr. (1860)

Thallus foliose-squamulose, rarely partly terete-nodulose, red- or olive-brown, often forming a ± continuous crust; lobes 0.05–0.3 mm wide; surface matt, rather roughened. Apothecia to 0.5 mm diam., dark red-brown, convex, with an often paler true exciple which soon becomes excluded; epithecium thin, brown; hymenium 85–105 μm tall, colourless. Paraphyses 1–2 μm wide, with apices to 5–7 μm wide. Asci 80–90 × 20 μm. Ascospores 40–80 × 4–5.5 μm, 7- to 14-septate, fusiform-elongate, with ± attenuated apices. Pycnidia *c.* 40 μm diam.; conidia 2–3 × 1 μm.

On mosses and hepatics on old trees and on a mossy boulder. Scotland (W. Sutherland, W. Ross and Easterness). Recorded last century also from Ireland (Galway, Kylemore). Fennoscandia.

May be confused with the more frequent *Massalongia carnosa* which occurs in similar habitats but has smaller, 11–27 × 4·5–8 μm, 1(-2)-septate ascospores, and a cortex 3–4 cells thick, not one cell thick as in *Arctomia delicatula*.

ARTHONIA Ach. (1806)

Thallus crustose, immersed or superficial, effuse or delimited (usually by brown lines), or absent; thallus hyphae usually I+ red or bluish and K/I+ blue. **Photobiont** *Trentepohlia*, some saprobic, weakly lichenized, or lichenicolous. **Ascomata** apothecia-like, variously shaped, rounded and flat or convex to elongate, or linear and ± branched to stellate, emerging apothecia on bark often with a thin rim of thallus tissue and bark cells. **Disc** red-brown to black, sometimes with reddish or white pruina. **Thalline exciple** absent. **True exciple** absent, rarely rudimentary and developed only laterally. **Epithecium** colourless to red- or dark brown. **Hymenium** usually I+ blue or red and K/I+ blue. **Hypothecium** ± absent to conspicuous, often poorly delimited from the hymenium. **Hamathecium** of few to numerous, sparingly to richly branched and anastomosed paraphysoids in a gel matrix, often more richly branched above and there often thickened by pigment (or intermixed with crystals); apices often with thin, well-defined, dark caps, 2–4 μm wide. **Asci** usually 8-spored, clavate, ellipsoid or subglobose, semi-fissitunicate, with large apical dome and usually distinct ocular chamber; apical dome K/I−, or K/I+ bluish in lower part near the apex of the ocular chamber (Fig. 33o). **Ascospores** mostly ovoid to oblong-ovoid, 1- to 7-septate, sometimes with an enlarged upper cell (Fig. 3a), often constricted at the mid septum, colourless and smooth; old spores often becoming brown and then usually warted; young spores often with a thin perispore (in most species <1 μm thick in K). **Conidiomata** pycnidial, often present but usually immersed, very small and inconspicuous. **Conidia** colourless, simple, mostly bacilliform, sometimes ellipsoid or thread-like. **Chemistry**: many without lichen products, a wide range of substances (including xanthone and several anthraquinone pigments) produced by others. **Ecology**: various, rarely found above 500 m.

Arthothelium differs only in the presence of muriform ascospores.

Arthonia patellulata Nyl. (1853) has been incorrectly reported from the British Isles, and *A. punctilliformis* Leighton (1876) is a non-lichenized fungus of uncertain identity.

LITERATURE: Alstrup & Hawksworth (1990), Coppins (1989a), Coppins & James (1978, 1979a), Hawksworth (1983), Jørgensen & Tønsberg (1988).

1	Lichenized thallus present or absent; not on other lichens	2
	Lichenized thallus absent; on apothecia or thalli of other lichens	30
2	Ascospores 1-septate	3
	Ascospores 2- or more septate	12
3(2)	Photobiont *Trentepohlia*	4
	Photobiont not *Trentepohlia* or absent	7

4(3) Apothecial sections with K+ purplish, violet or magenta pigments 5
Apothecial sections without K+ purplish, violet or magenta pigments ..... **10. *didyma***

5(4) Ascospores mostly over 11 × 4 μm, becoming brown when old 6
Ascospores 7–11 × 3–4 μm, remaining colourless ..... **35. *spadicea***

6(5) Apothecia 45–70(-100) μm tall, fleck-like, flat; thallus pale brown; mainly on smooth bark ..... **10. *didyma***
Apothecia 85–140 μm tall, convex (biatorine); thallus usually ± tinged or stained orange; mainly on rough bark ..... **39. *vinosa***

7(3) Apothecia <0.5 mm diam., rounded, distinctly convex; photobiont cells numerous; habitat various, rarely smoothish bark 8
Apothecia often >0.5 mm diam., *either* rounded and flat, *or* linear-elongate; photobiont cells absent or few (?adventitious); on smoothish bark 11

8(7) Hymenium I−, K/I− 9
Hymenium I+ red, K/I+ blue 10

9(8) Ascospores 17–22(-26) μm long ..... **27. *ligniaria***
Ascospores 10.5–14 μm long ..... **28. *ligniariella***

10(8) Ascospores 8–12(-14) × 2.3–4 μm; epithecium of 2–3 rows of periclinally arranged extensions of the paraphyses ..... **25. *leucodontis***
Ascospores 11–15(-18) × 4–7 μm; epithecium of ± erect, ± capitate apices of paraphyses ..... **24. *lapidicola***

11(7) Apothecia 0.16–1 × 0.14–0.7 mm, rounded to oblong; thallus chalky white; hymenium K+ violet ..... **16. *galactites***
Apothecia 0.3–1.5 × 0.1–0.14 mm, linear-elongate; thallus almost inconspicuous; hymenium K− ..... **14. *excipienda***

12(2) Apothecial sections K+ purple, violet or magenta 13
Apothecial sections K+ greenish, yellowish or ± unchanged 14

13(12) Ascospores (18-)20–28 × 7–9.5 μm; apothecia mostly irregularly rounded ..... **7. *cinnabarina***
Ascospores (12-)14–19 × 4.7–7 μm; apothecia mostly linear to stellate ..... **11. *elegans***

14(12) On rock 15
On bark or wood 19

15(14) Thallus C− 16
Thallus C+ red ..... **12. *endlicheri***

16(15) Hypothecium brown; ascospores <17 μm long 17
Hypothecium colourless or pale brownish; ascospores >17 μm long 18

17(16) Ascospores 11–17 × 4.5–5(-7) μm; photobiont cells in distinct filaments (*Trentepohlia*) ..... **4. *arthonioides***
Ascospores 10–13(-14) × 2.8–3.8 μm; photobiont cells single (not *Trentepohlia*) ..... **29. *myriocarpella***

18(16) Thallus brown, thin, smooth; apothecia round, not pruinose ..... **31. *phaeobaea***
Thallus pale to slate grey, thickish or arachnoid; apothecia rounded to substellate, often pruinose ..... **6. *atlantica***

19(14) Ascospores with enlarged upper cell 20
Ascospores without an enlarged upper cell 27

20(19) Hypothecium <50 μm tall, colourless or pale 21
Hypothecium >90 μm tall, dark brown ..... **4. *arthonioides***

21(20) Ascospores mostly <26 μm long × <10 μm wide 22
Ascospores 26–36 × 10–13 μm ..... **21. *ilicina***

22(21) Ascospores >17 μm long 23
Ascospores 12–14(-16) μm long ..... **3. *anombrophila***

23(22) Apothecia white-pruinose 24
Apothecia not pruinose 25

24(23) Thallus UV+ orange, always K− in squash preparations; ascospores 19–26 × 6–7(-9) μm, 3- or 4-septate .... **5. *astroidestra***
Thallus UV−, often K+ red crystals in squash preparations; ascospores 16–22(-24) × 5–7 μm, 3(-4)-septate .... **40. *zwackhii***

25(23) Apothecial sections C− 26
Apothecial sections C+ red .... **2. *anglica***

26(25) Apothecia irregularly rounded to elongate; old ascospores brown and ± smooth (× 1000) .... **19. *ilicinella***
Apothecia linear to stellate; old ascospores brown and distinctly warted (even at × 400) .... **36. *stellaris***

27(19) Thallus and apothecia C− 28
Thallus and apothecia C+ red .... **20. *impolita***

28(27) Hypothecium <30 μm tall, colourless or pale 29
Hypothecium >30 μm, dark brown .... **26. *leucopellaea***

29(28) Thallus usually distinct, with *Trentepohlia*; apothecia 70–100 μm tall, rounded to stellate .... **34. *radiata***
Thallus usually inconspicuous, without *Trentepohlia*; apothecia 40–60 μm tall, rounded to linear, rarely branched .... **33. *punctiformis***

30(1) Ascospores 1-septate 31
Ascospores mostly 2- or 3-septate 38

31(30) Apothecial sections not K+ purple 32
Apothecial sections K+ purple .... **9. *cohabitans***

32(31) Apothecia inhabiting thallus 33
Apothecia inhabiting hymenium .... **8. *clemens***

33(32) Ascospores remaining colourless and smooth 34
Ascospores soon brown and warted .... **32. *punctella***

34(33) Ascospores mostly >6 μm wide 35
Ascospores mostly <6 μm wide 36

35(34) Apothecia 0.1–0.35 mm diam.; on *Porpidia* .... **1. *amylospora***
Apothecia 0.3–1.5 mm diam.; on *Peltigera* .... **30. *peltigerae***

36(34) Paraphysoids <3 μm wide 37
Paraphysoids 4–6 μm wide; on *Placopsis* .... **17. *gelidae***

37(36) Pycnidia usually present; on *Peltigera* .... **15. *fuscopurpurea***
Pycnidia absent; on *Phaeophyscia*, *Physcia* or *Xanthoria* .... **13. *epiphyscia***

38(30) Apothecia inhabiting hymenium 39
Apothecia inhabiting thallus 40

39(38) Apothecia discrete; on *Lecanora rupicola* .... **38. *varians***
Apothecia diffuse, with asci and paraphyses intermixed with those of the host; on *Lecidella* .... **22. *intexta***

40(38) Apothecia rounded to elongate, rarely branched 41
Apothecia stellate-dendritic; on *Schismatomma quercicola* .... **23. *invadens***

41(40) Ascospores (13-)14–17 μm long; hypothecium colourless .... **18. *graphidicola***
Ascospores 11–14 μm long; hypothecium reddish brown .... **37. *thelotrematis***

1. A. amylospora Almq. (1880)

Thallus absent, lichenicolous. Apothecia 0.1–0.35 mm diam., immersed in host thallus, rarely emergent, usually lichenicolous in small clusters, black, rounded, flat to convex; epithecium dark brown, K+

olivaceous; hymenium 40-55 μm tall, pale brownish, I+ reddish; hypothecium medium to dark brown, K+ olivaceous. Paraphysoids 1.7–2.5 μm wide; apices with dark brown apical caps, 4–6 μm wide. Ascospores (13-)16–21 × (5-)6–8 μm, 1-septate, ellipsoid to oblong-ovoid, colourless, perispore distinct and K/I+ blue.

On thallus of *Porpidia glaucophaea*, commensalistic, scarcely damaging host. Scotland (Mid-Perth, Is. of Skye). Sweden.

2. A. anglica Coppins (1989)

Thallus immersed, delimited by a dark brown line, to 6 × 3 mm, grey-white to pale fawn. Photobiont *Trentepohlia*. Apothecia mostly irregularly star-shaped, arms 0.08–0.16 mm wide, pale to dark reddish brown, not pruinose, in section 80–130 μm tall; epithecium pale to dark reddish brown, K+ olivaceous; hymenium 50–70 μm tall, colourless; hypothecium 20–50 μm tall, colourless. Paraphysoids 1(-1.5) μm wide, sometimes thickened and 1.5–2.5 μm wide in the epithecium, a few with dark caps, 2–3 μm wide. Asci 50–55 × 21–24 μm. Ascospores (16-)17–24 × 6–7(-8) μm, ovoid to oblong-ovoid, (2-)3- or 4-septate, the upper cell enlarged; old ascospores brown and warted. Pycnidia absent. Thallus Pd−, K−, KC−, C− (apothecial sections C+ red, gyrophoric acid), UV−.

On smooth bark of *Fagus* and *Ilex* in ancient woodlands. S. England (Sussex; S. Hampshire, New Forest; Devon). Madeira.

Like *A. stellaris* **(36)**, but differing in the C+ red apothecial sections and the absence of the unidentified substances seen by t.l.c. in that species.

3. A. anombrophila Coppins & P. James (1989)

A. cinereopruinosa auct. brit., non Schaerer (1850)

Thallus effuse or delimited by brown line, immersed or (when on rough bark) partly superficial, to 70 μm thick, whitish to pale grey, matt. Photobiont *Trentepohlia*. Apothecia 0.15–0.3 mm diam., numerous and often crowded, irregularly rounded, or usually elongate, 0.2–1 × 0.05–0.2 mm, and often irregularly branched or substellate, reddish brown but usually white-pruinose, in section 45–100 μm tall; epithecium pale to dark red-brown, K+ greenish; hymenium 35–55 μm tall, colourless or pale reddish brown above; hypothecium 25–35 μm tall, colourless or pale reddish straw. Paraphysoids *c.* 1 μm wide, numerous, often pigmented in epithecium and to 2.5 μm wide, sometimes with brown apical caps. Ascospores 12–14(-16) × 4–5(-6) μm, ovoid, (1-)3-septate, the upper cell enlarged, old ascospores brown and finely warted. Pycnidia 40–80 μm diam., few, immersed; wall red-brown, K+ greenish; conidia 3–5 × 0.7–1 μm, bacilliform. Thallus Pd−, K−, KC−, C−, UV− (unidentified substance ?related to confluentic acid).

On dry, shaded, usually rough, bark towards bases of old trees (especially *Quercus*), more rarely on smooth bark (*Betula*, *Corylus*, *Sorbus*), in old woodland and parkland; rare. Mainly W. Britain northwards to W. Sutherland, N. Ireland (Fermanagh, Donegal). ?Endemic.

Distinguished from *A. impolita* **(20)**, *A. zwackhii* **(40)**, and (when not pruinose) *A. anglica* **(2)** and *A. stellaris* **(36)**, by the smaller ascospores, and chemistry.

4. A. arthonioides (Ach.) A. L. Sm. (1911)

A. aspersa Leighton (1856)

Thallus whitish (or pinkish when fresh), usually effuse; when on rock, usually thickish (to 1 mm), soft and scurfy, sometimes dispersed into soft verrucae to 0.8 mm diam.; when on bark, mostly immersed; hyphae K/I−, loosely entwined around distinct filaments of *Trentepohlia*. Apothecia 0.15–0.6 mm diam., rounded and ± convex, or (especially on bark) somewhat elongate to 1.2 × 0.2–0.5 mm, black, not pruinose; in section 140–350 μm tall; epithecium dark brown, K+ greenish; hymenium 30–40 μm tall, pale reddish brown; hypothecium 100–300 μm tall, dark brown, K− or dulling. Paraphysoids 1–1.5 μm wide, thickened in epithecium by pigment, often with brown apical caps. Ascospores 11–17 × 4–5(-7) μm, (1-)3(-4)-septate, ovoid-oblong, the upper cell enlarged. Pycnidia not found. Thallus Pd−, K− or K+ faint yellowish, KC−, C− (lichen products not detected by t.l.c.).

On acid rocks and roots in dry underhangs, also dry undersides of trees (*Betula*, *Ilex*, *Quercus*, *Sorbus*, *Pinus*); local. Devon, S. Wales, N.E. England, Scotland, S.W. & N. Ireland. Norway, C. European mtns, Canada.

Morphs of this species on bark have until recently been treated as *A. aspersa*. Diminutive morphs on rocks have been confused with *A. myriocarpella* **(29)**.

5. A. astroidestra Nyl. (1874)

A. stellaris auct. p.p., non Krempelh. (1861), *A. zwackhii* var. *macrospora* P. James (1978)

Thallus immersed, usually delimited by a brown line, creamy white to cream-yellow. Photobiont *Trentepohlia*. Apothecia linear (to 1.2 mm long), irregularly branched or stellate (to 1.2 mm diam.) with arms 0.06–0.2 mm wide, pale to dark brown but usually thickly white-pruinose and in the herbarium often developing a waxy bloom of minute, needle-shaped crystals (× 50); in section 45–85 μm tall; epithecium colourless to red-brown (K+ greenish), often with dense minute granular crystals dissolving in K; hymenium 35–50 μm tall, colourless; hypothecium indistinct or to 25 μm tall, colourless. Paraphysoids 1(-1.5) μm wide, numerous, sometimes pigmented in epithecium and 1.5–2.5 μm wide, sometimes with apical caps. Ascospores (17-)18–24(-27) × 6–7 μm, (3-)4-septate, oblong-ovoid, the upper cell enlarged; old spores brown and warted. Pycnidia not seen. Thallus Pd−, K−, KC−, C−, UV+ orange; pruinose apothecia UV+ bright orange-yellow (lichexanthone).
On smooth bark, especially *Ilex*, in ancient woodlands. S.W. England (S. Hampshire, New Forest), Wales, S.W. to N.W. Ireland (Donegal). France (Bretagne).

Distinguished from *A. anglica* **(2)** and *A. stellaris* **(36)** by the pruinose apothecia, UV+ orange reaction, and chemistry (t.l.c.); poor specimens, especially from dead bark, can be difficult to identify with certainty.

6. A. atlantica P. James (1970)

incl. var. *positiva* P. James (1978)

Thallus effuse or forming mosaics with other species, white to pale grey, matt, often to 500 μm thick and cracked-areolate, sometimes thinly arachnoid in places; medulla whitish, densely packed with minute crystals. Photobiont *Trentepohlia*. Apothecia ± immersed, irregularly rounded (to 0.2 mm diam.) but more usually elongate (to 1.2 × 0.08–0.16 mm) or ± stellate, often joining to form an irregular reticulum, brown black but often thinly white-pruinose; epithecium reddish brown, K+ grey-brown; hymenium 40–50 μm tall, ± colourless; hypothecium sometimes rooting, pale reddish straw or sometimes with reddish brown patches (especially at outer edge). Paraphysoids 1–1.8 μm wide; apices with dark caps. Ascospores 16–24 × 6–7 μm, (2-)3- to 4-septate, oblong-ovoid, the upper cell enlarged; old ascospores brown and warted. Pycnidia not seen. Thallus Pd− or Pd+ yellowish or reddish, K+ yellow, KC−, C− (confluentic acid, and either atranorin with one or two unidentified compounds [var. *atlantica*] or stictic acid and unidentified compound [var. *positiva*]).

On hard, acid rocks in dry underhangs and sheltered sides of old walls, usually near the coast; rare. W. Scotland (Coll), S.W. and N.W. Ireland (Killarney, Mayo). ?Endemic.

Superficially like *Enterographa*, but distinguished by the shape of the ascospores. *Sclerophyton circumscriptum* differs in the Pd+ deep yellow thallus and dark brown, 4- to 7-septate ascospores.

7. A. cinnabarina (DC.) Wallr. (1831)

A. tumidula (Ach.) Ach. (1806)

Thallus immersed, often delimited by brown line and often mosaic-forming, whitish, pale grey, pale fawn or sometimes tinged dull orange in parts, surface smooth or (especially on thick bark) finely powdery. Photobiont *Trentepohlia*. Apothecia (0.2-)0.3–1 × 0.2–0.5(-0.7) mm, flat to slightly convex, roundish, ± polygonal or linear, occasionally branched but rarely stellate (to 1.5 mm diam.); white-pruinose with (usually) marginal, rose-red to cinnabar-red pruina, old apothecia often not pruinose and dark purplish brown; in section 110–140 μm tall; epithecium brown with blotches of purple-red pigment (K+ red or purple, dissolving) especially at the outer edge, often, also with numerous minute granules (pruina); hymenium 60–70 μm tall, colourless or pale orange-red; hypothecium 25–40 μm tall, ± colourless. Paraphysoids 1–1.5 μm wide, but often brown-walled and to 2.5 μm wide in epithecium, a few with apical caps. Ascospores (18-)20–28 × 7–9.5 μm, (2-)4- to 5(-6)-septate, oblong-ovoid (Fig. 3a); old ascospores brown, wrinkled but not warted. Pycnidia immersed, *c.* 60 μm diam., the walls red-brown, K+ olive-grey; conidia 3.5–5.5 × 0.8 μm, bacilliform. At least three unidentified anthraquinones (orange, lilac and red) by t.l.c.

On smooth bark of small (e.g. *Corylus*) or young trees, sometimes on roughish bark of large trees (especially *Fagus* and *Fraxinus*), in sheltered situations, mainly in old woodlands; common. W. British Is., extending locally to unpolluted eastern areas. Temperate Europe, North America.

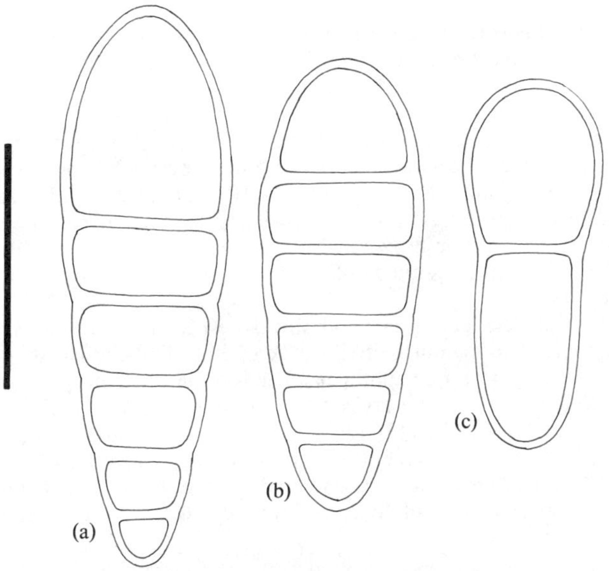

Fig. 3. *Arthonia* ascospores. (a) *A. cinnabarina*; (b) *A. impolita*; (c) *A. lapidicola*. Scale - 10 μm.

Often confused with *A. elegans* **(11)** but distinguished by generally larger, more rounded apothecia, taller hymenium, and larger ascospores.

8. A. clemens (Tul.) Th. Fr. (1867)

Thallus absent, lichenicolous. Apothecia developing in the host hymenium and blackening the disc, 0.1–0.4 mm diam., black, rounded, flat to slightly convex; epithecium dark brown, K+ olivaceous; hymenium 40–50 μm tall, colourless, I+ blue; hypothecium colourless or very pale brownish. Paraphysoids 1.5–2 μm wide, apices with dark pigmented caps, 2–4 μm wide. Ascospores 9–15 × 3–5 μm, 1-septate, narrowly ovoid to slipper-shaped, colourless.

In apothecia of *Lecanora dispersa* group; common. Almost throughout British Is. Europe, N. Africa, China, Greenland.

Like *A. varians* **(38)** but ascospores narrower and 1-septate. A similar, unidentified species in the apothecia of *Lecanora leptacina* from the Cairngorms differs in having a dark brown, K+ olivaceous hypothecium. There is also a similar collection on *Diploicia canescens* from Co. Kerry.

9. A. cohabitans Coppins (1989)

Thallus absent, lichenicolous. Apothecia scattered to clustered, developing below the surface of the bark but soon erumpent, fleck-like, brown-black, not pruinose, rounded and 0.06–0.16 mm diam., or elongate and to 0.3(-0.4) × 0.08–1 mm; in section (50-)60–70 μm tall, K+ purple due to yellow-orange pigment; epithecium often discontinuous, of pale yellow-orange pigment mixed with the dark brown walls of the paraphyses; hymenium 35–38 μm tall, pale yellow-orange, I+ blue; hypothecium 16–23 μm tall, mottled dark brown with yellow-orange background. Paraphysoids rather scanty, simple or sparingly branched, 1–1.5 μm wide, apices more richly branched, brown-walled, 1.5–2 μm wide, sometimes with dark brown caps to 3.5 μm wide. Ascospores 10.5–14 × 4–5 μm, 1-septate, the upper cell broader and longer than the lower, oblong-ovoid to oblong-ellipsoid or slipper-shaped, at first colourless with a thin perispore, later becoming brown and warted.

On thallus of *Arthothelium reagens*, commensalistic or weakly parasitic, rarely forming necrotic patches. W. Scotland (Argyll). Endemic.

The non-lichenicolous *A. didyma* **(10)** is distinguished by the larger ascospores.

10. A. didyma Körber (1853)

Thallus immersed, effuse, wide-spreading or in tiny patches (especially around lenticels and scars), inconspicuous, pale, fawn or olive-grey, sometimes pinkish when fresh. Photobiont *Trentepohlia*. Apothecia 0.06–0.6 × 0.06–0.3 mm, rounded to shortly linear, flat, dark brown to blackish, matt, not pruinose; in section 45–70(-100) μm tall; epithecium red-brown or indistinct; hymenium 33–50 μm tall, colourless or pale orange-red-brown in parts; hypothecium 5–30 μm tall, red-brown; red-brown pigment is K+ grey or olivaceous, but epithecium, hymenium and sometimes hypothecium with additional orange-red, K+ purple-violet pigment. Paraphysoids 0.5–1(-1.5) μm wide, rather scanty, or (especially in epithecium) brown-walled and to 2.5 μm wide, a few with apical caps. Ascospores (12-)14–17 × 4.7–7 μm, 1-septate, ovoid, colourless but soon brown and warted. Pycnidia 40–60 μm diam., rare, immersed, the wall red-brown, K+ olive-grey; conidia 3.5–4 × 0.7 μm, bacilliform. No lichen products or anthraquinone(s) by t.l.c.

On usually smooth bark of deciduous trees and shrubs, especially *Corylus*, mainly in woodlands; common in unpolluted areas. Throughout British Is. Temperate Europe, North America.

Usually recognized by the tiny fleck-like apothecia. The K+ purplish morph is separated from small morphs of *A. spadicea* **(35)** by its larger ascospores, and from *A. vinosa* **(39)** by the flat, shallower apothecia. The lichenicolous *A. cohabitans* **(9)**, on *Arthothelium reagens* has ascospores, 10.5–14 × 4–5 μm. Morphs without any K+ purplish pigment have been recognized as *A. aspersella* Leighton (1872), but both types of apothecia are often found on the same thallus. *A. excipienda* **(14)** differs from the K– form in its more elongated apothecia, well-developed lateral 'exciple', and absence of *Trentepohlia*.

11. A. elegans (Ach.) Almq. (1880)

Thallus immersed, whitish to dull ochre-yellow, or dull reddish- to purplish brown, usually delimited by pale orange to dark brown line and often mosaic-forming. Photobiont *Trentepohlia*. Apothecia 0.3–1.2(-1.6) × 0.1–0.2 mm, linear with pointed apices, but often branched or stellate and to 1.5 mm diam., dark purplish brown, not pruinose or with reddish pruina along the edges; in section 60–85 μm tall; epithecium reddish brown or purplish in parts (especially laterally), K+ purple-red and/or blue-violet (quickly dissolving); hymenium 35–45 μm tall, ± colourless or pale yellow-reddish; hypothecium 10–25 μm tall, pale yellow-reddish. Paraphysoids 1–1.5 μm wide, brown-walled in epithecium and to 2.5 μm wide, many with apical caps. Ascospores (12-)14–19 × 4.5–7 μm, (2-)3–4(-5)-septate, upper cell enlarged; old ascospores brown and faintly warted (× 1000). Pycnidia not seen. At least three unidentified anthraquinones (orange, lilac and red) by t.l.c.

On smooth bark, especially of *Corylus*, less often on rough bark, in sheltered woodland. Common in W. Britain, confined to old woodlands further east. Temperate Europe.

Often confused with *A. cinnabarina* **(7)**; other species with similarly shaped apothecia and upper cell enlarged ascospores lack K+ purplish or violet pigments.

12. A. endlicheri (Garov.) Oxner (1956)

Thallus 0.3–0.5(-2) mm thick, tartareous, irregularly cracked, very uneven, often forming bullate ridges or lobe-like proliferations, whitish pink when fresh becoming pale brownish grey in the herbarium, matt to finely scabrous; in places with erose soralia, soredia 50–100 μm diam., internally, densely packed with minute crystals. Photobiont *Trentpohlia*. Apothecia to 0.8 mm diam., very rare, pale, angular to rounded, or shortly elongate, white-pruinose; epithecium red-brown; hymenium 50–60 μm tall; hypothecium 50–70(-100) μm tall, ± colourless. Ascospores 14–20 × 5–7 μm, 3- to 5-septate, ovoid, upper cell not enlarged. Pycnidia 120–200 μm diam., frequent but immersed and inconspicuous, the wall colourless; conidia 15–26 × 1 μm, short thread-like, often curved or sigmoid. Thallus Pd–, K–, KC+ red, C+ red, UV– (lecanoric acid).

On acid rocks in dry, sheltered underhangs, especially near the coast, very local. S.W. England, Channel Is., Wales (Pembroke). ?S. Europe.

Dirina massiliensis f. *sorediata* on acid rocks can usually be distinguished by the generally thinner thallus, ± delimited, convex soredia with finer soralia (30–50 μm diam.) and the presence of erythrin; difficult specimens require t.l.c.

13. A. epiphyscia Nyl. (1875)

Thallus absent, lichenicolous. Apothecia 0.1–0.35 mm diam., often in clusters, black, rounded, convex; epithecium brown, K ± olivaceous; hymenium 30–35 μm tall, I+ red; hypothecium colourless (K−) to dark reddish-brown (K+ olivaceous). Paraphysoids 1.5–2.5 μm wide, apices thickened with dark pigment, to 4 μm wide. Ascospores 10–14 × (3-)3.5–5 μm, 1-septate, oblong to slipper-shaped, colourless.

On thalli of *Physcia adscendens*, *P. dubia*, *Phaeophyscia orbicularis* and *Xanthoria parietina*, commensalistic or weakly parasitic. Throughout British Is. Europe, Greenland, South America, S. Africa, Subantarctic Is.

More than one species may be included here; collections with a colourless or very pale hypothecium possibly belong to *A. molendoi* (Heufl. ex Frauenf.) R. Sant. (1986).

14. A. excipienda (Nyl.) Leighton (1871)

A. dispersa auct. brit. p.p., non (Schrader) Nyl. (1861), *A. hibernica* Nyl. (1876)

Thallus immersed, often scarcely apparent, effuse or sometimes delimited from adjacent crusts by a brown line. Photobiont absent, but a few adventitious chlorococcoid or *Trentepohlia* cells sometimes present. Apothecia 0.3–1.5 × 0.1–0.14 mm, often curved or flexuose, occasionally a few branched; 'margins' slightly raised towards a slit-like disc that sometimes expands in old apothecia. In section 40–50 μm tall, with well-developed, lateral, dark brown (K+ greenish) exciple-like zone 12–18 μm wide; epithecium pale brown or olivaceous; hymenium 30–40 μm tall, colourless; hypothecium indistinct, rarely over 5 μm tall, colourless. Paraphysoids 0.08–1.5 μm wide, numerous, only rarely thickened by pigment above, apical caps not seen. Ascospores 15–17(-18) × 5–7 μm, 1-septate, ovoid, colourless. Pycnidia not found.

On smooth bark (e.g. *Corylus*) or smooth plateau surface of fissured bark; ?rare. W. Scotland, W. Ireland. W. Europe.

Distinguished from superficially similar *Arthonia* species by the well-developed 'exciple'-like margin; easily overlooked for *A. punctiformis* **(33)**, which has 3-septate ascospores. *Melaspilea atroides* has a definite black true exciple and smaller, dark brown ascospores.

15. A. fuscopurpurea (Tul.) R. Sant. (1960)

Thallus absent, lichenicolous. Apothecia (0.2-)0.5–1.5(-2) mm diam., red-brown to brown-black, often surrounded by a ring of white, chlorotic host tissue, rounded, flat to convex, often with central part convex and surrounded by a flat, maculate zone of individual patches of developing young hymenium and sometimes also pycnidia; epithecium orange- to reddish brown, K+ dull brown, sometimes with olivaceous tinge; hymenium 25–30 μm tall, pale brown-orange, K+ dull, I+ red; hypothecium 40–100 μm tall, pale brown-orange, ± with red-brown mottling, K+ dull. Paraphysoids 1.5–2.5(-3) μm wide, extending periclinally above the asci to form the epithecium in which the colourless apical walls are embedded in a pigmented matrix. Ascospores 9–15 × 3–5 μm, 1-septate, ovoid to oblong-ovoid or slipper-shaped, upper cell shorter and broader than the lower, colourless. Pycnidia grouped in apothecia-like clusters, or in the outer edge of the apothecia, 25–40 μm diam., red-brown, wall reddish brown, K+ dull; conidia 3.5–5 × 0.8–1 μm, bacilliform.

On thallus of *Petigera lactucifolia*, probably parasitic; rather common. Mostly in W. British Is. Europe, Greenland.

Can be confused with *A. peltigerae* **(30)**.

16. A. galactites (DC.) Dufour (1818)

Thallus to 80 μm thick, chalky-white, usually in ± circular patches without a dark border, immersed, of loose bark cells, scattered hyphae and abundant, minute crystals. Photobiont absent. Apothecia 0.16–0.8(-1) × 0.14–0.6(-0.7) mm, flat, black, not pruinose; in section 35–55 μm tall; epithecium dark brown; hymenium 20–35 μm tall, pale greenish yellow; hypothecium 5–10 μm tall, greenish yellow or brown; brown pigment K−, greenish yellow pigment K+ violet. Paraphysoids 1–2 μm wide, brown-walled and to 3 μm in epithecium, a few with dark apical caps. Ascospores 12–14 × 4–5 μm, 1-septate, ovoid, colourless. Pycnidia 40–80 μm diam., often numerous, black, wall brown, K− or K+ olivaceous; conidia 10–14 × 0.5 μm, curved. Thallus K+ faint yellow, KC−, C− (lichen products not detected by t.l.c.).

On smooth bark of *Populus*; ?extinct. S. England. S. Europe.

17. A. gelidae R. Sant. (1986)

Thallus absent, lichenicolous. Apothecia 0.2–0.5 mm diam., black, round or slightly elongate, erumpent., convex to almost hemispherical; epithecium dark brown; hymenium colourless, I+ red; hypothecium colourless. Paraphysoids with apices 4–6 μm wide, dark brown. Ascospores 11.5–16 × (4-)5–6 μm, 1-septate, oblong-ovoid, colourless. Pycnidia immersed, *c.* 30 μm diam., the wall colourless; conidia *c.* 5 × 1 μm, bacilliform.

On thallus of *Placopsis gelida*, commensalistic. S. Wales, Scotland. N.W. Europe, Iceland, Canary Is., Tanzania.

18. A. graphidicola Coppins (1989)

Thallus absent, lichenicolous. Apothecia scattered, developing below the surface of the bark but soon erumpent, fleck-like, rounded or ± polygonal and 0.06–0.16 mm diam., or elongate ± with pointed ends and to 0.42 × 0.06–0.1 mm, reddish to dark brown, not pruinose; in section 60–85 μm tall; epithecium indistinct or pale reddish brown, K+ dull olivaceous; hymenium 40–50 μm tall, pale reddish brown, K+ pale olivaceous, I+ red: hypothecium 10–28 μm tall, colourless, I+ blue. Paraphysoids rather scanty, sparingly branched, 1–1.5 μm wide, apices more richly branched, brown-walled, 1.5–2 μm wide, occasionally with dark brown caps to 3.5 μm wide. Ascospores (13-)14–17 × 4.5–5.5 μm, 2- to 3-septate, the upper cell enlarged, oblong-ovoid, at first colourless and smooth but often with a thin perispore, later covered in dark brown, granular warts. Pycnidia rare, immersed, *c.* 40 μm diam., the wall reddish brown, K+ greenish; conidia colourless, bacilliform, 4.5–5 × 0.8 μm.

On thallus of *Graphis scripta*, commensalistic. S.W. England (N. Cornwall), W. Scotland, N.W. Ireland. ?Endemic.

Similar to *A. thelotrematis* **(37)** but ascospores longer and hypothecium paler.

19. A. ilicinella Nyl. (1867)

Thallus immersed, effuse or delimited by brown line, whitish grey or creamy-white. Photobiont *Trentepohlia*. Apothecia 0.1–0.6 mm diam., or elongated to 0.8 × 0.15–0.3 mm, irregularly rounded, ± flat, black, not pruinose; in section 80–95 μm tall; epithecium red-brown, K+ greenish; hymenium 40–60 μm tall, pale red-brownish, K+ pale greenish; hypothecium 10–25 μm tall, concolorous with hymenium. Paraphysoids *c.* 1 μm wide, but brown-walled and 1.5–2 μm wide in epithecium, sometimes a few with apical caps. Ascospores (16-)18–23 × 7–9 μm, oblong-ovoid, 3- to 4(-5)-septate, the upper cell enlarged; when old becoming brown but smooth or very faintly warted (× 1000). Pycnidia often present but usually few, 60–80 μm diam., the wall reddish brown, K+ greenish; conidia 7–9 × 1 μm. Lichen products not detected by t.l.c.

On smooth bark, especially *Corylus*; rare. W. Scotland (Argyll to Skye), W. Ireland. ?Endemic.

Closely related to *A. ilicina* **(21)**, but distinguished by the smaller apothecia and ascospores. *A. stellaris* **(36)** has ± stellate apothecia, and ascospores that become distinctly warted (× 400) when brown.

20. A. impolita (Hoffm.) Borrer (1831)

Thallus effuse and often wide-spreading, whitish to pale grey or light brown, matt or powdery, rimose-cracked and 0.1–0.3 mm thick but sometimes warty and to 1 mm thick. Photobiont *Trentepohlia*. Apothecia to 1 mm diam., rounded or angular, sometimes somewhat elongate or substellate, immersed, pinkish to dark brown but usually thickly white-pruinose and hence inconspicuous when dry; in section 100–250(-250) μm tall; epithecium red-brown, K+ grey or greenish; hymenium 40–60 μm tall, colourless; hypothecium 45–85(-170) μm tall, colourless. Paraphysoids numerous, 1–1.8 μm wide, in epithecium often with brown walls and to 2.5 μm wide. Ascospores 13–22 × 4.5–7 μm, (3-)4- or 5-septate, oblong-ovoid, the upper cell not enlarged, colourless (Fig. 3b), old ascospores apparently never brown. Pycnidia 100–200 μm diam., rare, ± immersed, pale or red-brown, internally convoluted, the wall colourless or pale reddish brown (K+ greenish); conidia 10–16 × 0.5 μm, short-thread-like, usually curved. Thallus and apothecia pruina Pd−, K−, C+ red, UV− (arthoniaic acid).

On dry bark of tree trunks, especially *Acer* and *Quercus*, often dominating the side of the tree, rarely on wooden boards; frequent. S. Britain extending locally to S. Aberdeenshire. Temperate Europe, (?)N. Africa, North America.

Distinguished from other *Arthonia* species on bark with pruinose apothecia by the C+ red reaction. *A. anglica* **(2)** apothecia are not pruinose are C+ red in section, and the ascospores have enlarged upper cells. Recalls *Schismatomma decolorans* in the field, which is always sterile in Britain, sorediate and C−.

21. A. ilicina Taylor (1836)

Arthothelium ilicinum (Taylor) P. James (1965)

Thallus immersed, effuse or delimited by brown line, whitish to creamy yellow. Photobiont *Trentepohlia*. Apothecia irregularly rounded and 0.2–1 mm diam., or a few elongated to 2 × 0.2–0.6 mm, flat to slightly convex, dark reddish brown to black, not epruinose; in section 95–104 μm tall; epithecium red-brown, K+ greenish; hymenium 60–85 μm tall, colourless, or pale brownish red, K+ greenish; hypothecium 20–30 μm tall, concolorous with hymenium or darker in places. Paraphysoids numerous, *c.* 1 μm wide, brown-walled and 1.5–2 μm wide in epithecium. Ascospores 26–36 × 10–13 μm, ovoid-oblong, often slightly curved, (4-)5- to 6(-7)-septate, the upper cell enlarged, old spores brown but smooth or very faintly warted (× 1000). Pycnidia rare, 60–80 μm diam., brown, the wall red-brown, K+ greenish; conidia 7–9 × 1 μm, bacilliform. No lichen products detected by t.l.c.

On smooth, more rarely rough, bark of deciduous trees in old woodlands; locally common. S. & W. British Isles. Oceanic districts from S. Norway to Portugal, Macaronesia, Canada, S. Africa, Tasmania.

A. ilicinella **(19)** has smaller ascospores and generally smaller apothecia. Similar, but muriform, ascospores occur in *Arthothelium dictyosporum* and *A. reagens*.

22. A. intexta Almq. (1880)

Thallus absent, lichenicolous. Apothecia developing in the hymenium of the host, poorly delimited, with asci and paraphysoids intermixed with those of the host. Paraphysoids 1.5–3 μm wide; apices thickened by olive-brown pigment to 5.5 μm wide. Ascospores (11-)12.5–17(-20) × (3-)3.5–5(-6) μm, (1-)2(-3)-septate, oblong-ovoid, colourless. Pycnidia immersed in host hymenium, 50–70(-90) μm diam., the wall olive brownish above, becoming colourless below; conidia 5–5.5 × 1 μm.

In apothecia of saxicolous *Lecidella* species; scattered. Throughout Britain (W. Cornwall, Shropshire, N.E. Yorkshire, Shetland). Europe, N. Africa, North and South America, Australia.

23. A. invadens Coppins (1989)

Thallus absent, lichenicolous. Apothecia evenly and mostly densely scattered, developing below the surface of the bark but soon erumpent, dark brown, not pruinose, linear to more usually stellate or stellate-dendritic, 0.2–0.8(-1) mm diam., linear apothecia or 'arms' very narrow, 0.03–0.8(-0.1) mm wide; in section 54–60 μm tall, I+ blue; epithecium 7–12 μm tall, red-brown, K+ olivaceous; hymenium 30–40 μm tall, colourless, or pale reddish brown in places; hypothecium indistinct, less than 10 μm tall, colourless. Paraphysoids rather scanty, branched, 1(-1.5) μm wide; apices often brown-walled and to 2 μm wide, without dark caps. Ascospores 11–15(-17) × (4-)4.5–5.5(-6) μm, 2- to 3- septate, the upper cell enlarged, oblong-ovoid, at first colourless and often with a thin perispore, later covered in brown, granular warts. Pycnidia few, immersed, 40–50 μm diam., the wall reddish brown, K+ greenish; conidia bacilliform, 3.8–5 × 0.8 μm.

On thallus of *Schismatomma quercicola* in ancient woodland; commensalistic or weakly parasitic. S. England, S.W. Ireland (W. Cork).

The Pd+ red reaction of the host's soralia distinguish this species from *A. anombrophila* **(2)** and *A. stellaris* **(36)**.

24. A. lapidicola (Taylor) Branth & Rostrup (1869)

Catillaria ooliticola W. Watson (1935)

Thallus thin and partly immersed to irregularly scurfy granular-verrucose, verrucae to 150(-300) μm diam., pale fawn to dull grey or olive-brown. Photobiont cells 6–17(-21) μm diam., green (not *Trentepohlia*), mostly ± globose, but rarely irregularly globose to ellipsoid and 4–10 × 4–6 μm. Apothecia (0.1-)0.2–0.4(-0.5) mm diam., convex, dark brown to black; in section 70–190 μm tall; epithecium to 15 μm tall, ± reddish brown, K+ dull- or olivaceous brown; hymenium 35–60 μm tall, colourless or pale greenish in upper part; hypothecium dark, ± reddish brown, K+ dull- or olivaceous brown. Paraphysoids (1-)1.5–2 μm wide, numerous, apices mostly swollen to 3.5 μm with dark apical caps. Ascospores 11–15(-18) × 4–7 μm, 1-septate, ovoid, or constricted at septum and sole-shaped with upper cell ± rounded and elongated; lower cell narrower, colourless (Fig. 3c). Pycnidia 40–50 μm diam., few to numerous, immersed, the wall reddish brown, K+ dull brown; conidia 4.5–6 × 1 μm, bacilliform. Lichen products not detected by t.l.c. (or a greenish spot at A & G:4 before developing).

On calcareous rocks and mortar, including loose stones, mostly in open habitats, also ± basic window sills of churches below metal gratings; widespread. Throughout British Is. Temperate Europe, North America.

Easily mistaken in the field for a small member of Lecideaceae s. lat. Morphs of *A. leucodontis* **(25)** on rock differ in the smaller ascospores, scanty, narrower paraphyses, and epithecium of compacted periclinal hyphae.

25. A. leucodontis (Poelt & Döbb.) Coppins (1989)

Bryostigma leucodontis Poelt & Döbb. (1979), *Arthonia exilis* auct. brit., non (Flörke) Anzi (1860)

Thallus effuse, usually forming small patches (<5 mm wide) among other crusts, dull olive-grey, olive-brown or grey-brown, thin, minutely scurfy-granular; granules when discrete 20–50 μm diam. Photobiont 6–14(-17) μm diam.; cells chlorococcoid (not *Trentpohlia*). Apothecia 0.05–0.25(-0.3) mm diam., rounded, convex, black, not pruinose; in section 60–100 μm tall; epithecium 5–7 μm tall, greenish- or reddish brown, K+ greenish, composed of 2–3 rows of periclinally arranged, dark-walled hyphae 1.5–2 μm wide; hymenium 25–30 μm tall, colourless or pale greenish; hypothecium 30–70 μm tall, reddish brown, K+ dull olive-brown. Paraphysoids 0.5–1 μm wide, very scanty. Ascospores 8–12(-14) × 2.5–4 μm, 1-septate, ovoid, upper cell broader and shorter than the lower but often with a slightly pointed apex. Pycnidia rare, *c.* 40 μm diam., the wall brownish, K+ olivaceous; conidia bacilliform, 3–4 × 0.5 μm. Lichen products not detected by t.l.c.

On twigs, especially of *Sambucus*, or tree trunks (e.g. *Acer*, *Ulmus*), sometimes over bryophytes, less often on fence-posts and shaded, ± acid rocks; apparently common but overlooked and tolerant of suburban conditions. Throughout British Is. Lowland W. Europe.

Morphs on rock need to be carefully distinguished from *A. lapidicola* **(24)**. Often confused with the non-British *A. apatetica* (Massal.) Th. Fr. (1866) and *A. exilis* in the past. The correct name of this species appears to be ***A. muscigena*** Th.Fr. (1865).

26. A. leucopellaea (Ach.) Almq. (1880)

Thallus mostly immersed, sometimes thinly scurfy or arachnoid, effuse, whitish but often with pink tinge when fresh; thallus hyphae K/I−, loosely woven around distinct filaments of *Trentepohlia*. Apothecia irregularly rounded to bluntly stellate or lobate, dark brown to black but usually thinly white-pruinose and with whitish arachnoid margin, centre adpressed but margin often ascending; in section 70–90 μm tall; epithecium brown, K+ yellow greenish tinge; hymenium 25–36 μm tall, colourless or pale brownish; hypothecium 35–70 μm tall, dark brown, K+ yellow greenish tinge. Paraphysoids 1–1.5 μm wide, scanty, brown- walled and to 2.5 μm in the epithecium. Ascospores 10–15(-17) × 3.5–5 μm, (1-)3- to 4-septate, oblong-ovoid, the upper cell not enlarged, colourless but old ascospores brown and warted. Pycnidia not found. Thallus Pd−, K− or K+ faint yellowish, KC−, C− (two unidentified substances).

On acid bark of mature trees (especially *Betula, Ilex, Quercus, Pinus*), also on old *Calluna* stems; locally frequent in ancient birch-pine and oak woodlands. S. & S.W. England (Hampshire, New Forest), Wales, Scotland (Highlands), S.W. Ireland. N.W. Europe, C. European mtns.

27. A. ligniaria Hellbom (1884)

Thallus inconspicuous, immersed, or apothecia associated with subgelatinous, mixed-algal film. Apothecia rounded, flat to convex, 0.1–0.3 mm diam., black; in section 95–100 μm tall, all parts K/I−; epithecium 7–10 μm tall, dark brown, K+ olivaceous brown; hymenium 40–50 μm tall, colourless to pale yellow-brown, K+ pale yellow-olive; hypothecium 45–55 μm tall, pale to dark brown, K+ olivaceous tinge. Paraphysoids numerous, dense, much-branched, 1.7–2 μm wide; apices branched and knobby, to 5 μm wide, with dark caps. Ascospores 17–22(-26) × 6.5–9.5 μm, 1-septate, ellipsoid to oblong-ovoid, constricted at septum, colourless or becoming brown when old. Pycnidia not found.

On bark of mature *Quercus*, rotting wood, and on ± acid soil of coastal landslip; rare. N. & S.W. England (N.E. Yorkshire, Devon, W. Scotland). Sweden.

Distinguished from *A. lapidicola* **(24)** and *A. leucodontis* **(25)** by large ascospores and K/I− tissues; *A. ligniariella* **(28)** has smaller ascospores.

28. A. ligniariella Coppins (1991).

Like **27** in appearance, general anatomy and pigmentation. Hymenium 25–30 μm tall. Ascospores 10.5–14 × 3–3.8 μm, 1-septate, narrowly clavate, constricted at septum, colourless; upper cell usually shorter and broader than the lower cell.

Mostly on rotting wood of stumps of fallen trees, also on *Quercus* bark, weathered hardboard and moribund *Leucobryum* cushions. Throughout British Is. Sweden.

29. A. myriocarpella Nyl. (1863)

Thallus immersed or minutely granular, whitish. Photobiont 6–12 μm diam.; cells green, not *Trentepohlia*, ± globose. Apothecia 0.08–0.3 mm diam., convex, black; in section 65–90 μm tall; epithecium dark brown, K−; hymenium *c.* 30 μm tall, colourless; hypothecium 35–55 μm tall, brown, K−. Paraphysoids 0.5–1 μm wide, numerous, with conspicuous capitate, brown-capped apices, 3–6 μm wide. Ascospores 10–13(-14) × 3–4 μm, oblong to ovoid-oblong, evenly 3-septate or 2-septate, upper cell ± enlarged. Pycnidia 40–60 μm diam., sessile, black; conidia 3.5–4 × 0.5–1 μm, bacilliform.

On calcareous schist, possibly lichenicolous; known only from the type collection. N. Scotland (Inverness, Aviemore).

Has been confused with *A. arthonioides* **(4)**.

30. A. peltigerae Th. Fr. (1866)

Thallus absent, lichenicolous. Apothecia 0.3–1.5 mm diam., black, convex to hemispherical, sometimes with a flat outer rim; epithecium brown, K+ olivaceous; hymenium 45–50 μm tall, pale brown ± with olive tinge, K+ olivaceous, I+ red; hypothecium to 250 μm tall, dark mottled red-brown, K+ olivaceous. Paraphyses 1.5–2.5 μm wide, branched but erect and ± parallel; apices ± wider, to 3.5 μm, erect, often coated with dark pigment. Ascospores 15–20 × (5-)6–7 μm, 1-septate, oblong to bluntly slipper-shaped, upper cell slightly broader than the lower, colourless.

On thallus of *Peltigera rufescens* in coastal heath; weakly parasitic. N. Scotland (E. Sutherland). Scandinavia.

Differs from *A. fuscopurpurea* **(15)** in the darker coloured apothecia and hypothecium, larger ascospores, and erect, paraphysoid-apices that do not extend periclinally above the asci.

31. A. phaeobaea (Norman) Norman (1869)

Thallus red-brown or brown with faint mauve tinge, thin, smooth or slightly cracked, usually forming delimited patches but without any bordering line, to 100 μm thick, cortex of ± globose, brown-walled cells 3–4 μm diam. Photobiont cells 7–12 μm diam., green, not *Trentepohlia*. Apothecia 0.16–0.3 mm diam., rounded, flat to slightly convex, brown-black or black-pruinose; in section 90–120 μm tall; epithecium dark reddish brown, K−; hymenium 35–50 μm tall, colourless; hypothecium 48–85 μm tall, pale straw-brown or colourless, K−. Paraphysoids 1(-1.5) μm wide; apices ± capitate, brown-capped, coherent. Ascospores 17–30 × 5–7 μm, 3- to 5-septate, oblong, constricted at mid-septum and upper half slightly broader than lower, colourless; perispore usually distinct, *c.* 2 μm thick in K. Pycnidia 40–80 μm diam., numerous, black, immersed; wall brown K+ olivaceous brown; conidia 4–5.5 × 1.5–2 μm, oblong. Lichen products not detected by t.l.c.

On hard, siliceous maritime rocks, mesic-supralittoral; common. N. & W. British Is. Norway, North America.

Easily overlooked, especially when sterile, but then recognized by the characteristic thallus colour and the numerous punctiform pycnidia.

32. A. punctella Nyl. (1859)

Thallus absent, lichenicolous. Apothecia 0.07–0.22 mm diam., black, rounded, convex; epithecium dark reddish brown, K+ dull or ± olivaceous; hymenium 30–60 μm tall., colourless, I+ blue; hypothecium 30–70 μm tall, dark reddish brown, K+ dull or ± olivaceous. Paraphysoids 1.5–3 μm wide; apices thickened with pigment, to 4(-5) μm wide. Ascospores 12–17 × 5–6.5(-7.5) μm, 1-septate, ovoid to oblong-ellipsoid, colourless at first but soon brown and warted.

On thalli of *Diplotomma* species on rocks and walls; commensalistic. S. England (W. Kent), Scotland (Fife, Caithness), S. Ireland (Cork, Carlow). ?Endemic.

33. A. punctiformis Ach. (1808)

Thallus immersed, usually effuse, indistinct or slightly 'bleaching' the bark. Photobiont absent. Apothecia 0.2–1.4 × 0.1–0.4 mm, roundish, elliptic to linear, rarely branched or substellate, black, not pruinose, usually immersed then erumpent and often with a thin 'margin' of bark tissue; in section 40–60 μm tall; epithecium brown or olive-brown, K+ greenish; hymenium 26–50 μm tall, colourless; hypothecium indistinct, to 10 μm thick, colourless. Paraphysoids 1–1.5 μm wide, numerous, apices thickened to 3.5 μm wide with dark brown walls. Ascospores 13–23 × (4-)5–7 μm, 3(-5)-septate, oblong-ovoid to oblong; apical cell not enlarged, colourless. Pycnidia not found. Lichen products not detected by t.l.c.

On smooth bark of a wide variety of trees and shrubs; often a primary colonizer of twigs along with *Arthopyrenia punctiformis*; common except in heavily polluted areas. Throughout British Is. Europe, North America.

Distinguished from *A. radiata* **(34)** by the less conspicuous, apparently non-lichenized thallus, and rarely branched, shallower apothecia.

34. A. radiata (Pers.) Ach. (1808)

Thallus immersed, usually delimited by brown line and often mosaic-forming, whitish, pale grey, sometimes with brownish or olivaceous tinge. Photobiont *Trentepohlia*. Apothecia variably shaped, rounded to substellate and 0.15–0.8 mm diam., linear or sparingly branched and to 1.4 × 0.1–0.2 mm, or stellate and to 1.5(-2.2) mm diam., flat or slightly convex, black, not pruinose; in section 70–100 μm tall; epithecium brown or olive-brown, K+ greenish; hymenium 35–50 μm tall, colourless; hypothecium 10–25 μm tall, colourless to pale olive-brown, K+ greenish. Paraphysoids numerous, 1–2 μm wide, often to 3 μm wide and brown-walled in the epithecium and the often with dark apical caps. Asci (Fig. 33o). Ascospores 15–20 × 4.5–6 μm, oblong-ovoid to oblong, 3-septate; apical cell not enlarged. Pycnidia 60–85 μm diam., rare, immersed, black, the wall brown, K+ greenish; conidia 4–5 × 1 μm, bacilliform. Lichen products not detected by t.l.c.

On mostly smooth bark of many trees and shrubs; widespread and common except in polluted areas. Throughout British Is. Europe, North America.

A common and very variable species, often confused with *A. punctiformis* **(33)**.

35 A. spadicea Leighton (1854)

Thallus immersed, occasionally superficial to 35 μm thick, effuse, pale to dark grey, green-grey, or grey-green; hyphae K/I−. Photobiont *Trentepohlia*. Apothecia (Fig. 28c), (0.1-)0.2–1.5 mm diam., rounded, flat to slightly convex, red-brown to blackish, often slightly glossy, resembling spots of tar; in section 55–70 μm tall, ± uniformly orange-brown, K+ dull purplish; epithecium indistinct; hymenium 25–35 μm tall; hypothecium 30–40 μm tall. Paraphysoids 0.5–1(-1.5) μm wide, not swollen or pigmented at apices; apices often embedded in a thin, clear gel-like layer. Ascospores 7–11 × 3–4 μm, 1-septate, ovoid or sole-shaped, colourless. Pycnidia frequent, 80–100 μm diam., brown, the wall orange- to red-brown, K+ purplish; conidia 3–4.5 × 0.5–1 μm, bacilliform to narrowly ellipsoid, straight or slightly curved. Lichen products not detected by t.l.c.

On very shaded bark of trunks (usually at or near base) of deciduous trees in moist woodland; rather common except in very polluted areas, rare in N. Scotland. Throughout British Is. Temperate Europe, ?North America, South Africa.

Distinguished from *A. vinosa* **(39)** by the usually less convex apothecia, smaller ascospores, and apothecial pigment which turns purplish, but does not dissolve, in K.

36. A. stellaris Krempelh. (1861)

Thallus immersed, usually delimited by a brown line and 1–2 cm diam., sometimes larger on tree trunks, whitish with green or grey tinge, or pale fawn. Photobiont *Trentepohlia*. Apothecia linear or more usually irregularly branched to stellate and to 2 mm diam., with 'arms'; 0.08–0.15 mm wide, reddish brown to blackish, not pruinose but sometimes covered by a thin layer of bark cells; in section 60–95 μm tall; epithecium reddish brown, K+ greenish; hymenium 40–50 μm tall, colourless or pale red-brownish above; hypothecium indistinct or to 20 μm tall, colourless or pale red brownish, K+ greenish in patches. Paraphysoids 1(-1.5) μm wide, brown-walled and to *c.* 2 μm wide in the epithecium, sometimes with dark apical caps. Ascospores (13-)16–22(-24) × (5-)6–7(-9) μm, ovoid to oblong-ovoid, (2-)3- or 4-septate;

upper cell enlarged; old ascospores brown and warted. Pycnidia rare, immersed, 40–60 μm diam., the wall red-brown, K+ greenish; conidia 4.5–5.5 × 0.5–1 μm. Thallus Pd−, K−, KC−, C−, UV− (unidentified substance in t.l.c.).

On smooth bark, especially of *Corylus*, *Ilex*, *Quercus*, and *Sorbus*, in old woodland. W. Britain, locally frequent in Scotland (E. Perth), rare elsewhere. Temperate Europe in oceanic or montane areas.

Often confused with other species with ascospores with an enlarged upper cell: *A. anglica* **(2)**, *A. anombrophila* **(3)**, *A. astroidestra* **(5)**, *A. ilicinella* **(19)**, *A. zwackhii* **(40)**, weakly pigmented morphs of *A. elegans* **(11)**, and the lichenicolous *A. invadens* **(23)**.

37. A. thelotrematis Coppins (1989)

Thallus absent, lichenicolous forming circular to elliptic necrotic patches 3–7 × 1.5–5 mm. Apothecia crowded, developing below the surface of the bark but soon erumpent, fleck-like, rounded to ± polygonal and 0.06–0.2 mm diam., or elongate and to 0.4 × 0.06–0.12 mm, sometimes slightly branched or bluntly stellate, dark reddish brown to black, not pruinose; in section 55–60 μm tall; epithecium 12–15 μm tall, reddish brown, K+ olivaceous; hymenium 33–35 μm tall, pale reddish brown, K+ greenish, I+ red; hypothecium 5–15 μm tall, reddish brown, K+ olivaceous, I+ blue. Paraphysoids rather scanty, sparingly branched, *c.* 1 μm wide; apices often branched, brown-walled, 1.5–2.5 μm wide, a few sometimes with a dark brown cap 3–3.5 μm wide. Ascospores 11–14 × 4.5–5 μm, (2-)3-septate, upper cell enlarged, oblong-ovoid, at first colourless and often with a thin perispore, later covered in dark brown, granular warts.

On thallus of *Thelotrema lepadinum* on *Corylus*, parasitic. Devon, W. Scotland, N. Ireland. Azores. Similar to *A. graphidicola* **(18)**.

38. A. varians (Davies) Nyl. (1861)

A. glaucomaria (Nyl.) Nyl. (1861)

Thallus absent, lichenicolous. Apothecia developing in the hymenium of the host and blackening the disc, 0.3–1 mm diam., black, level with the host disc or slightly more convex; epithecium brown to olive-brown, K+ greenish; hymenium 45–60 μm tall, colourless, I+ blue; hypothecium colourless. Paraphysoids 1.5–2(-2.5) μm; apices cemented together by dark pigment. Ascospores 11–17 × 5–6.5 μm, (1-)2- to 3-septate, ovoid to oblong-ellipsoid, colourless.

In apothecia of *Lecanora rupicola*, rather common in most of the host's localities. Throughout British Is. Europe, N. Africa, North America.

The very similar ***A. subfuscicola*** (Lindsay) Triebel (1991), occurring in the hymenium of *Lecanora carpinea*, is reported from S. Aberdeen.

39. A. vinosa Leighton (1856)

Thallus immersed or thinly powdery to scurfy-granular, effuse, whitish to pale fawn, or more usually stained orange-yellow, sometimes pinkish when fresh, often with dull yellowish to orange-yellow blotches, K+ purple. Photobiont *Trentepohlia*. Apothecia 0.2–0.5(-0.6) mm diam., rounded, convex, orange-red-brown to brown-black, usually matt, not pruinose; in section 85–140 μm tall, yellow-orange-red throughout or brown-red in upper hypothecium, K+ magenta and K+ purple with pigments dissolving; epithecium indistinct; hymenium 33–40 μm tall; hypothecium 45–100 μm tall. Paraphysoids 0.5–1(-1.5) μm wide, mostly neither swollen nor dark-walled above, a few sometimes with dark walls and apical caps. Ascospores 11–15 × 4–5 μm, ovoid-ellipsoid to sole-shaped, 1-septate, colourless but old ascospores brown and warted. Pycnidia frequent, 40–60 μm diam., the wall red-brown, K+ purple; conidia (3.5-)4–6 × *c.* 1 μm, bacilliform or slightly curved. Thallus Pd−, K− or K+ purplish (pigment), C−, (unidentified orange anthraquinone).

On bark or exposed wood of old trees, especially *Quercus* but often also *Alnus*, confined to old woodland and ancient parklands; common. S., W. & N. Britain. Europe, ?North America.

40. A. zwackhii Sandst. (1903)

Thallus to 80 μm thick, effuse, whitish to pale grey, matt, slightly rimose. Photobiont *Trentepohlia*. Apothecia numerous and often crowded, rounded to polygonal and 0.2–0.3 mm diam., or linear (usually along edges of cracks in the bark) and to 1 × 0.01–0.15 mm, or stellate and to 1.2 mm diam., reddish brown but mostly thinly white-pruinose; in section 65–95 μm tall; epithecium red-brown, K+ greenish;

hymenium 40–50 μm tall, colourless, or pale red-brown above; hypothecium 15–35 μm tall, colourless to pale straw. Paraphysoids 1–2 μm wide, but brown-walled and to 2.5 μm wide in the epithecium, a few with apical caps. Ascospores 16–22(-24) × 5–7 μm, oblong-ovoid, 3- to 4-septate, the upper cell enlarged; old ascospores brown and warted. Pycnidia not seen. Thallus Pd− or Pd+ yellow in places, K− or K+ yellow to red (crystals; reaction patchy), KC−, C− (unidentified substance 'A' of Coppins & James, 1978; ± norstictic acid).

On dry bark of mature *Fagus*, *Fraxinus* and *Quercus* in ancient woodlands; rare. S. England (Dorset; S. Hampshire, New Forest), Scotland (E. Perthshire). Germany, Denmark, S. Sweden.

Distinguished from *A. impolita* **(20)** by the C− thallus and apothecia, and ascospores with upper cells enlarged, and from *A. anombrophila* **(2)** by the longer ascospores and lichen products.

ARTHOPYRENIA Massal. (1852)

Thallus immersed in bark, inconspicuous or causing discoloration of bark, sometimes blackish due to copious development of brown hyphae; lichenized thalli usually whitish to pale brown, often pinkish when fresh, effuse or occasionally delimited in lichenized species. **Photobiont** absent or *Trentepohlia*. **Ascomata** perithecia, circular to ellipsoid in surface view, in section comprising a dark, often laterally spreading, clypeate involucrellum composed of compacted hyphae and bark cells, and a thin, usually colourless true exciple that surrounds the centrum; hyphae dark brown, K− or K+ greenish. **Hamathacium** usually of branched and anastomosed, ± moniliform or slender pseudoparaphyses, I−, rarely without interascal filaments and then often I+ bluish, downwardly projecting periphysoids also present. **Asci** with two functional wall layers, fissitunicate, obpyriform, obclavate or ± cylindrical, 8-spored, I−. **Ascospores** clavate, oblong or fusiform, 1- or 3-septate, usually medianly constricted, sole-shaped, colourless, smooth, but occasionally brownish and finely warted when old; perispore (in K) inconspicuous to distinct. **Conidiomata** pycnidia, blackish, wall concolorous with involucrellum of ascomata. **Conidiogenous cells** unbranched, enteroblastic, often percurrently proliferating, ± cylindric, lageniform or ± globose. **Conidia** variously bacilliform, oblong, ovoid or thread-like, 0(-1–3)-septate, some species have two anamorphs. **Chemistry**: no substances detected by t.l.c. **Ecology**: on, usually smooth, bark.

Species on rocks, earth and mosses, with *Nostoc* as photobiont, previously included in this genus, are treated under *Pyrenocollema*. *Mycomicrothelia* has a similar morphology and habitat ecology to *Arthopyrenia* but has dark-coloured and usually ornamented ascospores. Other superficially similar genera on bark, but with different ascomatal, ascus, ascospore, or hamathecium structures, are *Acrocordia, Anisomeridium, Leptorhaphis, Porina, Strigula* and *Tomasellia*.

The status of several species reported from the British Is. is currently not clear and these are not considered in the following account: *Arthopyrenia allogena* (Nyl.) Arnold (1870) (on *Rhizocarpon concentricum*; new genus probably required), *A. aphorisa* (Stirton) A. L. Sm. (on *Ulmus*, W. Inverness), *A. desistens* (Nyl.) Sm. (1911) (?lichenicolous), *A. lomnitzensis* B. Stein (1879) (on *Hymenelia lacustris*), *A. platypyrenia* (Nyl.) Arnold, *A. socialis* Körber (1865) (records dubious) and *A. spilobola* (Nyl.) Arnold (1891) (records dubious).

LITERATURE: Coppins (1988a).

1	Ascospores 1-septate (a few old spores sometimes 3-septate)	2
	Ascospores soon 3-septate	11
2(1)	Involucrellum K+ green	3
	Involucrellum remaining brown in K	6
3(2)	Ascospores with a distinct gelatinous perispore, *c.* 2 μm wide in K	4
	Ascospores without or with a very thin perispore, <1 μm thick in K	5

4(3) Both cells of mature ascospores with a distinct median constriction; ascomata 0.2–0.45 mm wide .... **5. *cinereopruinosa***
Cells of mature ascospores without a median constriction, except sometimes the lower cell; ascomata (0.3-)0.4–0.6 mm wide .... **7. *lapponina***
5(3) Ascospores with a short upper cell and an elongated lower cell .... **10. *ranunculospora***
Ascospores with cells of ± equal length .... **13. *viridescens***

6(2) Ascospores <27 μm long or 8 μm wide, colourless 7
Ascospores 27–40 × 8–13 μm, often becoming brownish .... **1. *antecellans***

7(6) Pseudoparaphyses persistent, but sometimes ± moniliform 8
Pseudoparaphyses absent, but periphysoids arising from upper wall (section) .... **11. *salicis***

8(7) Ascomata <0.25 mm diam.; asci <58 μm tall 9
Ascomata 0.25–0.5 mm diam.; asci 60–100 μm tall .... **6. *fraxini***

9(8) Thallus lichenized with *Trentepohlia*; ascospores never with distinct perispore 10
Thallus non-lichenized (without *Trentepohlia*); ascospores often with thin but distinct perispore .... **8. *punctiformis***

10(9) Ascospores (18-)19–26(-29) × 4.5–5(-6) μm; asci 40–60 μm tall; thallus whitish to pale buff .... **2. *nitescens***
Ascospores (14-)15–19(-21) × 3–4.5 μm; asci 31–45 μm tall; thallus pale fawn to darkish brown .... **3. *carneobrunneola***

11(1) Involucrellum remaining brown in K; ascomata 0.1–0.25 mm wide 12
Involucrellum K+ green; ascomata 0.3–0.5 mm wide .... **4. *cerasi***

12(11) Thallus not discolouring the bark; ascospores 22–31 × 7–9 μm .... **12. *subcerasi***
Thallus a blackish stain; ascospores 14–20 × 4.7–6 μm .... **9. *rhyponta***

1. A. antecellans (Nyl.) Arnold (1870)

Thallus whitish, greenish white or pale brown, probably non-lichenized. Ascomata 0.2–0.4 × 0.2–0.3 mm, circular or ellipsoid; involucrellum yellowish brown in K. Pseudoparaphyses 1.5–2.5 μm wide, persistent, much branched, cells 7–16 μm long. Asci 65–105 × (20-)25–35(-40) μm, obpyriform. Ascospores 27–40 × 8–13 μm, 1(-3)-septate, often becoming brownish and finely warted, apices ± pointed; perispore indistinct. Pycnidia *c.* 100 μm diam., usually numerous; conidia 4–6 × 0.8 μm, bacilliform.

On smooth bark, especially of *Corylus, Fagus, Ilex* and *Sorbus*, mostly in old woodlands. Widely distributed in W. British Is. Europe, North America.

Usually recognized in the field by the abundant pycnidia contrasting with the intermixed, larger ascomata and easily confirmed by the large ascospores that often become brownish; the ascospores of *Mycomicrothelia* species become dark brown at an earlier stage.

2. A. nitescens (Salwey) Mudd (1861)

Thallus whitish to pale buff, usually pinkish when fresh, lichenized with *Trentepohlia*. Ascomata 0.15–0.2 × 0.12–0.17 μm, circular to ellipsoid, mostly scattered; involucrellum brown in K. Pseudoparaphyses 1.5–3 μm wide, persistent, much branched, ± moniliform, the cells 3–8 μm long. Asci 40–60 × (16-)19–23 μm, obpyriform or obclavate. Ascospores (18-)19–26(-29) × 4.5–5(-6) μm, 1-septate, scarcely constricted at septum; cells ± equal, each usually biguttulate; perispore indistinct. Pycnidia 30–40 μm diam., frequent; conidia 2.8–3.8 × 0.8 μm, bacilliform.

On smooth bark, mainly *Corylus*, in old woodlands. W. British Is. Endemic?

Apt to be confused with *A. punctiformis* **(8)**, but the ascospores never have a distinct perispore, and the species grows in mature, species-rich communities (i.e. with *Arthonia, Pyrenula, Thelotrema* species, etc.). Closely related to *A. carneobrunneola* **(3)**.

3. A. carneobrunneola Coppins (1988)

Like **2**, but thallus usually darker (brownish); ascomata slightly smaller, 0.1–0.2 mm diam., and often confluent; asci 31–45 × 14–17(-21) μm; ascospores (14-)15–19(-21) × 3–4.5 μm.

On smooth bark, mainly *Corylus* (often with **2**), in old woodlands. W. England (rare), W. Scotland, W. Ireland. Endemic?

4. A. cerasi (Schrader) Massal. (1852)

Thallus inapparent or slightly bleaching the bark, non-lichenized. Ascomata 0.3–0.5 × 0.25–0.4 mm, often ellipsoid; involucrellum brown, K+ greenish. Pseudoparaphyses 1.5–2 μm wide, persistent, slender, cells *c.* 8–14 μm long. Asci 60–75 × 14–18 μm, cylindric-clavate. Ascospores 17–22 × 5–7 μm, clavate with rounded apices, 3-septate, constricted at septa especially the median one; perispore distinct, *c.* 2 μm thick in K. Pycnidia 80–120 μm, with either macro- or microconidia; macroconidia 11–13 × 2–2.5 μm, oblong, 3-septate; microconidia 9–14 × 0.8 μm.

On smooth bark, especially *Corylus*. Scotland, England (Lake District); local. Europe.

Close to *A. lapponina* **(7)** but differing in the smaller ascomata, 3-septate ascospores, and conidial characters.

5. A. cinereopruinosa (Schaerer) Massal. (1855)

Thallus inconspicuous or slightly bleaching the bark, non-lichenized. Ascomata (0.2-)0.3–0.45 × 0.2–0.3 mm, circular or ellipsoid, often covered by a thin layer of bark cells giving a whitish pruinose appearance. Pseudoparaphyses 1–1.5 μm wide, persistent, slender, cells 8–16 μm long. Asci 55–77 × 12–19 μm, cylindrical-clavate. Ascospores 14.5–22 × 5–8 μm, clavate, with rounded apices, 1-septate, much constricted at the septum and each cell with a median constriction; perispore distinct, *c.* 2 μm thick in K. Pycnidia 60–100 μm diam., with either macro- or microconidia; macroconidia 8–11 × 1.8–2.2 μm, bacilliform; microconidia 4.7–6 × 1 μm, bacilliform.

On smooth bark, especially young *Fraxinus*. N. & W. British Is. Europe, North America.

Close to *A. lapponina* **(7)**, but ascomata somewhat smaller, ascospores with a distinct median constriction in both cells, slightly narrower pseudoparaphyses and different conidia.

6. A. fraxini Massal. (1852)

Thallus inconspicuous, or visible as a pale fawn or pale brown stain, non-lichenized. Ascomata 0.25–0.5 × 0.2–0.4 mm, circular or ellipsoid; involucrellum remaining brown in K. Pseudoparaphyses 1–1.5 μm wide, persistent, richly branched, cells 6–12 μm long. Asci 60–90(-100) × 16–25 μm, narrowly obpyriform. Ascospores (17-)19–26 × (4.5-)5–7 μm, 1(-3)-septate, ± constricted at the septum, the apices ± rounded; perispore distinct, 1–1.5 μm thick in K. Pycnidia *c.* 100 μm diam.; conidia 4–6 × 0.8 μm, bacilliform.

On smooth bark, usually on branches, of *Fraxinus*, also on *Alnus*, *Betula*, *Corylus*, *Ilex* and *Quercus*. Throughout Scotland, probably elsewhere in Great Britain. Europe, North America.

Superficially, easily mistaken for *A. lapponina* **(7)**, but distinguished by its K− involucrellum, more richly branched pseudoparaphyses, ± obpyriform asci and shorter conidia. The internal features of *A. fraxini* more closely resemble *A. punctiformis* **(8)**, which has smaller ascomata, asci and ascospores and ± moniliform pseudoparaphyses.

7. A. lapponina Anzi (1864)

Thallus inconspicuous, not discolouring bark, non-lichenized. Ascomata (0.28-)0.4–0.6 × 0.22–0.55 mm, usually ellipsoid and with flat border formed by the spreading involucrellum; involucrellum brown, K+ greenish. Pseudoparaphyses 1.5–2 μm wide, persistent, slender, distantly branched, cells 8–15(-18) μm long. Asci 65–100 × 12–16(-20) μm., ± cylindrical. Ascospores 15–23 × 5.5–8(-9) μm, 1-septate (a few old spores occasionally 3-septate), clavate with rounded apices and markedly constricted at septum, each cell often biguttulate but without a median constriction (except sometimes faintly so in the lower cell); perispore distinct and *c.* 2 μm thick in K. Pycnidia, when present, 60–100 μm diam.; conidia 7–11 × 0.8–1 μm, thread-like, ± straight.

On smooth bark of many trees, often as an early colonizer on thin branches; common. Throughout British Is. Europe, North America.

Usually recognized by the rather large, ± ellipsoid ascomata in an inapparent thallus (without *Trentepohlia*), but several less common species have a similar appearance: *A. cerasi* **(4)**, *A. cinereopruninosa* **(5)**, *A. fraxini* **(6)**, *Julella sericea* and *Pyrenula coryli*.

8. A. punctiformis Massal. (1852)

A. cembrina (Anzi) Grumman ex D. Hawksw. (1980), *A. laburni* Arnold (1861), *?A. padi* Rabenh. (1858)

Thallus inconspicuous, or bark darkened (due to abundant brown hyphae), non-lichenized. Ascomata 0.1–0.23 mm diam., ± circular, scattered, or clustered; involucrellum remaining brown in K. Pseudoparaphyses (1-)1.5–2.5(-3) μm wide with cells 3–8 μm long, persistent, much branched, ± moniliform. Asci 40–55 × 15–22 μm, obpyriform. Ascospores 16–20(-22) × 4.5–5 μm, 1-septate and colourless, (sometimes yellowish brown and 3-septate when old), slightly constricted at the septum, cells often biguttulate but without median constrictions; perispore usually distinct, *c.* 1 μm thick in K. Pycnidia not detected.

On smooth bark of many trees and shrubs, usually as an early colonizer of twigs and thin branches and often accompanied by *Arthonia punctiformis*; common, (except in very polluted areas). Throughout British Is. Europe, North America.

A rather broad concept of the species is used here to include extreme morphs with a blackish thallus (i.e. *A. laburni*) or with irregularly clustered ascomata (i.e. *A. cembrina*); further critical studies on this complex are needed. Usually easily recognized in the field by the small ascomata and habitat ecology, but can be confused with *A. salicis* **(11)**, which has slightly larger ascomata and different hamathecial characteristics. Other species with small ascomata and 1-septate ascospores are *A. carneobrunneola* **(3)**, *A. nitescens* **(2)** and *A. viridescens* **(13)**, but they all have a clearly lichenized thallus containing *Trentepohlia*. A similar species, but with 3-septate ascospores, is *A. rhyponta* **(9)**.

9. A. rhyponta (Ach.) Massal. (1852)

Like morphs of **8** with a blackish thallus (due to abundant brown hyphae), but ascospores becoming 3-septate at an early stage.

On smooth bark, especially branches; local. From S. England (Sussex) to Scotland (Highlands). Europe, North America.

Most material referred to this species belongs either to *A. punctiformis* **(8)** or to *Stigmidium microspilum* (Körber) D. Hawksw. (1975), a lichenicolous fungus forming circular grey patches on *Graphis scripta* and with ascospores 14–15(-19) × 3–5 μm. The doubtfully distinct *A. grisea* (Schleicher ex Schaerer) Körber (1855), which does not have a darkened thallus, is incorrectly reported from Great Britain.

10. A. ranunculospora Coppins & P. James (1988)

Thallus immersed, white to fawn, often pinkish when fresh, lichenized with abundant *Trentepohlia*. Ascomata 0.22–0.36 × 0.2–0.3 mm, circular or ellipsoid, often matt or ± pruinose due to thin covering of bark cells; involucrellum brown, K+ greenish. Pseudoparaphyses 0.5–1.5 μm wide, persistent, much branched, cells 4–10 μm long. Asci 35–42 × 12–17 μm, obpyriform. Ascospores 14–16 × 4.5–5.7 μm, 1-septate and constricted at septum, with short, uniguttulate upper cell and an elongate, biguttulate lower cell, with a median constriction; perispore indistinct. Pycnidia frequent, of two types: (a) 80–120 μm diam., the macroconidia 7.6–10(-12) × 3.5–4.3 μm, oblong-ellipsoid, 0(-1)-septate; (b) 30–60 μm diam., the microconidia 3.5–4 × 0.5–0.8 μm, bacilliform.

Usually on smooth areas of fissured bark of tree trunks, especially *Fraxinus* and *Quercus*, sometimes on smooth bark of trunks or main stems of e.g. *Corylus*, *Ilex* and *Sorbus*, also *Calluna*; widely distributed in sheltered, old woodlands. W. British Is. S.W. Norway and Denmark.

Easily identified by its 'tadpole-shaped' ascospores, K+ green involucrellum, and well-developed thallus; also, it is the only British *Arthopyrenia* found on rough bark. It is occasionally found only with macroconidia.

11. A. salicis Massal. (1852)

Thallus often inconspicuous, pale fawn brown, sometimes pinkish when fresh, more rarely dark brown (due to abundant brown hyphae). Ascomata 0.15–0.3 × 0.15–0.2 mm, often with depressed ostiole, circular or ellipsoid; involucrellum remaining brown in K. Pseudoparaphyses absent (or gelatinized before asci are mature); periphysoids 7–15 × 1–1.5 μm, unbranched, 0- to 2-septate; hymenial gel sometimes I+ pale blue. Asci 33–45 × 14–20 μm, obpyriform. Ascospores (12-)14–17(-21) × 4–5.5 μm, 1-septate, constricted at the septum, each cell biguttulate, the lower cell often with a median constriction, the apices rather blunt; perispore indistinct. Pycnidia *c.* 60 μm diam.; conidia 3–3.7 × 1 μm, ± bacilliform.

On smooth bark, especially of *Corylus*; rather common. Throughout British Is., especially frequent in the north and west. Europe, North America.

Apparently facultatively lichenized, as thalli vary from having no detectable to abundant *Trentepohlia* cells. Differs from *A. punctiformis* **(8)** and all other small perithecioid species in lacking pseudoparaphyses and possessing periphysoids.

12. A. subcerasi (Vainio) Zahlbr. (1921)

Thallus inconspicuous, non-lichenized. Ascomata 0.2–0.26 × 0.14–0.2 mm, ± ellipsoid; involucrellum remaining brown in K. Pseudoparaphyses 1.5–2.5 μm wide, persistent, much-branched; cells 4–10 μm long. Asci 55–80 × 19–30 μm, obpyriform. Ascospores 22–31 × 7–9 μm, 3-septate, colourless, sometimes brownish and warted when old; perispore very thin or indistinct. Pycnidia 40–60 μm diam; conidia 4–5 × 0.8–1 μm, ± bacilliform.

On smooth bark of *Betula* trunks in native pinewoods; rare. Scotland. Scandinavia.

Most likely to be confused with *A. antecellans* **(1)**, but differs in the generally smaller ascomata and smaller ascospores, which become 3-septate very early and only rarely become brownish. *A. cerasi* **(4)** differs in its larger ascomata, ± cylindric asci, shorter ascospores, K+ green involucrellum and different conidia.

13. A. viridescens Coppins (1988)

Thallus a whitish or pale fawn stain on bark, sometimes pinkish when fresh, lichenized with *Trentepohlia*. Ascomata 0.14–0.2 × 0.1–0.15 mm, circular to ellipsoid, mostly scattered; involucrellum brown, K+ green. Pseudoparaphyses 0.8–1 μm thick, persistent, much-branched; cells 4–10 μm long. Asci 31–49 × 11–17 μm, obclavate to short-cylindrical. Ascospores 12–17(-19) × 3.5–4.7 μm, 1-septate, usually constricted at the septum; upper cell usually distinctly wider, the cells usually biguttulate, without a median constriction, colourless or sometimes brownish and faintly warted when old; perispore indistinct. Pycnidia 40–50 μm diam., with either: (a) macroconidia 9–10.5 × 2–2.5 μm, oblong, 2- to 3-guttulate; or (b) microconidia 4–5 × 1–1.3 μm, ovoid-oblong to bacilliform, eguttulate.

On smooth bark of *Corylus* in old woodlands, W. England, Wales, W. Scotland, N. Ireland. ?Endemic.

Occurs in similar communities to *A. carneobrunneola* **(3)**. Differs from other species with very small ascomata in having a K+ green involucrellum and thin, slender paraphyses. Probably related to *A. cinereopruinosa* **(5)**, which has larger ascomata and larger ascospores, which also have a distinct perispore.

ARTHOTHELIUM Massal. (1852)

Thallus crustose, immersed or superficial, effuse or delimited. **Photobiont** *Trentepohlia*, chlorococcoid or absent. **Ascomata** apothecia-like, variously shaped, flat to convex, elongate or stellate. **Disc** red-brown to black, sometimes pruinose. **Thalline exciple** absent. **True exciple** absent. **Epithecium** colourless to red-brown or dark brown. **Hymenium** usually I+ blue. **Hypothecium** poorly developed to well developed and dark red-brown to dark brown. **Hamathecium** of sparsely to richly branched and anastomosed paraphysoids, the apices usually swollen and red-brown, often capitate. **Asci** *Arthonia*-type. **Ascospores** ovoid to ellipsoid, colourless, the upper part sometimes enlarged, muriform. **Conidiomata** *Arthonia*-like. **Chemistry**: lichen products absent, or with various substances (incl. xanthones and anthraquinones). **Ecology**: on bark, especially when smooth, in humid, little-disturbed habitats.

Separated from *Arthonia* only by the muriform, not transversely septate, ascospores and retained only for convenience pending further studies of inter-relationships in the family.

LITERATURE: Coppins (1989), Coppins & James (1979a).

1	Ascospores with an undivided, enlarged upper cell	2
	Ascospores without an undivided, enlarged upper cell	3

2(1) Apothecial sections K+ purple-violet or magenta .... **2. *reagens***
Apothecial sections K+ greenish .... **1. *dictyosporum***

3(1) Photobiont absent or chlorococcoid 4
Photobiont *Trentepohlia* 5

4(3) Ascospores oblong-ovoid, the transverse cells with 0–1 longitudinal septa .... **3. *lirellans***
Ascospores ovoid-ellipsoid, the four central transverse cells with 2–3 longitudinal septa .... **5. *orbilliferum***

5(3) Ascospores >25 × 10 μm 6
Ascospores (15-)17–25(-26) × 7–9.5(-10.5) μm .... **6. *ruanum***

6(5) Apothecia mostly 0.3–0.6 mm at widest or longest point; epithecium reddish brown, K+ green; paraphysoid apices with dark caps .... **4. *norvegicum***
Apothecia mostly 0.5–1(-2) mm at widest or longest point; epithecium red-brown, K+ reddish or K− ; paraphysoid apices without dark caps .... **7. *spectabile***

1. A. dictyosporum (Coppins & P. James) Coppins (1989)

A. ilicinum var. *dictyosporum* Coppins & P. James (1979)

Thallus immersed, often delimited by brown line, whitish, brown-grey or fawn. Photobiont *Trentepohlia*. Apothecia 0.3–0.8 × 0.14–0.3 mm, irregularly rounded, polygonal, or shortly elongate, flat or slightly

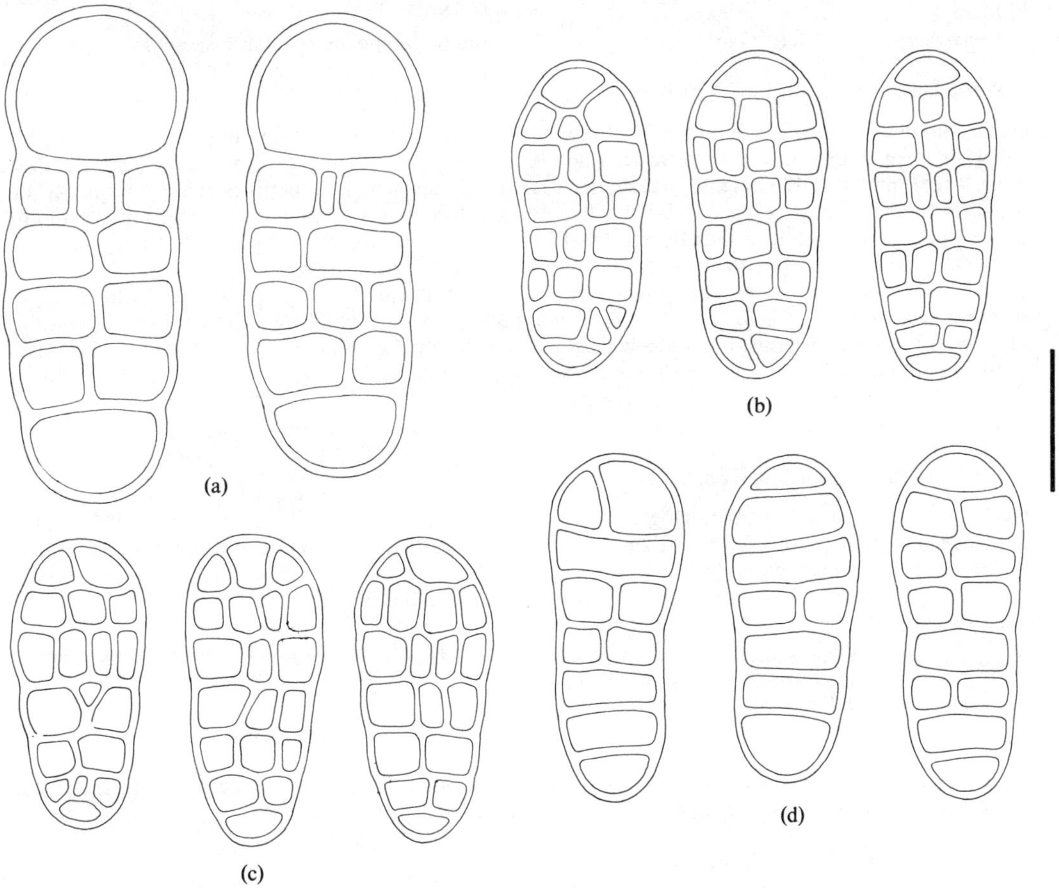

Fig. 4. *Arthothelium* ascospores. (a) *A. reagens*; (b) *A. ruanum*; (c) *A. orbilliferum*; (d) *A. lirellans*. Scale = 10 μm.

convex, sometimes disintegrating at centre, blackish, not pruinose; in section 80–190 μm tall; epithecium brown, K+ greenish; hymenium 60–90 μm tall, colourless; hypothecium 10–85 μm tall, colourless, or pale brownish (K+ greenish) in parts. Paraphysoids 1–1.5(-1.8) μm wide, numerous, mostly thickened by pigment to 2.5 μm wide in epithecium and a few with apical caps. Ascospores 28–33 × 9–12 μm, ovoid-oblong, muriform, the upper cell enlarged, the lower part with 5–7 transverse septa and most cells with 1 or 2 longitudinal septa, when old brown and warted. Pycnidia 50–60 μm diam., few, the wall reddish brown, K+ greenish; conidia 4.5–6.5 × 0.6–0.8 μm, bacilliform. Lichen products not detected by t.l.c.

On smooth (*Corylus*, *Ilex* and *Sorbus*) or roughish bark (*Salix*) in old woodlands; rare. Scotland (W. & C. Highlands). ?Endemic.

Similar in appearance to *Arthonia ilicina*, but ascospores with a muriform 'tail' and distinctly warted when brown, and to *Arthothelium reagens* **(2)** which has K+ purplish and magenta pigments.

2. A. reagens (Coppins & P. James) Coppins & P. James (1981)

A. ilicinum var. *reagens* Coppins & P. James (1979)

Like **1**, but apothecia with brownish purple tinge; thallus around apothecia often reddish orange; epithecium purplish, K+ purple-violet; epithecium, hymenium and hypothecium with orange-red, oily inclusions that are K+ magenta. Ascospores 20–38 × 10–14 μm, the lower part with 4–6 transverse septa and at maturity most transverse cells with 1–2 longitudinal septa, when old brown and distinctly warted (Fig. 4a).

On smooth bark of *Corylus* and *Fraxinus* in old hazelwoods; rare. W. Scotland (Argyll to Skye). ?Endemic.

A. macounii (G.K. Merr.) Noble (1987), from W. Canada, could be the same species.

3. A. lirellans (Almq.) Coppins (1979)

Thallus immersed, usually delimited by brown line, indistinct but discolouring bark whitish, dull yellowish, fawn, or pale grey to grey-brown. Photobiont absent. Apothecia to 0.8 mm diam., or 0.3–1.2 × 0.1–0.3(-0.5) mm, shortly elongate, irregularly rounded or polygonal, sometimes slightly branched or bluntly stellate, in section 45–60 μm tall; epithecium green-brown, K+ greenish; hymenium 35–40 μm tall, colourless; hypothecium indistinct or to 15 μm tall, colourless. Paraphysoids 1–2 μm wide, but pigmented and to 3 μm wide in epithecium, apices often with dark cap. Ascospores 17–27(-30) × 8–12 μm, oblong-ovoid, muriform with (4-)5–8 transverse septa and 0–4 transverse cells with a single longitudinal septum (Fig. 4d). Pycnidia 60–80 μm diam., rare, the wall brown, K+ greenish; conidia 3.5–5 × 0.5–0.7 μm, bacilliform. Lichen products not detected by t.l.c.

On smooth bark (*Corylus, Ilex, Quercus, Sorbus*), often on branches, in old woodlands (especially in ravines). W. Britain, most common in Scotland. ?Endemic.

Easily overlooked in the field for *Arthonia punctiformis*. Close to *A. orbilliferum* **(5)**.

4. A. norvegicum Coppins & Tønsb. (1984)

Thallus immersed, white or greenish grey, usually forming small (<3 cm diam.) patches. Photobiont *Trentepohlia*. Apothecia 0.3–0.6 mm diam., ± rounded, slightly convex, black, not pruinose, in section 90–130 μm tall; epithecium reddish brown, K+ greenish; hymenium 70–100 μm tall, colourless or pale straw, K+ greenish yellow; hypothecium 10–30 μm tall, colourless or patchily reddish-brown (K+ greenish). Paraphysoids 0.7–1.5 μm wide, numerous, but pigmented and to 2.5 μm wide in epithecium; apices often with dark hood or cap. Ascospores 29–36 × 12–15 μm, ovoid-ellipsoid to ovoid-oblong, muriform with 7–8(-9) transverse septa, at maturity most transverse cells with 2–3 longitudinal septa, when old brown and warted. Lichen products not detected by t.l.c.

On *Calluna* stems in coastal heath and on *Sorbus*; rare. W. Scotland (Coll), W. Ireland (Mayo, Donegal). Coastal C. Norway, ?W. Canada.

The Norwegian material has ascospores 31–42 μm, with mostly 9 transverse septa. Distinguished from *A. spectabile* **(7)** by the smaller apothecia and K+ greenish epithecium.

5. A. orbilliferum (Almq.) Hasse (1913)

Thallus immersed, grey-white or creamy-white, usually delimited by brown line. Photobiont absent. Apothecia to 0.7 mm diam., rounded, rarely elongate or bluntly stellate, usually soon disintegrating at the centre, black, not pruinose, in section 45–65 μm tall; epithecium greenish brown, K+ greenish; hymenium 30–50 μm tall, colourless or pale greenish; hypothecium to 10 μm tall, indistinct, colourless. Paraphysoids 1–1.5 μm wide, but pigmented and to 3 μm wide in the epithecium; apices often with dark hood or cap. Ascospores (16-)19–26 × (9-)10–14 μm, ovoid-ellipsoid, muriform, with 5–6 transverse septa, the 4 central cells with 2–3 longitudinal septa (Fig. 4c). Pycnidia 50–80 μm diam., rare, black, the wall greenish in K; conidia 3.7–4.7 × 0.5–0.8 μm. Lichen products not detected by t.l.c.

On smooth bark (*Corylus, Ilex, Quercus, Sorbus*) in old (especially ravine) woodlands; rare. W. Scotland, W. Ireland. France (Bretagne).

Very close to *A. lirellans* **(3)** but thallus usually whiter, apothecia more rounded and with a greater tendency to disintegrate at the centre, and ascospores relatively broader and with more longitudinal septa.

6. A. ruanum (Massal.) Körber (1861)

Thallus immersed, effuse or delimited by brown line, creamy white, pale grey, brown-grey or olive-grey. Photobiont *Trentepohlia*. Apothecia to 1.6(-2) mm diam., irregularly rounded to bluntly stellate, often disintegrating in places and regenerating to give the appearance of a crowded swarm of punctiform apothecia, black, not pruinose (but often long remaining covered by bark-cells), in section 70–95 μm tall; epithecium dark red-brown, K+ green; hymenium 35–50 μm tall, colourless or pale brown (K+ greenish); hypothecium 10–30 mm tall, dark red-brown, K+ dark green. Paraphysoids 1–1.5 μm wide, numerous; apices obscured by dense pigment. Ascospores (15-)17–24(-26) × 7–9.5(-10.5) μm, ovoid-oblong, muriform, with 5–8 transverse septa and (2-)3–7 transverse cells with 1–3 longitudinal septa, brown and warted when old (Fig. 4b). Pycnidia 60–80 μm diam., scarce, the wall red-brown, K+ green; conidia 4–6 × *c.* 1 μm, bacilliform. Lichen products not detected by t.l.c.

On smooth bark (*Castanea, Corylus, Fraxinus, Sorbus*), usually by streams in sheltered woodland; rare. S. England, W. England (Lake District), Wales, S. Scotland. Temperate Europe, though absent from highly oceanic areas.

7. A. spectabile Flotow ex Massal. (1852)

Thallus immersed, creamy- or grey-white, usually delimited by brown line. Photobiont *Trentepohlia*. Apothecia to 1.2 mm diam., or somewhat elongate to 2 mm long, irregularly rounded or polygonal, black, not pruinose, in section 120–150 μm tall; epithecium red-brown, K+ reddish; hymenium and hypothecium pale reddish brown, K− or K+ faintly greenish. Paraphysoids 0.5–1 μm wide, numerous; apices usually not pigmented and without caps. Ascospores 26–36 × 12–15 μm, ellipsoid to oblong-ellipsoid, muriform, with 5–7 transverse septa, most transverse cells with 1–3 longitudinal septa. Lichen products not detected by t.l.c.

On ± smooth bark; ?extinct (not reported this century). S.W. England, Wales, S.W. Ireland. C. & E. Europe, North & South America, E. Asia.

ARTHRORHAPHIS Th. Fr. (1860)

Thallus crustose, without a distinct prothallus, enclosed within the thallus of the host and grey to grey-green, or free at some stage and soon verrucose-squamulose, the squamules strongly convex, bright greenish yellow to whitish grey, ecorticate or with a colourless epinecral layer. **Soredia** occasionally present. **Photobiont** chlorococcoid, forming a thick layer. **Ascomata** apothecia, sessile or located between the squamules, black, urceolate or discoid, mostly ± filled with brownish green, crumbling, drop-like concretions. **True exciple** poorly developed, of lax hyphae with strongly swollen walls. **Hamathecium** of paraphyses, free, thin, branched and net-like, anastomosing, the apices not or only weakly thickened. **Hymenium** containing oil droplets. **Asci** 8-spored, clavate, K/I−, the wall scarcely thickened at the apex, with an ocular chamber. **Ascospores** cylindrical to mostly acicular, 7- to 12-septate. **Conidiomata** not known. **Ecology**: restricted to acidic substrata, either parasitic on *Baeomyces* species or, at least later, free-living on soil amongst mosses or on weathered rock surfaces in cool regions.

LITERATURE: Hawksworth (1983), Poelt & Vězda (1977).

1 Thallus bright greenish yellow; not parasitic; contains rhizocarpic acid 2
Thallus whitish grey or greenish grey, never yellow; parasitic on *Baeomyces*; rhizocarpic acid absent 3

2(1) Thallus ± dissolved in fine granular soredia; epinecral layer absent .... **2. *citrinella***
Thallus not sorediate, of convex squamules; epinecral layer present .... **1. *alpina***

3(1) Ascospores 20–70 × 2–2.5 μm; commensalistic on *Baeomyces rufus* .... **4. *grisea***
Ascospores 60–82 × 3–4 μm; commensalistic on *Baeomyces roseus* .... **3. *fuscireagens***

1. A. alpina (Schaerer) R. Sant. (1980)

Thallus of strongly convex or bullate squamules, to 1 mm diam., either remaining ± discrete or becoming confluent and forming small, ± compact irregular patches to *c.* 1 cm diam.; marginal lobes indistinct, not becoming elongate; upper surface bright yellow-green, matt, the surface often roughened and appearing ± pruinose; soredia absent; in section with a colourless, continuous epinecral layer 7–10 μm thick. Photobiont cells 7–14 μm diam., ± globose. Apothecia infrequent, located between squamules, at first slightly concave, becoming flat, the margin excluded or becoming so; disc to 1.5 mm diam.; true exciple and upper part of thecium blackish grey-green. Ascospores 35–60 × 2–5 μm, 8- to 11-septate.

On soil and decaying mosses, often in crevices on ± vertical siliceous rock faces, montane; rare. Scotland. N. Europe, C. European mtns.

When sterile may be confused with *Catolechia wahlenbergii*, which is very rare but may occur in similar habitats, and which has a larger thallus with distinct, elongate marginal lobes and a smooth, ± shiny upper surface.

2. A. citrinella (Ach.) Poelt (1969)

Like **1**, but of flat to slightly convex areoles to 0.5 mm diam., often rather thin; areoles fragile (readily broken when touched with a needle), soon becoming at least partly sorediate; without a distinct, continuous epinecral layer; ascospores (35-)50–70(-100) × 2–4(-5) μm.

Similar habitats to **(1)**, upland areas; often frequent. Throughout British Is. Bipolar, at high altitudes in tropical and subtropical regions.

Said by some central European authors to be at first parasitic on *Baeomyces rufus*, later forming an independent powdery thallus; this has not been observed in British material. Possibly the sorediate counterpart of *A. alpina* **(1)**.

3. A. fuscireagens (Vainio) Poelt (1969)

Thallus verrucose, whitish grey or greenish grey, often rather inconspicuous. Apothecia 0.2–0.3 mm diam., top-shaped, distinctly constricted at the base, at first concave with a protruding thin margin, later flat and almost immarginate. Ascospores 60–82 × 3–4 μm, 7- to 12-septate.

Commensalistic on *Baeomyces roseus*; on heavy metal-rich shingle gravels in Wales and near the summits of Scottish mtns. C. Wales (Dyfed, Ystwyth), N.W. Scotland (Sutherland, Ross). N. Scandinavia.

4. A. grisea Th. Fr. (1860)

Thallus immersed in modified thalli of *Baeomyces rufus*, whitish grey to dark grey-green, associated with abundant algae. Apothecia sessile, scarcely constricted at the base, at first strongly urceolate to almost closed with a thick margin, later disc-shaped with an often persistent, thick, rough margin; disc to 0.5 mm diam. Ascospores (20-)30–50(-70) × 2–2.5(-4) μm, (8-)12- to 15-septate.

With the host on acid soils and weathered siliceous rocks; local. Throughout British Is.

ASPICILIA Massal. (1852)

Thallus crustose, weakly cracked to distinctly areolate, areoles contiguous to scattered, some species placodioid, with radiating marginal lobes; chalky white, greyish, greenish or brownish. **Isidia** and **soredia** present in some species. **Cortex** pseudoparenchymatous.

Photobiont chlorococcoid, cells globose to ellipsoid. **Medulla** I−. **Ascomata** apothecia, aspicilioid (crypotolecanorine), mostly immersed, occasionally becoming emergent, more rarely lecanorine and soon emergent and ± sessile. **Thalline exciple** poorly developed and becoming excluded, more rarely well developed and persistent, often little more than a slightly raised rim of thallus tissue, but well differentiated in some species. **True exciple** usually colourless and poorly developed, lateral to and below the hymenium, sometimes broadening above and concolorous with the epithecium. **Disc** flat to concave, black-brown, sometimes white-pruinose. **Epithecium** pigment brown to green, N− or N+ green, K− or K+ fading to brown ('Aspicilia green'). **Hymenium** usually more than 100 μm tall, colourless, I+ greenish or blue. **Hypothecium** colourless or very pale brownish. **Hamathecium** of paraphyses, simple to sparingly branched, frequently anastomosed, short-celled and moniliform above, the apices cohering to form a well-defined epithecium. **Asci** 4- to 8-spored, cylindrical to clavate, outer coat K/I+ blue, wall and apical dome K/I−. **Ascospores** simple, ellipsoid to globose, colourless, thin-walled, sometimes with a thin perispore, I−. **Conidiomata** pycnidia, immersed, elongately flask-shaped to almost globose, singular or in clusters; wall colourless, or at least in upper part brownish or greenish (pigment as in epithecium). **Conidiogenous cells** sessile or on short condiophores, subcylindrical. **Conidia** acrogenous, simple, bacilliform to shortly thread-like and ± straight, colourless. **Chemistry**: some species with fatty acids (especially aspicilin), β-orcinol depsidones (especially norstictic and stictic acids), or unidentified terpenoids; thalli and apothecia of all species are C−. **Ecology**: on calcareous and acidic rocks; *c.* 150 species, mainly temperate and arctic.

Distinguished from *Lecanora* by the K/I− ascus apical dome and moniliform paraphyses. Species of *Bellemerea* and *Koerberiella* differ in having *Porpidia*-type asci, and superficially similar species of *Pertusaria* differ in their larger, often thick-walled ascospores, and thinner, not moniliform, paraphyses. Close to *Hymenelia* which has a generally thinner, non-corticate or weakly corticate thallus, and pale-coloured apothecial discs; also to *Ionaspis*, which should perhaps be combined with *Hymenelia*, but which has *Trentepohlia* as photobiont. *Megaspora* is distinguished by the very large (over 30 μm long), thick-walled ascospores and in growing on mosses.

A difficult genus containing many confusing species subject to environmental modification. Collections from Great Britain and Ireland continue to present many unresolved taxonomic problems and the genus requires an in-depth study. In particular, the status in the British Is. of *A. cupreogrisea* (Th. Fr.) Hue (1912), *A. gibbosa* (Ach.) Körber (1855), and *A. verrucigera* Hue (1912) is uncertain, and reliable descriptions are unavailable. Most British material named as *A. gibbosa* is referable to *A. caesiocinerea*. It is important that the colour should be accurately recorded on collection as this may rapidly alter in the herbarium.

LITERATURE: Clauzade & Roux (1984), Creveld (1981), Hermann *et al.* (1973), Laundon (1986), Magnusson (1939).

1 Thallus with radiating lobe margins 2
Thallus without radiating lobe margins 3

2(1) Thallus K+ red .... **12. *subcircinata***
Thallus K− .... **10. *melanaspis***

3(1) Thallus K+ red [crystals] 4
Thallus K− or yellow-red [crystals absent] 7

4(3) Isidia or soredia absent 5
Isidia or soredia present .... **6. *grisea***

5(4) Ascospores (15-)20–30 × 11–15(-20) μm; conidia to >15 μm or shorter than 12 μm long 6
Ascospores 12–22 × 6–13 μm; conidia 11–16 × 1 μm .... **4. *cinerea***

6(5) Conidia 7–12 μm long .... **7. *intermutans***
Conidia 15–28 μm long .... **5. *epiglypta***

7(3) Thallus without papillae, isidia, or soralia; apothecia usually present 8
Thallus effusely isidiate, or granular sorediate; apothecia usually absent ..... **9. *leprosescens***

8(7) Ascospores 12–35 × 8–29 μm; algal layer absent below hypothecium 9
Ascospores 9–14 × 7–9 μm; algal layer present below hypothecium ..... **11. *recedens***

9(8) Asci (6-)8-spored, clavate; ascospores 12–35 × 8–15 μm, ellipsoid, uni- or biseriate 10
Asci (3-)4(-6)-spored, shortly cylindrical; ascospores 18–34 × 13–29 μm, globose to broadly ellipsoid, uniseriate 11

10(9) Medulla Pd+ orange, K+ yellow; thallus weakly cracked, without discrete areoles **8. *laevata***
Medulla Pd−, K−; thallus usually with discrete areoles ..... **1. *caesiocinerea***

11(9) Thallus of scattered, rounded areoles or granules, brownish grey or chalk-white 12
Thallus cracked-areolate, chalk- or grey-white ..... **2. *calcarea***

12(11) Thallus of flattened, rounded, often white-pruinose areoles; apothecia ± immersed; asci 4-spored ..... **3. *contorta***
Thallus of semiglobose, smooth, not pruinose granules; apothecia sessile; asci mostly 8-spored ..... **14. *tuberculosa***

1. A. caesiocinerea (Nyl. ex Malbr.) Arnold (1886)

A. gibbosa auct. brit. p.p., non (Ach.) Körber (1855)

Thallus to 20 cm diam., rather thick, generally wide-spreading, cracked to warted-areolate; areoles continuous, slightly concave to flat or ± convex, occasionally ± subsquamulose in the centre of the thallus, irregular to rounded, 0.3–1(-2) mm wide, often uneven with distinct edges and a rough, matt surface, pale bluish, brownish or dark grey; prothallus when present grey, delimiting. Apothecia 0.2–0.8 mm diam., almost non-pruinose, at first crater-like, sometimes becoming emergent; thalline exciple evident from an early stage, thin, entire, but rarely prominent; disc black, becoming expanded. Asci (4-)6- to 8-spored. Ascospores 14–30 × 7–16 μm, broadly ellipsoid to ± globose, often poorly developed. Conidia 6–12 × 1 μm. Cortex and medulla Pd−, K− (aspicilin).

On nutrient-rich rocks, especially by lakes and seashores and bird-perching stones, also on walls and memorials; locally common. Widespread in British Is., especially in N. & W. Europe, North America.

2. A. calcarea (L.) Mudd (1861)

Thallus to 30 cm diam. or more, rather thick, forming ± circular patches, continuous or usually cracked-areolate; areoles mostly radially orientated at least in outer part of thallus, chalky or grey-white, more infrequently greyish, particularly towards centre, rarely rust-coloured; prothallus usually delimiting, darkish grey, ± zoned. Apothecia 0.2–1 mm diam., ± immersed, rounded or angular; thalline exciple slightly raised; disc black, not or slightly white-pruinose. Asci 4-spored. Ascospores 18–30 × 14–27 μm, broadly ellipsoid to subglobose, with a thin perispore. Cortex and medulla Pd−, K− (aspicilin).

On calcareous rocks and stonework, especially hard limestones, also memorials; very common. Widespread in British Is. Europe, North America.

See also *A. contorta* **(3)**. *A. lilliei* B. de Lesd. (1906), described from N.E. Scotland, is a doubtful species said to closely resemble *A. calcarea* but apparently differing in being yellow internally and in having much narrower ascospores 13–25 × 5–6 μm.

3. A. contorta (Hoffm.) Krempelh. (1861)

Like **2**, but thallus reduced to scattered (rarely crowded), ± rounded areoles, 0.2–1.2(-1.8) mm diam., often convex, grey to chalky-white; prothallus indistinct, never delimiting. Apothecia 0.2–0.6(-0.8) mm diam., mostly 1(-3) per areole.

On calcareous rocks and hard limestones, more often than *A. calcarea* on mortar, concrete and impure limestones; common. Widespread in British Is. Europe, North America.

When the areoles are crowded they may become ± angular, but unlike *A. calcarea* **(2)**, they are never radially orientated and the thallus is not delimited by a distinct prothallus. Somewhat intermediate specimens may occur, and it has been suggested that hybridization may account for this.

4. A. cinerea (L.) Körber (1855)

Thallus cracked-areolate, becoming warted or papillate in nutrient-rich environments, smooth, whitish grey, grey-brown or at times rust-coloured or greenish; prothallus sometimes dark grey, delimiting. Apothecia (0.2-)0.4–1.2(-2) mm diam., concave to flat, at first immersed, later sometimes becoming sessile, round or irregular; thalline exciple thick, persistent; disc matt, black. Asci (6-)8-spored. Ascospores 12–22 × 6–13 μm, oblong- to broadly ellipsoid. Conidia 11–16 × 1 μm. Thallus Pd+ orange, K+ red (norstictic acid).

On exposed, siliceous rocks; locally frequent. Widespread in N. & W. British Is., rare elsewhere. Europe.

It is likely that several species are included here; see also *A. epiglypta* **(5)** and *A. intermutans* **(7)**.

5. A. epiglypta (Norrlin ex Nyl.) Hue (1912)

Thallus coarse, markedly areolate, areoles rounded-warted, at times ± roughened, usually greenish pale brown-grey to pale yellowish brown or dark grey. Apothecia 0.5–1.5 mm diam.; true exciple thin, indented, at times segmented or convoluted, persistent; disc dark brown-black, generally rough and appearing wrinkled. Conidia 15–28 × 1 μm. Ascospores 20–25 × 12–15 μm. Cortex and medulla Pd+ yellow-orange, K+ red (norstictic acid).

On coastal granitic rocks, often in nutrient-rich situations; rare. Channel Is., Is. of Scilly, S. Wales (Pembroke). Europe, except Mediterranean.

Other material reported on schistose rocks in montane regions of C. Scotland (Perth, Westerness) may represent a different species.

6. A. grisea Arnold (1891)

A. insolata (Manusson) Havaas (1944)

Thallus thin, cracked-areolate, often poorly delimited, pale to dark grey; isidia scattered, often infrequent, paler or almost white, sometimes becoming eroded and soredia-like. Apothecia 0.3–0.6 mm diam., infrequent, sessile; true exciple well developed; disc black. Asci 8-spored. Ascospores 18–22 × 9–12 μm. Thallus Pd+ orange, K+ red (norstictic acid).

On well-lit siliceous rocks and walls; scarce. Widespread, especially in W. British Is. Europe.

Some collections from Scotland may belong to *A. simoensis* Räsänen (1925).

7. A. intermutans (Nyl.) Arnold (1887)

Like **4**, but ascospores (20-)22–28 × (11-)12–14 μm and conidia 7–11 × 1 μm.

On flat, low-lying rocks in a field and an outcrop; very rare. Scotland (Outer Hebrides, Tiree; E. Lothian). Europe, Asia.

A poorly understood species within the *A. cinerea* **(4)** complex characterized by the larger ascospores and shorter conidia.

8. A. laevata (Ach.) Arnold (1887)

?*A. sylvatica* Arnold (1884)

Thallus continuous or slightly cracked, without discrete areoles, rather smooth, thin, pale to dark green-grey, often glossy; prothallus delimiting, blackish or inconspicuous. Apothecia 0.2–0.5 mm diam., numerous and usually crowded, crater-like to emergent; thalline exciple absent; true exciple prominent; disc black, flat to concave. Asci 8-spored. Ascospores 13–20 × 9–13 μm, broadly ellipsoid. Conidia 18–25 × 0.8 μm. Medulla Pd+ orange, K− or K± yellow (stictic and ± norstictic acids, ± an unidentified terpenoid).

On rocks and boulders in streams or woodlands, especially in shaded situations; frequent. W. & N.W. British Is. Europe, North America.

9. A. leprosescens (Sandst.) Havaas (1918)

Thallus cracked-areolate, scurfy, of easily detached, ± papillate or flattened granules, at times later becoming ± granular-sorediate, pale grey-blue to dark bluish grey; prothallus sometimes evident, conspicuous, dark green-grey, delimiting. Apothecia 0.3–1 mm diam., infrequent, at first urceolate, later emergent; thalline exciple often granular, mostly paler than thallus. Asci (4-)8-spored. Ascospores 14–30 × 7–16 μm, subglobose to ellipsoid. Thallus Pd−, K− (aspicilin).

On nutrient-rich siliceous rocks, birds' perching rocks and associated seepage tracks in the xeric-supralittoral; locally frequent. Widespread in coastal N. & W. Great Britain & Ireland, very rarely inland (Wilts, Avebury). W. Europe, North America.

10. A. melanaspis (Ach.) Poelt & Leuckert (1973)

Thallus whitish grey or glaucous, greening strongly when wet, with elongate, radiating marginal lobes; lobes ± convex, loosely adpressed and readily removed from the substratum, 0.4–1 mm wide; pale below. Apothecia to 1.8 mm diam.; thalline exciple prominent; disc flat to ± convex, red-brown to black-brown; epithecium red-brown, N−; hymenium 60–75 μm tall. Ascospores 10–13 × (6-)8–10 μm. Conidia 4.5–6 × 1 μm. Cortex and medulla Pd−, K−.

On siliceous rocks, often near lakes; rare. N. Scotland (Argyll, W. Sutherland). European mtns., Scandinavia, North America.

The loosely attached lobes and ± sessile apothecia recall an *Anaptychia* or *Phaeophyscia*.

11. A. recedens (Taylor) Arnold (1896)

Thallus thick, coarsely warted to cracked-areolate, dark grey. Apothecia 0.3–1.0 mm diam., round, numerous, crowded; thalline exciple thin, persistent; true exciple pseudoparenchymatous, cells 5–8 × 3–5 μm; disc dark brown-black, matt, flat, not pruinose; hypothecium with an algal layer below. Ascospores (9-)12–14 × 7–9 μm. Conidia 3–5 × 1 μm. Cortex and medulla Pd−, K− (unidentified compound by t.l.c.).

On rocks by the seashore; very rare. N.W. Scotland (Is. of Skye); recorded also last century from Wales (Merioneth, Barmouth, Llyn Bodlyn), Scotland (Perth, Aberfoyle; Roxburgh, Kelso). Europe.

12. A. subcircinata (Nyl.) Coppins (1980)

Thallus rather thin, grey, cracked-areolate towards centre, lobate at margins, forming ± round, closely adpressed patches; surface matt, ± pruinose especially towards ends of lobes; marginal lobes to 0.4–1(-1.5) mm wide, narrow, usually paler, contiguous, flat, or slightly convex. Apothecia abundant, becoming crowded towards thallus centre, ± immersed and concave when young, later becoming flatter; thalline exciple entire; disc 0.2–1 mm diam., brown-black. Ascospores 10–15 × 6–9 μm, broadly ellipsoid. Conidia 4–6 × 1 μm. Thallus Pd+ orange, K+ red, C− (norstictic acid).

On calcareous rocks and memorials, also on flints; local. Mainly S. & E. Great Britain, extending to N. Scotland (E. Perth, Angus). ?Endemic.

Possibly a synonym of *A. radiosa* (Hoffm.) Poelt & Leuckert (1973), which is widely distributed in Europe, North America.

13. A. subdepressa auct. brit., non (Nyl.) Arnold (1869)

Thallus rather thin, flat, irregularly cracked-areolate, smooth or somewhat wrinkled, grey. Apothecia rather small, scattered, immersed then emergent; thalline exciple excluded; true exciple and disc black; hymenial gelatine I+ faintly blue to quickly wine-red. Paraphyses coherent, often flexuose, dark brown above. Ascospores 18–32 × 11–15 μm, ellipsoid. Conidia 9–15 × 0.5 μm. Thallus Pd+ orange or −, K+ yellow to red or K− (± norstictic and stictic acids).

On schistose rocks, montane. N. Wales, Scotland (Highlands), S.W. & W. Ireland. Europe.

A poorly understood species; material under this name almost certainly comprises more than one taxon.

14. A. tuberculosa (Ach.) Laundon (1986)

Thallus to 0.5(-0.7) mm diam., of separate, scattered to crowded, regular, semi-globose, brownish grey granules, at times resembling isidia, often sterile; prothallus usually present, grey. Apothecia *c.* 1 mm diam., sessile, globose; thalline exciple swollen; disc black, often pruinose. Asci (4-)8-spored. Ascospores 22–28 × 20–26 μm, globose. Cortex C+ white; cortex and medulla Pd−, K− (aspicilin).

On flint nodules, on the coast and on chalk downs; ?extinct, not collected since *c.* 1830. S. England (Sussex, Is. of Wight). ?Endemic.

Similar to *A. caesiocinerea* **(1)** which has a more continuous thallus and narrower ascospores.

BACIDIA de Not. (1846)

Thallus crustose, smooth, cracked, warted, granular or sorediate, usually whitish, pale greenish or pale grey or fawn. **Photobiont** green; cells globose or broadly ellipsoid. **Ascomata** apothecia, strongly convex or ± flat, ± sessile, variously coloured, mostly to 1(-1.3) mm diam. **Thalline exciple** absent. **True exciple** always present, but sometimes reflexed and excluded, composed of ± coherent, radiating hyphae that often have swollen lumina at least in outer part. **Hymenium** I+ blue. **Hypothecium** colourless or variously coloured. **Hamathecium** of paraphyses, few to numerous, simple or rarely branched; apices often swollen. **Asci** clavate to cylindrical-clavate, 8-spored; *Bacidia*- type (Fig. 33a), apical dome K/I+ dark blue with a pale, ± conical apical cushion or in a few species with a dark, tubular ring-structure, wall pale, outer layer dark blue. **Ascospores** colourless, mostly 3- or more-septate at maturity, oblong, fusiform, bacilliform, acicular, or worm-like, mostly without a distinctive perispore [except in *B. sabuletorum* **(41)**]. **Conidiomata** pycnidia, immersed to sessile, variously shaped and coloured. **Conidia** colourless, variously shaped, sometimes transversely septate. **Chemistry**: thalli all Pd− K−, KC−, C−, and I−, mostly also [except *B. carneoglauca* **(34)** and *B. viridifarinosa* **(35)**] UV−; a wide range of pigments in apothecia and pycnidia. **Ecology**: various, but few species found on very acidic substrata.

Although many species formerly placed here are now referred to other genera (e.g. *Byssoloma*, *Catinaria*, *Fuscidea*, *Lecania*, *Micarea*, *Scoliciosporum*, *Stereocaulon*, *Toninia*, *Vezdaea*), *Bacidia* remains heterogeneous, and several species require relocation.

Many species are very variable, especially in the pigmentation of apothecia and the colour and extent of thallus development, particularly in specimens in shaded habitats. Many unidentified specimens of this difficult genus await further study. The final part of the key is to species commonly occurring without apothecia. The members of the *B. arnoldiana* group cannot be identified with certainty without apothecia, although *B. arnoldiana* **(2)** is probably the commonest species in the group to occur sterile.

LITERATURE: Coppins *et al.* (1992), Vainio (1922), Vězda (1961).

1 Apothecia present 2
Apothecia absent 49

2(1) Ascospores bacilliform or acicular, length/breadth ratio 10 or more 3
Ascospores oblong or fusiform, length/breadth ratio 9 or less 35

3(2) Epithecium K+ green 4
Epithecium never K+ green 11

4(3) On rocks or soil, sometimes on bryophytes 5
On bark or wood, sometimes on bryophytes 7

5(4) Hypothecium colourless, or upper part dark and K ± olivaceous brown 6
Hypothecium (upper part) K+ dark red- or purplish brown .... **25. *bagliettoana***

6(5) Hypothecium brown in upper part; thallus granules 15–60(-80) μm diam. .... **5. *egenula***
Hypothecium colourless throughout; thallus granules 80–120 μm diam. .... **6. *viridescens***

7(4) Hypothecium dark brown, at least in upper part 8
Hypothecium colourless throughout or pale 9

8(7) Upper hypothecium and inner exciple K+ dark red- or purple-brown; ascospores ± bacilliform .... **23. *subincompta***
Upper hypothecium K+ fuscous or ± olivaceous brown; ascospores acicular .... **5. *egenula***

9(7) Ascospores acicular, mostly curved or flexuose, tapering towards lower end 10
Ascospores ± bacilliform, straight, not markedly tapering at lower end .... **20. *circumspecta***

10(9) Ascospores 45–80 μm long, 7- to 16-septate; epithecium violet or blue-violet, K+ green, rarely green to begin with; hymenium 65-90 μm tall .... **13. *absistens***

Ascospores 30–50(-65) μm long, 3- to 7-septate; epithecium green; hymenium 40–65 μm tall ..... **14. *friesiana***

11(3) Epithecium and hymenium not greenish in water 12
Epithecium or hymenium greenish in water, K+ violet, C+ violet ..... **27. *beckhausii***

12(11) Paraphyses simple, or a few forked above, slender, 1–2 μm wide 13
Paraphyses branched and ± anastomosed, <1 μm wide ..... **28. *vezdae***

13(12) Not on coastal rocks; if on rocks with granular thallus then inner exciple and hypothecium K− (without reddish or orange tints) 14
On coastal rocks in the xeric-supralittoral; thallus usually thick and warty granular or granular-sorediate, rarely thin and cracked; inner exciple and upper hypothecium K+ pale to dark reddish- to orange-brown ..... **19. *scopulicola***

14(13) Hypothecium dark coloured, at least in upper part 15
Hypothecium colourless or pale straw, or very pale coloured 18

15(14) Ascospores mostly >30 μm long, acicular 16
Ascospores 15–25 μm long, bacilliform ..... **24. *incompta***

16(15) Thallus smooth, cracked or granular-warted, never with distinct soredia-like granules 17
Thallus granular; granules soredia-like, 20–40 μm diam. ..... **2. *arnoldiana***

17(16) Apothecia red-brown, most remaining flat and with a raised thalline exciple; upper hypothecium K± orange-brown ..... **26. *herbarum***
Apothecia pale to dark brown but without distinct red tinge, usually soon convex; hypothecium K ± fuscous or dull yellow-brown ..... **1. *arceutina***

18(14) Ascospores 14–28 × 2–3 μm, narrowly fusiform, on rocks or soil (on bryophytes, plant debris), never on bark or wood 47
Ascospores mostly >28 μm long, acicular, if bacilliform or narrowly fusiform then on bark or wood 19

19(18) On bark, or wood ± with blackish apothecia 20
On rocks, soil or mosses, if on bark then apothecia not blackish 24

20(19) Epithecium or exciple edge brown or purplish brown, K+ purplish tinge 21
Epithecium and exciple edge brown, K± yellowish tinge, never purplish ..... **1. *arceutina***

21(20) Ascospores <40 μm long, 3- to 7-septate; hymenium <60 μm tall; apothecia to 0.5 mm diam. 22
Ascospores mostly >45 μm long, 7- to 16(-22)-septate; hymenium >60 μm tall; apothecia mostly >0.5 mm diam. 23

22(21) Ascospores 21–40 × 2–3(-4) μm, strongly curved, sigmoid, worm-like, or a few short-acicular; conidia 7–9 × 0.5–1 μm, ± bacilliform, ... **21. *hegetschweileri***
Ascospores 18–28 × 1.5–2 μm, short-acicular; conidia 20–38 × 1 μm, curved ..... **22. *subcircumspecta***

23(21) Thallus granular or granular-isidiate ..... **11. *biatorina***
Thallus smooth to warted, often cracked, never with distinct granules ..... **10. *laurocerasi***

24(19) Exciple edge and epithecium colourless, or if coloured, then K+ orange, reddish or purplish 25
Exciple edge and epithecium pale to dark fuscous brown, K± yellowish tinge . **1. *arceutina***

25(24) Without above combination of characters; if on bryophytes or plant debris or calcareous substrata, the thallus finely granular with discrete granules 26
Apothecia (0.2-)0.3–0.7 mm diam., red-brown; thallus varnish-like or minutely granular-warted; on bryophytes or plant debris on calcareous substrata ..... **26. *herbarum***

26(25) Thallus finely granular-sorediate, with discrete granules 20–50 μm diam. 27
Thallus not granular-sorediate, if granular-isidiate then granules 60–120 μm diam. 28

27(26) Apotheca pale to dark pinkish or reddish brown; upper exciple edge pinkish brown, K+ purplish .... **3. *caligans***
Apothecia whitish or pale; upper exciple colourless, K− .... **4. *delicata***

28(26) Apothecia 0.1–0.3(-0.45) mm diam. 29
Apothecia mostly >0.3 mm diam. at maturity 30

29(28) Apothecia whitish or pinkish orange; exciple $\pm$ entirely composed of hyphae with $\pm$ globose lumina to 5 μm diam. in K .... **8. *chloroticula***
Apothecia pale to dark brown, rarely pale; exciple of narrow hyphae that abruptly swell to form large ovoid or globose vesicle-like cells to 20 $\times$ 12 μm in K .... **9. *saxenii***

30(28) On bark or wood, not aquatic 31
On rocks in streams or lake margins .... **7. *inundata***

31(30) Apothecia pink- or red-brown; dimensions of apothecia, hymenium and ascospores often greater than below 32
Apothecia pale, beige or pale orange-pink; 0.2–0.5 mm diam., hymenium 35–50 μm tall; ascospores 29–45 $\times$ 1.5–2 μm .... **15. *phacodes***

32(31) Thallus $\pm$ smooth to warted, never with isidia-like granules 33
Thallus thinly to richly granular-isidiate .... **16. *rubella***

33(32) Hypothecium and exciple K− or K+ yellowish, or outer part of exciple brownish and K$\pm$ purplish 34
Hypothecium and exciple (and often upper hymenium) pale yellow or reddish yellow, K+ mauve .... **18. *polychroa***

34(33) Apothecia red-brown or brick-red; hymenium 45–60 μm tall; paraphyses rarely <3 μm wide, only slightly swollen at apices; ascospores (30-)42–54 $\times$ 1.5–2 μm, 3- to 7-septate .... **17. *assulata***
Apothecia pinkish brown; hymenium 65–90 μm tall; paraphyses often distinctly swollen at apices, <5 μm wide; ascospores 34–70(-86) $\times$ 2.5–4 μm, 7- to 16-septate **10. *laurocerasi***

35(2) Hypothecium dark brown 36
Hypothecium colourless or pale 41

36(35) Ascospores <4.5 μm wide, without warted perispore 37
Ascospores 5–8 μm wide, often with warted perispore .... **37. *sabuletorum***

37(36) Ascospores usually <26 μm long, but if longer, then narrower and paraphyses mostly simple and slender 38
Ascospores 30–35(-42) $\times$ 3–4.5 μm; paraphyses much branched and $\pm$ anastomosed .... **28. *vezdae***

38(37) On bark or wood 39
On rocks .... **33. *trachona***

39(37) Ascospores >15 μm long 40
Ascospores 9–13 μm long .... **32. *subturgidula***

40(39) Epithecium green .... **23. *subincompta***
Epithecium colourless or pale reddish .... **24. *incompta***

41(35) Hymenium colourless or variously coloured, if greenish then K− or K+ green intensifying 42
Hymenium or epithecium dull greenish, K+ violet .... **27. *beckhausii***

42(41) On bark or mosses over bark 43
On rocks or soil, if on mosses then not over bark 45

43(42) Thallus thinly granular-warted, or smooth and $\pm$ cracked; ascospores (0-)3(-5) or 3- to 7(-9)-septate 44
Thallus thickly and entirely granular; ascospores (3-)5- to 6(-7)-septate .... **12. *auerswaldii***

44(43) Hymenium 70–120 μm tall; apothecia soon markedly convex; paraphyses 2–3 μm wide .... **37. *sabuletorum***
Hymenium <70 μm tall; apothecia flat to shallow-convex; paraphyses 1–1.5 μm wide .... **31. *naegelii***

45(42) Pycnidia absent or immersed in thallus; apothecial margin not pubescent; thallus UV− 46
Pycnidia 0.2–0.4 mm diam., always abundant, prominent, white, pubescent; apothecia whitish, with pubescent margin; thallus UV+ pink .... **34. *carneoglauca***

46(45) Ascospores 2–3 μm wide 47
Ascospores 4–7 μm wide 48

47(46) Apices of paraphyses <3 μm wide, colourless; conidia 10–20 μm long .... **29. *cuprea***
Apices of paraphyses <5 μm wide, often dark coloured; conidia 31–55 μm long **30. *subfuscula***

48(46) Ascospores <25 μm long, 1- to 3(-5)-septate, without warted perispore; paraphyses 1–2 μm wide .... **36. *fuscoviridis***
Ascospores 18–50 μm long, 3- to 7(-9)-septate, often with warted perispore; paraphyses 2–3 μm wide .... **37. *sabuletorum***

49(1) Pycnidia present 50
Pycnidia absent 57

50(49) Pycnidia whitish 51
Pycnidia brown or blackish 53

51(50) Conidia <6 μm long, bacilliform; thallus UV+ pink 52
Conidia >15 μm long, thread-like; thallus UV− .... **2. *arnoldiana***
[or *B. caligans, B. delicata* or *B. egenula*]

52(51) Pycnidia 200–400 μm diam.; thallus without farinose soredia .... **34. *carneoglauca***
Pycnidia 150–260 μm diam.; thallus with effuse patches of yellowish green soredia .... **35. *viridifarinosa***

53(50) Conidia <10 μm long, ellipsoid to bacilliform 54
Conidia >15 μm long, thread-like 55

54(53) Pycnidia 100–300 μm diam., sessile; conidia 3–5 × 1–1.5 μm .... **33. *trachona***
Pycnidia <100 μm diam., immersed; conidia 5–9 × 1.8–2.5 μm .... **24. *incompta***

55(53) Pycnidia immersed, in a thickish thallus or hidden among thallus granules; conidia <30 μm long 56
Pycnidia sessile, on a thin, scurfy thallus; conidia (20-)30–40 μm long .... **28. *vezdae***

56(55) On bark; conidia 16–24 × 0.5 μm .... **16. *rubella***
On rocks; conidia 18–30 × 0.7 μm .... **19. *scopulicola***

57(49) On bark or, if on rocks, then thin or composed of fine granules <40 μm diam. 58
On maritime rocks; thallus thick, warty to coarsely granular, ± with short isidia-like protuberances or soredia-like granules 40–80 μm diam. .... **19. *scopulicola***

58(57) Thallus granules mostly >40 μm diam.; outer hyphae of granules pigmented or not 59
Thallus finely granular; granules 20–40 μm diam.; outer hyphae of granules not pigmented .... **2. *arnoldiana***

59(58) Thallus granules 40–80(-100) μm diam., ± leprose, never glossy, often brownish grey on the upper surface (outer hyphae pigmented) .... **11. *biatorina***
Thallus granules 60–120 μm diam.; the surface smooth, sometimes slightly glossy, never brownish (outer hyphae colourless) .... **16. *rubella***

1. B. arceutina (Ach.) Arnold (1869)

B. leightoniana (Larbal. ex Leighton) H. Olivier (1911)

Thallus whitish, pale green-grey or grey, ± immersed, or thin, ± smooth or rimose, sometimes minutely granular-warted. Photobiont cells 5–12 μm diam. Apothecia 0.2–0.6(-0.8) mm diam., pale brown to brown-black, at first flat but often becoming convex; true exciple raised at first, colourless but outer edge usually brown, of thin (*c.* 1 μm) radiating hyphae, bound in a gel matrix that swells in K, but usually with a single row of swollen, rounded lumina (to 6 μm wide) at outer edge; epithecium yellow-brown to dark brown, K−, N−; hymenium 45–65 μm tall, colourless; hypothecium colourless or upper part (subhymenium) straw or yellow-brown, K± intensifying yellow. Paraphyses 1–1.5 μm wide, simple or a few irregularly forked above; apices slightly widening to 2.5(-4) μm wide, colourless. Ascospores (32-)35–55(-67) × 1.5–2(-2.5) μm, thinly 3- to 7-septate, acicular.

On trunks and branches of trees and shrubs with a ± basic or nutrient-rich bark, more rarely on sheltered, often slightly calcareous, rocks or on mosses on them; rare in polluted areas. Throughout British Is. Europe, North America.

2. B. arnoldiana Körber (1860)

Thallus dull yellowish green to fawn, finely granular; granules (goniocysts) 20–40 μm diam. Photobiont cells 5–10(-12) μm diam. Apothecia 0.35–0.7(-1.2) mm diam., usually few and scattered, or absent, flat, whitish grey to grey-brown; true exciple raised, usually paler and often ± white-pruinose; inner part concolorous with hypothecium, the outer part colourless; hyphae towards outer part with ellipsoid lumina 3–7 × 1.5–5 μm; hymenium (35-)40–50 μm tall, colourless; hypothecium dark reddish brown, K+ dull olivaceous brown, in upper part, ± colourless below. Paraphyses 1–2 μm wide, simple, or a few forked above; apices often swollen to 5 μm. Ascospores (19-)24–46 × 1–2 μm, thinly 1- to 3-septate, acicular. Pycnidia 100–200 μm diam., white, ± immersed; conidia 26–47 × 1–1.5 μm., curved.

On shaded, damp calcareous rocks or stonework, or shaded bark (especially *Sambucus*); tolerant of suburban conditions. Throughout Great Britain, especially frequent in E. & S.E. England, but rare in Scotland. Europe, North America.

Related species with a granular thallus in similar habitats are *B. caligans* **(3)**, *B. delicata* **(4)** and *B. egenula* **(5)**; all four are mainly separated by the internal pigmentation of their apothecia. Specimens of this group without apothecia (but often with pycnidia) cannot be determined with certainty, but *B. arnoldiana* is probably the most common.

3. B. caligans (Nyl.) A. L. Sm. (1911)

Thallus dull green, grey-grey or fawn-brown, finely scurfy granular; granules (goniocysts) 20–50 μm diam. Photobiont cells 5–13 μm diam. Apothecia 0.2–0.6(-1) mm diam., flat or later irregularly convex, reddish brown to brown-black, margin often darker than disc; true exciple apparent at first, becoming excluded, colourless except for reddish brown, K+ purplish, outer edge; with cellular lumina (to 7 μm wide) towards outer edge and in lower part; hymenium 45–55 μm tall, colourless, or pale purplish brown above, K+ intensifying purple; hypothecium colourless. Paraphyses 1–2 μm wide, simple; apices often swollen to 5 μm. Ascospores 24–48(-63) × 1.5–2(-2.5) μm, thinly 3- to 7-septate, acicular. Pycnidia immersed, whitish or pink-brown around the ostiole; conidia 29–35 × 1 μm, curved.

On sheltered, calcareous rocks and stonework, asbestos-cement, and mosses on crumbling mortar, rarely on bark (e.g. *Sambucus, Ulmus*); locally frequent. England and Wales, extending sporadically to E. Scotland. Belgium, Luxembourg, Netherlands, Poland.

4. B. delicata (Larbal. ex Leighton) Coppins (1980)

Thallus pale green to fawn, finely granular; granules (goniocysts) (15-)20–35(-50) μm diam. Photobiont cells 5–10(-12) μm diam. Apothecia (0.2-)0.3–0.6(-0.7) mm diam., flat, whitish, beige or orange-pink; true exciple persistent, colourless, with cellular lumina (to 7 μm wide) towards outer edge and in lower part; hymenium 35–50(-55) μm tall, colourless; hypothecium colourless. Paraphyses 1.5–2 μm wide, simple, or a few forked above; apices often swollen, to 5 μm. Ascospores (21-)24–48 × 1–1.5(-2) μm, thinly 3- to 7-septate, acicular. Pycnidia ± immersed, white; conidia 21–40 × 1 μm, curved.

On shaded substrata, including ± calcareous rock and stonework (often on mosses) and basic bark. Throughout British Is. Denmark, Italy, S. Sweden.

Distinguished from other members of *B. arnoldiana* **(2)** group by the absence of pigments in the apothecia, and from *B. phacodes* **(15)** by the distinctly granular thallus, larger cellular lumina in the outer exciple, and curved conidia. Morphs with small apothecia differ from *B. chloroticula* **(8)** in the more numerous paraphyses, and the cellular lumina in exciple being mostly confined to the outer parts.

5. B. egenula (Nyl.) Arnold (1870)

Thallus dull yellowish green, fawn or fawn-brown, finely granular; granules (goniocysts) 15–60(-90) μm diam. Photobiont cells 5–12 μm diam. Apothecia (0.15-)0.3–0.6(-0.75) mm diam., grey-brown, blue-grey or black, flat; true exciple persistent, colourless except upper outer edge which is olive-brown, K− or K+ purplish in part (especially above), cellular lumina to 7(-10) μm wide towards outer edge and in lower part; epithecium pale to dark olivaceous to blue-green, K−, N+ red-violet (often with blue crystals); hymenium (35-)40–55 μm tall, colourless; hypothecium reddish brown, K+ dull (± olivaceous) brown in upper part, colourless below. Paraphyses 1–1.5 μm wide, simple; apices swollen to 6 μm, and often pigmented. Ascospores 25–40(-45) × 1.5–2(-2.5) μm, thinly 3(-7)-septate, acicular. Pycnidia 100–200 μm wide, rare, ± immersed, white; conidia 20–35 × 1 μm, 0- to 3-septate, curved-sigmoid.

On shaded rocks, stonework (e.g. red-brick, sandstone), flints, sometimes on mosses, rarely on shaded bark (*Salix*), also on rabbit droppings in sand dunes. Throughout British Is. Europe, North America.

Distinguished from *B. arnoldiana* **(2)** by the greenish epithecium, and from *B. viridescens* **(6)** in the dark upper hypothecium and more finely granular thallus.

6. B. viridescens (Massal.) Norman (1868)

Thallus grey- or green-white or pale fawn, finely granular-warted; warts 80–120 μm diam. Photobiont cells 6–12 μm diam. Apothecia (0.2-)0.3–0.6(-0.8) mm diam., pale brown, grey or black, flat or becoming convex; true exciple evident at first, becoming excluded, purple-brown, K+ intensifying purple in outer and uppermost part, sometimes also with greenish pigment in upper parts, ± colourless below, cellular lumina to (7 μm wide) towards outer edge; hymenium 30–50 μm tall, olivaceous or green above, sometimes discoloured brownish; hypothecium colourless. Paraphyses simple, or rarely branched above, 1.5–2 μm wide, apices often swollen, to 5 μm, and pigmented. Ascospores 25–50(-58) × 1.5–2 μm, 3- to 7-septate, acicular. Pycnidia rare, ± immersed, blackish, the wall brown, K+ purplish and olivaceous in parts; conidia 15–25 × 1–1.3 μm, curved.

Mostly on soil (on bryophytes, plant debris or rabbit droppings) in lead mine spoil, in sand dunes or over limestone, rarely directly on soft limestone stonework or crumbling mortar. Throughout much of British Is. Finland, Netherlands, Italy.

Resembles *B. egenula* **(5)**. Most British specimens are poorly developed.

7. B. inundata (Fr.) Körber (1855)

Thallus grey-green or greenish fawn, rimose, the surface granular-warted, usually with a conspicuous, white bordering prothallus. Photobiont cells 5–12(-17) × 5–12 μm. Apothecia 0.2–0.6(-1) mm diam., flat to convex, pale to dark reddish brown, margin often paler; true exciple evident at first, becoming excluded, colourless or pale straw or brownish, but usually dark brown above, cellular lumina, 2.5–5(-7) μm wide in outer and lower parts; hymenium 45–55 μm tall, colourless or pale pink- or purplish brown, K± purplish; hypothecium colourless or pale straw, or pinkish brown in upper part. Paraphyses 1–1.5(-2) μm wide, simple, or a few forked above; apices often swollen to 3.5(-5) μm wide. Ascospores 24–43 × 2–2.5 μm, thinly 3(-7)-septate, acicular. Pycnidia ± immersed, of two types: (a) 100–200 μm diam., whitish, conidia 23–47(-55) × 0.5–1 μm, faintly 3-septate curved; (b) dark brown, wall brown (± purplish), conidia 11–15(-19) × *c.* 1 μm, curved.

On hard, siliceous rocks (rarely bark or wood) in streams and rivers, more rarely lake margins, often where shaded by overhanging trees; frequent in upland areas. S.W. to N. England, Scotland. Europe, North America.

A distinctive species, often seen as orbicular patches surrounded by a white prothallus; the only British *Bacidia* known to have two conidial types. *B. arnoldiana* **(2)** can occur in similar habitats on calcareous rocks; it has a more granular thallus and a darker hypothecium.

8. B. chloroticula (Nyl.) A. L. Sm. (1911)

Thallus dull green or grey-green, thin, scurfy granular. Photobiont cells 5–12(-14) μm diam. Apothecia 0.08–0.2 mm diam., often crowded and confluent, flat; disc pinkish to pale grey brown; true exciple persistent, usually paler than disc, colourless, ± entirely composed of hyphae with elliptical to globose lumina to 5(-7) μm wide; hymenium 35–42 μm tall, colourless, or pale brownish above; hypothecium colourless. Paraphyses 1.5–2 μm wide, few, simple; apices often swollen to 4 μm, colourless. Ascospores

21–38 × 1–1.8 μm, thinly 0- to 3-septate, narrowly acicular. Pycnidia whitish, immersed; conidia 30–43 × 0.5–1 μm, curved.

On various shaded substrata, such as tree bases and exposed roots, tops of cut stumps and fence posts, cement and brick rubble, and tarmacadam; rare. E. England, S. Wales, Ireland (Galway, Londonderry). Netherlands, Norway, N. Germany, Poland.

Recognized by the small, pale apothecia; pale morphs of *B. saxenii* **(9)** are distinguished by the large, vesicle-like cells in the true exciple.

9. B. saxenii Erichsen (1941)

Thallus inconspicuous or thin, green-grey or fawn, scurfy granular. Photobiont cells 7–14(-19) μm diam. Apothecia 0.1–0.3(-0.45) mm diam., flat or convex, pale or more usually pale to dark brown; true exciple indistinct, often paler than the disc, colourless, except for dark brown K+ purplish upper part in dark apothecia, hyphae 1.5–2 μm wide, lax in K, terminating in large ovoid or ± globose, vesicle-like cells to 20 × 12 μm; hymenium 40–45 μm tall, pale brown, K− or K+ purplish; hypothecium colourless. Paraphyses 1.5–2 μm wide, simple, or forked above; apices often capitate, to 5 μm. Ascospores 26–40(-45) × 1.5–2.3 μm, 3-septate, acicular. Pycnidia 50–100 μm diam., rare, immersed wall colourless; conidia 33–48 × 0.5–1 μm, curved-thread-like.

On shaded, calcareous or dust-enriched substrata, including bones, stones and boulders by farm tracks, decaying mosses on top of fence posts, and old leather in sand dunes, also on old iron railings and lead mine spoil. E. & N. England, Scotland, E. Ireland (Dublin). N. Germany.

Easily identified microscopically by the remarkable exciple structure with its large vesicle-like cells.

10. B. laurocerasi (Delise ex Duby) Zahlbr. (1926)

Thallus pale grey or green-grey, usually thin, smooth or rimose, or ± warted. Photobiont cells 5–12(-14) μm diam. Apothecia (0.2-)0.4–1(-1.2) mm diam, usually numerous, flat to convex, pinkish brown (in shade) to black; true exciple purplish brown, K+ intensifying purple, at outer edge, internally pale red-brown but often darker and sometimes with a purple tinge in the upper part; epithecium pale grey-brown (K−) to dark (± purplish) brown, K+ intensifying purple, N+ red; hymenium 65–90 μm tall, colourless; hypothecium colourless or pale straw, but lower part often pale orange- or red-brown towards the exciple - the coloration rarely extending to the centre. Paraphyses 1–1.5 μm wide, simple or branched above, apices often pigmented and swollen to 5 μm. Ascospores 34–70(-86) × 2.5–4 μm, 7- to 16-septate at maturity, acicular. Pycnidia immersed, the walls colourless; conidia 13–17 × 1 μm, curved.

On ± basic or nutrient-rich bark of trunks and branches of trees and shrubs, usually in open situations or secondary woodland; frequent. S. England, W. Wales, very rare and local in Scotland, Ireland. Europe, North America, Australia, New Zealand.

Closely related to *B. biatorina* **(11)**, but more often confused with *B. absistens* **(13)** which has a purple-violet, K+ green, epithecium.

11. B. biatorina (Körber) Vainio (1922)

Thallus pale green to grey-green, granular or granular-isidiate, the granules often brownish or grey on their upper surface. Photobiont cells 5–12 μm diam. Apothecia very like those of **10**, but larger, (0.5-)0.7–1.3 mm diam., and ascospores mostly longer (43-)60–75(-88) × 2.5–3(-4) μm, and 9- to 16(-22)-septate.

On trunks of mature trees (mainly *Quercus*) in ancient parkland or woodland. S. & W. England, Wales, extending northwards in W. Scotland, rare in Ireland. C. & W. Europe.

Often sparingly fertile or sterile; can be confused with sterile *B. rubella* **(16)**, although the isidiose granules of the latter are usually coarser and never dark-coloured on their upper surface.

12. B. auerswaldii (Hepp ex Stizenb.) Mig. (1929)

Like **11**, but differs in having ascospores 20–26 × 4–5 μm, (3-)5- to 6(-7)-septate, oblong-fusiform.

On trunk of *Ulmus* (? in parkland); not recorded since 1927–37. S.W. England (Somerset, near Taunton). Europe, Canary Is.

13. B. absistens (Nyl.) Arnold (1870)

Thallus whitish or grey-grey, thin, granular-warted. Photobiont cells 5–12(-14) μm diam. Apothecia 0.3–0.8(-1) mm diam., flat, black; true exciple persistent, dark greenish, K+ intensifying green at upper outer edge, internally pale reddish brown (K ± purplish) and with abundant, minute, colourless, granular inclusions (less than 1 μm diam.) dissolving in K; epithecium blue-violet or purple-brown (rarely green), K+ green, N+ purple-violet with blue crystals; hymenium 65–90 μm tall, colourless; hypothecium colourless or pale yellow-brown. Paraphyses 1–1.5 μm wide, simple, or forked above, apices often pigmented and swollen to 4 μm. Ascospores 45–80 × 2–3(-4) μm, 7- to 16-septate, acicular.

On ± acid bark of deciduous trees, mainly in old woodlands; common. Scotland (Highlands), extending locally elsewhere. N.W. Europe.

Often confused with *B. laurocerasi* **(10)** but easily separated by the epithecium pigmentation. Morphs with a green epithecium differ from *B. friesiana* **(14)** in the longer, more septate ascospores, and colour and inclusions in the inner exciple.

14. B. friesiana (Hepp) Körber (1860)

Thallus whitish to pale grey-green, partly immersed, mostly thin and minutely granular-warted. Photobiont cells 6–14 μm diam. Apothecia 0.2–0.6(-0.8) mm diam., flat to convex, blue-grey, grey-brown or black, but sometimes ± pale when deeply shaded; true exciple at first ± apparent, ± colourless within, outer edge (especially above) and upper parts usually dark brown, K+ purplish or in part, or wholly dark greenish and K+ intensifying green, hyphae with swollen cells to 6 μm wide towards outer edge; epithecium pale grey-green to dark blue-green, K−, N+ red or violet (often with blue granules); hymenium 60–65 μm tall, colourless or pale greenish; hypothecium colourless or yellowish brown to brown in upper part (subhymenium). Paraphyses 1–1.5 μm wide, simple, or a few forked above, apices often swollen to 5 μm. Ascospores 30–50(-65) × 3 μm, 3- to 7(-8)-septate, acicular.

On nutrient-rich bark of trees and shrubs (especially *Sambucus*); rather rare. Throughout British Is. Europe.

Rather variable but only likely to be confused with some morphs of *B. absistens* **(13)**.

15. B. phacodes Körber (1860)

Thallus whitish to grey-green, thin to irregularly warted or mealy, never with discrete granules. Photobiont cells 7–10 μm diam. Apothecia 0.2–0.5 mm diam., whitish, beige or pale orange-pink, the true exciple often paler than disc, flat to convex, often confluent and clustered; true exciple colourless, K ± pale yellowish, of radiating hyphae with sometimes ellipsoid (<2 μm wide) cells, but some cells to 4 μm at outer edge; hymenium 35–50 μm tall, colourless, or upper part pale yellow-brown, K ± yellowish; hypothecium colourless. Paraphyses simple, 1–1.5 μm wide; apices sometimes clavate, to 3 μm wide. Ascospores 29–45 × 1.5–2 μm, 3- to 7-septate, acicular. Pycnidia immersed, the walls colourless; conidia 28–50(-60) × 1–1.5 μm, thinly 3- to 7-septate, thread-like, ± straight.

On nutrient-rich bark of mostly mature trees (especially *Ulmus*), often in shaded, sheltered habitats; frequent. Throughout British Is., but very rare in the north. Europe, North America.

Distinguished from *B. rubella* **(16)** and *B. assulata* **(17)** by the paler apothecia, and from the former also by the non-granular thallus, and shorter hymenium and ascospores. *B. delicata* **(4)** has a finely granular thallus, outer exciple with large lumina (to 7 μm), and strongly curved conidia.

16. B. rubella (Hoffm.) Massal. (1852)

Thallus grey-green to yellow-green, thinly to richly granular-isidiate; granules 60–120 μm diam. Photobiont cells 5–17 μm diam. Apothecia (0.4-)0.7–1(-1.3) mm diam., when present, usually distinctly constricted below, flat, sometimes convex, pale to dark red-brown (testaceous); true exciple raised, sometimes white-pruinose, in section colourless but upper part pale yellow-orange or yellow-straw, sometimes (pruinose morphs) with radiating streaks of minute crystals, hyphae with lumina 1–2 μm wide or to 5 μm wide towards outer edge; hymenium 70–105 μm tall, colourless or faintly orange-red or yellowish in upper part; hypothecium colourless, or upper part pale yellowish- or orange-straw, K ± yellow intensifying. Paraphyses 1–1.5 μm thick, simple, or forked above, the apices often slightly swollen to 2.5 μm. Ascospores (35-)40–70(-75) × 2.5–3(-4) μm, acicular. Pycnidia pale pink to red-brown; conidia 16–24 × 0.5 μm, curved or sigmoid.

On trunks of mature trees with nutrient-rich bark (especially *Acer*, *Fraxinus*, *Ulmus*) in parklands, wayside situations and woodlands on rich soils, very rarely on sheltered gravestones in E. England; absent from polluted areas. Throughout British Is., rare in W. Scotland. Europe, North America.

Easily recognized when fertile, but sterile morphs can be confused with *B. biatorina* **(11)**.

17. B. assulata (Körber) Vězda (1967)

Thallus green-white or pale green-grey, thin or minutely warted. Photobiont cells 5–10 μm diam. Apothecia (0.2-)0.3–0.6 mm diam., flat, pale red-brown or brick-red; true exciple forming a distinct margin, colourless or very pale orange-brown, of radiating hyphae, the outer edge with single row of rounded lumina to 7 μm wide; hymenium 45–60 μm tall, colourless; hypothecium pale orange-brown (K −) above, pale-coloured or colourless below. Paraphyses 1–1.5 μm wide a few becoming clavate to 3 μm wide. Ascospores (30-)42–54 × 1.5–2 μm, 3- to 7-septate, acicular.

On bark of *Eucalyptus*, *Fraxinus* and *Populus* trunks; rare. S.W. England, Is. of Scilly, S.W. Ireland. Europe, North America.

The apothecia are similar to those of *B. rubella* **(16)** but are smaller, and the hymenium and ascospores are shorter, whilst the thallus is not granular. Distinguished from *B. phacodes* **(15)** by the darker apothecia and hypothecium.

18. B. polychroa (Th. Fr.) Körber (1860)

Thallus whitish or pale green-grey, thin, or thickish and irregularly warted. Photobiont cells 7–14 μm diam. Apothecia (0.3-)0.5–1(-1.5) mm diam., flat to sometimes convex, pale to dark reddish or red-brown; true exciple and hypothecium yellow or reddish yellow, K+ mauve in upper parts, ± colourless below, true exciple lumina 1.5–3.5 μm wide but 5(-6) μm at outer edge; hymenium 70–100 μm tall, colourless or upper part yellowish (K ± mauve). Paraphyses 1–1.5 μm, simple; apices to 2 μm wide, scarcely swollen. Ascospores (43-)50–70 × (2.5-)3–4(-4.5) μm, (3-)7- to 14-septate, acicular. Pycnidia immersed, reddish; conidia 18–12 × 0.5 μm, curved.

On mature trees; no modern records. S. England. Europe, North America.

Identified by the yellow, K+ mauve pigment in the exciple and hypothecium.

19. B. scopulicola (Nyl.) A. L. Sm. (1911)

Thallus pale olive-green to fawn-brown (sometimes dark green in narrow crevices), rarely thin and rimose, mostly thick, warted and coarsely granular, with short, coralloid, isidium-like protuberances, or extreme morphs with soredia-like, ± loose granules 40–80 μm diam. Photobiont cells 5–10 μm diam. Apothecia 0.4–1.3 mm diam., often few or absent, at first flat but soon convex, pale to dark brown, often with pinkish tinge, rarely black, often irregularly shaped or tuberculate; true exciple at first thick, usually darker than the disc, orange-brown (K −) in inner part, outer and sometimes lower parts colourless, of radiating, thick, densely gelatinized hyphae with narrow (to 2 μm) lumina, but lumina near outer edge often widening to 5(-7) μm; hymenium 45–60 μm tall, colourless or lower part pale red-orange; hypothecium colourless except upper part (subhymenium) which is usually pale red-orange. Paraphyses 1–1.5 μm wide, simple, often forked above; apices sometimes slightly swollen to 2(-4) μm. Ascospores (21-)29–45(-51) × 1.7–2 μm, 3- to 7-septate, acicular. Pycnidia reddish brown, ± immersed; conidia 18–30 × *c.* 0.7 μm, strongly curved.

On siliceous, maritime rocks in the xeric-supralittoral in crevices and underhangs, rarely on soil or decaying tufts of *Armeria*; locally abundant. W. & N. Britain, Ireland. S. Scandinavia, N.W. France, Azores.

Very variable in the thickness and texture of the thallus and colour of the apothecia, but the only *Bacidia* with acicular ascospores on seashores.

20. B. circumspecta (Norrlin & Nyl.) Malme (1895)

Thallus whitish or pale green-grey, thin, irregularly warted but without distinct granules. Photobiont cells 7–14(-16) μm diam. Apothecia 0.2–0.7(-1) mm diam., flat, black; true exciple with upper part and outer edge dark purplish brown, K+ intensifying purple, becoming pale-coloured to colourless below; epithecium blue-green, K −, N+ purple-violet, often with blue crystals; hymenium 45–55 μm tall, colourless or pale green above; hypothecium colourless or pale straw. Paraphyses 1–1.5 μm wide, simple, or forked above, the apices often pigmented and swollen to 4 μm. Ascospores (18-)20–30(-38) × (1.5-)2–2.5 μm, ± bacilliform. Pycnidia 100–160 μm diam., rare, sessile, ± globose, black, the wall with purple-brown and green pigments; conidia 7–9 × 1.5–2 μm, 0(-1)-septate, oblong.

On trunks of mature deciduous trees in old woodlands; rare. S. & N. (N. Yorkshire) England, Mid-Wales, E. Scotland. Europe, North America.

Distinguished from *B. subincompta* **(23)** by the non-granular thallus and pale hypothecium, and from both *B. hegetschweileri* **(21)** and *B. subcircumspecta* **(22)** by the bacilliform ascospores and green epithecium.

21. B. hegetschweileri (Hepp) Vainio (1922)

Thallus ± immersed to irregularly warted, white. Photobiont cells 5–14 μm diam. Apothecia 0.16–0.5 μm diam., flat, or occasionally convex, black; true exciple thin, brown-red or brown-purple, darker towards outer edge, K+ intensifying purple; epithecium brown-red to purple-brown, K+ intensifying purple; hymenium 35–50 μm tall, ± colourless; hypothecium colourless or pale red-brown. Paraphyses 1–1.5 μm wide, simple, upper 12–15 μm pigmented and apices clavate to 5 μm. Ascospores 21–40 × 2–3(-4) μm, 3- to 7-septate, short-acicular or worm-like. Pycnidia 40–50 μm diam., ± immersed, black, the wall brown-purple in K; conidia 7–9 × 0.5–1 μm, bacilliform or slightly curved.

On bark of *Betula*, *Juniperus* and *Ulmus* in old Caledonian pineforest areas. C.N. Scotland (Highlands, near Aviemore). C. & N.W. Europe, North America.

Similar to *B. subcircumspecta* **(22)** in pigmentation, but with different ascospores and conidia.

22. B. subcircumspecta Coppins (1992)

Thallus immersed or scurfy granular; granules, when discrete, 60–100 μm diam. Apothecia 0.2–0.5(-0.8) mm diam., flat to convex, sometimes tuberculate, dark purplish brown to black; true exciple thin, dark purple-brown, K+ intensifying purple along outer edge, pale purplish within, hyphae with lumina *c.* 1 μm wide except at outer edge where often expanded to 4 μm; hymenium 30–35 μm tall, pale brown-purple below, becoming dark purple brown above, K+ intensifying purple; hypothecium colourless. Paraphyses 1.5–1.8 μm wide, simple or forked above, the apices to 3(-4) μm, swollen. Ascospores 18–28 × 1.5–2 μm, (0-)3-septate, short-acicular. Pycnidia 60–80 μm diam., immersed, black, the wall mainly purple-brown but greenish around ostiole; conidia 20–38 × 0.5–1 μm, curved.

On hard wood, usually fence posts or rails. N. Scotland. ?Endemic.

Similar to *B. circumspecta* **(20)** and *B. hegetschweileri* **(21)**.

23. B. subincompta (Nyl.) Arnold (1870)

Thallus of whitish, scattered to confluent granules, not forming a thick granular crust, the granules 40–100 μm diam. Photobiont cells 7–14 μm diam. Apothecia 0.3–0.6(-0.9) mm diam., flat, later convex, sometimes tuberculate and to 1.2 mm diam., black, or rarely pale (albino morph); true exciple weakly developed, dark red-brown within, outer edge ± colourless; epithecium pale to dark green, K−, N+ purple-violet; hymenium 50–65 μm tall, colourless; hypothecium dark red-brown, K± purplish in upper part, pale red-brown below. Paraphyses 1–1.5 μm wide, simple, or rarely forked above; apices only slightly wider, to 3 μm wide. Ascospores 20–36(-40) × 2.3–3.5(-4) μm, 3- to 7-septate, bacilliform.

On trunks of mature trees (*Fraxinus*, *Quercus*, *Ulmus*) in ancient woodlands; rare. N.W. England (Cumbria), Mid-Wales, N. Scotland (C. & E. Highlands). Europe, North America.

Characterized by the thinly granular thallus, black apothecia with green epithecium and red-brown upper hypothecium (except the rare albino morph), and bacilliform ascospores; *B. incompta* **(24)** has a mealy granular thallus, a colourless or pale reddish epithecium, shorter ascospores and a totally dark red-brown hypothecium.

24. B. incompta (Borrer ex Hooker) Anzi (1860)

Thallus dull green, grey-green, whitish or fawn, mealy granular. Photobiont cells 5–12(-14) μm. Apothecia (0.2-)0.3–0.8(-1) mm diam., dark purple-brown to black, sometimes becoming convex; true exciple thin, paler than disc; exciple and hypothecium dark red-brown, K+ purplish, but outermost edge of exciple sometimes ± colourless; hymenium 35–45(-50) μm tall, colourless or pale reddish. Paraphyses 1.5–2(-2.5) μm wide, simple, or rarely forked above, not or only slightly swollen at apices (to 3 μm). Ascospores 15–25(27) × 2–2.5 μm, (0-)3(-5)-septate, bacilliform. Pycnidia ± immersed, blackish, walls red-brown, K+ purplish; conidia 5–9 × 2–2.5 μm, 0(-1)-septate, ellipsoid-oblong.

On trunks of trees with basic bark (especially *Ulmus*, sometimes *Fagus* or *Ilex*), often forming extensive dark green-grey, uneven, vertical streaks along rain or wound seepage tracks, sometimes on exposed roots of *Ulmus* on cliff faces; recently much declined owing to demise of the elm. Throughout S. England, extending north eastwards to Scotland (Aberdeen); rare in Ireland. Europe, North America.

Recognized by the mealy thallus and dark internal pigmentation of the apothecia. Occasionally found sterile, although the inconspicuous pycnidia are usually present.

25. B. bagliettoana (Massal. & de Not.) Jatta (1900)

B. muscorum (Swartz) Mudd (1861)

Thallus white or green-white, ± granular-warted. Photobiont cells 7–15 μm diam. Apothecia 0.3–1(-1.2) mm diam., flat, eventually ± convex, usually distinctly constricted below, black; true exciple red-brown in upper and outer parts, pale red-brown within but becoming ± colourless below, hyphae radiating, *c.* 3–5 μm wide (in K) with thick, gelatinized walls and lumina to less than 1.5 μm; epithecium olive- to blue-green, K−, N+ violet with blue granules; hymenium 40–55 μm tall; hypothecium red-brown and K+ intensifying red in upper 25–35 μm, pale reddish to colourless below. Paraphyses 1.5–2 μm wide, simple, or a few forked above, the apices swollen to 4(-5) μm. Ascospores 25–45 × 2–2.5(-3) μm, 3- to 7-septate, acicular or ± bacilliform.

Over mosses or plant debris on calcareous rocks (including old walls) or on the ground in chalk or limestone grassland or calcareous dunes; local. Throughout British Is. Europe, North America.

Often confused in the field with *B. sabuletorum* **(37)**, but easily separated by the ascospore shape and width and darker apothecia.

26. B. herbarum (Stizenb.) Arnold (1865)

Thallus whitish, thin and varnish-like, or thicker and minutely granular-warted. Photobiont cells 7–14 μm. Apothecia (0.2-)0.3–0.7 mm diam., red-brown, mostly remaining flat, distinctly constricted at base; true exciple persistent, ± colourless or straw, but often orange-or red-brown at outer edge, composed of radiating, thin hyphae but with 1–2 rows of swollen lumina (to 6 μm wide) at outer edge; hymenium 45–60 μm tall, colourless or very pale orange-brown above; hypothecium pale orange-brown and K− in upper part (subhymenium), lower part pale orange-brown, straw or ± colourless. Paraphyses 1–1.5 μm wide, simple, apices to 3(-4.5) μm, becoming clavate. Ascospores 36–60 × 2–2.5 μm, 3- to 7-septate, acicular.

On bryophytes, small perennial plants or plant debris on calcareous turf or amongst limestone rocks, from sea-level to *c.* 800 m; rare. Throughout British Is. Europe, North America.

27. B. beckhausii Körber (1860)

Thallus whitish or pale grey, ± immersed, or thin and varnish-like or slightly warted. Photobiont cells 8–14 μm diam. Apothecia 0.5–1.5 mm diam., at first ± flat, but soon convex, pale grey in shade to bluish black, often thinly white-pruinose, or with bluish 'bloom' when wet; true exciple thin, colourless or partly pale greenish, K+ violet; hyphae 1.5–2 μm wide, more distinct in K but not separating; epithecium usually with minute granules dissolving in K; hymenium 35–50 μm tall, upper part (and often vertical streaks) dull greenish K+ violet, N+ red; hypothecium colourless. Paraphyses 1–1.5 μm wide, simple or sparingly branched; apices not, or only slightly widening to 2 μm. Ascospores 17–26(-29) × 1.5–2.5 μm, (1-)3(-7)-septate, bacilliform or slightly curved. Pycnidia 50–100 μm diam., immersed, black, the wall olivaceous, K+ violet; conidia 3–3.5 × 1–1.5 μm, oblong-ellipsoid.

On bark or rarely wood of trunks of mature deciduous trees (especially *Fraxinus*, *Quercus*, *Ulmus*), mainly in old woodlands; frequent. N. Scotland (C. & E. Highlands), two 19th century records from N.E. England (Cleveland), S. England (New Forest). Europe, North America.

The large-celled photobiont, epithecial granules, and coherent exciple hyphae distinguish this species from *Micarea globulosella* and *M. synotheoides*.

28. B. vezdae Coppins & P. James (1978)

Thallus pale grey-green or pale grey, thin, scurfy. Photobiont cells 4–7 μm diam. Apothecia 0.2–0.4 mm diam., sometimes absent, at first ± top-shaped, some convex, pink- or grey-brown; true exciple clearly apparent, colourless, hyphae with ± ellipsoid lumina to 2.5 μm wide; epithecium red-brown, K−; hymenium 40–45 μm tall, colourless or pale brown; hypothecium darkish brown, K−. Paraphyses 0.5–1 μm wide, branched and anastomosed, netted. Ascospores 30–35(-42) × 3–4.5 μm, 3- to 7-septate, acicular-fusiform to acicular. Pycnidia 180–200 μm diam., to 300 μm tall, usually numerous, sessile, ± globose or elongated vertically, pink-brown to brown-black; conidia (20-)30–43 × 0.5–1 μm, curved.

Usually on shaded bark, sometimes on bryophytes or other lichens, on deciduous trees in woodland including coppices and plantations, occasionally on sheltered mossy rocks, especially near the coast, tolerant of moderately polluted conditions; common in S., rare in N. Throughout British Is. W.C. Europe.

A usually inconspicuous, but easily recognized species, often seen sterile as small groups of pycnidia.

29. B. cuprea (Massal.) Lettau (1912)

Catillaria umbraticula (Nyl.) P. James (1965)

Thallus pale grey or grey-green, often overgrown by green algae, thin or rather thick, uneven, minutely and irregularly granular-warted or scurfy granular, but granules never discrete. Photobiont cells 5–12 μm diam. Apothecia 0.2–0.4(-0.6) mm diam., flat or convex, pale beige, pale orange-pink, or pale to dark reddish brown; true exciple colourless, or upper outer part pale to dark pinkish brown and K+ purplish, hyphae narrow but often with expanded lumina, ellipsoid and to 5 μm wide towards outer edge; hymenium (35-)40–45(-50) μm tall, colourless, or pale grey and K± purplish in dark apothecia; hypothecium colourless or pale straw. Paraphyses 1–1.5 μm wide, the apices to 3 μm, colourless. Ascospores 14–28 × 2–3 μm, 3(-5)-septate, narrowly fusiform. Pycnidia immersed, whitish; conidia 10–19 × 0.8–1.2 μm, strongly curved.

On deeply shaded limestone rocks, mainly in dry underhangs and deep crevices. Throughout British Is.

Very variable in appearance. Most likely to be confused with *B. subfuscula* **(30)** and some *Lecania* species with 3-septate ascospores. Recently transferred to *Lecania* as ***L. cuprea*** (Massal.) v.d. Boom & Coppins (1992).

30. B. subfuscula (Nyl.) Th. Fr. (1867)

B. circumpallens (Nyl.) Arnold (1870)

Thallus granular, greenish white to pale buff, granules 40–200 μm diam. discrete or more usually coalescing to form a granular-warted crust. Photobiont cells 6–12(-16) × 6–12 μm. Apothecia 0.2–0.6 mm diam., at first flat but soon convex, pale, pinkish brown, grey-brown or blackish, margin sometimes remaining pale; true exciple colourless except for upper outer edge and apex which may be brown and K+ purplish, mostly 'cellular' with radiating ellipsoidal lumina 5–13 × 2.5–4.5(-7) μm; hymenium 40–50(-65) μm tall, colourless or upper part olivaceous grey to pinkish brown, K− or ± purplish, N+ red; hypothecium colourless or pale straw, K+ yellowish. Paraphyses 1.5–2 μm wide, the apices to 5 μm, swollen, often pigmented. Ascospores 14–28 × (2-)2.5–3 μm, 3(-5)-septate, narrowly fusiform to ± bacilliform. Pycnidia immersed, walls colourless; conidia 31–55 × 1–1.5 μm, mostly curved.

On rocks, turf or clayey soil in nutrient-rich situations (e.g. bird colonies), even on bird-lime; local. Mainly N. & W. Britain. Scandinavia, Iceland, N. North America.

Pale morphs on calcareous rocks are distinguished from *B. cuprea* **(29)** by the more swollen, often coloured, apices of the paraphyses and longer conidia; also *B. subfuscula* has never been found in dry underhangs.

31. B. naegelii (Hepp) Zahlbr. (1909)

Thallus thin, ± smooth, pale green-grey, usually forming small patches amongst other crusts. Photobiont cells 6–16 μm diam. Apothecia 0.2–0.6(-0.8) mm diam., often in small groups, flat to convex, variously coloured, ± whitish, blue-grey, grey-brown to blackish, often piebald; true exciple weakly developed, usually paler than the disc, colourless or with dark brown upper parts, the hyphae with expanded, ± ellipsoid lumina to 3 μm wide; epithecium colourless or more usually pale greenish, greenish grey or brown, to blue-green, K± intensifying green, N+ red, rarely partly or totally purplish brown and K± intensifying purple; hymenium 50–65 μm tall, colourless; hypothecium colourless. Paraphyses 1.5–2 μm wide, simple or forked above, the apices swollen to 5 μm, often coloured. Ascospores 13–24(-28) × 4–5(-6) μm, (0-)3(-5) septate, oblong to fusiform, often ± curved.

On nutrient-rich bark of trees and shrubs, usually in *Xanthorion* communities, very rarely on nutrient-rich rocks; common in unpolluted areas. Throughout British Is. Europe, North America.

32. B. subturgidula (Nyl.) Zahlbr. (1926)

Thallus immersed, whitish. Photobiont cells 6–12(-14) μm diam. Apothecia 0.16–0.4(-0.5) mm diam., flat to convex, bluish grey, grey-brown or brown-black, ± weakly pruinose; true exciple inconspicuous, yellowish to reddish brown, K+ yellow intensifying, but pale reddish yellow at outer edge, of radiating hyphae coherent in K, 1.5–4 μm wide with 1–1.5 μm wide lumina; hymenium 30–40 μm tall, pale brownish above due to minute granules that dissolve in K; hypothecium concolorous with inner exciple. Paraphyses 1–1.5 μm wide, simple or a few-branched, ± slightly swollen at apices. Ascospores 9–13 × 2.5–4 μm, (1-)3-septate, oblong to fusiform. Pycnidia 30–50 μm diam., numerous, immersed, brown, the wall reddish brown, K–; conidia 9–11 × 1–1.5 μm, bacilliform.

On wood in hollows in an old *Ilex*; ?extinct. S. England (Hampshire, New Forest). ?Endemic.

33. B. trachona (Ach.) Lettau (1912)

Thallus thin, rimose, pale grey-green or fawn, surface minutely granular-warted, rarely thicker and minutely leprose-granular. Photobiont cells 5–12 μm diam. Apothecia 0.2–0.5 mm diam., often absent, sessile, flat, black; true exciple persistent; epithecium dark green, K–, N+ red; hymenium 40–50 μm tall, pale reddish brown below, pale green above; hypothecium and true exciple dark red-brown, K+ purplish. Paraphyses 1–1.5 μm wide, contorted and often branched or anastomosing, the apices to 2 μm, not or only slightly swollen. Ascospores 11–17 × 3–4.5 μm, (0-)3-septate, fusiform. Pycnidia 100–300 μm diam., ± sessile, black; wall dark brown, K+ purplish; conidia 3–5 × 1–1.5 μm, short-bacilliform, often biguttulate.

On deeply shaded, hard, igneous rocks, rarely exposed large roots, in underhangs and caves. Throughout Great Britain, N. Ireland. Europe, North America.

Most often found sterile with its characteristic pycnidia, and known with apothecia only from S.W. England (Cornwall) and Channel Is.

34. B. carneoglauca (Nyl.) A. L. Sm. (1911)

Thallus thin, rimose and finely granular, often with a felt-like appearance, glaucous green or grey-green, developed on a white prothallus which often forms a white bordering margin. Photobiont cells 4–10 μm diam. Apothecia 0.4–0.6 mm diam., usually few or absent, sessile, the disc beige or pale pinkish; true exciple white pubescent, persistent, colourless, of thin-walled hyphae 1.5–3(-4) μm wide; hymenium 45–60 μm tall, colourless; hypothecium colourless. Paraphyses 1.5–2 μm wide, simple or forked above, not or only gradually widening to 3 μm at apices. Ascospores (14-)19–28 × 3–4 μm, 3- to 5-septate, fusiform to fusiform-acicular. Pycnidia 200–400 μm diam., ± sessile, white-pubescent; conidia 4–5.7 × (1-)1.5 μm, ± oblong, biguttulate, often with a median constriction. Thallus UV+ pink, 3 unidentified xanthones.

On hard siliceous rocks, rarely on exposed large roots, in dry shaded underhangs, usually near the coast or by large rivers or lakes; local. N. & W. Great Britain, W. Ireland. France.

Easily recognized, even when sterile, by the white-pubescent pycnidia and UV+ pink thallus. *B. viridifarinosa* **(35)** has an effusely sorediate thallus and, when present, smaller pycnidia. Often occurring with sterile *B. trachona* **(33)**, which has blackish pycnidia.

35. B. viridifarinosa Coppins & P. James (1992)

Like **34**, in pycnidia (when present), conidia, photobiont and chemistry, but apothecia unknown and differing in the effuse, irregular, often confluent patches of yellowish green, farinose soredia (soredia 10–30(-40) μm diam.), and slightly smaller (mostly 150–260 μm diam.), less frequent pycnidia.

On shaded, dry bark of mature deciduous trees, especially *Quercus* and *Ulmus*, in humid woodlands, rarely on shaded, siliceous or slightly basic rock-faces, old walls and monuments. Throughout Great Britain. Belgium, Luxembourg, W. Norway, S. Sweden.

36. B. fuscoviridis (Anzi) Lettau (1912)

Thallus whitish, irregularly cracked, surface ± smooth. Photobiont cells 6–12(-14) μm diam. Apothecia 0.3–1 mm diam., at first flat, but soon strongly convex, light red-brown; true exciple thick at first, colourless or upper and outer parts pale orange or pinkish, hyphae with ± rectangular lumina 7–12 × 2–4 μm; hymenium 80–95 μm tall, colourless or pale yellow-brown in upper part in places; hypothecium colourless or pale straw, K± yellowish. Paraphyses 1.5–2 μm wide, simple, the apices only slightly widening to 2.5(-3) μm. Ascospores 12–18 × 4–4.5 μm, (1-)3-septate, fusiform-ellipsoid to fusiform.

On partly calcareous mica-schist of old walls; not recorded this century. W. Ireland (Connemara). C. Europe.

Specimens overgrowing bryophytes on rocks in alpine situations should be compared with *Biatora carneoalbida* and *B. tetramera*.

37. B. sabuletorum (Schreber) Lettau (1912)

B. accedens (Arnold) Lettau (1912)

Thallus granular-warted or scurfy-granular, whitish or pale grey. Photobiont cells 7–13(-15) μm. Apothecia 0.3–0.8(-1) mm diam., mostly markedly convex but sometimes ± flat when young, pale pinkish brown, pale to dark brown, or black; true exciple thick when young, usually dark red-brown in upper parts, otherwise pale reddish brown to colourless, the hyphae 1.5–2.5(-3) mm wide, bound in dense gel matrix, the apical cells often clavate to 5 μm wide; hymenium 70–110 μm tall, colourless or upper part pale brown, grey, or olivaceous to blue-green, K−, N+ red; hypothecium 50–60 μm, dark (rarely pale) red-brown, K− or + reddish or purplish in upper, pale-coloured or colourless below. Paraphyses 2–3 μm wide, simple or a few forked above, apices to 5(-6) μm, often swollen. Ascospores (16-)18–40(-50) × 5–8(-9) μm, 3- to 7(-9)-septate, fusiform, rarely one cell with a longitudinal septum, often with finely warted perispore (best seen in K).

Usually on mosses or on plant debris over calcareous rocks, stonework or turf, rarely directly on rock, occasionally on shaded trunks of old trees (especially *Ulmus*); common, mainly at low altitudes but up to 850 m in Scotland (Highlands). Throughout British Is. Europe, North America, Australia, New Zealand.

Very variable in the colour of the apothecia; intensely pigmented forms with black apothecia have been referred to as *B. accedens*. *B. sabuletorum* is easily distinguished from other superficially species in similar habitats (e.g. *B. bagliettoana* **(25)**, *B. herbarum* **(26)**, *Biatora tetramera* and *Lecidea hypnorum* aggr. by its large, fusiform ascospores. Probably closely related to *Toninia lobulata*, which has a ± squamulose-areolate thallus and smaller, 3-septate ascospores.

BACTROSPORA Massal. (1852)

Thallus crustose, immersed or ± superficial, thin, continuous, scurfy, uniform or ± rimose; prothallus indistinct. **Cortex** absent. **Photobiont** *Trentepohlia*. **Ascomata** apothecia, mostly sessile, black or black-brown. **Thalline exciple** absent. **True exciple** persistent, dark red-brown at outer edge, paler within, pigment K+ blackish olive. **Epithecium** pale to dark red-brown. **Hymenium** colourless, I− or I+ blue. **Hypothecium** colourless. **Hamathecium** of branched and anastomosed paraphysoids; apices not swollen, richly branched. **Asci** 8-spored, appearing multispored, narrowly clavate or ± cylindrical, fissitunicate, the apex with a narrow K/I± pale blue apical dome penetrated by a small ocular chamber that is sometimes surrounded by a small K/I+ dark blue, ring-like zone. **Ascospores** ± acicular, colourless, multiseptate, ± separating into spherical, cuboid or rectangular part-spores. **Conidiomata** pycnidia; apical wall around ostiole dark brown, K+ green-black, pale below. **Conidia** simple, bacilliform or long-ellipsoid, with ± rounded ends. **Chemistry**: lichen products not detected by t.l.c. **Ecology**: dry recesses in bark of medium-aged to old broad-leaved trees.

Differs from *Lecanactis* primarily in the tendency to form part-spores which give the mature asci a multispored appearance; *Lecanactis homalotropa* possibly belongs to this genus.

LITERATURE: Sherwood (1977)

1 Ascospore part-spores ± rounded, oval, usually separate outside the ascus .... **1. *corticola***
Ascospores part-spores cuboid or rectangular, often in short chains of 2–4 outside the ascus .... **2. *dryina***

1. B. corticola (Fr.) Almq. (1869)

Thallus immersed or superficial, grey-white, scurfy, diffuse or weakly rimose, not sorediate. Apothecia 0.2–0.5(-1.0) mm diam., round, brown-black, deep red-brown when wet, at first ± flat, becoming ± convex, disc non-pruinose; true exciple pale within, ± dark red-brown, particularly towards the outer edge; epithecium dark brown, of congealed granules, K+ blackish olive, not dissolving (including the exciple); hymenium *c.* 80 μm tall, I−; hypothecium I−. Paraphysoids 2.2–2.5 μm wide, most densely branched at apices; apices not swollen. Asci 60–70 × 6–9 μm, conspicuous, ± cylindrical; walls *c.* 2 μm thick. Ascospores 50–100 × (1.5-)2(-2.5) μm, at first indistinctly long and very narrow, faintly spiralled, soon fragmenting into numerous, rounded individual segments, 2–3.5 × (1.5-)2(-2.5) μm. Pycnidia 0.1–

0.2(-0.3) mm diam, frequent, though few in thalli with abundant apothecia, brown-black, ostiole often gaping and torn; conidia 3.5–4 × 1–1.5 μm, simple, bacilliform or long ellipsoid, straight or slightly curved.

In deep fissures on the sides of old oaks in rather sheltered situations in ancient parklands; rare and scattered. Throughout Britain and N. Ireland. S. Sweden, Denmark.

Occasionally found as a sterile crust with pycnidia only.

2. B. dryina (Ach.) Massal. (1852)

Like **1**, but apothecia more irregular, often oval or sub-lobulate, convex and ± shining. Hymenium 80–90 μm tall, I+ blue; hypothecium I+ blue. Asci 70–80 × 11–12 μm, long-clavate, broadest near apex. Ascospores 70–77 × 2 μm, ± retaining original acicular form and less inclined to completely fragmented into individual part-spores as in **1**; part-spores 2–3 × 2 μm, notably rectangular or cuboid.

On oak bark, very rare. W.C. Scotland (Argyll, Loch Sunart). Sweden, C. Europe.

BAEOMYCES Pers. (1794)

Thallus crustose, granular, squamulose or almost foliose. **Photobiont** *Coccomyxa* or *Elliptochloris*. **Cortex** pseudoparenchymatous or of interwoven hyphae running ± parallel to the upper surface, but absent where sorediate. **Soralia** sometimes present. **Schizidia** discoid, detachable. **Ascomata** apothecia, pink or brown, single or several on short stipes, which may or may not be green (podetia) and corticate below or throughout, sometimes almost sessile. **Hamathecium** of paraphyses, slender, simple or sparingly branched above; hymenial gel I+ blue or I−. **Asci** 8-spored, cylindrical, thin-walled, the apex truncated, with a single functional wall layer, K/I− or with a thin K/I+ blue internal apical cap. **Ascospores** colourless, fusiform or ellipsoid, simple or to 3-septate. **Conidiomata** pycnidia, immersed in small warts on the thallus. **Conidiogenous cells** catenate, barrel-shaped, compacted. **Conidia** arising laterally and apically, short, bacilliform. **Chemistry**: β-orcinol depsides and depsidones. **Ecology**: characteristic of temporary and recently disturbed sites on acid gravels, roadsides, earth banks, mine-spoil heaps, heathland.

The ascus structure clearly separates the genus from the Cladoniaceae, with which it has often been placed.

LITERATURE: Frey (1933), Thomson (1967a)

1 Thallus entirely crustose or of minutely compacted squamules, not radiating at the margin 2
Thallus conspicuously and coarsely squamulose, with ± radiating margins .. **1. *placophyllus***

2(1) Apothecia rose-pink, ± domed when mature; thallus crustose, grey; schizidia scattered, coarse, soft, rounded, white or pale pink .... **2. *roseus***
Apothecia pale to deep brown, mostly flattened; thallus crustose or congested-squamulose, dull green-grey; schizidia absent or scattered, flattened, disc-like .... **3. *rufus***

1. B. placophyllus Ach. (1803)

Thallus forming compact partial or complete rosettes, rather thick, ridged-rugose or squamulose in centre, with distinct, flattened, smooth, contiguous lobes, with slightly upturned, paler margins, glaucous green-grey occasionally brownish, becoming bright green when wet, to 5 mm wide; schizidia peltate, often light-coloured, fragile, flat to ± convex, to 0.2 mm across, especially towards centre and lobe-margins, leaving shallow scars. Photobiont cells 6–13(-14) μm diam., globose, or a few ellipsoid, <14 × 12 μm diam. Apothecia to 2 mm diam., rare, flattened, red-brown, on short, furrowed, solid stipes, ± corticate below, often minutely squamulose. Ascospores 8–14 × 2–4 μm, simple or indistinctly 1-septate. Pycnidia not seen. Thallus Pd+ orange, K+ yellow, KC+ orange, C−, UV+ bright orange, rarely dull red (stictic acid).

On peaty and humus-rich gravels in upland areas, sometimes in association with heavy metals on mine spoil heaps; rather local. N.W. England, Wales, Scotland, Ireland (very rare). Europe, North America.

2. B. roseus Pers. (1794)

Thallus in small patches, or becoming more wide-spreading, smooth, ± shiny, pale grey to dull green-grey, sometimes whitish; schizidia scattered, coarse, rounded, or flattened, pink-white, shiny, varying in texture and size, 0.1–0.3 mm on fertile thalli, to 1 mm on sterile thalli, sometimes farinose-sorediate; margin ± determinate, without marginal lobes or squamules. Photobiont cells 5–8 μm diam., mostly globose. Apothecia to 3 mm diam., occasional, conspicuously stalked, bright rose-pink, domed; stipe white or pink, ecorticate, ± hollow; hymenium I+ blue. Asci with a thin K/I+ blue apical cap. Ascospores 10–26 × 2–3 μm, simple or 1-septate. Pycnidia *c.* 200 μm diam.; wall colourless or brown above. Conidia 3.8–5 × 0.8–1 μm. Thallus Pd+ orange, K+ orange, KC+ orange, C−, UV+ orange (mainly apothecia and schizidia) (baeomycesic, ± squamatic acids and atranorin).

Mineral or peaty soil on acid heaths and moors, frequently sterile. Widespread in N. England, Wales, Scotland, Ireland; rare and very local elsewhere. Europe, North America.

Sometimes confused with *Icmadophila ericetorum*, which has larger, flatter, sessile apothecia, a thallus which contains perlatolic and thamnolic acids, and ellipsoid photobiont cells; most records of that species from E. England refer to *Baeomyces roseus*.

3. B. rufus (Huds.) Rebent. (1804)

Thallus green to dull grey-green, sometimes brownish, discrete or ± wide-spreading, continuously crustose, nodulose, or of minute, overlapping, compacted squamules to 1 mm wide, ± delimited, non-sorediate or with diffuse, coalescing, irregular, greenish soralia; schizidia occasionally present, <0.2 mm diam., disc-like. Photobiont cells 6–13(-14) μm diam., globose or a few ellipsoid, <14 × 12 μm. Apothecia almost sessile, to 1 cm tall, dark red-brown to pale dull pink, ± translucent when wet, flattened, often incurved at base, to 2 mm diam., single or occasionally several together on a short, white, furrowed, solid stipe, mostly ecorticate, sometimes greenish and corticate towards the base; hymenial gel and asci I−, K/I−. Ascospores simple or indistinctly 1-septate. Pycnidia not seen. Thallus Pd+ orange, K+ yellow, KC+ yellow, UV− (stictic, ± norstictic, ± constictic acids).

On gravelly or peaty acid soils, particularly in recently disturbed or burnt heathland, often spreading over stones, mosses and peat overhangs in damp locations, including stream-sides and shaded earth banks, more rarely on sheltered walls and tree trunks; locally abundant. Throughout British Is.; local in Midlands and E. of England, and Ireland. Circumpolar in temperate and boreal zones. Europe, North America.

The differences between this species and *B. roseus* **(2)** suggest that two genera are involved; *B. roseus* **(2)** seems to have a strong affinity with *Icmadophila*.

BELLEMEREA Hafellner & Roux (1984)

Thallus crustose, cracked- to warted-areolate; areoles dispersed to contiguous; prothallus black. **Photobiont** chlorococcoid, many cells ellipsoid, at times algal layer continuous beneath hypothecium. **Medulla** I+ blue. **Ascomata** apothecia, ± immersed, at times occupying virtually the entire surface of areole; disc concave to flat. **Thalline exciple** thin, not raised, scarcely distinguishable from the thallus (cryptolecanorine). **True exciple** very thin, or absent. **Hamathecium** of paraphyses, ± branched and anastomosed; apices slightly swollen, often with a thin, dark brown cap. **Epithecium** brown or green-brown, N− or N+ purple. **Hypothecium** colourless. **Asci** 8-spored, clavate, *Porpidia*-type. **Ascospores** simple, occasionally spuriously 1-septate, ± ellipsoid, colourless, with a distinct I+ blue perispore. **Conidiomata** pycnidia, immersed. **Conidiogenous cells** in short chains. **Conidia** acrogenous or pleurogenous, short-bacilliform, simple, colourless. **Chemistry**: norstictic acid present in some species. **Ecology**: on montane rocks; a genus of *c.* 6 species.

Separated from *Aspicilia* by the *Porpidia*-type ascus. Distinguished from *Amygdalaria* and *Porpidia* species with immersed apothecia by the colourless (or absent) exciple, colourless hypothecium and ascospores with an I+ blue perispore.

LITERATURE: Clauzade & Roux (1984, 1985).

1. B. alpina (Sommerf.) Clauz. & Roux (1984)

Thallus whitish to blue-grey; prothallus black. Apothecia 0.2–1.5 mm diam., dull brown when dry, bright red-brown when wet; epithecium brown, N−; algal layer discontinuous beneath hypothecium. Ascospores 8–23 × 7–13 μm. Thallus Pd+ yellow-orange, K+ red, KC+ red, C− (norstictic acid).

On quartzite boulders near late snow patch. N. Scotland (Cairngorms); also recorded last century from schists on Ben Lawers. Scandinavia, Alps, North America.

B. cinereorufescens (Ach.) Clauz. & Roux (1984) [? *B. diamarta* (Ach.) Hafellner & Roux (1984)] differs in having a brown-black apothecial disc when dry (red-brown when wet), the algal layer continuing beneath hypothecium and ascospores 7–16 × 4–9 μm. It was recorded last century from N. Scotland (Lochan nan Cat, Ben Lawers and walls at Glen Fender, Blair Atholl) but material on which these records are based may have originated from outside British Is.

BELONIA Körber ex Nyl. (1852)

Thallus crustose, superficial, thin or immersed within substratum. **Cortex** absent. **Photobiont** *Trentepohlia* or *Trebouxia*. **Ascomata** apothecia, resembling perithecia but immersed in thalline warts or emergent and enclosed by pale true and thalline exciple. **Hymenium** exposed at surface via a pore-like opening, ± widening at maturity, I+ blue, containing small, yellow-orange oil droplets. **Hamathecium** of paraphyses, thin, to 1.5 μm wide, simple, septate, lax, not forming distinct epithecium but filling up cavity formed by ascoma walls. **Asci** (4-)8-spored, narrowly cylindrical, thin-walled. **Ascospores** colourless, mostly needle-shaped and septate, rarely muriform with attenuated apices. **Conidiomata** unknown. **Chemistry:** lichen products not detected by t.l.c.

LITERATURE: Jørgensen *et. al.* (1983), Swinscow (1961, 1964), Vězda (1959).

1 Ascospores septate 2
Ascospores muriform .... **3. *nidarosiensis***

2(1) On rocks; apothecia entirely covered with a thalline exciple 3
On soil, mosses or algal mats; apothecia without thalline exciple in upper part **2. *incarnata***

3(2) Thallus superficial; ascospores (50-)75–125 μm .... **4. *russula***
Thallus immersed; ascospores 50–75 μm .... **1. *calcicola***

1. B. calcicola W. Watson (1935)

Thallus mostly immersed, scarcely apparent, a few granules on the surface of the rock, effuse, pink-white. Photobiont *Trentepohlia*. Apothecia discrete or 1- to 3-contiguous, ¼- to ½- immersed, in fertile warts 0.2–0.25 mm diam.; involucrellum absent; true exciple colourless to pale yellow-brown or pink, naked above. Asci 8-spored. Ascospores 50–75 × 3–7 μm, colourless, acicular, 10- to 20-septate.

On limestone; rare. S.W. & N.W. England (N. Somerset, Goblin Coombe; ?Cumbria, Lake District, Buttermere). Endemic.

Closely related to *B. russula* **(4)** and perhaps conspecific. *B. russula* differs in being mainly superficial, the apothecia are enclosed in thalline warts, and the perithecia and ascospores are larger.

2. B. incarnata Th. Fr. & Graewe ex Th. Fr. (1865)

Thallus indistinct, ± gelatinous when wet, effuse, thin, varnish-like, greyish white or greenish grey. Photobiont *Trebouxia*. Apothecia ½- to ¼-immersed, irregularly scattered, single or in clumps, pinkish or yellowish grey. Apothecia 0.25–0.45 mm diam.; opening remaining pore-like; true exciple pale pink to reddish brown, thin; thalline exciple absent, at least from upper part of apothecium. Asci 8-spored. Ascospores 120–160 × 3–4 μm, acicular, multiseptate, colourless.

On mosses on ± basic soils; rare, probably overlooked. N.E. England (Teesdale), Scotland (Highlands), also on algal films overlying peat at mine spoil heaps in S.W. England (Dartmoor) and Wales (Ystwyth). Scandinavia.

3. B. nidarosiensis (Kindt) P. M. Jørg. & Vězda (1983)

Clathroporina calcarea W. Watson (1925)

Thallus ± superficial, forming an uneven, ± entirely powdery-sorediate, irregularly cracked, often wide-spreading, effuse, rather thin, pink-orange crust, fading to whitish cream in the herbarium. Photobiont *Trentepohlia*. Apothecia rather rare, scattered, pale pink, globose, with a pore ± opening at maturity to reveal a brownish disc, visible through a ± widening pore. Asci (4-)6- to 8-spored. Ascospores (36-)50–60(-82) × (8-)10–12(-14) μm, colourless, muriform, with (7-)12–15(-19) transverse septa, elongate, with attenuated apices and ± constricted at the middle septum.

On dry, vertical, sheltered surfaces and shaded overhangs on calcareous rocks, especially hard limestones, also common on N- and E-facing cement-mortar-plaster-stone walls; widely distributed and locally frequent. W. British Is., from Devon to Shetland Is., W. Ireland, extending to S.E. England in churchyards. Scandinavia, France, C. Europe.

Often forming extensive, sterile, sorediate, orange-pink sheets on vertical rock-faces and shaded, N-facing church walls. Some states of this species resemble strongly sorediate morphs of *Opegrapha gyrocarpa*, the thallus of which is, however, C+ orange-red (gyrophoric and schizopeltic acids).

4. B. russula Körber ex Nyl. (1857)

Thallus mostly superficial, thickly granular to inconspicuous, effuse, dark reddish grey to reddish brown. Photobiont *Trentepohlia*. Apothecia mostly discrete, rarely contiguous, immersed in prominent thalline warts (0.25-)0.4–0.6 mm diam., dull reddish brown, the apex pale grey or pink-grey, translucent when wet; involucrellum absent; true exciple colourless in outer part, the inner part pale yellow-brown, ± totally invested by thalline tissue. Asci 8-spored. Ascospores (50-)75–100(-125) × 3–6 μm, colourless, acicular, with 10–20 transverse septa, sometimes faintly constricted at septa.

On epidiorite, calcareous schists and limestones in moist, shaded overhangs, usually above 600 m; local. Scotland (Highlands). Mtns of N. & C. Europe.

For differences from *B. calcicola*, see **1**.

BIATORA Fr. (1817)

Biatora Fr. (1817), non Ach. (1809), *Mycobilimbia* Rehm (1890)

Thallus crustose, effuse, sometimes membranous in part but usually granular to granular-verrucose, creamy white, dull green, glaucous green or green-grey; usually ecorticate, sometimes sorediate; prothallus sometimes visible when the thalline granules are scattered, white, arachnoid. **Photobiont** chlorococcoid. **Ascomata** apothecia, biatorine, sessile, convex and immarginate, or at first ± top-shaped with a flat disc and shallow margin, occasionally tuberculate, whitish to dark brown but never black, not pruinose, sometimes with a reddish brown pigment (especially in hymenium and upper exciple), but never with purplish, greenish or very dark brown (carbonaceous) pigmentation. **True exciple** soon reflexed, well developed in sections, chondroid, composed of radiating hyphae, 2–2.5 μm wide, with terminal cells sometimes to 4 μm wide, tightly bound in a gel-matrix that does not dissolve or markedly swell in K; outer edge sometimes covered by a narrow gel layer. **Hymenium** 60–80(-100) μm tall, without a well-defined epithecium (sometimes with pale pigmentation in the upper part), granules or oil-droplets, I+ blue. **Subhymenium** distinct, somewhat opaque, often forming a straw- to brownish coloured zone between the hymenium and hypothecium. **Hypothecium** chondroid, of interwoven hyphae, 1–2 μm wide in a dense gel-matrix. **Hamathecium** of coherent (in K) paraphyses, 1.5–2 μm wide, simple or sparingly branched, occasionally anastomosed, the apices ± swollen (to 3 μm wide) but sometimes clavate (to 5 μm wide), not surmounted by distinct apical 'cap' or 'hood'. **Asci** 8-spored, cylindrical-clavate, lateral walls 0.7–1 μm wide, *Biatora*-type (Fig. 33b) with K/I+ blue apical dome penetrated from below by a narrow, K/I− apical cushion surrounded by a narrow, deeply K/I+ blue zone, wall K/I− but surrounded by a K/I+ blue outer layer, ocular chamber poorly developed or not apparent. **Ascospores** colourless, ellipsoid or oblong, to oblong-fusiform, simple to 3(-5)-septate, smooth, without a distinct perispore. **Conidiomata** not found. **Chemistry**: various, but most species without substances detectable by t.l.c. **Ecology**: usually mossy rocks or the trunks of old trees (often overgrowing bryophytes or plant debris). On somewhat acidic substrata but avoiding extreme acid or calcareous conditions and in relatively undisturbed habitats. The species growing on bark are all characteristic of ancient woodland.

When fertile, species of *Biatora* can be recognized by their wide-spreading thalli with medium-sized to rather large, convex, immarginate and pale apothecia. Similar taxa include: *Bacidia sabuletorum* and *Toninia lobulata*, which have thick-walled asci, broader paraphyses, ascospores with a warted perispore, and purplish and greenish pigments in the apothecia, and the *Lecidea hypnorum* group which has *Porpidia*-type asci.

LITERATURE: Hafellner (1984, 1989).

1 Thallus with effuse, pale greenish or yellowish soralia 2
Thallus without soralia 3

2(1) Soralia C+ red .... **3. *epixanthoidiza***
Soralia C− .... **2. *epixanthoides***

3(1) Ascospores mostly 0- to 1-septate (rarely a few 2- to 3-septate) 4
Ascospores mostly 3- septate 6

4(3) Thallus Pd− 5
Thallus Pd+ red .... **1. *cuprea***

5(4) Thallus whitish to green-white, granular-verrucose or with coherent granules 80–150 μm diam.; ascospores 0(-1)-septate .... **7. *vernalis***
Thallus grey-green, finely to mealy granular; granules 25–70 μm diam.; ascospores (0-)1(-3)-septate .... **4. *sphaeroides***

6(3) Apothecia grey- to dark (± reddish) brown, with reddish brown pigment in hymenium and/or upper exciple in section .... **5. *tetramera***
Apothecia whitish, without pigmentation in section .... **6. *carneoalbida***

1. B. cuprea (Sommerf.) Fr. (1831)

Lecidea cuprea Sommerf. (1826)

Thallus warted, whitish; warts 0.1–0.4 mm diam., smooth and ± corticate. Apothecia (0.3-)0.4–0.8 mm diam., usually immarginate, convex to globose, often tuberculate and to 1.5 mm diam., pale to darkish red-brown; true exciple pale to dark red-brown (at least in part), K− ; hymenium and hypothecium pale yellowish- or reddish brown. Ascospores 10–16(-18) × 3–5 μm, 0- to 1(-3)-septate, oblong- to fusiform-ellipsoid. Thallus Pd+ red, K−, KC−, C− (unidentified substance at A7 and G7, above argopsin and pannarin).

On soil at *c.* 1000 m. C. Scotland (Perthshire, Ben Lawers); no records this century. Scandinavia, North America.

Distinguished from other *Biatora* species by the Pd+ red thallus. *Lecidea limosa* has a similar Pd+ red thallus, but the apothecia are black.

2. B. epixanthoides (Nyl.) Diederich (1989)

Bacidia epixanthoides (Nyl.) Lettau (1912)

Thallus pale, grey-green or grey, immersed or thinly granular-verrucose, with irregular, effuse, often confluent, ± yellowish or pale buff, soralia. Soredia 20–60 μm diam. Apothecia often absent, similar to *B. tetramera* (**5**). Ascospores 17–25(-30) × 4.5–7 μm, 0- to 3(-5)-septate, fusiform. Thallus Pd−, K−, KC−, C− (lichen products not detected by t.l.c.).

Over bryophytes or on bark of mossy trunks of old deciduous trees in ancient woodlands, rarely on mossy rocks in wooded ravines; local. N. & W. Britain. Europe, North America.

3. B. epixanthoidiza auct., non (Nyl.) Räsänen (1939)

Lecidea epixanthoidiza auct., non Nyl. (1875)

Like **2**, but soralia C+ red (gyrophoric acid). Apothecia 0.4–0.8(-1.0) mm diam., whitish or pale, without pigmentation in sections. Ascospores (10-)12–17 × (4-)4.5–5(-6) µm, 0(-1)-septate.

Similar habitats to **2** but with more easterly distribution of N.E. England and C. & E. Scotland.

Sterile material of *B. epixanthoidiza* (Nyl.) Räsänen [syn. *Lecidea efflorescens* (Hedl.) Erichsen (1932)], has recently been discovered in N. Ireland (Fermanagh) on *Salix* bark. It is frequent in Scandinavia and differs in the smaller, more well-delimited soralia which are Pd+ red (argopsin). The soralia are larger (0.1–0.6 mm diam.) than those of *Halecania viridescens* (mostly 0.1–0.2 mm diam.) which has an identical chemistry. The current name for *B. epixanthoidiza* auct. is ***L. gyrophorica*** Tønsb. (1992).

4. B. sphaeroides (Dickson) Körber (1853)

Catillaria sphaeroides (Massal.) Schuler (1902)

Thallus grey-green, surface finely or mealy granular; granules 25–70 µm diam. Apothecia 0.4–0.8(-1.1) mm diam., at first hemispherical, later becoming ± globose, pale to dull orange-pink, or sometimes (especially when old) pale reddish brown, in sections without pigmentation. Ascospores (9.5-)11–17 × 3.5–5 µm, (0-)1(-3)-septate, fusiform-ellipsoid.

On sheltered, mature deciduous tree trunks (especially *Fraxinus* and *Quercus*), often around their bases, in old woodlands, more rarely on mossy rocks in wooded ravines or sheltered mountain gullies; local. N. & W. Britain, extending eastwards to S. England (E. Sussex). W. Europe, northwards to Scandinavia (Hordaland).

The more finely granular thallus and predominately 1-septate ascospores separate this species from *B. vernalis* **(7)**. Has been confused with *Lecania chlorotiza* and *L. cyrtellina*, which have smaller apothecia and narrower spores.

5. B. tetramera (de Not.) Coppins (1992)

Bacidia fusca (Massal.) Du Rietz (1929) , *B. obscurata* (Sommerf.) Zahlbr. (1905) , *B. tetramera* (de Not.) Coppins (1988), *Mycobilimbia obscurata* (Sommerf.) Rehm (1890)

Thallus whitish, green-white or pale grey, granular-verrucose, esorediate. Apothecia 0.4–1.2 mm diam., at first sessile and ± turbinate, later hemispherical to subglobose, grey- to red-brown; upper hymenium and upper true exciple pale brown, K−; hypothecium colourless or upper part (subhymenium) pale reddish brown. Ascospores 17–26 × 4.7–7 µm, (0-)3-septate, fusiform. Thallus Pd−, K−, KC−, C− (lichen products not detected by t.l.c.).

On bryophytes, small shrubby plants and plant debris on or amongst rocks in alpine situations above 1000 m; rare. Scotland (Highlands). Scandinavia, mtns of C. & S. Europe.

Distinguished from *B. epixanthoides* **(2)** by the esorediate thallus, and from *B. carneoalbida* **(6)** by the darker apothecia. Pale morphs of *Bacidia sabuletorum* (i.e. ± lacking greenish or purplish brown pigments) differ in their stouter paraphyses and often longer, 3- to 7(-9)-septate, ascospores with a finely warted perispore. *Bacidia fuscoviridis* has stouter hyphae in the true exciple, and grows on rock at low altitudes.

6. B. carneoalbida (Müll. Arg.) Coppins (1992)

Bacidia carneoalbida (Müll. Arg.) Coppins (1988) ["*carneopallida*"], *Biatora sphaeroides* auct. p.p., non (Dickson) Körber

Like **5**, but apothecia without internal pigmentation, ivory-white and ascospores slightly smaller, (12-)13–22 × 4–7 µm.

On mosses on rock in gully at 1160 m. C. Scotland (Perthshire, Ben Lawers), the only British locality. Scandinavia, mtns of C. & S. Europe, Canada.

Differs from *B. sphaeroides* **(4)** and *B. vernalis* **(7)** in the predominantly 3-septate ascospores. Has been confused with *Micarea alabastrites*, an oceanic species with C+ red apothecia and a very different anatomy.

7. B. vernalis (L.) Fr. (1831)

Lecidea vernalis (L.) Ach. (1803)

Thallus whitish or green-white, often with a varnish-like sheen, ± membranous in part but mostly granular verrucose or with discrete (but never loose) granules 80–150 μm diam. Apothecia 0.4–0.8 mm diam., at first top-shaped, later convex-adpressed to subglobose, pale to testaceous, without distinct pigmentation in section, except for a pale straw (K+ yellowish) coloured hypothecium. Ascospores (10-)12–19 × 3.5–5 μm, 0(-1)-septate. Thallus Pd−, K−, KC−, C− (lichen products not detected by t.l.c.).

In old woodlands on mossy tree trunks (especially *Quercus*), rarely on sheltered mossy rocks. N.W. England (Lake District), Scotland (W. & C. Highlands). Scandinavia, montane areas of W. & C. Europe.

Distinguished from *B. sphaeroides* **(4)** by its more coarsely granular thallus and mainly simple ascospores; it is also more light-demanding. Often sterile, but the absence of soralia then separates it from *B. epixanthoides* **(2)**, *B. epixanthoidiza* **(3)** and *Catillaria pulverea*, and the Pd− thallus separates it from *Phyllopsora rosei*. *Megalospora tuberculosa* has a softer (never varnish-like), usually greenish yellow thallus and contains usnic acid and zeorin.

BIATORELLA de Not. (1846)

Thallus crustose, usually thin and poorly developed, unstratified, continuous, effuse and often ill-defined; prothallus absent. **Photobiont** chlorococcoid. **Ascomata** apothecia, biatorine, yellowish or pale pink-yellow to pale reddish brown, usually convex to hemispherical, sometimes flat. **Thalline exciple** absent or thin, soon disappearing. **True exciple** usually very thin to almost absent, of parallel, colourless to pale yellow hyphae, I+ blue. **Epithecium** yellowish to reddish brown, K−. **Hymenium** colourless or pale straw-yellow, I+ deep blue. **Hypothecium** yellowish, I+ pale- to green-blue. **Hamathecium** of paraphyses, simple, becoming richly branched and entangled at the epithecium. **Asci** multispored, cylindrical to clavate, gelatinous I+ blue outer layer, with a definite apical dome (tholus), I− or I+ faint blue. **Ascospores** bacilliform, ellipsoid or globose, simple, thin-walled, colourless. **Conidiomata** not known.

Differs from *Strangospora* in that the asci have an I+ blue gelatinous outer layer and an apical dome (tholus) which is I− or I+ only faint blue. The epithecium comprises the solidly matted, richly branched tips of otherwise simple paraphyses, not set in a gelatinous matrix. Differs from *Sarea* Fr. (1825), the species of which grow on resin, in being lichenized and, more importantly, in lacking abundant orange or black pycnidia, which produce globose conidia.

LITERATURE: Hawksworth & Sherwood (1981), James (1971), Magnusson (1935).

1	Ascospores ellipsoid or bacilliform; not on bark	2
	Ascospores globose; on bark	**3. *monasteriensis***
2(1)	Ascospores 5–8 μm long; in montane habitats	**1. *hemisphaerica***
	Ascopsores 8–13 μm long; in lowland habitats	**2. *fossarum***

1. B. hemisphaerica Anzi (1860)

B. fossarum auct. brit., non (Dufour) Th. Fr. (1874)

Thallus pale grey, thin, fine-grained to mealy-arachnoid. Apothecia 1–2 mm diam., bright or dull orange-red, convex to hemispherical and immarginate from the start, immersed to adpressed, usually ± discrete; without a visible true exciple; epithecium pale to orange-yellow, sometimes with dark yellow-brown pruina; hymenium 150–200 μm tall, colourless; hypothecium (100-)150–300(-500) μm tall, pale grey-yellow, of densely packed hyphae 1 μm thick. Paraphyses *c.* 1 μm thick, thickened and clavate at the apices to 2.5–3.5 μm, clearly visible in water, conglutinate in K. Asci 135–150 × 15–18 μm, oblong-cylindrical, not symmetrical, 200- to 400-spored. Ascospores 5–8 × 2–3.5 μm, elongate-ellipsoid. Thallus K−, C−, I−.

On base-rich soils, particularly mica-schist, and on associated mosses, usually in shady rock crevices and ledges at high altitudes; very rare. C. Scotland (Perthshire, summit of Ben Lawers). N. & C. Europe, Alps, Carpathians.

2. B. fossarum (Duf.) Th. Fr. (1874)

Like **1,** but apothecia smaller, 0.4–1 mm diam., and ascospores longer, 8–13 × 2.3–3 μm, bacilliform. On sunny, compacted, calcareous, sandy soil in lowlands; very rare (single record). S.W. England (Somerset). S. Europe.

Has been confused with *B. hemisphaerica* **(1)**, which is a montane, arctic-alpine species. *Sarcosagium campestre* var. *macrospora* has similar ascospores, but smaller, marginate apothecia which are pink or red without an orange tinge.

3. B. monasteriensis (Lahm) Lahm (1883)

Thallus wide-spreading, pale grey-green when dry, ± bright green when moist, scurfy, effuse, continuous, ± minutely areolate. Apothecia 0.2–0.5 mm diam., yellowish, reddish yellow or pale pink to dull red-brown, translucent and paler when wet, flat to convex, partly immersed in the thallus when young, becoming sessile, numerous, scattered or ± contiguous; thalline exciple pale, thin, granular, soon excluded; true exciple rarely distinct, occasionally to 12 μm thick, increasing to 17 μm at the margin; epithecium translucent, pale red-brown, the colour contained within the apical cells of the paraphyses; hymenium 60–85 μm tall, colourless or pale straw yellow; hypothecium 25–85 μm tall, of densely packed fine hyphae, colourless or pale yellowish brown. Paraphyses 1.5 μm wide towards the base, swelling to 4–5 μm at the clavate apex, separating and clearly visible in K. Asci 60–65 × 12–17 μm, narrow-clavate, the walls 1 μm, increasing to 2–5 μm at apex, 100- to 150(-200)-spored. Ascospores 3–3.5 μm, globose. Thallus K−, C−, I−.

On trunks of broad-leaved trees, particularly *Ulmus*, *Fraxinus* and *Sambucus*, in sheltered habitats; rare. N.E. England (Cleveland), N. Wales (Merionethshire, Bala), S.W. & E.C. Scotland (Galloway, Newton Stewart; E. Perthshire). Scattered in C. Europe, particularly Germany and Switzerland.

BLARNEYA D. Hawksw., Coppins & P. James (1980)

Thallus crust-like, not corticate, of floccose hyphae, irregularly branched, colourless, their surfaces often encrusted with crystals, dissolving in C. **Photobiont** *Trentepohlia*; cells arranged in filaments or separating, invested by hyphae of the fungus without penetration of the protoplast. **Ascomata** not present. **Conidiomata** sporodochia, of ± vertically to irregularly orientated hyphae, very densely interspersed with crystals dissolving in C. **Conidiophores** semi-macronematous, mainly arranged in sporodochia, but occasionally arising directly from mycelium. **Conidiogenous** cells holo-, mono- or polyblastic, terminal or lateral, each conidium in turn acting as a conidiogenous cell prior to release from the conidiophores. **Conidia** simple to 1-septate, the apices rounded or truncate, guttulate, with wrinkled walls. **Ecology**: on lichens on bark.

One of only two lichenized hyphomycetes recognized in Europe.

LITERATURE: Hawksworth, Coppins & James (1980).

1. B. hibernica D. Hawksw., Coppins & P. James (1980)

Thallus pale pinkish white, fading to whitish in the herbarium, effuse, cotton-like, the margin clearly delimited with arachnoid-fasciculate threads radiating outwards on to the host thallus, the host thallus being killed in advance of the hyphal threads, persistent and perennial; hyphae 1.5–2 μm wide. Sporodochia present, often forming over effete apothecia or pycnidia of the host. Conidiophores 2–3 μm wide; conidia 8–13 × 3–4.5 μm. Thallus and sporodochia Pd−, K−, KC+ red or −, C+ red or −, UV+ glaucous (lecanoric and schizopeltic acids).

At first lichenicolous, spreading over *Enterographa* or *Lecanactis* thalli (chiefly *L. abietina*, also *L. premnea*) and apparently taking algae from them, finally independent, on bases of sheltered, dry sides of ancient *Ilex*, rarely *Quercus*, in ancient woodland; locally frequent. S.W. England, Wales (Cardigan), W. Ireland (Kerry, Waterford). Spanish Pyrénées (overgrowing sorediate *Belonia* sp.).

The occurrence of schizopeltic acid and the method of conidiogenesis suggest that *B. hibernica* is an anamorph with affinities to the Opegraphales.

BRIGANTIAEA Trevisan (1853)

Thallus crustose, verrucose. **Photobiont** chlorococcoid; cells 6–14 μm diam. **Ascomata** apothecia, with thick true exciple. **True exciple** of thick, radiating hyphae. **Hamathecium** of paraphyses not or slightly thickened at apex, indistinctly septate. **Asci** 1(-2)-spored, clavate-cylindrical, thick-walled, with distinct I+ blue tholus and both external and inner ascus walls. **Ascospores** muriform, thin-walled, colourless. **Conidiomata** not known. **Chemistry**: the apothecia contain parietin, K+ purple. **Ecology**: on soil, mosses and low-growing vegetation; montane.

Previously included in *Lopadium* which differs in lacking a distinct I+ blue tholus, has paraphyses which are distinctly enlarged at their apices and apothecia which lack parietin.

1. B. fuscolutea (Dickson) R. Sant. (1981)

Thallus smooth, granular-verrucose or minutely squamulose, grey-green to yellowish white. Apothecia to 2 mm diam., infrequent, sessile to subpedicellate, flat, becoming ± convex, golden yellow occasionally tinged olive-black, surface pruinose, margin often prominent, irregularly wavy and inflexed; epithecium with numerous yellow crystals, K+ purple; hymenium to 250 μm tall, I+ blue, paraphyses to 1.5 μm wide. Ascospores 48–100 × 24–55 μm, ellipsoid. Thallus Pd−, K+ yellow, KC+ yellow, C− (atranorin, chloratranorin); apothecia K+ purple (parietin).

Encrusting mosses and low-growing plants, occasionally on rocks or soil, mostly above 900 m., but descending to sea level; very local, but possibly overlooked when sterile. N. Scotland (Highlands, Sutherland, Shetland). Iceland, Fennoscandia, Faeroes.

BRYONORA Poelt (1983)

Thallus crustose. **Photobiont** chlorococcoid. **Ascomata** apothecia, sessile. **Thalline exciple** present but superficially often appearing biatorine. **True exciple** distinct, continuous, strongly conglutinated, swollen; outer layer consisting of anastomosing, net-like, randomly-orientated hyphae in a firm gelatinous matrix, the marginal hyphae with globose, ± swollen apices; inner, non swollen layer red-brown. **Epithecium** red-brown. **Hypothecium** strongly conglutinated, with photobiont cells below, pale brown. **Hamathecium** of paraphyses, simple, thin; apices thickened; inner wall layer brownish, strongly conglutinated. **Asci** 8-spored, with an I + blue outer wall layer and a distinctly I + blue tholus. **Ascospores** thick-walled, ellipsoid-elongate to spindle-shaped, simple or septate, colourless. **Conidiomata** pycnidia. **Conidia** bacilliform. **Chemistry**: lichen products not detected by t.l.c. **Ecology**: on mosses, montane.

Characterized by the clearly distinct conglutinated cortex of the exciple and by its ecology.

LITERATURE: Poelt (1983).

1. B. curvescens (Mudd) Poelt (1983)

Lecanora curvescens (Mudd) Nyl. (1865)

Thallus mostly within mosses, indistinct, grey- to black-brown, irregularly granular, nodulose, or ± continuous and film-like, effuse. Apothecia scattered to crowded, ± sessile, 0.5–1 mm diam.; disc dark brown to black-brown, initially ± grey-white-pruinose, at first ± rounded, concave, later becoming irregularly lobate, flat or often ± convex, to 5 mm diam., the margin grey to deep red-brown, mostly concolorous or somewhat paler than disc; true exciple, when evident, 0.03–0.05 mm wide. Paraphyses with apices to 4 μm wide; end cell red-brown. Ascospores (18-)24–35(-41) × 5–6.5(-8) μm, simple to 4(-6)-septate, ellipsoid-subcylindrical, rarely clavate, with rounded or weakly pointed apices, straight, curved or weakly S-shaped. Thallus Pd−, K−, KC−, C−, UV−.

On *Andreaea* and other mosses on inclined, preferably seepage, tracks on north-facing rock outcrops, sometimes with *Caloplaca nivalis*, mostly above 900 m; very rare. C. Scotland (Ben Lawers, Caenlochan, Ben Nevis range). Iceland, Fennoscandia, Czechoslovakia, Austria, Greenland, Switzerland, France (Corsica), Nepal.

The ascospores are characteristically arranged parallel to each other in the ascus and their septation is not clear in squash preparations in water.

BRYOPHAGUS Nitschke ex Arnold (1862)

Thallus crustose, not layered, of irregular algal colonies penetrated by thin, 0.7–1 μm wide, branched hyphae. **Photobiont** *Gloeocystis*-like, globose or elongate, compacted into colonies. **Ascomata** apothecia, at first immersed and closed, later ± emerging, becoming urceolate with a narrow opening. **True exciple** comprising outer part of septate, thin-walled, ± radiating hyphae, and an inner part of pseudoparenchymatous cells. **Hymenium** I+ blue to yellow-brown. **Hamathecium** of paraphyses, straight, unbranched, non-septate, conglutinate. **Asci** 8-spored, cylindrical, thin-walled, slightly thickened at the apex. **Ascospores** cylindrical-fusiform, colourless, septate. **Conidiomata** pycnidia, pyriform, immersed. **Conidia** shortly cylindrical. **Chemistry**: no lichen products detected by t.l.c. **Ecology**: on bryophytes and algal mats.

LITERATURE: Vězda (1969).

1. B. gloeocapsa Nitschke ex Arnold (1862)

Thallus membranous, ± shining, very thin when dry, pale fawn to greenish, becoming gelatinous when wet, effuse or vaguely delimited. Apothecia 0.2–0.5 mm wide, mostly scattered and discrete, rarely crowded, at first ± immersed and closed, becoming emergent and deeply concave; disc pale yellow-brown to orange-red, rarely dark brown; true exciple pale yellowish to pale brown or concolorous. Ascospores 20–30 × 1.5–2 μm, 3- to 4-septate.

On bryophytes and algal mats, especially on recently disturbed soil, on shaded road-cuttings and banks as well as mineralized, acid soil associated with old mine workings; frequent and widespread, but often overlooked. W. & N. British Is. C. & N. Europe, from Lapland to Alps and Carpathians, Madeira, Azores.

The thallus forms distinctive pale yellow-fawn necrotic patches on mosses and hepatics; the tiny pale urceolate apothecia are often ± concolorous with the thallus.

BRYORIA Brodo & D. Hawksw. (1977)

Thallus shrubby, erect, decumbent or pendent; branches ± terete and smooth, hair-like, occasionally becoming flattened and ± pitted towards the base, with or without lateral spinules; branching aniso- or isotomic dichotomous (Fig. 5). **Soralia** frequent, tuberculate or fissural (Fig. 6). **Pseudocyphellae** present or absent, usually inconspicuous. **Cortex** composed of periclinal, conglutinate hyphae immersed in moderate or relatively little matrix; medullary hyphae usually ornamented. **Photobiont** trebouxioid. **Ascomata** apothecia, lateral, very rare. **Thalline exciple** persistent to excluded, concolorous with thallus. **Asci** clavate, thick-walled, *Lecanora*-type, (6-)8-spored. **Ascospores** simple, ellipsoid, without distinctive perispore, colourless. **Conidiomata** pycnidia, usually rare. **Conidia** minute, cylindrical to fusiform. **Chemistry:** usnic acid absent, atranorin and chloratranorin frequent, β-orcinol depsidones predominating or lichen products not detected.

Distinguished from other alectorioid genera, even in the absence of apothecia, by the characteristic chemistry, cortical structure and the scarcity and inconspicuous nature of the pseudocyphellae.

LITERATURE: Brodo & Hawksworth (1977), Hawksworth (1972b), Holien (1989), Krog (1980).

1 Thallus erect or prostrate; lateral spinules usually present, often abundant ... 2
Thallus pendent or prostrate; lateral spinules usually absent or sparse ... 6

2(1) Apices usually paler than base of thallus ... 3
Apices ± concolorous with base of thallus ... 4

3(2) Soralia with tufts of spinules present; medulla Pd− ... **11. *smithii***
Soralia absent; medulla Pd+ red (fumarprotocetraric acid) ... **1. *bicolor***

4(2) Thallus Pd+ orange-red, K−, KC− (fumarprotocetraric acid) ... 5
Thallus Pd+ yellow, K+ yellow, KC+ red (barbatolic and alectorialic acids) **9. *nadvornikiana***

5(4) Soralia with tufts of spinules present; on trees ... **4. *furcellata***
Soralia absent; on the ground ... **10. *nitidula***

6(1) Thallus K+ yellow, KC+ red (barbatolic and alectorialic acids) ... 7
Thallus K+ orange-red or K−, KC− (barbatolic and alectorialic acids absent) ... 8

7(6) Lateral spinules frequent; soralia absent ... **9. *nadvornikiana***
Lateral spinules absent; soralia often present ... **2. *capillaris***

8(6) Thallus K− (norstictic acid absent); pseudocyphellae absent ... 9
Thallus K+ yellow to orange-red (norstictic acid); pseudocyphellae present, at least on basal branches ... **7. *implexa***

9(8) Main stems <0.5 mm diam.; soralia various; main stems ± terete, usually matt; medulla and soralia Pd+ red (fumarprotocetraric acid), occasionally Pd− ... 10
Main stems >0.5 mm diam.; soralia tuberculate; main stems often compressed, contorted and pitted, usually shiny; medulla usually Pd−; soralia Pd+ red (fumarprotocetraric acid) ... **3. *chalybeiformis***

10(9) Thallus pale to dark brown or black, the base paler than or concolorous with apices; soralia various ... 11
Thallus whitish grey to grey, sometimes pale brown, the base never paler than apices; soralia tuberculate ... **6. *subcana***

11(10) Main stems usually <0.25 mm diam., flexuose, dark brown to black, concolorous throughout; soralia rare, <0.2 mm diam., tuberculate ... **8. *lanestris***
Main stems usually >0.25 mm diam., brown to dark brown (never black), the base usually paler than apices; soralia rare or abundant, <0.75 mm diam., tuberculate or fissural ... **5. *fuscescens***

1. B. bicolor (Ehrh.) Brodo & D. Hawksw. (1977)

Thallus to 4 (-10) cm, erect, rarely decumbent, forming dark cushions; branches 0.2–0.7 mm diam., terete, occasionally compressed towards base; branching isotomic- to anisotomic-dichotomous (Fig. 5a); basal branches black or brownish black, the apices olive-grey to pale brown, shiny; lateral spinules 1.5–5 mm, numerous, with constricted bases, arising at right angles from main stems, often arcuate, usually simple; pseudocyphellae absent. Soralia absent; soredia occasionally present at ends of broken branches. Apothecia unknown in Britain. Medulla Pd+ red (often patchy), K−, KC−, C− (fumarprotocetraric acid).

Amongst mosses, sometimes on soil or rock in hilly areas; local and decreasing. N. & W. British Is. Europe, North America, Africa, Himalaya, Japan, Malaysia.

Distinguished from *B. smithii* **(11)** by the lack of soralia and associated clusters of spinules, and the Pd+ red medullary reaction, and from *Alectoria nigricans* by the smaller size, absence of conspicuous pseudocyphellae, presence of short, lateral, perpendicular, spinulose branches and by being K− and C−.

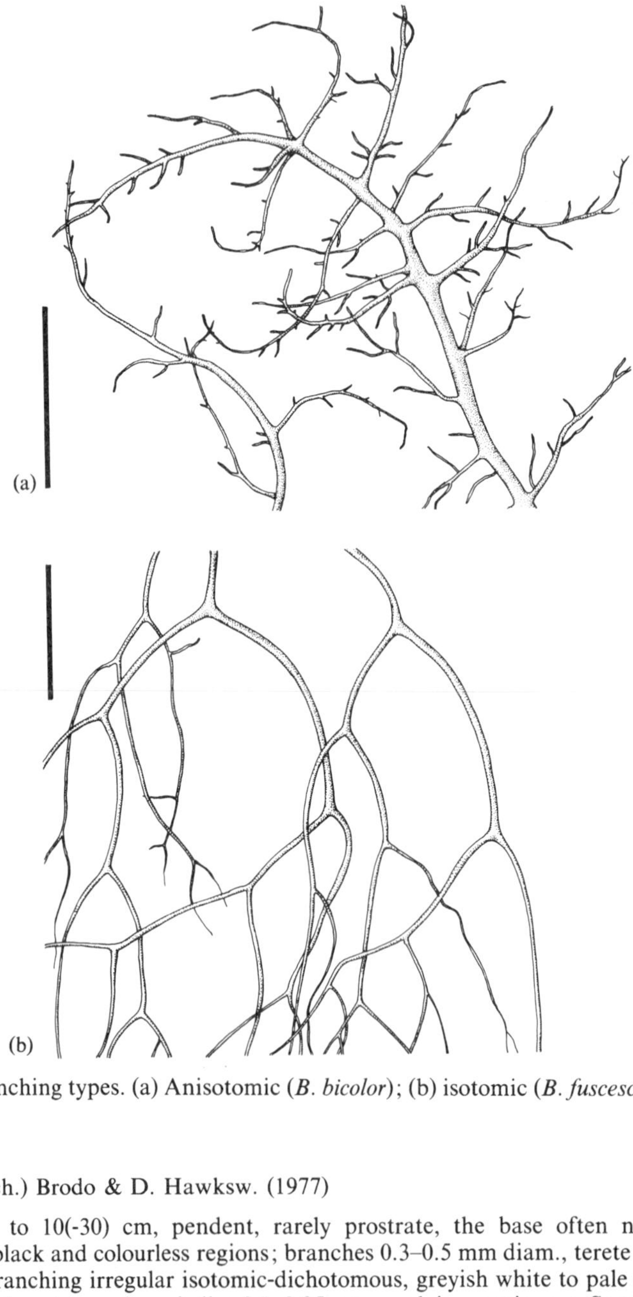

Fig. 5. *Bryoria* branching types. (a) Anisotomic (*B. bicolor*); (b) isotomic (*B. fuscescens*). Scale = 5 mm.

2. B. capillaris (Ach.) Brodo & D. Hawksw. (1977)

Thallus (Fig. 36j) to 10(-30) cm, pendent, rarely prostrate, the base often not persistent, often fragmenting with black and colourless regions; branches 0.3–0.5 mm diam., terete or ± compressed at axils, abundant; branching irregular isotomic-dichotomous, greyish white to pale to dark brown; true lateral spinules ± absent; pseudocyphellae 0.1–0.25 mm, oval, inconspicuous. Soralia to 0.5 mm diam., rare to abundant, or absent. Apothecia and pycnidia not observed in British material. Medulla Pd+ yellow, K+ yellow, KC+ red, C+ red or C− (barbatolic and alectorialic acids); soralia Pd+ orange-red (fumarprotocetraric acid).

On conifers and *Betula* in relict pinewoods; rare but locally abundant. C. & E. Scotland; formerly also in Ireland. Scandinavia, C. Europe, Pyrénées, Canary Is., North America, Japan.

Exhibits considerable variation in colour and abundance of soralia. Distinguished from all other British species (especially *B. subcana* **(6)**), apart from *B. nadvornikiana* **(9)**, which has spinulose lateral branches and dark main stems, by not being K− and KC−.

3. B. chalybeiformis (L.) Brodo & D. Hawksw. (1977)

Thallus to 10 cm, prostrate and entangled; branches 0.5–2 mm diam., terete or dorsiventrally compressed; branching sparse, irregularly isotomic-dichotomous, black or dark brown, usually ± concolorous, occasionally developing blackened areas or base slightly paler; lateral spinules and pseudocyphellae absent. Soralia to 1.5 mm diam., tuberculate, sparse or absent. Apothecia and pycnidia unknown. Cortex and medulla Pd−, K−, KC−, C− (lichen products not detected); soralia Pd+ red or Pd− (± fumarprotocetraric acid).

On non-calcareous, mossy rocks, rarely horizontal branches, and on the ground in hilly and montane districts; local and often wrongly recorded; rare, upland Britain. Europe, North America, Africa, ?Asia, ?Japan, Antarctica.

Distinguished from all other British taxa, apart from *B. fuscescens* var. *positiva* **(5)**, by the absence of lateral spinules and pseudocyphellae, its colour and chemical reactions. It is separated from the usually corticolous var. *positiva* by its thicker main stems, sparser branching and soralia, darker thallus colour, colour of basal parts, type of soralia and chemical reactions. The name is retained here in the sense of recent authors; a new typification is required to fix its application in this sense.

4. B. furcellata (Fr.) Brodo & D. Hawksw. (1977)

Thallus 3–5(-12) cm, tufted; branches 0.3–0.5 mm diam., terete, sometimes compressed towards base; branching regularly isotomic-dichotomous, pale to dark brown to black, ± shiny; lateral spinules usually present, occasionally abundant; pseudocyphellae absent. Soralia 0.3–1 mm, white, usually abundant, oval to fusiform, fissural, sometimes slightly raised, becoming covered with tufts of spinules, 0.5–2.5 mm. Cortex, medulla and soralia Pd+ red, K−, KC−, C− (fumarprotocetraric acid).

Known from decorticate *Pinus* trunks and *Betula* twigs in a few relict pine woodlands. Scotland (E. Inverness). C. & E. Europe, Fennoscandia, Madeira, North and Central America, Himalaya, Japan.

A well-delimited species, characterized by the uniform, pale brown, ± shiny, tufted thallus, the short, perpendicular lateral spinules and the soralia, which develop tufts of spinules. *B. smithii* **(11)**, which shares this last feature, has paler apices, lacks detectable lichen products, has a different ecology and is south-western in the British Is.

5. B. fuscescens (Gyelnik) Brodo & D. Hawksw. (1977)

Thallus 5–15(-30–65) cm, pendent or prostrate; branches to 0.5(-0.8) mm diam., terete, becoming compressed towards base and axils; branching irregularly isotomic-dichotomous (Fig. 5b), pale to dark brown, rarely blackish, the basal parts paler or concolorous with apices; black fragmentation regions occasionally present, ± matt; lateral spinules and pseudocyphellae absent. Soralia to 0.75 mm, abundant to scarce, rarely absent, tuberculate or fissural. Apothecia very rare; disc 1–1.5 mm diam., brown to dark brown. Thallus Pd+ red or Pd−; medulla Pd−; soralia Pd+ red, K−, KC−, C− (fumarprotocetric acid).

On conifers, acid-barked, broad-leaved trees, such as *Betula*, siliceous rocks, amongst mosses, on walls and worked timber; common in hilly areas. Throughout British Is., especially the north and east. Moderately sensitive to SO_2 pollution ($<55\ \mu g\ m^{-3}$). Europe, North America, Canary Isles, E. Africa.

The commonest species of the genus in Britain. Very variable, becoming contorted and spinulose in exposed situations, but generally easily recognizable by the tendency to have a pale base, fissure-like as well as tubercle-like soralia, the matt surface and smokey-brown colour; the soralia are always Pd+ red, but the thallus is Pd+ or Pd−. Most often confused with *B. chalybeiformis* **(3)** and *B. subcana* **(6)**.

var. *fuscescens*: Thallus pendent, pale brown, pale at base. Soralia abundant.

var. *positiva* (Gyelnik) Brodo & D. Hawksw. (1977): Thallus prostrate, dark brown, concolorous at the base. Soralia scarce or absent.

6. B. subcana (Nyl. ex Stitzenb.) Brodo & D. Hawksw. (1977)

Like **5**, but thallus much paler, grey-green to whitish grey, sometimes blackened towards base, and the soralia are always tuberculate.

On conifers and broad-leaved trees. N. & S. England, Wales, Scotland (locally frequent), Ireland (locally rare). W. Europe, W. North America.

Small specimens of *B. fuscescens* **(5)** may be difficult to distinguish from this species, but the paler colour, and rapid, intense Pd+ orange-red reaction of thallus, medulla and soralia (fumarprotocetraric acid) are usually diagnostic.

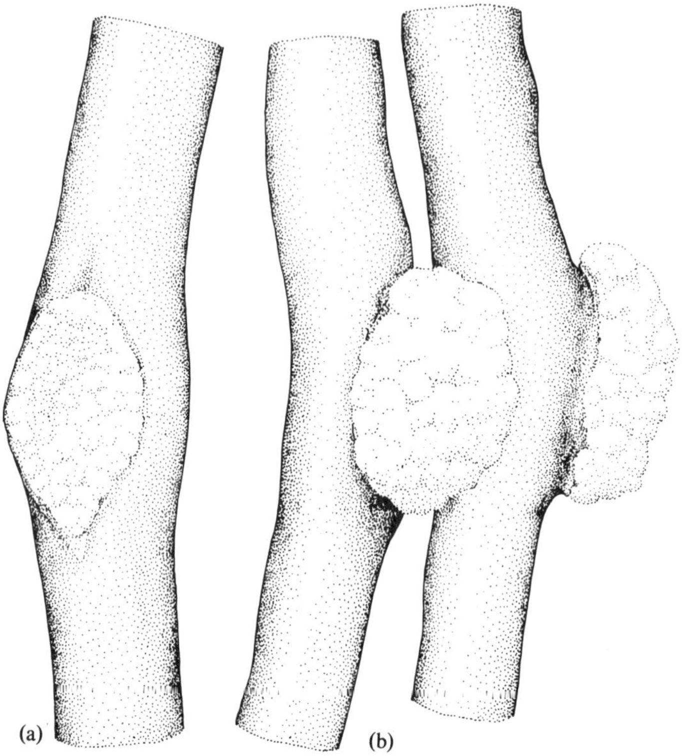

Fig. 6. *Bryoria* soralium types. (a) Fissural; (b) tuberculate.

7. B. implexa (Hoffm.) Brodo & D. Hawksw. (1977)

B. pseudofuscescens (Gyelnik) Brodo & D. Hawksw. (1977)

Thallus 5–10(-25) cm, pendent; branches 0.5–0.35 mm diam., terete; branching ± isotomic-dichotomous, pale brown to greyish, becoming dark brown or black, often olivaceous, especially towards base, dull or shiny. Soralia and true lateral spinules absent; pseudocyphellae to 0.5(-1.0) mm, usually abundant, white, short to elongate-fusiform. Apothecia and pycnidia unknown in British Is. Medulla Pd+ red, K+ yellow-red, KC−, C− (norstictic acid).

On coniferous trees. C. Scotland (Perthshire, Inverness, Angus). C. & N. Europe, North America.

Collected at three sites last century; possibly extinct in British Is. Distinguished from all other British species by the presence of norstictic acid. Recent research suggests that this species comprises several chemotypes; only the norstictic acid one (i.e. *B. pseudofuscescens*), not that with psoromic acid (i.e. *B. implexa* s.str.), has been found in the British Is.

8. B. lanestris (Ach.) Brodo & D. Hawksw. (1977)

Thallus 5–10(-15) cm, pendent; branches 0.1–0.25(-3.0) mm diam., thread-like, contorted, friable, irregular in diameter; branching irregularly isotomic-dichotomous, brownish black to black, usually concolorous; true lateral spinules and pseudocyphellae absent. Soralia to 0.2 mm diam., rare or absent, tuberculate type predominating. Apothecia and pycnidia unknown in British Is. Cortex and medulla Pd−, K−, KC−, C−; soralia Pd+ red (fumarprotocetraric acid).

On coniferous wood and palings; rare. N. England, Scotland. Europe, North America.

A distinctive species unlikely to be confused with other British *Bryoria*, readily separated from *B. fuscescens* **(5)** by its darker colour, wire-wool appearance (fragmenting in packets), friable texture, usually concolorous thallus, the absence of conspicuous fissural soralia and much narrower main stems and branches, which are characteristically uneven in diameter.

9. B. nadvornikiana (Gyelnik) Brodo & D. Hawksw. (1977)

Thallus 5–15 cm, erect, becoming pendent; branches 0.2–0.3 mm diam., terete or rarely ± compressed at axils; branching isotomic-dichotomous towards base, anisotomic-dichotomous towards apices; basal branches black; apical branches pale greyish green to pale brown or olivaceous brown; lateral spinules arising at right angles, frequent; pseudocyphellae absent. Soralia, apothecia and pycnidia not observed in British material. Medulla Pd+ orange-red, K+ yellow, KC+ red, C− or C+ pink (barbatolic and alectorialic acids).

Known only from a few sandstone crags, where it is associated with *Usnea filipendula*. N. England (Northumberland-Cumbria border). N. & C. Europe, North America, Hawaii, E. Africa, Himalaya, Japan.

Apart from *B. capillaris* **(2)**, which lacks lateral spinules and the basal darker branches, is the only species of *Bryoria* in Britain which reacts KC+ red and K+ yellow.

10. B. nitidula (Th.Fr.) Brodo & D. Hawksw. (1977)

Thallus to 5(-8) cm, erect, sometimes decumbent, dying from the base; branches 0.5–0.8 mm diam., terete; branching anisotomic-dichotomous, pale brown to black throughout, shiny; lateral spinules up to 2.5 mm, with slightly constricted bases, arising sparsely from main stems, concolorous with thallus; pseudocyphellae and soralia absent. Apothecia and pycnidia unknown. Cortex and medulla Pd+ orange-red, K−, KC−, C− (fumarprotocetraric acid).

Characteristic of arctic and subarctic heaths, particularly of 'snow-patch' communities. C. Scotland (Cairngorm Mtn). Collected on 2 separate occasions last century; probably extinct in British Is. Arctic and subarctic, absent from Alps and C. European mountains.

11. B. smithii (Du Rietz) Brodo & D. Hawksw. (1977)

Thallus 4–7 (-12) cm, decumbent; branches *c.* 1 mm diam., frequent, terete or slightly compressed towards base, frequently with depressions; branching isotomic-dichotomous, the basal branches dying, dark brown to black, the apices paler, olivaceous to pale brown, exceptionally concolorous with main stems; lateral spinules 2–7 mm, with slightly constricted bases, numerous, arising at right angles to main stems, frequently arcuate. Soralia present, arising from pseudocyphellae, fissural, concave to flat; soredia often sparse, mixed with short spinules arising in tufts from the soralia. Apothecia and pycnidia unknown. Medulla and soralia Pd−, K−, KC−, C− (lichen products not detected by t.l.c.).

On ancient trees of *Quercus robur* in sheltered, but rather well-lit situations; very rare. S.W. England (Devon, Dartmoor). Formerly collected in N. Wales (Merioneth) and S.C. Scotland (Perthshire) but probably extinct there. S. Scandinavia, C. Europe, Himalaya, China, Malaysia, Hawaii.

A well-defined species, characterized by its soralia mixed with short spinules, and its Pd− reaction; it most resembles *B. bicolor* **(1)**, which lacks soralia, has a medulla which is Pd+ red, and is rarely on trees.

BUELLIA de Not. (1846)

Hafellia Kalb, Mayrh. & Scheid. (1986)

Thallus crustose, smooth, rimose, areolate, granular, or slightly placodioid, sometimes immersed, whitish to grey, brown or yellowish, often delimited by a dark prothallus and mosaic-forming, corticate or not. **Photobiont** chlorococcoid. **Ascomata** apothecia, immersed, emergent or superficial from the beginning, black, sometimes white-pruinose. **Thalline exciple** absent, or in immersed apothecia not distinguishable from the thallus (cryptolecanorine). **True exciple** thin, pale and inconspicuous to well developed and dark-pigmented. **Epithecium** brown to olive-green, K−, N− or N+ reddish. **Hymenium** colourless, or greenish in upper part, I+ blue, with or without numerous oil droplets that do not dissolve in K. **Hypothecium** pale or more usually dark brown, sometimes olivaceous in part. **Hamathecium** of paraphyses, septate, simple or branched in upper third, apices swollen and pigmented, many with a dark brown cap. **Asci** (4-)8(-many)-spored, clavate, *Lecanora*-type. **Ascospores** brown, 1-septate, rarely 3-septate or submuriform, ellipsoid, oblong or fusiform, straight or curved; wall of uniform thickness, or thickened at the septum (rarely also subapically), sometimes thinner and paler at the apices, surface smooth or finely ornamented (use oil immersion), without a distinct perispore. **Conidiomata** pycnidia, immersed; wall dark brown at least in upper part. **Conidia** simple, colourless, ellipsoid to oblong, bacilliform, fusiform-cylindrical, or thread-like and curved. **Chemistry**: paradepsides, tridepsides, depsidones, xanthones and unidentified substances reported. **Ecology**: on a wide range of substrata, especially siliceous rocks and acid stonework, bark, wood, soil, bryophytes and lichens. A large, cosmopolitan genus of *c.* 400 species.

Rinodina species differ in having a distinct thalline exciple, although in some species this is concolorous with the disc and so evident only in section. Species of *Rinodina* have a pale (usually colourless) hypothecium and usually much pronounced wall thickenings in the ascospores. *Diplotomma* differs in that the ascospores are 3-septate to muriform and distoseptate as in *Rhizocarpon*; the apothecia are also often densely pruinose. *Diploicia* is distinguished by the distinctly placodioid thallus. Similar, but unrelated crustose genera containing species with discoid apothecia and brown, 1- to 3-septate or submuriform ascospores are *Arthonia*, *Catolechia*, *Epilichen*, *Melaspilea*, *Orphniospora* and *Rhizocarpon*, as well as the lichenicolous or non-lichenized *Abrothallus*, *Dactylospora* and *Rhizodiscina*; all these genera have different ascus structures.

British material of *B. geophila* (Flörke ex Sommerf.) Lynge (1937) is corticolous *Diplotomma alboatrum*. *B. succedens* (Nyl.) Arnold (1870), originally found on rocks in Scotland (Perth, Ben Lawers), is excluded as neither authentic material nor a reliable description of this species has been seen; it was reported to have a thin, white thallus and 1- to 3-septate ascospores, 9.5–18 × 4–6 μm.

Tentatively assigned here is the recently described sterile sorediate crust, ***B. violaceofusca*** Thor & Mhur (1991). It has a wide-spreading, ± immersed, white to pale grey thallus and numerous, slightly elevated, often confluent soralia which are dark brown with a ± violet tinge (greenish when abraded), and 0·15–0·6 mm diam. Thallus and soralia Pd –, K –, KC –, C –, UV – (lichen products not detected by t.l.c.). On dry, fissured bark of mature *Fraxinus* and *Quercus* at woodland edge or in pasture woodland, rare. Scotland (Peebles, E. Sutherland). C. Sweden.

LITERATURE: Schauer (1969), Scheidegger (1987), Sheard (1964).

1 Thallus not immersed in lichens or on crustose lichens 2
Thallus immersed in that of *Physconia* .... **16. *pulverulenta***

2(1) Thallus with diffuse soralia, or soredia absent 3
Thallus with delimited soralia .... **9. *griseovirens***

3(2) On bark, woody stems, wood, or old leather 4
On rock, soil, or on bryophytes over soil or rock 10

4(3) Thallus smooth, cracked or warty, KC – 5
Thallus powdery-granular, KC + red .... **15. *pulverea***

5(4) Hymenium with numerous oil droplets; ascospore wall thinner and paler at apices 6
Hymenium without numerous oil droplets; ascospore wall not thinner at apices 8

6(5) Ascospores 25–32 × (9-)10–14 μm; hymenium 100–130 μm tall 7
Ascospores (13-)17–26(-30) × (6.5-)7–10(-13) μm; hymenium less than 100 μm tall .... **6. *disciformis***

7(6) Thallus K + yellow; paraphyses apices 3–4.5 μm wide .... **3. *arnoldii***
Thallus K + yellowish red (crystals); paraphyses apices 2–3 μm wide .... **18. *sanguinolenta***

8(5) Thallus K – ; apothecia 0.2–0.6 mm diam. 9
Thallus K + yellow or K + yellowish red; apothecia 0.4–1.2 mm diam. .... **7. *erubescens***

9(8) Ascospores mostly <11 × 4.5 μm .... **20. *schaereri***
Ascospores mostly >11 × 5 μm .... **17. *punctata***

10(3) Apothecia remaining immersed 11
Apothecia superficial at maturity 16

11(10) Thallus whitish, grey or grey-brown, C− or C+ fleeting red 12
Thallus yellowish, C+ persistent orange .... **14. *ocellata***

12(11) Thallus C−; ascospore surface smooth or faintly warted 13
Thallus C+ red (in sections); ascospore surface striate .... **24. *uberior***

13(12) Thallus K+ yellowish red (crystals) 14
Thallus K− or K+ yellow 15

14(13) Apothecia not pruinose; ascospores >12 μm long, finely warted; medulla I+ blue or I− .... **2. *aethalea***
Apothecia pruinose; ascospores 8.5–10.5 μm long, smooth; medulla I− ...... **13. *meiosperma***

15(13) Thallus whitish, K+ yellow; medulla I−; ascospores (8-)10–13(-15) × 5–7(-8.5) μm .... **22. *stellulata***
Thallus whitish or pale to dark grey, K−; medulla I− or I+ blue; ascospores (9.5-)12–20 × (4.5-)6–10(-12) μm .... **2. *aethalea***

16(10) On soil, turf, or on bryophytes over rocks 17
On rock 18

17(16) Thallus with marginal lobes; asci 4-spored .... **4. *asterella***
Thallus without marginal lobes; asci 8-spored .... **10. *insignis***

18(16) Thallus C− 19
Thallus C+ red .... **19. *saxorum***

19(18) Thallus K− or K+ yellow 20
Thallus K+ yellowish red (crystals) .... **23. *subdisciformis***

20(19) Hymenium with numerous oil droplets 21
Hymenium without numerous oil droplets 22

21(20) Ascospores 14.5–23.5 × 6.5–11.5 μm, with subapical and median thickenings **12. *leptoclinoides***
Ascospores (10-)11.5–15 × 5–8 μm, with median thickening only .... **8. *excelsa***

22(20) Thallus well-developed and K−, or inconspicuous; medulla I- 23
Thallus well-developed, K+ yellow-orange; medulla I+ bluish........ **21. *spuria***

23(22) Ascospores >5.5 μm wide; thallus usually conspicuous 24
Ascospores 4.5–5.5 μm wide; thallus inconspicuous .... **1. *abstracta***

24(23) Thallus of scattered to often aggregated, brown areoles that are often warty or ± lobed; ascospores mostly 7.5–9 μm wide .... **5. *coniops***
Thallus smooth, cracked, warty, or of unlobed areoles, whitish, grey or brown; ascospores mostly 6–8 μm wide .... **17. *punctata***

1. B. abstracta (Nyl.) H. Olivier (1903)

Thallus immersed and inconspicuous; medulla I−. Apothecia 0.1–0.2 mm diam., contiguous, often forming lines or irregular clusters; disc flat to slightly convex; margin persistent, thin, raised and wavy-crenulate. Ascospores 11–13 × 4.5–5.5 μm, 1-septate, narrowly oblong, surface minutely pored or wrinkled. Pycnidia unknown. Chemistry not tested.

On granite pebbles and small stones; rare. Is of Scilly. Mediterranean.

Differs from *B. punctata* **(17)** in the inconspicuous thallus and narrower ascospores. Resembles *Lecidea sarcogynoides* in habit.

2. B. aethalea (Ach.) Th. Fr. (1874)

Thallus thin to rather thick, distinctly areolate, often less than 2 cm diam. and mosaic-forming; areoles 0.2–0.6(-1.2) mm wide, ± angular, flat, or rarely convex, pale to dark grey, sometimes discoloured brown; delimited by a black prothalline line, the blackish prothallus sometimes visible between the areoles; medulla I+ blue or I−. Apothecia 0.2–0.4(-0.7) mm diam., immersed, sometimes surrounded by a thalline collar; disc flat; true exciple thin or inconspicuous, usually dark olivaceous; epithecium brown to olivaceous, N± reddish; hypothecium pale to dark brown. Ascospores (9.5-)12–18(-20) × (4.5-)6–10(-12) μm, 1-septate, minutely warted. Pycnidia rare; conidia 5.5–8.5 × 1 μm, bacilliform. Thallus Pd± yellow-orange, K+ yellowish red (crystals), C− (norstictic and ± stictic acids), or Pd−, K−, C− (no lichen products detected by t.l.c.).

On acid rocks, pebbles and stonework, often a pioneer species, sometimes in nutrient-rich situations, rarely on timberwork; common. Throughout British Is. Europe, North America.

Specimens with an I− medulla are sometimes referred to as *B. soraria* Th. Fr. (1874), but a broad concept of *B. aethalea* is used here pending a critical revision. Specimens with a weak or negative K reaction are often mistaken for *B. stellulata* **(22)**, but that species has a different chemistry (t.l.c.), smaller ascospores, and usually a whiter thallus. Also of similar habit are *B. meiosperma* **(13)** — apothecia pruinose, ascospores smaller and smooth; *B. ocellata* **(14)** — thallus yellowish, C+ persistent orange; and *B. uberior* **(24)** — thallus C+ red.

3. B. arnoldii Servít (1931)

Thallus thin to moderately thick, smooth, wrinkled or rimose; medulla I−. Apothecia 0.4–1 mm diam.; disc flat; true exciple persistent, shiny; hymenium 100–130 μm tall, with numerous oil droplets. Paraphyses 1.5–2 μm wide; apices 4–4.5 μm wide. Ascospores 25–32 × (9-)12–15 μm, 1-septate, often strongly curved; apices pointed, thin-walled and pale; wall internally thickened at the septum. Conidia 3–3.5 × 1–1.5 μm, narrowly ellipsoid to oblong. Thallus Pd± yellowish, K+ yellow, C− (atranorin).

On smooth conifer bark. N. Scotland (Inverness, W. Ross). Sweden, Alps, North America.

Differs from *B. disciformis* **(6)** in the larger ascospores and shorter conidia, and from *B. sanguinolenta* **(18)** in the K reaction and broader paraphyses.

4. B. asterella Poelt & M. Sulzer (1974)

Thallus white, matt, containing numerous crystals, with short marginal lobes, to 2 mm long and 0.4–1 mm wide, forming rosettes when young but often becoming irregular and fragmented when older; medulla I−. Apothecia 0.4–0.8 mm diam., at first immersed with a thalline collar, later superficial; disc flat to strongly convex, ± pruinose when young; true exciple narrow or excluded; hymenium 50–70 μm tall, without oil droplets. Asci 4-spored, (rarely more). Ascospores 10–17 × 5–8 μm, 1-septate, finely warted, apices rounded. Thallus Pd+ yellow-orange or −, K+ yellowish red or K−, C− (± norstictic acid).

On dry, basic soil. Very local and decreasing. E. England (E. Anglia, Breckland). Norway, France, Germany, Switzerland.

5. B. coniops (Wahlenb. ex Ach.) Th. Fr. (1861)

Thallus areolate; areoles 0.1–0.3 mm diam., pale grey-brown to dark brown, convex, sometimes becoming shortly lobed, often aggregating to form a warty, secondarily rimose crust up to 1–2 mm thick in the thallus centre; a dark prothallus sometimes evident when areoles are dispersed; medulla I−. Apothecia 0.3–0.6 mm diam., superficial; disc flat to slightly convex; true exciple usually persistent; epithecium brown, N−; hymenium without oil droplets. Paraphyses *c.* 1.5 μm wide, apices up to 4–5 μm wide. Ascospores 13–18 × (7-)7.5–9(-10) μm, 1-septate, wall uniformly thickened, minutely warted. Conidia 13–20 × 0.8–1 μm, thread-like, curved. Thallus Pd−, K−, C− (no lichen products detected by t.l.c.).

On maritime rocks, especially those manured by birds. Mainly N. & W. Scotland, extending to S.W. England (Cornwall), W. Wales, Ireland. Scandinavia, Iceland, North America, and also in the Southern Hemisphere.

Distinguished from *B. punctata* **(17)** by the thicker, more warted thallus, and generally larger ascospores.

Superficially similar is *B. badia* (Fr.) Massal. (1853). It has an initially lichenicolous, dark brown, areolate to subsquamulose thallus, ascospores 10–15(-17) × 7–8.5 μm, and bacilliform conidia, 4.5–5 × 1 μm. Recently discovered on red-tiles, associated with *Parmelia mougeotii* and *P. verruculifera*. E. Suffolk and S. Wiltshire. Europe, N. Africa, Arabia.

6. B. disciformis (Fr.) Mudd (1861)

Thallus immersed or sometimes ± superficial, thin and almost smooth to thickish, rimose and somewhat warted, whitish to yellowish grey, often delimited by a blackish prothallus; medulla I−. Apothecia 0.3–1.3 mm diam., superficial; disc flat to slightly convex; true exciple usually persistent; epithecium brown, N−; hymenium with numerous oil droplets. Ascospores (13-)17–26(-30) × (6.5-)7–10(-13) μm, 1(-3)-septate, sometimes slightly curved, wall thinner and paler at the pointed apices. Conidia 5.5–8.5 × 0.7–1 μm, straight. Thallus Pd± yellowish, K+ yellow, C− (atranorin, traces of two or more unidentified lichen products).

On smooth bark; frequent. N. & W. Britain, Ireland; extending locally elsewhere. Circumpolar in Northern Hemisphere.

See also *B. arnoldii* **(3)**, *B. erubescens* **(7)**, and *B. sanguinolenta* **(18)**. Often confused with morphs of *Lecidella elaeochroma* with a white thallus. *B. arborea* with similar apothecia and a sorediate thallus is discussed under *B. griseovirens* **(9)**.

7. B. erubescens Arnold (1873)

Thallus immersed to thinly superficial, whitish to pale grey, smooth to rimose or slightly wrinkled, usually delimited by a blackish prothallus; medulla I−. Apothecia 0.4–1.6 mm diam., sessile; disc flat to slightly convex, usually pruinose, at least when young; margin prominent and usually persistent; hymenium without oil droplets. Ascospores 14–20 × 6–9 μm, 1-septate, apices rounded, wall evenly pigmented, smooth to finely warted. Conidia 6–8 × 0.5–1 μm, straight. Thallus Pd+ yellow-orange, K+ yellowish red (crystals), C− (atranorin, norstictic acid and traces of stictic acid complex), rarely Pd± yellowish, K+ yellow (lacking norstictic acid).

On ± smooth bark, often confined to ancient woodlands; local. Great Britain, S.W. Ireland. W. & C. Europe, Azores, North America.

Distinguished from *B. disciformis* **(6)** by the absence of hymenial oil droplets, the rounded, evenly pigmented ascospore apices, and the normally K+ red (crystals) thallus.

8. B. excelsa (Leighton) A. L. Smith (1911)

B. punctata var. *excelsa* (Leighton) Sheard (1964)

Thallus whitish, areolate; areoles variable in size, dispersed or forming a continuous, warted crust; prothallus inconspicuous; medulla I−. Apothecia superficial; disc flat to slightly convex; true exciple persistent; epithecium brown, N−; hymenium with numerous oil droplets. Ascospores (10-)11.5–15 × 5–8 μm, 1-septate; wall smooth to faintly roughened. Thallus Pd−, K+ yellowish, C− (atranorin).

On schistose rocks. W. Ireland (W. Galway, Mt. Doughruagh). Denmark.

Distinguished from *B. punctata* **(17)** by the whitish, K+ yellowish thallus and numerous hymenial oil droplets.

9. B. griseovirens (Turner & Borrer ex Sm.) Almb. (1952)

Thallus ± immersed and thin to thick, grey, smooth to wrinkled or rimose; a grey-black prothallus sometimes evident (especially on smooth bark). Soralia mostly discrete, and 0.15–0.4 mm diam., often crowded and few becoming confluent, pale green-grey to ash-grey but pale yellowish when abraded; soredia 15–26 μm diam., pigmented hyphae brown, K−, N−. Medulla I−. Apothecia very rare, 0.4–1 mm diam., superficial; disc flat to convex; hymenium 110–120 μm tall, without oil droplets. Ascospores (13-)15–28 × 7–13 μm, submuriform. Soralia Pd+ yellow or yellow-orange, K+ yellowish red (crystals), C− or C+ yellowish (atranorin, norstictic acid and traces of other substances of the stictic acid complex).

On ± smooth bark of deciduous trees and shrubs, less often on conifers, wood, and rarely on brick, usually in well-lit situations, tolerant of moderately polluted areas; frequent. Throughout British Is. Cosmopolitan in the N. & S. temperate zones.

A common and variable species. The K reaction is often weak and production of red crystals should be checked in squash preparations. Often confused with *Mycoblastus sterilis* and *Xylographa vitiligo*. A similar (when sterile) species with Pd− soralia (atranorin and placodialic acid) is known on wood and bark of old pines in the Scottish pinewoods and S. Norway; when fertile it has apothecial characters similar to *B. disciformis*; it has recently been described as ***B. arborea***, Coppins & Tønsb. (1992).

10. B. insignis (Naeg. ex Hepp) Th. Fr. (1860)

Thallus thin to moderately thick and somewhat warted; prothallus indistinct; medulla I−. Apothecia 0.5–1.1 mm diam., superficial; disc flat at first, later convex, ± pruinose when young; hymenium *c.* 100 μm tall, ± with oil droplets. Ascospores 21–32 × 8–13 μm, 1-septate, sometimes curved; wall paler and thinner at the apices, smooth to finely warted. Thallus Pd− or Pd+ yellowish, K+ yellow, C+ yellow (atranorin and unidentified substance).

Overgrowing mosses; very rare. N.C. Scotland (Perth, Ben Lawers). Scandinavia, Alps, North America.

11. B. leptocline (Flotow) Massal. (1854)

Thallus very thin, to thick and rimose-areolate, grey; areoles ± angular, 0.2–1 mm wide; prothallus indistinct; medulla I+ blue. Apothecia 0.4–1.2(-1.5) mm diam., superficial; disc flat to slightly convex; margin prominent and persisent; true exciple orange-brown, effusing an orange solution with K; epithecium brown, N−; hymenium 80–100 μm tall, without oil droplets. Ascospores 12–16(-18) × (6-)7–8.5(-11) μm, 1-septate; wall faintly warted, pigmentation often thickened at the septum. Pycnidia few; conidia 4 × 1 μm. Thallus Pd−, K+ yellowish, C− (atranorin).

On hard, exposed siliceous rocks, upland. N.W. England (Lake District), N. Wales, N. Scotland (Highlands). Europe, North America.

Easily mistaken superficially for species of *Lecidea* or *Porpidia*. *B. saxorum* **(19)** and *B. subdisciformis* **(23)** have different chemistries and usually more abundant pycnidia.

12. B. leptoclinoides (Nyl.) Steiner (1907)

Hafellia leptoclinoides (Nyl.) Scheid. & Mayrhofer (1986)

Thallus rimose, pale yellowish grey; medulla I−. Apothecia 0.45–0.8 mm diam., superficial; disc flat to slightly convex; epithecium brown, N−; hymenium with numerous oil droplets. Ascospores 14.5–23.5 × 6.5–11.5 μm, 1(-3)-septate, sometimes slightly curved, smooth to faintly warted; wall thinner and paler at apices. Thallus Pd+ yellowish, K+ yellowish, C− (atranorin, placodiolic acid).

On maritime rocks in the xeric-supralittoral, often in recesses; locally abundant. S.W. England, E. Ireland. Pyrénées.

13. B. meiosperma (Nyl.) Müll. Arg. (1887)

Thallus rimose to areolate, whitish; medulla I−. Apothecia 0.2–0.3 mm diam., immersed; disc pruinose; epithecium brown. Ascospores 8.5–10 × 5–6.5 μm, 1-septate, not constricted at the septum, wall smooth. Thallus Pd+ yellowish red, K+ yellowish red (crystals), C− (norstictic acid).

On maritime rocks, mostly sandstones. S. England, Channel Is, Scotland (Wigtown, Mull of Galloway). Spain, Italy.

Distinguished from *B. aethalea* **(2)** by the pruinose apothecia, and smaller, smooth-walled ascospores.

14. B. ocellata (Flotow) Körber (1855)

B. verruculosa auct., non (Sm.) Mudd (1861)

Thallus forming small patches 1–2(-5) cm diam., thin to thickish, yellowish grey, areolate (Fig. 36b), sometimes delimited by a black prothallus; areoles 0.3–1.3 mm wide, flat to convex; medulla I−. Apothecia 0.15–0.5 mm diam., immersed, sometimes surrounded by a thalline collar; disc concave to flat; true exciple thin or inconspicuous; epithecium brown to usually olivaceous, N+ reddish; hymenium without oil droplets; hypothecium brown, but greenish in upper part. Ascospores (11-)13–17(-21.5) × 6.5–9(-11.5) μm, 1-septate; apices rounded; wall uniformly thickened, finely warted. Conidia 4.5–5.5 × 0.8–1 μm. Thallus Pd−, K−, C+ persistent orange (xanthones, especially arthothelin).

On exposed acid rocks, pebbles and stonework. Throughout British Is. Europe, North America.

Similar in habit to, and often occurring with, *B. aethalea* **(2)**, but distinguished by the yellowish, C+ orange thallus.

15. B. pulverea Coppins & P. James (1978)

Thallus effuse, powdery granular, thin to thick, dull grey-green to dark brown-grey, often with irregular patches of soredia; soredia 12–22 μm diam., I−. Apothecia 0.2–0.4(-5) mm diam., usually at least a few present, immersed among the granules or sometimes emergent; disc flat to slightly convex; true exciple thin, ± persistent; epithecium brown, N−. Ascospores 12–14(-17) × (5-)6–7(-8) μm, 1-septate, straight or slightly curved; wall uniformly thickened, smooth. Thallus Pd+ yellow, K−, KC+ red, C+ pinkish (alectorialic acid).

On exposed, acid bark, wood and *Calluna* stems, often associated with *Lecanora conizaeoides* and *Scoliciosporum chlorococcum*, tolerant of sulphur dioxide and flouride pollution. N. & W. Britain. Belgium, Luxembourg.

Resembles *B. punctata* **(17)**, but differs in the powdery, sorediate thallus and chemical reactions; stains herbarium paper pink-brown in time.

16. B. pulverulenta (Anzi) Jatta (1900)

Thallus developed within the medulla of the host, with an independent cortex 10–20 μm tall and its own photobiont cells. Apothecia 0.3–0.5(-7) mm diam., superficial; hymenium 55–77 μm tall. Paraphyses 1.6–2.3 μm wide; apices to 5 μm wide. Ascospores (14.5-)16–21(-23) × 6.5–8.5 μm, 1- to 3-septate, finely warted.

On *Physconia distorta*; rare. N.C. Scotland (Mid-Perth). Europe, Iraq, North America.

A lichenicolous lichen living in the thallus of the *Physconia* which sometimes becomes ± bleached in infected areas. See also note regarding *B. badia* under *B. coniops* **(5)**.

17. B. punctata (Hoffm.) Massal. (1852)

Thallus thin or inconspicuous to thickish, rarely more than 0.5 mm thick in the centre, smooth to rimose and often warted, pale to dark grey, rarely brown; a dark or grey prothallus rarely present; medulla I−. Apothecia 0.2–0.6 mm diam., superficial; disc flat to slightly convex, rarely strongly so; true exciple thin, persistent, sometimes later excluded; epithecium brown, N−; hypothecium pale to dark brown; hymenium without oil droplets. Ascospores (8.5-)11.5–16(-19.5) × (4.5-)6–8(-10.5) μm, 1-septate, ± slightly curved; wall uniformly thickened, smooth to usually finely warted. Pycnidia sparse; conidia 14–20 × 0.5–1 μm, curved-thread-like. Thallus Pd−, K−, C− (no lichen products detected by t.l.c.).

On ± nutrient-rich or -enriched bark, wood, rocky substrata and debris including fence posts, birds' perching rocks; tolerant of sulphur dioxide pollution and inorganic fertilizers; abundant. Throughout British Is. Cosmopolitan.

Very variable. Many records on rock refer to *Rinodina lecideina*. Other similar, but rarer, species on rocks include *B. abstracta* **(1)**, *B. coniops* **(5)**, *B. excelsa* **(8)**, *B. spuria* **(21)**; and on bark or wood, *B. pulverea* **(15)** and *B. schaereri* **(20)**. *B. punctata, B. coniops* and *Rinodina lecideina* have long, curved conidia and should probably be assigned to a separate genus.

18. B. sanguinolenta Schauer (1965)

Thallus thin, ± immersed, whitish grey, sometimes delimited by a blackish prothallus; medulla I−. Apothecia (0.3-)0.4–0.8 mm diam., superficial; disc flat to convex; true exciple thin, mostly persistent; hymenium *c.* 120 μm tall, with numerous oil droplets. Paraphyses 1–1.4 μm wide; apices to 2–3 μm wide. Ascospores 25–32 × 10–14 μm, 1-septate, often curved, thin-walled and paler at the pointed apices; inner wall thickened at the septum. Pycnidia usually few; conidia 3.5–4 × 1–1.5 μm, oblong-ellipsoid. Thallus Pd+ yellow-orange, K+ yellowish red (crystals), C− (norstictic acid).

On smooth bark of *Sorbus*; very rare. N.C. Scotland (Mid-Perth). E. Alps.

Distinguished from *B. arnoldii* **(3)** and *B. disciformis* **(6)** by the K reaction, and from the latter, also by the larger ascospores.

19. B. saxorum Massal. (1852)

Thallus moderately thick, rimose, white to pale yellowish grey, delimited by a black prothallus and often mosaic-forming; medulla I+ blue. Apothecia 0.3–1 mm diam., superficial; disc flat to slightly convex; true exciple prominent; epithecium brown, N−; hymenium without oil droplets. Ascospores 10–18 × 5–9 μm, 1-septate, wall uniformly thickened, smooth. Pycnidia numerous; conidia 7–9 × *c.* 1 μm; fusiform-cylindrical. Thallus Pd± yellowish, K+ yellow, C+ red (atranorin, gyrophoric acid).

On exposed, hard, siliceous rock of standing stones and sarsens; local. S. England. Europe.
Similar to *B. subdisciformis* **(23)** but differing in chemistry and ecology. *B. leptocline* **(11)** is C−, has few pycnidia and shorter conidia.

20. B. schaereri de Not. (1846)

Thallus thin, continuous to slightly cracked or wrinkled, pale grey, effuse; medulla I−. Apothecia 0.2–0.3(-0.5) mm diam., superficial; disc flat to convex; true exciple thin to excluded; epithecium brown, N−; hymenium 35–60 μm tall, without oil droplets. Ascospores 6–10(-12) × 2.5–4(-4.5) μm, 1-septate; wall uniformly thin-walled, smooth. Pycnidia often numerous, 50–70 μm diam.; conidia 2–3 × 1–1.4 μm, shortly oblong to ellipsoid. Thallus Pd−, K−, C− (no lichen products detected by t.l.c.).
On acid bark and wood, especially of conifers and *Quercus*; local. N. & W. Britain, rarely elsewhere (S. to Devon). N. & C. Europe, North America.
Distinguished from *B. punctata* **(17)** by the smaller, thinner-walled ascospores and shorter, ± oblong conidia.

21. B. spuria (Schaerer) Anzi (1860)

Thallus moderately thick, granular or cracked-areolate, whitish; prothallus indistinct; medulla I+ faint bluish. Apothecia 0.4–1 mm diam., superficial; disc flat to convex; true exciple thin, mostly persistent; epithecium brown to olivaceous, N+ pale red; hymenium without oil droplets. Ascospores 9–15 × 4.5–6.5 μm, finely warted. Pycnidia usually few; conidia 4.5–5.5 × 1–1.5 μm. Thallus Pd+ yellow-orange, K+ yellow-orange, C− (atranorin, stictic acid, traces of substances of the stictic acid complex).
On sandstone. Wales. C. Europe.
Distinguished from *B. punctata* **(17)** by the whiter thallus, Pd and K+ positive reactions and different conidia. Closely resembles *Lecidella carpathica* in habit.

22. B. stellulata (Taylor) Mudd (1861)

Thallus thin, whitish to pale grey, areolate, mostly 1–2 cm diam., delimited by a black prothallus and often forming mosaics; areoles 0.2–0.5 mm wide, flat; medulla I−. Apothecia 0.2–0.36 mm diam., immersed; true exciple thin, evident at least when young; epithecium dark olivaceous to olive-brown, N+ reddish; hymenium without oil droplets. Ascospores (8-)10–13(-15) × 5–7(-8.5) μm, wall evenly thickened, minutely roughened. Pycnidia rare; conidia 3.5–5 × 1–1.5 μm. Thallus Pd± yellowish, K+ yellow, C− (atranorin, confluentic and 2'-*O*-methylperlatolic acids).
On hard rocks and pebbles, mainly maritime in the xeric-supralittoral. N. & W. Britain, Ireland. N. & C. Europe.
Has been much over-recorded, especially in inland localities, owing to confusion with *B. aethalea* **(2)**.

23. B. subdisciformis (Leighton) Vainio (1890)

Thallus thin to thick, rimose-areolate, areoles 0.2–1.2 mm wide; (Fig. 36c), sometimes becoming warted, pale to medium-grey, sometimes with a yellowish tinge, usually delimited by a blackish prothallus, often mosaic-forming; medulla I−. Apothecia 0.4–1(-1.8) mm diam., superficial; disc flat to slightly convex; true exciple prominent, mostly persistent; disc and margin grey-white pruinose at least when young; epithecium brown, N−; hymenium without oil droplets. Ascospores 9.5–19.5 × 5.5–10 μm, 1-septate; apices rounded; wall uniformly thickened, finely warted. Pycnidia numerous; conidia 6–9.5 × 0.8–1 μm, straight. Thallus Pd+ yellow-orange, K+ yellowish red (crystals), C− (atranorin, norstictic acid, traces of other substances of the stictic acid complex).
On hard, siliceous, mainly maritime rocks in the xeric-supralittoral; locally abundant. S.W. & W. Britain, Ireland.
Differs from *B. saxorum* **(19)** in the K+ red, C− thallus.

24. B. uberior Anzi (1866)

Thallus areolate, grey, mostly 1–5 cm diam.; areoles 0.35–0.8 mm diam.; medulla I+ blue. Apothecia 0.25–0.8(-1) mm diam., immersed; epithecium olivaceous to olive-brown, N+ reddish; hymenium without oil droplets. Ascospores 8–11.5 × 4.5–6.5 μm, striate. Conidia 3.5–4 × 1–1.5 μm, oblong. Thallus Pd− or + orange, K− or + yellow-orange, C+ red (gyrophoric and ± stictic acids).

On hard siliceous rocks. N.E. England (Northumberland), C. Scotland (Mid Perth). Scandinavia, Novaya Zemlya, C. Europe, S. France, Spain, U.S.A.

Distinguished from *B. aethalea* **(2)** by the C+ red thallus (in sections) and smaller ascospores with a striate ornamentation.

BYSSOLOMA Trevisan (1853)

Thallus crustose, effuse, cortex absent. **Photobiont** chlorococcoid. **Ascomata** apothecia, sessile, ± round. **Thalline exciple** absent. **True exciple** formed by loosely intricate hyphae, tomentose-arachnoid in some species. **Hymenium** I + blue. **Hypothecium** dark red-brown, K ± purple. **Hamathecium** of paraphyses, not or slightly thickened at apices, simple or branched. **Asci** 8-spored, clavate, in K/I with amyloid apical dome containing a darker blue tubular ring-structure, and an amyloid fuzzy coat. **Ascospores** 3-septate, colourless. **Conidiomata** pycnidia, sessile, ± globose, usually covered by a layer of loosely interwoven hyphae. **Conidiogenous cells** simple, with apically formed conidia. **Conidia** flask-shaped or ± ellipsoid and constricted at the middle. **Chemistry**: lichen products not detected by t.l.c. **Ecology**: predominantly epiphytic and on leaves. Mainly pantropical.

Characterized by the *Byssoloma*-type ascus structure, byssoid true exciple and pyriform conidia.

LITERATURE: James (1971), Santesson (1952), Vězda (1987).

1 True exciple arachnoid-woolly or, in well-developed specimens, spreading laterally as an arachnoid whitish network of hyphae 2
True exciple not arachnoid, with a ± swollen, smooth, shiny margin .... **2. *marginatum***

2(1) Thallus in small irregular patches, whitish to whitish green; prothallus hardly apparent; apothecial disc black, sometimes tinged bluish; apothecial margin usually white, rarely grey-white .... **3. *subdiscordans***
Thallus continuous, smooth or irregularly farinose, grey-green; prothallus sometimes conspicuous, brown to black; apothecial disc orange-brown to dark brown, never black; apothecial margin yellowish white to yellowish grey .... **1. *leucoblepharum***

1. B. leucoblepharum (Nyl.) Vainio (1926)

Thallus to 3 cm diam., continuous, grey to greenish grey, smooth or irregularly farinose, usually thin, sometimes rather thick and irregularly-cracked; prothallus sometimes conspicuous, brown to black, sometimes bluish. Apothecia 0.3–0.5 mm diam., flat or slightly convex, sometimes clustered in groups and becoming ± distorted; disc orange-brown to dark brown, sometimes with a bluish tinge, with a tomentose-arachnoid margin, yellowish white to yellowish grey. Ascospores 10–18 × 2.5–4 μm. Conidia ± pyriform.

On old *Calluna* stumps and on bark. S. & S.W. England (Cornwall, Lizard; Hampshire, New Forest), Is. of Scilly, Ireland (Offaly, Clonmacnois). Widely distributed in tropical and subtropical regions, spreading to temperate zones. Locally abundant in southern Europe, especially on *Buxus* leaves and on smooth (*Corylus*) or rough bark (old *Quercus*) trunks.

2. B. marginatum (Arnold) Sérusiaux (1991)

Thallus to 4–5 cm diam., usually smaller, almost always continuous, greenish white to greenish, irregularly farinose, rather thin, sometimes almost indistinct or made of dispersed granules; prothallus absent. Apothecia 0.3–0.7 mm diam., flat to strongly convex, sometimes confluent and forming lobate aggregates; disc brownish to almost black, usually with a bluish tinge, with a slightly prominent (in young apothecia), persistent true exciple always paler than the disc; base of apothecia sometimes with a whitish rim of outwardly directed hyphae appearing as anchorage of the apothecium to the substratum. Ascospores 12–18 × 3–5 μm. Conidiomata usually conspicuous. Conidia 3.7–4.5 × 1.3–1.7 μm, pyriform or slightly constricted at the middle.

Crevices of smooth bark (e.g. *Corylus*), rough bark or overgrowing mosses on bark (*Quercus*, *Salix*, etc.) in humid habitats; local. W. British Is. N.W. Spain (Galicia), Canary Is. (Tenerife). Formerly in C. Europe but apparently now extinct there.

3. **B. subdiscordans** (Nyl.) P. James (1971)

Thallus to 2 cm diam., of small irregular and dispersed patches, whitish to whitish green, the surface verrucose; prothallus present but hardly detectable. Apothecia 0.3–0.5(-0.6) mm diam., flat, sometimes clustered in groups of two to four and becoming ± distorted; disc black, sometimes with a bluish tinge, matt with a tomentose-arachnoid margin, usually white, rarely greyish white. Ascospores 10–17 × 3–5 μm. Conidia regularly pyriform.

On damp, shaded rocks and *Calluna* stems. W. & C. Ireland (Connemara, W. Mayo, Offaly). Widely distributed in tropical and sub-tropical regions, spreading to temperate zones, where it typically occurs on *Buxus* leaves. S.W. Europe, C. Europe (where now almost extinct).

CALICIUM Pers. (1794)

Thallus crustose, verrucose-granular or immersed, grey to grey-grey, pale yellow or dark green. **Photobiont** *Trebouxia*. **Ascomata** apothecia, mostly with a long, distinct stalk and globose to lens-shaped head (Fig. 28f), rarely sessile; stalk of sclerotized, blackish brown to greenish black, irregularly interwoven hyphae. **True exciple** well developed, supporting the dry spore mass. **Asci** cylindrical to clavate, formed singly from ascogenous hyphae with croziers, dissolving at an early age. **Ascospores** 1-septate, forming a well-developed, black, dry-spore mass; wall thick, dark brown, often with a distinctive ornamentation. **Conidiomata** pycnidia, sessile or slightly immersed, simple, globose, ostiole punctiform. **Conidiophores** branched. **Conidiogenous cells** subcylindrical, enteroblastic. **Conidia** broadly ellipsoid to short-cylindrical, colourless, non-septate. **Chemistry**: β-orcinol depsidone and *para*-depside, xanthones, and pulvinic acid derivatives may be present. **Ecology**: usually on dry bark and wood (*C. corynellum* **(3)** occurs on siliceous rocks) in rather humid and sheltered situations.

Resembles *Cyphelium* in thallus morphology, apothecial anatomy, chemistry and ecology but differs in the apothecia being stalked or, rarely, sessile but never immersed.

LITERATURE: Middelborg & Mattsson (1987), Tibell (1975, 1977).

1 Pruina on apothecia yellow 2
Pruina on apothecia absent or not yellow 4

2(1) Apothecia > 0.3 mm tall; head > 0.2 mm diam 3
Apothecia 0.1–0.2 mm tall; head <0.15 mm diam. .... **4. *diploellum***

3(2) Thallus granular, grey; asci clavate; ascospores with spirally arranged ridges . **2. *adspersum***
Thallus immersed, inapparent; asci cylindrical; ascospores with irregular cracks **6. *trabinellum***

4(1) Pruina absent, or white on exciple margin; apothecia 0.4–1.3 mm tall 5
Pruina brown, on lower side of exciple; apothecia 0.9–1.9 mm tall 6

5(4) Apothecia >0.3 mm tall; head mostly >0.15 mm diam. 7
Apothecia 0.1–0.2 mm tall; head <0.15 mm diam. .... **4. *diploellum***

6(4) Thallus granular, green; asci clavate; ascospores with irregular cracks .... **11. *viride***
Thallus immersed, a whitish stain; asci cylindrical; ascospores with spirally arranged ridges .... **10. *salicinum***

7(5) Thallus superficial, well developed; ascospore surface of spirally arranged ridges 8
Thallus immersed, inapparent or thinly superficial; ascospore surface warted or cracked-areolate 9

8(7) Thallus yellowish green, farinose-granular; apothecia very short, 0.3–0.5 mm diam. .... **3. *corynellum***
Thallus greyish green; verrucose-granular; apothecia medium, 0.7–1 mm .... **9. *quercinum***

9(7)	Apothecium stalk in squash I−	10
	Apothecium stalk in squash I+ blue	**7. *lenticulare***
10(9)	Asci cylindrical	11
	Asci clavate	**8. *parvum***
11(10)	Exciple margin usually with white pruina; asci 35–41 μm long; ascospores 9–13 μm long with surface of irregular cracks	**5. *glaucellum***
	Exciple without white pruina; asci 44–52 μm long; ascospores 11.5 -15 μm long, minutely warted to minutely cracked-areolate	**1. *abietinum***

1. C. abietinum Pers. (1797)

Thallus immersed. Apothecia 0.6–0.9 mm tall, 6–11 times as high as width of stalk, shiny black or dark brown; pruina absent, I−; head 0.21–0.28 mm diam., lens- or slightly bell-shaped; stalk 0.08–0.13 mm wide. Asci 44–52 × 4–5 μm, cylindrical. Ascospores 11.5–15 × 5–7 μm, uniseriately arranged, minutely warted to minutely cracked-areolate. Lichen products not detected by t.l.c.

On wood, especially oak and conifers, also decaying stumps and logs; local. S. England. C. Europe to C. Scandinavia, North and South America, Australia, New Zealand.

Characterized by the non-pruinose apothecia, the black to frequently brownish to olivaceous stalks and the large and minutely warted ascospores. Formerly often incorrectly recorded for *C. glaucellum* **(5)**, which has smaller, often faintly white-pruinose apothecia, shorter asci and smaller ascospores with minute, irregular ridges and cracks.

2. C. adspersum Pers. (1798)

Thallus granular, grey. Apothecia 0.6–1 mm tall, normally 3–5 times as high as width of stalk, black, I−, very thin, outermost part I+ blue; yellow pruina present on head and exciple margin; head 0.3–0.6 mm diam., broadly lenticular; stalk 0.1–0.3 mm wide. Asci clavate-cylindrical. Ascospores 13–16 × 5.5–6.5 μm, with spirally arranged ridges. Thallus Pd+ yellow-red, K+ red (norstictic acid); the pruina contains vulpinic acid.

On dry, old *Quercus* bark; rare. S.E. England (Oxford), W.C. Wales (Montgomery). C. Europe, S. Fennoscandia, New Zealand.

Characterized by the yellow pruina which is best developed on the heads of young apothecia. In *C. trabinellum* **(6)** the yellow pruina is restricted to the lower surface of the head.

3. C. corynellum (Ach.) Ach. (1803)

Thallus farinose-granular, often well developed, yellowish green. Apothecia ± sessile or short-stalked, 0.3–1.2 mm tall, 3–5 times as high as width of central part of stalk I−; disc to 2.5 mm diam. in sessile fruits; pruina white, faint, on the lower side of the exciple. Asci clavate. Ascospores 12–16 × 4–6 μm, with spirally arranged ridges when young, later becoming cracked. Thallus Pd−, K− (rhizocarpic and ? usnic acids).

On siliceous rock underhangs in very humid habitats, possibly lichenicolous on *Lepraria* species; very rare. N.E. England (Northumberland, Bywell). C. & N. Europe, Canada.

The only British *Calicium* species on rocks, characterized by the rather well-developed, granular thallus (perhaps of a host lichen) and short black apothecia, which are slightly pruinose on the lower part of the exciple.

4. C. diploellum Nyl. (1868)

Thallus inconspicuous, very thin, greyish white. Apothecia 0.1–0.2 mm tall, black, top-shaped; head 0.08–0.12 mm diam., often greenish yellow-pruinose when young. Ascospores 6–9 × 3–4(-4.5) μm, often remaining simple, with spirally arranged ridges.

On *Ilex* bark; very rare. N.W. Scotland (W. Inverness, Loch Sunart), S.W. Ireland (Killarney). ?Endemic.

A minute, poorly understood species; the ascospore ornamentation is particularly well developed.

5. C. glaucellum Ach. (1803)

Thallus usually immersed, rarely superficial with an indistinct, dark greyish green granular thallus. Apothecia 0.5–0.9 mm tall, 4–8 times as high as width of stalk, shiny black, usually with a faint white pruinose region at edge and below the head, I−; head 0.23–0.34 mm diam., obovoid to lens-shaped; stalk 0.11–0.17 mm wide. Asci 34–41 × 3.5–4.5 μm, cylindrical. Ascospores 9–13 × 5–6.5 μm, uniseriately arranged, with irregular cracks and ridges on the surface. Pycnidia frequent; conidia 4–5 × 0.8 μm, narrowly cylindrical. Thallus K ± dull yellow (unidentified substance).

On old stumps or standing or fallen wood of coniferous and deciduous trees, more rarely on bark; common. Throughout British Is., especially in the east. C. & N. Europe, North, Central & South America, Papua-New Guinea, Australia, New Zealand.

Characterized by the rather short-stalked, black apothecia that frequently have a whitish rim of pruina on the upper part of the exciple, the immersed or rarely superficial thallus, and the medium-sized ascospores that have distinctive irregular cracks and ridge fragments at maturity. Formerly often misidentified as *C. abietinum* **(1)**, which has black-brown apothecia, lacks white pruina and has longer asci and larger ascospores with an irregular warty surface. Also resembles *C. trabinellum* **(6)**.

6. C. trabinellum (Ach.) Ach. (1803)

Like **5**, but has apothecia with greenish yellow pruina on the lower surface of the head, and an ascospore ornamentation of irregular cracks. Pruina contains vulpinic acid.

On wood, e.g. old fence posts; extinct? E.C. Scotland (Perth). Europe, North America, N. Africa, Asia, Australia, New Zealand.

Characterized by the yellow pruina which is restricted to the lower surface of the head. In *C. adspersum* **(2)** the yellow pruina occurs over the head and the ascospores have spirally arranged ornamentation.

7. C. lenticulare Ach. (1816)

C. subquercinum Asah. (1932)

Thallus granular to verrucose or almost immersed, greenish to yellowish grey or with a bluish tinge. Apothecia 0.6–1.3 mm tall, 4–8 times as high as width of stalk, black, mostly with a distinct, white pruina on lower surface of head; stalk and exciple I+ dark blue (in squash); head 0.2–0.55 mm diam., obconical to lens-shaped; stalk 0.08–0.22 mm wide, black or with a reddish tinge. Asci narrowly clavate to ± cylindrical. Ascospores 9–11 × 4–5 μm, remaining non-septate while in the asci, then 1-septate, broadly ellipsoid, minutely verrucose to areolate. Thallus Pd−, K−, KC−, C−, UV− (lichen products not detected).

On bark or wood, often decaying stumps, especially of *Quercus*. W. England, Wales, W. Scotland, S.W. Ireland. Europe, North, Central & South America, Africa, Asia, Papua-New Guinea, Australia, New Zealand.

Recognized by the obconical, often slightly reddish brown apothecia, strong I+ blue reaction of the stalk and exciple, narrowly clavate asci, and the late formation of the ascospore septum. The *Calicium* species most tolerant of high humidity.

8. C. parvum Tibell (1975)

Thallus thin, grey to greenish grey, verrucose or ± immersed. Apothecia 0.5–0.8 mm, 6–8 times as high as width of stalk, black, shiny, I−, sometimes with a slight whitish pruina at the edge of the exciple; head 0.12–0.4 mm diam, lens-shaped; stalk 0.08–0.12 mm diam. Asci clavate. Ascospores 8–11 × 3.5–5 μm, not uniseriate, incised at septum and with irregular cracks on the surface. Pycnidia frequent; conidia 2–3 × 1–1.5 μm, ellipsoid. Thallus K+ dull yellow, Pd+ pale yellow, C− (diffractaic acid).

On wood and bark, mostly of *Pinus*; rare, ?overlooked. C. and E. Scotland in native pinewoods. Fennoscandia, North America, U.S.S.R.

Resembles *C. glaucellum* **(5)**, but distinguished from that species by the smaller apothecia, which are usually not pruinose, the superficial thallus, clavate asci, and smaller, narrower ascospores with ± polygonal warts.

9. C. quercinum Pers. (1797)

Thallus verrucose-granular, distinct, greyish green, thick. Apothecia 0.5–1 mm tall, 5–7 times as high as width of stalk, black, matt, with white pruina on lower surface of head, I−; head 0.16–0.54 mm diam., stalk 0.08–0.24 mm diam. Asci cylindrical. Ascospores 9–11 × 4–5 μm, uniseriately arranged, with spirally arranged ridges. Thallus Pd+ yellow, K+ yellow-red (?norstictic acid).

On oak trunks and timber; not recorded this century and possibly extinct. S. England, S.E. Scotland (Lanark, Falls of Clyde). Scandinavia, C. Europe.

Characterized by the well-developed thallus, pruinose apothecia, cylindrical asci and ascospores with spirally arranged ridges.

10. C. salicinum Pers. (1794)

Thallus immersed, often staining substratum grey. Apothecia 0.4–1.5 mm tall, 10–15 times as high as width of stalk, brown-black, with brown pruina (K+ purple-red crystals) on the lower side of the exciple and sometimes upper part of stalk, I– ; head 0.18–0.4(-0.6) mm diam., lens-shaped; stalk 0.05–0.12 mm wide. Asci cylindrical. Ascospores 8–11 × 4–5 μm, uniseriately arranged, ellipsoid, with spirally arranged ridges. Thallus Pd–, K– or ± Pd+ yellow-orange, K+ yellow-red (± norstictic acid).

On dry, acid bark and wood, especially on the dry sides of old *Quercus*. Throughout British Is., but rare in the west. Europe, North, Central & South America, N. Africa, Asia, Australia, New Zealand.

Characterized by the brown pruina on the lower side of the head, the small, spirally striated spores and the K+ yellow-red thallus. *C. viride* **(11)** has a bright green, rather than greyish thallus, clavate asci and larger ascospores. *C. lenticulare* **(7)** sometimes has a reddish brown tinge on the lower side of the head but differs in the head being obconical, the smooth to minutely dotted ascospores, I+ blue apothecia stalk and exciple and K– thallus.

11. C. viride Pers. (1794)

Thallus granular, usually well developed, bright green. Apothecia (Fig. 28f) (0.3-)0.6–1.9 mm tall, mostly 9–16 times as high as width of stalk, black, I– ; head 0.14–0.6 mm diam., often with a brown pruina on lower surface; stalk 0.04–0.16 mm diam. Asci clavate. Ascospores 11–13.5 × 6–7 μm, not uniseriately arranged, commonly incised at septum, with deep, irregular cracks. Contains rhizocarpic acid and epanorin.

On wood and dry, acid bark; very common generally, but rarer in the west. Throughout British Is. Europe, North & South America, Siberia.

Recognized by the bright green thallus. Some specimens with an immersed thallus and well-developed brown pruina on the lower side of the head may resemble *C. salicinum* **(10)**, which has cylindrical asci and smaller ascospores with spirally arranged ridges. The most resistant *Calicium* species to SO_2 air pollution.

CALOPLACA Th. Fr. (1860)

Thallus crustose, sometimes lobed towards the margins and placodioid, more rarely squamulose, yellow-orange, grey, or rarely black, closely adpressed or occasionally immersed. **Isidia** and **soredia** present or absent. **Photobiont** trebouxioid, chiefly as "*Pseudotrebouxia*". **Medulla** white. **Ascomata** apothecia; disc usually greenish-yellow to red, brown or black. **Thalline exciple** and/or **true exciple** present or absent, at times becoming excluded. **Epithecium** usually with numerous yellowish brown crystals. **Hymenium** usually colourless. **Hypothecium** colourless. **Hamathecium** of paraphyses, simple to branched, septate, apical cells frequently ± enlarged. **Asci** 8-spored, clavate, *Teloschistes*-type. **Ascospores** ellipsoid, polarilocular or rarely, plurilocular, or in a single instance (*C. nivalis* **(37)**) simple, colourless (Fig. 7). **Conidiomata** pycnidia, immersed, inconspicuous, *Xanthoria*-type. **Conidia** small, simple, ellipsoid, straight, colourless. **Chemistry**: species with ± orange thalli are K± dark violet-red; apothecia with orange to brown-coloured granules in the epithecium and thecium are K+ dissolving violet-red under the microscope (anthraquinones, especially physcion [= parietin]). **Ecology**: on well-lit, ± nutrient-rich or -enriched rocks and trees, rarely soil and wood.

Characterized by the crustose or placodioid thallus, typically orange apothecia (K+ violet-red) and colourless, polarilocular ascospores. Other closely related genera are *Leproplaca* (leprose; included in *Caloplaca* by some authors), *Xanthoria* (leafy), and *Teloschistes* (shrubby) - all members of the Teloschistaceae. The main taxonomic problems concern closely-related lichens growing on different substrata (e.g. *C. crenularia* and *C. ferruginea*, *C. cerina* and *C. stillicidiorum*), and it is unclear whether these should be regarded as vicarious species showing ecological differentiation, or merely habitat modifications of a single entity.

Caloplaca vitellinula (Nyl.) H. Olivier (1897) is excluded as all British specimens studied are poorly developed morphs of other species. Four species (*C. crenulatella*, *C. haematites*, *C. irrubescens* and *C. pollinii*) have not been seen this century and may be extinct.

LITERATURE: Hafellner & Poelt (1979), Kärnefelt (1990), Laundon (1976, 1992a), Magnusson (1944), Nordin (1972), Santesson (1970), Wade (1965), Wunder (1974).

1 Soredia or isidia present ... 2
Soredia and isidia absent ... 18

2(1) Soredia present; isidia absent ... 3
Soredia absent; isidia present which are often granular ... 13

3(2) Thallus conspicuously lobate, placodioid; orange ... 4
Thallus indistinctly lobate, or mostly crustose, not placodioid; grey or orange ... 5

4(3) Soralia laminal, bright yellow; lobes not pruinose, darker than the soralia, ± brown-orange ... **14. *cirrochroa***
Soralia terminal or marginal, orange; lobes ± densely pruinose, ± concolorous with soredia ... **20. *decipiens***

5(3) Thallus not forming scurfy patches, without rounded marginal lobes ... 6
Thallus of ± rounded, scurfy-mealy patches, with ill-defined, rounded marginal lobes ... **45. *teicholyta***

6(5) Soredia diffuse, often forming extensive areas, obscuring the thallus ... 7
Soredia ± delimited, scattered or ± contiguous but usually ± discrete, not obscuring thallus ... 8

7(6) Soredia pale grey-green (in shade) to bright yellow, K+ violet-red ... **15. *citrina***
Soredia greenish to black-grey, K− ... **50. *virescens***

8(6) Thallus grey to yellow, not areolate, ± continuous or inconspicuous ... 9
Thallus bright orange, ± areolate ... **38. *obliterans***

9(8) Soralia grey, green, yellow or brown-orange, K+ violet-red ... 10
Soralia bright orange, K+ violet-red ... **12. *chrysophthalma***

10(9) Soralia K− ... 11
Soralia K+ violet-red ... 12

11(10) Thallus with mostly crowded, papilla-like grey vesicles, bursting to form soralia, often with uneven lacerated margins, revealing yellow-green to bluish grey soredia; apothecia brown, K− ... **39. *obscurella***
Thallus with scattered, low, rounded vesicles, irreguarly bursting to reveal yellow-green soralia; apothecia orange, K+ violet-red ... **47. *ulcerosa***

12(10) Thallus grey, K−; soralia pale yellow to dull orange-brown, sunken and ulcerose **33. *lucifuga***
Thallus yellowish grey, K+ purple; soralia greenish to bright yellow, superficial **15. *citrina***

13(2) Thallus placodioid, rounded ... 14
Thallus crustose, irregular or sometimes rounded, or of small subsquamules ... 15

14(13) Lobes to 6 mm long ... **49. *verruculifera***
Lobes to 2(-3) mm long ... **25. *granulosa***

15(13) Thallus orange-yellow ... 16
Thallus grey; isidia usually obscuring the thallus ... 17

16(15) Thallus crustose; isidia ± elongate, coralloid, often forming a thin to thick and areolate crust ... **32. *littorea***
Thallus of minute squamules to 1 mm wide, with small, convex, granular lobes; isidia small, spherical, often forming distinct, convex clusters ... **7. *britannica***

17(15) Thallus pale grey, continuous; isidia well-differentiated, papillate, simple to coralloid; apothecia entirely orange-red .... **27. *herbidella***
Thallus dark grey, often with a bluish tinge, usually coarsely areolate, surface mealy; isidia minute, spherical, crowded, poorly differentiated; apothecial disc yellow orange with a persistent grey thalline exciple .... **30. *isidiigera***

18(1) Discs of apothecia black, sometimes with a red-brown tinge, often bluish pruinose 19
Discs of apothecia greenish to orange, red or brown, not pruinose 23

19(18) Discs with a red-brown tinge, without pruina; epithecium K+ violet-red; on bark or coastal siliceous rock 20
Discs often ± with a white-mauve pruina; epithecium K− to + pale violet; on limestone 21

20(19) Thallus whitish to immersed, not areolate; on bark .... **41. *pollinii***
Thallus dark grey to black, areolate; on rock .... **16. *concilians***

21(19) Thallus superficial, ± areolate; apothecia immersed in thallus or emergent 22
Thallus ± immersed, not areolate; apothecia sunk in shallow depressions in substratum .... **1. *alociza***

22(21) Apothecia *Aspicilia*-like, persistently immersed in thallus, several per areole; thallus bluish- or lead-grey .... **11. *chalybaea***
Apothecia emergent to superficial, one, rarely more, per areole; thallus dark grey **48. *variabilis***

23(18) Thallus placodioid, orange; lobes to 1–2 mm wide, usually extending to more than 3 mm beyond apothecia; ascospores swollen-orbicular-rhomboid to lemon-shaped 24
Thallus placodioid to crustose; if placodioid and orange, lobes less than 1 mm wide, not extending for more than 2 mm beyond apothecia; ascospores evenly ellipsoid 26

24(23) Lobe-ends convex, ± regularly furrowed or pleated, orange 25
Lobe-ends flattened, not regularly furrowed or pleated, bright orange-yellow ..... **6. *aurantia***

25(24) Lobe-ends divided, finger-like, shiny, bright orange; paraphyses ± branched towards the apices .... **46. *thallincola***
Lobe-ends palmately rounded, matt, orange; paraphyses simple at the apices . **22. *flavescens***

26(23) Thallus placodioid, forming rosettes 27
Thallus subsquamulose to crustose, not forming rosettes 29

27(26) Thallus yellow to orange, centre smooth or ± pruinose 28
Thallus greyish-white, lobes indistinct, rounded or absent, adpressed; centre scurfy-mealy .... **45. *teicholyta***

28(27) Apothecia constricted at base; thallus not pruinose .... **44. *scopularis***
Apothecia not constricted at base; thallus often ± densely pruinose .... **43. *saxicola***

29(26) Thallus yellow-green to orange 30
Thallus grey or immersed 39

30(29) Thallus crustose, not of scattered granules 31
Thallus of small, often scattered to contiguous, rounded or elongate, minute squamules, to 2 mm diam., convex .... **36. *microthallina***

31(30) Apothecia immersed in the thallus at first 32
Apothecia superficial from the first 33

32(31) Apothecia urceolate, becoming concave, to 0.5 mm diam.; thallus smooth, continuous, felted, not mosaic-forming; ascospores becoming 4-locular .... **40. *ochracea***
Apothecia flat, becoming subconvex, to 0.8 mm diam.; thallus ± finely rimose-cracked, mosaic-forming, frequently dissected by black lines; ascospores polarilocular **19. *dalmatica***

33(31) Apothecial disc ± flat from the begining 34
Apothecial disc deeply concave at first, with swollen persistent exciple, becoming flat .... **42. *ruderum***

34(33) Thallus well developed, ± continuous or of contiguous areoles 35
Thallus poorly developed, of scattered areoles 38

35(34) Thallus yellow to orange; apothecia concolorous with thallus 36
Thallus greenish yellow; apothecia orange 37

36(35) Thalline exciple swollen when young, not warted; apothecia constricted at base; ascospore septa $\frac{1}{3}$ length of ascospore .... **35. *marina***
Thalline exciple conspicuous, slightly warted; apothecia sessile; ascospore septa narrow, less than $\frac{1}{4}$ length of ascospore .... **18. *crenulatella***

37(35) Thallus continuous; apothecia to 3 mm diam.; on bark .... **23. *flavorubescens***
Thallus cracked-areolate; apothecia to 1.1 mm diam.; on rock .... **24. *flavovirescens***

38(34) Apothecia conspicuous, remaining flat; thalline exciple orange, persistent; ascospore septum *c.* 1 μm wide .... **2. *approximata***
Apothecia inconspicuous, becoming sub-convex; thalline exciple orange, becoming excluded; ascospore septum 3–5 μm wide .... **29. *irrubescens***

39(29) Apothecia orange, brownish-red or greenish, K+ violet-red; thallus lacking simple papillae and erose spots 40
Apothecia brown, constricted at base, flat, swollen and translucent when moist, soon becoming ± immarginate; epithecium K−; thallus with scattered, or ± crowded, small, simple papillae that burst leaving uneven, torn margins .... **39. *obscurella***

40(39) Thallus areolate 41
Thallus continuous or immersed, not areolate 45

41(40) True exciple prominent, orange to brownish-red 42
Thalline exciple prominent, grey 44

42(41) Thallus greyish 43
Thallus ± black .... **5. *atroflava***

43(42) True exciple glossy; disc red-brown to rust-coloured, unchanged when wet, not pruinose; without a photobiont layer beneath the apothecia .... **17. *crenularia***
True exciple matt; disc orange becoming dull, somewhat pellucid-greenish when wet, ± pruinose; with a thick photobiont layer beneath the apothecia .... **8. *ceracea***

44(41) Apothecia becoming constricted at base; on bark .... **26. *haematites***
Apothecia not becoming constricted at base; on coastal rock .... **3. *aractina***

45(40) Thalline exciple yellow to orange, at least in part; discs orange to red 46
Thalline exciple grey, persistent, elevated above the disc; disc orange to green-yellow **9. *cerina***

46(45) Asci 8-spored; apothecia 0.2–1 mm diam., scattered to crowded 47
Asci (8-)12 to 16-spored; apothecia to 0.3 mm diam., very crowded .... **10. *cerinella***

47(46) True exciple brownish red 48
True exciple orange 50

48(47) Apothecia not constricted at base 49
Apothecia constricted at base .... **13. *cinnamonea***

49(48) Thallus whitish-grey or immersed; apothecia 1–2 mm diam., flexuose-uneven; on bark .... **21. *ferruginea***
Thallus grey to ± immersed; apothecia to 1 mm diam., ± rounded; on rock **17. *crenularia***

50(47) Apothecia superficial from the first 51
Apothecia immersed at first .... **34. *luteoalba***

51(50) Thallus white to grey, or immersed 52
Thallus ± black, superficial .... **5. *atroflava***

52(51) Thallus on stone or wood, rarely bark; ascospores septate, polarilocular, ellipsoid 53
Thallus a grey membrane over mosses; ascospores ± simple, narrowly elongate-ellipsoid .... **37. *nivalis***

53(52) Ascospore septa 1–2 μm wide 54
Ascospore septa 2–5 μm wide 55

54(53) Ascospores 15–20 × 8–10 μm; apothecia often constricted at the base, urceolate at first .... **31. *lactea***
Ascospores 9–13 × 3–6 μm; apothecia sessile, flat, with prominent exciples **2. *approximata***

55(53) Ascospore septa 2.0–3.5 μm wide .... **4. *arenaria***
Ascospore septa 3–5 μm wide .... **28. *holocarpa***

1. C. alociza (Massal.) Migula (1925)

Thallus crustose, inconspicuous, immersed, sometimes forming a very thin, ± scurfy, dark grey film, often limited by a black prothallus. Apothecia to 0.5 mm diam., numerous, closely scattered, sunk in depressions, the top of disc level with the surface, or occasionally ± emergent, flat to slightly convex, black, sometimes thinly white-grey pruinose, leaving pale, shallow pits when shed; at first with a grey thalline exciple, becoming excluded; epithecium grey, with colourless crystals. Paraphyses with apical cells 2–5 μm wide. Ascospores 15–18 × 7–8 μm, ellipsoid, septum 1–3 μm wide, about $\frac{1}{6}$-$\frac{1}{9}$ of length of ascospore. Apothecia K−, epithecium and upper part of thecium K+ mauve-lilac.

On hard, sunny limestone, especially on or near the coast, more rare inland; rather scarce. Throughout British Is. Europe, especially in the S., extending to Iran and N. Africa.

Superficially resembles a small-fruited *Lecidea* s. lat., but the polarilocular ascospores, etc., indicate that it belongs to *Caloplaca*. The ± immersed thallus distinguishes it from other dark-fruited members of the genus.

2. C. approximata (Lynge) Magnusson (1946)

Thallus crustose, yellowish orange, always poorly developed and often ± reduced to a black prothallus. Apothecia to 0.5 mm diam., conspicuous, numerous, closely scattered, bright orange. persistently flat; with ± elevated, regular exciple. Paraphyses mostly simple, cells only slightly enlarged towards the tips.

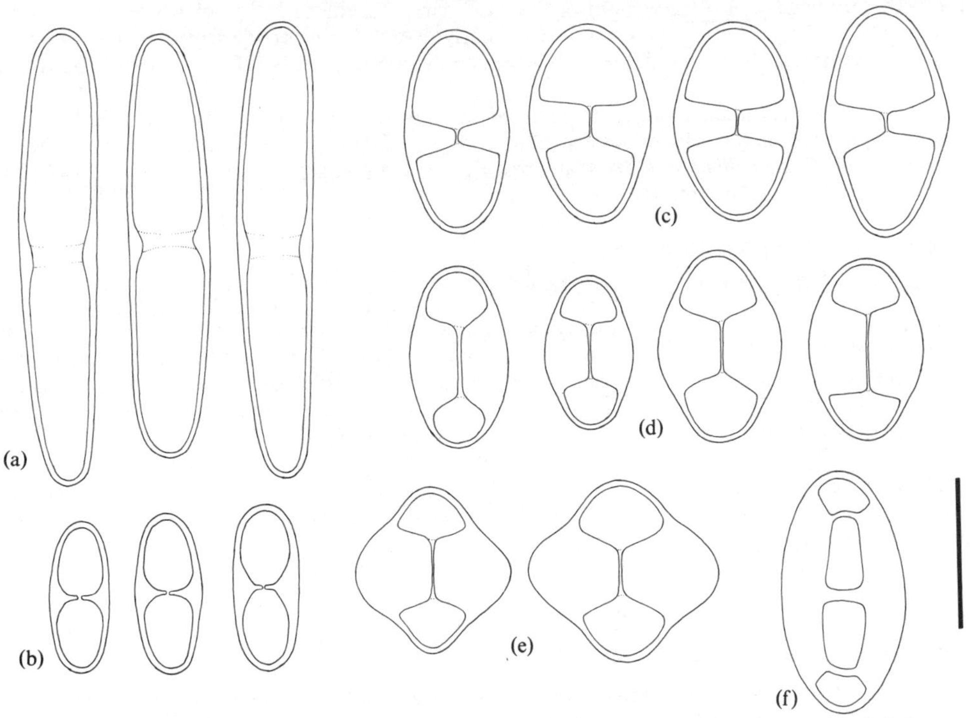

Fig. 7. *Caloplaca* ascospores in 10% K. (a) *C. nivalis*; (b) *C. approximata*; (c) *C. lactea*; (d) *C. holocarpa*; (e) *C. flavescens*; (f) *C. ochracea*. Scale = 10 μm.

Ascospores variable, mostly long and narrow, (9-)10(-13) × (3-)4(-6) μm, occasionally shorter and broader, 8 × 5 μm, septum 1(-2) μm wide (Fig. 7b). Apothecia K+ violet-red.

On calcareous schist at above 800 m; rare. C. Scotland (Mid-Perth, Ben Lawers and Angus, Caenlochan). Arctic, ± circumpolar.

Closely resembles *C. holocarpa* **(28)** but has ascospores which are mostly narrower with a distinctive septum only 1 μm wide, thus often appearing as if truly 1-septate and not polarilocular.

3. C. aractina (Fr.) Häyrén (1914)

Thallus wide-spreading, crustose, dark grey to black, the surface uneven but ± smooth, somewhat oily, ± distinctly cracked-areolate, often with a ± zoned margin bounded by a black prothallus. Apothecia to 1 mm diam., widely scattered, often scarce, flat, immersed soon becoming emergent to sessile; thalline exciple persistent, dark grey; discs brownish orange, roughened. Paraphyses with apical cells scarcely swollen. Ascospores 10–15 × 5–8 μm, broadly ellipsoid, septum 3–5 μm wide, *c.* ⅓ of the length of the ascospore. Thallus K−; discs of apothecia K+ violet-red.

On acid, sunny coastal rocks often near bird perches; rare. S.W. England (Cornwall, Lizard Peninsula) to S.W. Scotland (Strathclyde region). W. Europe, also inland S. Europe.

The dark, oily thallus resembles that of a marine *Verrucaria* or *Halecania ralfsii*, but bears scattered, ± sessile apothecia with brownish orange discs surrounded by a persistent, even, dark, thalline exciple. Specimens with a roughened thallus from inland localities in Europe are sometimes separated as *C. fuscoatra* (Decuill.) Zahlbr. (1894), but this is evidently only an ecological modification.

4. C. arenaria (Pers.) Müll. Arg. (1862)

C. subpallida Magnusson (1945)

Thallus crustose, inconspicuous, immersed, often overgrowing other lichens and then thinly uneven, pale to dark grey-brown. Apothecia to 0.5 mm diam., scattered to crowded, rounded to flexuose, flat; discs dark to dull orange; true exciple orange to reddish orange, prominent, ± elevated, persistent. Paraphyses broadening gradually, 2–3 μm wide at the apices. Ascospores 13–17 × 5–8 μm, ellipsoid, septum 2–3.5 μm, wide, *c.* ⅕ of the length of the ascospore. Apothecia K+ violet-red.

On acid rocks, especially slates and basalt, frequently overgrowing other crustose lichens; rare. N. England (Derbyshire), S.W. England (Devon), N. Wales, Scotland. Europe, North America, N. Africa, Asia.

Characterized by the small, reddish orange, flat apothecia with a persistent true exciple, on an immersed, inapparent thallus. Resembles a small morph of *C. holocarpa* **(28),** but the ascospores have a narrower septum; *C. subpallida* was originally distinguished by having paler apothecia; however, a continuum from orange to red-orange occurs.

5. C. atroflava (Turner) Monguillon (1914)

?incl. var. *submersa* (Nyl.) Magnusson (1944)

Thallus crustose, dark grey to black, thin, cracked-areolate when well developed; areoles *c.* 0.5 mm diam.; prothallus black, bordering thallus, ± fimbriate. Apothecia to 0.5 mm diam., scattered to crowded, rounded to flexuose, flat, constricted at base; thalline exciple absent; true exciple persistent, prominent, even, convex, glossy, bright orange; discs brownish orange. Paraphyses slender and flexuose, septate, not or only slightly broadened at the apices. Ascospores 13–16 × 9–10 μm, broadly ellipsoid, swollen, septum 5–7 μm wide, ⅓-½ (or more) of the length of the ascospore. Thallus K+ faintly violet-red in places; apothecia K+ violet-red.

On flint pebbles on chalk downland and on pebbles and boulders at the coast; rare. S. England. Europe, North America, Israel.

Resembles *C. holocarpa* **(28)** but the thallus is almost black. The broad ascospores with a broad septum are distinctive.

6. C. aurantia (Pers.) Steiner (1896)

Thallus placodioid, forming large, closely-adpressed, partially or entirely rounded thalli to 12 cm across, usually ± zoned or piebald; lobe-ends rounded and flat, without furrows, bright orange-yellow, the interior frequently paler or white due to loss of pigment, the central area ± cracked-areolate, in part brownish orange due to the apothecia. Apothecia to 1(-1.5) mm diam., confined to the central area, usually abundant and crowded, flat with a distinct orange-yellow thalline exciple, often becoming ± convex and ± immarginate when old; disc orange-brown, deeper coloured in strong light. Paraphyses 1–2 μm thick, slender and ± straight, septate, not or only slightly swollen at apices. Ascospores 10–13 × 8–10 μm, swollen, orbicular-rhomboid to lemon-shaped, the septum variable in width, to 5 μm wide. Thallus and apothecia K+ violet-red.

On well-lit limestones in nutrient-rich habitats, especially abundant on oolite walls of the Cotswolds; common. Throughout British Is. S. Europe, Mediterranean, E. to Iran, North America.

Closely related to *C. flavescens* **(22)**, with which it often grows and which is usually uniformly orange with convex, pleated, lobe-ends; *C. aurantia* is often ± concentrically zoned in colour with orange-yellow, flattened lobe-ends. *C. flavescens* may also sometimes become zoned by the death of the central part of the thallus, leaving it with either dark or white necrotic areas. The ascospores of *C. aurantia*, *C. flavescens* and *C. thallincola* **(46)** differ from all other British species in their lemon-shape.

7. C. britannica R. Sant. (1992)

Thallus entirely of small, swollen, convex, granular lobes or subsquamules, to 1 mm diam., orange-yellow, in places forming a thick, cracked-areolate thallus. Isidia to 0.1 mm diam., often numerous, coralloid, of groups of small, spherical granules which often form mounds on inner areas of the thallus. Apothecia to 0.5 mm diam., occasional, scattered, flat, orbicular to inflexed; thalline exciple persistent, orange-yellow; discs pale orange. Paraphyses lax, variable, some with narrow apical cells, others swollen, to 7 μm. Ascospores 11–14 × 4–6 μm, ellipsoid, septum 3–5 μm wide, about ⅓ of the length of the ascospore. Thallus and isidia K+ violet-red.

In crevices in the mesic-supralittoral zone on acid rock shores and also on sheltered birds' perching rocks; rare. W. Britain. ?Endemic.

The convex subsquamules, bearing numerous, minute, spherical granules of different sizes, are distinctive.

8. C. ceracea Laundon (1992)

C. caesiorufa auct. brit., non (Schrader) Flagey (1895)

Thallus crustose, wide-spreading, pale to dark grey, areolate, often thick and crumb-like, the surface roughened; prothallus pale or absent. Apothecia to 1 mm diam., usually crowded, ± contiguous, rounded to flexuose, flat, constricted at base; true exciple persistent, ± swollen, matt, orange, ± pruinose, yellowish orange and waxy when wet; thalline exciple present only in young apothecia, disappearing or occasionally persisting as a cushion on the lower surface of apothecia; disc orange-brown-pruinose. Algae forming a dense green layer of closely packed cells running beneath a colourless hypothecium (visible in sections). Paraphyses slightly swollen at the apices. Ascospores 13–15 × 8 μm, ellipsoid, septum 3–5 μm wide, about ⅓ of the length of the ascospore. Thallus K− ; apothecial exciple and disc K+ violet-red.

On acid, coastal rocks and in a few isolated inland localities; local. W. British Is. France, Norway.

Distinguished from *C. crenularia* **(17)** by the apothecial exciple which is matt and ± pruinose when dry, becoming dull greenish orange and waxy when wet and the thick layer of algae beneath and at the sides of the apothecium; in *C. crenularia* the true exciple is glossy and remains rusty or brownish red when wet; the algal layer is discontinuous below the apothecium.

9. C. cerina (Ehrh. ex Hedwig) Th. Fr. (1860)

C. stillicidiorum (Vahl) Lynge (1921)

Thallus crustose, pale to dark grey, often tinged glaucous or bluish, rather thick or thin, occasionally immersed and inconspicuous, continuous, somewhat waxy, the surface smooth or rarely ± warted; prothallus pale or absent. Apothecia to 1.5(-2.0) mm diam., scattered to contiguous, sometimes overlapping, often angular, ± sessile, constricted at base; thalline exciple persistent, even, swollen, raised, ± flexuose, grey; discs orange, orange-yellow or greenish, concave when young becoming ± flat when mature. Paraphyses broadening towards the tips, to 4 μm wide. Ascospores 12–15 × 8 μm, ellipsoid, septum 5–8 μm wide, about ⅓-½ of the length of the ascospore. Thallus and thalline exciple K− ; discs K+ violet-red.

On bark, more rarely wood, especially with a high pH (*Acer pseudoplatanus*, *Populus tremula*, *Fraxinus*, *Sambucus*, *Ulmus*), or where affected by organic nutrients, also overgrowing mosses and plant debris over basic rocks and earth, rarely, directly on calcareous rocks; formerly frequent, now decreasing and rather rare. Throughout unpolluted areas of British Is. Europe, extending to Iran, North America.

Readily identified by the frequently flexuose apothecia with orange to greenish yellow discs, concave when young, surrounded by a distinctive, persistent, grey, smooth, thalline exciple. Thalli overgrowing mosses were formerly referred to *C. stillicidiorum* but these are best considered as ecological modifications. *C. isidiigera* **(30)** and *C. virescens* **(50)** are closely related but reproduce asexually and are infrequently fertile.

10. C. cerinella (Nyl.) Flagey (1896)

Thallus crustose, white to grey, thin, smooth and non-corticate, in the herbarium developing crystals on the surface and appearing waxy, but often almost immersed, often in small patches among other lichens. Apothecia to 0.3 mm diam., closely grouped or ± contiguous; true exciple persistent, pale to yellow; discs flat, pale yellow to orange, rarely deeper orange. Paraphyses with broad tips, to 6 μm wide. Asci (8-)12- to 16-spored. Ascospores 10–13 × 6–7 μm, ellipsoid, septum 3–5 μm, wide, about $\frac{1}{4}$-$\frac{1}{2}$ of the length of the ascospore. Thallus K−; apothecial exciple and discs K+ violet-red.

On bark, especially twigs of *Sambucus nigra*, in nutrient-rich communities, especially *Xanthorion*; local. Throughout British Is. Europe.

An inconspicuous, but distinctive species in small patches, often with *Lecania cyrtella*. Distinguished by the closely grouped, minute apothecia with a pale yellow exciple and 12- to 16-spored asci.

11. C. chalybaea (Fr.) Müll. Arg. (1862)

Thallus distinct, smooth, even, markedly coarsely-rimose-areolate, blackish brown to lead-grey, sometimes with a bluish or violet tinge, well delimited by a ± conspicuous black prothallus, sometimes giving a ± zoned and radiate margin, thallus occasionally thinner, appearing spotted greyish on a dark prothallus. Apothecia to 0.5 mm diam., closely scattered, several per areole, deeply immersed in the thallus at first; thalline exciple ± distinct, often paler than the disc; epithecium grey, with colourless crystals; disc flat, ± level with areole surface, when mature, black, often bluish pruinose. Paraphyses broadening towards the tips, to 4 μm wide. Ascospores 10–16 × 6–8 μm, septum (1-)3–5 μm wide, to $\frac{1}{3}$ of the length of the ascospore. Thallus and apothecia K− or K+ pale violet; epithecium and upper part of the hymenium K+ mauve-lilac.

On hard limestones, on natural boulders, buildings (e.g. bridges), and chest tombs, apparently especially when ± nutrient-enriched; scattered. N. & W. Britain, very rare in the east. Europe, E. to Turkey, North Africa.

Superficially resembles an *Aspicilia* (e.g. *A. calcarea*) in habit but easily distinguished by the polarilocular ascospores. Differs from *C. variabilis* **(48)** in the thicker, distinctly areolate-cracked thallus and with persistently immersed apothecia.

12. C. chrysophthalma Degel. (1944)

Thallus immersed, inconspicuous, or a thin grey film; prothallus indistinct. Soredia bright orange-yellow, crowded, flat to ± convex, ± rounded, discrete, erose to superficial; soredia farinose to minutely granular. Apothecia rare, absent in Britain, to 1 mm diam., scattered, ± elevated, orbicular, concave at first but becoming flat; thalline exciple sometimes present, yellowish, sorediate; true exciple well developed; disc orange. Paraphyses 2 μm thick, simple, apex thickened to 3 μm. Ascospores (10-)13–20 × 7–10 μm, ellipsoid, septum 3–5 μm wide, about $\frac{1}{4}$ of the length of the ascospore. Soredia and apothecia K+ violet-red.

On nutrient-rich bark in open situations; rare. Is. of Scilly; there are also two doubtful 18th century records from S. England (Sussex). Europe, North America.

Characterized by the bright orange-yellow, delimited soralia contrasting with the deep orange apothecia and inconspicuous thallus; the British material is poorly developed.

13. C. cinnamomea (Th. Fr.) H. Olivier (1909)

Thallus crustose, white to grey, thin, uneven or granular. Apothecia to 1.5 mm diam., crowded, ± contiguous, rounded or somewhat angular or flexuose, flat at first, becoming ± convex, sessile; true exciple rust-red to brownish orange, irregular, somewhat swollen, ± uneven, ± greyish below, excluded in old apothecia; discs rust-red to brownish orange. Paraphyses flexuose, simple to forked, gradually broadening towards tips. Ascospores (12-)15–18 × 6–8 μm, elongate-ellipsoid, septum 3–7 μm wide, ¼–½ of the length of the ascospore. Thallus K−, apothecia K+ violet-red.

Encrusting mosses above 1000 m; rare. C. Scotland (Mid-Perth, Ben Lawers). Arctic, circumpolar.

Closely related to *C. ferruginea* **(21)**, differing in the sessile, generally rounded apothecia. *C. cinnamomea* has recently been included in *C. ammiospila* (Wahlenb. ex Ach.) H. Olivier (1909), but that species grows on wood and has apothecia with a distinct thalline exciple; their relationship is in need of further study.

14. C. cirrochroa (Ach.) Th. Fr. (1871)

Thallus to 5 cm across, placodioid, forming narrow, finger-like, closely adpressed, divided lobes in irregularly rounded or small scattered or ± contiguous thalli, dirty or brownish orange, the lobe-ends often ± pruinose and paler orange; lobe-ends 0.2–0.5 mm wide, rounded, forked, shallowly convex, elongated, ± contiguous, separated by almost parallel-aligned furrows. Soralia usually present, laminal, scattered, small, flat, ulcer-like, rounded, to 0.8 mm diam.; soredia farinose, lemon-yellow. Apothecia rare, to 0.5 mm diam., scattered, flat; thalline exciple persistent, orange; disc deeper orange. Paraphyses flexuose, mostly not swollen at the tips. Ascospores 10–15 × 5 μm, narrowly ellipsoid, septum 2–3 μm wide, under ⅓ of the length of the ascospore. Thallus and soredia K+ violet-red.

On inclined and vertical, hard, calcareous rocks, chiefly limestones, in sheltered and, often dry and shaded situations; rarely on medieval buildings and walls; local. W. & N. Britain, rarely in the east. Europe, North America.

The elongate finger-like lobes are reminiscent of *Xanthoria elegans*, but the paler brown-orange, not brown-red, lobes, and the contrasting punctiform, yellow soralia make this lichen unmistakable. Essentially a relict species occurring on, or near, hard limestone rocks which remained free of tree cover during the forest maximum.

15. C. citrina (Hoffm.) Th. Fr. (1860)

incl. f. *phlogina* (Ach.) D. Hawksw. (1969)

Thallus crustose, of very small yellowish green to bright orange, subsquamulose areoles or minutely scattered or continuously granular or ± immersed in the substratum and inconspicuous, diffuse. Soredia conspicuous, often covering the whole thallus, diffuse, not in distinct soralia, thin to thick, when thick, broken up into irregular coarse areoles, usually yellow but sometimes orange, yellowish grey in shade, farinose. Apothecia frequent, to 1.0 mm diam., scattered, immersed at first, becoming ± sessile, flat somewhat swollen when mature; thalline exciple usually persistent, even, smooth to almost completely sorediate, yellowish; disc orange. Paraphyses some with narrow tips, others ± clavate to 3 μm wide. Asci often long and narrow. Ascospores 10–15 × 5–6 μm, broadly ellispoid, septum *c.* 3 μm wide, about ⅓ of the length of the ascospore. Soredia K+ violet-red.

On limestone building stone, concrete, mortar, etc., especially on walls, usually in sunny situations, at times overgrowing mosses and dead plants (e.g. *Armeria*), frequent on nutrient-rich bark and wood, where the sorediate thallus is thin, less common on naturally calcareous and occasionally acid rock outcrops; very common. Throughout British Is. Cosmopolitan.

Readily recognized by the bright, diffuse soredia. The underlying thalline granules are variable. Paler sterile specimens in shade are occasionally difficult to distinguish from *Leproplaca* species, and the presence of corticate thalline granules is then important for the identification of *C. citrina*.

In E. England a similar, but possibly distinct, entity occurs with deeper orange, flattened, ± angular areoles with initially ± sorediate margins, the soredia eventually involving the whole surface of the squamules. It occurs with *C. citrina* on calcareous memorials and is easily recognized by the deeper colour.

16. C. concilians (Nyl.) H. Olivier (1909)

Thallus crustose, dark grey to black, thick, coarsely rimose-areolate, the surface irregular. Apothecia to 0.8 mm diam., crowded, discrete or contiguous, ± rounded, flat at first, becoming markedly convex; thalline exciple even but later excluded, dark rust-red, reddish black to black. Paraphyses branched, with narrow or clavate apices. Ascospores 12–17 × 6–9 μm, broadly ellipsoid, the septum about ⅓ of the length of the ascospore. Thallus K−, apothecia K+ violet-red.

On schistose coastal and montane rocks; no recent records. W. Scotland (Skye), N. Scotland (Ben Lawers), Ireland. Norway, Czechoslovakia.

A very rare lichen identified by the very dark, areolate thallus and convex apothecia which are reddish brown-black when dry, ± dark rusty red when wet.

17. C. crenularia (With.) Laundon (1984)

C. festiva (Ach.) Zwackh (1864)

Thallus crustose, dark grey, occasionally pale grey or rarely immersed and inconspicuous, usually forming a roughened, continuous or cracked-areolate surface. Apothecia to 1.0 mm diam., scattered to contiguous, ± rounded but occasionally slightly flexuose, flat to somewhat swollen; true exciple well-defined, glossy, brownish red; disc reddish brown. Paraphyses flexuose, the tips only slightly broadened, to 3.5 μm. Ascospores 12–14 × 6–8 μm, ellipsoid, septum *c.* 5 μm wide, to ½ of the length of the ascospore. Pycnidia 0.1–0.2 mm diam., brownish red. Thallus K−, apothecia and pycnidia K+ violet-red.

On acid rocks growing in sheltered, preferably rather damp, situations, also occasionally on walls and memorials; one of the most common species of *Caloplaca*. N. & S. Britain, absent or rare in the English Midlands and E. Anglia. Europe, North America, Israel.

Distinguished by the deep brownish red, almost rust-coloured, apothecia on a grey, roughened thallus. The ascospores have a particularly broad septum. Occasionally forms extensive thalli with only tiny, pimple-like, aborted apothecia or pycnidia; this morph is often characteristic of pollution-stressed sites. Closely related to the bark-inhabiting *C. ferruginea* **(21)**, but distinct in the darker, rough thallus and smaller apothecia.

18. C. crenulatella (Nyl.) H. Olivier (1909)

Thallus crustose, yellowish, corticate, non-pruinose, ± flat and rimose-areolate. Apothecia to 1 mm diam., becoming crowded, ± contiguous, ± rounded, flat; true exciple surrounded by a slightly warted thalline exciple, yellow. Paraphyses simple, broadened towards the tips, to 4 μm. Ascospores 18 × 8 μm, ellipsoid, septum 1.5–2.5 μm wide, under ¼ of the length of the ascospore. Thallus and apothecia K+ violet-red.

On inland limestone; very rare. N.W. England (Cumbria). ?Endemic.

A species needing further investigation known only from the type gathering; the narrow ascospore septum is distinctive.

19. C. dalmatica (Massal.) H. Olivier (1909)

C. schaereri (Arnold) Zahlbr. (1890), *C. velana* (Massal.) Du Rietz (1925)

Thallus crustose, brownish orange to yellowish white, occasionally in part whitish, fairly smooth to minutely roughened, continuous, sometimes pruinose, usually finely rimose-areolate; prothallus grey-black, mosaic-forming or absent. Apothecia to 0.8 mm diam., scattered or grouped, ± rounded, ± immersed at first, sometimes becoming ± sessile, later sub-convex; thalline exciple conspicuous, often becoming ± excluded in old apothecia; disc yellowish to light brownish orange, often concolorous with the exciple. Paraphyses ± straight, sparsely branched, most tips broadened, the apical cell 2–8 μm wide. Ascospores 9–13 × 5–9 μm, broadly ellipsoid, septum 3–5 μm wide, ¼–½ of the length of the ascospore. Thallus and apothecia K+ violet-red.

On hard, smooth limestones in sunny situations, and especially on walls, rarely hard mortar of church walls; local. England, chiefly in the S., S. & W. Ireland. Europe, especially Mediterranean, extending E. to Iran.

Readily identified by the yellowish to orange thallus, often with paler areas due to the browsing of molluscs, dissected by a black prothallus forming a mosaic, and the subconvex and immarginate apothecia. The type specimen of *C. dalmatica* differs from the British material in having a smooth, non-areolate, orange thallus with immersed apothecia. However, intermediates are frequent in this relatively variable species. In N. England (N. Yorkshire) thalli are often reduced and poorly developed. *C. ochracea* **(40)** is similar but has different ascospores.

20. C. decipiens (Arnold) Blomb. & Forss. (1880)

Thallus placodioid, of closely contiguous, adpressed lobes in ± rounded thalli to 3 cm across, greenish yellow (shade) to deep orange-yellow; lobe-ends to *c.* 1 mm wide, convex, ± densely pruinose, palmately branched. Soralia usually present, lip-shaped, on small inner lateral lobes, often spreading along the areoles; soredia minutely granular, concolorous with the thallus. Apothecia to 1.0 mm diam., occasional; thalline exciple swollen, persistent; disc concave at first, becoming rounded, constricted at base and flat when mature, orange. Paraphyses lax, simple, most tips broadened, apical cells 2–6 μm wide. Ascospores 10–15 × 5–8 μm, ellipsoid, septum 2–3 μm wide, about $\frac{1}{4}$–$\frac{1}{3}$ of the length of the ascospore. Thallus, apothecia and soredia K+ violet-red.

On calcareous walls, mortar and asbestos-cement, scarce on natural limstone rock outcrops where it is almost restricted to birds' perches at the coast, rarely on coastal timber and trees; frequent. E. England and Scotland becoming rarer in the W. and Ireland. C. Europe, North America, Iran, New Zealand.

The ± lip-shaped soralia of this lichen are distinctive; *C. cirrochroa* **(14)** has laminal, ulcer-like soralia. Yellow morphs closely resemble *Candelariella medians* which is distinguished by the K− thallus.

21. C. ferruginea (Huds.) Th. Fr. (1860)

Like *C. crenularia* **(17)**, but differing in the entire or ± immersed, usually smooth, whitish grey, rarely darker, thallus and apothecia, 1.0–2.0 mm diam., often with a markedly unevenly flexuose margin.

On smooth, well-illuminated bark of often young trees, less common on wood, usually as isolated thalli, now chiefly confined to unpolluted areas. S.W., W. & N.W. Britain, Ireland. Europe, North America, Hawaii, S. Africa, Australia.

22. C. flavescens (Huds.) Laundon (1984)

C. heppiana (Müll. Arg.) Zahlbr. (1890)

Thallus placodioid, forming large patches, closely adpressed, rounded or ± irregular, surface matt, pale to deep orange; lobe-ends palmate, rounded, convex, not or faintly pruinose, lobes contiguous, pleated, sometimes overlapping, centre ± coarsely areolate, uneven, of overlapping, sometimes contorted and congested lobes, sometimes whitish, non-pigmented, dotted with orange apothecia. Apothecia to 1.5 mm diam., mainly confined to the centre of the thallus, usually abundant and crowded, flat at first, becoming convex when old; thalline exciple present at first, ± concolorous, later excluded; disc orange to orange-brown. Paraphyses slender and ± straight to flexuose, not noticeably swollen at apices. Ascospores 12–15 × 8–10 μm, swollen, orbicular-rhomboid to lemon-shaped, septum variable in breadth, to 5 μm wide (Fig. 7e). Thallus and apothecia K+ violet-red.

On calcareous stone, especially limestone, walls, stonework, mortar and asbestos cement in ± nutrient-rich habitats, very rare on bark (usually trees near quarries); the most common placodioid *Caloplaca* in England. Throughout England, Wales and Ireland, extending locally to Scotland. Europe, especially in the S.

This variable lichen is closely related to *C. aurantia* **(6)** and *C. thallincola* **(46)**. *C. saxicola* **(43)** has a smaller thallus, is often densely pruinose, and has ellipsoid, not lemon-shaped, ascospores and tends to grow in more sheltered, drier situations.

23. C. flavorubescens (Huds.) Laundon (1976)

Thallus crustose, whitish, yellow-green, sometimes diffused grey, continuous, smooth and even to coarsely convex-granular-papillate, often delimited by a ± conspicuous, smooth, white-grey to blue-grey prothallus. Apothecia to 3.0 mm diam., scattered to crowded, rounded, flat, but often slightly convex, becoming irregular and distorted when old; thalline exciple thin, grey-yellow, crenulate, soon becoming excluded but retained on the underside of the apothecia; true exciple prominent, somewhat elevated, smooth, orange; disc orange to deep orange, ± flat. Lower part of hymenium and hypothecium with numerous oil droplets. Paraphyses mainly forked towards the tips, apices often broader, to 4 μm. Ascospores 15–18 × 6–10 μm, ellipsoid, septum 5–9 μm wide, to $\frac{1}{2}$ of the length of the ascospore. Thallus and apothecia K+ violet-red.

On wayside and parkland trees, especially isolated trees of *Fraxinus* in pastures, particularly on limestone, scarce on wood; formerly common and widespread before the industrial revolution but now rare and restricted. Mainly in W. Scotland, Ireland, now very rare in E. England (Suffolk). Europe, E. to Iran, North America, Australia.

Distinguished by the conspicuous orange apothecia on a greenish yellow-tinged, continuous thallus. The saxicolous *C. flavovirescens* **(24)** is closely related.

24. C. flavovirescens (Wulfen) Dalla Torre & Sarnth. (1902)

Like **23**, but differing in the rimose-areolate, not continuous, thallus, and smaller and darker apothecia, to 1.1 mm diam., and orange-brown, often darker and more convex discs. Thallus and apothecia K+ violet-red.

On ± calcareous rock outcrops and walls, or on acid rocks in close proximity to basic rocks or mortar, rare on pure limestone, chiefly in upland areas but also at sea-level near the coast; frequent. Throughout British Is. Europe, North America, Israel.

Usually occurs as large, conspicuous single thalli.

25. C. granulosa (Müll. Arg.) Jatta (1900)

Thallus placodioid, forming closely-adpressed ± rounded thalli, yellow to orange, the margin of short, convex lobes, centrally ± granular-areolate; lobes to 2(-3) × *c.* 0.5 mm, convex, ± contiguous, divided by furrows, palmately branched. Isidia abundant, variable in size, to 0.1 mm diam., spherical, usually ± obscuring the centre of the thallus as a granular-areolate crust. Apothecia to 1 mm diam., scattered, rounded, ± sessile, constricted at base; disc concave becoming convex; thalline exciple smooth, swollen, later granular-crenulate. Paraphyses broadened towards the tips, apical cell to 8 μm diam. Ascospores 11–13 × 7(-9) μm, broadly ellipsoid, septum, 3–6 μm wide, about $\frac{1}{4}$-$\frac{1}{2}$ of the length of the ascospore. Thallus K+ violet-red.

On sheltered limestone; very rare. S.W. England (Dorset, Somerset). S. Wales (Gower, Skomer) and N. Wales (Anglesey). S. & C. Europe, North America, N. Africa.

Distinguished by the isidiate, placodioid thallus. The isidia resemble those of *C. littorea* **(32)** which lacks marginal lobes and is coastal. *C. verruculifera* **(49)**, is larger and confined to nutrient-enriched birds' perching basic or acid rocks on or near the coast.

26. C. haematites (Chaub. ex St-Amans) Zwackh (1862)

Thallus crustose, grey, moderately thick, of convex granules, yet ± smooth towards the margin, unevenly areolate towards the centre, often surrounded by a broad, pale grey, blue-tinged prothallus. Apothecia to 1.0 mm diam., ± contiguous, moderate, rounded by mutual compression, constricted at base; thalline exciple grey, smooth and ± swollen at first, becoming thinner when old; disc brownish red, concave when young, becoming flat when mature. Paraphyses simple or forked, septate, broadening towards the tips, the apical cells 3 μm diam. Ascospores 11–14 × 5–8 μm, ellipsoid, septum 3–5 μm wide, about $\frac{1}{4}$-$\frac{1}{2}$ of the length of the ascospore. Thallus K–, apothecial discs K+ violet-red.

On smooth bark, especially twigs of fruit trees and *Populus*; not collected this century and probably extinct. S. England (S. Devon, Worcester, Cambridge). S. Europe, Israel.

A distinctive species, differing from *C. cerina* **(9)** in the brownish red apothecial discs; resembles members of the *Lecanora chlarotera* group.

27. C. herbidella (Hue) Magnusson (1932)

Thallus crustose, wide-spreading, whitish grey to grey, thin and continuous. Isidia numerous, crowded, usually ± obscuring the thallus, simple or branched, elongate, warted-papillate, nodulose, or coralloid, whitish grey, sometimes with a yellow tinge or suffused brownish at the tips. Apothecia to 1.3 mm diam., occasional, scattered, rounded to flexuose; true exciple concolorous, persistent, smooth; disc flat to slightly convex, rough, deep orange-red. Paraphyses simple or forked, septate, broadening towards the apices. Ascospores 10–14 × 6–8 μm, ellipsoid, septum *c.* 5 μm wide, $\frac{1}{3}$-$\frac{1}{2}$ of the length of the ascospore. Thallus and isidia K– or tips K+ faint purple, apothecia K+ violet-red.

On old trees, especially *Quercus*, rarely *Castanea*, in wood-pasture, royal forests and ancient deer parks; rare. England (Wiltshire, Savernake Forest; Sussex, Hampshire, Hereford; Westmorland, Lowther Park). Continental Europe, North America.

Distinguished by the numerous, whitish grey isidia which cover much of the thallus. The apothecia appear to be rather short-lived and of intermittant appearance, when sterile the thallus resembles that of delicate morphs of *Pertusaria coronata*.

28. C. holocarpa (Hoffm.) Wade (1965)

C. furfuracea sensu Wade (1965), *non* Magnusson (1944)

Thallus often immersed, inconspicuous or sometimes forming a thin, continuous, pale grey to black crust. Apothecia to 0.1–0.3(-0.8) mm diam., usually very crowded, contiguous, rounded to contorted; thalline exciple grey, not or poorly developed, usually inconspicous; true exciple well-delimited, often glossy, orange, ± excluded when old; disc orange to orange-brown, flat to ± convex. Paraphyses ± straight, broadened towards the tips, to 4–5 μm. Ascospores 10–15 × 5–10 μm, ellipsoid, septum 3–5 μm wide, *c.* $\frac{1}{3}$ to $\frac{1}{4}$ of the length of the ascospore (Fig. 7d). Thallus K−, apothecia K+ violet-red.

On calcareous stone, mortar, asbestos-cement, less frequently on wood, bark and acid stone, often in sunny, ± nutrient-enriched habitats; very common. Throughout British Is. Europe, E. to Iran, Saudi Arabia, North & South America, S. Africa, Australia.

Distinguished by the crowded apothecia of varying ages, many of which have a well-formed, corticate, often glossy, orange exciple. Separated from *C. arenaria* **(4)** and *C. lactea* **(31)** by the broad septa of the ascospores. *C. holocarpa*, as presently understood, is a variable species; on hard limestone the thallus may be thin, pale and inconspicuous with numerous, clustered or scattered orange apothecia, in urban sites the thallus becomes blackened and the numerous, small, deep orange apothecia stand out starkly in contrast.

29. C. irrubescens (Arnold) Zahlbr. (1898)

Thallus crustose, in small patches, orange, of small scattered to ± contiguous areoles, to 0.5 mm diam., flat, ± angular or irregularly rounded, on a blackish prothallus, often associated with cyanobacteria. Apothecia to 0.2–0.3(-0.5) mm diam., numerous, scattered, ± rounded, orange; true exciple becoming excluded with age; disc concolorous or brownish orange, flat to ± convex. Paraphyses simple, broadening towards the tips, the apical cells 2–3 μm diam. Ascospores 10–13 × 6–7 μm, broadly ellipsoid, septum 3–5 μm wide, about $\frac{1}{3}$-$\frac{1}{2}$ of the length of the ascospore. Thallus K+ violet-red.

On siliceous rock; known only from a single 19th century collection. C. Wales (Merioneth, Barmouth). S. Europe, North America, Saudi Arabia.

It is not not known whether the specimen was from a coastal or mountain site. The small, flat, rounded to irregular, orange-red areoles and small apothecia of a similar colour on a black background are distinctive; its affinities are uncertain.

30. C. isidiigera Vězda (1978)

C. cerina var. *cyanopolia* (Nyl.) Zahlbr. (1930) , ?*C. chlorina* (Flotow) H. Olivier (1909)

Thallus crustose, wide-spreading, medium to dark grey, often with a bluish tinge, thin to thickish, usually coarsely cracked-areolate, irregular, surface unevenly granular. Isidia ± covering the thallus, very small, fine (× 25), spherical, ± coralloid, usually poorly differentiated and giving the surface a mealy appearance. Apothecia to 1.5 mm diam., rather rare, scattered, often irregular, immersed in the granules; thalline exciple thick, grey, corticate, usually densely covered with coralloid isidia; disc yellow-orange to dull orange, flat to convex. Paraphyses simple, apical cell swollen to 5 μm. Ascospores 10–15 × 5–7 μm, septum 3–5 μm wide, about $\frac{1}{3}$ of the length of the ascospore. Thallus and isidia K+ pale violet, apothecial discs K+ violet-red.

On acid and especially damp, or shaded, ± basic rocks, often near water, rarely mortar, particularly on the horizontal tops of limestone tombs, rarely on nutrient-rich bark. Throughout British Is. Europe.

Distinguished by the mealy, isidiate, bluish-tinged thallus. The isidia are sometimes poorly differentiated and frequently pass unnoticed in the field. Some early records of *C. chlorina* (which could represent an earlier epithet for this species) belong here.

31. C. lactea (Massal.) Zahlbr. (1901)

Thallus thin, forming a whitish stain or immersed and inconspicuous, prothallus indistinct. Apothecia similar to *C. holocarpa* **(28)** but often more rounded, constricted at base and widely dispersed, sometimes forming shallow pits when shed; true exciple prominent, sometimes ± elevated. Ascospores 15–20 × 8–10 μm, septum 1–2 μm wide, less than $\frac{1}{8}$ of length of the ascospore (Fig. 7c). Thallus K−, apothecia K+ violet-red.

On limestone, chalk pebbles and shells, especially at the coast; local. Throughout Britain. Europe, North America, E. to Iran, New Zealand.

The ascospores need to be examined to confirm the identity of the species and specimens with small apothecia, resembling those of *C. holocarpa* **(28)**, on a whitish thallus in particular, need careful study. The apothecia of *C. lactea* often appear pellucid and the disc irregularly concave when moist.

32. C. littorea Tavares (1956)

Thallus crustose in small to extensive patches, dull orange, thin, continuous to minutely cracked-areolate, sometimes poorly formed, rarely weakly flattened-lobulate at the margins, prothallus pale or absent. Isidia usually very numerous, small, crowded, simple, peg-like to branched-coralloid, often forming a thin to thick and coarsely areolate, dull orange, isidiate crust. Apothecia to 0.8 mm diam., rare, often scattered and few, sometimes in groups, ± rounded, flat; true exciple orange, often becoming excluded; disc orange to deep orange. Paraphyses lax, branched, apices swollen to 4 μm wide. Ascospores 11–15 × 5–7 μm, ellipsoid, septum 1–4 μm wide, to about $\frac{1}{4}$ of the length of the ascospore. Thallus and isidia K+ violet-red.

In dry, sheltered recesses in the xeric-supralittoral zone on acid, rocky sea-shores; local. S. & S.W. England, to N. Scotland. Spain, Portugal.

Readily distinguished by the crowded, dull orange isidia on a crustose thallus in sheltered recesses of coastal rocks.

33. C. lucifuga Thor (1988)

Thallus immersed, inconspicuous, or a thin, grey film. Soralia numerous, pale yellow to dirty yellow-orange-brown, very crowded, often ± contiguous, flat and ulcer-like to convex, ± rounded; soredia farinose. Apothecia unknown. Thallus K−, soralia K+ violet-red.

In crevices of rough bark of old trees, especially *Quercus*, in wood-pasture; rare. W. & N. England (Hereford, Northumberland). W. Europe, from Italy, Spain to Sweden.

Distinguished by the dull-coloured, crowded, erose soralia. *C. ulcerosa* **(47)** has more scattered, discrete, K− soralia, and those of *C. chrysophthalma* **(12)** are more brightly coloured.

34. C. luteoalba (Turner) Th. Fr. (1860)

Thallus crustose, wide-spreading, grey-white, thin, continuous, uneven, without a cortex, developing a characteristic waxy pruinose 'tomentum' of crystals in the herbarium, prothallus pale or absent. Apothecia 0.3–0.5(-0.8) mm diam., numerous, crowded, often ± contiguous, rounded, immersed at first, becoming ± superficial; thalline exciple undeveloped, or sometimes thinly enveloping sides of young apothecia; true exciple orange, ± excluded; disc bright orange, deeply concave when young becoming flat to ± convex when mature. Paraphyses lax, apical cell often swollen, to 5 μm. Ascospores 8–12 × 3–6 μm, ellipsoid, septum *c.* 1 μm wide, about $\frac{1}{8}$ of length of ascospore. Pycnidia orange-red, ± emergent; conidia 3–4 × 1 μm, straight. Thallus K−, apothecial discs K+ violet-red.

On bark, almost exclusively on *Ulmus*, rare on *Acer pseudoplatanus* and *A. campestre*, especially in the wound tracks of injured, old trunks in parklands; very rare. Formerly widespread in lowland S. & E. Britain, N. to Scotland. Europe, North America, E. to Iran.

Much declined due to the loss of *Ulmus* due to Dutch Elm Disease, also rarely, but still persisting, on soft, calcareous stone (e.g. chalk pebbles) and mortar. Characterized by the congregate, rounded, bright orange apothecia which are immersed at first but become convex and superficial, the absence of a persistent thalline exciple, and narrowly septate ascospores. The thallus is unusual in becoming waxy and pruinose in the herbarium. The species often occurs with *Bacidia incompta*.

35. C. marina (Wedd.) Zahlbr. ex Du Rietz (1921)

Thallus crustose, orange to red-orange, sometimes paler, moderately thick, continuous and areolate, or of ± scattered, ± convex areoles, sometimes ± warted, usually growing in ± rounded patches, sometimes with a few, ill-defined, short, convex lobes at the margin, a pale prothallus is sometimes present. Apothecia to 0.8 mm diam., scattered to crowded, rounded, immersed soon becoming ± sessile; thalline exciple ± swollen becoming narrower when old; true exciple sometimes also conspicuous; disc concave at first but becoming subconvex, concolorous or deeper orange. Paraphyses broadened towards the tips, to 5 μm. Ascospores 10–18 × 5–8 μm, ellipsoid, septum 3–5 μm wide, about $\frac{1}{3}$ of the length of the ascospore. Thallus and apothecia K+ violet-red.

On acid and calcareous coastal rocks, forming the mesic supralittoral or sometimes within the upper part of the littoral zone above black *Verrucaria maura* and sea-spray zones; also on concrete sea defences, flints on shingle beaches, or ± inundated consolidated shingle and on wooden groynes, etc.; common. S.W., W. & N. British Is., rarer in the east. Europe, North America.

Distinguished by the areolate to warted thallus without marginal lobes, bearing rounded, ± flat apothecia, constricted at base with swollen exciples when young. *C. thallincola* **(46)**, which grows in the same seashore zone, has well-defined marginal lobes and a preference for more sheltered habitats. A uniformly yellow-orange, possible ecotype, occurs on sheltered, ± vertical, siliceous rock outcrops some distance (< 1 km) from the coast. It tends to have a more waxy aspect and lacks a white prothallus. The relationship of this seemingly distinct entity to the deeper-coloured *C. marina* is in need of critical evaluation.

36. C. microthallina (Wedd.) Zahlbr. (1931)

Thallus of contiguous or often ± scattered, small, individual microsquamules, rounded or ± elongate, 0.5–2 mm wide, convex, yellowish orange. Apothecia to 1 mm diam., scattered, sometimes isolated; flat but becoming convex; true exciple conspicuous, yellow-orange at first, reduced or ± crenulate when mature; disc orange. Paraphyses often with apical cell swollen, to 7 μm diam. Ascospores 12–15 × 6–8 μm, ellipsoid, septum (1-)3 μm wide, about ¼ of the length of the ascospore. Thallus and apothecia K+ violet-red.

On acid and calcareous coastal rocks, usually in the mesic supralittoral zone on *Verrucaria maura*, sometimes on bare rock, often in crevices; locally frequent. W. & N. British Is. Europe, North America.

The yellowish orange, scattered, minute squamules, usually on *Verrucaria maura*, are characteristic. The colour tends to be paler than that of *C. marina* **(35)** with which it often occurs.

37. C. nivalis (Körber) Th. Fr. (1871)

Thallus crustose, spreading over mosses, thin, skin-like or ± coarsely crowded-granular, pale to black-grey; prothallus conspicuous. Apothecia to 0.7 mm diam., numerous, crowded, ± contiguous, rounded, ± sessile, somewhat concave at first, becoming flat or ± convex; thalline exciple pale to dark grey; true exciple ± orange, persistent or ± excluded; disc roughened, orange to rust-coloured or blackish orange, pruinose. Paraphyses 1–3 μm wide, lax, simple to sparsely branched, of chains of obovate cells, submoniliform, apical cell slightly enlarged. Ascospores 25–30 × 3–5 μm, narrowly ellipsoid, sometimes with a median indentation and then slipper-shaped, straight to ± curved, simple, occasionally with signs of a pseudoseptum (Fig. 7a). Thallus K−, apothecia K+ violet-red.

Overgrowing mosses on acid rocks at altitudes exceeding 1000 m., apparently parasitic on *Andreaea* and *Grimmia*; very rare. N. Scotland (Highlands, near the summits of Ben Lawers and Ben Cruachan). Arctic-alpine, N. & C. Europe, North America.

Distinguished by the orange apothecia on mosses in arctic-alpine situations; the ascospores are long and narrowly ellipsoid, unlike any other species of *Caloplaca* (Fig. 7a). The ambivalent form of the ascospores and paraphyses in this species suggests that its inclusion in *Caloplaca* is in need of study.

38. C. obliterans (Nyl.) Blomb & Forss. (1880)

Thallus of contiguous or ± scattered areoles, to 0.3 mm wide, often elongate-irregular and somewhat sparingly branched, very closely adpressed, flat to ± convex, orange to brownish orange, prothallus sometimes present, pale, occasionally fimbriate. Soralia few and scattered to numerous and ± continuous, ± punctate or irregular in shape; soredia small-granular. Apothecia rare, not known in British material, to 0.5 mm diam., becoming convex. Ascospores 14 × 6 μm, the septum about ⅓ of length of ascospore. Thallus and soredia K+ violet-red.

On ± basic to acid, ± deeply shaded rocks, chiefly on cliffs, lowland and upland; rare and local. N. Scotland (C. & W. Highlands and islands). N. & C. Europe, North America, Israel.

Identified by the presence of orange soralia on an orange thallus of ± contiguous to scattered, irregular areoles; marginal lobes are absent. In very sheltered situations the areoles become very small, scattered, often with a white, partially fimbriate, interconnecting prothallus and very few soralia.

39. C. obscurella (Lahm ex Körber) Th. Fr. (1871)

C. sarcopisioides (Körber) Zahlbr. (1901)

Thallus crustose, initially thin, continuous, pale to dark grey, vivid green when wet, smooth to weakly areolate, with scattered to contiguous, minute, rounded, erose soralia, 0.1–0.3 mm diam., becoming uneven-scurfy, whitish to pale yellow-green or bluish grey, concave to flat, derived from the rupture of convex, pimple-like vesicles; prothallus ± present, grey or bluish black. Apothecia to 0.8 mm diam., rather rare, small, inconspicuous, scattered, rounded, flat or slightly swollen; thalline and a true exciple at first both distinct but soon excluded, dull to dark brown, pellucid when wet; epithecium pale greyish brown due to colour of tips of paraphyses, lacking the coloured crystals present in most other species of the genus. Paraphyses lax, simple to forked, septate, with dark globose apical cells 3–6 μm wide, Ascospores 10–13 × 6–8 μm, ellipsoid, septum 4–8 μm wide, about half (or more) of length of the ascospore. Thallus and apothecia K−.

On ± nutrient-rich, basic bark in damp situations; frequent. Throughout British Is., especially in the W. Europe.

Unusual in the apparent lack of physcion and corresponding K reactions. The scurfy texture of the older parts of the thallus is derived from the surplus of the numerous vesicles; in sheltered habitats the margins of the broken vesicles may remain intact. The apothecia are distinctive, being brown, rounded, flat or ± swollen, constricted at the base and soon immarginate, containing paraphyses with dark, globose, apical cells. Specimens are often sterile and these may be recognized by the rounded, erose, pale spots scattered over the grey, uneven thallus. The erose spots of *C. ulcerosa* **(47)** are similar, but usually more scattered and the soredia never turn bluish grey; the thallus in this species is pale and smooth.

40. C. ochracea (Schaerer) Flagey (1886)

Thallus crustose, white-grey, ± patchily tinged yellow, forming a thin, even film, rarely ± finely areolate, the surface non-corticate, of hyphae and crystals giving a 'felted', matt appearance, prothallus thin, black, sometimes mosaic-forming. Apothecia to 0.5 mm diam., frequent, discrete, rarely contiguous, rounded, urceolate, ± immersed at first, becoming ± sessile, rounded; true exciple swollen, waxy, orange; disc concave to flat, concolorous. Paraphyses apical cells only slightly broadened, 2–4 μm. Ascospores 12–15 × 5–7 μm, ellipsoid, internal locules dumb-bell-shaped, with the two end loculi connected by a wide canal which divides into two, becoming four-loculate when mature (Fig. 7f). Thallus and apothecia K+ violet-red.

On sunny, hard limestones; occasional. Chiefly in S. & S.W. England extending to N. England (Derbyshire, Peak District). S. Europe, N. Africa, Israel.

Distinguished by the unusual ascospores with a dumb-bell-shaped internal cavity when mature. The non-corticate, patchily yellow thallus, with small, regular, waxy, orange-brown, ± concave apothecia, immersed in the thallus at first, are useful field characters.

41. C. pollinii (Massal.) Jatta (1900)

C. phaeocarpella (Nyl.) Zahlbr. (1931)

Thallus crustose, wide-spreading, inconspicuous, immersed to superficial, uneven to warted, ± rimose-cracked, pale to dark grey; prothallus inconspicuous or pale. Apothecia to 1.0 mm diam., scattered to crowded-contiguous, rounded or ± flexuose, flat, becoming convex when mature; true exciple at first thick, excluded as the apothecium becomes convex, red-brown to black; epithecium and hymenium patchily yellowish brown; discs ± concolorous, the pigmentation variable, often darker towards the centre of the discs and sometimes with a reddish tinge, especially towards the exciple. Paraphyses branched, thickened to 3 μm towards the tips. Ascospores 13–15 × 7–10 μm, ellipsoid, septum 3–5 μm wide, $\frac{1}{5}$–$\frac{1}{3}$ of length of the thick-walled ascospore. Thallus K−, apothecia K+ violet-red.

On acid bark of trees; not collected this century and probably extinct. N. England (Yorkshire), S. Scotland (Grampian Region). C. Europe, North America.

Distinguished by the large, dark apothecia, which become convex when mature, the coloured epithecium and the thick-walled ascospores. *C. phaeocarpella* was separated on account of the dark brown apothecia, but falls within the variation of *C. pollinii* **(41)**.

42. C. ruderum (Malbr.) Laundon (1976)

Thallus crustose, yellow to yellowish white, thick, very uneven, of prominent, usually pruinose, markedly convex to spherical ± contiguous, irregularly-shaped, sometimes almost squamulose, granule-like areoles, areoles 0.5–1.5 mm diam. Apothecia to 1.5 mm diam., scattered to crowded, ± rounded but often distorted by compression when old; thalline exciple thick, swollen, persistent, pruinose, 0.2 mm thick; true exciple thinner; disc brownish orange, at first urceolate, becoming flat. Paraphyses simple, many with swollen apical cells to 5 μm wide. Ascospores 10–15 × 6–8 μm, ellipsoid, septum 2.5–5.0 μm wide, between $\frac{1}{5}$-$\frac{1}{3}$ of length of the ascospore. Thallus and apothecia K+ violet-red.

On soft, calcareous walls and mortar, especially on the S. side of medieval churches, rarely also on basic maritime rocks on W. coasts; frequent in S. & E. England, rare in W. British Is. W. part of N. European plain.

Characterized by the yellow (never orange), usually pruinose thallus of coarse, convex areoles and the urceolate young apothecia with a thick, swollen thalline exciple; the thallus has an overall hoary appearance.

There is a similar, unidentified species with smaller, more crowded, non-pruinose, ± warted, matt areoles of a deeper orange-red. Apothecia to 2 mm diam., numerous; disc flat or concave, deep orange-red; thalline exciple often also present, incomplete, ± granular; true exciple paler, flexuose. Ascospores 13–16 × 9–12 μm, septum 2–2.5 μm thick. On sunny, exposed, dry, hard limestones; rare. S.W. England (S. Devon, Berry Head), S. Wales (Pembroke, Bosherston), S.W. Scotland (Lismore Is.).

43. C. saxicola (Hoffm.) Nordin (1972)

C. arnoldii (Wedd.) Zahlbr. ex Ginzb. (1915), *C. miniatula* (Nyl.) Zahlbr. (1931)

Thallus placodioid, forming closely-adpressed, ± rounded, rosette-like thalli, yellow to deep orange, the margin of short, ± contiguous, finger-like to broad lobes, to only 2 mm long and 1.5 mm wide, convex, sometimes divided by furrows, often ± densely white-pruinose, the central area extensive, of convex granules with crowded apothecia. Apothecia to 1.0 mm diam., abundant and crowded, usually obscuring the central area of thallus; exciple conspicuous, yellow to orange, excluded when mature; disc orange to brownish orange, flat at first but convex at maturity. Paraphyses tips capitate, 6–8 μm diam. Ascospores 10–15 × 5–7(-8) μm, ellipsoid, septum 2–4 μm wide, about $\frac{1}{4}$-$\frac{1}{5}$ of length of ascospore. Thallus and apothecia K+ violet-red.

On nutrient-rich or -enriched, acid and calcareous rocks and walls, often in dry, sheltered situations; often the most common species of *Caloplaca* on building stone in the W., rare on bark and wood. Throughout British Is. Europe, North America, Asia.

Identified by the short often contiguous lobes and centrally crowded apothecia which become convex when mature. Specimens from overhangs lacking pruina named as *C. arnoldii* have ascospores, on average, smaller than typical *C. saxicola* (10–13 μm vs. 12–15 μm long); their affinities need to be studied in more detail.

44. C. scopularis (Nyl.) Lettau (1912)

Like **43**, but more robust and not pruinose, the apothecia constricted at the base and with swollen exciples when young.

On the tops of heavily entrophicated birds' perching rocks; locally frequent. N. Scotland (St Kilda, Thurso, Wick, Shetland), S. Wales (Pembroke, Skomer). N. Europe, North America.

45. C. teicholyta (Ach.) Steiner (1895)

C. albolutescens (Nyl.) H. Olivier (1909), *C. erythrocarpa* auct., non (Pers.) Zwackh (1862)

Thallus ± obscurely placodioid, forming closely adpressed rounded thalli, thin to thick, greyish white; lobe-ends rounded, obtuse, flat or slightly convex, without furrows, often poorly differentiated, forming a ± complete lobate zone around the edge of the thallus; ± uniform, plain and scurfy towards the centre, often with small granules, sometimes weakly areolate. Soredia farinose, whitish, poorly developed, formed by the erosion of the surface in the central area. Apothecia to 0.8 mm diam., uncommon but conspicuous when present, scattered, sometimes crowded, immersed in the thallus and deeply concave at first, becoming ± flat; thalline exciple sometimes also present, white, undulate; true exciple flexuose, swollen, bright orange; discs orange-red-brown, sometimes ± white-pruinose. Paraphyses slender, flexuose, branched, not noticeably swollen at the apices. Ascospores 15–18 × 7–10 μm, ellipsoid, septum often under 4 μm wide, up to $\frac{1}{4}$ of the length of the ascospore. Thallus K− or + pale violet, apothecia K+ violet red.

On calcareous stone, especially limestone and mortar, also ironstone and, more rarely, brick; mostly on buildings, walls and monuments, especially chest-tombs in churchyards; the natural habitat is ± basic, coastal rocks, from where it has spread inland; locally common. S. & E. Britain, rare in the W. extending to Sutherland on the coast. S. Europe, N. America, Israel.

The mealy, grey ± placodioid, rounded, patches of this lichen are very distinctive. It is usually sterile, but apothecia, of striking colour, do occasionally occur, especially on natural substrata. *C. erythrocarpa* has been recorded from the British Is. in error; the collections are of *C. teicholyta*.

46. C. thallincola (Wedd.) Du Rietz (1921)

Thallus placodioid (Fig. 36d), forming bright orange, closely adpressed, rounded or palmate thalli; lobe-ends 0.3–0.7 mm broad, convex, finger-like, contiguous, divided by frequently parallel furrows; the central area convex-areolate. Apothecia to 0.8 mm diam., mostly confined to the central area, usually scattered; thalline exciple orange, ± excluded when old; discs brownish orange, at first flat but becoming convex. Paraphyses lax, some simple but most forked towards the tips, which are slightly swollen. Ascospores 11–14 × 8–11 μm, swollen, ellipsoid to rhomboid to lemon-shaped, septum 4–5 μm wide, between $\frac{1}{3}$-$\frac{1}{2}$ of the length of the ascospore. Thallus and apothecia K+ violet-red.

On acid, rocky coasts; in the mesic to submesic supralittoral zones, with *Caloplaca marina* **(35)**, often overgrowing *Verrucaria maura*, more rarely on limestone cliffs near the sea, often in more sheltered situations than *C. marina*; common. S.W., W. & N. Britain. W. Europe, from France N. to Scandinavia, North America.

Closely related to *C. flavescens* **(22)** and distinguished by the divided, finger-like lobes and bright orange, almost shiny, as opposed to matt, orange colour. Most paraphyses of *C. thallincola* are branched, whereas in *C. flavescens* most are simple. *C. thallincola* generally inhabits different habitats from *C. flavescens*, but the latter may occasionally occur in similar sites, such as on sea-walls.

47. C. ulcerosa Coppins & P. James (1979)

Thallus crustose, wide-spreading, greyish white, forming a thin, continuous, non-areolate, film. Soralia rather scattered, discrete, rarely ± coalescing, very small, concave, rounded, to 0.3 mm diam., greyish to pale yellow-green, arising from low, bursting pustules, forming ulcer-like craters in the thallus; soredia coarsely farinose. Apothecia to 0.6 mm diam., usually present, usually scattered, soon emerging to sessile; thalline exciple thin, greyish white; true exciple orange, both exciples may become excluded; disc ± concave at first, soon becoming flat then convex, bright orange. Paraphyses tips swollen, to 6 μm wide. Ascospores 8–12 × 4–8 μm, broadly ellipsoid, septum 3–6 μm wide, between $\frac{1}{3}$-$\frac{1}{2}$ of the length of the ascospore. Thallus and soredia K−, apothecia K+ violet-red.

On nutrient-rich, basic bark of well-lit, mature trunks, *Ulmus*, *Acer pseudoplatanus*, *Fraxinus* in wayside sites; local. Throughout British Is. N. Spain, Germany (Schleswig Holstein), S. Sweden and Estonia.

When sterile, very similar to *C. obscurella* **(39)**.

48. C. variabilis (Pers.) Müll. Arg. (1862)

Thallus crustose, grey to deep grey or greyish brown, rather thin to thick, fairly smooth but markedly areolate, ± rounded, with the central area darker and the margin occasionally surrounded by a black prothallus. Apothecia to 1 mm diam., scattered to crowded, ± sessile, flat, becoming slightly convex; thalline exciple often present, ± dense, grey-bluish pruinose, persistent; disc black; epithecium pale. Paraphyses broadening and becoming rather stout towards the tips, to 3 μm diam. Ascospores 14–16(-21) × 7–9 μm, ellipsoid, septum 2–3(-5) μm wide, at times up to $\frac{1}{3}$ of the length of the ascospore. Thallus and apothecia K− to + pale violet.

On ± nutrient-rich limestone, on both natural boulders and stonework, especially the horizontal tops of chest-tombs and the copings of walls; local. Throughout England and Wales, most frequent in English Midlands, extending locally to Scotland and Ireland. Europe, North America, E. to Iran.

Distinguished by the dark, thick, areolate thallus, sessile, blackish apothecia with often pruinose exciples, stout paraphyses, and large ascospores. Old records often include *C. alociza* **(1)**.

49. C. verruculifera (Vainio) Zahlbr. (1931)

Like *C. granulosa* **(25)**, but differing in the larger size; lobes to 6 mm long, 0.5–1.0 mm wide, and the thallus often of a correspondingly greater diameter; ascospores are narrower with a narrower septum, typically narrowly ellipsoid with parallel sides, 12–14 × 5–6 μm, the septum only 1–3 μm wide.

On nutrient-enriched, acid and calcareous maritime rocks, rarely on timber of coastal fence posts; locally common. W. & N. British Is. N. Europe, North America.

The majority of old records of *C. granulosa* **(25)** belong here.

50. C. virescens (Sm.) Coppins (1980)

Thallus crustose, wide-spreading, pale to dark grey, often tinged bluish, green-grey when fresh, thick, coarsely areolate, unevenly coarsely granular, becoming mostly sorediate. Soredia granular, forming a thick crust of irregular thickness over the surface. Isidia corticate-granular or absent. Apothecia 0.4–1 mm diam., scarce, scattered or a few ± contiguous, rounded to ± irregular and contorted, flat or uneven; thalline exciple granular, grey, becoming ± excluded; disc pale orange. Paraphyses flexuose, simple to forked, some apical cells swollen to 6 μm diam. Ascospores 11–13 × 4–5 μm, narrowly ellipsoid, septum 1–3 μm wide, to ¼ of the length of the ascospore. Thallus K−, or K ± violet in squash preparations, apothecia K+ violet-red.

On the base of wayside, nutrient-enriched trunks, *Fraxinus*, *Acer pseudoplatanus*, and especially *Ulmus* often spreading extensively; probably formerly widespread, but now scarce. Lowland Britain. Belgium.

This little-known species has doubtless been much overlooked on account of the usually sterile condition. It is identified by the thick, greenish to dirty grey, often tinged blue, granular-sorediate thallus; it is most closely related to *C. isidiigera* **(30)** and belongs to the *C. cerina* **(9)** complex.

CANDELARIA Massal. (1852)

Thallus minutely fruticose or ± foliose, yellow-green to chrome-yellow, K−, both surfaces corticate. **Photobiont** chlorococcoid. **Upper and lower cortices** paraplectenchymatous. **Ascomata** apothecia, laminal, sessile. **Thalline exciple** persistent. **True exciple** rudimentary. **Epithecium** yellow-brown, granular. **Hymenium** colourless, I+ blue. **Hypothecium** colourless. **Hamathecium** of paraphyses, apices swollen, simple, nodular. **Asci** 20- to 50-spored, *Candelaria*-type, with apical dome I + blue only in the lower part and penetrated by a broad apical cushion (Fig. 33n). **Ascospores** colourless, ellipsoid, simple or thinly 1-septate, biguttulate. **Conidiomata** pycnidia, laminal, immersed within raised warts on the thallus surface; wall colourless. **Conidiogenous cells** enteroblastic, elongate-ampulliform, arising in groups on branched conidiophore cells. **Conidia** ellipsoid, simple, colourless. **Chemistry**: calycin and pulvinic dilactone. **Ecology**: on nutrient-rich bark, rarely rock.

LITERATURE: Hillmann (1936), Maas Geesteranus (1947).

1. C. concolor (Dickson) B. Stein (1879)

Thallus in small, discrete, suborbicular cushions, irregularly more wide-spreading, coalescing or in scattered fragments, ± closely adpressed, occasionally ± dissolved into coarse, ascending granules; lobes flattened, finely divided, *c.* 1 mm long, 0.1–0.5 mm wide, sometimes very narrow and almost hair-like, surface flat or ± wavy, often somewhat raised and fan-like, the margins distinctly crenulate, ± overlapping, entire, or becoming ± isidiate-granular or coarsely sorediate, bright yellowish green to chrome-yellow above, matt, almost white beneath; rhizines scattered, simple, white. Apothecia 0.4–1 mm diam., rare. Disc dull greenish orange or paler brownish yellow; thalline exciple smooth, becoming ± granular-uneven with age, concolorous. Ascospores 6–14 × 4–6 μm, biguttulate. Pycnidia *c.* 100 μm diam., rare; conidia 1.8–2.7 × 1 μm. Thallus Pd−, K−, KC−, C−, UV− or ± orange-black.

On nutrient-rich bark of well-lit, wayside broad-leaved trees, especially *Acer*, *Fraxinus*, *Tilia* and *Ulmus*, wooden fences, occasionally on ± nutrient-enriched rocks and walls; local, decreasing due to agricultural pollution. Throughout British Is. Cosmopolitan.

Forming small, delicate, yellow, intricate patches often in the crowns and rain tracks of tree trunks; somewhat resembles *Xanthoria candelaria* but in that species the thallus is less intricately branched, more orange and K+ crimson.

CANDELARIELLA Müll. Arg. (1894)

Thallus crustose, granular, nodular or warted, indistinctly or sometimes markedly placodioid and then lobed at circumference, citrine, yellow-green, yellow-orange. **Photobiont** chlorococcoid. **Cortex** pseudoparenchymatous. **Ascomata** apothecia, sessile. **Thalline exciple**

thin, persistent. **True exciple** indistinct. **Disc** bright yellow, usually flat. **Hypothecium** colourless. **Epithecium** yellow-brown, granular. **Hymenium** 50–75 μm tall, colourless, I+ blue. **Hamathecium** of paraphyses, the apices not or only slightly swollen, mostly simple a few sometimes sparingly branched. **Asci** clavate, 8- to 32-spored, *Candelaria*-type (Fig. 33n). **Ascospores** colourless, ellipsoid, ± cylindrical with rounded ends, straight or curved, or tear-drop-shaped, simple, often biguttulate, a few thinly 1-septate. **Conidiomata** pycnidia, immersed; wall colourless. **Conidiogenous cells** elongate-ampulliform, enteroblastic, acrogenous. **Conidia** ellipsoid or bacilliform. **Chemistry**: calycin, pulvinic dilactone, pulvinic acid; Pd−, K−, KC−, C+ orange. **Ecology**: varied, favoured by mild eutrophication and some species frequent in urban areas.

A yellow, crustose lichen with a K− thallus will belong to this genus. Occasionally markedly citrine-green-coloured chemotypes arise through the suppression of calycin; they are here recognized at the rank of form.

LITERATURE: Gilbert *et al.* (1981), Hakulinen (1954), Harris & Buck (1978), James (1971), Laundon (1970), Poelt & Vězda (1977).

1 Thallus crustose, minutely squamulose or indistinct, without marginal lobes 2
Thallus placodioid, ± orbicular, with closely adpressed marginal lobes .... **2. *medians***

2(1) On rocks 3
On bark, wood, soil or dead cushions of *Armeria* 5

3(2) Thallus distinct, coarsely granular to subsquamulose; apothecia absent to numerous, often contiguous; asci 12- to 32-spored; on neutral or siliceous, often nutrient-enriched rocks 4
Thallus inconspicuous; apothecia frequent, scattered; asci 8-spored; on basic rocks **1. *aurella***

4(3) Thallus bright golden-yellow, ± continuous, uniform or coarsely cracked, of contiguous, coarse, rounded granules .... **5. *coralliza***
Thallus dull yellow-orange, of scattered to compacted, ± flattened, somewhat extended granules in small discrete to contiguous clusters .... **4. *vitellina***

5(2) Soredia absent; thallus of corticate granules or subsquamulose 6
Soredia present, arising from margin of minute, ± flattened, ± crenate granules **3. *reflexa***

6(5) Thallus-granules rounded, 0.01–0.05 mm diam., forming a thin, continuous, uniform crust .... **6. *xanthostigma***
Thallus-granules, ± oval or flattened and somewhat extended, 0.1-0.5 mm diam., often clustered .... **4. *vitellina***

1. C. aurella (Hoffm.) Zahlbr. (1928)

Thallus of scattered, yellow, to green-yellow, convex granules, 0.5–1.5 mm diam., sometimes inconspicuous and disappearing; prothallus thin, continuous, dark grey to black, often extensive and dominant especially in urban areas. Apothecia frequent, 0.2–1.2 mm diam., discrete, ± regularly dispersed, or sometimes more crowded, yellow; thalline exciple ± entire. Asci 8-spored. Ascospores 10–18 × 5–6 μm, oblong, ellipsoid, straight or curved.

On man-made, basic substrata, concrete, mortar and asbestos-cement, rarely on hard, dust-impregnated wood, occasionally on natural limestone outcrops; abundant in urban areas, rare and local elsewhere. Throughout British Is. Widespread in Europe, North America.

The apothecia are frequently the most conspicuous feature on a thin, grey, black or inapparent prothallus.

A bright citrine-green chemotype, f. *smaragdula* Szat. (1931) [syn. f. *heidelbergensis* (Nyl.) P. James (1981)], occasionally occurs interspersed with normal-coloured thalli.

2. C. medians (Nyl.) A. L. Sm. (1918)

Thallus placodioid, to 3 cm diam., orbicular, radiating, the marginal lobes ± contiguous, yellow, citrine or grey-green yellow, surface ± scabrid or pruinose, the centre granular-areolate to minutely coralloid-isidiate, concolorous or commonly greyish compared with the margin; lobes 0.3–1 mm wide, mostly contiguous, discrete to overlapping, flattened to ± convex. Apothecia small, 0.3–1.2 mm diam., occasional, dull yellow, flat to slightly convex with a smooth to crenulate margin. Asci 8-spored. Ascospores 11–17 × 4–6 μm, simple, occasionally 1-septate, rather variable, ellipsoid, oblong, tear-drop-, or slipper-shaped.

On shaded to sunny, nutrient-rich, man-made calcareous substrata, mortar, asbestos cement, tops of vertical gravestones; rare in natural sites on nutrient-rich rock and calcareous birds' perching rocks. E. England (abundant), extending to W. & N.W. Britain (scarce). Mediterranean and C. Europe.

In shaded habitats the thallus is often pale green-grey with only a hint of yellow. A citrine-green chemotype, f. *steepholmensis* O. Gilbert (1981), occasionally occurs with normal-coloured thalli. *C. medians* resembles *Caloplaca decipiens*, which, however, is more orange in colour, K+ purple and has delimited soralia. The only British collection of another placodioid species, *C. arctica* (Körber) R. Sant. (1966), made last century, ostensibly from sea cliffs in E. Scotland (Kincardineshire, Portlethen), probably originated from outside the British Is.; this coastal species has strongly convex, finger-like lobes and multispored asci.

3. C. reflexa (Nyl.) Lettau (1912)

Thallus of dispersed or ± contiguous granules or minute squamules, ± sorediate, sometimes entirely finely sorediate; squamules to 1 mm diam., usually smaller, rounded or lobed to sub-stellate, mostly ± flattened-adpressed, dull yellow-green to citrine-yellow, the margins usually rapidly breaking down into fine-granular soredia; soredia 0.05–0.07 mm diam., spherical, bright yellow, often coalescing and covering the entire upper surface of the squamules, then tending to form a continuous, leprose crust. Apothecia 0.5–1 mm diam., very rare; margin smooth or partly sorediate; disc pale yellow to bright orange-yellow. Asci 8-spored. Ascospores 10–16 × 4.5–5.5 μm, oblong-ellipsoid, slightly curved, simple.

In nutrient-enriched habitats, such as the base of trees fouled by dogs, by hen-runs, rough-barked trees around farmsteads and villages, on sloping trunks of large *Salix*, also as tiny thalli amongst other lichens on slightly nutrient-enriched broad-leaved trees in woodland; shade- and moderately pollution-tolerant. Throughout British Is. Europe, Chile, Tasmania.

Resembles *Caloplaca citrina*, from which it can easily be separated by the K− reaction of the thallus. *C. xanthostigma* **(6)** lacks flattened squamules and the rounded, soredia-like but corticate, granules are larger, more constant in size, and have a dull orange-yellow colour; this species is also rarely found on nutrient-enriched substrata.

It has not been possible to refer any British specimens with certainty to *C. efflorescens* R. C. Harris & Buck (1978), a North American vicariant of *C. reflexa* with 32-spored asci.

4. C. vitellina (Hoffm.) Müll. Arg. (1894)

Thallus yellow, orange to brown-orange, continuous and coarsely cracked or in scattered patches, usually rather thick, composed of dispersed to crowded, small, convex, often somewhat flattened, nodular or subsquamulose granules, 0.5–2 mm wide. Apothecia frequent, 0.5–1.5 mm diam., flat; margin prominent, persistent, smooth to crenulate; disc greyish yellow, sometimes darkening when old. Thalline margin smooth, crenulate, nodular. Asci (12-)16- to 32-spored (Fig. 33n). Ascospores 9–15 × 3.0–6.5 μm, simple to weakly 1-septate. Pycnidia frequent, 80–100 μm diam.; conidia 2.5–3 × 1–1.5 μm, bacilliform.

On siliceous and non-calcareous rocks, walls, wood, bark, tarmacadam and occasionally soil, rusting iron or stained glass, particularly well developed in nutrient-enriched and dusty, man-made habitats; also birds' perching rocks; often abundant. Throughout British Is., in suburban and rural areas. Europe, North America, probably cosmopolitan.

A citrine-green chemotype, f. *flavovirella* (Nyl.) A. Henderson (1981) [*C. flavovirella* (Nyl.) Lettau (1912)], occasionally occurs in association with normal-coloured thalli.

A very variable species in which the thallus, especially when young, may be reduced to a thin, inconspicuous, ± continuous crust of scattered to contiguous, convex to spherical granules or is compacted into small, shortly lobulate clusters of often fertile, subsquamulose rosettes; the colour ranges from yellow to dull deep orange.

5. C. coralliza (Nyl.) Magnusson (1935)

Like **4**, but the thick, coarsely rimose-cracked thallus surface is entirely composed of ± continuous, or occasionally compacted, ± spherical, granular-coralloid granules, 0.05–0.3 mm diam. Apothecia 0.6–

1.5 mm wide, rare. Thalline margin granular. Ascospores 10–14 × 4.5–6 μm, oblong-ellipsoid, some ± curved. *C. coralliza* is consistently bright golden-yellow without any hint of orange.

C. coralliza may occur with *C. vitellina* but is not an urban species and has a much more restricted range of habitat confined to natural siliceous outcrops or isolated boulders frequently used as birds' perches in inland and, especially, coastal areas; frequent on sarsen stones and basalt. Throughout British Is., but local and scarce in many areas. Throughout C. and N. Europe.

6. C. xanthostigma (Ach.) Lettau (1912)

Thallus entirely of corticate, soredia-like, granules, 0.07–0.1 mm diam., which form a ± uniform, effuse, continuous, dull orange-yellow crust. Apothecia 0.2–0.9 mm diam., rare, concolorous, scattered. Asci 12- to 32-spored. Ascospores 9–12 × 4–5 μm, oblong or ovoid.

On vertical trunks of old, rough-barked, well-lit trees, especially *Acer*, *Fraxinus* and *Ulmus*, in parklands, glades or at the margins of woodland, not or only weakly nitrophilous. Widely distributed, but local throughout the British Is., rare in E. Europe, North America.

Characterized by the uniformly thin, minutely corticate-granular thallus, its absence from nutrient-enriched sites, and is duller in colour and lacks the true soredia of *C. reflexa* **(3)**.

CARBONEA (Hertel) Hertel (1983)

Thallus crustose, ± superficial, sometimes immersed and inapparent, or commensalistic without an independent thallus. **Photobiont** chlorococcoid. **Ascomata** apothecia, black, shining, subconcave to plane or subconvex; disc not pruinose. **Thalline exciple** absent. **True exciple** well developed, persistent, raised, black, opaque; hyphae dark-pigmented. **Epithecium** blue-green to emerald-green. **Hymenium** colourless in lower parts, dull blue, or ± vivid blue-green to bright emerald-green above. **Hypothecium** pale yellow-brown to blue-green in upper parts, dark red-brown to brown-black below, K−. **Hamathecium** of conglutinated, simple paraphyses, 1–2 μm thick, the apical cell to 3.5 μm wide, coherent in a gelatinous matrix. **Asci** 8-spored, *Lecanora*-type, clavate. **Ascospores** oblong-ellipsoid, simple, small, colourless; perispore absent. **Conidiomata** pycnidia, immersed. **Conidiogenous cells** elongate, ampulliform, enteroblastic, arising singly or in small groups. **Conidia** curved, thread-like, simple, colourless. **Chemistry**: lichen products not detected by t.l.c. **Ecology**: on siliceous rocks and walls in upland areas, or on lichens.

Distinguished by the vivid aeruginose-blue hymenium and opaque, black, true exciple.

LITERATURE: Hawksworth (1983), Hertel (1967, 1983, 1985), Hertel & Rambold (1985).

1 On lichens; pycnidia absent 2
 Not on lichens; pycnidia usually present .... **3. *vorticosa***

2(1) On *Candelariella vitellina*; ascospore ends rounded; epithecium smoky blue to emerald .... **2. *vitellinaria***
 On *Lecanora polytropa*; ascospore ends attenuate; epithecium clear pale blue . **1. *supersparsa***

1. C. supersparsa (Nyl.) Hertel (1983)

Thallus immersed in host. Epithecium pale bright blue or grey-blue. Ascospores 9–13 × 5–7 μm, ovoid to fusiform, the ends attenuate.

On thallus of *Lecanora polytropa*, perhaps also on *Rhizocarpon* and *Toninia*, parasitic. In upland and montane areas; very local. C. & W. Scotland, Is. of Skye.

2. C. vitellinaria (Nyl.) Hertel (1983)

Thallus immersed in host. Epithecium smoky blue, greenish blue or emerald. Ascospores 7–12(-13) × (4-)5–6(-7) μm, ellipsoid, the ends rounded.

On thallus of *Candelariella vitellina*, parasitic; widespread but local. S.W. England (Cornwall) northwards.

3. C. vorticosa (Flörke) Hertel (1983)

Thallus in irregular patches 5–15 mm wide, often evanescent, pale olivaceous or greyish to ± blackened, irregularly wrinkled-granular to subareolate, cracked-areolate, hummocky, subsquamulose to eroded-dispersed, matt or shining. Photobiont cells 6–19 μm diam. Apothecia 0.1–0.5(-0.7) mm diam., numerous, scattered, small, solitary or in clusters of 2–10, rounded to ± distorted through mutual pressure; disc black, matt or shining, flat or concave; true exciple ± slightly raised, entire to subcrenate; hymenium 40–55 μm tall; hypothecium dark brown. Ascospores (8-)10–12 × 3.5–4.5(-6) μm, oblong-ellipsoid, straight or slightly curved. Pycnidia *c.* 60 μm diam., frequent; wall green-brown in upper part, ± colourless below; conidia 17–30 × 0.7 μm. Pd−, K−, KC−, UV− (no lichen substances detected by t.l.c.).

On siliceous rocks and walls in upland to montane habitats; widespread, but often overlooked. N. Wales, N.W. England (Lake District, on metal-rich rocks), Scotland (Highlands). Arctic and N. Hemisphere mountain ranges, Himalayas, South America (Andes), New Zealand, subantarctic and antarctic islands.

CATAPYRENIUM Flotow (1850)

Thallus squamulose (Fig. 44b), attached to the substratum by intricate, colourless or brownish rhizoidal hyphae. **Upper cortex** paraplectenchymatous; **lower cortex** composed of rounded to angular cells, or absent. **Photobiont** chlorococcoid. **Ascomata** perithecia, immersed in the thallus or emergent between the squamules. **Involucrellum** present or absent. **True exciple** composed of tangentially arranged, ± elongated cells, pigmented around the ostiole, lower part pale or ± darkened. **Hamathecium** of periphyses; paraphyses absent. **Asci** 8-spored, thin-walled, clavate or cylindrical. **Ascospores** uniseriate or biseriate, colourless, simple, ellipsoid or ± ovoid-clavate in some species. **Conidiomata** pycnidia, laminal, immersed in the upper surface of the squamules or forming spherical projections at their margins. **Conidia** oblong or shortly bacilliform. **Chemistry**: no lichen products detected by t.l.c. **Ecology**: mostly on soil or over decaying bryophytes and plant debris, some species preferring calcareous substrata, very rarely on bark.

Closely related to *Placidiopsis* but with simple spores.

LITERATURE: Breuss (1990), Breuss & Hansen (1988).

1 Perithecia immersed in the squamules; involucrellum absent; thallus distinctly squamulose 2
Perithecia between the squamules, globose, black; involucrellum entire; thallus minutely squamulose and crust-like .... **10. *waltheri***

2(1) Rhizoidal hyphae brown, forming a dense, dark prothallus; asci clavate 3
Rhizoidal hyphae colourless; asci cylindrical 5

3(2) Squamules <4 mm across, irregularly rounded or ± elongated, especially the peripheral ones, lower cortex absent 4
Squamules rarely >2 mm, totally adpressed, finely notched to deeply incised, usually ± whitish pruinose; lower cortex present .... **2. *cinereum***

4(3) Usually on bark; distal parts of squamules often with grey or violet-grey pruina in defined patches .... **7. *psoromoides***
Usually on soil; pruina, if present, not in defined patches .... **3. *daedaleum***

5(2) Squamules usually loosely adpressed or partly ascending, crowded to overlapping, pycnidia marginal 6
Squamules ± firmly adpressed, not ascending, dispersed to contiguous; pycnidia laminal or unknown 8

6(5) Pycnidia less conspicuous; conidia 3–5 μm long, oblong 7
Pycnidia usually very conspicuous, globose, black and friable; conidia 5–7 μm long, bacilliform .... **4. *lachneum***

7(6) Ascospores 15–20 × 7.5–9.5 μm; squamules to 10 mm across, thick and leathery, ± wavy, with thick medullary tissue of intricated hyphae; colourless, hair-like hyphae always absent .... **8. *rufescens***
Ascospores 12–17(-19) × 5.5–7.5 μm; squamules to 6 mm across, thin; medullary tissue less developed; often with colourless, hair-like hyphae present on margins of younger squamules .... **6. *pilosellum***

8(5) True exciple persistently pale, the uppermost part around the ostiole pigmented 9
True exciple brownish throughout when mature .... **5. *michelii***

9(8) On mortar or calcareous rocks; squamules 2–4 mm across, often with dark margins; lower side black .... **1. *boccanum***
On soil; squamules 2–7 mm across, not dark-margined; lower side pale to brown, rarely blackish .... **9. *squamulosum***

1. C. boccanum (Servít) Breuss (1990)

Squamules 2–4 mm across, scattered to contiguous, firmly attached throughout or with slightly upturned margins, irregularly orbicular to somewhat lobed; upper surface brown, matt or slightly glossy; lower surface black up to the margin, giving a dark rim around each squamule; lower cortex of round to ± angular cells; rhizoidal hyphae colourless. Perithecia immersed; true exciple pale, except around ostiole. Asci cylindrical. Ascospores 12–16 × 5.5–7.5 μm, ellipsoid, uniseriate. Pycnidia laminal, immersed; conidia 3–4 μm long.

Mostly on mortar, rarely on calcareous rocks. England (S. Devon). Mediterranean region. Only once collected in Britain.

2. C. cinereum (Pers.) Körber (1855)

Thallus of small, ± densely aggregated squamules, to 0.3 mm thick, rarely more than 2 mm across, closely adpressed, finely incised, especially at the periphery, sometimes subgranular in central parts of thallus, usually whitish pruinose, rarely naked and brown, often with a darker margin; lower cortex paraplectenchymatious, blackish; hypothallus blackish, spongy, of intricate, rhizoidal hyphae *c.* 4 μm diam. Perithecia numerous, immersed in the squamules, the ostioles sometimes slightly elevated; true exciple pale when young, soon becoming brown or blackish throughout. Asci clavate. Ascospores 17–23 × 6–9 μm, usually ± clavate, biseriate.

On soil, humus or mosses, especially on calcareous ground; common in upland areas, also on consolidated soil on or near the coast. Mainly in N. and W. British Is. Throughout much of Europe, especially in northern and alpine regions.

3. C. daedaleum (Krempelh.) B. Stein (1879)

Squamules to 4 mm across, dispersed to crowded and then thalli often almost rosette-like, thickish, rounded or slightly elongated at the periphery, firmly adpressed or with slightly raised margins; upper surface matt, not pruinose, brown, or weakly pruinose and then with a greyish tinge; lower surface non-corticate; medullary tissue merging into the hypothallus of dark rhizoidal hyphae. Perithecia usually numerous, immersed; true exciple, except for ostiole, pale when young, later brownish throughout. Asci clavate. Ascospores 17–22 × 6–9 μm, oblong-ellipsoid to slightly clavate, biseriate.

On soil or humus or moribund bryophytes, confined to upland areas, rare. E. C. Scotland (Mid-Perth, Breadalbane Mtns.). C. & N. Europe.

4. C. lachneum (Ach.) R. Sant. (1980)

Squamules (Fig. 44b) to 8 mm across, scattered or more often crowded to imbricate, ± rounded or lobed, partly ascending; upper surface brown to dark red-brown, matt; lower surface black; margins with conspicuous, globular pycnidia up to 0.4 mm diam.; lower cortex paraplectenchymatous; rhizoidal hyphae colourless; medullary layer thick, of interwoven hyphae. Perithecia laminal, entirely immersed; true exciple, except for ostiole, colourless. Asci cylindrical. Ascospores 14–18 × 6–8 μm, ellipsoid, uniseriate. Conidia 5–7 μm long, bacilliform.

On soil and humus, fairly common in upland areas. Scotland, Wales. C. & N. Europe, Fennoscandia.

5. C. michelii (Massal.) R. Sant. (1980)

Squamules 2–4 mm diam., usually dispersed, rarely contiguous, firmly attached to the substratum, rounded to slightly incised, pale to medium brown above, dull; lower surface blackish, with a thin, paraplectenchymous cortex; rhizoidal hyphae colourless. Perithecia immersed; true exciple brown in mature ascomata. Asci cylindrical. Ascospores 11–15 × 5–6 μm, ellipsoid, uniseriate. Pycnidia unknown.

On sandy soils; very rare and local. S.E. England (Surrey). Alps, Sweden.

6. C. pilosellum Breuss (1990)

Squamules to 6 mm across, loosely attached or partly ascending, scattered to ± imbricate, rounded or lobed, pale brownish to reddish brown above, matt; younger squamules often with colourless, hair-like hyphal extensions on the margins. Perithecia immersed in the squamules; true exciple pale, except for ostiole. Asci cylindrical. Ascospores 12–17 × 5.5–7.5 μm, ellipsoid, uniseriate. Pycnidia marginal, usually few; conidia 3–4 μm long, oblong.

On soil or humus, often among mosses, preferring coastal areas. S.W. England (Somerset) to Scotland. Europe (sub-Atlantic-Mediterranean), Morocco, Australia.

7. C. psoromoides (Borr.) R. Sant. (1980)

Squamules to 4 mm across, dispersed to contiguous or partly imbricate, ± divided, often somewhat elongated, especially at the periphery, loosely attached or slightly ascending at the apices; upper surface dull greyish or greenish brown to fawn; distal parts of squamules often with grey-violet pruina in defined patches; lower surface non-corticate; medullary layer gradually merging into the dark, hypothalline tissue of interwoven, rhizoidal hyphae. Perithecia moderately frequent, fully immersed; true exciple pale or slightly brownish. Asci clavate. Ascospores 14–18 × 5–7 μm, ellispoid or ± attenuated at the apices, biseriate, sometimes pseudoseptate.

Mostly on bark or bryophytes thereon; very rare and local. E.C. Scotland (E. Perth); reported during last century from England (Sussex) and Wales (Gwynedd, Harlech). Europe, North America, E. Africa, New Zealand.

The only bark-inhabiting species of *Catapyrenium* in Europe, closely related to *C. daedaleum* **(3)**.

8. C. rufescens (Ach.) Breuss (1985)

Squamules to 10 mm diam., crowded to imbricate, ± lobate, thick, leathery, usually undulate; upper surface pale to dark brown, often with a reddish tinge, dull or somewhat glossy; lower cortex paraplectenchymous, of rounded to angular cells; rhizoidal hyphae colourless; medullary tissue well developed, of densely interwoven hyphae. Perithecia completely immersed; true exciple, except for ostiole, colourless. Asci cylindrical. Ascospores 15–20 × 7.5–9.5 μm, broadly ellipsoid, uniseriate. Pycnidia marginal, sparse to abundant; conidia 3–5 μm long, oblong.

On dry, calcareous rocks or on soil; widespread but local. Throughout much of British Is. (Westmorland, N.E. Yorkshire, Derbyshire, Mid Ebudes, Sligo). C. & S. Europe.

Similar to *C. lachneum* **(4)**, from which it can easily be distinguished by its larger ascospores and the different size and shape of the conidia. *C. pilosellum* **(6)** has markedly thinner squamules and smaller ascospores, while *C. squamulosum* **(9)** and *C. michelii* **(5)** both differ in smaller, ± adpressed squamules, smaller ascospores, and in lacking marginal pycnidia.

9. C. squamulosum (Ach.) Breuss (1985)

Dermatocarpon hepaticum auct., non (Ach.) Th. Fr. (1860)

Squamules 2–5 mm across, scattered to contiguous, attached by the entire underside or with slightly upturned margins, pale to dark brown above, matt, irregularly orbicular to somewhat lobed; lower cortex of rounded to ± angular cells; rhizoidal hyphae colourless. Perithecia fully immersed; true exciple pale, dark only around ostiole. Asci cylindrical. Ascospores 12–16 × 5.5–7.5 μm, ellipsoid, uniseriate. Pycnidia laminally immersed in the squamules; conidia 3–5 μm long.

On soil and humus, the most frequent species in British Is; widespread. England, Wales and Scotland (N. to Fife), Ireland. Europe, but rarer northwards.

10. C. waltheri (Krempelh.) Körber (1855)

Thallus minutely squamulose, squamules 0.5–1.5 mm wide and rarely more than 0.2 mm thick, firmly adpressed and confluent to form a continuous, crust-like cover over the substratum; upper surface fawn to dark brown, matt, not pruinose; rhizoidal hyphae brown, forming a dense, dark hypothallus. Perithecia between the squamules, often aggregated to groups or lines, black, spherical, semi-emergent; true exciple brown-black; involucrellum entire. Asci clavate. Ascospores 17–21 × 8–10 μm, ± ovoid to clavate, biseriate.

On soil and humus, confined to high altitudes; one collection. N.C. Scotland (Westerness, Ben Alder). C. & N. Europe.

CATILLARIA Massal. (1852)

Kiliasia Hafellner (1984)

Thallus crustose, immersed, evanescent, rimose, warted or areolate, whitish, greyish, greenish, brownish or blackish. **Cortex** absent or rudimentary. **Photobiont** chlorococcoid (e.g. *Dictyochloropsis*, *Myrmecia, Trebouxia*). **Ascomata** apothecia, variously coloured, usually epruinose. **Thalline exciple** absent. **True exciple** well developed in most species, usually evident at least when young, of coherent (in K), branched, radiating hyphae. **Hamathecium** of simple or sparingly branched, septate paraphyses, the apices abruptly swollen with a dark brown apical cap (species **1–5**), or the apices swollen or not and either colourless or surrounded by a pigmented hood. **Asci** 8(-16)-spored, subcylindrical to clavate, in K/I with a blue outer coat and uniformly blue apical dome (*Catillaria*-type, Fig. 33c; in species **1–5**, **11**, **12**) or with a K/I– conical apical cushion (*Bacidia*- or *Biatora*-types). **Ascospores** 1-septate, colourless. **Conidiomata** pycnidia, usually immersed and inconspicuous. **Conidiogenous cells** enteroblastic, arranged in chains with conidia borne terminally and laterally (in species **1–3**, **13**) or in a single layer and broadly ampulliform. **Conidia** ellipsoid to bacilliform, colourless, simple. **Chemistry**: lichen products not detected by t.l.c., reactions negative except in *C. laureri* **(15)** and *C. pulverea* **(17)**. **Ecology**: various.

As assembled here the genus is heterogeneous. Species **1–5** belong to *Catillaria* s. str. and the others will be transferred when problems concerning generic delimitation and nomenclature are resolved. *Catillaria* s. str. is characterized by the asci with uniformly K/I + blue apical dome, the abruptly swollen paraphyses apices with a dark brown cap, non-halonate ascospores and the conidiogenous cells in chains. *Halecania* shares these features except for the thick-walled ascospores. Species **6–10** belong to *'Kiliasia'*, which is probably referable to *Toninia* s. str. Other crustose genera with *Lecidea*-like apothecia and at least some species with colourless, 1-septate ascospores are *Absconditella*, *Arthonia*, *Bacidia* (immature), *Cliostomum*, *Dimerella*, *Fellhanera*, *Fuscidea*, *Lecania*, *Lecidea* (a few old spores), *Micarea* and *Tylothallia*.

LITERATURE: Coppins (1989b), Kilias (1981), Vainio (1922).

1 Soredia absent 2
Soredia present .... **17. *pulverea***

2(1) Apices of paraphyses abruptly swollen, with a well-defined, dark brown apical cap; asci *Catillaria*-type 3
Apices of paraphyses not to distinctly swollen, if pigmented then without a distinct cap; asci *Bacidia*- or *Biatora*-type 7

3(2) True exciple blackish throughout 4
True exciple pale, or dark-coloured only at outer edge 5

4(3) Apothecia 0.2–0.5 mm diam.; hypothecium dark brown .... **1. *chalybeia***
Apothecia 0.1–0.2 mm diam.; hypothecium colourless .... **2. *atomarioides***

5(3) Ascospores mostly 8–10 × 2.5–3.5 μm 6
Ascospores mostly 10–16 × 4.5–6 μm .... **5. *subviridis***

6(5) On rocks .... **3. *lenticularis***
On bark .... **4. *nigroclavata***

7(2) Hypothecium dark brown, K+ purplish; on rocks 8
Hypothecium colourless or pale-coloured, K− or, if K+ purplish, on bark 10

8(7) Epithecium indistinct, not greenish; not on lichens 9
Epithecium distinct, green-black; on *Aspicilia calcarea* .... **10. *episema***

9(8) Apothecia 0.2–0.4(-0.5) mm diam., adpressed; ascospores 8–11(-15) × (2.6-)3–4(-5) μm .... **8. *modesta***
Apothecia 0.3–0.8 mm diam., constricted below; ascospores (8-)12–16 × 3–4.5 μm **9. *picila***

10(7) Ascospores >3.5 μm; if narrower not on bark 11
Ascospores 1.5–3 μm wide; on bark .... **14. *globulosa***

11(10) Apothecia dark brown to blackish 12
Apothecia pale, pale reddish brown or reddish orange .... **16. *minuta***

12(11) On rocks, mosses over rocks, or on soil 13
On bark, mosses over bark, or on wood 15

13(12) On calcareous rocks 14
On mosses or plant debris over rocks or on the ground; in 'acidic' habitats .. **13. *contristans***

14(13) Ascospores (8-)9–11(-13) × 3.5–4.5 μm .... **6. *aphana***
Ascospores (10-)12–17(-19) × 4.5–5 μm .... **7. *scotinodes***

15(12) Thallus pale to dark grey-brown, or evanescent; K− 16
Thallus whitish to grey-green; K+ yellowish .... **15. *laureri***

16(15) Asci 8-spored .... **11. *atropurpurea***
Asci 12- to 16-spored .... **12. *neuschildii***

1. C. chalybeia (Borrer) Massal. (1852)

Thallus effuse or delimited, sometimes mosaic-forming, ± evanescent or thin and rimose to verrucose-areolate, with areoles 0.1–0.4 mm diam., beige to more usually dark olivaceous to blackish grey or dark grey-brown, matt or slightly glossy; prothallus black, limiting. Photobiont *Myrmecia*; cells 7–17 μm diam. Apothecia (0.15-)0.2–0.5(-1) mm diam., scattered or a few clustered, mostly flat, sometimes becoming convex, black (rarely dark brown), matt; true exciple thin, slightly raised, greenish black, usually concolorous with disc but sometimes paler; epithecium dark brown to green-black; hymenium 40–60 μm tall, colourless or usually pale blue-green (at least in lower part); hypothecium dark brown, K−; green pigment in epithecium, hymenium and exciple K− (or green intensifying), N+ purple-red. Paraphyses 1.5–2 μm wide, simple or rarely forked; apices capitate to 6 μm wide, with dark brown cap. Asci *Catillaria*-type. Ascospores (7.5-)9–12(-15) × 2.5–4 μm, oblong-ovoid to oblong-ellipsoid. Pycnidia 50–100 μm diam., the wall dark green, the outer cells with dark brown caps; conidia 1.8–3.5 × 0.5–0.8 μm. Lichen products not detected by t.l.c.

On wide range of acidic or, more usually, slightly basic or enriched rocks (incl. sea-shore) and stonework, rare on highly calcareous substrata; occasionally on dust-impregnated timberwork and tree trunks; common except above 600 m. Throughout British Is. Europe, N. Africa, North & South America.

Very variable, especially in thallus appearance. Morphs with paler, grey- to dark brown apothecia occur especially on rocks in or by lakes and rivers; these lack any green pigment in the hymenium and are referred to var. *chloropoliza* (Nyl.) Kilias (1981).

2. C. atomarioides (Müll. Arg.) Kilias (1981)

Like **1**, but apothecia 0.1–0.2 mm diam.; hymenium 30–40 μm tall, colourless; hypothecium colourless; ascospores 8.5–13.5 × 3–4 μm.

On hard, acid rocks (including river shingle and slate) and brick. S.W. England (Shropshire), S. Wales, C. Scotland (Perthshire). Sweden, France, Germany, Austria, S. Africa.

Easily mistaken for a diminutive form of *C. chalybeia* **(1)**.

3. C. lenticularis (Ach) Th. Fr. (1874)

Lecania actea (Nyl.) B. de Lesd. (1910)

Thallus usually effuse, ± immersed to thin and rimose, whitish, beige or pale brown, sometimes dark brown or olivaceous due to invasive cyanobacteria. Photobiont *Dictyochloropsis*; cells 7–15 μm diam. Apothecia 0.15–0.4 mm diam., numerous, scattered or in small groups, often partly immersed in thallus or substratum, reddish to dark brown, rarely black (then brownish when wet), flat to convex; true exciple sometimes paler than disc, pale to dark brown at outer edge (many hyphae with brown apical caps), pale brown to colourless within; epithecium pale to dark brown, K−, N−; hymenium 35–50(-70) μm tall, colourless; hypothecium colourless or pale straw. Paraphyses and asci as in *C. chalybeia* **(1)**. Ascospores 7–10(-12) × (2-)2.5–3.5(-4) μm, oblong or oblong-ellipsoid. Pycnidia 50–70 μm diam., the wall brown, the outer cells with dark caps; conidia 2–3.5 × 0.7–1.3 μm. Lichen products not detected by t.l.c.

On limestones and other highly calcareous rocks and building materials; widespread and rather common, especially in areas with abundant limestone. Throughout British Is. Europe, North America.

Often confused with *C. chalybeia* **(1)**, which has a blackish exciple, dark brown hypothecium, usually greenish pigment in the hymenium and is relatively rare on highly calcareous substrata. The dark apical caps of the paraphyses help to distinguish it from small *Lecania* species occurring in similar habitats.

4. C. nigroclavata (Nyl.) Schuler (1902)

Like 3, but differing mainly in growing on bark and the pale to dark brown hypothecium (at least in upper part); thallus thin and often immersed, pale to dark grey or grey-brown; apothecia 0.15–0.3 mm diam, dark brown to black; hymenium 30–40 μm tall; ascospores 8–10 × (2-)2.5–3.5(-4) μm; pycnidia not seen.

On smooth or rough bark of deciduous trees (e.g. *Alnus*, *Corylus*, *Quercus*, *Ulmus*); scattered. N. & W. Britain, Ireland. Europe, North America.

Easily mistaken in the field for such common species as *Buellia punctata*. *C. chalybeia* **(1)** occasionally occurs on bark, but has slightly larger apothecia with a blackish exciple (in section) and usually a green pigment in the hymenium.

5. C. subviridis (Nyl.) Zahlbr. (1926)

Thallus effuse, thin and rimose, beige to green-grey or grey-brown. Photobiont cells 8–12 μm, ellipsoid, ovoid or ± oblong. Apothecia black; true exciple dark brown at outer edge (ends of hyphae with dark brown caps), ± colourless within; hypothecium colourless. Ascospores mostly 10–16 × 4.5–6 μm. Pycnidia not seen. Lichen products not detected by t.l.c.

On nutrient-enriched (e.g. by birds) rocks; very rare. Channel Is., Is. of Scilly. ?Endemic.

Distinguished from *C. chalybeia* **(1)** and *C. atomarioides* **(2)** by the pale inner true exciple, and from *C. lenticularis* **(3)** by the black apothecia and larger ascospores.

6. C. aphana (Nyl.) Coppins (1989)

Lecidea aphana Nyl. (1867)

Thallus effuse, whitish to pale grey, thin to ± verrucose or rimose-areolate, rarely immersed; areoles *c.* 0.1–1.0 mm wide. Photobiont cells 7–13 μm diam. Apothecia 0.2–0.5 mm diam., scattered or a few confluent, dark brown to black; disc concave to flat, later convex; true exciple shallow, becoming excluded; outer and upper parts mostly dark brown, K+ purplish or violaceous, and upper parts dark grey-brown, K+ greenish grey, hyphae coherent in K, with *c.* 1–2 μm wide lumina and thick, gelatinized walls, overall 5–8 μm wide in K; inner parts of exciple colourless to pale straw-yellow; epithecium dark reddish brown, K+ purplish or violaceous, with some grey, K+ greenish grey pigment; hymenium 43–55 μm tall, ± colourless below, pale reddish brown above; hypothecium colourless or pale straw-yellow. Paraphyses 1.5–2 μm wide, coherent in K, simple or occasionally branched; apices clavate, to 4.5 μm wide, most surrounded by dense dark brown or grey pigment. Asci *Bacidia*-type. Ascospores (8-)9–11(-13) × 3.5–4.5 μm, 0- to 1(-3)-septate, ellipsoid to oblong-ellipsoid. Pycnidia not found. Lichen products not detected by t.l.c.

On calcareous rocks (schists, calcite, chalk) on vertical, dry, ± well-lit, coastal and inland cliff faces below 300 m; local. S. England (Dorset), C. Wales, W. Scotland, W. &. N. Ireland. ?Endemic.

Closely related to *C. scotinodes* **(7)**. Morphs with a well-developed thallus can resemble *Diplotomma* species and immersed forms of *Catillaria lenticularis* **(3)**; the latter has paraphyses lax in K, with discrete, dark brown apical caps.

7. C. scotinodes (Nyl.) Coppins (1989)

Kiliasia scotinodes (Nyl.) Coppins (1988), *Lecidea scotinodes* Nyl. (1873), *L. confusior* Nyl. (1874), *Catillaria confusior* (Nyl.) Zahlbr. (1926)

Like **6**, but outer exciple, upper hymenium and epithecium mostly dark green, a brownish, K+ purplish, pigment sometimes additionally present; ascospores larger, (10-)12–17(-19) × 4.5–5 μm.

On calcareous mica-schist and calcareous sandstone on ± vertical, overhung cliff-faces; rare. N.E. England (N.E. Yorkshire), C. Scotland (Angus & Perthshire). ?Endemic.

8. C. modesta (Müll. Arg.) Coppins (1989)

Lecidea modesta Müll. Arg. (1871)

Thallus effuse, pale fawn to ochraceous, thin, rimose or slightly warted. Photobiont cells 5–14 μm diam. Apothecia 0.2–0.4 mm diam., flat to convex; true exciple thin, becoming excluded, the outer edge pale brownish or colourless, the hyphae coherent in K, 2–4(-6) μm wide, the inner part dark reddish brown; epithecium indistinct; hymenium 45–55 mm tall, colourless, or pale reddish brown below (or in vertical streaks); hypothecium dark reddish brown, K+ purplish. Paraphyses 1.3–2 μm wide, simple or branched above, apices swollen to 3(-4) μm wide, colourless. Asci *Bacidia*-type. Ascospores (8-)9–12(-15) × 3.5–4.5 μm, 0- to 1-septate, ellipsoid or ovoid-ellipsoid to oblong. Pycnidia not found. Lichen products not detected by t.l.c.

On limestone. Scotland (West Ross, Loch Maree), Ireland (W. Donegal). Switzerland.

Most British records refer to *Micarea lutulata*. Closely related to *C. aphana* **(6)** and *C. scotinodes* **(7)** but with a dark hypothecium and inner exciple also seen in *C. picila* **(9)**. The *Bacidia*-type ascus structure distinguishes this species and *C. picila* **(9)** from the common *Clauzadea monticola* with *Porpidia*-type asci.

9. C. picila (Massal.) Coppins (1989)

Catillaria anomaloides auct., non (Massal.) Lettau (1912).

Like **8**, but apothecia 0.3–0.8 mm diam., mostly constricted at base or shortly stalked; ascospores (8-)12–16 × 3–4.5 μm.

On calcareous schist and limestone. Scotland (Mid-Perthshire), Ireland (Galway); no modern records. C. & S. Europe.

10. C. episema (Nyl.) H. Olivier (1905)

Kiliasia episema (Nyl.) Hafellner (1984), *Scutula episema* (Nyl.) Zopf (1896)

Thallus immersed in host, inapparent. Apothecia 0.2–0.4(-0.6) diam., mostly in small clusters, at first ± immersed, soon emergent, black; true exciple well developed, raised, becoming excluded, brown-black, K+ purplish; disc concave to convex; epithecium green-black, K+ green intensifying, N+ purple-red plus formation of blue granules; hymenium 35–40 μm tall, colourless or pale greenish above; hypothecium pale reddish brown, K+ purplish, N+ orange-red. Paraphyses simple, 1.7–2 μm wide; apices swollen to 5 μm, most coated in dark pigment. Asci *Bacidia*-type. Ascospores (9-)10–14 × 3.5–4.5 μm, ellipsoid.

On thallus of *Aspicilia calcarea* on hard limestones, commensalistic; local. N. & W. British Is. S. & W. Europe, Algeria.

Recently transferred to *Toninia* as ***T. episema*** (Nyl.) Timdal (1991).

11. C. atropurpurea (Schaerer) Th. Fr. (1874)

Biatorina atropurpurea (Schaerer) Massal. (1852), *Catinaria atropurpurea* (Schaerer) Vězda & Poelt (1981).

Thallus effuse, evanescent, thin, pale to dark grey-brown, often minutely granular; granules 15–70 μm diam., ± immersed, often scattered, minutely speckled appearance when on light-coloured bark. Photobiont cells 5–9 μm diam. Apothecia 0.2–0.6(-0.8) mm diam., reddish brown to dull black; true exciple dark, at first prominent, later sometimes excluded, dark brown at edge, colourless to pale brown within, the hyphae coherent in K, 1.5–2.5 μm wide but thickened by pigment at outer edge; disc at first concave, later flat or sometimes convex; epithecium yellowish to dark brown, K –, N –; hymenium 60–75 µm tall, colourless; hypothecium colourless to pale brownish. Paraphyses, 0.8–1 µm wide, simple or sparsely branched above; apices ± swollen, to 2 µm wide, often with dark brown hood 3(-4) µm wide. Asci clavate, with ± uniformly K/I + blue apical dome. Ascospores 10–15 × 5–7 µm, ellipsoid, with smooth walls to 1.5 µm thick. Pycnidia not seen. Lichen products not detected by t.l.c.

On bark or bryophytes on trunks of mature trees (especially *Fraxinus*, *Quercus*), sometimes on mossy branches, rarely on wood of fallen trees; mainly in *Lobarion* communities in ancient woodlands. Widespread in British Is., especially the north and west. Europe, North America.

Usually easily recognized in the field by the reddish brown to black apothecia, but can be confused with *Lecidea hypnorum* agg., which often occurs in similar habitats. The sterile thallus resembles that of *Zamenhofia coralloidea*. Some specimens from Ireland (Donegal) and C. Europe have ascospores with thinner but warted walls, and may belong to a distinct species. A related, undescribed species with warted, 3-septate ascospores is known from S.E. England (Sussex) and C. Scotland (Perthshire).

12. C. neuschildii (Körber) Th. Fr. (1874)

Biatorina neuschildii Körber (1860), *Catinaria neuschildii* (Körber) P. James (1965)

Like **11**, but for 12- to 16-spored asci, and ascospores 9–13 × 5–6 µm.

On *Juniperus* bark and old *Salix* wood; very rare. C.N. Scotland (E. Easterness, Abernethy Forest; Banff, Cairngorms). N. Europe.

13. C. contristans (Nyl.) Zahlbr. (1926)

Thallus warty-granular, sometimes with discrete ± globose areoles 0.05–0.3(-0.5) mm diam., whitish or usually pale to dark grey, or grey-brown. Photobiont cells 9–16 µm diam. Apothecia 0.2–0.6(-1) mm diam., convex to subglobose, black, often glossy, sometimes (especially when young) with a white, byssoid collar at the base; true exciple reflexed, inapparent from above, colourless or pale greenish or brownish in part, the hyphae coherent in K, 2–2.6 µm wide, occasionally some extending as loose hyphae to form a byssoid collar; epithecium dark green, K –, N + red; hymenium 35–45 µm tall, olivaceous to aeruginose in upper part, pale green, colourless or pale brown in lower part; hypothecium colourless or pale fuscous brown (especially upper part), K –. Paraphyses 1.7–2.3 µm wide, simple or branched (many with short lateral branches), coherent in K, each surrounded by a gel coat; apices not or only slightly swollen, colourless or with an external dark green hood. Asci *Biatora*-type. Ascospores 9–16.5 × 2.5–4.5 µm, (0-)1-septate, oblong or ovoid-oblong, ± constricted at septum. Pycnidia black, ± immersed, the wall dark green; conidiogenous cells in chains; conidia 3.5–4 × 0.7–1 µm, bacilliform. Lichen products not detected by t.l.c.

On bryophytes or plant debris on rocks or on the ground, in 'acid' habitats, especially above 500 m; locally frequent. England (N. Pennines), Scotland (Highlands). Scandinavia, Iceland, Austria and Tasmania.

Similar to *Micarea lignaria* and small morphs of *Lecidea limosa*, which both occur in the same habitats, but can be separated microscopically.

14. C. globulosa (Flörke) Th. Fr. (1874)

Bacidia globulosa (Flörke) Hafellner & V. Wirth (1987)

Thallus effuse, whitish, thin, smooth to powdery, ± rimose. Photobiont cells 7–14(-18) µm diam. Apothecia 0.15–0.4 mm diam., at first ± flat, soon convex to subglobose, usually grey-brown or grey- to greenish black, rarely pale in shade morphs; true exciple indistinct from above, colourless except upper edge which is often concolorous with the epithecium, the hyphae coherent in K, lumina ellipsoid, 2–4 µm wide; epithecium mostly slate-grey to dark olive-green, K + green intensifying, N + red; hymenium 30–40 µm tall, colourless; hypothecium colourless. Paraphyses 1.2–2 µm wide, simple or forked above; apical 1–2 cells often grey- or green-walled and swollen to 6 µm wide. Asci *Bacidia*- or *Biatora*-type. Ascospores 7–12(-16) × 2–2.5 µm, 0- to 1-septate, oblong-fusiform or bacilliform. Pycnidia inconspicuous, partly immersed, 40–100 µm diam., black; the wall dark green; conidiogenous cells in a single layer, short-ampulliform; conidia of two types (a) bacilliform or ± curved, 3.5–5(-6) × 0.5–0.7 µm (commoner), or (b) ellipsoid, 2–2.8 × 1–1.5 µm; the two types not in the same pycnidium.

On rough bark of *Quercus* trunks in woodlands or other sheltered situations; local. Scotland (C. & E. Highlands). Europe.

Often difficult to distinguish in the field from *Bacidia beckhausii*, with which it often grows.

15. C. laureri Hepp ex Th. Fr. (1874)

Catinaria laureri (Hepp ex Th. Fr.) Degel. (1941).

Thallus whitish to grey-green, rimose, thin and smooth, or to 150 μm thick with a granular-warted surface; prothallus black, delimiting. Photobiont cells 6–14 μm diam. Apothecia 0.4–1.1 mm diam.; disc flat to slightly convex; true exciple shallow, often paler than the disc, ± colourless, sometimes brownish (concolorous with epithecium) at upper outer edge, at base of apothecium often concolorous with hypothecium, swelling in K, the hyphae distinctly radiating, branched and anastomosing, 1.5–2 μm wide, without swollen ends; epithecium dark purple-brown, K+ purplish, N+ red, sometimes with additional patches of greenish, K+ green, N+ red pigment; hymenium 65–80 μm tall, colourless to pale purple brownish; hypothecium purple-red, K+ purple in upper part (subhymenium); lower part pale golden-brown, K−. Paraphyses 1.3–1.7 μm wide, simple; apices clavate to 3 μm wide, cemented by adhering pigment. Asci *Bacidia*-type, with distinct apical cushion. Ascospores 12–18 × 5–7 μm, ellipsoid to fusiform-ellipsoid. Pycnidia not found. Thallus Pd−, K+ pale yellow, KC−, C−, UV−.

On trunks of *Fagus*; very rare. S. England (Hampshire, New Forest). W. Europe, from Portugal to S. Scandinavia, E. North America.

Superficially resembles some morphs of the common *Lecidella elaeochroma*.

16. C. minuta (Massal.) Lettau (1912)

Thallus effuse, ± immersed, whitish. Apothecia 0.2–0.3(-0.4) mm diam., light red-brown to reddish orange, at first with concave disc, later flat or convex; true exciple at first prominent, later shallow or excluded, often paler than disc, outer part pale yellow, K−, inner part colourless; epithecium pale yellow, K−; hymenium 50–55 μm tall, colourless; hypothecium colourless. Paraphyses simple, 1–1.5 μm wide; apices ± swollen to 2.5 μm wide. Asci probably *Bacidia*- or *Biatora*-type. Ascospores 12–17 × 4–6 μm, ellipsoid- to ovoid-fusiform, ± constricted at septum; old spores with warted surface. Pycnidia not seen.

In shaded crevices on hard limestone; no modern records. W. Scotland (Argyll), W. Ireland (Connemara). C. Europe, Sweden.

The K− apothecia distinguish them from similarly coloured apothecia of *Caloplaca* and *Protoblastenia* species.

17. C. pulverea (Borrer) Lettau (1912)

Catinaria pulverea (Borrer) Vězda & Poelt (1981)

Thallus effuse, thin, ± evanescent; soralia large, effuse, grey-white to glaucous, granular, sometimes coalescing to form an almost continuous, sorediate crust; soredia 40–100 μm diam. Photobiont cells 5–14 μm diam. Apothecia 0.5–2.2 mm diam., rare, flat, sometimes convex and contorted, black; true exciple distinct, colourless, or pale brownish in upper part, sometimes concolorous with epithecium at upper outer edge, the hyphae coherent in K, 2–3 μm wide, each with a thick gel coat; epithecium dark green, K+ green intensifying, N+ red, sometimes with additional, violet, K+ green patches; hymenium 60–75 μm tall, colourless; hypothecium (and exciple) with minute, colourless crystals that do not dissolve in K or N, colourless, or pale brown in uppermost and/or lowermost parts. Asci *Bacidia*- or *Biatora*-type, the apical cushion narrow, in K/I ± surrounded by thin, dark blue zone. Ascospores 10–16(-19) × 4.5–6.5 μm, broadly to fusiform-ellipsoid, the walls rather thick; perispore absent. Pycnidia not seen. Thallus Pd+ red, rarely Pd− or faint yellow, K+ yellow, KC− (atranorin, zeorin, and usually fumarprotocetraric acid).

On mossy trunks (rarely branches) in old, especially moist woodlands, occasionally on mossy rocks or soil (especially near the coast), and on old *Calluna*. W. Britain. Oceanic N.W. Europe, Alps, Macaronesia, Tasmania, Chile.

Usually easily recognized, even when sterile, but can be confused with *Megalospora tuberculosa*, *Phyllopsora rosei* or thickly sorediate morphs of *Mycoblastus sterilis*.

CATOLECHIA Flotow (1850)

Thallus squamulose to ± placodioid, thick, ridged, lobed at the margins, yellow-green. **Photobiont** unidentified coccoid green algae. **Medulla** well developed, I−, dark brown beneath, with rhizinose strands extending into the substratum; prothallus absent. **Ascomata** apothecia, black, arising between the lobes, immersed at first but becoming superficial. **True exciple** soon excluded, dark reddish brown; cells with scarcely distinct lumina. **Thalline exciple** absent. **Epithecium** dark. **Hymenium** I+ blue. **Hypothecium** dark, of ± radiating hyphae. **Hamathecium** of densely branched and anastomosing paraphyses, not markedly swollen at the tips. **Asci** elongate-clavate, the outer coat I+ blue, ± thickened at the apex, with an internal I+, deep blue, meniscus-like cap, 8-spored. **Ascospores** ellipsoid, rounded at the ends, brown, 0- to 1(-3)-septate, smooth, lacking a distinct perispore, the central septum with an internal torus. **Conidiomata** unknown. **Chemistry**: pulvinic acid derivatives. **Ecology**: on soil, mosses or rock crevices, arctic-alpine.

This monotypic genus resembles *Epilichen* in both ascus structure and the nature of the hamathecial filaments but differs in thallus type and the origin of the apothecia.

LITERATURE: Hafellner (1978), Sheard (1964).

1. C. wahlenbergii (Ach.) Körber (1855)

Thallus bright yellow-green, sometimes stained brownish in parts, puckered and ± folded in ridges, the surface interspersed with yellowish crystals. Apothecia 2 mm diam., arising between the squamules; adjacent cortical tissues dark red in section; epithecium dark olive-green to ± black, K−, N+ rose; hymenium 100–120 μm tall; hypothecium dark brown, reddish in parts. Asci 50–70 × 14–17 μm. Ascospores (12-)13–17(-18) × 7–10 μm, (0-)1(-3)-septate. Thallus Pd+ yellow-orange (medulla), K−, KC−, C− (rhizocarpic acid and ?argopsin).

On acid solifluction soils, mosses and rock crevices above 900 m; very rare. Scotland (Westerness, Ben Nevis, Aonach Beag; Is. of Skye). Arctic-alpine, C. European mtns, Scandinavia, North America.

See also *Arthrorhaphis alpina*.

CAVERNULARIA Degel. (1938)

Thallus foliose, dorsiventrally flattened, or almost tubular, corticate on both sides, directly attached by parts of lower surface, lacking rhizines. **Lower surface** with numerous, small depressions formed by indentation of lower cortex. **Cortex** pseudoparenchymatous. **Photobiont** trebouxioid. **Soralia** present. **Ascomata** apothecia. **Thalline exciple** present. **Asci** 8-spored. **Ascospores** 3–4 μm, colourless, simple, globose. **Chemistry:** cortex contains atranorin; medulla with physodic acid.

Differs from *Hypogymnia*, *Parmeliopsis* and *Foraminella* by the numerous, small, pore-like cavities on the lower surface; *Menegazzia* has true pores in the upper surface.

LITERATURE: Ahti & Henssen (1965), Swinscow (1960).

1. C. hultenii Degel. (1938)

Thallus small, to 2 cm, forming ± orbicular rosettes or fragmentary; lobes to 1 mm wide, ± palmate, radiating, contiguous or overlapping towards centre, ± discrete and horizontally spreading; upper surface somewhat roughened, grey, becoming green-grey when wet. Soralia frequent, on or near lobe-ends, convex, occasionally becoming confluent; soredia finely granular, pale greenish white. Lower surface dark brown-black, without rhizines, with numerous, almost contiguous depressions, 0.1–0.2 mm diam. Apothecia very rare, shortly stalked, with a thick thalline exciple; disc brown. Cortex K± yellow (atranorin); medulla Pd−, K−, KC+ red, C− (physodic acid).

On *Pinus sylvestris, Betula, Sorbus aucuparia* and *Ilex*, mostly in moist, sheltered areas in relict Caledonian pine forests and ravines; rather rare but locally frequent. N.W. Scotland. Scandinavia, W. North America.

Distinguished from *Hypogymnia* by the abundant, perforation-like depressions on the lower surface; *C. lophyrea* (Ach.) Degel. (1938), the fertile counterpart, is confined to N.W. North America. The superficially rather similar *Foraminella hyperopta* and *Parmeliopsis aleurites* have rhizines and lack pore-like depressions on the lower surface.

CELOTHELIUM Massal. (1860)

Thallus crustose, immersed in bark. **Photobiont** *Trentepohlia*-like, the filaments of which are mixed with the bark cells below the surface of the substratum. **Ascomata** perithecia-like, circular to elongated in surface view, black, immersed to superficial, dimidiate and flattened, arising singly or in groups, covered by a common involucrellum with a separate ostiole for each ascoma, clypeate. **Hamathecium** a branched and anastomosing net of paraphysoids; periphyses absent; centrum not turning deep blue in iodine. **Asci** 8-spored, arising from the base or angles of the ascomata, narrowly elongate-cylindrical, bitunicate, sometimes with a small internal apical beak, generally capped by a meniscus visible in Congo Red and Cotton Blue, discharge presumably fissitunicate. **Ascospores** very long and filiform, colourless, multiseptate, the apices pointed, fasciculate or spiralled in the ascus. **Conidiomata**, when present, pycnidial or stromatic, immersed to superficial, dark brown to black; wall clypeate. **Conidiogenous cells** ± elongate-ampulliform, lining the conidiomatal cavity. **Conidia** thread-like, colourless, multiseptate.

This genus has been confused with both *Tomasellia* and *Leptorhaphis*, both genera with cellular pseudoparaphyses and not paraphysoids in the hamathecium. The very long and thread-like ascospores and conidia and the often grouped ascomata, set it apart from *Leptorhaphis*. The genus is probably mainly tropical; seven species are currently known, of which only one occurs in the British Is.

LITERATURE: Aguirre-Hudson (1991), Swinscow (1965).

1. C. ischnobelum (Nyl.) Aguirre (1991)

Leptorhaphis ischnobela (Nyl.) Coppins (1980)

Thallus greenish white, creamy when dry, smooth, sometimes delimited by a dark brown prothallus. Ascomata 0.45–1 mm wide, in groups; involucrellum pitted with ostiolar openings. Asci 80–110 × 6–10 μm. Ascospores 80–110 × 1.5–2.5 μm, at least 8-septate, with refringent particles between the septa. Conidiomata 0.4–0.6 mm long, circular to elongated; conidiogenous cells 7–10 × 1.5–2 μm; conidia 60–100 × 1.45–2 μm.

On smooth bark, especially of *Corylus*, but also known on *Betula, Fraxinus, Ilex* and *Quercus*, in humid situations. S. & W. Great Britain, S. & W. Ireland. Italy (Tuscany).

There has been some uncertainly as to whether this species is lichenized, but microtome reactions reveal a thin *Trentepohlia* layer associated with hyphae of the mycobiont below the bark surface. *Rhaphidicyrtis trichosporella* differs in having smaller, ± globose ascomata, an I + blue centrum, and slightly broader ascospores.

CETRARIA Ach. (1803)

Thallus dorsiventral, either foliose, ± loosely adpressed, the lobe-margins often ascending and rarely forming rosettes or shrubby, erect, the lobes often channelled, sometimes ± tubular, rarely flat, forming tufts. **Cortex** often 2-layered, cells thick-walled, pseudoparenchymatous to prosoplectenchymatous. **Pseudocyphellae** often present on lower surface or marginal. **Soralia** occasionally present. **Photobiont** trebouxioid. **Ascomata** apothecia, ± marginal, often attached obliquely. **Thalline exciple** present, often incurved. **Epithecium** red-brown to dark brown. **Hymenium** I+ blue. **Hypothecium** colourless. **Hamathecium** of paraphyses, branched, sparsely anastomosed; apices often swollen and brownish. **Asci** 8-spored, clavate to elongate-clavate, *Lecanora-type*. **Ascospores** 4–10 × 2.5–6 μm, colourless, simple, ellipsoid or globose. **Conidiomata** pycnidia; ostiole blackened, occasionally present at ends of marginal projections or stalked. **Conidia** 3.5–8 × 0.5–1 μm, colourless, cylindrical, dumb-bell-shaped or fusiform. **Chemistry:** higher fatty acids, depsides and depsidones.

The foliose species **(1–5)** superficially resemble *Parmelia* s. lat. but are characterized by the marginal, almost always inclined, apothecia and the pycnidia which are never sunken in the thallus. All species except *C. islandica* **(10)** and related species **(9, 11)** will eventually require transfer to at least one, and perhaps two, other new genera. *Tuckermannopsis* Gyelnik (1933) has been used by some authors for these, but we have not followed this here as all are certainly not congeneric with the type species of that genus.

LITERATURE: Hillman (1936), Kärnefelt (1979), Maas Geesteranus (1947).

1 Thallus ± shrubby, ascending; lobes erect, intertwined; dying from the base, loosely attached 2
Thallus ± foliose, spreading; lobes horizontal, ± free, apices and margins ascending; base persistent, firmly attached 7

2(1) Thallus straw-yellow to pale yellow-white 3
Thallus pale to dark reddish or greenish brown 4

3(2) Lobes slightly channelled, ± flat, distinctly wrinkled and ridged; base bright yellow **7. *nivalis***
Lobe-margins strongly inrolled, ± tubular, ± smooth; base bright purple ..... **8. *cucullata***

4(2) Upper surface colour ± uniform or reddish towards base; apices not finely divided; medulla C− 5
Upper surface pale yellowish or greenish brown towards base; apices finely divided into small lobules; medulla C+ red ..... **9. *delisei***

5(4) Laminal pseudocyphellae rare and indistinct; lobes ± subtubular, with incurved margins, usually 1–5 mm wide 6
Laminal pseudocyphellae usually distinct; lobes flattened, usually 4–10 mm wide **10. *islandica***

6(5) Medulla Pd+ red; surface ridged and pitted; marginal pseudocyphellae ± absent **10. *islandica***
Medulla Pd−; surface ± smooth; marginal pseudocyphellae in marginal lines **11. *ericetorum***

7(1) Thallus bright yellow 8
Thallus brown, green-brown or brown-black 9

8(7) Soralia present on margins; pycnidia rare; apothecia absent ..... **5. *pinastri***
Soralia absent; marginal stalked pycnidia frequent; apothecia frequent ..... **6. *juniperina***

9(7) Thallus dark brown to brown-black; without soredia; apothecia abundant 10
Thallus brown when dry, dull olive-green when wet; with marginal soredia; apothecia rare ..... **1. *chlorophylla***

10(9) Thallus to 20 cm diam.; lobes closely adpressed; apothecia not covering entire surface .. 11
Thallus to 2 cm diam.; lobes raised; apothecia numerous, often obscuring the thallus **4. *sepincola***

11(10) Lower surface pale brownish white; medulla Pd− ..... **2. *commixta***
Lower surface dark brown-black; medulla Pd+ yellow-orange ..... **3. *hepatizon***

1. C. chlorophylla (Willd.) Vainio (1896)

Thallus 1–6 cm diam., foliose, loosely tufted, rarely rosette-forming; lobes few to numerous, 1–3 × 0.2–1 cm, flat to channelled, wavy, crinkled, notched or often deeply incised, entirely ascending or ± raised at margins; upper surface olive-brown or brown, olive-green when wet, smooth or somewhat wrinkled; soredia frequent, greyish white, entirely marginal; lower surface pale brown, wrinkled; rhizines sparse, towards the centre, pale. Apothecia rare, on lobe-margins and apices, sessile; disc to 2 mm diam.; thalline exciple sorediate. Conidia not seen. Medulla Pd−, K−, KC−, C− (protolichesterinic acid).

On twigs, branches of wayside and woodland trees, shrubs, fence posts, also rarely on siliceous rocks. Throughout Great Britain, extending locally to Ireland where it is rare, possibly currently spreading as a result of bark acidification. Europe, North America, Chile, S. Argentina, New Zealand.

More-or-less brown specimens of *Platismatia glauca* from exposed situations remain unchanged in colour when wet; the underside of *C. chlorophylla* also becomes paler towards the centre, that of *Platismatia glauca* becomes darker.

2. C. commixta (Nyl.) Th. Fr. (1871)

Thallus 1–5(-15) cm diam., foliose, often forming mat-like rosettes; lobes 5–20 × 1–2 mm, irregularly divided, ± channelled, often overlapping, the axils angular or rounded; upper surface dark brown-black, ± shiny; lower surface pale brown, yellowish brown towards margin; rhizines to 2 mm, whitish brown, simple or branched, scattered. Apothecia 1–5(-8) mm diam., frequent; disc concolorous with thallus; thalline exciple often crenate. Pycnidia *c.* 0.1 mm diam., frequent, black, ± globose, sessile, on margins of lobes and apothecial margins, rarely occurring on upper surface. Conidia 3.5–5 × 1 μm, rod-shaped. Medulla Pd−, K−, KC+ pink or KC−, C−, UV± ice blue (± α-collatolic and/or alectoronic acids; no lichen substances usually detected by t.l.c.).

On siliceous rocks in exposed, montane and hilly situations. N. & S.W. England (Lake District, Pennines, Dartmoor), Scotland (Highlands). Europe, North America.

3. C. hepatizon (Ach.) Vainio (1899)

Like **2**, but distinguished by the dark brown to black lower surface with scattered, dark rhizines, conidia which are 3.5–5.5 × 1 μm, dumb-bell-shaped, and a different chemistry; medulla Pd+ yellow, K+ yellow, KC+ yellow-orange, C−, UV− (stictic and ± norstictic acids).

Siliceous rocks in uplands. S.W. England (Devon, Dartmoor), N.W. England (Cumbria), N. Wales, Scotland (Highlands). Circumboreal.

4. C. sepincola (Ehrh.) Ach. (1803)

Thallus 1–2 cm diam., 0.5–1 cm tall, foliose, forming neat, round, cushion-like tufts, ± monophyllous, irregularly incised-lobate; upper surface dark grey-green, brownish or olive-brown, matt, ± wrinkled; lower surface white or pale brown, weakly veined or wrinkled; rhizines pale, simple or branched, scattered. Apothecia abundant, often covering entire thallus, at ends of short lobes; disc 1–4 mm diam., red-brown, flat, smooth or ± convex with a thin thalline exciple. Conidia 3.5–5 × 1 μm, bacilliform. Medulla Pd−, K−, KC−, C−, UV− (protolichesterinic acid).

On twigs, predominantly of *Betula*, more rarely *Salix*, *Larix* and *Juniperus*, often with *Cetraria chlorophylla* **(1)** and *Parmelia septentrionalis* in well-lit situations, rarely on fence posts. Mainly N. & E. Britain, locally abundant in Scotland (Highlands), extending to and rare in N. England, C. Wales. C. & N. Europe, North America.

Separated from *C. chlorophylla* **(1)** by a tendency to become covered with apothecia, the darker thallus and absence of soredia.

5. C. pinastri (Scop.) Gray (1821)

Thallus 1–2 cm diam., foliose, forming irregular rosettes; lobes 1–5 mm wide with rounded apices, margins ascending; upper surface pale yellow or yellow-grey, greenish when wet; soredia abundant along lobe-margins, often of a brighter yellow colour, spreading to the upper surface; medulla yellow; lower surface pale yellowish or brownish white, veined; rhizines brownish white, squarrose. Apothecia not seen in British material. Medulla Pd−, K−, KC−, C−, UV− or blackish red (usnic, vulpinic and pinastric acids).

Chiefly on bark of *Pinus* and *Betula*, also on *Vaccinium* stems, rarely on old fence posts and pebbles; very rare. Scotland (Highlands), extending locally to England (Durham, Northumberland, formerly E. Anglia and Somerset). C. & N. Europe, North America.

Easily recognized by the conspicuous, bright yellow soralia, which form a fringe on ascending lobe-margins.

6. C. juniperina (L.) Ach. (1803)

Like **5**, but the lobe-margins lack soredia and have abundant, black, stalked pycnidia. It is also frequently fertile; disc 2–6 mm diam., red-brown with a thin thalline exciple.

In ancient Caledonian pine forests. N.E. Scotland (Grampian mtns), also N. England; ?extinct. Europe, North America, Japan.

7. C. nivalis (L.) Ach. (1803)

Thallus 2–4 (-6) cm tall, shrubby, erect, rigid, forming tufts; branches to 1 cm wide, irregularly branched, with truncate, deeply indented apices; upper and lower surfaces pale straw yellow to yellow-grey, bright yellow to yellow-brown at base; surface with depressions, ridged and wrinkled, often slightly channelled. Apothecia, not seen in British material, 3–20 mm diam., yellow-brown. Medulla Pd−, K−, KC−, C−, UV− (protolichesterinic acid).

Characteristic of lichen-moss heaths in *Calluna - Vaccinium uliginosum - Empetrum - Salix herbacea* communities, avoiding patches with late snow-lie, usually at >900 m alt. Scotland (Highlands, E. Cairngorm mtns), extending locally to north and west Scotland. Europe, North America, Arctic.

8. C. cucullata (Bellardi) Ach. (1803)

Like **7**, but branches narrower (to 4 mm wide), markedly channelled, with strongly incurved margins, appearing ± tubular, smooth and without depressions. The base has a characteristic deep red-purple tinge.

N.E. Scotland (Grampian mtns); reported in 18th and 19th centuries, but these collections could have originated outside Britain. Europe, North America, Arctic.

9. C. delisei (Bory ex Schaerer) Nyl. (1866)

Thallus 2–4(-5) cm tall, shrubby, erect, forming rigid, ± interwoven spiky tufts; lobes narrow, 2–5 mm wide, often channelled, smooth, ± matt, richly branched and dissected towards apices with abundant, short, fimbriate prolongations, to 0.5 mm, with stalked pycnidia along margins; upper surface dark brown-black, lighter olive- to yellow-brown towards base; lower surface concolorous or paler; pseudocyphellae frequent, white, elongate, laminal and marginal. Apothecia unknown in British material. Conidia 3.5–7 × 0.5 µm. Medulla and pseudocyphellae Pd−, K−, KC+ pink-red, C+ pink-red, UV− (hiascinic and ± gyrophoric acids).

In small, wet, grit-lined depressions near areas of late snow-lie, usually above 900 m; rare, but locally frequent. N.E. Scotland (Grampian mtns). Arctic, Alps, S. Argentina, Falkland Islands, New Zealand.

Separated from *C. islandica* **(10)** by the greater degree of branching in the upper parts of the thallus, the tendency to develop numerous, short, marginal prolongations towards the apices, the paler base of the thallus and the Pd−, KC+ red medulla.

10. C. islandica (L.) Ach. (1803)

Thallus 2–6 cm tall, shrubby, erect or decumbent, forming tufts; lobes 1–10(-30) mm wide, ± channelled, wider lobes flatter with a distinct marginal edge, moderately branched; upper surface dark grey-brown, red-brown or grey-green, basal parts often reddish, smooth or sometimes pitted, shiny or dull; marginal projections 0.1–1 mm, often frequent, rarely absent; lower surface usually concolorous or paler with numerous, conspicuous, white pseudocyphellae, also present on margins. Apothecia rare on lower surface of wide lobe ends; disc 2–20 mm, dark brown,; thalline exciple often crenulate. Pycnidia at ends of marginal projections, dark brown; conidia 6–7 × 1 µm. Medulla and pseudocyphellae Pd+ orange, K−, KC−, C−, UV− (fumarprotocetraric, and ± protocetraric and protolichesterinic acids).

In lichen-rich heaths, especially hilly and montane districts; locally dominant. Scotland (Highlands), extending locally to N. England (Pennines) and lowland E.England (Yorkshire, Lincolnshire, Norfolk). Europe, North and South America, Arctic, Asia, Australia, New Zealand.

Subsp. *crispiformis* (Räsänen) Kärnef. (1979) is distinguished from subsp. *islandica* by the very small, indistinct, laminal pseudocyphellae, the often rather narrower lobes which often become ± tubular and which lack a distinct marginal edge, the few, short lateral branches and the more pitted and ridged surface.

11. C. ericetorum Opiz (1852)

Like **10**, but with lobes 0.5–1 mm wide, strongly channelled, strongly incurved, appearing ± tubular, with linear pseudocyphellae on margins of lower surface, particularly distinct in upper parts; apical branches usually more branched than in *C. islandica* **(10)**. Medulla and pseudocyphellae Pd− (protolichesterinic acid).

In lichen-rich heaths in montane situations above 900 m; rare. C. (Ochil Hills) & N.E. Scotland (Grampian mtns). N. Europe, North America, S. Argentina.

CETRELIA Culb. & C. Culb. (1968)

Thallus foliose, with rounded lobes, without marginal rhizines. **Upper surface** with punctiform pseudocyphellae and marginal soredia. **Lower surface** black, with rhizines. **Cortex**

prosoplectenchymatous. **Photobiont** *Trebouxia*. **Medulla** I−. **Ascomata** apothecia, laminal, occasionally marginal, usually perforate. **Thalline exciple** persistent. **Hymenium** colourless. **Asci** *Parmelia*-type, 8-spored. **Ascospores** ellipsoid, simple, colourless. **Conidiomata** pycnidia. **Conidia** bacilliform, colourless, simple. **Chemistry:** atranorin (K+ yellow) in cortex; medulla with a range of orcinol depsides and depsidones.

LITERATURE: Culberson & Culberson (1968, 1979), Jørgensen & Ryvarden (1970)

1. C. olivetorum (Nyl.) Culb. & C. Culb. (1968)

Thallus to 10 cm diam., foliose, forming wavy, wide-spreading patches; lobes 0.5–2 cm wide, contiguous, often overlapping, rounded at apices and with wavy, crisped, raised margins; upper surface glaucous-grey, sometimes tinged brown, dotted with minute, scattered, white, punctiform pseudocyphellae (× 10 lens); margins of older parts of lobes, ± continuously sorediate; lower surface black, somewhat wrinkled; rhizines scattered, simple, with a wide, often paler, rhizine-free zone towards the margin. Apothecia not seen in British material, not perforate; hymenium 60–110 μm tall. Ascospores 12–15 × 7–10 μm. Pycnidia unknown. Three chemotypes occur in the British Is.: (a) Pd−, K−, KC− or KC+ pink, C− (± perlatolic and imbricaric acids) [*C. cetrarioides* (Del. ex Duby) Culb. & C. Culb. (1968) incl. *C. monochorum* (Zahlbr.) Culb. & C. Culb (1968)]; (b) Pd−, K−, KC+ red, C+ red (olivetoric acid) [*C. olivetorum* s.str.]; (c) Pd−, K−, KC+ pink, C− (α-collatolic and alectoronic acids) [*C. chichitae* Culb. & C. Culb. (1968)]. (a) is predominant and (b) and (c) are very rare. In all specimens the soralia and medulla are UV+ white, less intense in the case of (b).

Amongst mosses on broad-leaved trees, or less frequently rocks, in well-lit, but moist or boggy, sheltered, often long-established, woodlands and *Salix* carr; often locally frequent. W. British Is. Europe, S. E. Asia (chemotype (c) also in the Pacific and E. North America).

Characterized by the large lobes with minute, white, laminal pseudocyphellae and marginal soralia. Differs from *Platismatia glauca* in the presence of pseudocyphellae, the ± continuously rounded margins, the I− medulla and distinctive chemistry. May also be mistaken for large *Parmelia* species (e.g. *P. perlata*, *P. arnoldii*).

CHAENOTHECA (Th. Fr.) Th. Fr. (1860)

Cybebe Tibell (1984)

Thallus crustose, superficial or immersed, farinose, granular, verrucose or squamulose, grey, greenish grey, golden yellow or brownish. **Photobiont** *Dictyochloropsis*, *Stichococcus*, *Trebouxia*, or *Trentepohlia*. **Ascomata** apothecia, stalked; stalk short to long and slender, consisting of periclinally arranged brown hyphae; head globose to obconical. **Thalline exciple** absent. **True exciple** ± well developed, formed as a continuation of the stalk tissue. **Hamathecium** of paraphyses, sparingly branched, continuing to grow into the ascospore mass. **Asci** cylindrical, ellipsoid or irregular, dissolving at an early stage, formed from ascogenous hyphae, with or without croziers, either singly or in chains. **Ascospores** forming a dry brown spore mass, globose, brown to pale brown, not or rarely 1- to 5-septate, ellipsoid-cylindrical; spore wall ± thick, smooth or with an ornamentation of irregular cracks (Fig. 8). **Conidiomata**: absent. **Chemistry**: pulvinic acid derivatives occur as a pruina and more rarely within the thallus; sometimes also contain unidentified Pd+ yellow-red depsides or depsidones. **Ecology**: mainly on bark and wood, rarely on soil and rocks, mostly in very sheltered situations with high humidity and low light levels.

Calicium differs in the black ascospore mass and exclusively ellipsoid ascospores; it also always has *Trebouxia* as photobiont. In *Microcalicium* the ascospore mass is greenish black or dark brown, and all the species are saprobes or grow on lichens.

LITERATURE: Middelborg & Mattsson (1987), Tibell (1980).

1 Ascospores ellipsoid; pruina on apothecia yellow-green 2
 Ascospores spherical; pruina on apothecia white, yellow-green or absent 4

2(1) Thallus olive-yellow to grey-green or immersed; ascospores 5–8 × 3–3.5 μm 3
Thallus bright yellow, of rounded, coarse granules; ascospores 6–9 × 4–5 μm **4. *chrysocephala***

3(2) Photobiont cells ellipsoid or rectangular, <10 μm wide, in short chains (*Stichococcus*); thallus immersed or ± of olive-yellow-green granules .... **3. *chlorella***
Photobiont cells rounded, 10–15 μm diam., separate or in clusters (chlorococcoid); thallus granular-verrucose, grey-green to green .... **9. *laevigata***

4(1) Photobiont cells with bright green contents, the walls thinner 5
Photobiont cells with red-green contents, the walls thick (*Trentepohlia*); ascospores 5–7.5 μm diam. .... **8. *hispidula***

5(4) Photobiont cells elongate or rectangular, in short chains, transversely septate, the cells <10 μm wide (*Stichococcus*) 6
Photobiont cells rounded, clustered or solitary, not transversely septate, the cells 10–15 μm diam. (chlorococcoid) 11

6(5) Pruina on apothecia bright yellow-green 7
Pruina on apothecia white or absent 8

7(6) Thallus inconspicuous, immersed; ascospores 3–4 μm diam. .... **1. *brachypoda***
Thallus conspicuous, leprose, bright yellow-green; ascospores 2–3 μm diam. ... **6. *furfuracea***

8(6) Thallus immersed or thin and leprose 9
Thallus superficial, distinctly squamulose or coarsely granular; ascospores 4–5 μm diam. .... **12. *trichialis***

9(8) Thallus thin, farinose 10
Thallus immersed .... **13. *xyloxena***

10(9) Ascospores 2.5–3 μm diam. .... **7. *gracilenta***
Ascospores 3.5–5 μm diam. .... **11. *stemonea***

11(5) Thallus without rusty spotting; ascospores 3.5–4.5 μm diam., rarely larger 12
Thallus grey with yellow-orange to rusty spots and patches; ascospores 5.5–7.5 μm diam. .... **5. *ferruginea***

12(11) Thallus superficial; apothecia with yellow pruina; ascospores 5.5–7 μm diam. **10. *phaeocephala***
Thallus immersed; apothecia ± white-pruinose or pruina absent; ascospores 3.5–4.5 μm .... **2. *brunneola***

1. C. brachypoda (Ach.) Tibell (1987)

C. sulphurea (Retz) Midd. & Mattsson (1987) *nom. inval., Coniocybe sulphurea* (Retz.) Nyl. (1886)

Thallus immersed. Photobiont *Stichococcus*. Apothecia 0.4–1.4 mm tall; head and stalk covered in a dense yellow-green pruina; stalk black beneath; true exciple poorly developed; ascospore mass dark brown, covered in yellowish green pruina. Ascospores 3–3.5 diam., globose, with a slightly cracked surface. Pruina with vulpinic and pulvinic acids.

On dry bark in crevices of deciduous trees and wood of hollow trunks. E. British Is, rarer in W. and N.W, W. Ireland. Europe, North America.

Characterized by the immersed thallus and rather short, bright yellow-green stalks. *C. furfuracea* **(6)** has a distinct, leprose thallus, longer stalks and smaller ascospores.

2. C. brunneola (Ach.) Müll. Arg. (1862)

Thallus immersed, more rarely poorly developed on surface and then finely granular-verrucose, ecorticate. Photobiont *Dictyochloropsis* or *Trebouxia*. Apothecia 0.6–1.6 mm tall, 9–23 times as high as width of stalk; stalk 0.05–0.1 mm wide, shining black, without pruina, often branched so that one stalk carries 2–5 heads; head ± globose; true exciple poorly developed, without pruina, but lower side sometimes appearing pruinose from pale hyphae projecting from the surface. Asci of variable shape, produced in chains, with 1- to 2-seriately arranged ascospores. Ascospores 3.5–4.5 μm diam., smooth or with irregular fissures (Fig. 8a). Thallus Pd+ yellow, K−, KC−, C− (baeomycesic and squamatic acids).

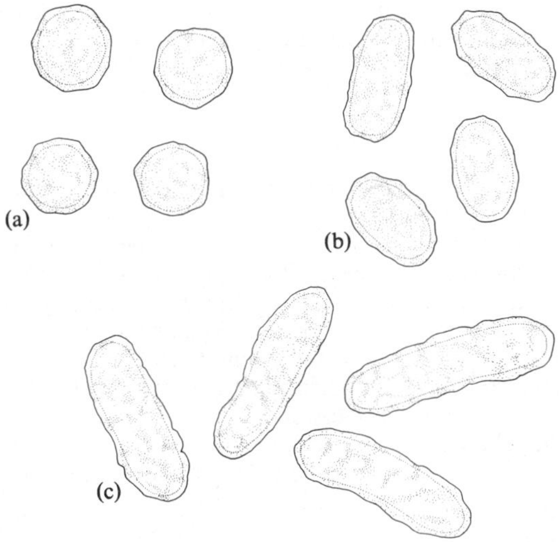

Fig. 8. *Chaenotheca* ascospores. (a) *C. brunneola*; (b) *C. chlorella*; (c) *C. laevigata*. Scale = 10 μm.

On wood of both coniferous and deciduous trees, rarely on bark of old trees, especially *Quercus*, rarely on old fruiting bodies of *Polyporus*. Throughout British Is. Europe, North and C. America, Australia, New Zealand.

Characterized by the globose head with usually strongly reduced exciple, the shiny, black, epruinose stalks which are often branched and the ± poorly differentiated, Pd+ thallus. When farinose, it may resemble *C. stemonea* **(11)**, which has *Stichococcus* as photobiont and a thick hyphal web on the lower side of the head. It may resemble *C. ferruginea* **(5)**, which has a greyish thallus and frequently yellow to red, K+ deep red spots and larger ascospores and *C. trichialis* **(12)**, which has a Pd− thallus and contains *Stichococcus*.

3. C. chlorella (Ach.) Müll. Arg. (1862)

C. carthusiae (Harm.) Lettau (1912)

Thallus usually distinct, covering surface, granular-verrucose, ecorticate, greyish green to green. Photobiont *Stichococcus*. Apothecia 0.4–1.6 mm tall, 8–17 times as long as width of stalk; stalk 0.04–0.08 mm wide, black, with a dense, yellowish green pruina in the upper part; head broadly obovoid to lens-shaped; true exciple mostly well developed, with a dense, yellowish green pruina on the lower side. Asci cylindrical, not produced in chains. Ascospores 5–7.5 × 2.5–3.5 μm, ellipsoid, with a coarse and irregular ornamentation when mature (Fig. 8b). The pruina contain vulpinic acid.

On dry undersides and inside hollow trunks of deciduous trees, rare. Throughout Great Britain, from Hereford, Hertford and Wiltshire to Fife and Inverness. Europe, North America, U.S.S.R., Australia, New Zealand.

Usually recognized by the short apothecia with greenish yellow pruina, the granular, greyish green thallus, and the ellipsoid ascospores with a coarse and irregular surface ornamentation. Specimens with a fully immersed thallus may resemble *C. chrysocephala* **(4)**, which contains *Trebouxia*.

4. C. chrysocephala (Turner ex Ach.) Th. Fr. (1860)

Thallus usually well developed, forming a continuous crust of rather small, ecorticate, ± globose, bright yellow granules, more rarely immersed. Photobiont *Trebouxia*. Apothecia 0.6–1.3 mm tall, 10–22 times as long as width of the stalk, developing from lower side of thalline granules; stalk 0.04–0.08 mm wide, with a dense yellow pruina in the upper part, the lower part black to pale brown; head obovoid or broadly obconical; true exciple well developed, with a dense, yellowish green pruina on the lower side. Asci cylindrical, with uniseriately arranged ascospores, produced singly. Ascospores 6–9 × 4–5 μm, shortly ellipsoid or ± globose, with a coarse and irregular ornamentation. Thallus Pd−, K−, KC−, C− (thallus and pruina with vulpinic acid).

On bark and wood of conifers, *Betula* and *Quercus*; common. Widely distributed throughout British Is., but especially in E. Scotland. Europe, North & C. America, Asia, Australia, New Zealand.

The only *Chaenotheca* with a well-developed, bright greenish yellow, granular thallus. Specimens with a poorly developed thallus may resemble *C. chlorella* **(3)**, which contains *Stichococcus*.

5. C. ferruginea (Turner ex Ach.) Mig. (1930)

Thallus usually covering surface, ecorticate, granular-verrucose, pale whitish grey, often with limited yellow to rusty reddish patches, rarely immersed. Photobiont *Trebouxia*. Apothecia 0.8–1.7 mm tall, 7–15 times as long as width of stalk; stalk 0.07–0.15 mm wide, shiny, black, without pruina; head broadly to narrowly obconical; true exciple well developed, without pruina; ascospore mass pale brown, often with a yellow tinge. Asci cylindrical to narrowly clavate. Ascospores 5–7.5 μm, globose, when mature coarsely and irregularly fissured, 1- to 2-seriately arranged in asci. Yellow-red, quinonoid pigment in thallus, K+ deep red; ascospore mass with yellowish K+ red pigment.

On acid bark and wood; tolerant of SO_2 air pollution. Throughout British Is., but becoming rarer in the west. Europe, North & C. America, Asia, Australia, New Zealand.

Usually recognized by the yellow-red, K+ deep red spots on the grey, often sterile thallus. When the pigment is absent may be mistaken for *C. trichialis* **(12)**, but differs in having larger ascospores and ± globose photobiont cells.

6. C. furfuracea (L.) Tibell (1984)

Coniocybe furfuracea (L.) Ach. (1816)

Thallus leprose, well developed, bright yellow-green. Photobiont *Stichococcus*. Apothecia 1.6–2.7 mm tall, slender; head and stalk covered in a yellowish green pruina, black beneath; true exciple indistinct; ascospore mass at first also covered in pruina, becoming pale brown. Ascospores 2–3 μm, globose. Conidiomata hyphomycetous, of dispersed, ± erect, colourless macronematous conidiophores; conidia ellipsoid, adhering in chains, colourless, simple. Thallus and pruina with vulpinic acid, pulvinic acid and pulvinic acid dilactone.

In shady, humid crevices amongst tree roots and in rock underhangs; most frequent in upland situations. Throughout British Is. Europe, North America.

Characterized by the yellow-green, leprose thallus with very slender apothecia. When sterile, it may resemble *Psilolechia lucida*, which normally has *Trebouxia* and contains rhizocarpic acid.

7. C. gracilenta (Ach.) Mattsson & Middelborg (1987)

Coniocybe gracilenta (Ach.) Ach. (1816)

Thallus thin, farinose, greyish green. Photobiont *Stichococcus*. Apothecia 2.5–3.5 mm tall, with flexuose and slender stalks; stalks black, with greyish, granular pruina; lower surface of head and upper stalk with a faint, greyish white pruina; true exciple poorly developed. Asci ellipsoid, formed in chains. Ascospores 2.5–3 μm diam., globose, very pale brown to colourless, with a slightly cracked surface. Unidentified substances in t.l.c.

On dry bases of old *Acer pseudoplatanus* and *Ulmus glabra*. N. England (Yorkshire) and Scotland (Angus). N. & C. Europe, Macaronesia, North America.

Probably the most shade tolerant species in the Caliciales. Included in the monotypic *Cybebe* on the basis of the absence of croziers and the pale pigmentation of ascospore wall.

8. C. hispidula (Ach.) Zahlbr. (1922)

Thallus immersed, inapparent. Photobiont *Trentepohlia*. Apothecia 0.5–1 mm tall, 8–16 times as long as width of stalk; stalk 0.05–0.08 mm wide, black in lower part with a thick yellowish green pruina in upper part; head obconical-obovoid; true exciple well developed, with dense yellowish green or rarely ochraceous pruina on lower side. Asci cylindrical, either without a stalk and produced in chains or with a well-developed stalk and formed singly. Ascospores 5–7.5 μm diam., globose, when mature with an ornamentation formed by irregular, reticulate fissures surrounding small, polygonal areas.

On dry, ± acid bark and wood, especially tree bases. Throughout British Is. Europe, North America, U.S.S.R., Nepal, Australia, New Zealand.

Characterized by the short-stalked apothecia, which have a greenish yellow pruina on the lower side of the exciple and by the large, globose ascospores; the only *Chaenotheca* sp. in the British Is. with *Trentepohlia*.

9. C. laevigata Nádv. (1934)

Thallus immersed or, when well developed, covering the surface in globose, olivaceous-yellow granules with a distinct cortex, 8–15 μm thick. Photobiont chlorococcoid. Apothecia 0.9–1.9 mm tall, 13–27 times as long as width of stalk; stalk 0.05–0.08 mm wide, usually covered by a dense, greenish yellow pruina; head obovoid-broadly obconical; true exciple well developed, with dense, yellowish green pruina on the lower side. Asci cylindrical, produced singly. Ascospores 5.5–6.5 × 3.2–3.5 μm, ellipsoid, with a roughly cracked surface (Fig. 8c), 1- to 2-seriately arranged in the ascus. Pruina on the head and stalk contain vulpinic acid.

On bark in crevices of *Fraxinus* and *Ulmus*; very rare. E. Scotland. C. Europe, Scandinavia, North America, U.S.S.R..

Characterized by the greenish yellow pruina on the lower side of the exciple, the ellipsoid ascospores and chlorococcoid photobiont.

10. C. phaeocephala (Turner) Th. Fr. (1860)

Thallus normally thick, squamulose, with large, convex, uneven verrucae, or thinner, and then verrucose-granular, rarely fully immersed, dark grey to greenish brown. Photobiont *Trebouxia.* Apothecia 0.5–1.2 mm tall, 6–15 times as long as width of stalk, stout; stalk 0.06–0.1 mm wide, pale to dark brown, not pruinose or distinctly thinly yellowish green-pruinose over uppermost part; head deep purple-brown, ± brown-orange when abraded, broadly obconical-lenticular; true exciple well-developed. Asci cylindrical to narrowly clavate, produced singly from branched, ascogenous hyphae. Ascospores 5.5–7 μm diam., globose, with a surface ornamentation of irregular, polygonal areas, 1- to 2-seriately arranged in asci. Pruina on lower side of head contains vulpinic acid.

On old worked timber, forming extensive patches particularly on uncreosoted barns; a single recent record (1991). S. England (W. Sussex, Parham). Europe, North America, U.S.S.R. Japan.

Characterized by the well-developed exciple, which is greenish yellow-pruinose on the lower side, the rather robust apothecia, usually well-developed verrucose-squamulose thallus and the globose ascospores.

11. C. stemonea (Ach.) Müll. Arg. (1862)

Thallus thin, farinose, green to bluish green. Photobiont *Stichococcus.* Apothecia 0.8–1.4 mm tall, 10–18 times as long as width of stalk; stalk 0.06–0.1 mm wide, the upper part whitish dull brown, the lower part brown-black; head globose; true exciple poorly developed; lower side of exciple and upper part of stalk covered by a loose, whitish pruina-like hyphal collar. Asci cylindrical or irregular, produced in chains. Ascospores 3.5–5 μm diam., globose, smooth, with some irregular cracks, 1- to 2-seriately arranged in asci. Thallus Pd+ yellowish red, K−, KC−, C− (contains 2 unidentified compounds).

On dry bark and wood, particularly *Alnus, Betula, Quercus* and *Pinus,* often in deeply shaded situations, including tree bases. N. England, S. England (Somerset), C. & E. Scotland. Europe, North America, U.S.S.R., Asia, Nepal, Australia, New Zealand.

Characterized by the farinose thallus and almost globose heads which have a whitish brown, pruina-like web on the lower side. *C. trichialis* **(12)** has a well developed exciple without a hyphal web on the lower side and never has a farinose thallus, though it may be minutely granular.

12. C. trichialis (Ach.) Th. Fr. (1860)

Thallus granular-squamulose, often rather glossy, more rarely finely granular, greenish grey, forming a crust or restricted to small patches. Photobiont *Stichococcus.* Apothecia 1–1.9 mm tall, 13–24 times as long as width of stalk; stalk 0.05–0.1 mm wide, the upper part often with a faint whitish pruina, the lower part black; head obconical- to lens-shaped; true exciple well developed, sometimes with a faint whitish pruina below. Asci shortly cylindrical or irregular, formed in chains. Ascospores 4–5 μm diam., globose, smooth, with irregular, narrow cracks, 1-(2-)seriately arranged in the asci. Thallus Pd−.

In dry, acid bark crevices and on wood. Throughout British Is., especially the east. Europe, U.S.S.R., Nepal, China, North, C. & South America, Australia, New Zealand.

Morphologically variable but usually has a distinct Pd−, verrucose-squamulose thallus, medium-sized apothecia and a faint white pruina on the lower side of the head. The thallus may be ± completely immersed, the apothecia vary in size and shape, the pruina may be absent and the stalk is sometimes brown rather than black. *C. xyloxena* **(13)**, differs in the totally immersed thallus, and more slender apothecia with a thick, white pruina. The thallus of *C. brunneola* **(2)** is Pd+ yellow and contains trebouxioid algae.

13. C. xyloxena Nádv. (1934)

Thallus immersed, sometimes visible as a greenish grey tinge. Photobiont *Stichococcus*. Apothecia 0.9–1.5 mm tall, 15–25 times as long as width of stalk; stalk 0.04–0.08 mm wide, the upper part usually with a thick, white pruina, the lower part black; head obconical; true exciple well developed, cushion-like, usually with a thick, white pruina below. Ascospores 3–4 μm diam., globose, smooth, with irregular, narrow cracks, 1- to 2-seriately arranged. Lichen products not detected by t.l.c.

On wood of old *Alnus, Betula, Quercus* and *Pinus* trees. Scotland (Highlands). Europe, North America, U.S.S.R., Asia, New Zealand.

Usually distinguished by the completely immersed thallus, and slender apothecia which are densely pruinose on the lower side of the head and upper part of the stalk. The thickened and cushion-like outer part of the exciple distinguish it from *C. trichialis* **(12)**.

CHAENOTHECOPSIS Vainio (1927)

Thallus usually indistinct, ± loosely associated with algal colonies, lichenicolous or saprobic. **Photobionts**, when present, chlorococcoid or *Trentepohlia*. **Ascomata** apothecia, short- to long-stalked, black or stalk pale below; head obovoid to lens-shaped. **True exciple** of a few rows of periclinally arranged or interwoven hyphae; stalk ± colourless in central part, usually brown, reddish or greenish in outer part; hyphae periclinally to irregularly arranged. **Hamathecium** absent. **Asci** 8-spored, ± cylindrical, with thickened apex penetrated by a narrow canal; persisting until ascospores mature. **Ascospores** 5–10 × 2–4 μm, simple or 1-sepate, ellipsoid to oblong- or fusiform-ellipsoid, brown; septum colourless to dark brown; surface smooth to minutely warted; not aggregated into a dry spore-mass. **Anamorphs** pycnidia, walls of outwardly pointed cells (*Asterophoma* species), amongst the spore-masses of host caliciales, or sometimes borne on long, branched stalks. **Conidiogenous cells** narrowly ampulliform to subcylindrical, arising singly, enteroblastic, acrogenous. **Conidia** rod-shaped to oblong-ellipsoid, simple, colourless. **Chemistry**: ascomata with various pigments. **Ecology**: mostly on dry bark or wood of old, hollowed or dead trees; also on *Cladonia* squamules, rotting wood polypores and in rock underhangs.

Distinguished from *Calicium, Chaenotheca* and *Microcalicium* by the ascospores not forming a dry spore-mass. *Mycocalicium, Phaeocalicium* and *Stenocybe* differ in having a uniformly thickened ascus apex and the last two genera have larger ascospores. The species of *Chaenothecopsis* are generally difficult to identify and the genus awaits a thorough revision. The better known, unidentified species are referred to here but many additional specimens also await critical study.

LITERATURE: Hawksworth (1978, 1980), Schmidt (1970), Tibell (1975, 1981, 1982, 1987).

1	Ascospores simple	2
	Ascospores 1-septate	3
2(1)	Apothecia stalk white-pruinose; ascospores 6–8(-11) μm long	**10. *viridialba***
	Apothecia stalk epruinose; ascospores 4.5–6 μm long	**8. *savonica***
3(1)	Apothecia ± scattered; ascospores mostly <10 μm long and 3.5 μm wide	4
	Apothecia in tufts; ascospores 9–14 × 3–4.5 μm	**1. *caespitosa***
4(3)	Ascospore-septum darker than outer wall	5
	Ascospore-septum paler than or of equal shade to outer wall	6
5(4)	Stalk (in squash) reddish brown; on *Chaenotheca trichialis*	**3. *epithallina***
	Stalk (in squash) greenish; on *Stichococcus*	**4. *nigra***

6(4) On wood, *Calicium* species, *Chaenotheca* species, or *Trentepohlia* 7
On *Cladonia* species .... **5. *parasitaster***

7(6) Ascospore-septum colourless or brown, but paler or thinner than outer wall; associated with algae 8
Ascospore-septum ± equal to outer wall in colour and thickness; not associated with algae .... **2. *debilis***

8(7) Head (in squash) with effuse, yellowish pigment, K+ purple .... **7. *pusiola***
Head (in squash) without effuse pigment, K− or only slightly reddish or green 9

9(8) Reddish pigment (if any) in apothecia K− or dulling 10
Reddish pigment in apothecia K+ green or blue-green .... **11. *viridireagens***

10(9) Ascospores mostly 6–8 μm long; associated with *Trentepohlia* .... **9. *vainioana***
Ascospores mostly 5–7 μm long; not associated with *Trentepohlia* .... **6. *pusilla***

1. C. caespitosa (Phillips) D. Hawksw. (1980)

Apothecia in compact tufts, 2–4 mm tall, black; heads 0.15–0.35 mm wide. Ascospores 9–14 × 3–4.5 μm, 1-septate.

On decaying polypore and rotten *Taxus* bark. S.W. England (Gloucester, Hereford). ?Endemic.

Not associated with algae. Easily distinguished by its large apothecia which occur in tufts.

2. C. debilis (Turner & Borrer ex Sm.) Tibell (1975)

Apothecia 0.4–1.3 mm tall, black; heads 0.14–0.4 mm wide; stalk 0.04–0.1 mm wide; epithecium and true exciple reddish brown, K−, N−; hypothecium greenish, N+ green intensifying; stalk pale wine-red, K− or K+ purplish red, N+ violet-red. Ascospores 5.5–7.5(-9.8) × 2–2.3(-3.3) μm, 1-septate; septum of ± same contrast as outer wall; surface smooth or minutely cracked.

On dry, weathered timber of a building; no modern records. S. England (Sussex). Temperate N. Hemisphere, Australia, New Zealand.

Usually not associated with algae. Distinguished from *C. pusilla* **(6)** by the internal pigmentation and darker ascospore septum.

An undescribed, similar species, also not associated with algae, has apothecia 0.3–0.6 mm tall, ascospores (4.5-)5–7(-7.5) × 2.0–2.5(-3) μm, and internally olivaceous, K−, N−, pigmentation. It occurs on wood of barkless or hollowed trunks of *Alnus* and *Quercus* in Scotland (Highlands).

3. C. epithallina Tibell (1975)

Apothecia 0.5–1 mm tall, entirely black or stalk brown below; heads 0.1–0.2 mm wide; stalk 0.05–0.08 mm wide; internally brown or reddish brown, K−, or hypothecium greenish, K+ brownish; all parts N-. Ascospores 4.8–6.7(-7.6) × (1.5-)1.9–2.3 μm, 1-septate; septum darker than outer wall.

On thallus of *Chaenotheca trichialis* which contains *Stichococcus*, parasitic; on decorticate trees in native pinewoods. N. Scotland (Easterness). Scandinavia, Canada.

4. C. nigra Tibell (1987)

Apothecia (0.5-) 0.7–1.2(-2) μm tall, black; heads 0.14–0.3 mm wide; stalk 0.04–0.08 mm wide; epithecium and hymenium brown, K−, N−; hypothecium and stalk greenish, K− or dulling, N−. Ascospores 4.8–7 × 1.5–2.2(-2.5) μm, 1-septate; septum much darker than outer wall.

Usually associated with *Stichococcus*, on wood of *Pinus*, *Alnus*, *Betula*, *Corylus* and *Quercus*; local. Scotland (Highlands). Tasmania, New Zealand.

5. C. parasitaster (Bagl. & Car.) D. Hawksw. (1978)

Apothecia 0.34–0.7 mm tall, black; head 0.12–0.24 mm wide; stalk 0.04–0.075 mm wide; epithecium and hymenium reddish brown; stalk greenish; all parts K−, N−. Ascospores (5-)6–9.5(-14) × 2.3–3(-3.5) μm, 1-septate; septum brown but thinner than outer wall; surface minutely warted.

On *Cladonia digitata*, *C. macilenta* and *C. polydactyla*, parasitic (infected parts discoloured brownish), on dry sides of large stumps and peaty overhangs in pinewoods. Scotland. Temperate Europe.

6. C. pusilla (Ach.) A. Schmidt (1970)

C. subpusilla (Vainio) Tibell (1975)

Apothecia 0.4–1 mm tall, black or more usually with stalk greenish or grey-brown below; head 0.12–0.28 mm wide; stalk 0.04–0.08 mm wide; internally dull greenish or brownish, K−, N−. Ascospores 5–7(-8.5) × 2–2.5(-3) μm, 1-septate; septum only tardily coloured, much paler than outer wall; surface ± smooth.

On wood of various trees (e.g. *Alnus, Betula, Castanea, Quercus, Pinus*); associated with globose-celled algae, sometimes (?commensalistic) on *Calicium glaucellum*. S.W. England (Hereford), Scotland (Highlands). Widely distributed in temperate and boreal regions of N. & S. Hemispheres, also in montane tropics.

Some unidentified collections are like *C. pusilla*, but have entirely black stalks and are associated with *Stichococcus*. On wood of *Alnus* and *Pinus*. Scotland (C. & E. Highlands).

7. C. pusiola (Ach.) Vainio (1927)

C. lignicola (Nádv.) A. Schmidt (1970)

Apothecia 0.4–0.8 mm tall, black or stalk rarely brown below; head 0.15–0.38 mm wide; stalk 0.04–0.1 mm wide; internally all parts with diffuse, greenish yellow pigment, K+ purple (dissolving). Ascospores 5.4–7.8(-9.5) × 2.2–2.7(-3) μm, 1-septate, the septum colourless or pale brown and much thinner than outer wall, the surface smooth.

On wood of *Pinus* (rarely *Betula*), usually on thallus of *Chaenotheca brunneola*, sometimes associated with *Stichococcus*, in native pinewoods; local. Scotland (C. & E. Highlands). Cool temperate N. Hemisphere, New Zealand.

8. C. savonica (Räsänen) Tibell (1984)

Apothecia 0.4–0.8 mm tall, black; head 0.1–0.2 mm wide; stalk 0.03–0.07 mm wide; internally dull or reddish brown, K−, N−, or partly greenish (K+ dull brownish, N± green intensifying). Ascospores 4.5–6 × 2–2.7(-3) μm, simple, ellipsoid with rounded ends, the surface minutely warted.

Associated with *Stichococcus* on dry bark of old *Quercus*. England (Hereford) and W. Scotland. Temperate N. Hemisphere, Australia, New Zealand.

Two unidentified species also have epruinose stalks and simple ascospores, although in both the ascospores are smooth and longer. The first has short-stalked apothecia, 0.2–0.3 mm tall, with ascospores (4.8-)6–7.5(-9) × 2–2.5 μm, and a leprose thallus, with globose algal cells 8–12(-16) μm diam.; it occurs in bark crevices of *Alnus* and *Quercus* and on mosses in a rock underhang, in N.E. England and E. Scotland. The second is parasitic on *Arthonia vinosa* (see **9**).

9. C. vainioana (Nádv.) Tibell (1979)

Apothecia 0.4–0.7 μm tall, black or stalk dark brown; head 0.15–0.3(-0.36) mm wide; stalk 0.04–0.07 mm wide; exciple and outer part of stalk brown; hymenium and hypothecium greenish; all parts K+ dull brown, N−. Ascospores (5-)6–8(-11) × 2–2.6(-3.3) μm, (0-)1-septate, septum paler than outer wall.

Associated with *Trentepohlia* or *Trentepohlia*-containing lichens (*Arthonia leucopellaea, Lecanactis abietina*), on bark or wood of *Pinus* and *Ilex*. Scotland (Highlands). Temperate Europe.

C. epithallina **(3)** and *C. pusilla* **(6)** are similar but have generally more robust apothecia, shorter ascospores, and are not associated with *Trentepohlia*.

Two unidentified species are also associated with *Trentepohlia*. The first, which may be *C. rubescens* Vainio (1927), has simple ascospores, (5-)5.5–7(-9.5) × 1.8–2.7(-3) μm, and grows on *Arthonia vinosa* in old woodlands in C. & E. Scotland. The second has darkly septate ascospores, 5–6.3 × 2 μm, and grows on *Opegrapha vulgata* on old *Quercus* in S.E. England.

10. C. viridialba (Krempelh.) A. Schmidt (1970)

Apothecia 0.5–2 mm tall, black; head 0.15–0.5 mm wide; stalk to 0.2 mm wide, white-pruinose; hypothecium and stalk reddish brown, K+ green. Ascospores 6–8(-11) × (2-)3–3.5 μm, simple.

Apparently not associated with algae, on barkless part of *Quercus* trunk by a stream. W. Scotland (Argyll). Circumboreal.

11. C. viridireagens (Nádv.) A. Schmidt (1970)

Apothecia 0.5–1.2 mm tall, black; head 0.1–0.24 mm wide; stalk 0.04–0.07 mm wide, the upper part ± with reddish tinge; epithecium and stalk mainly brown or olive-brown, but stalk sometimes reddish in parts; hypothecium brown-red or orange-red; reddish parts K+ olive-green or aeruginose. Ascospores (5-)6–7 × 2–2.7 μm, 1-septate; septum paler than outer wall.

On thalli of *Chaenotheca stemonea* and *C. trichialis* (both with *Stichococcus* photobiont), parasitic, on wood; rare. N.E. England, E. Scotland. Widely distributed in N. Hemisphere, Australia, New Zealand.

C. epithallina **(3)** lacks the reddish, K+ green pigment.

CHIODECTON Ach. (1814)

Thallus crustose, thin or thick, whitish, granular to cracked-areolate, often powdery. **Photobiont** *Trentepohlia*. **Ascomata** apothecia, elongate (lirellate), many immersed together in rounded or elongate, raised cushion-like stromata, black below. **Thalline exciple** absent. **True exciple** not or weakly developed. **Disc** often grey-pruinose. **Hypothecium** well developed, black or brown-black. **Hamathecium** of paraphysoids, slender, branched and anastomosed at apices. **Asci** 8-spored, clavate, *Opegrapha*-type. **Ascospores** (2-)3- to multiseptate, fusiform to needle-shaped or wavy, colourless. **Conidiomata** pycnidia. **Conidia** thread-like, straight or curved. **Chemistry**: atranorin and various depsidones. **Ecology**: on coastal rocks and bark.

A mainly tropical genus of *c.* 175 species distinguished from *Enterographa* and *Sclerophyton* by the deep black, dense stroma-like hypothecium which often connects the apothecia at the base.

LITERATURE: James (1970).

1. C. myrticola Fée (1825)

Thallus wide-spreading, chalk- or grey-white to grey, continuous, smooth, warted or ± dispersed in coarse, irregular patches, continuous or ± rimose-cracked, uneven, with convex, randomly dispersed, ± rounded, warts starkly white-pruinose or with an outline of a ± faint mosaic of lirellae with a brain-like appearance; lirellae pale grey to black, often ± entirely obscured by a dense ± white pruina; epithecium of densely interwoven hyphae in upper part, remaining uncoloured or pale brown, very densely interspersed with crystals mostly disappearing in K, scattered, elongate, simple or clustered crystals remaining; hymenium 100–120 μm tall, colourless; hypothecium brown-black, becoming blackish olive in K. Asci 80–95 × 15–19 μm, elongate-clavate, uniformly thick-walled. Ascospores (30-)35- 45(-50) × (4-)4.5–5 μm, 3-septate, straight or slightly curved. Thallus Pd+ rust-red, K± yellow or yellow to red, C− (protocetraric acid, atranorin, ± norstictic acid).

On smooth, dry bark and on rocks in dry, sheltered underhangs, also *Armeria* tufts and vegetable debris in sheltered situations; rare. S.W. England, Channel Is., Is. of Scilly, S.W. Ireland. Macaronesia.

Distinguished by the very uneven, often nodulose, thallus with ± elevated, concolorous or paler, pruinose, convex stroma with a closely packed mosaic of lirellae.

C. petraeum Delise ex Duby (1830), reported from dry rock crevices in the Channel Is. and Is. of Scilly, has been said to differ in the thicker thallus, rounded ascomata, and ascospores 30–40 × 3 μm. However, the status of this species and of the British records is currently unclear and requires a critical re-evaluation.

CHROMATOCHLAMYS Trev. (1860)

Thallus crustose, verrucose, membranous to cracked, dirty whitish or pale grey to pale brownish. **Photobiont** trebouxioid. **Ascomata** perithecia, ± immersed, thick-walled, pale to brownish, wall composed of oval, elongate-narrow cells below. **Hamathecium** of paraphysoids, persistent, delicate, branched and/or anastomosed, especially towards the base and at the apices; conglutinate periphysoids towards the inner side of the ostiole. **Asci** 2- to 8-spored,

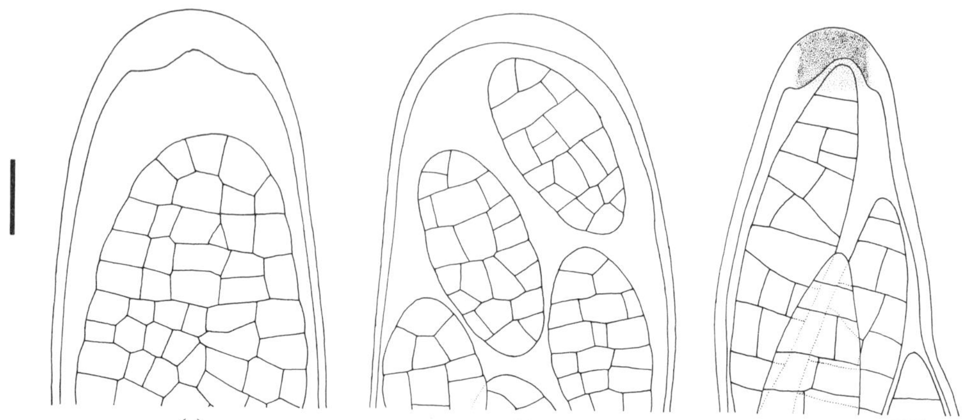

Fig. 9. *Chromatochlamys*, *Protothelenella* and *Thelenella* ascus apices. (a) *C. muscorum*; (b) *T. modesta*; (c) *P. sphinctrinoides*. Scale = 10 μm.

with two functional wall layers, fissitunicate, I−, the inner wall layer thin and the outer thick; apex rounded and sinuous (Fig. 9a). **Ascospores** submuriform to strongly muriform, the outer wall scarcely thicker than the septa, colourless to brownish.

Characterized by the fissitunicate, very thick-walled asci with an internal beak, the muriform, thin-walled ascospores, the persistent, branched-anastomosed paraphysoids and the presence of periphysoids.

LITERATURE: Mayrhofer & Poelt (1985)

1 Ascospores 10–14 μm wide, submuriform to muriform; on damp rocks, at first on *Verrucaria*; .... **1. *larbalestieri***
Ascospores (15-)20–27 μm wide, muriform; overgrowing mosses and plant remains or on soil .... **2. *muscorum***

1. C. larbalestieri (A. L. Sm.) Mayrh. & Poelt (1985)

Microglaena larbalestieri A. L. Sm. (1911)

Thallus membranous, thin, continuous or ± finely rimose-cracked, forming small patches covering *Verrucaria* species, olive-brown or brownish. Perithecia 0.3–0.5 mm diam., scattered, $\frac{1}{4}$-$\frac{1}{2}$-immersed and partly covered by thallus, blackened where exposed, scattered; wall rounded or somewhat pyriform, darkened and opaque above, paler below. Paraphysoids 1–1.5 μm wide, filament-like, not richly branched, rarely anastomosed. Asci (6-)8-spored, 1- to 2-seriate, cylindrical; wall 2–3 μm thick. Ascospores 44–60 × 10–14 μm, colourless, submuriform to muriform, elongate-ellipsoid, ± fusiform-tapered at one end.

On shaded, inundated, mica-schist, quartzite and dolerite rocks and pebbles in or by non-polluted streams, also on shaded, damp stones in woodlands; rare. W. Scotland (Loch Moidart and Is. of Mull), W. Ireland (Connemara). ? Endemic.

An inconspicuous species recognized by the characteristic ascospores, persistent paraphysoids and semi-aquatic habit.

2. C. muscorum (Fr.) Mayrh. & Poelt (1985)

Microglaena muscorum (Fr.) Th. Fr. (1860)

Thallus ± membranous, effuse, often mixed with scattered, pale granules, whitish or pale brown. Perithecia 0.3–0.6 mm diam., rounded or broadly pyriform, ± immersed amongst moss-leaves and stems with only ostioles obvious; wall 50–70 μm thick, brownish above, pale or colourless below. Paraphysoids 0.5–1.5 μm thick, thread-like, branched and anastomosed. Asci 2- to 4-spored, cylindrical; wall 3–6 μm thick (Fig. 9a). Ascospores 60–110 × 20–27 μm, colourless, becoming straw-coloured or brownish when old, markedly muriform, elongate-ellipsoid or subcylindrical.

In open situations encrusting pleurocarpous mosses on basic-barked trees (*Ulmus*, *Fraxinus*, *Corylus* and *Acer*), also on ± basic dunes or overgrowing mosses on limestones, serpentines, mica-schist, epidiorite, often in exposed situations. Widespread. W. British Is. Austria, Bulgaria. Finland, France, W. Germany, Italy, Jugoslavia, Norway, Romania, Spain, Sweden, Switzerland, U.S.S.R., Tunisia.

var. *octospora* (Nyl.) Mayrh. & Poelt (1985) has (6-)8-spored asci and smaller, ellipsoid-elongate ascospores 40–60 × 15–20 μm. N. Wales & N. Scotland (Argyll, Easterness & Sutherland). W. Germany.

The thallus is often only apparent as a thin, skin-like membrane with deeply immersed, pale perithecia. *C. muscorum* is a distinctive species with large, pale brown ascospores and persistent paraphysoids.

CHRYSOTHRIX Mont. (1852)

Thallus leprose, conspicuous and stain-like, bright yellow to bright yellowish green, non-corticate, unstratified and coloured throughout, comprising a mass of powdery, globose granules and anastomosing hyphae, often interspersed with granular crystals, the thallus margin effuse or ± determinate, without lobes; prothallus not apparent. **Photobiont** chlorococcoid. **Ascomata** apothecia, to 2.0 mm diam., immersed or superficial, concave-globose, surrounded by a thin, non-corticate margin or immarginate; disc green-yellow to light orange to brownish. **True exciple** poorly developed, of anastomosing hyphae. **Hymenium**, including epithecium, to 70 μm tall, colourless, ± granular above. **Hamathecium** of paraphyses, 1–1.5 μm, thick, septate, richly anastomosed above, the apices not swollen. **Asci** 20–35 × 8–13 μm, 8-spored, the inner wall extending to form a papilla, which is I+ blue. **Ascospores** 9–14 × 2.5–3 μm, narrowly obovoid to narrowly ellipsoid, colourless, (1- to 2-)3-septate. **Conidiomata** unknown. **Chemistry**: pulvinic acid derivatives. **Ecology**: on trees, wood, and shaded acid rocks.

LITERATURE: Laundon (1981).

1 Thallus leprose; apothecia absent or, if present, not yellow-green-brown 2
Thallus immersed; apothecia present, yellow-green-brown .... **3. *chrysophthalma***

2(1) Thallus bright yellow, rarely greenish yellow; apothecia scarce or unknown, when present ± flat; rhizocarpic acid absent 3
Thallus yellowish-green; apothecia occasionally present, yellow-green-brown, globose; rhizocarpic acid present .... **3. *chrysophthalma***

3(2) Thallus bright yellow, thick; granules 0.1–0.2 mm diam.; calycin and vulpinic acids present .... **2. *chlorina***
Thallus yellow, often slightly orange or greenish tinged; granules 0.01–0.1(-0.2) mm diam.; calycin and/or pinastric acids present .... **1. *candelaris***

1. C. candelaris (L.) Laundon (1981)

Thallus leprose, diffuse (rarely delimited), of scattered granules or producing a fairly thin, ± continuous, non-areolate, vivid, bright yellow crust, often with a golden or greenish tinge; granules 0.01–0.1(-0.2) mm diam., minutely convex to globose. Apothecia not known from British Is., minute, to 0.5 mm diam., rare, ± superficial, rounded-distorted angular, flat to slightly convex, at first surrounded by a very thin, non-corticate margin; disc smooth, pale orange, often yellow-pruinose; true exciple to 0.01 mm thick, entire, smooth, concolorous and sometimes obscured by the thallus, excluded as the apothecia mature. Thallus Pd−, or Pd+ orange, K and KC− or + orange, darkening to red-black, C− (calycin or pinastric acid, rarely both).

On the dry, shaded sides of rough-barked trees, more rarely on shaded acid rocks and walls. Throughout British Is. Cosmopolitan, from alpine areas to the tropics; absent from arctic-alpine and desert areas.

Characterized by its bright yellow, diffuse, leprose thallus.

2. C. chlorina (Ach.) Laundon (1981)

Thallus leprose, diffuse, forming scattered granules or a thick, continuous, non-areolate to strongly cracked crust, bright yellow; granules 0.1–0.2 mm diam. Apothecia unknown. Thallus Pd−, K−, or + faintly orange, KC− or + red, C− (calycin and vulpinic acid).

On acid rocks in very shaded, dry habitats beneath overhangs and in crevices, where it forms the species-poor association *Leprarietum chlorinae*, very rarely on the highly acid trunks of conifers and on man-made substrata; uplands. Wales, Scotland. N. Italy to Scandinavia, Himalayas, North America.

Characterized by its thick, bright yellow, leprose thallus of rather coarse granules growing on rock. May be difficult to separate from *C. candelaris* **(1)**, which has a thinner, yellow to orange-yellow thallus, usually of minute granules, grows on bark and has a different chemistry. *Psilolechia lucida* has a yellowish green, completely powdery thallus, which is slightly less brightly coloured, due to the presence of rhizocarpic acid.

3. C. chrysophthalma (P. James) P. James & Laundon (1981)

Thallus immersed, diffuse, often appearing as a pale grey stain. Soredia up to 25 μm, sometimes absent, scattered to continuous, often in extensive patches, thin to thick and cracked, diffuse, superficial, farinose, dull to bright yellowish green. Apothecia to 0.9 mm diam., rather rare, scattered, superficial, ± round, occasionally somewhat tuberculate, convex to globose, immarginate; discs smooth, soft, yellow-green to brown. Soredia Pd−, K−, KC−, C− (rhizocarpic acid and an unidentified substance).

On well-lit, rather dry, non-nutrient-rich wood and bark of mature, broad-leaved and coniferous trees; frequent, ?increasing. Throughout British Is. W. Europe, Greece (Rhodes), Azores.

Easily identified by the yellow-green-brown, globose, immarginate apothecia, and, when sterile, forming extensive, effuse, sorediate sheets on wood or bark of a wide variety of trees. Closely resembles the predominantly rock-growing *Psilolechia lucida,* which may occasionally occur on tree bases and exposed roots, but which lacks the distinctive unidentified substance always present in *C. chrysophthalma*. Two species may be present: *C. chrysophthalma* s.str. which is richly fertile with a ± inconspicuous and immersed thallus without soredia and an unnamed entity which is more superficial and ± continuously sorediate and only rarely fertile. The latter is very widespread and increasing on rather dry, sheltered sides of broad-leaved trees and conifers; *C. chrysophthalma* s.str. is mainly confined to Scotland on conifer wood.

CLADONIA Hill ex Browne (1756)

Primary thallus (basal) squamulose (subg. *Cladonia*), or crustose (subg. *Cladina*) and soon disappearing; **basal squamules** (subg. *Cladonia*) closely adpressed or ascending to erect, rounded to elongate, entire or variously indented, occasionally sorediate, often persistent, occasionally dominant. **Upper surface** usually corticate, the cortex dense, of ± vertical hyphae. **Lower surface** without rhizines, decorticate. **Photobiont** trebouxioid. **Medulla** usually 2-layered, the outer white, of ± loose hyphae, containing algal glomerules, the inner is glassy or white, cartilaginous, with conglutinate hyphae surrounding central canal. **Secondary thallus** (erect) consisting of usually hollow podetia, growing from margins or upper surface of primary thallus, blunt, pointed or cup-forming, simple or branched; upper surface corticate (lacking in subg. *Cladina* and some other species), often with ± decorticate areas, corticate granules, soredia, or squamules which may be peeling or not, occasionally areolate; apices, cups and axils closed or perforate. **Ascomata** apothecia, at apices of podetia, on cup rims or rarely sessile on basal squamules. **Thalline exciple** absent. **True exciple** often becoming excluded. **Hymenium** red, pale or dark brown. **Asci** elongate-clavate, thickened at the apex, 8-spored, with a I+ blue tholus and I+ blue outer gelatinous sheath. **Ascospores** simple, colourless, fusiform, oblong or ovoid. **Conidiomata** pycnidia, at apices of podetia, cup rims or on basal squamules, sessile or short-stalked, cylindrical to top-like. **Conidia** 5–14 × 0.5–1 μm, cylindrical, thread-like, curved, rarely straight. **Chemistry:** depsides, depsidones, dibenzofurans, terpenes and quinone pigments. **Ecology**: on soil, over mosses, on bark, wood, or soil in rock crevices, several species often forming complex, extensive communities, as on heathlands; mainly on acid humus-rich substrata, a few species on calcareous substrata.

The "reindeer lichens", with a disappearing, granular, primary thallus, have sometimes been assigned to a separate genus, *Cladina*, but recent studies have confirmed these are more appropriately treated as a subgenus within *Cladonia*. The British records of *Cladonia acuminata* (Ach.) Norrlin (1875), *C. amaurocraea* (Flörke) Schaerer (1823) and *C. turgida* Hoffm. (1796) are based on material collected last century which originated from outside the British Is.; only *C. amaurocraea* is considered likely to occur here.

Note: UV tests should be made on cut or abraded surfaces as cortical pigments (e.g. usnic acid) can mask the UV + white reactions.

LITERATURE: Ahti (1961, 1966, 1978, 1980, 1984), Hennipman (1978), Laundon (1965), Ruoss (1987), Ruoss 1989, Ruoss & Ahti (1988), Thomson (1968).

1 Some squamules present, either scattered or forming a ± continuous, persistent, basal thallus and/or dispersed on podetia; podetia rarely with continuous growth, dying at base 2
 Squamules present at extreme juvenile stages only or reduced to coarse areoles; podetia with continuous growth, often dying at base 93

2(1) Apothecia red, K+ violet-purple 3
 Apothecia pink to dark brown, K−, or absent 21

3(2) Squamules densely sorediate below 4
 Squamules not, or very sparingly sorediate below 5

4(3) Squamules ± horizontal, margins ± entire, with necrotic basal parts tinged orange below; apothecia on podetia ..... **12. *digitata***
 Squamules ± erect, the margins crenate, not tinged orange below; apothecia sessile on squamules ..... **18. *incrassata***

5(3) Basal squamules dominant; apothecia often ± sessile on squamules or on short, curved podetia 6
 Podetia dominant, taller than squamules; apothecia terminal on podetia 8

6(5) Lower surface of squamules white or pale grey, never cottony-arachnoid 7
 Lower surface of squamules pale yellow, with a thick cottony-arachnoid layer (× 10 lens) ..... **19. *luteoalba***

7(6) Thallus grey, Pd+ yellow-orange, K+ yellow-orange, UV−; lower surface of squamules tinged orange ..... **14. *macilenta***
 Thallus yellow-grey, Pd−, K−, UV+ white; lower surface of squamules not tinged orange ..... **18. *incrassata***

8(5) At least some podetia terminating in well-developed cups 9
 All podetia with pointed apices; cups absent 15

9(8) Podetia tinged green or yellow-green (usnic acid present) 10
 Podetia tinged grey or blue-grey (usnic acid absent) ..... **15. *polydactyla***

10(9) Podetia mostly <2 cm; cups wide, shallow, rarely hidden by apothecia; pointed podetia absent 11
 Podetia mostly >1.5 cm; cups narrow, often hidden by apothecia; pointed podetia often present 13

11(10) Podetia ± coarsely corticate-granular, or areolate-corticate, with or without crystals on herbarium specimens 12
 Podetia finely granular-sorediate, with fine, needle-shaped crystals on surface in herbarium specimens ..... **9. *pleurota***

12(11) Cups gradually tapered towards base; podetial squamules sparse or absent; medulla UV− ..... **8. *coccifera***
 Cups abruptly tapered towards base; podetial squamules frequent; medulla UV+ white ..... **10. *metacorallifera***

13(10) Most parts of podetia usually devoid of squamules; sorediate 14
Podetia ± densely peeling-squamulose; not sorediate .... **11. *bellidiflora***

14(13) Podetia with regular cups, often tinged grey within, without longitudinal fissures; medulla UV−, with fine, needle-shaped crystals on surface of herbarium specimens .... **17. *deformis***
Podetia irregular, with deformed cups, the inner and outer surface concolorous, often longitudinally split near apices; medulla UV+ white, lacking crystals on surface of herbarium specimens .... **16. *sulphurina***

15(8) Podetia ± uniformly farinose-sorediate 16
Podetia not sorediate, but often with few to numerous coarse, corticate granules 20

16(15) Thallus Pd+ yellow-orange, K+ yellow-orange 17
Thallus Pd−, K− 18

17(16) Basal squamules 5–15 mm wide, ± horizontally spreading; margins ascending, mostly entire .... **12. *digitata***
Basal squamules 2–8 mm wide, ± erect; margins incised .... **14. *macilenta***

18(16) Podetia with yellow or green-yellow soredia 19
Podetia with grey or pale grey-white soredia .... **14. *macilenta***

19(18) Apices of podetia fissured; UV+ white .... **16. *sulphurina***
Apices of podetia not fissured; UV− .... **17. *deformis***

20(15) Podetia <1 cm tall, pale to dark grey, without a yellow tinge; usually fertile **13. *floerkeana***
Podetia >1 cm tall, with a distinct yellow tinge; often sterile .... **11. *bellidiflora***

21(2) Basal squamules dominant; podetia absent or poorly developed and often inconspicuous amongst the squamules 22
Basal squamules not dominant, numerous to very sparse or restricted to the lower part of the podetia only; podetia dominant 39

22(21) Squamules C− 23
Squamules C+ green .... **44. *strepsilis***

23(22) Squamules not or sparsely sorediate below 24
Squamules densely farinose-sorediate below .... **12. *digitata***

24(23) Squamules pale yellow-green below 25
Squamules white, grey or mauve below 27

25(24) Squamules smooth below; Pd+ red 26
Squamules ± cottony arachnoid-tomentose below; Pd− .... **19. *luteoalba***

26(25) Squamules 4–10 × 1–3 mm .... **47. *foliacea***
Squamules 15–25 × 2–8 mm .... **48. *convoluta***

27(24) Squamule margins not coralloid-sorediate 28
Squamule margins ± coralloid-sorediate .... **30. *parasitica***

28(27) Apothecia absent or, if present, on corticate or sorediate podetia 29
Apothecia sessile or on very short, colourless, decorticate podetia .... **28. *caespiticia***

29(28) Squamules K+ yellow, orange or red 30
Squamules K− 33

30(29) Squamules white below 31
Squamules pale to deep grey-violet below .... **49. *firma***

31(30) Squamules to 0.5 cm tall, sometimes tinged yellow-ochre, without black necrotic bases .. 32
Squamules 0.5–2.0 cm tall, becoming grey towards the base on lower surface, with persistent black necrotic bases .... **46. *subcervicornis***

32(31) Squamules blue-grey above; lower surface sometimes tinged yellow-ochre at the base; K+ yellow (no crystals); on acid humus .... **14. *macilenta***

Squamules green above, starkly white below, K+ (red crystals), on calcareous soils in rock clefts and limestone .... **43. *symphycarpa***

33(29) Squamules Pd− 34
Squamules Pd+ red 35

34(33) Squamules UV+ white; cushion-forming or dispersed, not noticeably fragile . **59. *squamosa***
Squamules UV+ bluish white; sward-forming, very fragile and easily broken **29. *fragilissima***

35(33) On mosses on soil or rocks or directly on soil 36
On bark or wood .... **35. *coniocraea***

36(35) Squamules mostly ascending or erect, tufted or scattered, not rosette-forming 37
Squamules mostly spreading horizontally, rosette-forming, ± imbricate .... **21. *pocillum***

37(36) Squamules thick, often tough, rarely veined below, curved at apices when dry; lower surface white-grey-mauve 38
Squamules thin, often delicate and brittle, not recurved at apices when dry; the lower surface white .... **42. *ramulosa***

38(37) Squamules green above; lower surface with ± distinct pink veins .... **50. *cyathomorpha***
Squamules grey-green above; the lower surface white-grey-mauve, not veined **45. *cervicornis***

39(21) Podetia and squamules C− 40
Podetia and squamules C+ green .... **44. *strepsilis***

40(39) Cups present, mostly as wide as or wider than podetial stalk 41
Cups absent or, if present, not wider than podetial stalk 68

41(40) Podetia yellow-tinged 42
Podetia not yellow-tinged 46

42(41) Podetia mostly or entirely farinose-sorediate 43
Podetia not sorediate, with coarse, corticate granules or ± squamulose 45

43(42) Podetia to 4 cm tall, with a short, corticate zone at base, densely sorediate above; cups narrow, shallow; apothecia absent or, if present, red 44
Podetia 0.5–1.5 cm tall, entirely farinose-sorediate; cups goblet-shaped, rarely proliferating; apothecia present, pale brown to pale flesh-coloured .... **25. *carneola***

44(43) Cups irregular; podetia often with gaping, longitudinal fissures towards apices; UV+ white .... **16. *sulphurina***
Cups regular; podetia rarely with fissures; UV− .... **17. *deformis***

45(42) Cups wide, rather regular; podetia <1 cm tall, with coarse, corticate granules sometimes developing into non-peeling squamules towards the base .... **8. *coccifera***
Cups narrow or apices pointed, often irregular; podetia 1–4 cm tall, with abundant peeling squamules .... **11. *bellidiflora***

46(41) Podetia without granules or soredia 47
Podetia entirely or partly sorediate or with delimited, convex, corticate granules 57

47(46) Podetia not proliferating or else proliferating from margins of cups 48
Podetia regularly proliferating from centre of cups, forming tiers .... **45. *cervicornis***

48(47) Cups >3 mm, wide; podetial squamules absent; K− 49
Cups <1.5 mm, wide; podetial squamules present; K+ yellow .... **60. *stricta***

49(48) Cups mostly conspicuous, very rarely centrally perforate; UV− or ± white 50
Cups inconspicuous, often reduced to 3–6 points surrounding a central rounded, gaping, perforation; UV+ white .... **56. *crispata***

50(49) Cups not centrally perforate; UV− 51
Cups conspicuous, centrally perforate, with long, subulate proliferations; UV+ white .... **56. *crispata***

51(49) Podetia <6 cm; basal and podetial squamules often very sparse or absent 52
Podetia <2 cm; basal and podetial squamules usually numerous and well developed 53

52(51) Podetia rather swollen, 2–3 mm wide, blue-green, brownish at apices and yellowish at base, never with squamules .... **58. *maxima***
Podetia mostly slender, 0.5–1.5 mm wide, olive-brown, dark brown-black at base, ± sparingly squamulose .... **57. *gracilis***

53(51) Bases of older podetia marbled, with ± dispersed, white or grey areoles on a black decorticate surface 54
Podetia with a uniform cortex, or if broken, never becoming blackened at the base 55

54(53) Thallus Pd+ red; cups irregular, the surface often matt, pale grey, subarachnoid, frequently proliferating from margins to form intricate clusters of podetia **61. *phyllophora***
Thallus Pd−; cups ± regular, the surface not subarachnoid, not proliferating at margins except to form short, ± coralloid extensions bearing apothecia .... **29. *fragilissima***

55(53) Podetia not longitudinally fissured; lower surface of basal squamules tinged grey or mauve; K− or K+ yellow (some morphs of *C. chlorophaea* **(24)** may key out here) ... 56
Podetia often ± longitudinally fissured; lower surface of basal squamules white; K+ red (crystals) .... **43. *symphycarpa***

56(55) Podetia with wide, shallow, mostly regular cups; basal squamules tinged mauve below; K− .... **45. *cervicornis***
Podetia irregular, often turgid-coralloid; lower surface of basal squamules tinged grey-black towards base; K+ yellow .... **46. *subcervicornis***

57(46) Podetia and cups with coarse, corticate granules or peeling squamules 58
Podetia and cups with soredia 61

58(57) Cups not perforate; podetia with corticate granules and squamules derived as outgrowths from cortex; Pd+ orange-red 59
Cups perforate, often irregular; podetia with numerous squamules derived from peeling of cortex; Pd− or yellow .... **59. *squamosa***

59(58) Cups wide, ± regular, not occluded when fertile 60
Cups narrow, ± irregular, usually occluded when fertile .... **42. *ramulosa***

60(59) Basal squamules well developed, thick, chalky inside and below, forming ± imbricate or contiguous, horizontally spreading rosettes; calciphile .... **21. *pocillum***
Basal squamules sparse to abundant, thin, not chalky, contiguous or dispersed, becoming erect, not rosette-forming; calcifuge .... **20. *pyxidata***

61(57) Cups not perforate at base; UV− 62
Cups all perforate at base; UV+ white .... **41. *cenotea***

62(61) Thallus Pd+ red or rarely Pd−, K− or yellow, ... 63
Thallus Pd+ yellow-orange, K+ yellow-orange .... **15. *polydactyla***

63(62) Podetia corticate for 1–3 mm, rarely more at base 64
Podetia entirely sorediate, not or very slightly corticate at base 67

64(63) Cups narrow to wide, with a well-developed stalk (1–3.5 cm); sometimes proliferating from margins; K− 65
Cups wide, subsessile, or with a short (<1 cm), ± corticate stalk; not proliferating from margins; K+ yellow .... **22. *humilis***

65(64) Cups narrow in relation to length of podetial stalk 66
Cups wide in relation to length of podetial stalk .... **24. *chlorophaea***

66(65) Cups ± regular, abruptly expanded; soredia greenish white; cortex green; Pd+ orange-red, UV− .... **36. *ochrochlora***
Cups irregular, gradually tapered to podetial stalk; soredia and cortex brownish; Pd− or slowly orange, UV+ white .... **38. *rei***

67(63) Podetia green to grey, goblet-shaped; cups wider than podetial stalk, rather regular with denticulate, little-proliferating margins .... **23. *fimbriata***
Podetia ash grey, antler-like, mostly with pointed apices; cups narrow, irregular, often with marginal proliferations .... **39. *subulata***

68(40) Podetia ± longitudinally lacerate and fenestrate 69
Podetia entire, not obviously longitudinally lacerate 71

69(68) Podetia 1–3 cm, conspicuously lacerate; margins of squamules crenate-incised; apothecia present or absent 70
Podetia 0.3–1 cm, sparingly lacerate; margins of basal squamules entire; apothecia always present .... **27. *peziziformis***

70(69) Podetia verrucose or coarsely corticate-granular or with numerous peltate squamules; Pd+ yellow, K− .... **34. *macrophylla***
Podetia with scattered areoles; Pd+ yellow, K+ yellow-orange .... **26. *cariosa***

71(68) Podetia entirely or partly sorediate 72
Podetia not sorediate, smooth, ± scabrid or granular-corticate 78

72(71) Lower half of podetia corticate 73
Podetia entirely sorediate, or corticate for only a few mm at the base 74

73(72) Podetia-cortex smooth, ± finely mosaic-areolate; soredia farinose; UV− .... **40. *cornuta***
Podetia-cortex verrucose-granular; soredia ± granular; UV+ white .... **38. *rei***

74(72) Thallus Pd+ orange or red 75
Thallus Pd− 77

75(74) Podetia Pd+ red, K− or faintly yellow 76
Podetia Pd+ orange, K+ yellow-orange .... **14. *macilenta***

76(75) Podetia <2 cm, often curved, the apices mostly pointed and unbranched .... **35. *coniocraea***
Podetia 2–6 cm, often branched towards apices and becoming antler-like .... **39. *subulata***

77(74) Podetia often with an obscure, single, longitudinal fissure at one side, seen as a shallow depression (× 10 lens), often ± branched; UV+ blue .... **37. *glauca***
Podetia without a longitudinal fissure (× 10 lens), often unbranched; UV− .... **14. *macilenta***

78(71) Podetia <10 mm; apothecia frequent 79
Podetia >10 mm; apothecia infrequent 82

79(78) Podetia grey; apothecia dark brown or red 80
Podetia yellowish green; apothecia pale cream to pale brown .... **31. *botrytes***

80(79) Podetia rough, not fissured, often with granules or squamules; apothecia brown or red 81
Podetia smooth, weakly fissured, without granules or attached squamules; apothecia brown .... **27. *peziziformis***

81(80) Podetial surface verrucose-scabrid; apothecia red; Pd− .... **13. *floerkeana***
Podetial surface areolate-squamulose; apothecia brown; Pd+ red .... **42. *ramulosa***

82(78) Podetial surface rough; squamules derived from peeling cortical scales 83
Podetial surface smooth; squamules, if present, derived from outgrowths of cortex 86

83(82) Podetia mainly unbranched or with irregular branching; Pd− or ± yellow 84
Podetia ± dichotomously branched; Pd+ rust-red .... **53. *scabriuscula***

84(83) Thallus Pd+ orange, K+ orange, UV− 85
Thallus Pd−, K−, UV+ white [if yellowish (usnic acid) (see *C. bellidiflora* (**11**)] **59. *squamosa***

85(84) Podetia not perforate, <0.5 cm; squamules a ± continuous sward, compacted, much dissected .... **30. *parasitica***
Podetia perforate, <8 cm; squamules sparse to numerous, not dissected .... **59. *squamosa***

86(82) Podetia perforate at apices 87
Podetia not perforate at apices 88

87(86) Terminal perforation surrounded by 3–6 'points', one of which may proliferate; UV+ white .... **56. *crispata***
Terminal perforation surrounded by 'points', which do not proliferate; UV+ bluish .... **29. *fragilissima***

88(86) Podetia mostly branched towards base; incipient cups sometimes present 89
Podetia branching at all levels; cups absent 90

89(88) Podetia K+ yellow, yellowish at base .... **58. *maxima***
Podetia K−, brownish white at base .... **57. *gracilis***

90(88) Podetia ± ascending, without white concretions 91
Podetia ± prostrate, often with scattered, white concretions or nodules . **52. *subrangiformis***

91(90) Podetia hollow 92
Podetia solid .... **54. *stereoclada***

92(91) Branching markedly divergent; algae confined to dispersed, raised, green areoles (× 10 lens); usually Pd−, K+ yellow, occasionally red .... **55. *rangiformis***
Branching weakly divergent; algae ± continuous or forming a mosaic intersected by thin, white, decorticate lines or dispersed in scattered (not raised) clusters (× 10 lens); Pd+ rust-red, K− .... **51. *furcata***

93(1) Podetia moderately dichotomously branched, rarely tri- or tetrachotomous; branches abruptly tapered at apices; thallus stiff and spikey 94
Podetia very richly branched, the branching sympodial; branches not abruptly tapered at apices; forming extensive and elegant tufts 95

94(93) Podetia blue-green or grey-green, with fine crystals at branch apices developing in the herbarium; surface opaque, not truly corticate, verrucose; thallus decumbent, the branches widely divergent .... **33. *zopfii***
Podetia yellow-green to pale grey-green, without crystals towards apices; surface smoothly corticate, somewhat glossy; thallus ± erect, the branches moderately to closely divergent .... **32. *uncialis***

95(93) Podetia forming rounded 'heads', the branched-apices divergent, not orientated in one direction 96
Podetia not forming rounded 'heads', the branch-apices ± orientated in one direction, often in ± drooping clusters 98

96(95) Branching dense to lax, anisotomic, tri- or dichotomous, the ultimate branches 2(-3), unorientated or clustered and curved in various directions 97
Branching very dense, isotomic, polytomous, the apices of (3-) 4–6 branchlets in a whorl around a central, open axil .... **7. *stellaris***

97(96) Branching trichotomous; apical branches very thin, ± curved .... **5. *portentosa***
Branching dichotomous; apical branches ± stout, not curved .... **6. *mediterranea***

98(95) Branchlets and apices predominately tri-or tetrachotomous 99
Branchlets and apices mainly dichotomous .... **4. *ciliata***

99(98) Podetia with a green-yellow tinge (usnic acid); K− 100
Podetia white, without a green-yellow tinge (usnic acid absent); K+ yellow . **1. *rangiferina***

100(99) Podetia Pd+ red .... **2. *arbuscula***
Podetia Pd− .... **3. *mitis***

Subgen. CLADINA (Nyl.) Leighton (1871)

Primary thallus crustose, soon disappearing; **secondary thallus** consisting of erect, richly branched podetia, always ecorticate and not squamulose, lacking cups; pycnidia terminal.

1. C. rangiferina (L.) Weber ex Wigg. (1780)

Podetia 4–10 cm tall, grey-white, often brown and with a ± purple or bluish grey tinge towards apices, the surface uniformly, finely cottony-arachnoid (× 10), richly branched; branches distinctly orientated in one direction, the branching predominantly tetrachotomous, more rarely trichotomous at apices; young apices to 2 mm diam., blunt. Apothecia brown, inconspicuous, rare. Pycnidia containing colourless jelly. Thallus Pd+ red, K+ yellow, KC−, C−, UV− (fumarprotocetraric acid, atranorin).

On exposed, moss-lichen heaths and in pockets of soil on cliff ledges and rock outcrops in montane and maritime areas; rather local. N. England, N. Wales, Scotland, Ireland. Europe, North & South America, Asia, Antarctica.

Characterized by the even, cottony-arachnoid surface, white (not yellow-green) colour, tetrachotomous branching and the presence of atranorin (K+ yellow). *C. ciliata* var. *ciliata* **(4)** differs in the roughened surface, thinner branches, dichotomous branching and pycnidia containing red jelly. *C. arbuscula* **(2)** and *C. ciliata* var. *tenuis* **(4)** are yellowish green due to the presence of usnic acid.

The similar ***C. stygia*** (Fr.) Ruoss (1985) from Scandinavia, has red pycnidial jelly and the bases of the podetia are blackened with scattered, white areoles; it has recently been identified amongst collections of '*C. rangiferina*' made last century from Scottish Highlands (Braemar) and Wales (Snowdonia) and, more recently, from Wales (Merioneth, Borth Bog).

2. C. arbuscula (Wallr.) Flotow (1839)

Cladonia arbuscula subsp. *squarrosa* (Wallr.) Ruoss (1987)

Podetia 4–10 cm tall, yellow-green, surface ± roughened, richly branched; terminal branches strongly orientated in one direction, branching predominantly trichotomous (Fig. 10b) and tetrachotomous at apices; young apices to 2 mm diam., blunt. Apothecia brown, inconspicuous, rather rare. Pycnidial jelly colourless. Thallus Pd+ red, K−, KC+ yellow, C−, UV− (fumarprotocetraric and usnic acids).

In montane and lowland heathlands and acid dunes; rather local. Throughout British Is. Europe, North & South America, Asia.

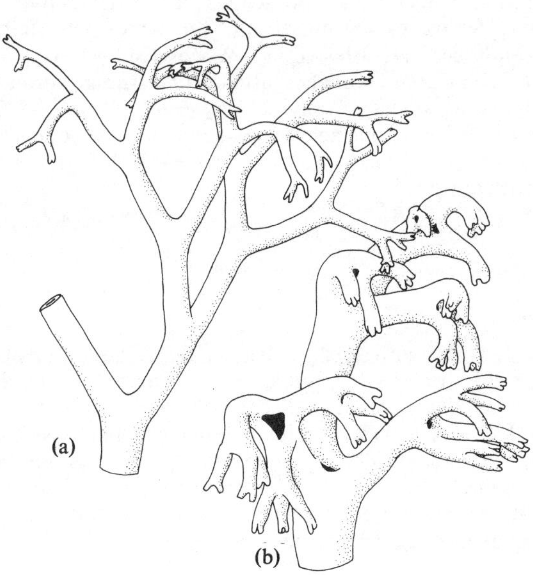

Fig. 10. *Cladonia* branching types. (a) Dichotomous (*C. ciliata*); (b) trichotomous (*C. arbuscula*).

Characterized by the yellow-green podetia, trichotomous and tetrachotomous branching and markedly orientated apices. The similarly coloured *C. portentosa* **(5)** has terminal branches which diverge in all directions; *C. ciliata* var. *tenuis* **(4)** differs in the dichotomous branching, more slender branches, and more delicate overall appearance. For differences from *C. mitis* see **(3)**, but some difficulty may be experienced in separating certain specimens of this species from the more common *C. arbuscula*.

3. C. mitis Sandst. (1918)

Cladina mitis (Sandst.) Hustich (1951)

Like **2**, but more slender, paler grey, apices less branched, the terminal branches are also less, or not at all, strongly orientated in one direction. Thallus Pd− (rarely Pd+ red), K−, KC+ yellow, C−, UV− (usnic and rangiformic acids, rarely also fumarprotocetraric acid).

On acid dunes; very rare. S.E. England, E. Scotland. Europe, Arctic, North & South America, Asia, Australia, New Zealand, Antarctica.

This taxon may be better treated as *C. arbuscula* subsp. *mitis* (Sandst.) Ruoss (1987).

4. C. ciliata Stirton (1888)

Podetia 3–6 cm tall, grey-white [var. *ciliata*], often with a distict purple-brown tinge towards apices, richly-branched; branches distinctly orientated in one direction, the branching predominantly dichotomous at the apices (Fig. 10a); young apices slender, to 0.8 mm diam. Apothecia brown, inconspicuous, rather rare. Pycnidial jelly red. Thallus Pd+ red, K−, KC−, C−, UV−.

On montane and lowland heathlands and acid dunes; local. Throughout British Is., Europe, North America, Turkey, Himalaya, E. Asia.

Two chemotypes are recognized at varietal level: var. *ciliata* has a grey-white thallus containing furmarprotocetraric and occasionally traces of usnic acids. **Var. *tenuis*** (Flörke) Ahti (1977) has a yellowish green thallus (usnic acid); it is much more frequent in the British Is. than var. *ciliata*. *C. ciliata* var. *ciliata* differs from *C. rangiferina* **(1)** in the predominantly dichotomous branching, more delicate habit, red pycnidial jelly, and chemistry.

Var. *tenuis* is very variable; large, coarse specimens with terminal branches not strongly orientated may resemble *C. portentosa* **(5)** which is, however, Pd−, UV+; *C. arbuscula* **(2)** is always coarser with a strongly emphasized orientation of the terminal branchlets and has tri and tetrachotomous branching. ***C. azorica*** Ahti (1961), characterized by the presence of fumarprotocetraric, perlatolic, usnic and ± protocetraric acids, has recently been recorded from S.W. England (Devon), Wales (Glamorgan) and S.W. Ireland; thallus pale yellow-grey, branches usually not strongly orientated in one direction, branching di- to trichotomous at apices; pycnidial jelly colouréless. Additionally, a grey morph, containing fumarprotocetraric and perlatolic acids, but lacking usnic acid, has been identified amongst material named as *C. ciliata* subsp. *ciliata*. Apparently frequent in W. British Is. but the status of this material requires further critical study.

5. C. portentosa (Dufour) Coem. (1865)

incl. f. *subimpexa* (Duvign.) Ahti (1978)

Podetia 4–10 cm tall, grey-green often with a yellowish green tinge, the surface often roughened, areolate, particularly towards base, richly branched; branches not orientated in one direction, the branching predominantly trichotomous at apices, often with perforate axils. Pycnidial jelly colourless. Thallus Pd−, K−, KC+ yellow or KC−, C−, UV+ white in decorticate parts (perlatolic and ± usnic acids).

On heathlands and dunes; abundant. Throughout British Is. W. Europe, W. North America.

Characterized by the unorientated terminal branches and distinctive chemistry. Very variable, often forming neat tufts amongst low vegetation, sometimes with distorted main branches; inflated and grossly modified morphs occur in boggy situations.

6. C. mediterranea Duvign. & des Abb. (1947)

Like **5**, but tends to form low, straggling clumps of a rather open, divergent habit, the podetia have longer, extended internodes, the branching is predominantly dichotomous and the surface is more smooth and matt.

On somewhat basic soils associated with serpentine. S.W. England (Lizard Peninsula). Mediterranean region, W. France, Yugoslavia, Turkey, N. Africa, Canary Islands.

British material differs somewhat from continental specimens which form ascending, rounded tufts with shorter internodes.

7. C. stellaris (Opiz) Pouzar & Vezda (1971)

Like **5**, but forming very compact, ascending, bouffant tufts of richly branched thalli with a rounded outline. Pycnidial jelly red.

Amongst *Rhacomitrium*; not recorded this century. N.E. Scotland (E. Grampian mtns). Europe, U.S.S.R., Greenland, North America, Asia.

The reindeer lichen; an important source of food, collectively with *C. mitis* **(3)**, *C. rangiferina* **(1)** and *C. arbuscula* **(2)**, for reindeer and caribou. Used in the manufacture of wreaths and for modelling as miniature trees.

Subgen. CLADONIA

Primary thallus persistent or sometimes disappearing; **secondary thallus** consisting of erect, unbranched to richly branched, usually corticate, often squamulose podetia which may bear cups; pycnidia basal on squamules or terminal on podetia.

8. C. coccifera (L.) Willd. (1787)

Podetia 1–2 cm tall, yellow-grey to yellow-green; cups wide, gradually tapering to base with only a short stalk (*C. pyxidata*-type), regular or ± deformed, rarely proliferating from margins; surface smooth or uneven, ± corticate throughout, with conspicuous, coarse, smooth, corticate granules, especially within the cups, sometimes ± squamulose at the base; soredia absent. Basal squamules up to 2 mm diam., very variable, ± rounded and sparingly indented; lower surface often yellow-brown towards base, ± dispersed or occasionally forming ± compact cushions. Apothecia and pycnidia red, frequent, on margin of cups. Thallus Pd−, K−, KC+ yellow, C−, UV−. Two chemotypes occur: (a) usnic acid, zeorin and ± porphyrilic acid; widespread. (b) usnic acid and barbatic acid; rare.

Predominantly on acidic, sandy or peaty soils, occasionally on mossy rocks, wall tops and rotting wood, in montane, moorland and coastal sites; common. Throughout British Is. Europe, Arctic, North & South America, Asia, New Zealand?, Antarctica.

Sterile specimens are separated from *C. pyxidata* **(20)** by their yellow colour, Pd− reaction and chemistry. The barbatic acid chemotype (b) has been recognized as *C. borealis* Stenr. (1989) and is distinguished also by the smooth to shallowly areolate-corticate surface of the podetia compared with the elevated, areolate-corticate cortex with bullate and scaly plates as vegetative propagules in *C. coccifera* s.str. *C. diversa* Asperges (1983) has been segregated from chemotype (a) to include material with podetia which are densely squamulose at the base and with granules, microsquamules and scaly plates on the upper part and within the cups; common in W. Europe and more frequent than *C. coccifera* s.str. in the British Is. *C. coccifera* s.str. appears to be confined to montane localities in the Scottish Highlands.

9. C. pleurota (Flörke) Schaerer (1850)

Like **8**, but with coarse, distinctly granular soredia on the upper part of the podetia and within the cups. The podetia are patchily corticate below, particularly towards the base, where they may become ± squamulose. Contains usnic, ± iso-usnic, porphyrilic acids and zeorin.

On *Rhacomitrium* heath near areas of late snow lie, also heathlands; very rare. N. Wales, N.E. Scotland (Cairngorm mtns and elsewhere). Europe, Arctic, North & South America, Asia, Australia.

10. C. metacorallifera Asah. (1939)

Like **8**, but the cups are more abruptly tapered (*C. chlorophaea*-like) and covered in numerous, minute, descending squamules which give the podetium a scabrid appearance. Parts of the podetia, particularly towards the margin of the cups, have blackened, decorticate areas. Contains usnic, didymic and squamatic acids, UV+ white.

On mossy boulders in ± sheltered situations; very rare. S.W. England (Shropshire), C.W. & N.E. Scotland (Argyll, Cairngorm mtns). Europe, North America, Japan.

11. C. bellidiflora (Ach.) Schaerer (1823)

Podetia 3–4 (-5) cm tall, grey-green to yellow-green, straight or somewhat twisted, pointed at apices, unbranched, often entirely densely peeling-squamulose, especially towards the base, becoming corticate-granular or ± decorticate and paler towards apices, never sorediate. Basal squamules small, often inconspicuous, deeply indented, yellowish brown (K+ brown-purple) towards base on lower surface. Apothecia frequent, red, sessile or on short projections at apex of podetia, often becoming clustered. Thallus Pd−, K−, KC+ yellow, C−, UV+ white (usnic and squamatic acids).

Amongst *Calluna* and mosses between rocks, particularly on stabilized scree on moorland and montane areas. Scotland, extending locally to N. England, Wales and Ireland. Europe, Arctic, North and South America, Asia, New Zealand, Antarctica.

A distinctive species which, though often stunted in southern localities, is characterized by the simple, pointed podetia, the yellow-green colour, the dense, peeling squamules and the red apothecia and pycnidia. When sterile it is distinguished from *C. squamosa* **(59)** by the yellow colour, absence of perforations and yellowish, K+ purplish pigment on lower side of basal squamules; entirely grey specimens, devoid of usnic acid, are rarely encountered.

12. C. digitata (L.) Hoffm. (1796)

Podetia to 1 cm tall, often ± curved-decumbent, pointed or with irregular cups which occasionally proliferate from margins, inside of cup corticate, otherwise entirely farinose-sorediate or with a few corticate patches towards the base. Basal squamules to 1 cm diam., dominant, ± horizontally spreading, often overlapping, yellowish green-grey, ± rounded and entire, scarcely indented, densely white farinose-sorediate on lower surface and on the upturned margins, often with a distinct orange tinge on the lower surface towards point of attachment to substratum. Apothecia and pycnidia red, occasional, at ends of pointed apices or on short projections from margins of cups. Thallus Pd+ orange, K+ yellow, KC−, C− (thamnolic acid).

On rotting wood, especially old tree stumps, more rarely on peaty soils, especially in mature woodlands; often locally common. Throughout British Is. Europe, North America, E. Africa, Asia.

The only species of *Cladonia* in Europe which is truly densely farinose-sorediate on the lower surface of basal squamules. The squamules in *C. incrassata* **(18)** and *C. parasitica* **(30)** are often coarsely granular-sorediate, but are much smaller, deeply incised and compacted. In other species, e.g. *C. polydactyla* **(15)**, the squamules may be sparsely sorediate, often the result of the shedding of soredia from associated podetia; *C. polydactyla* **(15)** has smaller, more incised basal squamules and usually more numerous, erect podetia.

13. C. floerkeana (Fr.) Flörke (1828)

Podetia to 1.5 (-3) cm tall, often shorter, pale to dark grey, ± tinged brown in exposed sites, without cups, simple or occasionally sparingly branched near apices, ± entirely corticate or partially decorticate, especially towards apices, rarely granular-sorediate in patches but very frequently with coarse granules, or podetia ± entirely densely squamulose. Basal squamules small, inconspicuous, often tinged orange on lower surface. Apothecia commonly abundant; hymenium and pycnidia red, terminal on podetia, single or clustered. Thallus Pd−, K+ yellow or K− (orange pigment reacts K+ purple), KC−, C−, UV− or UV+ blue (barbatic acid and, rarely, thamnolic acid).

On heathland and moorland, also on mossy scree, well-rotted wood, fence posts and tree stumps in open situations; common. Throughout British Is. Europe, North America, Asia, Australia.

The "Bengal match" lichen. Usually a well-fruited, and then conspicuous, species, but the podetia are variable and may be almost entirely decorticate except for a few scattered corticate granules, or, conversely, densely squamulose throughout. *C. macilenta* **(14)** which may occur in similar habitats, is always partly to wholly finely sorediate and K+ yellow.

14. C. macilenta Hoffm. (1796)

C. bacillaris (Leighton) Arnold (1880)

Podetia to 1.5(-3) cm tall, often less, grey-blue, often tinged orange where abraded, pointed at apices, without cups, usually simple or rarely sparingly branched near apices, often twisted and ± decumbent, usually ± entirely farinose to coarsely granular-sorediate, at times patchily decorticate, rarely bearing squamules, particularly near the base. Basal squamules small, ± elongate, much indented, rarely sparingly sorediate below. Apothecia and pycnidia red, at tips of podetia, single or clustered, rather frequent. Thallus Pd+ orange, K+ yellow, KC−, C−, UV− (thamnolic, ± barbatic, and ± didymic acids).

On soil, bark and wood, and amongst mosses in ± acid woodlands and heathlands; common. Throughout British Is. Europe, North & South America, Africa, Asia, Australia, New Zealand.

Squamulose morphs of *C. macilenta* may be difficult to separate from juvenile or pollution-stressed morphs of *C. polydactyla* **(15)** which may only develop distinct cups at a late stage of development. In more polluted sites, *C. macilenta* often grows with *C. coniocraea* **(35)**, when it may be identified by its characteristic grey-blue colour (unchanged when wet) in contrast to the dark green colour of *C. coniocraea* (brighter green when wet).

Specimens reacting Pd−, K−, KC+ pale yellow or KC−, C−, UV+ pale blue or UV− (barbatic and ± didymic acids) have been named as *C. bacillaris* but are best considered as a chemotype of *C. macilenta*; they typically occur in sandy heaths and dunes and, when sterile, may superficially resemble *C. glauca* **(37)** which, however, contains squamatic acid (UV+ white) and usually has a single obscure longitudinal fissure ± midway down the sides of the podetia. Thalli with finely divided basal squamules, often with a reddish medulla and superficial red spots, K+ purple, due to mite infestation, with barbatic and 4-*O*-methylbarbatic acids (UV+ bluish white) are ***C. norvegica*** Tønsb. & Holien (1984); recently recorded from Somerset (Hawkridge) on mossy rocks in old woodland.

15. C. polydactyla (Flörke) Sprengel (1827)

Podetia 1–2(-3) cm tall, pale grey to whitish or greenish grey, pointed at first, later with narrow, usually irregular, abruptly tapering cups (to 5 mm wide), often variously proliferating from margins or remaining dentate, usually simple, occasionally sparingly branched near apices, ± entirely farinose- to coarsely granular-sorediate and mixed with squamules which are particularly frequent near the base, at times patchily decorticate, especially above. Basal squamules small, incised, occasionally ± thinly sorediate below and with orange pigmented regions at base. Apothecia and pycnidia red, at apices of podetia, single or clustered, rather frequent. Thallus Pd+ orange, K+ yellow, KC−, C−, UV− (thamnolic acid).

On rotting wood, also on soil and amongst mosses in woodlands and heathlands in hilly and low-lying areas; often common. Throughout British Is. W. Europe, ?North America, Asia.

Very variable, characterized by podetia which bear narrow, irregular, proliferating cups, often with intermixed soredia and squamules. Specimens which are young or from polluted sites are difficult to separate from *C. macilenta* **(14)** which lacks cups. *C. digitata* **(12)**, which may occur in similar habitats, has large, rounded squamules with conspicuously sorediate lower surfaces. In polluted sites *C. polydactyla* often grows with *C. coniocraea* **(35)**, which is dark green and contains fumarprotocetraric acid and is Pd+ rust red.

C. umbricola Tønsb. & Ahti (1980), has a thallus which is Pd−, K−, KC−, C−, UV+ white (squamatic acid). It has been recorded on decaying wood and bark and amongst mosses on moist rocks, usually in deep shade in C. & N. Scotland (Stirling, Cairngorm mtns), Norway and W. North America and is probably a chemotype of *C. polydactyla*.

16. C. sulphurina (Michaux) Fr. (1831)

Podetia to 5 cm tall, yellow to grey- or green-yellow, without cups, irregular, with ± pointed apices, unbranched, usually conspicuously longitudinally fissured, particularly in upper part, ± farinose-sorediate throughout, with 1–2 mm corticate area at base, which is often ± squamulose. Basal squamules often large, 2–8 mm wide, ± rounded, the lower surface white, often yellowish brown towards base, occasionally sorediate, horizontally spreading or ± ascending. Apothecia and pycnidia red, at tips of podetia, rarely fertile. Thallus Pd−, K−, KC+ yellow, C−, UV+ white (usnic and squamatic acids).

On montane and less frequently on lowland heaths, also rarely on decaying thatch. Much of Scotland, extending locally to England and Wales. Europe, North & South America, Asia, New Zealand.

17. C. deformis (L.) Hoffm. (1796)

Like **16**, but the podetia lack terminal fissures and are often surmounted by ± regular cups with even margins. Thallus Pd−, K−, KC+ yellow, C−, UV− (usnic acid and zeorin).

On peat and decaying wood in montane areas; very rare. Scotland, especially Cairngorm mtns. Europe, Arctic, North America, Asia, New Zealand.

18. C. incrassata Flörke (1826)

Podetia to 0.2–0.5 cm tall, infrequent, often deformed or produced from an extension of squamule, ± flattened with a partly corticate-granular and sorediate surface. Basal squamules to 4 mm, dominant, numerous, yellowish grey-green, often contorted, crenulate-incised, usually conspicuously farinose- to granular-sorediate below, occasionally also above, ± erect, not tinged orange at point of attachment. Apothecia red, frequent, usually sessile on upper surface and extended margins of the squamules, more rarely on short, sometimes flattened, irregular podetia. Thallus Pd−, K−, KC+ yellow, C−, UV+ white (three chemotypes occur: (a) usnic, squamatic and didymic acids; (b) usnic, squamatic and barbatic acids; (c) squamatic, didymic and thamnolic acids).

On rotting wood, more rarely on acid soil in woodland or on old cut surfaces of peat; local. From S. England and Wales to N. Scotland, N.W. Ireland (Donegal). Europe, E. North America, Asia.

The small, sessile, red apothecia, yellowish thallus and chemistry are diagnostic. Chemotype (c) is very rare.

19. C. luteoalba Wheldon & A. Wilson (1907)

Podetia to 1 cm, rare, decorticate and blackened when old, often hidden among squamules. Basal squamules 2–10 mm diam., greenish yellow above, ± rounded and entire, scarcely indented, sometimes ± elongate, the margins ascending, when dry recurved and exposing pale yellow, distinctly cottony-arachnoid lower surface (× 10 lens). Apothecia and pycnidia red. Apothecia rare, on margins of cups, or sessile on squamules. Thallus Pd−, K−, KC+ yellow, C−, UV+ white or UV− (three chemotypes occur: (a) usnic acid, zeorin and ± porphyrilic acid (widespread); (b) usnic, and barbatic acids (rare, Is. of Mull); (c) usnic, squamatic and didymic acids (rare, Cairngorm Mtns).

Often overgrowing other species of *Cladonia* on peaty soils, mossy boulders and on rock ledges and walls in exposed heathlands and on mountain summits, rarely on bases of old trees. N. England (Lake District, Pennines) extending to S.W. England (Cornwall) and N.E. Scotland (Caithness). Europe, Alps, Greenland, Iceland, Spitsbergen, Alaska, Tierra del Fuego, Falkland Is., China.

Recognized by the conspicuous, large, curled squamules, pale yellow and cottony arachnoid below, a feature which distinguishes it from other species with well-developed basal squamules. The similarly coloured *C. foliacea* **(47)** has deeply incised squamules, is not cottony-arachnoid below, contains fumarprotocetraric acid, and occurs in different habitats.

20. C. pyxidata (L.) Hoffm. (1796)

Podetia 0.5–1.5(-3) cm tall, grey, occasionally ± brownish; cups 3–10 mm diam., gradually tapered towards base, ± regular, rarely proliferating from rim; surface uneven with coarse, corticate granules, particularly well developed within the cups, often with scattered, partially decorticate, pale areas; soredia absent. Basal squamules often rather small, rounded or elongate, with few indentations, sparse or often abundant, becoming ± erect. Apothecia and pycnidia brown, common, on cup rim; apothecia often shortly stalked. Thallus Pd+ red, K−, KC−, C−, UV− (fumarprotocetraric acid).

On mossy rocks, walls, tree trunks and acid soil, chiefly in rather dry habitats; common. Throughout British Is. Europe, North & South America, Africa, Asia, Australia, New Zealand.

Old woodland morphs usually have well-developed basal squamules. *C. chlorophaea* **(24)** differs in having more slender, gradually tapering cups with a short stalk and has granular soredia in the upper part of the podetium and within cups, but poorly developed specimens may be impossible to distinguish from *C. pyxidata*.

21. C. pocillum (Ach.) O. J. Rich. (1878)

C. pyxidata subsp. *pocillum* (Ach.) Dahl (1950)

Like **20**, but with particularly well-developed, ± horizontally spreading, contiguous and overlapping basal squamules, which tend to radiate from the margins and form ± imbricate, compact rosettes. The podetia are coarsely granular-corticate as in *C. pyxidata*. Thallus Pd+ red, K−, KC−, C−, UV− (fumarprotocetraric, ± atranorin).

Mainly on sandy, calcareous soils, old mortar on walls, calcareous dunes and soils overlying calcareous rocks; common. Throughout British Is. Europe, North & South America, Africa, Asia, New Zealand, Antarctica.

Possibly a calcicole ecotype of *C. pyxidata* **(20)** with apparent intermediates. This species and *C. rangiformis* **(55)** are the chief hosts of *Diploschistes muscorum*.

22. C. humilis (With.) Laundon (1984)

C. conoidea Ahti (1980)

Podetia (Fig. 36h) to 5(-7) mm tall, grey or green-grey; cups up to 6 mm diam., even, regular, not proliferating from rim, abruptly tapered to a short stalk, often subsessile, continuously corticate for a 1–2 mm zone at base, entirely farinose-sorediate above and within cups. Basal squamules often rather large, to 5 mm diam., ± rounded, sparingly incised, often well developed and forming a continuous crust. Apothecia and pycnidia brown, rare, on cup rim. Thallus Pd+ red, K+ yellow, KC−, C− (atranorin and fumarprotocetraric acid).

Mainly on dry or sandy ground especially on recently disturbed soils, on roadsides, sea cliffs and in suburban gardens, absent from montane areas. Throughout British Is. Europe, North America, Asia, New Zealand.

Characterized by the subsessile, short, regular cups with farinose soredia and a corticate base. The only brown-fruited, sorediate species which contains atranorin in Europe. The very similar *C. conista* A.W. Evans (1930), containing fumarprotocetraric and bourgeanic acids, may eventually be found in the British Is.; it is known from Norway and the Netherlands.

23. C. fimbriata (L.) Fr. (1831)

Podetia 0.5–1.5 cm tall, greyish to rarely brownish green; cups to 0.5 cm diam., ± regular, sometimes dentate at rim, not proliferating, abruptly joined to long stalk and appearing ± goblet-shaped, farinose-sorediate throughout, including base and within cups. Basal squamules often rather small and inconspicuous, ± elongate and often incised. Apothecia and pycnidia brown, infrequent, on cup margins; apothecia sessile or stalked. Thallus Pd+ orange-red, K−, KC−, C−, UV− (fumarprotocetraric acid).

Particularly on recently disturbed sites, gardens and old walls, occasionally in heathlands and dunes, absent from damp montane areas; common. Throughout British Is. Europe, North & South America, Asia, New Zealand.

Characterized by the abruptly expanded, goblet-shaped cups, the absence of cortication and entirely farinose-sorediate podetia. *C. chlorophaea* **(24)** has more gradually tapering cups, ± corticate or non-corticate podetia and coarse, granular soredia.

24. C. chlorophaea (Flörke ex Sommerf.) Sprengel (1827)

Podetia 0.5–2.5 (-4) cm tall, greyish to brownish green or dark brown; cups 3–6 mm diam., gradually tapered to stalk, sometimes ± regular, but often proliferating from margins, patchily corticate, becoming granular-sorediate above as well as within cups, sometimes almost totally corticate, except within the cups. Basal squamules often small, ± erect, sometimes incised. Apothecia and pycnidia brown, frequent; apothecia stalked or sessile on cup margins. Thallus Pd+ red, K−, KC−, C−, UV− (fumarprotocetraric acid; but also see below).

On peat, rotting logs and among rocks, in dry to rather moist, open heathland or moorland habitats; common. Throughout British Is. Europe, North & South America, Asia, Australia, New Zealand, Antarctica.

Additional chemotypes occur, often given specific status since there is some, but not unequivocal, correlation between chemistry and morphology: (a) *C. grayi* G. K. Merrill ex Sandst. (1929), thallus Pd− or red, K−, KC−, C−, UV+ blue-white (grayanic and ± fumarprotocetraric acids), rare; (b) *C. cryptochlorophaea* Asah. (1940), thallus Pd+ orange-red, K+ yellow-red, KC+ red, C− or C+ yellow, UV+ pale yellowish (cryptochlorophaeic and fumarprotocetraric acids), frequent; (c) *C. merochlorophaea* Asah. (1940), (i) thallus Pd− or Pd+ orange-red, K−, KC+ red, C+ red, UV+ faint blue (merochlorophaeic, 4-*O*-methylcryptochlorophaeic and ± fumarprotocetraric acids), frequent, (ii) var. *novochlorophaea* Sipman (1973), thallus Pd− or Pd+ orange-red, K−, KC−, C+ yellow, UV+ white (sekikaic and homosekikaic acids and ± fumarprotocetraric acid), rare. *C. asahinae* J. W. Thomson (1976), containing rangiformic, norrangiformic and fumarprotocetraric acids, has recently been found in N. Wales. W. Norway. *C. merochlorophaea* is commonly entirely esorediate.

Very variable and not easily resolved in morphological terms, *C. chlorophaea* s.lat. is most frequently confused with *C. fimbriata* **(23)**, which has more abruptly tapering, goblet-shaped cups and is entirely farinose-sorediate. Similar to *C. pyxidata* **(20)**, which, however, has gradually tapered cups almost to the base with distinctly coarser, corticate granules, present even in juvenile material.

25. C. carneola (Fr.) Fr. (1831)

Podetia 0.5–1.5 (-2) cm tall, pale yellow-green, appearing ± translucent when wet; cups 3–6 mm diam., abruptly tapered to a long, narrow stalk, ± regular with finely denticulate ± even rims which rarely proliferate; surface entirely farinose-sorediate. Basal squamules often rather small and inconspicuous. Apothecia pale yellow-brown or pale flesh-coloured, rather rare; pycnidia dark brown; both apothecia and pycnidia on rim of cups. Thallus Pd −, K −, KC+ yellow, C − (usnic and ± barbatic acids, zeorin; all British material contains barbatic acid, but its presence in this species is less frequent elsewhere).

On rotting conifer stumps, more rarely peat and *Calluna* stems, in ancient Caledonian pine woods, more rarely on blanket bog above 600 m; locally frequent. Scotland (especially Cairngorm mtns and Cheviot hills). Europe, Arctic, North & South America, Asia.

In the absence of apothecia, may be mistaken for the morphologically similar *C. fimbriata* **(23)** which is Pd+ red and is not yellowish green; the chemically similar *C. pleurota* **(9)** lacks soredia on the stalks of the podetia.

26. C. cariosa (Ach.) Sprengel (1827)

Podetia 0.5–1.5 (-2) cm tall, grey-green, sometimes tinged brownish; cups absent, often sparingly branched above, the surface partly decorticate, ± covered with numerous small corticate granules, sometimes with scattered squamules, distinctly longitudinally fissured with large, gaping cracks, often net-like; soredia absent. Basal squamules mostly rather small, ± erect, the margin entire or incised, with dark brown pycnidia on upper surface. Apothecia usually abundant, dark brown at apices of podetia and branches, usually larger than width of branches, often clustered and ± corymbose. Thallus Pd+ yellow, K+ yellow, KC −, C − (atranorin, ± norstictic acid), rarely Pd+ rust-red, K+ yellow, KC −, C − (atranorin, fumarprotocetraric acid).

In sandy situations, especially moderately calcareous heaths and mine spoil- heaps; rare. Throughout British Is. Europe, North & South America, N. Africa, Asia.

Readily recognized by the longitudinally fissured podetia and large, clustered, terminal brown apothecia. Norstictic acid, apparently unique to British material, is only sparingly produced and often only trace amounts are detected by t.l.c. Similar morphs, but containing atranorin and fumarprotocetraric acid (the usual acid in non-British material), are known from mine spoil heaps in Cornwall and ± calcareous dune heathlands in E. Scotland; British species of the *C. cariosa* group are in need of critical evaluation. For differences from *C. peziziformis* see **(27)**.

27. C. peziziformis (With.) Laundon (1984)

C. capitata (Michaux) Sprengel (1827)

Podetia to 1 cm tall, brownish grey-green, usually simple, ± granular-corticate with partly decorticate areas, occasionally fissured and bearing a few squamules; cups and soredia absent. Basal squamules small and granule-like or becoming ± rounded or ear-shaped, rather thick, with dark brown-black pycnidia on upper surface, scattered when mostly horizontal, or forming a ± continuous crust and then ± erect. Apothecia frequent, dark brown, produced mostly singly at apex and characteristically exceeding width of podetia, mostly simple, but secondary apothecia on short, lateral branches are sometimes produced which may fuse with the main apothecium. Thallus Pd+ red, K −, KC −, C − (fumarprotocetraric acid, ± ursolic acid).

On peaty soil, mainly in undisturbed coastal to montane heathlands; very rare, few recent records. Throughout British Is. Europe, E. North America, South America.

Characterized by the short, longitudinally fissured podetia bearing a single large terminal apothecium. *C. cariosa* **(26)** is larger, usually has several, proportionately smaller, apothecia in a ± corymbose arrangement on the podetia, and contains atranorin.

28. C. caespiticia (Pers.) Flörke (1828)

Podetia to 0.3 cm tall, decorticate. Basal squamules dominant, 2–7 mm long and to 1.5 mm wide, often irregularly incised and ± erect, often well developed and forming low cushions; upper surface greyish or brownish green; lower surface white, esorediate. Apothecia usually present, inconspicuous, pale brown, single or tuberculate, sessile or on short, broad stalks lacking algae and ± translucent. Pycnidia brown-black, on upper surface of basal squamules. Thallus Pd+ red, K −, KC −, C − (fumarprotocetraric acid).

On mossy tree trunks, rarely on earth banks in sheltered situations, mainly in long-established, mature woodlands; rather local. S. & W. British Is. W. Europe, E. North America, S. & E. Africa, Asia.

Characterized by the tendency to form compact cushions, the small, incised squamules bearing pycnidia as well as apothecia, the latter either sessile or on very short, alga-free, ± translucent stalks. Similar to *C. parasitica* **(30)** in habit, which has crowded basal squamules, disintegrating into coralloid-sorediate granules at their margins, conspicuous corticate podetia and contains thamnolic acid.

29. C. fragilissima Østh. & P. James (1977)

Podetia to 1.5 cm tall, irregular, ± simple but often branched at apices, smooth or wrinkled, ± completely corticate or with a mosaic of areoles on a white, decorticate surface, with scattered, mostly inconspicuous, fissures, becoming more decorticate and blackened when older, especially at base. Basal squamules 1–3(-5) mm long, 0.5–1(-2) mm wide, ± elongate, erect, incised, sometimes recurved, scattered or forming ± continuous mats, very brittle; upper surface grey-green, often bronze- or brown-tinged. Apothecia *c.* 0.2 mm diam., brown, contiguous in terminal clusters towards and at apex of podetia, sometimes on short projections from the squamule surface. Pycnidia dark brown, on basal squamules. Thallus Pd−, K−, KC−, C−, UV+ blue-white (grayanic acid).

On acid, sometimes waterlogged, peaty soils and at edges of peat hags in lowland and montane heathlands, also on mine spoil heaps; rather local. N. Britain from Cheshire and Yorkshire to Argyll and Outer Hebrides. W. Europe.

Often occurs only as basal squamules when the olive or bronze-green colour may be reminiscent of small morphs of *C. strepsilis* **(44)**, which is C+ green (strepsilin) and has tougher, more coriaceous squamules. May resemble morphs of *C. ramulosa* **(42)** in lowland habitats which may also have brittle squamules; these are Pd+ red (fumarprotocetraric acid).

30. C. parasitica (Hoffm.) Hoffm. (1796)

Podetia to 0.5(-2) cm tall, rather rare, often conspicuously elevated above the basal squamules, irregular, deformed, sometimes flattened and very irregularly branched, often ± covered in small, isidia-like, granules and partly decorticate, with scattered or numerous squamules and granules, fissured with gaping holes when well developed. Basal squamules up to 0.5 cm long, 0.2–1 mm wide, numerous, elongate, greatly dissected, often coralloid-branched, ± ascending, often coarsely granular-sorediate, forming crowded, compact, ± continuous, spreading swards; upper surface pale to dark grey; pycnidia dark brown-black, on upper surface of basal squamules; the lower surface white. Apothecia small, dark brown, at tips of podetia, often on extended projections, mostly clustered. Thallus Pd+ yellow, K+ yellow, KC−, C− (thamnolic and ± barbatic acids).

On hard wood of branches, trunks and logs of *Quercus* and *Pinus*; locally frequent in ancient woodlands. Throughout British Is., but rare in Ireland. Europe, North America, Asia.

Resembles *C. incrassata* **(18)** in habit, which has contorted, flattened yellowish squamules with podetia-like contortions or extensions bearing red apothecia and which contain usnic acid. *Cladonia squamosa* **(59)** has larger, less dissected squamules, and larger podetia, which are often perforate, with peeling squamules.

31. C. botrytes (Hagen) Willd. (1728)

Podetia to 0.5 cm tall, yellow-green, without cups, usually simple or occasionally sparingly branched towards apices when well developed; surface smooth and corticate to the base, becoming ± verrucose-areolate and sometimes longitudinally fissured. Basal squamules small, scattered, rather inconspicuous, ± erect and incised. Apothecia pale brown to pinkish, terminal on podetia even when these are young, sometimes on small, lateral branches near apices. Pycnidia often present on basal squamules. Thallus Pd−, K+ yellow or K−, KC+ pale yellow or KC−, C− (usnic and barbatic acids).

On relatively recently cut surfaces of *Pinus* stumps, more rarely dead *Calluna* stems; very rare. N. Scotland (Cairngorm mtns). Europe, North America, Asia.

The small size, yellow-green colour, absence of cups, and pale brown to pinkish apothecia are diagnostic. *C. carneola* **(25)** has similarly coloured apothecia, but always has well-developed cups.

32. C. uncialis (L.) Weber ex Wigg. (1780)

Podetia to 6 cm tall, without cups, yellow-green or greyish green, often brownish towards the pointed apices, forming ± compact, spiky tufts, the branching divergent, the axils mostly perforate, corticate throughout, the surface at times marbled; internal surface smooth or only weakly striate. Basal squamules absent. Apothecia and pycnidia brown, terminal. Thallus Pd−, K−, KC+ yellow, C−, surface of central canal UV+ white or UV−.

Two subspecies are recognized:

Subsp. *biuncialis* (Hoffm.) M. Choisy (1951): Podetia ± regularly dichotomously branched; surface of central canal white-powdery. Thallus with usnic and usually squamatic acids. On damp or dry acid soils, particularly on moorlands, at all altitudes; common. Throughout British Is. Europe, E. North America. Apothecia very rare in this subspecies. Characterized by the spiky, entangled tufts, yellow colour, absence of squamules and soredia, and UV+ reaction. In boggy situations podetia often become grossly inflated (to 10 mm diam.) and irregular. Dry heathland morphs have narrow (1.5 mm wide) rather richly-branched, ± decumbent podetia.

Subsp. *uncialis*: Podetia predominantly tri- to tetra-. or polychotomously branched, forming more regular tufts or cushions; surface of central canal not powdery. Thallus with usnic, rarely also squamatic acids. Sand dunes; rare. N.E. Scotland. Europe, North America, Asia; predominantly continental. Pycnidia and apothecia are more frequent in this subspecies.

C. amaurocraea (Flörke) Schaerer (1823), distinguished from **32** by the podetia which can terminate in cups, 1–3 mm wide, often with several small, brown, marginal projections; with strongly browned podetial apices and the thallus UV− (barbatic and usnic acids), was recorded from Scotland (Cairngorm mtns) last century, but the material probably originated from outside British Is. It is known from alpine fell-field and high altitude boulder woodlands in Europe, North America and Asia.

33. C. zopfii Vainio (1920)

Like **32** but usually more decumbent, a duller, darker green, and developing a thin, grey-violet "bloom" composed of long, fine, needle-shaped crystals (? terpenoid) in the herbarium. Internal surface of podetia fibrillose-striate. Contains usnic acid and is UV−.

On gravelly soil especially in montane regions and on acid dunes in N.E. Scotland, Ireland (Donegal); rather rare. Throughout British Is. Europe.

34. C. macrophylla (Schaerer) Stenhammar (1865)

Podetia 2–4(-5.5) cm tall, grey-green to brown, often contorted, simple or branched, particularly in upper half, without cups, the apices ± blunt, the surface often fissured and gaping, with numerous peltate squamules and grey or black decorticate areas, especially at base. Basal squamules 3–6 mm diam., often well developed, round or elongate, ± incised, the lower surface white. Apothecia and pycnidia brown, frequent, terminal on podetia; pycnidia also on peltate squamules of podetia. Thallus Pd+ yellow, K−, KC−, C−, UV− (psoromic acid).

Montane, often associated with mineral spoil heaps, also on acid peat, especially on scree and amongst mossy boulders; rather rare. Scotland (Highlands). Europe, Arctic, North America, Asia.

Characterized by the fissured, partly decorticate, podetia with scattered to crowded peltate squamules; the presence of psoromic acid is diagnostic.

35. C. coniocraea (Flörke) Sprengel (1827)

Podetia 1–2.5(-4) cm tall, grey-green, mostly pointed at apices, unbranched, often curved, occasionally bearing small, narrow cups not exceeding width of podetium and hardly even corticate inside, farinose-sorediate above, with a 1–2 mm corticate zone at base, which often bears squamules. Basal squamules dull green above, pale white below, very variable, often sorediate. Apothecia brown, rare, on tips of podetia. Thallus Pd+ red, K−, KC−, C− (fumarprotocetraric acid).

Chiefly on acid bark of living, or recently dead, trees and wood, less often on peaty soils; common. Throughout British Is. Europe, North America, Asia, Australia, ?New Zealand.

Tolerant to SO_2 pollution, often occurring with *C. macilenta* **(14)**, which may be recognized by its overall blue-grey tinge (± unchanged when wet) and K+ yellow reaction.

The name is currently misapplied, but used here in the traditional sense pending clarification of the application through a different typification. See note on *C. norvegica* under *C. macilenta* **(14)**.

36. C. ochrochlora Flörke (1828)

Like **35**, but distinguished by the more robust podetia, 2–4(-5) cm tall, with small, abruptly expanded cups, which are wider than the podetia at the apices; cups corticate inside; podetia with corticate areas at base sometimes extending upwards to the tips. Thallus Pd+ red, K−, KC−, UV− (fumarprotocetraric acid).

Predominantly on rotting tree stumps, but also heathlands over peat; common. Throughout British Is. Europe, North & South America, Africa, Asia.

British material is possibly a well-developed morph of *C. coniocraea* **(35)**, with juvenile and poorly developed specimens often being difficult to distinguish from that species. *C. rei* **(38)** may also have narrow cups but is UV+ white due to homosekikaic acid.

37. C. glauca Flörke (1828)

Podetia 1–5 cm tall, pale grey with pointed apices, without cups, mostly simple or 2- to 3-branched towards the apices, each with a single inconspicuous longitudinal fissure (in branched specimens the axils may also be perforated) often densely squamulose in lower part, sorediate in upper part, sometimes also with scattered squamules extending upwards. Basal squamules small, often inconspicuous, ± elongate and incised, greyish green above, white below. Apothecia brown, rare. Pycnidia brown, on apices of podetia, frequent. Thallus Pd−, K−, KC−, C−, medulla UV+ bluish white (squamatic acid).

On old rotting tree stumps, also on peat in heathlands. C. & E. Scotland extending to E. & S. England. W. Europe, E. North America, Asia.

When sterile may be confused with a chemotype of the red-fruited *C. macilenta* **(14)**, which has no lateral furrow and is Pd−, K− (barbatic acid). *C. subulata* **(39)**, when well developed, has antler-like branching, lacks a lateral furrow and is Pd+ red (fumarprotocetraric acid). For differences from *C. rei* see **(38)**.

38. C. rei Schaerer (1823)

Like **37**, but dirty or drab green-brown, occasionally with deformed apical cups, the margins rarely proliferating, often ± entirely farinose- or coarsely granular-sorediate, with a corticate and sometimes squamulose area at the base. Thallus Pd+ slowly yellow-orange or Pd−, K−, KC−, C−, UV+ white (homosekikaic, and ± fumarprotocetraric, acids).

On earth-banks in woodlands and on waste ground; rare. S.E. England. Europe, North America, E. Africa, Asia, Australia.

Distinguished by the UV ± white reaction (homosekikaic acid). *C. ochrochlora* **(36)** has small, abruptly expanded, ± regular cups which are wider than the podetia and is UV− and Pd+ quickly red. *C. glauca* **(37)** never has cups, is perforate, and contains squamatic acid.

39. C. subulata (L.) Weber ex Wigg. (1780)

Podetia 1–6 cm tall, dull to bright grey-green, sometimes tinged brownish grey, thin, with pointed apices, often irregularly branched or furcate towards apices and appearing antler-like, occasionally with irregular cups which often proliferate from margins, usually entirely farinose-sorediate throughout, but sometimes ± corticate at base which may be ± squamulose. Basal squamules inconspicuous or absent, elongate and deeply incised. Apothecia brown, rare, at apices of podetia. Thallus Pd+ red, K−, KC−, C−, UV− (fumarprotocetraric acid).

In well-drained, sandy heathlands and acid dunes, also on earth banks and wall tops, often amongst *Calluna*; often abundant. Throughout British Is. Europe, North & South America, Asia.

C. glauca **(37)** is more glaucous-grey, contains squamatic acid (Pd−, UV+), and has a narrow, thin, furrow at the side of the podetium. *C. rei* **(38)** has less branched podetia, often with deformed cups, which rarely proliferate, and homosekikaic acid (UV+ white).

40. C. cornuta (L.) Hoffm. (1791)

Podetia 1–4 cm tall, brown-green, smooth and entirely evenly corticate in lower half, grey-green to grey-white and farinose-sorediate in upper half, the apices pointed, unbranched; squamules occasionally present towards base. Basal squamules rather small, ± rounded, scarcely indented. Apothecia brown, very rare, at apices of podetia. Thallus Pd+ red, K−, KC−, C− (fumarprotocetraric acid).

On peat and rotting wood in moorlands, especially montane; common. Scotland; local and rare elsewhere. Europe, North & South America, Asia, Australia, New Zealand.

The half-corticate, half-sorediate, pointed, simple podetia are diagnostic.

41. C. cenotea (Ach.) Schaerer (1823)

Podetia 1–2.5 cm tall, greyish or greenish brown, unbranched or sparingly branched above, with narrow, markedly perforate cups which proliferate from margins, farinose-sorediate throughout, sometimes with small squamules, particularly at base. Basal squamules small, indented, often rather inconspicuous. Apothecia brown, rare, at apices of podetia. Thallus Pd−, K−, KC−, C−, UV+ white (squamatic acid).

On rotting stumps and humus amongst *Calluna* in ancient Caledonian pine woodlands. C. & E. Scotland. Europe, North & South America, Asia.

Distinguished by the perforated cups, a feature absent in other brown-fruited, sorediate, ± unbranched species with cups such as *C. fimbriata* **(23)** and *C. chlorophaea* **(24)**.

42. C. ramulosa (With.) Laundon (1984)

C. anomaea (Ach.) Ahti & P. James (1980)

Podetia 1–3.5 cm tall, green-brown, very variable, with pointed apices or irregular, poorly developed, narrow cups, not or sparingly branched towards apices; surface variously squamulose, in part granular or at times almost entirely decorticate, rarely with few, scattered, granular soredia. Basal squamules small, elongate and indented, sometimes tufted, fragile. Apothecia frequent, pale brown, often turgid when wet and forming convex clusters at apices of podetia. Thallus Pd+ red, K−, KC−, C− (fumarprotocetraric acid).

On inland and coastal sandy heathlands, moorlands, rotting tree stumps, fence posts, earth banks, wall tops, thatch and decaying *Armeria* tufts in coastal sites; frequent. Throughout British Is. Europe, North & South America, Africa, Asia, Australia, New Zealand.

Very variable, but distinguished by the irregular, often inconspicuous podetia, which are often partly granular, rarely sorediate, mixed with brittle squamules, often with ± decorticate areas and turgid, ± contiguous fruits, forming blunt apices.

43. C. symphycarpa (Ach.) Fr. (1826)

Podetia very rare, to 1 cm tall with irregular cups, ± continuously corticate, sometimes furrowed. Basal squamules dominant, 2–3 mm tall, scattered to contiguous, ± horizontally spreading, with recurved margins when dry; upper surface pale grey-green; lower surface starkly white. Apothecia brown, rare. Pycnidia dark brown-black, frequent on basal squamules. Thallus Pd+ yellow, K+ yellow-red, KC−, C− (norstictic acid and atranorin).

On coastal limestones, basic dune systems and montane epidiorite; rare. Throughout Great Britain (from Somerset to Sutherland). Europe, North America, Asia.

Easily overlooked in the absence of podetia, this calcicole is distinguished by the presence of norstictic acid and the occurrence of pycnidia on the basal squamules.

44. C. strepsilis (Ach.) Vainio (1894)

Podetia rare, with irregular cups or turgid-branched, ± uniformly corticate, often squamulose and fissured, sometimes areolate, unbranched. Basal squamules dominant, to 4 mm long, rounded or ± elongate, indented, often contiguous, forming cushions; upper surface bronze-green, white below. Apothecia dark brown, rare. Pycnidia brown, frequent on basal squamules. Thallus Pd+ yellow, K−, KC−, C+ bright emerald green, UV+ whitish (baeomycesic acid, squamatic acid and strepsilin).

On poorly drained, wet, acid sites in open, montane situations, less frequent in low altitude heathlands; common, W. & N. British Is., extending south but local and rare. Europe, E. North America, West Indies, E. Asia.

Distinguished by the C+ green reaction and tendency to grow in wetter habitats than most other species. Sorediate specimens with stunted podetia and conspicuous yellow-green soralia have recently been collected from heathland in the New Forest (S. England, Hants) and erroneously assigned to *C. brevis* (Sandst.) Sandst. (1922).

45. C. cervicornis (Ach.) Flotow (1849)

Podetia to 1 cm tall, with ± regular cups, corticate, areolate, often sparingly squamulose and proliferating from centre. Basal squamules dominant, often well developed, 2–3(-5) mm long, ± rounded, the margins indented, reflexed when dry, mostly contiguous and forming cushions; upper surface grey-green to grey-brown; lower surface white, often tinged mauve or grey; soredia absent. Apothecia rare, brown, on cup margins. Pycnidia frequent, brown, on cup margins. Two chemotypes occur: (a) thallus Pd+ red, K−, KC−, C− (fumarprotocetraric acid); (b) thallus Pd+ yellow, K−, KC−, C− (psoromic acid).

On acid and basic soils, particularly on coastal dunes; common. Throughout British Is. Europe, North America.

Distinguished from *C. subcervicornis* **(46)** by the smaller, more rounded, grey-green basal squamules, which are not conspicuously blackened towards the base, and the K− reaction. The psoromic acid - chemotype of *C. cervicornis*, sometimes treated as *C. cervicornis* subsp. *pulvinata* (Sandst.) Ahti (1983), is known only from N. England (Cleveland, Yorkshire) and C. Scotland (Perth); it was formerly erroneously ascribed to *C. rappii*, A. W. Evans (1952), which does not occur in Europe.

The above description refers to subsp. *cervicornis*. Subsp. *verticillata* (Hoffm.) Ahti (1980) has less well-developed basal squamules and the podetia have more regular cups, to 5 cm, which are ± abruptly tapered to the stalk and form proliferating tiers, in 2–3(-5) levels from centre of cup. It is characteristic of heathlands and is known from Europe, North & South America, Asia, Australia and New Zealand.

46. C. subcervicornis (Vainio) Kernst. (1900)

Podetia to 1.5 cm tall, with irregular or deformed turgid cups, flared from the base, bearing squamules, often proliferating from rim. Basal squamules dominant, elongate, 0.5–2 cm long, ± erect, with indented margins, mostly contiguous and forming ± compact cushions; upper surface bluish or lead-grey; lower surface white, oftened blackened towards base. Apothecia small, dark brown, on rim of cups, often becoming clustered, rather frequent. Pycnidia dark brown on rim of cups. Thallus Pd+ red, K+ yellow, KC−, C− (fumarprotocetraric acid, atranorin).

On humus, particularly pockets in rock crevices on moorlands, woodlands and montane sites; often common. N. & W. British Is., extending locally eastwards. Europe, Greenland.

Distinguished from *C. cervicornis* **(45)** by the larger, more elongate, often lead-grey basal squamules which are often blackened on the underside towards the base, the K+ yellow cortex and the irregular, often turgid, podetia.

47. C. foliacea (Huds.) Willd. (1787)

Podetia to 1 cm tall, inconspicuous, often absent, smooth, corticate, with a ± regular cup. Basal squamules predominant, elongate, 0.4–1(-2) × 0.1–0.3 cm, indented, often recurved, forming compact mats; margins with occasional tufts of black hairs; upper surface greyish yellow to yellow-green; lower surface pale yellow. Apothecia pale brown, on cup rims, very rare. Pycnidia dark brown, frequent on basal squamules. Thallus Pd+ red, K−, KC+ yellow, C− (fumarprotocetraric and usnic acids).

In moss-lichen swards on well-drained, calcareous, coastal soils and humus-stabilized sand dunes and shingle, in wind-swept, sunny situations, sometimes in rock crevices, usually coastal, rarely inland (E. England, Breckland). Throughout British Is. Europe, N. Africa, S.W. Asia.

Distinguished from *C. cervicornis* **(45)** by the yellow-green lobes. The red-fruited *C. luteoalba* **(19)** has ± rounded, scarcely indented lobes with a pale yellow, cottony-arachnoid lower surface.

48. C. convoluta (Lam.) Cout. (1913)

Like **47**, but has much larger squamules, 1.5–2.5 × 0.2–0.8 cm, also with occasional black tufts of marginal hairs; it forms loose, scattered, ± straggling clusters.

On soil associated with warm, sunny limestones or base-rich coastal slopes; rare and very local. S. & S.W. England (Sussex, Somerset, S. Devon), Channel Is. Europe, N. Africa, S.W. Asia.

49. C. firma (Nyl.) Nyl. (1861)

Basal squamules dominant, 0.4–1 cm tall, ascending, ± elongate, indented, recurved at apices when dry, contiguous, forming cushions; upper surface grey-green; lower surface pale to dark grey-violet.

Apothecia absent. Thallus Pd+ red, K+ yellow, KC−, C− (fumarprotocetraric acid, atranorin and ± rangiformic acid).

On ± basic soil, especially on or near the coast; locally common. W. Britain, Channel Is. and Is. of Scilly, northwards to Shetland Is. Mediterranean Europe, N. Africa.

Differs from *C. foliacea* **(47)**, with which it often grows, by the characteristic grey-violet colour of the lower surface of the basal squamules, the presence of atranorin (K+ yellow) and absence of usnic acid; *C. cervicornis* **(45)** has smaller squamules, lacks atranorin and is not restricted to basic soils.

50. C. cyathomorpha Stirton ex W. Watson (1935)

Podetia to 0.5 cm tall, gradually tapering to a funnel-shaped cup, coarsely granular within; marginal proliferations absent. Basal squamules dominant, 5–10 mm diam., ± rounded, indented, sometimes flat, often recurved at margins and becoming ± erect; upper surface grey-green; lower surface white with raised, pale pink- or yellow-brown veins. Apothecia and pycnidia brown, on cup rims. Apothecia rare; pycnidia frequent. Thallus Pd+ red, K+ yellow, KC−, C− (fumarprotocetraric acid and an unidentified compound).

On vertical faces of mossy rocks in hilly and montane districts; rare. N.W. England (Lake district), N. Wales (Pembroke), W.C. Scotland (Is. of Mull). West Europe, Canary Is.

Characterized by the large squamules with pronounced, ± raised, pale pink venation on the lower surface. Differs from *C. pyxidata* **(20)** in having larger, veined, basal squamules and an additional unidentified compound with fumarprotocetraric acid.

51. C. furcata (Huds.) Schrader (1794)

Podetia (Fig. 36i) 2–5(-8) cm tall, bluish green to dark grey-green (or brownish in open situations), slender, hollow, dichotomously branched, forming ± spiky, irregular tufts, the apices markedly pointed, without cups, the axils sometimes perforate; upper surface corticate throughout, smooth, often faintly marbled, occasionally with scattered squamules and cracks. Basal squamules disappearing, rounded, up to 3 mm diam.; lower surface white. Apothecia brown, on short lateral branches towards apices, ± corymbose, rather frequent. Pycnidia dark brown, terminal, abundant. Thallus Pd+ red, K− or K+ yellow, KC−, C− (fumarprotocetraric acid, rarely traces of atranorin).

On heathlands, dunes, mossy rocks in woodlands and mountain summits, preferring ± acid habitats. Throughout British Is. Europe, North & South America, Asia, Australia, New Zealand.

Very variable, but characterized by the dichotomously-branched podetia with pointed apices and tendency to form irregular tufts. *C. gracilis* **(57)** is less branched and has small cups, but juvenile stages of these two species are often very difficult to separate. *C. rangiformis* **(55)** is more richly branched, the branches diverging at a wider angle, and has a cortex which tends to become more mosaic-like with raised green areoles on a pale decorticate background. For differences from *C. furcata* **(51)**, *C. subrangiformis* **(52)**, *C. scabriuscula* **(53)** and *C. stereoclada* **(54)**, see under those taxa.

52. C. subrangiformis Sandst. (1924)

C. furcata subsp. *subrangiformis* (Scriba ex Sandst.) Pišùt (1961)

Like **51**, but often with raised, white, ± circular concretion-like spots towards the base of the dark brown or olive-brown, often shiny, podetia formed by local rupturing of the cortex, a cortex which is often cracked and brown, the often prostrate and little-branched habit and robust appearance, and the presence of spinose lateral branches which often arise perpendicularly from podetia. Thallus Pd+ red, K− or pale yellow, KC−, C− (atranorin and fumarprotocetraric acids).

On calcareous soils in short turf, particularly on chalk; local. S.E. England. Europe, S.W. Asia.

Possibly a morph of *C. furcata* deformed by excessive production of calcium oxalate crystals.

53. C. scabriuscula (Delise) Nyl. (1876)

Like **51**, but with a more slender habit, paler greenish white colour and the tendency for the cortex to peel, become partly decorticate and scabrid with small, peeling squamules and scattered, coarse soredia, particularly at apices of the podetia.

On peat and mossy rocks in moorlands and montane situations; rather local. Throughout British Is. Europe, North & South America, Asia, Australia, New Zealand.

Sometimes difficult to distinguish from *C. furcata* **(51)** but specimens of *C. scabriuscula* usually have, in part, a scabrid surface especially towards the tips due to an incipient peeling of the cortex. *C. glauca* **(37)**, is characterized by a single, inconspicuous furrow along the length of most podetia and is UV+ white.

54. C. stereoclada des Abb. (1946)

Like **51**, but with very thin (0.3–0.7 mm diam.), sparingly branched, often entangled, flexuous, solid podetia with blunt apices. Thallus Pd+ red (fumarprotocetraric acid).

On pockets of soil in acid rock faces; known in the British Is. only from a single collection. S.W. Ireland (Sybil Head). Macaronesia.

55. C. rangiformis Hoffm. (1796)

Podetia 2–6 cm tall, grey-white to grey-green, richly branched, the branches diverging at wide angles, forming spiky tufts with pointed apices, without cups, axils usually closed; upper surface corticate, distinctly areolate; attached squamules rather rare; soredia absent. Basal squamules disappearing, small; lower surface white. Apothecia very rare, brown, at apices of podetia, sometimes ± corymbose. Pycnidia brown, at apices of branches, frequent. Thallus Pd− (in *c.* 70% of collections) or Pd+ red, K+ yellow, KC−, C− (usually with atranorin and rangiformic acid, more rarely also with fumarprotocetraric acid).

On neutral or basic grassland, chalk-flint grassland, cliffs and dunes, especially along the coast; often common. Throughout British Is. Cosmopolitan.

Characterized by the pale, distinctly areolate, richly branched thallus, the widely divergent branches with pointed apices, and the K+ yellow reaction. *C. furcata* **(51)** is often darker, the surface not so obviously areolate (best seen when wet), and has fewer, less divergent, branches and is K−.

56. C. crispata (Delise) Vainio (1887)

Podetia 2–6 cm tall, pale to dark brown-grey to brown-green, irregularly branched, particularly in upper parts, initially terminating in a cup-like structure consisting of a single perforation surrounded by short spines, later one of these spines proliferating to form an additional proliferational podetia; upper surface ± smooth, corticate or ± areolate, sometimes partly squamulose. Basal squamules 1–2 mm, indented, often disappearing. Apothecia to 1 mm diam., pale to dark brown, on terminal spines, ± corymbose, rather frequent. Pycnidia dark brown, at apices of podetia, frequent. Thallus Pd−, K−, KC−, C−, UV+ white (squamatic, ± barbatic acids).

In *Calluna* heathland in moorland and montane areas; frequent. Throughout British Is. Europe, North & South America, Asia, Australia, New Zealand.

Characterized by the dark colour (paler in shade morphs), irregular branching, podetia terminating in a single perforation surrounded by short spines and in being Pd− and UV+ white. The rather more delicate and sparingly branched *C. gracilis* **(57)** lacks perforations, has a more continuous cortex and is Pd+ red and UV− (squamatic acid absent).

The above description applies to **var. *cetrariiformis*** (Delise) Vainio (1887). Var. *crispata* differs in having obvious, open cups with shorter, subulate proliferations and more squat, sturdy podetia.

In coastal grey dune systems; very rare. N. England (Yorkshire), E. Scotland. Europe, North & South America, Asia, Africa.

57. C. gracilis (L.) Willd. (1787)

Podetia 1–6 cm high, slender, 0.5–1.5 mm diam., greenish grey to dark brown, becoming dark brown-black towards base, unbranched or sparingly branched above, pointed at apices or with a ± regular shallow cup, 1–2 mm diam., not perforate, often with dentate margins, sometimes proliferating from rim; upper surface smooth, corticate, usually conspicuously areolate, attached squamules usually absent or scarce. Basal squamules up to 2 mm, indented, mostly disappearing. Apothecia and pycnidia dark brown, at apices of podetia; apothecia rather rare; pycnidia frequent. Thallus Pd+ red, K−, KC−, C− (fumarprotocetraric acid).

On sandy or acid soil, amongst rocks and, more rarely, on rotting wood in heathlands, dunes, and in montane situations; frequent. Throughout British Is. Europe, North & South America, Asia, Antarctica.

Characterized by the slender habit and smooth podetia, which are either pointed or terminate in small, non-perforate cups. *C. furcata* **(51)** lacks cups and has perforate axils. *C. crispata* var. *cetrariiformis* **(56)** has perforate cups, is more branched and is Pd−, UV+ white (squamatic acid).

58. C. maxima (Asah.) Ahti (1978)

Like **57**, but differs in its more robust podetia, to 3 mm wide, which are 1- to 2-forked towards apex, the glaucous tinge, brownish towards apex and pale yellow towards base. Podetial and basal squamules absent. Thallus Pd+ red, K+ yellow, KC−, C− (fumarprotocetraric acid and often atranorin).

On boggy alluvial grassland above 900 m, often in vicinity of late snow-line; rare. N. Scotland (Cairngorm mtns; W. Inverness, Aonoch Mór). Europe, North America, E. Asia.

59. C. squamosa (Scop.) Hoffm. (1796)

Podetia 2–5 cm tall, green to grey-brown, branching irregular, the apices pointed or with an irregular, perforate, narrow (to 2 mm wide) cup which is often dentate and proliferates from rim, the surface scabrid, covered in numerous squamules, particularly at base, derived from peeling of the cortex, often having extensive decorticate areas. Basal squamules to 2 mm, indented, often forming compact mats. Apothecia and pycnidia brown, at apices of podetia; apothecia rather small, rare, often in corymbose clusters. Thallus Pd−, K−, KC−, C−, UV+ (squamatic and ± barbatic acids).

On earth banks, rotting wood, old tree stumps, tree trunks, and mossy rocks by streams; common. Throughout British Is. Europe, North & N.W. South America, Africa, Asia, Australia, New Zealand.

Very variable but characterized by the perforate podetia, numerous peeling squamules and absence of soredia. In some morphs the podetia have a ± intact cortex and a semblance of cups, in which case they can resemble *C. crispata* **(56)** or *C. cervicornis* **(45)**. In these cases the surface of the podetia is nearly always finely roughened and scabrous, a feature absent in the other two species. The podetia of *C. squamosa* are nearly always ± gaping perforate either at the apex or sides.

The var. *subsquamosa* (Nyl. & Leighton) Vainio (1881) differs in the often more robust thallus and consistently more thickly squamulose thallus, which reacts Pd+ orange, K+ yellow, KC−, C−, UV− (thamnolic, and ± barbatic acids). It is common in the British Is. and has the same distribution as given above, with the exception of Asia and the addition of Antarctica. It is probably a chemotype of *C. squamosa* and probably not worthy of formal recognition since thamnolic and squamatic acids are closely related biosynthetically.

60. C. stricta (Nyl.) Nyl. (1869)

Podetia 0.5–2 cm tall, grey to grey-green, blackened towards base, sparingly branched above, terminating in narrow cups, to 1.5 mm wide, or with pointed apices, occasionally proliferating from centre of cup, the surface corticate, ± smooth with scattered squamules; soredia absent. Basal squamules to 8 mm, indented. Apothecia brown, not observed in British material. Pycnidia dark brown on rim of cups, frequent. Thallus Pd+ red, K+ yellow, KC−, C− (fumarprotocetraric acid and atranorin).

Amongst boulders near a snow patch at 1300 m; very rare. N. Scotland (Cairngorm mtns). Europe, Arctic, North America, Asia.

Characterized by the grey colour, central proliferations, the narrow, indistinct cups, the blackened base and K+ yellow reaction. *C. cervicornis* subsp. *verticillata* **(45)** differs in the wider cups and unbranched podetia, while *C. phyllophora* **(61)** has marginal proliferations; both these species are K−.

61. C. phyllophora Hoffm. (1796)

Podetia 1–5 cm tall, the surface subarachnoid, matt, pale grey-brown, areolate towards base, with blackened decorticate areas contrasting with the pale areoles; cups irregular, with often extended, occasionally dentate margins, which often proliferate to form extensive tiers; cups and lower parts of podetia variously squamulose. Basal squamules small, ± rounded, pale below. Apothecia rare; apothecia and pycnidia terminal on podetia, brown. Thallus Pd+ red, K−, KC−, C−, UV− (fumarprotocetraric acid).

On mossy rocks in montane situations; rare and local. S.W. & N. Scotland (Cairngorm mtns). Europe, North America, Asia, Tierra del Fuego, subAntarctic islands.

A polymorphic species, usually forming compact, turgid, interlocked clusters of irregular podetia, the lower part with pale areoles on a black, necrotic medulla. *C. gracilis* **(57)** has more regular cups, a smooth, non-arachnoid cortex and the podetia have uniformly brownish black bases; *C. ramulosa* **(42)** is not darkened below.

CLAUROUXIA D. Hawksw. (1988)

Pseudolecidea Clauz. & Roux (1984), non N. March. (1896)

Thallus crustose, superficial, areolate, on rock; prothallus black. **Photobiont** trebouxioid. **Ascomata** apothecia, black, immersed to erumpent; disc ± concave to flat. **Thalline exciple** absent. **True exciple** well developed, raised, black, carbonaceous and easily fragmenting. **Epithecium** blue-green. **Hymenium** I+ blue. **Hypothecium** purple-black to black, ± continous with the exciple. **Hamathecium** of mainly unbranched paraphyses, ± moniliform with several swollen cells towards the tip, the apical cell more strongly swollen with a brown apical cap. **Asci** 8-spored, elongate-clavate, strongly thickened at the apex, with a densely I+ blue colour subtending a conspicuous I− apical cushion. **Ascospores** broadly ellipsoid to subglobose, simple, colourless, smooth, without a distinct perispore. **Conidiomata** not known. **Chemistry**: no lichen products reported, but with an unidentified K+ reddish purple pigment in the true exciple and hypothecium. **Ecology**: on hard, siliceous rocks. Europe.

Claurouxia is distinguished amongst the lecideoid lichens by the ascus-structure, moniliform paraphyses, and particularly the thick, black and friable exciple and hypothecium which are continuous and extend below the whole apothecium. It shows some affinity in ascus structure to *Candelariella*.

LITERATURE: Clauzade & Roux (1984), Eriksson & Hawksworth (1988).

1. C. chalybeioides (Nyl.) D. Hawksw. (1988)

Lecidea chalybeioides Nyl. (1875), *Pseudolecidea chalybeioides* (Nyl.) Clauz. & Roux (1984)

Thallus black, often mottled greenish grey, areoles angular, 0.1–0.2 mm diam., medulla I−. Apothecia 0.2–0.4 mm diam.; hymenium 50–60 μm tall; hypothecium K+ reddish purple in parts. Paraphyses to 3 μm wide at the apices. Ascospores (4-)5–8(-9) × (3-)4–5 μm. Thallus Pd−, K−, C−, KC−.

On hard, siliceous rocks in damp or sporadically inundated situations; rare. S.W. England (Devon), C. Wales, E. & C. Scotland (Angus & mid-Perth). Finland, France.

Resembles *Clauzadeana macula* but very different microscopically.

CLAUZADEA Hafellner & Bellem. (1984)

Thallus crustose, generally immersed; when superficial grey or brown, granular or ± areolate; prothallus, when present, ± black. **Photobiont** *Trebouxia*. **Ascomata** apothecia, red-brown to black, sessile or immersed in rock and sometimes leaving pits, naked or ± pruinose. **Thalline exciple** absent. **True exciple** generally persistent and black. **Epithecium** red-brown to brown. **Hymenium** I+ pale blue in part. **Hypothecium** ± colourless to red-brown or orange-brown. **Hamathecium** of branched and occasionally anastomosing paraphyses, septate, scarcely or markedly swollen and pigmented at the apices. **Asci** 8-spored, elongate-clavate, *Porpidia*-type. **Ascospores** ellipsoid, simple, ± colourless, smooth, with a gelatinous perispore when young. **Conidiomata**, when present, pycnidial, ± immersed. **Conidiogenous cells** in chains. **Conidia** borne laterally, bacilliform, simple. **Chemistry**: no lichen products detected by t.l.c. **Ecology**: on calcareous rocks.

This genus is close to both *Amygdalaria* and *Porpidia*, but differs in the often immersed thallus, the strictly calcicolous habitat and, particularly, the conidial state. The species with immersed apothecia superficially resemble *Verrucaria* species of the *V. hochstetteri* group in the field, especially when dry, but are easily distinguished when wet by the different types of ascomata. *Lecidea advertens* (q.v.) may also prove to belong to *Clauzadea*.

LITERATURE: Clauzade & Roux (1985), Hafellner (1984).

1 Apothecia ± immersed; disc ± plane, in deep or shallow pits; thallus generally immersed 2
Apothecia sessile; disc becoming convex, not in pits; thallus generally superficial .... **3. *monticola***

2(1) Disc ± pruinose; apothecia mainly less than 0.5 mm diam., in deep pits, rather regularly arranged; ascospores thin-walled, 10–15(-19) × (6-)7–8(-9) μm .... **1. *immersa***
Disc not pruinose; apothecia mainly exceeding 0.5 mm diam., in shallow pits, randomly arranged; ascospores thick-walled, (15-)16–20(-28) × (6-)8–10(-12) μm .... **2. *metzleri***

1. C. immersa (Hoffm.) Hafellner & Bellem. (1984)

Lecidea immersa (Hoffm.) Ach. (1803)

Thallus white to grey, immersed; prothallus absent or black. Apothecia in pits, ± regularly arranged; disc (0.2-)0.3–0.5(-0.8) mm diam., ± level with the rock surface or slightly depressed, dark red brown to black, red-brown and ± translucent when wet, often white-grey-pruinose; true exciple black to white-grey, pruinose, persistent; epithecium orange-brown to brown; hymenium 45–60 μm tall; hypothecium ± colourless to orange-brown. Asci 55–60 × 12–20 μm. Ascospores 10–15(-19) × (6-)7–8(-9) μm; perispore up to 1 μm thick and ± granular in old spores. Pycnidia often frequent. Conidia 3–4 × 0.7–1 μm.

On hard, calcareous rocks and walls, especially limestones, in sunny and somewhat sheltered situations; locally frequent. Throughout British Is. Europe, particularly mountains close to the Mediterranean.

2. C. metzleri (Körber) Clauz. & Roux ex D. Hawksw. (1992)

Lecidea metzleri (Körber) Th. Fr. (1874)

Thallus white to pale grey, ± immersed; prothallus absent. Apothecia in shallow pits, scattered or occasionally in contiguous lines along crevices, $\frac{1}{2}$-$\frac{3}{4}$ immersed; disc (0.3-)0.5–0.8 mm diam., flat to slightly convex, red-brown to black, red-brown with black margins when wet, not pruinose; true exciple black, not pruinose, persistent, slightly raised; epithecium reddish brown; hymenium 75–100 μm tall; hypothecium ± colourless to reddish brown. Asci 50–60 × 12–16 μm. Ascospores (15-)16–20(-28) × (6-)8–10(-12) μm, the perispore up to 2 μm thick and not or scarcely granular.

On calcareous rocks, walls, and pebbles mainly in chalk and limestone areas, often in somewhat sheltered and moist situations; local. Throughout England and Wales, rare in Scotland and Ireland. Europe, from the Mediterranean to the Alps, Scandinavia.

Distinguished from *C. immersa* **(1)** by the generally larger dimensions, lack of pruina and less immersed, irregularly distributed apothecia.

3. C. monticola (Ach.) Hafellner & Bellem. (1984)

Lecidea monticola (Ach.) Schaerer (1833)

Thallus brownish grey to ochre or grey, granular to almost rimose and superficial, grey and scurfy or ± immersed; prothallus absent. Apothecia sessile, not immersed, constricted at the base, not forming pits; disc (0.2-)0.3–0.8(-1.5) mm diam., flat to convex, reddish brown to dark brown or black, red-brown when wet, not pruinose; true exciple black, slightly raised, generally persistent but sometimes finally excluded, not pruinose; epithecium orange-brown to reddish brown; hymenium 50–70 μm tall; hypothecium deep red-brown. Asci 45–60 × 12–19. Ascospores (6-)8–12(-14) × 4–6(-9) μm, the perispore to 0.5 μm thick and not becoming granular.

On sunny or sheltered, calcareous rocks, walls, cement, mortar and asbestos cement; rarely abundant. Throughout British Is. Europe, North America.

The commonest British species of the genus, superficially not unlike certain *Sarcogyne* species (e.g. *S. privigna*) when fresh but easily separated by the 8-(not multi-)spored asci. *C. monticola* is frequently confused with *Lecidella stigmatea*, which often occurs in similar habitats; in the latter the hypothecium is colourless, the spores longer and there is no trace of the red-brown pigmentation within the apothecia, so characteristic of *Clauzadea*.

CLAUZADEANA Roux (1984)

Thallus crustose, superficial; prothallus black. **Photobiont** trebouxioid. **Ascomata** apothecia, immersed in thallus areoles, aspicilioid, the algal cells extending below the hymenium and into the excipular tissues. **Thalline exciple** not raised, ± colourless. **Epithecium** brownish to bluish green, N+ purple red. **Hymenium** I+ blue. **Hypothecium** ± colourless. **Hamathecium** of richly branched and anastomosing paraphyses, composed of short, irregularly shaped cells, almost pseudoparaphysis-like in parts; not swollen or capitate. **Asci** broadly clavate, *Lecanora*-type. **Ascospores** ellipsoid, simple, colourless, without a distinct perispore. **Conidiomata** unknown. **Chemistry**: argopsin. **Ecology**: on hard, siliceous rocks. Europe.

The aspicilioid ascomata have lead to this taxon being subsumed under *Aspicilia*, but it differs in the nature of the hamathecial tissues and ascus-structure. Aspicilioid apothecia also occur in *Amygdalaria* and *Schaereria*, but *Clauzadeana* differs from both these genera on the basis of ascus structure, and further by the presence of algal cells in the exciple and under the hymenial tissues.

LITERATURE: Rambold (1989), Roux (1984).

1. C. macula (Taylor) Coppins & Rambold (1989)

Lecidea macula Taylor (1836), *Lecidea pissodes* Stirton (1877), *Lecidea instratula* Nyl. (1878), *Aspicilia morioides* Blomb. ex Arnold (1881), *Clauzadeana instratula* (Nyl.) Roux (1984)

Thallus composed of regular, reddish brown, angular and ± shiny areoles, 0.2–0.8 mm diam., dispersed on a well-developed, black prothallus which is often dominant, giving the whole a black colour without study with a hand lens. Apothecia 0.1–0.4 mm diam.; disc flat to slightly convex; thalline exciple not raised; hymenium 40–60 μm tall. Asci 30–40 × 15 μm. Ascospores (7-)8–10(-14) × (3-)3.5–5(-6) μm. Medulla Pd+ red, K−, KC−, C− (argopsin).

On exposed, hard, siliceous rocks, especially granites, upland; local. S.W. England (Devon), Wales, C. & N. Scotland (Caithness, Inverness, Kincardine, Colonsay), Ireland (Galway, Mayo). France, Norway, Sweden; reports from Japan and North America are dubious.

The habit and thallus colour superficially recall *Schaereria fuscocinerea*, with which the species not infrequently occurs, but that species differs in the blackish areoles and in the ascus apices being thin-walled and without any I+ blue, internal apical structures.

CLIOSTOMUM Fr. (1825)

Thallus crustose, ± smooth and rimose or areolate to granular-verrucose, whitish, pale grey or straw-coloured, effuse or bordered by a dark prothalline line. **Photobiont** trebouxioid. **Ascomata** apothecia; disc concave to flat, rarely convex, black, sometimes faintly pruinose. **Thalline exciple** absent. **True exciple** well developed, persistent, ± colourless but appearing straw-coloured through densely interspersed, minute granules; hyphae 1.7–2 μm wide, branched and radiating. **Epithecium** straw-coloured to dark brown, often minutely granular, granules dissolving in K. **Hymenium** 35–50 μm tall, I+ blue. **Hypothecium** colourless, I−. **Hamathecium** of numerous, mostly simple, paraphyses, the apices not thickened, clavate or capitate, sometimes pigmented. **Asci** 8-spored, clavate-cylindrical, colourless, *Biatora*-type. **Ascospores** colourless, smooth, mostly 1-septate, without a distinct perispore. **Conidiomata** pycnidia, usually numerous, rather large and conspicuous, stromatic; wall dark, purplish-brown (K+ purple intensifying) above, becoming thinner and ± colourless below. **Conidiogenous cells**, cylindrical, enteroblastic. **Conidia** colourless, smooth, simple, ovoid to ovoid-cylindrical, acrogenous. **Chemistry**: atranorin, usnic acid and fatty acids. **Ecology**: on bark or wood, rarely rocks.

Distinguished from other species of *Catillaria* s. lat and *Tylothallia* by the numerous pycnidia with dark brown (K+ purplish) walls. Both species often occur without apothecia, but usually have pycnidia.

LITERATURE: Galloway (1985).

1 Apothecia pinkish brown to black, occasionally pruinose; pycnidia 0.1–0.2 mm diam. .... **1. *griffithii***
Apothecia pale to creamy white; pycnidia 0.2–0.5 mm diam. .... **2. *corrugatum***

1. C. griffithii (Sm.) Coppins (1980)

Thallus whitish to pale grey, matt or slightly glossy, granular, often warted-areolate. Apothecia 0.2–0.6(-0.8) mm diam.; disc concave-flat or slightly convex, often thinly white-pruinose, pink-brown to dark brown-grey or blackish; true exciple thin, paler or concolorous, in section colourless or brown (K+ purplish tinge) at the upper outer edge. Ascospores 8–16 × 2.5–3.5 μm, (0-)1(-3)-septate, narrowly ellipsoid-oblong. Pycnidia 0.1–0.2 mm diam., usually present as black dots, the wall K+ purple. Conidia 3.5–4 × 1.5–2 μm. Thallus K+ yellow, (atranorin and roccellic acid).

On moderately acid bark and wood, often in rather dry situations, more rarely on sheltered, ± vertical rock faces or walls, *Calluna* stems and low shrubs, also decaying *Armeria* tufts; abundant. Throughout British Is. Europe, North America, Tasmania, New Zealand.

Very variable. The thallus is commonly smooth and thin, but is sometimes thick, irregularly warted or corrugate; the latter morphs are often confusing but usually bear the characteristic pycnidia, the walls of which are K+ purple. Apothecia are often variable in colour on the same thallus. When on rocks it differs from *Tylothallia biformigera* in the pigment of the pycnidia, the generally thinner thallus and chemistry.

2. C. corrugatum (Ach.) Fr. (1831)

C. graniforme (Hagen) Coppins (1980)

Like **1**, but the apothecia pale to creamy white, and lack pigment, and the pycnidia are larger, 0.2–0.5 mm diam. When well developed it has a more coarsely verrucose thallus. Atranorin in thallus; usnic acid in apothecia.

On wood of old farm buildings, coastal jetties and fence rails; very rare. E. England (E. Anglia); formerly Hereford and Sussex (extinct). Europe and North America.

All modern collections lack apothecia.

COCCOTREMA Müll. Arg. (1888)

Thallus crustose, granular-verrucose, when crustose with a pale, ± conspicuous prothallus. **Soredia** or **isidia** sometimes present. **Cephalodia** usually present, laminal, containing *Nostoc* or *Stigonema*. **Photobiont** green, ?*Myrmecia*. **Ascomata** apothecia but perithecia-like, ± globose, within thalline warts with a terminal ostiole. **True exciple** pale, semi-translucent in section. **Hamathecium** of paraphyses, thin, simple or usually branched and anastomosed; periphyses also present. **Asci** 4- to 8-spored; walls thin, sometimes evanescent. **Ascospores** simple, colourless, with uniformly thickened, non-layered wall. **Conidiomata** pycnidia, in small, pale thalline warts; wall pale. **Conidia** short, bacilliform. **Ecology**: usually on bark or rock, less often on soil or overgrowing mosses, widely distributed in temperate oceanic areas.

This genus, only recently recorded for Europe, was formerly allied to *Pertusaria* by some authors. However, the perithecia-like ascomata, the presence of periphyses and cephalodia and the thin-walled spores indicate little affinity with that genus.

LITERATURE: Brodo (1973), Coppins *et al.* (1992).

1. C. citrinescens P. James & Coppins (1992)

Thallus wide-spreading, thin to rather thick, smooth to unevenly verrucose, occasionally ± granular, continous or ± irregularly and unevenly cracked-areolate, pale green-grey, at times with a ± well-developed, pale grey, filmy prothallus. Soralia 0.5–1 mm diam., numerous, at first punctiform, bursting from low, nodular elevations on areoles, remaining discrete or becoming ± efflorescent and then ±

contiguous, pale greenish yellow or ± concolorous with the thallus; soredia coarsely to finely granular. Cephalodia scattered, at times very sparse, inconspicuous, scale-like and partly immersed in areoles or more conspicuous, irregularly peltate, ± partly convex, pale grey to pinkish yellow-brown, ± pellucid, containing *Stigonema* or *Nostoc*. Apothecia 0.6–0.8(-1) mm diam., perithecia-like, rather rare, often restricted to a limited area on the thallus, ± spherical with a flattened apex and ± entirely enveloped in a thin thalline cuff; true exciple 40–50 μm wide at base, 100–140 μm towards apex, pale, of elongate periclinal hyphal cells *c.* 2 μm wide. Ascospores (25-)30–50(-60) × (19-)21–28(-33) μm, with a uniform 2–3 μm thick wall. Warts containing pycnidia *c.* 0.3 mm diam.; pycnidia *c.* 100 μm diam.; conidia 3.8–5 × 0.8–1 μm. Soralia Pd+ red-orange, K+ bright golden yellow, C−, I−, UV− (stictic and cryptostictic acids).

On ± sheltered and humid, often ± vertical, rock faces, especially granite and basalt; widespread but local, especially near the coast, rarely inland. W. Scotland from Dunoon to Unst (Shetland), N.W. Ireland (Mayo, Donegal). W. Norway (Leka Island).

The vivid K+ golden yellow reaction of the soralia is diagnostic. The soralia may resemble those of *Lecanora jamesii*, which may also rarely occur on rocks but in that species the thalli are less extensive and the soralia are K−.

COELOCAULON Link (1833)

Thallus yellowish brown to dark brown, shrubby, erect to subpendent, forming richly-branched tufts on the ground, sometimes unattached; branches numerous, terete to flattened, cartilaginous, forked, bearing lateral and apical spinules. Pseudocyphellae usually present. **Cortex** of densely packed, pigmented, isodiametric cells overlying a layer of gelatinized, periclinal hyphae. **Photobiont** chlorococcoid. **Medulla** of dense or loose hyphae, which is in places so poorly formed as to give rise to a cavity. **Soralia** rare. **Ascomata** apothecia, lecanorine, on main branches; disc glossy, brownish, contorted when mature. **Thalline exciple** thin, spinulose. **Hamathecium** of paraphyses, septate, capitate, branched. **Asci** *Lecanora*-type, 8-spored. **Ascospores** colourless, simple, ellipsoid. **Conidiomata** pycnidia, occurring at ends of spinules. **Conidia** straight. **Chemistry:** fatty acids. **Ecology:** mainly acid heathlands.

Easily recognized by the dark, glossy, wiry thallus.

LITERATURE: Kärnefelt (1986).

1 Main branches flattened, uneven, furrowed, pitted, to *c.* 1 mm diam.; pseudocyphellae concave, excavate, elongate; branching open and coarse .... **1. *aculeatum***
 Main branches rounded, even, delicate, to *c.* 0.5 mm diam.; pseudocyphellae flat, circular; branching dense and spinulose .... **2. *muricatum***

1. C. aculeatum (Schreber) Link (1833)

Thallus forming shrubby tufts 2–4 cm high, of open, coarse, spinulose branches, matt to glossy brown; main branches to *c.* 1 mm diam., flattened, uneven, furrowed, pitted, the minor branches more rounded and even. Pseudocyphellae concave, excavate, and elongate, usually present in some pits on the main branches. Soredia very rare; present only in damaged material. Apothecia scarce; disc 2–5 mm diam., when mature. Thallus Pd−, K−, C− (lichesterinic and protolichesterinic acids).

Characteristic of acid soil amongst *Calluna*. Widespread and common on heaths, often forming ± extensive patches, from sea-level to *c.* 1000 m, locally frequent on coastal sand dunes; mossy scree slopes; rare on mossy rocks, tree trunks and stumps. Throughout British Is. Widespread in N. Hemisphere, on mountains in Africa, South America, Australia, New Zealand, extending to the Antarctic Peninsula.

Abnormal morphs with gross, efflorescent, pale "soralia" are occasionally encountered; the cause of this malformation is obscure.

Bryocaulon divergens (Ach.) Kärnefelt (1986) [syn. *Coelocaulon divergens* (Ach.) R. Howe (1912)] is similar, but is stouter, more divergently branched, with abundant, ± convex pseudocyphellae which, together with the medulla, react C+ red (olivetoric acid). Recorded last century on the ground amongst mosses on Cairngorm Mtns, but that material probably originated outside the British Is.

2. C. muricatum (Ach.) Laundon (1984)

Like **1**, and sometimes difficult to separate, but smaller, 1–3 cm high, more delicate, and more densely and evenly branched, usually with numerous, small, lateral spinules; these differences are especially apparent when the two species grow together.

On the ground, scattered on acid heaths from sea-level to *c.* 600 m., often with *C. aculeatum* **(1)**, but generally less abundant, occasionally on mossy rocks, old tree trunks and mossy scree slopes; locally abundant. Throughout British Is., particularly in upland Britain. Widespread in N. Hemisphere, mountains in Africa and South America, extending to Antarctica.

COLLEMA Weber ex Wigg. (1780)

Thallus mainly foliose, rarely crustose or shrubby, often swelling when wet, upper surface dark olive-green to brown-black, rarely grey-blue. **Upper** and **lower cortex** absent, attached to substratum by hapters, or, more rarely, by white rhizines. **Isidia** frequent, **soredia** absent. **Photobiont** blue-green, *Nostoc*; cells distributed through the thallus and not forming a discrete layer. **Ascomata** apothecia, with a pale brown, red-brown or brown-black disc. **Thalline exciple** present, usually persistent. **True exciple** plectenchymatous to pseudoparenchymatous. **Epithecium** indistinct, colourless or pigmented. **Hymenium** colourless, I+ blue. **Hypothecium** ± colourless. **Hamathecium** of paraphyses, separating in K, simple or branched, often anastomosed, especially near the apices, apices often clavate or subglobose, sometimes yellowish to reddish brown. **Asci** clavate, apex strongly thickened, apical dome I+ blue, with a downwardly projecting I+ blue annulus and apical cap, (2-)8(-16)-spored. **Ascospores** septate, submuriform or muriform, sometimes spirally arranged. **Conidiomata** pycnidia, innate, marginal or laminal, with a pale ostiole. **Conidia** (3-)4–6(-8) μm, oblong, not or ± enlarged at both ends. **Chemistry:** lichen products not detected by t.l.c. **Ecology:** on ± basic rocks, soil, trees, in exposed, dry habitats to moist, semi-innundated sites.

Distinguished from the closely related genus *Leptogium* by the absence of a one-cell thick, upper and lower cortex. The thallus colour in most *Leptogium* species is deep red-brown or bluish or glaucous in contrast to *Collema* in which it is commonly greenish black. Most of the commoner species of *Collema* show considerable variation in their morphology and are thus confusing and difficult to identify. Particular attention should be given to overall appearance and particular features of the lobes, the presence, shape and distribution of isidia, spore-shape, size and septation, as well as the substratum and specific habitat information.

C. ryssoleum (Tuck.) A. Schneider (1898) is omitted as it has not been correctly reported from the British Is.

LITERATURE: Degelius (1954, 1974).

1 On bark, or on mosses on bark 2
On rock, or on mosses on rock, or on soil 10

2(1) Thallus to 5 mm diam., or adjacent thalli fusing to cover larger areas, overall crustose or with minute lobes 3
Thallus above 5 mm diam., usually distinctly foliose or forming swollen, rounded tufts with indistinct lobes 5

3(2) Ascospores 8.5–17 μm wide, submuriform; thallus not forming cushions covered with apothecia 4
Ascospores 3–4.5 μm wide, 1 (-3)-septate; thallus forming small cushions covered with apothecia .... **4. *conglomeratum***

4(3) Ascospores cuboid-oblong with obtuse angles, 13–22 × 9–15 μm; thallus crustose, granular or with minute, ± scattered lobes .... **19. *occultatum***

	Ascospores ± ellipsoid, 16–30 × 8.5–17 μm; thallus foliose, ± rounded, lobes distinct, flattened, often channelled, arranged in rosettes	**12. *fragrans***
5(2)	Thallus, when wet, not cushion-forming; lobes ± horizontally spreading, usually distinct; apothecia present or absent; ascospores neither worm-like nor spiralled in the ascus	6
	Thallus, when wet, forming rounded, swollen cushions to 1 cm thick, of erect, wrinkled, often indistinct lobes; apothecia frequent, often covering thalli; ascospores worm-like and spiralled in the ascus	**8. *fasciculare***
6(5)	Thallus ± smooth, without longitudinal ridges	7
	Thallus with well-developed ridges or folds (bats' wing-like)	8
7(6)	Isidia at first globose, becoming flattened and squamule-like	**9. *flaccidum***
	Isidia globose to cylindrical, never becoming squamule-like	**10. *subflaccidum***
8(6)	Thallus with ± distinct coarse globose isidia, or isidia absent; apothecia frequent	9
	Thallus with cylindrical, simple or branched isidia, especially on the ridges; apothecia rare	**14. *furfuraceum***
9(8)	Ascospores 3–4.5 μm wide, acicular to bacillar; isidia often present	**22. *nigrescens***
	Ascospores 6–6.5 μm wide, irregularly clavate; isidia absent	**21. *subnigrescens***
10(1)	On soil (see also *C. crispum* and *C. auriforme*)	11
	On rocks or amongst mosses on the ground	13
11(10)	Thallus ± foliose; asci 8-spored; ascospores various	12
	Thallus crustose, thin, membrane-like, without distinct lobes; asci 4-spored; ascospores muriform	**17. *limosum***
12(11)	Thalline exciple ± smooth, even; ascospores 6–10 μm wide, colourless; pycnidia usually present; conidia 4–6 × 1–2 μm	**23. *tenax***
	Thalline exciple coarsely crenulate; ascospores *c.* 13 μm wide, usually pale yellow-brown; pycnidia absent; conidia 11–13 × 2–3 μm	**24. *bachmanianum***
13(10)	Thallus with well-developed longitudinal ridges; (bats' wing-like)	14
	Thallus smooth or wrinkled, without longitudinal ridges	16
14(13)	Thallus with globose isidia or isidia absent	15
	Thallus with cylindrical, simple or branched isidia	**13. *furfuraceum***
15(14)	Ascospores 3–4.5 μm wide, acicular to bacillar; isidia often present	**22. *nigrescens***
	Ascospores 6–6.5 μm wide, acicular to fusiform, often irregularly clavate and thicker towards one end; isidia absent	**21. *subnigrescens***
16(13)	Thallus isidiate, usually sparingly or not fertile	17
	Thallus without isidia, usually abundantly fertile when mature	27
17(16)	Isidia becoming flattened and squamule-like	18
	Isidia remaining globose or cylindrical	20
18(17)	Lobes numerous, ± extended and deeply divided into lobules	19
	Lobes rather few, large, wavy, sparingly and not deeply divided	**9. *flaccidum***
19(18)	Lobules corrugate-bullate, with markedly wavy margins; ascospores 13–16(-18) × 10–12 μm, ± cuboid	**16. *latzelii***
	Lobules smooth, without wavy margins; ascospores 26–34 × 13–15 μm, ellipsoid	**5. *crispum***
20(17)	Thallus <2 cm diam.; lobes <1 mm wide	21
	Thallus >2 cm diam.; lobes >2 mm wide	22
21(20)	Lobes 0.4–0.8 mm wide, white-pubescent at least at apices	**11. *fragile***
	Lobes 0.1–0.3 mm wide, not white-pubescent	**26. *parvum***
22(20)	Lobes ± elongate and deeply divided	21
	Lobes ± rounded, large, few, sparingly and not deeply divided	**10. *subflaccidum***

23(22) Lobes flat, or if concave, without lobulate margins 24
Lobes concave, often with elevated nodulose or lobulate margins .... **6. *cristatum***

24(23) Isidia ± globose, rarely forming a continuous crust 25
Isidia cylindrical, simple or branched, often forming a continuous crust ..... **15. *glebulentum***

25(24) Thallus usually paler, frequently dispersed in discrete lobes, pustulate with laminal isidia, the lobe margins flat or often wavy, down-turned when dry; ascospores submuriform 26
Thallus very dark (often blackish), forming ± continuous rosettes of channelled lobes, with isidia usually confined to the wavy ascendant margins; ascospores 3-septate .... **27. *undulatum***

26(25) Thallus not strongly swollen when wet, often pustulate-bullate; isidia small, scurfy when wet .... **14. *fuscovirens***
Thallus swollen and pulpy when wet, upper surface minutely striate when dry; isidia, rounded, bead-like when wet .... **1. *auriforme***

27(16) Thallus without swollen and ± channelled lobes; apothecia not confined to ascending lobes; ascospores submuriform or muriform or, if 3-septate, dimensions not as above 28
Thallus with (at least in the apices) swollen and ± channelled lobes; apothecia on ascending lobes; ascospores 3-septate, 18–28 × 6.5–8.5 μm .... **20. *polycarpon***

28(27) Thallus >0.5 cm diam., foliose or ± fruticose, broadly attached below 29
Thallus <0.5 cm diam., ± crustose, attached to substratum only at the centre **2. *callopismum***

29(28) Thallus distinctly foliose, the lobes ± extended and deeply divided, not forming cushions 30
Thallus appearing ± shrubby, with ± erect lobes and terete, marginal lobules, forming cushions .... **3. *ceraniscum***

30(29) Lobes concave, channelled or with ± ascending, uneven margins 31
Lobes flat or convex, not channelled, the margins not or little raised, often fan-shaped .. 32

31(30) Thallus blackish, convoluted but not obviously channelled; thallus margin level with or below the apothecial disc when mature; asocospores 3-septate, 6.5–9 μm wide **25. *undulatum***
Thallus dark olive-green-black, channelled, with lobulate or nodulose margins; thalline margin level with or elevated above the apothecial disc; ascospores submuriform (3- to 6-celled), 8–13 μm wide .... **6. *cristatum***

32(30) Thallus blackish olive-green to brown-black, minutely striate; ascospores 4.5–6.5 μm wide; on dry limestone rocks .... **18. *multipartitum***
Thallus bright to dark green, not striate; ascospores 8.5–13 μm wide; in streams **7. *dichotomum***

1. C. auriforme (With.) Coppins & Laundon (1984)

C. auriculatum Hoffm. (1796)

Thallus 2–4(-10) cm diam., foliose, ± rounded, ± loosely attached and partly ascending, ± deeply and irregularly lobed, rather thick, swollen and pulpy when wet; lobes to 10 mm wide, few, often rounded, ± striate or wrinkled when dry, with an entire or sometimes indented margin, occasionally bearing folioles; upper surface dark olive-green to brownish black, rarely blue-grey. Isidia often numerous, crowded, globose or clavate, rarely branched, usually particularly conspicuous when wet. Apothecia rather rare, often ± immersed when young; disc 2–3 mm diam., with entire, granular-isidiate thalline exciple. Ascospores 26–36 × 8.5–13 μm, ellipsoid to oval, submuriform.

Amongst mosses and on highly calcareous rocks, mortar and soil, chalk paths, shell-sand dunes, in rather moist, mostly shaded situations; often common. Throughout British Is. Europe, Tunisia, N. Africa, U.S.A.

Similar to *C. fuscovirens* **(14)** but has a thicker, generally brownish, non-olivaceous, non-bullate, but ± wrinkled thallus; *C. fuscovirens* generally grows directly on rocks. When wet the thallus of *C. auriforme* is thicker, more pulpy and the globose isidia larger. *C. auriforme* has a wider ecological amplitude, growing on a wide variety of basic substrata; it prefers moister habitats than *C. fuscovirens* and often occurs on or amongst mosses.

2. C. callopismum Massal. (1856)

Thallus to 0.5 cm diam., crustose, becoming ± tufted, rounded or angular in coarsely areolate patches, ± deeply lobed, fixed to substratum only at the centre; lobes to 0.2 mm wide, of equal length, usually repeatedly branched, ± contiguous, flat or swollen, at times partly ascending and crowded; upper surface dark olive-green or blackish, smooth or with very small isidia-like granules. Apothecia infrequent; disc 0.3–0.7 mm diam., flat or concave, with a thin thalline exciple, smooth or slightly lobed. Ascospores 17–26 × 8.5–10.5 μm, ± ellipsoid, submuriform, rarely 3-septate.

On limestone. N.W. England (Cumbria). Europe, Greenland, North America.

Rather variable in the degree of thallus-lobation and areolation. The var. *rhyparodes* (Nyl.) Degel. (1954) differs in the larger ascospores, 26–43 × 12–15 mm. On mica-schist and epidiorites in montane situations; very rare. C. Scotland (Perthshire, Ben Lawers, Ben Alder).

The thallus is often reduced to a poorly defined crust lacking distinct lobes.

3. C. ceraniscum Nyl. (1865)

Thallus to 3 cm diam., ± rounded, compact, forming intricate cushions; lobes 0.5–2 mm wide, ± erect, densely packed and of about equal length, richly branched, the lower part flattened, the terminal branches ± terete and isidia-like; upper surface black, matt, smooth or somewhat verrucose, without isidia. Apothecia rare in Britain, numerous and crowded when present; disc 0.4–0.8 mm diam., urceolate, with a ± thick, entire or lobulate thalline margin. Asci (2-)4-spored. Ascospores 20–36 × 13–22 μm, ± oval or subglobose, muriform.

On soil, often growing on living and decaying carpets of lichens and bryophytes on calcareous and schistose rocks above *c.* 1000 m alt.; rare. C.S. Scotland (Perthshire). Scandinavia, Novaya Zemlya, Siberia.

Distinguished by the very dark, richly branched, thin, ascending lobes with terete apices, which form compact clusters on decaying mosses.

4. C. conglomeratum Hoffm. (1796)

Thallus 2–10 mm diam., foliose or ± crustose forming rounded cushions, often attached by a central point; lobes 0.5–1.5 mm wide, few, free or overlapping, ± flattened, distinctly swollen with smooth or sometimes verrucose surface and lobulate margins. Apothecia mostly numerous, crowded and predominating, together often forming a ball, sessile; disc 0.5–1.5(-2.0) mm diam., flat to convex, with a rather thin, entire thalline exciple. Ascospores 15–24 × 3–4.5 μm, fusiform with acute apices, 1-(3)-septate, never with longitudinal septum.

On bark, (*Ulmus*, *Fraxinus*) in wayside, nutrient-rich sites. S. England (Sussex), C.S. Scotland (Perthshire); collected last century, no recent records. S. Europe, North America.

Differs from other small *Collema* species in the narrow, fusiform, mostly 1-septate ascospores.

5. C. crispum (Huds.) Weber ex Wigg. (1780)

Thallus usually small, sometimes forming colonies to 5 cm diam. or more, foliose, rather thin, deeply lobed, often forming rosettes, slightly swollen when moist; lobes 0.5 to 6 mm wide, often rather few, ear-like, ± rounded at apices, concave, often overlapping with ± often ± ascending, wavy and sometimes convoluted margins; margins entire or crenate, occasionally somewhat lacrate, but never swollen; upper surface olive-green-brown to black, smooth or isidiate; lower surface often with white rhizines, sometimes forming ± scattered hapters. Isidia at first globular, soon flattened and squamule-like. Apothecia infrequent, adpressed; disc 1–2(-2.5) mm diam., flat; thalline exciple often bearing lobules. Ascospores 26–34 × 13–15 μm, 3-septate or submuriform, rarely 4- or 5-septate, subellipsoid or ± oval, with rounded ends.

Calcareous rocks and walls, particularly on old crumbling mortar, often in damp, shaded places, less often on calcareous soils and shell-sand, rarely on dust-impregnated bark; abundant. Throughout British Is. Europe, Iceland, Palestine, N. Africa, U.S.A.

Distinguished by the numerous, ear-like, often overlapping, lobes and flattened, squamule-like isidia which can resemble small lobes. The isidia are sometimes so numerous that they obscure the thallus. In very shaded habitats the thallus can become glaucous-emerald-green tinged. A diminutive morph, to 3 mm diam., with an almost crustose thallus of scattered, ± overlapping lobes, to 0.5 mm wide. has been referred to var. *metzleri* (Arnold) Degel. (1954). This morph is chiefly associated with shaded chalk outcrops; intermediates are frequent.

6. C. cristatum (L.) Weber ex Wigg. (1780)

Thallus 2–5(-10) cm, foliose, rounded, semi-circular, or irregular, dying away in centre when old, closely adpressed, deeply lobed, rather thin; lobes narrow, notably concave, radiating, irregularly branched, contiguous or discrete, the margins elevated, wavy, sometimes convoluted, ± entire or ± incised and often distinctly small-lobulate, not swollen; upper surface dark olive-green-brown to black, with (var. *marginale*) or without (var. *cristatum*) wart-like isidia; lower surface with rounded hapters forming large white tufts. Apothecia often densely crowded, sometimes absent, ± marginal, sessile or stalked; disc to 5 mm diam., flat or convex, with an even ± thalline margin. Asci 4- to 6(-8)-spored. Ascospores 18–32 × 8–13 μm, ellipsoid, with ± acute ends, submuriform (Fig. 11d).

On calcareous rocks and soils. Throughout British Is. Europe, North America, Arctic, Africa, Madeira, Morocco, Algeria, Asia, Turkestan.

Very variable, morphs are recognized at the varietal level, though intermediates do occur: (a) var. *cristatum* has rather short, wide and irregularly branched, often ascending lobes; margin ± wavy and coarsely incised and appearing crisp. Apothecia to 5 mm diam., often rather few. Isidia absent. N. and montane regions. (b) **var. *marginale*** (Huds.) Degel. (1954) has extended, furcate lobes which are less incised. Apothecia to 2 mm diam., mostly numerous. Often isidiate; isidia ± globose or club-shaped, rarely flattened. More southern than (a). Densely isidiate and wide-lobed morphs of var. *marginale* may resemble *C. auriforme* **(1)** but differ in the darker, non-striate thallus with distinctly branched, usually narrow lobes, and the abundance of apothecia. *C. cristatum* is usually firmly attached to rocks forming distinctive, complete or incomplate, radially spreading, intricate rosettes, however morphs occur which are detached and often collect in ± moist declivities in rock where they form irregular, rather spiky, contorted thalli sometimes with narrow, attenuated, curled lobes. Though a very variable species, *C. cristatum* is recognized by the narrow, persistently concave lobes.

7. C. dichotomum (With.) Coppins & Laundon (1984)

C. fluviatile (Huds.) Steudel (1824), *Leptogium rivulare* auct brit. non (Ach.) Mont (1846)

Thallus to 2.5 cm diam., but sometimes forming more extensive patches, foliose, rather thin, adpressed or in part ascending, deeply lobed; lobes to 2 mm wide, few, ± fan-shaped, repeatedly branched, with discrete or overlapping, flattened lobules, 0.5–1.5 mm wide, the ends at times ascending, ± broadened and lip-like; upper surface bright-dark green, sometimes with a glaucous tinge, roughened, without isidia. Apothecia mostly sparse and scattered, rarely numerous, superficial; disc 0.7–1 mm diam., flat, with a thin, entire, smooth thalline exciple. Ascospores 20–30 × 8.5–13 μm, 3-septate or submuriform, with one longitudinal septum. Pycnidia 300–550 μm diam., rather frequent, superficial, often similar to young urceolate apothecia, with a small 'disc' and a thick thalline margin, reddish yellow with blackened apex, appearing as swollen low tubercles with a prominent opening; conidia 5–6.5(-8) × 1.2–1.5 μm.

On mostly permanently immersed rocks in unpolluted upland streams; rare and decreasing. Wales, N. England, Scotland, N. Ireland. Fennoscandia, France, Germany, U.S.S.R.

A distinctive species with rather small thalli with a characteristic repeatedly forked branching pattern, small apothecia and unusually large pycnidia. The only *Collema* which occurs in ± permanently inundated habitats. Very sensitive to pollution in streams, this species has disappeared from many of its former sites in recent years.

8. C. fasciculare (L.) Weber ex Wigg. (1780)

Thallus to 1 cm diam., crustose or small-foliose, very swollen, pulpy, to 1 cm deep when moist, ± coarsely nodulose-wrinkled and indistinctly lobed, forming rounded cushions. Lobes, when distinct, short, flattened, adpressed, overlapping, sometimes distinctly lobulate, the lobules developed from wrinkles; upper surface dark olive-green to brown-black, paler when moist, often wrinkled and ridged, without isidia. Apothecia numerous, crowded, overlapping, generally predominant; disc 0.8–2 mm diam., flat, often with a thick, ± wrinkled thalline margin. Asci (6-)8-spored. Ascospores 50–95 × 4.5–5 μm, 9- to 16-septate, worm-like and spiralled, often attenuated towards one or both ends.

Frequently associated with mosses on trunks of usually old trees, especially *Corylus*, in rather moist, shady, sheltered sites, characteristic of the *Lobarion*, rarely on mossy rocks; locally frequent. W. & N. Britain, W. Ireland. Europe, N. Africa, W. Asia, Tasmania, New Zealand.

Characterized by the mainly crustose thallus, which becomes remarkably swollen, spherical and pulpy when wet, abundant apothecia and the characteristic worm-like spores.

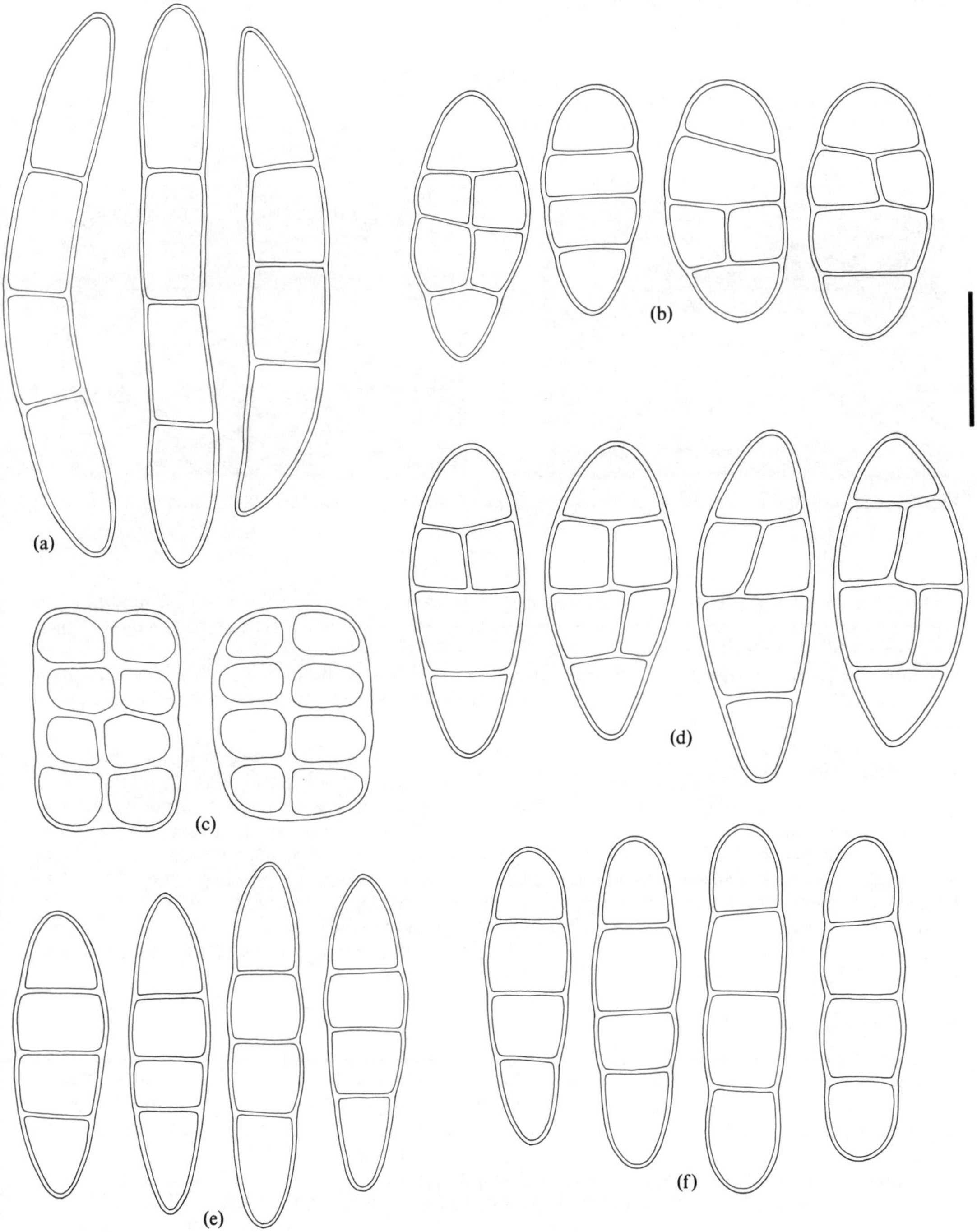

Fig. 11. *Collema* ascospores. (a) *C. multipartitum*; (b) *C. tenax*; (c) *C. occultatum*; (d) *C. cristatum*; (e) *C. polycarpon*; (f) *C. undulatum*. Scale 10 μm.

9. C. flaccidum (Ach.) Ach. (1810)

Thallus 3–6 cm diam., foliose, membrane-like, often forming extensive colonies, ± deeply and very irregularly lobate; lobes 0.5–1.5 cm wide, few, thin, ± rounded, loosely attached, often partly ascending, irregularly folded, rumpled and wavy; margin entire or rarely coarsely lacerate, sometimes ± revolute; upper surface dark olive-green to black, with numerous, often crowded isidia on surface and margins. Isidia concolorous, globose when young, becoming flattened and squamule-like, ± horizontal or ascending (Fig. 12b). Apothecia very rare, laminal; disc 1.5–2.5 mm, flat; thalline margin smooth or striate. Ascospores 26–34 × 6–6.5 μm, 3- to 5-septate, ellipsoid to fusiform.

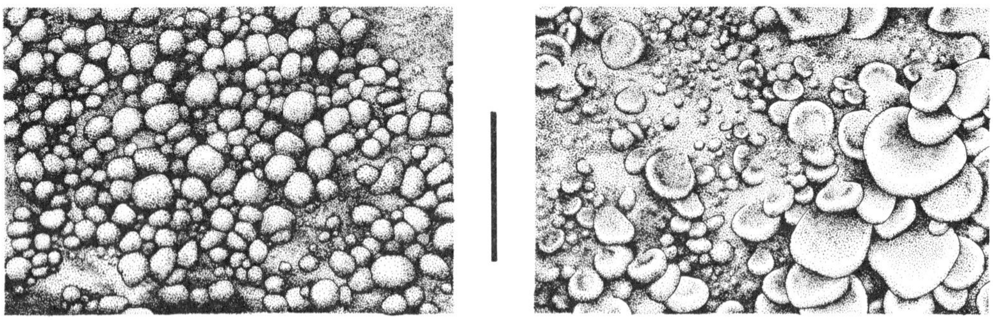

Fig. 12. *Collema* thalli surfaces with isidia. (a) *C. subflaccidum*; (b) *C. flaccidum*. Scale = 0.5 mm.

On sheltered, damp siliceous rocks, more rarely calcareous substrata, often amongst mosses or in sheltered seepage tracks, more rarely on bark; frequent. N. & W. British Is. Europe, Iceland, North America, Tunisia, Asia Minor, Siberia.

Distinguished by the large, rather ragged, membrane-like thallus with squamule-like isidia. *Leptogium cyanescens* has similar, flattened isidia but is blue-grey and has a well-developed, one cell thick, upper cortex.

10. C. subflaccidum Degel. (1974)

Like **9**, but with smaller, densely distributed, laminal isidia, which are mostly globular or cylindrical, very rarely becoming flattened and squamule-like (Fig. 12a). The thallus is often tinged brown. The numerous isidia give a scurfy appearance to the upper surface. Apothecia very rare, laminal, to 2 mm diam. with a persistent, densely isidiate thalline exciple; disc red-brown, flat. Ascospores 42–55 × 4.5–6.5 μm, 5- to 7-septate, narrowly fusiform or ± acicular, straight or slightly curved.

On bark, especially old *Fraxinus* in relatively moist, shady places, more rarely on rocks; rather frequent. N. & W. British Is., extending to S. England (Hampshire, New Forest). Europe, North America, Africa, Asia, Hawaii.

11. C. fragile Taylor (1836)

Thallus to 2 cm diam., of ± radiating, crowded, convex, nodulose-contorted closely attached lobes, 0.4–0.8 mm wide, which are ± white-pubescent, at least at their apices, furcate, forming a dispersed or ± confluent crust with a ± radiating margin; upper surface dark green-brown to black, matt with few to numerous globose isidia 0.1–0.2 mm diam. Apothecia very rare; disc 0.5–0.8 mm diam., at first with rather thick, thalline margin, later disappearing. Ascospores 26–30 × 13–17 μm, ellipsoid-ovoid, submuriform with 4 transverse septa and 1 or 2 longitudinal septa.

Directly on shady, often steeply inclined, limestone rocks and walls; rather rare. W. British Is. S.W. Europe.

Distinguished by the closely attached, often partly fragmentary, thallus directly on hard limestones, sometimes forming extensive colonies in seepage tracks. Only the tips of actively growing lobes are finely pubescent. This species has the aspect of a small *Leptogium* but lacks a cellular upper cortex. *C. parvum* **(26)** has smaller, more adpressed lobes without white pubescence.

12. C. fragrans (Sm.) Ach. (1814)

Thallus to 0.5 cm diam., but thalli often numerous and crowded together, foliose, deeply lobed forming ± rounded rosettes, usually fixed to substratum in centre; lobes 0.3–1.5 mm wide, free or imbricate, ± flattened, not swollen, often crenate, ± flat, channelled, adpressed or ascending to erect or somewhat spreading; upper surface dark olive-green to blackish, smooth or with globose, isidia-like papillae or terete to flattened lobules on margins and on upper surface. Apothecia often numerous and crowded in centre of thallus; disc 0.4–0.7 mm diam., ± immersed at first, flat to concave with a thin to moderately thick thalline exciple, entire or papillate. Asci (4-) 8-spored. Ascospores 16–30 × 8.5–17 μm, submuriform, with 2–5 transverse and 2–5 longitudinal septa, oval or ± ellipsoid.

On nutrient-enriched bark, especially of *Ulmus*, in old parkland, very rarely on rocks. S. England, C. Scotland. W., C. & S. Europe, Morocco, U.S.A.

This species is characterized by forming small, scattered or ± contiguous, neat rosettes of ± stellately arranged lobes, often with abundant apothecia towards thallus-centre. *C. fragrans* has decreased dramatically in recent years due to the demise of *Ulmus*; its centre of distribution now seems to be S. England (Hampshire, New Forest), where it colonizes wound seepage tracks on mature *Fagus*.

13. C. furfuraceum (Arnold) Du Rietz (1929)

Thallus 3–6(-10) cm diam., foliose, membrane-like, thin, closely adpressed, conspicuously lobed; lobes 0.5–1 cm wide, few, rounded or extended, ± overlapping; upper surface dark olive-green to black, paler and ± transparent when moist, markedly ridged, ridges radiating, in young parts of thallus short, sometimes ± rounded, becoming long, narrow and flexuous, 0.1–0.3 mm wide, 1.5 mm tall, simple or branched. Isidia present on ridges, abundant, terete, style-like, to 0.3 mm long, simple, becoming branched when old. Apothecia very rare; disc 0.5–1.5 mm diam., flat, with a densely isidiate thalline exciple. Ascospores 40–80 × 3–6.5 μm, 4- to 5-septate, fusiform to acicular, often curved.

On bark and occasionally rock in humid, well-lit situations; rather frequent but rare and decreasing in the south. N. & W. British Is. extending to S. England (Dorset). Europe, Mediterranean region, North and South America, Asia, N. Africa.

Similar to *C. nigrescens* **(22)** and *C. subnigrescens* **(21)** in its membrane-like ("bats' wing"-like) thallus; however, the characteristic ridges of *C. furfuraceum* become ± densely isidiate. *C. subflaccidum* **(10)** is also ± densely isidiate but completely lacks ridges.

14. C. fuscovirens (With.) Laundon (1984)

C. tuniforme (Ach.) Ach. (1810)

Thallus 3–5 cm diam., foliose, ± rounded or irregular, deeply lobed, adpressed; lobes 2–5 mm wide, rather few, extended, often repeatedly branched, channelled or ± rounded with ± ascending, wavy margins, never swollen; upper surface dark olive-green to nearly black, paler and ± transparent when moist; lower surface paler than above, sometimes greyish or bluish. Isidia on surface and margins, globular, often numerous and partly covering the thallus to give a scurfy appearance, concolorous with thallus. Apothecia numerous or sparse; disc 0.5–1.5 mm diam., flat or with a ± thick thalline exciple, smooth or isidiate. Ascospores 15–24 × 6.5–13 μm, submuriform, often with 3 transverse and 1 longitudinal septa, oval, ellipsoid or ± globose.

On hard, sunny and exposed calcareous rocks or rocks covered in calcareous dust; local. Throughout British Is. Europe, Iceland, U.S.A., Greenland, U.S.S.R., Turkestan, N. Africa.

Closely related to *C. auriforme* **(1)** but has a thinner, darker thallus which is often pustulate, not noticeably swollen when wet, with narrow, more wavy lobes and smaller isidia; when dry, the minute, crowded isidia give the thallus a scabrid, matt appearance.

15. C. glebulentum (Nyl. ex Crombie) Degel. (1952)

Thallus 3–6 cm diam., foliose, membrane-like, rounded or irregular, closely adpressed, ± deeply lobate; lobes to 5(-10) mm wide, often numerous, rounded or extended, ± overlapping or with contiguous, ascending margins; upper surface dark olive-green to black, folded and conspicuously wavy, margin entire or at times crenate or isidiate, never swollen. Isidia *c.* 0.1 mm wide, to 1.5 mm long, terete, ± branched, globose when young, rarely becoming flattened, usually numerous, often forming a dense, coarsely areolate crust in older parts of thallus. Apothecia unknown.

On rocks in upland situations; on siliceous, rarely ± calcareous rocks in river and lake shores which are periodically inundated; rare in N. Wales, N.W. England, more frequent in Scotland (Highlands), W. Ireland. Fennoscandia, Germany, Switzerland, Iceland, Greenland, Novaya Zemlya.

An often overlooked species close to *C. flaccidum* **(9)** and *C. subflaccidum* **(10)**, but distinguished by the rosette-forming, more markedly adpressed and divided thallus and rather large, terete or coralloid, finger-like isidia. *C. furfuraceum* **(13)** occurs predominantly on bark and is pustulate and ridged with smaller, terete isidia.

16. C. latzelii Zahlbr. (1909)

Thallus to 1.5 cm diam., foliose, forming contorted, rounded or irregular cushions, deeply lobed; lobes 1–3 mm wide, radiating, ± extended, repeatedly branched, contiguous, often overlapping with ascending, ± entire, markedly wavy margins, never swollen; upper surface dark olive-green to black, ± smooth, matt. Isidia, when present, as low warts to ± coarse globose or squamule-like. Apothecia not frequent, scattered or crowded, inconspicuous often submarginal; disc 0.3–0.8 mm diam., flat to concave. Ascospores 13–16(-18) × 10–12 μm, oval-subglobose or ± cuboid, submuriform.

On vertical, sunny, coastal serpentine rock faces. S.W. England (W. Cornwall, Lizard). Portugal, Spain, Italy, Yugoslavia, Morocco.

The contorted, plicate-lobed thallus with coarse, granular isidia resembles *Leptogium plicatile*, which is more red-brown. The often quadrately divided, oval to subglobose spores are diagnostic.

17. C. limosum (Ach.) Ach. (1810)

Thallus thin, membrane-like, without distinct lobes, forming a dark olive-green, blue-grey to brown-black crust, the surface smooth or verrucose, shrinking and almost disappearing when dry. Apothecia often numerous, immersed or sessile; disc 1–3 mm diam., flat, with a smooth or ± verrucose thalline exciple. Asci (2-) 4-spored. Ascospores 26–34 × 10–15 μm, ovoid-ellipsoid, muriform.

On damp clay or sandy soils, especially pathways; rather rare. S. & W. British Is., E. Scotland (Fife). Europe, North America, Africa, China.

Differs from *C. tenax* **(23)** by the 2(-4)-spored asci, larger ascospores, and the poorly developed, crustose thallus without distinct lobes.

18. C. multipartitum Sm. (1814)

Thallus to 2–3 cm diam., foliose, rounded or irregular, deeply lobate, loosely adpressed to partly ascending, very fragile; lobes 0.5–1.5 mm wide, often remaining partly separate, often fan-shaped, ± contorted, often nodular and irregular, convex, often sub-cylindrical, repeatedly branched, not ascending; upper surface ± dark olive-green to brown-black, matt, without isidia, often minutely longitudinally striate, ± swollen and semi-transparent when moist. Apothecia rather common, scattered; disc 1–2 mm diam., flat with ± thick, entire, crenulate, or lobulate thalline exciple. Ascospores 26–43 × 4.5–6.5 μm, 3-septate, ellipsoid-fusiform, sometimes curved (Fig. 11a).

On hard, sunny and exposed limestones, rarely on siliceous rock associated with shell-sand; local. Throughout W. British Is. Europe, Morocco, Canada.

Readily distinguished from all other *Collema* species by the much-branched thallus with narrow, often partially discrete, notably convex-rounded, minutely striate lobes, and the characteristic ellipsoid-fusiform ascospores.

19. C. occultatum Bagl. (1861)

Thallus small, crustose, of widely scattered to ± contiguous granules, ± covering rather large areas, sometimes forming a thin or ± thick, often minute, effuse crust; granules 50–100 μm diam., ± globose, black to green-black, rarely with very small somewhat flattened lobes. Apothecia rather frequent, globose or perithecia-like when young; disc 0.1–0.3 mm diam., red-brown to blackish, somewhat exposed when mature. Ascospores 13–22 × 9–15 μm, ± cubic-oblong with rounded angles or ± globose, submuriform (Fig. 11c).

On ± basic bark, especially of *Acer*, *Fraxinus*, *Ulmus* and *Sambucus*; local, most frequent in the north. S.W. & N. Britain. Europe, Mediterranean region, Algeria, North America.

Readily distinguished by the poorly developed thallus of minute granules, the small perithecia-like apothecia and ± cuboid, submuriform spores.

20. C. polycarpon Hoffm. (1796)

Thallus 2–6 cm diam., or the smaller rosette-like cushions, foliose, deeply lobed, rather thick; lobes 1–2.5 mm wide, numerous, ± radiating, ± flattened, contiguous, often ascending at the centre of the thallus, often channelled towards apices with raised margins, swollen, somewhat contorted; upper surface dark olive-green to black, without isidia. Apothecia numerous, terminal ± elevated, crowded and frequently covering thallus except outermost parts of lobes, mostly appearing stalked; disc 0.5–1.5 mm diam., flat to convex, with rather thin, smooth thalline exciple. Ascospores 18–28 × 6.5–8.5 μm, 1- to (2-)3-septate, fusiform, with acute apices (Fig. 11e).

On hard, exposed limestones; local. N. England, S. Wales, Scotland, Ireland. Europe, Arctic, Africa, Asia.

The small, rosette-like cushions with ± raised, channelled lobes with numerous terminal apothecia are diagnostic.

21. C. subnigrescens Degel. (1954)

Thallus 2–6(-20) cm diam., foliose, membrane-like, thin, rounded, ± closely adpressed to substratum, deeply lobed; lobes 0.5–1.5 cm wide, few, rounded or ± extended, overlapping, often folded; upper surface dark olive-green to black, strongly ridged; ridges radiating, short, in young parts of thallus, in old parts becoming long and narrow, to 1.5 mm tall and 0.1–0.3 mm wide lacking isidia; lower surface with depressions corresponding to ridges, paler green. Apothecia often present; disc 0.8–1.5 mm diam., naked or rarely ± white-pruinose, flat with a narrow, smooth margin. Ascospores 40–75 × 6–6.5 μm, 4- to 5-septate, acicular-fusiform, usually curved.

On ± nutrient-rich bark and rocks; local. W. British Is. Europe, Madeira, Algeria, Tunisia.

22. C. nigrescens (Huds.) DC. (1805)

Like **21**, but usually smaller, to 10 cm diam., and often more stunted and convoluted, with lobes to 1 cm wide, densely pustulate and ridged (more than in **21**), dark olive-green to brown-black, coarse, nodular-granular isidia, to 0.2 mm diam., sometimes present on upper surface and margins. Apothecia usually very numerous, often covering most of the thallus; disc 0.4–1 mm wide, flat or convex, with a narrow, at times isidiate, thalline exciple. Ascospores 50–90 × 3–4.5 μm, (4)5- to 12-septate, acicular to bacilliform.

On ± nutrient-rich bark, occasionally seepage tracks on rocks, especially on the coast; rather local, decreasing. N. & W. British Is., eastwards to Dorset. Europe, North America, N. Africa.

23. C. tenax (Swartz) Ach. (1810)

Thallus 3–4(-10) cm diam., very variable, foliose, rounded in small rosettes, or irregular, closely or loosely attached or partly ascending, rather thick, much swollen when moist; lobes often numerous, mostly radiating from centre, with parallel sides or widening towards apices, contiguous and overlapping or discrete, entire or crenate, flattened or usually ± concave, smooth; upper surface dark olive-green to brownish black,. Isidia, when present, large, globose, recalling adventive lobes. Apothecia often present, sometimes predominating, on the surface or margins of lobes; disc to 3 mm diam., flat, with entire thalline exciple, becoming convex. Asci (4- or 6-) 8-spored. Ascospores 17–26 × 6.5–10.5 μm, 3-septate or submuriform, fusiform to ellipsoid with pointed or rounded ends (Fig. 11b). Pycnidia 0.1–0.2 mm diam., rather common, immersed, ± globose. Conidia 4–6 × 1–2 μm.

On basic clay, sandy and calcareous soils and mortar; abundant. Throughout British Is., particularly the south. W. Europe, Mediterranean region, North America, Iceland, Greenland, Novaya Zemlya, Africa, Asia.

The most frequent *Collema* and also the most variable: 3 morphs can be distinguished; these are: **Var. *ceranoides*** (Borrer) Degel. (1954) with ± compacted or loosely tufted thalli of simple or often branched, digitate, cylindrical lobes, to 1.5 cm tall and 1–2 mm diam.; rarely fertile; on loose sandy, basic, soils especially shell-sand, and is probably an adaptation to shifting sand and soil. Var. *corallinum* (Massal.) Degel. (1954), which is almost crustose with very few lobes and often forms a coralloid crown around the disc of the apothecia. Var. *vulgare* (Schreber) Degel. (1954), the common form on mortar, has chunky, short, swollen convex lobes in small confluent rosettes and is usually abundantly fertile.

Differs from *C. bachmanianum* **(24)** in having apothecia with a smooth thalline exciple, shorter conidia, and narrower spores. May be distinguished from *C. limosum* **(17)** by the 8-spored asci containing smaller spores, and the lobed thallus. *C. confertum* Arnold (1867), with ascospores 8–12 μm wide and a marked tendency to grow directly on rock, is sometimes regarded as a distinct species.

24. C. bachmanianum (Fink) Degel. (1954)

Like **23**, but the apothecia have a coarsely crenate, thalline exciple with warts or lobules and larger, broadly ellipsoid, submuriform ascospores, 26–36 × 10–15 μm, colourless, becoming straw-coloured or pale brown; pycnidia absent; internal conidia 10.5–13 × 2–3 μm.

On sandy ± basic soils and shell-sand dunes; rare but easily overlooked. Throughout British Is. Europe, Iceland, Greenland, Spitsbergen, North America, Siberia.

No pycnidia are developed, the conidia, few in number, are much larger than in other species of the genus. They develop from the ends and sides of hyphae deep in the thallus and remain on the conidiophores forming small groups or bundles of up to 15 conidia; similar internal conidia are also known in *C. multipartitum* **(18)**.

25. C. undulatum Laurer ex Flotow. (1850)

Thallus 3–6 cm diam., foliose, rounded or irregular, adpressed, deeply lobed; lobes 2–4 mm wide, channelled, with ascending, entire or sparingly incised, wavy margins, overlapping, never swollen; upper surface dark olive-green to black, smooth, not isidiate. Apothecia usually numerous, often densely crowded; disc 1–1.5 mm diam., flat or ± convex, with a ± thick thalline exciple. Ascospores 17–30 × 6.5–9 μm, 3-septate, fusiform-ellipsoid (Fig. 11f).

On hard, exposed upland limestone rocks; local. N. England, N. Wales, Scotland. Europe, Syria, Iceland, Greenland, Spitsbergen, Novaya Zemlya, Alaska.

Specimens with numerous globose, rarely terete or squamule-like isidia, 0.1–0.2 mm diam., have been referred to var. *granulosum* Degel. (1954).

Distinguished by the 3-septate spores, dark thallus with densely compacted, concave, markedly wavy lobes with apothecia and a thalline exciple level with, or below that of, the disc.

26. C. parvum Degel. (1954)

Thallus to 1.5 cm diam., of discrete to contiguous, plane to convex, adpressed, ± radiating lobes, 0.1–0.3 mm wide, forming entire or partial rings; upper surface dark brown to blackish, matt, not white-pubescent, with few to numerous irregularly globose isidia to 0.05(-1) mm diam. Apothecia unknown.

On bare limestone on top of exposed boulder at *c.* 800 m alt., recently discovered in E.C. Scotland (Angus, Caenlochan). W. Europe from Arctic to Mediterranean.

Resembles *Leptogium diffractum* in habit, but that species has ± glossy lobe ends and a well-defined upper cortex.

CORNICULARIA (Schreber) Hoffm. (1794)

Thallus shrubby, erect, bilaterally symmetrical, the lobes ± flattened. **Cortex** thick, of massive, periclinal hyphae. **Photobiont** *Trebouxia*. **Ascomata** apothecia. **Thalline exciple** entire, ± toothed or spinose. **Asci** *Lecanora*-type, 8-spored. **Ascopores** colourless, simple, ellipsoid. **Conidiomata** pycnidia, black, globose, half sunken, mostly at apices of branches. **Conidia** dumb-bell-shaped. **Chemistry**: no lichen products detected by t.l.c. **Ecology:** on hard, siliceous rocks.

Differs from *Coelocaulon* in being firmly attached, erect and sparsely branched, with usually abundant, apical apothecia.

1. C. normoerica (Gunn.) Du Rietz (1926)

Thallus 1–2 cm tall, forming ± erect tufts, firmly attached; main branches 0.3–1 mm wide, erect, stiff, tough, solid, flattened, sparingly branched, tapering abruptly towards apices, arising from a proliferating holdfast; surface black-brown, ± shiny. Apothecia 1–5 (-8) mm diam., frequent, apical or subapical; disc black and shiny, usually with a few, short, cylindrical extensions. Ascospores 5–6 × 3–4 μm.

Well-lit, coarse-grained, siliceous rocks, widespread in wind-swept, usually montane situations. N. England extending to the S.W. (Cornwall, Dartmoor), Wales, N. & W. Scotland, Ireland. Europe, North America, Japan.

CRYPTOLECHIA Massal. (1853)

Thallus crustose, superficial. **Photobiont** *Trentepohlia*. **Ascomata** apothecia, at first immersed, with a pale pseudothalline exciple as they emerge; disc flesh-coloured to yellowish, without attached cortical remnants. **Thalline exciple** absent. **True exciple** pale and indistinct. **Hymenium** I+ pale blue (mainly ascus walls). **Hamathecium** of mostly simple, septate paraphyses with ± swollen apices. Asci narrowly clavate, with a single functional wall layer, (8-)12(-16)-spored, wall K/I+ blue, apex not thickened, without internal apical apparatus. **Ascospores** 3(-4) transversely septate, colourless. **Conidiomata** pycnidia, immersed; wall colourless. **Conidiogenous cells** short-cylindrical to elongate-ampulliform, sometimes arising in groups on branched conidiophores, enteroblastic. **Chemistry**: no lichen products reported by t.l.c..

LITERATURE: Hawksworth & Dibben (1982), Vězda (1969a).

1. C. carneolutea (Turner) Massal. (1853)

Thallus white or glaucous-white, smooth, continuous to somewhat rimose. Apothecia 0.4–1 mm diam., often abundant, discrete or rarely 2- to 3-contiguous; disc pinkish orange or pale yellow, immersed at first, emerging with an irregularly striate to crenulate margin, becoming flat to convex and ± immarginate; hypothecium pale. Paraphyses 1.5–2 μm wide, the apices sometimes swollen to 4 μm wide. Ascospores 10–20 × 3.5–5 μm, oblong, the ends ± abruptly rounded. Pycnidia *c.* 200 μm diam., in swollen warts but inconspicuous; conidiogenous cells 4–6 × 2–3 μm; conidia 7.5–9 × 3.8–4.5 μm. Thallus Pd−, K−, KC−, C−.

A rare and decreasing species on the shaded trunks of old trees with nutrient-rich bark, especially *Ulmus*, also *Acer*, *Fagus*, *Fraxinus* and *Hedera*. Woodland, parkland, wayside trees, usually near the coast. S. & S.W. England, from Sussex to Cornwall, Is. of Scilly, Channel Is., formerly north to Morecambe Bay and Is. of Man, W. Ireland (Clare). S.W. Europe, Madeira, Azores.

The extensive, smooth, white thallus and pinkish apothecia, some of which, when young, are reminiscent of *Petractis clausa* in having radiately fissured margins, distinguish this species in the field.

CYPHELIUM Ach. (1815)

Thallus crustose, verrucose to ± areolate, thick or thin, superficial or immersed, grey to greenish grey, yellow, greenish or brownish yellow. **Cortex** 10–35 μm thick, formed by interwoven hyphae, sometimes interspersed with small yellow or colourless crystals. **Photobiont** *Trebouxia*. **Ascomata** apothecia, black, sessile to immersed. **True exciple** of variable thickness, blackish brown, sometimes white-pruinose, rarely reduced laterally. **Hamathecium** of paraphyses, simple. **Asci** formed singly from ascogenous hyphae with croziers, at first clavate but becoming cylindrical or broadly clavate, with a single functional wall layer, disintegrating at an early stage, the spores being released into a dry black powery mass (mazaedium). **Ascospores** 1-septate, rarely submuriform, broadly oblong or ± globose, with a slight or pronounced constriction at septum; wall thick, dark brown, smooth or with a distinctive ornamentation of irregular cracks or parallel ridges. **Chemistry**: pulvinic acid derivatives (yellow pigments), together with unidentified compounds. **Ecology**: on bark and wood.

Calicium is similar but has stalked apothecia.

LITERATURE: Bailey (1974), Coppins & James (1974), Tibell (1971).

1 On wood or bark 2
On *Pertusaria* species 4

2(1) Apothecia immersed; true exciple thin, rim-like; thallus yellowish green 3
Apothecia sessile; true exciple thickened at base; thallus grey .... **1. *inquinans***

3(2) Acospores 1-septate .... **5. *tigillare***
Ascospores submuriform .... **2. *notarisii***

4(1) Apothecia sessile; ascospores at first coarsely spirally striate, later with irregular ridges and cracks; on *Pertusaria coccodes* .... **3. *sessile***
Apothecia immersed; ascospores with a few shallow fissures; on *Pertusaria pseudocorallina* .... **4. *marcianum***

1. C. inquinans (Sm.) Trevisan (1862)

Thallus verrucose to cracked-areolate, pale to dark grey, usually rather thick, sometimes almost immersed. Apothecia 0.4–1.6(-2.5) mm diam., black, sessile; true exciple thick, strongly thickened at the base, frequently with faint white pruina. Asci cylindrical; ascospores 14–19(-21) × 8–11 μm, uniseriately arranged, broadly ellipsoid, dark brown, with a finely striated surface and a few irregular cracks when mature. Thallus Pd− or + pale yellow, K+ yellow to reddish brown, C− (3 unidentified lichen products, including a yellow pigment, by t.l.c.).

Frequent on wood and acid bark (*Pinus* and *Quercus*), particularly old fence posts, gates and rails. S.E., C. and N.E. England, Scotland (Berwick and Highlands). Scandinavia, C. Europe, North & South America, Australia, New Zealand.

Characterized by the usually well-developed pale grey thallus, the abundant, sessile apothecia with the true exciple strongly thickened at the base and the medium-sized ascospores with a minute ornamentation of parallel ridges. When on bark, distinguished from *C. sessile* **(3)** by its larger apothecia and larger ascospores. In the field it may be confused with black-fruited species of *Lecidea* and *Lecidella*, but can readily be distinguished by the loose spore mass which leaves a black trace on finger or paper. Relatively tolerant of SO_2 air pollution, often occurring with *Lecanora conizaeoides*.

2. C. notarisii (Tul.) Blomb. & Forss. (1880)

Thallus verrucose to almost totally immersed, bright yellowish green. Apothecia 0.2–0.5(-0.8) mm diam., produced singly on thalline warts, immersed, not pruinose; true exciple, thin, rim-like, not thickened at the base. Asci 2- to 3-seriate, bursting at very early stages. Ascospores 13–32 × 11–19 μm, irregular in shape, 1-septate to submuriform. Thallus Pd−, K−, KC−, C− (rhizocarpic acid).

On wood, particularly weathered fence rails and gates, especially near the sea; rare. S.E. England. S. Sweden, Finland, C. Europe, North America.

Characterized by the vivid yellow-green, verrucose thallus, immersed apothecia and submuriform ascospores. In external morphology it is identical with *C. tigillare* **(5)**, which has only 1-septate ascospores and is restricted to ancient woodlands. *C. notarisii* probably has the most easterly distribution of any British bark- or wood-inhabiting lichen.

3. C. sessile (Pers.) Trevisan (1862)

Apothecia 0.25–0.6 mm diam, sessile, not pruinose; true exciple thick throughout, distinctly thickened at the base. Ascospores 10–15 × 6–9 μm, slightly constricted at septum; surface coarsely spirally striated, finally cracking irregularly, to give a complicated pattern of spiral ridges and irregular cracks.

On *Pertusaria coccodes*, more rarely *P. amara* and *P. albescens*, commensalistic to mildly pathogenic, on trunks of old oaks; local. S. & E. British Is. W. Europe northwards to S. Scandinavia.

This species causes considerable modification of the host thallus; it is usually darker in colour and the production of isidia and soredia of the host is partly suppressed. Specimens of *C. inquinans* **(1)** on bark have larger apothecia and ascospores. For differences from *C. marcianum*, also parasitic on *Pertusaria*, see **4.**

4. C. marcianum B. de Lesd. (1908)

Like **3**, but with immersed apothecia; ascospores 11–13 × 6–8 μm, with fewer, shallower fissures.

On *Pertusaria pseudocorallina*; very rare. Scotland (Argyll), Ireland (Mayo). France, Spain.

5. C. tigillare (Ach.) Ach. (1815)

Thallus verrucose, of variable thickness, bright yellowish green. Apothecia 0.2–0.6(-0.8) mm diam., immersed, not pruinose; true exciple thin, rim-like. Ascospores 17–21 × 9–11 μm, 1-septate, broadly ellipsoid, the surface smooth, without cracks. Thallus Pd −, K −, KC −, C − (rhizocarpic acid, epanorin and 2 unidentified compounds).

On wood, especially of conifers, and posts and rails in old forests; very rare. S. England (Hampshire, New Forest), Scotland (E. Highlands). Fennoscandia, C. Europe, Balkans, U.S.S.R., North America.

Characterized by the yellowish green, verrucose thallus, immersed apothecia and 1-septate ascospores. Morphologically very similar to *C. notarisii* **(2)** which has submuriform ascospores.

CYSTOCOLEUS Thwaites (1849)

Thallus filamentous, of elongated but contorted hyphae closely surrounding and adhering to single filaments of the photobiont, dark brown to black. **Photobiont** *Trentepohlia*. **Ascomata** and **conidiomata** not known. **Chemistry**: no lichen products detected by t.l.c. **Ecology**: vertical or overhanging siliceous rock faces.

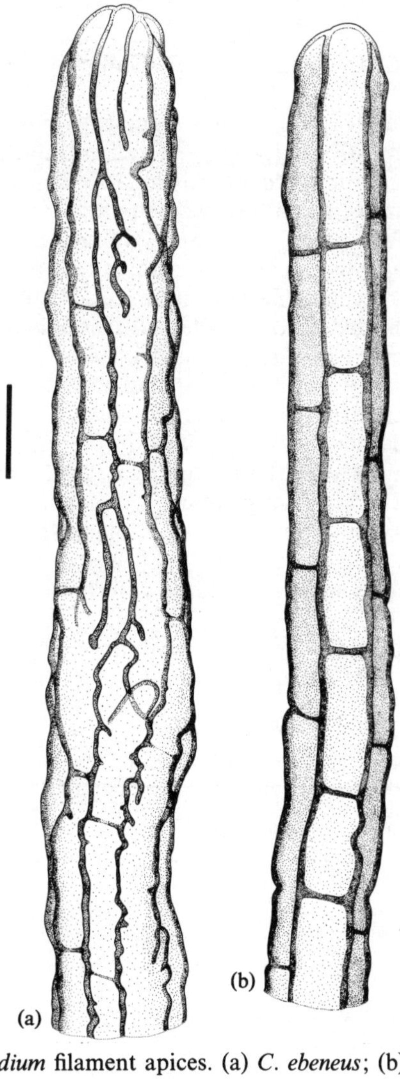

Fig. 13. *Cystocoleus* and *Racodium* filament apices. (a) *C. ebeneus*; (b) *R. rupestre*. Scale = 10 μm.

LITERATURE: Lindsay (1971, 1974), Santesson (1952), Schade (1932).

1. C. ebeneus (Dillwyn) Thwaites (1849)

C. niger (Huds.) Hariot (1890)

Thallus forming wefts or mats, often in small circular patches, the margin not delimited; filaments numerous, contorted, branched, the cells short and barrel-shaped, with strongly thickened walls (Fig. 13a).

On siliceous rock faces (rarely on tree trunks), in situations where the air is normally persistently humid but not directly wetted by rain; characteristic of the *Racodietum rupestris*, often growing intermixed with *Racodium rupestre*; mainly upland. S.W. & N. England (Dartmoor, Peak District, Pennines), Wales, W. Scotland, Ireland. Europe, North America, Chile, S. Georgia, S. Orkney Is., New Zealand, Antarctica.

Frequently confused with *Racodium rupestre* with which it may occur and only separated with certainty microscopically; differs in the contorted and narrow, nodular, shorter-celled hyphae which appear rather paler than those of *R. rupestre* under the microscope.

DEGELIA Arvidsson & D. Galloway (1981)

Thallus foliose, lobate or placodioid, ± round to irregular, ± loosely attached by a dense felt of rhizines; lobes to 1 cm wide, mostly contiguous, laterally sometimes ± imbricate, the apices rounded, entire or notched, sometimes secondarily lobate, schizidiate or isidiate, grey-blue to pale grey when dry. **Upper surface** smooth or with distinct longitudinal ridges, occasionally minutely scabrid. **Lower surface** pale, with dense, simple, entangled, blue-black rhizines, rarely extended as a prothallus. **Upper cortex** of compacted, periclinal or anticlinal hyphae. **Lower cortex** ± apparent or lax and merging into blue-black rhizines. **Photobiont** *Nostoc* (*Scytonema* in S. Hemisphere). **Medulla** of parallel, horizontally orientated hyphae. **Ascomata** apothecia, laminal, sessile, biatorine with a pale margin; disc concave at first, becoming flat or ± convex, matt, non-pruinose, ± red-brown, occasionally ± blackened. **True exciple** well developed, of radiating isodiametric paraplechtenchymatous cells. **Hymenium** yellow-brown above, I+ deep blue. **Hamathecium** of paraphyses, simple, not or slightly enlarged at apex. **Asci** elongate-clavate, 8-spored, clavate to subcylindrical, thickened at apex with a distinct I+ blue plug and occasionally an external amyloid cap. **Ascospores** simple, ellipsoid, colourless. **Conidiomata** pycnidia, laminal to marginal, wart-like, blackish. **Conidia** bacilliform, lateral or terminal on branching, short-celled conidiophores. **Chemistry**: lichen products not detected by t.l.c. **Ecology**: on bark and mossy rocks in oceanic environments; often a component of the *Lobarion*.

The genus has two main centres of world distribution, temperate S. Hemisphere (sect. *Degelia*) and amphi-Atlantic (sect. *Amphiloma* (Fr.) P. M. Jørg. & P. James).

LITERATURE: Jørgensen & James (1990)

1 Thallus with isidia or schizidia; apothecia absent or few 2
Thallus without isidia or schizidia; apothecia abundant .... **1. *plumbea***

2(1) With erect or unorientated, ± decumbent, ± dorsiventral, elongate, spathulate schizidia .... **3. *ligulata***
With knob-like, ± erect, terete, coralloid isidia .... **2. *atlantica***

1. D. plumbea (Lightf.) P. Jørg. & P. James (1990)

Parmeliella plumbea (Lightf.) Vainio (1890)

Thallus plate-like, monophyllous, thick and rigid, forming ± orbicular, placodioid patches 5–10 cm diam.; lobes mostly contiguous for entire length, rounded at apices, scarcely indented with thickened, ascending margins; upper surface blue-grey, occasionally brownish or olivaceous, with prominent longitudinal ridges and concentric, lateral, crescent-shaped curves, giving the lobes a scallop-like appearance; lower surface with a prominent, blue-black, beard-like hypothallus. Soredia and isidia absent, though occasionally ± knob-like lobules may develop. Apothecia 0.5–1 mm, usually numerous; disc red to dark brown with a paler true exciple, often becoming convex. Ascospores 17–25 × 7–10 μm, colourless, ellipsoid, smooth-walled; perispore absent. Conidiomata 0.1–0.2 mm diam. Thallus and medulla Pd−, K−, KC−, C− (lichen products not detected by t.l.c.).

On mossy, broad-leaved trees and rocks in ± open, relict woodlands and on maritime rocks. S. & W. British Is., locally abundant in N.W. Scotland and W. Ireland, rather rare elsewhere. Europe, North America, N. Africa, Macaronesia.

Readily recognized by the thick, monophyllous-placodioid, lead-blue, orbicular thallus. It may consist of small imbricate lobes which are occasionally rather knob-like, but these are more flattened than the isidia of *D. atlantica* **(2).**

2. D. atlantica (Degel.) P. Jørg. & P. James (1990)

Parmeliella atlantica Degel. (1935)

Like **1**, but knob-like or partially coralloid; marginal and laminal isidia, to 0.2 mm wide, occur on the upper surface, these may be very numerous and form a thick, irregular, cracked-areolate crust ± obscuring the thallus. Apothecia very rare, usually few per thallus when present.

On ± shaded, mossy, broad-leaved trees in cool, sheltered humid woodlands, more rarely on mossy rocks. W. British Is.; locally abundant in W. Scotland and Ireland, less frequent elsewhere.

3. D. ligulata P. M. Jørg. & P. James (1990)

Like **1**, but rarely fertile and with schizidia. Schizidia <1 mm, numerous, ± flattened, ligulate to spathulate, ± erect or randomly decumbent, usually blackened towards the apices and thus imparting an overall dark hue to the thallus. Apothecia <0.5 mm diam., very rare; disc brown ± blackened with age, with a paler true exciple.

In crevices in maritime rocks and at the bases of small trees and shrubs in dry but intermittently humid coastal lowlands, possibly dependent on onshore sea mists; rare. W. Scotland (Wigtown and Mid Ebudes), W. Ireland. Azores, Madeira.

A distinctive species differing from *D. atlantica* **(2)** in possessing elongate, spathulate schizidia. A newly recognized component of the Macaronesian element in the British flora, which also includes *Herteliana taylorii*, *Pseudocyphellaria lacerata* and *Nephroma tangeriense*.

DERMATOCARPON Eschw (1824)

Thallus foliose, single-lobed and attached to substratum by a usually central holdfast (peltate or umbilicate), or multi-lobed and attached by several, scattered holdfasts. **Upper surface** grey to dark brown, ± whitish pruinose, mostly smooth but with slightly raised, punctiform spots of the ostioles of ascomata or conidiomata. **Lower surface** smooth, verrucose or veined; rhizines absent (in W. European species); tomentum absent. **Upper cortex** pseudoparenchymatous, formed of adhering, vertically orientated hyphae with ± globose cells. **Lower cortex** of several rows of rounded to usually almost cuboid, vertically aligned, thick-walled cells. **Photobiont** chlorococcoid. **Medulla** of thin, filamentous hyphae. **Ascomata** perithecia, uniloculate, immersed, globose; involucrellum absent (Fig. 41a). **True exciple** colourless, reaching the surface by a short to rather long ostiolar neck, the neck often slightly protruding and brown pigmented. **Hamathecium** of numerous periphyses; hymenial gel I+ reddish or blue. **Asci** 8-spored, cyndrical-clavate to saccate, thick-walled, K/I−, *Verrucaria*-type; apical dome distinct in young asci and with an ocular chamber, but becoming reduced and almost absent as the ascus matures. **Ascospores** simple or rarely a few thinly 1- to 3-septate, globose to narrowly ellipsoid or ovoid, colourless, smooth, without a conspicuous perispore. **Conidiomata** pycnidia, immersed, similar in size to perithecia, wall colourless, internally pseudoparenchymatous with irregular locules lined with poorly differentiated conidiogenous cells. **Conidia** bacilliform, simple, colourless. **Chemistry**: lichen products not detected by t.l.c. **Ecology**: on hard acid or calcareous rocks, often by rivers or lakes or in seepage tracks; cosmopolitan, but mainly in cool temperate regions.

Species of *Catapyrenium* are distinguished by their lower surface having numerous rhizines or a tomental hyphal mat. The presence of one or more of the following characteristics will distinguish sterile specimens of *Lasallia* or *Umbilicaria* species from *Dermatocarpon*: lower surface with rhizines or black thalloconidia; upper surface or margin with isidia, erose sorediate areas or cilia; photobiont trebouxioid; detectable lichen products in the medulla (by spot tests or t.l.c.).

Care must be taken to distinguish a true pruina, which is not water-soluble, from clusters of water-soluble crystals often formed on the thallus surface in the herbarium.

The delimitation of many species in the genus is still unclear, and a modern monographic revision is required.

LITERATURE: Degelius (1934), Thomson (1984), Zschacke (1933).

1 Thallus attached by a solitary, central holdfast, single-lobed, or sometimes with secondary lobes which are not attached to the substratum directly 2
Thallus multi-lobed, attached by several, scattered, sometimes almost marginal holdfasts 6

2(1) Lobes 3–15(-23) mm wide 3
Lobes mostly >20 mm wide 4

3(2) Thallus blackish below; ascospores 10–18(-25) μm long, ellipsoid to narrowly ellipsoid .... **5. *meiophyllizum***
Thallus pale brown below; ascospores 6–7(-9) μm long, broadly ellipsoid to globose .... **3. *leptophyllum***

4(2) Upper surface not pruinose; lower surface with raised plicate ridges or veins 5
Upper surface whitish pruinose; lower surface smooth to verrucose, very rarely ridged or veined .... **6. *miniatum***

5(4) Thallus 0.4–0.7 mm thick; ascospores 13–15 μm long .... **1. *arnoldianum***
Thallus 0.1–0.3 mm thick; ascospores (13-)15–21(-23) μm long .... **7. *rivulorum***

6(1) Thallus bright green when wet; medulla Melzer's I+ brown-red .... **4. *luridum***
Thallus scarcely changing colour when wet (never bright green); medulla Melzer's I– .... **2. *intestiniforme***

1. D. arnoldianum Degel. (1934)

Thallus single-lobed, to *c.* 50 mm diam., often irregularly incised, rigid, 0.4 to 0.7 mm thick; upper surface greyish brown, not pruinose, green when wet; lower surface pale brown, roughened at first, later folded and veined; cells of upper cortex *c.* 8.5 μm diam. Perithecia 0.25–0.45 mm diam. Asci subcylindrical. Ascospores 13–15 × 4.5–6.5 μm, oblong-ellipsoid.

On semi-inundated boulders in streams and at margins of upland lakes. N.W. Scotland (Is. of Mull). N. & C. Europe, North America (Alaska).

Characterized by the non-pruinose, thick, single-lobed thallus which eventually has a conspicuously veined underside. *D. rivulorum* **(7)** has a thinner thallus and longer ascospores. The British material referred to these two species is, however, scant and merits critical re-evaluation.

2. D. intestiniforme (Körber) Hasse (1912)

Thallus multi-lobed; lobes crowded, mostly 3–10 mm wide, 0.15–0.3(-0.4) mm thick. Upper surface grey-brown to dark brown, little changed when wet, at least young lobes usually thinly blue-white-pruinose; lobe margin at first ascending but later becoming inrolled giving the centre of the thallus a contorted, intestine-like appearance; underside smooth, pale to more usually dark brown, often extending to give the margin of erect lobes a blackish rim; holdfasts scattered, in the middle of the lobes or frequently close to the margins. Perithecia 0.2–0.25 mm diam. Asci 35–50 × 13–16 μm, clavate. Ascospores 9–12(-15) × (4.5-)6–7(-9) μm, broadly ellipsoid, ellipsoid or ovoid.

On ± basic, steeply sloping rocks in seepage tracks or on rocks in the splash zone of lake shores [above the level of *D. luridum* **(4)**], sea-level to *c.* 800 m alt.; local. N.E. England (Teesdale, Cleveland), N. Scotland (Highlands). N. & C. Europe, North America.

Non-pruinose specimens can be distinguished from *D. luridum* **(4)** by the negative reaction of the medulla with Melzer's iodine. Distinguished from multi-lobed morphs of *D. miniatum* **(6)** by the more numerous holdfasts on the usually darker underside, the often conspicuously blackened rim of the erect lobes, and the slightly thinner thallus.

3. D. leptophyllum (Ach.) G. Lång (1912)

Thallus single-lobed, to 10 mm diam., basically circular but often notched-lobed, 0.2–0.4 mm thick, with a solitary central holdfast; upper surface pale grey to brownish, ± thinly pruinose. Ascospores 6–7(-9) × (5-)6–7 μm, globose to broadly ellipsoid.

On rocks and boulders in rivers and lakes; rare. N. England, C. Wales, Scotland. N. Europe.

The name *D. leptophyllum* **(3)** has often been used to include a range of small, single-lobed taxa, the most common of which is *D. meiophyllizum* **(5)**. Included in *D. leptophyllum* **(3)** here are specimens with a pale underside and short ascospores, but more than one species may be involved. A collection from N. England (Northumberland), on calcareous rocks, has a Melzer's I+ brown-red medulla. A species, possibly referable to *D. leptophylloides* (Nyl.) Zahlbr. (1922), growing in Mid-Wales (R. Wye) has lobes that become sinuately incised and overlapping, pale grey above and pale to mid-tan below, and longer ascospores, 13–19 × 4.5–6.5 μm.

4. D. luridum (With.) Laundon (1984)

D. weberi (Ach.) Mann (1825)

Thallus multi-lobed, flabby (especially when wet), mat; lobes mostly 5–15 mm wide, 0.15–0.4 mm thick, often convex, especially in the centre of the mat; upper surface greyish brown to dark brown, but bright green when wet, not pruinose but often developing clusters of white crystals in the herbarium; lower surface dark brown but sometimes with paler patches, mostly smooth. Perithecia 0.2–0.3 mm diam. Asci 60–70 × 14–17 μm, clavate. Ascospores (10-)12–19(-23) × 5–8 μm, ellipsoid to usually narrowly ellipsoid. Medulla Melzer's I+ brown-red.

On rocks and boulders in lakes, streams and rivers, or in seepage tracks, usually submerged or inundated for most of the year; locally common. N. & W. Britain, Ireland. Temperate to arctic N. Hemisphere, New Zealand.

Distinguished from other multi-lobed species by the bright green colour when wet, and the brown-red reaction of the medulla with Melzer's iodine. The former character, and its occurrence in wet habitats, can lead to it being mistaken for a thallose liverwort.

5. D. meiophyllizum Vainio (1921)

Thallus single-lobed, 3–12(-23) mm diam., 0.2–0.45 mm thick, attached by a solitary central holdfast; lobes often crowded or overlapping, with adpressed or rarely ascending margins, circular or often notched-lobed; upper surface pale grey-brown to dark brown, little changed when wet, not pruinose; lower surface dark brown to black, smooth. Perithecia 0.2–0.4 mm diam. Asci 55–75 × 12–16 μm, clavate to cylindrical. Ascospores 10–18(-25) × 5–8(-9.5) μm, narrowly ellipsoid.

On siliceous rocks and boulders in streams, rivers and lake shores, frequently inundated; rather rare. N. & W. British Is. Europe, N.W. Himalaya.

D. leptophyllum **(3)** has a pale underside and shorter ascospores. In Mid-Wales (R. Wye), it occurs with an unidentified species, also with long ascospores, but with a pale underside — see under *D. leptophyllum*.

6. D. miniatum (L.) Mann (1825)

Thallus mostly single-lobed, ± circular but margin often notched or torn, usually concave with raised margin, 10–70 mm diam., (0.2-)0.3–0.5(-0.7) mm thick, sometimes appearing multi-lobed [var. *complicatum* (Lightf.) Hellbom (1870)], due to formation of secondary (?regeneration) lobes, such clusters of lobes always attached by a solitary holdfast; upper surface reddish brown to grey-brown, grey-white pruinose; lower surface pale tan-brown, sometimes with dark grey-brown blotches, smooth or warted, very rarely ridged or veined. Perithecia 0.2–0.3 mm diam. (Fig. 41a). Asci 50–67 × 9–12 μm, cylindrical. Ascospores (8-)9–12(-14) × (4.5-)5–6(-7) μm, ellipsoid or ovoid-ellipsoid.

On ± vertical rock surfaces, usually on cliff and crag faces of hard limestones, occasionally on siliceous rocks near sea and lake shores; usually on dry rocks, sometimes subjected to periodic flushing or spraying; mostly at low altitudes, very rare above 500 m. W. Britain, Ireland. Europe, North America, W. Himalaya, Japan, Australia.

Usually easily recognized by the large, single-lobed thallus. Multi-lobed morphs need care to separate them from *D. intestiniforme* **(2)**; most montane records of the var. *complicatum* refer to that species.

7. D. rivulorum (Arnold) Dalla Torre & Sarnth. (1902)

Thallus single-lobed, 30–50 mm diam., 0.1–0.3 mm thick, attached by a central holdfast; upper surface dark grey-brown, olive-dark brownish when wet, not pruinose; lower surface dark grey-brown, slightly to distinctly ridged or veined; cells of upper cortex *c.* 6.5 μm diam. Perithecia 0.2–0.35 mm diam. Asci 50–60 × 10–15 μm, subcylindrical. Ascospores (13-)15–21(-23) × 6.5–8.5 μm, narrowly ellipsoid, sometimes appearing thinly 1- or 3-septate.

On siliceous rocks and boulders by streams; usually in snow-melt streams outside the British Is. N.C. Scotland (Perth, Ben Lawers). Scandinavia, C. Europe, Greenland, North America (Alaska).

See under *D. arnoldianum* **(1)**.

DICTYONEMA C. Ag. (1821)

Thallus absent. **Basidiomata** matt-like to foliose, often bracket-like; upper side blue- to grey-green, villose to fibrillose due to irregularly to radially arranged filaments of the cyanobacterial photobiont; hyphae without clamp connections; lower side ('hymenophore') white, cream or buff, smooth to ± tomentose, continuous or broken up. **Photobiont** filaments surrounded by a coating of longitudinally arranged hyphae. **Basidia** with 4, ± curved, slender sterigmata. **Basidiospores** colourless, simple, thin-walled, smooth, I−. **Chemistry**: no lichen products reported. **Ecology**: on trees, mossy rocks and on the ground; 6 species, mostly tropical and subtropical.

The only consistently lichenized bracket fungus known in the British Is.

LITERATURE: Coppins & James (1979c), Parmasto (1978).

1. D. interruptum (Carm. ex Hook.) Parm. (1978)

Basidiomata crustose or of ± bracket-shaped lobes to 1 cm wide. Photobiont *Scytonema*, filaments 14–17(-22) μm wide, each surrounded by a 3–5 μm thick coat of longitudinally arranged hyphae; hyphae in surface view 6–9 μm wide, with interlocking, sinuate edges. Free ('generative') hyphae 4–7.5 μm wide, thin- to thick-walled (wall to 1 μm) also present; subhymenial hyphae ± vertically orientated, mostly dichotomously branched, 4–8 μm wide; a few clamp connections present on generative and subhymenial hyphae. Basidia 12–20 × 5–7(-8) μm. Basidiospores subnavicular-obovoid, 6–7.7 × 3.5–4 μm.

Amongst bryophytes on sheltered tree trunks and rocks in moist, relict woodland; very rare. S.W. Ireland (N. Kerry), formerly also in S. England, S. Wales, W. Scotland (Argyll, Appin). Azores.

The thallus-like and well-formed brackets of the lichenized basidiome are rarely developed by *D. interruptum* which, in consequence, is most likely to be overlooked as flattened, interwoven tufts of free-living *Scytonema*.

DIMERELLA Trevisan (1880)

Thallus crustose, superficial, ecorticate, thin and often inconspicuous, membranous or scurfy granular, pale grey. **Photobiont** *Trentepohlia*. **Ascomata** apothecia, sessile, marginate at least when young, concave-urceolate to flat, whitish yellow to orange-red. **Thalline exciple** absent. **True exciple** colourless, mostly pseudoparenchymatous. **Hymenium** I+ blue. **Hamathecium** of unbranched, distinctly septate paraphyses with ± swollen apical cells. **Asci** 8-spored, narrowly cylindrical, wall K/I+ pale blue, thin, not thickened at apex and without internal apical structures. **Ascospores** fusiform to ellipsoid, 1-septate, colourless, smooth, without a perispore. **Conidiomata** pycnidia, ½- to ¾-immersed, pale dull yellowish; wall colourless. **Conidiogenous cells** 5–9 × 1.8–2.5 µm, arising singly or several together from branched supporting cells, cylindrical or lageniform. **Conidia** ± oblong or ellipsoid, simple, sometimes biguttulate, colourless. **Chemistry**: no lichen products detected by t.l.c. **Ecology**: in humid, shaded or sheltered habitats, mostly on acid bark, sometimes over mosses.

Widely distributed in moist temperate and tropical regions; many tropical species are foliicolous. The British species are commonly sterile, but with pycnidia.

1 Apothecia 0.4–2 mm diam.; disc orange-yellow; ascospores (6-)8–11 µm; conidia 3–4(-5) × 1.3–1.8 µm .... **1. *lutea***
Apothecia 0.2–0.4 mm diam.; disc whitish to dull yellowish or pinkish; ascospores 9–14 µm; conidia 6–7(-8) × 1.8–2.6 µm .... **2. *pineti***

1. D. lutea (Dickson) Trevisan (1880)

Apothecia 0.4–2 mm diam., distinctly constricted at the base; disc concave at first but soon flat or slightly convex, bright orange-yellow (rarely paler), with paler and often flexuous margin. Paraphyses 1.5–2.7 µm wide, the apical cell(s) to 4.5 µm. Ascospores (6-)8–11 × 2–3.5 µm. Pycnidia 150–260 µm diam.; conidia 3–4(-5) × 1.3–1.8 µm, ellipsoid.

Mostly on bryophytes on trees and over siliceous rocks and accumulated soil/debris, in humid, shaded situations, mostly in *Lobarion* communities; locally frequent. S. & W. British Is. Pantropical, extending into suboceanic temperate regions in both Hemispheres.

Problematical, immature or sterile specimens can be distinguished from *D. pineti* **(2)** by measurement of paraphyses or conidia.

2. D. pineti (Ach.) Vězda (1975)

D. diluta (Pers.) Trevisan (1880)

Apothecia 0.2–0.4 mm diam., sessile, ± contricted at base; disc concave to flat, whitish to dull yellow or pinkish yellow, with concolorous or paler exciple. Paraphyses 1–2 µm wide, the apical cell(s) to 4 µm. Ascospores 9–14 × 2–3.5 µm. Pycnidia 100–220 µm diam.; conidia 6–7(-8) × 1.8–2.6 µm, oblong, biguttulate, with ± clear median constriction.

On shaded, ± acid bark, at tree bases or in deep crevices, more rarely on wood, mosses, rocks or old leather, usually in woodlands; common, often overlooked in deeply shaded niches. Throughout British Is. Temperate N. Hemisphere.

Can be confused with immature *D. lutea* **(1)** and with *Absconditella* species which, however, have an I− (or reddish) hymenium, indistinctly septate paraphyses, a distinct apical dome in their asci, and lack *Trentepohlia* as the photobiont.

DIPLOICIA Massal. (1852)

Thallus crustose, placodioid, forming rosettes, glaucous or greyish white; prothallus absent. **Upper cortex** pseudoparenchymatous, packed with minute crystals, K−, not dissolving. **Photobiont** chlorococcoid. **Ascomata** apothecia, at first immersed within thallus, emergent; disc black or brown-black, not pruinose. **Thalline exciple** absent. **True** exciple thin, concolorous with disc, often occluded. **Epithecium** granular, brown-black. **Hamathecium** of simple paraphyses with swollen and darkened apices. **Hypothecium** brown-black. **Asci** 8-spored, clavate, *Lecanora*-type. **Ascospores** polarilocular, brown, ellipsoid. **Conidiomata** pycnidia, immersed in thallus; wall colourless but brown around the ostiole. **Conidiogenous cells** enteroblastic, intercalary, branched. **Conidia** arising acrogenously and pleurogenously, bacilliform, simple, colourless.

1. D. canescens (Dickson) Massal. (1852)

Thallus to 6 cm diam., often smaller and coalescing, ± crustose at the centre with radiating, closely contiguous, apparently pleated, convex marginal lobes, glaucous-grey-white, sometimes becoming darker, more verruculose and ± cracked and areolate towards the centre; lobes 0.5–1 mm wide, wider at margins, matt, densely white-pruinose, sorediate. Soralia scattered, coalescing to form erose patches, mainly laminal; soredia white or concolorous-granular, ± farinose. Apothecia 0.3–0.9 mm diam., rare; true exciple dark brown, opaque, with marginal zone of crystals. Ascospores (9-)11–13(-15) × (4.5-)5.5–6(-7.5) μm. Pycnidia rare; conidia 6–7 × 1 μm. Thallus Pd−, K ±yellow, KC−, C−, UV+ dull orange or UV− (atranorin, chloratranorin, diploicin and a xanthone).

On rocks, walls and trees in mainly dry, often ± sheltered, nutrient-enriched habitats, also on birds' perching rocks, in the xeric-supralittoral on sheltered coasts, limestones, and calcareous, man-made substrata; frequent especially in churchyards in the south, rarer and more restricted to dry crevices in coastal rocks and nutrient-rich trees in the north. Throughout lowland British Is. Atlantic - subatlantic temperate Europe, Mediterranean region, New Zealand.

Resembles *Solenopsora candicans* which is not sorediate, has apothecia with a thalline exciple, 1-septate, colourless ascospores, a different chemistry and is restricted to hard limestones.

DIPLOSCHISTES Norman (1853)

Thallus crustose, continuous to cracked-areolate, grey-white, yellowish in some non-British species, smooth to verrucose, often pruinose; medulla I+ bluish. **Photobiont** *Trebouxia*. **Ascomata** apothecia, at first usually ± perithecia-like, later urceolate, immersed; disc black, sometimes pruinose. **Thalline exciple** present, sometimes scarcely distinguished from the thallus or thick and pruinose. **True exciple** united with the thalline exciple, dark brown to black, of thick-walled, swollen to ± globose hyphae immersed in a matrix below, extended into pale brownish periphysoids on upper inside margin, remotely septate, not swollen apically, often appearing as a fringe in surface view, and presumed to close the ascomatal opening under xeric conditions. **Epithecium** ± colourless to black, sometimes with crystal inclusions. **Hymenium** colourless, I−. **Hypothecium** colourless to dark brown or blackish. **Hamathecium** of flexuous, ± unbranched paraphyses, remotely septate, the apices not swollen, sometimes brownish. **Asci** elongate-clavate to subcylindrical, the wall ± evenly thickened when mature, with a somewhat abrupt apical thickening with a thin, internal apical beak or a downward, convex swelling when young, lacking any apical apparatus, the contents I+ orange-red, the walls I−, not fissitunicate, (1-)4- to 8-spored. **Ascospores** broadly ellipsoid, brown to dark brown or purplish black, muriform, smooth, lacking a distinct perispore or gelatinous sheath, I− or I+ bluish. **Conidiomata** pycnidia, in slightly raised warts, black. **Conidiogenous cells** simple or branched at the base, elongate-ampulliform; conidia arising apically. **Conidia** bacilliform to elongate-ellipsoid, truncated to apiculate at the base, simple, colourless. **Chemistry**: *para*-depsides (e.g. lecanoric and diploschistesic acids); all British species have C+ red thalli; some species with gyrophoric, constictic or norstictic acids. **Ecology**: on rocks, soil, mosses or lichens. Cosmopolitan, especially in xeric situations, and including about 35 species.

The ascospores are at first colourless and only transversely septate, both the pigmentation and septation developing progressively as they mature.

Several species in this genus are of particular biological interest in that they are initially lichenicolous fungi. When the ascospores alight on a suitable host, as the fungus develops it parasitizes and kills the mycobiont in the existing "host" lichen, usurping its photobiont and using this to develop its own characteristic lichenized thallus on the site of the initial host lichen.

LITERATURE: Lettau (1932–7), Lumbsch (1988, 1989).

1 Apothecia urceolate, with ± open discs, >1 mm diam.; thallus uneven, often verrucose, areolate or not, white to grey 2

Apothecia ± perithecioid, the disc scarcely visible from above, 0.1–0.2 mm diam. in surface view; thallus ± smooth, regularly areolate, usually lead-grey to bluish grey .... **1. *caesioplumbeus***

2(1) Thallus whitish grey to pale or dark grey, not densely pruinose, K− or K+ yellow or red; hypothecium dark brown to black; on mosses or siliceous rocks 3
Thallus greyish white to chalk-white, densely pruinose, K−; hypothecium colourless; on calcareous rocks .... **2. *gypsaceus***

3(2) Thallus ± continuous, never with a yellow tinge, uneven, verrucose, K− or K+ yellow or red; asci 4-spored; ascospores with 5 transverse and 1–2 longitudinal septa; on *Cladonia* thalli, mosses or soil .... **3. *muscorum***
Thallus cracked-areolate, often with a yellow tinge, ± even to verrucose, K+ yellow; asci (4-)8-spored; ascospores with 5–7 transverse and 2–3 longitudinal septa; on siliceous rocks .... **4. *scruposus***

1. D. caesioplumbeus (Nyl.) Vainio (1921)

Thallus ± areolate, lead-grey to bluish grey, the areoles regular, smooth, ± flat, not pruinose, often ± shining. Apothecia 0.1–0.2 mm diam., ± perithecioid, immersed, the opening depressed, the marginal excipular hairs visible as a whitish fringe around the opening; hymenium 170–200 μm tall; hypothecium colourless. Asci 100–120 × 15–20 μm, (4-)8-spored. Ascospores (21-)30–45(-54) × 11–24 μm, with (4-)5–9 transverse and 2–5 longitudinal septa. Pycnidia immersed, recalling the apothecia; conidia 5–7 × 1–1.5 μm. Thallus Pd−, K−, KC+ red, C+ red, UV− (lecanoric acid).

On sunny, often ± nutrient-rich, siliceous rocks on seashores in the xeric-supralittoral; local. W. & S.W. Britain. W. Europe, Mediterranean region, Morocco.

Initially lichenicolous on *Lecanora gangaleoides*, killing the mycobiont and taking over its photobiont to produce an independent, lichenized thallus.

The thallus colour and immersed ascomata recall species of the *Aspicilia gibbosa* group which, in addition to the microscopic characters, differ in that the ascomatal opening lacks a whitish fringe and the thallus is C−. *D. caesioplumbeus* has sometimes been confused with the Mediterranean *D. actinostomus* (Pers.) Zahlbr. (1923), which has a consistently white-grey to greyish thallus, generally shorter spores, 18–32 μm in length, and is not known to be initially lichenicolous.

2. D. gypsaceus (Ach.) Zahlbr. (1892)

Thallus cracked to ± areolate, grey-to chalk-white, densely pruinose. Apothecia 2–5 mm diam., urceolate; disc pruinose; hymenium 120–190 μm tall; hypothecium colourless. Asci 90–130 × 20–30 μm, 4-spored. Ascospores 25–40 × 10–17 μm, with 4–7 transverse and 1–2 longitudinal septa. Thallus Pd−, K−, C+ red, UV− (lecanoric acid).

On hard, often vertical, calcareous rocks, especially limestones, in somewhat shaded and damp situations, associated with *Gyalecta jenensis*; rare. Almost throughout Great Britain, W. Ireland (Galway, Connemara). Eurasia, North and South America, North Africa, Australia, New Zealand.

3. D. muscorum (Scop.) R. Sant. (1980)

Thallus whitish to dark grey, ± continuous, not areolate, the surface uneven, ± coarsely verrucose, not pruinose. Apothecia 1–2 mm diam., urceolate; disc pruinose; thalline exciple not or slightly raised above the thallus; hymenium 75–120 μm tall; hypothecium dark brown. Asci 65–80 × 12–15(-20) μm, 4-spored. Ascospores (20-)25–35(-40) × (8-)12–15(-18) μm, with 5 transverse and 1–2 longitudinal septa. Thallus Pd−, K− or K± yellow or red (diploschistesic acid always detectable by t.l.c.), C+ red, UV− (diploschistesic and lecanoric acids).

Initially parasitic on *Cladonia* squamules and podetia, especially of *C. pocillum* and *C. rangiformis*, which at first are scarcely modified, apart from the presence of the apothecia and C+ red reaction imparted to the podetia or squamules; granules developing an independent thallus and spreading over adjacent mosses. Calcareous ground, wall tops or base-rich dunes, rarely trees; local. Throughout British Is. Europe, North and South America, N. & S. Africa, Asia, Australia, New Zealand.

4. D. scruposus (Schreber) Norman (1853)

Thallus pale to dark grey, often with a yellowish or yellow-green tinge when in shaded habitats, ± continuous to coarsely cracked-areolate, the surface ± uneven to verrucose, not pruinose. Apothecia urceolate, (1-)1.5–3 mm diam.; disc ± pruinose; thalline exciple ± raised above the thallus; hymenium 90–125 μm tall; hypothecium dark brown to black. Asci 90–110 × 20–25 μm, (4-)8-spored. Ascospores

(20-)22–40 × 10–18 μm, with (4-)5–7 transverse and (1-)2–3 longitudinal septa. Pycnidia crateriform; conidia bacilliform, 4–6 × 1 μm. Thallus Pd−, K− or K+ yellow, C+ red, UV− (diploschistesic and lecanoric acids, or lecanoric acid alone; both chemotypes occur in Great Britain).

Siliceous or somewhat basic rocks and walls, often in nutrient-enriched sites; frequent. Throughout British Is. Europe, North and South America, N. Africa, Asia, Australia, New Zealand.

Some British records refer to *D. muscorum* **(3)**, which, until recently, was often regarded as a variety of *D. scruposus* by many authors [i.e. as var. *bryophilus* (Ehrh. ex Ach.) Müll. Arg. (1892)].

DIPLOTOMMA Flotow (1849)

Thallus crustose, superficial, delimited, rarely effuse, pale to dark grey, continuous or ± rimose, immersed if on lichens. **Cortex** densely packed with minute crystals, not dissolving in K. **Photobiont** chlorococcoid. **Ascomata** apothecia, immersed at first, emergent. **Thalline exciple** sometimes present, pale, thin, crenate, spurious, concolourous with disc. Disc black, not or ± white-grey-pruinose. **True exciple** poorly developed, continuous with the hypothecium. **Epithecium** red-brown. **Hymenium** colourless, I+ blue. **Hypothecium** pale to dark brown, semi-opaque. **Hamathecium** of simple paraphyses with enlarged, pigmented apical cells. **Asci** 8-spored, ± ellipsoid-clavate, *Lecanora*-type. **Ascospores** septate or submuriform, with internally thickened walls (distoseptate), becoming red-brown, with minutely ornamented cells, the lumens rounded. **Conidiomata** pycnidia, pale to dark brown above, colourless below. **Conidia** ellipsoid, colourless, simple. **Ecology**: on basic or nutrient-enriched bark, also calcareous rocks, mortar and walls, more rarely asbestos-cement, or parasitic on lichens.

Differs from *Buellia* in the crystalline-granular cortex, immersed apothecia, septate to submuriform, red-brown spores, with rounded lumina, and the distoseptate ascospores. *Rhizocarpon* may have rather similar ascospores but the asci are of a distinctive type.

1 Thallus K− (no crystals), Pd− 2
Thallus K+ red (crystals), Pd+ orange .... **4. *chlorophaeum***

2(1) Not on lichens 3
On *Caloplaca teicholyta* .... **3. *murorum***

3(2) Ascospores soon becoming submuriform; on nutrient-rich bark and calcareous rocks .... **1. *alboatrum***
Ascospores septate, never submuriform; on calcareous rocks .... **2. *epipolium***

1. D. alboatrum (Hoffm.) Flotow (1849)

Thallus thin to thick, often wide-spreading, white or pale to dark grey, smooth or rimose-cracked or granular, ± determinate; dark prothallus absent. Apothecia 0.3–0.8(-1.5) mm diam.; disc at first flat, pruinose, later ± convex, and not pruinose; thalline exciple sometimes present, spurious, white, sometimes ± crenulate; true exciple inconspicuous; epithecium brown; hymenium 45–75 μm, tall, colourless; hypothecium brown. Asci (40-)55(-70) × (10.5-)17(-25) μm. Ascospores (11-)15–20(-30) × (5.5-)8–10(-17) μm, at first (1-)3-septate, mostly submuriform at maturity. Thallus Pd−, K−, KC−, C−, UV− (no lichen substances detected by t.l.c.).

On nutrient-rich bark, especially *Ulmus*, and *Fraxinus*, calcareous rocks and mortar. Throughout British Is. Europe, North America, Australia, New Zealand.

2. D. epipolium (Ach.) Arnold (1869)

Like **1**, but ascospores (1-)3-septate, lacking longitudinal septa, and more frequently curved.

On calcareous rocks; local. Throughout British Is. Europe, North America.

The taxonomic status of this species is uncertain; possibly only a morph of *D. alboatrum* **(1)**.

3. D. murorum (Massal.) Coppins (1980)

Like **1**, but thallus inapparent to ± continuous and powdery-superficial. Apothecia ± prominent, soon becoming convex, very pruinose. Ascospores 20–22 × 7.5–9 μm.

On *Caloplaca teicholyta*. Rare. S. England (Oxfordshire), W. Ireland (Galway). S. Europe.

4. D. chlorophaeum (Hepp ex Leighton) Szat. (1956)

Thallus thick, white to ochraceous, regularly rimose or warted, or thin and pale grey, cracked-areolate or disappearing, ± determinate; dark prothallus absent. Apothecia 0.2–0.6 mm diam.; disc flat, non-pruinose, later becoming convex; thalline exciple spurious, disappearing; true exciple indistinct; epithecium brown; hymenium (80-)105(-115) μm tall; hypothecium brown. Asci (40-)85(-80) × (13-)16(-20) μm. Ascospores (14.5-)18.5(-27.5) × (7.0-)10.0(-12.5) μm, 3-septate to submuriform. Medulla Pd+ yellow-orange, K+ yellow-red (crystals), KC−, C−, I± blue (norstictic acid).

On sheltered, slightly basic, maritime rocks, sea-walls, sandy mortar; locally frequent. W. British Is. Europe, Azores.

Distinguished by the reactions with Pd and K.

DIRINA Fr. (1825)

Thallus crustose, superficial, effuse or delimited, surface becoming cracked, often slightly verruculose or areolate, usually ± pruinose; prothallus present. **Upper cortex** with anticlinally arranged, colourless hyphae immersed in a pale yellow-brown, gelatinous matrix, often with crystals; epicortex absent. **Photobiont** *Trentepohlia*. **Ascomata** apothecia, discrete or aggregated, sessile. **Thalline exciple** present. **True exciple** thin or inconspicuous. **Epithecium** brownish. **Hymenium** colourless. **Hypothecium** black, opaque, or dark brown, clearly defined. **Hamathecium** of paraphysoids, branched and interwined, apices smooth or ± coloured, ornamented. **Asci** 8-spored, clavate, fissi- or semi-fissitunicate, the apex thickened with a small, internal I+ blue ring, *Opegrapha*-type. **Ascospores** 3-septate, colourless, fusiform, straight or ± curved. **Conidiomata** pycnidia, immersed or slightly elevated like black or dark brown dots. **Conidia** thread-like and sickle-shaped. **Chemistry**: erythrin, lecanoric acid and an unidentified substance. **Ecology:** dry underhangs in calcareous and siliceous environments.

LITERATURE: Tehler (1983).

1. D. massiliensis Durieu & Mont. (1847)

D. repanda (Fr.) Nyl.(1836)

Thallus thick, chalk-like, usually smooth, but sometimes uneven or verruculose or ± rimose-areolate, white to brown-grey, often with a violet tinge; cortex ± pruinose, densely packed with small crystals; prothallus brown or black-brown. Apothecia 0.5–3 mm diam., ± sessile; disc whitish dark grey, concolorous with the thallus, ± white-pruinose; thalline exciple smooth, at first regular, often becoming ± undulate, sometimes contorted; hymenium 50–140 μm tall. Paraphysoids 1–2 μm wide, apices 1–3 μm wide. Asci 70–120 × 12–18 μm. Ascospores 20–24 × 4–6 μm. Thallus (cortex) Pd−, K−, KC+ red, C+ red, UV−; medulla Pd−, K−, KC−, C−, UV+ pale yellow (erythrin, ± lecanoric acid, UV+ yellow and violet unknowns).

On dry, hard limestones, rarely siliceous rocks, often in underhangs, very rarely on mortared walls, on ± vertical cliffs and rocky outcrops on or near the coast; rare and local, very rarely inland. S. and S.W. Britain to N. Wales. W. Europe, Mediterranean region.

The **f.** ***sorediata*** (Müll. Arg.) Tehler (1983) [syn. *D. repanda* f. *stenhammarii* (Fr. ex Stenhammar) Clauz. & Roux (1975)] has soralia to 0.4 mm diam., usually paler than thallus, pale brown-grey, yellowish or whitish grey, at first punctiform, becoming confluent and spot-like, flat or convex; soredia 0.02–0.05 mm diam., coarsely granular. Cortex UV− or glaucous, the medulla UV+ glaucous or rarely ± yellowish. On dry, vertical, or overhanging, calcareous rocks, plaster or mortared walls, especially on the E. and N. sides of churches, also on coastal siliceous rocks; throughout the British Is.

On hard limestones, the thallus of f. *sorediata* is often thick, chalk-like, continuous and densely packed with calcium oxalate crystals, whereas on siliceous rocks it is often thin and ± dispersed; possibly more than a single species may be involved. Forma *sorediata* is sometimes given specific status on account of its different and more extensive distribution than *f. massiliensis*, as well as its wider ecological amplitude. Two UV+ substances are present in the medulla in both taxa; in *f. sorediata* the UV+ glaucous substance tends to predominate, whereas in *f. massiliensis* the UV+ yellow substance is usually dominant.

The *f. sorediata* resembles the rare, coastal *Arthonia endlicheri*, which is also C+ red but contains lecanoric acid and has a UV− medulla.

EIGLERA Hafellner (1984)

Thallus crustose. **Photobiont** chlorococcoid. **Ascomata** apothecia, immersed, *Aspicilia*-like. **True exciple** scarcely evident except in upper part where it is dark blue-green. **Hamathecium** of paraphyses, the apices sometimes broader or clavate, without pigmented caps or hoods, slender, septate, sparingly branched. **Asci** 8-spored, broadly clavate or ± ellipsoid, with a uniformly I+ blue apical dome and I+ blue outer fuzzy coat forming a thick cap over the apex. **Ascospores** simple, colourless, smooth, without a thickened perispore. **Conidiomata** pycnidia, immersed, blackish; wall dark blue-green. **Conidiogenous cells** arising in a single row, ampulliform, enteroblastic. **Conidia** simple, colourless, bacilliform. **Chemistry**: no lichen products reported by t.l.c. **Ecology**: on calcareous rocks.

Differs from *Aspicilia* and *Hymenelia* in having an I+ blue apical dome to the asci, and from *Bellemerea* in lacking an amyloid ring-structure in the ascus apex.

LITERATURE: Hafellner (1984).

1. E. flavida (Hepp) Hafellner (1984)

Aspicila flavida (Hepp) Rehm (1863)

Thallus crustose, effuse, superficial, grey to ochraceous-brown, matt, rimose, not corticate, to 200 μm thick but sometimes ± evanescent. Photobiont cells 7–12 μm diam. Apothecia 0.15–0.3 mm diam., immersed in thallus, aspicilioid, disc concave, blackish (tinged blue-green when wet); true exciple to 12 μm wide at the top; hymenium 60–70 μm tall, colourless below, pale to dark blue-green above, K−, N+ red, I+ dark blue; hypothecium colourless. Paraphyses (1-)1.5–1.7 μm wide, to 3 μm wide at apices. Ascospores 12–18 × 7–9 μm, ellipsoid. Pycnidia 30–50 μm diam.; conidia 3–4 × 0.8–1 μm. Thallus Pd−, K−, KC−, C−, UV−.

On calcareous rocks, stones, especially pebbles in coastal dunes or shingle ridges; rare. N.E. England (Cleveland; Northumberland, Holy Island), C. & N.W. Scotland (Angus at 700 m, Sutherland). Europe, North America.

ENDOCARPON Hedw. (1789)

Thallus squamulose (Fig. 36e) to rarely foliose, adpressed to ascending. **Upper cortex** pseudoparenchymatous; lower cortex occasionally present. **Photobiont** *Stichococcus*, present in both the thallus and the hymenium. **Ascomata** perithecia, immersed in the thallus, with a slightly projecting ostiole. **True exciple** dark throughout. **Hymenium** I+ reddish or bluish, K/I+ blue. **Hamathecium** without paraphyses; periphyses numerous, with 1- or 2-celled hymenial algae. **Asci** 1- to 2-spored, clavate or saccate, fissitunicate, thick-walled, the apex with a chitinoid plug staining red in Congo Red. **Ascospores** multiseptate to muriform, colourless, becoming yellowish brown. **Conidiomata** pycnidia, immersed, without involucrellum. **Conidiogenous cells** elongate-ampulliform, enteroblastic. **Conidia** rod-shaped, simple, colourless. **Chemistry**: no lichen products detected by t.l.c. **Ecology**: earth banks, calcareous soil, lakesides.

Differs from *Catapyrenium*, *Dermatocarpon* and *Placidiopsis* in having algal cells within the perithecial cavity.

LITERATURE: Zschacke (1934).

1 Squamules 0.7–1.5 mm, overlapping at edges; lower surface black with a pale border; asci 1-spored .... **1. *adscendens***
Squamules 0.5–3.5 mm, not, or only slightly, upturned at edges; lower surface pale, or if black then without a conspicuous pale border; asci 2-spored .... **2. *pusillum***

1. E. adscendens (Anzi) Müll. Arg. (1881)

Thallus nodulose-squamulose to almost foliose, the squamules overlapping, with ± radiating marginal lobes, forming a ± compact mat (Fig. 36e); squamules 0.7 × 1.5 mm, 0.2–0.26 mm thick when wet, often ± convex, not pruinose, the margins entire or ± indented-crenulate; upper surface fawn to light brown, dull greenish when wet, matt; lower surface black, with a broad, pale marginal zone. Perithecia 0.2–0.4 mm diam., 1–2 per squamule; upper part brown-black, conspicuous, ± elevated around the ostiole; ostiole depressed, globose; wall brown-black above, pale below; ± discrete crystal clusters few to frequent within perithecium. Hymenial algae 3.5–4.5 μm, spherical or shortly ellipsoid. Asci 1-spored. Ascospores (28-)32–47(-50) × 14–18 μm, elongate-ellipsoid. Pycnidia *c.* 100 μm diam., frequent; conidia 4–6 × 0.8 μm.

On periodically inundated granite, mica-schist, porphyry and sandstone, at the margins of rivers and lakes, often encrusting mosses, such as *Thamnobryum*; very rare. S. Wales (River Usk), W. Ireland (N. Kerry, Connemara). Europe, North America.

2. E. pusillum Hedw. (1789)

Thallus of scattered or ± contiguous squamules; squamules 0.8–3.5 mm, *c.* 0.18 mm thick, closely adpressed to substratum, the margins crenate, only slightly raised; upper surface pale to dark reddish brown, deep green when wet, matt; lower surface pale [var. *pallidum* (Ach.) Körber (1855)] or black [var. *pusillum*], with a few wart-like hairs. Perithecia 0.25–0.4 mm diam., globose. Hymenial algae 3–7 × 2–4 μm. Ascospores (1-)2 per ascus, (25-)45–55(-58) × 14–19(-23) μm.

On soil overlying chalk and on old walls; probably extinct. Recorded last century from S. & S.E. England (Norfolk, Thetford; Sussex, Rottingdean; Is. of Wight, Alum Bay). Europe, North America, Asia.

ENTEROGRAPHA Fée (1825)

Thallus crustose, not layered, smooth or cracked-areolate, white-grey to dark olive-green or brown, frequently forming mosaics; prothallus grey to black, separating the thalli. **Photobiont** *Trentepohlia*. **Ascomata** apothecia, immersed, punctiform, ellipsoid or ± lirellate, rarely stellate; disc flat, brown to black, not pruinose. **True exciple** thin, immersed, pale to brown. **Hamathecium** of paraphyses, very slender, anastomosing, becoming more branched above, conglutinated, without swollen tips. **Asci** 8-spored, cylindrical or cylindrical-clavate, fissituniate, apical dome with short ocular chamber surrounded by a minute, K/I+ dark blue ring; remaining part of apical dome and inner wall of ascus often K/I+ pale blue. **Ascospores** 3- to 16-septate, fusiform, colourless. **Conidiomata** pycnidia, immersed, visible as minute, pallid to brown specks; wall colourless, or pale brown above. **Conidiogenous cells** cylindrical, enteroblastic, arising singly. **Conidia** 5–6 × 0.8 μm, arising terminally, bacilliform, simple, colourless. **Chemistry**: confluentic, protocetraric, psoromic and conpsoromic acids and unidentified pigments, may be present. **Ecology**: mainly tropical genus of about 30 species.

LITERATURE: Coppins & James (1979a).

1 Thallus sorediate; abraded soralia and medulla Pd+ rust-red; apothecia very rare **4. *sorediata***
Thallus not sorediate; Pd+ yellow or Pd−; apothecia frequent 2

2(1) Ascospores (3-)5- to 7-septate; apothecia to 1 mm long; thallus Pd− 3
Ascospores 9- to 16-septate; apothecia 1–2 mm long; thallus Pd+ yellow-orange **2. *elaborata***

3(2) Apothecia 0.1–0.3 × 0.1 mm, punctiform and often arranged in lines or ± elongate-serpentine, not surrounded by a whitish rim; usually on trees .... **1. *crassa***
Apothecia 0.3–1.0 × 0.2–0.4 mm, comma-like or ± shortly serpentine, sometimes branched, often surrounded by a whitish rim; usually on rocks .... **3. *hutchinsiae***

1. E. crassa (DC.) Fée (1825)

Thallus thin to thick, continuous, smooth or faintly rimose-cracked, even or becoming coarsely and unevenly low-warted, often forming wide-spreading, complex mosaics, pale grey to pale brown, dark brown or dark olive-green, ± refractive with closely packed granules, K – ; prothallus intersecting the thalli, thin, persistent, dark brown to black. Apothecia usually very numerous, brown-black, deeply immersed, without a rim, minutely punctiform, 0.1 mm diam., or ± shortly ellipsoid, occasionally branched, <0.4 × 0.1 mm, sometimes in dotted or thread-like lines; epithecium pale to dark brown, semi-opaque, non-granular. Ascospores 20–35(-40) × 4–6(-)7 μm, 5- to 7-septate, straight to slightly curved. Pycnidia 40–50 μm diam., frequent. Conidia 4–6 × 1.5–2 μm. Thallus Pd –, K –, C – (confluentic acid).

On shaded bases and trunks of mature or senescent broad-leaved trees, especially in ancient woodlands, rarely on very shaded, acid rock, tolerant of low illumination and moderate levels of pollution; locally abundant. S. & S.W. England, extending to N. England and W. Scotland (Is. of Skye), S. & W. Ireland. Oceanic W. Europe from Iberia to S. & W. Scandinavia, rarely penetrating far inland.

The shape of the apothecia and thallus colour are both very variable.

2. E. elaborata (Lyell ex Leighton) Coppins & P. James (1979)

Like **1**, but thallus pale pink-brown to pallid brown, densely granular-refractive, K+ dissolving; apothecia <2 × 0.2 mm long, extended, more conspicuous, serpentine, often branched or stellate. Ascospores (42-)49–65 × 3.5–4.5 μm, 9- to 16-septate, often ± curved, with a thin perispore. Pycnidia not known. Thallus Pd+ yellow-orange, K – or faint yellow, C – (psoromic and 2'-*O*-demethylpsoromic acids).

Habitat similar to *E. crassa* **(1)**, but possibly ± confined to *Fagus*. N. Ireland (Fermanagh). Europe from Pyrénées to Denmark and S. Sweden, Azores, Madeira.

Only recently refound this century in British Is., formerly recorded from S.E. England (Hampshire, New Forest; Is. of Wight) and S.W. Ireland.

3. E. hutchinsiae (Leighton) Massal. (1860)

Thallus thin to moderately thick, continuous, smooth or finely rimose-cracked, even, often forming elaborate, wide-spreading mosaics, dark grey to dull or dark olive-green or -brown, densely packed with minute crystals, K – ; prothallus black, criss-crossing the thalli. Apothecia 0.3–1.0 × 0.2–0.4 mm, rarely rounded, ellipsoid, curved or comma-like, rarely bluntly stellate, discrete or 2- to 3-contiguous, immersed; disc fully exposed, flat or slightly convex (when wet), black or grey-brown-black, matt, slightly raised and sometimes surrounded by a whitish rim (pseudothalline exciple); epithecium brown to black-brown, semi-opaque, non-granular, K –. Ascospores (22-)24–28(-30) × 4.5 μm, 5- to 8-septate, straight or ± curved. Thallus Pd –, K –, KC –, C –, UV – (confluentic acid).

On shaded, ± permanently damp, often vertical, acid rock faces, especially in small, deep ravines or on shaded seepage tracks on rock faces on or near the coast; occasionally on the smooth bark of exposed tree roots and butresses by sheltered streams; local. W. British Is., extending to N.E. England. W. Mediterranean region to S. Scandinavia, Azores, Madeira.

More frequent in W. Scotland than *E. crassa* **(1)**, with which it is often confused. *E. crassa* has smaller, often inconspicuous apothecia and occurs predominantly on bark.

4. E. sorediata Coppins & P. James (1979)

Thallus thin, continuous, even, scurfy, ± coarsely rimose-areolate, grey or pale brown, sorediate, with packed, highly refractive crystals; soralia present, farinose, white-grey, concolorous or paler than thallus, rounded and discrete, to 0.5 mm diam. or becoming ± confluent forming larger, erose, sorediate patches, sometimes in lines along margins of cracks bordering the areoles. Apothecia similar to those in *E. crassa* **(1)** but paler. Ascospores 8–10 × 3–4 μm [?immature], (3-)4- to 5-septate, curved. Pycnidia not seen. Medulla and soralia Pd+ rust-red, K –, KC –, C –, UV – (protocetraric acid).

Mainly on the dry side of trunks of *Quercus* not directly wetted by rain in ancient woodland; rare or overlooked. S. & E. England (Hampshire, New Forest; East Anglia). ?Endemic.

The Pd+ rust-red reaction distinguishes this from *Schismatomma decolorans* (Pd−) and *S. niveum* (Pd+ yellow-orange); *S. quercicola* is Pd+ red but contains fumarprotocetraric acid (t.l.c.) and has rounded soralia.

EOPYRENULA R.C. Harris (1973)

Thallus immersed, inconspicuous or whitish, effuse. **Photobiont** *Trentepohlia*, or ?non-lichenized. **Ascomata** perithecia, ± globose or hemispherical; with a dark brown involucrellum, which either clasps the sides of the centrum or spreads outwardly. **True exciple** pale brown, colourless or almost inapparent below; dark tissue K−. **Hamathecium** of numerous, simple or sparingly branched or anastomosed paraphysoids; short periphyses, sometimes evident near the ostiole; gel matrix I+ yellowish or + pale bluish (especially around the asci). **Asci** subcylindrical, without apical ring-structure or (usually) distinct ocular chamber, the wall K/I− or + blue (especially below), 8-spored but often up to 6 spores aborting. **Ascospores** (1-)3- or more septate, brown, the end cells often paler, without conspicuous perispore or ornamentation. **Conidiomata** pycnidia, black. **Conidiogenous cells** ± cylindrical or narrowly lageniform, enteroblastic. **Conidia** of two types: macroconidia ellipsoid, oblong to cylindrical, brown, 1- or more septate; macroconidia bacilliform to filiform, colourless, simple. **Chemistry**: lichen products not detected by t.l.c. **Ecology**: on bark in woodland.

A small genus with 4 European species. Distinguished from *Pyrenula* by ascospores without lenticular lumina, straight microconidia, and brown, septate macroconidia.

LITERATURE: Coppins *et al.* (1992), Harris (1973).

1 Involucrellum clasping sides of centrum; spores 3- or more septate, mostly > 17 μm long; macroconidia with 1 pale septum or (5-)7 dark septa 2
Involucrellum outwardly spreading; spores 3-septate, < 17 μm long; macroconidia with 3 pale septa and rounded locules .... **1. *avellanae***

2(1) Ascospores mostly 18–26 × 6.5–10 μm and 3- to 5-septate; macroconidia with 1 pale septum; microconidia 6–10 μm long .... **2. *leucoplaca***
Ascospores mostly 21–34 × 9–10 μm and 5- to 7-septate; macroconidia with (5-)7 dark septa; microconidia 10–16.5 μm long .... **3. *septemseptata***

1. E. avellanae Coppins (1992)

Thallus inconspicuous. Photobiont filaments sometimes present. True exciple indistinct. Perithecia 0.2–0.28 × 0.16–0.22 mm, broadly ellipsoid, with distinct ostiolar depression, ± hemispherical in section; involucrellum outwardly spreading. Ascospores 10.5–16 × (5-)6–7 μm, ellipsoid, (1-)3-septate; septa pale or mid septum becoming dark. Macroconidia 9.5–13 × 4.5–6 μm, 3-septate, with rounded locules and pale septa; microconidia 2.8–4 × 0.7–1 μm, bacilliform.

On smooth bark of *Corylus*; local. W. Britain, N. Ireland. ?Endemic.

Easily mistaken in the field for *Arthopyrenia* species or *Pyrenula coryli*; the latter has ascospores with markedly lens-shaped lumina and curved microconidia (macroconidia not known). ***E. grandicula*** Coppins (1992) which has larger macroconidia, 14–19 × 6–7 μm, is common in S.W. England, W. Scotland and W. Ireland.

2. E. leucoplaca (Wallr.) R. C. Harris (1973)

Thallus whitish. Photobiont present. Perithecia 0.2–0.35 mm diam., ± globose. Ascospores (13-)18–26 × 6.5–10(-12) μm, ellipsoid to clavate-fusiform, (1-)3- to 5(-7)-septate; septa dark; middle 2 cells darkest and the outer cells progressively paler. Macroconidia 6.5–9.5 × 3–4 μm, ellipsoid, ovoid or pyriform, (0-)1-septate, the septum pale; microconidia 6–10 × 0.5–0.7 μm, filiform.

On fissured bark of *Fraxinus* in old woodland. N. Scotland (Inverness); the only British record but perhaps overlooked. Local in W. & C. Europe.

Could be mistaken in the field for a morph of an *Anisomeridium*.

3. E. septemseptata Coppins (1992)

Like **2**, but ascospores 21–34(-38) × 9–10 μm, 5- to 7(-8)-septate; macroconidia 18–26 × 4–5 μm, (5-)7-septate; microconidia (10-)13–16.5 μm long.

On *Corylus* stems in old woodland. W. Scotland (Argyll, Westerness). S.W. Norway.

EPHEBE Fr. (1825)

Thallus fruticose, filamentous, in decumbent, intricately branched, brownish black tufts, attached by a holdfast, without a cellular cortex; hyphae external to the algae when young, later also forming a central strand (Fig. 44a). **Photobiont** *Stigonema*. **Ascomata** developing from pycnidia, apothecia, with a true exciple, solitary or aggregated, immersed in lateral swellings on the branches; discs punctiform. **Hymenium** gelatinous, the upper part brown, I+ blue-green, K/I+ blue. **Hypothecium** dense. **Hamathecium of** septate paraphyses, sparsely branched, the apices thickened. **Asci** 8- to 16-spored, cylindrical to obclavate, thin-walled; wall with K/I+ blue outer layer; apical dome absent. **Ascospores** simple or occasionally 1- to 2-septate, ellipsoid, colourless. **Conidiomata** lateral pycnidia. **Conidiophores** branched. **Conidiogenous cells** enteroblastic, elongate-ampulliform. **Conidia** ellipsoid to bacilliform, simple, colourless, acrogenous. **Chemistry**: no lichen products detected by t.l.c. **Ecology**: on damp siliceous rocks, often in seepage tracks.

LITERATURE: Henssen (1963a).

1 Thallus 5–10(-30 mm) wide, with numerous short, spine-like lateral branchlets; asci <16-spored .... **2. *hispidula***
Thallus 20–30 mm wide, with few short, spine-like lateral branchlets; asci 8-spored **1. *lanata***

1. E. lanata (L.) Vainio (1888)

Thallus 20–30 mm wide, consisting of fine, cylindrical or angular, soft, slender much-branched, decumbent, sometimes matted filaments with occasional, short, spine-like branches, dark-green to brownish black, somewhat shiny. Ascomata to 0.25 mm, rare, occurring as minute, ± concolorous structures in swellings on the branches. Asci 8-spored. Ascospores 11–18 × 3.5–6 μm, 0(-1)-septate. Conidia 3–4.5 × 1.2–1.7 μm.

Seepage tracks on siliceous rocks and boulders, often on inundated rocks by streams; locally frequent. Upland British Is. Circumpolar in the Northern Hemisphere, extending north to Greenland.

2. E. hispidula (Ach.) Horwood (1913)

Like **1**, but thallus 5–10(-30) mm wide, the filaments with numerous, short, spine-like lateral branches. Ascomata to 0.25 mm diam., rare, more prominent, becoming subglobose, with an open disc. Asci 8- to 16-spored. Ascospores 7–9.5 × 4–5 μm. Conidia 2–3 × 1 μm.

On damp, siliceous rocks in mountains; scarce and very local. Scotland, Ireland. Probably circumpolar in the Northern Hemisphere.

Normally sterile, and distinguished from *E. lanata* **(1)** by its smaller size and numerous, spine-like branches, which give it a distinctly hairy appearance. Occasionally sterile specimens may be difficult to determine.

EPIGLOEA Zukal (1890)

Vorarlbergia Grumm. (1969)

Thallus inconspicuous, immersed in thin, gelatinous algal films. **Photobiont** *Coccomyxa*-type. **Ascomata** perithecia, gelatinous, dark-green to black, superficial, the wall without algal cells, consisting of unthickened, pigmented, periclinally-orientated hyphae; ostiole 15–50 μm diam. **Hamathecium** of paraphyses, thin, colourless, simple, the apices not capitate. **Asci** 8- to multispored, club-shaped to cylindrical, moderately thick-walled, non-fissitunicate (dehiscing bivalvally), the entire ascus wall I+ blue. **Ascospores** 1- to 5-septate, sometimes with an appendage at both ends, ovate-ellipsoid, colourless, thin-walled. **Conidiomata** pycnidia, globose, usually longer than wide, black, with a distinct ostiole, filled with gelatin-like substance. **Conidia** terminal, colourless, bacilliform to narrowly or broadly ellipsoid. **Chemistry**: lichen products not detected by t.l.c. **Ecology**: on algal films on rocks, mosses, lichens and soil in wet places.

LITERATURE: Döbbeler (1984).

1	Ascospores with apical appendages	1
	Ascopores without apical appendages	**3. *soleiformis***
2(1)	Asci 32-spored; ascospores 1-septate	**1. *grummannii***
	Asci 8-spored; ascospores 3-septate	**2. *medioincrassata***

1. E. grummannii Döbb. (1984)

E. bactrospora auct. brit., non Zukal (1889)

Perithecia 60–110(-130) μm diam., globose, pale brown, surrounded by a thin, colourless, gelatinous coat; ostiole 12–30(-60) μm, visible as a paler spot. Asci (35-)40–55(-67) × (7-)8–10(-11) μm, 32-spored, club-shaped. Ascospores 13–16.5(-18) × 1.5–2(-2.5) μm (with appendages), 1-septate, spindle-shaped, with tapering, thread-like appendages (to 3 μm) at both ends, usually with 2 oil bodies per cell, irregularly arranged in the ascus. Pycnidia not known.

Ascomata partly or completely immersed in algal films, usually on *Grimmia* and *Hypnum*. C.W. Scotland (Argyll). Germany, Italy (Tirol).

2. E. medioincrassata (Grumm.) Döbb. (1984)

Vorarlbergia medioincrassata Grumm. (1969)

Perithecia 100–190(-220) μm diam., ± globose, with a flattened apex, pale to dark brown, sometimes becoming black, with a very thin, colourless, gelatinous layer; apex with a circular swelling 20 μm thick. Asci (35-)47–70(-86) × (10-)11–15(-16) μm, 8-spored. Ascospores (18-)24–33(-38) × 3.5–5 μm (excluding appendages), 3-septate, ellipsoid to spindle-shaped, with thread-like appendages, 2–4(-8) μm, at either end.

Ascomata on, or immersed in, algal films on dead or dying bryophytes. S.W. Scotland (Dumfries). Austria, Italy, Germany.

3. E. soleiformis Döbb. (1984)

Perithecia (70-)90–140(-150) μm diam., globose, with a well-developed depression around the ostiole, green-black, sometimes grey or grey-black, with a colourless, gelatinous coat 4–15 μm thick. Asci (34-)40–55(-60) × 8–10.5(-12) μm, 8-spored. Ascospores (8.5-)9.5–12.5(-14) × 3.5–4.5(-5) μm, 1-septate, ellipsoid, with a slight constriction in the middle, without appendages.

Ascomata sunk into algal films on decomposing mosses (*Andreaea*, *Hypnum*, *Rhacomitrium* and *Gymnomitrion*) or on squamules of *Cladonia* species, rarely on raw humus, rotten wood or acid rocks; local. Throughout British Is. Germany, Austria, Italy, Marion Island (South Atlantic).

British material was formerly misidentified as *E. renitens* (Grumm.) Döbb. (1984), which differs in the larger perithecia, 105–200 μm, and ascospores 11.5–16 × 4.5–5.5 μm, not constricted at the septum.

EPILICHEN Clem. (1909)

Thallus crustose, areolate to subsquamulose, or absent; prothallus absent; distinct cortex lacking. **Photobiont** coccoid green algae. **Ascomata** apothecia. **Thalline exciple** absent. **True**

exciple black. **Epithecium** brownish. **Hypothecium** thick, dark brown. **Hamathecium** a branched and anastomosing net of paraphysoid-like filaments, not distinctly capitate. **Asci** elongate-clavate, the apex of the tholus and the hymenial gelatin turning blue in iodine. **Ascospores** brown, 1-septate, the septum with a torus, smooth. **Conidiomata** not known. **Chemistry**: ± pulvinic acid derivatives. **Ecology**: lichenicolous.

Formerly included in *Buellia*, but that genus differs in having broader, true paraphyses which are branched only at the swollen and deeply pigmented apices as well as *Lecanora*-type asci. *Catolechia* is distinguished by the different thallus type, ecology and, more significantly, the early stages of development of the ascomata.

LITERATURE: Hafellner (1978, 1979).

1. E. scabrosus (Ach.) Clem. (1909)

Thallus crustose, areolate to subsquamulose, yellow to yellowish green. Apothecia 0.2–0.8 mm diam., arising on the areoles, black, becoming convex, not pruinose; hymenium 70–100 μm tall, colourless to greenish; epithecium olive-brown or brown. Asci 50–70 × 14–17 μm, 8-spored. Ascospores (9-)11–17 × 6–10 μm, broadly ellipsoid, usually slightly constricted at the septum. Thallus Pd+ yellow-orange, K−, C−, UV+ orange (unidentified pulvinic acid derivatives).

On the thallus of *Baeomyces rufus*, especially around old mine-workings and on spoil heaps. N. England (Northumberland, N.W. Yorkshire), Wales, Scotland, W. & S.W. Ireland; probably often overlooked. Circumpolar in the Northern Hemisphere.

E. glaucinigellus (Nyl.) Hafellner (1978), which also grows on *Baeomyces* thalli elsewhere in N. Europe, completely lacks a thallus and is a commensal. The lichenicolous fungus *Dactylospora athallina* (Müll. Arg.) Hafellner (1979), which rarely occurs on *B. rufus* in N. Britain, has sparsely branched, not netted, hamathecial filaments, lacks a discrete thallus, ascospores only to 10 μm wide and asci in which only the outer coat turns blue in iodine.

EUOPSIS Nyl. (1875)

Thallus crustose, granular to minutely squamulose, dark reddish brown, gelatinous when moist. **Cortex** not differentiated; hyphae forming a reticulate pattern around the cyanobacterial photobiont, but pseudoparenchymatous in parts with a green algal photobiont. **Photobiont** *Gloeocapsa*, sometimes also with *Trebouxia*; cells or cell-clusters of *Gloeocapsa* near the surface with red-brown, K+ purplish gelatinous sheaths. **Ascomata** apothecia, with a glossy, brown disc. **Thalline exciple** well developed. **True exciple** narrow, of parallel, compacted paraphysis-like hyphae. **Epithecium** pale brown. **Hymenium** colourless, I−. **Hypothecium** colourless or pale brownish, of interwoven hyphae. **Hamathecium** of sparingly branched paraphyses, slender, not moniliform. **Asci** cylindrical, thick-walled, 8-spored; discharge rostrate; inner wall of ascus I+ blue, apical dome I−. **Ascospores** simple, ellipsoid, colourless, without a distinct perispore. **Conidiomata** pycnidia, immersed. **Conidiogenous cells** catenate. **Conidia** bacilliform, acrogenous and pleurogenous. **Chemistry**: no lichen products reported. **Ecology**: on moist, acid rocks, sometimes over mosses, soil or peaty debris. **Distribution**: 2 species in cool temperate to arctic regions of the Northern Hemisphere.

Distinguished from *Pyrenopsis* by the generally more disciform apothecia with glossy discs, I− apical dome of the ascus, and more slender paraphyses. *Psorotichia* has asci which are completely I−.

LITERATURE: Henssen *et al.* (1987)

1 Thalline exciple uneven, whitish, containing green algae; thallus mainly granular, with whitish patches .... **1. *granatina***
Thalline exciple crenulate, red-brown, containing cyanobacteria; thallus minutely squamulose, uniformly red-brown .... **2. *pulvinata***

1. E. granatina (Sommerf.) Nyl. (1875)

Pyrenopsis granatina (Sommerf.) Nyl. (1866)

Thallus reddish brown, with small, whitish patches, granular, often broken up into nodulose clumps to 1 cm diam. Photobiont *Gloeocapsa* but also with *Trebouxia* in the thalline exciple and in patches in the thallus. Apothecia to 0.5 μm diam., disc flat to convex; thalline exciple irregular, whitish, to 100 μm wide, pseudoparenchymatous, containing green algae; true exciple 10–25 μm wide; hymenium 70–80 μm tall; hypothecium 60–70 μm tall. Asci 50–60 × 10–12 μm. Ascospores 9–12 × 5–7 μm.

On moist, siliceous or slightly basic rocks; rare. S.W. England (S. Devon, Dartmoor), Scotland. Fennoscandia, North America.

Records of this species require verification, as many appear to refer to *E. pulvinata* **(2)**.

2. E. pulvinata (Schaerer) Vainio (1881)

Pyrenopsis pulvinata (Schaerer) Th.Fr. (1866)

Thallus coarsely granular-areolate to squamulose, forming pulvinate patches to 2 cm diam., uniformly reddish brown. Photobiont *Gloeocapsa*. Apothecia to 1 mm diam.; disc flat to convex; thalline exciple crenulate due to granular or subsquamulose outgrowths, concolorous with thallus, sometimes becoming excluded, to 150 μm wide, without green algae; true exciple 10–25 μm wide; hymenium 100–125 μm tall; hypothecium 70–80 μm tall. Asci 80–100 × 10–15 μm. Ascospores 10–15(-17) × 5–8 μm.

On moist, acid rocks (especially mica-schist), often over dead or dying mosses or small accumulations of soil; rare. N. Scotland (C. Highlands), C.W. Ireland (Connemara). Europe, North America.

EVERNIA Ach. (1810)

Thallus shrubby-foliose, ± pendent and somewhat flaccid, attached by a ± spreading basal holdfast; lobes few to numerous, dorsiventral, flattened, strap-shaped, corticate on both sides, lower surface paler than upper. **Photobiont** chlorococcoid. **Ascomata** apothecia, ± stoutly and shortly stalked, marginal. Disc red-brown. **Thalline exciple** irregular, crenate. **Asci** 8-spored, *Lecanora*-type. **Ascospores** colourless, ellipsoid, simple. **Conidiomata** pycnidia, laminal and marginal, immersed, rounded, blackened around the ostiole, rare. **Conidia** needle-shaped. **Chemistry:** cortex K+ yellow (atranorin, ± usnic acid); medulla with evernic acid or (in non-British species) divaricatic acid.

Differs from *Pseudevernia* by the absence of isidia and a ± black lower surface. Resembles *Ramalina* but is dorsiventral, with the photobiont cells mainly confined to the upperside thus accounting for the green upper and white lower surfaces; *Ramalina* species are not dorsiventral, green above and below and have 1-septate ascospores.

LITERATURE: Keissler (1958–60).

1. E. prunastri (L.) Ach. (1810)

Thallus (1-)2–6(-10) cm, foliose; lobes rather soft, few to numerous, strap-shaped, to 5 mm wide ± palmately branched, often twisted. Upper surface green-grey to pale greenish yellow, often with an incomplete network of elongate ridges centrally spreading towards the margins; lower surface white, occasionally dotted greenish especially toward the tips, broadly channelled. Soralia marginal and/or laminal, at first rounded and often confined to ridges and/or margins, later spreading and confluent, paler or concolorous with the upper surface. Apothecia very rare, 0.2–0.5(-1.5) cm diam. Ascospores 7–11 × 4–6 μm. Pycnidia *c.* 0.3 mm diam.; conidia 6–7 × 0.5 μm. Medulla Pd−, K−, KC−, C−, UV− (usnic and evernic acids, atranorin).

Widespread and often abundant with a wide ecological amplitude, principally on sunny, often windswept, neutral to acid-barked trunks or canopy of wayside and parkland trees, hedgerows, also on fence-posts, old stems of *Calluna* or on the ground, more rarely in sheltered woodland and boggy sites; occasionally on nutrient-rich siliceous rocks, gravestones, stabilized shingle, short turf and brick walls. Throughout British Is., but absent or rare in much of the English Midlands where mean SO_2 levels are above 60 μg m^{-3}. Europe, W. North America, N. Africa, Japan.

The "oak moss" of the perfume industry. Material from shaded habitats usually has fewer, paler lobes and is more elongate. Specimens from polluted sites are often reduced in size with contorted or down-turned lobes which are often markedly eroded and sorediate. The grey, usnic acid-deficient race has been accorded the status of form, *E. prunastri* f. *herinii* (Duvign.) D. Hawksw. (1980), which can grow with the yellowish f. *prunastri*, when it retains its identity. The presence of both atranorin and usnic acid is demonstrated on t.l.c. plates by the purple coloration produced on charring where the spots of the two substances overlap.

FARNOLDIA Hertel (1983)

Thallus crustose, immersed or superficial, continuous; prothallus absent; medulla, when developed, I+ violet. **Photobiont** trebouxioid. **Ascomata** apothecia, sessile, ± constricted below; disc black, sometimes pruinose, convex. **True exciple** well developed, raised, of strongly compacted, dark brown hyphae, black and friable. **Epithecium** greenish to brownish. **Hymenium** colourless or greenish, I+ blue. **Hypothecium** colourless, greenish to dark brown. **Hamathecium** of branched and anastomosed, septate paraphyses, apices not markedly swollen. **Asci** 8-spored, elongate-clavate, with a thickened tholus I+ blue, containing a more densely amyloid, tubular pore, *Porpidia*-like. **Ascopores** ellipsoid, simple, smooth, with a thick, gelatinous perispore. **Conidiomata** pycnidia, black, scattered on the surface of the substratum or immersed in the thallus. **Conidiogenous cells** ± cylindrical. **Conidia** bacilliform. **Chemistry**: lichen products not detected by t.l.c. **Ecology**: on calcareous rocks, mainly upland. Cosmopolitan.

Farnoldia should be compared with *Clauzadea*. Its separation from *Melanolecia* and *Tremolecia* is discussed under the latter.

LITERATURE: Clauzade & Roux (1985), Hertel (1967, 1983).

1. F. jurana (Schaerer) Hertel (1983)

Tremolecia jurana (Schaerer) Hertel (1977), *Melanolecia jurana* (Schaerer) Hertel (1981)

Thallus whitish grey, usually immersed and scarcely apparent, where present generally I+ violet. Apothecia (0.2-)0.5–1.5(-2.5) mm diam., numerous, generally convex, not pruinose; hymenium 70–130 μm tall; hypothecium to 120 μm or more tall, mostly dark brown, often intensely greenish in upper part, K± purplish in section (as is the true exciple). Ascospores (11-)13–28(-33.5) × (5.5-)7–14(-20) μm; perispore to 3 μm thick. Pycnidia 60–100 μm diam., often present; wall green above, colourless below; conidia 3.5–6 × 0.5–1 μm. Thallus Pd−, K−, KC−, C−.

On hard and dolomitic limestones, often in somewhat damp situations, a member of the *Gyalectetum jenensis* association; local. Throughout British Is. Europe, N. North America, Turkey.

Only likely to be confused in its habitat with *Clauzadea monticola*, which is most quickly separated by the red-brown hypothecium and smaller ascospores. The apothecia are exceptionally hard, internally conglutinate and extremely difficult to squash successfully.

FELLHANERA Vězda (1986)

Thallus crustose, farinose to scurfy granular, sometimes irregularly shallow-verrucose, often rimose, ecorticate, whitish to grey-green or dull ochraceous. **Photobiont** chlorococcoid. **Ascomata** apothecia, sessile, flat or later convex, or thinly white-pruinose, whitish, beige or yellowish, colourless in sections. **Thalline exciple** absent. **True exciple** of ellipsoid to globose cells, becoming excluded. **Hymenium** I+ blue. **Hypothecium** colourless or dull orange-brown. **Hamathecium** of numerous paraphyses, which are mostly unbranched but often branched and widening above. **Asci** 8-spored, clavate, in K/I with amyloid apical dome containing a darker blue tubular ring-structure, and an amyloid fuzzy coat. **Ascospores** 1- to 3-septate, colourless, ovoid, oblong-ovoid to fusiform, smooth, not or sometimes halonate. **Conidiomata** pycnidia, often numerous, ± immersed in thallus, whitish or brown-orange; ostiole often gaping. **Conidia** 3–4 × 1.3–1.7 μm, pyriform. **Chemistry**: Pd−, K−, KC−, C−, UV−; lichen products not detected by t.l.c.

Byssoloma differs in having an outer true exciple of thick-walled, loose (at least in K) hyphae, and a dark hypothecium. *Bacidia vezdae* has been transferred to this genus as *F. vezdae* (Coppins & P. James) V. Wirth (1987), but has thread-like conidia and a brown hypothecium.

LITERATURE: Vězda (1986).

1 Ascospores 1-septate .... **1. *bouteillei***
Ascospores (1-)3-septate .... **2. *subtilis***

1. F. bouteillei (Desm.) Vězda (1986)

Catillaria bouteillei (Desm.) Zahlbr. (1902)

Apothecia 0.1–0.3 mm diam; true exciple cells 2–5 μm wide; hymenium 30–50 μm, tall. Paraphyses 0.5–1 μm wide below, widening to 2 μm at the tips. Ascospores 9.5–14(-16) × 3.5(-7) μm, 1-septate, ovoid to ovoid-oblong, often constricted at septum and sometimes ± sole-shaped. Pycnidia 80–150 μm diam.

In sheltered situations on evergreen leaves and small twigs (especially *Buxus*), damp siliceous rocks or stones; local. Throughout British Is. Cosmopolitan.

Some records may refer to *F. subtilis* **(2)**.

2. F. subtilis (Vězda) Diederich & Sérusiaux (1991)

Bacidia subtilis Vězda (1961)

Like **1**, but apothecia 0.15–0.4 mm diam.; ascospores 11–16 × 2.5–4.5 μm, (1-)3-septate, oblong, fusiform or clavate-fusiform.

In sheltered situations, on living stems of *Vaccinium myrtillus*, *Buxus* twigs and leaves, *Larix* twigs and top of old fence posts; local, probably overlooked. S.W. England (Gloucestershire), C. Wales, C. Scotland (Fife, Midlothian). Cool-temperate Europe north to C. Sweden.

A pollution-tolerant species which may be increasing in many parts of Europe. An unidentified species with orange-brown apothecia, dull brown-orange hypothecium, 3-septate ascospores (10–14 × 2.3–3.5 μm), and brown-orange pycnidia is known from sheltered trunks of *Acer campestre, Carpinus, Fraxinus* and *Ulmus* in England (Cheshire, Essex) and S.W. Scotland (Wigtown). It often occurs with pycnidia only, but its pyriform conidia distinguish it from *Bacidia vezdae*.

FORAMINELLA Fricke Meyer (1982)

Thallus foliose, small, rosette-forming, closely adpressed; lobes ± radiating, discrete or contiguous and overlapping. **Cortices** pseudoparenchymatous; upper surface with a pored epicortex, sorediate. **Soralia** laminal. **Photobiont** chlorococcoid. **Medulla** of densely interwoven hyphae. **Lower surface** brown-black with dark, simple rhizines. **Ascomata** apothecia, laminal, sessile or somewhat elevated. **Thalline exciple** present, concolorous with the thallus; disc red- to dark brown. **Hamathecium** of paraphyses, simple. **Asci** 8-spored. **Ascospores** 7–12 × 3–4 μm, simple, kidney-shaped to curved, colourless. **Conidiomata** pycnidia, black, laminal, immersed. **Conidia** sickle-shaped. **Chemistry:** cortex with atranorin or usnic acid; medulla with divaricatic acid (UV + white).

Parmeliopsis differs in the paler lower cortex, short, ampulliform, straight conidia, shortly ellipsoid ascospores, and the presence of thamnolic acid (K+ orange).

LITERATURE: Meyer (1982).

1 Thallus yellow or greenish yellow .... **1. *ambigua***
Thallus pale grey .... **2. *hyperopta***

1. F. ambigua (Wulfen) Fricke Meyer (1982)

Parmeliopsis ambigua (Wulfen) Nyl. (1866)

Thallus 1–3(-4) cm diam., closely adpressed, often forming neat rosettes, or of ± scattered, ± unorientated lobes, often contiguous with adjacent thalli; lobes 0.5–1 mm wide, sometimes much reduced, ± elongate, often radiating, flat or ± concave, contiguous or partly overlapping centrally, ± discrete at margins, sparingly indented; upper surface bright to dull yellowish green, matt; soralia frequent, laminal, yellowish, flat or ± convex, rarely concave, paler or concolorous with thallus, sometimes ± contiguous to form a ± continuous sorediate crust in older parts of thallus; lower surface brown-black. Apothecia rare, to 2 mm diam. Ascospores 7–11 × 2.5–3 μm. Conidia 12–18(-22) × 0.5–1 μm. Cortex K+ faintly yellow, KC+ yellow (usnic acid and atranorin); medulla Pd−, K−, KC−, C−, UV+ white (divaricatic acid).

Last century confined to conifer wood (fallen trees, stumps, palings), in C. and E. Scotland; now on acid-barked, wayside and parkland broad-leaved trees, rarely on worked timber or siliceous rocks (mainly sandstones) in moderately polluted areas (SO_2 60–70 μg m^{-3}); widespread and spreading. Throughout N., C. and S.E. England, rare in W. Britain and Ireland. Europe, North America, S.E. Asia, Australia.

Saxicolous morphs sometimes resemble *Parmelia mougeotii* which differs in the Pd+ orange, K+ orange, UV− (stictic acid) medulla. Close to *F. hyperopta* **(2)**, sometimes considered to be only a chemotype, with additional usnic acid, of that species. See p. 685.

2. F. hyperopta (Ach.) Fricke Meyer (1982)

Parmeliopsis hyperopta (Ach.) Arnold (1884)

Like **1**, but thallus pale grey or blue-grey and the upper surface has scattered to contiguous, laminal, white to blue-grey-white, ± convex soralia. Cortex K+ weakly yellow (atranorin), medulla Pd−, K−, KC−, C−, UV+ white (divaricatic acid).

Mainly restricted to bark of old conifers, also occasionally on acid-barked, broad-leaved trees (*Betula*) and wood and palings in open, dry woodlands; local. N. England, Scotland (Highlands), extending to S. & S.E. England. Europe, North America.

F. hyperopta has not undergone the same dramatic increase in distribution as *F. ambigua* **(1)**.

Parmeliopsis aleurites has a medulla reacting Pd+ orange, K+ orange, UV− (thamnolic acid), a paler lower cortex, ellipsoid ascospores and short, straight, partly swollen conidia. See page 685.

FULGENSIA Massal. & de Not. (1855)

Thallus squamulose, dispersed to placodioid, orange-yellow, often with schizidia. **Photobiont** chlorococcoid. **Ascomata** apothecia, sessile; disc orange-brown. **Thalline exciple** present. **Asci** 8-spored, *Teloschistes*-type. **Ascospores** colourless, ellipsoid to pyriform, simple or occasionally 1-septate, never polarilocular. **Conidiomata** pycnidia, rare, immersed, multilocular. **Conidiogenous cells** lining locules within the pycnidial cavity, ± subglobose to short-ampulliform, enteroblastic, acrogenous. **Conidia** bacilliform, simple, colourless. **Chemistry**: thallus K+ purple (physcion). **Ecology**: calcicole, usually on soil or mosses.

LITERATURE: Gilbert (1978).

1 Thallus placodioid, the marginal lobes clearly differentiated; inner part of thallus ± continuous; ascospores ellipsoid to pyriform, 3.5–5 μm wide .... **2. *fulgens***
  Thallus warty-scaly, the marginal lobes not, or only weakly, differentiated, composed of dispersed groups and islands of areoles; ascospores mostly broadly ellipsoid, 4–7 μm wide. .... **1. *bracteata***

1. F. bracteata (Hoffm.) Räsänen (1931)

Thallus 1–2 cm diam., golden yellow, roughened but sparingly pruinose, of dispersed or ± contiguous groups of ± convex areoles, the margin not, or indistinctly lobed, the lobes scarcely longer than broad, the centre strongly scaly-warty, often ± overlapping and forming congested clusters, occasionally becoming detached as schizidia. Apothecia 0.5–1(-1.2) mm diam., frequent; disc red-brown. Ascospores 9–13 × 4–7 μm, simple, broadly ellipsoid. Thallus K+ purple, UV ± salmon-orange (physcion), a smaller amount of fallacinol, and fallacinal (trace).

On rock, soil and mosses on a buttress of metamorphosed 'sugar limestone' above 1000 m; very rare. N. Scotland (E. Inverness, Ben Alder range). Arctic-temperate Europe, North America; possibly circumpolar.

A number of minor, infraspecific taxa have been described; the Scottish material matches var. *alpina* (Th. Fr.) Räsänen (1943).

2. F. fulgens (Swartz) Elenkin (1907)

Thallus 1–3 cm diam., rounded or often irregular and fragmented, ± placodioid; marginal lobes neat, clearly differentiated, longer than broad, mostly over 1 mm wide, remaining ± discrete or often overlapping especially towards the centre, which often becomes verrucose-bullate with age, orange-yellow to whitish, ± coarsely pale pruinose; schizidia variable, in most populations part of thallus - surface is given over to their production which when shed, exposes the white medulla. Apothecia often present, 0.5–1.5 diam., at first concave with a thalline exciple, becoming convex with a ± excluded margin, deep orange. Ascospores 9–12 × 3.5–5 μm, 0(-1)-septate, ellipsoid, pyriform or clavate. Thallus K+ purple, UV± dull pale orange (fragilin, caloploicin), apothecia K+ purple (physcion).

On well-consolidated, well-lit, freely drained, basic substrata, such as chalk-pebble grassland and shell-sand, in warm, sheltered, but open, sites, mostly adjacent to dunes and near sea level, often associated with the moss *Trichostomum crispulum*; rare. S. & E. England (N. Somerset, Is. of Wight, Channel Isles, Breckland); formerly also on Sussex coast. Widespread in warmer areas of the N. Hemisphere, Australia, New Zealand.

British populations have an unusually high proportion of pyriform spores. The formation of schizidia appears to be environmentally controlled.

FUSCIDEA V. Wirth & Vězda (1972)

Rhopalospora Massal. (1860)

Thallus crustose, rimose-cracked or areolate; prothallus commonly present. **Upper cortex** undifferentiated, sometimes with a layer of brownish surface cells. **Lower cortex** absent. **Photobiont** chlorococcoid. **Medulla** rarely I+ blue. **Ascomata** apothecia, immersed to sessile; disc red- to black-brown. **Thalline exciple** usually absent, at most poorly developed. **True exciple** ± distinct, persistent or occluded, concolorous with disc or paler; inner part colourless or pale brownish; outer part brown or black-brown. **Epithecium** brown. **Hymenium** colourless or sometimes pale brownish, I−. **Hypothecium** colourless or pale straw; hyphae erect. **Hamathecium** of paraphyses 1.5–2 μm wide, simple or sparsely branched, weakly conglutinated in water, especially at the apices, becoming ± free in K, a brown pigment surrounding the 1–2 uppermost cells; upper cell clavate, to 5 μm wide. **Asci** 8-spored, subcylindrical to clavate, with a thin external and internal K/I+ dark blue cap, surrounded by a thick, K/I+ pale blue, gelatinous apical cap (Fig. 33d). **Ascospores** colourless, sometimes brown when old, ± spherical to usually ellipsoid, rarely elongate, straight to curved or bean-shaped, simple to 7-septate. **Conidiomata** pycnidia, immersed in thalline warts at thallus-margin. **Conidia** cylindrical, with obtuse apices, or ellipsoid. **Chemistry**: orcinol *para*-depsides and tridepsides; also fumarprotocetraric, alectorialic and fatty acids. **Ecology**: characteristic of nutrient-deficient, acid, siliceous rocks and acid-barked trees.

Rhopalospora is included as, although it has septate ascospores, all other features are consistent with *Fuscidea*. Conservation of the latter name is being proposed.

LITERATURE: Coppins *et al.* (1992), Hertel (1977), Inoue (1981a, b), Poelt & Vězda (1981), Tønsberg (1992).

1 Soralia present; apothecia usually absent 2
Soralia absent; apothecia present 10

2(1) Soralia Pd+ orange or red; UV+ yellow or UV− 3
Soralia Pd−; UV+ bluish white 5

3(2) Thallus areolate; soralia numerous; apothecia usually absent 4
Thallus rimose-cracked; soralia sparse; apothecia usually present .... **3. *cyathoides***

4(3) Soralia C+ red; Pd+ yellow-orange; UV+ yellowish; on rocks .... **12. *praeruptorum***
Soralia C−; Pd+ rust-red; UV−; on trees .... **1. *arboricola***

5(2) On trees 6
On rocks 8

6(5) Thallus indistinct, ± immersed or of convex, scattered to confluent verrucae, 0.08–0.16 mm diam.; soralia efflorescent and irregular, rarely well developed and ulcerose; perlatolic or divaricatic acid 7
Thallus distinct, of smooth, corticate granules, 0.1–0.25mm diam., thick; soralia often well developed and ulcerose, especially towards edge; divaricatic acid .... **8. *lightfootii***

7(6) Perlatolic acid present .... **15. *viridis***
Divaricatic acid present .... **13. *pusilla***

8(5) Soralia C−; thallus grey or brown 9
Soralia C+ red; thallus often with a greenish or yellowish tinge .... **5. *hibernica***

9(8) Areoles discrete, flat to weakly convex; prothallus conspicuous, black, amongst the areoles .... **4. *gothoburgensis***
Areoles contiguous or fused, strongly convex; prothallus only visible as a border surrounding the thallus .... **14. *recensa***

10(1) Ascospores simple 11
Ascospores 5- to 7-septate .... **9. *lugubris***

11(10) Ascospores straight; medulla Pd−, UV+ or UV− 12
Ascospores bean-shaped; medulla Pd+ red, UV− .... **3. *cyathoides***

12(11) Apothecia immersed 13
Apothecia emergent 14

13(12) Apothecia irregularly rounded; thalline exciple absent; asci at several levels within apothecia .... **7. *kochiana***
Apothecia circular; thalline exciple distinct; asci at one level within apothecia **6. *intercincta***

14(12) True exciple persistent; divaricatic acid present at least in the apothecia 15
True exciple becoming excluded; thallus without divaricatic acid, UV− .... **10. *lygaea***

15(14) Thallus UV+; apothecia regularly rounded, 0.3–0.7 mm diam.; true exciple black **11. *mollis***
Thallus UV−; apothecia flexuose, 0.5–2 mm diam.; true exciple brown, often with some unpigmented parts .... **2. *austera***

1. F. arboricola Coppins & Tønsb. (1992)

Thallus areolate or minutely warted; areoles 0.2–0.3 mm diam., green-brown, rarely greyish, discrete or locally contiguous, circular to irregularly rounded, convex; soralia present, 0.2–0.5 mm diam., pale yellow-green, at times brown-tinged, at first discrete and delimited at edges, often becoming confluent and coalescing towards centre of thallus, ± rounded; prothallus usually distinct, brown, of radiating hyphae, visible between areoles and as a border surrounding thallus. Apothecia to 0.8 mm diam., very rare, sessile, rounded; disc dark brown-black, ± flat; true exciple often becoming flexuose, of radiating hyphae firmly conglutinated and with colourless crystals dissolving in K; hypothecium I+ faintly blue. Ascospores 7–9 × 4–5 μm, ± oblong, with obtuse apices, simple or 1-septate, with a median constriction. Soralia and exciple Pd+ red, K+ brownish yellow, C−, UV− (fumarprotocetraric and ± protocetraric acids).

On the bark of old *Pinus*, rarely on *Betula*. N. Scotland (Highlands). Scandinavia.

Like *F. lightfootii* **(8)** but with a different chemistry and (mostly) habitat.

2. F. austera (Nyl.) P. James (1980)

Thallus pale grey to chocolate-brown, rimose-cracked; areoles 0.2–0.5 mm wide, contiguous, thin, uniform to somewhat irregular, slightly convex or tuberculate; prothallus black, ± delimiting the thallus. Apothecia 0.5–2 mm diam., sessile, rounded to flexuose, discrete, or aggregated, then to 2 mm wide; disc dull black-brown becoming reddish brown when wet, flat, at times ± sparingly pruinose; true exciple brown, often paler than disc, at times partly piebald, persistent, becoming ± strongly tuberculate-lobate. Ascospores (6-)9–11(-12) × 6.5–8(-9) μm, broadly ellipsoid, simple. Pycnidia *c.* 0.15 mm diam., black-brown, around ostiole. Thallus Pd−, K−, KC−, C− (divaricatic acid confined to apothecia).

On sheltered, vertical, siliceous rocks, forming extensive mosaics; local. N. Scotland (Highlands), rare in England. S. Scandinavia, C. European mtns.

A morph with discrete and non-pruinose apothecia has been named *F. taeniarum* (Malme) V. Wirth & Vězda (1972), but is linked with typical *F. austera* through transitional stages.

Distinguished by the large, sessile, often conglomerate apothecia, which take on a distinctive red-brown hue when moistened. The ascospores are rounded and divaricatic acid, frequent in the genus, is here confined to the apothecia only.

3. F. cyathoides (Ach.) V. Wirth & Vězda (1972)

incl. var. *corticola* (Fr.) Kalb (1976)

Thallus very variable, pale grey-fawn to dark grey, usually with a ± strong brown tinge, rimose-cracked, rarely sorediate; areoles irregular, flat or, in part, uneven-convex to 0.6 mm wide; prothallus dark brown or black, bordering thallus, often mosaic-forming. Apothecia to 1.5 mm diam., immersed, with a cuff-like thalline exciple to sessile, rounded to strongly flexuose; disc brown-black, flat or ± convex; true exciple concolorous with, or often paler than the disc, persistent or ± disappearing. Ascospores 9.5–13(-14) × 5–5.5(-6) μm, bean-shaped. Thallus Pd+ rust-red, K+ brownish yellow, UV− (fumarprotocetraric acid).

On coarse-grained, nutrient-deficient, siliceous rocks, stabilized shingle, frequently forming extensive mosaics, rarely on bark; often abundant. Throughout British Is., but rare in E. England. Europe, North & South America, Africa, Japan.

A very variable species, especially in the colour and thickness of the thallus, characterized by the bean-shaped spores and the presence of fumarprotocetraric acid (Pd+ rust red). The var. *sorediata* (Magnusson) Poelt (1978) is distinguished from var. *cyathoides* by the presence of convex, rounded or oval soralia, 0.4–2 mm diam., discrete, occasional, pale yellow-brown to pale greenish, occasionally occurring with var. *cyathoides*. The var. *corticola* appears to be merely *F. cyathoides* on bark and so is not accepted here although it has a thicker thallus and restricted distribution; on *Betula* and *Alnus* in boggy sites in W. Scotland. *F. kochiana* **(7)** has a thicker thallus containing divaricatic acid (UV+) in the medulla and nearly spherical ascospores.

4. F. gothoburgensis (Magnusson) V. Wirth & Vězda (1972)

Thallus whitish, pale grey to grey-brown, areolate, sorediate; areoles 0.2–2.5 mm wide, discrete and scattered to locally contiguous, rounded to irregular, flat to slightly convex, ± cracked; soralia to 0.2–0.4 mm diam., usually concolorous with the thallus, sometimes greenish white, rounded, erose or delimited by a thin, elevated thallus margin, or effuse; prothallus black, well developed, between areoles and bordering thallus. Apothecia 0.3–1.7 mm diam., rare, sessile, at times flexuose; disc dull black-brown, flat to convex, at times with an umbo; true exciple black or concolorous with thallus, thin, distinct, persistent. Ascospores 7.5–11 × 5–7 μm, broadly ellipsoid, simple. Medulla and soralia UV+ bluish white (divaricatic acid).

On vertical, ± sheltered, siliceous rocks, often beneath overhangs; very local. S.W. England, Wales, Scotland. S. Sweden.

Very variable; characterized by the dispersed or clustered, sorediate areoles on a black prothallus and the UV+ blue medulla and soralia. A distinctive morph with large, greyish, erose soralia and rounded, ± discrete clusters of areoles on a black prothallus, has been named by some continental authors as f. *maculosa* (Magnusson) Poelt (1981); it intergrades with typical *F. gothoburgensis*.

5. F. hibernica P. James & Poelt (1992)

Thallus pale grey-green to dark chocolate-brown, areolate, sorediate; areoles to 1 mm diam., discrete and thinly dispersed or contiguous, often becoming confluent, irregularly rounded, flat to weakly convex;

soralia present, 0.2–0.3(-0.4) mm diam., pale ochre yellow to pale yellow-green, discrete or ± confluent, mostly irregularly rounded, weakly concave to flat; soredia farinose; prothallus dark brown, K+ fuscous brown, N−, discontinuous, irregular, forming somewhat glossy bands between or at the edge of the thallus. Apothecia and pycnidia unknown. Cortex and soralia KC+ red, C+ red, UV+ bluish white (gyrophoric and fatty acids).

On vertical, siliceous rock, often in crevices and gullies. W. Wales, W. Scotland, S.W. Ireland. Norway.

Distinguished by the pale yellow-ochre, somewhat erose, soralia and C+ red reaction.

6. F. **intercincta** (Nyl.) Poelt (1978)

F. atlantica (Magnusson) P.James & Poelt (1981)

Thallus pale to dark grey-brown, continuous or rather thick, cracked; areoles to 0.6 mm wide, irregular, flat to convex; prothallus black to brown-black, delimiting and ± in patches within thallus. Apothecia often numerous, 0.3–0.5(-0.6) mm diam., discrete or crowded, immersed or somewhat emergent, sessile, usually rounded, sometimes irregular; thalline exciple concolorous with thallus, very thin, halo-like and ± elevated, often whitish; true exciple brown-black, thick; disc black, flat to slightly convex, matt, rarely gnarled. Ascospores 9.5–13 × 6.5–8 μm, colourless, broadly ellipsoid. Medulla UV+ bluish white (divaricatic acid).

On moist, often steeply inclined, siliceous rocks in uplands; rare. Throughout British Is. Scandinavia, Portugal.

Distinguished by the small, often numerous apothecia with a pale, thin, halo-like thalline exciple and a thicker, dark true exciple and disc. The apothecia superficially resemble those of *Schismatomma graphidioides* and *Opegrapha rufescens*.

A morph with irregularly rounded apothecia, a thin, indistinct thalline exciple, and a thin thallus, named *F. atlantica*, is of no taxonomic significance. A specimen [?*F. oculata* Oberholl. & V. Wirth (1984)] from N. Wales (Llanberis) has numerous, crowded or scattered, fleck-like, yellowish white soralia, 0.1–0.3 mm diam. but the apothecia concur with *F. intercincta* in all aspects.

7. F. **kochiana** (Hepp) V. Wirth & Vězda (1972)

Thallus pale to medium grey often tinged brown, thick, deeply cracked; areoles 0.5–2.5 mm wide, coarse, irregular, flat; prothallus usually well developed, brown-black, broadly delimiting the thallus. Apothecia usually numerous, to 3 mm wide, immersed; disc brown-black, flat, usually irregular and ± angular; true exciple excluded, pale brown, narrow, with an epinecral layer. Asci commonly at varying levels in the apothecium. Ascospores 8–11 × 5–7 μm, broadly ellipsoid, becoming red-brown. Medulla UV+ bluish white (divaricatic acid).

On exposed, coarse-grained, siliceous and sandstone rocks and boulders in open, upland situations; locally frequent. Throughout British Is.

Differs from *F. cyathoides* **(3)** in the usually more immersed apothecia lacking a raised exciple, the thicker thallus, rounded, not bean-shaped, ascospores, asci at varying levels in the hymenium and the UV+ blue medulla.

8. F. **lightfootii** (Sm.) Coppins & P. James (1978)

Thallus olive to dull grey-green or suffused brownish, finely to coarsely verrucose-areolate, occasionally ± papillate; usually sorediate; soralia pale or yellowish green, becoming confluent, often bursting from apices of areoles, soredia granular, sometimes farinose; prothallus pale grey to dark brown. Apothecia 0.5–1 mm diam., frequent, dark grey-brown to black, shining; true exciple paler or concolorous, flexuose, irregular or sublobate, at times surrounded by thalline granules and appearing to have a thalline exciple. Ascospores 8–11 × 3.5–4.5 μm, with obtuse ends, medianly constricted. Thallus UV+ white (divaricatic acid).

On ± horizontal boughs and twigs, particularly *Fraxinus* and *Salix*, usually near boggy sites and overhanging streams and rivers; locally abundant. N. & W. British Is. W. Europe.

Variably sorediate; the thallus also varies from coarse, papilla-like granules to ± continuously small cracked-areolate. Closely resembles *F. arboricola* **(1)** which, however, is Pd+ red (fumarprotocetraric acid) and has a different ascospore shape; their shape in *F. lightfootii* is diagnostic.

9. F. lugubris (Sommerf.) P. James & Purvis (1992)

Rhopalospora lugubris (Sommerf.) Poelt (1980)

Thallus thick, dark grey-brown or black-brown, often purple-tinged, finely cracked-areolate; areoles somewhat verrucose, ± convex. Apothecia to 0.5–1.5 mm diam., sessile, black, discrete or often ± crowded; disc matt, flat to ± convex; true exciple ± occluded, thick, concolorous with the thallus, of strongly conglutinated, faintly brownish hyphae, with dark brown spots, marginally dark brown; hypothecium pale straw to red-brown, semi-opaque. Ascospores 17–48 × 4–7 μm, narrowly clavate, often attenuated at one end, often curved, 6- to 7-septate. Thallus Pd−, K−, KC−, C− (unidentified substance by t.l.c.).

On exposed, vertical, siliceous, acidic rock outcrops; rare. C. & N.W. Scotland (Argyll, E. Ross, mid-Perth). N. Europe, Alps (very rare).

Distinguished by the septate, attenuate-clavate ascospores.

10. F. lygaea (Ach.) V. Wirth & Vězda (1972)

F. periplaca (Nyl.) V. Wirth & Vězda (1972), *F. tenebrica* (Nyl.) V. Wirth & Vězda (1972)

Thallus pale to dark grey-brown or dark brown, often with a purplish tinge, thin, continuous, or ± incompletely cracked; areoles to 0.7 mm wide, often smaller, irregular, flat; prothallus dark brown to black, thin, bordering thallus. Apothecia 0.2–0.6 mm diam., emergent to adpressed, sessile, black, rounded; disc flat to markedly convex; true exciple at first thin, soon occluded, usually indistinct. Ascospores 5–9.5 × 5–6.5 μm, spherical to broadly ellipsoid. Lichen products not detected by t.l.c.

On hard, siliceous rocks; common in montane west and north. Upland British Is. S. & C. Europe.

There is considerable variation in the thickness of thalli and the cracking of the upper surface. The species is characterized by the thin, rather even, sometimes minutely areolate, thallus with convex, immarginate apothecia. The rounded ascospores distinguish it from poorly developed specimens of *F. cyathoides* **(3)** with which it often grows.

11. F. mollis (Wahlenb.) V. Wirth & Vězda (1972)

Thallus pale grey-brown to dark grey, deeply and regularly cracked, thick; areoles 0.3–1 mm wide, irregular, flat; prothallus dark brown-black, delimiting. Apothecia 0.3–1 mm diam., sessile, rounded to slightly flexuose, discrete; disc black, flat to slightly convex; true exciple black, elevated. Ascospores 8–9.5 × 5.5–6.5 μm, broadly ellipsoid. Medulla UV+ bluish white (divaricatic acid).

On damp, serpentine rocks; rare. N. Scotland (Shetland Is., Balta Sound). N. Scandinavia.

Resembling *F. cyathoides* **(3)**, but distinguished by the presence of divaricatic acid and broadly ellipsoid ascospores.

12. F. praeruptorum (Du Rietz & Magnusson) V. Wirth & Vězda (1972)

Thallus pale to dark brown, areolate; areoles to 0.4 mm diam., contiguous to discrete, uneven and delimited by a black prothallus, rounded, tuberculate; soralia 0.2–0.5 mm diam., at first pale greenish or ochre- or cream-coloured, punctiform, rounded, soon becoming brownish and ± confluent. Apothecia to 0.8 mm diam., very rare, sessile, black; disc slightly convex; true exciple narrow. Ascospores 9.5–12 × 3–4 μm, bean-shaped. Soralia Pd+ yellow, KC+ red, C+ red, UV+ faintly yellowish (alectorialic acid).

On ± vertical, shaded, siliceous rocks, often under overhangs, in uplands; rather rare. Upland Great Britain. W. Europe, Scandinavia.

The thallus is coarsely uneven-verrucose, with rather small, irregular soralia; in the herbarium becoming a pinkish hue in time. *Opegrapha gyrocarpa* has a smoother thallus and *Trentepohlia* as photobiont; it is also C+ red but contains gyrophoric and schizopeltic acids.

13. F. pusilla Tønsb. (1992)

Thallus usually forming minute rosettes to 10 mm diam., areolate; areoles to 0.4 mm diam., greyish green to green, discrete to contiguous, irregularly rounded, markedly convex; soralia green to pale yellow, irregular, becoming confluent in specimens with contiguous areoles; soredia farinose; prothallus distinct, of pale to dark brown, ramifying hyphae, visible between areoles and/or as a thallus border.

Apothecia and pycnidia unknown. Medulla and soralia UV+ bluish white (divaricatic and ± nordivaricatic acids).

On *Alnus*; a single British collection. Scotland (Glen Strathfarrar). Norway.

14. F. recensa (Stirton) Hertel, V. Wirth & Vězda (1972)

Thallus pale grey or brownish grey, areolate; areoles 0.1–0.5 mm diam., usually contiguous, discrete, rounded and uneven-convex or tuberculate; soralia 0.2–0.7 mm diam., at first whitish, soon becoming brownish and confluent, uneven, sometimes convex; prothallus dark brown, thin, ± visible between areoles and delimiting the thallus. Apothecia 0.4–0.8 mm diam., infrequent, sessile, black, rounded; disc flat to slightly convex; true exciple thin, conspicuous. Ascospores 9.5–12 × 3.5–4 μm, curved-kidney-shaped. Medulla and soralia UV+ bluish white (divaricatic and ± nordivaricatic acids).

On hard, often ± vertical, siliceous rocks; overlooked. Upland British Is. Europe, E. North America, Indonesia.

This sterile species is identified by the rather uneven, crowded, greyish soralia and UV+ blue (divaricatic acid) response.

15. F. viridis Tønsb. (1984)

Thallus greyish green, indistinct, ± immersed or areolate; areoles to 0.2 mm diam., matt, discrete or ± contiguous, rounded, convex; soralia green, yellowish green or straw-coloured, effuse, irregular, often becoming confluent, soredia farinose; prothallus ± distinct, brown. Apothecia 0.2–0.4 mm diam., very rare, sparse, sessile, black; disc flat; true exciple persistent, brown. Asci 12- to 16-spored. Ascospores 22–44 × 2–3.5 μm, narrowly clavate, attenuate, often curved, 5- to 6- septate. Thallus and medulla Pd−, K−, KC−, C−, UV+ bluish white (perlatolic, hyperperlatolic acids, and related substances).

On damp woodland trees in boggy sites. N.E. England (N.E. Yorkshire), N. Wales (Merioneth), Scotland. Norway.

The soralia are typically diffuse, efflorescent and irregular, and only rarely well-delimited and ulcerose as usual for *F. arboricola* **(1)** and *F. lightfootii* **(8)**.

GOMPHILLUS Nyl. (1853)

Thallus crustose, semi-transparent and sub-gelatinous when wet, effuse or ± delimited. **Photobiont** chlorococcoid. **Ascomata** apothecia, top-shaped, black-brown above, pale to colourless below. **Thalline exciple** absent. **True exciple** well developed, extended below into a stalk-like extension, pale brown at the edge, pale within, of thin, anastomosing hyphae in a gelatinous matrix, not carbonized. **Epithecium** red-brown, non-granular. **Hamathecium** of numerous, anastomosing paraphyses, <1 μm wide, the apices not swollen. **Asci** 8-spored, narrowly cylindrical, with an apical tholus, I−, the discharge rostrate. **Ascospores** very long, thread-like, multiseptate, colourless. **Conidiomata** pycnidia, black above, pale below. **Conidia** minute, ellipsoid. **Ecology**: on mosses, liverworts, and sometimes, lichens; oceanic.

LITERATURE: Frey (1933), Hafellner (1984), Vězda (1979).

1. G. calycioides (Del.) Nyl. (1853)

Thallus thin, membrane-like, sometimes almost disappearing; hyphae penetrating moss and hepatic leaf cells. Photobiont cells (6.5-)9–11 μm, round. Apothecia few to numerous, scattered, sessile but appearing shortly stalked, simple, rarely 2- to 4- per stalk, lower part of "stalk" ± furrowed; epithecium red-brown. K ± brown. Asci 280–300 × 7–10 μm. Ascospores 160–200 × 1.5–2(-3) μm, straight or ± curved, the apices rounded, 25–65(-100)-septate. Conidia 2–4 × 0.5–0.7 μm. Thallus Pd−, K−, KC−, C−, UV− (no lichen products detected by t.l.c.).

Overgrowing mosses and hepatics on mossy trees and rocks, more rarely on the ground, mainly in sheltered, well-wooded sites or isolated trees in sheltered valleys, often by streams and lakesides; local, frequently overlooked. N. Wales, S.W., W. & N. Scotland, S. to N. Ireland. W. Europe from Portugal to Scotland. Macaronesia, Hawaii, Chile.

Distinguished by the top-shaped, black-brown apothecia, thin, skin-like thallus and hair-like, multiseptate spores; the thallus of *Chromatochlamys muscorum* appears similar, but has pale, immersed perithecia.

GRAPHINA Müll. Arg. (1880)

Thallus crustose, immersed or partly superficial, effuse. **Photobiont** *Trentepohlia*. **Ascomata** apothecia, immersed to prominent, lirelliform, simple or branched; disc slit-like or little-exposed, black. **Thalline exciple** absent. **True exciple** black, opaque. **Epithecium** colourless or brown. **Hymenium** colourless, I−. **Hypothecium** pale or dark. **Hamathecium** of simple paraphyses. **Asci** 8-spored, ± cylindrical, with a single functional wall layer, the apex thickened, I− (Fig. 35c). **Ascospores** colourless, septate, becoming sub-muriform or muriform, I+ violet. **Conidiomata** pycnidia, black. **Conidia** ellipsoid, colourless. **Chemistry**: Pd−, K−, KC−, C−, UV− or patchy pale or dull orange (unidentified xanthones), depsidones, zeorin, atranorin and lichexanthone reported from non-European species. **Ecology**: on bark.

Closely related to *Graphis* which differs in having septate, not sub-muriform to muriform ascospores.

LITERATURE: Coppins & James (1978).

1 Ascospores 30–50(-70) × 10–18 μm, more than 30-celled 2
Ascospores 13–15 × 5.5–6 μm, less than 6-celled .... **2. pauciloculata**

2(1) Apothecia erumpent to ± superficial, mostly unbranched, ovoid to oblong; exciple prominent; disc slit-like .... **3. ruiziana**
Apothecia semi-immersed, branched; exciple often inconspicuous; disc usually expanded and whitish-pruinose .... **1. anguina**

1. G. anguina (Mont.) Müll. Arg. (1882)

Thallus ± immersed, smooth or faintly cracked, grey, ± delimited. Apothecia ± immersed in thallus, flat or slightly elevated, usually branched, the branches ± straight, curved or serpentine; true exciple thin and often inconspicuous, not furrowed, black, pale below; disc narrow and slit-like or ± narrowly exposed and ± white-pruinose; hypothecium pale. Ascospores (25-)30–50(-70) × (10-)14–18(-20) μm, with more than 30 cells, with 7–11 transverse septa.

On smooth bark of broad-leaved trees in open or sheltered habitats, especially in coppice, old woodlands and *Salix* carr. Widespread in British Is., but absent from most of N. Scotland and C. & E. England. N.W. Germany.

The ± immersed apothecia more closely resemble some morphs of *Graphis scripta* than those of the two following species.

2. G. pauciloculata Coppins & P. James (1978)

Thallus partly immersed, thin, smooth, pale grey or grey-green, continuous. Apothecia 0.5–1.5 × 0.3–0.5 mm, erumpent to ± superficial, simple or branched, straight or occasionally curved or sigmoid, frequently tightly clustered, gnarled, shortly stellate or in irregular clusters; true exciple well developed, not furrowed, black, shiny; disc a narrow slit; hypothecium pale. Ascospores 13–15 × 5–5.6 μm, to 6-celled, with 3–4 transverse and 1–2 longitudinal septa. Pycnidia numerous; conidia 4–6 × 1–1.5 μm.

Smooth bark of *Corylus*, ancient *Ilex*, and young *Quercus* in old, moist, open woodland; very rare. S.W. England (E. Cornwall, S. Devon), S.W. Ireland (E. Cork). Endemic.

Differs from *G. ruiziana* **(3)** in the smaller, often tightly clustered, gnarled apothecia, and from both *G. anguina* **(1)** and *G. ruiziana* **(3)** in the smaller, few-celled ascospores. The tightly compacted, clustered apothecia have the superficial aspect of an *Arthonia* (e.g. *A. ilicinum*) or *Arthothelium*.

3. G. ruiziana (Fée) Müll. Arg (1887)

Thallus largely immersed, light grey-green. Apothecia 0.5–2.2(-3.2) × 0.3–0.5 mm, markedly elevated, scattered, rarely contiguous, simple, rarely branched, straight, curved, or somewhat flexuose; true exciple well developed, sometimes weakly furrowed, or smooth, shining, occasionally ± white-pruinose adjacent to slit; disc thin and slit-like; hypothecium dark. Ascus (Fig. 35c). Ascospores 28–45 × 12–18 μm, with more than 30 cells, with (6-)7(-8) transverse septa. Pycnidia frequent; conidia 2.5–3.5 × 1 μm, ellipsoid.

On smooth, leached bark, especially *Ilex*, young *Quercus* and *Sorbus aucuparia*, in mild, humid valleys, rarely on siliceous rocks in damp woods. S.W. England, N.W. England (Lake District), W. Wales, W. Scotland, S.W. to N.W. Ireland. France (Brittany), Africa, S.E. Asia, Pacific Is., South America.

Distinguished by the short, ± straight, elevated, lirelliform apothecia, with a depressed, narrow slit. Occasionally mistaken for the superficially similar, widespread and non-lichenized *Gloniopsis praelonga* (Schwein.) Zogg (1962), which has I− ascospores, 17–27 × 6–10 μm, and 3–7 transverse and 1–2 longitudinal septa; it is especially frequent on old *Rubus fruticosus* aggr. canes.

GRAPHIS Adans. (1763)

Thallus crustose, effuse, immersed to superficial. **Photobiont** *Trentepohlia*. **Ascomata** apothecia, immersed to superficial, usually elongate (lirelliform), simple, branched or stellate. **Thalline exciple** absent. **True exciple** usually black, opaque, prominent and strongly developed, sometimes longitudinally furrowed, rarely pale. **Hymenium** I−. **Hypothecium** thin, pale. **Hamathecium** of simple paraphyses. **Asci** to 8-spored, I−, clavate to subcylindrical, non-fissitunicate, dehiscence by an apical split. **Ascospores** transversely septate, I+ violet, lumina rounded to lens-shaped, colourless. **Conidiomata** pycnidia. **Conidia** rod-shaped, colourless. **Chemistry**: no lichen products or with norstictic or psoromic acids; old apothecia of some species sometimes with a K+ purple anthraquinone. **Ecology**: on bark.

Phaeographis differs in the brown ascospores and, in the case of the British species, flat lirellae with inconspicuous, scarcely raised margins and more broadly expanded discs; the ascospores of *Graphis* are colourless at maturity but may be brownish when over mature. Superficially resembling *Opegrapha*, species of which have fissitunicate asci, I− ascospores and anastomosed hamathecial elements.

LITERATURE: Coppins *et al.* (1992), Redinger (1938).

1	Apothecia black; ascospores at least 5-septate	2
	Apothecia pale; ascospores 3-septate	**1. *alboscripta***
2(1)	Margins of apothecia furrowed; thallus K+ (yellow to red)	**2. *elegans***
	Margins of apothecia not furrowed; thallus K− (brownish)	**3. *scripta***

1. G. alboscripta Coppins & P. James (1992)

Thallus smooth, continuous to faintly rugose or scurfy-pruinose, often mosaic-forming, white-grey to grey, containing numerous, blunt-edged crystal clusters, K−. Apothecia *c.* 2 × 0.3–0.5 mm, numerous, single or 2- to 3-branched, rarely stellate, when abundant forming a closed reticulum, elongate, serpentine, the margin white, becoming ± tumid; disc slit-like, often scurfy and not or little-exposed; true exciple not blackened, the upper edge with minute, brown crystals dissolving in K. Ascospores (15-)16–19 × (6-)7–8(-9) μm, 3-septate, broadly ellipsoid to slightly clavate. Thallus Pd+ pale yellow, K+ pale yellow, KC−, C−, UV−; apothecia Pd+ deep orange-yellow (psoromic, 2'-*O*-demethylpsoromic acids).

On smooth bark of *Corylus*, *Fraxinus* and *Ilex*. Coastal W. Scotland (Argyll to Is. of Skye). ?Endemic.

2. G. elegans (Borrer ex Sm.) Ach. (1814)

Thallus smooth, membrane-like, continuous verrucose to uneven and cracked, pale to dark grey, densely packed with coarse crystal clusters, K−. Apothecia simple or branched, short or elongate, straight or curved; true exciple black, each side with 1–6 longitudinal furrows, disc narrow and slit-like; old apothecia occasionally with orange, K+ purple pruina. Ascospores 32–55 × 6–12 μm, (8-)10- to 12-septate, fusiform-cylindrical. Thallus Pd+ yellow becoming slowly red, often in parts, K+ reddish (crystals) (norstictic acid).

On moderately shaded, smooth bark on a wide range of trees, rarely on moist, shaded, siliceous rocks; often frequent. Throughout British Is., but now absent from most of C. England. Oceanic Europe, Dominican Republic, New Zealand; perhaps cosmopolitan but many records doubtful.

The furrowed apothecial margins are usually conspicuous, but sometimes the margin may have only a single furrow and immature apothecia may lack them.

3. G. scripta (L.) Ach. (1809)

Thallus smooth or uneven, whitish to grey, rarely orange-brown, with ± scattered crystal clusters, K−. Apothecia branched or not, often elongate; true exciple black, the sides not furrowed; disc narrow or expanded, white-pruinose or not. Ascospores 25–70 × 6–10 μm, 5- to 15-septate, fusiform-cylindrical. Pycnidia occasional; conidia 2–5 × 1 μm. Thallus Pd−, K− (lichen products not detected by t.l.c.).

On moderately to deeply shaded smooth bark on a wide range of trees. Throughout British Is., but now absent from most of C. England. Europe.

Not distinguishable with certainty in the field from *Graphina anguina* which, however, has muriform ascospores. The apothecia of *G. elegans* **(2)** and *G. scripta* are extremely variable in shape, from short, deeply immersed and simple, to well developed and serpentine or in stellate clusters; on young bark the apothecia tend to be ± parallel to the circumference of the branches, growing most rapidly in the direction the bark itself is expanding.

GYALECTA Ach. (1808)

Thallus crustose, smooth, continuous, sometimes ± cracked and scurfy, effuse, or inconspicuous. **Photobiont** *Trentepohlia*. **Ascomata** apothecia, disc ± flat to strongly concave or urceolate, frequently waxy-looking, pallid, flesh-coloured, orange or brown, translucent and semi-opaque when wet. **Thalline exciple** absent. **True exciple** well developed, persistent, white to grey or concolorous with the disc. **Asci** thin-walled, lacking an apical apparatus or tholus, contents and wall I+ blue, rarely I− (Fig. 33q). **Ascospores** 8 per ascus, ellipsoid to fusiform, trans-septate, to muriform, not halonate, not indented at the septa, colourless. **Conidiomata** pycnidia, colourless to brown. **Conidia** rod-shaped, short. **Chemistry**: no lichen products detected by t.l.c. **Ecology**: on soil, mosses, bark and rock in ± nutrient- or ± base-rich substrata.

Old ascomata frequently lose their hymenium, leaving the shining, pale, sterile base and empty thalline exciple of the hypothecium.

LITERATURE: Lettau (1932–37), Vězda (1958).

1 Ascospores transversely septate only 2
 Ascospores muriform, at least with a single longitudinal septum 5

2(1) Ascospores 3-septate 3
 Ascospores 5- to 13-septate ..... **1. *derivata***

3(2) True exciple not pruinose, even, not crenulate 4
 True exciple white-pruinose, ± crenulate; disc pinkish brown; ascospores 15–23 × 5–8 μm ..... **7. *ulmi***

4(3) Apothecia 0.2–0.5(-0.6) mm diam.; ascospores 9–14(-16) × 4–7 μm, with at least a few oblique septa ..... **4. *geoica***
 Apothecia (0.4-)0.6–0.8(-1.0) mm; ascospores (10-)12–16(-21) × 5–6(-7) μm, the septa consistently transverse ..... **3. *foveolaris***

5(1) Ascospores muriform; septa oblique, rarely transverse 6
 Ascospores 5–9(-13)-septate, with an occasional longitudinal septum ..... **1. *biformis***

6(5) On trees 7
 On rocks and mortar ..... **5. *jenensis***

7(6) Ascospores 9–14(-17) × 6–9(-11) μm, 6 or fewer cells visible in optical view; all septa oblique .... **2. *flotowii***
Ascospores 14–28(-31) × 5–9 μm, 6 or more cells visible in optical view; transverse and oblique septa visible .... **6. *truncigena***

1. G. derivata (Nyl.) H. Olivier (1911)

Thallus thin, or somewhat scurfy, smooth or immersed and then inconspicuous, pale grey or green-grey. Apothecia 0.2–0.3 mm diam., scattered or confluent, ± immersed at first, becoming emergent; disc 0.2–0.4 mm diam., urceolate, concave, with well-developed, raised, smooth and pale or concolorous true exciple, buff to orange. Ascospores 21–31(-33) × (3-)4–6 μm, (5-)11- to 13-septate, oblong-fusiform, with occasional single longitudinal septum (Fig. 14a).

On bark, of wayside, broad-leaved trees, particularly nutrient-rich or wounded trunks of *Ulmus*, *Fraxinus*, *Acer*, *Sambucus* and *Quercus*; local. W. British Is., rare in E. Britain. C. & S. Europe.

Pachyphiale fagicola is similar, but has 16- or more-spored asci. *G. derivata* is considered as a variety of *G. truncigena* **(6)** by some authors.

G. biformis (Körber) H. Olivier (1911) differs only in its shorter, no more than 7-septate ascospores and saxicolous habit. On crumbling serpentine or ± calcareous rocks, especially near the coast. W. Wales (Merioneth), W. Scotland (Dumbarton, Argyll, Westerness).

2. G. flotowii Körber (1855)

Thallus thin, superficial to immersed, inconspicuous, smooth, continuous to somewhat cracked, grey to grey-green. Apothecia 0.2–0.4 mm diam., usually numerous, scattered or confluent, small, sessile to immersed; disc to 0.2 mm diam., concave, with paler or concolorous, prominent, smooth to occasionally crenate, true exciple, cream to orange. Ascospores 9–14(-17) × 6–9(-11) μm, ± spherical to shortly ellipsoid, muriform, most of the cross walls obliquely arranged, with 6 or fewer cells in optical view (Fig. 14c).

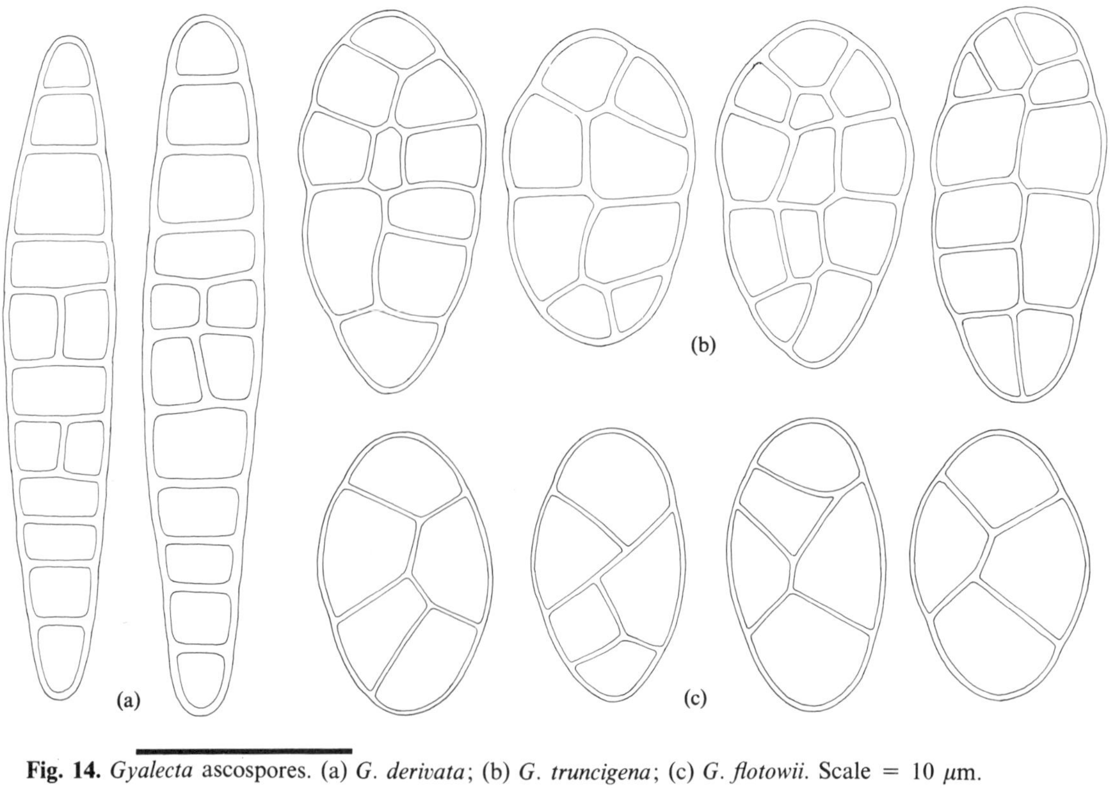

Fig. 14. *Gyalecta* ascospores. (a) *G. derivata*; (b) *G. truncigena*; (c) *G. flotowii*. Scale = 10 μm.

On base-rich bark and nutrient-tracks of *Acer*, *Ulmus* and *Fraxinus*, local. Throughout British Is. Europe, North America.

Differs from the more frequent *G. truncigena* **(6)** in the smaller, usually nearly spherical, fewer-celled ascospores.

3. G. foveolaris (Ach.) Schaerer (1836)

Thallus thin to thick, ± even, becoming coarsely verrucose-granular, grey-white, the granules sometimes tumid and intricately convoluted; surface matt to ± shining, often pruinose, not cracked or areolate. Apothecia (0.4-)0.6–0.8(-2) mm diam., frequent, sessile, scattered to crowded-confluent; disc shallowly to deeply concave-urceolate, flesh-pink to orange; true exciple well developed, smooth to crenate, sometimes ± tumid, white to flesh-pink. Ascospores (10-)12–16(-21) × 5–6(-7) μm, (2-)3(-4)-septate, ellipsoid.

Overgrowing mosses and soil associated with basic and serpentine rocks, mica-schists, epidiorite, in uplands, rarely coastal areas; rare, very local. N. England, Scotland. Europe, North America.

Differs from *G. geoica* **(4)** in the paler thallus, larger apothecia and ascospores with consistently transverse, not oblique, septa.

4. G. geoica (Wahlenb.) Ach. (1808)

Thallus uneven, occasionally somewhat gelatinous, surface thin, matt, pale green-grey, and without convolutions, effuse. Apothecia (0.2-)0.3–0.4(-0.6) mm diam., numerous, ± immersed to sessile, scattered or confluent with well-developed, smooth raised, even, rarely slightly crenate, grey to flesh-pink true exciple; disc 0.2–0.3 mm diam., shallow to deeply concave, yellow-pink when young, often becoming orange to brown. Ascospores 9–14(-16) × 4–7 μm, 3-septate, some oblique, ellipsoid.

On ± calcareous or serpentine soils, crumbling rocks, humus, and associated mosses, also very rarely on base-rich bark (*Ulmus*); rare and local. Scotland (Highlands), extending to scattered localities in England. Europe, North America.

More widespread than *G. foveolaris* **(3)**, which has a better developed, nodulose-verrucose thallus, larger apothecia, and ascospores with consistently transverse, not oblique, septa.

5. G. jenensis (Batsch) Zahlbr. (1924)

Thallus thin or rather thick and then uneven, sometimes cracked or effuse, inconspicuous, grey pinkish or orange-pink, fading in the herbarium. Apothecia 0.3–1 mm diam., usually numerous, sessile, scattered or occasionally confluent, with well-developed, smooth, even to ± coarsely radiate-striate, crenate, pinkish true exciple; disc at first pore-like, then concave, often shiny, pale to deep orange. Ascospores (11-)13–23(-25) × 6–10 μm, 3-septate, muriform when mature, ellipsoid.

On damp, often sheltered and shaded, calcareous rocks as well as mortar, sometimes spreading to adjacent soil and mosses. Throughout British Is. Europe, North America.

The **var. *macrospora*** Vězda (1969) differs in the larger ascospores, (27-)30–38(-40) × (10-)12–16 μm. It occurs on shaded, siliceous rocks, mainly granite, with seepage from adjacent shell-sand and mortar. Very local, S.W. England (Cornwall; Channel Is. Jersey, Guernsey; Is. of Scilly). Endemic.

The immature apothecia are ± spherical prior to the development of the disc and then resemble pale pink perithecia. The thallus is sometimes associated with bright orange tufts of the photobiont (*Trentepohlia*).

6. G. truncigena (Ach.) Hepp (1853)

Thallus inconspicuous, thin to scurfy, encrusting bark, glaucous, greenish or grey. Apothecia 0.2–0.5 mm diam., immersed, rarely superficial, in bark, scattered to confluent, pale orange, flesh-coloured or pale brown; true exciple thick, smooth, even, rarely uneven-wavy, cream to brownish or yellowish pink; disc to 0.25 mm diam., deeply concave, pale to deep orange or brownish. Ascus (Fig. 33q). Ascospores 14–28(-31) × 5–9 μm, ellipsoid, ovoid or oblong-fusiform, sparingly muriform, with 6 or more cells in optical view; septa oblique, rarely regularly transverse (Fig. 14b).

On nutrient-rich bark of trees such as *Ulmus*, *Fraxinus* and *Quercus*, often in well-lit situations; common. Throughout British Is. Europe, North America.

7. G. ulmi (Swartz) Zahlbr. (1905)

Thallus thin, smooth or ± scurfy to finely verrucose-granular, thick white to white-grey or occasionally grey-green, smooth, rarely cracked and following irregularities of the substratum. Apothecia 0.6–1.6 mm diam., numerous, scattered or frequently contiguous, emerging to sessile, scattered to confluent; true exciple pale, ± densely white-pruinose, smooth or often crenate; disc covered at first, 0.2–1.0 mm diam., concave, orange-brown, ± thinly white-pruinose especially when young, sometimes persisting, becoming chestnut-brown with age. Ascospores (12-)15–20(-23) × 5–7(-8) μm, 3-septate, occasionally one, ± oblique, broadly ellipsoid.

On calcareous substrata, such as soil and mosses in limestone areas, and at the base of *Ulmus* in humid and sheltered sites; rare and decreasing. S. & N. England, Scotland. Europe.

Distinguished by its distinctive apothecia, with a chestnut-coloured disc, and a prominent, white-pruinose, ± crenate true exciple.

GYALIDEA Lettau (1937)

Thallus crustose, inconspicuous to superficial, smooth or ± sparingly cracked, sometimes ± gelatinous when moist, effuse. **Prothallus** absent. **Photobiont** *Cystococcus*- or *Leptosira*-like. **Ascomata** apothecia, gyalectoid, urceolate to flat, pale cream, olive-brown, black-brown, rarely pink, translucent when wet. **True exciple** well developed, elevated, generally darker than the disc. **Hymenium** colourless, I+ yellow or red-brown. **Hamathecium** of abundant, septate, mostly simple, paraphyses, 1–1.5 μm wide, rarely thickened at the apices, bound with the asci in a gelatinous matrix. **Asci** (4-)8-spored, subspherical, cylindrical or clavate, thin-walled except for a slightly thickened apical tholus, ± with a small ocular chamber. **Ascospores** (1-)3-septate to muriform, the septa ± constricted, ellipsoid to fusiform, colourless; often with a thin perispore. **Conidiomata** pycnidia, black, immersed in the thallus. **Conidia** bacilliform. **Chemistry**: lichen products not detected by t.l.c. **Ecology**: on soil, mosses and rocks, often in specialized habitats, e.g. on mica-schist, pebbles in or near streams, mine spoil heaps.

Gyalecta differs in having *Trentepohlia* as photobiont, lacking a thin perispore around the ascospores which are also without constricted septa and has a different apical structure of the ascus.

LITERATURE: Vězda (1966).

1 Ascospores 1- to 3-septate 2
Ascospores muriform 3

2(1) Ascospores 1(-3)-septate; disc dark when dry, lilac-grey-opalescent when moist; apothecia 0.2–0.4 mm diam. .... **1. *diaphana***
Ascospores 3-septate; disc pale orange to brown when dry, translucent when moist; apothecia 0.4–1 mm diam. .... **3. *hyalinescens***

3(1) Ascospores mostly >22 μm long; on rocks 4
Ascospores mostly <22 μm long; on soil or mosses .... **6. *subscutellaris***

4(3) Epithecium pale green-brown; true exciple dark brown 5
Epithecium colourless; true exciple pale pink- or yellowish brown to cream .... **5. *roseola***

5(4) Hymenium brown-green above; apothecia regular; on ± damp, siliceous rocks .... **2. *fritzei***
Hymenium dark brown-green above, green-brown colour extending down from epithecium into upper hymenium; apothecia regular or ± contorted; on ± dry, calcareous rocks .... **4. *lecideopsis***

1. G. diaphana (Körber ex Nyl.) Vězda (1978)

Thallus thin, blackish, cracked, effuse ± gelatinous when wet. Apothecia 0.2–0.4 mm diam., superficial; disc flat, dark brown to black when dry, violet- grey-opalescent when wet; true exciple thin, smooth, dark brown to black, unchanged when wet; hymenium 70–80 μm, colourless. Ascospores 8–12(-15) × 5–6 μm, 1(-3)-septate, ellipsoid to fusiform.

On micaceous rocks in streams; very rare. E.C. Scotland (Angus, Glen Clova and E. Perth, Cairngorm). Poland-Czechoslovakia border (Sudeten Mtns.).

G. fritzei **(2)** is similar in habitat preference but has larger, muriform ascospores.

2. G. fritzei (Stein) Vězda (1966)

Thallus thin, superficial or ± immersed, smooth, rarely ± cracked, sometimes ± rusty red, often ± disappearing, effuse, white. Apothecia 0.2–0.6 mm diam., 0.1–0.2 mm high, emergent, sessile and scattered or 2- to 3-clustered; disc persistently deeply concave, perithecioid when young, pale orange to dark brown; true exciple well developed, markedly fuscous brown, sometimes fissured, sometimes appearing spuriously pruinose with loose rock fragments; epithecium pale green-brown; hymenium (80-)120–160 µm tall; hypothecium very pale brown. Asci (76-)100–120 × (25-)30–45 µm, (6-)8-spored. Ascospores (18-)24–35(-43) × (10-)12–16(-21) µm, ellipsoid, subspherical or ellipsoid, occasionally ± attenuated at one end, muriform.

On somewhat basic rocks and pebbles, e.g. basalt and mica-schist in upland areas, often in seepage tracks and spasmodically wet situations; rare but widespread. C. & N. Scotland (Mid-Perth, Argyll, Westerness). Sweden, Czechoslovakia.

Resembles *G. lecideopsis* **(4)**, which occurs on dry limestones and has the upper part of the hymenium darker. A specimen with apothecia with a regular, deeply fissured margin, and smaller, well-muriform spores, 16–18 × 7–9 µm, collected on galena in mid-Wales, may be a new species closely related to *G. fritzei*.

3. G. hyalinescens (Nyl.) Vězda (1966)

Thallus thin, membranous, smooth or ± cracked, effuse, grey to brown-grey. Photobiont *Leptosira*-like. Apothecia to 1 mm diam., dispersed or 2- to 5-confluent, ± immersed or superficial; disc flat to concave, orange to brown, translucent when wet, with a persistent, ± raised, pale grey to brownish margin; hymenium 70–85(-120) µm tall, colourless. Asci 70–80 × 10–14 µm. Ascospores 12–16(-24) × 5–7 µm, 3-septate, cylindric-ellipsoid.

On moist basalt or soft, siliceous rocks in damp ravines; rare. N. Scotland (Argyll), W. Ireland. French Pyrénées, Germany, W. Canada, Azores.

Superficially resembles some morphs of *Hymenelia lacustris* but differs in the well-developed exciple and septate, not simple, ascospores.

4. G. lecideopsis (Massal.) Lettau ex Vězda (1966)

Thallus immersed or thin to inconspicuous, cracked-scurfy, effuse, a pale or greyish stain. Apothecia 0.3–0.6 mm, ± immersed, scattered to confluent; disc deeply concave, whitish to grey-brown, the margin somewhat darker, well developed, smooth to crenate; hymenium 90–130 µm tall, dark green-brown to black-brown in upper part, remainder colourless. Asci 60–120 × 15–30 µm, 4- to 8-spored. Ascospores (17-)22–30(-36) × 10–12(-15) µm, ellipsoid to oblong-ellipsoid, muriform.

On silt-encrusted limestone amongst heavy-metal-rich shingle by river; very rare. N.E. England (N. Northumberland), also on chalk nodules in a downland site (W. Sussex). C. Europe.

Very similar to *G. fritzei* **(2)**, which apparently has a paler brown-green upper hymenium and does not occur on limestone; the delimitation of these taxa is uncertain.

5. G. roseola (Arnold) Lettau ex Vězda (1966)

Thallus thin, smooth or uneven-scurfy, effuse, white. Apothecia 0.3–0.7 mm diam. superficial; disc concave; true exciple well developed, elevated, smooth, concolorous, cream to pinkish or yellowish brown; hymenium 90–120 µm tall, colourless. Asci 70–100 × 15–25 µm. Ascospores (20-)29–34(-36) × (8-)12–14 µm, ellipsoid, muriform.

On shaded limestone at entrance to an old lead mine; very rare. N. Scotland (Argyll, Strontian). N. Sweden, Austria.

Differs from other species in the genus in the flesh-coloured apothecia.

6. G. subscutellaris (Vězda) Vězda (1966)

Thallus continuous, thin, green-brown, ± gelatinous when wet, effuse, becoming grey-brown or blackish or inconspicuous when dry, encrusting soil and mosses and merging imperceptibly into adjacent algal colonies. Apothecia numerous, emergent to superficial, scattered or confluent; true exciple well developed, brown; disc at first urceolate, especially when dry, later becoming ± concave to flat, dull yellow to fawn when young and moist, darkening to brown with age or when dry; hymenium 90–120 μm high, colourless. Ascospores (15-)17–20(-22) μm × 7–10 μm, ellipsoid, muriform.

Overgrowing mosses, such as *Weissia* species, on ± basic, heavy metal-rich soil at three old copper and lead/zinc mines. Very rare, S.W. Wales (Cardigan). Czechoslovakia.

GYALIDEOPSIS Vězda (1972)

Thallus crustose, often filmy, smooth or rarely verrucose, effuse, often with hyphophores; hyphophores erect, spine-like, top-shaped or scale-like (Fig. 15). **Photobiont** trebouxioid. **Ascomata** apothecia, orbicular, collapsed, red-brown to almost black, becoming plump and translucent when wet, with a raised margin. **Thalline exciple** absent. **True exciple** and **hymemium** of a distinctive, loosely anastomosing network of thin hyphae in a gelatinous matrix. **Asci** 2- to 8-spored, cylindrical-clavate to ovate, thickened at the apices, the contents I+ wine-red. **Ascospores** transversely septate to densely muriform, colourless, with a thin to thick perispore. **Conidiomata** pycnidia unknown, but hyphophores may have parallel roles. **Chemistry**: no lichen substances detected by t.l.c. **Ecology**: rocks, trees and associated bryophytes in humid woodland situations, also on montane schist and epidiorite soils.

LITERATURE: James (1975), Vězda (1979).

1 Ascospores muriform; hyphophores grey, green or brown, spine- or top-like 2
Ascospores transversely 15- to 20-septate; hyphophores black, scale-like, with a serrated edge .... **2. *muscicola***

2(1) Ascospores (6-)8 per ascus, < 35 μm long; hyphophores glassy-white, spine-like **1. *anastomosans***
Ascospores (1-)2(-3) per ascus, > 35 μm long; hyphophores brown, top-shaped .... **3. *scotica***

1. G. anastomosans P. James & Vězda (1972)

Thallus filmy, often thin, continuous, skin-like, smooth or ± verrucose, glaucous to pale grey, in patches to 1 cm diam., occasionally coalescing; prothallus often delimiting, pale grey or white. Hyphophores thin, *c.* 0.1 mm long, usually abundant and crowded, sometimes absent, often flattened and translucent at the pointed apices (Fig. 15c). Apothecia frequent, 0.2–0.4 mm diam., scattered or occasionally crowded and contiguous, orange-red to black-brown, round or angular, with margin smooth or irregularly uneven, pale, translucent, when wet with a dark brown margin. Asci (6-)8-spored. Ascospores 20–30 × 6–13 μm, ellipsoid or subclavate, muriform or submuriform, < 30-celled.

On smooth bark, especially *Salix* and *Fagus*, in moist, sheltered woods, copses and carrs, rarer on wood, mosses among boulder scree and on rock underhangs; pollution tolerant; locally frequent, ?increasing. Throughout British Is. France (Brittany), Belgium, Norway, ?Azores.

2. G. muscicola P. James & Vězda (1972)

Thallus very thin, continuous, inconspicuous, forming a pale grey to glaucous grey film over mosses, the hyphae penetrating moss cells. Hyphophores minute, black, hooked forming an apical flange with lateral projections on either side holding a clear, spherical droplet (Fig. 15a). Apothecia frequent, 0.3–0.6 mm diam., brown-black becoming red-brown, single or in groups of 2–4. Asci 8-spored. Ascospores 48–55 × 2–3 μm, cylindrical, 15- to 20-septate.

Over mosses on trunks, rock faces and boulders in very damp situations preferring the 'drip tassels' of pendent mosses, particularly *Isothecium myosuroides*, especially on *Fraxinus*: scattered throughout western oceanic woodlands. S.W. England (Devon), N. Wales, W. Scotland, N.W. Ireland. Crimea, France, Norway.

Superficially resembles *Gomphillus calycioides*, which occurs in similar habitats.

3. G. scotica P. James (1975)

Thallus varnish-like, ± shining and translucent when wet, grey-white, mostly effuse and ill-defined, to *c.* 1.5 cm diam. Hyphophores 0.1–0.3 mm tall, 0.1–0.2 mm wide, creamy brown to red-brown, numerous,

Fig. 15. *Gyalideopsis* hyphophores. (a) *G. muscicola* (on moss); (b) *G. scotica* (on dead hepatic); (c) *G. anastomosans* (on bark). Scale = 1 mm.

evenly scattered, rounded and squat, peg- or top-like (Fig. 15b). Apothecia rare, scattered in groups of 2 to 3, sessile, but often hidden. Asci (1-)2(-3)-spored, thick-walled, ovoid-ellipsoid. Ascospores 40–63 × 22–30 μm, densely muriform.

On decomposing liverworts and mosses over base-rich (mica-schist, epidiorite) mountain soils; very rare. N. England (Lake District), C. Scotland (Ben Alder, Ben Hope and Ben Lawers). Endemic.

Probably overlooked; the peg-like hyphophores are diagnostic.

HAEMATOMMA Massal. (1852)

Loxospora Massal. (1852)

Thallus crustose, weakly corticate, usually partly or ± entirely sorediate. **Photobiont** trebouxioid. **Ascomata** apothecia, sessile, but sometimes ± immersed in soredia; disc red or brown. **Thalline exciple** concolorous with thallus, often sorediate or excluded. **True exciple**, if evident, red or brown. **Hymenium** I+ blue. **Hypothecium** colourless. **Hamathecium** of branched and anastomosed, or rarely simple, paraphyses, not or only slightly swollen at the apices. **Asci** 8-spored, clavate, *Lecanora*-type (mostly poorly developed), or with a uniformly amyloid apical dome. **Ascospores** fusiform to broadly acicular, transversely multiseptate, colourless. **Conidiomata** pycnidia, immersed. **Conidiophores** branched. **Conidiogenous cells** terminal and intercalary, ampulliform, enteroblastic. **Conidia** curved-filiform or bacilliform, simple, colourless. **Chemistry**: various; all species containing at least atranorin in the thallus and anthraquinone pigment(s) in apothecia and pycnidia. **Ecology**: on bark or rocks, rarely on wood.

A widespread genus in tropical and temperate regions of both hemispheres. All British species are sorediate and rarely fertile.

LITERATURE: Rogers & Hafellner (1988).

1 Thallus, at least in younger parts, with delimited soralia; prothallus absent or, if present, then not white or cottony 2
Thallus ± entirely leprose- or farinose-sorediate; usually delimited by a white, cottony prothallus .... **4. *ochroleucum***

2(1) Soralia K+ and Pd+ pale yellowish, or K− and Pd− 3
Soralia K+ bright yellow, Pd+ yellow to orange .... **2. *elatinum***

3(2) Soralia whitish or bluish grey, Pd−, K−, UV+ white; prothallus often conspicuous, blue-grey .... **1. *caesium***
Soralia whitish, Pd+ and K+ pale yellowish, UV−; prothallus absent or thin and grey .... **3. *leprarioides***

1. H. caesium Coppins & P. James (1978)

Thallus effuse and wide-spreading, or forming small, poorly delimited patches amongst other species, thin, ± continuous, pale grey and rather scabrid or (especially on smooth bark) with scattered areoles on a dark, thin, blue-grey prothallus; soralia numerous, efflorescent, discrete, 0.2–1 mm diam., or often confluent, whitish to bluish grey; soredia 30–80(-100) μm diam., external hyphae often blue-grey, N+ red. Apothecia and pycnidia not known. Thallus, medulla and soralia Pd−, K−, KC−, UV+ white (perlatolic acid).

On smooth and rough acid bark and, occasionally, wood, rarely on shaded acid rocks (vertical surfaces), mainly in woodland; common in the west, very rare elsewhere. N. & W. Great Britain, extending locally to S.E. England. Norway, North America.

On smooth bark, easily recognized by its conspicuous, blue-grey prothallus, but on rough bark and wood often confused with *H. elatinum* **(2)** or *Mycoblastus sterilis* which, however, have K+ yellow, Pd+ yellow-orange or red, UV− soralia. *Buellia griseovirens* often has a bluish tinge, but its soralia are smaller, excavate, K+ and Pd+ yellow or reddish, and UV−. *Fuscidea pusilla*, *F. viridis*, *Lecidea botryosa* and *L. nylanderi* all occur on acid bark or wood and have poorly delimited soralia with identical reactions, due to divaricatic or perlatolic acids, but none has a blue-grey pigment in the prothallus or soralia. Recently transferred to *Mycoblastus* as ***M. caesius*** (Coppins & P. James) Tønsb. (1992).

2. H. elatinum (Ach.) Massal. (1854)

Loxospora elatina (Ach.) Massal. (1852)

Thallus effuse, thin, smooth or minutely verrucose, pale grey; soralia 0.6–2 mm diam., convex, pale yellowish to yellow-grey, sometimes bluish grey, often becoming confluent; soredia 40–80(-100) μm diam., the external hyphae colourless. Apothecia 0.4–1.5 mm diam., very rare; disc flat, reddish brown, ± thinly white-pruinose; thalline exciple thin and sorediate, eventually excluded; epithecium straw-brown, with dense crystals, K+ dissolving yellow leaving a pale reddish brown pigment. Paraphyses mostly simple. Asci with uniformly K/I+ blue apical dome. Ascospores 37–49 × 4–5 μm, 3- to 5-septate. Pycnidia rare, immersed in thallus or as small thalline warts, *c.* 200 μm diam.; wall colourless; conidia 3–7-4.5 × 0.8 μm, bacilliform. Thallus and especially soralia Pd+ yellow to orange, K+ bright yellow, UV− (thamnolic and elatinic acids).

On acid bark (especially *Betula* and *Quercus*), less often on wood, in old or boggy woodland; common. N. & W. British Is., extending locally to E. England. N. & C. Europe, Azores.

The K+ immediately bright yellow soralia distinguish this from similar species, e.g. *H. caesium* **(1)**, *Biatora epixanthoides*, *B. epixanthoidiza*, *Ochrolechia inversa* and *Pertusaria hemisphaerica*.

3. H. leprarioides (Vainio) Vainio (1934)

Thallus thin, continuous or weakly rimose, sometimes patchy, pale grey; prothallus grey-black; soralia 0.1–0.3(-0.7) mm diam., rounded, minutely punctiform or fleck-like, discrete or occasionally contiguous, flat to ± convex; soredia farinose to finely granular. Apothecia 0.2–0.4(-0.7) mm diam., rare, often poorly developed, ± emergent to sessile at maturity; disc scarlet, K+ purple-black; thalline exciple persistent, even or ± crenulate, not or partially sorediate; cortex of thallus and edge of thalline exciple densely finely granular, granules dissolving in K+, large coarse crystal clusters below and within photobiont layer and inner part of thalline exciple, not dissolving in K; hymenium 50–60(-70) μm tall; epithecium bright orange-red, finely orange granular, K+ dissolving fleeting violet. Paraphyses simple, apices not swollen. Ascospores (50-)60–80 × 4–5(-5.5) μm, (8-)9–12(-16)-septate, elongate, one end ± attenuated. Thallus and soralia Pd+ pale yellow, K+ pale yellow, KC+ yellow, C−, UV− (atranorin).

On smooth bark and branches of young trees (*Salix*) in boggy, wooded carrs; very rare. S.W. Ireland (Killarney). W. Brittany, Portugal.

This easily overlooked species somewhat resembles underdeveloped forms of *Pertusaria pupillaris*, the soralia of which, however, are Pd+ rust-red (fumarprotocetraric acid); the crystal formation in the thallus and apothecia is diagnostic for *H. leprarioides*.

4. H. ochroleucum (Necker) Laundon (1970)

Thallus soft, leprose or farinose, often wide-spreading (especially on rocks); prothallus usually conspicuous, white cottony. Apothecia 0.4–1.5(-2.5) mm diam., uncommon, at first hidden by thallus granules, later emergent; disc scarlet, K+ purplish; thalline exciple sorediate, often excluded; epithecium scarlet, K+ magenta. Paraphyses richly branched, especially in the upper hymenium. Asci *Lecanora*-type. Ascospores 30–60 × 5–7 μm, 3–7-septate. Pycnidia 150–200 μm diam., sometimes present, immersed but upper part scarlet (K+ magenta); conidia 12–20 × 0.5–0.7 μm, curved-filiform. Thallus Pd− pale yellow, K+ pale yellow, C−, UV− (atranorin, zeorin, porphyrilic acid and ± usnic acid).

On acid or slightly calcareous, relatively dry, vertical rock-faces, church walls, gravestones, often forming patches 1 m or more across, also on slightly enriched tree trunks, rarely on bark (mainly var. *ochroleucum*). Almost throughout the British Is., common in the W. W. & N. Europe, North America.

Two varieties are recognized: var. *porphyrium* (Pers.) Laundon (1970), with whitish to pale greenish grey thallus (lacking usnic acid), is the commoner, especially in S.E. England; var. *ochroleucum*, with pale yellow-green or yellow-grey thallus (usnic acid), occurs in E. England and E. Scotland. The two varieties retain their identities when growing together.

When sterile, distinguished from other similar species (e.g. *Lepraria* and *Pertusaria* species) by the conspicuous, white cottony prothallus. *Phlyctis argena* can be separated by the K+ yellow to red thallus.

HALECANIA M. Mayrh. (1987)

Thallus crustose, ± effuse, rimose-cracked or granular to warted, mostly whitish to bluish- or brownish grey; **cortex** usually thin or indistinct, the terminal cells often with a brown cap; medulla I−. **Photobiont** chlorococcoid. **Ascomata** apothecia, sessile. **Thalline exciple** soon receding with a broad algal zone and a broad to indistinct cortex; outer edge ± with brown layer of apically capped cells. **True exciple** thin laterally, usually expanding above with a surface layer of apically brown-capped cells. **Hymenium** I+ blue. **Hypothecium** colourless. **Hamathecium** of simple, or occasionally forked, septate paraphyses, with swollen, mostly dark brown, capitate apices. **Asci** 8-spored, clavate to subcylindrical, *Catillaria*-type. **Ascospores** colourless, 1-septate, ellipsoid or ovoid- to oblong-ellipsoid, sometimes ± sole-shaped, perispore swelling markedly in K (not apparent in over-mature spores). **Conidiomata** pycnidia, usually present, immersed; wall colourless except around the ostiole. **Conidiogenous cells** enteroblastic, arranged in branched chains, pleurogenous. **Conidia** simple, bacilliform, colourless. **Chemistry**: argopsin, ± zeorin; surfaces often with brownish (K−, N−) or greenish (K−, N+ reddish) pigments.

Similarities in the structure of asci, paraphyses and conidiomata suggest affinities with *Catillaria s.str.*, but *Halecania* differs in having a thalline exciple and ascospores with a distinct perispore. *Lecania s.str.* differs from *Halecania* in the *Bacidia*-type asci, acrogenous, sickle-shaped or curved-filiform conidia, and lack of dark apical caps on its surface cells (incl. paraphyses); also, the epithecial pigment (when present) is reddish brown and K+ purplish.

LITERATURE: Coppins (1989b), M. Mayrhofer (1987).

1 On rocks; thallus not sorediate, but sometimes granular 2
On bark; thallus sorediate .... **5. *viridescens***

2(1) Thallus granular or warted, matt, whitish, brown-grey or blackish; not on sea-shore 3
Thallus rimose-areolate, smooth, oily, bluish to slate-grey; on sea-shore (mesic-supralittoral) rocks .... **2. *ralfsii***

3(2) Epithecium brown, N−; montane 4
Epithecium greenish, N+ reddish; lowland .... **4. *spodomela***

4(3) Thallus granular- to warted, pale to dark brown-grey; thalline exciple persistent **1. *alpivaga***
Thallus surface minutely granular, blackish; thalline exciple receding and inconspicuous .... **3. *rhypodiza***

1. H. alpivaga (Th. Fr.) M. Mayrh. (1987)

Lecania alpivaga Th. Fr. (1871)

Thallus of discrete to clustered, granular warts, to *c.* 1 mm thick, matt, pale to dark brown-grey. Apothecia 0.3–0.8 mm diam., at first ± globose with pore-like disc, expanding to reveal a flat to slightly convex, brown-black disc; thalline exciple mostly entire and persistent, occasionally receding in old apothecia, the cortex distinct, without crystalline inclusions; epithecium and outer edge of true exciple brown, K−, N−. Ascospores 14–17 μm × 5–7 μm (in water, incl. perispore) [× 4.5–5 μm in K, excl. perispore]. Pycnidia usually present; upper wall brown, N−.

On calcareous rocks above 900 m; very rare. C. Scotland (Ben Alder, Ben Lawers). Arctic-alpine; C. Europe, Scandinavia, Spitzbergen, Greenland.

2. H. ralfsii (Salwey) M. Mayrh. (1987)

Lecania ralfsii (Salwey) A. L. Smith (1918) , *Catillaria jejuna* (Nyl.) P. James (1965)

Thallus ± effuse, rimose-cracked, surface ± smooth, waxy or oily in appearance, bluish- to slate-grey; cortex with brownish (pigment caps) and greenish (N+ reddish) pigments. Apothecia 0.2–0.5 mm diam.; disc mostly flat, dark brown to black, paler when wet; thalline exciple concolorous with thallus, soon receding; true exciple blackish, slightly raised when wet; epithecium brown, K−, N−, adjacent upper edge of true exciple usually greenish brown (N+ reddish). Ascospores (13-)15–17(-20) × 6–9 μm (in water, incl. perispore) [× 4.5–6 μm wide in K, excl. perispore] (Fig. 16). Pycnidia numerous; upper wall green-brown, N+ reddish. Thallus Pd+ red, K− or K+ yellowish (argopsin and zeorin).

On hard, siliceous, sea-shore rocks, usually associated with *Lecanora actophila* in the mesic-supralittoral zone. Frequent in W. Britain and Ireland; E. Scotland (rare). N.W. Europe from British Is. to Norway (Sogn og Fjordane).

Dark morphs of *Lecanora helicopis* differ in the Pd− thallus, *Lecanora*-type asci, predominantly simple, shorter ascospores (mostly 9–12 μm long) and thin, curved conidia (12–17 × 0.7 μm).

3. H. rhypodiza (Nyl.) Coppins (1989)

Catillaria rhypodiza (Nyl.) Zahlbr. (1926)

Thallus effuse, to *c.* 200 μm thick, rimose, surface minutely granular, blackish, often invaded by cyanobacteria. Apothecia 0.3–0.6(-0.8) mm diam.; disc black brown when wet, flat; thalline exciple soon

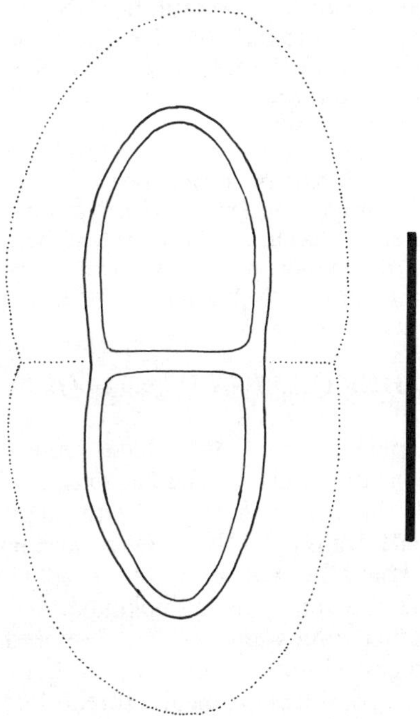

Fig. 16. *Halecania ralfsii* ascospore in 10% K. Scale = 10 μm.

receding and obscured by thalline granules; true exciple thin; epithecium and outer edge of true exciple brown, K−, N−. Ascospores 12–15 × 4.5–6 μm (in water). Pycnidia with brown (N−) upper wall.

Calcareous rocks. C. Scotland (Perth, Ben Lawers and Angus, Caenlochan). Endemic.

The black, granular thallus recalls *Placynthium*. Distinguished from *H. alpivaga* **(1)** by the thallus and inconspicuous thalline exciple; the lowland *H. spodomela* **(4)** differs from both these species in the greenish (N+ reddish) epithecium.

4. H. spodomela (Nyl.) M. Mayrh. (1987)

Lecania spodomela (Nyl.) A. L. Smith (1918), *Lecidea nigrificans* Nyl. (1876)

Thallus effuse, warted-areolate, whitish to pale grey, appearing blackish owing to a scurfy-granular covering of a *Gloeocapsa*-like cyanobacterium (cells 20–26 μm diam.) and sometimes goniocysts (15–40 μm diam.), dark brown hyphae surrounding chlorococcoid algal cells 7–10(-12) μm diam. Apothecia 0.3–0.6 mm diam., flat to slightly convex; disc dark brown to black; thalline exciple grey (but often coated with black granules), persistent or receding, without a distinct cortex; epithecium dull olive-green, K−, N+ reddish, but adjacent edge of true exciple brown, N−. Ascospores 10–13(-15) × 5–7.5 μm (in water, incl. perispore) [× 4–5 μm in K, excl. perispore]. Pycnidia with greenish (N+ reddish) upper wall.

On acid, but slightly enriched, siliceous rocks (e.g. granite), usually near the coast but always above the supralittoral zone; rare or overlooked. S.W. England (Cornwall, Devon), S. Wales (Pembroke), W. Ireland (Galway, Mayo). Sweden, Italy (Piedmont), Romania.

Distinguished by the greenish (N+ reddish) epithecium. The apparent obligate relationships of this species with the blackish cyanobacterium and the goniocysts (? an unidentified parasitic lichen) deserve closer study.

5. H. viridescens Coppins & P. James (1989)

Thallus effuse, minutely warted-areolate, the areoles 0.05–0.1 mm diam., fragile (disintegrating when touched with a needle), pale green to greenish brown, matt, dissolving to form pale, bright green or blue-green soralia; soralia (0.1-)0.2(-0.3) mm diam., punctiform to confluent, farinose; soredia 12–20 μm diam., the surface hyphae with grey- to dark brown walls (K −, N −). Apothecia 0.2–0.4 mm diam., few or absent; disc flat to convex, watery grey to dark brown, not pruinose, sometimes with narrow; thalline margin sorediate, decorticate, often receded; true exciple brown (K −, N −) at upper edge, often thinly white-pruinose; epithecium pale to brown, K −, N −. Ascospores 12–17(-20) × 4–6 μm (in water, incl. perispore) [× 3.5–4.5 μm in K, excl. perispore], fusiform to clavate-fusiform, upper cell slightly wider. Pycnidia not found. Thallus and soralia Pd+ red, K −, KC −, C −, UV − (argopsin).

On slightly enriched bark on upper side of ± prostrate branches of *Salix* and *Sambucus*; rare or overlooked. W. Britain from Devon north to Dumbarton. Norway, Luxembourg.

The only sorediate member of the genus on bark. When sterile, easily mistaken for *Rinodina efflorescens* which has larger (0.1–0.5 mm diam.), less fragile (not distintegrating when touched with a needle) areoles, larger soralia (0.1–0.4 mm diam.), and contains pannarin.

HERTELIANA P. James (1980)

Thallus crustose, superficial, pale-coloured. **Photobiont** chloroccocoid. **Ascomata** apothecia, immersed to sessile, red-brown; disc convex. **Thalline exciple** absent. **True exciple** persistent, pseudoparenchymatous, pale brown to orange. **Epithecium** orange-brown, ± granular. **Hymenium** ± colourless to pale orange, I+ blue. **Hypothecium** well developed, root-like, ± colourless to yellow-orange. **Hamathecium** of simple or sparsely branched paraphyses, the apices swollen. **Asci** elongate-clavate, with a thickened, I+ blue tholus including a less intensely staining ocular chamber, almost *Bacidia*-like, 8-spored. **Ascospores** broadly fusiform, somewhat pointed at the apices, simple, colourless, smooth, lacking a distinct perispore. **Conidiomata** pycnidia, immersed, black. **Conidia** thread-like, falcate, simple, colourless, arising singly from the apices of the conidiogenous cells. **Chemistry**: β-orcinol (atranorin) and orcinol (confluentic acid) *para*-depsides, and unidentified products. **Ecology**: on damp, siliceous rocks in oceanic situations.

This monotypic genus is close to *Biatora*, which differs in the often anastomosing hamathecial filaments, which are not swollen at the apices and in that the ascospores become 1- or more septate; distinctive falcate conidia are also produced. The root-like hypothecium, best seen in strongly convex apothecia, and yellow-orange pigmentation, are particularly characteristic.

LITERATURE: Hafellner (1984), Poelt & Vězda (1981).

1. H. taylorii (Salwey) P. James (1980)

Thallus creamy white, pale greenish white or glaucous to bluish green, thick, cracked; prothallus ± distinct, black, marginal. Apothecia 0.5–0.7 mm diam., immersed at first, then sessile, red-brown, strongly convex to tuberculate; true exciple not raised. Paraphyses 1–2 μm wide, the apices to 3–4.5 μm. Ascospores (16-)18–22(-25) × 8–10(-11) μm. Conidia 17–20 × 1 μm. Thallus Pd −, K+ yellow, KC+ yellow, C −, UV − (atranorin, confluentic acid, 2 unidentified substances).

On siliceous rocks in damp overhangs, or damp, exposed rocks in sheltered woodlands, sometimes with *Gyalidea hyalinescens* and *Lecidea phaeops*, in mainly oceanic situations; local, more frequent in W. Ireland. S.W. England (Cornwall, Devon) to N.W. Scotland (Shetland Is., Highlands), N.E. England (Northumberland), S. & W. Ireland. Madeira, Azores.

A morph with clusters of granular soredia abundantly developed from the surface of the areoles occurs in W. Cornwall (Lizard). The thallus of *H. taylorii* often forms extensive, continuous, sterile, glaucous-green patches minutely dotted with incipient apothecia and pycnidia. *Lecidea phaeops* has a similar aspect but then the apothecia are persistently immersed, *Aspicilia*-like, and the thallus chemistry is different.

HETERODERMIA Trevisan (1868)

Thallus foliose, in rosettes or irregular, sometimes combining to form extensive, radiating mats; marginal lobes discrete or ± contiguous and closely adpressed or ribbon-like, ascending and loosely attached, white to greyish. **Soredia**, lobules or isidia usually present. **Upper cortex** of periclinal hyphae. **Photobiont** chlorococcoid. **Lower surface** with or without a cortex, when present, of periclinal hyphae and with simple or branched rhizines, marginal, also black or pale, hair-like cilia often also present. **Ascomata** apothecia, very rare; disc brown-black. **Thalline exciple** thick, incurved, subhymenium pale. **Asci** 8-spored, *Lecanora*-type. **Ascospores** very thick-walled, 1-septate, often with additional small locules between main locules and spore wall, brown, ellipsoid-oblong, surface smooth. **Chemistry:** atranorin (upper cortex), zeorin and other terpenes, depsidones and pigments occasionally also present (medulla). **Ecology**: on bark or ground amongst low vegetation.

Distinguished from *Anaptychia* by the smooth, thick-walled spores, the more complex chemistry, including atranorin, and the generally paler coloured thallus.

LITERATURE: Kurokawa (1962, 1973); Culberson (1966).

1 Thallus firmly attached; lobes rosette-forming, small or absent, lacking marginal cilia; isidia, soralia or lobules present 2
Thallus ± loosely attached; lobes entangled, lobes ascending, with long, black marginal cilia; soredia absent or in a central furrow on the underside of lobes ..... **2. *leucomelos***

2(1) Soredia absent, but with lobules or isidia towards the centre of thallus; underside pale, K− or yellow 3
Soralia lip-shaped, on the underside of broadened lobe tips; underside yellow-orange, K+ purple ..... **3. *obscurata***

3(2) Medulla K+ red; lobules present, ± flattened, compacted, overlapping, never erect, sometimes coarsely granular; lobes shortly spreading ..... **4. *propagulifera***
Medulla K−; isidia present, spathulate or ± cylindrical; lobes mostly erect . **1. *isidiophora***

1. H. isidiophora (Vainio) Awasthi (1973)

Thallus 1–2 cm diam., ± rosette-forming; lobes 0.5–1 mm wide, ± flat, contiguous, not ascending, sparingly branched; upper surface grey-white, ± entirely covered in isidia, except at lobe margins; isidia at first rounded, ± dorsiventral, ± flattened or erect and cylindrical, simple or sparingly branched; lower surface corticate, white, with white to dark brown-black rhizines extending to the lobe margins, to 0.8 mm. Apothecia not observed in British material. Medulla Pd−, K± yellow, KC−, C− (zeorin).

On mosses amongst serpentine rocks; very rare. S.W. England (Cornwall, Lizard peninsula). Central & South America, Africa, Asia.

The British collection has reduced marginal lobes but otherwise resembles extra-European material.

2. H. leucomelos (L.) Poelt (1965)

Thallus 5–15 cm across, often in loose rosettes or forming ± entangled mats, ± loosely attached; lobes 0.5–3 mm wide, elongate, mostly dichotomously branched, entangled, ± ascending at apices, sometimes reflexed, with conspicuous, long, grey or black, simple or sparsely branched marginal cilia, 5–9 mm long; upper surface ivory-white, smooth; lower surface white, channelled, central part arachnoid or powdery and ± sorediate; lower cortex not developed. Apothecia not observed. Medulla Pd+ yellow, K+ yellow-red, KC+ yellow-red, C− (zeorin and salazinic acid).

On mossy rocks or moss-lichen turf on sunny, exposed, coastal cliffs, very rarely on trunks and branches of wayside, broad-leaved trees; rare, local and declining. S.W. England (Cornwall, Is. of Scilly, formerly in Devon, Dorset and Sussex), Channel Is., Wales (Anglesey), S.W. Ireland. Widespread in tropics and subtropics.

The white, ascending, extended lobes and long, marginal, often intertwined, dark cilia are characteristic. The development of soredia is variable, in some specimens they are absent and the lower cortex remains arachnoid but frequently at least the distal part of the underside is sorediate and then ± strongly reflexed and exposed by an upward curling of the lobes.

3. H. obscurata (Nyl.) Trevisan (1869)

Thallus 3–8 cm diam., ± forming neat rosettes or more rarely irregularly spreading; lobes 0.7–2 mm wide, ± flat, contiguous, not ascending or ± raised at the tips, ± palmately branched; upper surface smooth, ivory white with spreading, lip-shaped soralia on recurved apices of lobes and black, marginal cilia; lower surface white, not corticate, with orange-brown hyphae overlying the medulla, K+ purple. Apothecia very rare; disc brown-black with a coarsely crenulate, sometimes sorediate, thalline exciple. Ascospores (23-)25–30(-32) × 15–18 μm. Upper surface Pd−, K± yellow, KC−, C− (zeorin), lower surface K+ purple (unidentified pigments).

On mossy trunks, short turf and on mossy rocks in ancient parklands, woodlands and exposed coastal heaths; rare. S. & W. Britain, W. Scotland, C. & W. Ireland. Europe, North and South America, Africa, Asia, Australia, New Zealand.

The British localities represent the most northern limit of its predominantly southern-oceanic range in Europe.

4. H. propagulifera (Vainio) Dey (1977)

Like **3**, but differs in the smaller size, the presence of minutely dissected, often densely crowded, ± granular, nodulose and eroded, dorsiventral lobules on the upper surface, concentrated at the centre of the thallus and medulla reacting Pd+ orange, K+ red (crystals), KC+ red, C−, UV− (norstictic acid, zeorin).

On peaty soil with *Calluna* on low, exposed, coastal cliffs; very rare. Is. of Scilly (Tresco). Europe, Africa, Asia, North and South America, Hawaii.

H. isidiophora **(1)** differs in having ± cylindrical, true isidia, not minute, congested overlapping lobules. The K+ red reaction separates *H. propagulifera* from the other species.

HYMENELIA Krempelh. (1852)

Thallus crustose, immersed to superficial. **Cortex** poorly differentiated. **Photobiont** chlorococcoid. **Ascomata** apothecia, aspicilioid, immersed within thallus or substratum; disc concave to flat. **True exciple** thin, ± prominent, without photobiont cells, narrow but often broadening above to form the margin, colourless or pale-pigmented (especially at upper surface). **Hypothecium** colourless to pale brownish. **Hamathecium** of paraphyses, simple, or ± branched above, becoming shorter celled above and ± constricted at the septa, apical cells ± strongly conglutinate, swollen, without well-defined pigmented hoods or caps. **Asci** 8-spored, cylindrical-clavate; outer coat I+ blue, but inner walls and apical dome I−, K/I−. **Ascospores** globose to ellipsoid, simple, colourless, without a distinct perispore. **Conidiomata** pycnidia, immersed. **Conidiogenous cells** elongate-ampulliform, enteroblastic, acrogenous. **Conidia** bacilliform, simple, colourless. **Chemistry**: no lichen products detected by t.l.c. **Ecology**: on rocks, usually on periodically inundated or moist rocks; *c.* 9 species, temperate Northern Hemisphere.

The two species treated have pale apothecial discs which distinguish them from *Aspicilia*, whose species have blackish discs, as well as a usually thicker thallus with a well-differentiated pseudoparenchymatous cortex. *Hymenelia* is doubtfully distinct from *Ionaspis*, which differs in possessing *Trentepohlia* as photobiont.

LITERATURE: Poelt & Vězda (1981).

1	Thallus superficial, well developed; on acid rocks	**1. *lacustris***
	Thallus immersed or superficial and poorly developed; on hard calcareous rocks	**2. *prevostii***

1. H. lacustris (With.) M. Choisy (1949)

Thallus to 0.4 mm thick, smooth, ± even, continuous to cracked (especially around the apothecia), pale whitish cream to deep rusty red, effuse or mosaic-forming and then delimited by red-brown hypothalline lines. Apothecia 0.15–0.4(-0.6) mm diam., often crowded, ± immersed, or with a slightly raised true exciple, rounded to ± irregular, pale orange- to red-brown, but pale pink to bright orange when wet; true exciple ± colourless, the uppermost and outer parts pale brown to red-brown, the outer edge granular; epithecium pale orange to dark red-brown, interspersed with minute granules not dissolving in K;

hymenium 90–105 µm tall. Ascospores 13–20 × 6–11 µm, broadly ellipsoid to ± globose. Pycnidia 50–80 µm diam., red-brown; conidia 4.5–6.5 × 1 µm.

On siliceous rocks by rivers and lakes, often forming an extensive zone of mosaics above or with aquatic *Verrucaria* species (especially *V. aethiobola*), also in seepage tracks; common in upland areas. Throughout British Is. N. & C. Europe, Mediterranean mountain region, North America.

Rhizocarpon lavatum occurs in similar habitats, often with this species, and has a similarly coloured thallus, but has immersed black apothecia and muriform ascospores.

2. H. prevostii (Duby) Krempelh. (1852)

Thallus immersed or thin and scurfy, or to 0.3 mm thick and cracked-areolate, ivory, fawn or pale brown, effuse. Apothecia 0.1–0.4(-0.5) mm diam., scattered to continuous, immersed in shallow to deep pits in the substratum or thallus, when dry separated from wall of pit by a gaping fissure; disc concave, rarely flat, pale pink to deep pinkish brown, ± transparent when wet; true exciple colourless; epithecium colourless to pale red-brown; hymenium 120–160 µm tall. Ascospores globose to broadly ellipsoid, 10–24(-26) × 7–17(-19) µm. Pycnidia not seen.

On hard limestones, especially on ± horizontal surfaces, including flat chest-tombs, often in slightly sheltered or moist situations; local. Throughout Britain, N.W. Ireland (Fermanagh and Sligo). Scandinavia, Alps, Jura.

Recalls *Ionaspis epulotica*, which has a yellowish to pinkish grey thallus containing *Trentepohlia* (cells 20–40 µm diam.) and not chlorococcoid algae (cells 7–15 µm diam., rarely ellipsoid to 20 × 15 µm). Collections from England and Wales have large, mostly ellipsoid, ascospores, 17–24 × 12–17 µm, whereas those from Scotland have smaller, often globose, ascospores, 10–17 × 7–12 µm; the possibility that two distinct taxa are involved merits further study.

HYPERPHYSCIA Müll. Arg. (1894)

Physciopsis M. Choisy (1950)

Thallus foliose, lobate, very closely adpressed. **Lobes** ± radiating, pale brownish grey to dark brown, epruinose, matt; underside pale, with a few, inconspicuous, very short rhizines. **Upper cortex** pseudoparenchymatous, the cells with lumina 3–7 µm diam.; **lower cortex** colourless, prosoplectenchymatous, the cells with lumina *c.* 2 µm wide. **Soralia** present. **Photobiont** trebouxioid. **Ascomata** apothecia, laminal, sessile; disc brown, not pruinose. **Thalline exciple** present. **Epithecium** pale brown. **Hymenium** colourless, I+ blue. **Hamathecium** of paraphyses, often branched above; apices clavate, light brown with a thin, dark brown cap. **Asci** 8-spored, cylindrical-clavate, *Lecanora*-type. **Ascospores** brown, 1-septate, thick-walled. **Conidiomata** pycnidia, immersed; walls colourless except for brown zone around ostiole. **Conidiogenous** cells arising in branched chains, barrel-shaped to subcylindrical, enteroblastic, acrogenous and pleurogenous. **Conidia** thread-like. **Chemistry:** lichen products not detected by t.l.c. **Ecology**: on bark or rarely rock.

LITERATURE: Hafellner *et. al.* (1979), Maas Geesteranus (1952), Moberg (1977).

1. H. adglutinata (Flörke) Mayrh. & Poelt (1979)

Physciopsis adglutinata (Flörke) M. Choisy (1950)

Thallus to 2 cm diam., ± orbicular, or irregular and confluent with others; lobes 0.2–0.5 mm wide, separate or ± overlapping; soralia laminal, *c.* 0.15–0.3 mm diam., spot-like, sometimes ± confluent at centre of thallus, concolorous or paler. Apothecia rare; thalline exciple ± smooth. Ascospores 13–19 × 7–9 µm, but often poorly formed. Pycnidia rare; conidia 11–21 × *c.* 0.5 µm.

On nutrient-rich or -enriched (especially dust impregnated) bark of tree trunks in parkland and wayside situations, rarely also on shaded, vertical surface of calcareous or nutrient-rich rocks and walls; commonest in south. Throughout lowland British Is. S. Europe to S. Scandinavia, Saudi Arabia, East Africa, Central, North & South America, Australia, New Zealand.

Resembles diminutive morphs of *Phaeophyscia orbicularis* but is distinguished by the very adpressed lobes, difficult to detach, and colourless, prosoplectenchymatous lower cortex.

HYPOCENOMYCE M. Choisy (1951)

Thallus crustose to squamulose, pale grey to dark brown. **Upper cortex** usually of 2 layers, the upper of colourless, ± periclinally oriented hyphae, the lower [in *anthracophila* **(1)**, *friesii* **(3)** and *scalaris* **(4)**] of randomly arranged or ± anticlinally orientated hyphae. **Soredia** sometimes present. **Photobiont** chlorococcoid. **Medulla**, when present, of loosely interwoven, randomly orientated, ± thin-walled hyphae. **Ascomata** apothecia, sessile, brown-black, persistently flat to convex; disc frequently gyrose and margin flexuose. **Thalline exciple** absent. **True exciple** raised, persistent or becoming excluded, of radiating hyphae, colourless to dark brown in inner part, rim green or brown. **Epithecium** brown-green. **Hymenium** I+ blue. **Hypothecium** colourless to brown. **Hamathecium** of paraphyses, septate, sparingly branched, weakly conglutinate. **Asci** 8-spored, clavate to cylindrical; apical dome K/I+ blue but poorly developed in some species. **Ascospores** simple or 1(-3)-septate, ellipsoid-fusiform. **Conidiomata** pycnidia, sessile, globose-ovate, black. **Conidia** subglobose to filiform, colourless. **Chemistry**: fumarprotocetraric, protocetraric or lecanoric acids, or unidentified substances. **Ecology**: almost exclusively on wood or bark, often on burnt wood, rarely on rocks and walls.

Distinguished from *Lecidea* by the squamulose thallus, the weakly conglutinate paraphyses, the ± fusiform and sometimes septate ascospores, the sessile pycnidia, and preference for charred wood.

LITERATURE: Timdal (1984).

1 Thallus squamulose, Pd− or Pd+ red, K− 2
Thallus crustose, Pd+ yellow, K+ yellow .... **5. *xanthococca***

2(1) Squamules C− 3
Squamules C+ red .... **4. *scalaris***

3(2) Squamules adpressed, bullate or irregularly ascending; thallus Pd−; apothecia flat, black, marginate 4
Squamules ascending, sorediate; thallus Pd+ red; apothecia strongly convex, brown, immarginate .... **1. *anthracophila***

4(3) Squamules bullate or irregularly ascending, dull; ascospores ellipsoid-fusiform, simple to 1(-3)-septate .... **2. *caradocensis***
Squamules normally adpressed, ± shiny; ascospores ellipsoid, simple .... **3. *friesii***

1. H. anthracophila (Nyl.) P. James & G. Schneider (1980)

Thallus squamulose; squamules to 0.8(-1.3) mm diam., ± ascending, at first concave, becoming convex, the margin slightly upturned, entire, with lip-shaped, farinose soralia; upper surface green-brown, shiny; lower surface pale brown, with grey soralia. Apothecia to 0.8(-1.3) mm diam., infrequent, attached marginally or laminally to lower surface of squamule, simple or in groups, becoming convex; true exciple becoming excluded; epithecium brown, N−. Ascospores (7-)9.9(-13) × (1.5-)2.1(-2.5) μm, narrowly ellipsoid-fusiform, 0(-1)-septate. Pycnidia rare, the wall brown, K−, N−; conidia 6.5–12.5 × *c*. 1 μm, bacilliform to shortly thread-like. Thallus Pd+ red, K−, KC−, C−; soralia and medulla UV+ bluish white (fumarprotocetraric and protocetraric acids, unidentified UV+ substances).

On *Pinus* stumps in relict woodland. Scotland (Glen Strathfarrar). Boreal Fennoscandia, Corsica, Greece, North America, Japan.

Characterized by the ascending squamules, with sorediate margins and lower surface, and the Pd+ red (fumarprotocetraric acid) reaction.

2. H. caradocensis (Leighton ex Nyl.) P. James & G. Schneider (1980)

Thallus squamulose; squamules to 1(-1.5) mm diam., round or irregular, often crowded, bullate or irregularly ascending, esorediate, margin crenulate; upper surface greyish green to dark brown, dull. Apothecia to 0.5(-0.8) mm diam., infrequent, attached laminally to squamules, flat and marginate, black, epruinose. Ascospores (6.5-)10.6(-14) × (2-)3.4(-4) μm, 0 to 1(-3)-septate, narrowly ellipsoid to fusiform.

Pycnidia rare, the wall brown, K−, N−; conidia *c.* 2.5–4.5 × 1.2 μm, oblong-ellipsoid. Thallus Pd−, K−, KC−, C−; medulla UV+ white (unidentified substances).

On bark and especially wood, tolerating atmospheric pollution and occasionally growing with *Lecanora conizaeoides*. E. & C. England, Scotland. Atlantic-subatlantic Europe, Alps.

Characterized by the bullate, frequently deformed and partly flattened squamules. Closely related to *H. friesii* **(3)**; young, sterile specimens may be indeterminable, though older specimens are distinguished by the squamule-shape, the dull upper surface and the partly fusiform, septate spores.

3. H. friesii (Ach.) P. James & G. Schneider (1980)

Thallus squamulose; squamules to 1.0(-1.5) mm diam., adpressed or rarely ± ascending, discrete, flat or convex, esorediate, margin entire, crenulate or incised, not upturned; upper surface greyish green to dark brown, shiny, margin concolorous. Apothecia to 1.0(-1.4) mm diam., normally abundant, attached laminally to squamules, flat and marginate, black, not pruinose; epithecium brown, K−. Ascospores (4.5-)6.2(-7.5) × (2.5-)3.0(-3.5) μm, simple, ellipsoid. Pycnidia frequent, the wall dark brown, K−, N−; conidia 2.5–5 × 1 μm, bacilliform. Thallus Pd−, K−, KC−, C−; medulla UV+ white (unidentified substances).

On coniferous bark and wood, especially burnt, in relict pine woodlands, more rarely on old *Quercus*. Scotland. Fennoscandia, Germany, Austria , France, Italy, U.S.A.

In exposed sites squamules are dark brown, ± flat, with an entire margin, in more sheltered sites they are paler, ± convex, with a crenulate, deeply incised margin and may be difficult to distinguish from *H. caradocensis* **(2)** when young.

4. H. scalaris (Ach. ex Lilj.) M. Choisy (1951)

Thallus squamulose; squamules to 1.2(-2.0) mm diam., ascending, ± remaining discrete, often overlapping, the margin slightly upturned, entire or crenulate with lip-shaped, farinose soralia; upper surface greyish green, olive-brown, to dark brown, dull; lower surface white-sorediate along margin, yellowish brown or greenish. Apothecia to 1.5(-2.5) mm diam., infrequent, attached marginally to the base of the squamules, flat and marginate, black with bluish white pruina; epithecium greenish, N+ reddish. Ascospores 7–8 × 3–4 μm, narrowly ellipsoid to fusiform, simple. Thallus Pd−, K−, KC+ red, C+ red; medulla and soralia UV+ white (lecanoric acid and unidentified substances).

On acid bark and wood, also occasionally siliceous rockwork, memorials and brick; common. Throughout British Is. Europe, U.S.S.R., North America, Australia.

Characterized by the ascending, usually overlapping, marginally sorediate, C+ red squamules, which vary from greyish green to dark brown and which may have an entire or crenulate margin. Fertile specimens are usually only sparingly sorediate.

5. H. xanthococca (Sommerf.) P. James & G. Schneider (1980)

Thallus crustose, areolate; areoles to 0.7(-1.2) mm diam., not sorediate; surface pale grey to yellowish brown, dull. Apothecia to 0.8(-1.2) mm diam., often present, attached to areoles, flat, black, not pruinose; true exciple persistent, raised; epithecium greenish, sometimes brownish, K+ violet, N+ red. Ascospores (7-)7.9(-13) × (3-)4–6(-7) μm, simple, ellipsoid. Pycnidia usually numerous, the wall greenish, N+ reddish; conidia 2.5–4 × 1.5–3 μm, subglobose. Thallus Pd+ yellow, K+ yellow, KC+ red, C+ red, UV− (alectorialic acid and unidentified substance).

On wood in native pinewood. Scotland (E. Highlands). Fennoscandia.

Characterized by the crustose, areolate thallus, which reacts C+ red, and the flat, black apothecia with a persistent, raised exciple. Stains herbarium paper pink-brown in time.

The recently recorded *H. sorophora* (Vain.) P. James & Poelt (1981) is similar, but the areoles burst into diffuse, greenish or yellowish brown, farinose soralia. The conidia are 3.5-5 × 1.5–2.5 μm, ellipsoid or bacilliform (more so than in *H. xanthococca*). Chemistry similar. On wood of fallen *Quercus*, Hants (New Forest).

H. leucococca R. Sant (1986) has scattered, white, subsquamulose areoles with marginal, ± lip-shaped soralia, and a similar chemistry; apothecia unknown; it has been collected once on pine bark in N.W. Scotland (W. Ross, Coulin Forest).

HYPOGYMNIA (Nyl.) Nyl. (1896)

Thallus foliose, attached by adhesive discs below, or rarely by most of the lower cortex. **Lobes** inflated, often ± tubular, solid or ± hollow. **Upper surface** corticate, grey or ± suffused brown. **Soredia** often present. **Photobiont** chlorococcoid. **Medulla** white or ± hollow, hyphae loosely woven, arachnoid or ± absent. **Lower surface** black, corticate, shiny, wrinkled, without rhizines, sometimes with gaping perforations. **Ascomata** apothecia, sessile or ± prominently stalked. **Thalline exciple** well developed, persistent, concolorous. **Disc** shining, concave to flat, red- to yellow-brown. **Asci** 8-spored. **Ascospores** simple, colourless, ± globose. **Conidiomata** pycnidia, immersed, wall blackened around the ostiole, pale below. **Conidia** 5–8 × 1 μm, bifusiform. **Chemistry**: upper cortex with atranorin, the medulla a range of depsidones (including physodic, physodalic and fumarprotocetraric acids).

Parmelia and related genera have flattened lobes with rhizines; physodic and physodalic acids are never present. *Menegazzia* has perforations on the upper surface of the lobes, a different chemistry and larger, ellipsoid, thick-walled ascospores. *Cavernularia* has a smaller thallus with numerous perforation-like depressions below.

LITERATURE: Elix (1979), Krog (1974), Maas Geesteranus (1947).

1 Medulla hollow or hyphae lax, arachnoid; soredia present 2
Medulla solid, compact; soredia absent .... **2. *intestiniformis***

2(1) Soralia discrete, confined to ends of the lobes, Pd+ or Pd− 3
Soralia diffuse, laminal, irregularly spreading, Pd− .... **1. *bitteriana***

3(2) Soralia capitate on ends of lobes; medulla Pd− .... **4. *tubulosa***
Soralia lip-shaped on ruptured, ± upturned ends of lobes; medulla Pd+ orange to rust-red .... **3. *physodes***

1. H. bitteriana (Zahlbr.) Räsänen (1947)

Thallus to 5 cm diam., closely adpressed, ± forming rosettes or irregularly spreading; lobes 1–3 mm wide, hollow, ± convex but flattened towards the ends, contiguous and overlapping, sometimes intricately convoluted at centre, often ± discrete towards margins; upper surface grey, wrinkled; soralia spreading, laminal; lower surface black. Apothecia not observed in British material, to 6.5 mm diam.; disc red-brown. Ascospores 6–7.5 × 3–4.5 μm. Medulla and soralia Pd−, K−, KC+ red, C−, UV+ pale violet-blue (physodic acid).

On trunks of *Pinus* in relict Caledonian forests, once on *Fagus*; rare. Scotland (Highlands, Midlothian). Europe.

Distinguished by the diffuse, laminal, Pd− soralia; this species may be a recent arrival and starting to spread.

2. H. intestiniformis (Vill.) Räsänen (1943)

Thallus to 5 cm diam., loosely attached, forming ± complete rosettes or irregularly spreading; lobes narrow, to 1 mm, ± convex, ± nodular-uneven, contiguous, overlapping, often narrower and ± ascending towards the centre; upper surface pale grey, ± tinged brown especially towards the apices, ± shiny; lower surface black. Apothecia rare, to 8 mm diam.; thalline margin often irregular and sub-lobulate. Pycnidia numerous. Medulla Pd+ rust-red, K−, KC−, C−, UV− (fumarprotocetraric acid).

On exposed, siliceous boulders above 900 m; local. Scotland (Highlands). Europe.

The centre of the thallus often has somewhat flattened, decumbent, branching, ± unorientated, finger-like lobes. Two closely related European species, *H. atrofusca* (Schaerer) Räsänen (1943) and *H. oroarctica* Krog (1974), contain physodic acid (medulla UV+ blue-white); these taxa could occur in the Scottish Highlands.

3. H. physodes (L.) Nyl. (1896)

Thallus to 10 cm diam., ± loosely attached, forming rosettes or irregularly spreading, often forming large patches; lobes 2–3 mm wide, ± hollow, often ascending towards the tips; upper surface grey, smooth, ±

shining; lower surface black, pale brown towards margin, wrinkled; soralia lip-shaped, developing from the rupture of the underside tips of the lobes, becoming upturned and fan-like. Apothecia rather rare, with short, thick stalks; disc red-brown; thalline exciple thin, persistant. Medulla and soralia Pd+ orange to red, K−, KC+ red, C−, UV+ pale violet-blue (physodic, physodalic, 3-hydroxyphysodic and ± protocetraric acids).

On siliceous rocks, trees, *Calluna* stems and other acidic (*c.* pH 4.5) substrata, supralittoral to montane; abundant. Throughout British Is. Europe, North America, East Africa, Himalaya.

A polymorphic species, ranging from contorted, plate-like, thalli with coarse, overlapping and swollen lobes, to thin, finely finger-like thalli with discrete lobes; the latter is often characteristic of moist, sheltered habitats, e.g. amongst *Calluna*. Characterized by the Pd positive reaction and the presence of spreading, upturned, lip-shaped soralia at the ends of lobes. One of the most ubiquitous macrolichens in the British Is. and absent only where mean SO_2 levels exceed 100 $\mu g\ m^{-3}$.

4. H. tubulosa (Schaerer) Havaas (1918)

Like **3**, but the lobes more tubular and ascending and the upper surface more roughened, matt; the soralia are rounded, capitate, not ruptured, each involving the entire tip of a lobe. Medulla and soralia Pd−, K−, KC+ red, C−, UV+ pale violet-blue (physodic and 3-hydroxyphysodic acids).

On similar substrata as *H. physodes* **(3)**, and often with it, but not as frequent. Throughout British Is. Europe, North America, East Africa, Himalaya.

ICMADOPHILA Trevisan (1852)

Thallus crustose, superficial, delimited, not corticate; prothallus white or absent. **Photobiont** *Coccomyxa*. **Ascomata** apothecia, rounded or sub-lobulate, pale, sessile or slightly elevated, appearing very shortly stalked. **Thalline exciple** soon excluded. **True exciple** well developed, persistent or partially excluded, extended below, of intricately interwoven hyphae. **Hymenium** I+ blue. **Hamathecium** of simple, slender paraphyses, apical cells often swollen. **Asci** (6-)8-spored, cylindrical, K/I− except for a thin, dark blue cap in the apical wall; apical dome with a concave depression at the base, without an ocular chamber; not fissitunicate. **Ascospores** ellipsoid-fusiform, 1- to 3-septate, colourless, without a perispore. **Conidiomata** pycnidia, immersed; wall colourless. **Conidiogenous cells** enteroblastic, in chains, short-cylindrical. **Conidia** arising laterally and terminally, bacilliform, simple, colourless. **Ecology**: on humus, moorlands, rotting wood.

Icmadophila is most closely related to certain disparate elements of *Baeomyces*, e.g. *B. roseus*.

LITERATURE: Frey (1933).

1. I. ericetorum (L.) Zahlbr. (1895)

Thallus rather soft, of compacted, irregular granules forming a continuous very uneven crust, pale green, glaucous-green or whitish-grey, K+ yellow. Photobiont cells 6–10 × 3.5–5 μm. Apothecia 1–2(-3) mm diam., pink or pale orange-pink; true exciple thin or excluded, paler or concolorous with the disc, even, smooth or irregularly crenulate or nodulose; disc smooth or ± corrugate-wrinkled, naked or faintly pruinose; epithecium red-brown, densely packed with minute crystals, K+ dissolving, orange; hymenium 120–140 μm tall; hypothecium colourless, central part of interwoven hyphae with scattered, interspersed clusters of coarse crystals, K− not dissolving. Paraphyses *c.* 1 μm wide, the apices to 5 μm. Ascospores 13–27 × 4–6 μm. Conidia 3.5–4 × 0.5–1 μm. Thallus Pd+ orange, K+ orange, KC+ orange, C−, UV+ glaucous (apothecia UV−) (thamnolic, perlatolic acids).

On damp ground, peat and rotted wood, especially exposed moorland; usually in uplands, very rare and decreasing in lowlands. Widespread in N. England, Scotland and Ireland, very local elsewhere. Widespread in N. Hemisphere, New Zealand.

Often confused with *Baeomyces roseus*; all lowland records need careful checking against this species.

IONASPIS Th. Fr. (1871)

Thallus crustose, immersed or superficial, smooth to cracked-areolate, effuse or ± delimited by a whitish prothallus, yellow-white to dark grey but with a ± pink tinge when fresh, not corticate; algal layer and medulla not differentiated. **Photobiont** *Trentepohlia*. **Ascomata** apothecia, aspicilioid, immersed, rarely becoming ± sessile, pallid to blackish; disc concave at first, later ± flat. **True exciple** thin, ± conspicuous except towards the surface where it broadens to form a distinct margin, colourless or pale brownish to green above. **Epithecium** not well differentiated. **Hymenium** colourless, or pale brownish to green or blue-green in upper part, I+ reddish, greenish or bluish. **Hypothecium** colourless. **Hamathecium** of paraphyses, simple or ± branched above, shorter-celled above and ± constricted at the septa; apices without well-defined pigmented hoods or caps. **Asci** 8-spored, clavate to cylindrical, the outer coat I+ blue, the inner walls and apical dome I−. **Ascospores** globose to ellipsoid, simple, colourless, thin-walled, without a distinct perispore. **Conidiomata** pycnidia, immersed. **Conidiogenous cells** enteroblastic, elongate-ampulliform, acrogenous. **Conidia** 3–8 × 0.7 μm, bacilliform, simple. **Chemistry**: lichen products not detected by t.l.c. **Ecology**: on ± moist, calcareous or siliceous rocks in upland habitats. **Distribution**: northern latitudes and montane regions of Northern Hemisphere.

Seems to be identical to *Hymenelia* except for the trentepohlioid photobiont.

LITERATURE: Jørgensen (1989), Magnusson (1933).

1 Apothecia pale pink or pale brown; hymenium colourless or pale brownish to olivaceous above 2
Apothecia dark bluish or blackish; hymenium blue-green above 3

2(1) Upper parts of hymenium, true exciple and ± thallus K+ violet; on acid rocks .. **5. *odora***
All parts K−; on calcareous rocks .... **2. *epulotica***

3(1) Upper hymenium N+ red 4
Upper hymenium N− .... **6. *suaveolens***

4(3) Ascospores 6–12(-13) μm diam. or long; thallus superficial 5
Ascospores (10-)15–17(-20) μm long; thallus ± immersed .... **4. *melanocarpa***

5(4) Thallus smooth or faintly cracked; on acid rocks .... **1. *cyanocarpa***
Thallus cracked-areolate or powdery; on calcareous rocks .... **3. *heteromorpha***

1. I. cyanocarpa (Anzi) Jatta (1910)

Thallus thin, smooth or faintly cracked, whitish grey or pinkish. Apothecia 0.2–0.5 mm diam., immersed, exciple ± prominent; disc blue-black, true exciple concolorous or paler; hymenium 85–100 μm tall, the upper part blue-green, K−, N+ reddish. Ascospores 10–12(-14) × 5–8.5 μm. Pycnidia 50–80 μm diam., numerous, blackish, the wall green.

On acid rocks by streams or on flushed cliff faces, over 700 m. Scotland (Highlands). N. and montane Europe, Siberia.

Differs from *I. suaveolens* **(6)** in the N+ reddish hymenium and usually numerous pycnidia.

2. I. epulotica (Ach.) Arnold (1871)

I. rhodopsis (Sommerf.) Blomb. & Forss. (1880)

Thallus immersed to cracked-areolate, yellowish to pinkish grey. Apothecia (0.2-)0.3–0.5(0.7) μm diam., immersed, when dry often separated from thallus by a circular fissure, occasionally emergent, pink (especially when wet) to pale brown; disc concave to flat; hymenium 80–140 μm thick, colourless, K−, N−. Ascospores 13–22 × 5–12 μm. Pycnidia 60–100 μm diam., indistinct, pallid, the wall colourless.

On moist, calcareous rocks, more rarely acid rocks subjected to basic flushing. Mainly in W. & N. British Is. N. and montane Europe, Siberia, North America.

Morphs with a superficial thallus are sometimes distinguished as var. *crustosa* Magnusson (1933) or *I. rhodopsis*, but intermediates occur in British material. The trentepohlioid photobiont distinguishes this species from *Hymenelia prevostii*.

3. I. heteromorpha (Krempelh.) Arnold (1874)

Thallus white to yellow-grey, cracked-areolate, the areoles 0.1–0.5 mm wide, rarely powdery or ± immersed. Apothecia 0.2–0.35 mm diam., immersed, often separated from the thallus by a circular fissure; hymenium, 60–100 μm tall, the upper part blue-green, K−, N+ reddish. Ascospores 7–12 × 7–10 μm, globose to broadly ellipsoid. Pycnidia 80–100 μm diam., the wall green.

On hard limestones, upland, above 300 m. N. England (Pennines), E.C. Scotland (Perth). N. and montane Europe, Siberia, Alaska.

Differs from *I. melanocarpa* **(4)** in the smaller ascospores and usually better developed thallus.

4. I. melanocarpa (Krempelh.) Arnold (1887)

Thallus immersed or thinly farinose, never areolate, whitish to yellow-grey. Apothecia 0.2–0.5 mm diam., immersed, blackish; hymenium 120–150 μm tall, the upper part blue-green, N+ reddish. Ascospores (10-)15–17(-20) × 8–11 μm. Pycnidia *c.* 100 μm diam., the wall green.

On calcareous rocks; very rare. N.W. Scotland (W. Sutherland, Inchnadamph). N. and montane Europe, Siberia, Canada.

5. I. odora (Ach.) B. Stein (1879)

Thallus superficial, yellowish- to olive-grey, thin and smooth, sometimes thicker and cracked-areolate. Apothecia 0.1–0.4 μm diam., pale pink to brownish; hymenium 50–70 μm tall, the upper part (and adjacent exciple and thallus surface) K+ violet, N−. Ascospores 9–12 × 5–6 μm. Pycnidia 50–80 μm diam., the wall greenish, K+ violet.

On damp siliceous rocks, usually by streams or in seepage zones, mainly above 700 m. Scotland (Highlands). N. and montane Europe, Siberia, Canada.

Easily identified by the K+ violet reactions.

6. I. suaveolens (Schaerer) Arnold (1874)

Thallus cracked-areolate. rarely thin and almost smooth, dark green to olive-brown. Apothecia 0.2–0.4 mm diam., blackish; hymenium 50–60 μm tall, the upper part blue-green, K−, N− or fading. Ascospores 6–8(-11) × 5–7 μm, globose to broadly ellipsoid. Pycnidia unknown.

On siliceous rocks in the splash zone of streams, uplands, rare. Great Britain. N. and montane Europe, Alaska.

British distribution uncertain owing to earlier confusion with *I. cyanocarpa* **(1)** and *I. odora* **(5)**; these three species are separated by testing their hymenia with K and N.

JULELLA Fabre (1879)

Polyblastiopsis Zahlbr. (1903)

Thallus crustose. **Photobiont** (if present) *Trentepohlia*, not or only facultatively lichenized. **Ascomata** perithecia, circular to ellipsoid in outline. **Involucrellum** present, dark, usually laterally spreading, composed of compacted hyphae and bark cells; wall pigment greenish or brown, K+ greenish, N−. **True exciple** thin, pale or colourless, surrounding the centrum. **Hymenium** I−. **Hamathecium** of abundant, richly branched and anastomosing, slender paraphyses. **Asci** (2-)8-spored, cylindrical to clavate, I−, the inner wall thickened towards apex, ± with a shallow, broad ocular chamber. **Ascospores** ellipsoid, submuriform to muriform, colourless, with a distinct, gelatinous perispore. **Conidiomata** pycnidia, black. **Conidia** bacilliform to shortly thread-like, simple, colourless. **Chemistry**: no lichen products detected by t.l.c. **Ecology**: on bark. **Distribution**: cosmopolitan in temperate to tropical regions, *c.* 13 species.

Distinguished from *Mycoglaena* by the halonate ascospores, less cylindrical asci, more richly branched, hamathecial filaments, and the N− ascomatal pigments.

LITERATURE: Barr (1985), Harris (1973), Morgan-Jones & Swinscow (1965).

1. J. sericea (Massal.) Coppins (1992)

Polyblastiopsis sericea (Massal.) Zahlbr. (1903)

Thallus effuse, pale grey or yellowish grey, often inconspicuous. True exciple 200–400 μm diam. Perithecia 0.3–0.6 mm diam., flattened-hemispherical, ½- to ⅕- immersed, blackish, but covered by a thin layer of bark cells; involucrellum dark brown. Ascospores 15–25 × 10–15 μm (excluding gelatinous perispore), muriform, with 5–7 transverse septa and 1–2 longitudinal septa. Pycnidia not seen.

On smooth bark of branches of *Corylus* and *Quercus*, very rare. S.W. Ireland (Glengarriff). Italy.

Closely resembles *Arthopyrenia lapponina* when examined under a hand lens but that species has 1-septate ascospores. The more widely recorded (C. Europe, U.S.A., Canada) *J. fallaciosa* (Stizenb. ex Arnold) R. C. Harris (1987) is possibly a synonym.

KOERBERIELLA B. Stein (1879)

Thallus crustose, superficial, areolate, areoles contiguous or dispersed. **Prothallus** indistinct, or blackish and mainly between the areoles. **Isidia** usually present, cylindrical to slightly clavate. **Cortex** ± pseudoparenchymatous, outer cells rounded, without a layer of dead tissue (epinecral layer). **Photobiont** trebouxioid. **Medulla** I−. **Ascomata** apothecia, sessile, developing from isidia-like protuberances. **Thalline exciple** persistent, ± pseudoparenchymatous, outer zone brownish. **True exciple** not clearly differentiated. **Disc** flat to slightly convex, red-brown to dark brown, not pruinose. **Epithecium** brownish to brown, not granular. **Hymenium** colourless, I+ bluish brown or bluish brown to orange. **Hypothecium** colourless or occasionally brownish below. **Hamathecium** of paraphyses, branched and anastomosed, apices thickened. **Asci** 8-spored, elongate-clavate, *Porpidia*-type. **Ascospores** ellipsoid, colourless, simple, with a strongly thickened perispore when young, wall I−. **Conidiomata** pycnidia, immersed in the tips of the isidia, pycnidial walls brown at the apex and colourless below. **Conidiogenous cells** arranged in chains, enteroblastic. **Conidia** arising mainly pleurogenously, bacilliform, colourless, simple. **Chemistry**: gyrophoric acid and unidentified substances. **Ecology**: on rock.

This monotypic genus is distinguished from other genera of the Porpidiaceae, apart from *Bellemerea*, by the presence of a thalline exciple; it is currently separated from the closely related *Bellemerea* primarily by the I−, not I+ blue, ascospore walls.

LITERATURE: Rambold *et al.* (1990).

1. K. wimmeriana (Körber) B. Stein (1879)

Aspicilia leucophyma (Leighton) Hue (1912)

Thallus areoles 0.3–0.7 mm diam., smooth, rounded or contorted, sometimes cracked-areolate, pale grey to grey-pink; isidia usually present, mostly one per areole, to 0.75(-1.25) × (0.3-)0.5–0.6 mm; cortex *c.* 20 μm thick, outer cells (4.5-)6–6.5(-7.5) × (4.5-)5–5.5(-6.5) μm. Apothecia 0.5–0.8(-1.5) mm diam., rare; disc pale to dark red-brown; hymenium 110–140 μm tall. Paraphyses 1.5–2 μm wide, apices 2.5–5 μm thick. Asci (80-)105–115 × 20–25 μm. Ascospores 17–30 × 9–15 μm. Conidia 3.5–5 × 1–1.5 μm. Thallus Pd−, K−, KC−, C− (gyrophoric acid trace, and an unidentified substance).

On wet, periodically inundated, calcareous schists and epidiorites, alpine; local. N. England (Westmorland, Lake District), Scotland (Highlands). C. & E. Europe, Scandinavia, U.S.S.R.

Specimens vary from almost completely non-isidiate to densely so. Non-isidiate specimens in continental Europe are frequently infected by lichenicolous fungi.

LASALLIA Mérat (1821)

Thallus foliose, monophyllous, plate-like, dorsiventral, attached by a stout, ± central umbilicus. **Upper surface** densely pustulate, with raised, rounded or oval, ± convex pustules. **Isidia** present. **Upper cortex** paraplectenchymatous. **Photobiont** chlorococcoid. **Lower cortex** scleroplectenchymatous. **Lower surface** with broad, excavate depressions corresponding to pustules on the upper surface, without rhizines. **Ascomata** apothecia, sessile or ± stalked. **Thalline exciple** absent. **True exciple** present, persistent. **Disc** black, flat, smooth to roughened. **Asci** 1(-2) spored, *Umbilicaria*-type. **Ascospores** muriform, pale to dark brown. **Conidiomata** pycnidia, immersed, with ± pale brown wall. **Conidia** bacilliform.

Separated from *Umbilicaria* which differs in having a non-pustulate thallus, 8-spored asci, and smaller, simple or sparingly muriform ascospores.

LITERATURE: Llano (1950).

1. L. pustulata (L.) Mérat (1821)

Thallus 3–6 (-15) cm diam., monophyllous, dull black-brown, often densely white, ± scabrid-pruinose when dry, green-brown when wet; upper surface with numerous, crowded, conspicuous, convex, oval pustules, especially towards the centre of thallus, flattening towards the margins which become ± eroded, and lacerated, rarely fenestrate; lower surface grey, brown or black, roughened. Apothecia very rare, 1–3 mm diam.; disc black, smooth; true exciple smooth or partly isidiate. Ascospores 28–70 × 18–34 µm. Pycnidia 150–180 µm diam.; conidia 2.5–3 × 1 µm. Medulla Pd−, K−, KC+ red, C+ red, UV− (gyrophoric acid).

On nutrient-, and often mineral-rich, siliceous rocks, standing stones, birds' perching rocks and boulders, sometimes forming spectacular swards of large numbers of thalli in nutrient-enriched seepage tracks on steep rock faces; rather local. W. Britain (Dartmoor, Lake District, N. Wales, Scotland), extending to E. England (rare). Europe, Canary Is., North America, Asia.

A very distinctive species characterized by the conspicuously pustulate upper surface with black, coralloid clusters of isidia. Locally called rock tripe.

LECANACTIS Fr. (1825)

Thallus crustose, thin, continuous, scurfy or cracked, to irregularly areolate or verrucose, rarely leprose, effuse or delimited by a brown-black hypothallus; cortex absent or poorly differentiated. **Photobiont** *Trentepohlia*. **Ascomata** apothecia, disc-like to elongate (lirelliform), mostly sessile, black, but often densely white-pruinose. **Thalline exciple** absent. **True exciple** well developed, often persistent, raised, dark brown to black. **Epithecium** pale to dark brown, exciple and hypothecium dark brown; pigment usually K+ greenish. **Hymenium** I+ reddish or blue. **Hamathecium** of branched paraphyses (paraphysoids) with ± swollen apices. **Asci** 8-spored, narrowly clavate to cylindrical, fissitunicate, apex with narrow, I± pale blue apical dome penetrated by a small ocular chamber that is sometimes surrounded by a small, I+ dark blue, ring-like zone. **Ascospores** narrowly clavate, fusiform or acicular-fusiform, mostly 3- or more septate, sometimes with a thick perispore, colourless, but old spores becoming brown-warted in *L. abscondita* **(2)**. **Conidiomata** pycnidia, inconspicuous and immersed or large, prominent, cylindrical, and white-pruinose; walls dark brown, usually K+ greenish. **Conidiogenous cells** elongate, ampulliform to subcylindrical, arising in a single layer, enteroblastic, acrogenous. **Conidia** colourless, simple, bacilliform (? rarely curved). **Chemistry**: including confluentic, lecanoric, lepraric and schizopeltic acids, erythrin, and several unidentified substances. **Ecology**: mostly on humid, yet dry, shaded rocks and dry recesses of bark; cosmopolitan, with *c.* 50 species.

Traditionally separated from *Opegrapha* by the disc-like, not elongate, ascomata. This distinction is untenable and a reassessment of characters and generic limits is required. The same is true for *Bactrospora*, with thread-like ascospores breaking up into part-spores, and *Schismatomma*, with a thalline exciple and poorly developed true exciple. *Lecanactis abscondita* **(2)**, *L. grumulosa* **(5)** and *L. lyncea* **(9)** are often included in *Opegrapha*.

LITERATURE: Coppins & James (1979a), Gilenstam (1969), James & Coppins (1979), Jørgensen & Tønsberg (1988), Lettau (1932).

1 On bark or wood ... 2
On rock or mosses on rock ... 10

2(1) Apothecia present ... 3
Apothecia absent, often with pycnidia ... 7

3(2) Apothecia circular, black and not pruinose or whitish to yellow-grey-pruinose ... 4
Apothecia mostly elongated, oval to elliptical, ± lirelliform, black but with dense brown-grey (later whitish) pruina on the disc ... **9. *lyncea***

4(3) Apothecia ± pruinose (except when old); ascospores 3- to 5-septate, 14–40 × 2–7 μm ... 5
Apothecia black, not pruinose; ascospores multiseptate, 130–140 × 3–5 μm **7. *homalotropa***

5(4) Apothecial discs not or only thinly white- or grey-pruinose; ascospores fusiform, < 30 μm long; cylindrical pycnidia absent ... 6
Apothecial discs thickly yellow-grey pruinose; ascospores fusiform-acicular, 28–40 μm long; cylindrical pycnidia with white-pruinose tips (C+ red) usually present .. **1. *abietina***

6(5) Apothecia to 2 mm wide, not or thinly grey- to green-pruinose; ascospores (4-)5-septate, 18–25 × 5–7 μm; thallus thin, grey ... **10. *premnea***
Apothecia to 0.5 mm wide, white-pruinose; ascospores 3- or 4-septate, 14–27 × 2–3 μm; thallus thick, chalk-white, speckled with tiny, brown, granular flecks ... **3. *amylacea***

7(2) Thallus thin, greyish, with conspicuous, cylindrical or globular, pruinose-tipped pycnidia ... 8
Thallus thick, whitish, wholly powdery, without conspicuous pycnidia ... 9

8(7) Pruinose pycnidial tips C+ red, K−; conidia 12–17 μm long ... **1. *abietina***
Pruinose pycnidial tips C−, K+ yellow; conidia 3.7–5 μm long ... **11. *subabietina***

9(7) Thallus wholly chalky white; confluentic acid present ... **9. *lyncea***
Thallus white, with evenly dispersed, tiny, brown, granular flecks; confluentic acid absent ... **3. *amylacea***

10(1) Thallus always without apothecia or pycnidia, forming cushions; lepraric acid present ... **8. *latebrarum***
Thallus with apothecia and/or pycnidia, not forming cushions; lepraric acid absent ... 11

11(10) Pycnidia absent or inconspicuous, less than 0.1 mm diam., not pruinose ... 12
Pycnidia conspicuous, cylindrical, 0.2–0.3 mm diam., with white-pruinose tips ... **1. *abietina***

12(11) Thallus and apothecial pruina C+ red ... 13
Thallus and apothecial pruina C− ... 15

13(12) Ascospores to 3–4.5(-5) μm wide, not becoming brown or warted ... 14
Ascospores 5–7(-8) μm wide, many becoming brown-warted ... **2. *abscondita***

14(13) Apothecia rounded to elliptic or lirelliform, with persistent margin; hypothecium K+ greenish ... **5. *grumulosa***
Apothecia rounded, with margin soon excluded; hypothecium K− ... **6. *hemisphaerica***

15(12) Thallus whitish, conspicuous, Pd+ yellow-orange; apothecial discs mostly white-pruinose; ascospores 3-septate ... **4. *dilleniana***
Thallus dull greyish, thin or immersed, Pd−; apothecial disc naked or thinly green-pruinose; ascospores (4-)5-septate ... **10. *premnea***

1. L. abietina (Ach.) Körber (1855)

Thallus thin, effuse, mauve-grey to whitish grey. Apothecia (not always present) 0.7–2 mm diam., circular, sessile or ± adpressed, thickly yellow-grey-pruinose over brown-black disc, which has a prominent, persistent true exciple. Ascospores fusiform to acicular, 3(-4)-septate, ± curved, 28–40 × 3–6 μm. Pycnidia 0.2–0.3 mm diam., numerous and evenly scattered, cylindrical, knob-like, with white-pruinose apices; walls 25–50 μm thick, dark brown in water mount, green-black in K;

conidiogenous cells 8–12 × 1.9–2.4 µm; conidia cylindrical, 12–17 × 2–3 µm. Pycnidial pruina C+ red, K−; apothecial pruina and thallus Pd−, K−, KC−, C−; medulla UV+ yellow- or glaucous grey (lecanoric and schizopeltic acids, unknown UV+ substances).

On dry, often shaded, acid bark and exposed wood in medium-aged to mature woodland on deciduous and, sometimes, coniferous trees; best developed in vertical crevices on N. or N.E. sides of trunks; occasionally in underhangs on acid rocks, bryophytes or plant debris; locally common. Throughout Great Britain, except in extremely oceanic situations, throughout Ireland (rare). Europe, North America, Chile.

When sterile, distinguished from *L. subabietina* **(11)** and *Opegrapha vermicellifera* by the C+ red pycnidial pruina and large conidia.

2. L. abscondita (Th. Fr.) Lojka (1874)

Thallus thin, ± continuous, or partly cracked to verrucose-areolate; areoles sometimes in small, scattered groups, yellowish or greenish grey; medulla I+ blue. Apothecia 0.3–0.8 mm diam., rounded, sessile; disc ± flat, black, mostly bluish white-pruinose; true exciple ± prominent, entire, not pruinose; exciple, hypothecium and epithecium K− (not greenish). Ascospores (15-)19–23 × 5–7(-8) µm, often with a distinct perispore, 3- to 5-septate, sub-clavate, ± constricted at septa, at first distinctly halonate, later becoming brown-warted. Pycnidia 80–120 µm diam., black, conidia 5.5–6.5 × 1 µm, bacilliform. Thallus and apothecial pruina Pd−, K−, C+ red (erythrin and lecanoric acid).

On dry, sheltered rock underhangs on slightly calcareous rocks. C. Scotland (Perth; Aberdeen, Mar Forest; W. Ross). Scandinavia, arctic-alpine.

Like *L. dilleniana* **(4)** and *L. grumulosa* **(5)** in appearance and habitat, but differs from both in I+ blue thallus hyphae, broader, often brown-warted ascospores, and K− (not greenish) pigment in the apothecia; further differs from *L. dilleniana* **(4)** in the Pd− thallus.

3. L. amylacea (Ehrh. ex Pers.) Arnold (1880)

Thallus thick, chalk-white, white-powdery, with numerous, evenly scattered, pale brown, granular flecks (apparently ascomatal initials) to 0.2(-0.3) mm diam. Apothecia usually absent, 0.2–0.5 mm diam., disc white-pruinose, concave; true exciple black. Ascospores 14–27 × 2–3 µm, 3(-5)-septate. Pycnidia not seen. Apothecial pruina and thallus Pd−, K−, KC−, C−; medulla UV+ citrine-glaucous (unidentified substance "A" and an unsaturated fatty acid, UV+ blue after charring).

On dry bark of ancient *Quercus*, usually low down, often in concavities between root buttresses, always where not directly wetted by rain; in ancient woodlands or parklands; rare. Great Britain, from Cornwall and Hampshire (New Forest) to Skye, Moray and Nairn. Europe, from Denmark to Tuscany, North America.

T.l.c. is recommended for confirmation of sterile specimens.

4. L. dilleniana (Ach.) Körber (1855)

Thallus pale white-grey, tinged mauve or orange-pink (greyish in the herbarium), thin, effuse, ± tomentose-scurfy, areolate to verrucose, non-sorediate. Apothecia (0.3-)0.5–1.5(-2) mm diam., rounded, a few becoming angular or ± elongate, mostly emergent or sessile; disc ± exposed at maturity, naked or usually ± grey-pruinose; true exciple usually not pruinose. Ascospores 18–32(-40) × 4–5 µm, fusiform, straight, 3-septate. Pycnidia to 0.1 mm diam., rare, semi-immersed, blackish; conidia 5–8 × 1–1.5 µm, bacilliform. Thallus and apothecial pruina Pd+ yellow-orange, K−, KC−, C− ; medulla UV± yellow-grey, rarely glaucous (psoromic, 2'-*O*-demethyl psoromic and ± schizopeltic acids).

In dry, shaded rock crevices and overhangs; rather rare. W. & N. British Is. Scandinavia, subarctic and boreal Europe.

Distinguished from *L. abscondita* **(2)**, *L. grumulosa* **(5)** and *L. hemisphaerica* **(6)** by the Pd+ yellow-orange thallus, and longer ascospores.

5. L. grumulosa (Dufour) Fr. (1831)

L. delimis (Nyl.) A. L. Sm. (1911), *L. mirifica* (Stirton) A. L. Sm. (1926), *L. monstrosa* Bagl. (1857), *L. nothiza* (Nyl.) P. James (1980)

Thallus usually thick, rarely thin, greyish white to pale brownish grey, ± continuous, farinose or cracked-areolate. Apothecia black, at first immersed then sessile, roundish to angular, 0.2–1.1 mm diam., or ellipsoid to elongate to 2 mm long; disc flat, usually white or bluish white-pruinose; true exciple thin, often prominent. Ascospores 14–23 × 3–4(-5) μm, 2–3(-5)-septate, oblong-fusiform. Pycnidia not found; conidia said to be curved, *c.* 5–7 × 1 μm. Thallus and apothecial pruina Pd−, K−, C+ red (lecanoric acid and erythrin).

On dry, ± calcareous rocks and mortar, often on sheltered underhangs and shaded walls; rather rare. S. & W. British Is. Europe.

Most likely to be confused with *L. abscondita* **(2)**, *L. dilleniana* **(4)** or *L. hemisphaerica* **(6)**.

6. L. hemisphaerica Laundon (1976)

Thallus thick and areolate, delimited, chalk-like, continuous, uneven, forming irregular, convex warts between the areoles where well developed, pale grey, but ± flat and wrinkled where poorly developed; prothallus not apparent. Apothecia to 0.4–1.2 mm diam., scattered or crowded, superficial, at first flat, soon becoming convex and semiglobose, usually orbicular, but occasionally a little irregular; disc pale to dark grey, pruinose; true exciple very thin, soon becoming excluded; exciple and hypothecium K− (not greenish); epithecium colourless to dark brown (K+ greenish), with abundant, small, granular crystals. Ascospores 13–25 × 3–4.5 μm, (2-)3(-7)-septate, narrowly ovate to fusiform. Pycnidia unknown. Thallus and apothecial pruina Pd−, K−, KC+ orange or KC−, C+ orange or C− (lecanoric acid, erythrin).

On plaster, rarely mortar, on shaded, vertical walls of churches, coastal; rare. S. & E. England (Kent, Sussex, Suffolk, Norfolk, Dorset). Italy.

Similar species with a C+ red thallus, *L. abscondita* **(2)** and *L. grumulosa* **(5)**, have apothecia with persistent, prominent true exciples; the apothecia of *Dirina massiliensis* have a persistent thalline exciple.

7. L. homalotropa (Nyl.) Gilenstam (1969)

Thallus thin, smooth, greyish white. Apothecia 0.2–0.6 mm diam., disciform, rounded, black, roughened, not pruinose, at first immersed, then sessile and raised 0.1–0.2 mm above the thallus, with a thick, often ragged true exciple and concave to flat disc. Asci cylindrical, with 8 parallel ascospores. Ascospores to *c.* 50-septate, cylindrical-acicular, blunt-ended, 130–140 × 3–5 μm. Pycnidia unknown. Thallus Pd−, K−, C−, UV+ in part orange- to yellow-grey; no substances detected by t.l.c.

On mainly smooth bark of *Fraxinus*, *Corylus*, *Ilex*, *Salix* and *Sorbus*, in humid, sheltered oceanic woodlands, often with *Thelotrema subtile*, etc., in a *Graphidion* community; locally frequent. N. Scotland (Highlands and Islands N. to Ross), W. Ireland. Azores.

Easily identified by the multiseptate, acicular-cylindrical ascospores. Perhaps related to *Bactrospora*, the ascospores of which, however, break up into part-spores at maturity.

8. L. latebrarum (Ach.) Arnold (1885)

Thallus greyish white, tinged pink-violet, becoming paler in the herbarium, often forming conspicuous, semiglobular, spongy cushions to 5 mm tall and 10 mm diam.; prothallus when developed, whitish or pale brown, forming a lax, felted mat; thallus of photobiont cells interwoven by coarse hyphae heavily encrusted by crystals. Apothecia and pycnidia unknown. Thallus Pd−, K−, KC−, C−; medulla UV+ ice-blue (lepraric, and ± roccellic acids).

On very shaded, acidic rock underhangs in humid conditions; overlooked. W. British Isles. Norway, W. Sweden, France, European mtns.

Easily mistaken for a species of *Lepraria* or *Leproloma* which, however, do not have *Trentepohlia* as photobiont. May also be confused with sterile thalli of *Arthonia arthonioides* and *Lecanactis abietina* **(1)**; the former has no lichen products detectable by t.l.c., the latter is pale grey and contains schizopeltic acid.

9. L. lyncea (Sm.) Fr. (1831)

Opegrapha lyncea (Sm.) Borrer ex Hook. (1833)

Thallus thick, chalk-white, ecorticate and wholly powdery. Apothecia rarely rounded, 0.2–0.4 mm diam., mostly elliptic to shortly lirelliform, occasionally elongate to 1.6(-2) × 0.2–0.3(-0.5) mm; disc thinly to densely whitish- to brown-grey-pruinose; true exciple prominent, mostly not or only thinly pruinose. Ascospores 6- to 9-septate, 22–35 × 3–4.5 μm, elongate-fusiform. Pycnidia unknown. Thallus Pd−, K−, KC−, C−, UV+ ice-blue (confluentic and 2'-*O*-methylmicrophyllinic acids).

On dry, rough, usually well-lit bark of ancient *Quercus* or (rarely) *Fagus*, particularly on underhangs below large branches, in ancient, usually open, parklands or forests; locally frequent. S. & E. England, S. & C. Wales, N. England (extinct), very rare in W. Scotland (Loch Sunart) and C. Ireland (Laois). W. Europe north to Denmark and S. Sweden, N. Africa.

Often grows with, and apt to be confused with, *Arthonia impolita* which, however, has C+ red apothecial pruina.

10. L. premnea (Ach.) Arnold (1861)

L. plocina auct., non (Ach.) Massal. (1860)

Thallus thin, effuse and greyish, or evanescent. Apothecia 0.4–1.5(-2) mm diam. disciform, large, black, sessile; disc naked or thinly grey-green-pruinose, flat; true exciple persistent, prominent, ± crenate. Ascospores 18–25 × 4.5–6(-7) μm, cylindrical, (4-)5-septate, often curved, with pointed ends. Pycnidia occasional, mostly arranged in lines or groups, black, ½-immersed to sessile, 60–100 μm diam.; conidia bacilliform, 3.5–4.6 × 0.8 μm. Thallus and apothecial pruina Pd−, K−, C−, UV− (lichen products not detected by t.l.c.).

On dry, rough, usually well-lit bark of ancient *Quercus*, *Taxus* and (rarely) *Fagus*; the dominant species of the post-climax *Lecanactidetum premneae* association on old trees, often with *L. lyncea* **(9)**, in ancient, usually open, parklands or forests, more rarely on hard, acid rocks in dry underhangs; most frequent in S., rare in N. England, W. Scotland (on rocks only), Wales, Ireland. W. Europe, North America.

Specimens on rocks have been distinguished as *L. plocina*, but no differences other than substratum are apparent.

11. L. subabietina Coppins & P. James (1979)

Thallus 20–80 μm thick, effuse, thin, whitish grey, ± continuous, scurfy, cracked. Apothecia unknown. Pycnidia 0.18–0.5 mm diam, numerous, crowded, ± sessile, globose, then cylindrical or widened above, black (K+ greenish) but densely white-pruinose and often with a pale yellowish conidial mass extruded from the wide ostiole; conidia oblong to ellipsoid, 3.7–5 × 1.2–1.7 μm. Pycnidial pruina K+ lemon-yellow, C−; thallus Pd−, K−, KC−, C− , UV+ bluish or yellowish white (confluentic, 2'-*O*-methylmicrophyllinic and lepraric (in pycnidia) acids).

On dry, usually shaded, acid bark of (mostly) old *Quercus*, less often *Betula* and *Pinus*, usually low down on trunk; on *Hedera* stems and decaying fern fronds in dry underhangs, also *Calluna* stems. S. & W. Britain from Hampshire and Cornwall, north N.E. Yorkshire and Isle of Skye, S.W. Ireland. France (Brittany), N. Spain, Madeira, Azores.

Can be confused with sterile morphs of *L. abietina* **(1)** which has larger conidia and K−, C+ red pycnidial pruina and with *Opegrapha vermicellifera*, which has K− pycnidial pruina, K− (not greenish) pycnidial walls and smaller conidia.

LECANIA Massal. (1853)

Thallus crustose, thin to thick, continuous or of scattered granules, warted or areolate, rarely papillate or somewhat lobed. **Upper surface** smooth to warted-papillate, occasionally with soralia or minute, granule-like outgrowths ("blastidia"), grey-white, white-yellow, to brown-black, sometimes ± pruinose. **Upper cortex** three to four cells thick, of anticlinal hyphae, occasionally densely crystalline or covered by an amorphous layer. **Photobiont** chlorococcoid. **Ascomata** apothecia, sessile, at first mostly flat, often becoming ± convex; disc 0.4–0.6(-1) mm diam., pale- to black-brown, to orange, sometimes pruinose. **Thalline exciple** usually present, rarely excluded, ± cellular or of hyphae arranged in a gelatinous matrix. **True exciple** sometimes developed, the cells at outer edge, rounded, thick-walled, elongate, thin-walled below. **Hymenium** colourless, K/I+ blue. **Hypothecium** colourless or pale. **Hamathecium** of paraphyses, thick, simple, conglutinated, sometimes sub-moniliform or of 1 to 2 terminal cells with dark-pigmented cap. **Asci** 8(-16)-spored, *Bacidia*-type. **Ascospores** (0-)1- to 3(-7)-septate, oval to elongate-ellipsoid, thin-walled, colourless, without a thickened perispore. **Conidiomata** pycnidia, frequent in species on trees, very rare in rock species.

Chemistry: most species lack a distinctive chemistry; atranorin chloratranorin, gangaleoidin and unidentified terpenes sometimes present. **Ecology**: most frequently associated with nutrient-rich, or -enriched, ± basic substrata in inland and coastal sites, especially old, mortared walls, ragstone and *Sambucus*.

Distinguished from *Bacidia*, *Halecania*, *Lecanora* and *Catillaria* mainly by the different ascus-type, ascospores, epithecium and paraphyses; the presence or absence of a thalline exciple is of less importance. Reference should also be made to *Lecanora tenera*. In *Halecania* the ascus has an I+ blue apical dome lacking a 'masse axiale', and the ascospores have a more thickened outer layer. *Lecania dubitans* (Nyl.) A. L. Sm. (1928) has been incorrectly reported from the British Is. Some of the species at present recognized can be difficult to separate and may prove to be variants of a single species. Thallus morphology, presence of blastidia, and structure of the thalline exciple provide important diagnostic features.

LITERATURE: M. Mayrhofer (1988), v.d. Boom (1992).

1 On trees or wood 2
On rocks, mortar, or asbestos-cement 5

2(1) Ascospores simple or 1-septate 3
Ascospores 3-septate .... **10. *fuscella***

3(2) Thallus pale green or dull greenish grey, disappearing, or of contiguous, smooth, low warts; apothecia pale to dark red-brown, emergent to sessile 4
Thallus bright green, scurfy; apothecia bright pink or ± piebald-brownish, semi-immersed .... **4. *chlorotiza***

4(3) Ascospores 10–16 × (3-)4–5 μm; on nutrient-rich bark in *Xanthorion* .... **5. *cyrtella***
Ascospores 8–12(-15) × 2–3(-4) μm; on basic bark, not associated with *Xanthorion* **6. *cyrtellina***

5(1) Thallus and apothecial margin with minute granule-like outgrowths ("blastidia") or ± sorediate 6
Thallus without blastidia, not sorediate, ± smooth, ± areolate-cracked, sometimes warted or papillate 8

6(5) With scattered, clustered, or thickly distributed blastidia 7
With abundant, large, diffuse, uniformly or speckled indigo-blue to glaucous soralia, often coalescing .... **3. *baeomma***

7(6) Ascospores 9–14 × 3–5 μm, 1-septate; paraphyses not submoniliform .... **9. *erysibe***
Ascospores 19–25 × 4–4.5 μm, (1-)3-septate; paraphyses ± submoniliform **7. *coeruleorubella***

8(5) Ascospores 0- to 1-septate 9
Ascospores 2- to 7-septate 16

9(8) Thallus ± even to tumid-warted, not coarsely papillate, ± cracked-areolate or roughened, not obscurely lobed at the margin, naked to densely white-pruinose 10
Thallus coarsely knobbly-warted to papillate overall, ± obscurely lobed at the margin, not pruinose .... **1. *aipospila***

10(9) Thalline exciple soon excluded, without algal cells; disc markedly convex 11
Thalline exciple ± persistent, with algal cells; disc (concave-) flat to ± convex 12

11(10) Thallus superficial, wide-spreading, continuous, smooth or ± cracked-areolate; apothecia pink-yellow to dark brown, at times piebald, not translucent when wet; on non-calcareous rocks .... **11. *hutchinsiae***
Thallus immersed or thin and disappearing, granular-uneven; apothecia orange to dark brown, translucent when wet; on calcareous rocks and pebbles .... **12. *sylvestris***

12(10) Thallus white or dirty white, often ± pruinose (see also *Lecanora tenera*) 13
Thallus dark brown, grey-brown to black, not pruinose 15

13(12) Apothecia not pruinose; thalline exciple with a well-differentiated cortical zone; epithecial pigment diffuse only sporadically concentrated around tips of paraphyses ... 14

Apothecia ± densely white-pruinose; thalline exciple without well-differentiated cortical zone; dark pigment concentrated around tips of paraphyses .... **18. *turicensis***

14(13) Thallus white to pale grey, cracked-areolate; disc roughened, dark brown to black, crowded, when wet minutely dotted (×20) due to fusion of clusters of tips of paraphyses .... **2. *atrynoides***
Thallus yellow-white, warted; disc smooth, orange- to red-brown, not dotted due to fusion of tips of paraphyses .... **15. *olivacella***

15(12) Areoles smooth, rather regular and angular, not minutely papillose .... **16. *rabenhorstii***
Areoles poorly developed, nodular, minutely papillose .... **13. *inundata***

16(8) Ascospores 3-septate 17
Ascospores 4- to 6-(-7)-septate .... **8. *coerulescens***

17(16) Thalline exciple excluded, cortical zone of thalline exciple of thin-walled, isodiametric, pseudoparenchymatous cells .... **17. *suavis***
Thalline exciple persistent, ± tumid, whitish cortical zone of thalline exciple of thick-walled compacted hyphae with narrow net-like lumina .... **14. *nylanderiana***

1. L. aipospila (Wahlenb.) Th. Fr. (1867)

Thallus often forming extensive, coalescing patches, pale brown to grey-brown, or in part purplish or indigo-black, of scattered or ± contiguous, low to ± elevated, coarse, knobbly papillae and warts, often indistinctly lobed at the margin, prothallus sometimes ± distinct, pale. Apothecia 0.5–1 mm diam., usually frequent, sessile or somewhat immersed, often apical on the warts or papillae; disc red-brown, black-brown or black, reddish brown when wet, ± concave to weakly convex, not pruinose; thalline exciple concolorous with thallus, persistent, or occasionally eventually excluded, the outer edge anatomically similar to cortex of the papillae, with a colourless, superficial, amorphous layer above a single outer layer of isodiametric cells with brown-pigmented, thickened walls, within composed of compact, short-celled, somewhat radiating pseudoparenchyma intermixed with photobiont cells; epithecium brown, red-brown to orange, without granules; hymenium 65–90 μm tall. Paraphyses slightly swollen and short-celled towards apices, sometimes almost moniliform. Ascospores 9–14(-16) × 4–6 μm, 1-septate. Contains ± 2 unidentified terpenes.

On weakly basic, neutral or acidic rock outcrops and cliffs, usually on or near nutrient-rich sites and associated with seepage tracks in sheltered to exposed, rocky coastal areas; locally frequent. Channel Is. (Jersey), throughout British Is. Norway, Sweden, France.

Very variable, distinguished by the generally coarsely warted to papillate, often indistinctly lobed, thallus. In shade the thallus is pale fawn, and in exposed sites more intensely coloured and often tinged deep bluish.

2. L. atrynoides Knowles (1913)

Thallus even or ± granular, ± rimose-areolate, flat, sometimes almost disappearing, white to pale brown-grey, often partly darkened by the invasive hyphae of other commensalistic fungi. Apothecia 0.3–0.7 mm diam., scattered or in groups, often crowded, sometimes irregular in shape; disc dark brown to black, ± roughened, pale brown, speckled with darker brown when wet (× 20), flat or ± convex, not pruinose; thalline exciple persistent, or sometimes excluded, concolorous or darker than disc, in section with a narrow cortex, occasionally also with a poorly developed, inner true exciple with thicker, brown-walled cells at the external edge; epithecium brown or black-green; hymenium (60-)70–90 μm tall. Paraphyses rather sparse, simple, gradually widening to 3–5 μm at apices and there ± conglutinated. Ascospores 10–14(-16) × 4.5–6 μm.

On neutral to acid rocks in ± exposed situations on the coast; rather local. Throughout British Is. France, Italy.

L. turicensis **(18)** differs in the darker-coloured thallus and often ± pruinose apothecia.

3. L. baeomma (Nyl.) P. James & Laundon (1980)

Thallus wide-spreading, forming extensive patches, mostly indeterminate, white or yellow-white, thin to medium thick, scaly or scurfy, ± irregularly, coarsely, rimose-cracked-areolate, surface entirely dissolved in, or with effuse, scattered to ± contiguous, diffuse or ± delimited, irregular, erose or pustular soralia. Soralia speckled indigo-blue (× 20) on the surface, uniformly pale within. Apothecia 0.9–1.5 mm diam., sometimes deformed, occasional, scattered, rarely numerous and to 3 contiguous; disc dark brown to black, ± densely mauve-grey-pruinose, concave or flat to somewhat convex; thalline exciple persistent, prominent, raised, ± sorediate, concolorous or paler than the thallus; epithecium black-brown, sometimes with an olivaceous tinge, densely granular; hymenium 60–100 μm tall. Paraphyses 2–2.5 μm thick, with apices not or slightly thickened. Ascospores 12–16(-18) × 4–5.5 μm, 1-septate. Contains atranorin, chloratranorin and gangaleodin.

On shaded, sheltered, often north- or east-facing outcrops of neutral to acid rocks, predominantly coastal; locally frequent. Channel Is., Great Britain northwards to Scotland (Shetland), S. & W. Ireland. S.W. Norway.

Distinguished by the abundant, diffuse, indigo-speckled soralia and pruinose apothecia with an eroded-sorediate thalline exciple; its inclusion in *Lecania* needs investigation.

4. L. chlorotiza (Nyl.) P. James (1992)

Catillaria chlorotiza (Nyl.) P. James (1965)

Thallus scurfy-leprose, bright to glaucous-green, effuse, ± continuous, often wide-spreading. Apothecia 0.1–0.3 mm diam., occasional, semi-immersed to ± sessile, ± convex, rounded or ± tuberculate, bright pink or ± piebald brownish; thalline exciple ± excluded; hymenium 25–40 μm tall. Ascospores (9-)10–12(-18) × 2–3 μm, 0- to 1-septate. Pycnidia: (a) minute, to 50 μm diam., microconidia 7–10 × 0.5 μm, curved or hooked; (b) 70–160 μm diam., pale with gaping ostioles, macroconidia 3–6 × 1–2 μm, cylindrical.

On very shaded, ± basic bark and inside hollow trees, especially *Ulmus*, *Fraxinus* and *Salix* in sheltered, wayside and woodland sites and by water; rare. England. France, Denmark, Norway.

5. L. cyrtella (Ach.) Th. Fr. (1871)

L. sambucina (Körber) Arnold (1884)

Thallus very thin, rather smooth, granular-uneven, occasionally finely scurfy-effuse or immersed, white to pale grey. Apothecia 0.25–0.5 mm diam., very numerous to crowded, emergent to sessile, yellowish, pale pink, red-brown or blackish, ± translucent when wet; thalline exciple very thin, smooth to crenulate, often becoming incomplete and excluded, white or pale grey; disc flat to markedly convex. Asci 8- to 12(-16)-spored. Ascospores 10–16 × (3-)4–5 μm, narrowly ellipsoid, 1(-3)-septate, generally curved when mature. Microconidia 17–20 × 0.5 μm, curved.

On nutrient-rich bark in the *Xanthorion*, especially on *Sambucus*, *Fraxinus* and *Acer*; often common. Throughout British Is. Europe, North America, Australia, New Zealand.

The name *L. sambucina* has been applied to specimens with predominately 12- to 16-spored asci.

6. L. cyrtellina (Nyl.) Sandst. (1912)

Thallus inconspicuous or continuous, rather smooth, pale green to dull greenish grey. Apothecia pale pink to brownish or piebald; hymenium 30–35 μm tall. Ascospores 8–12(-15) × 2–3(-4) μm. Microconidia (12-)14–16(-18) × 0.5 μm, curved; macroconidia 10–15(-18) × 1.5(-2) μm, 0- to 1-septate, curved, crescent-shaped.

On basic bark, but not associated with *Xanthorion*; local. S. England (Hampshire, New Forest), N. England, Scotland. Europe, North America.

Very similar to the more frequent *L. cyrtella* **(5)** which has consistently broader ascospores and is mostly confined to *Xanthorion* communities.

7. L. coeruleorubella (Mudd) M. Mayrh. (1988)

Thallus thick, determinate, coarsely rimose-cracked, thickly and densely blastidiate, yellow-brown to brown-grey. Apothecia 0.5–1.2 mm diam., sessile, or sometimes ± partially obscured by blastidia, single or in small clusters, often deformed; disc red-brown to black, ± densely grey-pruinose, convex to flat; thalline exciple well developed, irregular, swollen, irregularly cracked to incomplete and partially hidden, surface densely and minutely granular with blastidia; epithecium brown, densely granular; hymenium 60–70 μm tall. Paraphyses short-celled towards the apices, becoming sub-moniliform, cells often noticeably vacuolate. Ascospores 19–25 × 4–4.5 μm, 3-septate.

On old, crumbling mortar; not recorded since last century. N.E. England (N.E. Yorshire, Ayton). W. Germany, Czechoslovakia, France, Austria.

This species forms a thick thallus of abundant blastidia and may be extremely rare fertile and consequently overlooked.

8. L. coerulescens Mudd (1861)

Thallus thin to thick, cracked-arolate, grey-brown to black-brown. Apothecia crowded, broadly sessile; disc black-brown, red-brown when wet, flat to weakly convex, irregular, disc and thalline exciple densely pruinose; thalline exciple swollen, irregularly crenate, concolorous with thallus, with a ± cartilaginous cortex with a radiating network of hyphae with elongated, narrow cells and much thickened walls; true exciple absent; epithecium black-brown to green-black; hymenium 65–95 μm tall. Paraphyses mostly simple, 1- to 2-celled at apices. Ascospores 19–29 × (4-)4.5–6 μm, (1-)3(-7)-septate, elongate, ± slightly curved, ends somewhat attenuated to ± abruptly rounded. Conidia 12–13 × 0.7 μm, curved, thread-like, simple.

On ± calcareous sandstone of walls; N.E. England (Yorkshire, Cleveland) and E. England (E. Norfolk). Endemic.

Distinguished by the 3- or more septate ascospores and the unique, netted structure of the hyphae in the thalline exciple.

9. L. erysibe (Ach.) Mudd (1861)

Incl. f. *sorediata* Laundon (1967)

Thallus thin, scurfy, ± cracked-areolate, areoles sharply angled, scattered or ± contiguous, delimited, ± densely blastidiate, greenish yellow to dirty greenish grey-brown. Apothecia to 0.4 mm diam., usually present, adpressed to sessile, single and scattered or ± clustered; disc orange- to black-brown, moderately to strongly convex; cell walls thick, brownish towards external edge. Thalline exciple lacking a cortical layer, developing granule-like blastidia; true exciple often present; epithecium yellow-brown, greenish dark brown to black-brown; hymenium 50–70 μm tall. Paraphyses mostly not widened or slightly clavate at apices. Ascospores 9–15 × 3–5 μm, 1-septate, ellipsoid.

In nutrient-enriched habitats, often near domestic habitation in suburban sites, e.g. on brick, damp, ± basic rocks, asbestos-cement, and farm-polluted limestones, monuments, very rarely decaying wood; often abundant. Throughout British Is. W. Germany, Denmark, France, North America, Australia.

Although frequently fertile, *L. erysibe* can form extensive, sterile colonies (e.g. on gravestones) and is then distinguished by the minutely blastidiate thallus; some morphs have pale, ± discrete soralia. The hymenium and ascospores are similar to those in *L. inundata* **(13)** and *L. rabenhorstii* **(16)**.

10. L. fuscella (Schaerer) Körber (1855)

Thallus thin, granular, effuse, pale grey-white, containing numerous, coarse crystals clusters. Apothecia 0.5–0.8 mm diam., ± sessile; disc pale brown to brown-black or bluish grey, naked or ± pachily white or bluish pruinose; thalline exciple thin, persistent or sometimes excluded; epithecium ± pale to deep red-brown, K− or K+ deep purplish brown, N−; hymenium (45-)65–70 μm tall. Paraphyses 2–2.5 μm wide, simple or forked, apices ± slightly swollen. Asci 8- to 16-spored. Ascospores 12–22 × 4–6 μm, 3-septate, straight or ± curved, fusiform-ellipsoid with abruptly rounded ends.

On smooth, rarely roughened, bark, especially *Acer campestre* and *Populus tremula*; very rare and not recorded recently, apparently extinct. S. England (Gloucester, Kent). Mediterranean, North America.

This lichen has the aspect of a small-fruited species of the *Lecanora chlarotera* group.

11. L. hutchinsiae (Nyl.) A. L. Sm. (1918)

Catillaria littorella auct. brit., non (Nyl.) Zahlbr. (1926)

Thallus thin, wide-spreading, continuous to ± rimose-cracked-areolate, greenish grey to grey tinged brown. Apothecia 0.3–0.6 mm diam., sessile, mostly single and discrete or in small groups; disc pale pink-brown to dark brown, at times piebald or paler in shaded habitats, strongly convex even when young, margin excluded; thalline exciple medulla with few algal cells and cortex ± restricted to the lower edge of the apothecium; true exciple chiefly wide, almost colourless; epithecium yellow-brown to colourless; hymenium 40–70 μm tall. Ascospores (9-)10–16 × 3–4.5 μm, ellipsoid.

On neutral to acid rocks or scattered boulders, often in moderate to deep shade in old woodlands or scrub especially near the coast; locally frequent. S.W. England, extending locally to the S.E., Wales, Scotland, W. Ireland. W. Germany, Roumania.

Very variable and notable for the very convex apothecia, the soon excluded, well-developed true exciple and much reduced thalline exciple. See also *L. sylvestris* **(12)**.

12. L. sylvestris (Arnold) Arnold (1884)

Like **11**, but thallus mostly immersed or thin and inconspicuous, granular-uneven or as a few, scattered areoles. Apothecia orange to dark brown, translucent with a darker excipular zone apparent when wet.

On shaded, chalk pebbles and limestones. S. England, W. Ireland. W. & C. Europe.

13. L. inundata (Hepp ex Körber) M. Mayrh. (1987)

Thallus thin, with a few thicker nodules or areoles, sometimes uniformly compact, rarely crumbling into individual nodules; surface with small, pale papillae, more conspicuous when moist, yellow-brown to grey-brown and dark brown, at times becoming greenish. Apothecia to 0.8 mm diam., sessile, often clustered in small groups, rarely crowded and deformed; disc orange-brown to black-brown, becoming translucent when wet, flat to moderately convex, at times strongly convex when old, rarely white-pruinose; thalline exciple almost always present, pale grey to pale brown, in young apothecia almost swollen, later somewhat receding, often incompletely fissured, with an algal-rich medulla and algal-free cortical zone particularly well developed towards the base of the apothecia, sometimes transformed into an epinecral layer; true exciple sometimes present in places, the outer cell walls of which may be concolorous with the epithecium; epithecium pale orange, red-brown to dark brown, rarely black-brown, spotted. Paraphyses hardly or slightly swollen towards apices, but at least some usually strongly swollen, the apical cell distinct, externally brown-pigmented. Ascospores 12–18 × 5.5–6(-7.5) μm.

On limestones, ragstones and calcareous rocks. S. England, northwards to S. Scotland. C. & S. Europe, N. Africa.

Characterized by the small papillae on the surface of the nodules and areoles. The individual papillae are larger than blastidia, have a smooth upper surface and well-developed, algal-free zone which may in part be transformed into a layer of dead cells.

14. L. nylanderiana Massal. (1856)

Thallus wide-spreading, thin, areolate, areoles sharply angled, whitish grey to pale brown. Apothecia sessile, rarely single, mostly crowded and becoming deformed by mutual pressure; disc black, when moist ± translucent with a dark-pigmented edge, pruinose, mostly flat, rarely convex; thalline exciple pale grey, swollen in flat apothecia, appearing narrower in ± convex apothecia, inner part algal-rich, and a broad cortical zone of netted, ± radiating, thread-like hyphae in a gelatinous matrix, particularly well developed on the lower part and sides of the exciple; true exciple narrow to rather robust, the cell walls of the outermost hyphal cells slightly to strongly brown-pigmented. Paraphyses swollen towards apices, the terminal 2–3 cells with dark pigmented walls, some with a cap-like deposit of a dark brown pigment. Ascospores 12–16(-18) × 4–5 μm, 3-septate, the ends abruptly rounded, thin-walled.

On maritime, mostly vertical or overhanging calcareous rocks; rare. Throughout British Is. C. Europe, North America.

Characterized by the flat, black, pruinose apothecia, with a pale, persistent margin and the rather broad, dumpy ascospores. *L. suavis* **(17)** also has 3-septate ascospores but differs in having a pseudoparenchymatous outer part to the thalline exciple and narrower, 3–4 μm wide, ascospores.

15. L. olivacella (Nyl.) Zahlbr. (1928)

Thallus of scattered scales or areoles, dirty white to yellow-brown. Apothecia to 0.6 mm diam., sessile, single to numerous; disc orange-red to dark brown, when moist becoming paler to cinnamon-brown, a dark-pigmented edge often persists, flat to moderately convex; thalline exciple narrow, regular, whitish, with an algal-rich medulla and cortical zone of isodiametric, pseudoparenchymatous cells, overlain by a narrow layer of dead cells; true exciple narrow, upper part with dark-pigmented walls; hymenium 55–65 μm tall. Paraphyses mostly unbranched, the apices slightly swollen and pigmented. Ascospores 11.5–14 × 4.5–6 μm, 1-septate, thin-walled.

On rocks; only collected last century. S.W. England (Devon, Torpoint). Scattered throughout Europe, N. Africa.

Morphologically similar to *L. hutchinsiae* **(11)**, but thallus ± thicker, scaly-areolate, and the apothecia have a persistent thalline exciple.

16. L. rabenhorstii (Hepp) Arnold (1848)

Thallus thin to thick, usually of contiguous, angular areoles, the edges often curling upwards; surface yellowish brown-grey, at times with small, scattered, angular or lobe-like fragments; cortex of pseudoparenchymatous, isodiametric hyphal cells, overlain by a well-developed layer of dead cells. Apothecia to 0.7 mm diam., sessile, in parts numerous and covering areoles, rarely deformed; disc orange, brown-red to dark brown, pruinose, paler when moist, ± or rarely strongly convex; thalline exciple yellowish white, distinct, not swollen, becoming excluded only in strongly convex apothecia, algal-rich, inner zone and a cortical zone of pseudoparenchymatous, isodiametric hyphal cells, the walls of the outermost cells thick, overlain by a layer of dead cells; true exciple poorly developed, narrow, upper part brownish; epithecium mottled pale brown-orange to dark brown; hymenium 65–85 μm tall. Paraphyses mostly not or slightly swollen towards the apices, a few strongly swollen, to 6 μm wide, mostly colourless, a few with an external, yellow-brown deposit on the outer cell wall of the terminal 1–2 cells. Ascospores 11–13(-16) × 4–7 μm, 1-septate, thin-walled.

On ± basic rocks, particularly in coastal areas; probably common. Throughout British Is. S. & C. Europe to N. Mediterranean, N. Africa.

Closely resembles *L. inundata* **(13)** in anatomy, but that species has more bullate areoles and the apothecia are usually grouped and less numerous.

17. L. suavis (Müll. Arg.) Mig. (1926)

Thallus of large and small, flat to convex areoles or warts, rarely continuous; upper surface of warts smooth, yellowish white to red-brown or dark brown, becoming paler when moist, at times with a dark pigmented edge, flat, convex to semi-globose, irregular, at times pruinose; thalline exciple in part swollen, becoming excluded in strongly convex apothecia, ± concolorous with disc, with an algal-containing medulla and well-differentiated cortex of pseudoparenchymatous, isodiametric hyphal cells, the outer layers with ± thicker, slightly swollen margins, overlain by a layer of dead cells, cortex particularly well developed at sides of apothecia; true exciple of varying thickness, of elongate, thickly conglutinate hyphal cells, the cell walls deeply pigmented including those deep within the apothecium; epithecium orange-brown to dark brown, slightly mottled; hymenium 45–65 μm tall, at times ± brownish. Paraphyses strongly swollen towards apices, to 8–10 μm thick, the terminal 2–4 cells with brown-pigmented walls. Ascospores 13–20 × (3-)4–4.5 μm, 3-septate, thin-walled, lens-shaped.

On calcareous, or at least nutrient-rich, inclined rock surfaces and underhangs; local. Throughout British Is. to Shetland. N. Europe to northern S. Europe.

The commonest *Lecania* with 3-septate ascospores, typically of rounded warts with apothecia varying from being moderately convex, marginate with a distinct exciple to strongly convex with the exciple excluded.

18. L. turicensis (Hepp) Müll. Arg. (1862)

L. albariella (Nyl.) Müll. Arg. (1880)

Thallus thin, granular to areolate, white, reddish white or white-grey, mostly K−; cortical zone of isodiametric cells overlain by a layer of dead cells sometimes formed. Apothecia to 0.8 mm diam., sessile, numerous to thickly crowded and deformed; disc red-brown to black, when moist becoming paler, with a dark-pigmented edge, pale grey-white pruinose; thalline exciple narrow, white, often becoming excluded and disappearing completely, with an algal-containing medulla and in part, a narrow, algal-free cortical zone, overlain by layer of dead cells, at times overlain by a thick crystalline layer; hymenium 55–80 μm tall, occasionally partly pigmented as epithecium. Paraphyses, unbranched, strongly swollen towards apices, the terminal 1–3 cells to 6 μm wide with an external pigmentation. Ascospores 10.5–13 × 4.5–6 μm, 1-septate, thin-walled, occasionally thinly pigmented.

On calcareous and non-calcareous rocks, also mortar. Throughout British Is. C., W. & S. Europe, North America, N. Africa.

The commonest *Lecania* on rocks. On limestone the thallus is mostly continuous, areolate, with a pruinose upper surface; on non-calcareous rocks the areoles are often more scattered. Well-developed morphs have a tumid-small-lobulate thallus often densely white-pruinose, resembling a diminutive *Solenopsora* sp. (e.g. *S. vulturiensis*).

LECANORA Ach. (1810)

Thallus crustose, granular, areolate, or placodioid, rarely immersed; sorediate or not, isidia and cephalodia absent. **Photobiont** *Trebouxia* and perhaps other chlorococcoid algae. **Ascomata** apothecia, sessile or shortly stipitate, rarely immersed. **Thalline exciple** present, generally conspicuous and concolorous with the thallus, in some species disappearing at an early stage or becoming excluded. **True exciple** poorly differentiated in most species. **Epithecium** greenish brown to dark brown or reddish brown, sometimes granular. **Hymenium** ± colourless, I+ blue. **Hypothecium** colourless or pale. **Hamathecium** of paraphyses, simple, sparsely branched or anastomosed, septate, apices slightly swollen to capitate, often pigmented (never "Aspicilia-green"). **Asci** elongate-clavate, *Lecanora*-type (Fig. 33e) [or *Bacidia*-type (in **10** and **70**)], apex strongly thickened with an I+ blue apical dome and a broad I− apical cushion, outermost gelatinous coat I+ blue, 8(-32)-spored. **Ascospores** ellipsoid to subglobose, usually rounded at the apices, rarely apiculate, colourless, simple, sometimes with a central protoplasmic strand, walls thin or thick, lacking a distinct perispore, smooth-walled. **Conidiomata** pycnidia, immersed, walls colourless but brown around the ostiole. **Conidiogenous cells** sessile or borne on short, branched conidiophores, elongate-ampulliform, enteroblastic with broad collarettes. **Conidia** acrogenous, bacilliform, filamentous, arc-like or broadly falcate, colourless, simple. **Chemistry:** atranorin often in the cortex, a wide range of depsides and depsidones, terpenoids or xanthones in the medulla. **Ecology:** on a wide range of natural and man-made substrata. **Distribution:** Cosmopolitan.

This generic name has traditionally been applied to a wide range of lichens whose apothecia have thalline exciples and which produce simple and colourless ascospores. This has led to disparate taxa being subsumed under a single generic name, as has also been the case with *Lecidea s.lat.*, and a modern reappraisal of the circumscription of the genus is required. A few of the more distinctive segregates have been recognized here, notably *Aspicilia, Eiglera, Protoparmelia,* and *Tephromela*. Others will also certainly have to be accepted, for example, *Glaucomaria* M. Choisy (1929) for *L. rupicola*, and *Lecanoropsis* M. Choisy (1949) for *L. quercicola* and *L. saligna*.

Species referred to this genus have received scant attention from most European authors and species concepts are far from clear in several species groups, for example those around *L. cenisia, L. chlarotera, L. dispersa,* and *L. polytropa*. Determination in many of these groups remains problematic. *L. pacifica* Tuck. (1882), included in Cannon *et al.* (1985), appears to be restricted to the west coast of North America; Scottish material previously referred to that species most probably represents a new taxon close to *L. intumescens*.

LITERATURE: Brodo (1984), Eigler (1969), Imshaug & Brodo (1966), Leuckert & Poelt (1989), Poelt (1958, 1969).

1 Thallus placodioid, margins lobe-like in surface view 2
Thallus not placodioid, margins not lobe-like in surface view 5

2(1) Thallus C+ orange 3
Thallus C− or weakly yellowish 4

3(2) Thallus surface granular-pruinose; lobes not strongly convex; on calcareous rocks **50. *pruinosa***
Thallus surface not pruinose; lobes swollen and convex; on nutrient-enriched siliceous rocks .... **61. *straminea***

4(2) Thallus grey-green to yellow-brown; lobes flat to almost concave, closely adpressed, not or scarcely overlapping; apothecia yellow-brown to red-brown .... **42. *muralis***
Thallus pale yellow-green to greyish white; lobes flat to slightly convex, often loosely attached and overlapping; apothecia brown-black .... **1. *achariana***

5(1) Soredia present 6
Soredia absent 22

6(5) Soredia yellowish or yellow-green 7
Soredia grey-green, white or grey 16

7(6) Soredia C− 8
Soredia C+ orange 15

8(7) Soredia K+ strongly yellow 9
Soredia K− or sometimes faintly yellow 11

9(8) Thallus white to grey or brownish 10
Thallus yellow-green; soredia arising from the margins of the areoles .... **43. *orosthea***

10(9) Areoles convex; soredia arising over the surface; on siliceous rocks and walls **11. *caesiosora***
Areoles ± flat; soredia arising in mounds which may later coalesce; on wood or bark .... **27. *farinaria***

11(8) Thallus areolate, yellow-green, yellowish brown, greenish grey or blackish green; soredia in convex mounds or not; usually on rocks 12
Thallus not areolate, grey; soredia in discrete, strongly convex soralia; usually on bark .... **38. *jamesii***

12(11) Soralia Pd+ yellow, orange or red 13
Soralia Pd− 14

13(12) Soredia arising on the margins of the areoles; Pd+ orange or red .... **63. *subaurea***
Soredia arising on the surface of the areoles; Pd+ yellow or orange .... **24. *epanora***

14(12) Areoles rounded and warted, not subsquamulose, margins entire .... **24. *epanora***
Areoles flattened, subsquamulose, margins often crenulate .... **59. *soralifera***

15(7) Thallus areolate; areoles brownish yellow, poorly defined; soredia at first in clearly delimited convex soralia .... **6. *alboflavida***
Thallus effuse and granular, rarely cracked-areolate; areoles yellowish green; soredia forming a ± continuous crust .... **26. *expallens***

16(6) Thallus whitish grey, grey, or brownish grey, K+ yellow, Pd− or yellowish 17
Thallus grey-green, K− or weakly yellow, Pd− or + orange-red 21

17(16) Soredia arising superficially; on rocks 18
Soredia arising in excavate soralia; on peat, mosses or decaying plant materials **25. *epibryon***

18(17) Apothecia sessile, discs red-brown to dark brown or black 19
Apothecia immersed, discs pinkish to brownish .... **54. *rupicola***

19(18) Apothecial discs red-brown; thallus granular or areolate, surface lacking brown granules; soredia white or grey 20
Apothecial discs dark brown or black; thallus areolate, surface often with brown granules; soredia white or bluish to yellowish white .... **11. *caesiosora***

20(19) Soredia white to greenish white; on limestone or dust-contaminated bark ..... **12. *campestris***
Soredia whitish grey; on siliceous rocks .... **14. *cenisia***

21(16) Thallus KC−, Pd+ orange-red; soredia arising over the whole thallus surface **21. *conizaeoides***
Thallus KC+ yellow, Pd−; soredia mainly arising on the margins of the areoles **33. *handelii***

22(5) On bark, wood, decaying plant material, mosses, or lichenicolous 23
On rocks 60

23(22) Thallus yellow, yellow-green to grey-green 24
Thallus white, grey, orange or inapparent 32

24(23) Thallus C+ orange 25
Thallus C− 26

25(24) Thallus yellow-green to grey-green, K−; apothecial discs pale yellow to greenish, when young the exciple crenulate and containing algae .... **19. *confusa***
Thallus yellowish white to greenish black, K+ yellowish; apothecial discs pinkish brown to greenish black, when young the exciple smooth and lacking algae **69. *symmicta***

26(24) On bark or wood 27
On mosses or decaying plant material .... **40. *leptacina***

27(26) Thallus Pd− 28
Thallus Pd+ yellow .... **71. *varia***

28(27) Thallus areolate, areoles often dispersed, yellow-green to grey-green 29
Thallus not clearly areolate, ± continuous or granular, yellowish white to yellow-green or green-black 31

29(28) Apothecia yellowish green or green-black to brownish 30
Apothecia pinkish brown .... **66. *sublivescens***

30(29) Thallus of ± contiguous areoles with crenulate margins or minutely lobed; discs often becoming blackish .... **36. *intricata***
Thallus inconspicuous or of dispersed granular or convex areoles, not lobed or crenulate; discs usually remaining yellowish or yellow-green .... **47. *polytropa***

31(28) Thalline exciple persistent, often crenulate; thallus non-corticate, with projecting crystals after storage .... **62. *strobilina***
Thalline exciple soon excluded, not crenulate; thallus corticate, lacking projecting crystals after storage .... **69. *symmicta***

32(23) Thallus and (or) thalline exciple or discs Pd+ deep yellow to orange or red 33
Thallus, exciple and discs Pd− 40

33(32) Ascospores not exceeding 15 μm in length 34
Ascospores exceeding 15 μm in length 39

34(33) Ascospores exceeding 4 μm in width 35
Ascospores not exceeding 4 μm in width 38

35(34) Apothecial discs orange-brown to red-brown, dark brown or black, not or weakly pruinose; thalline exciple persistent 36
Apothecial discs yellowish to pinkish brown, densely pruinose; thalline exciple becoming excluded .... **44. *pallida***

36(35) Thalline exciple corticate, not pruinose; apothecia <1 mm wide 37
Thalline exciple not corticate, densely pruinose; apothecia >1 mm wide .... **37. *intumescens***

37(36) Epithecium granules not dissolving in K, Pd+ orange-red (with crystals) ... **17. *cinereofusca***
Epithecium granules dissolving in K, Pd− (no crystals formed) .... **51. *pulicaris***

38(34) Thalline exciple soon excluded, Pd+ orange .... **10. *cadubriae***
Thalline exciple persistent, Pd− .... **45. *piniperda***

39(33) Thalline exciple corticate, not pruinose, Pd+ yellow .... **14. *cenisia***
Thalline exciple not corticate, densely pruinose, Pd+ orange .... **37. *intumescens***

40(32) Apothecial disc red-brown, orange-red, or white-pruinose 41
Apothecial disc pale brown, yellowish brown or brown to black 54

41(40) On decaying plant materials or lichenicolous 42
On bark or wood 43

42(41) Ascospores (5-)8–9.5 μm wide, ends rounded; on decaying plant materials; thallus whitish to pale grey .... **25. *epibryon***
Ascospores 4–5 μm wide, ends pointed; on *Lecanora epanora, L. handelii* or *L. subaurea*; thallus absent, orange-grey or grey-brown .... **32. *gisleriana***

43(41) Thallus K+ yellow 44
Thallus K− 48

44(43) Thallus shades of grey 45
Thallus yellowish or greenish white .... **8. *argentata***

45(44) Apothecial disc C− 46
Apothecial disc C+ orange-yellow .... **13. *carpinea***

46(45) Thalline exciple with large crystals inside 47
Thalline exciple lacking large crystals inside .... **12. *campestris***

47(46) Thallus ± continuous; thalline exciple not or slightly warted; apothecial discs usually not pruinose .... **15. *chlarotera***
Thallus granular, usually dispersed; thalline exciple warted; apothecial discs usually pruinose .... **53. *rugosella***

48(43) Pycnidia abundant; conidia broadly falcate, over 1.5 μm wide 49
Pycnidia sparse or absent; conidia filiform or narrowly arc-like, not exceeding 1.5 μm wide 50

49(48) Thalline margin excluded; apothecial disc pruinose; macroconidia 8–12 μm long **52. *quercicola***
Thalline margin persistent; apothecial disc ± not pruinose; macroconidia 7–8(-9) μm long .... **55. *saligna***

50(48) Ascospores mainly less than 6 μm wide 51
Ascospores 6–9 μm wide .... **35. *horiza***

51(50) Asci 8-spored 52
Asci (8-)16–32-spored .... **57. *sambuci***

52(51) Ascospores not exceeding 4 μm wide 53
Ascospores 4–6 μm wide .... **39. *laevis***

53(52) Thallus of dispersed granules or inconspicuous .... **45. *piniperda***
Thallus of irregular granules often aggregated into mounds .... **58. *sarcopisioides***

54(40) Apothecial disc dark brown to black even when young 55
Apothecial disc yellowish or pale brown when young, sometimes darkening when old 57

55(52) Thallus K+ yellow; thalline exciple persistent 56
Thallus K−; thalline exciple becoming excluded .... **4. *aitema***

56(55) Apothecial disc dark brown; lower parts of thallus K− .... **30. *fuscescens***
Apothecial disc black; lower parts of thallus K+ purple .... **31. *gangaleoides***

57(54) Ascospores mainly exceeding 4 μm wide 58
Ascospores 2.5–4.5 μm wide .... **65. *subintricata***

58(57) Apothecial discs ± flat or concave 59
Apothecial discs convex when mature .... **69. *symmicta***

59(58) Thallus forming distinct patches; apothecia immersed, at first ± urceolate ... **48. *populicola***
Thallus inconspicuous or of widely scattered granules; apothecia sessile to raised **23. *dispersa***

60(22) On calcareous or other base-rich rocks 61
On siliceous rocks 71

61(60) Thallus K+ yellow or red 62
Thallus K− 64

62(61) Thalline exciple persistent, true exciple not visible; thallus white or grey 63
Thalline exciple soon excluded, true exciple well developed and persistent; thallus usually lemon-yellow .... **41. *marginata***

63(62) Thallus K+ yellow changing to red, Pd+ yellow to orange; apothecia pinkish, pruinose .... **64. *subcarnea***
Thallus K+ yellow, Pd− or weakly yellowish; apothecia red-brown to red-black, not pruinose .... **12. *campestris***

64(61) Thallus C+ yellow, orange or red 65
Thallus C− 67

65(64) Thallus and exciple C+ orange or red; thalline exciple persistent 66
Thallus and exciple C+ yellow; thalline exciple sometimes excluded .... **18. *conferta***

66(65) Thallus margins somewhat lobed and pruinose, C+ orange .... **50. *pruinosa***
Thallus margins not or slightly lobed, not pruinose, C+ red .... **20. *congesta***

67(64) Apothecia pinkish brown to red brown; epithecium pale to dark brown; ascospores often exceeding 10 μm in length 68
Apothecia black; epithecium blue-green; ascospores not exceeding 10 μm in length .... **3. *agardhiana***

68(67) Thalline margins entire or irregularly crenulate; apothecial discs not or lightly pruinose 69
Thalline margins regularly deeply crenulate; apothecial discs densely white to blue-grey pruinose .... **22. *crenulata***

69(68) Thallus granular or inconspicuous 70
Thallus composed of swollen, white areoles .... **5. *albescens***

70(69) Apothecial discs pinkish, pale olivaceous, or pale greenish grey; thallus usually inconspicuous; paraphyses 1.5–2 μm thick .... **23. *dispersa***
Apothecial discs pale brown to red-brown; thallus of rounded granules; paraphyses 2–2.5 μm thick .... **56. *salina***

71(60) Thallus yellowish white, white, grey, or inconspicuous 72
Thallus yellow-green, sulphur-yellow, or greyish green 85

72(71) Thallus and (or) apothecial disc C+ yellow or orange to red 73
Thallus and apothecial discs C− 77

73(71) Thallus C+ yellow or orange 74
Thallus C− (apothecial discs C+ yellow to orange) 76

74(73) Thallus granular to crustose, not areolate, K− or weakly yellowish; apothecia not or weakly pruinose 75
Thallus cracked-areolate, K+ deep yellow; apothecia densely pruinose .... **68. *swartzii***

75(74) Thallus white or inconspicuous; apothecial discs pale brown to red-brown .... **7. *andrewii***
Thallus yellowish brown or tawny brown; apothecial discs yellowish to reddish brown .... **29. *fugiens***

76(73) Thallus usually continuous; apothecia sessile from the start, 0.5–1(-1.5) mm diam., disc pale reddish brown; very rare on rock .... **13. *carpinea***
Thallus cracked-areolate throughout; apothecia immersed at first, 0.5–2(-3) mm diam., disc pinkish brown; exclusively on rocks .... **54. *rupicola***

77(72) Thallus K+ yellow or red 78
Thallus K− 82

78(77) Epithecium lacking coarse granules; thallus K+ yellow 79
Epithecium with coarse granules; thallus K+ yellow or yellow to red 80

79(78) Apothecial discs black; epithecium green or brownish green, N+ purple-red **31. *gangaleoides***
Apothecial discs red-brown to red-black; epithecium pale orange to reddish brown, N− .... **12. *campestris***

80(78) Apothecia usually exceeding 0.5 mm diam.; ascospores exceeding 4 μm wide 81
Apothecia 0.1–0.2(-0.5) mm diam; ascospores 1.5–2(-2.5) μm wide .... **70. *tenera***

81(80) Thallus coarsely areolate, continuous; apothecial discs pale brown to dark brown .... **49. *praepostera***
Thallus granular or of dispersed areoles; apothecial discs red-brown to black or blackish green or yellowish brown .... **14. *cenisia***

82(77) Apothecial discs pinkish, olivaceous or red-brown 83
Apothecial discs greenish black 84

83(82) Thallus yellowish white .... **2. *actophila***
Thallus grey .... **34. *helicopis***

84(82) Thallus areolate, areoles papillate; apothecial disc red-brown .... **46. *poliophaea***
Thallus of dispersed granules or inconspicuous; apothecial disc pinkish to olivaceous .... **23. *dispersa***

85(71) Thallus C− 86
Thallus C+ yellow-orange .... **9. *atrosulphurea***

86(85) Areoles flat to slightly convex, mainly under 1 mm diam. 87
Areoles strongly convex, to 1–1.5 mm diam. .... **28. *frustulosa***

87(86) Thalline exciple soon excluded 88
Thalline exciple persistent 89

88(87) Areoles forming a ± continuous crust, ± flat; apothecia at first immersed, not constricted below, discs yellowish to pinkish brown, blue-green or green-black **67. *sulphurea***
Areoles usually dispersed, convex and warted; apothecia sessile from the first, constricted below, discs black .... **16. *chlorophaeodes***

89(87) Thallus granular or of dispersed areoles with entire margins; disc usually remaining yellowish or yellow-green 90
Thallus of ± contiguous areoles with crenulate margins or minutely lobed; discs often becoming blackish .... **36. *intricata***

90(89) Ascospores (4.5-)5–6.5(-7) μm wide .... **47. *polytropa***
Ascospores 3–4 μm wide .... **60. *stenotropa***

1. L. achariana A. L. Sm. (1918)

Thallus placodioid, forming rosettes or cushions, pale yellow to yellow-green or grey-white, central lobes often piled-up and overlapping, marginal lobes 1–2 mm broad, white below, not marginate, flat to slightly convex. Apothecia generally abundant, sessile to short-stalked; thalline exciple becoming crenulate and wavy; discs rose-brown to brown, not pruinose. Ascospores 10–15(-16) × 4–6 μm. Medulla Pd−, K− (?rarely K+ yellow to red), C− (zeorin, usnic acid, unidentified triterpenoids).

On upland, siliceous rocks; very rare. N.W. England (Lake District), formerly also in Wales (Merioneth). C. Europe N. to Norway and Sweden.

2. L. actophila Wedd. (1875)

Thallus forming ± circular or continuous crusts, yellow-white to white-green, smooth, cracked-areolate; prothallus white or bluish. Apothecia 0.3–0.6(-1) mm diam., sessile, sometimes constricted below; thalline exciple persistent raised, rarely slightly excluded, sometimes crenulate or flexuose; discs finally black with a distinct, greenish blue tinge, not distinctly pruinose but often with a bluish bloom when wet; epithecium pale blue to yellow-brown or olivaceous, N+ purple-red, sometimes with yellow-brown granules dissolving in K; hymenium 45–50 μm tall. Paraphyses 1.5–2.5 μm wide, conglutinate, sparsely branched, ± simple above, apices to 3.5 μm wide, swollen, bluish to olivaceous, weakly capitate. Ascospores 8–14 × 5–6 μm, sometimes guttulate and then appearing 1-septate. Pycnidia 50–100 μm diam., immersed, blue-black; conidia 15–20 × 0.5–0.75 μm, arc-like. Thallus and medulla Pd−, K−, KC+ weakly yellowish, C− (usnic acid, unidentified pigments).

On siliceous, maritime rocks, mesic-supralittoral. Widely distributed on rocky coasts throughout British Is. W. Europe, Canada (New Brunswick).

L. helicopis **(34)**, which occurs in similar or more sheltered situations, has a predominantly grey thallus and the apothecial disc is not pruinose and lacks a greenish blue tinge, especially when wet.

3. L. agardhiana Ach. (1814)

Thallus thin, ± immersed, white to bluish white or inconspicuous, indeterminate; prothallus absent. Apothecia 0.2–0.5 mm diam., forming in pits in the rock, often irregular in shape; thalline exciple thin, white, at first raised, crenulate or flexuose, finally excluded; disc black to brown-black, concave to strongly convex, blue-grey-pruinose; epithecium blue-green or olivaceous, K+ intensifying blue-green, N+ purple-red; hymenium 30–45 μm tall. Paraphyses 1–2 μm wide, sparsely branched, apices to 2.5 μm wide, pale bluish green, not or slightly swollen. Asci 25–38 × 8–12 μm. Ascospores 8–10 × 3.5–5 μm, ellipsoid. Thallus Pd−, K− or slightly yellowish, KC−, C−.

On hard, calcareous rocks, especially limestones on rocky shores; rare. Wales (Caernarvon), Scotland (Wigtown), Ireland (Antrim). C. Europe.

Most likely to be confused with *L. crenulata* **(22)**, which may occur in similar habitats but that species has sessile apothecia, neatly crenulate thalline exciples and lacks the distinctive blue-green epithecium.

4. L. aitema (Ach.) Hepp (1853)

Thallus in small patches, dispersed to ± continuous, uneven to granular or irregularly areolate, creamy white; prothallus inconspicuous or whitish brown. Apothecia 0.2–0.5(-0.6) mm diam., sessile; thalline exciple ± distinct in young apothecia, soon excluded; discs red-brown to black, sometimes mottled, strongly convex, not pruinose; epithecium yellowish brown; hymenium 55–75 μm tall. Paraphyses 2–2.5 μm wide, branched and anastomosed, short-celled, apices to 3.5 μm wide, not or slightly swollen, yellowish brown, with granules dissolving in K. Asci 35–45 × 10–15 μm. Ascospores (10-)12–17 × 4.5–5.5 μm, narrowly ellipsoid. Thallus Pd−, K−, KC− or pinkish, C−.

On acidic bark, predominantly twigs, and commonly wood and worked timber, especially upland areas; local. Widespread in British Is. C. Europe, Fennoscandia.

This characteristic, upland member of the poorly understood *L. symmicta* **(69)** complex is primarily distinguished by the dark discs of the apothecia. The early exclusion of the thalline exciple has led to this species often being treated in *Lecidea*. Many early records from the British Is. referred to this species, particularly as var. *saepincola* (Ach.) Hedl. (1892), do not appear to belong here and require further study.

5. L. albescens (Hoffm.) Branth & Rostrup (1869)

Thallus to 1 cm wide, in compact rosettes, convex, of swollen areoles, sometimes almost lobe-like and notched at the margins, brilliant white or rarely greyish white, surface slightly rough to granular. Apothecia 0.1–0.7(-1.5) mm diam., abundant, generally compacted in the centre, sessile, broad at the base; thalline exciple persistent, even to crenulate to wavy; discs pale pinkish brown to yellowish brown or olivaceous, white-pruinose or not; epithecium brownish or ± colourless, N−; hymenium usually N+ purple-red. Paraphyses to 2 μm wide, apices 1.5–3 μm thick. Ascospores (7-)11(-16) × (3-)5–6 μm, broadly ellipsoid. Conidia 20–25 × 1 μm. Thallus Pd−, K−, KC−, C−.

On hard calcareous rocks, mortar and memorials in well-lit situations; common. Widespread in C. & E. British Is. Europe, North America.

Distinguished from most of the *L. dispersa* **(23)** complex by the white, well-developed, almost placodioid, thallus; *L. andrewii* **(7)**, *L. conferta* **(18)**, *L. congesta* **(20)** and *L. pruinosa* **(50)** have C+ yellow, orange or red thalli.

6. L. alboflavida Taylor (1836)

Thallus of dispersed to aggregated areoles, pale greenish yellow, areoles brownish yellow, convex, margins entire, surface granular and composed of low warts or papillae; soredia arising on the surface of the areoles, initially from circular soralia but becoming confluent and spreading over the surface, paler than the areoles, yellow-green to greenish white; prothallus inconspicuous to black. Apothecia, when present, to 1.2 mm diam.; thalline exciple becoming contorted and crenate to irregularly lobed; discs pale reddish brown, shiny, not pruinose. Thallus Pd−, K− or weakly yellow, C+ orange (atranorin, arthothelin, unidentified UV+ orange pigment).

On siliceous rocks; very rare. Ireland (Kerry, Killarney). ?Endemic.

The material referred to this taxon strongly recalls and has identical chemical reactions to that described for the primarily corticolous *Ochrolechia inversa*; the relationship between these taxa requires re-examination.

7. L. andrewii B. de Lesd. (1913)

Thallus of coarse, scattered granules, often inconspicuous and immersed, white, granules rounded and unequal in size; prothallus inconspicuous. Apothecia 0.5–1.5 mm diam., aggregated, sessile; thalline exciple well developed, persistent, raised, coarsely crenulate to almost lobulate or contorted; discs pale brown to red-brown, flat to concave, sometimes white-pruinose; epithecium interspersed with yellowish brown granules; hymenium 50–65 μm tall; hypothecium yellowish. Paraphyses 1.5–2.5 μm wide, mainly simple, sparsely anastomosed below and sometimes branched above, apices 3–4.5(-5) μm wide, swollen, capitate, yellowish. Asci 35–50 × 12–15 μm. Ascospores 9–12(-13) × 5–6 μm, often guttulate. Thallus Pd+ orange-red, K− or weakly yellow, KC+ red, C+ orange-red (arthothelin, 2,7-dichlorolichexanthone).

On siliceous, maritime rocks, xeric-supralittoral, exceptionally inland, especially on quartzite and schists; rare. S.W. England (Dorset, Portland), N. & S. Wales, Scotland, W. Ireland. Norway, Sweden.

Resembling the *L. dispersa* **(23)** group but distinguished from that, and *L. congesta* **(20)** which may be found in similar habitats, by the chemical reactions. The apothecia also tend to be larger and to have better-developed thalline, often very flexuous, exciples and more red-brown discs than *L. dispersa* **(23)**. *L. fugiens* **(29)** differs in the yellowish brown to yellowish green thallus and more reddish brown discs.

8. L. argentata (Ach.) Malme (1897)

L. subfuscata Magnusson (1932)

Thallus continuous, yellowish to greenish white, smooth at the margins but uneven to warted centrally; prothallus usually black and well developed. Apothecia 0.4–0.8(-1) mm diam., sessile, dispersed or aggregated, constricted below; thalline exciple persistent, smooth to crenulate, containing massive crystals not soluble in K; discs reddish or dark brown, sometimes piebald; epithecium reddish brown, not granular; hymenium (60-)75–90(-100) μm tall. Paraphyses 1.5–2 μm wide, sparsely branched and anastomosed, apices to 3 μm wide, slightly swollen, yellowish. Asci 45–55 × 18–22 μm, broadly clavate. Ascospores (10.5-)11.5–14.5(-17.5) × (5.5-)6–8.5 μm, broadly ellipsoid, walls 0.5–1 μm thick. Thallus Pd− or weakly yellow, K+ yellow, C− (atranorin, gangaleoidin, usually traces of californin).

On bark of deciduous trees; scarce or overlooked. British Is. mainly in the N. C. Europe, Fennoscandia, E. North America.

A distinctive member of the *L. subfusca* group distinguished from *L. chlarotera* **(15)** and allied species by the absence of granules in the epithecium.

9. L. atrosulphurea (Wahlenb.) Ach. (1814)

Thallus irregularly cracked-areolate, the margin often diffuse and the centre compacted, pale sulphur-yellow to greenish yellow or greenish grey, surface smooth or granular; prothallus indistinct. Apothecia 0.5–2(-2.5) mm diam., at first ± immersed, finally sessile and somewhat constricted at the base; thalline exciple at first thin and crenulate, scarcely protruding above the level of the disc and soon becoming excluded; discs becoming strongly convex, often uneven, olivaceous black to greenish black or black; epithecium brown-black or bluish brown, K+ green, N+ purple-red; hymenium 50–75 μm tall, blue-brown or dark brown above; hypothecium to 0.2 mm tall, becoming increasingly thick with age. Paraphyses 1–2 μm wide, branched and anastomosed, some apices expanded to 3–5 μm wide and others scarcely so, blue-brown to black. Asci elongate-clavate, long-stalked, apices only moderately thickened, 50–60 × 11–14 μm. Ascospores (8-)11–15(-16.5) × 4–5(-6) μm, narrowly ellipsoid. Pycnidia 0.1–0.15 mm diam., frequent, immersed, black; conidia 22–28 × 0.5–0.75 μm, thread-like, curved to sigmoid. Thallus Pd−, K+ yellow, C+ yellow-orange to red (usnic acid, zeorin, ± arthothelin, ± norstictic acid, ± thiophanic acid, ? skyrin).

On exposed, montane siliceous rocks, especially epidiorite and schists at over 1000 m. Scotland (Highlands). Arctic, Fennoscandia, North America, probably circumpolar in the Northern Hemisphere.

L. gangaleoides **(31)** differs in the primarily grey thallus and its chemistry, and in the apothecia where the thalline exciple is not excluded and the disc ± flat.

10. L. cadubriae (Massal.) Hedl. (1892)

Lecidea cadubriae (Massal.) Th.Fr. (1874)

Thallus granular, sometimes dispersed or scale-like, white to grey-white, sometimes with a creamy tinge; prothallus white, sometimes well developed and ± fibrous. Apothecia 0.2–0.6(-0.8) mm diam.; thalline exciple at first thin and entire, often becoming ± excluded with age; discs flat to strongly convex, pinkish to pale brown or brownish black, often weakly pruinose; epithecium brown, not distinctly granular; hymenium 30–45 μm tall. Paraphyses 1.5–2 μm wide, simple or sparsely branched above, apices to 4 μm wide, swollen, brown-capitate. Asci 35–45 × 10–14 μm, clavate, *Bacidia*-type. Ascospores (7-)8–11 × 3–4 μm, narrowly ellipsoid. Pycnidia 60–100 μm diam., frequent, immersed to somewhat erumpent, black, sometimes gaping; conidia (3-)4–5(-6) × 1–1.5 μm, bacilliform. Thallus and thalline exciple Pd+ orange, K+ yellow changing to orange or red, C− (stictic acid).

On bark and wood of aged *Pinus*, often associated with *Lecidea hypopta*; very local. Scotland. Montane C. Europe, Scandinavia, North America.

The ascus type and form of the conidia clearly indicate that this species is not satisfactorily placed in either *Lecanora* or *Lecidea s. str*. It may be related to *Lecidea turgidula* which occurs in similar habitats but is K− and Pd− and lacks a thalline exciple. A norstictic acid chemotype of *L. cadubriae* has been reported from North America.

11. L. caesiosora Poelt (1966)

Thallus areolate, composed of almost contiguous, convex areoles, surface roughened, whitish to grey or sometimes brownish with minute dark brown granules; soredia arising from soralia on the surface of the areoles, soralia mainly 1–2 mm diam., convex, the whole surface sometimes sorediate, white or bluish to yellowish white, always paler than the thallus. Apothecia rare in the British Isles, 0.5–0.7 mm diam., ± sessile; thalline exciple persistent; discs dark brown to black, often slightly pruinose; epithecium greenish brown, coarsely granular; hymenium 45–50 μm tall. Paraphyses 1.5 μm wide, conglutinate, apices to 2 μm, slightly swollen. Asci 40–45 × 15 μm, broadly clavate. Ascospores 12–14.5 × (6-)7.5–8 μm, ellipsoid. Pycnidia *c.* 100 μm diam.; conidia 20–25 × 0.5–1 μm, arc-like. Thallus Pd ± yellow, K+ yellow, C− (atranorin, psoromic and usnic acids).

On vertical and often well-lit faces of siliceous walls and rocks; local. N. England, N. Wales, Scotland; specimens with apothecia found only in Scotland (Angus, Perth). N. & C. Europe.

Material without apothecia and lacking the characteristic dark brownish granules on the areoles may be difficult to separate from sorediate morphotypes of *L. campestris* **(12)** and *L. cenisia* **(14)** without t.l.c.

12. L. campestris (Schaerer) Hue (1888)

Thallus generally forming ± circular patches, composed of dispersed to more usually contiguous areoles, surface rough to warted with small rounded protruberances, light to dark grey; prothallus often conspicuous, fimbriate, white; soredia [**subsp. *dolomitica*** O. L. Gilbert (1984)] arising in convex soralia bursting through the upper cortex, coalescing to cover the thallus surface with age, white to greenish white. Apothecia 0.5–1.5(-2) mm diam., usually abundant, often crowded over most of the thallus area, sessile, slightly constricted at the base; thalline exciple persistent, raised, entire or variously flexuose or contorted, with small, irregular crystals in the medulla; discs flat or sometimes slightly convex, red-brown to red-black; epithecium pale orange to reddish brown, not granular; hymenium (60-)70–85(-90) μm tall. Paraphyses 1.5–2.5 μm wide, sparsely branched, especially towards the apices, apices 3–4.5 μm wide, slightly swollen, reddish brown. Asci 50–60 × 12–21 μm, broadly clavate. Ascospores 10–14(-17) × 6–8.5(-9) μm, broadly ellipsoid. Conidia 12–20(-30) × 1 μm, thread-like to arc-like. Thallus Pd− or weakly yellow, K+ yellow, C−, UV− (atranorin, zeorin, and 1–3 unidentified compounds).

On mortar, asbestos-cement, calcareous and nutrient-enriched siliceous rocks, and rarely, worked wood; very common. Widespread in the British Is., becoming scarcer in N. & N.W. Scotland. Europe, North America.

This species superficially recalls the corticolous *L. chlarotera* **(15)** which differs in the large, irregular crystals in the medulla of the thalline exciple as well as in the chemical components. Apothecia are scarce in subsp. *dolomitica* which is known only from magnesian limestone outcrops, man-made structures and dust-contaminated bark in N. England (Peak District).

13. L. carpinea (L.) Vainio (1888)

Thallus forming discrete patches, generally continuous and smooth, thin, light grey to white, tending to become greyer and somewhat cracked in the centre with age; prothallus white. Apothecia 0.5–1(-1.5) mm diam., sessile, constricted at the base, often crowded; thalline exciple well developed, persistent, entire, sometimes becoming almost excluded with age; pale reddish brown to creamy or purplish, flat to somewhat convex, densely white-grey-pruinose; epithecium granular, pale yellowish brown, the granules dissolving in K; hymenium 45–65 μm tall. Paraphyses 1–2 μm wide, sparsely branched and anastomosed, apices not or slightly swollen. Asci 55–70 × 14–18 μm, clavate. Ascospores (9-)10–12.5(-14) × (5-)6–8 μm., subglobose to ellipsoid. Conidia 15–16 × 1 μm, thread-like to arc-like. Thallus Pd −, K+ yellow, C−, the apothecial disc C+ yellow or orange (atranorin, chloratranorin, sordidone).

On smooth bark of deciduous trees, particularly on twigs and small branches, exceptionally on stonework; local. Widespread, especially in S. & E. British Is. Europe, North America, Asia.

L. pallida **(44)** has a similarly pale-coloured thallus, pruinose apothecia, and occurs in the same habitat, but the apothecial discs react Pd+ red and C−, and the thallus is Pd+ yellow changing to red.

14. L. cenisia Ach. (1810)

Thallus a ± continuous, granular crust, or composed of dispersed to aggregated warted areoles, whitish grey to pale grey, areoles ± flat with entire to crenate margins; prothallus when evident white or bluish; soredia [**var. *sorediians*** Suza (1928)] rare. Apothecia 1–2(-2.5) mm diam., sessile, strongly constricted below and somewhat elevated; thalline exciple well developed, persistent, entire to crenulate or flexuose, with large crystals in clumps in the medulla; discs reddish brown to black, yellow-brown, grey or greenish black, sometimes yellowish white-pruinose; epithecium brown to olivaceous brown, with coarse granules dissolving in K, K+ greenish, N+ purple-red; hymenium 60–90 μm tall. Paraphyses 1.5–2 μm wide, sparsely branched, apices slightly swollen, olivaceous. Asci 45–50 × 7–9 μm. Ascospores (7-)9–15(-19) × (4.5-)6–8.5 μm. Thallus Pd+ yellow or yellow to orange-red, K+ yellow or yellow to red (with crystals), C−, thalline exciple Pd+ yellow (atranorin; five chemotypes may occur, with roccellic acid, gangaleoidin, norstictic acid, an unidentified fatty acid, or no additional compound).

On coarse, siliceous rocks, especially in the xeric-supralittoral but also inland. Channel Islands, Wales, N. Scotland. Europe, North America.

Part of a complex of species not well understood in the British Is. and requiring a modern revision. Specimens with black apothecial discs and sparse pruina [var. *atrynea* (Ach.) Clauz. & Roux (1985)] recall *L. gangaleoides* **(31)**, in which the epithecium lacks coarse granules. Reports from the British Is. on wood appear to be errors for other species. The var. *sorediians* is only known in the British Is. from inland sites in Scotland (Angus, Banff, Lothian).

15. L. chlarotera Nyl. (1872)

Thallus continuous, forming neatly delimited patches, smooth to roughened or warty, pale grey or sometimes yellowish grey; prothallus absent. Apothecia 0.4–0.8(-1.5) mm diam., sessile, often aggregated; thalline exciple well developed, persistent, entire to irregularly crenulate or slightly warted, the medulla with massive irregular crystals not dissolving in K; disc flat, pale dull brown to rose-, orange- or red-brown, not or rarely slightly pruinose; epithecium colourless to greyish brown, interspersed with coarse granules soluble in K and N; hymenium 70–95 μm tall. Paraphyses 1.5–2 μm wide, simple or sparsely branched, apices 2.5–4 μm wide, slightly swollen and yellowish brown, not capitate. Asci 65–80 × 10–15 μm, subcylindrical to elongate-clavate. Ascospores (9-)11–13(-15) × 6.5–7.5(-8.5) μm, broadly ellipsoid. Conidia 10–14 μm long, cylindrical. Thallus Pd−, K+ yellow, C−, (atranorin, gangaleoidin, ± roccellic acid, ± californin).

On bark and wood, mainly of deciduous trees and also worked timber; common. Throughout Great Britain & Ireland, except in the most polluted areas. Widespread in Europe, North America.

The commonest member of the '*L. subfusca*' group in the region. Superficially similar to the primarily saxicolous *L. campestris* **(12)**; specimens on timber require t.l.c. to separate them with certainty. Black pycnidia of *Vouauxiella* species often occur at the junction between the exciple and disc; infected apothecia are often discoloured. *L. pulicaris* **(51)** differs in the Pd+ red exciple and the red to orange-brown epithecium with fine granules, not soluble in N; *L. rugosella* **(53)** more commonly has white-pruinose discs but is further separated by the strongly warted thalline exciple and the more granular to warted thallus. Species separations in this complex are increasingly based on detailed studies of the anatomy of the exciple; see Brodo (1984) for further information.

16. L. chlorophaeodes Nyl. (1873)

Thallus areolate, areoles dispersed to ± contiguous, convex to aggregated, pale yellow-green to greenish grey, surface warted; prothallus absent. Apothecia 0.4–1(-1.2) mm diam., sessile, constricted at the base, scattered; thalline exciple entire to crenulate and slightly raised at first but soon becoming ± excluded; discs black, often somewhat shiny, becoming strongly convex and uneven; epithecium bluish green, interspersed with yellowish brown granules dissolving in K, N+ weakly purple-red; hymenium 60–75 μm tall; hypothecium becoming massively developed. Paraphyses 1–2 μm wide, simple or sparsely branched, occasionally anastomosed below, apices to 2.5–3 μm wide, not or only slightly thickened. Asci 45–55 × 12–18 μm, broadly clavate, long-stalked. Ascospores (9-)10–12(-13) × 4–5 μm, occasionally 1-septate, narrowly ellipsoid. Conidia 18–27(-30) × 1–1.5 μm, flexuose to arc-like or sigmoid. Thallus Pd−, K+ yellow, KC+ yellowish, C− or weakly yellowish (usnic acid, zeorin, fatty acids).

On exposed siliceous (syenite) rock, montane; very rare. N. Scotland (Sutherland, Ben Loyal). Fennoscandia.

Resembling *L. leptacina* **(40)** but with a less yellow thallus and different chemical components. Greyer specimens recall *L. gangaleoides* **(31)** which has broader ascospores. Earlier reports from the British Isles refer to *L. fugiens* **(29)** differing in the paler and C+ orange thallus.

17. L. cinereofusca Magnusson (1932)

Thallus continuous or irregularly cracked-areolate, pale grey to grey, smooth to warted; prothallus absent or blackish; soralia rarely present in North American material. Apothecia 0.7–1.5 mm diam., immersed at first, finally sometimes sessile; thalline exciple well developed, persistent, coarsely crenate to almost dentate, papillate or flexuose with age, white or a paler grey than the thallus, medulla with very large crystals not soluble in K; disc flat to slightly convex, orange-brown or pale reddish brown; epithecium red-brown, interspersed with coarse granules also present on the surface, not dissolving in K, Pd+ orange (with the formation of crystals); hymenium 60–85 μm tall. Paraphyses 2–3 μm wide, sparsely branched and anastomosed, apices not or slightly swollen. Ascospores (7.5-)10–14.5 × (6-)7–8.5(-9.5) μm, broadly ellipsoid. Conidia 10–14 μm long, cylindrical. Thallus Pd−, K+ yellow, C−, thalline exciple and epithecium Pd+ yellow-orange (atranorin, pannarin, ± placodialic acid, ± roccellic acid).

On *Salix*; very rare. Scotland (Argyll, Loch Sunart). C. European Alps, Scandinavia, North America (on a variety of deciduous trees), E. Asia.

L. cinereofusca can often be separated from *L. pulicaris* **(51)** by the more orange apothecial discs, but examination by t.l.c. is necessary for accurate identification.

18. L. conferta (Duby ex Fr.) Grognot (1863)

Thallus ± absent to thick and areolate, yellowish ochre to pale grey, surface smooth to warted. Apothecia 0.3–0.8 mm diam., sessile, constricted at the base, generally crowded and becoming angular by compression; thalline exciple yellowish or yellowish grey, at first entire or rarely crenulate, sometimes excluded when mature; disc pale brown, greyish brown or olivaceous, flat to slightly convex; epithecium colourless, yellowish or slightly brownish, granules absent; hymenium 45–55 μm tall. Paraphyses 1.5–2 μm wide, sometimes branched, apices slightly swollen. Asci 36 × 12 μm. Ascospores (8-)9–12(-15) × 4–7 μm. Conidia 10–15 × 0.5 μm. Thallus K−, C+ yellow (also thalline exciple), apothecial discs C+ yellow-orange.

On hard limestones; very rare. England (S. Devon, Yorkshire), N. Wales, N.E. Scotland; there are old records from Ireland. Europe.

A member of the poorly understood *L. dispersa* **(23)** group distinguished primarily by the C+ yellow reaction. The name *L. conferta* has been misapplied and the correct name as well as the taxonomy of this species needs further study.

19. L. confusa Almb. (1955)

Thallus granular to granular-areolate, forming small discrete patches, greenish grey to yellow-grey, smooth. Apothecia sessile, slightly constricted below, in small groups; thalline exciple generally slightly paler than the thallus, at first crenulate but gradually becoming excluded and represented by a few irregular granules; discs flat to somewhat convex, pale yellow-green, more rarely pinkish or brownish; epithecium brownish, with coarse granules; hymenium 65–85 μm tall. Paraphyses 1–2 μm wide, branched and anastomosed, apices not swollen. Asci 32–45 × 11–15 μm, broadly clavate, long-stalked. Ascospores (8-)10–14(-16) × 4–5 μm, narrowly ellipsoid. Thallus Pd−, K−, C+ orange-red.

On bark, wood and worked timber of deciduous trees, particularly on twigs; locally abundant, especially near the coast. S.W. England, Wales, Lowland Scotland, E. Ireland. W. Europe N. to S. Scandinavia, North America.

Separated from the eastern *L. varia* **(71)** and sparsely sorediate specimens of *L. conizaeoides* **(21)** by the more yellow colour of the thallus, the gradually excluded exciple, and the C+ thallus reaction. *L. expallens* **(26)** has soredia, and *L. symmicta* **(69)** an overall grey, usually K+ yellowish thallus and paler immarginate apothecial discs. See also *L. strobilina* **(62)**.

20. L. congesta Clauz. & Vězda (1969)

Thallus to 1 cm diam., forming patches, areolate, to 1 mm thick, the centre often disintegrating, the margins not or slightly lobed, whitish, not pruinose; prothallus absent. Apothecia 0.8–1.2 mm diam., sessile, constricted below and slightly raised, often crowded and becoming angular by mutual compression; thalline exciple well developed, persistent, ± wavy; disc pale brown-grey, flat, not pruinose; epithecium brownish, interspersed with minute, brown granules dissolving in K, K−, N−; hymenium 60–70 μm tall. Paraphyses 1.5–2 μm, simple or sparingly branched, apices slightly thickened. Ascospores (9.5-)10–12 × 5–6.5(-7) μm. Thallus Pd−, K−, C+ pink to red (gyrophoric acid).

On steep to vertical, north-facing, hard limestones on sea cliffs; not recorded for over 130 years and perhaps extinct. N.E. Ireland (Down). W. Mediterranean, North America.

The thallus recalls *L. albescens* **(5)** in which the apothecial discs are paler and the thallus C−. In *L. pruinosa* **(50)** the margins of the thallus are somewhat lobed, pruinose and C+ orange.

21. L. conizaeoides Nyl. ex Crombie (1885)

Thallus coarsely granular, thick, ± continuous, grey-green; soredia forming over the whole thallus surface, grey-green, coarsely granular, often predominating. Apothecia 0.4–1(-1.5) mm diam., frequent, sessile, narrowed at the base, often at or beneath the level of the soredia; thalline exciple persistent, crenulate to granular, sorediate or irregularly flexuose; discs pale green-grey to grey-brown, sometimes pinkish, flat or concave, rarely slightly convex; epithecium colourless to pale yellow-brown, generally with granules dissolving in K−, not granular; hymenium (45-)55–70(-85) μm tall. Paraphyses 1.5–2.5 μm wide, branched and anastomosed, most frequently branched towards the apices, apices to 3 μm wide, not or slightly swollen. Asci 40–50 × 12–18 μm, broadly clavate, apical dome I+ blue with an I− internal cone, recalling the *Bacidia*-type. Ascospores 10–14(-16) × 4–6(-7) μm. Conidia 12–22 × 0.5–1 μm, arc-like to thread-like. Thallus Pd+ yellow changing to red, K− or weakly yellowish, C− (fumarprotocetraric acid).

On a wide range of low pH substrata including bark, wood, siliceous rocks, walls, decaying plant debris, peaty soils, and rubberized materials; the commonest lichen on trees throughout Lowland Britain where air pollution restricts the occurrence of other species, becoming rarer and mainly restricted to worked timber in S.W. England, Wales (Pembroke), N.W. Scotland. Widespread in Great Britain, introduced into Ireland and there restricted to eastern coastal areas. Widespread in lowland areas of Europe subjected to air pollution, introduced into several eastern North American cities, also near sulphur springs in Iceland.

Colonies are commonly infected by the parasitic fungus *Athelia arachnoidea* causing pale brown circular lesions with felted, white margins. Apothecial discs infected by *Lichenoconium lecanorae* turn black. In the most polluted areas thalli are reduced to a sorediate crust, while in cleaner areas apothecia can predominate and soredia become sparse; the latter specimens recall *L. confusa* **(19)** and *L. varia* **(71)** which respectively differ in the Pd− and Pd+ yellow (not changing to red) thalli.

22. L. crenulata Hook. (1833)

Thallus usually immersed and inconspicuous, more rarely areolate, pale grey, surface smooth to finely granular. Apothecia (0.1-)0.2–0.6(-0.9) mm diam., constricted at the base, slightly raised, often in small groups and then becoming angular by compression; thalline exciple well developed, persistent, white, regularly deeply crenate with 5–8 segments, rarely non-crenulate; discs rather variable in colour, reddish brown to yellowish or brownish-black, grey or blue-grey-pruinose; epithecium brownish or blue, interspersed with fine granules, N− or N+ faintly pink; hymenium 55–70 μm tall, yellow-brown above. Paraphyses *c.* 2 μm wide, sparsely branched, apices capitate, the terminal cell brownish and swollen to about 3 μm wide. Ascospores 6–10(-15.5) × (4-)4.5–6(-7) μm. Conidia 11–15 × 0.5 μm. Thallus Pd−, K−, C−.

On hard, calcareous rocks and mortar; scattered. Commonest in S.E. & E. England (East Anglia). Widespread in the Northern Hemisphere.

In *L. albescens* **(5)** and *L. dispersa* **(23)** which occur in similar habitats, the exciple is never regularly deeply crenate and the apothecial discs are not or lightly pruinose and usually paler in colour.

23. L. dispersa (L.) Sommerf. (1826)

Thallus immersed or sometimes consisting of scattered granules, white to pale grey. Apothecia (0.15-)0.2–1(-3) mm diam., sessile, constricted below, scattered or aggregated into dense groups; thalline exciple well developed, persistent, entire to crenulate to contorted or flexuose, often pruinose when young; discs very variable in colour, pinkish brown to olivaceous brown, or pale yellowish or greenish grey, sometimes white-pruinose (especially when young or on highly calcareous substrata); epithecium pale yellowish brown or brownish, densely interspersed with granules not dissolving in K, N− or N+ paler; hymenium 70–100 μm tall; hypothecium sometimes brownish. Paraphyses 1.5–2 μm thick, branched and frequently anastomosed, apices to 3.5 μm wide, slightly swollen and brownish. Asci 50–65 × 12–18 μm, broadly clavate, short-stalked. Ascospores (7-)8.5–14 × (3-)4–7 μm. Thallus Pd−, K−, C− (β-sitosterol).

On a wide range of calcareous substrata, including rocks, walls, mortar, and many man-made substrata and articles, also on nutrient-enriched or dust-contaminated bark; extremely common. Throughout Great Britain & Ireland, becoming less frequent in upland regions. The most air pollution tolerant lichen known. Widespread through the Northern Hemisphere.

In heavily soot-contaminated areas the thallus may incorporate soot particles to form a blackish stain on concrete buildings. Some corticolous specimens could belong to the species called *L. hageni* (Ach.) Ach. (1810) or *L. umbrina* (Ehrh.) Massal. (1852) by some European authors. The group is badly in need of a monographic study. The separation from *L. albescens* **(5)** and *L. crenulata* **(22)** is discussed under those species.

24. L. epanora (Ach.) Ach. (1810)

Thallus coarsely granular or forming a ± continuous, areolate crust, yellow-green or sometimes blackish green, areoles strongly warted and convex; prothallus when visible black; soralia arising on the upper surface of the areoles, soredia dull yellow to lemon-yellow, usually paler than but sometimes concolorous with the thallus, 0.2–0.5 mm diam., but often coalescing to form an almost continuous, sorediate crust. Apothecia 0.5–1.5(-2) mm diam., rare, sessile; thalline exciple well developed, entire to crenate or coarsely lobed, often flexuose or sorediate, raised; disc yellow-brown to pale red-brown, flat to slightly convex; epithecium yellowish to greenish or brownish, interspersed with yellowish granules not dissolving in K; hymenium 70–80 μm tall. Paraphyses 1–2 μm wide, branched and anastomosed, apices not or slightly swollen. Ascospores 8–12(-13) × 5–6.5 μm. Thallus and soredia Pd− or occasionally Pd+ yellow or orange, K−, C− (epanorin, rhizocarpic acid, zeorin).

In shaded overhangs and dry crevices in iron sulphide-rich rocks. N. England (Peak District, S. Pennines, Northumberland), Isle of Man, Wales, Scotland. C. Europe, Canada (British Columbia).

In *L. subaurea* **(63)** the soralia mainly arise on the margins not on the surface of the areoles and rarely coalesce to form a sorediate crust; thallus and soralia of that species are also consistently Pd+ orange or red due to pannarin.

25. L. epibryon (Ach.) Ach. (1810)

Thallus weft-like or of dispersed, convex, wart-like granules, whitish to pale grey; soralia when present excavate, soredia coarsely granular, greyish white. Apothecia 1–2(-3) mm diam., sessile, constricted at the base, often short-stalked, frequently in dense groups; thalline exciple persistent, entire to crenulate or flexuose, containing small, irregular crystals in the cortex and sometimes the medulla not dissolving in K; disc dark reddish brown, shiny or sometimes with small clumps of pruina, ± flat; epithecium reddish brown, sometimes with small oil drops; hymenium 60–85 μm tall. Paraphyses 1.5–2 μm wide, simple or sparsely branched, apices to 3.5 μm, somewhat swollen, reddish brown. Asci 48–55 × 12–18 μm, broadly clavate, short-stalked. Ascospores (11-)13.5–18(-19) × (5-)8–9.5 μm, narrowly ellipsoid. Thallus Pd−, K+ yellow, C− (atranorin, ± zeorin).

On peat, mosses and other plant remains, arctic-alpine, 800–1000 m; very rare. N. England (W. Yorkshire), Scotland (Highlands). Circumpolar in the Northern Hemisphere.

Resembling a terricolous and bryophilous *L. campestris* **(12)**; their relationship merits careful study.

26. L. expallens Ach. (1810)

Thallus effuse and granular to cracked-areolate, usually forming circular to elongate patches, yellowish green, sometimes with radially orientated crystals evident on the surface; soredia initially arising from discrete, somewhat excavate, 0.1–0.3 mm, lens-shaped soralia, soon becoming confluent to make a continuous, sorediate crust, farinose, pale yellow to lemon yellow; prothallus inconspicuous or when on smooth bark, bluish grey. Apothecia 0.3–0.8(-1.5) mm diam., when present sessile, not constricted below, sometimes angular by compression; thalline exciple at first crenulate but becoming excluded, sometimes sorediate; disc pale yellow, pinkish or brownish, flat to slightly convex, sometimes pruinose; epithecium colourless to pale yellowish; hymenium 35–55 μm tall. Paraphyses 1.5–2 μm thick, sparsely branched and anastomosed, apices not or slightly thickened. Ascospores 10–16(-17) × 4–7(-7.5) μm, broadly ellipsoid. Thallus Pd −, K+ yellow, C+ deep yellow or orange to reddish (thiophanic and usnic acids, zeorin).

On bark, wood and worked timber, generally in well-lit or other xeric situations, occasionally also on dry, siliceous rock faces and vertical memorials or walls; common. Throughout Great Britain & Ireland. Reported as widespread in the North Temperate zone, but North American records are mainly *L. strobilina* **(62)**.

When sterile this species may be confused with *Pyrrhospora quernea*, which has a somewhat duller yellow, rather than a yellow-green or citrine, thallus and rather coarser soredia; it also contains isoarthothelin and lacks usnic acid and zeorin.

27. L. farinaria Borrer (1834)

Thallus of contiguous areoles, grey-white, sometimes slightly yellowish, greenish or brownish, smooth; soredia arising in small mounds which may coalesce and cover the whole thallus surface, farinose to granular, yellowish or whitish. Apothecia rare, 0.4–1(-1.5) mm diam., dispersed or aggregated, at first immersed amongst the soredia but finally erumpent; thalline exciple at first well developed, later becoming excluded, sorediate when young, the medulla with very large crystals not dissolving in K; disc black or sometimes reddish brown, sometimes partly pallid, finally convex; epithecium yellow-brown to greenish brown, interspersed with granular crystals dissolving in K but not N; hymenium 50–75 μm tall. Paraphyses *c.* 1 μm thick, anastomosed below, sparsely branched above, apices greenish brown, not swollen. Ascospores (10-)14–17(-20) × 7–15 μm, broadly ellipsoid to subglobose. Thallus and soralia Pd+ weakly yellow, K+ yellow, C− (atranorin, chloratranorin, roccellic acid).

On wood (and occasionally adjacent loose bark), rarely on bark of healthy trees e.g. *Salix* in carr; rare, scattered. N. England, Scotland, N. Ireland (Fermanagh, Donegal); formerly also in S. England (W. Sussex). N.W. Europe, North America.

Most easily separated from other sorediate taxa on wood when fertile as the blackish apothecial discs are surrounded by a characteristic thick, sorediate thalline exciple.

28. L. frustulosa (Dickson) Ach. (1810)

Thallus of scattered to ± contiguous, strongly convex areoles, areoles 1–1.5 mm diam., the outermost areole margins crenulate to almost palmate, pale yellow to yellow or yellow-green, surface minutely roughened; prothallus well developed, extensive, black. Apothecia 0.4–2(-3) mm diam., at first immersed in or between the areoles, finally ± sessile; thalline exciple at first well developed and entire, becoming crenate, flexuose or irregular, often excluded with age; discs brown to red-brown, flat to convex; epithecium reddish brown, not granular; hymenium 60–90 μm tall. Paraphyses 1.5–2.5 μm wide, mainly simple or sparsely branched above, apices 3.5–5 μm wide, capitate, reddish brown or brownish. Ascospores 10–15(-18) × 5–7(-9) μm. Thallus Pd+ orange-red, K+ yellow, C− (usnic acid, unidentified fatty acids; ± stictic, ± norstictic and ± cryptostictic acids, ± atranorin, ± epanorin, zeorin, ± unidentified fatty acids).

On siliceous rocks in montane areas; rare. N. England (Lake District), Scotland (Highlands, Ben Lawers). Circumpolar in the Northern Hemisphere, essentially arctic-alpine but extending S. to North America (California).

The strongly convex areoles separate this species from all other yellow and yellow-green species of the genus in Great Britain. Of the various chemotypes, that with fatty and usnic acids is the commonest in the British Is., although the stictic acid-group chemosyndrome also occurs.

29. L. fugiens Nyl. (1873)

Thallus immersed or of scattered to loosely aggregated, rounded granules, exceptionally irregularly areolate, pale yellow to yellowish brown. Apothecia 0.3–0.9(-3) mm diam., often arising each on a single granule, dispersed or aggregated, constricted at the base; thalline exciple persistent, entire, becoming crenulate or sometimes flexuose by compression; disc yellowish to reddish brown, flat or slightly convex, sometimes slightly white-grey-pruinose; epithecium pale red-brown, interspersed with minute, granular crystals; hymenium 45–60 μm tall. Paraphyses 1.5 μm wide, unbranched, apices to 2 μm wide, swollen. Asci 40–50 × 6–10 μm, broadly clavate, short-stalked. Ascospores (8-)9–12(-14) × (4-)5–6 μm, narrowly ellipsoid. Thallus K+ faintly yellow, KC+ red to orange-red, C+ orange to orange-red, thallus, exciple, disc and epithecium Pd+ orange to orange-red (arthothelin).

On maritime rocks, xeric-supralittoral, rarely on decaying *Armeria maritima*; local. Channel Is., W. coasts British Is., N.E. Scotland. W. Europe.

The distinctive Pd reaction, best demonstrated on the apothecia, separates this from colour morphs of the *L. dispersa* **(23)** group; *L. andrewii* **(7)** has a white thalline exciple. There appear to be two or three other closely related, but undescribed taxa similar to *L. fugiens* and the complex is in need of critical evaluation.

30. L. fuscescens (Sommerf.) Nyl. (1871)

Lecidea fuscescens Sommerf. (1824)

Thallus effuse, granular, granules 0.05–0.3(-0.5) mm diam., poorly delimited, white or greyish; prothallus sparsely developed, blackish. Apothecia (0.2-)0.4–0.6(-0.8) mm diam., sessile; thalline exciple at first entire, becoming excluded; disc pale to dark brown; hymenium 35–50 μm tall. Paraphyses lax in K, apices swollen and brown to reddish brown. Ascospores (5-)6–10(-13) × 4–7(-8) μm, broadly ellipsoid to subglobose. Conidia needle-like. Thallus K+ yellow.

On bark and wood; reported last century, ?extinct. Scotland (Highlands). C. European Alps, Poland, Scandinavia, North America.

The status of this species in Great Britain requires further study. Records made in recent decades mainly refer to *L. cadubriae* **(10)** which has narrowly ellipsoid ascospores 3–4 μm wide and a Pd+ orange thallus.

31. L. gangaleoides Nyl. (1872)

Thallus irregularly areolate, delimited, separate thalli often coalescing, areoles warted to strongly convex, grey, sometimes with a greenish tinge; prothallus white. Apothecia (Fig. 28b) to 1.5(-2) mm diam., sessile, slightly constricted at the base; thalline exciple persistent, entire or rarely crenate, the medulla with large crystals not dissolving in K; disc black, flat; epithecium green or brownish green, not granular, K+ green and N+ purple-red; hymenium 70–90(-100) μm tall, often with an olivaceous green tinge; hypothecium colourless to pale pinkish brown. Paraphyses 1–1.5 μm wide, infrequently branched, anastomosed, apices not swollen or capitate. Asci 55–70 × 11–15 μm, broadly cylindrical. Ascospores (10-)12–15(-18) × (5-)6–8(-9) μm. Thallus Pd± yellow or red, K+ yellow, C− (atranorin, chloratranorin, californin, gangaleoidin, and norstictic, stictic and thiophanic acids), older, lower parts of the thallus with an orange, K+ purple anthraquinone.

On hard siliceous rocks, commonest on somewhat shaded, maritime rocks in the xeric-supralittoral, exceptionally on worked timber, also locally abundant inland. Widespread in British Is. but absent from large parts of C. & S.E. England. Europe, North America.

Most commonly confused with *Tephromela atra*, which prefers more basic rocks, and generally has a rather paler grey thallus without any greenish tinge, a purple-red hymenium (often demonstrable in the field by hand sections), a dark red-brown hypothecium, and never forms orange anthraquinones in the lower parts of the thallus (extracted and easily observed in acetone). The two species may occur intermixed in the xeric-supralittoral. Greenish thalli should be compared with *L. atrosulphurea* **(9)** in which the exciple becomes ± excluded. The colour of the epithecium distinguishes *L. gangaleoides* from the poorly understood *L. cenisia* **(14)** group.

32. L. gisleriana Müll. Arg. (1874)

L. gisleri Poelt & Ullrich (1964)

Thallus absent or represented by small, orange-grey to grey-brown lobules intermixed with the host lichen. Apothecia 0.2–0.7(-1) mm diam., dispersed; thalline exciple persistent, ± entire to irregularly crenulate or crenate, pale grey; disc orange-brown to reddish brown, flat to somewhat convex; epithecium dark red-brown; hymenium 40–55 μm tall. Paraphyses 1–2 μm wide, sparsely branched, apices 2–3 μm wide, slightly swollen, not capitate. Asci 40–48 × 12–15 μm, broadly clavate, outer coat and apical dome I+ blue, with a broad I− apical cushion, discharge rostrate, ?*Lecanora*-type. Ascospores (7-)8.5–11 × 4–5 μm, broadly fusiform, apices pointed.

Lichenicolous lichen, on thalli of *L. epanora* **(24)**, *L. handelii* **(33)** and *L. subaurea* **(63)**, on metal-rich spoil heap; very rare. S.W. England (North Somerset, Mendip Forest) and S. Wales (Ystwyth). C. Europe, N. Sweden.

Distinguished from other lichenicolous taxa by the pointed ascospores.

33. L. handelii Steiner (1909)

Thallus areolate, dispersed or forming a ± continuous crust, areoles flattened and subsquamulose or more commonly convex, sometimes strongly so, grey-green to grey; soredia arising from soralia on the margins of the areoles, more rarely from their surface, concolorous with the thallus or whitish to bluish green or slightly yellowish; prothallus absent. Apothecia 0.2–0.5 mm diam., often absent, sessile; thalline exciple persistent, not or scarely raised, entire; disc pale brown to pinkish brown, flat, sometimes with bluish black pruina; epithecium colourless to pale yellowish, not granular; hymenium 50–65 μm tall. Paraphyses 1.5–2.5 μm wide, sparsely branched and anastomosed, apices not or slightly swollen. Ascospores not found in British material. Thallus Pd−, K−, KC+ yellow, C− (usnic acid, zeorin).

On metal-rich siliceous rocks, particularly spoil heaps from copper mines; rare. England (Lake District, E. Cornwall, Somerset), Wales (Pembroke). C. Europe, Alps, Scandinavia.

L. subaurea **(63)** has a much brighter yellow-green thallus and contains rhizocarpic acid and pannarin in addition to zeorin. *L. handelii* belongs to the *L. polytropa* **(47)** group and is related to *L. soralifera* **(59)** which has distinctly yellow-green soredia produced from the start on the surface of flattened and darker areoles; in *L. handelii* the soralia are initiated from the margins of the areoles.

34. L. helicopis (Wahlenb.) Ach. (1814)

Lecania helicopis (Wahlenb.) Clauz. & Roux (1985)

Thallus cracked-areolate, forming ± continuous, discrete patches, grey to dark grey, surface smooth and somewhat glossy to slightly roughened; prothallus often conspicuous, white, dark blue or black. Apothecia 0.2–0.5(-0.7) mm diam., at first ± immersed, becoming sessile, not strongly constricted below; thalline exciple at first entire and concolorous with the thallus, becoming brown to black, finally almost excluded; disc black or sometimes pinkish brown; epithecium yellowish or brownish, K+ yellow, N−; hymenium 40–45 μm tall, colourless or greenish grey below and greenish to olivaceous above; hypothecium grey. Paraphyses 2–3 μm thick, sparsely branched, apices to 3.5–4 μm wide, swollen, bluish green to dark brown. Asci 28–35 × 12–20 μm, broadly clavate, *Lecanora*-type. Ascospores 9–15 × 4–6 μm, appearing 1-septate, often guttulate. Pycnidia to 0.06 mm diam., often numerous, black; conidia 18–25 × 0.5–1 μm, thread-like to arc-like, sigmoid or irregularly contorted. Thallus Pd−, K−, C−.

On siliceous, maritime rocks in the mesic-supralittoral; common. Widespread in British Is. Coastal W. & N. Europe, North America.

For separation from *L. actophila* **(2)** see under that name. May also be confused with *Halecania ralfsii* which has truly 1-septate ascospores, upturned, whitish margins to the areoles and a Pd+ red medulla (argopsin). Although *L. helicopis* has been referred to *Lecania*, its asci are not of the *Bacidia*-type so it is retained here.

35. L. horiza (Ach.) Lindsay (1869)

L. parisiensis Nyl. (1866)

Thallus continuous, in small patches, white to yellowish white, rough to slightly warted, rarely smooth; prothallus when evident white. Apothecia 0.5–1.2(-2.5) mm diam., sessile, somewhat constricted at the base, often densely crowded; thalline exciple well developed, persistent, entire or sometimes flexuose, raised, medulla with small, irregular crystals not dissolving in K; disc orange, reddish orange or reddish brown, sometimes white-pruinose when young; epithecium reddish brown, interspersed with fine but lacking coarse crystals; hymenium 65–80 μm tall. Paraphyses 1.5–2.5 μm wide, simple to sparsely branched, rarely anastomosed, apices 2.5–4 μm wide, yellowish brown, slighty swollen. Ascospores (11-)12–15(-17) × (5.5-)6.5–8.5(-9.5) μm. Thallus Pd−, K− or sometimes weakly yellowish, C− (atranorin, 1–3 unidentified substances).

On bark of deciduous trees, especially mature *Acer, Fraxinus* and *Ulmus*; local. Mainly in S. & E. British Is. France, U.S.A. (California), Mexico.

Frequently overlooked for *L. chlarotera* **(15)** which has a more granular epithecium, particularly large crystals in the medulla of the exciple, and a distinct K+ yellow thallus. *L. laevis* **(39)** is doubtfully distinct.

36. L. intricata (Ach.) Ach. (1810)

Thallus of subsquamulose, flattened areoles, each 0.2–0.5 mm wide, contiguous or sometimes dispersed, areole margins crenulate (jig-saw-like), grey-green or yellow-green, surface smooth or finely wrinkled; prothallus when present black. Apothecia (0.4-)0.8–1(-1.5) mm diam., immersed or sessile, not constricted at the base, 1–2 per areole; thalline exciple at first entire and slightly raised but becoming flexuose and often ± excluded; discs blackish green or greenish brown, concave to slightly convex; epithecium green-brown or brown, interspersed with granules dissolving in K; hymenium 60–70 μm tall. Paraphyses simple or sparsely branched, apices slightly swollen to 2.5 μm wide. Ascospores (8-)10–14(-15) × (4.5-)5–7 μm. Conidia 23–25 × 0.5–1 μm, arc-like. Thallus Pd−, K− or K+ yellow, KC+ yellow, C− (usnic acid, zeorin).

On siliceous rocks and walls, more rarely worked timber, xeric-supralittoral to upland. Widespread in British Is., rare in Lowland Britain; mainly coastal in Ireland. Circumpolar in the Northern Hemisphere.

Distinguished from *L. polytropa* **(47)**, with which it has sometimes been subsumed, by the crenulate margins of the areoles, patterned surface, mainly immersed apothecia with darker discs that do not become strongly convex, and the generally more continuous thallus.

37. L. intumescens (Rebent.) Rabenh. (1845)

Thallus continuous, smooth or minutely cracked, only becoming areolate in old, central parts of the thallus, often poorly delimited, pale grey; prothallus absent or black. Apothecia (0.8-)1–2.5(-3) mm diam., sessile, slightly constricted at the base, dispersed or rarely aggregated when they may be angular by compression; thalline exciple well developed, persistent, entire to almost crenate or flexuose, raised, white-grey, often pruinose, without a differentiated cortex; discs orange-brown, red brown or dark brown, pruinose when young but becoming convex and non-pruinose with age; epithecium interspersed with coarse, yellowish granules dissolving in K; hymenium (70-)80–110 μm tall. Paraphyses 1.5–2.5 μm wide, simple or sparsely branched and anastomosed, apices to 3.5 μm wide, slightly swollen, colourless to yellowish brown. Ascospores 11.5–18 × 5–8 μm. Thallus K+ yellow, C−, exciple Pd+ deep yellow or orange (atranorin, chloratranorin, psoromic and 2'-*O*-demethylpsoromic acids, ± lichexanthone, zeorin).

On smooth bark of deciduous trees in wayside and woodland situations; widespread but rather scarce. C. & S. England, Wales, Scotland, Ireland. C. Europe N. to Scandinavia, North America.

Older records often refer to *L. chlarotera* **(15)**, distinguished by the non-pruinose and generally smaller apothecial discs, the presence of large irregular crystals in the non-pruinose exciple, and the chemical components. The bloated, white margin of the apothecia and presence of psoromic acid (Pd+ deep yelow-orange) are diagnostic.

38. L. jamesii Laundon (1963)

Thallus to 2–3 cm diam., forming continuous, smooth to granular patches, grey; prothallus when present black; soredia arising in ± circular, convex soralia, to 1 mm diam., farinose, pale yellow (to whitish on storage). Apothecia to 0.8 mm diam., usually absent, dispersed, immersed or sessile; thalline exciple entire to delicately crenulate, becoming ± excluded, including large crystal clusters; discs pale greenish brown, pinkish brown or rarely almost black, flat; epithecium pale yellowish brown interspersed with small, granular crystals; hymenium 35–60 μm tall. Paraphyses sparsely branched, apices not swollen or pigmented. Asci 45–50 × 10–15 μm. Ascospores (7-)10–14 × (4-)6–8 μm. Thallus Pd−, K−, C−, soralia sometimes weakly K+ yellowish (atranorin, gangaleoidin).

On smooth bark of deciduous trees, most frequently *Salix* in damp situations, rarely on wood or siliceous rocks; locally abundant. S.W. & W. Great Britain, scattered in Ireland. France (Brittany), N. Spain, Chile.

39. L. laevis Poelt (1952)

Thallus continuous, not to distinctly cracked-areolate or warted, forming delimited patches, almost white to grey, containing small crystals; prothallus indistinct or whitish. Apothecia 0.2–0.6(-0.7) mm diam., sessile, slightly constricted below; thalline exciple persistent, entire or slightly crenulate, medulla with small crystals not dissolving in K; discs red-brown, dull, flat or finally slightly convex; epithecium red-brown, not granular; hymenium 60–70 μm tall; hypothecium with abundant, small crystals. Paraphyses 1.5–2.5 μm wide, simple or sparsely branched, apices not swollen. Ascospores 13–15 × (4-)5-8(-9) μm, ellipsoid. Thallus Pd−, K− or weakly yellowish, C− (atranorin).

On bark of deciduous trees; rare. N. Wales, Scotland, Ireland (Sligo). Atlantic-Mediterranean, also drier parts of C. Europe, Africa, Saudi Arabia, Indian Subcontinent.

This belongs to a poorly understood southern European group within the *L. subfusca* complex which is in need of a thorough revision. The relationship to *L. horiza* **(35)** merits a particularly critical examination; the two could prove to be synonyms.

40. L. leptacina Sommerf. (1826)

Thallus of dispersed to aggregated, convex to subglobose squamules, the margins somewhat crenulate, tending to form an uneven, areolate thallus with somewhat flattened and radiating margins, yellow to yellow-green; prothallus black. Apothecia (0.4-)1–1.5 mm diam., at first ± immersed, becoming sessile but not or slightly constricted below; thalline exciple persistent, well-developed, entire, becoming flexuose, crenate and bead-like with age; discs olivaceous to green-black or black, uneven to slightly convex, matt, often thinly white-blue-pruinose; epithecium yellowish brown, ± granular, K+ blue-green; hymenium 60–80 μm tall, yellow-brown above. Paraphyses 1.5–2 μm wide, sparsely branched and anastomosed, apices to 4.5 μm wide, slightly thickened and brownish. Ascospores (9-)10–13(-15) × (4.5-)6–9 μm, ellipsoid or narrowly ellipsoid, walls *c.* 1 μm thick. Thallus Pd− or yellowish, K−, C− (atranorin, psoromic and 2′-*O*-demethylpsoromic acids).

On mosses (especially *Andreaea*) and decaying plant materials, usually in areas of late snow-lie, montane, usually only over 650 m; very rare. Scotland (Highlands). C. Europe (Tatra Mountains, Alps), Greenland, Fennoscandia, arctic North America, circumpolar in the Northern Hemisphere.

41. L. marginata (Schaerer) Hertel & Rambold (1985)

Thallus well developed, continuous or cracked-areolate, rarely dispersed, lemon-yellow to yellowish white or rarely deep yellow; prothallus prominent, white or bluish grey. Apothecia 0.5–1.5(-2.5) mm diam., immersed, eventually becoming almost sessile; thalline exciple soon becoming excluded; true exciple persistent, black; discs flat to somewhat convex, not pruinose; epithecium yellow-green to greenish black; hymenium 55–75 μm tall. Paraphyses 1.5–2.5 μm wide, simple or sparsely branched and anastomosed, apices 3.5–5(-6) μm wide, swollen, capitate, brown. Ascospores (8-)10–14 × (4-)4.5–6(-7.5) μm, narrowly ellipsoid. Pycnidia frequent; conidia 11–15 × 0.5–1 μm, vermiform to arc-like. Thallus Pd− or weakly orange, K+ yellow, KC+ yellow, C− (usnic acid, atranorin also reported).

On hard, calcareous rocks or other base-rich rocks, montane; very rare. Scotland (Highlands). C. Europe, Scandinavia.

The only species of the genus in the British Isles with a predominantly yellow thallus occurring on calcareous or other base-rich rocks. *L. subcarnea* **(64)** has a much more greenish, usually K− thallus and lacks the large, black apothecia; a new generic name may be required for these and some allied species.

42. L. muralis (Schreber) Rabenh. (1845)

Thallus to 10 cm diam., forming placodioid, circular patches or rosettes, usually closely adpressed but becoming uneven when old, marginal lobes flat to concave, corticate below, thallus centre sometimes areolate, in extreme habitats reduced to dispersed granules or areoles, greenish yellow to yellow brown, tending to be darker centrally, ± shiny to slightly pruinose, the undersurface and edges of the marginal lobes whitish. Apothecia 0.5–1.5(-2) mm diam. sessile, densely aggregated in the centre of the thallus;

thalline exciple well developed, entire to crenulate or flexuose, often angular by compression, pruinose, generally persistent, sometimes almost excluded in old apothecia; discs yellow-brown to reddish brown, flat to slightly convex, not pruinose; epithecium pale yellowish or brownish, with granular crystals on the surface not dissolving in K; hymenium (50-)65–75(-80) μm tall. Paraphyses 2–3 μm wide, mainly simple or sparsely branched above, apices not swollen or capitate. Asci 30–40 × 8–12 μm. Ascospores 9–15(-16) × (4-)5–7 μm. Conidia 20–25 × 0.75–1 μm, thread-like to arc-like. Thallus Pd+ yellowish or Pd−, K−, KC+ yellowish, C− (atranorin, murolic, psoromic and usnic acids, zeorin; ± fumarprotocetraric acid in North America).

On nutrient-rich, -enriched and calcareous rocks, common on a wide variety of man-made substrata (concrete, asbestos-cement, tarmac, paving slabs, tiles, etc.), also dust and nutrient-rich bark, wood and worked timber; air pollution tolerant. Extremely common in Lowland Britain, but becoming rare in the S.W., W., Scotland. Europe, Mediterranean (including North Africa), North America.

Only subsp. *muralis* is known from Great Britain and Ireland. The massive expansion on man-made substrata seen here does not occur in much of the world's range of the species; the material actively spreading in lowland, man-made habitats may represent a special ecotype distinct from upland, bird-perch populations. Unlikely to be confused with any British species apart from *L. straminea* **(61)** which has a C+ orange thallus with convex lobes; specimens with a much reduced thallus are sometimes confused with *L. polytropa* **(47)**.

43. L. orosthea (Ach.) Ach. (1810)

Lecidea orosthea (Ach.) Ach. (1803)

Thallus continuous, often poorly delimited, areolate, areoles flat, irregular in shape, yellowish green or yellow-green; soredia arising from soralia at the margins of the areoles, spreading to cover the whole thallus surface, yellow-green; prothallus white or bluish. Apothecia 0.3–0.7(-0.8) mm diam., sparse or absent, dispersed, sessile, scarcely exceeding the level of the soredia, thalline exciple evident only in young apothecia, soon excluded; discs pinkish brown to yellow-green or grey-black, somewhat convex, not or slightly pruinose; epithecium green to brownish, interspersed with granules dissolving in K; hymenium 50–60(-70) μm tall. Paraphyses mainly unbranched, slightly swollen at the apices. Ascospores (8-)9–16(-20) × (3-)4.5–6(-7) μm. Conidia 13–20 × 1 μm, needle-like. Thallus Pd−, K+ yellow-brown, KC+ yellow, C− (usnic acid, ± skyrin, ± zeorin).

On dry, vertical and in recesses and overhangs of siliceous rocks, occasionally spreading to the upper, exposed surface, rarely on gravestones and old walls, wood or bark (particularly of *Fagus*); frequent. Widespread in N. & W. British Is., rare in E. England. Europe, North America.

Densely sorediate specimens in which the areoles are obscured recall *L. expallens* **(26)**, distinguished by the C+ orange thallus and soredia. The finely isidiate *Pertusaria flavida* is also C+ orange. *L. orosthea* belongs to the same group as *L. sublivescens* **(66)** and *L. sulphurea* **(67)**, both of which lack soredia.

44. L. pallida (Schreber) Rabenh. (1845), non Chev. (1826)

Thallus ± continuous, becoming cracked-areolate in the centre, clearly delimited and forming small patches, whitish grey to grey, surface smooth to slightly granular. Apothecia 0.5–1.5 mm diam., dispersed to closely aggregated, sessile, slightly constricted at the base; thalline exciple well developed, usually entire, conspicuous at first but becoming ± excluded, medulla with granular crystals dissolving in K; discs pinkish to rose or occasionally light brown, flat to convex, densely white-pruinose; epithecium yellowish brown, interspersed with coarse granules dissolving in K; hymenium 60–85 μm tall. Paraphyses 1.5–2.5 μm wide, sparsely branched and anastomosed, apices not or only slightly swollen. Ascospores (9-)10–12(-15) × (5-)6–8(-9) μm, broadly ellipsoid. Thallus K+ yellow, C−, exciple and discs Pd+ yellow to red (atranorin, chloratranorin, protocetraric acid).

On bark of deciduous trees, particularly twigs or when slightly nutrient-rich. Frequent in S. England but becoming scarce northwards, very rare in Scotland (Highlands), scattered in Ireland. Europe N. to S. Fennoscandinavia, North America, Mexico, India.

The separation from *L. carpinea* **(13)** is discussed under that species. Soredia produced in small, raised soralia have been reported from *L. pallida* in France. A K+ red (norstictic acid) variety [var. *rubescens* Imsh. & Brodo (1966)] is known in North America. The epithet *pallida* is retained here in the interests of nomenclatural stability in preference to the unfamiliar *L. albella* (Pers.) Ach. (1810) which could be taken up under the current Code.

45. L. piniperda Körber (1859)

Thallus of scattered granules or ± inconspicuous, not clearly delimited, grey, yellowish or greenish grey. Apothecia 0.2–0.4(-0.8) mm diam., scattered or in small groups, sessile, somewhat constricted at the base; thalline exciple entire to slightly crenate, ± white to brownish grey, generally persistent but becoming thinner and often ± excluded with age; discs pinkish red to reddish or blackish brown, flat, often white or blue-grey-pruinose; epithecium brownish, interspersed with granules; hymenium 45–75 μm tall. Paraphyses 1.5–2 μm wide, branched and anastomosed, apices 2.5–3.5 μm wide, yellowish brown, slightly swollen, capitate. Asci 30–40 × 9–11 μm, broadly clavate. Ascospores (7-)8–12(-16) × 2.5–4 μm, narrowly ellipsoid. Conidia of two types: macroconidia 7–10(-14) × 1–2μm, curved; microconidia 3–5 × 1 μm, bacilliform. Thallus Pd− or + red in parts, K− or weakly yellowish, C−.

On wood and worked timber, also bark on *Pinus sylvestris*; scattered. S.W. England, Wales, Scotland (Highlands), very rare or overlooked in Ireland. Probably circumboreal in cool temperate to subboreal zones in the Northern Hemisphere.

A well-characterized but often overlooked species distinguished by the inconspicuous or dispersed granular thallus and the minute and generally pale blue-grey pruinose apothecia.

46. L. poliophaea (Walhenb.) Ach. (1810)

Thallus continuous, irregularly cracked-areolate, grey to brownish grey or bluish grey, surface densely granular to warted or papillate, papillae rounded to flattened or branched; prothallus often prominent, fimbriate, with alternating white and greenish to black zones. Apothecia to 1 mm diam., often aggregated, constricted at the base, scarcely exceeding the level of the papillae; thalline exciple crenulate to strongly crenate, sometimes papillate, persistent; discs dull red-brown, flat or slightly convex; epithecium yellowish brown, not granular; hymenium 45–70 μm tall. Paraphyses 2–3 μm wide, mainly simple, some sparsely branched, apices 3.5–5 μm wide, pale yellowish brown, markedly swollen, capitate. Asci 35–45 × 12–15 μm, clavate. Ascospores (7-)9–13(-15) × (4-)5–7 μm, ellipsoid. Thallus Pd−, K−, C− (no substances detected by t.l.c.).

On ± flat, nutrient-enriched maritime and lakeside siliceous rocks, especially those used by birds, most frequent in the mesic-supralittoral in coastal areas; local. Wales, Scotland, Ireland, rarely elsewhere. N. Europe, Scandinavia.

47. L. polytropa (Hoffm.) Rabenh. (1845)

Thallus immersed and inconspicuous, or of dispersed granules or areoles, sometimes forming a ± continuous, areolate crust, areoles when present ± rounded, surface flat and with ± even and not crenulate margins, yellow-green to grey-green or brownish, surface ± smooth; prothallus when evident black. Apothecia 0.3–1(-1.5) mm diam., dispersed, sessile, constricted below, tending to arise singly on areoles which are then obscured; thalline exciple initially well developed and entire but becoming crenulate and excluded; discs pale yellow to pale yellow-green, occasionally greenish brown or pale brown, finally convex, glaucous; epithecium colourless to yellowish or reddish brown, interspersed with granular crystals dissolving in K; hymenium 45–60 μm tall. Paraphyses 1.5–2(-2.5) μm wide, sparsely branched, apices slightly thickened. Asci 32–50 × 12–17 μm. Ascospores (9-)10–14(-15) × (4.5-)5–6.5(-7) μm. Conidia (12-)18–22(-25) × 1 μm, needle-like to arc-like. Thallus Pd−, K+ yellow, KC+ yellow, C− (rangiformic and usnic acids, eulecanoral, zeorin).

On siliceous rocks and walls, rarely on worked timber; common. Throughout British Is., particularly in maritime areas. Circumpolar in temperate to arctic areas of the Northern Hemisphere.

Often confused with or used to include *L. intricata* **(36)**, which has darker, emerald green-brown discs, is always areolate and the areoles are usually contiguous and have crenate, jig-saw-like margins and a wrinkled surface. See also *L. stenotropa* **(60)**.

48. L. populicola (DC.) Duby (1830)

Thallus continuous, becoming areolate in the centre, forming clearly delimited patches, concentrically zoned, grey; prothallus fibrous, grey. Apothecia to 1.5 mm diam., aggregated in small groups, ± immersed in the thallus; thalline exciple entire to irregularly contorted, persistent, often pruinose and including numerous crystals; discs light brown to pale yellow-brown, flat, greyish white-pruinose when young; epithecium colourless to brownish, interspersed with numerous granules; hymenium 45–75 μm tall; hypothecium with small granules. Paraphyses 1–2 μm wide, sparsely branched, apices not swollen. Ascospores 10–16 × 6–9 μm. Thallus Pd−, K−, KC−, C− (2,5-dichlorolichexanthone).

On nutrient-enriched *Populus* bark; not recorded for 150 years and apparently extinct. England (E. Norfolk, Coltishall). Europe N. to Scandinavia, U.S.S.R. (Siberia), North America.

49. L. praepostera Nyl. (1873)

Thallus cracked-areolate, often uneven, effuse, areoles convex and often raised in clusters, white, pale green to yellowish white; prothallus zoned white and blue. Apothecia (0.3-)0.5–1.5(-1.8) mm diam., dispersed or in small groups, sessile, somewhat constricted below; thalline exciple well developed, crenulate to crenate, persistent, greenish grey or as the thallus; discs pale brown to dark brown or black, often piebald, flat to slightly convex, sometimes white-grey-pruinose; epithecium colourless or olivaceous, ± interspersed with granules, K+ red (with abundant, red, needle-like crystals); hymenium 70–90 μm tall. Paraphyses 1.5–2 μm wide, simple to sparsely branched, anastomosed below, apices 2–3 μm wide, slightly swollen. Asci 50–65 × 16–18 μm, elongate-clavate. Ascospores (8-)11–15 × (5-)6–8 μm, ellipsoid. Conidia 22–31 × 1 μm, thread-like to arc-like. Thallus Pd+ orange-red, K+ yellow to red (crystals), C− (atranorin, norstictic acid).

On ± vertical, somewhat sheltered, maritime, siliceous rocks; local. S.W. England, Wales, Channel Is., Is.of Scilly. France, Corsica, Canary Islands.

Distinguished from *L. rupicola* **(54)** and *L. swartzii* **(68)** by the K+ yellow to red (not yellow) thallus, from *L. swartzii* also by the C− thallus, and from *L. rupicola* further by the C− apothecial discs and the sessile and not sunken apothecia. Some or all of the British material referred to *L. praepostera* may belong to *L. schistina* (Nyl.) Nyl. ex Crombie (1894); the separation from some morphotypes and chemotypes of *L. cenisia* **(14)** is also unclear.

50. L. pruinosa Chaub. (1821)

Thallus 1.5–4 cm diam., placodioid, forming circular patches, ± areolate in the centre, marginal lobes 0.4–0.8 mm broad, radiating, ± flattened to slightly convex, white to yellowish white or pale yellow-green, surface granular-pruinose; prothallus absent. Apothecia 0.3–0.8(-1.5) mm diam., sessile, slightly constricted below; thalline exciple well developed, persistent, entire to crenulate, raised, pruinose; discs pinkish brown to dark red-brown, white-grey-pruinose; epithecium yellowish brown, interspersed with coarse, angular crystals not dissolving in K; hymenium 55–80 μm tall. Paraphyses 1.5–2.5 μm wide, simple or sparsely branched, apices 2.5–3.5 μm, slightly swollen, yellowish brown, capitate. Asci elongate-clavate, apical dome ± evenly I+ blue with only a shallow, meniscus-like I− apical dome, 45–52 × 10–14 μm. Ascospores 8–13 × 4–7 μm. Thallus Pd−, K− or yellowish, C+ orange (arthothelin, 2,7-dichloronorlichexanthone).

On limestone; not seen this century and presumed extinct. S. England (Somerset, Cleeve Hill). S. Europe, particularly Mediterranean, Carpathians, U.S.A. (Oklahoma).

Unlikely to be confused with any other British species. The placodioid *L. straminea* **(61)** on nutrient-enriched, siliceous rocks is also C+ orange but has swollen, convex lobes lacking pruina.

51. L. pulicaris (Pers.) Ach. (1814)

Thallus continuous, generally clearly delimited, yellowish white to grey, surface smooth to somewhat roughened or warted; soralia very rare, 0.1–0.2 mm diam., when present greenish-white; prothallus white, blue-black or absent. Apothecia 0.3–1(-1.5) mm diam., sessile, somewhat constricted at the base; thalline exciple well developed, entire to slightly warted, persistent, medulla with large crystals not dissolving in K; discs reddish brown or black, ± flat; epithecium red to orange-brown, with granules on the surface and interspersed between paraphyses dissolving in K but not N; hymenium 60–85 μm tall. Paraphyses 1.5–2.5 μm wide, simple or sparsely branched, apices to 3.5 μm wide, not or slightly swollen. Ascospores (9-)11–15(-16) × (6.5-)7.5–9.5(-11) μm, broadly ellipsoid. Thallus K+ yellow or weakly yellow, KC−, C−, thallus and exciple normally Pd+ orange red (atranorin, usually fumarprotocetraric acid, ± roccellic acid, UV+ blue unidentified substance).

On bark and wood of coniferous and deciduous trees, most often on branches and twigs. Widely distributed in British Is., most common in N.E. England & Scotland. Circumboreal in the Northern Hemisphere.

Discussed further under *L. chlarotera* **(15)**.

52. L. quercicola Coppins & P.James (1979)

Thallus dispersed, or composed of contiguous patches, then ± continuous, pale grey to pale green-grey, surface finely warted or granular; prothallus indistinct. Apothecia 0.3–0.6(-0.7) mm diam., usually forming a ± continuous, compacted sward, more rarely dispersed, ± immersed at first but becoming sessile with slightly constricted bases; thalline exciple at first irregularly crenate, becoming excluded; discs pale brown to dull red-brown, rarely pinkish brown, flat to slightly convex, ± whitish or yellowish pruinose; epithecium pale yellow or reddish brown; hymenium (40-)50–65 μm tall. Paraphyses *c.* 1 μm wide, mainly simple, unbranched, apices not or slightly expanded, capitate. Asci 40–45 × 6–8 μm. Ascospores 8–11(-12) × 4–5 μm, narrowly ellipsoid. Pycnidia *c.* 70 μm diam., common, ± immersed to erumpent, black; macroconidia 8–12 × 3–4 μm, broadly fusiform, curved, truncated at the base; microconidia 8–13 × 0.5 μm, arc-like, pointed. Thallus Pd−, K+ weakly yellowish, C−, UV+ pale grey-white (1–3 unidentified pigments).

On rough bark of ancient deciduous trees, especially *Quercus*, in open woodland and parkland. Mainly S.E. & E. England. France, Italy, N. Spain.

L. saligna **(55)** differs in the persistent thalline exciple, non-pruinose apothecial discs, the smaller macroconidia, less regular conidiogenous cells, and usually a single pigment visible on t.l.c. plates.

53. L. rugosella Zahlbr. (1928)

Thallus continuous to dispersed, warted, granular or of scattered areoles, determinate, yellowish white to grey; prothallus absent. Apothecia (0.5-)0.7–1.3(-2) mm diam., dispersed or loosely aggregated, somewhat elevated or sessile, constricted below; thalline exciple well developed, warted, rarely thin, medulla with large, irregular crystals not dissolving in K; discs pale brown to pink-brown or reddish orange, ± flat or concave, often lightly pruinose with crystalline granules; epithecium colourless to reddish brown; hymenium 60–85 μm. Paraphyses sparsely branched, apices capitate. Ascospores (10-)12–17(-18) × (6-)7–9.5(-11.5) μm. Thallus Pd−, K+ yellow, C− (atranorin, gangaleoidin, ± roccellic acid and californin).

On bark, particularly of coniferous trees, and also wood; scattered. Mainly N. British Is. Europe, North America.

An infrequently recognized member of the *L. chlarotera* **(15)** group distinguished primarily by the well-developed, warted thalline exciple and the coarsely granular thallus.

54. L. rupicola (L.) Zahlbr. (1928)

Aspicilia lindsayi W. Watson (1939)

Thallus forming continuous patches, clearly delimited, cracked-areolate, often thick and uneven, areoles flat to slightly convex, whitish grey or sometimes deeper grey to greenish or brownish grey, surface ± smooth; soredia usually absent, when present [**var. *efflorens*** Leuckert & Poelt (1989); syn. f. *sorediata* (Flotow) Zahlbr. (1928)] arising in discrete, convex soralia, creamy white, tending to coalesce; prothallus black. Apothecia 0.5–2(-3) mm diam., immersed to slightly raised, arising singly or juxtaposed; thalline exciple entire and crenulate to flexuose or convoluted, not extending above the thallus surface; discs pinkish to brownish, flat to convex, densely pale grey to blue-grey-pruinose; epithecium pale brown to dark olivaceous brown, interspersed with crystals not dissolving in K; hymenium 80–90 μm tall. Paraphyses 2–3 μm wide, sparsely branched, apices to 3.5 μm, not or slightly thickened, colourless. Ascospores (8-)9–14.5(-15) × 5.5–7(-7.5) μm. Conidia 14–25 × 1 μm, thread-like to arc-like. Thallus K+ yellow, C−, apothecial disc Pd−, C+ yellow (atranorin, chloratranorin, sordidone, lecanoric, norstictic, roccellic, thiophanic and variolaric acids).

On hard, siliceous rocks, frequent in the xeric-supralittoral on coastal rocks, but also locally frequent inland. Widespread in British Is. The var. *efflorens* is known only from the dolerite and basalts in N. Northumberland, Montgomery, Midlothian. Cicumpolar from temperate to arctic areas in the Northern Hemisphere.

The apothecia are often blackened by the lichenicolous *Arthonia varians* (see under *Arthonia*). The species varies considerably throughout its range and several subspecies have been recognized; the material from Great Britain & Ireland conforms to subsp. *rupicola*. *L. swartzii* **(68)** differs in the C+ orange-red thallus and white prothallus, but also occurs in rock underhangs. *L. praepostera* **(49)** also has a C− thallus but the apothecia are sessile, and shades of brown to black. The vivid C+ yellow reaction of the apothecial disc is diagnostic.

55. L. saligna (Schrader) Zahlbr. (1928)

Thallus to 5 cm or more across, of dispersed to ± contiguous granules, forming poorly delimited but often extensive patches, yellowish grey to grey; prothallus indistinct. Apothecia 0.2–0.8 mm diam., sessile, often densely aggregated and then becoming angular by compression, scarcely exceeding the thallus granules; thalline exciple persistent, entire to crenulate but becoming inconspicuous with age; discs pale red-brown, flat to slightly convex, sometimes slightly pruinose at first; epithecium olivaceous green to brown; hymenium 45–65 μm tall. Paraphyses 1.5–3 μm wide, sparsely branched, especially above, apices 2.5–4 μm wide, olivaceous brown, capitate. Asci 50–60 × 13–18 μm. Ascospores 7–10(-13) × 4–6(-7) μm, ellipsoid. Pycnidia frequent, black; conidia 7–8(-9) × 1.5–2.5(-3) μm, broadly fusiform to arc-like, truncated at the base. Thallus Pd−, K−, KC−, C− (atranorin, one unidentified pigment).

On wood and worked timber, more rarely dead bark, generally in exposed situations; locally abundant. England, extending to C. & E. Scotland. Mediterranean N. to S. Scandinavia, Turkey, North America.

The separation from *L. quercicola* **(52)** is discussed under that species.

56. L. salina Magnusson (1926)

Thallus *c.* 0.25 mm diam., of dispersed, rounded granules, more rarely aggregated with the larger somewhat lobed, not clearly delimited, whitish grey to grey or yellowish grey; prothallus indistinct. Apothecia 0.2–0.6(-0.8) mm diam., dispersed or aggregated, sessile, constricted below; thalline exciple entire or irregularly crenulate to somewhat flexuose, persistent, medulla lacking crystals; discs pale brown to red-brown or yellow-grey, flat to slightly convex, sometimes weakly pruinose when young; epithecium colourless to pale brownish yellow, not granular; hymenium 45–55(-65) μm tall. Paraphyses 1.5–2.5 μm wide, richly branched, apices 2.5–3.5 μm wide, not or slightly thickened and capitate, yellowish brown. Ascospores 8–11(-15) × 4.5–6.5(-7) μm. Thallus Pd−, K−, KC−, C− (2,5-dichlorolichexanthone).

On weathered brick wall; very rare. England (East Suffolk, Frostenden). Scandinavia, Poland, North America.

A little understood member of the the *L. dispersa* **(23)** group distinguished from *L. albescens* **(5)** by the granular thallus, finally non-pruinose and smaller apothecia, and the shorter hymenium, and from *L. dispersa* **(23)** also by the colour of the apothecial discs and broader paraphyses. It also recalls a diminutive *L. campestris* **(12)** but doubtful specimens can be separated by the smaller ascospores.

57. L. sambuci (Pers.) Nyl. (1861)

Thallus of clusters of minute granules *c.* 1 mm diam., more rarely ± continuous and irregularly cracked, poorly delimited, white to grey; prothallus not conspicuous. Apothecia 0.2–0.4(-0.8) mm diam., dispersed or aggregated, sessile, arising amongst the thallus granules, slightly constricted at the base; thalline exciple entire or slightly crenulate at first but becoming irregular and finally almost excluded; discs pinkish brown to red-brown, flat to slightly convex; epithecium colourless to pale reddish brown, not or finely granular; hymenium (40-)55–65(-70) μm tall. Paraphyses 1.5–2 μm wide, simple to sparsely branched, apices 2.5–3.5 μm wide, colourless to brownish, the final 2 cells capitate. Ascospores (6-)8–12 × 4–6(-7) μm, (8-)16–32 per ascus. Pycnidia 0.05–0.1 mm diam., frequent, black, erumpent; conidia 10–12 × 1 μm, irregularly arc-like or curved, ± straight. Thallus Pd−, K−, KC−, C−.

On nutrient-enriched bark, especially twig axils of *Sambucus*, in areas of low air pollution; rare. Mainly in S.W., W. & N.W. British Is. Europe, North America.

The only member of the genus occurring in Great Britain & Ireland which has more than 8-spored asci.

58. L. sarcopisioides (Massal.) A.L.Sm. (1918)

Thallus of irregular granules, sometimes aggregated into convex groups *c.* 2 mm wide, poorly delimited or forming extensive patches, creamy to grey; prothallus absent. Apothecia 0.2–0.6(-1.2) mm diam., dispersed or aggregated into small groups, constricted at the base; thalline exciple at first paler than the thallus but later concolorous, entire to crenulate, persistent; discs red-brown, flat to slightly convex; epithecium interspersed with granules; hymenium 35–40 μm tall. Paraphyses 1–1.5 μm wide, sparsely branched, apices not or slightly thickened. Ascospores 7–12 × 3–3.5(-4) μm. Conidia 3–7 × 1–1.5 μm, ± bacilliform. Thallus Pd−, K−, KC−, C−.

On bark, especially of coniferous trees, and wood or worked timber; local. N.W. England (Lake District), Scotland (Highlands). C. Europe, Scandinavia.

Poorly understood in the British Isles. It differs from *L. saligna* **(55)** in the absence of the distinctive, broadly falcate macroconidia. *L. piniperda* **(45)** has smaller and generally pruinose apothecia. The Pd+ orange reaction and the exciple structure separates *L. cadubriae* **(10)**.

59. L. soralifera (Suza) Räsänen (1931)

Thallus to 2mm diam., of aggregated or more rarely dispersed areoles, areoles flattened, subsquamulose, and often with crenulate margins, greenish grey to yellowish green or greenish brown to greenish black, often not clearly delimited; soredia mainly 1–2 mm diam., abundant, arising from soralia on the surface of the areoles, usually discrete and circular but sometimes becoming ± confluent to form a continuous crust, pale yellow; prothallus inconspicuous. Apothecia to 0.5 mm diam., usually absent, arising on single areoles, sessile and somewhat constricted below; thalline exciple well developed, entire to crenulate, becoming excluded with age, greenish to pinkish brown; discs finally slightly convex, yellowish, pinkish brown or brownish black; epithecium brown, crystalline inclusions absent. Thallus Pd−, K−, KC+ yellow, C− (usnic acid, zeorin).

On siliceous rocks and walls, often in exposed situations, more rarely on wood or worked timber, moderately air pollution tolerant; locally frequent. Upland Britain, scattered in Ireland. C. Europe, Scandinavia.

Allied to the non-sorediate *L. intricata* **(36)**, of which it has been considered a variety. The thallus recalls *L. handelii* **(33)**, with which it can occur, but in that species it is grey-green to grey and the soredia originate from the margins of the areoles.

60. L. stenotropa Nyl. (1872)

Thallus of dispersed to somewhat coalescing, rounded granules, sometimes forming a ± continuous areolate crust, brownish green to green-grey; prothallus white or absent. Apothecia (0.2-)0.4–0.8(-1) mm diam., sessile, dispersed to densely aggregated and then angularly compressed; thalline exciple entire to flexuose, whitish grey, scarcely raised and becoming almost excluded; discs pale brown to grey-brown, flat to slightly convex, not pruinose; epithecium colourless, not granular; hymenium 45–60 μm tall. Paraphyses 1.5–2 μm wide, simple or sparsely branched, apices slightly swollen. Ascospores 8–12 × 3–4 μm, narrowly ellipsoid. Thallus Pd−, K−, KC− or weakly yellowish, C− (rangiformic and usnic acids, zeorin).

On coarse, siliceous rocks, particularly sandstone walls and memorials where there is alkaline run-off from mortar, rarely on worked timber, or directly on basic substrata; especially in polluted areas. Widespread in British Is. Scandinavia.

Easily mistaken for *L. polytropa* **(47)** which has a more distinctly yellow-green to grey-green thallus, generally more greenish apothecial discs and ascospores (4.5-)5–6.5(-7) μm wide.

61. L. straminea Wahlenb. ex Ach. (1810)

Thallus to *c.* 5 cm diam., of placodioid, circular patches or rosettes, the margins of radiating, convex lobes 3–6 × 1.5 mm, the centre becoming uneven and irregularly areolate, pale yellow-green, the centre darkest, surface smooth to slightly warted, not or sometimes white-pruinose. Apothecia 2–3 mm diam., sessile, finally constricted at the base; thalline exciple well developed, swollen, striate to crenulate, flexuose or granular, scarcely exceeding the disc and sometimes ± excluded; discs flat to slightly concave, red-brown to dark brown, not or slightly pruinose; epithecium well developed, yellowish brown, slightly granular; hymenium 80–90 μm tall. Paraphyses 1.5–3 μm wide, simple or branched, especially above, ± contorted, apices to 3–4 μm wide, irregularly thickened. Ascospores 8–12(-14) × 4.5–6(-6.5) μm. Thallus Pd−, K−, C+ orange or red (atranorin(?), arthothelin, thiophanic acid, 2,7-dichloronorlichexanthone, three other xanthones).

On nutrient-enriched rocks, especially the tops of boulders frequented by birds in coastal situations; rare and local. Scotland (Outer Hebrides, Shetland Is., St Kilda). Arctic-maritime, Fennoscandia, North America; probably circumpolar in the Northern Hemisphere.

L. muralis **(42)** is distinguished by the flattened and broader lobes and C− thallus.

62. L. strobilina (Sprengel) Kieffer (1895)

Thallus continuous or dispersed, granular to irregularly subareolate, ± undelimited, uneven, whitish to whitish yellow or pale yellowish green, surface not corticate and developing a pruinose coating of projecting crystals to 0.1 mm in length after storage; prothallus inconspicuous. Apothecia to 1 mm diam., aggregated, sessile, the base slightly constricted; thalline exciple entire to crenulate, finally ± excluded, not corticate; discs flat to somewhat convex, grey-yellow to pale red-brown or grey-brown; epithecium colourless, often interspersed with colourless, granular crystals; hymenium 35–65 μm tall; hypothecium with small, granular crystals. Paraphyses 1–1.5 μm wide, simple or sparsely branched, apices not swollen. Asci 35–45 × 10–17 μm. Ascospores 10–15(-16) × (3.5-)4–6 μm, narrowly ellipsoid and often somewhat reniform. Conidia 25 × 1 μm, arc-like. Thallus Pd−, K+ yellowish to brown, KC+ yellowish or −, C− (usnic acid, ± zeorin).

On bark, mainly of coniferous trees and worked timber; very rare. Channel Is., Is. of Scilly, N. Wales. S. Europe, E. North America, Mexico, N. Africa.

Most likely to be mistaken for C− chemotypes of *L. symmicta* **(69)** separable by the corticate thallus and exciple, the less yellow colour (usnic acid absent), and the hymenium exceeding 70 μm tall.

63. L. subaurea Zahlbr. (1928)

Thallus areolate, forming ± continuous crusts or more rarely somewhat dispersed, areoles flat to convex or almost squamulose, grey-green to yellow-green; soredia arising on the margins of the areoles, occasionally coalescing to form a sorediate crust, bright yellow or pale yellow-green, paler than the thallus; prothallus absent. Apothecia not known in British material, 0.3–0.6 mm diam., sessile, ± immersed amongst the soredia; thalline exciple well developed, persistent, often sorediate; discs concave to ± flat, red-brown to olivaceous brown; epithecium red-brown to olivaceous, sparsely interspersed with granules; hymenium 50–60 μm tall; hypothecium with oil drops. Paraphyses unbranched, apices red-brown and slightly expanded. Ascospores 7–10 × 6–8 μm, broadly ellipsoid to subglobose. Thallus and soredia Pd+ orange to red, K−, KC−, C− (pannarin, rhizocarpic acid, zeorin).

On ± metal-rich, siliceous rocks, especially sandstones and siltstones in well-lit situations; local. N. England (Derby, Peak District; S.W. Yorkshire), E. Scotland (Fife). C. Europe (Silesia), Scandinavia.

See also *L. epanora* **(24)**, which occurs in similar, but generally more shaded habitats.

64. L. subcarnea (Liljebl.) Ach. (1810)

Thallus continuous and finely cracked to areolate, clearly delimited, areoles often uneven in height, ± flat, margins crenulate, white or pale grey, surface coarsely granular to warted; prothallus white, fibrous, generally well-developed. Apothecia (0.2-)0.4–1(-1.7) mm diam., aggregated, immersed to sessile and finally constricted below; thalline exciple entire, tending to be paler than the thallus, becoming excluded; discs pinkish to pale grey-brown, convex to almost subglobose, usually densely white or bluish white pruinose; epithecium brownish, granular, K+ red (with needle-like crystals); hymenium 60–80(-100) μm tall. Paraphyses 1.5–2 μm wide, simple or sparsely branched, especially above, apices to 3.5 μm wide, generally with the terminal cell subglobose. Ascospores (8)10–14 × (5-)6–8 μm. Thallus Pd+ orange to red, K+ yellow to red, C− (atranorin, chloratranorin, protocetraric, norstictic and connorstictic acids).

In dry recesses of base-rich maritime rocks in the terrestrial-halophobic, by lakes, or rarely on calcareous rocks or in woodlands; local. England (N. Devon, Westmorland), Channel Is., Wales (Pembroke, Merioneth), C. & E. Scotland. W. Europe, U.S.A. (California).

Unlikely to be confused with other species in similar habitats because of the strongly convex pruinose apothecia and chemical reactions.

65. L. subintricata (Nyl.) Th.Fr. (1871)

Thallus immersed and inconspicuous or of dispersed fine granules, white, yellowish grey or grey, poorly delimited; prothallus inconspicuous. Apothecia (0.2-)0.4–0.6(-1) mm diam., dispersed, sessile, usually constricted below; thalline exciple entire but becoming excluded; discs flat to convex, yellowish, olive-brown or brownish black, sometimes pruinose, with a bluish bloom when moist; epithecium colourless to olive-grey, interspersed with granules; hymenium 40–65 μm tall. Paraphyses 1–2 μm wide, sparsely branched, anastomosed, apices to 3.5 μm wide, swollen, capitate, brown. Ascospores (5-)6–10(-12) × 2.5–4(-4.5) μm, narrowly ellipsoid. Conidia 3–7 × 1–1.5 μm, bacilliform. Thallus Pd−, K− or weakly yellow, KC−, C−.

On bark, wood, and especially worked timber; rare. Scotland (Highlands). Coniferous forests in upland areas of C. Europe, Scandinavia, North America.

Differing from C− chemotypes of the *L. symmicta* **(69)** group in the shorter ascospores and conidia.

66. L. sublivescens (Nyl.) A. L. Sm. (1918)

Lecidea sublivescens (Nyl. ex Crombie) P. James (1965)

Thallus areolate, dispersed or forming a ± continuous crust, areoles flattened with somewhat irregular margins, pale yellowish green, surface ± smooth; prothallus inconspicuous. Apothecia 0.2–0.6 mm diam., dispersed to compacted, arising on single areoles, at first ± immersed but becoming almost sessile; thalline exciple visible only when young, soon excluded; disc concave or more commonly convex to undulate, often convoluted, pinkish brown; epithecium ± colourless, sparsely interspersed with brownish granules; hymenium 55–65 μm tall. Paraphyses 1–2 μm wide, branched and anastomosed, apices not thickened. Ascospores 10–14(-15) × 4–5 μm, narrowly ellipsoid. Thallus Pd –, K – or weakly yellowish, KC+ yellowish, C –.

On dry bark of ancient *Fagus* and *Quercus* trees; rare. S. England, Wales. S. Sweden.

Recalling a pale and poorly developed *L. sulphurea* **(67)**, of which it could be a corticolous morphotype.

67. L. sulphurea (Hoffm.) Ach. (1810)

Lecidea sulphurea (Hoffm.) Wahlenb. (1812)

Thallus cracked-areolate, rather thick, forming ± continuous crusts, yellow-green to yellow-grey or greenish yellow, areoles flat to slightly convex, surface smooth to granular or roughened; prothallus green-black. Apothecia 1–1.5(-2.5) mm diam., initialy immersed, scarcely extending above the thallus surface; thalline exciple entire, irregular, becoming ± excluded; discs yellowish to pinkish brown, green-brown, blue-green or greenish black, flat to strongly convex, densely grey-pruinose; epithecium greenish or bluish grey, K+ green-black, interspersed with granular crystals dissolving in K; hymenium 55–90 μm tall. Paraphyses 2–2.5 μm wide, sparsely branched, apices slightly thickened. Ascospores (9-)9.5–12.5(-16) × 4.5–5.5(-6) μm. Conidia 14–20 × 0.5–1 μm, thread-like to arc-like. Thallus Pd –, K – or + yellowish brown, KC+ yellow, C – (atranorin, gangaleoidin, α-collatolic and usnic acids, zeorin).

On exposed, somewhat nutrient-rich, siliceous rocks and walls, especially in the xeric-supralittoral but also inland; common. Widespread in British Is. Circumboreal in the Northern Hemisphere.

Separated from other yellowish-green saxicolous non-sorediate species by chemical tests. See also the corticolous *L. sublivescens* **(66)**.

68. L. swartzii (Ach.) Ach. (1810)

L. subradiosa Nyl. (1872)

Thallus continuous at the margins, cracked-areolate in the centre, uneven or wrinkled, grey-brown, surface smooth to roughened; prothallus white. Apothecia 0.3–1(-2) mm diam., at first immersed, becoming sessile, base strongly constricted and apothecia easily detached; thalline exciple entire, flexuose, becoming inconspicuous and excluded; discs yellow-brown, becoming strongly convex, densely blue-grey pruinose; epithecium colourless to olivaceous green, ± interspersed with granules; hymenium 60–85 μm tall. Paraphyses 2–2.5 μm wide, mainly simple, apices slightly thickened. Ascospores (9-)10–12(-14) × 5–6(-7) μm, ellipsoid. Thallus Pd –, K+ yellow, C+ orange-red.

In underhangs of siliceous rocks, upland; rare. Scotland (Banff, Glen Avon; Moray, Perth). C. European mtns, Scandinavia.

The British material conforms to subsp. *swartzii* [syn. *L. radiosa* var. *karelica* Räsänen (1934)]. Separated from *L. rupicola* **(54)**, which occurs in more exposed situations, by the reaction with C and the sessile apothecia strongly constricted below. *L. praepostera* **(49)**, which occurs only in maritime situations, has K+ yellow to red (crystals) and C – thalli.

69. L. symmicta (Ach.) Ach. (1814)

Thallus of dispersed granules to irregularly areolate, often forming discrete patches but sometimes almost leprose, very variable in colour, whitish to pale yellowish green or greenish grey; prothallus inconspicuous. Apothecia 0.3–0.8(-1) mm diam., dispersed to aggregated, sessile, scarcely constricted below; thalline exciple poorly developed, entire, soon excluded, often devoid of algal cells and composed of radiating narrow hyphae; discs variable in colour, creamy to pinkish, pale orange or brownish, flat to strongly convex; epithecium colourless or yellowish brown to olivaceous, interspersed with yellow-brown granules dissolving in K; hymenium 40–70(-100) μm tall. Paraphyses 2–2.5 μm wide, branched and anastomosed, apices 2.5–3.5 μm wide, not or slightly thickened. Ascospores 9–15.5(-16) × 4–5(-6) μm. Conidia 18–25 × 0.5–1 μm, thread-like to arc-like. Thallus Pd –, K –, KC – or yellowish, C – or + orange or red (± thiophanic acid in the apothecia, usnic acid, xanthone, zeorin, three unidentified compounds, ? atranorin).

On bark, wood and worked timber, particularly on twigs or in other exposed situations; frequent. Widespread in British Is. Throughout the North Temperate zone.

Part of a complex of badly understood species which vary in many features, particularly colour and chemical reactions. *L. aitema* **(4)**, often included within this species, is primarily separated by the darker apothecial discs. A related unnamed species with 3-septate ascospores occurs on *Juniperus* in the Speyside native pinewoods of the Scottish Highlands. The arctic-boreal sorediate *L. orae-frigidae* R. Sant. (1984), which also has thinner medullary hyphae and paraphyses, has not yet been found in the British Isles.

70. L. tenera (Nyl.) Crombie (1874)

Thallus of minute scattered to ± contiguous irregular areoles, areoles flat to somewhat warted with age, creamy, whitish grey or green-grey, paler towards the margins; prothallus, when evident, white. Apothecia 0.1–0.2(-0.5) mm diam., dispersed, immersed to sessile, the base slightly constricted, inconspicuous; thalline exciple poorly developed, irregularly crenulate, often excluded; discs white to pale green, pinkish brown or pale brown, becoming strongly convex, not or slightly pruinose; epithecium ± colourless, interspersed with coarse granules; hymenium 25–40 μm tall. Paraphyses 1.5–2 μm wide, simple or sparsely branched, apices to 5 μm wide, not swollen or expanded. Asci 30 × 10 μm, broadly clavate, *Bacidia*-type. Ascospores 7–10(-15) × (1.5-)2–3 μm, rarely appearing 1- to 3-septate, fusiform-ellipsoid, apices pointed, often curved. Pycnidia frequent, immersed; conidiogenous cells sessile; conidia 4–5.5 × 1–1.5 μm, narrowly ellipsoid to bacilliform. Thallus Pd + red, K + yellow, C – (atranorin, chloratranorin, stictic and ± norstictic acids, unidentified compounds).

In dry sheltered crevices of maritime siliceous rocks in the xeric-supralittoral; locally abundant. W. British Is., decreasing northwards. Europe N. to Scandinavia.

The ascus type suggests a placement in *Lecania*, but the chemistry and conidia are distinct from the recognized species of that genus.

71. L. varia (Hoffm.) Ach. (1810)

Thallus of dispersed to contiguous rounded granules, sometimes thickening into a ± areolate crust, yellow-grey to dark green-grey, surface somewhat shiny when areolate; prothallus inconspicuous. Apothecia 0.4–1(-1.5) mm diam., dispersed to closely aggregated, not compacted, sessile, slightly constricted at the base; thalline exciple well developed, entire to crenulate or convoluted, persistent, raised; discs pinkish brown to greenish brown, ± flat; epithecium ± colourless, interspersed with minute granules; hymenium (60-)70–80(-90) μm tall. Paraphyses 1–2 μm wide, sparsely branched and anastomosed, apices slightly thickened. Asci 35–60 × 14–18 μm. Ascospores (7-)9–11.5(-13) × 5–7(-8) μm. Conidia 12–22 × 0.5–1 μm, thread-like to arc-like. Thallus Pd + yellow, K – or weakly yellowish, KC –, C – (psoromic and usnic acids).

On wood and worked timber, rarely on bark of living trees; locally frequent. Predominantly in E. Great Britain, scattered in Ireland. Probably throughout the North Temperate zone, most frequent in continental C. Europe, North America.

In the past much confused with abundantly fertile and scarcely sorediate specimens of *L. conizaeoides* **(21)**, especially by continental authors; uncertain specimens can be distinguished by applying Pd. North American material with protocetraric rather than psoromic acid merits further study. Often occurring with *L. symmicta* **(69)** which has a shallow exciple often without algae and a Pd – often whiter thallus.

LECIDEA Ach. (1803)

Thallus crustose, superficial, or immersed, continuous to areolate, usually grey; prothallus present or absent.. **Photobiont** chlorococcoid green algae, mostly *Trebouxia* (incl. *Pseudotrebouxia*), but also *Chlorosarcina* and perhaps other genera. **Ascomata** apothecia, immersed to sessile, usually persistently marginate. **Thalline exciple** absent. **True exciple** of irregular chains of swollen cells, outer layer brown to dark brown, internally ± colourless. **Epithecium** often pigmented, olivaceous or brown to blackish or green-black. **Hymenium** usually I+ blue. **Hypothecium** colourless to brown or black. **Hamathecium** of simple or sparsely branched paraphyses, anastomoses absent or sparse, the apical cells usually pigmented. **Asci** elongate-clavate, *Lecidea*-type (Fig. 33f), outer coat I+ blue, apex thickened, apical dome I+ pale blue, usually with a distinct I+ blue meniscus-like, shallow subapical ring but sometimes prolonged into a tube. **Ascospores** ellipsoid to oblong, simple, with a central plasma bridge, smooth, without a perispore, colourless. **Conidiomata** when present pycnidial, immersed, black. **Conidia** elongate-bacilliform, simple, colourless. **Chemistry**: ± orcinol *para*-depsides and *β*-orcinol depsidones, rarely with traces of atranorin. **Ecology**: on rocks, especially common on hard siliceous substrata. Cosmopolitan, centred in humid and cool regions of both Hemispheres.

The description above is mainly applicable to those species of *Lecidea* s.str. (**6–8**, ?**16**, **21**, **22**, **27**, **31–33**, **36**, **37**, **39**, **42**, **44**, **48**, **49**). The remaining species may deviate in several respects, especially regarding their hamathecia, ascus-structure, conidia, chemistry and ecology.

This generic name has been applied historically to a wide variety of unrelated crustose lichens which have in common simple, colourless ascospores formed in apothecioid ascomata lacking a thalline exciple. As a result of more critical studies, especially within the last 15 years, more homogeneous groups have been recognized and referred to a wide variety of families and genera: *Adelolecia*, *Amygdalaria*, *Biatora*, *Carbonea*, *Claurouxia*, *Clauzadea*, *Clauzadeana*, *Farnoldia*, *Fuscidea*, *Herteliana*, *Hypocenomyce*, *Lecanora*, *Lecidella*, *Melanolecia*, *Micarea*, *Miriquidica*, *Nesolechia*, *Porpidia*, *Psilolechia*, *Psora*, *Pyrrhospora*, *Rimularia*, *Schaereria*, *Tephromela*, *Trapelia*, *Trapeliopsis* and *Tylothallia*. The structure of the ascomata, especially the nature of the hamathecial tissues, exciple and asci (as observed when stained in iodine), have played major roles in these regroupings.

In determining lecideoid lichens particular attention must be paid to microscopical features, especially colours and reactions in sections.

The following names, all based on material growing on rocks from Scotland, are excluded as of uncertain application, since none of the type material has been traced: *L. callista* Stirton (1874), *L. calpodes* Stirton (1876) and *L. stirtoniana* Zahlbr. (1925) [syn. *L. relicta* Stirton (1876)]. *L. sincerula* Nyl. (1876) described from the Antarctic, is a most dubious British record and so is not treated here. British material named as *L. meiocarpa* Nyl. (1876) was found to be *Lecania cyrtellina* and that name is also therefore omitted here. The current placement of the numerous other taxa now excluded from the genus can be traced through the Index.

LITERATURE: Coppins *et al.* (1992), Hertel (1967, 1968, 1969a, 1970a-b, 1971, 1973, 1975a-b, 1977, 1981, 1984), Hertel & Rambold (1990),Magnusson (1952), Rambold (1989), Rambold & Triebel (1989), Schwab (1986), Tønsberg (1990), Triebel & Rambold (1988), Vainio (1922).

1	On bark or wood, mosses or soil, or on lichens	2
	On rocks or pebbles	26
2(1)	On bark or wood	3
	On mosses or soil, or on lichens	14
3(2)	Discrete soralia absent; apothecia reddish brown to black; thallus Pd−	4
	Discrete soralia present; apothecia deep orange; thallus and soralia Pd+ red	**13. *cinnabarina***
4(3)	Pycnidia absent or immersed	5
	Pycnidia sessile, barrel-shaped	**15. *doliiformis***
5(4)	Ascospores globose	6
	Ascospores ellipsoid, fusiform, or ovoid	7
6(5)	Thallus superficial, granular-sorediate, UV+ white	**40. *nylanderi***
	Thallus immersed in substratum, UV−	**3. *antiloga***

7(5) Hypothecium dark brown 8
Hypothecium ± colourless, sometimes tinged yellowish or pale orange 11

8(7) Thallus superficial and cracked-areolate, or immersed and ± inapparent; apothecia arising singly 9
Thallus superficial, well developed, coarsely granular to ± coralloid; apothecia in dense botryose clusters .... **10. *botryosa***

9(8) Thallus very thin or ± inapparent, C−; apothecia mostly <0.5 mm diam. 10
Thallus cracked-areolate, C+ red; apothecia mostly >0.5 mm diam. .... **21. *fuscoatra***

10(9) Hymenium 20–30 μm tall; epithecium dark brown to black; pycnidia absent **41. *paraclitica***
Hymenium 35–50 μm tall; epithecium dark greenish blue to greenish brown; pycnidia abundant .... **17. *erratica***

11(7) Ascospores <7 μm wide 12
Ascospores 8.5–10 μm wide .... **12. *carrollii***

12(11) Epithecium red-brown, N− 13
Epithecium colourless, brown or greenish, N+ reddish .... **51. *turgidula***

13(12) Apothecia mainly <0.5 mm diam.; exciple at first well developed; ascospores oblong-ellipsoid, (8-)9–12 μm long .... **24. *hypopta***
Apothecia mainly >0.5 mm diam.; exciple thin and soon excluded; ascospores fusiform-ellipsoid, (8-)9–14(-18) μm long .... **18. *erythrophaea***

14(2) On other lichens 15
On bryophytes, plant debris or soil 21

15(14) Independent thallus absent (lichenicolous fungi) 16
Independent thallus present (lichenicolous lichens) 18

16(15) Ascospores ellipsoid to subcylindrical or fusiform 17
Ascospores subglobose .... **4. *associata***

17(16) Apothecia 0.3–0.4 mm diam.; ascospores 4–7 μm wide; hypothecium brown .. **28. *inquinans***
Apothecia (0.1-)0.15–0.2 mm diam.; ascospores 2–3 μm wide; hypothecium reddish violet to purplish brown .... **46. *punctum***

18(15) Thallus whitish to brown or olivaceous, surface smooth to granular-areolate, never isidiate, C− 19
Thallus dark brown to blackish, surface minutely granular or isidiate, usually C+ pink .... **20. *furvella***

19(18) Apothecia lacking a central umbo; asci *Bacidia* or *Rimularia*-type 20
Apothecia with a central umbo; asci *Lecidea*-type .... **52. *umbonella***

20(19) Thallus grey to brown or olivaceous brown, areolate; apothecia ± immersed between the areoles; epithecium brown, N−; asci *Rimularia*-type .... **30. *insularis***
Thallus greenish brown, granular; apothecia ± superficial; epithecium dark greenish, N+ purple-red; asci *Bacidia*-type .... **29. *insidiosa***

21(14) Thallus white-grey, grey or blackish; ascospores <10 μm wide, the walls <1.5 μm thick 22
Thallus brown; ascospores (8-)12–15 μm wide, the wall 2–3.5 μm thick .... **50. *tornoensis***

22(21) Apothecia immarginate; epithecium greenish, N+ red or purplish 23
Apothecia marginate, at least when young; epithecium colourless or pale brownish, N− 24

23(22) Thallus white-grey to grey, Pd+ orange, KC−; apothecia not pruinose; ascospores often attenuated at apices, (8-)10–14(-15) × 3–4.5(-5) μm .... **35. *limosa***
Thallus grey to dark grey or blackish, Pd−, KC+ orange; apothecia ± pruinose; ascospores not attenuated at apices, (12-)15–19(-24) × 5–7(-9) μm .... **11. *caesioatra***

24(22) Thallus grey or tinged brown or dull green, thin and membranous; apices of paraphyses colourless, <3 μm wide 25

Thallus white, tartareous (mostly *c.* 200 μm thick); apices of paraphyses clavate to capitate, ± brown-walled, <6 μm wide .... **9. *berengeriana***

25(24) Apothecia dark brown to black, often persistently marginate; ascospores 4.5–6(-7) μm wide, often thinly 1(-3)-septate and finely warted .... **25. *hypnorum***
Apothecia pale to dark brown, usually soon immarginate; ascospores 3–4.5(-4.8) μm wide, always simple and smooth .... **26. *sanguineoatra***

26(1) On siliceous rocks, boulders, walls or pebbles N+ 27
On chalk pebbles, limestones or other highly calcareous rocks 61

27(26) Thallus orange, ochre or rust-coloured 28
Thallus whitish, grey, green, brown or inapparent, if rust-coloured only partly so 29

28(27) Thallus with soralia; medulla I−; medulla K+ yellow .... **5. *atrofulva***
Thallus lacking soralia; medulla I+ blue; medulla K− .... **49. *silacea***

29(27) Thallus K+ yellow or K−, C− 30
Thallus K−, C+ red .... **21. *fuscoatra***

30(29) Thallus K+ yellow changing to red (red needle-like crystals) 31
Thallus K− or yellow (no red needle-like crystals) 32

31(30) Thallus whitish grey to rust-coloured in parts; areoles even; medulla I+ blue .... **32. *lactea***
Thallus grey-brown; areoles convex to tuberculate or botryose; medulla I− .... **38. *mullensis***

32(30) Hypothecium ± colourless 33
Hypothecium brown to black or orange or red-brown 35

33(32) Apothecia mostly >0.2 mm diam. 34
Apothecia <0.2 mm diam. .... **45. *polycarpella***

34(33) Epithecium green-black to black, N+ red or purplish .... **44. *plana***
Epithecium olive-brown, N− .... **36. *lithophila***

35(32) Apothecia remaining immersed in the areoles 36
Apothecia arising on the surfaces of the areoles, or ± directly on the rock, sessile 40

36(35) Medulla I+ blue 37
Medulla I− 38

37(36) Thallus greyish or whitish, sometimes partly rust-coloured; hypothecium pale (rarely dark) brown .... **31. *lapicida***
Thallus brown to reddish brown; hypothecium dark brown .... **22. *paupercula***

38(36) Thallus dark grey to grey-brown; ascospores less than 12 μm long 39
Thallus white to pale grey or greenish white; ascospores 12–18(-23) μm long .... **43. *phaeops***

39(38) Apothecia 0.2–0.5 mm diam.; epithecium brown, N− .... **23. *gyrizans***
Apothecia 0.5–1.1 mm diam.; epithecium dark greenish, N+ red or purplish **39. *nigrogrisea***

40(35) Ascospores <5 μm wide 41
Ascospores >5 μm wide 45

41(40) Hymenium ± greenish or reddish violet 42
Hymenium ± colourless or slightly brownish 44

42(41) Hymenium ± greenish 43
Hymenium reddish violet .... **48. *sarcogynoides***

43(42) Ascospores narrowly ellipsoid, mainly >8 μm long 44
Ascospores broadly ellipsoid to subglobose, <8 μm long .... **8. *brachyspora***

44(43) Exciple C+ carmine-red in section .... **7. *diducens***
Exciple C− in section .... **6. *auriculata***

45(40) Ascospores usually <10 μm long 46
Ascospores usually >10 μm long 49

46(45) Thallus whitish grey to dark grey, cracked-areolate to granular-warted, never ± squamulose ... 47
Thallus red brown to dark brown, of convex to subglobose, ± squamulose, wart-like areoles ... **19. *fuliginosa***

47(46) Epithecium greenish, N+ red or purplish ... 48
Epithecium brown, N− ... **23. *gyrizans***

48(47) Apothecia >1 mm diam.; pycnidia with elevated margins ... **27. *inops***
Apothecia <0.5 mm diam.; pycnidia lacking elevated margins ... **17. *erratica***

49(45) Hymenium >60 μm tall ... 50
Hymenium usually <60 μm tall ... 55

50(49) Apothecia black; thallus areolate or of scattered granules ... 51
Apothecia reddish brown at first, later brown-black; thallus ± smooth and continuous ... **2. *ahlesii***

51(50) Hymenium usually >70 μm tall ... 52
Hymenium usually <70 μm tall ... 53

52(51) Medulla I+ blue; epithecium greenish, N+ red or purplish ... **33. *confluens***
Medulla I−; epithecium brown, N− ... **23. *gyrizans***

53(51) Medulla I− ... 54
Medulla I+ blue ... **31. *lapicida***

54(53) Thallus grey-brown, of ± flat to convex areoles; apothecia 0.5–1.1 mm diam.; ascospores broadly ellipsoid ... **39. *nigrogrisea***
Thallus greenish white, of scattered bullate granules; apothecia 0.2–0.4 mm diam.; ascospores narrowly ellipsoid ... **16. *endomeleana***

55(49) Thallus K+ yellow, Pd− yellow ... 56
Thallus K−, Pd− ... 57

56(55) Thallus cracked-areolate; apothecia mainly >0.5 mm diam., arising between the areoles, only occasionally compacted, not on a common stroma ... **31. *lapicida***
Thallus of dispersed verrucae or areoles; apothecia <0.4 mm diam., usually in botryose clusters of up to 25 on a common black stroma ... **47. *pycnocarpa***

57(55) Medulla I+ blue ... 58
Medulla I− ... 59

58(57) Ascospores (8-)9–12 × (2.5-)3(-4) μm ... **27. *inops***
Ascospores (8-)10–15 × (4.5-)5–8 μm ... **31. *lapicida***

59(57) Ascospores narrowly ellipsoid or ellipsoid, not dumb-bell-like, mainly >4 μm wide ... 60
Ascospores ellipsoid to ± dumb-bell-like, <4 μm wide ... **14. *commaculans***

60(59) Thallus ± absent, prothallus predominant; apothecial disc often ± pore-like ... **37. *matildae***
Thallus dark grey, of scattered scurfy granules, prothallus not dominating; apothecial disc open, not pore-like ... **42. *pernigra***

61(26) Thallus superficial; epithecium dull greenish; hypothecium red-brown to black; ascospores 11–14 × 7–9 μm ... **1. *advertens***
Thallus ± immersed; epithecium red-brown; hypothecium orange-brown; ascospores (14-)16–19 × (5-)6–8 μm ... **34. *lichenicola***

1. L. advertens Nyl. (1866)

Thallus superficial, granular, scurfy, black, dispersed; prothallus absent. Apothecia 0.2–0.3 mm diam., flat to somewhat convex, black; true exciple becoming excluded, black; epithecium somewhat granular, dull greenish; hymenium 75–90 μm tall; thalline exciple and hypothecium dark red-brown, K−. Paraphyses *c.* 1 μm wide, very delicate, not swollen or capitate at the tips, branched and anastomosed.

Asci not seen undischarged, outer coat at least I+ blue. Ascospores 11–14 × 7–9 μm, ellipsoid. Thallus K−, C−.

On ± calcareous rocks, associated with cyanobacteria. N. Wales (Merioneth), W. Ireland (Mayo); not collected this century. ?Endemic.

The dark brown thalline exciple and hypothecium and calcicolous habitat suggest this could be a *Clauzadea*, but the greenish epithecium and thin paraphyses indicate otherwise; healthy asci are required to ascertain the true affinities of this species.

2. L. ahlesii (Körber) Nyl. (1872)

L. valentior Nyl. (1877)

Thallus effuse, ± shiny, thin, continuous or irregularly rimose, pale grey or grey-green; medulla I−. Apothecia (0.2-)0.3–0.6(-0.8) mm diam., sessile, reddish brown to brown-black; disc concave to slightly convex; true exciple usually prominent and persistent, the outer edge and hypothecium dark reddish brown; inner exciple pale reddish brown or almost colourless; epithecium colourless or yellowish to pale reddish brown, K−; hymenium 65–80(-90) μm tall; hymenium and hypothecium usually with blue-violet (K+ green) granules. Paraphyses *c.* 1.5 μm wide, apices slightly widened to 2.5–3 μm, simple or sparingly branched and anastomosed, without apical cap or hood. Asci *Porpidia*-type. Ascospores 12–17 × (5-)6–7(-9) μm. Thallus Pd−, K−, KC−, C− (no lichen products detected by t.l.c.).

On periodically inundated, shaded siliceous rocks; local. S.W. England (N. Devon), N.W. England (Lancashire), Wales, Scotland (Highlands), S.W. Ireland. C. Europe.

Close to *L. sanguineoatra* **(26)** but distinguished by its wider ascospores and habitat on rocks.

3. L. antiloga Stirton (1877)

Thallus immersed, hardly visible, not or scarcely discolouring the wood; prothallus absent. Apothecia (0.1-)0.2–0.4(-0.5) mm diam.; disc ± flat to somewhat concave, black, sometimes ellipsoidal with the wood grain; true exciple persistent, black, shiny, ± colourless and granular within, the surface cells expanded, brown capitate (K+ greenish), 4–6 μm wide; epithecium granular, brown, K+ greenish olivaceous; hymenium 30–60 μm tall; hypothecium colourless. Paraphyses generally branched only near the apices, with dark brown caps to 3.5 μm wide. Asci 30–40 × 8–10 μm, broadly clavate, tholus well developed, *Bacidia*-type. Ascospores (3.5-)4–5.5(-6) μm, sometimes ± uniseriate, globose. Thallus Pd−, K−, KC−, C−.

On hard, exposed, conifer palings and on dead *Pinus* cones and loose or detached bark; rare. N. Scotland (Inverness, Aviemore; Moray, Culbin Forest). France (Provence), Sweden, Finland.

The spherical ascospores are distinctive.

4. L. associata Th. Fr. (1867)

Thallus absent, lichenicolous. Apothecia 0.2–0.35 mm diam., black; true exciple persistent, of ± parallel hyphae, granular; epithecium reddish brown; hymenium I−. Paraphyses filamentous, simple, not swollen at the apices. Asci 50–60 × 6–8 μm, long stalked, apex not strongly thickened, I−. Ascospores 7–8(-9) × 5–6 μm, subglobose.

On thalli of *Ochrolechia frigida*, *O. tartarea* and *Thamnolia vermicularis*; rare. W. & N. (Highlands) Scotland. Probably circumpolar in the arctic and northern boreal zones of the Northern Hemisphere.

A new generic name, *Geltingia* Alstrup & D. Hawksw. (1990), has recently been introduced for this, as ***G. associata*** (Th.Fr.) Alstrup & D. Hawksw. (1990), and some allied species. It has often been included in *Nesolechia* but is distinct from the type species of that genus, *N. oxyspora* (Tul.) Massal. (1856), primarily a pathogen of *Parmelia saxatilis* and *P. sulcata*.

5. L. atrofulva Sommerf. (1826)

Thallus areolate, ochre to dull orange-yellow or rusty yellow-brown, the areoles mainly 0.2–1 mm diam., often dispersed, uneven to strongly convex, medulla I−; prothallus grey or black, with crateriform to tuberculate dark blue-grey to blackish soralia, 0.1–0.2 mm diam. Apothecia rare, 0.2–0.6 mm diam., brown-black; true exciple dark brown at outer edge, colourless within, of radiating hyphae with short swollen cells; epithecium olive-brown to pale brown; hymenium 50–60 μm tall; hypothecium colourless. Paraphyses mainly simple, apices not markedly swollen but often with brown caps. Asci 55–65 × 14–18 μm, *Lecanora*-type. Ascospores 8–12 × (5-)5.5–7.5(-8) μm, broadly ellipsoid. Conidia 8–22 × 0.5–1 μm, thread-like. Thallus, medulla and soralia Pd± yellow, K+ yellow, KC−, C− (stictic, cryptostictic, ± norstictic acids; ± atranorin and ± confluentic acid are reported in Norway).

On metal-rich, siliceous rocks, often with *Acarospora sinopica*, *Lecidea silacea* **(49)** and *Tremolecia atrata*. N.W. England (Lake District), Scotland (Highlands). Austria, France, Germany, arctic and subarctic areas of the Northern Hemisphere.

In the shade the soralia can predominate and the thallus is paler yellow. Easily recognized by the distinctive thallus colour, which is always much more yellow than that of *Rhizocarpon oederi* or *Tremolecia atrata*, neither of which are sorediate. Recently transferred to *Miriquidica* as ***M. atrofulva*** (Sommerf) Rambold & Schwab (1990).

6. L. auriculata Th. Fr. (1861)

Thallus grey, areolate; areoles ± flat to somewhat convex, often dispersed, with rock crystals visible between, often ± immersed; medulla I± blue (best seen in squash mounts); prothallus black, fimbriate, often predominating. Apothecia (0.4-)0.6–1.8(-4) mm diam., sessile, black, disc concave to flat, strongly constricted below, single or closely aggregated; true exciple persistent, raised, sometimes flexuose, exciple red-black at outer edge, internally reddish, intensifying in K, C− ; epithecium green, blue-black to green-black; hymenium 30–50 μm tall, ± greenish; hypothecium massively developed, brown to dark brown, to 200 μm or more tall in old apothecia. Paraphyses 1.3–1.8 μm wide, mainly simple, occasionally anastomosed, apical cells clavate to 3.5 μm wide, with a dark green-brown hood to 5 μm wide. Asci 27–40 × 8–12 μm, the apex strongly thickened, the outer coat I+ blue. Ascospores (5-)8–11(-12) × (1.5-)2–3(-4) μm, narrowly ellipsoid, abruptly truncated at the ends. Pycnidia immersed, not marginate; conidia 10–15 × 0.8 μm, bacilliform. Thallus and medulla Pd−, K−, KC−, C− (confluentic acid).

On coarse-grained, siliceous rocks, especially granites, often in fine cracks. N. Wales, Scotland (Highlands), Ireland. Mainly arctic and subarctic in Europe, rare in the central European mountains, North America, Himalayas.

This species, in which the reddish tone of the exciple is especially characteristic, is most likely to be confused with *L. brachyspora* **(8)**, which differs in the short, broad ascospores, *L. diducens* **(7)** and *L. matildae* **(37)**, which have a C+ rose exciple, and *L. inops* **(27)**, which has marginate pycnidia.

7. L. diducens Nyl. (1865)

Like **6**, but differing in the exciple, which is C+ carmine-red [2'-*O*-methylanziaic acid (confluentic acid absent)] in section.

On siliceous rocks and pebbles, particularly near the coast; not uncommmon, especially in N. British Is. S.W. & N.W. England (Lake District), S. Wales (Pembroke), Scotland (especially Shetland), W. Ireland. Circumpolar, atlantic and alpine, Europe, North America, Nepal, Australia.

This species may be a chemotype of *L. auriculata* **(6)** and is sometimes treated as *L. auriculata* var. *diducens* (Nyl.) Th. Fr. (1874) by continental authors. Many specimens under this name in older collections refer to *L. auriculata* **(6)**.

8. L. brachyspora (Th. Fr.) Nyl. (1887)

Like *L. auriculata* **(6)**, but thallus areolate, the areoles flat to somewhat convex, often ± immersed, whitish grey; prothallus often inconspicuous. Apothecia similar in all details. Ascospores 4.5–7(-8) × 3.5–4.5(-5) μm, broadly ellipsoid to subglobose.

On coarse-grained siliceous rocks, especially pebbles, coastal; local. W. & N. Scotland (Orkney). Arctic, in Europe south to Denmark, North America, Nepal, ?Venezuela.

Mainly distinguished from *L. auriculata* **(6)** by the ascospore-shape. The length:breadth ratio of the ascospores in *L. auriculata* **(6)** is 2.4–3.7, while in this species it is only 1.1–2.1.

9. L. berengeriana (Massal.) Nyl. (1866)

Thallus mostly *c.* 200 μm thick, tartareous, white, obscuring the form of the underlying bryophytes or plant remains, surface of contiguous, granular warts, 0.1–0.3(-0.5) mm diam.; medulla I−. Apothecia 0.3–0.9 mm diam., flat and ± thinly marginate when young but soon convex and immarginate, medium to dark brown or blackish; exciple and hypothecium dark reddish brown, but outer edge of exciple colourless, giving the appearance of a thalline exciple; epithecium yellowish brown; hymenium 60–65 μm tall; hymenium and hypothecium with scattered blue-violet (K+ green) granules. Paraphyses 1.5–2(-2.5) μm wide, becoming clavate to capitate and brown-walled at apices and to 6 μm wide, mostly simple. Asci *Porpidia*-type. Ascospores (9.5-)11–16(-19) × 4–5(-6) μm, fusiform-ellipsoid, perispore absent. Thallus Pd −, K −, KC −, C − (no lichen substances detected by t.l.c.).

On bryophytes over ± calcareous rocks (especially mica-schist) or on exposed turf of mountain ridges or summits. W.C. England (Hereford, Downton Castle Park), N. Scotland (Highlands, Shetland), Circumpolar in mountains of N. Hemisphere.

Close to *L. hypnorum* **(25)** and *L. sanguineoatra* **(26)** but distinguished by its tartareous thallus and much broader, brown-walled apices of the paraphyses.

10. L. botryosa (Fr.) Th. Fr. (1874)

Thallus effuse, rather thickly granular-sorediate, grey to grey-brown; soredia 30–80 μm diam., at first arising from convex areoles (0.1–0.2 mm diam.) which soon become hidden or completely dissolved, external hyphae often brown-walled. Photobiont cells 5–10 μm diam. Apothecia (0.15-)0.2–0.3(-0.4) mm diam.; disc ± flat, arranged in dense botryose clusters, constricted below and often ± stipitate, dark reddish brown to black; true exciple very thin, whitish in young apothecia but later brown, pale brownish within, edge and epithecium brown or yellowish brown, K −; hymenium 30–55 μm tall, colourless to yellowish brown; hypothecium dark brown. Paraphyses 1.5–2 μm wide, the apices often with a dark brown cap, to 4 μm wide. Asci 35–45 × 9–11 μm, *Biatora*-type, in K/I with narrow (often indistinct) apical cushion bordered by a dark blue zone contrasting with the pale blue flanks of the apical dome. Ascospores 7–12(-13) × (2.5-)3–4 μm, 0(-1)-septate. Thallus Pd −, K −, KC −, C −, UV + white (perlatolic acid).

On wood and exposed old bark of *Pinus*; rare. N. Scotland (C. & E. Highlands). Probably circumboreal in the coniferous forest zone, N. and montane Europe, North America.

Closely related to *Toninia cumulata*, which has a verrucose-areolate thallus and mainly 1-septate ascospores and occurs on soil. T.l.c. is recommended to separate sterile specimens from *L. nylanderi* **(40)**, which occurs in similar habitats, but contains divaricatic acid.

11. L. caesioatra Schaerer (1818)

Thallus rather thick, of dense, subglobose, ± isidioid granules 0.1–0.2 mm diam., grey to dark grey or almost black, whitish when eroded; prothallus absent. Photobiont cells 6–12 μm diam. Apothecia sessile, or partly hidden by thallus granules (0.3-)0.5–1(-1.2) mm diam., strongly convex, sometimes tuberculate, bluish black with blue-grey bloom especially when wet; true exciple reflexed, colourless to pale dull yellowish, of radiating conglutinated hyphae (in K), 1.5–2 μm wide; epithecium and upper part of the hymenium bright blue-green, K −; hymenium 50–65(-75) μm tall; hypothecium ± colourless to reddish brown to almost violet, K + reddish orange. Paraphyses 1.3–1.8 μm wide, simple or sparsely branched; apices not markedly swollen but with individual, colourless gel-coats. Asci 50–60 × 15–20 μm, ± *Lecanora*-type with a broad apical cushion. Ascospores (12-)15–19(-24) × 5–7(-9) μm, ellipsoid. Thallus Pd −, K ± weakly yellowish, KC + orange, C − (sphaerophorin).

On mosses, particularly *Andreaea*, *Grimmia* and *Rhacomitrium*, over soil or schistose boulders; locally abundant over 700 m. N.W. England (Lake District), Scotland (especially the Highlands), W.C. Ireland (Galway). Arctic-alpine, C. Europe, Scandinavia, North America.

Often mistaken for *Micarea assimilata* and *M. incrassata*, which have apothecia without a bluish bloom and smaller photobiont cells (4–7 μm diam.).

12. L. carrollii Coppins & P. James (1979)

L. rufofusca auct. brit., non (Anzi) Nyl. (1869)

Thallus immersed to superficial, thin, whitish; prothallus black, thin. Apothecia 0.2–0.5 mm diam., flat, pale to dark red brown; true exciple persistent, colourless within, edge and epithecium pale red-brown, K + dull brown, N −: hymenium 60–70 μm; hypothecium 20–45 μm tall, colourless, I + blue. Paraphyses branched below, occasionally anastomosed, apices slightly thickened with brown caps. Asci 55–65 × 15–20(-28) μm, ± *Lecanora*-type with wide apical cushion. Ascospores 13–17(-19) × 8.5–10 μm, ellipsoid, thick-walled. Thallus Pd −, K − or K + faint yellow, KC −, C −, I −, UV − (atranorin).

On ± smooth bark of young deciduous trees or twigs, especially in moist carrs. W. Britain, from Jersey to Is. of Skye, W. Ireland. France (Brittany).

This species and *L. erythrophaea* **(18)** have previously been misidentified as *L. tenebricosa* (Ach.) Nyl. (1861) by British authors. *L. erythrophaea* differs in its narrower, thin-walled ascospores. *L. carrollii* is closely related to *L. tornoensis* **(51)** and the genus *Japewia* Tønsb. (1990) has recently been introduced to accommodate it [as ***J. carrollii*** (Coppins & P. James) Tønsb. (1990)] and *L. tornoensis*. A further, sorediate species, ***J. subaurifera*** Muhr & Tønsb. (1990) has been recorded from Scotland (Mid Perth, Easterness). This has a green-brown to brown, ± rimose thallus with discrete to ± confluent, spotted brown-yellow soralia, yellow internally. Apothecia rare, <0.4 mm diam., chestnut-brown. Ascospores 11–20 × (8-)12–13(-14) μm, with a thick wall, 1.5–3.0 μm. It is UV ± bluish white and contains unidentified pigments and ± lobaric acid. *L. exigua* Chaub. (1821) resembles *L. carrollii* but has smaller apothecia (0.15–0.2 mm diam.), smaller, thin-walled ascospores (9–10.5 × 4.8–6 μm) and a KC+ orange thallus. Recently found on *Corylus* in sheltered valley woodland in S. Devon. S.W. Europe.

13. L. cinnabarina Sommerf. (1823)

Thallus superficial, ± continuous or of ± dispersed, rounded, convex areoles, creamy to pale fawn; soralia arising from the areoles 0.3–0.8 mm diam., ± urceolate to slightly raised, discrete, white or cream; prothallus absent. Apothecia 0.4–0.8(-1.2) mm diam., becoming strongly convex, deep orange, K+ crimson; true exciple soon excluded; epithecium interspersed with granules, orange, K+ crimson; hymenium 35–55 μm tall; hypothecium well developed, colourless to pale yellow-orange. Paraphyses sparingly branched, becoming somewhat moniliform towards the tips, not capitate. Asci 30–40 × 10–12 μm; outer layer I+ blue; inner layers and tholus I−, perhaps except for an I+ blue apical tube. Ascospores (8-)9–10(-12) × 2–4 μm, narrowly ellipsoid. Thallus Pd+ red, K+ yellow, KC−, C−, UV+ cream (atranorin, chloratranorin and fumarprotocetraric acid).

On *Betula* trunks. N.E. Scotland (S. Aberdeen, Mar Forest). The single collection is doubtfully British. Europe, North America.

The generic position of this species is obscure and it has been referred to *Protoblastenia* by some continental authors. When fertile its apothecia could be mistaken for those of a *Caloplaca*, but it can be distinguished by the simple ascospores, soralia, and thallus reactions. Most records from the British Is. are misidentifications for *Pertusaria pupillaris* and *Schismatomma quercicola*.

14. L. commaculans Nyl. (1868)

Thallus ± inconspicuous, sometimes present as scattered, whitish grey areoles, medulla I−; prothallus if present black. Apothecia (0.2-)0.4–0.6(-1.3) mm diam., superficial, strongly convex, often in groups and then angular in shape, not or slightly constricted below, black; true exciple indistinct, soon excluded, greenish black at outer edge, ± colourless within; epithecium dark greenish black; hymenium (30-)40–60 μm tall, slightly brownish; hypothecium deep red, K+ crimson purple. Paraphyses branched and sometimes anastomosed, with a thick gelatinous sheath, the apices with green-black hood, to 4 μm wide. Asci 35–45 × 7–12 μm, *Catillaria*-type. Ascospores (7-)9–12 × (2-)3–3.5 μm, sometimes tending to be 1-septate, narrowly ellipsoid to somewhat dumb-bell-like. Conidia 9–20 × 0.5–1 μm, bacilliform.

On siliceous rocks, montane. N. Scotland (S. Aberdeen, Morrone, Lochnagar); no recent records. Arctic-alpine, Alps, Finland, Iceland, Spitzbergen.

The deep red hypothecium, intensifying in K, is particularly diagnostic for this species. *Micarea submoestula* has a red-brown hypothecium, a better developed thallus, and shorter and never dumb-bell-like ascospores. Allied to *L. limosa* **(35)**, which grows on soil and perhaps referable to *Austrolecia* Hertel.

15. L. doliiformis Coppins & P. James (1992)

Thallus effuse, often wide-spreading, finely granular but granules often proliferating to form a thickish crust, dull green-grey or whitish grey; granules 30–60(-80) μm diam. Photobiont cells 7–14 μm diam., trebouxioid. Apothecia 0.2–0.5 mm diam., rare, convex to subglobose, pale pinkish brown with whitish pruina; true exciple absent; epithecium pale brownish due to dense minute crystals that dissolve in K; hymenium 45–50 μm tall, colourless or with pale brown vertical streaks; hypothecium reddish brown, K+ yellowish brown. Paraphyses 0.7–1.5 μm wide, but widening to 2.5(-3) μm in the epithecium, branched and anastomosed, apices colourless. Asci 40–45 × 9 μm, cylindrical-clavate, ± *Porpidia*-type, with an I− apical dome penetrated by darker tube-like structure. Ascospores (7-)8–11 × 2.7–3.7 μm, ovoid- to oblong-ellipsoid. Pycnidia 80–160 μm diam., 140–300 μm tall, numerous, sessile or ± stalked, somewhat barrel-shaped, grey-brown to dark grey, usually thinly white-tomentose, ostiole often gaping; wall dull brown, K−, N+ reddish brown, but green (K+ intensifying, N+ red) around the ostiole;

conidia 3.5–4.7 × 1.5–2 μm, shortly oblong or wider at proximal end, mostly biguttulate. Thallus Pd−, K−, KC−, C−, UV− (no lichen substances by t.l.c.).

On dry, often sheltered, rough bark or long-exposed wood (especially *Quercus*) in woodland or sheltered parkland; local. S. and W. Britain. ?Endemic.

Easily recognized by its conspicuous pycnidia and its distinctive ecology. *Micarea* species with sessile to stalked pynidia all have a smaller-celled photobiont (cells 4–8 μm diam.).

16. L. endomelaena Leighton (1876)

Thallus of scattered, strongly convex to bullate granules, 0.2–0.5 mm diam., greenish white, medulla I−; prothallus poorly developed, black. Apothecia 0.2–0.4 mm diam., sessile, ± flat to slightly convex, black; true exciple persistent, black, brown in section but greenish at outer edge; epithecium olivaceous brown, becoming bluish green in K; hymenium 60–70 μm tall; hypothecium dark brown. Paraphyses 1.5–2 μm wide, mainly simple; apices clavate with dark greenish walls, to 3 μm wide. Asci not seen mature. Ascospores (9-)11–12(-15) × 4–5 μm, narrowly ellipsoid. Thallus and medulla Pd−, K+ weakly yellow, KC−, C−.

On old schist walls and in basalt crevices; rare. N. Wales (Caernarvon, no recent records), W.C. Scotland (Mull). ?Endemic.

17. L. erratica Körber (1861)

L. tephrizans Leighton (1876)

Thallus grey, continuous or of scurfy scattered fragments, if on wood ± immersed, hardly discolouring weathered grey wood, medulla I−; prothallus often predominant, fimbriate and black. Apothecia (0.15-)0.2–0.4(-0.7) mm diam., sessile, constricted at the base, disc concave to ± flat, rarely somewhat convex, black; true exciple persistent, raised, rarely finally excluded, blue-black at outer edge, pale brownish (K+ purplish) within; epithecium dark greenish blue to greenish brown, K+ green intensifying, N+ purple-red; hymenium 35–50 μm tall; hypothecium dark brown. Paraphyses 1.5–2.5(-3) μm wide, sparsely branched and anastomosed, not strongly swollen or capitate at the tips. Asci 25–30 × 7–10 μm, with K/I+ blue apical dome penetrated by a darker tube-like structure. Ascospores (6-)7–9(-10) × (2-)3–4(-5) μm, narrowly ellipsoid, rather abruptly truncated at the ends. Pycnidia 80–150 μm diam., often abundant, sometimes only those present, black, scattered on the prothallus or partly immersed in the thallus; conidia 3–5 × 1.5 μm, ± cylindrical. Thallus and medulla Pd−, K−, KC−, C−.

On pebbles, especially flints and stable shingle, rarely also on wood. S. & E. England, extending locally to Scotland; rare in Ireland. C. Europe, North America.

A characteristic species of this specialized habitat, where it can be confused with saxicolous morphs of *Buellia punctata* on rocks or *Catillaria atomarioides* and *C. chalybeia* without microscopic examination. It has been transferred to *Micarea* as ***M. erratica*** (Körber) Hertel, Rambold & Pietschm. (1989).

18. L. erythrophaea Flörke (1826)

L. hyalinella (Körber) Jatta (1889)

Thallus immersed or ± superficial, thin, whitish; prothallus absent. Apothecia 0.3–0.8(-1) mm diam., flat to convex, dark reddish brown to brown-black; true exciple thin to excluded, colourless within, the hyphal apices swollen at the outer edge, the edge and epithecium red-brown K−, N−; hymenium 35–45(-70) μm tall, I+ blue; hypothecium colourless to weakly yellowish. Paraphyses simple or rarely branched; apices to 5 μm wide, capitate, usually with a brown hood. Asci 30–40 × 9–12 μm, *Bacidia*-type. Ascospores (8-)9–14(-18) × 3–4.5(-5) μm, fusiform-ellipsoid. Thallus Pd−, K−, KC−, C−, I− (no substances detected by t.l.c.).

On bark of deciduous trees, especially young *Fraxinus*, in sheltered woodlands; rare. E. England, Scotland. C. & N. Europe, Scandinavia.

Most easily confused with *L. hypopta* **(24)**, which is best separated by the more raised exciple in young apothecia. Easily overlooked in the field for *Bacidia arceutina* with which it often grows.

19. L. fuliginosa Taylor (1836)

Thallus of convex to subglobose ± squamulose warts, (0.5-)1–2(-3) mm wide, rather loosely attached to the rock, constricted below, red-brown to dark brown, often shiny, surface uneven, medulla I−;

prothallus absent. Apothecia 0.3–0.5(-1) mm diam., sessile, ± flat to somewhat convex, black; true exciple black, becoming excluded, of dark brown radiating hyphae expanded to 6–8 μm at the surface; epithecium dark brown, K – ; hymenium 30–40 μm tall; hypothecium dark brown, massively developed and sometimes ± stalked. Paraphyses sparsely branched and anastomosed, apical cells with dark brown caps, to 6 μm wide. Asci 30–35 × 10–12 μm, *Catillaria*-type. Ascospores (5-)7–10(-12) × 4–5(-6) μm, ellipsoid, ± attenuated at the ends. Conidia reported to be 12–17 × 0.5 μm, thread-like and arcuate. Thallus and medulla Pd –, K –, KC –, C – (no lichen products detected by t.l.c.).

On exposed siliceous rocks; rare. S.W. England (Cornwall, Lizard), S. Wales (Pembroke, Tycanol), N.W. England, Scotland, W. Ireland (Kerry to Galway, but not recently). Europe, North America.

This species has a distinctive *Acarospora*-like habit. Its current generic disposition needs further study.

20. L. furvella Nyl. ex Mudd (1861)

Rimularia furvella (Nyl. ex Mudd) Hertel & Rambold (1987)

Lichenicolous lichen. Thallus to 0.8 mm thick, effuse, rimose-areolate, dark brown to blackish; surface minutely isidioid or granular (granules 30–50 μm diam.). Photobiont cells 6–12 μm diam. Apothecia 0.3–0.8(-1.5) mm diam., usually few or absent, black; disc flat to convex; true exciple thin, exciple and hypothecium dark brown; epithecium brown or olive-brown; hymenium 60–75 μm tall. Paraphyses branched and anastomosed, the apices not strongly capitate. Asci 50–60 × 15–18 μm. Ascospores 12–16(-20) × 6.5–9(-10) μm. Thallus usually either Pd –, K –, C + pink, or Pd + yellow, K + yellow to red (crystals), C – (with varying proportions of gyrophoric and norstictic acids; other reported substances are probably contaminants from host lichens).

Overgrowing various crustose lichens on exposed siliceous rocks; local. N. & W. Britain, N.E. Ireland (Down). C. & N. Europe, Canada.

Perhaps closest to *Rimularia*, but see under that genus. Sterile morphs of *Placynthiella icmalea* resemble this species, but have smaller photobiont cells (5–9 μm diam.) and occur in shady, humid situations, not on sunny, exposed rocks.

21. L. fuscoatra (L.) Ach. (1803)

Thallus ± continuous, cracked-areolate, whitish grey to pale yellowish brown or grey-brown, the areoles to 3 mm diam., flat to ± convex, medulla I – ; prothallus black. Apothecia 0.5–2(-3) mm diam., ± immersed, sunken into or between the areoles, flat to slightly convex, black to densely grey-pruinose; true exciple persistent, slightly raised, of irregular chains of subglobose cells, the outer brown and to 5 μm wide, ± colourless within, C + pink; epithecium olive-green to olive-brown, olivaceous in K; hymenium 40–60 μm tall; hypothecium dark brown to black, well developed and stipe-like. Paraphyses simple or sparsely branched near the apices, *c.* 1.5–2 μm wide, not swollen or gradually widening to 3(-5) μm at apices, the apices sometimes with olive-brown pigment. Asci 45–55 × 8–15 μm, *Lecidea*-type. Ascospores (7-)9–14(-17) × (3.5-)4–7(-10) μm, ellipsoid to narrowly ellipsoid. Conidia 7–10(-12) × 0.5–0.8 μm, bacilliform. Thallus Pd –, K –, C + red (gyrophoric and ± lecanoric acids).

On rather smooth, slightly nutrient-rich siliceous rocks, walls and brick, rarely on worked timber; common. Throughout British Is. Europe, from the Mediterranean to the Alps, Madeira, North America, Nepal, S.E. China.

The thallus colour is particularly variable, morphs with grey thalli ± lacking a brown tinge are sometimes treated as var. *grisella* (Flörke ex Schaerer) Nyl. (1856). The C + red reaction of the cortical surface is a striking feature.

22. L. paupercula Th. Fr. (1874)

Like **21**, but thallus a darker, ± reddish brown, and medulla I + blue, Pd + orange, K + yellow, KC –, C – (stictic acid).

On tops of large siliceous boulders above 500 m alt. C. & N. Scotland (E. Ross; Perth, Ben Lawers) Circumpolar in montane regions of N. Hemisphere.

L. lapicida **(31)** has the same medullary reactions, but has a paler thallus and, usually, hypothecium.

23. L. gyrizans Nyl. (1861)

L. fuscocinerea auct., non Nyl. (1852)

Thallus crustose, areolate, grey-brown to somewhat reddish brown; areoles 0.2–1 mm diam., strongly convex; medulla I –. Apothecia 0.2–0.5 mm diam., to 0.8 mm if tuberculate, immersed or a few ±

sessile, black; disc flat to slightly convex, minutely wrinkled; true exciple thin, often greyish, often angular, convolute or ± gyrose; exciple and hypothecium dark brown, K−; epithecium brown, K−; hymenium 80–120 μm tall. Paraphyses 1.5–1.8 μm wide, richly branched and anastomosed, short-celled and often ± moniliform above, with some cells to 4 μm wide. Asci 50–60 × 18–25 μm, clavate, *Rimularia*-type. Ascospores (8-)9–11 × 4.7–7(-8) μm, subglobose to ellipsoid. Pycnidia 60–150 μm diam., immersed; wall brown above, becoming colourless below; conidia 5–8.5 × 0.8–1 μm, bacilliform, simple, acrogenous. Cortex C−, medulla Pd± orange, K± yellow, KC−, C− (stictic and ± norstictic acids).

On schists, primarily montane, but known from sea level in S. Wales (Pembroke); locally rare. W. & N. Great Britain. Montane Europe, North America.

The type of *L. fuscocinerea* has recently been shown to be identical to *Schaereria tenebrosa* (= *S. fuscocinerea*), necessitating a name change to *Lecidea gyrizans* Nyl. However, this species has recently been transferred to *Rimularia* as *R. gyrizans* (Nyl.) Hertel & Rambold (1990).

24. L. hypopta Ach. (1803)

Thallus ± immersed, or superficial and then white, ± continuous; prothallus absent. Apothecia (0.2-)0.3–0.5(-0.7) mm diam., deep reddish brown, to dark brown or almost black, strongly convex to almost tuberculate; true exciple pale fawn, soon becoming excluded, giving the apothecia a piebald appearance in this process, exciple colourless within, outer edge ± pale brown; epithecium reddish brown, K± olivaceous brown; hymenium 25–40 μm tall; hypothecium colourless. Paraphyses simple or sparsely branched; apical cells brown and swollen to 3–4 μm wide. Asci 30–40 × 10–12 μm, walls ± evenly thickened throughout; outer layer and the shallow apical dome K/I+ uniformly blue, ± *Catillaria*-type. Ascospores (8-)9–12 × (2.5-)3–4(-5) μm, oblong-ellipsoid to slightly curved, often pale greyish brown or spuriously 1-septate. Conidia 14–18 × 4–5 μm, arcuate. Thallus Pd−, K−, KC−, C−.

On dry wood, or more rarely bark, of *Betula*, *Quercus* and, especially, conifers; local. S.W. (N. Somerset) & N. (S. Pennines, Northumberland) England, C. & N. Wales, N. Scotland (Highlands). Probably circumboreal, C. & N. Europe, North America.

This species has been included in *Lecanora* by some continental authors, but a thalline exciple is absent and the asci are not characteristic for that genus; its affinities require further study.

25. L. hypnorum Lib. (1853)

Thallus effuse, thin, membranous, whitish or pale grey, sometimes with brownish tinge; I−. Apothecia 0.3–1.2 mm diam. (sometimes forming larger tuberculate clusters), dark brown to black, sessile; disc flat to convex; true exciple thin and often flexuose, usually long-persistent; exciple and hypothecium dark reddish brown, K−; epithecium pale brown or almost colourless; hymenium 60–70 μm tall; hymenium and hypothecium usually with scattered blue-violet (K+ green) granules. Paraphyses 1.5–1.7 μm wide, mostly simple, the apices gradually widening to 2.5 μm, colourless. Asci *Porpidia*-type. Ascospores 10–16(-19) × 4.5–6(-7) μm, ellipsoid, often thinly 1(-3)-septate, mostly with finely warted epispore. Thallus Pd−, K−, KC−, C− (no lichen products detected by t.l.c.).

On bryophytes and plant debris over limestones and calcareous schists or their derived soils, very rarely on old tree trunks (*Fraxinus*); to over 1000 m in Scotland. N. & W. British Is. N. Europe, North America.

This species, together with *L. ahlesii* **(2)**, *L. berengeriana* **(9)** and *L. sanguineoatra* **(26)**, forms a distinctive group having affinities with *Clauzadea*.

26. L. sanguineoatra auct., non (Wulfen) Ach. (1803)

Like **25**, but apothecia usually a paler brown (rarely blackish), with a more readily excluded true exciple, and narrower ascospores, (8-)9–14(-15) × 3–4.5(-4.8) μm, which are smooth and never septate.

On bryophytes over acid rocks or bark (especially *Quercus*, *Betula*, *Sorbus*) in woodlands or sheltered valleys, very rarely directly on rock; below 600 m. N. & W. British Is. N. Europe, North America.

Rare morphs on rocks can be separated from *L. ahlesii* **(2)** by their narrower ascospores. *L. berengeriana* **(9)** has a thicker, tartareous thallus and stouter apices to the paraphyses. Many British records of *L. hypnorum* **(25)** belong here.

27. L. inops Th. Fr. (1874)

Thallus cracked-areolate, uneven, continuous or occurring as dispersed convex granules, whitish grey to dark grey, medulla I+ violet-blue; prothallus indistinct. Apothecia 1–3 mm diam., sessile, black, occasionally pruinose, usually strongly convex and often irregular in shape; true exciple soon excluded, rarely persistent and flexuose, olive-brown at outer edge, colourless within; epithecium olivaceous brown, K+ olive-green, N+ purple-red; hymenium 45–55 μm tall; hypothecium dark brown to ± black. Paraphyses simple, apices slightly swollen to 2.5–3 μm wide, brownish. Asci 25–45 × 6–9 μm, *Lecidea*-type. Ascospores 8–11(-13) × (2.5-)3(-4) μm, oblong-ellipsoid. Pycnidia *c.* 0.5 mm diam., usually abundant, immersed to slightly raised, with ± elevated margins, conidia 9–15 × 0.8 μm, bacilliform. Thallus and exciple Pd−, K−, KC−, C−, UV+ whitish (perlatolic acid and unidentified substance in the exciple).

On copper-rich rocks. N.W. England (Lake Distict, Coniston and Dale Head). France, Czechoslovakia, Norway, Sweden.

This species is like *L. diducens* (**7**), with which it has sometimes been confused, but differs in the less developed and soon excluded exciple, darker hypothecium and, especially, the marginate pycnidia and different chemistry. The bright blue colour of copper oxalate is often visible through cracks in the thallus.

28. L. inquinans (Tul.) Nyl. (1856)

Thallus absent, lichenicolous. Apothecia 0.3–0.4 mm diam., black, strongly convex, often aggregated; true exciple soon excluded; epithecium and hypothecium red-brown to dark brown; hymenium 45–60 μm tall, pale red-brown. Paraphyses to 3 μm thick, frequently branched; apices swollen and red-brown. Asci 40–50 × 12–18 μm, clavate, the apical dome and outer fuzzy coat I+ blue. Ascospores 8–13 × 4–7 μm, broadly ellipsoid to subglobose, sometimes ± uniseriate.

On thalli of *Baeomyces roseus*. N. Scotland (E. Inverness). France, Germany.

29. L. insidiosa Th. Fr. (1868)

Thallus at first granular, becoming almost areolate, greenish brown; hyphae I+ violet; lichenicolous lichen. Apothecia 0.2–0.5(-0.7) mm diam., black, ± with bluish bloom when wet, convex; true exciple soon excluded, dark green-brown at outer edge, paler within and hyphae I+ violet; epithecium dark green, N+ purple-red, surface layer brownish, of minute crystals that dissolve in K; hymenium 40–60 μm tall; hypothecium colourless or pale brownish or greenish. Paraphyses 1.3–1.8 μm wide, simple or sparingly branched; apices swollen to 3 μm and often with a dark green hood or brown cap. Asci *Bacidia*-type, with a broadly conical apical cushion. Ascospores (8-)9–12 × 4.5–5.5(-6) μm, ellipsoid. Thallus Pd−, K+ yellow, KC+ yellow, C−.

On *Lecanora varia*. S. & W. England (E. Kent, Shropshire), S.E. Scotland (Berwick). Austria, Finland, Germany, Sweden.

This is a lichenicolous lichen, the thallus of which is formed by its usurping of the photobiont from the host *Lecanora*. The apothecia of the host are colonized first.

30. L. insularis Nyl. (1852)

Rimularia insularis (Nyl.) Rambold & Hertel (1985)

Thallus areolate, grey-brown to brown or somewhat olivaceous; medulla I−; prothallus well developed, black; lichenicolous lichen. Apothecia 0.1–0.6 mm diam., at first sunken between the areoles, later ± sessile, flat to convex; true exciple raised ± flexuose, black; true exciple, epithecium and hypothecium dark brown, K−; hymenium 50–60 μm tall. Paraphyses *c.* 1–2 μm wide, branched and anastomosing, becoming moniliform above with swollen cells to 4 μm wide. Asci 40–55 × 12–16 μm, *Rimularia*-type. Ascospores (8-)8.5–14 × 4.5–7 μm, ellipsoid. Thallus Pd−, K± yellow, KC−, C−.

Forming islets in thalli of *Lecanora rupicola*; locally abundant. Mainly in W. & N. British Is. Europe, Macaronesia, North America, Tierra del Fuego, Australia, New Zealand.

31. L. lapicida (Ach.) Ach. (1803)

L. contiguella Nyl. (1873), *L. luteoatra* Nyl. (1873), *L. subcinerascens* auct. brit., non Nyl. (1877)

Thallus irregularly cracked-areolate, whitish to grey, often unevenly rust-coloured; medulla I+ blue-violet; prothallus often conspicuous, black. Apothecia (0.2-)0.5–1.2(-1.8) mm diam., immersed to ±

sessile, black, arising between the areoles and often compacted together with angular margins; true exciple persistent, raised, dark green at outer edge, ± colourless within, K+ yellowish; epithecium greenish black, K+ bright blue-green, N+ reddish purple; hymenium 50–70(-90) μm tall; hypothecium pale to dark brown, K−. Paraphyses *c.* 1.5–2 μm wide, simple or sparsely branched and anastomosed, apices expanding to 3 μm wide or with a green-brown hood and 3–4(-5) μm wide. Asci (40-)50–60(-65) × (12-)15–20 μm, *Lecidea*-type. Ascospores (8-)10–15(-16) × (4.5-)5–8 μm, broadly ellipsoid. Conidia 8–12(-20) × 1 μm, bacilliform. Medulla and exciple Pd− or yellowish, K± yellow (flooding out in squash mounts), KC−, C− (stictic and constictic acids, unidentified substances).

On siliceous, often iron-rich, rocks in upland areas. Mainly in W. & N. Britain, S.W. & N.E. Ireland. Widespread in N. & S. Hemispheres.

For differences from *L. lactea* (see **32**).

32. L. lactea Flörke ex Schaerer (1828)

L. demolita auct. brit., non Vainio (1934), *L. lapicida* var. *lactea* (Flörke ex Schaerer) V. Wirth (1987)

Like **31**, but differs in the predominance of norstictic acid (thallus K+ yellow to red, crystals), and when growing with *L. lapicida* by its usually thicker, whiter thallus with more neatly arranged, angular areoles. *L. lapicida* contains stictic with only traces of norstictic and reacts K+ yellow.

In similar habitats to **31**, but also coastal, local. W. & N. Britain, Ireland.

33. L. confluens (Web.) Ach. (1803)

Like **31**, but exciple blackish at outer edge, ± colourless within and epithecium green to olive-green, brownish green or black. Thallus and medulla Pd−, K−, KC−, C− (confluentic acid, 2'-*O*-methylperlatolic acid, and often trace of stictic acid).

On siliceous rocks. Scotland extending to S.W. England. Europe, North America.

This name has been widely misapplied to various large-fruited *Lecidea*-like species, particularly *L. lapicida* (**31**); its status in the British Is. merits re-assessment.

34. L. lichenicola (A. L. Sm. & Ramsb.) D. Hawksw. (1978)

? *L. obsoleta* Nyl. (1865)

Thallus inconspicuous, immersed, sometimes granular and greenish white due to superficial algae; prothallus absent. Apothecia sessile, sometimes immersed in shallow pits, pinkish to reddish brown, finally sometimes ± black, ± closed and perithecioid at first but finally expanded with a ± flat disc, 0.1–0.2(-0.35) mm diam.; true exciple persistent, colourless to creamy, becoming irregularly dentate; epithecium red-brown, K−; hymenium to 150 μm tall; hypothecium orange-brown. Paraphyses simple to sparsely branched, the apices not swollen or capitate. Asci 50–70 × 12–14 μm, obpyriform, somewhat attenuated towards the apex; apex thickened, I−; outer coat I+ blue. Ascospores (14-)16–19 × (5-)6–8 μm, elongate-ellipsoid, with conspicuous oil drops, sometimes pseudoseptate, the contents I+ yellow-orange.

On chalk pebbles in disturbed habitats (e.g. around rabbit burrows); local. S. & E. England (Is. of Wight, Sussex, Surrey, Hertford, E. Yorkshire), C. Scotland (Perth). ?Endemic.

The original material of *L. obsoleta*, described from calcareous soil in S.E. England (Sussex, near Lewes), could not be located, but according to the original description may well belong here. This species should probably be placed in *Trapelia*, from which it differs only in the immersed thallus and calcareous habit.

35. L. limosa Ach. (1810)

L. stenotera auct. brit., non (Nyl.) Nyl. (1866)

Thallus granular-verrucose, of convex granular areoles (0.1-)0.2–0.6(-1) mm diam., white-grey to grey; prothallus white or absent. Apothecia (0.3-)0.5–0.8(-1) mm diam. sessile, black, strongly convex, adpressed; true exciple not apparent, reflexed, ± colourless, of branched, radiating hyphae 1–2 μm wide; epithecium and hymenium deep bluish green, K−; hymenium 45–65 μm tall; hypothecium reddish brown, K+ orange. Paraphyses 1.3–2 μm wide, sparsely branched and anastomosed, sparsely septate, the apices not swollen or pigmented. Asci 35–45 × 9.5–11 μm, *Catillaria*-type but sometimes with an indistinct ocular chamber. Ascospores (8-)10–14(-15) × 3–4.5(-5) μm, narrowly ellipsoid to subfusiform, one or both apices often attenuated. Conidia 6–7 × 1.5 μm, reported as bacilliform. Thallus Pd+ orange-red, often patchily, K−, KC−, C− (pannarin and unidentified triterpenoids).

On soil, over mosses or humus; rare. N. Scotland (Highlands). C. & N. Europe, North America.

This species recalls *L. caesioatra* **(11)**, which occurs in similar situations, but differs in the less granular and thinner thallus, non-pruinose apothecia, the darker hypothecium, smaller asci with uniformly I+ blue apical dome, the more pointed and narrower ascospores, and in the thallus reactions. *L. limosa* is probably referable to *Austrolecia* Hertel.

36. L. lithophila (Ach.) Ach. (1814)

Thallus ± continuous to irregularly cracked-areolate, surface smooth to slightly warted, pale grey to blue-grey, often rust-coloured in parts, medulla I−; prothallus generally indistinct. Apothecia (0.2-)0.5–1.5(-2.5) mm diam., dark brown to black, disc brown when moistened, ± sessile, partly immersed only in the early stages, sometimes angular by compaction; disc ± flat to slightly convex, not constricted below; true exciple persistent, raised, outer edge dark brown ± with greenish tinge, internally pale brown to ± colourless, K−; epithecium brown, K−; hymenium (45-)50–80 μm tall, sometimes brownish in parts; hypothecium colourless, composed of ± vertically orientated hyphae. Paraphyses sparsely branched and anastomosed, the apices brownish, to 4.5 μm wide, often with granules adhering. Asci (45-)50–60 × 12–15 μm, elongate-clavate, *Lecidea*-type. Ascospores (9-)11–15 × (4-)5–6(-7) μm, narrowly ellipsoid. Conidia 12–18 × 0.5–1.5 μm, cylindrical. Thallus and medulla Pd−, K−, KC−, C− (4-*O*-demethylplanaic and ± planaic acids).

On exposed siliceous rocks, stones and pebbles, especially those rich in iron. Throughout N. & W. Britain, Ireland (rare). C. & N. Europe, North America, S.E. China.

The separation of the species from the closely related *L. plana* **(44)**, with which it has often been confused, is discussed there.

37. L. matildae Magnusson (1956)

Thallus ± immersed, loosely associated with algae, no cortex developed; prothallus dominant, black, irregularly spreading around rock crystals. Apothecia 0.15–0.25(-0.35) mm diam., black, shiny, sessile; disc concave to flat and ± pore-like; true exciple persistent, to 50 μm thick, the outer layer dark brown, cellular, K−, C+ weakly reddish; hymenium (45-)50–60 μm tall; epithecium olivaceous brown interspersed with granules; hypothecium brown. Paraphyses branched, particularly towards the apices, the apices brown, swollen to 4 μm wide. Asci 35–45 × (8-)10–15 μm, *Lecidea*-type. Ascospores (9-)10–12(-14) × 5–6(-6.5) μm, ellipsoid.

On granite. S.E. Ireland (Wicklow, Lugnaquilla mtns); one collection. Endemic.

This species is close to *L. diducens* **(7)**, differing primarily in the relatively broad ascospores.

38. L. mullensis Stirton (1877)

Thallus grey-brown, of strongly convex to tuberculate or ± botryose areoles, medulla I−; prothallus black, well developed. Apothecia (0.2-)0.3–0.5 mm diam., black, immersed; disc ± concave to flat, sometimes angular by compression, sometimes umbonate; true exciple persistent, slightly raised and forming a distinct rim to the sunken disc, dark brown; epithecium golden brown to dark reddish brown; hymenium 70–100(-110) μm tall, I+ blue, changing to red in parts; hypothecium brownish to deep reddish brown. Paraphyses 1.5–2.3 μm wide, branched and anastomosed, ± moniliform above with cells to 4.5 μm wide, the apices brown, not distinctly capitate. Asci 45–60 × 10–15 μm, *Rimularia*-type. Ascospores (6-)8–11 × 4–7 μm, subglobose to ellipsoid. Thallus and medulla Pd+ yellow changing to red, K+ yellow to red (with abundant crystals), KC+ red, C− (norstictic acid).

On exposed siliceous rocks. S.W. England (Devon), N. Scotland (Mull, S. Aberdeen). ?Endemic.

Very close to *L. gyrizans* **(23)**, differing mainly in the K+ red (crystals) reaction and possibly only a chemical race. Probably referable to *Rimularia*.

39. L. nigrogrisea Nyl. (1879)

Thallus grey-brown, areolate, areoles ± flat to convex, ± shiny, medulla I−; prothallus black. Apothecia 0.5–1.1 mm diam., black, arising between the areoles and finally appearing ± sessile, convex, sometimes angular by compression; true exciple tending to become excluded, poorly developed, ± continuous with the hypothecium in section, K−; epithecium green-black, intensifying in K, N+ deep reddish purple; hymenium 60–75 μm tall; hypothecium dark brown, deep reddish brown in K. Paraphyses mainly simple, rarely branched towards the apices and occasionally anastomosed, the apices slightly swollen, to 3 μm wide, greenish black, interspersed with crystalline granules. Asci 40–55 × 12–17 μm, *Lecidea*-type. Ascospores (7-)9.5–11 × 4–5.5(-6) μm, broadly ellipsoid. Conidia 6–8 × 0.6 μm. Medulla Pd−, K+ fainly yellow, KC−, C−.

On schists. C. & N. Scotland (Perth, Inverness). Montane regions of Europe, but distribution uncertain.

Perhaps identical to the montane European *L. obluridata* Nyl. (1873), but differing in the more strongly areolate and greyer thallus, and perhaps also the somewhat shorter hymenium and ascospores. Should also be compared with *L. plana* **(44)**, which differs primarily in the paler and often scarcely developed thallus and paler hypothecium.

40. L. nylanderi (Anzi) Th. Fr. (1874)

Thallus effuse, grey-white, usually tinged brownish in part, entirely granular-sorediate; granules 20–60(-70) μm diam.; external hyphae sometimes brown. Photobiont cells 5–12(-14) μm diam. Apothecia very rare, 0.2–0.5(-0.9) mm diam., pale to dark red-brown, flat to slightly convex; true exciple thin, pale brown, the outer part sometimes darker; epithecium brown, K−; hymenium 35–40 μm tall, colourless; hypothecium colourless. Paraphyses simple; apices abruptly swollen with dark brown caps to 4 μm wide. Asci *Bacidia*-type. Ascospores 6–7 μm diam., globose. Thallus Pd−, K−, KC−, C−, UV+ white (divarcatic acid).

On trunks of old *Pinus*, *Betula* and *Juniperus* in old oak-birch or pine-birch woodlands. C. & N. Scotland (Perth, S. Aberdeen, E. Inverness). Scandinavia, C. Europe, N. North America.

Difficult to separate from sterile *L. botryosa* **(10)** except by t.l.c. (both species are UV+ white), although the latter usually has a darker grey-brown thallus. *Haematomma caesium* contains perlatolic acid, and has blue-grey hyphae in some soredia and a usually conspicuous grey-blue-black prothallus.

41. L. paraclitica Nyl. (1872)

Thallus immersed, ± inconspicuous, effuse. Apothecia 0.15–0.3(-0.4) mm diam., black, flat to convex; true exciple soon excluded, of compacted, ± radiating, densely pigmented hyphae, 2.5–3 μm wide, with terminal cells to 5 μm wide; true exciple, epithecium and hypothecium dark brown to blackish, K−, N−; hymenium 20–30 μm tall, colourless. Paraphyses 1.7–3 μm diam., mostly simple, apices capitate, with dark brown caps 3–6 μm diam. Asci 20–25 × 8–12 μm, the outer coat and apical dome K/I+ blue, the latter with a narrow apical cushion surrounded by a dark blue border. Ascospores (5-)6–10 × 2.5–3.5 μm, narrowly ellipsoid. Pycnidia unknown. Thallus Pd−, K−, C−.

On dry, weathered conifer wood; rare. N. Scotland (Highlands). S. Finland, Sweden.

Has been confused with *L. erratica* **(17)**, but differs in the shallower hymenium, brown, N− epithecium, capitate paraphyses and the absence of pycnidia.

42. L. pernigra Hertel (1975)

Thallus dark grey, of ± scattered scurfy granules, or scarcely apparent; medulla I−; prothallus black. Apothecia 0.2–0.4(-0.5) mm diam., sessile, strongly constricted below, black, disc flat to convex; true exciple persistent, raised and somewhat swollen, of ± radially arranged swollen hyphae, the outer layer dark brown to black, internally reddish brown; epithecium dark brown, N+ reddish brown; hymenium 40–55 μm tall, I−; hypothecium dark brown. Paraphyses mainly branched only towards the apex, the apices brown, capitate and to 5.5 μm wide. Asci 45–55 × 12–15 μm; *Lecidea*-type. Ascospores (7-)10–15(-16.5) × (3-)4–5(-6) μm, narrowly ellipsoid, ± attenuated at the ends, often 2-guttulate. Thallus Pd−, K−, KC−, C−.

On coarse-grained sandstone. N. England (Derbyshire, Peak District; S. Lancashire). Sweden.

Material on stones in acid dunes in S.W. Scotland (Wigtown, Torrs Warren) is perhaps this species but has scattered white areoles.

43. L. phaeops Nyl.(1858)

Thallus white, pale greenish white or whitish grey, irregularly cracked to ± areolate, smooth, medulla I−; prothallus white, often well developed, sometimes replaced by dark contact zones with other thalli. Apothecia 0.2–0.5(-0.6) mm diam., black, immersed, aspicilioid, irregular in shape, sometimes coalescing; disc flat to concave; true exciple scarcely apparent in surface view, the outer edge red-brown where exposed, internally pale; epithecium red-brown, somewhat olivaceous in K; hymenium (95-)100–150(-160) μm tall; hypothecium dark red-brown. Paraphyses ± simple, occasionally branched towards the tips, apices reddish brown, scarcely swollen to 3 μm wide. Asci 60–80 × 12–15 μm, *Biatora*-type. Ascospores 12–18(-23) × 5–6(-7) μm, narrowly ellipsoid, somewhat inequilateral and plano-convex; apices often attenuated. Thallus Pd+ yellow, K+ yellow, KC+ yellow, C− (atranorin and psoromic acid).

In damp shaded crevices of siliceous rock outcrops, often near the ground, occasionally by streams; local. W. Britain (Devon to Shetland), S.W. & W. Ireland. Apparently oceanic in Europe, Azores, North America.

Amygdalaria pelobotryon, which occasionally occurs in similar habitats, can be distinguished by the C+ red thallus, the anastomosed paraphyses, ascus-apical apparatus, and larger ascospores. The continuous, ± glaucous thallus is very reminiscent of *Herteliana taylorii*, but the apothecia in that species are emergent to ± sessile and the thallus chemistry is different.

44. L. plana (Lahm) Nyl. (1872)

Thallus pale grey, granular to cracked and irregularly areolate, often dispersed and often inconspicuous, medulla I−; prothallus indistinct. Apothecia 0.2–0.5(-1.5) mm diam., black, remaining so when moistened, sessile, disc flat to slightly convex; true exciple persistent, raised, green-black at outer edge, internally yellowish grey or colourless; epithecium dark green black to black, intensified in K, N+ reddish purple; hymenium 35–50(-55) μm tall; hypothecium ± colourless to pale yellowish brown, of ± vertically orientated hyphae. Paraphyses sparsely branched and anastomosed, apices with a dark green-hood, to 4 μm wide. Asci 30–40 × 8–14 μm, *Lecidea*-type. Ascospores (7-)8–11(-13) × (2.5-)3.5–5(-6) μm, ellipsoid. Conidia 10–13 × 0.5–0.6 μm. Thallus Pd−, K−, KC−, C− (planaic and 4-*O*-demethylplanaic acids).

On siliceous, especially coarse-grained, upland rocks. N. England, S. Wales, Scotland, N. Ireland. Europe, North America, Japan, Australia.

Differs from *L. lapicida* **(31)** in the yellow-grey exciple and I− medulla, from *L. lithophila* **(36)** in the mainly immersed thallus, the small ascospores and greenish epithecium, and the disc remaining black when moist, and from *L. nigrogrisea* **(39)** in the less developed and paler thallus.

45. L. polycarpella Erichsen (1929)

Thallus effuse, thin, continuous or of slightly convex areoles *c.* 0.1 mm diam., pale greyish but often darkened by extraneous material. Photobiont cells (6-)9–21(-28) μm diam. Apothecia 0.08–0.15(-0.2) mm diam., immersed and flat to ± sessile and convex, black and immarginate when dry; true exciple appearing as a dark rim encircling a translucent greenish disc when wet 7–12 μm wide, dark greenish, the hyphae coherent in K; hymenium 35–50 μm tall, pale greenish above, I+ blue; hypothecium colourless or pale greenish. Paraphyses 0.8–2 μm wide, few, simple or sparingly branched, ± moniliform above; apices colourless. Asci 30–45 × 11–14 μm, clavate, in K/I apical dome pale blue with a dark blue tubular structure. Ascospores 8–12 × 4–5 μm, ellipsoid. Pycnidia 40–50 μm diam., immersed, black, the wall greenish; conidiogenous cells subglobose to short ampulliform, lining the pycnidial cavity; conidia 4.5–5.5 × 1.5–2 μm, bacilliform. Thallus Pd−, K−, C−, I−.

On loose siliceous stones in urban waste land. N. England (S. Yorkshire, Sheffield). N. Germany, Austria.

An inconspicuous species with some similarities to the *Micarea sylvicola* group and *Psilolechia* species, but differing in having a true exciple of coherent hyphae (in K).

46. L. punctum (Massal.) Jatta (1900)

?*L. cladoniaria* Nyl. (1857)

Thallus absent; lichenicolous fungus. Apothecia (0.1-)0.15–0.2 mm diam., ± immersed to erumpent, black; disc concave or flat; true exciple thin, dark brown, K+ purplish brown; epithecium dark brown; hymenium 45–55 μm tall, I−; hypothecium poorly developed, reddish violet to purplish brown. Paraphyses simple or sparsely branched, easily separating; apices not swollen. Asci 35–45 × 7–8 μm, ± cylindrical, the walls ± even, only slightly thickened at the apex, the outer coat I+ pale blue; apex I−. Ascospores 6–10 × 2–3 μm.

On *Cladonia* species, especially squamules of *C. coniocraea*, perhaps also *Cetraria islandica*. N. Scotland (Aberdeen, Perth, Is. of Skye), E. Ireland (Dublin). France.

The material on *Cetraria* may merit recognition as a separate species, *L. cetrariicola* Lindsay (1869), but is in need of a critical re-appraisal. Recently transferred to *Phaeopyxis* as ***P. punctum*** (Massal.) Rambold, Triebel & Coppins (1989). ***P. varia*** Coppins, Rambold & Triebel (1989) is a similar parasite on *Trapeliopsis gelatinosa*; Mid-Wales, Scotland.

47. L. pycnocarpa (Körber) Ohlert (1870)

L. segregans Nyl. (1866), *L. tabidula* Nyl. (1867)

Thallus white to pale grey, of dispersed ± shiny verrucae or areoles, sometimes scarce, the medulla I+ blue in parts; prothallus black, fimbriate, sometimes dominant. Apothecia 0.15–0.25(-0.35) mm diam., black, mostly arising in compact groups of to 25 on a common black stroma, ± botryose, sessile; disc ± flat to slightly convex; true exciple at first raised, later excluded; epithecium aeruginose to greenish-brown, intensifying in K, N+ reddish purple; hymenium 30–50 μm tall, often pigmented as the epithecium; hypothecium dark red-brown. Paraphyses branched and occasionally anastomosed, the apices olivaceous and swollen to 5 μm wide. Asci 40–50 × 12–15 μm, *Bacidia*- or *Biatora*-type with narrow apical cushion. Ascospores (9-)10–18(-22) × (3-)4–6 μm, elongate-ellipsoid. Thallus Pd+ weakly yellow, K+ yellow, KC+ yellow, C− (atranorin and unidentified substance).

On coarse siliceous rocks, especially granites, in upland areas. N. England, Wales, Scotland; old reports also from N.W. Ireland (Sligo, Leitrim). C. & N. Europe, North America.

The characteristic botryose clusters of minute apothecia are otherwise consistently seen only in *L. botryosa* **(10)**, which grows on bark. A mainly sterile, sorediate form, *L. pycnocarpa* **f. *sorediata*** Coppins & Fryday (1992), is found in the mountains of N. England and Scotland; it has convex, whitish areoles with black soralia and resembles *Miriquidica nigroleprosa* but differs in its K+ yellow cortex due to atranorin.

48. L. sarcogynoides Körber (1855)

Thallus usually ± absent, of dispersed greyish white areoles, the medulla I−; prothallus well developed, fimbriate, often predominant, black. Apothecia (0.3-)0.5–1.5(-2.2) mm diam., black, sometimes pruinose, sessile, strongly constricted at the base, often in groups; disc slightly concave to flat; true exciple persistent, raised, ± shiny, often flexuose, purple-black at outer edge, pale reddish within, K+ purplish; epithecium dark olivaceous brown to brownish black; hymenium 40–50 μm tall, pale reddish to violet, K+ reddish purple in parts; hypothecium brown-black. Paraphyses simple to sparsely branched; apices to 4 μm wide, olivaceous, capitate. Asci 35–45 × 11–12 μm, *Lecidea*-type. Ascospores (7-)10–12(-14) × (2-)2.5–3.5(-4) μm, narrowly ellipsoid; apices rounded to somewhat attenuated. Thallus Pd−, K−, KC−, C− (terpenoid traces).

On exposed granite rocks and stones. S.W. Britain (Channel Is., Is. of Scilly). Europe, from Portugal to S. Sweden, N. & S. Africa, S. America, Australia, New Zealand.

Similar to the *L. auriculata* **(6)** group, but distinguished by the colour of the exciple and hymenium, *Sarcogyne*-like apothecia, and less abruptly truncated ascospores. Sometimes confused with *Adelolecia pilati*.

49. L. silacea (Ach.) Ach. (1803)

Thallus ochre to rust-red, areolate; areoles to 1.5 mm diam., convex, the cortex deep rust in section (iron oxides), medulla I+ blue; prothallus indistinct. Apothecia 0.5–1.5 mm diam., black, arising between the areoles, at first appearing ± immersed but finally ± sessile, flat to strongly convex; true exciple becoming excluded, green-black at outer edge, internally pale; epithecium green to blue-green or black-green, K+ aeruginose, N+ reddish purple; hymenium 40–60 μm tall, colourless to greenish in parts; hypothecium dark brown, massively developed and almost stipe-like in section. Paraphyses sparsely branched and anastomosed, the 2–4 final cells swollen; apices to 5 μm wide, greenish. Asci 40–55 × 8–12 μm, *Lecidea*-type. Ascospores (8-)10–15(-17.5) × 4.5–6(-7) μm, ellipsoid to broadly ellipsoid. Conidia 8–12 × 1 μm, bacilliform. Thallus Pd−, K−, KC−, C− (± porphyrilic acid).

On upland metal-rich siliceous rocks, often with *Acarospora sinopica*, *Rhizocarpon oederi* and *Tremolecia atrata*; local. N.W. England (Lake District), Scotland (Highlands). Cool temperate and arctic Europe, N. America, Nepal, ?Tierra del Fuego.

50. L. tornoensis Nyl. (1859)

Thallus dark brown, uneven, or evanescent. Apothecia 0.3–0.5(-0.8) mm diam., strongly convex, deep red-brown; true exciple excluded; epithecium red-brown to brown; hymenium 50–80 µm tall; hypothecium ± colourless. Paraphyses branched and anastomosed, the apices swollen, capitate, red-brown, to 3.5 µm wide. Asci 55–70 × 24–37 µm, broadly clavate, the apex thickened with an internal apical beak, *Lecanora*-type. Ascospores (15-)17–20(-24) × (8-)12–15 µm, broadly ellipsoid to subglobose, with strongly thickened (2–3.5 µm thick) walls. Thallus Pd−, K−, KC−, C− (no lichen products detected by t.l.c.).

On bryophytes in sheltered rock crevices above 800 m. C. Scotland (Perth, Breadalbane Mtns). Circumpolar in Northern Hemisphere.

This species, distinguished by the massively thickened ascospore walls, has sometimes been placed in *Mycoblastus*, as *M. tornoensis* (Nyl.) R. Anderson (1974), but belongs neither to that genus nor to *Lecidea* s.str., and the new genus *Japewia* Tønsb. (1990) has been introduced to accommodate it [as ***J. tornoensis*** (Nyl.) Tønsb. (1990)] and the allied *L. carrollii* **(12)**.

51. L. turgidula Fr. (1824)

Thallus effuse, immersed or thinly granular, whitish to grey, often inconspicuous. Apothecia 0.2–0.5(-0.6) mm diam., black, with bluish bloom when wet, sometimes thinly white-pruinose, sessile or (on wood) partly immersed, flat to convex; true exciple excluded, usually of coherent (in K) radiating hyphae, colourless to darkish greenish or brownish, I ± pale violet; epithecium brownish to blue-green, K− or K+ green intensifying, also with minute pale granules that dissolve in K; hymenium 30–40(-45) µm tall, streaked violaceous (K+ green) or brownish to greenish (K+ intensifying); hypothecium colourless. Paraphyses 1.3–2 µm wide, sparingly branched and anastomosed, the apices scarcely swollen but sometimes with a dark brownish hood to 5 µm wide. Asci 25–30 × 10–13 µm, *Bacidia*-type. Ascospores 6–10(-12) × 2.5–4.5(-6) µm, 0(-1)-septate, narrowly ellipsoid to oblong, sometimes slightly curved. Pycnidia *c.* 100 µm diam., black, ± immersed in substratum, the wall green-brown, K+ green intensifying; conidiogenous cells catenate, subcylindrical; conidia arising laterally and terminally, 3–3.5 × 1.5–1.8 µm, bacilliform. Thallus Pd−, K−, KC−, C−.

On acid bark and wood of (mostly) conifers; locally common. Scotland (Highlands), extending locally to S.W. England (rare).

Often mistaken for a *Micarea*, but distinguished from that genus by the exciple of coherent hyphae, and minute granules in the epithecium that give the apothecia a pruinose appearance (bluish bloom when wet).

52. L. umbonella Nyl. (1866)

L. umbonata auct. brit., non (Hepp) Mudd (1861)

Lichenicolous lichen, often forming gall-like swellings on the host thallus. Thallus whitish; medulla I+ violet. Apothecia 0.1–0.6 mm diam., erumpent, black; disc with central umbo; true exciple prominent, brown-black; epithecium brown or greenish brown; hymenium 70–110 µm tall; hypothecium brown to dark brown. Paraphyses branched and anastomosed; apices scarcely swollen. Asci 50–80 × 12–18 µm, *Lecidea*-type. Ascospores 11–15(-16) × (6-)6.5–8(-10) µm. Conidia 5–9 × 1 µm, bacilliform. Thallus Pd+ yellow, K+ yellow to red (norstictic acid, ? from host).

On *Lecidea lactea*. S.W. & N.W. England (W. Cornwall, Lizard Peninsula; Lake District), N. Wales (Merioneth), N. Scotland (Highlands), W. Ireland (Galway). C. & N. Europe, Greenland, North America (Colorado).

The new generic name *Cecidonia* Triebel & Rambold (1988) has been introduced to accommodate this [as ***C. umbonella*** (Nyl.) Triebel & Rambold (1988)] and one related species. Similar collections on *Porpidia* species have smaller ascospores, (9-)10–12(-13) × 5–6.5(-7.5) µm, an I− and usually a K− medulla; these have been recognized as ***Cecidonia xenophana*** (Körber) Triebel & Rambold (1988) [syn. *L. umbonella* var. *alumnula* (Nyl.) Hertel (1970)]. This species also occurs widely in N. & W. British Is.; N.W. & C. Europe, Canary Is.

LECIDELLA Körber (1855)

Thallus crustose, immersed or superficial, continuous to verrucose-areolate. **Photobiont** chlorococcoid. **Medullary hyphae** I− ; prothallus absent or black to blue-black. **Ascomata**

apothecia, usually constricted at base, sessile, ± black or red-brown in shade. **Thalline exciple** absent. **True exciple** of radiating hyphae, dark green, blue-black or brown, at times crystalline, not black and friable. **Epithecium** green-black-brown. **Hymenium** colourless to pigmented, I+ blue. **Hypothecium** colourless, yellowish or brown to red-brown. **Hamathecium** of paraphyses, usually lax (especially in K), simple, the apices scarcely swollen, rarely capitate. **Asci** 8-spored, clavate, *Lecanora*-type. **Ascospores** simple or rarely 1-septate, colourless, ellipsoid, rather uniformly thick-walled, lacking a distinct perispore. **Conidiomata** pycnidia, immersed. **Conidia** thread-like. **Chemistry**: atranorin, chloratranorin, zeorin, diploicin and a range of xanthones frequent; psoromic acid present in one species. **Ecology**: widely distributed, especially in nutrient-enriched habitats.

There is considerable variation in the epithecial and excipular pigments, and also in the presence of crystals within these areas; their taxonomic significance is not always well understood.

LITERATURE: Hertel (1977), Knoph (1990), Hertel & Leuckert (1969), Leuckert *et al.* (1989).

1	On rocks	2
	On mosses, bark or wood	9
2(1)	Thallus sorediate or scurfy-blastidiate	3
	Thallus not sorediate, or blastidiate, smooth to verrucose-squamulose	4
3(2)	Thallus thin, with scattered granular, yellowish white to grey-green, round soralia, often paler when rubbed	**7. *scabra***
	Thallus thick, of crowded blastidia, unchanged when rubbed	**6. *prasinula***
4(2)	Hypothecium yellowish to red-brown, K+ orange-brown	5
	Hypothecium colourless or brownish when old, K−	6
5(4)	Thallus C+ orange	**9. *subincongrua***
	Thallus C−	**3. *carpathica***
6(4)	True exciple dark green, blue-black to dirty brownish at edge, pale and transparent within	7
	True exciple ± uniformly dark brown, entirely opaque	**1. *anomaloides***
7(6)	Thallus C−	8
	Thallus C+ orange	**11. *viridans***
8(7)	Thallus and/or apothecia Pd+ deep yellow	**2. *bullata***
	Thallus and/or apothecia Pd−	**8. *stigmatea***
9(1)	On bark and wood	10
	On mosses	**10. *wulfenii***
10(9)	Thallus sorediate or ± uniformly blastidiate	11
	Thallus without soralia or blastidia	**4. *elaeochroma***
11(10)	Thallus uniformly scurfy-sorediate or with diffuse, ± concolorous soralia; with or without apothecia	12
	Thallus with rounded, discrete, green-yellow soralia; apothecia numerous	**4. *elaeochroma***
12(11)	Thallus uniformly sorediate-scurfy (blastidiate); unchanged when rubbed	**5. *pulveracea***
	Thallus with irregularly scattered or ± contiguous concolorous soralia; often paler when rubbed	**7. *scabra***

1. L. anomaloides (Massal.) Hertel & Kilias (1980)

Thallus thin, ± scurfy, indistinctly areolate to ± granular, continuous, wide-spreading or a thin, film-like covering, pale to dark grey or greenish. Apothecia 0.4–0.7 mm diam., sessile, flat to slightly convex; true exciple well developed, often wavy, persistent, 50–70 μm thick, mostly opaque, dark brown, paler brown towards hypothecium; epithecium green-black to brownish, non-crystalline; hypothecium colourless or pale brown, red-brown when old, often K+ faintly reddish. Ascospores 9–15 × 7–8 μm.

On shaded, often vertical, ± damp, siliceous rocks, often in wooded situations or in sheltered gullies; local. Throughout W. British Is. Europe northwards from the Mediterranean, North America.

2. L. bullata Körber (1861)

Thallus moderately thick, white to grey-white, verrucose, of irregular, convex, occasionally bullate granules, to *c.* 1.5 mm diam.; surface ± eroded, at times pleated. Apothecia to 1.5 mm diam., sessile to almost immersed, at first flat, later convex, occasionally sublobulate, to 3 mm diam.; disc usually black, greyish pruinose; true exciple at first prominent, black, later excluded, dark brown in section, usually extending beneath hypothecium; epithecium crystalline, the crystals disappearing in K; hymenium 40–65(-90) μm, colourless, without crystals; hypothecium 2-layered, lower part densely crystalline, crystals disappearing in K. Ascospores 7–12 × 5–8 μm. Thallus and/or apothecia Pd+ deep yellow or rarely Pd−, K+ yellow, KC+ yellow, C− (atranorin, zeorin, ± psoromic acid, 2'-*O*-demethylpsoromic acid).

On vertical, non-calcareous, often metal-rich, rocks, also beneath underhangs, montane. Scotland (Highlands). Alps, Czechoslovakia, Fennoscandia, North America.

3. L. carpathica Körber (1865)

Thallus verrucose, to incipiently or unevenly coarsely granular, usually well developed, white or pale to dark grey. Apothecia 0.5–1 mm diam., ± immersed, at first flat, later becoming ± convex; true exciple thin, wavy, shiny, later excluded, in section greenish or blackish blue at edge, brown to red-brown within; epithecium partly greenish black, brownish tinged, crystals absent; hypothecium semi-opaque, bright red-brown, K+ bright orange-brown. Ascospores 10–16 × 6–8.5 μm. Thallus K+ yellow, KC+ yellow, C− (atranorin, diploicin, thuringione and chloratranorin).

On weakly basic, nutrient-enriched rocks and walls, also slate and asbestos cement roofing etc. Throughout British Is. C. Europe, North America, New Zealand.

L. stigmatea **(8)** differs in the pale hypothecium, only the outer edge of the true exciple coloured, and the usually less developed thallus.

4. L. elaeochroma (Ach.) M. Choisy (1950)

Thallus rather smooth, even and continuous or ± granular-verrucose, yellow-grey to yellow-green in sunny, exposed sites, becoming grey-green in shade, at times forming mosaics; prothallus black or bluish black. Apothecia to 1 mm diam., closely adpressed, rounded to irregular, at first flat, later often convex; disc black in exposed, well-lit situations, pale blue-black, brownish red or ± piebald with a darker true exciple in shade; true exciple persistent, flexuose, finally ± excluded; epithecium and edge of true exciple bluish green to dull grey-blue, crystalline, crystals dissolving in K; hymenium 40–70 μm tall, at times with inclusions towards lower part of thecium and hypothecium; hypothecium ± brownish orange, rarely colourless, K± bright red-brown. Ascospores 10–17 × 6–9 μm. Thallus K+ yellowish, KC+ yellow, C+ orange (reactions often faint) (arthothelin, granulosin and ± 4,5, dichloro-norlichexanthone).

On well-lit, smooth bark, especially twigs and small branches and wood, often forming mosaics, moderately tolerant of SO_2 air pollution; common. Throughout British Is. Europe, North America, New Zealand.

The pigmentation in the apothecia and thallus is much reduced in shaded situations. The species can, however, be distinguished by the lax, simple paraphyses, the brownish orange hypothecium intensifying in K, and the C+ orange thallus. Forma *soralifera* (Erichsen) D. Hawksw. (1972) is sorediate, with soralia 0.5–1 mm diam., bright yellow-green, punctiform to efflorescent, ± discrete, ± scattered over, or, at times, restricted to only part of thallus; the production of soralia does not depress the formation of apothecia. Soralia contain lichexanthone. It occurs in similar habitats to, and usually with, f. *elaeochroma*. Mainly W. British Is. C. & N. Europe.

The mainly continental and Asian *L. euphorea* (Flörke) Hertel (1980) [*Lecidea dolosa* (Nyl.) Vainio (1893)], distinguished by the K+ yellow, C− (?atranorin) reactions, is poorly understood and is probably not correctly recorded from the British Is.

See under *Lecidea carrollii* for note on *L. exigua* which has a KC+ orange thallus.

5. L. pulveracea (Schaerer) H. Sydow (1887)

Like **4**, but with a thin to thick, fine, powdery thallus dissolved into small blastidia to 0.1 mm, coarsely areolate, forming a continuous, wide-spreading crust.

On wood, confined to worked timber of barns, and fence rails, especially near farms, formerly widespread. England (Midlands and the south-east); not seen since 1879 and perhaps extinct. C. Europe, N. to Scandinavia, North America.

The identity of this species in unclear and shows affinities with *Pyrrhospora quernea*.

6. L. prasinula auct., non (Wedd.) Hertel (1980)

Thallus thin to thick, of numerous aggregations of proliferating granules (blastidia), dispersed in scattered clusters or ± continuous, forming a coarsely cracked-areolate, minutely granular crust, pale dull ochre to grey-ochre in shade. Apothecia to 1 mm diam., infrequent, rarely crowded, often intermingled with blastidia; disc flat or wavy, rarely convex, matt; true exciple thin, shiny, flexuose, persistent dark black-brown, paler within, strongly crystalline; epithecium dark blue-green, non-crystalline; hymenium 60–90 μm, crystalline, crystals not dissolving in K; hypothecium brown to dark red-brown. Ascospores 10–20 × 6–9 μm. Thallus C+ orange (often faint).

On non-calcareous, maritime rocks, rarely on decorticated wood and decaying *Armeria* tufts, coastal; rather rare. S. & W. British Is. (Channel Is. to Shetland). S. Sweden.

7. L. scabra (Taylor) Hertel & Leuckert (1969)

Thallus very thin to coarsely granular, becoming ± cracked-areolate, pale to dark green-grey; soralia scattered or ± confluent, concolorous to yellowish white, efflorescent, round. Apothecia sessile, concave, to somewhat irregular or convex, often convoluted, occasionally subgyrose; true exciple well developed, shiny, becoming excluded, dark brown at edge, paler within, K+ brighter red-brown, crystalline; epithecium brown-black to blue-black; hypothecium dull yellow- to red-brown. Ascospores 10–15 × 6–8 μm. Thallus C+ orange (atranorin, arthothelin, ± thuringion).

On siliceous or slightly basic rocks, often on walls and memorials, occasionally on dry and often dust-impregnated wood, worked timber, or bark. Almost throughout British Is. C. & N. Europe, North America.

The number and extent of soralia are very variable. When sterile, the thallus becomes ± continuously sorediate, when fertile, soralia are usually few and scattered. The surfaces of the soralia are concolorous with the thallus but become pale greenish when rubbed.

8. L. stigmatea (Ach.) Hertel & Leuckert (1969)

Thallus white to dirty grey-green, dark grey-brown, blackish or rust-red to brown, immersed or superficial and then continuous, faintly rimose-cracked or granular-verrucose. Apothecia to 1.5 mm diam., flat or rarely convex; true exciple well developed when young, the edge blue-black, indigo to green-black, colourless within, non-crystalline, K ± unchanged; epithecium brown to purplish brown, K+ purple intensifying; hymenium 60–80 μm tall, not interspersed with crystals; hypothecium colourless to pale yellow-brown. Ascospores 10–16 × 6–9 μm. Thallus K+ yellow or K−, UV− (three chemotypes: atranorin and zeorin; lichexanthone and zeorin; or zeorin).

On ± basic and calcareous rocks, walls and mortar or adjacent siliceous rocks; often very common. Throughout British Is. Europe, North America.

Very variable in the shape of apothecia and thickness of thallus. The contrasting brown to purplish brown epithecium and the thin blackish or indigo external part of the true exciple, with the colourless or pale hypothecium, are diagnostic. This species, *Caloplaca flavovirescens* and *Protoblastenia rupestris* are useful indicators of weakly calcareous rocks. See also *Clauzadea monticola*.

9. L. subincongrua (Nyl.) Hertel & Leuckert (1969)

L. asema (Nyl.) Knoph & Hertel (1990); incl. *L. effugiens* (B. Nils.) Knoph & Hertel (1990), *L. elaeochromoides* (Nyl) Knoph & Hertel (1990), *L. polyantha* Taylor ex Leighton (1879)

Thallus coarsely granular-papillate, rather thick, often strongly verrucose, coarsely cracked-areolate, pale ochre to yellow-green or rarely pale grey in shade. Apothecia 0.5–1 mm diam., black, at first flat, later becoming convex, often shiny, at least when young; true exciple becoming excluded, grey-green to grey-blue at edge, pale brown to red-brown within, inner part crystalline; epithecium blue-green; hymenium ± crystalline, the crystals dissolving in K; hypothecium yellowish to deep red-brown, K+ intensifying. Ascospores 10–13(-20) × 6–9 μm. Thallus K+ yellow, C+ orange (xanthones, ± atranorin).

On maritime, non-calcareous rocks in the xeric-supralittoral, occasionally on decaying *Armeria* tufts, widespread on exposed rocky shores, rarely also on sheltered montane rocks. Throughout British Is. W. Europe, Mediterranean region, North America.

Very variable. Characterized by the pale ochre to yellow-green, strongly verrucose thallus with jet-black apothecia and the maritime habitat. Recently divided into 3 species, based mainly on major xanthones present: *L. asema* (asemon, thiophanic acid), *L. effugiens* (isoarthothelin, 2,5,7-3-*O*-methylnorlichexanthone), *L. elaeochromoides* (arthothelin, ± thuringione or ± granulosin).

10. L. wulfenii (Hepp) Körber (1861)

Thallus like **9**; true exciple dirty drab black at edge, reddish within; epithecium dark brown-black or blue-green, crystalline; hypothecium pale red-brown above, darker below. Ascospores 7–16 × 7–8 μm, ellipsoid. Thallus C+ orange.

On mosses overgrowing calcareous substrata, 700–1100 m. C. Scotland (Ben Lawers, Caenlochan Glen). Arctic-alpine, particularly in wind-swept situations, southwards to mountains of Balkan peninsula, North America.

11. L. viridans (Flotow) Körber (1855)

Thallus thin, scurfy or finely granular, usually small, often wide-spreading in cracks and fissures, greyish to greenish yellow. Apothecia 0.2–0.3 mm diam., scattered to crowded, ± flat; true exciple persistent, often grey-green-pruinose, brown to grey-green at edge, paler within; epithecium dark brown, brown-green or green-grey, granular, crystals dissolving in K; hymenium 50–60 μm tall, not interspersed with granules; hypothecium colourless to pale straw. Paraphyses clavate, apices brown. Ascospores 9.5–12 × 5–7 μm. Thallus C+ orange (arthothelin and thiophaninic acid).

On hard, non-calcareous rocks. S. & W. England (Shropshire, Channel Is.), Wales. Warmer regions of C. Europe, Mediterranean region, North America.

L. stigmatea **(8)** has a blue-black edge to the true exciple, paraphyses which lack clavate apices, a C− thallus, and a different chemistry.

LECIDOMA G. Schneider & Hertel (1981)

Thallus ± crustose to squamulose-areolate; prothallus when present black. **Upper cortex** plectenchymatous. **Lower cortex** undifferentiated, but the lowest medullary hyphae deep brown, attached directly to substratum. **Photobiont** chlorococcoid; dividing into 2–4 daughter cells. **Medulla** I−, the hyphae 3–4 μm thick, thin-walled. **Ascomata** apothecia; disc ± immersed, flat or convex. **True exciple** with outer part pale brown, the inner part, colourless. **Epithecium** 12–18 μm tall, red-brown. **Hymenium** colourless, I+ blue. **Hypothecium** colourless. **Hamathecium** of paraphyses, straight, thin-walled, strongly conglutinate, only slightly branched and anastomosed, apical cells swollen with a thin, dark brown cap. **Asci** thin-walled, clavate, with a well-developed, K/I+ pale blue apical dome and a K/I+ dark blue tubular structure, *Porpidia*-like. **Ascopores** simple, ellipsoid or ovoid, thin-walled, smooth, without a perispore. **Conidiomata** not seen. **Chemistry**: lichen products not detected by t.l.c.

LITERATURE: Schneider (1980).

1. L. demissum (Rutström) G. Schneider & Hertel (1981)

Thallus coarsely areolate to subsquamulose, forming small to spreading lumpy mounds, to 7 cm diam., dark brown, rarely grey to grey brown (shade morphs), ± shiny; areoles 1–2(-4) mm across, flat to convex, polygonal or somewhat subsquamulose, tightly packed, ± turgid, often contiguous and ± coalescing; lower surface deep brown-black. Apothecia 0.5–1(-1.5) mm diam., discrete or becoming confluent, ± immersed or adnate; true exciple thin when young, becoming excluded when mature; disc round, dull brown-black, red-brown when moist; hymenium 60–70 μm tall. Paraphyses 2–3 μm wide, the apices to 4 μm. Asci 50–55 × 10–12 μm. Ascospores 12–16 × 5.5–7 μm. Thallus Pd−, K−, KC−, C−.

On peaty soils and acid gravels, rarely on bare rock on exposed mountain sides and summits. Scotland (Highlands), extending locally to N.W. England (Lake District), N. Wales, Ireland (Wicklow). C. Europe, Fennoscandia, North America.

When well developed the thalli resemble small cow-pats in colour and surface appearance.

LEMMOPSIS (Vainio) Zahlbr. (1906)

Thallus minute, thin, granular-crustose, black, pseudoparenchymatous, gelatinous when moist, with an incomplete, uneven outer layer of fungal cells. **Photobiont** cyanobacterial (not *Nostoc*); cells globose to slightly ovate, in clumps. **Ascomata** apothecia, sessile, urceolate-discoid. **Thalline exciple** present but poorly developed. **True exciple** robust, broad-rimmed, of anastomosing hyphae. **Hymenium** colourless, I+ blue. **Hypothecium** narrow, pale golden-brown. **Hamathecium** of paraphyses, slender, septate, slightly conglutinate. **Asci** narrowly clavate, thin-walled, 8-spored. **Ascospores** simple, ovate, thick-walled, colourless. **Conidiomata** unknown. **Chemistry**: lichen products not detected. **Ecology**: on calcareous rocks and clay soil.

Psorotichia has a weak or virtually absent true exciple with a narrow rim, lax hyphae and a prominent thalline exciple. *Lempholemma* has a minute, fruticose growth habit with algae randomly scattered throughout the thallus and immersed apothecia with an inconspicuous true exciple. *Collema* and *Leptogium* differ in possessing a true exciple composed of small isodiametric cells, and do not produce unicellular or round-ended ascospores.

LITERATURE: Ellis (1981).

1. L. arnoldiana (Hepp) Zahlbr. (1906)

Thallus of rounded to subrectangular areoles, to 0.6 mm diam., scattered or aggregated, forming a thin, effuse, dark-coloured crust, gelatinous when wet. Apothecia 0.4–0.5(-0.7) mm, numerous, prominent, sessile; true exciple prominent, dull red, sometimes dull yellow-brown, the base thin, of isodiametric cells from which arise densely anastomosing hyphae running parallel to the sides of the hymenium. Ascospores 12.5–25 × 7.5–12.5 μm, simple, ovate to broadly ellipsoid.

On calcareous rocks in shaded, humid sites such as chalk pebbles on woodland floors, base of tufa stones in grassland, crevices in limestone cliffs, when well developed spreading to moss; very local. S. England, extending to N. England and Wales. Europe.

Resembles a small crustose *Leptogium* sp. but has simple ascospores.

L. oblongans (Nyl. ex Crombie) A. L. Sm. (1918), known only from the original gathering made 120 years ago on clay soil in Cumbria, is said to have ascospores longer in relation to their width (15–25(-27.5) × 5–7.5 μm) and more densely conglutinate paraphyses. It is unknown outside the British Is.

LEMPHOLEMMA Körber (1855)

Thallus crustose-warted, squamulose, foliose or minutely shrubby, blackish, or dark blue-green to olivaceous especially when wet, gelatinous; not layered, without a well-defined cortex. **Hormocystangia** present or not (Fig. 17). **Photobiont** *Nostoc*; cells in chains, occasionally compressed and clustered. **Ascomata** apothecia, mainly laminal, ± globose, ± immersed to sessile. **Thalline exciple** conspicuous. **Disc** pore-like, sometimes later expanding. **True exciple** usually thin and inconspicuous, rarely more than 20 μm wide at the sides, colourless, of compacted, parallel hyphae. **Hypothecium** shallow, colourless. **Hymenium** colourless, without a distinct epithecium. **Hamathecium** of scanty to numerous, simple or sparingly branched paraphyses without swollen apices; gel matrix I+ reddish, K/I+ blue. **Asci** 8-spored, clavate to cylindrical, without a distinct apical apparatus; wall K/I−, or blue in outer part (? adhering hymenial gel). **Ascospores** ellipsoid to globose, simple, colourless, thin-walled but often with a distinct, gelatinous perispore that disperses in K. **Conidiomata** pycnidia, mostly laminal, globose, immersed. **Conidiogenous cells** slender, cylindrical. **Conidia** acrogenous, bacilliform, or with a swelling at one or both ends, or fusiform, straight or slightly curved, simple, colourless. **Chemistry**: no lichen products reported. **Ecology**: on rocks, mortar, bryophytes or soil, usually on substrata that are calcareous or subjected to basic flushing; a cosmopolitan genus of *c.* 30 species.

The genus is heterogeneous and should be divided; the circumscription of most of the species is also uncertain and a critical revision is required. Apart from detailed anatomical aspects, the genus can be distinguished from most other gelatinous cyanobacterial lichens with a similar appearance by its *Nostoc* photobiont. Similar species of *Collema* have septate ascospores, as do those of *Leptogium*, which also have a well-defined cellular cortex.

The apothecia and pycnidia of *Lempholemma* species are often inconspicuous and are best observed in moistened thalli.

No British specimens or localized records have been traced for *L. isidioides* (Nyl. ex Arnold) Magnusson (1939).

LITERATURE: Henssen (1968).

1 Thallus without strap-like lobes, if lobes minutely shrubby then 0.15–3 mm wide; isidia absent 2
Thallus with strap-like lobes, 5–15 mm long and 0.2–0.6 mm wide; isidia clustered, globose to clavate ..... **6. *radiatum***

2(1) Thallus minutely shrubby; lobes 1–5 mm long and 0.15–0.3 mm wide 3
Thallus crustose, squamulose or foliose, or cushion-like; lobes less than 0.5 mm long 4

3(2) Ends of some lobes swollen, with globular to cup-shaped hormocystangia ..... **2. *cladodes***
Ends of lobes not swollen, hormocystangia absent ..... **3. *intricatum***

4(2) Thallus forming small cushions or tufts to 5 mm diam., with minute flattened squamules or short cylindrical lobes to 0.5 mm long; apothecia and pycnidia absent 5
Thallus forming a spreading granular, nodular or membranous crust, sometimes with rounded marginal lobes to 3 mm wide; apothecia and/or pycnidia usually present 6

5(4) Ends of some lobes swollen, with globular to cup-shaped hormocystangia ..... **2. *cladodes***
Ends of lobes scarcely swollen, hormocystangia absent ..... **1. *botryosum***

6(4) Ascospores 9–14 μm diam. or 9–16(-20) μm long, broadly ellipsoid to globose **4. *myriococcum***
Ascospores 20–33 μm long, ellipsoid ..... **5. *chalazanun***

1. L. botryosum (Massal.) Zahlbr. (1924)

Thallus of aggregated, short, minutely shrubby, ± imbricate cylindrical lobes forming small tufts or umblicate cushions 1.5–3 mm diam.; lobes to 0.3 mm long, 0.15–0.2 mm wide, unbranched or forked, surface ± minutely longitudinally striate, apices slightly swollen. Apothecia not seen in British Is., reported as half-immersed with a pore-like disc. Ascospores 7–9 × 5–7 μm.

On periodically wet, exposed, hard calcareous rocks, especially Carboniferous limestone, often in water-filled depressions in limestone pavement. N. England, N. Scotland (S. Aberdeen), N. Ireland (Fermanagh, Mayo). Europe.

Some of the material referred below to *L. intricatum* **(3)** may be a luxuriant morph of *L. botryosum*. *L. cladodes* **(2)** differs in producing hormocystangia. Sometimes confused with *Collema fragile* and *Synalissa symphorea*.

2. L. cladodes (Tuck.) Zahlbr. (1924)

L. albonigrum Magnusson (1946)

Thallus cushion- or tuft-like to 5 mm diam., at first a rosette of short, flattened squamules which eventually elongate into cylindrical, ± dichotomously branched, minutely wrinkled lobes to 2 mm long and 0.1–0.2 mm wide; lobe ends swelling to produce globose hormocystangia, 0.15–0.3 mm diam., which eventually burst and become cup-like (Fig. 17). Apothecia unknown in Europe, in American material

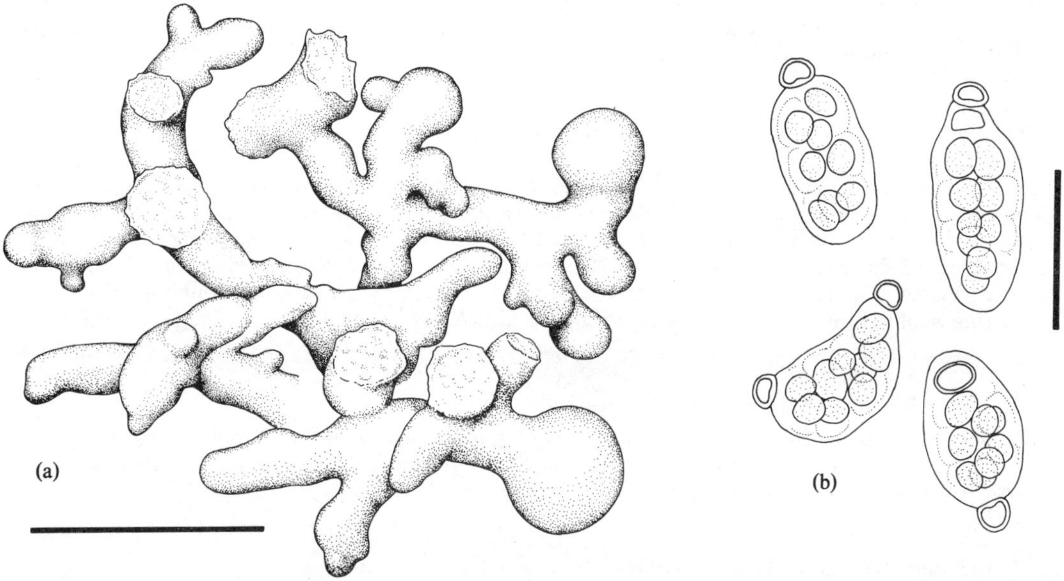

Fig. 17. *Lempholemma cladodes*. (a) Thallus branches (moist) with old and with developing hormocystangia; (b) hormocysts (not lichenized). Scale a = 1 mm, b = 30 μm.

reported as terminal or lateral, 0.1–0.4 mm diam., with blackish convex disc. Ascospores 15–20 μm diam., globose.

On low, flat, acid rock outcrops, affected by blown shell-sand in machair or on calcareous rock by a waterfall; rare. S.W. & N.W. Scotland. Norway, Sweden, North America.

Resembles *L. botryosum* **(1)** or in more luxuriant morphs of *L. intricatum* **(3)**, but differs from both in producing hormocystangia.

3. L. intricatum (Arnold) Zahlbr. (1924)

L. fennicum (Räsänen) Degel. (1946)

Thallus shrubby to filamentous, forming low cushions or matts to 3 cm across; lobes cylindrical, repeatedly dichotomously branched, decumbent, entangled, blackish, sometimes faintly longitudinally wrinkled, to 5 mm long and 0.12–0.2 mm wide. Apothecia rare, 0.15–0.2 mm diam., laminal or rarely terminal, ± globose, at first pore-like, later sometimes with slightly expanded, red-brown disc. Paraphyses scanty, 1–2 μm wide, sparingly branched. Asci clavate, sometimes narrowing to a papilla-like apex. Ascospores 10–14 × 8–11 μm, broadly ellipsoid to globose. Pycnidia *c.* 100 μm diam., immersed, laminal or subterminal; conidiogenous cells cylindrical, 15–23 × 1.5–2 μm; conidia 4–5 × 1–1.5 μm, fusiform.

On flushed, slightly basic, granite slabs or mica-schist at 300–400 m alt.; very rare. S.W. (Kirkcudbright, Black Craig) and C. Scotland (Grampian mtns.).

Of the British material, only that from Black Craig has pycnidia and apothecia, it also has slightly more robust and less distinctly wrinkled lobes than that from the Grampian mtns.; the latter may be a different species, and is perhaps closer to *L. botryosum* **(1)** or *L. cladodes* **(2)**.

4. L. myriococcum (Ach.) Th. Fr. (1861)

L. chalazanellum (Nyl.) Zahlbr. (1924), *L. chalazanodes* (Nyl.) Zahlbr. (1924)

Thallus dark olivaceous to blackish, nodular granular, or spreading to 5 cm diam. or more and then often with irregularly rounded lobes (to 3 mm wide) at the edge, centre of the thallus membranous but with irregular ridges which are often warted, dark olivaceous to blackish. Apothecia 0.2–0.3 mm diam., often aggregated along the ridges or lobe ends, ± globose, almost fully to ¼-immersed, with red-brown pore-like to expanded disc; hymenium 100–130 μm tall. Paraphyses numerous, 1.5–2 μm wide. Asci cylindric to cylindrical-clavate, the ascospores uniseriate at least in upper part of ascus. Ascospores (8-)9–16(-20) × 8–12(-15) μm, globose to ellipsoid, with a gelatinous perispore that disperses in K. Pycnidia frequent; conidia 2–3 × 0.5–1 μm, bacilliform.

On limestone and mortar, usually over mosses; common. Throughout British Is. Europe, North America.

A broad concept is used here pending a critical revision. Typical *L. myriococcum* has a spreading, ± lobate thallus (resembling moribund *Collema auriforme*) and mainly globose ascospores, 9–14 μm diam. Specimens formerly referred to *L. chalazanodes* have a similar habit but mainly broadly ellipsoid ascospores, 12–20 × 8–12(-15) μm, while those referred to *L. chalazanellum* have mainly ellipsoid ascospores but a much reduced, nodular-granular thallus growing directly on crumbly mortar and occurring mainly in low-rainfall, eastern areas. See also *L. chalazanum* **(5)**.

5. L. chalazanum (Ach.) B. de Lesd. (1910)

Like **4**, but ascospores 20–33 × 10–13 μm, ellipsoid (mostly at least twice as long as broad).

On the crumbly mortar of wall tops, occasionally on bare calcareous soil; rare. S. Britain. Europe, North America.

6. L. radiatum (Sommerf.) Henssen (1968)

Thallus forming rosettes, to 2–3 cm diam., or in larger mats, consisting of elongated, strap-like, ± channelled, repeatedly dichotomously branched lobes, 5–15 mm long and 0.2–0.6 mm wide; upper surface of lobes brown to blackish, with clusters of globose to clavate isidia or lobules, sometimes also with globular hormocystangia; lower surface pale brown or greenish, and often with a central, ± winged, nerve-like structure and bundles of white attachment hyphae; lobe ends usually with an elongate-clavate swelling. Apothecia not known in Europe, reported from North America as lateral, half-immersed, 0.3 mm wide, with a pore-like disc, and ellipsoid ascospores, 13.5–23 × 8–12 μm. Pycnidia frequent, ± immersed; conidia 3–5 × 1 μm, bacilliform or swollen at one or both ends, sometimes slightly curved.

On basic rocks, especially calcareous mica-schists, soil or overgrowing moribund bryophytes on moist ledges or periodically flushed situations, mostly at over 700 m alt. Scotland (Highlands). Scandinavia, Alaska, Canada (Alberta).

The prostrate, separated, strap-like, canaliculate, isidiate lobes, make this an easily recognized species.

LEPRARIA Ach. (1803)

Thallus leprose to granular, usually forming a powdery crust, whitish to dark grey to greenish, thick to thin, margin diffuse to delimited, composed of a mass of powdery or non-powdery spherical granules, covered by entangled hyphae which often project outwards (× 25). **Prothallus** scarce; hypothallus scarce. **Photobiont** species of the Chlorophyta. **Medulla** white or unstratified. **Ascomata** absent. **Conidiomata** absent. **Chemistry**: chiefly aliphatic acids, benzyl esters, depsides, depsidones, and triterpenoids. **Ecology**: frequently dominant in shaded habitats.

Characterized by the powdery sterile thallus comprising discordant species for which ascomata are unknown that have a common life-form. Several natural groups (e.g. *Chrysothrix*, *Leproloma*, *Leproplaca*) have been separated from genus *Lepraria*, and this process is expected to continue; *Leproloma* contains dibenzofurans. *Lepraria* and *Leproloma* species are often difficult to separate in the field and t.l.c. is usually required for confirmation.

LITERATURE: Laundon (1992b).

1 Thallus powdery; in shade 2
 Thallus granular, pale grey; on sunny, mossy, acid rocks 8

2(1) Thallus yellowish green to bluish grey to whitish; hyphae not forming a weft over the surface 3
Thallus verdigris green, becoming thick; hyphae forming a weft over the surface **5. *lesdainii***

3(2) Thallus bright greenish grey to whitish 4
Thallus dull greenish grey, usually with a bluish tinge; medulla undifferentiated .. **4. *incana***

4(3) Thallus almost lobed at margins, delimited, whitish; on vertical hard limestone 5
Thallus diffuse, bright greenish grey to yellowish green; not on hard limestone 7

5(4) Pd+ orange (atranorin and other acids) 6
Pd+ yellow (alectorialic acid) .... **3. *frigida***

6(5) Contains protocetraric and/or fumarprotocetraric acids and atranorin; Pd+ reddish orange .... **8. *nivalis***
Contains stictic acid and atranorin and ± constictic acid and zeorin; Pd+ orange **6. *lobificans***

7(4) Thallus KC− or + yellow; contains stictic acid and atranorin and ± constictic acid and zeorin; thallus with greenish tinge; common .... **6. *lobificans***
Thallus KC+ reddish orange; contains alectorialic and protocetraric acids; thallus with yellowish to greenish tinge; rare .... **2. *eburnea***

8(1) Thallus KC+ reddish orange, Pd+ lemon yellow (alectorialic acid) .... **7. *neglecta***
Thallus KC− or + yellow, Pd+ reddish orange or + yellow or − (fumarprotocetraric and fatty acids [if with atranorin and ± porphyrilic acid and fumarprotocetraric acid absent see *Leproloma cacuminum*] .... **1. *caesioalba***

1. L. caesioalba (B. de Lesd.) Laundon (1992)

L. zonata Brodo (1968)

Thallus a mass of convex granules 0.05–0.2 mm diam., usually without projecting hyphae, pale grey; delimited, the granules often lighter and clustered towards the margin, sometimes giving the thallus a zoned appearance, without lobes but often growing in dainty irregular rosettes; medulla greyish white, rarely exposed. Thallus Pd−, + yellow or + reddish orange, K− or + yellow, C− (atranorin usually present, fumarprotocetraric acid and either rangiformic or roccellic acids).

On moss cushions, lichens and cyanobacteria on sunny rocks, rarely on vertical rocks; locally common. Throughout upland Britain. Europe, North America.

Morphologically identical to the less common *L. neglecta* **(7)** which is C+ reddish orange and morphs of *Leproloma cacuminum* which have a different chemistry.

2. L. eburnea Laundon (1992)

Thallus leprose, of a mass of powdery spherical granules to 0.2 mm diam., with projecting hyphae, when well developed forming a thick powdery crust, bright whitish grey with a greenish to yellowish tinge, diffuse, without marginal lobes, medulla white. Thallus Pd+ orange, K− or + yellow, KC+ reddish orange, C− (alectorialic and protocetraric acids).

On bricks, mortar and mosses of brick walls in slight shade, rock crevices and tree trunks, lowland; rare. Throughout Britain. Norway, France, North America.

Characterized by its bright yellowish grey, powdery thallus which is KC+ reddish orange and Pd+ orange.

3. L. frigida Laundon (1992)

Thallus leprose, of a mass of powdery convex granules, to 0.6 mm diam., with projecting hyphae, which are often eroded to leave a leprose membrane, delimited by an obscurely lobed margin, thick, whitish, often with a yellow tinge; medulla white. Thallus Pd+ yellow, K− or + yellow, KC+ reddish orange, C− (alectorialic acid).

Over mosses in damp crevices on limestone cliffs. N. England, Scotland. Arctic-alpine, Norway, Austria, North America, Greenland.

Resembles *Leproloma vouauxii*, but has a quite different chemistry. Characterized by the presence of alectorialic acid with distinct colour reactions with KC and Pd.

4. L. incana (L.) Ach. (1803)

Thallus leprose, of a mass of powdery spherical granules, to 0.1 mm diam., with projecting hyphae, when well developed forming a thick, frequently areolate, powdery crust, dull greenish grey, usually with a bluish tinge, diffuse, without marginal lobes; medulla ± undifferentiated. Thallus Pd−, K−, C−, UV+ white (divaricatic acid, zeorin, and ± atranorin).

On acid, shaded bark, and on acid, shaded walls and rocks, especially on bare surfaces but also growing over mosses; common and pollution-tolerant, extending into towns. E. England, extending locally westwards. N. Europe, North America.

Identified by the bluish-tinged dull thallus with an undifferentiated medulla. *L. umbricola* Tønsb. is similar; it is K+ yellow, Pd+ deep orange (thamnolic acid) and often has a dull green thallus.

5. L. lesdainii (Hue) R. Harris (1987)

Thallus leprose, of a mass of powdery irregular granules, to 0.1 mm diam. but often much smaller, with numerous projecting hyphae which tend to form a weft over the surface, when well developed forming a thick powdery crust, green, but becoming dull greenish grey in the herbarium; forming ± irregular orbicular patches in the early stages, without marginal lobes; medulla not differentiated. Thallus Pd−, K−, C− (unidentified triterpene).

On rock and over mosses in limestone caves and clefts, and on limestone cliffs shaded by trees; confined to damp, heavily shaded habitats, in deeper shade than *Leproplaca chrysodeta* with which it often grows; local. Throughout British Is. N.W. Europe, North America.

The verdigris green colour and hyphae tending to form a weft over the surface are distinctive.

6. L. lobificans Nyl. (1873)

L. finkii (B. de Lesd. ex Hue) R. Harris (1985)

Thallus leprose, of a mass of powdery convex granules of variable shape and size, to 0.3 mm diam., with projecting hyphae which sometimes form a weft on the surface, or the granules becoming eroded to leave a leprose membrane; usually forming a thick powdery crust, bright greenish grey to whitish grey, the latter colour especially when eroded; usually diffuse, without marginal lobes but in eroded specimens the medulla often develops into delimited sublobes; medulla white. Thallus Pd+ orange, K− or + yellow, C− (atranorin, stictic acid, and ± constictic acid and zeorin).

On shaded bark and on shaded acid and calcareous rock, growing directly on the surface and especially over mosses; also on shaded earth; common. Throughout British Is. Europe, Asia, Africa, North America, Australia, New Zealand.

Morphologically variable, but with a uniform chemistry. The bright, pale green colour is typical and distinctive, but whitish morphs can closely resemble both *Lepraria nivalis* **(8)** and *Leproloma vouauxii*, and require separation by t.l.c.

7. L. neglecta (Nyl.) Lettau (1958)

?Lecidea pedatula Nyl. (1876)

Thallus granular, of a mass of convex granules, *c.* 0.1–0.2 mm diam., usually without projecting hyphae, pale grey, delimited, the granules lighter and clustered towards the margin, without lobes, but often growing in dainty irregular rosettes; medulla greyish white, rarely exposed. Thallus Pd+ yellow, K− or + yellow, KC+ reddish orange, C− or + reddish orange (alectorialic acid, ± atranorin).

On moss cushions on sunny acid rocks and on stony ground, chiefly on mountain summits, arctic-alpine. N. England, Scotland (Highlands). N. & C. Europe, North America.

This species is easily recognized by the granular, non-powdery, thallus growing in sunny sites; other British species of *Lepraria* are powdery and grow in shade. *L. neglecta* s. str. is identified by the presence of alectorialic acid; many old records belong to the morphologically similar *L. caesioalba* **(1)** or *Leproloma cacuminum*.

8. L. nivalis Laundon (1992)

L. crassissima auct. p.p., non (Hue) Lettau (1958)

Thallus leprose, forming an uneven non-corticate membrane on which occur powdery convex granules to 0.4 mm diam., with projecting hyphae, when well developed producing an extensive, thickish, powdery, folded crust, delimited, margins almost lobe-like, white to pale grey, sometimes with a bluish grey tinge; medulla white, ± undifferentiated. Thallus Pd+ reddish-orange, K− or + yellow, C− (atranorin, protocetraric or fumarprotocetraric acids; rangiformic or roccellic acids as uncommon accessories).

On bare rock and over mosses and lichens on shaded limestone, often dominant, also on siliceous, coastal rocks; locally common. Throughout British Is. N.W. Europe to Mediterranean region, North America, Himalayas.

Usually easily distinguished by the white to pale grey membranous thallus which is almost lobed-like. Some morphs of *L. lobificans* may look similar, when t.l.c. is needed for their separation.

LEPROCAULON Nyl. ex Lamy (1878)

Primary thallus ± persistent, densely powdery-leprose, diffuse, unstructured, giving rise to a secondary thallus of small, delicate slender, white, cartilaginous, ± terete, branching, interwoven pseudopodetia, ± clothed with leprose-sorediate, floccose-tomentose granules. **Photobiont** trebouxioid. **Ascomata** unknown. **Conidiomata** unknown. **Chemistry**: complex, including a range of depsides, depsidones, phloroglucinal derivatives, triterpenoids, fatty acids. **Ecology:** ± sheltered, rather dry crevices on siliceous rocks.

Formerly included in *Stereocaulon*, but distinguished by the lack of ascomata, the absence of corticate phyllocladia, cephalodia, pycnidia, and the different chemistry. The primary thallus resembles that of a *Lepraria*.

LITERATURE: Lamb & Ward (1974).

1. L. microscopicum (Vill.) Gams ex D. Hawksw. (1974)

Primary thallus a leprose-granular crust, persistent, blue-green to yellow-grey; secondary thallus of delicate, white, simple or sparingly branched, ± erect pseudopodetia, 2–4 mm tall, which, like the primary thallus, are covered in numerous, small, leprose-arachnoid granules. Thallus Pd−, K−, KC−, C−, UV− (variable; ± combination of atranorin, usnic, rangiformic acids, zeorin, and unidentified substances and fatty acids).

On thin soil, especially in crevices associated with ± dry sides of cliff faces and walls, rarely on the trunks of old broad-leaved trees and peaty soils; frequent in coastal sites, less common inland. S. & W. British Is. Europe, Asia Minor, N. Africa, Macaronesia, Central and W. North America, Australia.

Close examination will reveal the minute, white, ± erect, shrubby pseudopodetia which distinguish it from *Lepraria* species in the field.

LEPROLOMA Nyl. ex Crombie (1894)

Thallus leprose, rarely granular, ecorticate, of usually powdery, convex granules or eroded to leave a leprose membrane, often with powdery lobes best developed at the margin of the thallus; pale, often tinged yellow, sometimes grey; lobes (when present) and granules covered by loosely entangled hyphae, which often project outwards (× 25). **Medulla** white. **Hypothallus** primitive, whitish grey to brownish, of a continuous weft of loosely entangled, branched hyphae, these usually thick and brownish, in contrast to the narrow, colourless hyphae of the thallus. **Photobiont** species of Chlorophyta. **Ascomata** apothecia, very rare, small, superficial, orbicular, urceolate. **Thalline exciple,** tumid, sorediate, concolorous with the thallus. **Discs** brownish. **Hymenium** short, colourless. **Hypothecium** brown. **Hamathecium** of paraphyses straight, simple, septate, the cells slightly constricted at the septa, the tips slightly swollen. **Asci** clavate. **Ascospores** simple, ellipsoid, colourless. **Conidiomata** unknown. **Chemistry**: dibenzofurans, atranorin and roccellic acid.

Resembles *Lepraria*, but differs in the presence of dibenzofurans. Fertile material of the type species, *Leproloma membranaceum* **(3)**, was found in the 18th century, but has not been seen subsequently. Species remaining in *Lepraria* have never been found with ascomata. Specimens from C. Scotland (Perth) with a whitish thallus, protruding hyphae and only atranorin and fatty acids may belong to a further species of this genus described as *Crocynia rigidula* B. de Lesd. ex Hue (1924); similar specimens occur in lowland coastal and sheltered inland river gorges in Norway.

LITERATURE: Laundon (1989).

1	Thallus whitish, greyish, or yellowish white	2
	Thallus greenish yellow	**2. *diffusum***
2(1)	Thallus obscurely or not lobed, yellowish white to grey	3
	Thallus with distinct lobes, yellowish white	**3. *membranaceum***
3(2)	Thallus without lobes	4
	Thallus obscurely lobed	**4. *vouauxii***
4(3)	Thallus Pd+ reddish orange; thallus powdery	**2. *diffusum***
	Thallus Pd− or +yellow; thallus powdery to granular	**1. *cacuminum***

1. L. cacuminum (Massal.) Laundon (1992)

L. angardianum (Øvstedal) Laundon (1989)

Thallus crustose-leprose, diffuse, forming a thick, non-areolate crust of powdery to non-powdery granules, yellowish-white to grey; surface not corticate, a mass of coarse, spherical granules to 0.4 mm diam., or eroded to leave a leprose membrane, the margin absent, diffuse, undelimited to delimited, without lobes; medulla white; underside with a weft of loosely entanged hyphae forming the early development of a hypothallus, whitish grey. Thallus Pd− or + yellow, K+ yellow, C− (atranorin, ± porphyrilic acid and fatty acids).

On especially acid, mossy rocks, also acid bark, soil and other lichens. Temperate-arctic, N. & S. Hemispheres.

This species is variable, some morphs being whitish and powdery, whilst others are grey and granular; the latter morph is similar in appearance to *Lepraria neglecta* which has a different chemistry.

2. L. diffusum Laundon (1989)

Thallus (Fig. 36a) crustose-leprose, diffuse, forming a thick, non-areolate crust of powdery granules, either greenish white [var. *diffusum*] or greenish yellow [**var. *chrysodetoides*** Laundon (1989)], surface not corticate, a mass of powdery, spherical granules to 0.3 mm diam., not marginate, diffuse, usually undelimited but occasionally delimited, without lobes; medulla white, sometimes exposed in places; underside with a weft of loosely entangled hyphae forming the early development of a hypothallus, whitish grey to brownish. Thallus C+ yellow (var. *diffusum*) or C− (var. *chrysodetoides*), K− to + yellow (var. *diffusum*), or K+ slowly orange (var. *chrysodetoides*), Pd+ brownish red (both varieties have oxypannaric acid, sometimes with a trace of pannaric acid).

On mosses over calcareous and acid stone in shaded to moderately illuminated situations; rare. S.E. & N. England, Wales, Scotland, N. Ireland. Europe, W. North America. Var. *chrysodetoides* is known from mosses on rocks in Scotland and Wales, and from moderately calcareous rocks in Sweden.

The identity of var. *diffusum* should always be confirmed by t.l.c., but var. *chrysodetoides* is easily distinguished by the greenish yellow colour, which becomes orange-yellow in places, distinct from the ± uncoloured var. *diffusum*.

3. L. membranaceum (Dickson) Vainio (1899)

Lepraria membranacea (Dickson) Vainio (1921)

Thallus of powdery lobes, forming irregular rosettes, yellowish white; apices of lobes rounded, ± flat, surface not corticate, of numerous powdery convex granules, the margin either flat or raised, 1–17 mm wide, older lobes often disintegrate to powdery convex granules to 0.5 mm diam., especially at the centre of the thallus, or eroded to leave a leprose membrane; medulla whitish; underside a continuous, well-developed weft of branched hyphae forming a hypothallus, whitish grey to brownish. Apothecia to 2 mm diam., very rare, the sorediate thalline exciple 0.5 mm thick. Thallus Pd+ reddish orange, K− to + yellow, C− (pannaric and roccellic acids, atranorin occurring in about half the populations).

On shaded, vertical, acid rocks and walls, on the surface and over mosses, occasionally colonizing mosses on shaded tree trunks, especially *Quercus*; acidophilous. Upland Britain, extending locally to C. & W. England (Leicester, Charnwood Forest; Worcester, Malvern Hills). Upland Europe, mountains of E. Africa, North and South America, Australia.

Young specimens are recognized by their rounded, powdery squamules with a rolled ridge at the margin, recalling *Normandina*. The yellowish white colour is characteristic. Thalli with apothecia were last collected in the 18th century.

4. L. vouauxii (Hue) Laundon (1989)

Thallus of powdery lobes, forming irregular rosettes, to irregularly delimited, small, leprose crusts, whitish to pale yellowish to greenish grey; surface not corticate, a mass of powdery, convex granules, which are sometimes eroded to leave a leprose membrane, the margin obscurely lobed; medulla white, exposed in places on most specimens; underside a weft of loosely entangled hyphae forming the early development of a hypothallus, whitish grey to brownish. Thallus Pd− to + reddish orange, K−, C− (pannaric acid 6-methyl ester, ± pannaric acid, rarely atranorin).

On stone and bark, in shaded situations, favouring trees such as *Fraxinus* and *Ulmus* with a nutrient-rich bark, in sheltered woodland and parks, on bare surfaces, and mosses and liverworts, also commonly on soil abroad; common, but scarce on mountain rocks. Mainly lowland Britain. Europe, Greenland, E. & S. Africa, South America, Australia, New Zealand.

Variable in morphology and colour, resembling a poorly developed morph of *L. membranaceum* **(3)**, for which it has often been mistaken, but that species is distinctly lobed, the medulla is inconspicuous, the hypothallus is well developed, and always contains pannaric and roccellic acids.

LEPROPLACA (Nyl.) Harm. (1900)

Thallus leprose, non-corticate, sometimes with ± distinct marginal lobes. Thallus surface of powdery, yellowish or dull orange, convex granules. **Medulla** white. **Prothallus** and **hypothallus** absent. **Photobiont** species of the Chlorophyta. **Conidiomata** and **ascomata** absent. **Chemistry**: anthraquinones, parietin; K+ dark violet-red. **Ecology**: on shaded calcareous rocks, especially limestone.

The genus is derived from *Caloplaca* by reduction, in which the cortex and apothecia have been lost during evolution and adaptation to shaded habitats. Apothecia have reportedly been discovered in *L. xantholyta* and the genus has therefore been reunited with *Caloplaca*. The species resemble *Chrysothrix*, but differ in their chemistry. Often forms extensive patches on vertical surfaces.

LITERATURE: Laundon (1974).

1 Thallus dull brownish yellow, without marginal lobes .... **1. *chrysodeta***
Thallus bright yellow, with ± distinct marginal lobes .... **2. *xantholyta***

1. L. chrysodeta (Vainio ex Räsänen) Laundon (1974)

Thallus of powdery globose granules, *c.* 0.1 mm diam.; when well developed producing a thick, areolate crust, dull orange-, brownish to greyish yellow, effuse, without marginal lobes.

Limestone cliffs and other dry, shaded, calcareous rocks, often overgrowing mosses, occasionally extending to the mortar of shaded walls; common. Throughout British Is. N. Europe, Middle East (Jordan), N. & S. America, New Zealand.

Easily identified by the dull orange-yellow, powdery, globose granules in shaded, calcareous habitats. Sterile *Caloplaca citrina* is sometimes similar, but has brighter, convex (not globose) soredia, and grows on smooth, illuminated calcareous surfaces. The correct name in *Caloplaca* for this species is *C. chrysodeta* (Vainio ex Räsänen) Dombr. (1970)

2. L. xantholyta (Nyl.) Harm. (1900)

Thallus leprose, ± orbicular with ± distinct marginal lobes when well developed, with a flat powdery surface or of convex granules, yellow; medulla white, often exposed in places; delimited by radiating, flat to slightly convex and non-corticate marginal lobes.

On damp, limestone cliffs and limestone walls; locally abundant in uplands, rare in lowlands. Throughout Great Britain. Europe, Middle East (Israel), New Zealand.

Readily distinguished by its yellow, powdery thallus with marginal lobes which react K+ violet-red. The correct name for this species in *Caloplaca* is *C. xantholyta* (Nyl.) Jatta (1902).

LEPTOGIUM (Ach.) Gray (1821)

Thallus crustose, squamulose, foliose or minutely shrubby, ± gelatinous, blue-grey to brown or blackish: not layered in section, cortex of a single row of isodiametric to shortly rectangular cells; inner part of thallus with loosely interwoven hyphae, or (in sect. *Homodium*) pseudoparenchymatous throughout. **Upper surface** smooth to wrinkled or ridged, often glossy, rarely arachnoid. **Lower surface** smooth, arachnoid or hairy, sometimes with scattered hapters or groups of white rhizines. **Isidia** often present. **Photobiont** *Nostoc*, cells mostly arranged in distinct chains. **Ascomata** apothecia, sessile to shortly stalked, mainly laminal. **Thalline exciple** usually becoming excluded. **True exciple** raised, cupular, usually pseudoparenchymatous, colourless, ± reddish brown. **Disc** concave to flat. **Epithecium** colourless to reddish brown, K−, N−. **Hymenium** colourless, I+ blue. **Hypothecium** shallow, colourless or pale yellowish. **Hamathecium** of paraphyses, numerous, conglutinate, simple, apices ± swollen. **Asci** (4-)8-spored, clavate; K/I wall + blue, and apical dome + pale blue with a dark blue axial tube. **Ascospores** ellipsoid, ovoid or fusiform, often apiculate at one or both ends, septate, usually submuriform to muriform, colourless, without any distinct surface ornamentation or perispore. **Conidiomata** pycnidia, laminal or marginal, globose, ± immersed; wall colourless. **Conidiogenous cells** slender, cylindric. **Conidia** bacilliform, simple, colourless. **Chemistry:** no lichen substances detected by t.l.c., unidentified substances reported by g.l.c. **Ecology**: on various, often moist or periodically inundated or flushed, mostly basic to highly calcareous rocks and bark; *c.* 130 species, cosmopolitan.

The corticate thallus will usually distinguish species of this genus from *Collema* and sterile specimens of *Lempholemma*.

The name *Lempholemma minutissimum* (Flörke) Fr. (1846) has been used for several smaller species, and its correct application is at present uncertain.

With the exception of the array of morphs here included under *L. gelatinosum*, most of the larger species are well circumscribed. However, a critical revision of the small species is required. These comprise two groups: (1) sect. *Collemodium* - *L. massiliense*, *L. plicatile*, *L. schraderi* and *L. turgidum*; (2) sect. *Homodium* - *L. byssinum* s.lat., *L. diffractum*, *L. subtile*, *L. tenuissimum* and *L. teretiusculum*.

LITERATURE: Jørgensen & James (1983), Sierk (1964).

1 Thallus crustose, minutely squamulose, minutely foliose, or minutely shrubby; lobes <1 mm wide 2
  Thallus distinctly foliose; lobes >2 mm wide 10

2(1) Thallus squamulose, foliose or minutely shrubby 3
  Thallus crustose, granular or granular-areolate .... **4. *byssinum***

3(2) Thallus lobes either not radiating, or if radiating then >0.5 mm wide or not on hard limestone 4
  Thallus margin with radiating, contiguous, adpressed lobes, <0.5 mm wide, on hard limestone .... **8. *diffractum***

4(3) Lobes or squamules distinctly wrinkled when dry; centre of thallus (in section) with loosely interwoven hyphae 5
Lobes or squamules ± smooth; centre of thallus (in section) pseudoparenchymatous throughout 8

5(4) Lobes entirely cylindrical 6
Lobes flattened or nodulose, occasionally with short cylindrical extensions 7

6(5) Lobes horizontally spreading, adpressed .... **14. *massiliense***
Lobes erect, crowded, forming cushions .... **17. *schraderi***

7(5) Lobes distinct, 1–2(-3) mm wide .... **15. *plicatile***
Lobes indistinct, nodulose and crowded, <1 mm wide .... **21. *turgidum***

8(4) Thallus of rounded or deeply divided squamules or lobes; apothecia usually present 9
Thallus mainly of erect, crowded, minute cylindrical to coralloid branchlets; marginal squamules elongate and adpressed or absent; apothecia rarely present ... **20. *teretiusculum***

9(8) Lobes to 0.3 mm wide, often forming a rosette around each apothecium, not forming cushions; apothecia 0.2–0.5 mm diam. .... **18. *subtile***
Lobes to 2 mm wide, mostly imbricate and deeply divided, and forming cushions; apothecia to 0.8 mm diam. .... **19. *tenuissimum***

10(1) Upper or lower surface of lobes with fine white-grey tomentum 11
Upper or lower surface of lobes without tomentum 15

11(10) Upper surface not tomentose; lower surface with a fine, pubescent tomentum, occasionally forming long, coarse rhizines 12
Upper surface with a grey-white, arachnoid tomentum; lower surface ± naked but with localized tomentum in areas of attachment to surface .... **12. *juressianum***

12(11) Isidia present 13
Isidia absent 14

13(12) Upper surface dark olive-black, smooth; uniformly isidiate .... **16. *saturninum***
Upper surface bluish grey, strongly wrinkled-striate; patchily isidiate .... **10. *hibernicum***

14(12) Thallus forming compact, multi-lobed rosettes, often ± convex when mature; apothecia with abundant marginal folioles .... **3. *burgessii***
Thallus forming loose, wide-spreading rounded patches; apothecia without marginal folioles .... **11. *hildenbrandii***

15(10) Lobes only slightly swollen when wet, smooth or striate when dry 16
Lobes markedly swollen when wet, thin, markedly ridged and wrinkled when dry **1. *brebissonii***

16(15) Lobes thin, paper-like; isidia absent, or (if present) flattened or squamule-like 17
Lobes thick, tough, cartilaginous; a few coarse granular laminal isidia often present **15. *plicatile***

17(16) Margin of lobes ± lacerate-fimbriate; or surface with flattened isidia 18
Margins of lobes entire; isidia absent 21

18(17) Lobes often >2 mm wide; centre of thallus (in section) with interwoven hyphae 19
Lobes <2 mm wide; centre of thallus (in section) pseudoparenchymatous throughout .... **19. *tenuissimum***

19(18) Thallus brown-grey, ± wrinkled, the lobe margins deeply divided; laminal isidia few or absent 20
Thallus blue-grey, smooth with laminal, terete, or flattened and often overlapping isidia .... **7. *cyanescens***

20(19) Lobe margins with branched mostly cylindrical, isidia-like outgrowths which are sometimes also on the surface; lobe surface, especially below, with raised, narrow, ± vertically orientated rib-like wrinkles. Apothecia margins isidiate .... **13. *lichenoides***
Lobes margins deeply divided with flattened outgrowths which are never laminal; lobe surface smooth to wrinkled, rarely with a few raised ribs. Apothecia margins smooth .... **9. *gelatinosum***

21(17) Lobes narrow, <5 mm diam. 22
Lobes wide, >5 mm diam. 23

22(21) Lobes ± erect, the margins revolute forming tube-like structures; apothecia absent .... **6. *corniculatum***
Lobes decumbent with ascending margins which do not form tube-like structures; apothecia frequent .... **9. *gelatinosum***

23(21) Lobes matt, pale grey, very thin, transversely wrinkled; apothecia rare with short, pale stalks .... **2. *britannicum***
Lobes shiny, dark grey to greenish blue, rather thick, striate; apothecia frequent, ± sessile .... **5. *cochleatum***

1. L. brebissonii Mont. (1840)

Thallus forming conspicuous tufts 1–5 cm wide, when moist pulpose, with markedly ridged and swollen, indistinct lobes, shrinking to an irregular, flattened, often markedly ridged thallus when dry; upper surface dark green-black when wet, becoming grey-black when dry, strongly rugose and uneven; lower surface similar but paler, without tomentum; lobes partly fenestrate, ridges towards margin ± isidiate; isidia cylindrical or flattened. Apothecia not observed in British material.

On deciduous trees and mossy rocks; local. W. British Is., especially W. Scotland, W. Ireland. W. Europe, Macaronesia, E. Africa, New Zealand.

When well developed easily identified by the dark swollen thallus, the lobes with ridges and isidia. *L. hibernicum* **(10)** usually has better defined lobes which are finely tomentose below.

2. L. britannicum P. Jørg. & P. James (1983)

Thallus to 10 cm diam., in loose tufts or as irregularly dispersed lobes; lobes (0.3-)0.5–1.0(-1.5)mm wide × 1.0–1.5(-3.5) mm long, irregular, crinkled, very thin, 50–70 μm thick, tissue-paper-like, often ± ascending, often contorted, ± revolute at the margins; upper surface pale grey when wet, becoming pale grey-blue when dry, wavy, transversely wrinkled, bullate-honeycombed, convoluted, matt, ± smooth, finely striate; lower surface slightly paler than upper; rhizines and tomentum absent. Apothecia very rare, laminal; disc (0.5-)0.8–1.2(-1.5) mm diam., ± top-shaped with a distinct, short, pale, longitudinally furrowed stalk; disc at first concave, later becoming flat or convex, red-brown or brown; thalline exciple distinct, whitish brown to pale buff, soon becoming ± occluded; true exciple of narrow, periclinal hyphae. Ascospores (18-)20–30(-34) × (5-)6–7(-8) μm, 3- to 4-septate, or submuriform with 1 longitudinal septum, elongate-clavate to ellipsoid, apices ± pointed.

On ± sheltered coasts, in pockets of bare earth, short grass and low vegetation amongst rocks, wall tops; local. W. British Is., Channel Is. France (Brittany), Faeroe Is.

Characterized by the very thin lobes, absence of isidia, the rare occurrence of apothecia and restriction to coastal areas. Rather similar to *L. cyanescens* **(7)** which is, however, isidiate.

3. L. burgessii (L.) Mont. (1840)

Thallus to 10 cm diam., forming rounded tufts or rosettes of numerous, rounded or wavy lobes, often raised in centre; lobes 3–5(-8) mm wide, much convoluted, overlapping, crisped, with abundant folioles arising from margins and surface; upper surface blackish brown or blackish olive-green; lower surface uniformly shortly grey tomentose, paler than upper surface. Apothecia fairly frequent; disc 1–3 mm diam., red-brown to blackish brown, concave; thalline exciple thick with abundant folioles. Ascospores 30–40 × 13–17 μm, 5- to 7-septate, ellipsoid or ± fusiform, muriform.

On mossy trunks, especially of *Corylus* and *Fraxinus*, also on mossy rocks, in sheltered moist woodlands, often by lake-sides or on mossy rocks in valley bottoms; locally abundant. W. British Is. W. Europe, Azores, E. United States, Ecuador, E. Africa, S. India, New Zealand.

The number of secondary folioles is variable. Distinguished from *L. cyanescens* **(7)** and *L. gelatinosum* **(9)** by the thin, even, pubescent tomentum on the lower surface and by the thalline exciple with abundant folioles.

4. L. byssinum (Hoffm.) Zwackh ex Nyl. (1856)

L. biatorinum (Nyl.) Leighton (1879), *L. cretaceum* (Sm.) Nyl. (1855)

Thallus 0.5–5 cm diam., blue-grey to brown-black, initially of minute, rosette-like, adpressed squamules, 0.3–0.5(-0.8) mm diam., with crenulate to granular margins; granules 0.06–0.1 mm diam., often abundantly produced, soon obscuring the squamules and producing a greenish brown to blackish, areolate, granular crust; centre of squamules and granules pseudoparenchymatous throughout. Apothecia 0.2–0.6 mm diam., frequent, immersed among the thallus granules or becoming sessile; thalline margin smooth or with a few granules. Ascospores (18-)24–36 × (8-)12–16(-17) μm, muriform, with (3-)4–5(-7) transverse septa.

On calcareous or clayey soils, or on stones of soft limestone (especially chalk and oolite), rarely on old mortar; scarce. S. England, extending locally to S.E. Scotland (Angus). Sweden, C. & S. Europe, Siberia, North America.

An inconspicuous species distinguished by the granular thallus; the primary squamules are often obscured, especially when growing on soil. More than one species may be involved and a critical revision of this species group is required. Careful anatomical comparisons are required to separate sterile material from that of *Moelleropsis nebulosa.*

5. L. cochleatum (Dickson) P. Jørg. & P. James (1983)

Thallus forming extensive patches, to 10 cm diam.; lobes to 1 cm wide, 100–150 μm thick, numerous, wavy, overlapping, margins ± ascending, rounded, entire, intricately folded, particularly towards centre; upper surface dark green-grey, matt, ± distinctly finely striate, not wrinkled. Apothecia to 3 mm diam., usually present, sessile; thalline exciple thin, persistent, pale, ± wrinkled; true exciple parenchymatous. Ascospores (20-)25–30(-37) × (12-)14–16(-17) μm, faintly ornamented, muriform.

On sheltered trunks of old deciduous trees, particularly *Fraxinus* and *Corylus*, especially near streams, also on mossy rocks; local and rare. N.W. England (Cumbria), W. Scotland, S.W. Ireland. W. Europe, Azores, E. Africa, Indian subcontinent.

When well developed this species forms elegant, spreading patches with numerous, overlapping lobes with raised, entire margins and studded with very regular, sessile apothecia. When juvenile or less well developed, it forms more compacted, sterile tufts of ± ascending, intermingled, curled lobes; isidia are never present.

6. L. corniculatum (Hoffm.) Minks (1873)

L. palmatum (Huds.) Mont. (1846)

Like *L. gelatinosum* **(9)**, but lobes larger, to 5 mm wide, the margins characteristically revolute forming tube-like structures; upper surface blackish olive-green, without a reddish tinge, smooth or wrinkled. Apothecia unknown in British Is.

On mosses amongst boulders and on the ground in old dunes, rarely on tree trunks; scattered and rare. Mainly W. & N. British Is. Coastal Europe, W. North America, Japan.

Characterized by the revolute margins of the lobes which curl inwards becoming ± tubular, especially towards the apices.

7. L. cyanescens (Rabenh.) Körber (1855)

Thallus to 10 cm diam., forming irregularly spreading patches or small tufts; lobes to 1 cm diam., ± overlapping, rather thin; isidia abundant, laminal and sometimes also marginal, cylindrical or flattened; folioles occasionally present; upper surface pale bluish grey, smooth; lower surface pale grey, naked. Apothecia rare, sessile from the beginning; disc to 2 mm diam., red-brown; true exciple paler, of narrow, periclinal hyphae. Ascospores (15-)19–23(-25) × (6-)7–9(-10) μm, submuriform.

In old woodlands, on mossy trees and rocks or directly on rock when particularly near water; locally frequent. W. Britain, especially W. Scotland, W. Ireland. Cosmopolitan in temperate and subtropical regions.

Separated from *L. cochleatum* **(5)** by the thinner thallus with abundant, flattened or cylindrical isidia, the scarcity of apothecia, and the pale bluish grey colour. Young specimens may be distinguished from *L. burgessii* **(3)** by the presence of isidia and the smooth lower surface.

8. L. diffractum Krempelh. ex Körber (1865)

Thallus 0.5–1.0 cm diam., or confluent and forming larger patches, with adpressed, radiating, marginal lobes; lobes 0.4–1.1 mm long and 0.2–0.5 mm wide, convex or flat, often of about equal length, ± contiguous for most of their length; surface matt but never pubescent, often ± wrinkled and ± glossy at the lobe ends, brown-olive; middle of the thallus with convex areoles or granules, 0.1–0.2(-0.3) mm diam., or middle part falling away leaving areas of radiating marginal lobes. Apothecia rare; disc 0.2–0.5 mm diam. Ascospores 15–30 × 8–12 μm, ellipsoid, ± muriform.

On hard limestones. S.W. England, C. Wales. Sweden, Italy.

The radiating marginal lobes are reminiscent of several *Placynthium* species, also of *Collema parvum*.

9. L. gelatinosum (With.) Laundon (1984)

L. sinuatum (Huds.) Massal. (1853)

Thallus very variable, of numerous, overlapping or ± erect, convoluted, rounded lobes, forming compact tufts to 8 cm diam.; lobes 1–3(-5) mm wide, with ± erect, entire, crenate or divided margins; upper surface dark brown or reddish brown, becoming greyish in sheltered situations, coarsely ridged, pleated and roughened. Apothecia frequent; disc to 2 mm diam., smooth, concave or flat; true exciple raised, concolorous, non-isidiate. Ascospores 22–42 × 11–17 μm, muriform, with 5–7(-9) transverse septa.

Amongst mosses on ± basic to highly calcareous substrata including limestone soils, mortar, rarely at base of old trees and on mosses on flat chest tombs; locally abundant. W. British Is., extending locally eastwards. Europe, North America, India.

Probably the commonest British species of *Leptogium*. Very polymorphic and much confused with *L. lichenoides* **(13)**, particularly where the lobes are small and very richly divided. The marginal outgrowths or extensions are always flattened, never cylindrical as in *L. lichenoides*, and the lower surface of the lobes is usually less markedly ridged.

10. L. hibernicum Mitchell ex P. Jørg. (1973)

Thallus to 5 cm diam., becoming markedly swollen when wet; lobes 0.5–1 cm, overlapping with ± raised, rounded or partly dissected margins; upper surface brownish or bluish grey, ± transversely wrinkled-striate; isidia nodular, in patches, coarse, or lobules developing along margins and on the surface; lower surface paler, uniformly finely pubescent tomentose; tomental hairs *c.* 20 μm long with globose cells. Apothecia not known in British material.

On trunks, especially of *Corylus* and *Fraxinus*, in sheltered, moist, old woodlands, more rarely on wayside trees in sheltered valleys; local. W. Scotland, W. Ireland. W. Europe, E. & S. Africa, S. Chile.

Characterized by the coarse striations often running in one direction, the warted upper surface with nodular isidia and tomentose lower surface.

11. L. hildenbrandii (Garov.) Nyl. (1856)

Thallus large, to 6 cm diam., often forming rounded patches; lobes rounded, to 0.6 mm, partly overlapping, rather thick; upper surface deep blue- or black-grey, wrinkled, minutely irregularly striate; lower surface with a thick, white tomentum and tassels of rhizines. Apothecia laminal, often numerous and contiguous, ± immersed when young, becoming emergent; thalline exciple thin; disc 2–3 mm diam., red-brown.

On deciduous trees; apparently extinct. C. Scotland (E. Perth). C. & S. Europe, China, Japan.

Differs from *L. hibernicum* **(10)** and *L. saturninum* **(16)** which are isidiate, and rarely fertile. *L. burgessii* **(3)** has numerous, thinner lobes and the exciple conspicuous folioles.

12. L. juressianum C. Tav. (1950)

Thallus to 4 cm diam., thin, forming cushions, loosely attached; lobes to 5 mm diam., rounded, ± spreading, margins distinctly revolute, mostly entire; upper surface blue-grey, often brown-tinged, matt, partially covered by thin, dense, grey-white, arachnoid tomental hyphae with cylindrical cells; isidia marginal or laminal, granular to sublobulate; lower surface paler, ± naked but with a localized tomentum in areas of attachment to surface. Apothecia and pycnidia unknown.

On thick moss cushions, particularly those of *Isothecium myosuroides* on very shaded deciduous trees in old woodlands, rarely on mossy rocks; rare. S.W. Ireland (Glengarrif, Killarney). Portugal, Iberia, Kenya, Chile.

Characterized by the revolute margin, granular isidia and grey-white tomentum of the upper surface. Differs from *L. lichenoides* **(13)**, in which the lobes have ± ascending margins which are ± finely denticulate to fimbriate, and the upper surface is shiny.

13. L. lichenoides (L.) Zahlbr. (1924)

Like *L. gelatinosum* **(9)**, but the margins and sometimes the surface of lobes with small, simple or branched, cylindrical isidia-like extensions which may be very densely produced giving the appearance of fibrillose margins; upper surface smooth to wrinkled or with a few ridges; lower surface usually conspicuously ridged, often ± vertically orientated ribs, sometimes arachnoid towards the base. Apothecia 0.3–1 mm diam., rare ; thalline exciple isidiate. Ascospores (26-)33–45(-50) × 12–15 μm, muriform, with (5-)7(-9) transverse septa.

Amongst mosses particularly at the base of old trees, especially *Fraxinus*, also on rocks, walls and soil in rather damp situations, predominantly in old woodlands. Throughout British Is. Europe, North America.

Less frequent than *L. gelatinosum* **(9)** and much rarer in highly calcareous situations. Often confused with morphs of *L. gelatinosum* which have much divided lobe margins.

14. L. massiliense Nyl. (1879)

Thallus to 1 cm diam., forming ± circular patches, or irregularly spreading, closely adpressed; lobes 0.1–0.2(-0.3) mm wide, cylindrical and ± horizontally radiating, variously branched, often with nodular isidia-like extensions along lobes; upper surface pale grey-brown, often furrowed and irregularly ridged. Apothecia not observed.

On moist or shaded limestone, including loose stones, or on flat siliceous rocks at edge of mesotrophic rivers; rare. W. England (Somerset, Cumbria). C. & S. Europe.

Distinguished from *L. schraderi* **(17)** by the fan-like, radiating and closely adpressed thallus.

15. L. plicatile (Ach.) Leighton (1879)

Thallus to 5 cm diam., very tough, rigid, forming closely adpressed rosettes; lobes to 2(-3) mm wide, with ascending, twisted, often divided and ± pleated margins; upper surface dark brown-black with a reddish tinge, uneven, often ridged, sometimes with coarse, clustered, isidia-like warts; lower surface paler. Apothecia marginal or laminal, markedly concave when young; thalline exciple persistent, raised; disc 1–1.5 mm diam. Ascospores 18–30 × 8–16 μm.

On hard limestone rocks, usually in damp situations, especially in seepage tracks, more rarely on siliceous rocks influenced by calcareous seepage or by mesotrophic rivers, occasionally on chest tombs and butresses in churchyards. Throughout British Is. Europe, U.S.A., New Zealand.

Typically consists of rather ragged and scattered tufts with raised, convoluted and ± cartilaginous lobes firmly attached to rocks. May be confused with *Collema auriforme*, which lacks a cellular cortex, does not remain tough when wet and is blackish olive-green in colour. Small and immature specimens can be difficult to distinguish from *L. turgidum* **(21)**.

16. L. saturninum (Dickson) Nyl. (1856)

Like *L. hibernicum* **(10)** but lobe ends more regularly rounded, surface even, not wrinkled and densely covered in small globose, cylindrical or branched isidia; thallus to 8 cm across, black-brown; lobes 3–10(-15) mm wide; lower surface with white-grey tomentum with a few long white rhizines; tomental hairs to *c.* 100 μm long with cylindrical cells, 6–12 × 4–4.5 μm. Apothecia unknown in Britain.

On bark of old trees and on mossy, calcareous rocks; rare. C. & N. Scotland. Europe, North America, N. Asia.

Collema subflaccidum is similar, but has a smooth lower surface and lacks a distinct upper cortex.

17. L. schraderi (Ach.) Nyl. (1856)

Thallus to 5 mm tall, forming small, upright tufts; lobes to (0.1)0.3–0.6 mm wide, cylindrical, ascending, often divided towards base, distinctly wrinkled and pleated, glossy, a few globular isidia ± present. Apothecia occasional; disc to 1.5 mm diam. Ascospores 23–33 × 10–12 μm.

On mosses or soil in calcareous habitats, particularly on old mortar walls, over limestones and in short, calcareous turf; locally abundant. Throughout British Is. Baltic Is., C. & S. Europe.

Apparently intermediate morphs exist with *L. turgidum* **(21)**, the latter having more irregularly shaped lobes and often more numerous, irregular, globular isidia.

18. L. subtile (Schrader) Torss. (1843)

Thallus minutely foliose, blue-grey to brown; lobes very small, 0.1–0.3 mm wide, ± horizontal, often arranged in rosettes, 0.5–1.5 mm diam., below a single or small group of apothecia, apices of lobes often dividing into short, flattened to terete extensions, to 0.3 mm long and 0.06–0.1 mm wide; centre of thallus pseudoparenchymatous throughout. Apothecia 0.2–0.5 mm diam., usually abundant; thalline exciple mostly persistent, smooth to nodulose. Ascospores 20–31 × 10–12 μm, muriform, with (3-)5(-7) transverse septa.

On basic bark of old, usually fallen, trees, especially *Fraxinus* and *Ulmus*, old stumps, and on old polypores and plant debris on the ground; local. Throughout British Is. Europe.

Differs from *L. teretiusculum* **(20)**, in the shorter, rosette-forming lobes with only a few or shorter marginal extensions, and numerous apothecia. *L. tenuissimum* **(19)** has larger lobes, and scattered, larger apothecia.

19. L. tenuissimum (Dickson) Körber (1855)

L. humosum Nyl. (1858)

Thallus of spreading to overlapping lobes, forming blue-grey to brown cushions or mats; lobes to 2 mm wide, thin, elongate, deeply incised or crenate, with clustered coralloid outgrowths, surface ± smooth, olive- or brown-green; centre of thallus pseudoparenchymatous throughout. Apothecia to 0.8 mm diam., frequent; disc concave to flat, reddish brown; thalline exciple thick, often with attached lobules. Ascospores 20–35 × 9–12 μm, muriform, with 3–5(-7) transverse septa.

Amongst mosses on the ground; local. Throughout British Is. Scandinavia, Arctic, North America.

Very inconspicuous, rarely collected, and much over-recorded owing to confusion with diminutive morphs of *L. gelatinosum* **(9)** or *L. lichenoides* **(13)**; both these species have interwoven hyphae in the centre of their thalli. The larger lobes and apothecia distinguish this species from *L. subtile* **(18)**.

20. L. teretiusculum (Wallr.) Arnold (1892)

Thallus brown or (in shade) grey, at first with small, radiating, narrow, ± branched, flat, adpressed lobes 0.3–1 × 0.1–0.2 mm, which at their margins and apices develop long, cylindrical, often coralloid isidia-like extensions, 30–70(-100) μm diam. and to 0.6(-1) mm long, that become crowded to form dense clusters or cushions; at maturity, the primary squamules are mostly obscured except sometimes at the edge of the cushions; thallus pseudoparenchymatous throughout. Apothecia very rare, red-brown, urceolate; thalline exciple thick. Ascospores 20–25 × 10 μm, muriform with 3–5 transverse septa.

On rough, basic bark of old trees, especially *Acer*, *Fraxinus* and *Ulmus*, more rarely on basic rocks, old walls, and coastal turf, frequent but often overlooked. Throughout British Is., especially the S. & S.W.

Close to *L. subtile* **(18)** which has less elongated lobes with few, if any, isidia-like extensions, and numerous apothecia. Morphs with very long 'isidia' can resemble the cyanobacterial morphotype of *Lobaria amplissima*.

21. L. turgidum (Ach.) Crombie (1870)

Thallus thickish, lobate; lobes small, (0.1-)0.2–1 mm wide and to 2.5 mm long, turgid, sometimes ascending and appearing shrubby; upper surface dark olive- or red-black, wrinkled to plicate, concave at the circumference, often densely granular-isidiate; isidia 0.06–0.1 mm diam. Apothecia to 0.5–3 mm diam., absent to numerous; disc concave, becoming ± flat; thalline exciple swollen, entire, or sometimes slightly granular. Ascospores 23–32 × 10–12 μm, 3-septate to muriform.

On calcareous walls, especially crumbly mortar and soils; local. Mainly S. & E. Britain. Europe.

The status of the specimens included here requires critical study as apparently intermediate morphs exist between *L. plicatile* **(15)**, with larger lobes, and *L. schraderi* **(17)**, with more regular cylindrical lobes.

LEPTORHAPHIS Körber (1855)

Thallus absent, or immersed, whitish grey to greenish, smooth, continuous, not well-delimited. **Photobiont** absent; sometimes loosely associated with trentepohlioid algae. **Ascomata**

perithecioid, scattered singly, sometimes confluent, circular to ellipsoidal, immersed but becoming superficial, hemispherical to mammiform, ostiolate. **Involucrellum** dark brown to greenish black, the hyphae not changing colour in K, clypeate, forming a *textura epidermoidea* to *intricata*; radiating hyphae sometimes extending to form a basal fringe. **True exciple** pseudoparenchymatous, colourless to pale brown, developed within the involucrellum, inconspicuous or not continuous below the asci. **Hamathecium** of cellular pseudoparaphyses, branched and anastomosed; periphyses and periphysoids absent; hymenial gelatin I+ yellowish orange to bluish green, never deep blue. **Asci** cylindrical-clavate, bitunicate, with a broad and truncate, exceptionally shortly pointed, apex, fissitunicate, 8(-16)-spored. **Ascospores** acicular-fusiform, arcuate or sigmoid, colourless, 1- to 3-septate (in British species), smooth, thin-walled. **Conidiomata** pycnidial, immersed to semi-superficial, hemispherical to spherical. **Conidiogenous cells** enteroblastic, ampulliform to lageniform. **Macroconidia** colourless, acicular-fusiform, arcuate, smooth, thin-walled, somewhat reminiscent of the ascospores. **Microconidia** colourless, bacilliform. **Chemistry**: no lichen products detected by t.l.c. **Ecology**: 12 species on bark of deciduous trees in the Northern Hemisphere; some ± host specific.

Essentially bark saprobes, but algal cells are occasionally seen on the surface of the bark near the ascomata in fresh collections. Exceptionally, as in *L. atomaria* **(1)**, differentiated thalli with intimate fungal-algal associations are observed on the surface and in vertical section. Distinguished from other pyrenocarpous genera by the acicular-fusiform ascospores. *Celothelium* and *Rhaphidicyrtis*, previously included in *Leptorhaphis*, have thread-like ascospores, differ in ascomatal wall structure, hamathecial tissues and ascus morphology.

LITERATURE: Aguirre-Hudson (1991), Aguirre & Hawksworth (1987).

1 Ascomata 0.13–0.3 mm diam., somewhat circular, not surrounded by a basal fringe; hymenial gelatin I− or bluish; at least some ascospores 3-septate when mature 2
Ascomata 0.3–0.45 mm diam., ellipsoidal, surrounded by a dark basal fringe, *c.* 100 μm diam.; hymenial gelatin I+ bluish; ascospores 1-septate when mature .... **2. *epidermidis***

2(1) Thallus inconspicuous; hymenial gelatin I+ yellowish; ascospores (25-)30–45 × 1.5–2.5 μm, the apices attenuated; pycnidia present .... **3. *maggiana***
Thallus whitish-grey; hymenial gelatin I+ bluish; ascospores (20-)25–35 × 2–3.5 μm, the apices rounded; pycnidia absent .... **1. *atomaria***

1. L. atomaria (Ach.) Szat. (1928)

Thallus immersed, whitish grey, smooth, continuous, not well-delimited, the upper levels associated with trentepohlioid algae. Perithecia 0.13–0.2 mm diam., blackish, circular, scattered singly, superficial, sometimes dark brown below; ostiole central; involucrellum greenish brown, of leptodermatous cells; true exciple colourless to dark brown, continuing below the hymenium; hymenial gelatin I+ bluish. Asci (35-)40–55 × 10–15 μm, short-clavate, broadly stalked, (6-)8-spored, the apex broad, truncate. Ascospores (20-)25–32 × 2–3.5 μm, fusiform, curved, sometimes sigmoid, 1- to 3-septate, the apices rounded. Conidiomata unknown.

On bark of *Fraxinus* and, especially, *Populus*, often in nutrient-rich communities, probably only partly lichenized; rare. S. Wales (Dyfed), Scotland (Highlands). Temperate and boreal Europe.

Two other species occur on *Populus* bark outside the British Is.: *L. lucida* Körber (1865), with ascospores 45–70 × 2.5–3 μm and 8-spored asci, and *L. tremulae* Körber (1855), with ascospores 13–25.5 × 3–4.5 μm, and 8- to 16-spored asci.

2. L. epidermidis (Ach.) Th. Fr. (1860)

Thallus immersed, generally absent, occasionally with trentepohlioid algae in patches around the ascomata. Perithecia 0.3–0.45 mm diam., circular to ellipsoid, immersed but becoming superficial; ostiole central; involucrellum dark brown, extending to form a dark basal fringe; true exciple colourless to pale brown, continuing below the ascoma; hymenial gelatin I+ orange-amber. Asci 40–65(-70) × 10.5–15.5 μm, short-clavate, broadly stalked, 8-spored, the apex broad and truncate. Ascospores 24.5–35.5 × 2–3.5 μm, fusiform, curved, 1-septate, the apices pointed. Conidiomata *c.* 75 μm diam., scattered amongst the perithecia, immersed; conidiogenous cells 7.5 × 1.5–2 μm; macroconidia 24–5 × 1.5 μm, fusiform, arcuate, non-septate; microconidia to 4 μm in length, bacilliform.

On bark of *Betula*, probably not lichenized but sometimes loosely associated with *Trentepohlia*; locally abundant, especially in cooler regions of the north and west. Throughout British Is. Circumboreal in cool-temperate to sub-boreal zones of the Northern Hemisphere.

3. L. maggiana (Massal.) Körber (1865)

Thallus immersed, generally absent, trentepohlioid algae occasionally seen on the surface. Perithecia 0.13–0.3 mm diam., circular to slightly ellipsoid, immersed but becoming superficial; ostiole central in a small depression; involucrellum dark brown to greenish black, sometimes extending as a small basal fringe; true exciple colourless to greenish black; hymenial gelatin I+ yellowish. Asci 50–55 × 9–11 μm, cylindrical clavate, shortly stalked, 8-spored, the apex broad, truncate. Ascospores (25-)30–45 × 1.5–2.5 μm, thread-like to arcuate or sigmoid, 1(-3)-septate, the apices attenuated. Conidiomata 0.1–0.15 mm diam., scattered amongst the perithecia, immersed; conidiogenous cells 5–10 × 1.5–2.5 μm; macroconidia 20–25 × 1–1.5 μm, thread-like, curvate, 1-septate; microconidia 3.5–4 × 1–1.5 μm, bacilliform.

On smooth bark of *Corylus* (also on young *Quercus* bark in C. Europe), not lichenized but superficial algae fortuitously present; local S. England (Suffolk, Devon, Somerset), Scotland (Argyll, E. Perth), N. Ireland (Londonderry). Austria, Germany, Italy, Switzerland.

LEUCOCARPIA Vězda (1969)

Thallus crustose, granular, effuse. **Photobiont** *Trebouxia*. **Ascomata** perithecia, globose, colourless to pale, immersed in the substratum. **True exciple** continuous below the asci, colourless. **Hamathecium** at maturity composed only of periphyses, short-celled, gelatin I+ pale green to reddish. **Asci** 8-spored, obclavate, *Verrucaria*-type. **Ascospores** ellipsoid, colourless, muriform. **Conidiomata** unknown. **Chemistry**: lichen products not detected by t.l.c. **Ecology**: on soil in association with mosses.

LITERATURE: Vězda (1969b).

1. L. biatorella (Arnold) Vězda (1969)

Thallus thinly granular-verrucose, greenish to yellowish white. Perithecia immersed, only the pinkish or yellowish ostiolar region exposed, 0.4–0.6(-0.8) mm diam., globose or with a short ostiolar neck. Periphyses 60–70 × 1 μm. Asci 50–70 × 20–30 μm. Ascospores (25-)28–38(-45) × (10-)11–14 μm, at first 3-septate, soon becoming muriform with up to 9 transverse septa.

On bryophyte mats over mica-schist at 1000 m. alt. Scotland (Perth, Ben Lawers). C. Europe, Austria, Czechoslovakia.

Most likely to be confused with *Chromatochlamys muscorum* which, however, has numerous paraphyses and larger ascospores. *Polyblastia gelatinosa* has similar ascospores and grows on mosses, but has a dark coloured thallus and black perithecia.

LEUCOCARPOPSIS G. Salisb. (1975)

Thallus crustose, composed of ± densely packed, pseudoparenchymatous cells. **Photobiont** an unidentified green alga, the cells often ± vertically aligned in the thallus. **Ascomata** perithecia, ± globose to obpyriform, immersed, with a short neck. **True exciple** colourless, composed of interwoven thick-walled hyphae. **Involucrellum** absent. **Hamathecium** of periphyses lining the ostiolar canal; pseudoparaphyses gelatinized at an early stage and absent in mature ascomata; hymenial gelatin I+ blue. **Asci** broadly clavate to broadly saccate, short stalked, bitunicate in structure, with a broad internal apical beak, I−, 8-spored. **Ascospores** elongate-ellipsoid to broadly fusiform, colourless, thin-walled, smooth, non-septate.

Conidiomata not known. **Chemistry**: no characteristic lichen substances reported. **Ecology**: on siliceous rocks ± permanently submerged in freshwater.

A monotypic genus closely related to *Verrucaria*, from which it is distinguished by the colourless exciple and the absence of an involucrellum, and any other dark brown pigmented tissues.

LITERATURE: Salisbury (1975).

1. L. devensis G. Salisb. (1975)

Thallus greenish orange to yellowish green, smooth, ± continuous, not areolate, somewhat subgelatinous when wet, prothallus, if present, white. Perithecia 0.2–0.3(-0.4) mm diam., immersed, in somewhat raised, wart-like portions of the thallus, colourless to pale orange, the hymenial gelatin swelling when wet and the whole then translucent orange. Asci 45–50 × 15–22 μm. Ascospores (16-)18–24 × (7-)8–9(-10) μm, apices rounded to somewhat attenuated, sometimes spuriously 1-septate, with conspicuous oil drops.

On ± permanently submerged, hard, siliceous rocks in rivers, associated with aquatic *Verrucaria* species and *Staurothele fissa*; only known from the original collection. N. Wales (Merioneth, River Dee).

The ascospore shape and size recall *Verrucaria praetermissa*, one of the species mentioned as associated with the original collection, and it is possible *L. devensis* represents an unusual genotype of *V. praetermissa* having lost the ability to make all dark brown tissues; a similar phenomenom is known in *Acrocordia gemmata* f. *carnea*.

LICHENOTHELIA D. Hawksw. (1981)

Thallus on rocks or lichens, crustose, dark brown to black, forming convex cushions or ± areolate, sometimes placodioid at the margins, of ± angular, thick-walled, pseudoparenchymatous cells in ± vertical rows, especially below; some species with elongated stolons producing "microthallus" propagules; prothallus black. **Photobiont** absent; coccoid green algae or, particularly, cyanobacteria sometimes fortuitously(?) present on the thallus surface, not integrated into the thallus. **Ascomata** at first cushion-like, finally ± apothecia-like with a concave disc, black, sessile or ± stalked. **Thalline exciple** absent. **True exciple** and hypothecium recalling the thallus in structure. **Epithecium** poorly differentiated. **Hymenium** I− or I+ blue. **Hamathecium** of pseudoparenchymatous cells or elongated cellular pseudoparaphyses, irregularly branched and anastomosed. **Asci** clavate to subcylindrical, thick-walled, especially at the apex with an internal apical beak, fissitunicate; outer layer, I± blue; apex I−, 8-spored. **Ascospores** ellipsoid to soleiform, thick-walled, smooth or verrucose, 1- to 3-septate to submuriform, brown, often with a conspicuous gelatinous perispore swelling in K. **Conidiomata** either hyphomycetous, arising from the thallus surface, forming ± stalked, brown and ± verrucose, multicellular, subglobose macroconidia; or pycnidial, immersed, and bearing colourless, simple, bacilliform microconidia. **Chemistry**: no characteristic lichen products reported. **Ecology**: on rocks and, occasionally, lichens, mainly in exposed situations and extreme environments (e.g. hot and cold deserts); cosmopolitan.

The genus comprises at least 20 species, which appear to be essentially rock-surface dwelling, non-lichenized fungi. However, the thalli are often mixed amongst those of lichens and are superficially lichen-like. The variation in ascus iodine reaction and the nature of the hamathecial filaments are of particular phylogenetic and ontogenetic interest. The single species known in the British Is. belongs to a complex meriting more critical study and additional species are to be expected.

LITERATURE: Hawksworth (1981b), Henssen (1987).

1. L. convexa Henssen (1987)

Thallus of scattered or aggregated areoles, 0.1–0.4 mm diam., black, ± convex; ± stolons producing "microthalli". Ascomata cushion-like, to 0.2 mm diam. Hamathecium pseudoparenchymatous, I+ blue in parts. Asci 30–35 × 12–16 μm. Ascospores 11–13.5(-14.5) × 5.5–7(-7.5) μm, (1-)2–3-septate, verrucose. Macroconidia 10–15(-22) μm diam.

On slate, intermixed with *Rhizocarpon geographicum* thalli, and on sarsens. S. England (Wiltshire), N. Wales. Germany, Sweden, North America.

LICHINA Agardh (1817)

Thallus shrubby, ± erect, forming tufts, attached to rocks by an indistinct holdfast; lobes small, rounded or flattened, gelatinous when wet, dark olive-brown to black. **Cortex** indistinct, ± gelatinous. **Photobiont** cyanobacterial, *Calothrix* or *Dichothrix*. **Ascomata** apothecia, immersed in apices of branches; disc ± closed. **Hamathecium** of paraphyses, branched and anastomosed, becoming septate, with capitate apices. **Asci** 8-spored, thin-walled, deliquescing, not thickened at the apex; outer coat I+ blue. **Ascospores** colourless, simple, oblong-ellipsoid, thin-walled. **Conidiomata** pycnidia, with single ostiole or chambered. **Conidia** round or cylindrical. **Chemistry:** lichen products not reported. **Ecology**: marine, on rocks at or below high tide mark.

LITERATURE: Fletcher (1975).

1 Thallus forming erect, compact tufts to 0.5 cm tall; lobes to 0.3 mm thick, rounded **1. *confinis***
Thallus forming loose tufts of ± palmately divided branches to 1cm tall, lobes to 2 mm thick, ± flattened .... **2. *pygmaea***

1. L. confinis (Müller) Agardh (1821)

Thallus to 0.5 cm tall, erect, forming small, interwoven, compact tufts 5–10 mm diam., sometimes coalescing to form swards; lobes to 0.3 mm thick, terminal branches to 3 mm, dull, olive-brown to black, richly branched, branches rounded throughout. Apothecia to 0.5 mm diam. Ascospores 12–18 × 10–15 μm.

On coastal rocks at about high tide mark, particularly abundant on sunny, sheltered shores, small estuaries and creeks, becoming restricted to shade and crevices on exposed shores where it may occur well above high tide mark. N. & W. British Is. N. Europe, Madeira, North America.

Unlikely to be confused with any other lichen except for *L. pygmaea* (**2**) which is larger, has ± flattened lobes and is often mistaken for a small seaweed.

A morph with larger terminal branches, to 0.8 mm, which is weakly attached, decumbent and radially spreading, and pale grey to yellowish at the base, occurs in sheltered sea-lochs overgrowing *Verrucaria maura*, in W. Scotland and S.W. Ireland.

2. L. pygmaea (Lightf.) Agardh (1821)

Like **1**, but thallus to 1 cm tall and coarser in appearance, forming loose tufts of ± flattened and palmately divided branches to 2 mm thick, which may become rounded towards the apices; the tufts are often widespreading, forming swards several metres across. Apothecia to 2 mm diam. Ascospores 22–29 × 11–16 μm.

On sunny, ± exposed shores, where it occurs further down the seashore than *L. confinis*, and is associated with fucoid algae and barnacles, and is frequently inundated by sea water; locally abundant. Throughout British Is. Portugal, Madeira, Azores to S. Norway.

Resembles a small dark brown seaweed, especially *Catenella opuntia*.

LITHOGRAPHA Nyl. (1857)

Thallus crustose, areolate; prothallus present. **Photobiont** chlorococcoid. **Ascomata** apothecia, lirellate to ± round or angular, disc slit-like. **Thalline exciple** absent. **True exciple** opaque, of dark brown, conglutinated, anastomosing and radiating hyphae. **Hamathecium** of paraphyses, branched, soon disappearing. **Hymenium** colourless, I+ blue. **Hypothecium** colourless. **Asci** 8-spored, clavate; *Rimularia*-type. **Ascospores** simple, ellipsoid. **Conidiomata** unknown. **Ecology**: on rocks.

Wadeana has periphysoid filaments on the exciple, which is less strongly developed, red-brown pigmentation in the ascomata, multispored asci and *Trentepohlia* as photobiont.

LITERATURE: Hertel & Rambold (1990), Redinger (1937).

1. L. tesserata (DC.) Nyl. (1857)

Thallus whitish to pale grey-brown, clearly areolate; areoles to 0.4(-1.0) mm, scattered or ± contiguous, flat to convex, ± angular; cortex densely packed with small crystals, K+ dissolving; prothallus black, often visible between areoles and around the thallus. Photobiont cells 7–14 μm diam., globose, but many ellipsoid, to 20(-26) × 14(-16) μm. Apothecia numerous, 0.2–1.5 × 0.2–0.3 mm, black, lirellate, almost rounded or short, simple, straight or ± curved, sometimes divided, solitary or crowded and congested-gnarled, sessile, shiny, adpressed, ± emergent, constricted at base; disc slit-like, not opening; true exciple well developed, elevated and curved inwards, covering the disc. Paraphyses to 1 μm wide. Ascospores 9–12(-15) × 5–8 μm. Thallus Pd+ orange, K+ red (crystals) medulla only, KC+ red, C−, UV− (norstictic acid).

On ± moist, siliceous or ± basic rocks, especially basalt, in coastal and upland sites; rare and very local. N. & W. Britain. C. & N. Europe, Fennoscandia.

Distinguished by the areolate thallus, crowded, shortly lirellate apothecia, simple spores, and K+ red medulla.

British material referred to as *L. tesserata* **var. *petraea*** (Ach.) Redinger (1937) is distinct from *L. tesserata*; it has a scant, deep red-brown thallus, mainly around short apothecia which have swollen margins and a slit-like disc and are often gnarled and contorted. The asci contain numerous ascospores, 2 × 1 μm, the true exciple is well developed and gives an orange exudate in squashes in K; its affinity may be with *Polysporina*, possibly *P. dubia*. Reported from the Channel Is., Is. of Scilly, and W. Ireland on siliceous (often granite) coastal rocks.

LOBARIA (Schreber) Hoffm. (1796)

Thallus foliose, dorsiventral, irregularly spreading; lobes branched, the apices rounded at apices or truncate, often ± incised. **Upper surface** smooth, flat or wrinkled, sometimes with prominent depressions with a network of ridges. **Soredia** and **isidia** occasionally present. **Cortex** pseudoparenchymatous. **Photobiont** green, *Myrmecia* or *Trebouxia* or blue-green (*Nostoc* or *Scytonema*), the blue-green sometimes in internal or as shrubby cephalodia (Fig. 30d). **Lower surface** tomentose; rhizines present, simple to squarrose. **Pseudocyphellae** and **cyphellae** absent. **Ascomata** apothecia, rather rare. **Thalline exciple** present. **Asci** 8-spored, *Peltigera*-type. **Ascospores** colourless or pale brown, 1- to 7- septate, fusiform to elongate. **Conidiomata** pycnidia, immersed, pimple-like, ostiole black. **Conidia** ± cylindrical-elongated, slightly swollen at both ends. **Chemistry:** depsides, depsidones. **Ecology:** characteristic of ancient woodlands and parklands, especially in humid, sheltered areas.

The genus contains some of the largest British lichens, now seriously threatened in many areas owing to their extreme sensitivity to SO_2 pollution (<25 μg m^{-3}), acid rain and to changes in woodland management. The closely related genera *Sticta* and *Pseudocyphellaria* are frequent associates in the *Lobarion* community.

LITERATURE: James & Henssen (1976), Yoshimura (1971).

1 Upper surface ridged, with distinct depressions; ridges often sorediate or isidiate; underside with large, convex, tomentum-free areas 2

Upper surface smooth or partly wrinkled, distinct depressions and reticulate ridges absent; not sorediate or isidiate; underside uniformly tomentose 3

2(1) Upper surface matt and yellow-buff when dry, deep blue-grey when wet; photobiont blue-green .... **3. *scrobiculata***
Upper surface shiny, pale brown or greenish brown when dry, bright green when wet; photobiont green .... **2. *pulmonaria***

3(1) Upper surface pale green-brown when dry, without coralloid cephalodia on upper surface .... **4. *virens***
Upper surface pale grey-white when dry, often with dark brown, coralloid cephalodia on upper surface .... **1. *amplissima***

1. L. amplissima (Scop.) Forss. (1883)

Thallus often exceeding 30 cm diam., forming large patches, ± forming rosettes or irregularly branched; lobes 5–10 (-15) mm wide, with rounded apices, sinuate-indented, wavy, ± discrete at apices, contiguous and overlapping towards centre; upper surface grey-white when dry, green-grey when wet, thick, wrinkled or smooth, often with scattered pycnidia in conical warts to 0.7 mm diam.; cephalodia to 2 cm diam., occasionally absent, ± shrubby, dark green-brown to brown-black (Fig. 30d); lower surface tomentose, pale brown, paler towards margin, without cyphellae or pseudocyphellae. Photobiont green in thallus and blue-green in cephalodia. Apothecia occasional; disc red-brown. Ascospores 40–60 × 6–7 μm, 1- to 3-septate. Pycnidia frequent; conidia 5 × 1 μm. Medulla Pd−, K+ yellow or K−, KC± deep pink, C− (± scrobiculin).

Chiefly on trunks and main branches of *Fraxinus, Ulmus, Acer pseudoplatanus* and *Quercus*, occasionally on rocks. S. & N.W. England, N. Wales, C. & W. Scotland, S.W. Ireland. Europe, Spain to S. Scandinavia and east to Greece.

This distinctive, large species is characterized by its pale thallus with neat, rounded, notched lobe-margins, green photobiont and often the presence of densely branched, shrub-like cephalodia, which may occasionally occur independently and have thus been named *Dendriscocaulon umhausense* (Auersw.) Degel. (1942). The cephalodia represent the response of the mycobiont to cyanobacteria; very rarely additional leaflets of the green morphotype may secondarily arise directly from the cephalodia on the main thallus.

2. L. pulmonaria (L.) Hoffm. (1796)

Thallus wide-spreading, often exceeding 30 cm diam., ± loosely attached at one end, at times ± hanging and strap-shaped; lobes 1–3 cm wide, ± dichotomously divided, often rather elongate, sinuate-indented with truncate apices; upper surface pale green-brown when dry, bright green when wet, strongly reticulately ridged with marked depressions, often with coarse soredia or isidia along ridges; lower surface tomentose, brown, with naked, paler, convex areas corresponding to depressions in the upper surface. Photobiont green, often with blue-green internal cephalodia in the medulla. Apothecia occasional, on ridges and towards the margin; disc 2–4 mm diam., reddish brown, with a thin thalline margin. Ascospores 18–30 × 5–9 μm, colourless or pale brown, fusiform, (1-)3- septate. Pycnidia on the ridges, immersed; conidia 5 × 1 μm, bacilliform. Medulla Pd+ orange, K+ yellow-orange, KC+ yellow-orange, C− (norstictic acid, stictic and constictic acids).

On broad-leaved trees, low scrub, *Calluna* and mossy rocks; locally abundant but rare and decreasing in parts of the south. S.W., S.E. & N.W. England, N.W. & S. Wales, W. Scotland, W. Ireland. Oceanic and montane Europe, N. & S. Africa, North America, Asia.

Distinctive; most individuals are isidiate, occasionally specimens may be markedly sorediate.

3. L. scrobiculata (Scop.) DC. (1805)

Thallus to 10 cm wide, irregularly lobate; lobes 1–3 cm wide, rounded, often concave, the margin entire or slightly crenulate; upper surface dark bluish green or deep greyish blue when wet, yellowish green when dry, pitted with large, shallow depressions, slightly ridged but ridges not forming distinct network; soredia blue-grey, at first punctiform, becoming irregular and spreading along ridges and lobe margins; lower surface pale brown, tomentose, reticulate, convex areas corresponding to depressions in the upper surface, whitish, without tomentum. Photobiont blue-green. Pycnidia not seen. Apothecia infrequent, disc 1–2 mm diam., red-brown with thick, inflexed, sometimes sorediate thalline exciple. Ascospores 50–80 × 6–7 μm, 3- to 7- septate, colourless, elongate-fusiform. Medulla Pd+ orange, K+ yellow, KC+ deep orange-pink, C−, UV+ violet-glaucous (± usnic, stictic, ± norstictic and constictic acids, scrobiculin).

On old broad-leaved trees, often at the edges of old woodland, more widespread and also on wayside trees in Scotland, also *Calluna* and mossy rocks in coastal areas. Scotland (W, N. & C. Highlands), extending locally to S.W. Scotland, N. Wales, S.W. England and S.W. Ireland. Europe, Australia, New Zealand.

Declining, particularly in the S. and S.W.

4. L. virens (With.) Laundon (1984)

Lobaria laetevirens Zahlbr. (1906)

Thallus to 10 cm diam., ± closely adpressed, often forming rosettes; lobes 3–10 mm wide, ± rounded at apices, indented, wavy, occasionally with marginal folioles, often overlapping and becoming imbricate towards centre; upper surface pale grey-brown to grey-green when dry, pale green when wet, often smooth and ± shiny, more rarely becoming transversely wrinkled and ridged in older parts; lower surface pale brownish white, evenly tomentose without tomentum-free, naked patches. Photobiont green. Apothecia frequent, disc 1–3 mm diam., pink- or orange- red with thick, inflexed thalline exciple. Ascospores 25–45 × 8–11 μm, colourless or pale brown, fusiform, 1-septate. Pycnidia frequent; conidia 4–5 × 1–1.5 μm. Medulla Pd−, K+ pale yellow or K−, KC± pink, C− (± scrobiculin).

On *Quercus, Fraxinus, Ulmus* and *Fagus*, mostly confined to ancient woodlands, occasionally on sheltered rocks in woods and on western sea-cliffs; locally common. W. Britain southwards to S. & S.W. England (Cornwall, Dorset, Hampshire-New Forest), Ireland. Europe.

Characterized by the dull grey-green colour, its mostly closely adpressed thallus and tendency to tolerate more shaded sites than other members of the genus.

LOPADIUM Körber (1855)

Thallus crustose, granular to minutely squamulose, effuse. **Cortex** absent, but cartilaginous layer present. **Photobiont** chlorococcoid; cells 8–17 × 8–12 μm, globose or broadly ellipsoid. **Ascomata** apothecia, sessile, the base distinctly constricted, round, sessile, with a protruding margin; disc concave to flat. **Thalline exciple** absent. **True exciple** pseudoparenchymatous, dark red-brown. **Hamathecium** of paraphyses, to 2.5 μm, simple or forked, straight, terminal cells with swollen apices 4–7 μm wide and dark brown pigmented caps to 10 μm tall. **Hymenium** I+ blue. **Hypothecium** brown-black, K−. **Asci** 1-spored, ellipsoid to clavate-cylindrical, thick-walled, the outer layer I+ blue, without a distinct apical dome, with a thin K/I+ dark blue zone between the ascus wall and ascus contents in the ascus apex. **Ascospores** strongly muriform, oblong-ellipsoid, colourless or pale yellow-brown. **Conidiomata** not seen.

Characterized by the asci which lack a distinct apical dome and the thick, ± simple paraphyses with dark brown conical caps. *Schadonia* has thin, branched paraphyses and the asci have a distinct I+ blue tholus.

LITERATURE: Poelt & Vězda (1981), Santesson (1952).

1 On mossy, acid-barked trees .... **1. *disciforme***
On mosses and plant remains on the ground .... **2. *pezizoideum***

1. L. disciforme (Flot.) Kullhem (1870)

Thallus granular- or minutely-squamulose; squamules 0.1–0.4 mm diam., sometimes producing marginal isidia-like granules 0.05–0.1 mm diam., green-brown or brown, matt, often with paler or white margins. Apothecia 0.4–0.8 mm diam., black. Ascospores 70–115 × 23–45 μm.

On mossy, acid-barked trees, especially *Quercus*, but also *Fraxinus, Alnus* and *Corylus*. N.W. England (Cumbria), S. Wales (Brecon), Scotland. Alps, Carpathians, Scandinavia.

When sterile may resemble *Protoparmelia ochrococca*, which has uniform brown, shiny granules and occurs on *Pinus* and *Larix* trunks.

2. L. pezizoideum (Ach.) Körber (1855)

Thallus granular-warted, dark brown-black. Apothecia 0.7–1 mm diam., the disc and margin black; edge of true exciple red- to black-brown, ± shiny. Ascospores 70–115 × 23–45 μm.

On mosses and plant remains on the ground, montane (above 1000 m alt.). C. Scotland (Perth, Ben Lawers; W. Ross, Beinn Dearg). Arctic-alpine.

MACENTINA Vězda (1973)

Thallus crustose or minutely filamentous; hyphae smooth or papillose. **Photobiont** chlorococcoid. **Ascomata** perithecia, pale brown; surface naked or rarely with a ring-shaped collar. **Involucrellum** absent. **True exciple** pale, in section formed of periclinally elongated cells, sometimes with a surface layer of isodiametric cells. **Hamathecium** of periphyses, paraphyses absent; gel I+ red. **Asci** 8-spored, fissitunicate, clavate-cylindrical, thickened at apex when young, I−. **Ascospores** 12–26 × 3.5–5 μm, transversely (1-)3-septate to submuriform, elongate-ellipsoid, smooth, without a perispore, colourless. **Conidiomata** unknown. **Chemistry**: lichen products not detected by t.l.c. **Ecology**: on bark, leaves or on mosses and liverworts overgrowing leaves.

Distinguished from *Thelidium* by the pale, non-carbonaceous true exciple and different ecology. *M. aurantiaca* McCarthy & Vězda (1985) is a synonym of the non-lichenized *Nectria pseudopeziza* (Desm.) Rossman (1979).

LITERATURE: Coppins & Vězda (1977), Orange (1989).

1 Thallus minutely filamentous, often disintegrating into soredia-like granules; cortical cells papillose; perithecia 200–380 μm wide; ascospores 3- to 5-septate ... **2. *stigonemoides***
Thallus immersed or thin and leprose; cortical cells not papillose; perithecia 80–120 μm wide; ascospores 1- to 3-septate .... **1. *abscondita***

1. M. abscondita Coppins & Vězda (1977)

Thallus superficial, very thin, leprose, dull green. Perithecia 80–120 μm wide, pale brown when dry, almost translucent and colourless when wet; true exciple 10–12 μm wide, of colourless or pale yellowish, elongated cells. Asci clavate-cylindrical or sometimes pyriform. Ascospores 12–20 × 3.5–4.5 μm, 1- to 3-septate.

In fissures of bark of *Sambucus* or on stems of shaded *Juniperus* in coastal scrub. S.E. England (W. Kent, East Sussex), N.E. Scotland (E. Ross). ?Endemic.

2. M. stigonemoides A. Orange (1989)

Thallus 50–400(-600) μm tall, minutely fruticose, densely branched, branches 12–35 μm wide, ± prostrate or, more frequently, ascending, cylindrical; cortical cells with papillae *c.* 1 μm high; thallus locally disintegrating into soredia-like granules; granules 15–40 μm diam.; superficial cells often prominent and bearing one or more papillae. Perithecia 200–380 μm wide, pale brown, ovoid or ± obpyriform; true exciple 30–80 μm thick, the median layer of elongated cells, the outer layer of isodiametric cells. Asci fusiform-cylindrical. Ascospores (13-)16–21 × (4-)5–6 μm, 3- to 4(-5) septate.

On shaded bark in species-poor, bryophyte-dominated communities, mainly on *Sambucus* and *Ulmus*, in humid situations; frequent. Throughout British Is. France.

Frequently sterile and easily overlooked as an alga or moss protonemata. The cylindrical, papillose branches are, however, readily identifiable under the microscope. Sterile granular morphs have bulging papillose cells on the surface of the granules, and frequently short lengths of branches intermixed with the granules.

MASSALONGIA Körber (1855)

Thallus small, foliose; lobes squamulose or elongate forming rosettes, ± isidiate at margins. **Lower surface** of densely interwoven, longitudinally oriented hyphae. **Soredia** absent, isidia and folioles present. **Upper cortex** pseudoparenchmatous. **Lower cortex** not developed. **Photobiont** *Nostoc*. **Medulla** of loosely interwoven hyphae. **Ascomata** apothecia, hemiangiocarpic, laminal or marginal. **True exciple** present. **Hymenium** brown above, K−, colourless below, I+ blue. **Hamathecium** of coherent paraphyses, septate, simple, ± swollen and pigmented at the apices. **Asci** cylindrical, 8-spored, ?*Peltigera*-type. **Ascospores** colourless, ellipsoid to fusiform, 2(-4)-celled, sometimes constricted at the septa. **Conidiomata** pycnidia, brown above, paler below. **Conidia** bacilliform or slightly dumb-bell-shaped. **Chemistry**: lichen products not reported by t.l.c.

LITERATURE: Henssen (1963b).

1. M. carnosa (Dickson) Körber (1855)

Thallus widely scattered or aggregated, often forming rosettes 1–3 cm diam.; lobes flattened, elongate, to 10 mm long, 0.5–1.5 mm wide, horizontal or ± ascending, irregularly branched and overlapping; upper surface red-brown when dry, pale brown to deep green when wet, smooth, margins smooth, irregularly divided, nodulose or with coarse, ± branched, cylindrical or ± globose isidia; lower surface brownish white with sparse brown rhizines at the base of the lobes. Apothecia to 2 mm diam., rather rare, often ± stalked; true exciple 100–135 μm wide, of ± rounded cells, the marginal cells with short hairs; disc flat, red-brown with a smooth, paler true margin. Paraphyses 3–5 μm wide. Ascospores 11–23(-32) × 4.5–7(-8.5) μm, 1(-2)-septate. Pycnidia to 0.6 mm diam.; conidia 4–6 × 1 μm.

Among mosses and on peat, especially over rocks, particularly on boulders in or near streams, rarely on soil; rather rare. W. British Is. N. Europe, North America, South America, Australia, New Zealand.

Distinguished by the small, often overlapping, red-brown squamules, with coarse, nodular isidia and the 1-septate spores. See also *Arctomia*.

MEGALARIA Hafellner (1984)

Thallus crustose, without a distinct cortex. **Photobiont** *Trentepohlia,* but filaments and orange colour rarely evident. **Ascomata** apothecia, sessile, black, rather large. **Thalline exciple** absent. **True exciple** persistent, raised, black, of anticlinally arranged hyphae, each with a thick gel coat. **Hymenium** ± colourless, I+ pale blue. **True exciple**, epithecium and hypothecium with dark greenish pigment. **Hamathecium** of numerous slender paraphyses, simple or sparingly branched; apices clavate, without dark apical caps. **Asci** cylindrical clavate, (2-)8-spored, *Lecanora*-type with a broad, non-amyloid apical cushion and a shallow ocular chamber. **Ascospores** ± ellipsoid, 1-septate, colourless, thick-walled, smooth, lacking a distinct outer sheath. **Conidiomata** pycnidia. **Conidia** ellipsoid to oblong, simple, colourless. **Chemistry:** lichen products not detected by t.l.c. **Ecology:** on bark.

A monotypic genus distinguished from *Catillaria s. lat.* (including *Catinaria* Vainio) by the rather large apothecia, large, thick-walled ascospores and, especially, the *Lecanora*-type asci.

LITERATURE: Hafellner (1984).

1. M. grossa (Pers. ex Nyl.) Hafellner (1984)

Catinaria grossa (Pers. ex Nyl.) Vainio (1922)

Thallus effuse, ± smooth or slightly verrucose, often rimose, greenish white to pale grey. Apothecia 0.4–1.5(-2) mm diam.; disc flat to convex, matt; true exciple distinct, glossy; inner part colourless; epithecium, hypothecium and outer exciple bluish to greenish black, K+ green intensifying, N+ purple-red; hymenium 120–140 μm tall, usually colourless but this and hypothecium sometimes (especially in old apothecia) with additional reddish (K+ purple-red) pigment. Paraphyses 1.5–2 μm wide. Ascospores 20–30 × 10–15 μm, ellipsoid to oblong-ellipsoid, often slightly constricted at septum, wall 1–2 μm thick. Pycnidia 100–120 μm diam., inconspicuous, immersed; wall colourless or pale greenish in upper parts; conidiogenous cells 10–15 × 1–1.5 μm, ± cylindrical; conidia 3–4 × 1.5–2 μm. Thallus Pd−, K−, KC−, C−, UV−.

On nutrient-rich bark of usually mature trees (especially *Fraxinus* and *Acer campestre*), sometimes on younger or small trees (e.g. *Corylus*), usually in old, humid woodlands, very rarely on basic rocks; rare and local. Mainly in N. & W. British Is., much declined in England. W. Europe, North America, Australia, New Zealand.

Usually recognizable in the field by the contrasting pale thallus and rather large, black, sessile apothecia with a minutely roughened disc and shiny, raised exciple.

MEGALOSPORA Meyen (1843)

Thallus crustose, superficial, continuous, unstructured. Soredia present, in ± contiguous soralia. **Photobiont** chlorococcoid (? *Dictyochloropsis*). **Ascomata** apothecia, sessile, concave to flattened; disc brown to black-brown, roughened, ± shining. **Thalline exciple** absent. **True exciple** ± prominent, at first raised, 3-layered, inner zone of thin-walled, radiating hyphae, median zone ± lax-arachnoid; outer edge of dense, conglutinated hyphae. **Epithecium** pigmented pale to deep red-brown. **Hymenium** I− or I+ blue, with oil droplets. **Hypothecium** flat, pellucid. **Hamathecium** of paraphyses *c.* 1 μm wide, ± simple below, sparsely anastomosed, particularly above, apices slightly thickened. **Asci** *Lecanora*-type, ovoid-ellipsoid. **Ascospores** elongate, straight, septate, colourless; epispore smooth. **Conidiomata** pycnidia. **Conidiogenous cells** enteroblastic, cylindrical to elongate-ampulliform, acrogenous. **Conidia** bacilliform. **Chemistry**: zeorin, ± usnic acid, ± pannarin, ± lichexanthone. **Ecology** on mosses on siliceous rocks and trees; a pantropical-temperate genus of about 30 species.

LITERATURE: Sipman (1983).

1. M. tuberculosa (Fée) Sipman (1983)

Bombyliospora pachycarpa auct., non (Delise ex Duby) Massal. (1852)

Thallus thick, effuse, soft, powdery-granular, yellowish or greenish grey; prothallus absent or smooth, blue-black; thalline granules aggregated, partly corticate, glaucous blue-grey to yellowish grey, breaking down into irregular clusters of yellow-green to grey, *Lepraria*-like, soralia, which may spread over most of the thallus. Apothecia extremely rare (recently seen only from Somerset), rounded, flat; disc 0.5–2.0 mm diam., chestnut-brown, shining; true exciple thick, persistent, paler or concolorous with the disc; epithecium densely granular, the granules not dissolving in K; hymenium 130–180 μm tall, without oil droplets. Asci 1-spored. Ascospores 70–140 × 25–35 μm, 5- to 11-septate, the wall *c.* 2 μm thick and septa *c.* 0.5 thick. Thallus Pd−, K+yellow, KC+yellow, C−, UV− (usnic acid, zeorin, unknown UV+ blue).

On bark of broad-leaved trees, especially *Quercus*, *Fagus* or *Fraxinus*, in sheltered, humid but well-lit glades in ancient woodlands, also on mossy rocks and on the ground in coastal sites; rather local. S. & W. Britain, S and N.W. Ireland. Pantropical-temperate.

Usually sterile, and then readily confused with *Phyllopsora rosei*, which is Pd+ orange-red, K− and has scale-like, flattened, minute, pubescent granules on an arachnoid prothallus. Old herbarium specimens of *M. tuberculosa* become more glaucous and are covered in minute, needle-shaped crystals.

MEGASPORA (Clauz. & Roux) Hafellner & V. Wirth (1987)

Pachyospora auct., non Massal. (1852)

Thallus crustose, non-lobate, ± pruinose. **Photobiont** chlorococcoid. **Ascomata** apothecia, deeply immersed in thalline verrucae. **Thalline exciple** present. **True exciple** thin, colourless or pale straw, of strongly conglutinated vertically aligned hyphae. **Epithecium** brownish green, N+ bright green. **Hymenium** colourless, I+ blue. **Hypothecium** colourless. **Hamathecium** of paraphyses, richly branched and anastomosing, not swollen at apices, strongly conglutinate. **Asci** (4-)8-spored, clavate or cylindrical-clavate, thin-walled except towards the apex, with a K/I pale apical dome. **Ascospores** large, simple, globose to shortly ellipsoid, colourless, wall uniformly thickened. **Conidiomata** not seen. **Chemistry**: lichen products not detected by t.l.c. **Ecology**: montane.

LITERATURE: Clauzade & Roux (1984), Hertel (1971), Wirth (1987).

1. M. verrucosa (Ach.) Hafellner & Wirth (1987)

Pachyospora verrucosa (Ach.) Massal. (1852)

Thallus ± tartareous, continuous or ± dispersed in irregular patches, of ± compacted or loosely co-ordinated, coarse, confluent, inter-connecting granules; granules white to grey-white, somewhat mottled, swollen, convex or unevenly flattened, surface smooth, ± roughened or in part (especially near the apothecia) finely scabrid-areolate, ± densely white-pruinose. Apothecia 0.5–1.5 mm diam., 1(-2) in coarse thalline warts; disc 0.2–0.4(-0.7) mm diam., poriform or ± expanded, grey-black or black, surface roughened but not pruinose, the disc colour extending outwards over the inner part of the thalline exciple and true exciple, forming a dark halo around the disc; thalline exciple like the cortex densely packed with small granules; photobiont zone without granules; the medulla and below apothecia also densely granular, not dissolving in K; epithecium ± grey-black in part, K+ brownish, non-granular; hymenium 200–250 μm tall. Asci 200–230 × 45–50 μm. Ascospores (30-)35–50(-60) × (21-)25–39(-42) μm, wall 1.5–2.5 μm thick. Thallus Pd−, K−, KC−, C−.

On soil or, more frequently, over mosses and plant remains on calcareous rocks (epidiorite, mica-schist, limestone), usually in mountains, very rarely on consolidated dunes rich in shell-sand; widely distributed but very local. W. Britain, from S.W. England (Somerset) and N. Wales to N. & W. Scotland. Arctic to Mediterranean, Venezuela.

The position of this species, the only one of the genus, is unclear. The poriform apothecia are reminiscent of *Pertusaria* but, with subglobose ascospores with uniformly thickened walls and thin, numerous, anastomosing paraphyses, this species seems most closely related to *Ochrolechia*.

MELASPILEA Nyl. (1856)

Thallus crustose, immersed, often evanescent or scarcely apparent. **Photobiont** usually *Trentepohlia*. **Ascomata** apothecia, immersed to superficial, ± rounded to elongated and lirelliform, sometimes branched; disc exposed or slit-like, black. **Thalline exciple** absent. **True exciple** prosoplectenchymatous, sometimes indistinct to well developed and continuous below the hymenium or stipe-like, dark brown to black. **Epithecium** usually dark brown. **Hymenium** colourless or brown-striate; I− or + blue. **Hypothecium** colourless to brown. **Hamathecium** of paraphyses, thread-like, sparsely branched to anastomosed. **Asci** elongate-clavate, generally thickened at the apex and with an internal ocular chamber, usually I−. **Ascospores** ellipsoid to sole-shaped, 1-septate, colourless at first but generally brown at maturity, smooth-walled to warted. **Conidiomata** unknown. **Chemistry**: thallus reactions mostly negative, rarely with an unidentified pigment by t.l.c. **Ecology**: on bark, more rarely rock, or other lichens; cosmopolitan.

This generic name has been traditionally employed for crustose lichens with lirelliform apothecia and brown, 1-septate ascospores. It is heterogenous and badly in need of a modern revision. The generally accepted type species, *M. arthonioides* (Fée) Nyl. (1856), appears to be close to *Buellia*.

LITERATURE: Coppins *et al.* (1992), Miles (1980), Redinger (1938).

1 Growing on lichens 2
Not growing on lichens 3

2(1) Ascospores 10–13.5(-16) × 5–7.5 μm; on *Phaeographis* .... **8. *lentiginosa***
Ascospores 18–22 × 9–12 μm; on *Verrucaria* .... **7. *leciographoides***

3(1) On rock 4
On bark 5

4(3) Apothecia rounded to shortly lirellate, 0.1–0.4 mm long; ascospores (9.5-)11–14 μm long .... **5. *granitophila***
Apothecia lirellate, 0.4–1(-1.4) mm long; ascospores 17–23 μm long .... **6. *interjecta***

5(4) Apothecia elongate; disc slit-like 6
Apothecia ± rounded or angular; disc finally exposed, not slit-like 9

6(5) Ascospores <15 μm long 7
Ascospores mainly >15 μm long 8

7(6) Hymenium colourless, I–; ascospores smooth, 11–15 μm long .... **3. *constrictella***
Hymenium with brown bands, I+ blue; ascospores verrucose, 8.5–11 μm long .. **2. *atroides***

8(6) Ascospores 19.5–27(-32) μm long; paraphyses not anastomosing .... **4. *diplasiospora***
Ascospores 14–21(-23) μm long; paraphyses anastomosing .... **9. *lentiginosula***

9(5) Apothecia <0.75 mm diam.; true exciple thick; ascus apex I– 10
Apothecia <1.5 mm diam.; true exciple thin; ascus apex I+ blue .... **1. *amota***

10(9) Apothecia 0.1–0.2(-0.3) mm diam.; hymenium I+ blue .... **11. *proximella***
Apothecia (0.2-)0.3–0.75 mm diam.; hymenium I– .... **10. *ochrothalamia***

1. M. amota Nyl. (1867)

Thallus evanescent, grey to cream. Apothecia to 1.5 mm diam., rounded or angular, erumpent, disc flat; true exciple thin; hymenium colourless to brownish in parts, I+ blue; hypothecium dark brown. Paraphyses branched and anastomosing. Asci 45–75 × 12–25 μm, (4-)8-spored, clavate, with an internal ocular chamber, apex I+ blue. Ascospores 14.5–18.5 × 7.5–12.5 μm, sole-shaped, smooth, remaining colourless or pale brown.

On smooth bark of old *Quercus*. S.W. Ireland (S. Kerry, Killarney). Endemic.

Distinguished from *M. ochrothalamia* **(10)** and *M. proximella* **(11)**, which also have a dark hypothecium, by the I+ blue hymenium and ascus-apices, and from *M. atroides* **(2)** in the shape of the apothecia.

2. M. atroides Coppins (1992)

Thallus evanescent, white to cream or yellow tinged near the apothecia. Apothecia 0.2–2 × 0.1–0.2 mm, oblong to ellipsoid, sometimes branched or curved, disc slit-like; true exciple inflexed and raised; hymenium with brownish pigmented bands, I+ blue; hypothecium dark brown. Paraphyses anastomosing, the apices with brown granules. Asci 22–30 × 11–15 μm, 8-spored, clavate, with an internal ocular chamber; tip I–. Ascospores 8–5.11 × 3.5–5 μm, sole-shaped, warted, brown. Thallus K± orange; no lichen products detected by t.l.c.

On *Corylus*, associated with *Graphis*, *Pyrenula* and *Thelotrema* species. W. Scotland (Kintyre to Skye). W. Pyrénées.

Easily confused with *Opegrapha atra*, which has 3-septate ascospores.

3. M. constrictella (Stirton) A. L. Sm. (1911)

Thallus ± evanescent, cream. Apothecia 0.1–1 × 0.2–0.4 mm, elongate, branched, often tri-radiate, erumpent to ± superficial; disc slit-like but expanded with age; true exciple thin, strongly inflexed at first; hymenium ± colourless, I–; hypothecium colourless. Paraphyses anastomosed. Asci 40–65 × 12–17 μm, 8-spored, clavate, with an internal ocular chamber, I–. Ascospores 10–17 × 4–7 μm, sole-shaped, colourless, becoming pale brown.

On bark of a coniferous tree (? *Pinus*). Reported from Scotland (Argyll) last century but the specimen could be from Australia or New Zealand.

The origin of the only specimen known requires further investigation. *M. lentiginosa* **(8)** has similar apothecia but differs in the mainly simple paraphyses and also grows on other lichens.

4. M. diplasiospora (Nyl.) Müll. Arg. (1887)

Thallus evanescent, cream. Apothecia 0.3–1.5 × 0.2–0.5 mm, elongate, unbranched, occasionally curved, partly immersed to superficial; disc slit-like; true exciple well developed, raised; thecium colourless, I+ bluish. Paraphyses sparsely branched. Asci 75–105 × 28–37 μm, 8-spored, clavate, with an internal ocular chamber, I−. Ascospores 19.5–27(-32) × 9.5–13.5(-16) μm, sole-shaped, smooth, brown.

On smooth bark of *Ilex* and on *Tilia*. Wales (Merioneth), S.W. Ireland (Kerry, Killarney); not recorded recently but easily overlooked. Endemic.

Old apothecia fall away to leave a black elongate outline. Distinguished from other British species by the longer ascospores.

5. M. granitophila (Th. Fr.) Coppins (1989)

M. subarenacea J. Nowak & Kiszka (1974), *Arthonia granitophila* Th.Fr. (1865)

Thallus evanescent, dull green- or yellowish brown, often apparent only near the apothecia. Apothecia 0.1–0.3(-0.4) mm diam., immersed to sessile, rounded to somewhat elongate, rarely branched, sometimes curved; disc exposed; true exciple well developed, raised, extended below into a stipe-like structure; hymenium pale brown, I+ blue; hypothecium dark brown. Paraphyses unbranched or branched and anastomosed below. Asci 25–45 × 14–20 μm, 8-spored, clavate, with an internal ocular chamber, I−. Ascospores (9.5-)11–14(-19) × (3.5-)4.5–5.5(-6.5) μm, ovoid-oblong to slipper-shaped, colourless and smooth at first, becoming pale brown and verrucose.

On shaded siliceous rocks in upland areas, often associated with *Opegrapha gyrocarpa*, *O. zonata* or *Porina lectissima*. Scotland, Wales. Norway, Sweden, Poland, S.W. Germany.

6. M. interjecta (Leighton) A.L. Sm. (1911)

Encephalographa interjecta (Leighton) Th. Fr. (1874)

Thallus evanescent or thinly scurfy, pale green-grey to pale fawn. Photobiont cells 5–16 μm, diam., trebouxioid. Apothecia 0.4–1(-1.4) × 0.2–0.3(-0.5) mm, lirellate, simple or forked, at first with a slit-like disc and a tumid margin, later developing additional slits or becoming subgyrose-contorted; true exciple well developed, reddish brown-black, K ± olivaceous; epithecium pale to dark brown, K ± olivaceous; hymenium 70–85 μm tall, colourless, I+ yellowish, K/I+ blue; hypothecium concolorous with exciple. Paraphyses 1.5–2 μm wide, richly branched and anastomosed. Asci 47–53 × 21–26 μm, 8-spored, I−; apical dome with narrow ocular chamber. Ascospores 17–23 × 7–9 μm, 1-septate, slipper-shaped, much constricted and easily broken at septum, colourless but old spores brown and finely warted.

On vertical surfaces of slate and mica-schist rocks; rare. Wales (Brecon, Cardigan, Merioneth), Scotland (Stirling). Endemic.

Has the appearance of an *Opegrapha*, although specimens with subgyrose apothecia are likely to be overlooked as *Polysporina simplex*. Formerly considered a synonym of *Poeltinula cerebrina*, a species of hard limestone with a very different thallus, ascus structure and ascospore pigmentation.

7. M. leciographoides Vouaux (1913)

Thallus immersed in host lichen. Apothecia 0.15–0.5 × 0.1–0.2 mm, rounded at first but becoming elongate, sometimes in groups of 2–6, superficial; disc exposed; true exciple thin, raised; hymenium colourless, I+ blue; hypothecium brownish. Paraphyses branched; apices swollen to 3.5 μm wide and capitate. Asci 65–80 × 23–28 μm, 8-spored, clavate. Ascospores 18–22 × 9–12 μm, finally pale brown, sole-shaped.

On an unidentified *Verrucaria*. N. England (S. Lancashire); no recent records. France.

8. M. lentiginosa (Lyell ex Leighton) Müll. Arg. (1887)

Thallus immersed in host lichen. Apothecia 0.1–0.5 × 0.1–0.2 mm, oblong to ellipsoid, occasionally branched, solitary or irregularly aggregated, superficial; disc slit-like; true exciple thick, incurved, continuous below the hymenium, becoming stipe-like below and extending into the host; hymenium colourless to pale brown, I+ blue; hypothecium colourless, delimited below by the extended exciple. Paraphyses mainly simple, with periphysoid-like filaments at the margin of the exciple. Asci 25–35 × 8–12.5 μm, (4-)8-spored, clavate, with an internal ocular chamber. Ascospores 10–13.5(-16) × 5–7.5 μm, sole-shaped, smooth, brown.

On *Phaeographis dendritica*, mainly in ancient woodland. S. England (Hampshire, New Forest; Devon, Sussex), S.W. Ireland (Cork). France.

The minute apothecia are intermixed with those of the host, which is scarcely affected by their presence; presumably a commensal. A stipe-like structure is otherwise seen in the British Isles only in *M. granitophila* **(5)**.

9. M. lentiginosula (Nyl.) A. L. Sm. (1911)

Thallus inconspicuous. Apothecia 0.2–0.8 × 0.1–0.2 mm, oblong to ellipsoid, branched, usually curved, somtimes aggregated, superficial; disc slit-like; true exciple thick, raised, inflexed; hymenium colourless, I+ blue; hypothecium colourless and in contact with the substratum. Paraphyses sparsely branched. Asci 37–50 × 14–26 μm, (4-)8-spored, clavate, with an internal ocular chamber, I−. Ascospores 14–21(-23) × 6–11 μm, sole-shaped, ± warted, brown.

On smooth plateau surfaces of mature *Pinus* bark. N.W. England (Westmorland), N. Scotland (Perth, West Inverness). ?Endemic.

M. proximella **(11)** differs in the consistently rounded apothecia, the generally coloured hypothecium and hymenium, the latter also being I−.

10. M. ochrothalamia Nyl. (1865)

Thallus usually evanescent, white to grey-green or orange to rust-coloured, especially near the apothecia. Apothecia (0.2-)0.3–0.75 mm diam., rounded to angular by compression, solitary or in groups of 2–3, superficial; disc exposed; true exciple well developed, sometimes orange-tinged; epithecium green to orange-brown; hymenium pale brown, I−; hypothecium pale brown. Paraphyses to 4 μm wide, sparsely branched; apices capitate, pale green to brown. Asci 54–70 × 15–30 μm, 8-spored, clavate, with an internal ocular chamber, I−. Ascospores 17–21(-28) × 6–8.5(-12) μm, sole-shaped; apices sometimes attenuated, smooth (roughened, under SEM), pale brown.

On smooth bark, especially of *Quercus*, also *Castanea*, *Fagus*, *Salix* and *Sorbus*, in moist woodlands. S.W. & S. England, N. Scotland (Perth, West Inverness), Ireland (Cork, Wicklow). Belgium, France, Spain (Pyrénées).

The rounded to angular apothecia, often with orange-brown pigments, are particularly characteristic. *M. proximella* **(11)** differs in the I+ blue hymenium and shorter apothecia, with a conspicuous raised true exciple. The species may be referable to *Schrakia* Hafellner.

11. M. proximella Nyl. ex Norrlin (1873)

Thallus inconspicuous. Apothecia 0.1–0.2(-0.3) mm diam., ± rounded to angular, solitary or in groups, superficial; disc sometimes obscured by the inflexed exciple; exciple raised; epithecium granular; hymenium colourless to pale brown, I+ pale blue; hypothecium colourless to dark brown. Paraphyses sparsely branched and anastomosed; apices swollen to 4 μm with external pigmentation. Asci 36–55 × 14–21 μm, (4-)8-spored, clavate, with an internal ocular chamber, I−. Ascospores 14–19(-22) × 7–9 μm, sole-shaped, smooth, remaining colourless, pale brown only at maturity. Thallus reactions negative; no substances by t.l.c.

On *Juniperus* bark. Scotland (Highlands). Europe.

Most British records refer to *M. ochrothalamia* **(10)**, the separation from which is discussed under that species.

MENEGAZZIA Massal. (1854)

Thallus foliose, dorsiventral, lobate, ± radiate, often forming rosettes or irregularly spreading, rather closely attached throughout; lobes ± inflated, tubular, hollow. **Photobiont** chlorococcoid. **Upper surface** smooth, with rounded perforations. **Lower surface** corticate, naked, uneven, blackened, without rhizines or perforations; walls of internal cavity white above, black at the base. **Soredia** present. **Ascomata** apothecia, laminal, rounded, cup-shaped, sessile to ± stalked. **Thalline exciple** well developed, thin. **Disc** concave, ± shining. **Epithecium** brown, non-granular. **Hymenium** colourless, I+ blue. **Hypothecium** of thick-walled, conglutinated hyphae. **Hamathecium** of paraphyses, richly branched and anastomosed, apical cells not capitate, ± pigmented. **Asci** 2-spored (some 8-spored species in

S. Hemisphere). **Ascospores** colourless, simple, ellipsoid, thick-walled. **Conidiomata** pycnidia, immersed, punctiform, laminal; apex pigmented. **Conidia** bacilliform. **Chemistry**: atranorin in the cortex and various depsidones in the medulla. **Ecology**: on acid bark and siliceous rocks in oceanic woodlands.

Primarily a temperate Southern Hemisphere genus of *c.* 50 species, of which only *M. terebrata* is widespread in the Northern Hemisphere. Similar to *Hypogymnia*, but that genus lacks perforations on the upper surface, contains physodic acid and has smaller ascospores.

1. M. terebrata (Hoffm.) Massal. (1854)

Thallus to 5(-10) cm diam., often forming neat, closely adpressed rosettes or sometimes irregularly spreading; lobes 0.5–1.5 mm wide, ± rounded at apices, indented, branched, radiating, contiguous for most of their length, more rarely partially discrete, sometimes ± overlapping in older parts of the thallus; upper surface grey or green-grey, often with brownish margins, smooth, shiny, with scattered, round perforations, <1 mm diam.; soralia scattered, or sometimes contiguous, ± convex, occasionally ± elevated; lower surface black, wrinkled, without rhizines. Apothecia very rare, disc pale or red-brown. Ascospores 50–68 × 30–36 μm, ellipsoid; wall *c.* 5 μm thick. Pycnidia black, rare. Conidia 4–5 × 1 μm, bacilliform. Cortex K+ yellow (atranorin), medulla Pd+ yellow-orange, K+ yellow, KC−, C−, UV− (stictic, constictic, menegazziaic, ± norstictic acids).

On leached acid bark of broad-leaved trees, such as *Betula*, *Alnus* and *Quercus*, more rarely conifers, also on mossy siliceous rocks in humid woodlands, rarely in humid coastal sites; locally frequent. W. British Is. W. Europe, Madeira, Azores, North and Central America, China, U.S.S.R., Japan.

Resembles *Hypogymnia physodes* in colour and inflated lobes, but distinguished by the scattered, surface perforations, globose soralia on the upper surface and Pd+ orange medulla.

MICAREA Fr. (1825)

Thallus crustose or immersed, effuse, of ± spherical granules (goniocysts), convex to ± globose areoles, or a rimose or scurfy crust, the areoles sometimes dissolving into soredia; prothallus indistinct, ± ecorticate but sometimes with a colourless amorphous outer layer. **Photobiont** green, cells usually thin-walled, 4–7 μm diam., and often paired ('micareoid'); a few species with cephalodia containing *Nostoc* or *Stigonema*. **Ascomata** apothecia mostly under 1 mm diam., whitish, grey, bluish, brown or black, not pruinose, usually immarginate, adpressed, sessile or rarely ± stipitate. **Thalline exciple** absent. **True exciple** absent to well developed, composed of branched radiating hyphae, lax in K or bound by dense pigment. **Hymenium** I+ blue. **Hypothecium** colourless or pigmented. **Hamathecium** of mostly branched paraphyses, some species also with stouter unbranched paraphyses; apices not or slightly swollen, without a dark brown apical 'cap'. **Asci** 8-spored, clavate to cylindrical-clavate; in K/I with blue outer layer and apical dome and unstained wall, the apical dome with an apical cushion that in some species is surrounded by a faint to distinct dark cylindrical 'ring-structure'. **Ascospores** colourless, smooth, without perispore, simple to transversely septate, ellipsoid, ovoid, fusiform or acicular. **Conidiomata** pycnidia, or rarely sporodochia; pycnidia immersed, sessile or stalked. **Conidiogenous cells** enteroblastic, short-ampulliform to cylindrical, acrogenous. **Conidia** colourless, of 3 types: (a) *macroconidia*, curved, thread-like or helicoid, often septate, rarely ellipsoid; (b) *mesoconidia*, cylindrical, ellipsoid or ovoid, simple, often biguttulate, mostly 1–2 μm wide; (c) *microconidia*, ± cylindrical, simple, eguttulate, to 1 μm wide. **Chemistry**: no lichen products detected by t.l.c., or various, several species with gyrophoric acid. **Ecology**: on wide range of acidic substrata; rarely tolerant of nutrient enrichment.

A cosmopolitan genus with *c.* 50 species.

LITERATURE: Coppins (1983, 1985, 1988b, 1989d), Tønsberg (1992).

1	Ascopores needle-like, or 3- or more septate	2
	Ascopores ellipsoid, ovoid, oblong or fusiform, mainly 0- to 1-septate	17

2(1) Hypothecium colourless, pale or mottled brown or greenish 3
Hypothecium dark purple-brown .... **13. *melaena***

3(2) Upper hymenium dull greenish or brownish, K+ violet 4
Upper hymenium colourless, greenish or brownish, K− 6

4(3) Apothecial sections C+ red, the K+ violet pigment also C+ violet 5
Apothecial sections C− .... **7. *synotheoides***

5(4) Ascopores fusiform, 10–17(-19) × 2.5–3(-3.5) μm .... **5. *nitschkeana***
Ascospores acicular, 13–26 × 1.5–2.5(-3) μm .... **6. *globulosella***

6(3) Apothecial sections C+ red 7
Apothecial sections C− 11

7(6) Ascospores fusiform, more than 3 μm wide; pycnidia ± immersed or absent 8
Ascospores needle-like to sigmoid, less than 3 μm wide; short-stalked pycnidia present .... **22. *pycnidiophora***

8(7) Thallus of firm areoles or granules, Pd− 9
Thallus of fragile granules, often leprose-granular, Pd+ red .... **19. *leprosula***

9(8) Ascospores 3(-5)-septate, mostly <26 μm long; macroconidia 20–50 × 1μm 10
Ascospores (3-)5- to 7-septate, (19-)23–34(-38) μm long; macroconidia 50–110 × 1 μm .... **18. *cinerea***

10(9) Apothecia whitish .... **17. *alabastrites***
Apothecia pale grey to black (rarely whitish in extreme shade) .... **16. *peliocarpa***

11(6) Ascospores fusiform, >3 μm wide; apothecia grey to black; pycnidia ± immersed or absent 12
Ascospores needle-like to sigmoid, <3 μm wide; apothecia whitish; stalked pycnidia present .... **21. *stipitata***

12(11) Thallus Pd+ red or deep yellow 13
Thallus Pd− 14

13(12) Thallus of fragile granules, often becoming leprose-granular, C+ red, Pd+ deep yellow .... **20. *granulans***
Thallus of firm granular areoles (or sometimes ± immersed when on wood), Pd+ red, C− .... **14. *lignaria***

14(12) Thallus blackish or inconspicuous, C− 15
Thallus of yellowish white granular areoles, C+ persistent orange .... **13. *lignaria***

15(14) Hypothecium colourless to pale greenish, K− 16
Hypothecium mottled reddish brown, K− or + orange-brown .... **28. *turfosa***

16(15) Ascospores 0- to 1(-3)-septate; true exciple inconspicuous in section .... **29. *viridiatra***
Ascospores (0-)3-septate; true exciple distinct in section .... **15. *ternaria***

17(1) Upper hymenium greenish or dull brown, K+ violet 18
Upper hymenium colourless, green or brown, K− (rarely purple-brown, K+ purple intensifying in *M. crassipes*) 23

18(17) Hypothecium ± colourless 19
Hypothecium dark purple-brown .... **25. *subviolascens***

19(18) Pycnidia stalked 20
Pycnidia immersed to ± emergent, or absent 21

20(19) Pycnidia brown, thinly tomentose; thallus superficial, dull green, of 20–40 μm diam. granules; apothecia brown .... **2. *hedlundii***
Pycnidia black, not tomentose; thallus immersed, rarely with pale greyish, 60–120 μm diam., granular areoles; apothecia black .... **8. *misella***

21(19) Thallus immersed or formed of greenish grey to grey granular areoles; apothecial sections and thallus often C+ red 22

Thallus superficial, of minute, pale to dark greenish, scattered to densely aggregated granules; apothecial sections C−, but hymenial pigment is C+ violet; thallus C−, or rarely C+ persistent orange .... **1. *prasina***

22(21) Ascospores mostly 1-septate, often curved, (7-)9–16(-18) × 2–3(-3.5) μm; paraphyses numerous, 1–1.5 μm wide; apothecial and thallus sections usually C+ red ... **4. *denigrata***
Ascospores mostly simple, never curved, 6.5–9.5 × 2–3(-3.7) μm; paraphyses scanty, 0.5–1 μm wide; apothecial and thallus sections usually C− .... **8. *misella***

23(17) Apothecia whitish to pale reddish brown, without distinct pigmentation in section 24
Apothecia coloured, with obvious pigmentation in section 27

24(23) Thallus ± smooth, rimose, scurfy or granular-areolate, C−, without detectable substances by t.l.c. 25
Thallus of minute, scattered to densely aggregated granules, C− or rarely C+ orange, usually with substances by t.l.c. .... **1. *prasina***

25(24) Apothecia convex-hemispherical from the beginning, if tuberculate then clusters less than 0.7 mm diam.; ascospores simple, 6.5–10.5 × 2.5–4 μm; sporodochia absent 26
Apothecia at first ± flat and adpressed, later convex and often in large nodulose clusters to 2 mm across; ascospores 1-septate, 9–16 × 3–5 μm; sporodochia usually present **3. *adnata***

26(25) Photobiont cells 5–12 μm diam., or ellipsoid <15 × 10 μm; hypothecium colourless to pale greenish in upper part .... **30. *bauschiana***
Photobiont cells 4–8 μm diam.; hypothecium colourless or straw-yellow to pale orange-brown .... **34. *lithinella***

27(23) Hypothecium colourless to pale yellow-straw, or pale greenish in uppermost part 28
Hypothecium brownish, greenish or purplish 31

28(27) Ascospores simple, 6.5–10.5 × 2.5–4 μm 29
Ascospores 0- to 1-septate, 9–14 × 4–6 μm 30

29(28) True exciple absent (sections); photobiont cells 5–12 μm diam. .... **28. *bauschiana***
True exciple distinct in sections; photobiont cells 4–7 μm diam. .... **33. *excipulata***

30(29) Upper hymenium brown; photobiont cells 4–8 μm diam.; pycnidia usually present .... **38. *subnigrata***
Upper hymenium green; photobiont cells 7–21 μm diam.; pycnidia absent .... **39. *intrusa***

31(27) Hypothecium fuscous or reddish brown, without distinct purple or green tinges; N− or N+ orange-brown 32
Hypothecium with distinct purple or green tinges; N+ purple-red 37

32(31) Pycnidia inconspicuous and immersed, or absent 33
Pycnidia conspicuous, sessile to stalked, black .... **11. *botryoides***

33(32) Ascospores <10 μm in length 34
Ascospores >10 μm in length 36

34(33) Apothecia 0.2–0.8 mm diam.; hypothecium dark reddish brown (blackish in thick sections); hymenium colourless or greenish 35
Apothecia 0.1–0.25 mm diam.; hypothecium orange- or reddish brown (not blackish in thick sections); hymenium colourless or tinged orange-brown .... **37. *myriocarpa***

35(34) Thallus smooth to scurfy granular; paraphyses 0.8–1(-1.5) μm wide .... **35. *lutulata***
Thallus of convex, rounded areoles; paraphyses 1.5–2 μm wide .... **26. *submoestula***

36(33) Thallus of whitish to grey-brown, convex areoles; often with brown cephalodia; apothecia 0.3–1 mm diam. .... **23. *incrassata***
Thallus thin, scurfy, blackish; without cephalodia; apothecia 0.15–0.3 mm diam. **28. *turfosa***

37(31) Apothecia convex to subglobose from the beginning, sessile or partly immersed 38
Apothecia marginate at least when young, markedly constricted below, appearing turbinate or short-stalked .... **27. *crassipes***

38(37) Photobiont cells 4–8 μm diam. 39
Photobiont cells 5–12 μm diam., or ellipsoid <15 × 10 μm 46

39(38) Hypothecium green in water and K– 40
Hypothecium purplish brown in water, K+ purple intensifying or ± green in part 42

40(39) Hypothecium dark green or purplish; ascospores 7–13(-14)μm in length 41
Hypothecium pale green; ascospores (10-)12–17(-19) μm in length .... **29. *viridiatra***

41(40) Paraphyses scanty, 0.6–0.8 μm wide when not pigmented; hypothecium green or purple (often intermixed); ascospores ovoid to oblong-ovoid, 3–4.5 μm wide .... **12. *contexta***
Paraphyses numerous, 1–1.5 μm wide when not pigmented; hypothecium green (never purple); ascospores oblong to ovoid-oblong, 2.5–3.5 μm wide .... **10. *olivacea***

42(39) Pycnidia ± immersed in thallus or absent; ascospores mostly >9 μm in length, 0–1(-3)-septate 43
Pycnidia sessile to stipitate; ascospores <8.5 μm in length, simple .... **9. *nigella***

43(42) Thallus scurfy or granular-areolate, the granules mostly <0.2 mm diam.; cephalodia absent 44
Thallus of whitish, warted areoles, 0.1–0.4 mm diam., often with brownish cephalodia .... **24. *assimilata***

44(43) Ascospores 1(-3)-septate 45
Ascospores simple .... **36. *pseudomarginata***

45(44) Paraphyses numerous, monomorphic, <1 μm wide; microconidia 4.5–7 × 0.5–1 μm; immature forms of .... **13. *melaena***
Paraphyses scanty, dimorphic, 0.5–1 or 1.5–2 μm wide; microconidia 4–5 × 0.5–1 μm .... **12. *contexta***

46(38) Ascospores 0(-1)-septate, ellipsoid or ovoid-ellipsoid, (6-)7–10 × (2.5-)3–4.5 μm wide; apothecia 0.2–0.5 mm diam., to 1.2 mm if tuberculate; conidia 3.5–6(-6.5) × 1–2 μm .... **31. *sylvicola***
Ascospores 0- to 1-septate, oblong-ovoid or oblong-ellipsoid, 5.5–8(-9) × 1.5–2.5 μm wide; apothecia 0.1–0.3 mm diam., to 0.55 mm if tuberculate; conidia 3–4.5 × 1–1.5 μm .... **32. *tuberculata***

1. M. prasina Fr. (1825)

M. subviridescens (Nyl.) Hedl. (1892)

Thallus light green, pale to dark grey-green, often ± gelatinous when wet, of thinly scattered to densely aggregated, ± globose goniocysts; goniocysts 12–40(-60) mm diam., the outer hyphae often coated by greenish, K+ violet pigment. Photobiont cells 4–7 μm diam. Apothecia 0.1–0.4 mm diam., to 0.6 mm if tuberculate, shallow-convex to subglobose or tuberculate, whitish, pale to dark grey, brownish grey, or blackish; true exciple poorly developed; epithecium colourless, or dull greenish and K+ violet; hymenium *c.* 28–50 μm tall; hypothecium colourless to dull yellowish. Paraphyses 0.5–1 μm wide, to 1.5 μm towards apices, numerous. Asci 26–40(-50) × 8–12 μm. Ascospores 8–14(-17) × 2.3–4(-5) μm, ovoid-ellipsoid to oblong, 0- to 1(-3)-septate. Pycnidia white, or greyish around ostiole (due to greenish, K+ violet pigment): (a) 50–120 μm diam., emergent to sessile, the ostiole often gaping, with ± cylindrical to narrowly obpyriform mesoconidia, (3.5-)4–6 × 1–1.5 μm; (b)30–60(-100) μm diam., usually immersed, with cylindrical or narrowly fusiform microconidia, (5-)5.5–8 × 0.5–1 μm. Thallus and apothecia Pd–, K–, C– (except for the greenish K+ violet, C+ violet, N+ red pigment), rarely thallus C+ persistent orange (four chemotypes: methyloxymicareic acid; micareic acid; prasinic acid; unidentified xanthone(s) [thallus C+ orange]; ± trace gyrophoric acid-?contaminant).

On bark, stumps, fallen debris (especially in conifer plantations), soil or debris in rock crevices in coastal districts, rarely on shaded sandstone rocks; usually in shaded situations or niches; common and widespread, even close to centres of large conurbations. Throughout British Is. Temperate Europe, North America; related morphs also in Africa, South & Central America, Australia, New Zealand.

Extremely variable and frequently sterile. Within the British Isles, chemotypes show differing tendencies: methyloxymicareic acid - widespread and 'weedy' on various substrata; micareic acid - on wood and tree trunks in ancient woodlands; prasinic acid - on coastal turf in S.W. England; xanthones on stumps in native pinewoods of Scotland. Usually recognizable by the minutely granular thallus, but often confused with *M. adnata* **(3)**, *M. denigrata* **(4)**, *M. hedlundii* **(2)** and *M. misella* **(8)**, and also *Lecania cyrtella*, which has simple paraphyses with swollen apices.

2. M. hedlundii Coppins (1983)

Like **1**, but with numerous stalked, brown, thinly white-tomentose pycnidia, 0.1–1 mm tall, 0.07–0.14 mm diam., the stalk often branched, with up to 5 pycnidia; mesoconidia (4-)4.5–5.5(-6) × 1.5–2.0 μm, oblong-ellipsoid or oblong ovoid. The goniocysts often contain a yellowish K+ purple-violet pigment (as oily droplets). Ascospores 6.5–10(-12) × 2.5–4 μm, 0(-1)-septate. ± traces of micareic acid (?contaminant).

On tree stumps in ancient *Quercus* or *Pinus* woods; rare. S. Wales, Scotland (Highlands), N. Ireland. Norway, Sweden, Bavaria, Austria, North America.

3. M. adnata Coppins (1983)

Thallus ± smooth or scurfy, pale grey-green, ± waxy. Photobiont cells 4–7 μm diam. Apothecia pale to pale reddish brown, at first ± flat with adpressed, white rim, later convex, 0.2–0.4(-0.6) mm diam. or in nodulose clusters to 2 mm across; true exciple indistinct but evident in young apothecia, colourless; hymenium 35–40 μm tall, colourless; hypothecium colourless. Paraphyses 1–1.5 μm wide, numerous. Asci 30–35 × 10–12 μm. Ascospores 9–16 × 3–5 μm, ellipsoid or ovoid to oblong, (0-)1-septate. Conidiomata: (a) 0.08–0.25 mm diam., whitish, convex sporodochia; macroconidia 6.5–9.5 × 2.3–3 μm, ellipsoid to oblong, simple; (b) 40–60 μm diam., immersed, white; mesoconidia 4–5.5 × 1–1.5 μm, cylindrical. Lichen products not detected.

On wood and debris of old stumps, or on loose bark of old trees, in old *Betula*, *Quercus* or *Pinus* woodlands; rarely on dry peat hags. N.W. England, Wales, W. Scotland, N. Ireland. Sweden, S. Germany, Austria, Switzerland, Madeira.

Distinguished from *M. prasina* **(1)**, by ± smooth, waxy thallus (no goniocysts), apothecia-like sporodochia, and the absence of K+ violet pigment.

4. M. denigrata (Fr.) Hedl. (1892)

Thallus usually of greenish white to pale grey, convex areoles, 60–200 μm diam., sometimes scurfy and blackish (when parasitized), or immersed; areoles not corticate, the outer part ± with greenish, K+ violet pigment. Photobiont cells 4–7 μm diam. Apothecia 0.15–0.5 mm diam. or tuberculate and to 0.6 mm., dark grey to black, pale to brown in shade morphs, convex to subglobose; true exciple usually not evident at maturity; hymenium 25–40 μm tall, the upper part dull greenish, K+ violet; hypothecium colourless. Paraphyses 1–1.5(-1.7) μm, numerous. Asci 28–36 × 9–12 μm. Ascospores (7-)9–16(-18) × 2–3.3(-3.5) μm, oblong-ellipsoid, oblong-ovoid, fusiform or bacilliform, often ± curved, (0-)1-septate. Pycnidia immersed to ± emergent, whitish to black, usually with greenish, K+ violet pigment: (a) 60–150 μm diam., the ostiole often gaping, the macroconidia 12–24 × 1 μm, curved, (1-)3-septate; (b) 80–160 μm diam., the ostiole often gaping, mesoconidia 2.5–4.5(-5) × 1–2 μm, short-cylindrical or obovoid; (c) 30–50 μm diam., ostiole not gaping; microconidia, (4.5-)5–7.5 × 0.5–1 μm, cylindrical. Except for the greenish, Pd−, K+ violet, C+ violet pigment all parts K−, usually C+ orange-red (gyrophoric acid).

Usually on wood, on fallen trunks and stumps; an early colonizer on a wide variety of worked timber; more rarely on other substrata, e.g. shaded sandstone rocks and brickwork, dead bark, and moribund mosses in acid dunes; common. Throughout British Is.; common and widespread even in polluted areas. Throughout Europe, North America; rare to absent in many arctic and lowland mediterranean areas.

Polymorphic, usually identified by the C+ orange-red reaction of apothecial sections, and narrow, 1-septate ascospores. Often sterile on worked timber, with numerous pycnidia extruding mesoconidia as white blobs. Often confused with *M. nitschkeana* **(5)** (ascospores 3-septate), *M. prasina* **(1)** (thallus of goniocysts, gyrophoric acid absent) and *M. misella* **(8)** (ascospores mostly simple, paraphyses less than 1 μm wide, stalked pycnidia usually present).

5. M. nitschkeana (Lahm ex Rabenh.) Harm. (1899)

Like **4**, but the ascospores 10–17(-19) × 2.5–3(-3.5) μm, fusiform, often curved, (1-)3(-4)-septate.

Usually on twigs or small branches of trees and shrubs (especially *Calluna*), often in the *Bacidietum chlorococcae*; less often on wood and, rarely, stones in heathland; common. Throughout British Is. (especially common in E. Britain). Throughout most of Europe, North America.

6. M. globulosella (Nyl.) Coppins (1983)

M. bacidiella sensu Vězda, non (Vainio) Vězda (1976)

Like **4** and **5**, but distinguished by the bacilliform or ± needle-like curved, 0- to 3(-6)-septate ascospores, 13–26 × (1.5-)2–2.5)-3) μm, shorter microconidia, 3.5–5 × 0.5–1 μm, and the apparent absence of macroconidia.

On fallen decorticate *Quercus* branch in Hereford (Croft Park) and on top of a gate in a wooded valley in Wales. Europe, with scattered records on conifer bark, old polypores and wood, Canada (Quebec), Japan.

M. synotheoides **(7)** differs in the dark green granular areoles (± gelatinous when wet), often longer ascospores, and lack of gyrophoric acid. *Bacidia beckhausii* has a larger-celled photobiont, ± pruinose apothecia, excipular hyphae coherent (in K), and occurs on less acid bark or wood.

7. M. synotheoides (Nyl.) Coppins (1982)

Thallus grey-green to dark olivaceous, ± gelatinous when wet, of scattered to aggregated granular areoles 20–70 μm diam.; areoles not corticate, the outer hyphae often greenish or brownish, K+ violet. Photobiont cells 4–7 μm diam. Apothecia 0.1–0.3 mm diam., grey- or brown-black, convex to ± globose, or tuberculate to 0.5 mm diam.; true exciple indistinct; hymenium 35–45 μm tall, the upper part dull greenish, K+ violet; hypothecium colourless, or pale greenish and K+ weakly violet. Paraphyses 0.5–1 μm wide, sometimes to 1.5 μm towards apices. Asci 30–40 × 9.5–12 μm. Ascospores 14–35(-43) × 1.8–2.5(-3) μm, bacilliform to curved-acicular, 1- to 7(-11)-septate. Pycnidia immersed, emergent to sessile, whitish to grey-black; walls dull greenish, K+ violet: (a) 60–120 μm diam., the ostioles often gaping; mesoconidia 4.5–6 × 1–1.5 μm cylindrical to fusiform; (b) 30–40 μm diam.; microconidia 3.5–4.0 × 0.5–1 μm, short-cylindrical. Lichen products not detected.

On trunks of *Betula*, *Quercus* and *Pinus* etc. in communities allied to the *Parmelion laevigatae*, more rarely on wood. N. Wales, W. Scotland, W. Ireland. N. Spain, Macaronesia, Japan.

Easily confused with *M. prasina* **(1)** or *M. melaena* **(13)**, but microscopically more like *M. globulosella* **(6)** or *Bacidia beckhausii*.

8. M. misella (Nyl.) Hedl. (1892)

Thallus immersed in British specimens, inconspicuous. Photobiont cells 4–7 μm diam. Apothecia 0.1–0.3 mm diam., black, convex to ± globose; true exciple poorly developed; hymenium 25–36 μm tall, the upper part greenish, K+ violet; hypothecium colourless, or pale greenish and K ± violet. Paraphyses scanty, 0.5–1 μm wide, the apices sometimes to 1.5 μm. Asci 25–35 × 7–10 μm. Ascospores (6.5-)7–9.5 × 2–3(-3.7) μm, ellipsoid, ovoid or oblong-ovoid, never curved, 0(-1)-septate. Pycnidia black; walls green-brown, K+ violet: (a) sessile or on unbranched, non-tomentose stalks, 70–320 μm tall and 50–100 μm wide, the mesoconidia 3.5–5 × 1–1.5(-1.7) μm, short-cylindrical; (b) ± immersed, *c.* 40 μm diam., microconidia 3.5–6 × 0.5–1 μm, narrowly cylindrical. Thallus, apothecia and pycnidia K−, C− (except for K+ and C+ violet, greenish pigment).

On wood of fallen trees or stumps in *Betula* or *Pinus* woodlands and conifer plantations; local in England, more frequent in Scotland. Europe, especially in coniferous forests, Canada, Brazil.

Easily recognized, even when sterile, by the immersed thallus and numerous, usually stalked, black pycnidia with greenish, K+ violet pigments. Non-British morphs, with a superficial thallus (± gyrophoric acid) and few, if any, stalked pycnidia can be difficult to distinguish from immature *M. denigrata* **(4)**. Dark-coloured stalked pycnidia occur also in *M. botryoides* **(11)**, *M. hedlundii* **(2)** and *M. nigella* **(9)**.

9. M. nigella Coppins (1983)

Thallus immersed and inconspicuous, or thin and pale greenish grey. Photobiont cells 4–7 μm diam. Apothecia 0.1–0.3 mm diam., black, subglobose or tuberculate; hymenium 25–30 μm tall, colourless or with purple-brown streaks; upper part purple-brown, K+ dull green; hypothecium mottled, dark purplish brown, K+ dull green. Paraphyses scanty, dimorphic: 0.5–1 μm wide, the apices sometimes to 2 μm, colourless, or 2–3 μm wide and coated throughout by dark pigment. Asci 22–27 × 10–11 μm. Ascospores 6.5–12 × 2.5–4 μm, ellipsoid to oblong-ovoid, simple. Pycnidia 60–300 μm tall, 40–80 μm diam., numerous, black, sessile or stalked; wall dark purplish brown, K+ green; mesoconidia 3.4–4.3 × 1.2–1.6 μm, ellipsoid or short cylindrical. Lichen products not detected.

On ± soft wood of conifer and *Betula* stumps in shaded woodlands; rare. N.E. England, Scotland, N. Ireland. Denmark.

Distinguished from other species with stalked pycnidia by the purplish brown, K+ green pigmentation.

10. M. olivacea Coppins (1983)

Thallus immersed, weakly areolate, or scurfy, inconspicuous or pale grey to green-grey. Photobiont cells 4–7 μm diam. Apothecia 0.1–0.25 mm diam., black, convex to subglobose, or tuberculate to 0.4 mm; true exciple indistinct; hymenium 30–35 μm tall, pale olivaceous with darker streaks, K+ green intensifying; hypothecium dark olivaceous or olive-brown, K+ green intensifying. Paraphyses numerous, of two types: 1–1.5 μm wide, sometimes to 2 μm at apices, colourless, branched; 2–3 μm wide, ± simple and often surrounded by dark pigment. Asci 33–35 × 9–11 μm. Ascospores (7-)9–12.3 × 2.5–3.5 μm, oblong to ovoid-oblong, (0-)1-septate. Pycnidia 25–50 μm diam., ± immersed, the walls sordid greenish, K+ green intensifying; conidiogenous cells 4–5.5 μm long; mesoconidia 3–4.5 × 1–1.5 μm, cylindrical. No lichen products detected by t.l.c.

On hard wood of conifer stumps and shaded rocks in woodland; very rare, known from two sites. Scotland; ?endemic.

Specimens may be confused with non-pruinose morphs of *Lecidea turgidula*, which has a true exciple of conglutinate hyphae (in K), a paler hypothecium, and photobiont cells, 12–16 μm diam. *M. tuberculata* **(32)** is distinguished by the longer conidiogenous cells (5–10 μm).

11. M. botryoides (Nyl.) Coppins (1980)

Thallus usually thin, ± inconspicuous when on mosses, pale to dark dull green or blackish, rarely pale buff in deep shade, scurfy-granular, granules 15–50 μm diam. Photobiont cells 4–7 μm diam. Apothecia 0.1–0.25 mm diam., or tuberculate and to 0.5 mm diam., black, matt, convex to globose; true exciple indistinct, reflexed, reddish brown; hymenium 25–35 μm tall, colourless or partly tinged brown or olivaceous but always with dark brown (K+ olivaceous) vertical steaks; hypothecium dark reddish brown, K– or dulling, N– or red tinge intensifying. Paraphyses rather scanty, of two types: 0.5–1 mm wide, to 1.5 μm at apices, colourless, often branched; 2–2.5 μm wide, to 3.5 μm, at apices, fasciculate, coated by brownish pigment. Asci 20–35 × 7–9 μm. Ascospores 8–13(-16) × 2.3–4 μm, ovoid to oblong-ellipsoid, sometimes ± curved, 0- to 1(-3)-septate. Pycnidia to 400 μm tall and 40–90 μm wide, numerous, stalked, black; stalk often branched; wall dark greenish brown, K± green intensifying, N+ red, but stalk tissue ± reddish brown, K–, N– or red tinge intensifying. Conidia 3.5–5 × 1–1.5 μm, ± cylindrical, often biguttulate. Lichen products not detected by t.l.c.

Usually in shaded underhangs on rock, roots, soil and bryophytes, rarely on soft wood of stumps and, in polluted areas, on tree bases in sheltered woodland; common. Mainly W. and upland Britain. N.W. Europe, Azores.

Commonly sterile and most often seen as tiny black projections (pycnidia) from moribund mats of bryophytes in underhangs. Sterile morphs on wood differ from *M. misella* **(8)** and *M. nigella* **(9)** in pigmentation. On rock, *M. lutulata* **(35)** has similar apothecia but immersed pycnidia, smaller, simple ascospores and a larger-celled photobiont.

12. M. contexta Hedl. (1892)

Thallus immersed, ± inconspicuous. Photobiont cells 4–7 μm diam. Apothecia 0.1–0.2(-0.3) mm diam., ± globose ot tuberculate, black; true exciple indistinct; hymenium 33–45 μm tall, dark green (N+ red) above, paler below, ± with purple-violet (K+ blue-green) granules; hypothecium dark green or purple, the pigments often intermixed, K+ green, N+ red. Paraphyses rather scanty, of two types: less than 1 μm wide, with apices ± thickened to 2 μm wide, branched, colourless; some simple, ± in small fascicles, pigmented throughout, 1.5–2 μm wide or to 3 μm at apices. Asci 35–40 × 10–14 μm. Ascospores 7–13(-14) × 3–4.6 μm, ovoid to oblong-ovoid, 1-septate. Pycnidia immersed to sessile, black; walls greenish, K+ intensifying: (a) *c.* 40 μm diam., the mesoconidia 3.5–4.5 × 1.5–2 μm, oblong; (b) 20–40 μm diam., the microconidia 4–5 × 0.5–1 μm, bacilliform. Thallus C–.

On rotting wood of fallen *Pinus* branches. N.E. Scotland (E. Inverness; S. Aberdeen). C. Sweden.

Specimens of *M. melaena* **(13)** with immature ascospores differ in the usually better developed thallus, more numerous paraphyses, and longer microconidia. *M. olivacea* **(10)** has different paraphyses, narrower, ± oblong ascospores, and lacks purple pigmentation.

13. M. melaena (Nyl.) Hedl. (1892)

Thallus of scattered, pale buff, dull green or dark grey-green, granular areoles, 20–100 μm diam., often blackish and scurfy due to disruption by dark brown mycelial fungi and foreign algae. Photobiont cells 4–7 μm diam. Apothecia 0.1–0.4 mm diam. or tuberculate and to 0.5 mm, black, convex to ± globose; true exciple indistinct; hymenium 30–45 μm tall, usually blue-green (K ± intensifying), sometimes purplish (K+ intensifying or K+ green) below or rarely throughout; hypothecium dark, variously pigmented, usually purple-brown (K ± intensifying purple), the outer part or ± whole often K+ green. Paraphyses 0.5–1 μm wide, numerous, mostly branched, sometimes a few 1.5–2 μm wide, simple, coated with dense pigment. Asci 30–40 × 12–15 μm. Ascospores 12–21 × 4–5.5 μm., ovoid-oblong or oblong, straight, (1-)3(-5)-septate, the apices obtuse. Pycnidia rare, partly immersed to sessile, the walls purple-brown or greenish: (a) 100–140 μm diam., with macroconidia 18–33 × 1–1.5 μm, curved, 0- to 7-septate; (b) 40–60 μm diam., with microconidia 4.5–7 × 0.5–1 μm, fusiform-cylindrical. Thallus Pd−, K−, C+ red or C−; apothecial sections C− (± gyrophoric acid).

Usually on wood of stumps, fallen trees, and decaying timberwork, but also acid bark and peaty turf, rarely on rock in woodlands; common. Upland areas of N. & W. Britain, extending locally to S. England and Ireland. Europe, Azores, North America.

Variable in appearance and internal pigmentation, usually easily identified microscopically; *M. turfosa* **(28)** and *M. viridiatra* **(29)** are superficially similar, but restricted to turf in areas over *c.* 1000 m, where *M. melaena* is rare. Easily overlooked as a black or charred stain when on the tops of stumps or on peat (cf. also *Placynthiella*); on bark it can be confused in the field with *M. synotheoides* **(7)** or dark morphs of *M. prasina* **(1)**.

14. M. lignaria (Ach.) Hedl. (1892)

Thallus of scattered to confluent, ± globose, whitish, grey-green or bluish grey, glossy or matt areoles 80–300 μm diam., sometimes ± immersed. Photobiont cells 4–7 μm diam. Apothecia 0.15–0.6(-0.9) mm diam., black or blue-black, convex to ± globose, rarely with stalk to 1 mm tall [**f. *gomphillacea*** (Nyl.) Hedl. (1892)]; true exciple ± distinct in young apothecia, soon reflexed, brownish in outer parts; hymenium 50–75 μm tall, colourless, or upper and sometimes lower part olivaceous to blue-green, K−, rarely a few scattered, minute, violaceous (K+ blue-green) granules present; hypothecium pale blue-green or olivaceous, K−, but central part often ± colourless or pale brownish. Paraphyses 1.3–1.8 μm wide, numerous simple, or sparingly branched above; apices often ± swollen and coated in green or greenish brown pigment. Asci 45–50 × 11–19 μm. Ascospores 16–36(-38) × 4–6(-7) μm, fusiform, straight or slightly curved, 3- to 7-septate. Pycnidia ± immersed, the walls green (K-) in upper parts: (a) *c.* 100 μm diam., with macroconidia 16–22 × 1 μm, curved, 0- to 3-septate; (b) 100–140 μm diam., with mesoconidia 4–7 × 1.5–2 μm, ± cylindrical to obovoid; (c) 40–50 μm diam., with microconidia 5–7 × 0.8 μm, narrowly-cylindrical; only (c) is common. Two chemically separated varieties occur: var. *lignaria*, thallus Pd+ red, K−, KC−, C− (argopsin); **var. *endoleuca*** (Leighton) Coppins (1983), thallus Pd−, K−, KC+ and C+ persistent orange (unidentified xanthones). Apothecial sections C−.

On acidic substrata, often over mosses and plant debris; widespread and common, especially in upland districts. Throughout British Is. Cool temperate Europe, North America, extending to Arctic regions.

Var. *endoleuca* is less common; W. Britain, Ireland. ?Endemic.

M. cinerea **(18)** and *M. peliocarpa* **(16)** differ in the C+ red apothecial sections (gyrophoric acid). *Catillaria contristans* has a Pd−, C− thallus and 1-septate ascospores.

15. M. ternaria (Nyl.) Vězda (1970)

Like **14**, but ascospores 13–22 × 3.5–5 μm, never more than 3-septate, and no lichen products detected by t.l.c. Also, the apothecia have a more flattened appearance, with a more well-defined true exciple in section.

On coastal, hard siliceous rocks. N.W. Scotland (Shetland, Fair Is.), W. Ireland.

British collections have a ± obsolete thallus and the identification is tentative; the usual morph on mosses and debris in arctic Europe and Alaska has a well-developed thallus of convex to subglobose areoles.

16. M. peliocarpa (Anzi) Coppins & R. Sant. (1979)

Like **14**, but areoles 40–200 μm diam., usually less glossy. Apothecia 0.14–0.4(-0.6) mm diam., to 1 mm if tuberculate, adpressed, flat to convex, whitish (shade) or blue-grey to black; true exciple well developed, colourless; hymenium 40–55 μm tall, usually greenish to blue-green (K −, N + red) in upper part; hypothecium ± colourless. Paraphyses 1–1.5 μm wide, numerous, branched especially above; apices often thickened to 2.5 μm by green pigment. Asci 40–55 × 12–17 μm, clavate, without ring-structure. Ascospores (11-)15–23(-24) × 3–5(-6) μm, fusiform or oblong, ± curved, (1-)3(-5)-septate. Pycnidia frequent (especially on bark or wood): (a) 140–200 μm diam., immersed, white or greenish around often gaping ostioles, the macroconidia 21–40(-5) × 1–1.5 μm, curved or sigmoid, often faintly 1- to 5-septate; (b) 50–80 μm diam., ± sessile, whitish, the microconidia 6–7 × 0.5–1 μm, fusiform-cylindrical. Thallus, apothecia and pycnidia Pd −, K −, C + red (gyrophoric and 5-*O*-methylhiasic acids).

On a wide range of acid substrata; widespread and frequent, especially in uplands (but rare above 500 m). Britain. W. Europe, extending E. in montane areas, North America, Madagascar, Australia, New Zealand.

Very variable, and often confused with *M. lignaria* **(14)**, which has C − apothecial sections. *M. nitschkeana* **(5)** has smaller ascospores and a K + violet epithecium. *M. leprosula* **(19)** has similar apothecia but a Pd + red, sorediate thallus. See also *M. alabastrites* **(17)** and *M. cinerea* **(18)**. *Bacidia naegelii* has unbranched paraphyses, is C − and occurs on nutrient-enriched bark.

The recently described ***M. coppinsii*** Tønsb. (1992) has similar apothecia and an identical chemistry, but is usually sterile with numerous, discrete green soralia, resembling *Trapelia corticola*; on acid bark of *Betula*, *Calluna*, *Myrica* and *Sorbus*, also on acid rocks, in Scotland (Fife, W. Perth, Argyll and Is. of Skye). W. Norway.

17. M. alabastrites (Nyl.) Coppins (1982)

Like **16**, but apothecia and pycnidia not pigmented, cream- or ivory-white; apothecia 0.2–0.7 mm diam., to 1 mm when tuberculate; ascospores (16-)18–26(-29) × (4-)4.5–5(-6) μm, 3(-7)-septate.

Usually on bark (often over bryophytes), occasionally on wood and rarely on rocks; in oceanic woodlands. W. Britain, Ireland. Norway, Belgium, Macaronesia.

Extreme shade morphs of *M. peliocarpa* **(16)**, with whitish apothecia, are usually sparingly fertile, whereas the apothecia of *M. alabastrites* are almost invariably numerous and crowded. Distinguished from extreme shade morphs of *M. cinerea* **(18)** by the mostly 3-septate ascospores and shorter, curved or sigmoid macroconidia.

18. M. cinerea (Schaerer) Hedl. (1892)

Bacidia carneoalbens (Nyl.) A. L. Sm. (1911), *Hastifera tenuispora* D. Hawksw. & Poelt (1986) [anamorph]

Like **16**, but ascospores (19-)23–34(-38) × (2.5-)4–6 μm, (3-)7-septate; apothecia 0.2–0.7 mm diam., to 1.3 mm when tuberculate; hymenium 55–70 μm tall; macroconidia 50–110 × 1 μm, long, flexuose.

On trunks (often over bryophytes), especially *Betula* and *Quercus*, in woodlands, rarely on rocks; occasionally on exposed turf in the mountains and then mostly sterile with only the macroconidial anamorph (*Hastifera tenuispora*), which has pycnidia with widely gaping ostioles. Mainly W. Britain. Norway, montane C. & S. Europe.

19. M. leprosula (Th.Fr.) Coppins & A. Fletcher (1975)

Thallus blue-grey, more rarely grey-brown, of scattered to usually coalescing convex to globose areoles, 100–400 μm diam., often proliferating to form a thick crust; areoles matt, fragile, eroding to form sorediate patches, the soredia 20–50 μm diam.; outer hyphae greenish, K −, N + red. Photobiont cells 4–7 μm. Apothecia similar to *M. peliocarpa* **(16)**, rare. Ascospores 14–26(-29) × 4–5.5 μm, (1-)3-septate. Pycnidia unknown. Thallus Pd + red, K −, C + red; apothecial sections C + red; argopsin and gyrophoric acid.

On bryophytes on acid rocks, rarely bark or wood; mainly in upland areas. N. & W. Britain. Europe, North America.

20. M. granulans (Vainio) Timdal (1991)

M. subleprosula (Vězda) Vězda (1976)

Thallus like **19**, but Pd+ yellow, K−, C+ red (alectorialic acid, 2 accessory substances). Apothecia ± globose or tuberculate, grey-brown to black. Ascospores (35-)40–50(-60) × 5–6.5 μm, 3- to 7(-9)-septate.

Over mosses on rocks above 750 m. N. Wales, C.S. Scotland (Perth). Sweden, Austria, Czechoslovakia.

This species and *M. leprosula* **(19)** can be confused with sterile *Trapeliopsis granulosa*, which has a thallus Pd−, C+ red (gyrophoric acid).

21. M. stipitata Coppins & P. James (1979)

Thallus dull or grey-green, thin and uneven or with flattened convex areoles 40–100 μm diam. Photobiont cells 4–7 μm diam. Apothecia 0.1–0.3 mm diam., convex to globose or tuberculate, whitish, often absent; true exciple indistinct; hymenium 35–50 μm tall, colourless. Paraphyses 1–1.5 μm wide. Asci 30–35 × 10–12 μm. Ascospores (14-)21–34 × 2–2.5(-3) μm, short-acicular. Pycnidia numerous, whitish, on simple or branched stalks 400–800 μm tall and 60–100 μm wide.; conidia 6–8 × 1–1.8 μm, bacilliform to ellipsoid. All parts Pd−, K−, C− (no substances detected by t.l.c.).

On leached bark, especially *Betula* and *Quercus*, often over bryophytes, rarely on mossy rocks; oceanic woodlands. W. British Is. N.W. France, N. Spain, Macaronesia.

22. M. pycnidiophora Coppins & P. James (1979)

Like **21**, but all parts C+ red (gyrophoric acid); pycnidia less than 300 μm tall, and conidia 3.8–6 × 1–1.5 μm.

On ± smooth bark, especially *Fagus* and *Ilex*. S. England (especially Hampshire, New Forest). France (Bretagne, Pyrénées), Canary Is., U.S.A. (Louisiana).

23. M. incrassata Hedl. (1892)

Thallus dull grey-white, grey-brown to dark grey, matt, of confluent convex-verrucose areoles 80–300 μm diam. Photobiont cells 4–7 μm. Cephalodia usually present, areole-like, brown, scattered, with *Nostoc* (cells 3–5 μm diam.); clusters of *Stigonema* often amongst the areoles. Apothecia 0.3–0.8(-1) mm diam., convex, usually partly immersed by surrounding areoles, black; true exciple indistinct; hymenium 45–50 μm tall, the uppermost part dark greenish, K−, N+ red; hypothecium 150–400 μm tall, dark red-brown (no purple tinge) K−, N+ bright orange-brown. Paraphyses (1-)1.5–2 μm wide, numerous, sparingly branched, the apices sometimes to 3 μm. Asci 45–48 × 11–14 μm. Ascospores (10-)12–17 × 4–4.8 μm, ellipsoid to oblong-ovoid or oblong-fusiform, 0- to 1(-2)-septate. Pycnidia 30–60 μm, rare, immersed, blackish; conidia 6–9 × 1–1.3 μm, ± bacilliform. All tissues Pd−, K−, KC−, C− (lichen products not detected by t.l.c.).

On soil of heaths on mountain summits, or on acid coastal heaths. Scotland (Highlands). Scandinavia, C. European mtns, U.S.A. (Rocky Mtns), Kerguelen Is.

M. assimilata **(24)** differs in the purple-brown hypothecium, as does *M. subviolascens* **(25)**, which also has a green, K+ violet upper hymenium; *M. submoestula* **(26)** has a similar hypothecium, but smaller ascospores and conidia. *Lecidea limosa*, which occurs in similar habitats, has a Pd+ red thallus and pale hypothecium, while *L. caesioatra* has dark blue-grey apothecia with a pale bluish grey bloom when wet and photobiont cells 8–12 μm diam.

24. M. assimilata (Nyl.) Coppins (1983)

Like **23**, but areoles usually whitish; apothecia usually more prominent; hypothecium purple-brown, K+ purple intensifying or (especially above) K+ dark green, N+ purple-red; ascospores (10-)12–16(-19) × 3–5 μm, 0(-1)-septate.

Similar habitats to *M. incrassata* **(23)**; two sites above 1000 m. N. Scotland. Scandinavia, North America, Australia.

Sometimes confused with *Lecidea berengeriana* and *L. hypnorum*.

25. M. subviolascens (Magnusson) Coppins (1983)

Like **24**, but on rocks; hymenium greenish, K+ violet; *Nostoc*-containing cephalodia not known, but *Stigonema* clusters usually present between areoles; ascospores 9–16(-17) × 4–5 μm, simple; pycnidia unknown.

On small stones in turf around late snow-beds at above 1000 m. C. & N. Scotland (Perth, Ross, Westerness). Norway.

26. M. submoestula (Nyl.) Coppins (1992)

Lecidea submoestula Nyl. (1876)

Thallus dark grey, at first of coherent granular areoles 0.1–0.3 mm diam., later cracked into 'islands' *c.* 0.4–1 mm across. Photobiont cells 4–7 μm diam., also with partly lichenized clusters of *Stigonema* appearing as a black 'prothallus'-like structure amongst the areoles. Apothecia (0.3-)0.4–0.7 mm diam., convex, sessile, black; true exciple soon excluded, reflexed, dark green-black in outer part, ± colourless within; epithecium dark green, K+ intensifying, N+ red; hymeniun 28–40 μm tall; hypothecium dark reddish brown, K−, N−. Paraphyses 1.5–2 μm wide, numerous, sparingly branched, ± anastomosed, the apices ± slightly swollen. Asci 25–35 × 10–12 μm. Ascospores (6-)7–9.5(-10) × 2.5–3.5 μm, ellipsoid to oblong-ellipsoid, simple, some 1-septate when old. Pycnidia 80–100 μm diam., immersed; wall green; conidia 4–6 × 1–1.6 μm, bacilliform. Thallus Pd−, K−, C−.

On acid rocks. W. Scotland, W. Ireland (Galway). ?Endemic.

No recent records. Differs from *M. incrassata* **(23)**, in the smaller ascospores, shorter conidia, and in growing on rocks. Could be confused with *Lecidea commaculans*.

27. M. crassipes (Th.Fr.) Coppins (1983)

Helocarpon crassipes Th. Fr. (1860)

Thallus whitish, pale grey-brown or ash-grey in part; areoles 70–200 μm diam., which often produce secondary granules. Photobiont cells 4–7 μm diam. Apothecia 0.2–0.6 mm diam., sessile, turbinate or short-stalked, at first flat but later convex, rarely tuberculate, to 1.4 mm diam; true exciple well developed, especially in young apothecia, becoming excluded, the inner part concolorous with hypothecium, the outer part greenish (K+ intensifying); hymenium 45–50 μm tall, greenish (K+ intensifying) or purple-brown (K+ purple intensifying) above, pale purplish (K+ purple intensifying or K+ greenish) below; hypothecium dark purple-brown (K+ intensifying). Paraphyses numerous, simple or sparingly branched above, 1–1.5 μm wide, to 2.5 μm at apices. Asci 40–48 × 11–14 μm. Ascospores (9-)10–17(-21) × (2.5-)3–4.5 μm, 0(-1)-septate, fusiform-ellipsoid. Pycnidia 65–150 μm diam., black, sessile, rare; conidia 4.5–5.7 × 1.2–1.5 μm, bacilliform. Lichen products not detected by t.l.c.

Over mosses on summit rocks. C. Scotland (Ben Lawers range). Arctic-alpine, Scandinavia, C. European mtns, Alaska.

Distinguished from *M. assimilata* **(24)** by the finely granular thallus, turbinate or ± stalked apothecia and absence of cephalodia.

28. M. turfosa (Massal.) Du Rietz (1923)

Thallus blackish, thin, without distinct granules or areoles. Photobiont cells 4–7 μm diam. Apothecia 0.15–0.3(-0.4) mm diam., convex or ± globose; true exciple reflexed, reddish brown; hymenium 35–50 μm tall, greenish (K−, N+ red) or brownish in part (especially base); hypothecium mottled reddish brown, K−. Paraphyses numerous, branched, 1–1.5 μm wide. Asci 35–50 × 11–12 μm. Ascospores (10-)12–21(-25) × (3.5-)4–5 μm, 0- to 3-septate, oblong-ellipsoid to fusiform. Pycnidia 35–40 μm diam., immersed; conidia 3.5–4.5 × 1 μm, ± bacilliform. Lichen products not detected by t.l.c.

On exposed turf, usually on bryophytes on mountain summits. Scotland. Scandinavia, Czechoslovakia, Greenland, Alaska.

Forming a black stain on exposed turf. Distinguished microscopically from *M. melaena* **(13)** and *Placynthiella uliginosa*, but see *M. viridiatra* **(29)**.

29. M. viridiatra Coppins (1985)

Like **28**, but thallus and apothecia greenish tinged (especially when wet), often with granular areoles 0.06–0.12 mm diam.; true exciple absent; hymenium vivid blue-green; hypothecium pale blue-green; ascospores (10-)12–17(-19) × (3.5-)4–5 μm, 0- to 1(-3)-septate; conidia 4.5–6(-6.5) × 0.5–1 μm.

On exposed turf, usually on bryophytes, late snow beds. Scotland (Cairngorm mtns and the Ben Nevis range, Westerness).

30. M. bauschiana (Körber) V. Wirth, & Vězda (1976)

Lecidea rusticella Nyl. (1878)

Thallus continuous, cracked or scurfy granular, greenish grey, grey-green, ± tinged brownish or ferruginous. Photobiont cells 5–12 μm diam. Apothecia 0.2–0.3 mm diam., or tuberculate to 0.7 mm, convex to ± globose, pale bluish grey to black, sometimes brown (infected by *Bispora* sp.) or pale (in shade); true exciple absent; hymenium 35–60 μm tall, colourless, or greenish, K−, N+ red); hypothecium colourless, or pale greenish in upper part. Paraphyses rather scanty, of two types: mostly 0.5–1.5 μm wide, sparingly branched; some 1.5–2 μm wide, unbranched, often in fascicles. Asci 35–50 × 9–12 μm. Ascospores 6.5–10.5 × 2.5–4 μm, simple, ellipsoid to ovoid. Pycnidia 40–80 μm diam., immersed; conidia 4–6 × 0.5–1 μm, bacilliform. Lichen products not detected by t.l.c.

In dry but humid situations, especially underhangs, on rock, stones, roots and consolidated soil; tolerant of iron and copper rich rocks; common. Mainly N. & W. Britain, W. Ireland. Europe, Macaronesia, N.E. North America.

The commonest member of the *M. sylvicola* group but care is needed to separate it from species **31–34** and *Psilolechia clavulifera.*

31. M. sylvicola (Flotow) Vězda & V. Wirth (1976)

Like **30**, but thallus darker, often bluish grey, often ± areolate; apothecia 0.2–0.5 mm diam., or to 1.2 mm when tuberculate, black; hymenium 40–60(-70) μm tall; hypothecium blackish green (K− or + green intensifying) but at least lower part dark purple brown (K+ purple intensifying); ascospores (6-)7–10 × (2.5-)3.5–4 μm, 0(-1)-septate; pycnidia 40–200 μm diam., immersed; conidia 3.5–6(-7) × 1–2 μm, bacilliform to oblong-ovoid.

Usually in similar habitats to *M. bauschiana* **(30)**, but rarely found on old wooden fence posts; local. N. & W. Britain. N.W. & C. Europe, E. Canada, U.S.A.

Specimens with immature ascospores need care to separate them from *M. tuberculata* **(32)**; can also be confused with *M. pseudomarginata* **(36)**.

32. M. tuberculata (Sommerf.) R. Anderson (1974)

Like **31** in the dark green hypothecium, but with ascospores 5.5–8(-9) × 1.5–2.5 μm, 0- to 1-septate, oblong-ovoid or oblong-ellipsoid; hymenium 30–35 μm; conidia 3–4.5 × 1–1.5 μm. The thallus is pale buff to greenish, never blue-grey, and the apothecia are usually only 0.16–0.3 mm diam., or to 0.55 mm if tuberculate.

On siliceous rocks, stones and roots in dry underhangs; rare. N. & W. British Is. Scandinavia, Alps, Czechoslovakia.

Psilolechia clavulifera differs in the smaller, simple ascospores, paler hymenium, *Stichococcus* photobiont and a hyphomycetous anamorph. See also *M. olivacea* **(10)**.

33. M. excipulata Coppins (1988)

Thallus thin and discontinuous, ± scurfy, not forming discrete areoles or goniocysts, pale buff or greenish. Photobiont cells 7–17(-20) μm wide, ± globose. Apothecia 0.15–0.4 mm diam., flat or convex, ± weakly marginate, dark grey to black; true exciple well developed, the outer edge olivaceous, colourless within; epithecium olive-green, K−, N+ red; hymenium 38–45 μm tall; hypothecium pale olivaceous. Paraphyses 1–1.5 μm wide or to 2 μm at apices, rather sparse, ± branched. Asci 35–41 × 9–12 μm. Ascospores (6-)7–10 × 2.8–4(-4.5) μm, ellipsoid to ovoid, simple. Pycnidia 30–80 diam., ± immersed; wall dark green; conidia 3–5 × 0.5–1.5 μm, bacilliform. Thallus Pd−, K−, C−; apothecial sections C−.

. On cardboard drum in urban waste ground. N. England (S. Yorkshire, Sheffield). Austria (on loose siliceous stones).

Differs from *M. bauschiana* **(30)** and *M. lithinella* **(34)** in the well-developed true exciple, and from the latter also by the green epithecium and larger-celled photobiont.

34. M. lithinella (Nyl.) Hedl. (1892)

Thallus thin, often indistinct, sometimes with whitish convex areoles 40–100 μm diam. Photobiont cells 4–7 μm diam. Apothecia (0.1-)0.15–0.4 mm diam., convex, pale, dull yellow-orange to reddish brown; true exciple absent; hymenium 35–50 μm tall, colourless; hypothecium yellow-straw to pale orange-brown. Paraphyses 0.5–1 μm wide; apices to 1.5 μm, sparse, simple or irregularly forked above. Asci 35–50 × 8–13 μm. Ascospores 6.5–9.5 × 2.5–4 μm, simple, ovoid to fusiform-ellipsoid. Pycnidia *c*. 40 μm diam., immersed; conidia 4–5.5 × 0.5–1 μm, bacilliform. Lichen products not detected by t.l.c.

On shaded stones, often with *Trapelia* species, on ground by woodland tracks, old quarries, etc.; widespread, but overlooked. Throughout British Is. Sweden, Germany, Luxembourg, Spain.

Similar to *M. bauschiana* **(30)** but apothecia and pycnidia never with a green pigment, the paraphyses of one type, the hypothecium yellowish, and the photobiont small-celled; it never occurs in dry underhangs.

35. M. lutulata (Nyl.) Coppins (1980)

Lecidea botryiza Nyl. ex Stirton (1873)

Thallus smooth or cracked, or scurfy granular, lacking discrete areoles or granules, pale, grey, greenish or buff, sometimes ferruginous. Photobiont cells 5–12(-15) μm diam. Apothecia 0.2–0.4 mm diam., or tuberculate, to 0.8 mm, convex to ± globose, grey-brown to black; true exciple absent; hymenium 30–40 μm tall, colourless, pale brown olivaceous or blue-green, the pigment often vertically streaked; hypothecium dark fuscous or reddish brown, K – or + red intensifying (not purplish or greenish), N –. Paraphyses rather sparse, dimorphic: mostly 0.5–1(-1.5) μm wide, often branched; some 1.5–2 μm wide, unbranched and often in fascicles. Asci 30–40 × 7–10 μm. Ascospores 6–9 × 2–3(-4) μm, simple, ellipsoid or ovoid. Pycnidia 80–200 μm diam., immersed, black; conidia (3.5-)4–5(-5.5) × 1–1.5 μm., bacilliform. Lichen products not detected by t.l.c.

On rocks and roots in dry but humid underhangs in woods and small valleys. N. & W. British Is. Temperate Europe, N.E. North America.

M. sylvicola **(31)** differs in hypothecium pigmentation; *M. pseudomarginata* **(36)** differs in having a distinct true exciple, larger ascospores, and a small-celled photobiont.

36. M. pseudomarginata Coppins (1988)

Thallus scurfy-granular to granular-areolate, pale to dark grey or grey-green, matt, sometimes partly ferruginous; areoles (if discrete) 80–240 μm diam. Photobiont cells 4–7 μm diam. Apothecia 0.2–0.4 mm diam., convex or tuberculate to 0.8 mm, black; true exciple distinct, soon reflexed, dark brown, colour as hypothecium but outer part K + greenish; hymenium 40–60 μm tall, colourless, with dark brown vertical streaks; hypothecium dark brown, K – or K ± purplish, N + red. Paraphyses numerous, of two types: mostly 1–1.5 μm wide, sparingly branched; some 2–3 μm wide, unbranched and in small fascicles surrounded by dark pigment. Asci 35–55 × 9.5–12(-14) μm. Ascospores 9–12(-13) × 3.5–5(-5.5) μm, simple, ellipsoid, ovoid or oblong-ellipsoid. Pycnidia 80–140 μm diam., immersed to ± emergent, black; conidia 6–9 × 1.5–2 μm, ± bacilliform. Lichen products not detected by t.l.c.

On vertical rock surfaces in open but sheltered situations; local. Upland Britain. ?Endemic.

Distinguished from *M. lutulata* **(35)** and *M. sylvicola* **(31)** by the true exciple, absence of greenish pigment in the hymenium, thallus-type, small-celled photobiont and longer conidia. *Lecidea pycnocarpa* has a ± glossy, white, K + yellow thallus, paraphyses with dark apical caps and larger photobiont cells.

37. M. myriocarpa V. Wirth & Vězda ex Coppins (1983)

Thallus scurfy farinose-granular, pale greenish or buff. Photobiont cells 4–7 μm diam. Apothecia 0.1–0.15 mm diam., or tuberculate to 0.25 mm diam., immarginate, convex to globose, pale to dark reddish brown; true exciple absent; hymenium 25–35 μm tall, colourless or brownish streaked; hypothecium reddish- or orange-brown, K –, N –. Paraphyses rather sparse, of two types: mostly 0.5–1 μm wide, often branched; some 2–3 μm wide, unbranched, usually in brown fascicles. Asci 21–25 × 5–7 μm. Ascospores 5.5–8.5 × 1.5–2.5 μm, 0- to 1-septate, oblong or oblong-ovoid. Pycnidia 25–30 μm diam., ± sessile, doliiform; conidia 2.5–3 × 1–1.5 μm, shortly bacilliform. Lichen products not detected by t.l.c.

In dry underhangs on rock, stones, roots, bryophytes etc.; frequent. N. & W. British Is. Temperate and boreal Europe.

Distinguished from other *Micarea* species of dry underhangs by the small apothecia and pigmentation; confused with *Psilolechia clavulifera* and *P. leprosa*.

38. M. subnigrata (Nyl.) Coppins & Kilias (1980)

Catillaria subnigrata (Nyl.) Herre (1910), *Micarea confusula* (Nyl.) Hedl. (1892)

Thallus dark brown, of scattered to confluent, convex areoles 80–450 μm diam. Photobiont cells 4–7 diam. Apothecia 0.2–0.6 mm diam., or tuberculate to 1 mm, convex, dark brown to black; true exciple reflexed, pale brown; epithecium brown, K−, HNO_3−; hymenium 35–38 μm tall, colourless; hypothecium colourless. Paraphyses (1-)1.5–2 μm wide, to 2.5 μm near apices, numerous, branched. Asci 33–35 × 10–14 μm. Ascospores 8–12 × 4–5 μm, (0-)1-septate, ellipsoid. Pycnidia frequent, immersed: (a) 80–200 μm diam., the macroconidia 7–10 × 5–6 μm, helicoid; (b) 40–100 μm diam., the microconidia 4.5–6 × 0.5–1 μm, bacilliform. Lichen products not detected by t.l.c.

On ± exposed hard siliceous rocks, often forming small patches with other crustose lichens; local. N. & W. Britain. Scandinavia, E. France, S. Germany.

The helicoid macroconidia are diagnostic.

39. M. intrusa (Th. Fr.) Coppins & Kilias (1983)

Catillaria intrusa (Th. Fr.) Th. Fr. (1874), *Lecidea aphanoides* Nyl. (1868), *L. melaphana* Nyl. (1869)

Thallus effuse or forming small patches to 1 cm diam., dark olivaceous or grey, granular-verrucose or scurfy. Photobiont cells 7–21 μm diam. Apothecia 0.14–0.4 mm diam., convex, black; true exciple reflexed, colourless or pale brown, or outer edge dark brown; hymenium 40–50 μm tall, the upper part green K−, N+ red; hypothecium colourless or tinged faint olivaceous, sometimes dull orange (K+ red) in upper part. Paraphyses 1–1.5(-1.7) μm, the apices not swollen, numerous, branched. Asci 35–45 × 14–15 μm, contents (also of ascospores) sometimes with orange, K+ red pigment. Ascospores (7-)9–14(-17) × 4–6 μm, 0- to 1(-3)-septate (septum often eccentric), ellipsoid or ovoid-ellipsoid, rarely oblong to ± fusiform and ± curved. Anamorphs unknown. Thallus Pd−, K−, C−; apothecial sections C−.

On exposed, hard acid rocks; rare. Scotland (Highlands). Scandinavia, E. France, S.W. Germany.

Mistaken for *Scoliciosporum umbrinum*, which has worm-like ascospores.

MICROCALICIUM Vainio (1927)

Thallus immersed in lichens, amongst free-living algae or saprobic. **Photobiont** absent. **Ascomata** apothecia, black, sessile or stalked. **True exciple** well developed, dark, of paraplectenchymatous or thick-walled hyphae. **Hamathecium** of sparse, thick-walled paraphyses. **Asci** broadly ellipsoid, formed in chains from ascogenous hyphae without croziers, with a single functional wall layer, deliquescing at maturity. **Ascospores** 1- to 3(-7)-septate, broadly ellipsoid to cylindrical, blue-green, a distinctive ornamentation of spirally arranged ridges developing after the asci breakdown; forming a dry green-black ascospore mass (mazaedium). **Conidiomata** pycnidia, frequent in some species, ± globose and sessile, black. **Conidiogenous cells** ± cylindrical, arising singly, enteroblastic with distinct broad collarettes. **Conidia** terminal, formed singly, ellipsoid to subglobose, colourless, simple, extruded as a slimy drop. **Chemistry**: ascospores, true exciple and stalk contain unidentified pigments which are generally green in acidic and red-brown in alkaline solutions. **Ecology**: parasitic on lichens, algae or saprobic, on bark, wood and rocks in humid and shaded sites.

Calicium is similar, but lichenized, with a black ascospore mass.

LITERATURE: Hawksworth (1981b), Tibell (1978, 1981, 1984).

1 Apothecia stalked; ascospores 6–8 μm long; pycnidia absent 2
Apothecia sessile; ascospores (9-)11.5–15 μm long; pycnidia frequent .... **3. *disseminatum***

2(1) On wood; ascospore mass green-black, with sclerotized hyphae .... **1. *ahlneri***
On algae or lichens on rocks, or roots; ascospore mass dark brown or greenish black, without sclerotized hyphae .... **2. *arenarium***

1. M. ahlneri Tibell (1978)

Thallus immersed, inapparent. Apothecia 0.36–1.1 mm tall, stalks occasionally branched, black; ascospore mass not taller than width of head, blue-green, with sclerotized hyphae, pruina absent; stalk

0.06–0.12 mm wide, dull black with a coarsely granular surface; true exciple in section reddish brown, poorly developed. Asci cylindrical. Ascospores 5.7–7 × 2.2–2.8 μm, 1-septate, with spiral ridges.

On wood of ancient *Quercus* and of *Pinus* trunks, parasitic or saprobic; rare. S. England (Somerset, Hampshire), Scotland (C. and E. Highlands). Europe, North America.

M. arenarium **(2)** differs in always being long-stalked, in having a well-developed true exciple, and lacking persistent, sclerotized hyphae in the ascospore mass.

2. M. arenarium (Hampe ex Massal.) Tibell (1978)

Thallus immersed, inapparent. Apothecia 0.6–1.8(-2.5) mm tall, black, head 0.14–0.28 mm diam.; ascospore mass not higher than width of head, dark brown or greenish black, without sclerotized hyphae, pruina absent; stalk 0.08–0.12 mm diam., dull black with a coarsely granular surface; true exciple in section reddish brown, N+ greenish. Asci broadly ellipsoid. Ascospores 6.7–8.2 × 2.3–2.9 μm, ellipsoid, 1-septate, with spiral ridges.

On *Stichococcus* or other green algae, *Psilolechia lucida* and *P. clavulifera*, parasitic or commensalistic, in shaded and humid localities, often under overhanging, siliceous rocks, or on exposed roots, mainly upland; rather rare. N. & W. Britain. Europe, North America, Siberia.

3. M. disseminatum (Ach.) Vainio (1927)

M. subpedicillatum (Schaerer) Tibell (1978)

Thallus immersed, inapparent. Apothecia 0.12–0.35 mm diam., 0.14–0.2 mm tall, ± sessile; ascospore mass greenish black, protruding, with sclerotized hyphae; true exciple in section green, K+ brownish red. Asci broadly ellipsoid. Ascospores (9-)11.5–15 × 3.4–4.2 μm, 1- to 3(-7)-septate, with spiral ridges. Pycnidia frequent, erumpent, 40–60 μm diam., the walls ± blue-green and of thick-walled cells; conidiogenous cells 6–9 × 2–3 μm; conidia subglobose, 2–3 × 1.5–2 μm.

On free-living green algae or growing on other Caliciales, parasitic or commensalistic, sometimes saprobic directly on bark or wood of old *Pinus* in localities rich in Caliciales, in humid conditions; rare. S. England (Hereford, Leicester), Scotland (Highlands). Europe, North America, Siberia.

The pycnidia appear first, up to 15 per host areole, and infected lichen thalli are decolourized or brownish with the formation of extensive lesions.

MIRIQUIDICA Hertel & Rambold (1987)

Thallus crustose, of continuous to scattered areoles; areoles corticate, often with a distinct epinecral layer; medulla I−; black prothallus sometimes present. **Photobiont** chlorococcoid. **Ascomata** apothecia, black or dark brown, remaining immersed or becoming sessile. **Thalline exciple** usually absent. **True exciple** of radially orientated hyphae 5–7 μm wide at outer edge (in water mounts), reduced in species with immersed apothecia, greenish or brownish, at least at outer edge, never black and brittle; photobiont cells usually absent, occasionally present in the inner part of the exciple in some specimens. **Epithecium** greenish (N+ red) or brownish (N-). **Hymenium** colourless, I+ blue. **Hypothecium** colourless. **Hamathecium** of paraphyses, ± branched and anastomosed; apices ± clavate and surrounded by a closely adhering pigmented hood. **Asci** clavate, 8-spored, probably *Lecanora*-type but apical cushion often weakly defined and ocular chamber scarcely developed. **Ascospores** simple, occasionally with a false cytoplasmic septum, rarely 1-septate when old, ellipsoid to usually oblong-ellipsoid. **Conidiomata** pycnidia, immersed in areoles. **Conidia** 15–28 × 0.5–1 μm, curved-thread-like, simple, colourless. **Chemistry**: miriquidic acid in most species, ± other lichen products; miriquidic acid is easily identified on t.l.c. plates as a ± peacock blue spot. **Ecology**: on acidic rocks, rarely on wood. **Distribution**: about 13 species in boreal or montane regions of the Northern Hemisphere.

LITERATURE: Hertel & Rambold (1987).

1 No areoles sorediate; apothecia usually numerous 2
Some areoles sorediate; apothecia few or absent .... **6. *nigroleprosa***

2(1)	Areoles K− or K+ yellowish (no crystals)	3
	Areoles K+ yellow to red (crystals)	**5. *lulensis***
3(2)	Apothecia becoming emergent and sessile	4
	Apothecia remaining immersed	**1. *complanata***
4(3)	Areoles pale grey and ± glossy, or dark brown- or bluish-grey and matt; medulla Pd− (?rarely Pd+ orange)	5
	Areoles dark brown, glossy; medulla Pd+ orange	**2. *garovaglii***
5(4)	Areoles grey-white to brownish-grey, often glossy, rarely minutely lobed	**4. *leucophaea***
	Areoles grey-brown to dark bluish-grey, matt, often minutely lobed	**3. *griseoatra***

1. M. complanata (Körber) Hertel & Rambold (1987)

Aspicilia complanata (Körber) B. Stein (1879), *Aspicilia superiuscula* (Nyl.) Hue (1912)

Thallus of scattered, more rarely contiguous, areoles on a conspicuous black prothallus; areoles to *c.* 1 mm diam., rounded, ± minutely lobed, flat to slightly convex, pale- to brown-grey, matt; superficial epinecral layer indistinct. Apothecia 0.2–0.7 mm diam., immersed, mostly 1 per areole; disc brown to black, usually immarginate; true exciple reduced, the outer edge dark brown to greenish; epithecium brown. Ascospores 10–16(-18) × (4-)5–6(-7) μm. Thallus Pd−, K−, KC−, C− (miriquidic acid).

On moist, schistose rocks, mostly above 700 m; local. Scotland (Highlands). N. Europe.

This species has the aspect of an *Aspicilia*.

2. M. garovaglii (Schaerer) Hertel & Rambold (1987)

Thallus areolate; areoles 0.2–0.8 mm wide, contiguous, angular to rounded, flat to often convex, brown, glossy; superficial epinecral layer distinct. Apothecia 0.3–1(-1.3) mm diam., sessile, the base ± constricted, black; disc flat to convex; true exciple thin and usually persistent, olive-brown to dark green at outer edge, pale straw within; epithecium olive-brown to green. Ascospores 11–16.5 × 5–6.5 μm. Medulla (especially upper part) Pd+ orange, K+ yellow, KC−, C− (miriquidic and stictic acids).

On schistose rocks on SW-facing cliffs. N. Scotland (W. Ross, near Loch Maree). Europe, North America, Asian U.S.S.R. (Krasnojarski Kraj).

Characterized by the darkish brown, glossy thallus; *Lecidea paupercula* is of similar appearance, but has flatter areoles, an I+ blue medulla, a dark hypothecium and lacks miriquidic acid.

3. M. griseoatra (Flotow) Hertel & Rambold (1987)

Lecidea griseoatra (Flotow) Schaerer (1850), ? *Lecidea subplumbea* Anzi (1868)

Thallus of scattered to contiguous areoles; areoles *c.* 1 mm diam., rounded, convex, often minutely lobate and becoming verrucose, grey-brown to dark bluish-grey, matt; superficial epinecral layer absent or indistinct. Apothecia to 0.3–1.2 mm diam., sessile, the base ± constricted, black, flat to convex; true exciple usually persistent, greenish throughout or inner part brownish; epithecium olive-green. Ascospores 9–13(-14) × (4-)5–7 μm.

On siliceous rocks and boulders by late snow patches; rare. N. Scotland (Highlands). Montane Europe, North America.

Differs from *M. leucophaea* **(4)** in the darker coloured, always matt, and more frequently lobate areoles. *Lecidea subplumbea*, also recorded from the Scottish Highlands, is probably conspecific but has a violet-brown inner exciple, narrow ascospores (5–5.5 μm), and sometimes additionally contains stictic acid.

4. M. leucophaea (Rabenh.) Hertel & Rambold (1987)

Lecidea leucophaea (Rabenh.) Nyl. (1870)

Thallus areolate; areoles 0.2–0.8 mm diam., mostly contiguous, flat to convex or verrucose, rarely a few minutely lobate, grey-white, grey, or brownish grey, often glossy; superficial epinecral layer usually distinct. Apothecia 0.3–1.2 mm diam., immersed at first but usually soon becoming sessile, adpressed or ± constricted at the base; disc flat to convex, brown to black, ± glossy; true exciple thin, often persistent, concolorous or paler than the disc, brownish or green at the outer edge, pale straw within, the inner part sometimes containing algae; epithecium olive-brown, rarely brown. Ascospores 8–15 × 4–7 μm. Thallus Pd−, K−, KC± faintly pink, C−; exciple C± faint pink (miriquidic acid).

On siliceous and metal-rich rocks, rarely on wooden fencing. Throughout British Is., but rare in E. England. N. and montane Europe, North America, Australia.

The commonest member of the genus in the British Is.; very variable. The rarer species [e.g. *M. griseoatra* **(3)** and *M. lulensis* **(5)**] can easily be mistaken for it.

5. M. lulensis (Hellbom) Hertel & Rambold (1987)

Lecidea lulensis Hellbom (1866)

Like **4**, but thallus (especially upper medulla) K+ yellow to red (crystals), Pd+ orange (norstictic, ± connorstictic, and ± stictic acids).

On siliceous boulders in exposed boulder field at 980 m. N. Scotland (Perth, Ben Lawers). Scandinavia, Svalbard.

The only species of the genus in the British Is. to lack miriquidic acid (but see *Lecidea atrofulva*). The thallus reactions could cause confusion with *Lecidea lactea*, which usually has flatter, more angular areoles, and an I+ blue medulla.

6. M. nigroleprosa (Vainio) Hertel & Rambold (1987)

Lecidea nigroleprosa (Vainio) Magnusson (1927)

Thallus of scattered to contiguous areoles; areoles to 0.2–0.8 mm diam., ± rounded, convex, often minutely lobate, dark grey, matt, many sorediate; soralia 0.1–0.2 mm diam., usually 1 per areole, developing from the upper surface, dark blue-grey but abrading whitish; external cells of cortex and exposed soredia olive-green, N+ red; superficial epinecral layer distinct. Apothecia unknown in British material. Cortex K−; medulla Pd+ deep yellow, K−, KC−, C− (miriquidic and ± psoromic acids).

On exposed siliceous boulders. N.E. England (Northumberland, Colouring Crag), C. & N. Scotland.

Lecidea pycnocarpa f. *sorediata* has blackish soralia, and is locally common in similar habitats, in the Scottish Highlands and has been misidentified as *M. nigroleprosa*; it differs in having whitish areoles with a K+ yellow cortex (atranorin and an unidentified substance).

MNIACEA Boud. (1885)

Thallus inconspicuous, on leafy liverworts. **Photobiont** absent. **Ascomata** apothecia, sessile; disc concave to convex, matt; in section all parts I−, K/I−. **Thalline exciple** absent. **True exciple** distinct, of upwardly directed hyphae that meet the surface at a low angle; hyphae 3–4.5 μm wide with ± rectangular cells 7–10 μm long. **Hamathecium** of numerous, slender, septate paraphyses 1–1.8 μm wide, which are simple or forked above and have slightly swollen to clavate apices; gel weak or absent. **Asci** 8-spored, cylindrical-clavate, with thick, non-amyloid apical dome containing a cylindrical tube- or plug-like structure (observe in Congo Red). **Ascospores** ± ellipsoid, simple, smooth, without a distinct perispore, colourless. **Conidiomata** unknown. **Chemistry**: not known. **Ecology**: on leafy liverworts.

The biological status of the genus merits investigation.

1 Apothecia blue-green when moist .... **1. *jungermanniae***
Apothecia whitish or pinkish when moist .... **2. *nivea***

1. M. jungermanniae (Nees ex Fr.) Boud. (1906)

Apothecia 0.2–0.6 mm diam., blue-green when moist, drying green-black; epithecium and true exciple bright blue-green, K+ dull yellow-green, N−. hymenium 100–150 μm tall, colourless or pale blue-green above. Paraphyses with apices swollen, often clavate, green, to 7 μm wide. Ascospores (14-)16–23 × 8–12 μm.

On leafy liverworts (e.g. *Cephalozia*, *Diplophyllum*, *Nardia*) on acid turf, rocks or rotting wood; mostly upland. N. Britain. W. and montane Europe.

2. M. nivea (Crouan) Boud. (1907)

Like **1**, but apothecia whitish or pinkish (lacking blue-green pigment).

Habitats similar to **1**. Scotland (Highlands). France, Austria.

Apparently rare, probably overlooked as very inconspicuous when dry. Similar to some *Vezdaea* species which, however, have very different asci.

MOELLEROPSIS Gyelnik (1940)

Thallus crustose, granular, effuse. **Cortex** poorly differentiated. **Photobiont** *Nostoc*. **Ascomata** apothecia, sessile or sometimes ± sunken between thallus granules; disc brownish with an irregular, granular thalline exciple, concolorous with thallus. **True exciple** paraplectenchymatous, of radiating hyphae, often occluded by an irregular, granular blue-grey thalline exciple. **Hymenium** K/I+ blue-green. **Hypothecium** of intricately interwoven hyphae, pale brownish. **Hamathecium** of paraphyses, simple, straight, septate, conglutinate; upper cells scarcely enlarged, with external brown pigmentation. **Asci** 8-spored, narrowly cylindrical, thin-walled; apex with an I+ blue apical dome, best visible in young asci. **Ascospores** uniseriate or partly biseriate, colourless, ellipsoid, simple, smooth. **Conidiomata** unknown. **Chemistry**: lichen products not detected by t.l.c.

1. M. nebulosa (Hoffm.) Gyelnik (1940)

Thallus thin to thick, coarsely granular, irregularly cracked when well developed, the granules 0.03–0.1 mm diam., ± crowded, pale blue-grey to grey-blue-black, rounded or somewhat elongate; photobiont clusters interwoven and covered by short-celled hyphae. Apothecia to 1 mm diam.; disc flat to convex, reddish brown to brown; true exciple to 100 μm thick; hymenium to 100 μm tall; hypothecium to 100 μm wide. Asci 60–90 × 7.5–15 μm. Ascospores (11-)13–15(-20) × 6–8 μm, sometimes attenuated at one end, with a single, large oil droplet.

On light, often sandy and well-drained, ± bare, often recently disturbed, soils in open habitats, e.g. coastal cliffs, road cuttings, or earthy wall tops; mainly local and decreasing. W. British Is. Europe.

Pannaria pezizoides has a paler grey-brown, minutely flattened-squamulose thallus, apothecia to 2 mm diam., and ascospores with a ridged and warted epispore; it grows amongst mosses in damp situations.

A second, undescribed, species of *Moelleropsis* occurs in coastal areas in S.W. Ireland, with a dull brown-blue-black, coarser coralloid- and granular thallus. When young the thallus forms nodular extensions on a dark basal crust resembling that of *Lecania aipospila*. The apothecia often become secondarily lobed and proliferating, and the ascospores are 15–27 × 6–8 μm; it is also known from the Azores.

A misunderstood, ephemeral species, ***Biatora humida*** Kullh. (1871), should possibly be assigned to *Moelleropsis*. It has an olivaceous to dull grey-brown, granular thallus, resembling that of *Placynthiella uliginosa* and is not tinged blue. Apothecia 0.3–0.5 mm diam., red-brown, convex and immarginate; paraphyses 1–1.5 μm wide, branched, not swelling at apices; asci clavate with an extended 'foot' at base. Ascospores 14–19 × 7–9.5 μm, ellipsoid, thin-walled, smooth. On recently disturbed, acid soils and recently laid landscaping gravels; recently recorded from S. England (Hants) and Scotland (Stirling).

MOSIGIA Fr. (1845)

Non *Mosigia* Sprengel (1826)

Thallus crustose, superficial areolate, thin. **Prothallus** blackish. **Isidia**-like protuberances or soralia in some species. **Photobiont** chlorococcoid. **Ascomata** apothecia, immersed in the areoles, usually 1–2 per areole, rounded, aspicilioid, black with a wrinkled, often umbonate surface. **True exciple** dark-brown to black, concolorous and continuous with the hypothecium. **Epithecium** brown-black to black, of firmly conglutinated apices of the paraphyses, K−, N−.

Hymenium colourless to pale-brown, often divided by vertical columns of melanized sterile hyphae continuous with the epithecium and hypothecium, I+ blue. **Hypothecium** dark-brown. **Hamathecium** of paraphyses, branched and anastomosed, sparsely distributed in K, septate; apical cells hardly swollen but mostly dark pigmented. **Asci** clavate, *Rimularia*-type, 8-spored. **Ascospores** ovoid or ellipsoid, colourless, but ± brown when old, single, smooth-walled, lacking a distinct perispore. **Conidiomata** pycnidial; walls dark. **Conidia** bacilliform. **Chemistry**: gyrophoric acid and accessories (thallus C+ red), unidentified UV+ pink compound. **Ecology**: on siliceous rocks in upland areas.

A genus of 3 species in the N. Hemisphere, but together with some species included in *Lecidea*, it has recently been treated in *Rimularia* by Hertel & Rambold (1990). Separated from other 'aspicilioid' genera, with which it was formerly united by the C+ red medulla which also lacks the I+ blue reaction characteristic of *Bellemerea*; and ascus structure. The dark-brown hypothecium distinguishes it from *Clauzadeana*, *Hymenelia* and *Aspicilia*.

LITERATURE: Hertel & Rambold (1990), Magnusson (1930), Poelt & Vězda (1977).

1 Thallus areoles with a solitary isidium-like protuberance that sometimes abrades to produce a yellowish 'soralium'; apothecia absent .... **2. *intercedens***
Thallus areoles without isidium- or soralium-like structures; apothecia numerous .... **1. *illita***

1. M. illita (Nyl.) R. Sant. (1984)

Lecidea illita Nyl. (1879), *Lecanora contracta* (Th.Fr.) Zahlbr. (1928)

Thallus thin, yellowish to pinkish brown or dark brown, cracked-areolate; areoles 0.2–0.5 mm wide; prothallus dark brown to black. Apothecia mostly solitary; disc at first punctiform, then widening to 0.2–0.3 mm, black, concave to umbonate, not pruinose. Hymenium 70–100 μm tall. Paraphyses 1–1.5 μm wide; apices to 3 μm wide. Asci 50–70(-90) × 17–20 μm. Ascospores 10–17 × 7–10 μm. Pycnidia not seen.

On sunny, exposed, siliceous or ± basic rocks and boulders in humid areas. N. England, Wales (Snowdonia), N. Scotland (Argyll, Grampians). Scandanavia.

Recently transferred to *Rimularia* as ***R. badioatra*** (Krempelh.) Hertel & Rambold (1990).

2. M. intercedens (Magnusson) R. Sant. (1984)

Thallus thin, grey- to pinkish brown, cracked-areolate; prothallus black, often visible between the areoles; areoles mostly 0.2–0.5 mm wide, most with a single, central, isidium-like protuberance, 0.05–0.15 mm diam., that sometimes becomes detached or abraded to form a yellowish soralium-like structure. Apothecia very rare, similar to *M. illita* **(1)**, but 0.2–0.4(-0.5) mm diam. and ascospores 19–21 × 10–12 μm.

On sun-exposed, mostly fine-grained acidic rocks, boulders and pebbles, sea-level to 700 m alt.; local. N. England, Wales (Gwynedd), Scotland. Scandinavia.

Easily overlooked, especially when forming small thalli in species-rich mosaics, but readily identified with a hand-lens by the nipple-like, papillate areoles.

MYCOBLASTUS Norman (1852)

Thallus crustose, continuous or granular, sometimes irregularly cracked-areolate, very variable, whitish grey to grey. **Prothallus** sometimes present, grey or blackish. **Photobiont** chlorococcoid, possibly trebouxioid. **Soralia** often present. **Ascomata** apothecia, flat to convex or subglobose, sessile or adpressed, black. **Thalline exciple** absent. **True exciple** well developed, often excluded at maturity. **Epithecium** colourless or interspersed with violet granules, K+ blue-green. **Hymenium** colourless, blue-black or brownish, I+ blue. **Hypothecium** colourless, pale brown or red. **Hamathecium** of anastomosing paraphyses, intergrading imperceptibly into the hyphal network of the hypothecium. **Asci** 1- to 2(-3)-spored, cylindrical-clavate, very thick-walled; apical dome I+ blue, with a distinct ocular chamber (especially when young); outer layer thickened above, forming an I+ dark blue cap over the ascus apex. **Ascospores** medium to very large; walls strongly thickened, simple, oblong or ellipsoid, colourless, multinucleate. **Conidiomata** pycnidia, immersed in the thallus, visible as minute black dots; wall green in upper part, K−, N+ reddish. **Conidiogenous cells** short. **Conidia** bacilliform, simple, colourless. **Chemistry**: atranorin; fatty acids, depsides, depsidones, usnic acid and pigments. **Ecology**: on ± acidic substrata, predominantly on bark and wood, more rarely on rocks and associated mosses.

An anatomically distinct group of closely related species, unified by their apothecia, which lack a thalline exciple, ascospore morphology and anastomosing paraphyses.

LITERATURE: Coppins & James (1979b), James (1971).

1 Soralia absent; apothecia usually present 2
Soralia present; apothecia rarely present 4

2(1) Hypothecium and adjacent medullary tissue under apothecia colourless or pale straw brown to pale purple; asci (1-)2(-3)-spored; ascospores ellipsoid with ± pointed apices, 25–70(-100) × 15–42 μm 3
Hypothecium and adjacent medullary tissue under apothecia carmine-red; asci 1(-3)-spored; ascospores cylindrical with flattened apices, 65–100 × 32–50 μm **4. *sanguinarius***

3(2) Epithecium and ± hymenium blue-black or in part olive-brown, non-granular, K−; thallus Pd− .... **1. *affinis***
Epithecium and hymenium densely interspersed with violet granules, K+ dissolving blue-green; thallus Pd+ rust-red .... **3. *fucatus***

4(1) Soralia grey, punctiform to contiguous, Pd+ rust-red or Pd− 5
Soralia bright yellow or yellow-green, efflorescent, ± contiguous, Pd− .... **2. *alpinus***

5(4) Soralia Pd+ rust-red, scattered or frequently contiguous; apothecia rare, violet internally 6
Soralia Pd−, scattered, often few, rounded; apothecia often present; hypothecium carmine-red .... **4. *sanguinarius***

6(5) Prothallus well developed; soralia ± convex, efflorescent, frequently contiguous, sometimes continuous; on bark .... **5. *sterilis***
Prothallus inconspicuous; soralia flat or ± excavate, often scattered; on wood .... **3. *fucatus***

1. M. affinis (Schaerer) Schauer (1964)

Thallus rather thick, irregular, granular-warted, pale grey or green-grey. Soralia usually absent, scattered if present, white or very pale grey, granular. Apothecia 0.5–1.5 mm diam., matt, black, strongly convex even when young, ± sessile; epithecium ± opaque, blue-black, rarely in part pale brown or olive-brown; hymenium paler below, K−; hypothecium colourless, pale straw or pale brown, ± intensifying in K, or K+ red-brown. Asci (1-)2-spored. Ascospores (40-)47–70(-100) × (25-)30–42 μm, ellipsoid. Pycnidia often present. Cortex K+ yellow; medulla Pd−, K−, KC−, UV+ white (unidentified substances).

On acid bark, especially conifer or *Betula*, less frequently overgrowing mosses on rock, rarely directly on rock in well-wooded sites; local. S.W. England, Wales, C. & W. Scotland. N. Europe, mountains of C. Europe.

Differs from *M. sanguinarius* **(4)** in the pale base to the apothecia, the 2-spored asci and ellipsoid, not cylindrical, ascospores.

2. M. alpinus (Fr.) Kernst. (1893)

Thallus ± spreading, film-like, effuse, with scattered to contiguous, grey, low warts; prothallus grey to blackish; soralia at first scattered, punctiform, bright or citrine-yellow, bursting from low warts, often becoming efflorescent and forming irregular, granular-sorediate patches, ± covering the entire thallus. Apothecia very rare (unknown in Britain), similar to those in *M. sanguinarius* **(4)** but hypothecium and adjacent medulla colourless or pale straw-brown, K− or ± orange-brown (intensified). Thallus Pd−, K± yellow, KC± yellow, UV− (atranorin, planaic acid and usnic acid in soralia).

On fallen pine trunks in native pinewoods or vertical, acid rocks; rare. England (Derbyshire), N.E. & S.E. Scotland (Kirkcudbright). Europe.

3. M. fucatus (Stirton) Zahlbr. (1926)

Thallus effuse, immarginate, pale slate or dark grey, occasionally with a bluish tinge, rather thin, ± continuously or partly faintly rimose, surface smooth or partly uneven with low spreading warts, 0.1–0.3 mm diam., which often erode and become granular sorediate; soralia ± diffuse, sometimes well-defined, round or oval, to 1.5 mm diam, or irregular, excavate, enclosing pale yellowish green to deep blue-grey, granular soredia. Apothecia 0.5–1.5 mm diam., not common, often few, scattered, black, shiny, thin and plate-like, closely adpressed, flat or slightly concave; true exciple thin, often excluded at maturity; epithecium, hymenium and upper part of hypothecium ± densely interspersed with numerous, minute, vivid purple-violet granules which dissolve in K to a bright aeruginose-blue, lower part of hypothecium pale violet or colourless, rarely partly spotted red-brown (intensified in K). Asci (1-)2(-3) spored. Ascospores (25-)30–48(-52) × 15–21 μm, ellipsoid; wall 2.5–4 μm thick. Thallus Pd+ rust red, K− or + brownish, KC+ dull brown, C−, UV− (atranorin, chloratranorin, fumarprotocetraric acid).

On damp, often rotting, decorticate wood and fence posts, particularly in boggy, sheltered, well-wooded habitats; rare, very local, possibly overlooked. N. Wales, W. Scotland. N.W. Europe.

When fertile recognized by the large, simple ascospores and vivid purple colouring in the hymenium, which is K+ bright peacock-blue. Sterile specimens can resemble *Buellia griseovirens*, *Haematomma caesium*, *Pertusaria pupillaris* and *Xylographa vitiligo*, all of which also occur on decorticate wood. *Haematomma caesium* is Pd−, UV+ white, *B. griseovirens* is K+ red (crystals), *X. vitiligo* is Pd− and has a ± immersed thallus; the soralia are often oval in outline. *Pertusaria pupillaris*, like *Mycoblastus fucatus*, contains fumarprotocetraric acid, but in the former the soralia are rather scattered, punctiform and evenly distributed. *M. sterilis* **(5)** is a possible synonym.

4. M. sanguinarius (L.) Norman (1853)

Thallus pale to dark grey, often very irregular and uneven, thick and ± coarsely verrucose or papillose-warted, appearing marbled, continuous or ± cracked, more rarely thin, with scattered low warty papillae, a pale to dark grey prothallus sometimes present; soralia [f. *leprosus* Nádv. (1951)] rare, scattered, concolorous or paler than the thallus, convex, efflorescent. Apothecia 0.5–1.7(-3.0) mm diam., usually frequent, black, becoming convex or almost spherical, adpressed or sessile leaving a bright carmine-red spot when damaged or abraded; true exciple becoming excluded at maturity. Asci 1(-3)-spored. Ascospores 70–100 × (28-)35–45 μm, cylindrical; wall 6–7 μm thick. Pycnidia often present. Cortex Pd+ yellow, K+ yellow, C−; medulla K+ red (in part), UV− (atranorin, chloratranorin, rhodocladonic acid (= mysaquinone), caperatic acid); hypothecium and tissue under the apothecium carmine to blood-red, K+ bright red, pigment diffusing into solution.

On trees, particularly *Betula*, occasionally on wood, hard siliceous rocks and over mosses; frequent. Upland areas of Scotland, extending locally to C. & N. England and Wales. Upland N. & C. Europe, rare in the Alps.

Usually distinguished by the carmine-red tissue under the apothecia; rarely, this pigmentation is absent, leading to confusion with *M. affinis* **(1)**.

5. M. sterilis Coppins & P. James (1979)

Thallus very variable, of ± scattered to contiguous, ± convex, pale to dark grey areoles or becoming a ± continuous, thick, granular-sorediate crust; prothallus extensive, blue-black; soralia numerous, excavate or usually ± convex, ± efflorescent, scattered, <0.5 mm diam., or coalescing to form extensive, granular patches, bluish grey, becoming pale green-grey or whitish when abraded, sometimes fleck-like. Thallus and soredia Pd+ red, K+ yellow, KC−, C−, I−, UV− (atranorin, fumarprotocetraric acid).

On smooth, acid to neutral bark of young trees, or the flat surfaces of rough bark, especially *Betula, Alnus*, as well as *Fraxinus* (in SO_2-polluted sites), in wayside and woodland habitats, rarely on wood; locally frequent, increasing, especially in areas with 30–50 μg m^{-3} mean SO_2 levels. Throughout British Is. N.W. Europe.

Superficially resembling *Buellia griseovirens*, *Fuscidea lightfootii* and *Haematomma caesium* but differing in the Pd+ red reaction of the soralia. The recent discovery of apothecia in Norway suggest it is a corticolous morph of *M. fucatus* (3). The thallus is often host to two lichenicolous fungi, *Tremella lichenicola* and *Skyttea gregaria*.

MYCOCALICIUM Vainio (1890)

Thallus inapparent. **Photobiont** absent. **Ascomata** apothecia; upper part obovoid to lens-shaped, stalked; stalk consisting of parallel hyphae, dark brown to blue-green. **Thalline exciple** absent. **True exciple** ± well developed, of dark brown, angular pseudoparenchymatous cells or of radiating hyphae. **Hamathecium** of paraphyses, simple to sparsely branched; tops forming a layer over the hymenium. **Asci** cylindrical, formed singly from ascogenous hyphae with croziers, with strongly, uniformly thickened apex, with a single functional wall layer, persisting until maturity of ascospores. **Ascospores** simple, uniseriately arranged in the asci, not released, or a dry spore mass, dark brown. **Conidiomata** not known. **Chemistry**: vulpinic and pinastric acids in some species. **Ecology**: saprobic, not lichenized, cool temperate to tropical.

LITERATURE: Tibell (1975, 1984, 1987).

1. M. subtile (Pers.) Szat. (1925)

Apothecia 0.4–0.8 mm tall, black, stalk shiny black or dull; upper part 0.16–0.32 mm diam., obconical to lens-shaped; stalk 0.05–0.12 mm diam. Ascospores 5.5–9 × 2.3–4 μm, ellipsoid to broadly subfusiform, smooth or with a minutely warted surface ornamentation.

On dry wood; scarce. S. England (Hampshire, New Forest), Scotland (Perthshire). Europe, North America, Asia, Australia, New Zealand.

Very variable, but characterized by the dark greenish to brownish stalk tissue. Easily confused with *Chaenothecopsis* species in which the ascus apex is penetrated by a narrow canal.

MYCOGLAENA Höhnel (1909)

Thallus on bark, immersed, inconspicuous, probably non-lichenized. **Photobiont**, evidently absent. **Ascomata** perithecium-like, 0.2–0.4 mm diam., circular or broadly ellipsoid in outline, flattened, greenish black but often with a thin whitish border. **Involucrellum** present, bright green, K−, N+ red, composed of dark hyphae mixed with bark cells. **Exciple** not distinct. **Hamathecium** of paraphyses to 1 μm thick, simple or sparingly branched and anastomosed; gelatin I−. **Asci** ± cylindrical, with a ± truncate apex, K/I−, the apex usually thickened internally but without an ocular chamber, 8-spored. **Ascospores** transversely septate or muriform, ovoid-fusiform or clavate, colourless; perispore not apparent. **Conidiomata** pycnidia, resembling ascomata but much smaller; conidia bacilliform. **Ecology**: on smooth bark of trees and shrubs.

Distinguished from *Arthopyrenia* and *Polyblastiopsis* by the bright green, N+ red ascoma wall.

LITERATURE: Harris (1973).

1	Ascospores muriform	**1. *acuminans***
	Ascospores 3-septate	**2. *myricae***

1. M. acuminans (Nyl.) Vainio (1921)

Asci 80–95 × 17–20 μm. Ascospores 28–31(-38) × 7–9(-10) μm, clavate with attenuated lower part, muriform with 7–10 transverse septa and all but lowest 1–2(-4) cells with 1–3 longitudinal septa. Pycnidia not seen.

On young twigs of *Pinus*; rare or overlooked. S. Scotland. Europe.

2. M. myricae (Nyl.) R. C. Harris (1973)

Asci 50–75 × 10–15 μm. Ascospores 16–20 × 4.5–7 μm, ovoid-fusiform, 3-septate, the second cell ± enlarged, constricted at septa. Pycnidia 40–60 μm diam.; conidia 4.5–5(-5.7) × 0.8 μm.

On woody stems of *Myrica*, sometimes on other hosts, e.g. *Alnus, Betula, Sorbus*; common. N. & W. British Is. Europe, North America.

MYCOMICROTHELIA Keissler (1936)

Thallus crustose, immersed, rarely superficial, silvery or cream, sometimes ± lacking, sometimes with a ± black limiting fringe. **Photobiont** absent in most species, when present *Trentepohlia*. **Ascomata** perithecia, ± globose to flattened, usually lacking a lower wall, black; involucrellum present, dark brown, often olivaceous in K, extending laterally and sometimes forming a distinct fringe around the perithecia, formed of hyphae and incorporated bark cells (clypeate in structure). **True exciple** ± globose to flattened, often poorly developed to absent below, of interwoven brown to colourless hyphae. **Hamathecium** of pseudoparaphyses, branched and anastomosed, septate; gelatin I− or I+ violet (not blue). **Asci** 8-spored, elongate-clavate or broadly clavate to subcylindrical, with two functional wall layers, discharge fissitunicate, with an internal ocular chamber, I−. **Ascospores** ellipsoid to slipper-shaped, 1(-3)- septate, the upper cell usually larger, brown, usually distinctly warted, often with a thin gelatinous sheath swelling in K. **Conidiomata** pycnidia, similar to the ascomata in structure. **Conidiogenous cells** enteroblastic, elongate-ampulliform, unbranched, acrogenous. **Conidia** bacilliform, simple, colourless, or ellipsoid, 1-septate and brown, or ellipsoid, simple, with centrally thickened walls and brown. **Chemistry**: lichen products not detected by t.l.c. **Ecology**: on bark, cosmopolitan but mainly in the tropics.

This genus includes 26 species world-wide, most of which are ± host-specific and not lichenized. It is close to *Arthopyrenia* s. str., from which it differs in that the ascospores turn brown within the ascus prior to discharge and also have warted walls. Old degenerating ascospores in some *Arthopyrenia* species can become brownish, but then their walls are not usually warted. Reports of *M. melanospora* (Hepp) D. Hawksw. (1982) from the British Is. are based on misidentifications; that species has ascospores (12.5-)13.5–15.5(-17) × (6-)6.5–7.5(-8) μm, cells ± equal in size and rounded apices, and elongate-ellipsoid, brown, 1-septate conidia 13–14.5 × 5.5–6.5 μm; it occurs on smooth bark of *Mespilus germanicus* and (?) *Crataegus* in C. Europe.

LITERATURE: Hawksworth (1985).

1 Basal fringe around the perithecia inconspicuous, <125 μm wide, ± circular 2
Basal fringe around the perithecia conspicuous, 125–300 μm wide, ellipsoid to broadly fusiform .... **3. *wallrothii***

2(1) Ascospores (12.5-)13.5–16(-17.5) × (5.5-)6.5–8(-9) μm; pycnidia present .... **1. *atlantica***
Ascospores (15.5-)17.5–21(-25) μm in length; pycnidia absent .... **2. *confusa***

1. M. atlantica D. Hawksw. & Coppins (1985)

Thallus silvery grey to ± indistinct, lacking a delimiting fringe, sometimes loosely associated with *Trentepohlia*, probably not lichenized. Perithecia (0.15-)0.2–0.3(-0.35) mm diam., 90–120 μm tall, with a purple-black fringe 30–50 μm wide; hymenial gelatin I+ pale violet when young. Asci 40–55 × 12–17 μm, clavate. Ascospores (12.5-)13.5–16(-17.5) × (5.5-)6.5–8(-9) μm, ellipsoid, 1-septate, scarcely constricted at the septum, the cells ± equal in size; apices rounded. Pycnidia 45–75 μm diam., scattered amongst the perithecia. Conidia 3–4 × 0.5–0.7 μm, bacilliform, colourless, simple.

On smooth bark of *Corylus*, generally in deep shade; very rare. W. Scotland (W. Inverness, Skye), W. Ireland (Kerry, Clare, Donegal). Endemic.

2. M. confusa D. Hawksw. (1985)

Thallus creamy or whitish, often with a delimiting purplish black fringe, loosely associated with algae, facultatively lichenized with *Trentepohlia*. Perithecia 0.2–0.3(-0.4) mm diam., basal fringe to 25 μm wide; hymenial gelatin I –. Asci 55–70 × 12–18 μm, elongate-clavate. Ascospores (15.5-)17.5–21(-25) × (6-)7–9(-10) μm, sole-shaped, 1-septate, markedly constricted at the septum, the upper cell generally larger and rounded, the lower generally attenuated. Pycnidia unknown.

On smooth bark, particularly of *Corylus*, but also *Betula*, *Fraxinus*, *Ilex*, *Quercus*, and *Sorbus*, in shaded, ± constantly humid situations; local. W. British Is. Norway, Spain (Cataluña).

Most British records of "*Microthelia micula* Körber (1855)" are misidentifications for this species. *Mycomicrothelia confusa* is distinguished from *M. atlantica* **(1)** by the larger and ± slipper-shaped ascospores, with an attenuated lower cell, and the absence of pycnidia.

3. M. wallrothii (Hepp) D. Hawksw. (1980)

Thallus inconspicuous, not delimited, with algal cells occasionally scattered on the surface, but not lichenized. Perithecia 0.15–0.25(-0.3) mm diam., rare, situated in an ellipsoid, dark grey basal fringe, to 0.3 mm wide; hymenium I –. Asci 35–60 × 14–16(-20) μm, elongate-clavate. Ascospores (12-)14–18(-20) × 6–8(-8.5) μm, ellipsoid, 1-septate, not to slightly constricted at the septum; apices rounded. Pycnidia 50–100(-120) μm diam., intermixed with the ascomata or present alone. Conidia (9-)11–12.5(-14) × 4–6.5 μm, ellipsoid to doliiform, brown, thickened centrally internally and sometimes ± distoseptate.

On trunks of *Betula*; very rare. Scotland (Highlands, E. Ross). Probably circumboreal in the Northern Hemisphere. Also known from *Populus* in Norway.

Superficially recalling *Leptorhaphis epidermidis*, which also occurs on *Betula* bark; microscopic examination is necessary to conclusively separate these taxa.

MYCOPORUM Flotow ex Nyl. (1855)

Thallus immersed, inconspicuous or slightly bleaching the substratum. **Photobiont** absent and probably non-lichenized. **Ascomata** circular or irregular in outline, with one to several perithecia-like locules, each with an ostiole. **Involucrellum** overlying the locules, dark brown, K –, clypeate, composed of compacted hyphae mixed with bark cells; outer and lower edges of locules without a distinct true exciple. **Hamathecium** of branched, short-celled, ± moniliform pseudoparaphyses, ± gelatinized and indistinct when asci are fully mature; I –. **Asci** 8-spored, fissitunicate, broadly ellipsoid, obpyriform or subglobose, I –. **Ascopores** transversely septate to mostly submuriform, ± ovoid-oblong, constricted at middle septum with the upper half wider and more rounded than the lower, colourless, perispore not apparent. **Conidiomata** not known. **Chemistry**: no lichen products detected by t.l.c. **Ecology**: on smooth bark of twigs and branches.

The above description refers to the British taxa, which may belong to a different genus from the type species, *M. elabens* Flotow ex Nyl. (1856) from C. Europe, which has larger, thick-walled ascomata and is lichenized with *Trentepohlia*.

LITERATURE: Coppins & James (1979a).

1 Ascomata unilocular; ascospores with 3(-4) transverse septa, 0–2 of which have a longitudinal septum .... **1. *hippocastani***
Ascomata mostly multilocular; ascospores with (3-)4(-5) transverse septa and (3-)4(-5) longitudinal septa .... **2. *quercus***

1. M. hippocastani (DC.) Coppins (1979)

Ascomata 0.1–0.2 mm diam., ± circular in outline, unilocular. Asci 40–45 × 17–20 μm. Ascospores 11–17 × 4–6.5 μm, with 3(-4) transverse septa, 0–2 transverse cells with a longitudinal septum.

On twigs and branches of *Populus tremula*; rare or overlooked. Scotland (Highlands). Europe.

Easily confused with *Arthopyrenia punctiformis* which has 1(-3)-septate ascospores.

2. M. quercus (Massal.) Müll. Arg. (1882)

Ascomata 0.1–0.4 × 0.1–0.2 mm, mostly irreulgar in outline, 1- to 6-locular. Asci 25–31 × 19–22 μm. Ascospores (10-)12–18 × 4.5–7 μm, with (3-)4(-5) transverse septa and 1–4 transverse cells with a longitudinal septum.

Mainly on smooth, thin twigs of *Quercus*, but occasionally on *Corylus* branches; common. Throughout British Is. Europe.

Material of *M. quercus* var. *ilicis* (Riedl) Coppins (1979), on *Ilex* from W. Ireland, has not been traced; it is said to have ascomata with (3-)7–10 locules.

NEPHROMA Ach. (1810)

Thallus foliose, dorsiventral, spreading, ± loosely attached, ± rosette-forming or rarely fragmentary; lobes rounded to slightly elongate, entire, indented or with folioles, often ascending at the margins. **Upper surface** ± smooth, matt or shining. **Soredia** sometimes present. **Cortex** pseudoparenchymatous. **Photobiont** green (*Coccomyxa*) or blue-green (*Nostoc*), cephalodia (containing *Nostoc*) internal in species with a green photobiont. **Medulla** white or yellow. **Lower surface** corticate, smooth to subpubescent or tomentose. **Ascomata** apothecia, sessile, rounded to reniform, on lower surface at apices of lobes; disc pale brown to red-brown-black. **Thalline exciple** present. **Asci** 8-spored, simplified '*Peltigera*-type'. **Ascospores** pale brown, fusiform, 3-septate. **Conidiomata** pycnidia, marginal, in semi-immersed warts; ostiole black, pimple-like. **Conidia** bacilliform. **Chemistry:** hopane triterpenoids, depsides, pigments. **Ecology:** on bark, rarely on rocks, oceanic to boreal-montane; often characteristic of ancient woodlands and sensitive to SO_2 pollution.

Close to *Peltigera* both morphologically and chemically, but characterized by the apothecia arising on the lower surface and the presence of a lower cortex.

LITERATURE: James & White (1987), Wetmore (1960).

1 Photobiont blue-green; fresh thalli ± blue-black when wet; cephalodia absent 2
  Photobiont green; fresh thalli yellow-green when wet; cephalodia present .... **1. *arcticum***

2(1) Soredia absent; apothecia usually frequent or thallus with numerous folioles 3
  Soredia present; apothecia rare; folioles absent .... **4. *parile***

3(2) Lower surface smooth; medulla yellow or pale yellow, K+ purple, (rarely white and K−) 4
  Lower surface pubescent or tomentose; medulla white, K- 5

4(3) Folioles absent; commonly fertile .... **3. *laevigatum***
  Folioles present, marginal and often laminal; thallus rarely fertile .... **6. *tangeriense***

5(3) Lower surface distinctly tomentose, with pale raised papillae; thalline exciple entire .... **5. *resupinatum***
  Lower surface densely pubescent, without papillae; thalline exciple and supporting lobe pectinate .... **2. *helveticum***

1. N. arcticum (L.) Torss. (1843)

Thallus 6–8 cm diam., forming rosettes or fragmented and irregularly spreading; lobes *c.* 15 mm wide, ± rounded, the margins entire, undulate, rarely ± crenulate; upper surface yellow-green when wet, straw-coloured when dry, smooth to partly wavy, glabrous, with convex cephalodial warts (Fig. 30c); medulla white; lower surface pale brown-white at margins, with a conspicuous brownish black tomentum towards the centre. Photobiont green in medulla, blue-green in cephalodia. Apothecia not observed in British material, 2(-3) cm wide, rufous-brown; thalline exciple ± raised. Ascospores 23–27 × 4–6 μm, 3-septate, subfusiform. Pycnidia rare, marginal; conidia 3–4 × 1–2 μm. Thallus Pd+ yellow, K ± yellowish, KC+ yellow, C−; medulla UV+ (nephroarctin, phenarctin, zeorin, ± usnic acid).

On lichen-moss heaths; very rare. Scotland (N.W. Highlands). Circumpolar, subarctic to boreal, North America, Asia including Japan.

Characterized by the yellow-green, large-lobed thallus, absence of soralia and the dark tomentose lower surface. The only British species with a green photobiont. In arctic regions thalli are often larger and abundantly fertile. Distinguished from *Lobaria scrobiculata* by the lack of soredia and from *Peltigera* species by the presence of a lower cortex and absence of a network of veins and rhizines.

2. N. helveticum Ach. (1810)

Thallus 3–4 cm diam., forming ± compact rosettes or irregularly spreading; lobes 2–3 mm wide, ± elongate; margins entire or with small, terete, isidia-like outgrowths and phyllidia which occasionally spread to the lamina, fertile lobes ± pectinate. Upper surface pale grey-brown to red- or dark brown, ± smooth, sometimes thinly pubescent; medulla white; lower surface dark brown to black, tomentose or densely pubescent. Photobiont blue-green. Apothecia frequent; thalline exciple conspicuously pectinate, the dorsal surface scabrid, often markedly honeycombed. Ascospores 21–27 × 6–8 μm, 3-septate. Pycnidia rare. Thallus and medulla Pd−, K−, KC−, C− (hopane triterpenoids: 7β acetoxyhopan -22-ol (T1), hopane -7β, 22- diol (T4); unidentified accessory substances).

On mossy rocks or trees. Scotland (Perth, Killin, Achmore Woods); probably extinct in the British Is. Extinct in Scandinavia, rare elsewhere in Europe. ± Cosmopolitan.

A boreal-temperate species, forming part of a world-wide species-complex, exhibiting much variation and a wide ecological amplitude. Characterized by the presence of isidia-like outgrowths and phyllidia resulting in the distinctive pectinate thalline exciple and subtending lobes. Distinguished from *N. resupinatum* **(5)** by the less tomentose lower surface and absence of pale papillae.

3. N. laevigatum Ach. (1814)

Thallus 3–8 cm diam., spreading, rarely fragmentary, the lobes 2–10 mm wide, robust, often coriaceous, the margins entire, decumbent or ± reflexed, rarely developing small folioles; upper surface grey-brown to deep red-brown, ± smooth or partly wavy; medulla yellow; lower surface smooth to longitudinally striate - ridged to wavy, pale at margins, becoming brown-black towards the centre, without a tomentum. Photobiont blue-green. Apothecia frequent; thalline exciple entire or rarely subcrenulate; dorsal surface conspicuously scabrid-areolate and ridged. Ascospores 17–20 × 5–7 μm. Pycnidia infrequent. Medulla K+ purple, rarely K− (anthraquinones, triterpenoid hopane-6α, 7β, 22- triol (T6)).

Primarily on bark in mossy, humid communities in markedly oceanic habitats in relict woodlands. W. British Is. Widespread in W. Europe, but now largely extinct in the north-central lowlands. N. America, Asia.

A rare morph with a white medulla occurs on the Is. of Scilly (Tresco) and may have been originally introduced from Madeira into Tresco Gardens; it should not be confused with *N. bellum* (Sprengel) Tuck. (1841), a more northern, boreal-montane to continental species which does not occur in the British Is.; *N. bellum*, which also has a subtomentose lower surface, has a different hopane triterpenoid pattern.

4. N. parile (Ach.) Ach. (1810)

Thallus up to *c.*8 cm diam., forming rosettes or fragmentary; lobes spreading, 3–8 mm wide, often thin; margins entire or subcrenulate, rarely ascending; upper surface bluish grey to dark red-brown with marginal pustular soralia which later spread on to the surface; medulla white; lower surface smooth to wrinkled, naked or in part subpubescent, rarely tomentose; soralia blue-grey to brown then partially corticate. Photobiont blue-green. Apothecia extremely rare, often small and immature; dorsal surface ± smooth, sorediate. Ascospores 18–20 × 6–7 μm. Pycnidia rare. Medulla K+ yellow or K− [hopane triterpenoids: 15α-acetoxyhopan-22-ol (T2), hopane-6α, 22-diol (zeorin) (T3)] and hopane-15α, 22-diol (T5) distinctive unidentified accessory substances).

On bark and mosses, characteristic of old woodlands, forming part of the *Lobarion* association. N. & W. British Is.; similar in distribution to the commoner *N. laevigatum* **(3)** but less oceanic and more boreal-montane. N. temperate to boreal, South America.

The only sorediate British species of the genus. Distinguished from *Peltigera collina* by the presence of a lower cortex and the smooth lower surface, absence of a network of veins and rhizines and a different hopane complement.

5. N. resupinatum (L.) Ach. (1810)

Thallus to 10 cm diam., forming rosettes or ± unorientated; lobes 5–17 mm wide, robust; margins entire, rounded or crenulate, or lobulate, sometimes ascending; upper surface grey-brown, ± wavy to ± faveolate, ± uniformly pubescent or finely subtomentose, occasionally with phyllidia arising from irregular verrucae; medulla white; lower surface pale brown-white, markedly tomentose, with scattered, raised, white papillae. Photobiont blue-green. Apothecia frequent, to 1.1(-1.5) cm, subtending lobe markedly recurved, dorsal surface tomentose, coarsely scabrid and foveolate-reticulate. Ascospores 21–24 × 4–6 μm. Pycnidia rare, marginal; conidia 4–5 × 1–2 μm. Thallus and medulla Pd−, K−, KC−, C− (no distinguishing medullary substances detected by t.l.c.).

On bark or rocks in cool, moist shady habitats; extinct. Reliably recorded only from Scotland, Aberdeen, nr. Invercauld (in 1792). Circumpolar boreal to low alpine, widespread in Scandinavia.

Easily recognized by the tomentose thallus with raised, white papillae on the lower surface and the absence of any chemistry.

6. N. tangeriense (Maheu & A. Gillet) Zahlbr. (1932)

Thallus 3–4(-9) cm diam., forming compact rosettes or dispersed and fragmentary; lobes 2–6 mm wide, fragile, thin; margins entire, often ascending and crisped, frequently developing horizontal or ascending folioles, which are scattered or form dense imbricate clusters; folioles dorsiventral, rounded or variously sublobulate and notched at the margins; upper surface ± smooth, with scattered crenulate folioles, *c.* 1–2 mm wide; medulla yellow; lower surface smooth to ridged-undulate, without a tomentum, pale at the margins, becoming brown-black towards the centre. Photobiont blue-green. Apothecia rare, similar to *N. laevigatum* but smaller, not observed in British material. Pycnidia not observed. Medulla K+ purple, with the same anthraquinones as *N. laevigatum,* but a wider range of hopane triterpenoids: ± 15α-acetoxyhopan-22-ol (T2), hopane-6α, 22-diol (zeorin) (T3), ± hopane -7β, 22-diol (T4), ± hopane -6α, 7β, 22-triol (T6).

On mossy rocks and on soil in exposed situations, as well as on trees and shrubs, part of the *Lobarion* community, with a preference for warmer, drier habitats than *N. laevigatum*; very local. S.W. England, (Lizard), Channel Is. (? now extinct), Is. of Scilly (Tresco), N. Wales (Cardigan, Lleyn Peninsula). W. Mediterranean, Macaronesia.

Distinguished from the usually fertile *N. laevigatum* **(3)** by the numerous, crenulate, dorsiventral folioles along the margins and on the lamina, which may ± obscure the lobes when well developed. The small lobules, which may be produced as part of the regeneration of damaged thalli in *N. laevigatum*, remain exclusively marginal, decumbent and are more uniformly rounded in that species.

NORMANDINA Nyl. (1855)

Thallus squamulose, layered. **Upper cortex** pseudoparenchymatous or evanescent, distinct lower cortex absent. **Photobiont** *Trebouxia*. **Medulla** with mostly globose or ellipsoid hyphal cells arranged in a net-like structure surrounding groups of algal cells. **Conidiomata** and **Ascomata** unknown. **Chemistry**: lichen products not detected by t.l.c. **Ecology**: on other lichens or bryophytes in localities with somewhat oceanic conditions.

LITERATURE: David & Hawksworth (1989), Henssen (1976).

1. N. pulchella (Borrer) Nyl. (1861)

Thallus squamulose, glaucous to pale grey or greenish grey; squamules 1–2 mm diam., shell- or ear-like, concentrically ridged with raised paler margins, scattered or aggregated to form colonies; soredia often present on surface and margins, green or concolorous; lower surface white-tomentose.

On mossy deciduous trees or rocks in woodland and parkland, where it is epiphytic on other lichens (especially *Parmeliella, Pannaria, Degelia*) or bryophytes, rarely on bare bark; frequent. S. & W. British Is. Cosmopolitan.

The perithecia of ***Lauderlindsaya borreri*** (Tul.) J. David & D. Hawksw. (1989), a lichenicolous fungus, are not uncommon on the squamules. At times individual squamules may be reduced to a sorediate cuff around single perithecium of the fungus; they have been erroneously interpreted as the fruit bodies of the lichen.

OCHROLECHIA Massal. (1852)

Thallus crustose, ± uneven, often thick, rarely with spines and then appearing minutely shrubby, continuous to uneven, warted, not distinctly areolate, occasionally reduced to scattered convex warts, pale to dark grey, often tinged greenish; prothallus grey, well developed or absent. **Upper surface** not corticate, or with a thin cortex of thin-walled hyphae. **Photobiont** chlorococcoid. **Ascomata** apothecia; disc expanded, yellowish or brownish pink, often white-pruinose. **Thalline exciple** prominent, well developed. **True exciple** narrow, also visible in some species. **Hymenium** 150–200 μm tall. **Hamathecium** of paraphyses, thin, densely branched and anastomosed. **Asci** 2- to 8-spored, uniformly thick-walled, *Pertusaria*-like. **Ascospores** large, simple, thick-walled, smooth, even. **Conidiomata** pycnidia. **Conidia** cylindrical-elongate, straight. **Chemistry**: some species have gyrophoric acid and ± lecanoric acid. **Ecology**: on ± acid bark and mainly siliceous rocks.

Distinguished from *Pertusaria* primarily by the less thick, uniform ascospore walls.

LITERATURE: Hanko *et al.* (1986), Howard (1970), Tønsberg (1992), Verseghy (1962).

1	Thallus C+ and KC+ red, orange-red or bright orange	2
	Thallus C−, KC− or C+ and KC+ pale yellow	8
2(1)	Soredia or isidia present	3
	Soredia and isidia absent	7
3(2)	Thallus isidiate, or diffusely granular-sorediate, well-defined soralia absent	4
	Thallus sorediate, with well-defined, yellowish white or yellowish green soredia	5
4(3)	Thallus Pd−; on trees	**7. *subviridis***
	Thallus Pd+ pale orange; on mosses, the ground or over rocks	**2. *inaequatula***
5(3)	Thallus C+ pink-orange (fleeting)	6
	Thallus C+ bright orange-red	**4. *inversa***
6(5)	Thallus granular, developing spine-like extensions	**3. *frigida***
	Thallus granular-warted, not developing spine-like extensions	**1. *androgyna***
7(2)	Thallus granular, developing small spine-like extensions	**3. *frigida***
	Thallus granular-verrucose, not developing spines	**8. *tartarea***
8(1)	Soredia absent	9
	Soredia present	**9. *turneri***
9(8)	Apothecial disc pruina C+ and KC+ red	**5. *parella***
	Apothecial disc pruina C+ yellow, KC−	**6. *pallescens***

1. O. androgyna (Hoffm.) Arnold (1885)

Thallus very variable, rather thick, coarsely, irregularly granular-verrucose, or thin, even, continuous, whitish to dark grey, occasionally tinged brown, often wide-spreading; prothallus often conspicuous, pale. Soralia to 2–3 mm diam., ± round and convex or becoming irregular, yellow-green to grey-green, scattered, contiguous or coalescing to form a ± continuous sorediate crust. Apothecia occasional, scattered, 2–4 mm diam.; disc pale pinkish or orange-brown, not-pruinose, at first ± concave, becoming flat; thalline exciple thick, entire, prominent. Ascospores 30–45 × 13–22 μm. Thallus Pd−, K−, KC+ orange-red, C+ orange-red, UV− (gyrophoric acid).

Very variable and often forming large patches on mosses overgrowing, or directly on, acid-barked trees and siliceous rocks; often abundant. N. & W. British Is., more local in east and Midlands. Circumboreal.

The C+ orange-red reaction of the thallus and yellow-green, ± convex soralia are diagnostic. Recent work in Norway has identified 4 apparently distinct taxa differing in chemical, morphological, distributional and/or ecological traits (Tønsberg 1992).

2. O. inaequatula (Nyl.) Zahlbr. (1913)

Like **1**, but thallus coarsely erose-sorediate, uniformly grey-white, delimited, without yellowish soralia, and Pd+ pale orange (unidentified substance). Apothecia not known in British material.

Forming ± orbicular patches over bryophyte mats on exposed mountain ridges and on tops of boulders; local. Scotland (Highlands, Southern Uplands). Scandinavia.

3. O. frigida (Swartz) Lynge (1928)

Thallus, grey- or yellowish white, at first a thin crust of granules which soon develop small, smooth spine-like extensions or become ± coralloid, often forming extensive patches over low vegetation; rarely with convex soralia [?*O. lapuensis* (Räsänen) Räsänen (1931)]. Apothecia to 5(-7) mm diam., sessile; disc concave-flat, pale brown or reddish brown, not pruinose; thalline exciple rather thin, smooth, entire. Ascospores 25–40 × 15–26 μm. Thallus Pd−, K−, KC+ red, C+ red, UV− (gyrophoric acid).

Amongst mosses and overgrowing low vegetation, on high moors or mountain summits, rarely coastal heaths; frequent. Scotland, extending to N. England and Wales. Arctic, subarctic, high mountains in temperate North America, Europe and Asia.

Easily distinguished by the pale, distinctive, spine-like extensions of the thallus granules.

4. O. inversa (Nyl.) Laundon (1963)

Thallus rather thin, continuous, minutely granular-warted or minutely areolate, greenish grey to grey-yellow, sorediate; prothallus grey-black. Soralia round, flat or convex, pale yellowish green or ochre-coloured, ± discrete at first, later becoming ± confluent, rarely forming a continuous, thick, areolate, powdery crust; soredia farinose, rarely minutely granular. Apothecia very rare, unknown mature, round or distorted, flat; thalline exciple, to 0.1 mm diam., entire, smooth, not sorediate, persistent, raised above level of disc; disc flat, ± pruinose, pale brown. Thallus Pd−, K−, KC+ orange, C+ orange, UV+ dull orange (atranorin, thiophaninic acid, arthothelin and additional xanthones).

On old *Quercus, Fagus, Salix* and *Pinus*, rarely slate or sandstone rocks, in moist, old woodlands, especially in sheltered, poorly-drained boggy sites; local. S. & W. British Is., commonest in Hampshire (New Forest) and W. Ireland. S.W. Norway, France (Morbihan), Macaronesia.

When on rock, closely resembles *Pertusaria flavicans* which has granular, not farinose, soredia, lacks atranorin and is UV+ bright orange. Perhaps a synonym of *Lecanora alboflavida* (q.v.).

5. O. parella (L.) Massal. (1852)

Thallus thick, smooth to granular-warted, often fissured, grey-white, ± orbicular, often concentrically ridged and zoned at circumference; surrounded by a white prothallus. Apothecia usually numerous, crowded, round or irregular, to 2 mm diam.; disc deeply convex when young, becoming flat, pale pink-brown, often thickly and coarsely grey-white-pruinose; thalline exciple rather broad, swollen, at times almost closing over disc, smooth, entire; epithecium granular, K+ partly dissolving. Ascospores 45–65(-88) × 25–40(-50) μm, broadly ellipsoid. Thallus Pd−, K−, KC+ yellow, C− or + yellow, UV− or slightly bluish (variolaric acid); apothecial disc pruina and true exciple Pd−, K−, KC+ red, C+ red, UV+ white (gyrophoric acid).

On siliceous or ± basic rocks, walls, brickwork, slates, occasionally trees; frequent. Abundant throughout British Is. Europe, Canada, U.S.A.

A very variable species with a wide ecological amplitude. Well-developed specimens have a corrugate-uneven thallus with numerous, contiguous angular apothecia, in shaded habitats the thallus is thin, smooth, the apothecia scattered and initially pimple-like; the pale prothallus is often fimbriate at the edge.

O. upsaliensis (L.) Massal. (1852), with a rather thin, smooth to granular-wrinkled or warted thallus without soredia, abundant, pale yellowish or greyish buff, strongly white-pruinose apothecia, 1–3 mm diam., and ascospores (30-)50–75 × (25-)30–40 μm, was reported last century as encrusting mosses near the summits of Morrone and Craig Calliach in Highland Scotland. Thallus, thalline exciple and apothecia disc pruina Pd−, K−, KC−, C± yellow (variolaric acid). It seems likely that the material of this alpine species, on which these records are based, originated outside the British Is.

6. O. pallescens (L.) Massal. (1853)

Like **5**, but thallus thin, smooth, with evenly scattered apothecia, the apothecial disc yellowish brown; epithecium crystalline, C+ yellow, crystals disappearing (variolaric acid), true exciple C−.

On *Salix*, *Alnus* and *Betula* in moist carrs; rather rare. W. British Is. W. Norway.

In the absence of proper typification the status of this species remains unclear; most material named as such appears to be a morph of *O. parella* growing on trees. However, other material from W. British Is. (also W. Norway) may be distinct and is possibly referable to *O. pallescens*.

7. O. subviridis (Høeg) Erichsen (1930)

Thallus usually thick, waxy, smooth to granular or warted, often densely covered with soft, branched or ± firm, coralloid, cylindrical isidia, to 0.5 mm diam., becoming confluent towards centre and forming a continuous, uniformly concolorous crust, often breaking down into granular soralia. Apothecia rare, ± sessile, deeply concave with a thick, densely rugose-isidiate thalline exciple; disc to 4 mm diam., pale pink-brown, roughened, scabrid-pruinose; epithecium finely granular, the granules dissolving in K. Ascospores 30–68 × 25–35 μm. Isidia Pd−, K−, KC+ red, C+ red, UV+ glaucous, often with a yellow tinge (gyrophoric, ± lecanoric acids).

On bark of woodland and wayside broad-leaved trees; frequent. Throughout British Is. Oceanic and suboceanic Europe, eastern U.S.S.R., North America, Japan, Korea.

Distinguished by the extensive, diffuse patches of soft isidia, which easily erode to form granular soredia.

8. O. tartarea (L.) Massal. (1852)

Thallus often very thick, to 3 mm or more, tartareous, often with numerous irregular warts forming an uneven, corrugate crust, sometimes with a paler zoned margin and a pale prothallus, pale to dark grey. Apothecia usually frequent, immersed and closed at first, rounded or irregular, scattered or crowded, sessile; thalline exciple thick, wavy; disc to 5(-8) mm diam., pale brown to dull orange-pink, concave to flat, not, or translucently, pruinose, the surface often scabrose-roughened; epithecium granular, the granules dissolving in K. Ascospores (35-)40–70 × 20–40 μm, broadly ellipsoid. Medulla Pd−, K+ pale yellow, KC+ red, C+ orange-red, UV−; apothecial disc Pd−, K−, KC+ red, C+ vermilion red (gyrophoric and ± lecanoric acids).

On siliceous boulders, leached, acid bark and montane moss-lichen heaths, upland, oceanic and montane areas in wind-swept areas; frequent. N. & W. British Is. Circumboreal, C. European mtns extending south to Corsica and Madeira.

Frequently confused with *O. androgyna* (**1**), which has pale to yellowish green soralia. The apothecia are frequently contorted, or have proliferating exciples; occasionally the disc does not expand and the apothecia take on the aspect of a *Pertusaria* (e.g. *P. xanthostoma*).

9. O. turneri (Sm.) Hasselrot (1945)

Thallus thin, smooth, or ± wrinkled-granular, somewhat waxy, white-grey, sorediate; prothallus sometimes present and well developed, pale or whitish grey. Soralia numerous, very variable in size, often excavate and in lines, becoming crowded and later confluent towards the centre of the thallus, sometimes forming a ± continuous, granular sorediate crust; soredia 0.04–0.1 mm diam., granular, sometimes resembling isidia, usually paler than the thallus, originating from small warts which burst to form granular, wart-like soralia. Apothecia to 3 mm diam., occasional, concave to flat with a thick, wrinkled and granular sorediate thalline exciple to 0.5 mm thick; disc ± pruinose, rough, pale yellow- or brown-pink. Thallus Pd−, K−, KC± pale yellow, C± yellow, UV+ glaucous with an orange tinge (variolaric acid, ± lichesterinic acids); apothecial disc C+ red (gyrophoric acid).

On bark and wood, occasionally on acid rocks, especially sandstones; common. Throughout British Is. Europe, North America.

The race containing lichesterinic acid (fatty acid) is found on more acidic substrata and is sometimes considered to be a separate species, *O. microstictoides* Räsänen (1936). *O. turneri* is distinguished by the grey, excavate, sometimes ± linear development of soredia and the C± pale yellow (not red) thallus.

OMPHALINA Quél. (1886)

Thallus granular ('*Botrydina*'), squamulose ('*Coriscium*'), or absent. **Photobiont** *Coccomyxa*, absent in lichenicolous species. **Basidiomata** agaricoid, the basidia borne on gills. **Cap** small, convex to flat, often umbilicate to funnel-shaped, thin-fleshed, with water-soaked appearance when wet, often translucent-striate, mostly ± smooth, dull or brightly coloured, **cuticle** little-differentiated, of ± radially arranged, filamentous hyphae, **hyphae** with pigment membranal, incrusting or intercellular, sometimes also vacuolar, **clamp connections** present or absent. **Gills** shortly to broadly decurrent, sometimes thick and distant, **cystidia** absent, but gill edge sometimes with sterile cells. **Stem** central, slender, without a veil. **Basidia** 2- to 4-spored. **Basidiospores** ovoid to narrow-ellipsoid, almond-shaped, drop-shaped, or almost triangular in side-view, smooth, thin-walled, colourless, not I−, **spore print** white or whitish, rarely rose. **Gill trama** weakly to strongly irregular. **Chemistry**: lichen products not detected. **Ecology**: on soil, plant remains, wood, *Sphagnum*, or lichenicolous.

The lichenized European species possess a thallus of one of two types: **(1)** *Botrydina*-type, crustose, dark green, of globose granules 0.1–0.2 mm diam., the granules composed of clusters of algae surrounded by a pseudoparenchymatous hyphal layer; **(2)** *Coriscium*-type, squamulose, the young squamules ± circular, becoming lobed, lobes 0.9–2.0 mm wide, rather thin, pale green-grey, becoming greyish in the herbarium, closely attached to the substratum by pale hyphae, the margin raised and inrolled, pale; squamules composed of clusters of algae enveloped by hyphae, bounded by an upper and lower cortex.

Some of the lichenized species have been placed in *Gerronema* Singer (1951). More recently all the lichenized species have been assigned to a separate genus *Phytoconis* Bory (1797), but that division is not supported by basidiomatal characters and is not followed here; *Omphalina* has been proposed for conservation over *Phytoconis*. More work is required on the taxonomy and nomenclature of the British species.

In the preparation of material for the herbarium, careful notes on the colour of the fresh fruit-body and on the size of the spores (obtained from a spore-print) should be made.

In the following account the *c.* 24 non-lichenized, non-lichenicolous species are omitted. Algae have also been reported in the stipe tissues of *O. chlorocyanea* (Pat.) Singer (1951).

LITERATURE: Clémençon (1982), Redhead & Kuyper (1987, 1988), Watling (1981).

1 Basidiomata present 2
  Basidiomata absent 9

2(1) Basidiomata associated with a thallus of *Botrydina-* or *Coriscium-* type; not on *Peltigera* ... 3
  Basidiomata not associated with a thallus; on dying parts of *Peltigera* thalli **1. *cupulatoides***

3(2) Thallus *Botrydina*-type; cap various; stem whitish to yellow, orange or brown, but without lilaceous tints 4
  Thallus *Coriscium*-type; cap cream to buff or pale orange-yellow; stem pale lilaceous at first and later whitish but retaining a pink flush at apex .... **4. *hudsoniana***

4(3) Basidiomata whitish, bright yellow or orange, or brown at first but soon becoming to yellow-brown or straw; clamp connections absent 5
  Basidiomata grey-brown to black-brown; clamps connections present or absent 7

5(4) Basidiomata whitish, yellowish brown or straw 6
  Basidiomata bright lemon-yellow to chrome-yellow or orange .... **5. *luteovitellina***

6(5) Basidiomata whitish to ivory; basidiospores 8–9.5(-10) × 4.5–6 μm; often on *Sphagnum* .... **7. *pseudoandrosacea***
  Basidiomata yellowish brown to straw; basidiospores 7–10 × 6–7 μm; habitat various, rarely on *Sphagnum* .... **2. *ericetorum***

7(4) Stem finely pubescent with erect hairs (strong lens); basidiospores less than 12 μm long and less than 6 μm wide; basidia 4.5–8.5 μm wide; clamp connections absent 8

Stem fibrillose at first, but not finely pubescent; basidiospores 8–14(-16) × 4.5–8 μm; basidia 10–12 μm wide; cap often somewhat irregular and excentric, the surface fibrillose; clamp connections present .... **3. *griseopallida***

8(7) Basidia 4-spored; basidiospores 7–11 × 2.8–3.5 μm; gills broadly decurrent .. **6. *pararustica***
Basidia predominantly 2-spored; basidiospores 6–8 × 3–4 μm; gills shortly decurrent **8. *velutina***

9(1) Thallus of *Botrydina*-type 10
Thallus of *Coriscium*-type .... **4. *hudsoniana***

10(9) Hyphae amongst granules (2.5-)3–4(-6) μm diam. 11
Hyphae amongst granules 2–3 μm diam., thin-walled; not separable when sterile .... **6. *pararustica*** or **8. *velutina***

11(10) Hyphal walls mostly <1–1.5 μm or more thick; cortical cells of granules with thickened walls .... **5. *luteovitellina***
Hyphal walls 0.5–1 μm thick; cortical cells of granules with ± unthickened walls; not separable when sterile .... **2. *ericetorum*** or **7. *pseudoandrosacea***

1. O. cupulatoides P. D. Orton (1977)

Cap 5–20 mm, at first convex, soon expanded-umbilicate then depressed or funnel-shaped, the margin long remaining incurved, greyish-brown, translucent-striate, when dry clay to buff-brown. Gills arcuate-decurrent, pinkish buff to grey-brown, rather distant, often forked, sometimes veined on sides, the edge thick, blunt and fertile. Hyphae with the pigment finely encrusting; clamp connections present. Stem 10–30 × 0.7–2 mm, concolorous with gills, drying white, finely pruinose-tomentose to ± entirely pubescent, hairs 10–140 × 4–8 μm, cylindrical, sometimes slightly swollen in lower part, often flexuose. Basidia 4-spored. Basidiospores (7-)8–11 × (4-)4.5–5 μm, cylindric- or elongate-ellipsoid.

Associated with moribund parts of *Peltigera* (*P. lactucifolia* only?), non-lichenized. S. England (Hampshire, New Forest), Wales, Scotland. Endemic.

2. O. ericetorum (Fr.) M. Lange ex H. Bigelow (1970)

?*O. fulvopallens* P.D. Orton (1984), *Botrydina vulgaris* Bréb. (1839)

Thallus *Botrydina*-type, cortical cells of granules with ± unthickened walls, hyphae amongst granules (2.5-)3–4(-6) μm diam. with walls 0.5–1.0 μm thick. Cap 6–20 mm diam., at first planoconvex, then depressed, brown when young but soon becoming yellowish brown or straw-coloured, often with darker centre and furrows; margin soon crenate, furrowed, translucent-striate; surface smooth to very finely scurfy. Gills whitish to yellowish, distant, broadly decurrent. Stem 10–30 × 1–2 mm, dark brown with a ± purplish tinge at apex, otherwise ± concolorous with cap, very finely white-pubescent, white tomentose at base. Hyphae with pigment intercellular, but soon deposited on the hyphal wall; clamp connections absent. Basidia 4-spored. Basidiospores 7–10 × 6–7 μm, ellipsoid.

On soil, rotting wood and tree trunks, especially in humid habitats, less commonly on *Sphagnum*; common. Throughout British Is. Europe, North America; possibly elsewhere but confused with other species.

O. fulvopallens has been shown to be lichenized and probably belongs here; the basidia are predominantly 2-spored, the basidiospores 7–9 × 4.5–6 μm, and the stem apex is paler. See also *O. luteovitellina* **(5)**.

3. O. griseopallida (Desm.) Quélet (1886)

Thallus, if present, *Botrydina*-type. Cap 8–17 mm diam., at first planoconvex, then funnel-shaped, often somewhat irregular and excentric; margin somewhat furrowed, translucent-striate, grey-brown to dark grey-brown, whitish-grey when dry, the surface fibrillose to almost tomentose. Gills distant, shallow, decurrent, ± concolorous with cap. Cystidia and marginal cells absent. Stem 10–15 × 1–2 mm, ± concolorous with cap, the surface fibrillose, soon glabrous. Hyphae with intraparietal to incrusting pigment. Clamp connections present. Basidia (2-)4-spored, 10–12 μm wide. Basidiospores 8–14(-16) × 4.5–8 μm, ellipsoid to ovoid.

On soil; rare. S.W. England (S. Devon, Slapton). Europe.

It is possible that the species is normally non-lichenized and only facultatively associated with a *Botrydina*-type thallus.

4. O. hudsoniana (Jenn.) Bigelow (1970)

O. luteolilacina (Favre) Henderson (1958), *Coriscium viride* (Ach.) Vainio (1890)

Thallus *Coriscium*-type. Cap 10–20 mm diam., convex-depressed, becoming funnel-shaped, cream to buff or pale orange-yellow, finely tomentose; margin lightly crenate, weakly transparent-striate. Gills ± concolorous with cap, shortly to strongly decurrent, moderately distant, veined. Stem to 20 × 2.5–3 mm, pale lilaceous at first, soon becoming whitish or cream but retaining a pink flush at apex, finely pubescent throughout; hairs cylindrical. Hyphal pigment intercellular; clamp connections absent. Basidia 4-spored. Basidiospores 8.5–10 × 4.5–5.2 μm, ellipsoid.

On peaty soil, less often on moss, decaying vegetation and rotting wood, mainly in upland areas; rather frequent. Throughout British Is. Europe, North America.

Specimens from North America lack lilac tints in the stem. Reports of this type of thallus associated with other types of basidiomata are due at least in part to confusion with other species.

5. O. luteovitellina (Pilát & Nannf.) M. Lange (1957)

Thallus *Botrydina*-type; cortical cells of granules with slightly thickened walls; hyphae amongst granules mostly with walls to 1–1.5 μm or more thick. Cap 5–15 mm diam., flat to planoconvex, bright lemon-yellow to chrome-yellow or orange; margin crenate, translucent-striate. Gills concolorous with cap or paler, thick, distant, veined and often forked. Stem 10–15 × 1.5–2.5 mm, concolorous with gills, finely pubescent throughout; hairs cylindrical. Hyphal pigment cytoplasmic and intercellular, partly deposited on the hyphae; clamp-connections absent. Basidia 4-spored. Basidiospores 6.5–9.5 × 3.5–4.5 μm, ellipsoid.

On peaty soil in upland areas; rather frequent. Throughout British Is. Europe, North America.

Distinguished from *O. ericetorum* **(2)** in the bright yellow to orange colour and the narrower basidiospores. In fading specimens, the stem retains the characteristic colours longer than the cap.

6. O. pararustica Clémençon (1982)

Thallus *Botrydina*-type, hyphae amongst granules 2–3 μm diam., thin-walled. Cap 4.5–7 mm diam., planoconvex to weakly depressed, dark brown, smooth to lightly scurfy; margin lightly crenate, furrowed, translucent-striate. Gills distant, broadly decurrent, concolorous with cap or slightly paler, not veined. Cystidia and marginal cells absent. Stem 9–11 × 0.5–1 mm, concolorous with cap, pubescent throughout, hairs irregularly cylindrical, brown. Hyphae with brown-incrusting pigment; clamp connections absent. Basidia 4-spored, 5–7 μm wide. Basidiospores 7–11 × 2.8–3.5 μm, cylindric-ellipsoid to somewhat ovoid.

On soil; rare. S. Wales (Glamorgan), C. Scotland (Mid-Perth, Ben Lawers). Switzerland.

The non-British *O. baeospora* Singer (1977), described from Czechoslovakia, has clamp connections and basidiospores 5–7 × 3.5–5 μm.

7. O. pseudoandrosacea (Bull.) Moser (1967)

Thallus *Botrydina*-type, like **2**. Cap 10–20 mm, becoming depressed, whitish to ivory with darker striae, drying cream. Gills concolorous with cap. Stem 10–22 × *c.* 1.5–2 mm, concolorous with cap, pubescent, hairs cylindrical. Clamp connections absent. Basidia 2(-4)-spored. Basidiospores 8–9.5(-10) × 4.5–6 μm.

On *Sphagnum*, less frequently amongst other mosses on wet soil; locally frequent. W. & N. Scotland. Europe.

Usually regarded as a morph of *O. ericetorum* **(2)**, but distinguished by the whitish basidiomata and mostly 2-spored basidia. The correct application of the name *O. pseudoandrosacea* is in dispute, and in any case Moser's combination does not appear to have been validly published; the species concept followed here is that adopted by Quélet.

8. O. velutina (Quélet) Quélet (1886)

O. grisella (Weinm.) Moser (1953)

Thallus *Botrydina*-type; hyphae amongst granules 2–3 μm diam. Cap 5–11 mm diam.; margin furrowed, crenate and translucent-striate; at first dark brown, then grey brown, light brown when dry. Gills whitish or greyish at first, contrasting with cap and stem, slowly later dark grey brown to brown but still paler than cap, broad, shortly decurrent. Cystidia and marginal cells absent, perhaps rarely with very isolated marginal cells difficult to distinguish from basidia. Stem 15–30 × 1–3 mm, ± concolorous with cap or paler, finely pubescent; hairs brown incrusted, at least at base, the apex mostly colourless. Hyphal pigment intraparietal to incrusting; clamp connections absent. Basidia mostly 2-spored, 5–6 μm wide. Basidiospores 6–8 × 3–4 μm, ellipsoid.

On soil. Throughout British Is. Europe.

The non-British *O. rustica* (Fr.) Quélet (1886) differs in the larger basidiospores, 8–11 × 4–5.5 μm, and broadly decurrent gills.

OPEGRAPHA Humb. (1793)

Thallus crustose, thin to thick, smooth or ± rimose, sometimes partially or entirely immersed, white, grey, mauve-grey, grey-brown, dark brown or olive-green; prothallus often present and mosaic-forming. **Soralia** rarely confluent, occasionally present. **Photobiont** *Trentepohlia*; a few species lichenicolous on lichens with other photobionts. **Ascomata** lirellate apothecia, elongate, rarely ± rounded, sometimes branched or stellate, discrete or confluent, rarely densely clustered and contiguous, ± sessile, rarely immersed. **Thalline exciple** absent, but a thin false thalline exciple rarely developed. **True exciple** black, continuous under the hypothecium, opaque, sometimes ± swollen, disc ± exposed or remaining slit-like, K− or red-brown or ± olive-green in section, occasionally yellow-green, orange or white-pruinose. **Epithecium** colourless in those species with a slit-like disc, or brown where exposed. **Hymenium** I−, ± reddish brown or ± blue. **Hypothecium** colourless to pale straw, K− or ± olive-green. **Hamathecium** of paraphysoids, septate, branched, often richly anastomosed, apices not or rarely slightly swollen. **Asci** 8-spored, clavate-cylindrical, except for a sometimes discernible, minute I+ blue apical ring; dehiscence fissitunicate (Fig. 34g). **Ascospores** multiseptate, fusiform or acicular, I−, sometimes with a thin perispore, colourless or becoming ornamented and red-brown when over-mature. **Conidiomata** pycnidia, immersed to sessile, flask-shaped, black, rarely white-pruinose. **Conidiogenous cells** mainly elongate-ampulliform, arising singly, enteroblastic, acrogenous. **Conidia** very variable, simple, rarely septate, ellipsoid, rod- or sickle-shaped, colourless (Fig. 18). **Chemistry**: variable, no lichen products, unidentified pigments, or in a few species gyrophoric acid. **Ecology**: on a wide range of young and mature broad-leaved trees, more rarely conifers and wood; also on often sheltered, acid, neutral or basic rocks, rarely on soil and dead herbs; a few species are lichenicolous.

Sometimes confused with *Graphis*, which differs in having a hamathecium of unbranched paraphyses, I+ purple ascospores with lens-like cells, and quite different non-fissitunicate asci. The delimitation of *Opegrapha* from *Lecanactis* and *Schismatomma* is currently imprecise and these genera should also be consulted in cases of difficulty. The non-lichenized genus *Hysterium* is sometimes mistaken for *Opegrapha*, but differs in the early browning of the ascospore walls before release from the asci.

The size and shape of the conidia vary considerably in *Opegrapha* and are an important diagnostic feature of many species. In section, the inner edge of the true exciple, and sometimes the epithecium and hymenium, are tinged green-olive or red-brown, both colours K± intensified, according to the species. In those species with red-brown pigmentation, the ascospores in degenerating asci often become red-brown with age due to the uneven (often granular) deposition of a red-brown pigment on their outer wall.

The number of septa of the spores is frequently diagnostic; in a few species, e.g. *O. mougeotii* **(11)** and *O. paraxanthodes* **(16)**, one of the middle cells is ± enlarged. A distinct perispore is present in a few species, e.g. *O. viridis* **(29)**, *O. prosodea* **(18)**.

LITERATURE: Coppins *et al.* (1992), Pentecost & Coppins (1983), Redinger (1937), Torrente & Egea (1989).

1 Photobiont present; not on other lichens (not lichenicolous) 2
Photobiont absent; on other lichens (lichenicolous) 35

2(1) Thallus without soredia; apothecia frequent 3
Thallus sorediate or with numerous pycnidia; soralia punctiform to effuse; apothecia rare or absent 26

3(2) Ascospores 3-septate 4
Ascospores 4- to 15-septate 16

4(3) On bark or wood or plant remains 5
On rock, soil, mortar 10

5(4) Exciple and disc not orange-pruinose; not K+ purplish in section 6
Exciple and disc ± orange-pruinose; pruina and exciple K+ dissolving magenta-purple in section .... **14. *ochrocheila***

6(5) Ascospores 5–8 μm wide; perispore thin and distinct; exciple red-brown, K− in section .... **9. *herbarum***
Ascospores 3–5 μm wide; perispore indistinct; exciple K+ olive-green or brown in section 7

7(6) Apothecia ± sessile or emergent; disc a slit or only partly exposed; exciple well developed; pseudothalline exciple absent; ascospores 13–18(-20) μm long, ± clavate 8
Apothecia immersed; disc exposed; exciple thin, sometimes with a pale pseudothalline exciple; ascospores (15-)17–27 μm long, ± curved .... **20. *rufescens***

8(7) Thallus pale or inconspicuous 9
Thallus dark brown .... **23. *saxigena***

9(8) Disc partially open; exciple brownish in section, K+ diffusing magenta; conidia 12–20 μm long .... **14. *ochrocheila***
Disc a persistent slit; exciple greenish in section, K± olive-green; conidia 4–5 μm long .... **2. *atra***

10(4) Ascospores 20–25(-30) μm long 11
Ascospores 14–18(-22) μm long 13

11(10) Apothecial disc ± exposed or a slit, not pruinose; exciple smooth or uneven, not pruinose 12
Apothecial disc widely exposed, densely grey-pruinose; exciple smooth, not pruinose .... **25. *subelevata***

12(11) Apothecial disc a slit or only narrowly exposed; apothecia often short, gnarled-gyrose; ascospores without a distinct perispore .... **22. *saxicola***
Apothecial disc ± widely exposed; apothecia ± elongate; ascospores with a distinct perispore .... **9. *herbarum***

13(10) Thallus pale or ± inconspicuous, not mosaic-forming 14
Thallus conspicuous, dark brown to chocolate-coloured, often mosaic-forming . **23. *saxigena***

14(13) Apothecia not white-pruinose; conidia 4–7 μm long 15
Apothecia densely white-pruinose; conidia 11–20 μm long .... **6. *demutata***

15(14) Exciple K+ diffusing magenta-purple in section; margin ± orange-pruinose; disc ± widely exposed .... **14. *ochrocheila***
Exciple K− or red-brown in section; margin not pruinose; disc a slit or occasionally slightly exposed .... **21. *saxatilis***

16(3) Ascospores 4- to 7(-8)-septate 17
Ascospores (8-)9- to 15-septate 25

17(16) On bark or wood 18
On rock, mortar 21

18(17) Ascospores 2.5–3.5(-4) μm wide 19
Ascospores (5-)6–9 μm wide .... **27. *varia***

19(18) Pycnidia black, not pruinose 20
Pycnidia grey-white, ± densely pruinose .... **28. *vermicellifera***

20(19) Conidia 4–9 μm long, short, curved, worm-like ..... **13. *niveoatra***
Conidia 9–15 μm long, curved, occasionally straight, bacilliform, 3–5 μm long .. **30. *vulgata***

21(17) Ascospores 5–9 μm wide 22
Ascospores 3–4(-5) μm wide 24

22(21) Ascospores (18-)20–37 μm long, 4- to 6-septate; ends of apothecia ± rounded 23
Ascospores (20-)25–33(-55) μm long, 5- to 7(-8)-septate; ends of apothecia acute **11. *mougeotii***

23(22) Ascospores (18-)20–25(-27) μm long, 4- to 5-septate; thallus inconspicuous **16. *paraxanthodes***
Ascospores 20–37 μm long, 5- to 6-septate; thallus distinct, grey, ± rimose-cracked **27. *varia***

24(21) Ascospores 5- to 7-septate, 20–35 μm long; apothecia long, thin, ± serpentine, 0.07–0.2 mm wide; conidia curved, 10–14 μm long ..... **10. *lithygra***
Ascospores 4- to 5-septate, 15–22 μm long; apothecia short, stout, 0.13–0.4 mm wide; conidia bacilliform 6–7.5 μm long ..... **4. *cesareensis***

25(16) Apothecia short, ± knot-like, 0.4–1 mm long; ascospores 25–60 μm long ..... **29. *viridis***
Apothecia elongate, serpentine, 1–3 mm long; ascospores 50–80 μm long ..... **18. *prosodea***

26(2) Soredia absent; pycnidia usually abundant 27
Soredia present; pycnidia usually absent 30

27(26) Pycnidia ± pruinose 28
Pycnidia not pruinose 29

28(27) Pycnidia ± densely white-pruinose; conidia 4–7 × 1 μm ..... **28. *vermicellifera***
Pycnidia thinly white- or greenish pruinose; conidia 3–5 × 0.5-1(-2) μm ..... **27. *varia***

29(27) Conidia 4–9 μm long ..... **13. *niveoatra***
Conidia 9–15 μm long ..... **30. *vulgata***

30(26) Soralia C+ red-orange (often fleeting) 31
Soralia C− 33

31(30) Thallus pale, often inconspicuous, not mosaic-forming; soralia small, punctiform or diffuse, UV− 32
Thallus brown, dark brown, rarely pale, mosaic-forming; soralia 0.4–1.5 mm diam., becoming confluent, UV+ glaucous to yellow; apothecia gyrose-contorted .. **8. *gyrocarpa***

32(31) Soralia bright uniformly orange when fresh, ± punctiform; soredia farinose, 12–26 μm diam. ..... **24. *sorediifera***
Soralia fawn dotted with brown (× 20 lens) when fresh, diffuse, inconspicuous; soredia finely granular, 20–50 μm diam. ..... **7. *fumosa***

33(30) Thallus pale to brown or inconspicuous, not mosaic-forming; soralia pale grey, ± tinged ochraceous or orange, discrete or becoming confluent; apothecia unknown 34
Thallus dark brown, mosaic-forming; soralia pale brown to lilac-white, punctiform, discrete, 0.2–1.2 mm diam.; apothecia sometimes present ..... **31. *zonata***

34(33) Soralia pale grey or ± ochre-coloured, 0.2–0.7 mm diam., often confluent, forming irregular erose patches, 2–3 mm diam. ..... **5. *corticola***
Soralia bright orange, minute, punctiform, 0.1–0.3 mm diam., remaining mostly discrete ..... **12. *multipuncta***

35(1) Exciple K+ greenish in section 36
Exciple K± brown in section 38

36(35) Ascospores (5-)6-septate, 4 per ascus, 21–28 × 4.5–6 μm; on *Pertusaria leioplaca* ..... **17. *pertusariicola***
Ascospores 3-septate, 4 or 8 per ascus, 13–18 × 4.5–6 μm; on *Thelotrema* species 37

37(36) Asci 4-spored; on *Thelotrema subtile* ..... **3. *brevis***
Asci 8-spored; on *Thelotrema lepadinum* and *T. monosporum* ..... **26. *thelotrematis***

38(35) Ascospores 14–22 × 5–8 μm; on Verrucariaceae, *Caloplaca cirrochroa* and *Xanthoria parietina* ..... **15. *parasitica***
Ascospores 19–26 × 6–8 μm; on *Dermatocarpon miniatum* aggr. ..... **19. *pulvinata***

1. O. areniseda Nyl. (1875)

O. actophila Nyl. (1880)

Thallus very thin, or inconspicuous or immersed, indeterminate. Apothecia 1–2 × 0.15–0.3 mm, elongate, sessile, occasionally furcate; exciple K−, enclosing; disc a slit; epithecium pale brown; hymenium 75–90 μm tall, I+ red. Ascospores (14-)20–30(-40) × (4-)5–6 μm, 3–5(-8)-septate, often with a thin perispore. Pycnidia 0.1–0.2 mm, numerous, sterile, often clustered; conidia 3–5 × 0.5–0.7 μm, bacilliform. Thallus Pd−, K−, KC−, C−, UV− (no lichen products detected by t.l.c.).

On consolidated soil and sand, rarely on decaying wood or friable rock on or near the coast; rare. Channel Is., Is. of Scilly, S.W. Wales, W. Ireland. France, Madeira.

Differs from *O. mougeotii* **(11)** in the smaller ascospores lacking an enlarged middle cell. *O. areniseda* belongs to the *O. varia* aggregate. It frequently occurs without apothecia and is then distinguished by the large, prominent, sessile pycnidia which are nearly always abundant.

2. O. atra Pers. (1794)

Thallus thin, smooth, effuse or ± delimited, cream- to silver-white, sometimes ± tinged olivaceous, smooth, often ± immersed, usually in small clearly defined patches, occasionally mosaic-forming. Apothecia (0.3-)0.7–1.5(-2) × (0.09-)0.12–0.2(-0.3) mm, 100–130 μm tall, abundant, randomly distributed or in parallel or ± stellate clusters, variable, simple or frequently branched and often contiguous, sometimes forming dense contiguous black patches, sessile; disc a slit, rarely opening; exciple K+ olive-green; hymenium 60–90 μm tall, I+ blue; epithecium brown or green-brown, K+ olive-green. Ascospores 13–18(-20) × 2.5–4(-5) μm, 3-septate, ellipsoid or often ± clavate. Conidia 4–5 × 0.7–1 μm, straight or slightly curved, ends blunt, rounded, bacilliform. Thallus Pd−, K−, KC−, C−, UV− (no lichen products detected by t.l.c.).

On smooth bark of young or old trees, more rarely on wood, especially on smooth bark of *Corylus*, *Fraxinus* and *Ilex*; often abundant. Throughout British Is. W. Europe, North America, New Zealand.

Though normally superficial, the lirellae may become almost totally immersed on *Sambucus niger*. This common species often forms small, neat thalli on smooth bark of young trees. It is distinguished by the numerous, often contiguous apothecia and the 3-septate ascospores and olivaceous tinge of the hymenium and edge of the exciple.

3. O. brevis Coppins (1987)

Apothecia 0.12–0.24 mm wide, 0.1–0.14 mm tall; exciple in section K+ greenish. Asci 4-spored. Ascospores 3-septate, (14-)15–18 × 4.5–5(-6) μm, pigment in perispore. Conidia 5–7 × 1 μm.

On *Thelotrema subtile* on *Corylus*; local. Scotland (W. Inverness, Argyll, Kintyre, S. Ebudes, W. Ross). Endemic.

4. O. cesareensis Nyl. (1868)

O. hysteriiformis Nyl. (1876)

Thallus thin, even and smooth, sometimes rimose, rarely inconspicuous, effuse or delimited, bleached to pale or deep lilac-brown or grey-lilac, often forming extensive mosaics; a thin grey-black prothallus often present. Apothecia 0.3–1(-2) × 0.15–0.3(-0.45) mm, 0.2–0.6 mm tall, scattered, usually simple, or infrequently forked, rarely contiguous, sessile, straight or curved, prominent, sometimes elevated on a small thalline cushion; disc a persistent slit; exciple thick, somewhat swollen, K− or K+ dull green; epithecium brown; hymenium 80–90 μm, I+ blue in upper part. Ascospores (13-)15–22(-26) × 4–5 μm, (3-)4- to 5(-6)-septate, the cells elongate, equal in size. Conidia 6–7.5 × 0.7–1 μm, bacilliform, straight. Thallus Pd−, K−, KC−, C−, UV− (no lichen products detected by t.l.c.).

On sheltered faces or underhangs of moist or dry siliceous rocks in the xeric-supralittoral zone; locally frequent. S.W. England (Devon), Wales, extending to N. Scotland (Shetland), Ireland. France, Germany.

The lilac tinge of the thallus and 4- to 5-septate ascospores are distinctive.

5. O. corticola Coppins & P. James (1979)

Thallus wide-spreading, effuse, continuous, thin, or ± immersed or inconspicuous, grey-green in shade, ± suffused dark or dull brown in more exposed situations; prothallus usually not apparent or pale.

Soralia 0.2–0.7 mm wide, sometimes becoming confluent, 2–3 mm wide, scattered or ± contiguous, initially punctiform, becoming irregular and erose, greenish fawn, pale grey-brown or ochraceous, fading to whitish grey in the herbarium. Apothecia and pycnidia unknown. Soralia Pd−, K−, KC−, C−, UV± pale fawn (rarely two UV+ red spots).

Most frequent on old, rough bark of *Ulmus*, but also found on a wide range of wayside and woodland deciduous trees, sometimes in deep shade, as on old trees in coppice woodland and on *Taxus* in churchyards; common. W. British Is. France (Pyrénées), Italy (Rome, Tuscany).

Resembles *O. gyrocarpa* **(8)**, which may occasionally occur on bark, but lacks gyrophoric and schizopeltic acids. *O. corticola*, which has never been recorded fertile, may have close affinities with *Belonia*.

6. O. demutata Nyl. (1879)

Thallus thin, effuse, or of small rounded areoles, cream-white. Apothecia (0.3-)0.4–0.7(-1) × 0.12–0.2 mm, 40–80 μm tall, semi-immersed to sessile, frequently branched, sometimes surrounded by a pseudothalline exciple; disc slit-like or partly open; disc and exciple white-pruinose; exciple K−; epithecium brown; hymenium 60–70 μm tall, I+ blue in parts, the remainder I+ red. Ascospores (12-)14–18 × 3–5 μm, 3-septate. Conidia 11–15(-20) × 2 μm, straight or slightly curved. Pd−, K−, KC−, C−, UV− (no lichen products detected by t.l.c.).

On brickwork and shaded flint. E. England (Lincoln, Gibraltar Point; Suffolk, Wickham Market). France, Germany.

Distinguished from *O. ochrocheila* **(14)** by the white, K− pruina.

7. O. fumosa Coppins & P. James (1992)

Thallus thin, wide-spreading, effuse, immersed, inconspicuous, grey-brown. Soralia numerous, pale fawn, minutely mottled brown (never orange even when fresh), sometimes erumpent-punctiform or elliptical, 0.2–1 × 0.2–0.6 mm, ± flat or regularly convex with loosely heaped soredia, often confluent and diffuse (but never forming a continuous leprose crust); soredia finely granular, 20–50 μm diam., pale fawn, those on the surface often brown. Apothecia and pycnidia unknown. Soralia Pd−, K−, KC+ pink, C+ red, UV− (gyrophoric acid).

On trunks of mature *Quercus* in sheltered sites in ancient woodland. S. England (Hampshire, New Forest; N. Somerset, Exmoor), Scotland (Argyll, Loch Sunart). ?Endemic.

Forming extensive patches, sometimes to several square metres, characterized by diffuse, grey-brown stains with inconspicuous, punctiform to erose, fawn-coloured, C+ red (fleeting) soralia.

8. O. gyrocarpa Flotow (1825)

Thallus thin to medium, effuse or ± finely rimose, often forming extensive mosaics, pale tan to deep chocolate-brown; prothallus dark, delimiting. Soralia 0.4–1.5(-2.5) mm diam,, scattered, often in patches or becoming ± confluent, dull yellow-grey or ochraceous, fading to pale buff-grey in the herbarium. Apothecia not uncommon, 0.7–1.6(-2) mm wide, knot-like or clustered-gyrose, rounded, sessile, scattered, rarely crowded and contiguous; exciple and hypothecium K+ olive-green; epithecium pale; hymenium 80–120 μm tall, I−. Ascospores (12-)17–25(-30) × (3-)4–6 μm, slightly clavate, 3-septate, straight. Conidia 3.8–6 × 0.5–0.7 μm, bacilliform. Soralia Pd−, K−, KC+ red, C+ red, UV± glaucous or pale yellow (gyrophoric, schizopeltic acids, unidentified xanthone and accessory substances).

On humid, often ± vertical or overhanging, shaded, neutral or acid rock faces, especially in old woodlands, rarely on old walls; scarce on deciduous or conifer bark and *Ilex*. Widespread in W. England, Wales, Scotland, W. Ireland, very scarce elsewhere. W. Europe, Madeira, Azores.

Can be confused with *Belonia nidarosiensis*, which has C− soralia and is mainly confined to ± calcareous rocks. *O. gyrocarpa* should probably be placed in *Lecanactis*.

9. O. herbarum Mont. (1833)

?*O. protuberans* Zahlbr. (1923)

Thallus thin, sometimes inconspicuous or immersed, smooth, ash-grey, brown or dull olive-green. Apothecia 0.5–1(-1.6) × 0.15–0.3 mm, 40–100 μm tall, scattered, simple or infrequently branched, sessile; disc at first a slit, soon becoming ± fully exposed, occasionally ± green-pruinose; exciple K−; epithecium brown, K− or ± red-brown; hymenium 70–90 μm tall, I+ red. Ascospores (16-)18–24(-26)

× (4-)5–7(-8) μm, 3-septate, somewhat clavate, ends rounded, with a thin but distinct perispore, 0.5–1 μm wide, colourless, becoming ± red-brown when over-mature (Fig. 18d). Pycnidia very rare; conidia 3–6 × 0.5–1 μm, bacilliform, straight. Thallus Pd−, K−, KC−, C−, UV− (no lichen products detected by t.l.c.).

On acid to basic, usually smooth bark, wood, rarely on dead herbs and bramble stems and sandstone, often in rather shaded situations; widespread. S. & W. England, Wales, Scotland, Ireland. W. Europe, North America, Australia.

Distinguished by the red-brown exciple and epithecium, and the presence of a distinct perispore. The lichenicolous *Laeviomyces opegraphae* D. Hawksw., with simple brown conidia can occur on the thallus. Often confused with non-pruinose morphs of *O. ochrocheila* **(14)**.

10. O. lithyrga Ach. (1810)

Thallus thin, film-like, smooth, membranous, pale to dull grey, deep olive, rarely rusty brown. Apothecia 0.1–1.5(-2) × 0.07–0.2(-0.25) mm, 50–80 μm tall, scattered or becoming ± crowded-contiguous, sessile, delicate, serpentine-sinuate, simple, rarely branched or stellate, rarely contiguous; disc a persistent slit, rarely surrounded by a pale pseudothalline exciple; exciple in section K+ dull green; hymenium 60–80 μm tall, I+ red. Ascospores 20–35 × 3–5 μm, 5- to 7-septate, slender, acicular, ends tapered. Conidia of

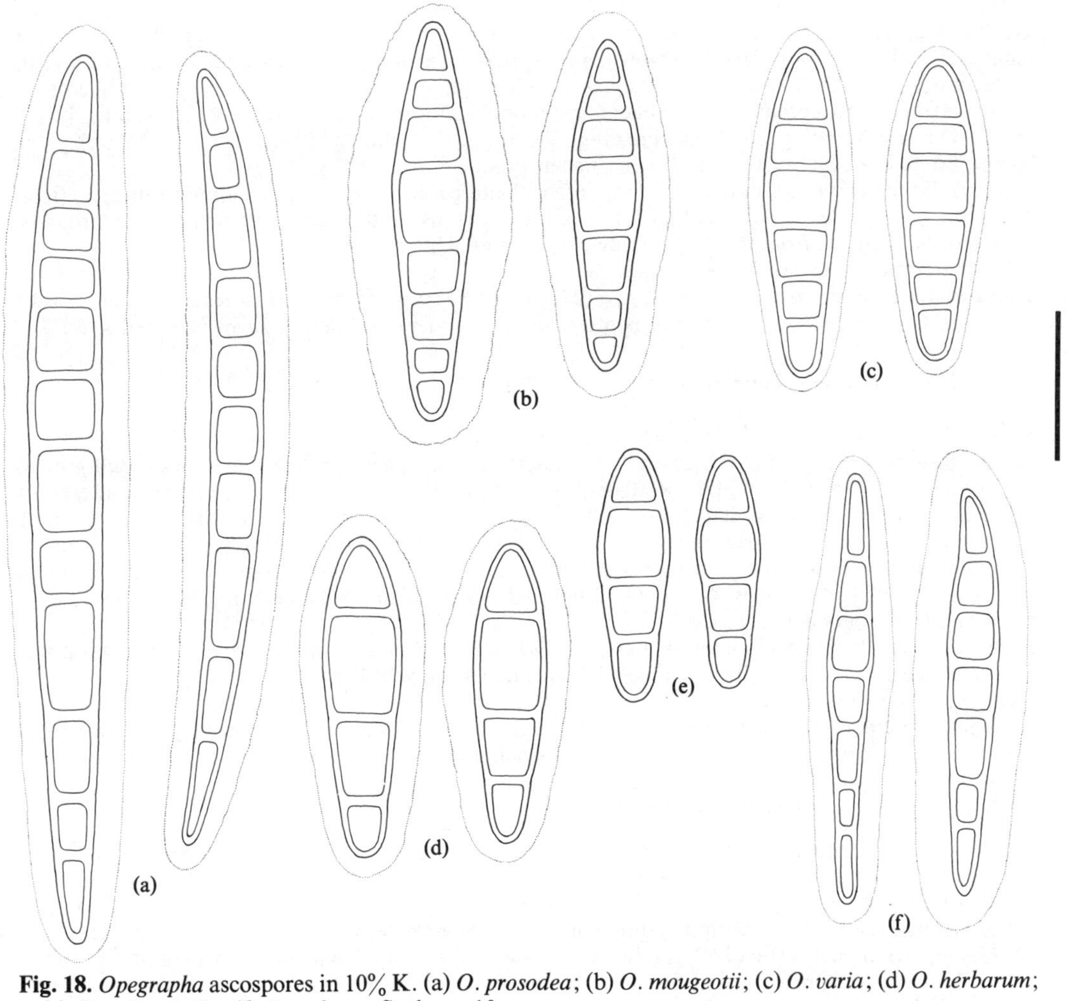

Fig. 18. *Opegrapha* ascospores in 10% K. (a) *O. prosodea*; (b) *O. mougeotii*; (c) *O. varia*; (d) *O. herbarum*; (e) *O. ochrocheila*; (f) *O. vulgata*. Scale = 10 μm.

two types, 10 14 × 0.5–0.7 μm, curved and 3–5 × 1–1.2 μm, bacilliform. Thallus Pd−, K−, KC−, C−, UV− (no lichen products detected by t.l.c.).

On deeply shaded siliceous rocks, often in deep clefts, sometimes with *Enterographa hutchinsiae*; local. Throughout W. Britain and W. Ireland. Sweden, France, Italy, North America.

Closely related to *O. vulgata* **(30)** but differing in the substratum and more delicate apothecia.

11. O. mougeotii Massal. (1853)

Thallus thin, often ± immersed, smooth, scurfy or finely rimose, cream-white, buff, rarely pale green or ochraceous. Apothecia 0.6–1.5(-2) × 0.3–0.5 mm, 150–220 μm tall, scattered or contiguous, usually sessile and simple, straight or curved, ends acute; disc initially a slit, soon becoming ± widely exposed; exciple K−, sometimes ± yellow-green or grey-blue-pruinose; epithecium brown, K−; hymenium 70–110 μm, I+ red. Ascospores (20-)25–33(-55) × 5–8(-10) μm, 5- to 7(-8)-septate, usually with an enlarged middle cell, a thin perispore present (Fig. 18b). Conidia of two types, 5–8 × 0.7–1 μm, bacilliform and 3–5 × 1.5 μm, ellipsoid. Thallus Pd−, K−, KC−, C−, UV− (no lichen products detected by t.l.c.).

On steep, shaded limestone, mortared walls or calcareous sandstone; very local. Lowland Britain and Ireland. Sweden, France, Germany.

12. O. multipuncta Coppins & P. James (1992)

Thallus thin, wide-spreading, smooth or somewhat sparingly rimose, or scurfy, superficial or partly immersed, effuse, dull or dark red-brown, rarely paler, becoming ± grey and discoloured in the herbarium. Soralia 0.1–0.3 mm diam., small, rounded or sigmoid, ± punctiform, very numerous, erose or ± excavate, discrete, becoming ± confluent or forming a minute mosaic of tiny irregular soralia, bright orange when fresh, fading to pale dull cream-grey in the herbarium. Apothecia and pycnidia unknown. Soralia Pd−, K−, KC−, C−, UV− (no lichen products detected by t.l.c.).

On small branches of scrub-trees in damp, often sheltered or boggy sites, including *Crataegus*, *Hebe*, *Juniperus*, *Salix* species, and *Ulmus glabra*, also on shaded trunks and on deeply shaded rock underhangs; scattered. W. Britain, from the Is. of Scilly to Shetland (Unst). Norway.

Characterized by the wide-spreading thallus studded with numerous, tiny, bright orange soralia. Young material may resemble *O. sorediifera* **(24)**, which has a paler thallus and generally paler soralia which are C+ reddish. *O. multipuncta* is often locally frequent and dominant in favoured sites.

13. O. niveoatra (Borrer) Laundon (1963)

O. dubia Leighton ex Arnold (1861)

Thallus thin, superficial, minutely cracked or smooth, effuse, occasionally inconspicuous, dull grey to olive-brown. Apothecia 0.3–1(-1.5) × (0.08-)0.1–0.2 mm, 50–90 μm tall, small, sessile, scattered or contiguous, simple, curved-sinuate or ± stellate, occasionally branched; disc at first a slit, becoming partially exposed with age; exciple somewhat uneven, K± olivaceous in section, epithecium brown; hymenium 40–70 μm, I+ red. Ascospores 20–30 × 2.5–4 μm, 4- to 7(-8)-septate, ± acicular. Conidia short and curved, of two types, 4–8 × 1–1.5 μm and 7–9 × 0.7 μm. Thallus Pd−, K−, KC−, C−, UV− (no lichen products detected by t.l.c.).

On neutral and basic deciduous bark, particularly *Fraxinus*, *Acer* and *Ulmus*, in wayside and woodland situations, more rarely on conifers and wood; frequent. Throughout British Is. Europe, North America.

Often confused with *O. vulgata* **(30)** but the short, worm-like conidia are distinctive; the species shows a distinct preference for basic bark. Pycnidia of *Laeviomyces opegraphae*, producing brown simple conidia, often occurs on the thallus and in the hymenium.

14. O. ochrocheila Nyl. (1865)

Thallus effuse, spreading, very thin or immersed, inconspicuous, white-grey, rarely tinged brownish or olive-green. Apothecia 0.5–1.2(-2.5) × 0.25–0.4 mm, 60–100 μm tall, sessile, variable, sometimes ± irregularly gnarled, simple, sinuate, or often branched, rarely stellate, scattered or contiguous; exciple, and sometimes the disc, ± with a red-orange pruina of unknown composition, K+ magenta-red, comprising two pigments (by t.l.c.), rarely totally absent; disc a slit, frequently ± widening to a broad disc with a thin margin; epithecium reddish-brown; hymenium 50–60 μm tall, I+ blue. Ascospores (12-)14–16(-22) × 3–4.5(-5) μm, ellipsoid or somewhat clavate (Fig. 18e). Conidia (10-)12–20 × 0.7–1.7

μm, sometimes 1- to 3-septate, bacilliform, straight or only weakly curved. Thallus Pd−, K± purplish, KC−, C−, UV± orange or glaucous (± 2 UV+ pigments).

On old, dry, shaded deciduous bark, especially *Quercus* in well-wooded sites, often on wood, more rarely on sheltered siliceous rocks; occasional. Throughout Great Britain, rare in Ireland. W. Europe, Canada.

Morphs with non-pruinose apothecia are easily mistaken for *O. herbarum* **(9)**, which has broader ascospores with a distinctly swollen second cell and shorter conidia.

A superficially similar species, ***O.* cf. *rubefacta*** Räsänen (1946), but with 5- to 8-septate spores, 16–26 × 4–6 μm, occurs on decorticated wood in N. Wales. The exciple has an orange pruina, K+ crimson-magenta, and short, stumpy lirellae, 0.5–1.0 × 0.2–0.4 mm, the thallus is pale green or inconspicuous.

15. O. parasitica (Massal.) H. Olivier (1906)

O. persoonii (Ach.) Chev. (1822) nom. illegit.

Apothecia 0.2–0.3 mm wide, 0.15–0.21 mm tall; exciple in section K+ brown-red. Asci 8-spored. Ascospores 3-septate, 14–22 × 5–8 μm, pigment in perispore. Conidia 3.5–5 × 0.8 μm.

On crustose Verrucariaceae and *Caloplaca cirrochroa* and on hard limestones or *Xanthoria parietina* on trees or rocks; local. Scattered in British Is. C. & N. Europe, Fennoscandia.

The fungus on *X. parietina* is best segregated as *O. physciaria* (Nyl.) D. Hawksw. & Coppins (1992).

16. O. paraxanthodes Nyl. (1879)

Thallus thin, finely cracked, inconspciuous or immersed, pale greenish yellow. Apothecia (0.4-)0.5–0.9(-1.3) × 0.2–0.3(-0.5) mm, 50–150 μm tall, sessile, short, scattered or ± clustered, occasionally branched; disc at first a slit, becoming ± widely exposed with age; exciple sometimes furrowed, K− in section; epithecium brown; hymenium 75–90 μm, I+ red. Ascospores 20–25(-27) × 5–8(-9) μm, 4- to 5-septate, with a narrow perispore, usually with 1–2, ± enlarged middle cells. Conidia 3.5–7 × 0.5–1 μm, straight. Thallus Pd−, K−, KC−, C−, UV− (no lichen products detected by t.l.c.).

In crevices of deeply shaded, base-rich rocks (sandstone, limestone, epidiorite) mainly in uplands; rare and very local. Throughout British Is. Endemic.

Resembles *O. mougeotii* **(11)** but the apothecia are consistently smaller and the thallus has a yellowish green tinge when fresh; the ascospores are shorter with more rounded ends and have fewer septa. The status of this species in relation to the rare collections of *O. varia* **(27)** on rock needs critical evaluation; in general *O. paraxanthodes* has persistently shorter ascospores with fewer septa, 4–5 against 5–6 in *O. varia*; the thallus in *O. varia*, a predominantly lowland species, is also consistently thicker.

17. O. pertusariicola Coppins & P. James (1979)

Apothecia 0.15–0.2 mm wide, 0.1–0.15 mm tall; exciple in section K+ greenish. Asci 4-spored. Ascospores (5-)6-septate, 21–28 × 4.5–6 μm, pigment in perispore. Conidia 3.5–4 × 1 μm.

On *Pertusaria leioplaca* on *Corylus*; local. N. Wales (Merioneth), Scotland, N.W. Ireland.

18. O. prosodea Ach.(1803)

Thallus thin, sometimes almost inconspicuous, smooth and somewhat membranous or minutely cracked, pale to dark grey, occasionally olive-green to ± dull olive-brown. Apothecia (1-)1.2–2.3(-3) × 0.2–0.35 mm, 100–200 μm tall, prominent, sessile, scattered or ± contiguous, sometimes massed and interconnected, straight or mostly curved or serpentine, infrequently branched; disc persistently slit-like; exciple swollen, K+ greenish olive in section; epithecium pale brown; hymenium 90–100 μm tall, I+ red. Ascospores (40-)50–70(-80) × 6–8 μm, 8- to 14-septate, elongate-fusiform, with a narrow perispore (Fig. 18a). Conidia 4–6 × 0.5–1 μm, straight. Thallus Pd−, K−, KC−, C−, UV− (no lichen products detected by t.l.c.).

On dry, shaded, rough bark of mature and over-mature trees, especially *Quercus*, also on old *Taxus* in churchyards, a characteristic old woodland and parkland species, often forming extensive patches; very local. S. England, S. Wales, S.W. Ireland, very rare elsewhere. S. France, Portugal, Spain, North America.

19. O. pulvinata Rehm (1869)

Apothecia 0.16–0.3 mm wide, 0.2–0.24 mm tall; exciple in section K+ brown-yellow or brown-red. Asci 8-spored. Ascospores 3-septate, (18-)20–24(-26) × 6–8 μm, brownish pigment in spore wall. Conidia 6.5–7 × 1–1.5 μm.

On *Dermatocarpon miniatum* aggr.; rare. W. Scotland (S. Ebudes, Colonsay; Argyll, Lismore Is.). Austria, W. Germany (Bavaria), N. Italy.

20. O. rufescens Pers. (1794)

Thallus thin, smooth or finely cracked, sometimes delimited by a dark prothallus and mosaic-forming, dull olive to red-brown, rarely grey. Apothecia (0.25-)0.32–0.5(-1) × 0.09–0.15(-0.3) mm, immersed, sometimes ± elevated on a thalline cushion, curved or substellate, usually numerous and often contiguous; disc fully exposed, and surrounded by a very thin exciple (K−) and sometimes a thin, persistent, pale, ± raised, pseudothalline exciple; exciple K−; epithecium brown; hymenium 50–60 μm tall, K−. Ascospores (15-)17–27 × 3–5 μm, fusiform, often ± curved. Conidia 4–8 × 0.8–2 μm, variable, straight or curved. Thallus Pd−, K−, KC−, C−, UV− (no lichen products detected by t.l.c.).

On smooth, shaded, nutrient-rich bark of broad-leaved trees, particularly *Acer*, *Corylus*, *Fraxinus* and *Ulmus*, in wayside and woodland sites; local. Throughout British Is., but most frequent in the south and west. W. Europe, North America.

Characterized by the small, immersed, often slightly raised apothecia with a fully exposed disc, thin true exciple, a surrounding, thin, ± whitish pseudothalline margin, and characteristic ascospores. *O. rufescens* resembles the rare *Schismatomma graphidioides*, but the true exciple of the former is always clearly apparent in sections. When sterile, distinguished from *O. niveoatra* **(13)** by the broader conidia.

21. O. saxatilis DC. (1805)

Thallus thin, finely cracked or smooth, sometimes immersed, white, grey, pale green-pink or rusty ochraceous. Apothecia very variable, 0.6–1.2(-2) × (0.1-)0.15–0.3 mm, 40–120 μm tall, sessile or rarely semi-immersed, scattered or contiguous, simple, sometimes branched or ± stellate; disc a slit, sometimes partially exposed with age; exciple K+ olive-green; epithecium brown, K+ greenish; hymenium 80–100 μm tall. Ascospores 16–20(-24) × 4–5(-6.5) μm, 3-septate, ± clavate. Conidia 4.2–7 × 0.5–1.4 μm, straight, bacilliform, or rarely ellipsoid, 4–5 × 1–1.5 μm in large pycnidia, to 0.3 mm diam. Thallus Pd−, K−, KC−, C−, UV− (no lichen products detected by t.l.c.).

On damp, shaded limestone, mortar, bone or base-enriched rock; frequent. Throughout British Is., particularly near the coast. Sweden, France, Algeria.

Similar to *O. saxicola* **(22)**, but with smaller ascospores and a greenish exciple, epithecium and hypothecium in K. The lichenicolous *O. parasitica* **(15)**, with rounded apothecia, also occurs on lichens on limestone.

The wide circumscription of *O. saxatilis* adopted here encompasses several ecotypes. The most distinctive of these have been named as: *O. calcarea* Turner ex Sm. (1807), which is widespread on sheltered, hard limestone rocks and has a ± pink, immersed thallus (fading in the herbarium) with abundant lipid-rich cells staining with Sudan black, 5–10 μm wide; the delicate lirellae are simple, serpentine or variously furcate or stellate. *O. conferta* Anzi (1862) has a thin, inapparent to evanescent, pale grey thallus, often confined to minute fissures in neutral to siliceous rocks, especially near the coast; the lirellae are very variable in form, sometimes scattered, irregularly nodular or deformed or often forming small stellate or knot-like, intertwined aggregations. *O. chevallieri* Leighton (1854) resembles *O. conferta* but has a conspicuous, superficial, pure white, chalk-like thallus with conspicuous, loose to compact clusters of lirellae, sometimes to 5 mm diam.; it is widespread on cement, mortar and plaster walls, especially in churchyards and on derelict buildings, etc. These ecotypes appear to intergrade with the typical state of the species and confusing intermediates occur; a detailed study of this complex is required. The name *O. saxatilis* also requires satisfactory typification, as does *O. saxicola* **(22)**, there being much confusion in the application of these names.

22. O. saxicola Ach. (1814)

Thallus thin or immersed, smooth or finely cracked, creamy white, grey, greenish to brown-red. Apothecia 0.5–1.8(-3) × 0.16–0.25(-0.35) mm, sessile to semi-immersed, scattered, infrequently furcate, sometimes elliptical and gnarled, sub-gyrose or, longitudinally furrowed; disc slit-like; exciple K−;

epithecium brown, K−; hymenium 90–120 μm tall. Ascospores (20-)22–26(-31) × 5–8 μm, 3-septate, ± clavate, usually with a thin perispore. Conidia 4–6 × 0.5–0.7 μm, bacilliform. Thallus Pd−, K−, KC−, C−, UV− (no lichen products detected by t.l.c.).

On sheltered moist limestone or calcareous schists (especially in underhangs), such as clints in limestone pavement, mortar or associated base-enriched siliceous rock; local. Throughout British Is. Sweden, Germany, Poland, North America.

Distinguished from *O. saxatilis* **(21)** by the K−, brown pigmentation of the apothecial tissues and longer ascospores, as well as the usually longitudinally furrowed or gnarled-subgyrose shape of the apothecia.

23. O. saxigena Tayl. (1836)

Thallus thin, continuous, smooth or partially cracked, superficial, surrounded by a thin, dark prothallus, often forming extensive mosaics, dark chocolate- to deep purplish brown, rarely partially grey-brown. Apothecia 0.5–1.2(-1.4) × (0.15-)0.2–0.35(-0.45) mm, 150–200 μm tall, scattered, rarely contiguous, short, sessile, infrequently branched, often sparse; disc a slit, rarely partially exposed; exciple occasionally furrowed, K+ olive-green; epithecium brown; hymenium 80–120 μm tall, I+ red or occasionally ± blue below. Ascospores (15-)16–20(-22) × 4–5(-6) μm, 3-septate, ellipsoid, rarely slightly clavate. Conidia 4–7 × 0.3–0.8 μm., bacilliform. Thallus Pd−, K−, KC−, C−, UV− (no lichen products detected by t.l.c.).

On deeply shaded, siliceous rock, especially in old woodlands, rarely on smooth bark (e.g. *Ilex*), often forming extensive communities with *O. gyrocarpa* **(8)** and *O. zonata* **(31)**; locally frequent. S.W. & W. England, Wales, W. Ireland. France, Spain, Portugal.

Distinguished from *O. gyrocarpa* **(8)** and *O. zonata* **(31)**, with which it often grows, by the absence of soralia and lack of lichen products.

24. O. sorediifera P. James (1962)

Thallus thin, smooth or somewhat scurfy, occasionally evanescent with a narrow, dark prothallus, pale to medium brown. Soralia 0.4–1.2 mm wide occasionally confluent, to 3 mm diam., punctiform, pale orange-yellow or buff when fresh, fading to cream in the herbarium; soredia farinose, 12–26 μm diam. Apothecia 0.3–0.6(-1.3) × 0.12–0.3 mm, 60–140 μm tall, not uncommon, scattered, semi-immersed to sessile, unbranched; disc a slit; exciple K+ olive-green; epithecium pale; hymenium 90–120 μm tall, I+ red. Ascospores 30–40(-58) × 4–5 μm, 10- to 14-septate, elongate-fusiform, with a distinct perispore. Conidia 4–6 × 0.6–0.8 μm, bacilliform. Soralia Pd−, K−, KC+ red, C+ pink-red, UV− (gyrophoric acid, ± unidentified UV+ red pigments).

On young, sometimes old, deciduous bark, especially *Acer* and *Salix*, in moist, shaded or open boggy woodland; frequent. S. & W. Britain, rare elsewhere, Ireland. Denmark, France.

Often sterile and then distinguished from *O. corticola* **(5)** by the much smaller, usually discrete, soralia and chemistry. Rare, juvenile morphs of *O. gyrocarpa* **(8)** on bark differ in chemistry. *O. multipuncta* **(12)** has a more extensive thallus with very numerous, fleck-like soralia; it is C−. See *O. viridis* **(29)**.

25. O. subelevata Nyl. (1888)

Thallus thin or rather thick or ± immersed, smooth to finely cracked and granular, ash- to yellow-grey. Apothecia 0.5–1.5(-2.5) × 0.12–0.3(-0.4) mm, scattered or massed, elongate, ± sessile, elevated, occasionally furcate, ends often acute; disc usually exposed, densely mauve-grey-pruinose; exciple thin, non-pruinose, K−; epithecium brown; hymenium 80–90(-100) μm, tall. Ascospores (16-)20–23(-27) × (5-)6–7(-8) μm, 3-septate, ± elongate-clavate, sometimes becoming brownish when old. Conidia 4–6 × 0.7–1 μm, bacilliform or slightly curved. Thallus Pd−, K−, KC−, C−, UV− (no lichen products detected by t.l.c.).

On steep limestone, mortar or ± basic sandstone or slate rocks; very rare. S.W. England (N. Devon), Channel Is. (Alderney). France, Italy, Algeria, New Zealand.

The prominent apothecia with ± open grey-blue-pruinose discs and black, naked margins are distinctive. Distinguished from *O. demutata* **(6)** by the larger apothecia, shorter conidia, and larger ascospores.

26. O. thelotrematis Coppins (1984)

Apothecia 0.1–0.14 mm wide, 0.1–0.12 mm tall; exciple in section K+ greenish. Asci (6-)8-spored. Ascospores 3-septate, 13.5–17 × (4-)4.5–5.5(-6) μm, pigment in perispore. Conidia 3.5–5 × 1 μm.

On *Thelotrema lepadinum* and *T. monosporum*, on *Corylus*; local. Scotland (W. Inverness, Argyll). Endemic.

27. O. varia Pers. (1794)

O. lichenoides Pers. (1794), *O. rimalis* Ach. (1809), *O. diaphora* (Ach.) Ach. (1803), *O. atrula* Nyl. (1877)

Thallus thin or evanescent, smooth or finely cracked, pale to dark grey, occasionally tinged dull brown. Apothecia 0.7–2.5(-4) × (0.1-)0.2–0.5 mm, 80–120 μm tall, mostly elongate, infrequently branched, rarely stellate, scattered or contiguous, sessile; disc very variable, a slit frequently becoming ± exposed with age or fully exposed from the beginning; exciple and/or disc sometimes ± grey- or green-pruinose; exciple K− in section; epithecium brown, K−; hymenium 60–80 μm, I+ red. Ascus (Fig. 34g). Ascospores (18-)20–37 × (5-)6–9 μm, 4- to 6-septate, somewhat clavate, the middle cells somewhat enlarged, with a thin perispore and rounded ends, becoming red-brown with age (Fig. 18c). Pycnidia often thinly white- or greenish-pruinose; conidia 3–5 × 0.5–1(-2) μm, bacilliform or slightly dumbell-shaped. Thallus Pd−, K−, KC−, C−, UV− (no lichen products detected).

On neutral, basic, rough, shaded bark, especially *Ulmus*, *Acer* and *Quercus*, occasionally on wood and decaying herbs (*Pteridium*), rare on sheltered limestone or mortar; frequent. Throughout British Is., though now absent from much of the English Midlands. France, Germany, U.S.S.R., North America.

O. varia is here considered as a very variable species which has previously been divided into several taxa based on minor differences in the apothecia, especially those relating to the degree of exposure of the disc, and also on conidia shape. Further studies are required to establish the validity of the many entities ascribed to *O. varia s.lat.* See also note under *O. paraxanthodes* **(16)**.

A similar, perhaps, undescribed, species is frequent in S. England and W. Scotland, mainly on dry bark of *Quercus* and *Taxus*. It differs in having narrower apothecia, 0.2–0.6 × 0.1–0.2 mm, with a ± persistent, slit-like disc, ascospores (13-)16–24 × 4.5–5 μm, and K+ green apothecial pigmentation.

28. O. vermicellifera (Kunze) Laundon (1963)

Thallus thin, smooth or ± finely cracked, often forming extensive patches, pale grey or grey. Apothecia (0.3-)0.6–1.2(-1.6) × 0.09–0.15(-0.25) mm, 20–50 μm tall, rather rare, often in restricted areas on the thallus, elongate, narrow, semi-immersed, serpentine, infrequently furcate, or ± in scattered, stellate clusters; disc a slit, occasionally partially exposed; exciple sometimes furrowed or ± grey-pruinose, K−; epithecium brown; hymenium 50–60 μm tall, I−. Ascospores 16–25(-30) × 3–4 μm, 4- to 7-septate, with rather pointed ends. Pycnidia prominent, white to pale grey-pruinose, (0.1-)0.15–0.25 mm wide and 0.15–0.2(-0.4) mm high; young pycnidia semi-immersed, chestnut-brown with a pale ostiole; conidia (3-)4–7 × 1–1.5 μm, straight, bacilliform. Thallus and pycnidia Pd−, K−, KC−, C−, UV− (no lichen products detected by t.l.c.).

On neutral to basic bark, particularly dry sides of *Ulmus*, *Acer* and *Fraxinus*. Throughout British Is., but scarce in the English Midlands and Ireland. France, Germany.

Lecanactis abietina and *L. subabietina* differ in having larger conidia, K+ green pycnidial walls, a different thallus and pycnidia chemistry, and are associated with more acid bark (*Betula, Quercus*, conifers); the pycnidia and thallus of *L. subabietina* are UV ± glaucous white, those of *Lecanactis abietina* UV+ glaucous with a ± orange tinge.

29. O. viridis (Ach.) Nyl. (1861)

Thallus very thin or inconspicuous, usually in small (2–4 cm) patches, smooth or somewhat scurfy, dull olive or brown, rarely pale. Apothecia 0.4–0.8(-1.0) × 0.12–0.35(-0.4) mm, 80–100 μm tall, sessile, initially semi-immersed, short, rounded, scattered, seldom shortly furcate, often elliptical or button-like; disc a narrow slit; exciple K+ olive green, reaction sometimes faint; epithecium pale; hymenium 70–80 μm tall, I+ red. Ascospores 23–60 × 6–9 μm, (8-)11- to 15-septate, with a perispore. Conidia 3.5–6 × 0.5–1 μm or (?)15–18 × 1 μm, bacilliform or slightly curved. Thallus Pd−, K−, KC−, C−, UV− (no lichen products detected by t.l.c.).

On smooth, young, rarely old, shaded bark, particularly *Acer*, *Corylus*, *Ilex*, *Quercus* and *Salix* in old woodland; very local. S.W. England, W. Scotland, S.W. Ireland. Sweden, France, Germany, North America, Tasmania.

Closely related to *O. sorediifera* **(24)**, which has C+ orange-pink soralia. *O. prosodea* **(18)** has similar ascospores but is easily distinguished by the larger, more extended, coarse lirellae and the preference for the dry sides of ancient broad-leaved trees and *Taxus*.

30. O. vulgata (Ach.) Ach. (1803)

O. cinerea Chev. (1822)

Thallus thin, sometimes effuse or conspicuous, smooth or ± rimose, white, grey, pale or deep brown, often with an olive tinge. Apothecia (0.5-)0.8–2(-3) × 0.15–0.25 mm, 50–100 μm tall, very variable, sessile or semi-immersed, elongate, serpentine-sinuate, simple or often furcate, sometimes stellate or forming interlinking network (Fig. 28e); disc persisting as a slit disc, only rarely becoming partially exposed in older apothecia; exciple K−; epithecium dull chocolate brown; hymenium 45–60 μm tall, I+ red. Ascospores (15-)20–30(-40) × 2.5–4(-4.5) μm, 4–7(-8)-septate, elongate-acicular, ± tapered at one end, straight or somewhat curved, thin-walled (Fig. 18f). Pycnidia often numerous, intermixed with the lirellae; conidia of three types, sickle-shaped and 9–18 × 1–1.5 μm, or 9–14(-17) × 0.5–0.7 μm, or bacilliform and 3–5 × 1–1(-1.8) μm. Thallus Pd−, K−, KC−, C−, UV− (no lichen products detected by t.l.c.).

On ± shaded, smooth, young bark or smooth areas on old trunks of a wide range of tree species; abundant. Throughout British Is., except the English Midlands. Europe, North America.

This widespread and very variable species is distinguished from *O. niveoatra* **(13)** by the characteristic, longer, curved conidia. The name *O. vulgata* is retained here in its traditional, but currently nomenclaturally incorrect, sense pending a typification with protected status; if not done this has implication for the genus as a whole as this is its conserved type species.

31. O. zonata Körber (1855)

Thallus thin to rather thick, smooth, even, or finely cracked, superficial, often delimited by a dark prothallus, often forming extensive mosaics, dark chocolate brown, often tinged mauve. Soralia (0.2-)0.5–1(-1.5) mm diam., small, punctiform, rarely confluent, dark brown to ± lilac-white, pale when abraded. Apothecia rather rare, 0.2–0.3(-0.4) × 0.1–0.3 mm, 80–120 μm tall, scattered, sessile, rounded or ellipsoid, unbranched; disc a slit; exciple K−; epithecium brown; hymenium 75–90 μm tall, I+ blue. Ascospores 24–28(-32) × 3.5–6 μm, 5- to 7(-11)-septate, with pointed ends. Conidia 5–6 × 0.5 μm, straight. Soralia Pd−, K−, KC−, C−, UV−, rarely ± deep blue (confluentic, 2'-*O*-methylmicrophyllinic and 2'-*O*-methylperlatolic acids).

On deeply shaded, siliceous rock overhangs in humid situations, rarely on shaded, smooth deciduous bark, e.g. *Fagus* and *Ilex*; locally frequent, often associated with *O. gyrocarpa* **(8)** and *O. saxigena* **(23)**. W. Britain, upland Ireland. Norway, France, Alps, North America.

Distinguished by dark thallus with rounded, ± dark or lilac-mauve-white soralia and the thallus chemistry; *O. zonata* has been placed in *Enterographa* as ***E. zonata*** (Körber) Källsten (1989).

OPHIOPARMA Norman (1853)

Thallus crustose, corticate. **Photobiont** trebouxioid. **Ascomata** apothecia, to 2 mm diam., round or somewhat irregular, sessile. **Thalline exciple** present or absent. **True exciple** thick, concolorous with the disc. **Hamathecium** of paraphyses slightly thickened at the tip, rarely branched or anastomosed. **Asci** 8-spored, clavate, with a shallow, uniformly K/I + blue apical dome, lacking a distinct ocular chamber or apical cushion. **Ascospores** ± spirally arranged in ascus, fusiform, transversely multiseptate, colourless. **Conidiomata** pycnidia, immersed, visible as black dots 100–400 μm diam.; wall colourless except for dark green (K−, N+ red) zone around the ostiole. **Conidiogenous cells** arranged in chains, sometimes branched, subcylindrical, enteroblastic, acro- to plurogenous. **Conidia** rod-shaped, simple colourless.

Haematomma, which also has a ± red apothecial disc and acicular transversely septate ascospores, differs in having a sorediate thallus (in British spp.), a thin true exciple, a *Lecanora*-type ascus, and thin, richly branched and anastomosed paraphyses not thickened at the tips.

LITERATURE: Rogers & Hafellner (1988).

1. O. ventosum (L.) Norman (1853)

Haematomma ventosum (L.) Massal. (1852)

Thallus wide-spreading or in large conspicuous patches, thick, very uneven, coarsely contiguous, verrucose-warted, areolate, sometimes ± low papillate, older parts ± coarsely incompletely rimose-cracked, pale to dark grey or ± intensely suffused yellow-green; prothallus distinct, pale, sometimes fimbriate. Apothecia 0.4–2.5 mm diam., frequent, scattered or in small contiguous groups, ± immersed to low-sessile, rounded or irregular by mutual pressure, or occasionally becoming sublobulate; thalline exciple ± well developed, extending under the apothecium, rugose; true exciple level with or slightly raised above the disc, ± concolorous with the disc or paler; disc flat or slightly convex, matt, orange- or blood-red or red-brown; hymenium 60–70 μm tall, merging by vertically aligned hyphae into a massive hypothecium; epithecium, hymenium, hypothecium and true exciple ± mottled-granular orange-red; epithecium K+ indigo-blue turning to patchily violet-blue; hymenium, hypothecium and true exciple K+ blue becoming intense magenta-pink to fiery orange-red. Paraphyses simple, rarely branched, *c.* 1 μm wide; apices not or little swollen. Asci 50–60 × 18–20 μm, ellipsoid-ovate. Ascospores (30-)40–50(-55) × 4.5–5 μm, indistinctly 3- to 7-septate, curved, the ends acute. Pycnidia frequent; conidia 7–10 × 1 μm. Cortex K+ yellow-orange-yellow, UV−; medulla Pd+ yellow-orange, K+ yellow-orange, KC+ yellow-orange, C−, UV± glaucous white (divaricatic, thamnolic, and sometimes usnic acids); the apothecia and parts of the medulla contain ventosin, an orange-red, K+ blue pigment.

On well-lit, nutrient-poor, especially coarse-grained, hard, siliceous rocks and scree in exposed situations, montane. S.W. England (Devon, Dartmoor), N. England, N. Wales (Snowdonia), Scotland, rare in Ireland. Europe.

The considerable variation in the intensity of the UV+ glaucous test of the medulla reflects the amount of divaricatic acid present. Two chemotypes, with or without usnic acid, occur, but show no ecological preferences and often occur side by side. A very distinctive species with a coarse thallus and orange-red-brown, flat apothecia.

O. lapponica (Räsänen) Hafellner & R. W. Rogers (1988) [syn. *Haematomma lapponica* Räsänen (1931)] is similar but has smaller (12–21 × 3–5 μm) ascospores, the cortex K− and medulla Pd−, K− (without thamnolic acid). Recorded last century from Scotland (Argyll, Ben Cruachan), but that specimen possibly originated from outside the British Is; this species is arctic-alpine, and circumpolar in the N. Hemisphere.

ORPHNIOSPORA Körber (1874)

Thallus crustose, areolate. **Prothallus** and **hypothallus** blackish. **Photobiont** chlorococcoid. **Ascomata** apothecia, black, not pruinose. **Thalline exciple** absent. **True exciple** and **hypothecium** dark brownish. **Hymenium** colourless or pale I+ blue. **Hamathecium** of simple or sparingly branched and anastomosed paraphyses, apices without pigmented caps. **Asci** clavate, 8-spored; apical dome narrow, not or weakly I+ blue but ascus apex surmounted with a strongly I+ blue, rather diffuse cap. **Ascospores** colourless or dark olive-brown, ellipsoid, thick-walled, simple or with an indistinct septum. **Conidiomata** pycnidia, immersed. **Conidiogenous cells** integrated in chains, cylindrical, acrogenous or pleurogenous. **Conidia** bacilliform, simple, colourless. **Chemistry**: no lichen products detected by t.l.c. **Ecology**: on siliceous rocks.

A genus of 2–3 species widely distributed in montane regions. *O. mosigii* (Körber) Hertel & Rambold (1988), of central Europe, has colourless ascospores. Possibly related to *Fuscidea*, which differs in the pale hypothecium and never containing any greenish or purplish pigments.

LITERATURE: Hafellner (1984), Hertel & Rambold (1988), Sheard (1964).

1. O. moriopsis (Massal.) D. Hawksw. (1982)

O. atrata (Sm.) Poelt (1972)

Thallus brownish black; areoles 0.3–1 mm diam., angular or rounded, flat to convex; outer cells of cortex greenish, N+ red; medulla brownish, I+ blue-violet. Photobiont cells 7–12(-15) μm diam. Apothecia

0.4–1 mm diam., at first immersed (later becoming ± sessile); true exciple at first shallow but later excluded and ± inapparent; epithecium olive-green, K+ green intensifying, N+ red, sometimes, with additional brownish, K+ purplish pigment; hymenium 80–110 μm tall, colourless, or pale brownish, K+ purplish; hypothecium dark brown, K+ purplish tinge. Paraphyses 1.5–2 μm wide; apices slightly widening to 3–4 μm. Ascospores 12–18 × 6.5–10 μm, dark olive-brown. Conidia 3–4 × 1 μm.

On hard, acid montane rocks; rare. N. England, Scotland. Europe, North America, Australia.

A morph with markedly convex areoles with pale brown apices, known from a single collection from Scotland (Perth, Craig Calliach), is sometimes recognized as var. *brunnea* (A. L. Sm.) D. Hawksw. (1982) [syn. *O. atrata* var. *brunnea* (A. L. Sm.) Poelt (1980)]. *O. moriopsis* is distinctive microscopically, but is easily mistaken in the field for *Schaereria fuscocinerea*.

PACHYPHIALE Lönnr. (1858)

Thallus crustose, superficial, whitish grey or green-grey, thin, effuse. **Photobiont** *Trentepohlia*. **Ascomata** apothecia, at first immersed, then emerging, becoming concave, with a conspicuous persistent true exciple and sometimes a thin thallus-like exciple; disc red-brown to orange, rather gelatinous in texture. **True exciple** well developed, of narrow, thin, compacted hyphae, the outer edge red-brown, pale within, often containing crystals. **Epithecium** well developed, fleshy, red-brown. **Hymenium** I+ pale blue (mainly ascus walls). **Hamathecium** of paraphyses, unbranched, thread-like, septate, ± hooked at the tips. **Asci** (8-)16 to 48-spored, elliptical; wall I+ blue, not or only slightly thickened at apex and without internal apical apparatus. **Ascospores** colourless, acicular to spindle-shaped or fusiform, multiseptate, lying parallel to one another or spirally arranged in ascus. **Conidiomata** pycnidia, *c.* 100 μm diam., reddish brown; wall pale red-brown above, colourless below. **Conidiogenous cells** elongate-ampulliform, arising in groups on short-branched conidiophores, enteroblastic, acrogenous. **Conidia** 12–20 × 1 μm, curved to thread-like, simple, colourless. **Chemistry**: no lichen products detected by t.l.c. **Ecology**: on bark of deciduous trees in ancient woodland or parkland.

LITERATURE: Lettau (1937).

1	Ascospores 38–85 × 3–5 μm, 9- to 15-septate	**1. *carneola***
	Ascospores 15–35 × 3.5–5 μm, 3- to 7-septate	**2. *fagicola***

1. P. carneola (Ach.) Arnold (1871)

P. cornea (Ach.) Poetsch (1872)

Apothecia numerous, scattered, rarely contiguous, gelatinous and ± translucent when wet, concave, red- or orange-brown; true exciple dark brown; at first sunk in thallus, then erumpent with a spurious ± crenulate pale thalline exciple, sometimes persistent. Asci (8-)16 to 48-spored, 75–100 μm long. Ascospores 38–80(-85) × 3–5(-6) μm, acicular, straight or somewhat curved or flexuous, with sharp-pointed ends, 9–15-septate. Pycnidia rare.

On bark of mature to old broad-leaved trees, mainly *Quercus*, *Fraxinus*, *Acer* and *Fagus*, sometimes *Corylus*, of pH > 5, mostly in ancient woodlands or parklands, mainly on sheltered, moderately well-lit, often mossy, bark in open patches in the *Lobarion*, and an ancient woodland indicator; locally frequent. Britain, from Kent and Cornwall northwards to Ross and Nairn, absent from N. & C. England and East Anglia, rare in S. & W. Ireland. Europe, from Norway to Portugal, eastwards to Poland and Yugoslavia, North and South America.

Thelopsis rubella, which grows in similar habitats, has fruits of similar width and colour; the apical pores of its perithecia are normally easily seen when wet (× 10).

2. P. fagicola (Hepp) Zwackh (1862)

Like **1**, but differs in the more immersed thallus, the often more crowded, rather duller, darker brown apothecia, often with a less widely exposed disc and shorter, 15–35 × 3.5–5 μm, 3–7-septate, fusiform ascospores, ends rounded or shortly acute (Fig. 19). Pycnidia occasional.

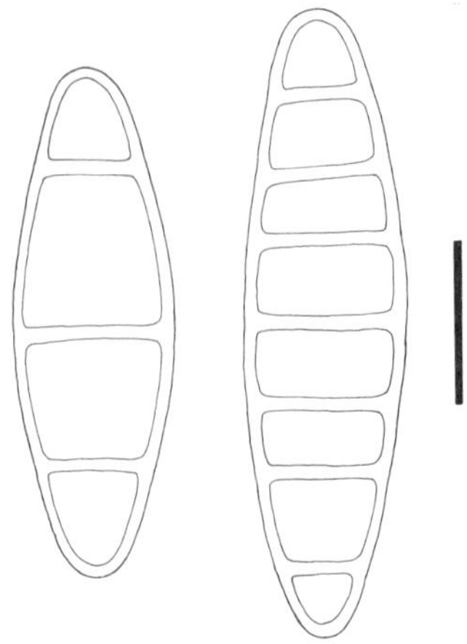

Fig. 19. *Pachyphiale fagicola* ascospores. Scale = 5 μm.

Chiefly on *Ulmus glabra* in old woodlands and on basic-barked, wayside trees; rare. W., C., & E. Scotland. Scattered and local in montane woods in Europe.

Gyalecta derivata is similar, but has 8-spored asci and ascospores with often more than 7 septa.

PANNARIA Delise (1828)

Thallus foliose, squamulose or crustose, often attached to lower surface by a blue to blue-black hypothallus which is sometimes visible from above. **Upper surface** grey-blue to brown-black. **Photobiont** blue-green, *Nostoc*. **Ascomata** apothecia, sessile, disc red-brown to black. **Thalline exciple** persistent, concolorous with thallus, outer part pseudoparenchymatous, inner loosely organized with densely packed photobiont cells. **True exciple** thin, pale; cells ± pseudoparenchymatous. **Asci** 8-spored, apical apparatus variable, a thin internal or external cap present in some species, or staining not apparent (e.g. *P. rubiginosa* group). **Ascospores** colourless, simple, ellipsoid, often apiculate at one or both ends and with a warted epi- or perispore. **Conidiomata** pycnidia. **Conidia** straight, bacilliform. **Chemistry:** lichen products not usually detected by t.l.c. (pannarin (Pd+ red) present in some species). **Ecology**: mostly on broad-leaved, mossy trees in *Lobarion* communities, more rarely on rocks, mainly in sheltered, humid, relict woodlands in W. British Is.

Closely resembles *Parmeliella*, which apparently differs only in the absence of a thalline exciple. The species of both genera are often sterile, and the key to *Parmeliella* should also be consulted for such material. Some species superficially resemble *Massalongia* and *Psoroma*; see also *Degelia* and *Moelleropsis*.

LITERATURE: Jørgensen (1978).

1 Soredia or isidia present; apothecia rarely present 2
  Soredia and isidia absent; apothecia usually present 4

2(1) Thallus uniform, without enlarged, radiating marginal lobes, Pd− 3
  Thallus with ± distinctly radiating marginal lobes, Pd+ orange-red .... **1. *conoplea***

3(2) Upper surface of squamules olive-brown; soralia dark grey-violet-blue .... **5. *mediterranea***
  Upper surface of squamules pale to dull red-brown; soralia cream to pale yellow **8. *sampaiana***

4(1) Thallus with enlarged ± radiating lobes, ± rosette-like 5
  Thallus without enlarged distinctly radiating lobes, not forming well-defined rosettes 6

5(4) Apothecial disc red-brown; thallus blue-grey or yellow-brown; surface ± finely scabrid pruinose; marginal lobes >2 mm wide .... **7. *rubiginosa***
Apothecial disc black; thallus pale to dark grey; surface not pruinose; marginal lobes <2 mm wide .... **2. *hookeri***

6(4) Apothecial disc pale red-brown to deep brown; thalline exciple ± conspicuous, granular or ± crenulate, sometimes becoming excluded 7
Apothecial disc dark brown-black; thalline exciple always ± inconspicuous, soon disappearing .... **4. *leucophaea***

7(6) Apothecia 0.3–0.9 mm diam., soon markedly convex; squamules often dispersed on an extensive blue-black hypothallus .... **3. *ignobilis***
Apothecia 0.9–2 mm diam., flat or slightly convex; squamules crowded, sometimes overlapping; hypothallus usually inconspicuous .... **6. *pezizoides***

1. P. conoplea (Ach.) Bory (1828)

Thallus foliose, <2–3 cm diam., rarely larger, ± forming conspicuous, irregular, entire or fragmentary rosettes; lobes 4–5 × 7–8 mm, deeply incised, mostly contiguous, fan-like or somewhat overlapping, ± concave, deeply indented with elongate, ± ascending margins; upper surface glaucous, blue-grey to ± pale brown in more exposed situations, often paler and ± scabrid-pruinose towards margins, with numerous, coarse marginal, at times abraded, knob- or squamule-like simple or coralloid isidia which may cover and obscure the thallus apart from the ± extended marginal lobes; hypothallus well developed, but rarely apparent as a blue-black zone surrounding the thallus. Apothecia 0.5–1.5 mm diam., very rare; disc brown-red; thalline exciple distinct, usually coarsely abraded-isidiate. Ascospores 15–19 × 9–10 μm, with perispore 20–24 × 10–12 μm, colourless, ellipsoid with ridged perispore, acuminate at one or both ends. Thallus Pd+ orange-red (pannarin).

On mossy bark of broad-leaved trees, rarely rocks, in humid, sheltered woodlands; locally abundant but declining in the S. S. & W. British Is. Europe, North & South America, Africa, Asia, Macaronesia.

The lobes may be much reduced, appearing ± crustose, the thallus ± reduced to a compacted mass of coarse bluish isidia. In exposed sites the thallus surface becomes distinctly pale brown, although the isidia always remain bluish in colour. *P. rubiginosa* **(7)** lacks isidia, is usually richly fertile, and forms more distinct rosettes; *Parmeliella jamesii* and *P. testacea* are Pd−.

2. P. hookeri (Borrer ex Sm.) Nyl. (1857)

Thallus in part crustose, in part minutely lobulate, in closely adpressed, ± incomplete rosettes, to 2–3 cm diam.; lobes to 2 × 3 mm wide, contiguous, convex, turgid, becoming fan-like, wider and elongate towards margins and verrucose-gnarled and fragmented towards centre; upper surface pale grey, sometimes tinged pale brownish, ± white striate-maculate; hypothallus thin, but often visible in between the squamules. Apothecia to 2 mm diam., abundant, often ± contiguous and angular by compression; disc black, rarely dark brown, roughened, matt, flat, conspicuous, irregularly crenulate; thalline exciple usually white- to pale blue-grey; epithecium green-black, opaque; hymenium 100–200 μm tall, colourless or in part ± red-brown. Ascospores 11–15 × 8–11 μm, colourless, ellipsoid to globular, smooth-walled, without a perispore. Thallus Pd± faint orange (traces of pannarin).

Predominantly on basic mica-schists and epidiorite or overgrowing associated mosses, often in temporary seepage tracks on mountains above *c.* 700 m; local. Scotland. Faroes, Scandinavia, Alps, Iceland, Greenland, Canada, Alaska, Siberia, Chile, Argentina, S. Georgia, New Zealand.

May be mistaken for a placodioid *Lecanora* because of the similar thalline appearance and black apothecial discs.

3. P. ignobilis Anzi (1862)

Thallus crustose, wide-spreading, of small, closely adpressed squamules, 0.5–1 mm; hypothallus well developed, conspicuous, indigo-black; squamules scattered or contiguous-contorted, ± rounded, or ± flattened or unevenly verrucose, pale grey, blue-grey, rarely olive-brown or brown, with a paler margin, slightly indented to irregularly crenulate, contiguous and overlapping or more often widely dispersed, appearing as rimose-cracked crust. Apothecia 0.3–0.9 mm diam., often numerous, single or in clusters; disc bright red-brown to chestnut brown, soon becoming markedly convex; thalline exciple ±

incomplete, grey, secondarily granular, soon excluded, often exposing a pale red-brown true exciple. Ascospores 10–14 × 8–9 μm, including perispore, 20–28 × 9–11 μm, colourless, ellipsoid with acute, attenuated apices, usually only at one end. Lichen products not detected by t.l.c.

On bark of old wayside trees, especially *Fraxinus* and *Ulmus glabra*, often in fissures; in drier habitats than other *Pannaria* species; local. C. Scotland (especially in Great Glen area). W. & S. Europe, W. Norway, Mediterranean, N.W. Africa.

Characterized by the grey or olive, subcrustose thallus and the numerous small, very convex, often red-brown apothecia with a granular margin which is soon excluded. The squamules are sometimes widely dispersed, rounded with minute marginal lobes on a blue-black hypothallus.

4. P. leucophaea (Vahl) P. Jørg. (1978)

Thallus crustose, of indented, unevenly crenulate ± rounded, adpressed, turgid squamules, to 2 mm diam., forming a rather thick, ± cracked-areolate crust; hypothallus usually obvious, black; squamules grey, grey-brown to black, discrete at margin, often becoming ± imbricate at centre, not enlarged at margins. Apothecia to 1 mm diam., frequent, often clustered; disc brown-black, rarely brownish in shade, ± convex; thalline exciple often indistinct, sometimes unevenly developed on one side of an apothecium. Ascospores 15–17 × 5–6 μm, colourless, outer surface smooth, without a perispore, inner side of spore wall uneven. Lichen products not detected by t.l.c.

On sheltered, ± basic, moist rocks, more rarely bases of trees in tree-lined, sheltered coastal sites or lake-sides. England, Wales, Scotland. Norway, widespread in Europe, North America, Macaronesia,

Very variable. In moist, shaded habitats the prothallus is often exceptionally well developed, the squamules remain discrete and widely dispersed, and the thallus and the apothecia are pale. In exposed sites the hypothallus is much reduced, the dark brown squamules become imbricate and the apothecia are dark brown-black. The thalline exciple is soon excluded and the species is then easily mistaken for a *Parmeliella*.

5. P. mediterranea C. Tav. (1965)

Thallus minutely squamulose, crust-like, forming small patches; hypothallus inconspicuous, thin; squamules 2–3 mm, often irregularly rounded, contorted, deeply indented; upper surface blue-grey to olive-brown, white felted-tomentose at margins. Soralia grey or violet-grey, granular, mostly on up-turned, ascending margins, occasionally spreading to upper surface, often appearing ± woolly (crystals × 50) in the herbarium. Apothecia very rare, unknown in British material; disc brown; thalline exciple sorediate. Ascospores 13–16 × 7–8 μm, with perispore 16–24 × 8–9 μm; perispore broadly attenuate. Containing steroids, often appearing as crystals on herbarium specimens.

On broad-leaved trees and rocks in relict woodlands; locally frequent. S.W. England, E. to Hampshire, Wales, C. & W. Scotland. Oceanic and mediterranean Europe, N. to Scandinavia, Macaronesia, Canada.

Characterized by the blue-violet colour of the granular soralia born on almost lip-shaped lobes of small, often inconspicuous, olive-grey squamules. The soralia are sometimes very abundant, the thallus then becoming an almost entirely grey-violet sorediate crust. See *Parmeliella jamesii*.

6. P. pezizoides (Weber) Trevisan (1869)

Thallus squamulose; squamules small, 0.5–1 mm, ± compacted, pale grey to dark brown, often with a red tinge, indented, sometimes finger-like flattened, ± overlapping, granular; hypothallus usually poorly developed and inconspicuous. Apothecia to 2 mm diam., flat, often very abundant; disc bright orange-brown to dark brown; thalline exciple persistent, crenulate, granular. Ascospores 19–25 × 8–10 μm, including perispore 25–30 × 9–12 μm, colourless, ellipsoid; epispore distinctly ridged and warted. Soredia and isidia absent. Lichen products not detected by t.l.c.

On damp mosses on rocks, walls, trees and the ground, usually in sheltered, undisturbed sites. Upland British Is. Europe, Azores, North America, Africa, Asia.

Characterized by the numerous, crowded, conspicuous, orange-brown apothecia with a persistent, granular thalline exciple. The thallus in moist habitats is blue-grey and the apothecial disc pale orange-brown; in exposed sites both thallus and apothecial disc are dark brown. It closely resembles *Psoroma hypnorum* but the presence of the *Nostoc* photobiont gives the thallus of *Pannaria pezizoides* a darker bluish colour when wet, in contrast to the brighter green of *Psoroma*. *Moelleropsis nebulosa* has a darker, more grey-blue, uniformly granular thallus, less frequent, smaller, 0.5–1 mm diam., red-brown discs, lacks a thalline exciple, has ascospores which lack a perispore, and grows on well-drained, often sandy, soils.

7. P. rubiginosa (Ach.) Bory (1828)

Thallus foliose, ± forming rosettes to 2–3 cm diam. with extended marginal lobes; upper surface whitish grey-blue to brown or olive; lobes 3–4 × 7–8 mm, deeply indented and mostly concave with thick, pale, ascending margins; surface smooth, ± faintly scabrid or occasionally thinly pruinose; hypothallus fibrous, well developed, obscure or sometimes extending as a blue-black zone surrounding the thallus. Apothecia 0.5–1.5 mm diam., frequent; disc red-brown; thalline exciple prominent, persistent, often crenulate. Ascospores 15–19 × 9–10 μm, with perispore 20–24 × 10–12 μm, colourless, ellipsoid; perispore uneven, acuminate at one or both ends. Thallus Pd+ orange-red (pannarin).

On mossy, broad-leaved trees, rarely on rocks in humid, sheltered woods; locally abundant. W. British Is., S. to Devon and Cornwall. W. & S. Europe, all continents except Australasia.

Readily recognized by the neat, foliose, bluish grey thallus, usually with numerous apothecia with bright red-brown discs with a pale crenulate thalline exciple. The lobes are sometimes rather narrow and the thallus at times distinctly squamulose, especially at the centre. In exposed sites the thallus becomes darker, often with an olivaceous tinge.

8. P. sampaiana C. Tav. (1950)

Thallus crustose, composed of thick, coarse, adpressed, ± rounded squamules, 2–3 mm diam., scattered or ± crowded, rarely with ± distinct, 3–4 mm, elongate, blunt marginal lobes; hypothallus blue-black, well developed at thallus margins and in cracks of thallus; areoles pale red-brown with paler margins and with gnarled, ± coarsely granular, cream-coloured marginal and laminal soralia. Apothecia unknown. Lichen products not detected by t.l.c.

On smooth or mossy bark of mature trees, especially *Fagus sylvatica*, *Ulmus glabra* and *Fraxinus excelsior*, rarely on mossy rocks in humid, relict woodlands, occasionally on wayside trees; locally frequent. S. & W. British Is. S. & W. Europe, Tunisia.

Characterized by the pale chestnut-brown colour of the closely adpressed squamules, the cream-yellow soralia and the pavement-like, almost crustose thallus on a black hypothallus. Occasionally the soredia may cover most of the thallus in a cream-yellow crust.

PARMELIA Ach. (1803)

Thallus foliose, loosely attached to closely adpressed, dorsiventrally flattened, corticate on both sides, attached by rhizines. **Rhizines** simple, forked, dichotomomously-branched or squarrose, few to numerous, covering the entire lower surface or ± restricted to centre. **Cortex** pseudoparenchymatous, often with pseudocyphellae or palisade plectenchymatous with overlying pored epicortex. **Photobiont** trebouxioid. **Soralia** and **isidia** frequent, papillae and folioles occasional. **Ascomata** apothecia, laminal, sessile, disc red-brown, concave or flat. **Thalline exciple** persistent, concolorous with thallus. **True exciple** seen only in section, colourless, pseudo- to plectenchymatous. **Epithecium** pale yellow-brown to red-brown. **Hymenium** colourless, I+ blue. **Hypothecium** colourless. **Hamathecium** of paraphyses, sparsely branched, apices somewhat swollen, often pigmented. **Asci** clavate, *Lecanora*-type, 8-spored. **Ascospores** colourless, simple, ellipsoidal to spherical, perispore absent. **Conidiomata** pycnidia, laminal, ± immersed in thallus, ostiole black. **Conidia** cylindrical, thread-like, unciform, or bifusiform. **Chemistry**: cortex: grey species contain atranorin, greenish yellow species contain usnic acid, brown species contain unidentified brown pigments; medulla: range of depsides and depsidones, pigments and fatty acids. **Ecology**: on bark, wood, siliceous rocks or sometimes on soil, cosmopolitan.

This large and disparate genus, with *c*. 1000 species world-wide, is in the process of being organized into smaller genera on the basis of cortical structure, ascospore size, conidial shape, types of cilia and rhizine branching, chemistry, and other characters. Several of these genera are still poorly delimited and described and their nomenclature and circumscription remains unstable. For this reason a broad concept of *Parmelia* is retained here and the latest segregate combinations cited as synonymys for those wishing to use them.

The cortex in the brown species belonging to *Neofuscelia* Essl. (1978) reacts N+ blue-green, those with a N− cortex belong to *Melanelia* Essl. (1978) if with a non-pored cortex, or to *Pleurosticta* Petrak (1931) with a pored epicortex. The predominantly grey species without epicortex belong to either *Parmelia* s.str., with linear, effigurate pseudocyphellae and squarrose rhizines, or to *Punctelia* Krog (1982) with orbicular pseudocyphellae and simple rhizines. The predominantly grey, epicorticate species include *Hypotrachyna* (Vainio) Hale (1974) with narrow, non-ciliate lobes and dichotomously branched rhizines, *Parmelina* Hale (1974) with adpressed, axillary ciliate lobes with a maculate surface and simple rhizines, *Parmelinopsis* Elix & Hale (1987) with narrow ciliate lobes and simple to forked rhizines, *Parmotrema* Massal. (1860) with wide, broadly rounded lobes and usually sparse rhizines with a bare marginal zone below, and *Flavoparmelia* Hale (1986) which includes two yellow-green species with narrow, non-ciliate lobes. Other yellow-green species have been assigned either to *Xanthoparmelia* (Vainio) Hale (1974) if lower surface is brown to black or to *Arctopamelia* Hale (1986) when the lower surface is white to pale brown. Numerous other segregates proposed do not have any British representatives. See also *Allantoparmelia, Cetrelia, Foraminella, Parmeliopsis*, and *Platismatia*.

LITERATURE: Elix & Hale (1987), Esslinger (1977, 1978), Hale (1965a, b, 1975, 1976, 1986, 1987, 1990), James (1982), Krog (1982), Lumbsch *et al.* (1988), Maas Geesteranus (1947).

1 Thallus green-grey, blue-grey to yellow-green, or yellow-grey, grey or yellow tone predominating ... 2
Thallus green-brown, grey-brown, red-brown or black-brown, brown tone predominating ... 30

2(1) Thallus green-grey to blue-grey, grey tone predominating ... 3
Thallus yellow-green to yellow-grey, yellow tone predominating ... 24

3(2) Isidia and soralia absent; surface without white pseudocyphellae ... 4
Isidia and/or soralia present; surface with or without white pseudocyphellae ... 6

4(3) Medulla Pd−, K−, C+ red; upper surface grey or blue-grey, not dark green when wet . 5
Medulla Pd+ orange, K+ red, C−; upper surface grey-green to brown-grey, dark oily green when wet ... **40. *acetabulum***

5(4) Thallus closely adpressed to substratum, lobes not overlapping at the margin; usually fertile, underside of apothecia ± densely rhizinose; cortex persistent, not peeling to expose a white medulla; rhizines simple ... **5. *quercina***
Thallus ± loosely attached to substratum, lobes overlapping at the margin; apothecia very rare, underside not rhizinose; cortex fragile, peeling to expose a white medulla; rhizines dichotomously branched ... **24. *taylorensis***

6(3) Isidia present ... 7
Soralia present ... 13

7(6) Upper surface ornamented with white pseudocyphellae, either in punctiform spots or as coarse, oval to elongate, ± raised ridges which may fuse to form a network ... 8
Upper surface devoid of pseudocyphellae ... 9

8(7) Upper surface with oval or elongate, ± raised pseudocyphellae which fuse to form a coarse, white network; isidia cylindrical; soredia absent; medulla K+ yellow, Pd+ yellow-orange ... **3. *saxatilis***
Upper surface with scattered, round pseudocyphellae; isidia globose, occasionally becoming coarsely sorediate or small- lobulate; medulla Pd−, K− ... **11. *reddenda***

9(7) Medulla C+ pink, red or orange-red; isidia not associated with scattered black cilia ... 10
Medulla C−; isidia mixed with scattered, fine, black cilia ... 12

10(9) Lobes >0.5 cm wide; cilia mostly axial; wayside trees and rocks ... 11
Lobes <0.5 cm wide; cilia apical and axial; ancient woodlands ... **9. *minarum***

11(10) Isidia pale brown to grey-brown or concolorous with thallus, cylindrical-clavate, simple or coralloid ... **6. *tiliacea***
Isidia blue-black, button-like with a short stalk, simple ... **7. *pastillifera***

12(9) Thallus large <10 cm diam., lobes <1.5 cm wide; isidia ± cylindrical; lower surface with wide, naked zone at the margin ... **27. *crinita***
Thallus small, <3 cm diam., lobes <0.5 cm wide; isidia ± flattened; lower surface densely rhizinose to the margin ... **8. *horrescens***

13(6) Upper surface with white pseudocyphellae, as punctiform spots or forming a network of coarse, ± raised lines ... 14
Upper surface smooth, devoid of white pseudocyphellae or with a faint mosaic of minute cracks ... 17

14(13) Upper surface with small, ± punctiform, rarely comma-like pseudocyphellae which may develop into soralia ... 15
Upper surface with oval or elongate pseudocyphellae which frequently fuse to form a coarse, white reticulum; soralia elongate, in lines, derived from pseudocyphellae **4. *sulcata***

15(14) Medulla C+ red; soredia fine, powdery, not forming folioles ... 16
Medulla C−; soredia coarse, granular, sometimes becoming corticate and forming ± flattened folioles ... **11. *reddenda***

16(15) Lower surface pale brown at the centre ... **10. *subrudecta***
Lower surface black at the centre ... **12. *borreri***

17(13) Soralia terminal, convex and delimited at ends of incised lobes; lobes ± elongate with markedly curved, sinuate-rounded axils; rhizines richly dichotomously branched; medulla C+ orange ... 18
Soralia marginal and/or laminal, mostly ± diffuse, if at lobe ends then on upper surface and only secondarily involving the apices; lobes ± rounded with indented margins and axils; rhizines unbranched or merely forked; medulla C− or C+ red ... 19

18(17) Medulla white ... **22. *laevigata***
Medulla pale yellow ... **23. *endochlora***

19(17) Upper surface smooth, without faint mosaic of pseudocyphellae or cracks; medulla K− or + yellow, never red ... 20
Upper surface with faint, close network of white lines, often developing into an intricate mosaic of hair-line cracks; medulla K+ blood red ... **29. *reticulata***

20(19) Medulla C+ red; thallus <6 cm diam.; lobes <5 mm wide, adpressed, without raised or wavy margins; marginal cilia absent ... 21
Medulla C−; thalli often >10 cm diam.; lobes 5–10 mm wide often with raised or wavy margins; marginal cilia usually present ... 22

21(20) Lobes often ± elongate, axils sinuate, ascending at margins; soredia granular, blue-black; upper surface often shiny ... **21. *britannica***
Lobes rounded, often indented, axils rounded but not markedly sinuose, down-turned at margins; soredia powdery, soft, concolorous or slightly darker than thallus; upper surface often matt ... **20. *revoluta***

22(20) Medulla and soralia K− ... 23
Medulla and soralia K+ orange-yellow ... **25. *perlata***

23(22) Medulla and soralia Pd−, KC+ pink, UV+ ice-blue ... **28. *arnoldii***
Medulla Pd+ rust-red, KC−, UV− ... **26. *robusta***

24(2) Lobes ± isidiate ... 25
Lobes with granular or farinose soralia ... 26

25(24) Isidia cylindrical becoming coralloid, often crowded but sometimes almost absent; medulla Pd+ orange, K+ yellow-orange .... **15. *conspersa***
Isidia globose or irregular, not branched; medulla Pd+ orange-red, K+ red .... **16. *tinctina***

26(24) Soralia laminal; axils not sinuate; rhizines simple or branched, not projecting beyond thallus margin 27
Soralia terminal, at ends of ± ascending, narrow lobes; axils sinuate; rhizines richly branched often projecting beyond thallus margin .... **19. *sinuosa***

27(26) Lobes 0.5–1.3 cm wide, rounded 28
Lobes <0.3 cm wide, elongate 29

28(27) Soredia coarse, granular; medulla Pd+ red, K+ yellow; lobes <1.3 cm wide **13. *caperata***
Soredia fine, farinose; medulla Pd+ orange, K+ yellow then red; lobes <0.7 cm wide .... **14. *soredians***

29(27) Lobes markedly convex; soralia on the ends of lobes within the thallus, convex; medulla KC+ pink, UV+ ice-blue .... **18. *incurva***
Lobes ± flat; soralia laminal, flat or excavate; medulla KC+ yellow-orange, UV− .... **17. *mougeotii***

30(1) Without soralia or isidia 31
With soralia and/or isidia 39

31(30) Upper surface with oval or elongate pseudocyphellae which frequently fuse to form a coarse, ± raised, white network 32
Upper surface devoid of white, ridge-like pseudocyphellae, no white network present 33

32(31) Medulla Pd+ orange, K+ yellow-orange .... **1. *omphalodes***
Medulla Pd+ red, K− .... **2. *discordans***

33(31) Lobes >2 mm wide, elongate or ± rounded, margins indented or entire; upper surface without pseudocyphellae 34
Lobes <2 mm wide, linear, margin irregularly nodulose; upper surface with minute psudocyphellae .... **38. *stygia***

34(33) Upper surface even or wavy, smooth, not transversely ridged; inland 35
Upper surface with transverse ridges and wrinkles; coastal 38

35(34) Marginal lobes usually present, well developed; folioles absent; apothecia often present 36
Marginal lobes ± absent; thallus of numerous, very small, overlapping folioles; apothecia absent .... **34. *laciniatula***

36(35) Thallus pale brown, red-brown, olive, or black-brown, ± brown when wet; lobes <1 cm wide, thin, often closely adpressed 37
Thallus grey-brown, dark oily green when wet; lobes <1.7 cm wide, coriaceous, not closely adpressed .... **40. *acetabulum***

37(36) Upper surface with numerous, ± evenly spaced, wart-like, conical papillae; apothecia usually present; thalline exciple swollen, ± papillate .... **30. *exasperata***
Upper surface smooth, without conical warts; apothecia crowded; thalline exciple very thin, smooth .... **37. *septentrionalis***

38(34) Upper surface pale grey-brown to light brown .... **42. *delisei***
Upper surface brown to black-brown .... **41. *pulla***

39(30) Thallus rarely pruinose; upper surface lacking hairs; margins of lobes adpressed 40
Thallus usually pruinose; upper surface with fine white hairs; margins of lobes usually ascending .... **33. *subargentifera***

40(39) Medulla C+ instantly bright carmine-red 41
Medulla C−, rarely C+ faint pink 42

41(40) Isidia ± globose, derived from soredia, leaving a pale yellow area when abraded; upper surface matt .... **31. *subaurifera***

Isidia cylindrical, becoming branched-coralloid, leaving white area when abraded; upper surface shiny .... **32. *glabratula***

42(40) Thallus lobes >0.2 cm wide; well defined marginal lobes present 43
Thallus of numerous, thin, flattened, overlapping folioles, <0.2 cm wide; marginal lobes poorly developed or absent .... **34. *laciniatula***

43(42) On rocks 44
On bark or wood 47

44(43) Isidia not inflated, ± solid, often branched, ± cylindrical 45
Isidia inflated, hollow, clavate or ± spathulate, often decumbent, orientated in all directions .... **36. *exasperatula***

45(44) Lobes <0.6 cm wide; pseudocyphellae absent 46
Lobes <0.15 cm wide; minute, white pseudocyphellae present .... **39. *disjuncta***

46(45) Isidia in ± scattered, coarse clusters, cauliflower-like; thallus yellowish to reddish brown; medulla KC+ orange-red, C− or C+ red .... **43. *loxodes***
Isidia diffuse, or ± in contiguous clusters; thallus brown to dark brown; medulla KC+ pink, C+ pink-red, ... **44. *verruculifera***

47(43) Upper surface with ± crowded, cylindrical isidia or flattened folioles; lower surface pale; apothecia very rare 48
Upper surface with numerous, evenly spaced, conical papillae; lower surface black; apothecia often present .... **30. *exasperata***

48(47) Folioles absent; isidia numerous, not resembling small lobes; marginal lobes always well developed 49
Thallus of numerous, small, thin, overlapping folioles; isidia absent; marginal lobes rarely conspicuous .... **34. *laciniatula***

49(48) Isidia clavate or ± spathulate, hollow, simple, often decumbent, orientated in all directions .... **36. *exasperatula***
Isidia cylindrical, ± erect, solid, often branched-coralloid .... **35. *elegantula***

1. P. omphalodes (L.) Ach. (1803)

Thallus to 20 cm, often forming extended patches, ± loosely attached; lobes to 4 mm wide, somewhat brittle, sublinear, discrete, contiguous or markedly crowded and overlapping, apices markedly truncated; upper surface shiny, flat or ± honeycombed, dark brown, partly greyish (in shade), with oval or elongate pseudocyphellae which frequently fuse to form a coarse, ± complete, raised, white network; lower surface black, rugose, with numerous, simple or forked, rarely branched, rhizines. Apothecia infrequent; disc to 10 mm diam., dark red-brown, for a long time cup-shaped; thalline exciple thin, roughened, paler. Ascospores 10–15 × 7–9 μm, ellipsoid. Conidia 5.5–6.5 × 1 μm, bacilliform. Medulla Pd+ orange, K+ yellow-red, KC+ orange, C−, UV± glaucous (salazinic and lobaric acids); cortex: K+ yellow (atranorin).

On nutrient-poor, siliceous rocks, especially in moorland and montane sites, rarely on acid bark. W. & N. British Is. Arctic-boreal, N. & C. Europe, Portugal, North America.

Distinguished from *P. saxatilis* **(3)** and *P. sulcata* **(4)** in thallus colour and the absence of isidia and soredia; shade specimens, which can be greyish, need to be checked carefully. See also *P. discordans* (**2**).

2. P. discordans Nyl. (1886)

Like **1**, but thallus uniformly dark brown, with less obvious, faint, white, linear pseudocyphellae, lobes 1–3 mm wide, generally narrower, consistently overlapping, seldom exceeding 1 mm wide. Medulla Pd+ rust-red, K−, KC− or + violet, C−, UV+ glaucous, often faintly so (protocetraric and lobaric acids).

In similar habitats to *P. omphalodes* **(1)**, scattered, rare. S.W. & N. England (Devon, Westmorland, Yorkshire), C. Wales, Scotland. Belgium, France, Fennoscandia, Spain.

3. P. saxatilis (L.) Ach. (1801)

Thallus 3–6(-20) cm diam., often forming complete or partial rosettes, ± loosely attached; lobes to 3 mm wide, slightly broadened and truncate at apices, discrete, or ± contiguous and overlapping; upper surface grey-white to grey-green, sometimes tinged brownish towards the lobe ends, with oval to elongate, ± raised, scattered, white pseudocyphellae which may fuse to form a ± complete, coarse, network. Isidia concolorous or tips tinged brownish, cylindrical, simple, becoming coralloid, at first arising from pseudocyphellae, later from the rest of upper surface, rather sparse or becoming numerous and ± obscuring the thallus. Lower surface black, brownish towards the margin; rhizines simple, occasionally forked. Apothecia to 1 cm diam., occasional; disc red-brown to dark brown, often isidiate at margin. Ascospores 16–18 × 9–11 μm. Conidia 6–7 × 1 μm, straight or slightly curved. Medulla Pd+ orange, K+ orange, KC+ orange, C−, UV− (salazinic acid and ± lobaric and norstictic acids).

On acid-barked trees and shrubs, *Calluna* stems as well as siliceous rocks, walls, memorials, roofing tiles, more rarely on the ground in shady to exposed situations, from the coast to mountain summits; often abundant. Relatively insensitive to SO_2 pollution (<70 μg m^{-3}). Throughout British Is. Widespread in Northern & Southern Hemispheres.

Very variable. The morphologically and chemically similar *P. sulcata* **(4)** has linear soralia, and *P. omphalodes* **(1)** is usually dark brown and lacks both soredia and isidia. *P. submontana* (see **4**) differs in habit, and in its isidia-like soredia in small, ± orbicular soralia.

4. P. sulcata Taylor (1836)

Thallus 5–10(-20) cm diam., often forming ± complete rosettes or randomly intricate, ± loosely attached, lobes to 0.5 cm wide, apices incised, discrete or ± contiguous and overlapping, especially at the centre of the thallus; upper surface flat to weakly honeycombed, grey-white to grey-green, sometimes partly white-pruinose, with oval or elongate, scattered, white pseudocyphellae which freqently fuse to form a conspicuous, coarse, ± incomplete network. Soralia elongate, laminal and marginal, derived from breakdown to cortex above the pseudocyphellae. Soredia granular, eroding. Lower surface black, brown towards the margin; rhizines simple or forked, some squarrosely branched. Apothecia occasional, disc red-brown to dark brown; thalline exciple partially sorediate. Ascospores 11–15 × 6–8 μm. Conidia 5–8 × 1 μm, rod-shaped, straight to slightly curved. Medulla and soralia Pd+ orange, K+ orange, KC+ orange, C−, UV− (salazinic acid).

On trees and rocks, occasionally on soil, from coastal regions to exposed mountain summits, perhaps more tolerant of nutrient-rich sites than *P. saxatilis* **(3)**; very common. Throughout British Is. Cosmopolitan, one of the most widely distributed lichens known.

Varies considerably in the size and degree of separation of the lobes, also in the amount of soralia development. *P. saxatilis* **(3)** differs only in being isidiate.

Parmelia submontana Nádv. ex Hale (1987) differs in its loosely attached, more elongate, little branched lobes with rolled down lateral margins and ± raised apices, small ± orbicular soralia with isidia-like soredia and fewer rhizines that are simple or only rarely forked. Recently discovered on exposed trunks of *Acer* and *Fagus* in S. Scotland (Dumfries, Lanarkshire). S. & C. European mtns, Sweden.

5. P. quercina (Willd.) Vainio (1899)

Parmelina quercina (Willd.) Hale (1974)

Thallus 2–5(-15) cm diam., closely adpressed, rosette-forming; lobes to 1 cm wide, ends rounded, ± indented or sometimes entire, continuous, ± overlapping at centre, not down-turned; upper surface blue-grey or ± grey, smooth, often ± shiny, faintly maculate. Lower surface dark brown to black, paler brown towards the lobe ends, rhizines simple. Apothecia usually present, thalline exciple smooth, even above, ± densely rhizinose at or towards the point of attachment. Ascospores (8-)9–11 × 6–8.5 μm, almost spherical to shortly ellipsoid. Conidia 4.5–6.5 × 0.5–1 μm, bacilliform, straight. Medulla Pd−, K−, KC+ red, C+ carmine-red, UV− (lecanoric acid).

On well-lit, nutrient-rich, hedgerow and parkland broad-leaved trees, frequently on ± horizontal boughs; very local. S. & S.W. England, extending N. to Wales (Merioneth). S. Europe, Alps, Canary Is., Asia.

This neat species is considered to be the fertile counterpart of the isidiate *P. pastillifera* **(7)**.

6. P. tiliacea (Hoffm.) Ach. (1803)

Parmelina tiliacea (Hoffm.) Hale (1974)

Thallus 4–8(-20) cm diam., sometimes forming elegant patches, rather unevenly attached, often wavy towards the centre with closely adpressed or ascending margins; lobes to 1 cm wide with short, rounded, often indented apices; upper surface whitish grey to grey, lacking a bluish tinge, sometimes weakly pruinose, faintly maculate, smooth. Isidia to 0.1 mm diam., laminal, pale brown to grey-brown or concolorous with thallus, sometimes brown-tipped, cylindrical to clavate, simple or coralloid, often crowded and forming a ± continuous crust. Lower surface black, brown towards lobe margins; rhizines to the lobe edges, simple or forked. Apothecia to 7 mm diam., rare; disc red-brown; dark rhizines arising from the underside of the thalline exciple near the point of attachment. Ascospores 9–10 × 6–9 μm, almost spherical to shortly ellipsoid. Conidia 4.5–1 μm, bacilliform, straight or slightly curved. Medulla Pd−, K−, KC+ red, C+ red, UV− (lecanoric acid).

On ± nutrient-rich bark of broad-leaved trees and siliceous rocks, roofing tiles and macadamized paths in sunny situations; local. Throughout British Is. Mainly in E. Europe, N. to C. Sweden.

Distinguished from *P. pastillifera* **(7)** in the form of the isidia.

7. P. pastillifera (Harm.) R. Schubert & Klem. (1966)

Parmelina pastillifera (Harm.) Hale (1976)

Like **6**, but thallus pale bluish grey, ± shiny, often entirely closely adpressed, with dark brown to blue-black, scattered, occasionally contiguous, isidia, to 0.2 mm diam. which are typically blue-black, flat-topped, knob- or button-like, sessile, or short-stalked.

On nutrient-rich bark, commonly *Fraxinus* and *Acer pseudoplatanus*, siliceous rocks, roofing tiles and memorials in well-lit situations. Widespread, especially in S. & W. British Is. W. Europe to Norway, S.C. Europe.

Distinguished by the distinctive, scattered, blue-black, knob-like isidia and the C+ red medulla. Considered an isidiate secondary species derived from *P. quercina* **(5)** which has an identical chemistry, apothecia, and thallus.

8. P. horrescens Taylor (1836)

Parmelinopsis horrescens (Taylor) Elix & Hale (1987)

Thallus to 3 cm diam.; forming small rosettes or ± irregular, lobes to 0.5 cm wide, flat or down-turned, ± overlapping, with frequently indented, ± incised ends; upper surface grey-white, matt or, in part, shiny. Isidia laminal, simple or ± branched, at first cylindrical, elongate, later flattened, decumbent and unorientated, often with brown-tipped ends, usually densely crowded and ± obscuring thallus, intermixed with scattered, fine, black cilia. Lower surface dark brown to black, becoming paler brown towards margins, rhizinose to margins; rhizines black, simple or forked. Apothecia very rare; disc red-brown with an irregular, coarsely crenate, isidiate-ciliate margin. Ascospores 12–15 × 9–11 μm, shortly ellipsoid. Medulla Pd−, K−, KC+ pink, C−, UV− (up to 8 compounds in the hiascic acid complex).

On neutral to acid-barked, old, broad-leaved trees in ancient woodlands; rare and very local. S. & S.W. England, Wales, S.W. Ireland. W. France, Portugal, Macaronesia.

Distinguished by cilia associated with isidia and the unusual chemistry; differing from *P. crinita* **(27)** also in the smaller size and chemistry. *P. minarum* **(9)** lacks the laminal cilia associated with isidia and has a C+ red medulla (gyrophoric acid).

9. P. minarum Vainio (1890)

Parmelinopsis minarum (Vainio) Elix & Hale (1987)

Thallus to 5 cm diam., often smaller, forming small, often irregular patches; lobes to 0.5 cm wide, flat, with much indented, ± overlapping, slightly rounded ends; upper surface grey-white. Isidia frequent, apices concolorous or becoming ± brown-tipped, cylindrical at first, simple, becoming ± richly coralloid-branched, remaining cylindrical, dense, often obscuring thallus, without associated cilia. Lower surface brown-black, paler towards margins, rhizinose to margins; rhizines simple or forked. Apothecia not seen. Medulla Pd−, K−, KC+ red, C+ red, UV− (gyrophoric acid).

On old trees, chiefly *Fagus*, in a few ancient woodlands; rare. S. England (Hampshire, New Forest), S.W. England (W. Cornwall). S.W. Europe, Macaronesia.

P. horrescens **(8)** has cilia associated with the isidia and a C− medulla; isidia in *P. minarum* are also finer and remain cylindrical.

10. P. subrudecta Nyl. (1888)

Punctelia subrudecta (Nyl.) Krog (1982)

Thallus 1–5(-6) cm diam., often rosette-forming; lobes to 1 cm wide, ± closely adpressed, margins contiguous or overlapping, rarely ascending, with rounded, sparingly indented margins and apices; upper surface pale grey, grey, or yellowish grey, with small, scattered, ± punctiform, rarely sigmoid pseudocyphellae, often developing into rounded, dot-like soralia. Soralia containing ± farinose soredia, 35–150 μm diam., more rarely soralia predominantly marginal, linear. Lower surface pale brown or partly white; rhizines pale, simple, not extending to margin. Apothecia very rare, to 5 mm diam.; disc red-brown; thalline exciple uneven-crenate, smooth or partially sorediate. Ascospores 14–17 × 12–15 μm, subglobose to ± ellipsoid. Conidia 5–7 × 1 μm, ± hooked at one end. Medulla and soralia Pd−, K−, KC+ red, C+ carmine-red, UV− (lecanoric acid).

On bark of broad-leaved trees, more rarely on wood, also amongst mosses on siliceous rocks, walls, roofing tiles, or standing stones in well-lit situations; often frequent. Throughout British Is. except N. Scotland and industrialized areas in C. & N. England. Cosmopolitan in tropical and temperate regions.

Characterized by the neat, closely adpressed thalli and punctiform pseudocyphellae which commonly become sorediate in older parts of the thallus. *P. borreri* **(12)** contains gyrophoric acid and has a jet-black lower surface; *P. reddenda* **(11)** is similar but has a black lower surface and a C− medulla. *Cetrelia olivetorum* has thalli to 20 cm diam., lobes to 2 cm wide, with raised, wavy margins, minute dot-like pseudocyphellae, and marginal soralia not arising from pseudocyphellae and a different chemistry.

11. P. reddenda Stirton (1877)

Punctelia reddenda (Stirton) Krog (1982)

Like **10**, but lower surface entirely black in the centre, and soredia coarser, often partially corticate, granular, 35–300 μm diam., in larger soralia, sometimes forming nodular or ± flattened folioles. Apothecia and pycnidia not seen. Medulla and soralia Pd−, K−, KC−, C−, UV− (up to 5 fatty acids).

On mossy trunks of broad-leaved trees, more rarely on mossy rocks, in sheltered, humid and often shaded situations; a member of the *Lobarion* and an old woodland indicator; rather local. S. & W. British Is. S. Scandinavia, France, North & South America, E. & S. Africa.

P. borreri **(12)** and *P. subrudecta* **(10)** both have a C+ red medulla.

12. P. borreri (Sm.) Turner (1807)

Punctelia borreri (Sm.) Krog (1982)

Like **10**, but lower surface jet-black and containing gyrophoric acid; soralia and medulla C+ orange-red. Generally more robust, more frequently fertile. The soralia, derived from pseudocyphellae are often more contiguous and coalescent and may form large, eroded patches; soredia 35–150 μm diam. Apothecia occasional, to 6 mm diam.; disc red-brown; pseudocyphellae on the thalline exciple sometimes ± sorediate. Ascospores 15–18 × 12–15 μm, subglobose to ± ellipsoid. Conidia 5–7 × 1 μm, hooked at one end.

On well-lit, ± nutrient-rich bark of broad-leaved trees in open situations, especially old orchards, less frequently on rocks; rare. S. & S.W. Britain, extending to S.W. Scotland (Solway Firth). ± Cosmopolitan in tropical and temperate regions.

May also be confused with *P. reddenda* **(11)** which has a C negative medulla.

13. P. caperata (L.) Ach. (1803)

Flavoparmelia caperata (L.) Hale (1986)

Thallus to 20 cm diam., conspicuous, often forming extensive patches, ± closely adpressed but centre often raised, wavy and uneven; lobes 0.5–1.3 cm wide, rounded, ± contiguous, ± overlapping, especially at the centre, rarely discrete, margins often indented; upper surface yellow to yellow-green-grey, occasionally green-grey (in shade), often wrinkled especially towards the centre. Soralia laminal, coarse, sometimes wart-like, occasionally gnarled, granular, derived from pustular, erose patches, at first punctiform, later coalescing to form more extensive, erose, spreading areas. Apothecia rare; disc to 0.8 cm diam., red-brown; thalline exciple ± sorediate. Ascospores 15–19(-22) ×

(8-)9–10 μm, ellipsoid. Medulla and soralia Pd+ orange-red, K−, KC± red, C−, UV− (protocetraric and caperatic acids); cortex KC+ yellow (usnic acid).

On well-lit, ± acid-barked, broad-leaved trees, a major species of the *Parmelietum revolutae*, also on conifers, fences, siliceous rocks, memorials, roofing tiles, coastal turf, and *Calluna* heath; common. Throughout S. & W. Britain in unpolluted habitats, actively recolonizing in N.E. England and C. & S. England where SO_2 levels are falling; rare and local in C. & N. Scotland. Worldwide.

P. soredians **(14)** has farinose soredia and a medulla which is K+ red.

14. P. soredians Nyl. (1872)

Flavoparmelia soredians (Nyl.) Hale (1986)

Like **13**, but thallus is smaller, tending to be more closely adpressed, lobes not exceeding 0.7 cm wide, soredia farinose, discrete at first, fusing to form finely sorediate patches. Mature apothecia not seen. Medulla Pd+ orange, K+ yellow to blood-red, KC+ blood-red, C−, UV− (salazinic acid).

On dry, neutral or ± acid-barked broad-leaved trees, fences, walls and monuments, siliceous rocks in sunny, exposed sites, especially near the coast, usually within the *Parmelietum revolutae*; local. Extreme S. & S.W. British Is., extending locally to S.E. Scotland (East Lothian). W. & S. France, S. Europe, South America, Macaronesia, C. & S. Africa, New Zealand.

15. P. conspersa (Ehrh. ex Ach.) Ach. (1803)

Xanthoparmelia conspersa (Ehrh. ex Ach.) Hale (1974)

Thallus 1–6(-10) cm diam., forming rosettes or coalescing into larger patches or as ± scattered lobes, closely adpressed and firmly attached; lobes 1–2 cm wide, slightly broader towards apices, discrete, contiguous or overlapping, margins often indented, not raised; upper surface yellow-grey, ± shining, smooth. Isidia often numerous, occasionally very sparse, laminal, cylindrical, simple or branched, coralloid, scattered or ± continuous, especially in old parts of thallus. Lower surface jet-black, with short, simple rhizines. Apothecia occasional; disc to 1 cm diam., red-brown with a thin, crenulate, ± inflexed, sometimes isidiate margin. Ascospores 6–10 × 4–5 μm. Conidia 4–5 × 1 μm, cylindrical. Medulla Pd+ orange, K+ yellow-orange, KC+ orange-red, C−, UV− (stictic, constictic and ± norstictic acids).

On sunny, siliceous rocks, walls, more rarely on roofing tiles, memorials, and hard wood in coastal and inland sites; common. S.W. & W. British Is., rarer eastwards. Cosmopolitan.

There is considerable variation in the number and shape of the isidia. The centre of the thallus may become ± obscured by a ± continuous, coarsely areolate crust of thin, crowded, coralloid, cylindrical isidia, alternatively, the isidia can be entirely missing and such specimens are often richly fertile. *P. tinctina* **(16)** has globose isidia and a different chemistry.

16. P. tinctina Maheu & A. Gillet (1925)

Xanthoparmelia tinctina (Maheu & A. Gillet) Hale (1974)

Like **15**, but with globose, somewhat inflated or irregular isidia and medulla Pd+ orange, K+ red (salazinic, and ± norstictic acids).

On sunny coastal boulders and slate roofs; very rare. S.W. England (Somerset, Norton Fitzwarren; Cornwall, Lizard Peninsula), Channel Is. (Jersey, Rozel). S. & W. Europe, often on serpentine, N. to Norway.

17. P. mougeotii Schaerer ex D. Dietr. (1846)

Xanthoparmelia mougeotii (Schaerer ex D. Dietr.) Hale (1974)

Thallus 0.5–1(-1.5) cm diam., ± forming rosettes or partially scattered, very closely adpressed, inner part sub-crustose, cracked; lobes narrow, to 3 mm wide, radiating, ± discrete and scattered, or contiguous, fan-like, not overlapping, elongate, incised, ± flat; upper surface yellow-grey, becoming ± darker grey-yellow at the centre. Soralia *c.* 1 mm diam., laminal, rounded, ± convex, flat or excavate, scattered. Lower surface black, rhizines short, stout, simple. Apothecia rare; disc to 2 mm diam., brown; thalline exciple entire, often sorediate. Ascospores 6–9 × 4–5 μm, ellipsoid. Conidia 4–5 × 1 μm, cylindrical. Medulla and soralia Pd+ orange, K+ orange, KC+ orange, C−, UV− (stictic acid).

On siliceous rocks, also on roofing tiles, as well as slate, memorials and quartzite chips in churchyards; locally common and increasing. Upland Britain, and S. & E. England on man-made substrata. W. Europe.

Rare saxicolous morphs of *Foraminella ambigua* have the medulla and soralia Pd −, K −, UV + starkly white (divaricatic acid). See also *P. incurva* **(18)**.

18. P. incurva (Pers.) Fr. (1826)

Arctoparmelia incurva (Pers.) Hale (1986)

Like **17**, but lobes markedly convex, intricate; bearing large, scattered, globular soralia, 2–4 mm diam., on the ends of smaller, inner lobes within the thallus. Lower surface pale. Medulla and soralia Pd − or ± rust-red, K −, KC + pink, C −, UV + vivid glaucous blue (alectoronic, ± α-collatolic and ± protocetraric acids); cortex KC + yellow (usnic acid).

On well-lit, siliceous rocks, very rarely on wood, pollution tolerant and increasing on man-made substrata; rather local. Upland Britain. Europe.

There is considerable variation in the protocetraric acid content; in some populations it is absent, in others it is predominant. *Parmelia centrifuga* (L.) Ach. (1803) [syn. *Arctoparmelia centrifuga* (L.) Hale (1986)], erroneously reported from Scotland, lacks soralia and usually has well-developed apothecia.

19. P. sinuosa (Sm.) Ach. (1814)

Hypotrachyna sinuosa (Sm.) Hale (1975)

Thallus 1–4(-6)cm diam., rather compact, sometimes forming extensive patches; lobes 0.2–0.5 cm wide, (1-)2–3(-4) mm wide, discrete or partly contiguous, dichotomously branched with wider, often ascending, truncate apices, axils markedly sinuate; upper surface yellow-grey to yellow-green, smooth, ± shiny. Soralia ± globose at ends of ± ascending lobes, soredia farinose. Lower surface jet-black, rhizines dichotomously branched, numerous, often projecting beyond the lobe margins. Apothecia not seen. Medulla and soralia Pd + orange, K + yellow-red, KC + red, C −, UV − (salazinic and constictic acids).

On siliceous rocks and well-lit, acid-barked trees, especially on twigs and young branches, occasionally on wood and *Calluna* stems; local. W. Scotland, extending locally to W. England (E. to Hampshire), Wales, W. & N. Ireland. Cosmopolitan.

20. P. revoluta (Flörke) Hale (1827)

Hypotrachyna revoluta (Flörke) Hale (1975)

Thallus 1–4(-6)cm diam., rather compact, sometimes forming extensive patches; lobes 0.2–0.5 cm wide, adpressed, ± overlapping towards the centre, rounded, often ± indented, axils rounded but not markedly sinuose, ends down-turned; upper surface grey to grey-green, smooth, ± matt; marginal cilia very sparse, usually absent. Soralia laminal, irregularly spreading, ± diffuse, erose, mainly on or near lobe ends, concolorous or slightly darker than thallus. Lower surface black, ± rhizinose to margins, rhizines simple, forked, or ± dichotomously branched. Apothecia rare; disc to 6 mm diam.; thalline exciple ± sorediate. Ascospores 11–16 × 8–10 μm, short-ellipsoid to almost spherical. Medulla and soralia Pd −, K −, KC + red, C + red, UV − (gyrophoric acid).

On wayside and woodland trees and shrubs, more rarely rocks and walls in inland and maritime situations; locally frequent. Throughout British Is. Europe, E. U.S.A., West Indies, Venezuela to Chile, South Africa, Japan, Indonesia.

The status of *P. afrorevoluta* Krog & Swinscow (1979) [syn. *Parmelinopsis afrorevoluta* (Krog & Swinscow) Elix & Hale (1987)] in the British Is. requires clarification; it is said to differ from *P. revoluta* in having laminal soredia, ± simple rhizines, short marginal cilia, ascospores 20–22 × 10–14 μm, and a tendency to produce paler soredia. See also *P. britannica* **(21)**.

21. P. britannica D. Hawksw. & P. James (1973)

Like **20**, but differing in the narrower, more elongate lobes with more sinuate axils; upper surface bluish grey, ± shiny; lobe ends, which are often ascending, developing irregular, erose, subterminal soralia with coarse, blue-black, granular, wart-like soredia; rhizines black, simple or tufted at ends. Apothecia not known.

On siliceous outcrops, boulders, memorials, and stone walls in ± sheltered, but well-lit sites, usually on or near the coast, rarely on the ground or on old *Calluna* stems and *Ulex*; local. W. British Is. France (Brittany), Spain, Portugal, Romania.

Distinguished by the extended, ± ascending lobes with indigo-blue to grey-black, coarse, semi-corticate, erose soralia and the largely coastal rock habitat. Has a superficial appearance of a reduced morph of *P. laevigata* **(22)**.

22. P. laevigata (Sm.) Ach. (1814)

Hypotrachyna laevigata (Sm.) Hale (1975)

Thallus wide-spreading, often forming extensive, elegant patches, to 15 cm diam., ± loosely attached, lobes 2–5 mm wide, ± elongate, branching ± divergently dichotomous, apices truncate, with markedly curved-sinuate axils; upper surface glaucous to bluish grey-white, ± shiny, smooth. Soralia mostly restricted to ends of lobes, rarely sublaminal, globose, concolorous or dark grey. Lower surface jet-black, rhizines numerous, dichotomously branched, often protruding beyond the margins (Fig. 39b). Apothecia rare, laminal, elevated; disc to 5 mm diam., dark brown; thalline exciple smooth or partially sorediate. Ascospores 18–21 × 9–13 μm. Conidia 4–6 × 1 μm, rod-shaped with a small swelling near both ends. Medulla and soralia Pd−, K−, KC+ yellow-orange, C+ orange, UV ± bluish-grey (barbatic, obtusatic, 4-*O*-demethylbarbatic, and norobtusatic acids).

On well-lit, acid-barked, leached, damp oak or birch woodlands and *Salix* carr, more rarely on mossy rocks and coastal cliffs, requiring rainfall in excess of 127 cm p.a. and with over 180 rain days p.a.; rather local but then often abundant. S. & W. British Is. Increasing in response to higher acid rain levels in W. Britain. W. Europe, E. U.S.A., C. & South America, E. Africa, Tasmania.

P. endochlora **(23)** differs in the pale yellow, not white, medulla.

23. *P. endochlora* Leighton (1871)

Hypotrachyna endochlora (Leighton) Hale (1975)

Like **22**, the medulla also C+ orange, but lobes usually smaller, 1.5–4 mm wide, the surface duller grey and medulla pale yellow. Medulla Pd− or + orange, KC+ orange, C+ orange, UV− or dull orange-red [barbatic, ± obtusatic, 4-*O*-demethylbarbatic, ± echinocarpic acids, secalonic acid A (pigment)].

In similar habitats to **22**, but more local; rare. W. & S.W. British Is. Portugal, W. France, C. & South America, Madagascar, E. & S. Africa, Azores, St. Helena, Hawaii.

24. P. taylorensis Mitchell (1961)

Hypotrachyna taylorensis (Mitchell) Hale (1974)

Thallus often forming extensive patches, to 15 cm diam., ± loosely attached; lobes to 0.8(-1.0) cm wide, overlapping, with ± tightly sinuate axils, margin entire or ± indented, lobe ends ± markedly down-turned; upper surface whitish grey, fragile, with occasional, irregular, white, abraded areas caused by a peeling of a fragile cortex. Soredia absent. Lower surface black, dark brown near the margins; rhizines richly dichotomously branched. Apothecia rare; thalline exciple coarsely crenulate, not rhizinose below. Ascospores 9–12 × 6–10 μm, almost spherical to shortly ellipsoid. Conidia 4–4.5 × 1 μm, bacilliform, with a small swelling near each end. Medulla Pd−, K−, KC+ red, C+ carmine-red, UV ± bluish (lecanoric and evernic acids).

Overgrowing mosses on trunks and rocks in ancient and relict woodlands, occasionally on mosses on rather exposed coastal rocks. S.W. & W. British Is., extending E. to S. Hampshire. Madeira, N. Spain, W. France (Pyrénées), N. Alps.

Young specimens may resemble *P. revoluta* **(20)** which is, however, always sorediate.

25. P. perlata (Huds.) Ach. (1803)

P. coniocarpa Laurer (1827), *Parmotrema chinense* (Osbeck) Hale and Ahti (1986)

Thallus 5–15(-20) cm diam., ± loosely attached; lobes to 1.5 cm wide, numerous, often overlapping, rounded, margin often raised and wavy, entire or sparingly indented, with black, marginal cilia, 1–2 mm in length; upper surface pale grey or grey, smooth. Soralia marginal, less frequently submarginal, discrete, ± globose, becoming ± contiguous, pale green-grey or whitish. Lower surface black with a broad, brown-black, naked zone towards the margin; rhizines few, simple, scattered. Apothecia rare; thalline exciple uneven, partially sorediate. Ascospores 21–28 × 13–15 μm. Medulla and soralia Pd+ orange, K+ yellow-orange, KC+ yellow-orange, C−, UV− (stictic and constictic acids, a range of accessory substances, including menegazziaic acid).

On well-lit, neutral to somewhat acid-barked, broad-leaved trees, also frequently on siliceous rocks and walls, sensitive to mean SO_2 levels >35 μg m^{-3}; locally abundant. Mainly S. & W. British Is. Cosmopolitan.

P. crinita **(27)** has laminal cilia associated with isidia and *P. arnoldii* **(28)**, a KC+ pink, UV+ ice-blue medulla; see also *P. robusta* **(26)**. The epithet *perlata* is retained in preference to the arguably nomenclaturally more correct *coniocarpa* as it is so well-known that the name change is undesirable and a protected status for *perlata* is pending.

26. P. robusta Degel. (1941)

Parmotrema robustum (Degel.) Hale (1974)

Like **25**, but differing in the larger, ± elaborately inrolled, very sparingly ciliate lobes, to 3 cm wide. Soralia rounded, rather sparse, often clustered on small extensions of the lobe margins. Medulla and soralia Pd+ rust-red, K−, KC−, C−, UV− (protocetraric acid).

On *Prunus spinosa* and mossy rocks on heathland near the coast; very rare. S.W. Ireland, (Cork, Baltimore). Portugal to W. France, Macaronesia, North and South America, Australia.

27. P. crinita Ach. (1814)

Parmotrema crinitum (Ach.) M. Choisy (1952)

Thallus often large, to 15 cm diam., sometimes forming extensive patches, ± loosely attached; lobes to 1.5 cm wide, contiguous or overlapping and intricate towards the centre, rounded at ends, frequently indented, uneven, surface wavy, margins partly ascending with occasional, scattered, black cilia, 1–3 mm; upper surface pale grey, matt, smooth or occasionally cracked. Isidia frequent, laminal and marginal, with rather coarse, concolorous or brown-tipped, ± cylindrical, often in ± proliferating coralloid clusters, intermixed with scattered, fine, black cilia, rarely becoming flattened, decumbent, and then forming small lobes. Lower surface black, with a broad, brown, naked zone towards the margin; rhizines simple (Fig. 39c). Apothecia very rare, thalline exciple isidiate. Ascospores 21–31 × 11–15 μm. Medulla Pd+ yellow-orange, K+ yellow-orange, KC+ yellow-orange, C−, UV− (stictic and constictic acids, and a range of accessory substances, including menegazziaic acid).

On ± mossy bark of broad-leaved trees and siliceous rock outcrops in sheltered and exposed situations, characteristic of well-lit *Lobarion* in undisturbed sites; local. S.W. & N.W. British Is., rare in Wales. Cosmopolitan.

A rather constant species but with sparse to numerous isidia and cilia. Resembles *P. perlata* **(25)**, which is sorediate. Intermixed isidia and cilia are only otherwise present in *P. horrescens* **(8)**, a smaller species with a K− medulla and a different chemistry.

28. P. arnoldii Du Rietz (1924)

Parmotrema arnoldii (Du Rietz) Hale (1974)

Thallus 5–15(-20) cm diam.; lobes 0.5 to 1.2(-2.0) cm wide, margins often ascending, wavy and indented; upper surface grey-white, sometimes faintly marbled, with often numerous and conspicuous, black, marginal cilia, to 5 mm long. Soredia ± marginal, globose. Lower surface almost entirely black, with an extensive, naked zone towards the margin; rhizines simple. Apothecia unknown in British material. Conidia 10–12(-14) × 0.5–1 μm, hair-like. Medulla and soralia Pd−, K−, KC+ pink-red, C−, UV+ ice-blue (alectoronic and α-collatolic acids).

Amongst mosses, especially on horizontal branches of trees and old shrubs in ± well-lit, mild and humid, undisturbed, woodlands, rarely in old *Calluna* heath; rare. W. Britain, eastwards to S. Hampshire (New Forest), S.W. Ireland. W. Europe, N. to S. Norway, C. European and Carpathian mtns, C., North & South America.

Distinguished by its Pd−, KC+ pink-red, UV+ intensely ice-blue medulla; in *P. perlata* **(25)** the medulla is Pd+ orange, KC+ orange, UV−.

29. P. reticulata Taylor (1836)

Parmotrema reticulatum (Taylor) M. Choisy (1952), *Rimelia reticulata* (Taylor) Hale & A. Fletcher (1990)

Thallus 5–15(-20) cm diam., ± loosely attached; lobes to 1.5 cm wide, margins often raised and wavy, frequently indented, apices often down-turned, especially when sorediate; marginal soredia absent; upper surface pale grey, with a faint, close network of white pseudocyphellae, often developing into a tight mosaic of hair-line cracks (× 10 lens). Soralia erose, irregular, occasionally coalescing, mainly on upper surface towards and at lobe-ends. Lower surface brown-black; rhizines simple, rarely forked, extending to the margin. Apothecia very rare; disc brown, sometimes perforate; thalline exciple ± sorediate. Ascospores 13–19 × 8–10 μm. Conidia (8-)12–16 μm, hair-like. Medulla and soralia Pd+ orange, K+ yellow then deep red, KC+ orange to deep red, C−, UV− (salazinic acid).

On sunny, wayside trees, sometimes in woodlands, and on exposed, sunny coastal rocks. S.E. England, extending N. to Scotland (Argyll, St. Kilda), S.W. & W. Ireland. Cosmopolitan.

Distinguished from *P. perlata* **(25)** by the minutely cracked upper surface and chemistry. Specimens in sheltered oceanic woodlands are often much larger with fewer lobes than those from open situations.

30. P. exasperata de Not. (1847)

Melanelia exasperata Essl. (1978)

Thallus to 2–3(-6) cm diam., entirely closely adpressed; lobes 1–4(-6) mm wide, uniformly thin, not coriaceous, discrete or commonly contiguous, or somewhat overlapping; upper surface red-brown, olive- or black-brown, ± green-brown when wet, roughened towards the centre, matt, with numerous, ± evenly scattered, conical papillae, 0.05–0.1 mm diam., with a pseudocyphella at the apex. Apothecia 3–5 mm diam., usually present; thalline exciple ± similarly papillate. Ascospores 9–12 × 5–6 μm. Conidia 7–8 × 1 μm, bacilliform. Medulla Pd−, K−, KC−, C−, UV− (no lichen substances detected by t.l.c.).

On ± acid-barked twigs and branchlets, almost entirely restricted to the pioneer *Lecanoretum subfuscae*, rarely fence posts and, exceptionally, rocks, very rarely on trunks of broad-leaved trees, in well-lit, but ± sheltered sites, usually as dispersed single plants; local, decreasing. W. British Is. Europe, North America, N. Africa.

Distinguished from all other *Parmelia* species by the presence of evenly spaced, pseudocyphellate papillae or warts on the upper surface and thalline exciple.

31. P. subaurifera Nyl. (1873)

Melanelia subaurifera (Nyl.) Essl. (1978)

Thallus 0.5–5(-10) cm diam., thin, closely adpressed, often ± rosette-forming; lobes to 0.5 cm wide, flat, radiating, contiguous, sometimes ± overlapping towards the centre; upper surface brown to green-brown, matt or rarely shiny in parts, smooth. Isidia globose, cylindrical or irregular, soft, secondarily derived from soredia by partial cortication, leaving a pale yellow area (subauriferin) where abraded. Lower surface dark brown to black; rhizines simple. Apothecia rare, to 2 mm diam., margin sorediate. Ascospores 9–12 × 5.5–7 μm. Conidia 5–5-7 × 1 μm. Medulla and soredia Pd−, K−, KC+ red, C+ carmine-red, UV− (lecanoric acid, subauriferin).

On smooth bark, especially branches and twigs of neutral to acid-barked trees, less frequently on trunks, rarely on rocks; moderately pollution tolerant, to *c.* 50 μg m^{-3} SO_2; frequent. Throughout British Is. Europe, North America.

P. glabratula **(32)** lacks soredia and has branched, true isidia, a more shiny upper surface, and a K+ violet-orange medullary pigment (rhodophyscin).

32. P. glabratula (Lamy) Nyl. (1883)

Melanelia glabratula (Lamy) Essl. (1978)

Thallus 1–5(-10) cm diam., thin, closely adpressed, rosette-forming or ± dispersed; lobes to (2-)3(-4.5) mm wide, flat, elongate, contiguous or overlapping; upper surface red-brown to olive-brown, shiny, especially towards lobe ends. Isidia cylindrical, smooth, becoming branched-coralloid, leaving a white area when abraded, usually numerous and often obscuring centre of thallus. Soredia absent. Lower surface black; rhizines simple. Apothecia to 5 mm diam., occasional; thalline exciple often becoming isidiate. Ascospores 10–14 × 5.5–8 μm. Conidia bacilliform, ± swollen towards each end. Medulla Pd−, K− or K+ violet, KC+ red, C+ red, UV− (lecanoric acid, rhodophyscin).

Chiefly on smooth bark or wood, occasionally on rocks; frequent. Throughout British Is. Europe, North America.

P. subaurifera **(31)** has isidia or partially corticate soralia, ± pale yellow when abraded and lacks the K+ violet medullary pigment rhodophyscin. See also *P. elegantula* **(35)**.

P. glabratula **subsp. *fuliginosa*** (Fr. ex Duby) Laundon (1965) differs in its consistently dark brown to black-brown colour and occurs primarily on siliceous rocks, tiles and memorials, and only rarely on broad-leaved trees. Mainly in Upland British Is. Europe, North America. It remains distinct when growing with subsp. *glabratula* and could merit specific status.

33. P. subargentifera Nyl. (1875)

Melanelia subargentifera (Nyl.) Essl. (1978)

Thallus *c.* 5 cm diam., loosely attached, margins usually ascending; lobes to 3 mm wide, flat; upper surface pale olive-brown to dark brown, often with a reddish or yellowish tinge, ± matt, usually pruinose, white cortical hairs present towards the lobe ends (× 20). Soralia laminal or marginal, usually confluent towards centre of thallus; soredia coarsely granular or isidioid. Medulla and soralia Pd−, K−, KC+ red, C+ carmine-red, UV− (lecanoric acid).

On nutrient-rich tree trunks, in *Xanthorion* communities; rare. N.E. Scotland (Kincardine). Europe, North America, Pakistan, N. India.

Distinguished from *P. subaurifera* **(31)** and P. *glabratula* **(32)** by the less adpressed thallus with ± reflexed lobes, the granular to isidioid, partially corticate soredia present usually in both laminal and marginal soralia, the lack of true isidia and the presence of few to many tiny cortical hairs.

34. P. laciniatula (H. Olivier) Zahlbr. (1916)

Melanelia laciniatula (H. Olivier) Essl. (1978)

Thallus to 5 cm diam., or thalli coalescing to form large patches to 15 cm diam., usually ± entirely composed of numerous, very small, thin, overlapping folioles, 5–8 mm long and 1–3 mm wide, occasionally with distinct, well-developed, closely adpressed, thin, marginal lobes; upper surface grey-brown to olive-brown, olive-green when wet; lower surface pale; rhizines simple, scattered. Apothecia unknown. Medulla Pd−, K−, KC−, C−, UV− (lichen products not detected by t.l.c.).

On well-lit, nutrient-rich trunks of broad-leaved trees, especially overhanging branches, particularly in parklands, occasionally on conifers, siliceous rocks and walls, increasing in relatively polluted (to 65 μg m^{-3} SO_2) sites; locally frequent. S.E. England, rarer in Scotland and Wales. Europe, N. Africa.

Very distinctive. Characterized by poorly developed marginal lobes, the numerous, small, thin, overlapping folioles, the C− medulla, absence of isidia, and the olive-green colour when wet.

35. P. elegantula (Zahlbr.) Szat. (1930)

Melanelia elegantula (Zahlbr.) Essl. (1978)

Thallus to 5 cm diam., ± closely adpressed; lobes to 2 mm wide, ± flat, contiguous, ± overlapping at centre; upper surface brown to green-brown, dark olive-green when wet, ± smooth, matt, occasionally shiny towards margins. Isidia solid, cylindrical, ± erect, often branched-coralloid, not inflated, ± smothering older parts of thallus, initials on newer lobes discrete, ± evenly spaced. Lower surface pale brown; rhizines simple, pale, scattered. Apothecia 2–3 mm diam., very rare; thalline exciple densely isidiate. Ascospores 8–11 × 4.5–6.5 μm, ellipsoid. Medulla Pd−, K−, KC−, C−, UV− (lichen products not detected by t.l.c.).

On ± nutrient-rich, acid-barked trees, more rarely on rocks including memorials, often in and increasing in relatively polluted areas (to <65 μg m^{-3} SO_2), frequently with *P. laciniatula* **(34)**. S.E. England, extending locally to C.N. Scotland (Perth). Europe, South America, N. Africa, Pakistan.

P. glabratula **(32)** has a dark lower surface, a thicker, C+ carmine-red medulla, and a tendency for the young initials of the isidia to be more randomly distributed. *P. exasperatula* **(36)** has coarser, ± spathulate, hollow, isidia.

36. P. exasperatula Nyl. (1873)

Melanelia exasperatula (Nyl.) Essl. (1978)

Thallus to 5 cm diam., ± closely adpressed at centre, often with raised, wavy marginal lobes, to 5 mm wide; upper surface pale olive-green to dark olive-brown or red-brown, semi-transparent when wet. Isidia clavate or ± spathulate, inflated, hollow, simple, swollen towards the middle, often decumbent and then orientated in all directions. Apothecia 2–3 mm diam., very rare; thalline exciple ± uneven, not isidiate. Ascospores 8–10 × 3.5–8 μm, ellipsoid. Medulla Pd −, K −, KC −, C −, UV − (lichen products not detected by t.l.c.).

On trunks and branches of nutrient-rich, wayside broad-leaved trees, occasionally saxicolous, especially on walls under *Acer pseudoplatanus*, often in relatively polluted areas (to 70 μg m^{-3} SO_2) and increasing; locally frequent. England, especially in the N. Pennines, rarer in Wales, Scotland, Ireland. Europe, Siberia, North America, Pakistan.

Readily recognized by the most distinctive isidia, which are clavate or ± spathulate and hollow.

37. P. septentrionalis (Lynge) Ahti (1966)

Melanelia septentrionalis (Lynge) Essl. (1978)

Thallus forming small rosettes, 1–3(-5) cm diam., closely adpressed; lobes to 2 mm wide, flat to somewhat convex, discrete to overlapping; upper surface red- or dark-brown, smooth, shiny; lower surface dark brown to black; rhizines simple, scattered. Apothecia abundant, often crowded and overlapping, covering most of the thallus apart from the extreme margin; thalline exciple thin, smooth with many pseudocyphellae. Ascospores 9–13 × 5–8 μm, ellipsoid. Conidia 5–6.5 × 1 μm, bacilliform. Medulla Pd + orange-red, K −, KC −, C −, UV − (fumarprotocetraric, ± protocetraric acids).

Mainly on twigs of *Betula* in well-lit, open situations, especially margins of woods, often with *Cetraria chlorophylla* and *C. sepincola*; very local. N. Scotland (E. Highlands). Europe, North America, Asia.

Forms small, neat, richly-fruited thalli. The occurrence on *Betula* in Britain is unusual as it is frequent on *Alnus incana* and *Salix* abroad. The closely related *P. olivacea* (L.) Ach. (1803), unknown in Britain, has a more wrinkled, matt thallus, a broader more distinct marginal zone without apothecia, and a crenulate thalline exciple; it has been incorrectly reported from Scotland.

38. P. stygia (L.) Ach. (1803)

Melanelia stygia (L.) Essl. (1978)

Thallus to 6 cm diam., firmly attached; lobes to 0.5–2 mm wide, very variable, ± convex, margin irregularly nodulose, elongate, linear, and overlapping, or ± entangled; upper surface dark brown-black, shiny, with small, inconspicuous, punctiform pseudocyphellae with a thick, ± raised, often darker, margin; lower surface black; rhizines simple or forked, scattered. Soredia and isidia absent. Apothecia to 5 mm diam., frequent; disc flat or convex, shiny; thalline exciple grossly warted, irregularly crenate and pseudocyphellate. Ascospores 7–10 × 4.5–6 μm. Pycnidia numerous, immersed, black, sometimes confluent, rounded or oval. Conidia 3.5–5.5 × 1 μm, bifusiform. Medulla Pd + orange-red or Pd −, K −, KC −, C −, UV − (± fumarprotocetraric and protocetraric acids).

On siliceous rocks, montane; rare. N. Scotland (Highlands), C. Wales (Radnor). Europe, Siberia, North America, Japan.

Cetraria hepatizon has marginal apothecia, marginal and markedly emergent pycnidia, strictly marginal pseudocyphellae, and its medulla is K + yellow-red (stictic and norstictic acid). *Allantoparmelia alpicola* has a matt surface, without pseudocyphellae, and a Pd + pale yellow, C + pink medulla.

39. P. disjuncta Erichsen (1939)

Melanelia disjuncta (Erichsen) Essl. (1978)

Thallus to 5 cm diam., closely adpressed; lobes to 0.5–1.5 mm wide, flat, short, often broadened towards the tips, bearing sub-marginal, white pseudocyphellae (× 10 lens); upper surface dark brown-black, shiny towards margins. Isidia very small, dark grey to black, peg-like, eroding into coarse soralia, scattered or aggregated into granular or wart-like clusters. Medulla Pd −, K −, KC + pink, C − or C + pink-red, UV + white (perlatolic, stenosporic and ± gyrophoric acids).

On hard, sunny, siliceous, commonly igneous, rocks, mainly upland, rarely on the coast; very local. C. & W. Britain, rarely in the E. on memorials. N. & C. Europe, North America.

P. verruculifera **(44)** has a similar dark brown thallus, but consistently broader lobes, no pseudocyphellae and a different chemistry.

40. P. acetabulum (Necker) Duby (1830)

Pleurosticta acetabulum (Necker) Elix & Lumbsch (1988)

Thallus 3–8(-30) cm diam.; lobes to 1.7 cm wide, ± closely adpressed, coriaceous, with wavy, sometimes crisped, margins and incised rounded ends, often wrinkled or warted towards centre, contiguous and overlapping; upper surface grey-green to brown-grey, occasionally ± grey-pruinose, dark oily green when wet; lower surface pale brown; rhizines simple. Apothecia 0.5–1.5 cm diam., often present; disc red-brown; thalline exciple ± unevenly crenulate, often strongly inflexed. Ascospores 14–17 × 7–8.5 μm. Conidia 6–7 × 1 μm, bacilliform. Medulla Pd+ orange, K+ red, KC−, C−, UV− (norstictic acid).

On nutrient-rich bark of trunks of broad-leaved trees, *Ulmus*, *Fraxinus*, *Acer pseudoplatanus*, and *Sambucus* in well-lit situations, very rarely on ragstone memorials; local and decreasing. S.E. England, extending locally to E.C. Scotland (Tayside) and Wales (Dolgellau). Europe.

41. P. pulla Ach. (1814)

Neofuscelia pulla (Ach.) Essl. (1978)

Thallus 3–12 cm diam., ± coalescing to form larger patches, ± closely adpressed; lobes to 5 mm wide, flat, elongate, rarely discrete, often contiguous, often overlapping and intermixed, radiating towards the margin; upper surface grey-brown to dark brown, unevenly rumpled and transversely wrinkled. Lower surface black; rhizines simple, frequent. Apothecia 2–7 mm diam., common, concentrated on older parts of thallus, concolorous with thallus; thalline exciple ± entire, occasionally lacerate. Ascospores 7–11 × 4–6 μm, ellipsoid. Conidia 5–7 × 1 μm, bacilliform. Medulla Pd−, K−, KC+ pink-red, C± pink-red, UV+ white (stenosporic and/or divaricatic and ± gyrophoric acids; both major substances are usually present together; specimens with stenosporic acid out-number those in which divaricatic acid predominates by 10:1, gyrophoric acid accounts for the occasional C+ pink reaction of the medulla in *c.* 25% of specimens).

On coastal siliceous rocks in the xeric-supralittoral, occasionally overgrowing mosses; locally abundant. Widespread. W. coasts of British Is. Europe, N. & S. Africa, Australia, New Zealand.

Resembles *P. delisei* **(42)** in all but the darker colour, and *P. verruculifera* **(44)** but lacks isidia.

42. P. delisei (Duby) Nyl. (1872)

Neofuscelia delisei (Duby) Essl. (1978)

Like **41**, but thallus paler brown with often broader lobes and a different chemistry (perlatolic, glomelliferic, glomellic acids and ± gyrophoric acids) although the reagent tests are the same.

On siliceous coastal rocks, xeric-supralittoral, rarely inland (Kent, Mereworth), often with *P. pulla* **(41)**. W. British Is. Europe, N. Africa.

The fertile counterpart of *P. loxodes* **(43).**

43. P. loxodes Nyl. (1872)

Neofuscelia loxodes (Nyl.) Essl. (1978)

Thallus wide-spreading, to 20 cm diam., ± loosely adpressed; lobes to 5 mm wide, flat, elongate, often overlapping, especially towards the centre, spreading marginally; upper surface pale yellowish grey-brown to reddish brown, uneven, frequently ± transversely wrinkled. Isidia frequent or sometimes scarce, coarse, cauliflower-like, clustered in pustulate outgrowths. Lower surface black; rhizines simple, frequent. Apothecia frequent; thalline exciple often isidiate. Ascospores (7-)8–10 × 4–5 μm. Medulla and abraded isidia Pd−, K−, KC+ orange-red, C− or C+ red, UV+ white (glomelliferic, glomellic, perlatolic and ± gyrophoric acid).

On maritime siliceous rocks, xeric-supralittoral, very rarely inland; often common in coastal areas. W. Britain, Ireland. Europe, W. North America, N. Africa, Israel.

The isidiate counterpart of *P. delisei* **(42)** and sparingly isidiate specimens are difficult to distinguish from the parent species. See also *P. verruculifera* **(44)**.

44. P. verruculifera Nyl. (1878)

Neofuscelia verruculifera (Nyl.) Essl. (1978)

Like **43**, but with thinner, darker dull brown to black-brown lobes with scattered, coarse isidia, which may become ± contiguous to form extensive, ill-defined patches on older parts of the thallus. Medulla and abraded isidia Pd−, K−, KC− or + pink, C− or C+ pink-red, UV+ white (divaricatic, and ± gyrophoric acid).

On siliceous rocks in maritime and inland areas, preferring well-lit xeric situations, memorials, standing stones, roadside walls, roofing tiles. E., S. & W. British Is. Europe, North America, Ethiopia, Saudi Arabia.

P. verruculifera has a much wider ecological amplitude than *P. loxodes* **(43)**, *P. pulla* **(41)** and *P. delisei* **(42)**, all of which are almost exclusively coastal. *P. disjuncta* **(39)** has a proportionally smaller thallus. *P. glabratula* subsp. *fuliginosa* **(32)** has a shining upper surface with thinner isidia.

PARMELIELLA Müll. Arg. (1862)

Thallus foliose, squamulose or ± crustose, often attached to lower surface by a blue to blue-black hypothallus which is sometimes visible from above. **Upper surface** grey-blue to brown-black. **Photobiont** blue-green, *Nostoc*. **Ascomata** apothecia, sessile; disc red-brown to black, with a thin, pale, true exciple. **Thalline exciple** absent. **True exciple** well developed, thick, pseudoparenchymatous; cells rounded to oblong, 15–20 μm diam. **Hamathecium** of simple paraphyses, not or slightly thickened at the apices, the pigment external. **Asci** 8-spored; apical apparatus a deeply blue-staining plug. **Ascospores** colourless, simple, ellipsoid, often apiculate at one or both ends and with a warted or ridged epispore. **Conidiomata** pycnidia, rare. **Conidia** bacilliform. **Chemistry:** lichen products not usually detected by t.l.c., some species when fresh have an antiseptic smell when moistened. **Ecology**: mostly on broad-leaved, mossy trees in *Lobarion* communities, more rarely on rocks, mainly in sheltered, humid, relict woodlands.

Closely resembles *Pannaria*, which apparently differs only in having a thalline exciple. Since the species of both genera are often sterile the key to *Pannaria* should also be consulted for such material. Some species superficially resemble *Massalongia* and *Psoroma*; see also *Degelia* and *Moelleropsis*.

LITERATURE: Jørgensen (1978)

1 Thallus ± uniform, without enlarged radiating marginal lobes 2
Thallus with enlarged, ± distinctly radiating marginal lobes .... **3. *testacea***

2(1) Soredia or isidia present; squamule margins ± concolorous 3
Soredia and isidia absent; squamule margins often white, the lobules with coarse knob-like protuberances .... **2. *praetermissa***

3(2) Soredia present, granular or ± powdery 4
Isidia present, cylindrical, branched .... **4. *triptophylla***

4(3) Squamule surface blue-grey, rarely tinged fawn .... **1. *jamesii***
Squamule surface pale to dull red-brown .... **3. *testacea***

1. *P. jamesii* S. Ahlner & P. Jørg. (1978)

Thallus squamulose; squamules 1–2 mm long, ± elongate and indented, crenulate, flat or concave, pale blue-grey, rarely tinged fawn, scattered or ± contiguous, forming small patches or a ± continuous crust-like cover on substratum; soralia marginal, coarsely granular-isidiate, often ± rounded and discrete, occasionally becoming confluent and transforming thallus into a sorediate crust. Apothecia to 0.5 mm diam., very rare; disc pale red-brown to blackish with a persistent paler true exciple, often becoming ± convex. Ascospores 15–18 × 6–7 μm, colourless, oblong, with a minutely warted epispore. Lichen products not detected by t.l.c.

On mossy, broad-leaved trees, occasionally conifers, more rarely on rocks and wood, often overgrowing mosses, particularly in sheltered moist woodlands and often close to streams; locally frequent. S. & W. Britain, Ireland. Scandinavia, Portugal, Italy, Macaronesia.

A small, inconspicuous and rather variable species. In exposed sites specimens are often tinged red-brown and have coarser, granular soralia, while the thalli of specimens from shaded habitats are pale blue-grey and usually have finer soralia which are often confluent, sometimes tending to form a continuous crust without marginal squamules. Resembles the rarer *Pannaria mediterranea* when sterile, which differs in having larger, (> 2 mm), ± rounded, more swollen, convex squamules, a dark blue-grey to olive-brown coloration and lilac-mauve-tinged soralia, which often appear woolly in the herbarium due to the gradual deposition of steroid crystals.

2. P. praetermissa (Nyl.) P. James (1965)

Thallus thick, squamulose; squamules 1–3 mm diam., ± rounded, indented, often growing densely compacted and overlapping, forming ± continuous, compact mats; hypothallus inconspicuous; upper surface brown-black, the margins often white; lobules and knob-like protuberances frequent, scattered. Apothecia unknown in British material; disc brown. Steroids are often present as needle-shaped crystals (× 20) on the thallus of herbarium specimens.

On mosses in sheltered recesses of ± basic, schistose rocks, above *c.* 750 m.; rare. Scotland (Grampian mtns, Ben Lawers range, Caenlochan). Iceland, Scandinavia, C. European mountains, Pyrénées, U.S.S.R., Japan, North America.

Characterized by the tendency to form thick, nodular, ± continuous crusts overgrowing mosses, the presence of numerous lobules and by the dark cyanobacterial layer which is readily observed when the lobules or squamules are broken. *Massalongia carnosa* has more flattened, elongate squamules, the margins of which are dissected by numerous flattened marginal accessory lobules or warts; this species occurs on non-calcareous soil or more frequently on moist, mossy, siliceous boulders often near streams.

This species might be better-placed in *Pannaria*.

3. P. testacea P. Jørg. (1978)

Thallus squamulose, ± forming rosettes, closely adpressed to substratum; squamules 2 × 2–3 mm, ± extended to 5 mm at margins; upper surface fragile, pale red-brown to smoky grey-brown, smooth or ± scabrid, matt; soralia predominantly marginal, sometimes also laminal, coarse, knob-like, at first concolorous, sometimes breaking down into a ± uniform surface of coralloid, ± blue-grey granules which may become coarsely cracked-areolate; hypothallus blue-black, fimbriate, sometimes apparent between cracks but seldom extending beyond squamules. Apothecia 1–1.5 mm, rare; disc red-brown, flat; true exciple rather thick, paler, persistent. Ascospores 16–20 × 9–10 μm, colourless, ellipsoid, without a warted epispore. Lichen products not detected by t.l.c.

On mossy broad-leaved trees, rarely on rocks in moist, sheltered woodlands; locally abundant. W. British Is. France, Portugal, Italy, Spain, Madeira, Tenerife.

Bruised fresh material smells strongly of antiseptic when wet. The size of soredia may vary considerably. Specimens with coarse, knob-like soralia may resemble *P. triptophylla* **(4)** which has consistently smaller squamules (to 1 mm), which are grey-brown above and produce cylindrical isidia. The similarly coloured *Pannaria sampaiana* has a more closely adpressed thallus, cream-yellow soralia, a distinct black hypothallus and lacks extended marginal lobes.

4. P. triptophylla (Ach.) Müll. Arg. (1862)

Thallus of minute, indented, rounded or ± elongate, flat to concave squamules, to 1 mm wide, scattered or ± contiguous, sometimes overlapping; hypothallus conspicuous, wide, smooth, black, giving the plant at a distance an overall dark appearance; upper surface of squamules grey-brown, ± shiny, with terete, finger-like, branched, erect or ± decumbent isidia along the margins, often interspersed with more flattened lobules, sometimes forming a ± continuous isidiate crust and often completely obscuring the underlying squamules. Apothecia to 1 mm diam., infrequent; disc red-brown to black; true exciple pale. Ascospores 10–17 × 5–8 μm, colourless, ellipsoid, without a warted epispore; outer spore wall smooth; inner wall uneven. Lichen products not detected by t.l.c.

On mossy trees, more rarely rocks, in damp, sheltered places; local. W. British Is., extending to S. England (Wiltshire, Dorset). Europe, Iceland, Greenland, Canada, Kamchatka, Madeira, Canary Is.

Characterized by the presence of a prominent dark smooth hypothallus and isidia which may be rather coarse and knobbly or occasionally very fine, finger-like and often branched. The squamules vary from grey to black-brown; when isidia are sparse there may be a close resemblance to *Pannaria leucophaea,* which has more rounded, convex, slightly swollen squamules and is usually fertile.

PARMELIOPSIS (Nyl.) Nyl. (1866)

Thallus foliose, small, rosette-forming, closely adpressed, lobes ± radiating, discrete or contiguous and overlapping. **Cortices** pseudoparenchymatous, upper surface with a pored epicortex. **Photobiont** chlorococcoid. **Lower surface** white to pale tan with pale, simple rhizines. **Ascomata** apothecia, laminal, sessile, thalline margin concolorous with the thallus; disc dark brown. **Hamathecium** of paraphyses, simple. **Asci** 8-spored, *Lecanora*-type. **Ascospores** colourless, ellipsoid. **Conidiomata** pycnidia, laminal and marginal, emergent, black. **Conidia** short, ampulliform, with 1(-2) swellings. **Chemistry:** atranorin and thamnolic acid.

Foraminella differs in having a darker lower cortex, kidney-shaped or long, curved conidia and chemistry.

LITERATURE: Meyer (1982).

1. P. aleurites (Ach.) Nyl. (1866)

Thallus 1–3(-5) cm diam., often forming rosettes or of ± scattered, unorientated lobes, closely adpressed to substratum. Lobes 1–3 mm wide, indented or sometimes branching towards the apices, discrete or contiguous and overlapping; upper surface grey-white, ± shining, older parts often ± obscured by a ± continuous crust of crowded, concolorous or slightly darker, ± cylindrical isidia which may become eroded; lower surface ± pale brown with simple rhizines. Apothecia infrequent; disc red-brown. Ascospores 6–9 × 5–6 μm. Conidia 3.5–4.5 × 1 μm. Thallus Pd+ yellow-orange, K+ yellow, C−, UV− (thamnolic acid and atranorin).

On acid bark and wood, especially conifer and *Betula*; formerly restricted to ancient Caledonian pine woodlands in Scotland, but now of scattered occurrence in C. Wales and N. England, and also frequent and spreading in S. England on artificial substrata such as worked oak and conifer wood such as fence rails; rarely directly on peat. Europe, North America.

Similar to *Foraminella hyperopta* which is, however, sorediate and has a medulla which is UV+ glaucous. The correct name is ***Imshaugia aleurites*** (Ach.) S. Meyer (1985).

PARMENTARIA Fée (1825)

Thallus immersed. **Cortex**-like layer present, consisting of bark cells compacted with numerous hyphae. **Photobiont** *Trentepohlia*. **Ascomata** perithecia, immersed, mostly radially arranged with lateral ostioles extending to the surface at the centre of the group. **Hamathecium** of paraphyses, septate, rarely branched; periphyses much-branched and anastomosed, highly gelatinized forming a plug with a narrow opening the length of the ostiole. **Asci** cylindrical-clavate, thick-walled, 1- to 8-spored. **Ascospores** colourless to brown, muriform. **Conidiomata** pycnidia, pyriform, divided into several chambers internally. **Conidia** colourless, thread-like, straight or curved. **Ecology**: on bark, *c.* 25 species mainly occurring in the tropics.

Characterized by the inclined or prostrate perithecia, usually radiating in small groups with centrally ± joined ostioles, and the brown, muriform ascospores with thickened walls. The genus is separated from *Pyrenula* mainly by the arrangement of the perithecia and the muriform spores and has been united with this genus by Harris (1989).

LITERATURE: Harris (1989), Imshaug & Harris (1969), Swinscow (1966).

1. P. chilensis Fée (1838)

Thallus immersed, pale olive-green or yellow-buff, sometimes with white dots, smooth, continuous and shiny, becoming rougher, rarely cracked and dull with age. Perithecia 1–1.2 mm diam., black, arising 1–1.5 mm beneath surface, at maturity often only end apices of ostiolar tubes visible at surface, occurring singly or joined by their ostioles in radially arranged groups of 2 to 5; outer wall black and friable; ostiole 0.2–0.3 mm diam., lateral at one end of perithecium, forming a pale yellow- or orange-brown, circular or oval depression at thallus surface. Paraphyses 1–2 μm wide. Asci 4- to 8-spored. Ascospores (57-)70–135(-140) × (24-)26–47(-55) μm, oblong to fusiform-ellipsoid, brown when mature. Pycnidia *c.* 2 mm diam.; conidia 24–27 × 1 μm. Thallus Pd−, K−, KC−, UV+ light yellow (lichen products not tested by t.l.c.).

On smooth bark of *Corylus* and *Ilex* in very sheltered, moist ravines at *c.* 100 m alt.; very rare. W.N. Scotland (Westerness, Loch Sunart), S.W. Ireland (Kerry, Killarney). Macaronesia, U.S.A. (Florida), West Indies, Juan Fernández, Colombia.

Distinguished from *Pyrenula* species by the arrangement of the perithecia and the large, muriform ascospores. Laterally developed ostioles are sometimes produced by *Acrocordia gemmata* and in an undescribed species of *Pyrenula* from S.W. Ireland and W. Scotland.

PELTIGERA Willd. (1787)

Thallus foliose, forming compact to very wide-spreading rosettes, lobes rounded or ± elongate, discrete, contiguous or overlapping, bright green, blue-grey, grey-brown or brown, flat, wavy or bullate, smooth, matt or shiny, ± scabrid, tomentose or pruinose, the margins entire or ± lobulate, or with folioles, schizidia or soralia. **Lower surface** non-corticate, densely arachnoid-tomentose or with anastomosing pale or dark veins, from which arise conspicuous rhizines (Fig. 20). **Cephalodia**, when present, external, containing *Nostoc*, on either upper (Fig. 30a) or lower surface of species with green photobiont. **Cortex** pseudoparenchymatous. **Photobiont** blue-green (*Nostoc*) and (or) chlorococcoid (*Coccomyxa*). **Medulla** white, of ± loosely interwoven hyphae. **Ascomata** apothecia, saddle-shaped, flattened or oval, red-brown to black, often with reflexed or crenulate margins, on horizontal or ± vertical, ascending lobules; development hemiangiocarpic. **Asci** 8-spored, cylindrical, fissitunicate, the apex of the endoascus with a K/I+ blue annulus, *Peltigera*-type (Fig. 34d). **Ascospores** narrowly fusiform to acicular, 3- to multiseptate, colourless to pale brown. **Chemistry**: 6 diagnostic hopane triterpenoids and tridepsides in some species. **Ecology**: on bare soil or in short turf, mossy lawns, bonfire sites or on mossy tree trunks and fallen logs.

Although a conspicuous and readily recognized genus, identification to species is not easy and depends on well-developed material in good condition. Particular attention must be given to the surface of the lobes, including the presence or absence of a tomentum, and also to the lower surface, including the colour and extent of veining and form of rhizines; the terpene patterns are diagnostic for those species without a tomentum on the upper surface and are essential when identifying fragmentary material.

LITERATURE: Poelt & Vězda (1981), Vitikainen (1982), White & James (1987).

1 Thallus bright green when moist (photobiont green, *Coccomyxa*); cephalodia present 2
Thallus pale to dark blue-grey to brown-green when moist (photoboint blue-green, *Nostoc*); cephalodia absent 4

2(1) Thallus >5 cm diam.; apothecia infrequent; discs saddle-shaped; cephalodia on upper surface; veins, ± anastomosing or inconspicuous 3

Thallus <2 cm diam.; apothecia frequent; discs flat, rounded; cephalodia on lower surface, on fan-shaped, dark brown veins .... **3. *venosa***

3(2) Cephalodia adpressed, convex and wart-like, never detaching; lower surface with a distinct network of pale or dark-coloured veins .... **1. *leucophlebia***
Cephalodia peltate, with free margins, flat or slightly concave, frequently detaching leaving white scars; lower surface without a distinct network of veins or veins absent .... **2. *britannica***

4(1) Isidia or soredia present 5
Isidia and soredia absent (flattened, marginal, regeneration lobules sometimes present) 8

5(4) Soredia present 6
Isidia present 7

6(5) Soralia laminal; thallus 1–2 cm diam., ± ascending; upper surface ± tomentose **9. *didactyla***
Soralia marginal; thallus >2 cm diam., radiating; upper surface not tomentose . **17. *collina***

7(5) Isidia laminal, peltate, darker than thallus, resembling cephalodia; lobes <1 cm wide .... **18. *lepidophora***
Isidia and schizidia marginal and laminal, squamulose and overlapping or coralloid, concolorous with thallus; lobes mostly >1 cm wide .... **6. *praetextata***

8(4) Upper surface tomentose at least at lobe margins 9
Upper surface never tomentose, entirely smooth or scabrid 14

9(8) Lower surface with distinct veins; upper surface with ± adpressed tomentum 10
Lower surface lacking veins or veins indistinct and only present towards lobe margins; upper surface with a thin, erect tomentum towards margins .... **10. *malacea***

10(9) Thallus more than 3 cm diam., with broad, spreading, sterile lobes, only ascending at ends when fertile 11
Thallus rarely over 3 cm diam., with narrow, ± soon erect, fertile lobes .... **9. *didactyla***

11(10) Lobes <1.5 cm wide; sterile lobes with up-turned, somewhat swollen, sometimes crisped margins 12
Lobes >2 cm wide; sterile lobes with thin, often down-turned, margins 13

12(11) Lower surface with dark, densely branched, fluffy rhizines, confluent at their bases and brown veins; upper surface often thickly pruinose or white-grey-incrusted towards centre .... **7. *rufescens***
Lower surface with pale, simple, usually unbranched rhizines and pale, raised veins; upper surface tomentose or glabrous towards centre .... **8. *ponojensis***

13(11) Veins raised, usually persistently downy; rhizines discrete, slender, simple, downy, or with numerous very short, squarrose outgrowths (bottle-brush-like) .... **4. *membranacea***
Veins flattened, smooth; rhizines irregular and fluffy, confluent at their base (like a shag-pile carpet) .... **5. *canina***

14(8) Veins dark, flattened at centre, or absent; lichen products present 15
Veins white to pale yellow-brown at centre, narrow and elevated; lichen products absent .... **15. *degenii***

15(14) Upper surface smooth, glossy or matt 16
Upper surface finely roughened .... **19. *scabrosa***

16(15) Lower surface with distinct veins; upper surface lacking schizidia, isidia or lobules 17
Lower surface without distinct veins, dark brown-black, felted with small, whitish, round spots; upper surface with marginal, flattened or nodular schizidia, sometimes also isidia or lobules .... **14. *elisabethae***

17(16) Lower surface of older parts of thallus with darker brown veins and rhizines; upper surface ± shiny 18
Lower surface of older parts of thallus with pale to deep ochre-coloured veins with pale rhizines; upper surface matt .... **11. *lactucifolia***

18(17) Lower surface veins varied but pale at margins; ascending lobes rarely becoming crisped-contorted or becoming squamulose 19
Lower surface with distinct, dark brown, anastomosing veins to the margins; ascending lobes often markedly crisped-contorted and becoming squamulose .... **13. *polydactyla***

19(18) Lower surface with few, wide veins, or veins almost absent; upper surface white-grey-pruinose at margins, often deeply cracked; apothecia saddle-shaped, dark brown-black, ascending .... **12. *neckeri***
Lower surface with a distinct network of dark brown-black, flat veins with elongate, white interstices; upper surface not pruinose, rarely cracked; apothecia flat, ± rounded, brown, horizontal .... **16. *horizontalis***

1. P. leucophlebia (Nyl.) Gyelnik (1926)

Thallus forming patches to 20 cm diam.; lobes 2–4 cm wide, rounded; margins often distinctly wavy, crisped, and irregularly divided; upper surface bright green when wet, pale grey or tinged brownish when dry, smooth or indented, with brown, adpressed, convex, wart-like cephalodia, not easily detached; lower surface with a distinct network of veins, brown-black towards the centre, paler at the margins, with correspondingly pale or dark simple rhizines. Apothecia rather rare, ± round or saddle-shaped, on short extensions from the lobe margins, fertile lobes only partly corticate or wholly decorticate on the underside beneath the apothecia. Ascospores 47–92 × 5–6 μm, 3- to 7-septate, colourless or pale brown. Tenuiorin, methyl gyrophorate, ± gyrophoric acid, 4 unidentified triterpenes (only one of which is common to *P. britannica*).

On mosses amongst rocks, often in ± calcareous habitats; widespread but local. Upland Britain. Widespread in N. Hemisphere.

Characterized by the bright green colour when fresh, large rounded lobes, the adpressed cephalodia on the upper surface which are difficult to remove, and the distinct network of dark brown veins on older parts of the thallus.

2. P. britannica (Gyelnik) Holt.-Hartw. & Tønsb. (1983)

Like **1**, but the lower surface usually without or with indistinct veins and the cephalodia on the upper surface (Fig. 30a) are concave-button- or saucer-shaped with free entire or somewhat crenulate, rarely ± proliferating margins, easily detached leaving white scars; rhizines often few, in scattered groups, bottle-brush-like. Tenuiorin, methyl gyrophorate, ± gyrophoric acid, phlebic acid A and B, and 2 unidentified terpenoids.

On bark or mossy rocks and walls in damp, sheltered habitats, often in woods avoiding calcareous sites; local. S.W. & N.W. England, N. Wales (Merioneth), Scotland. Norway, W. North America.

The more continental *P. aphthosa* (L.) Willd. (1787) has been incorrectly recorded for Britain and has cephalodia which are adpressed, convex, somewhat corrugate, and not readily detached from the surface, the lower surface is blackish towards the centre, woolly but without rhizines and, unlike *P. leucophlebia* (**1**), the apothecia are uniformly corticate on the underside. The relationship between *P. leucophlebia*, *P. aphthosa* and *P. britannica* needs further investigation.

A blue-green morphotype of *P. britannica*, which develops from the cephalodia and may become independent, has been recorded (Argyll, E. Ross, W. Sutherland). The specimens are to 3 cm across, have crenate or dentate margins with erect hairs; the upper surface is blue-grey with a characteristic network of pale interconnected spots, the underside is blackish, densely felted with scattered bundles of simple rhizines; secondary attached lobules of the green morphotype are usually present.

3. P. venosa (L.) Hoffm. (1789)

Thallus to 2 cm diam., of a single, rounded or mussel-shaped lobe or of many, discrete or overlapping fan-shaped lobes; lobes 0.5–2 cm wide, often thicker than 1 mm; margin entire or ± indented, rather regular; upper surface bright green when wet, grey-green when dry becoming pale to dark brown in the herbarium, smooth, ± shiny; lower surface with conspicuous, fan-shaped, pale to dark brown, radiating veins, without rhizines, tomentose; cephalodia scattered, associated with veins, superficial, dark brown or grey, rounded or irregular and wart-like. Apothecia usually present, 1–6 per lobe, marginal, oval, wider than long, black-brown, horizontal, flattened. Ascospores 30–47 × 7–8 μm, 3-septate, fusiform, colourless or pale brown. Tenuiorin, methyl gyrophorate, ± gyrophoric acid, phlebic acids A and B, 2 unknown terpenoids.

On calcareous schists, epidiorite, mine spoil heaps, more rarely on limestone, chiefly in upland sites; very local. Mainly N. England, Scotland. Widespread in N. Hemisphere.

A distinctive, neat, small, usually fertile, bright green species with conspicuous, fan-shaped, brown veins on a pale background on the lower surface. *P. britannica* **(2)** and *P. leucophlebia* **(1)** are much larger with cephalodia on the upper surface. A blue-green morphotype frequently occurs separately or sometimes with the green morphotype; this consists of small, ± ascending, overlapping, brown-black squamules, up to 1 mm long, semi-circular or ligulate, photobiont *Nostoc*, evenly distributed throughout inner part of lobe. This phase resembles extreme, compacted, reduced morphs of *Leptogium gelatinosum*.

4. P. membranacea (Ach.) Nyl. (1887)

Thallus often forming broadly radiating patches to 30 cm or more diam., grey or often brownish grey; lobes to 3 cm wide, discrete or usually ± contiguous, elongate or rounded, thin; margins uneven, down-turned, entire, ± indented or ± lacerate and radiating; upper surface ± thinly to thickly white-grey tomentose, at least towards and at the margins, rarely flat, frequently ± markedly bullate-ridged, without folioles but occasionally with swollen regenerative secondary lobules; lower surface uniformly whitish, with pale, raised veins and well-developed, simple, dispersed, downy to bottle-brush-like rhizines (Fig. 20a). Apothecia roundish, becoming saddle-shaped red-brown to brown. Ascospores 65–73 × 3.5–4 μm, 3- to 5-septate, colourless or pale brown. Lichen products not detected by t.l.c.

Overgrowing or amongst mosses on the ground, mossy rock faces, tree trunks and buttresses and rocks; common. Throughout British Is. Europe, Asia, Macaronesia, North America.

For differences from *P. canina* **(5)**, see below; for specimens with schizidia, see *P. praetextata* **(6)**.

5. P. canina (L.) Willd. (1787)

Like **4**, but with more flattened, less pronounced, smooth veins in the central parts of the thallus, more irregular rhizines, bushy or penicillately branched, floccose-woolly-fleecy (as in a shag-pile carpet), often confluent at the base (Fig. 20b).

On sandy or calcareous soils in oligotrophic habitats; rare. Scattered throughout British Is. Europe, North & South America, Asia, Africa.

Prefers more basic or calcareous conditions.

6. P. praetextata (Flörke ex Sommerf.) Zopf (1909)

Like **4**, but with few to numerous schizidia, scattered or in dense, overlapping clusters along margins of older parts of thallus and along fractures on thallus surface; rhizines as in *P. membranacea* but often remaining only downy (Fig 20c) and often becoming brown. Apothecia rounded, ± elongate or saddle-shaped.

On mossy tree trunks and bases, rotting logs, damp rocks and soil, often in sheltered situations; common. Throughout British Is. Widespread in the N. Hemisphere.

P. membranacea **(4)** and *P. praetextata* are probably a species-pair and are sometimes difficult to separate; the development of schizidia in the latter, although variable, is diagnostic.

7. P. rufescens (Weis) Humb. (1793)

Thallus rosette-forming or ± fragmented, to 20 cm diam.; lobes to 1–5 cm wide, ± radiating, contiguous or somewhat overlapping; margins even, crinkled or crisped, ± elevated, ± swollen; centre adpressed, often with adventitious lobules and small adpressed lobes; upper surface thickly tomentose especially towards the margins, brownish, ± white-grey-pruinose or sometimes patchily grey-white frosted in older parts, flat or wavy, rarely ± bullate when well developed; lower surface with numerous wide, flattened veins, not conspicuously hairy, rarely somewhat scabrid, usually darkening towards thallus centre, and with dark, branched and coalescing, ± fasciculate rhizines. Apothecia frequent, often large and conspicuous, elevated and recurved, on lateral ends of main lobes, usually saddle-shaped with an inrolled, coarsely crenulate or denticulate, uneven thalline margin. Ascospores 40–70 × 4–5 μm, 3- to 5(-6)-septate. No lichen products detected by t.l.c.

Preferring rather basic soils; rather common. Throughout British Is. Cosmopolitan.

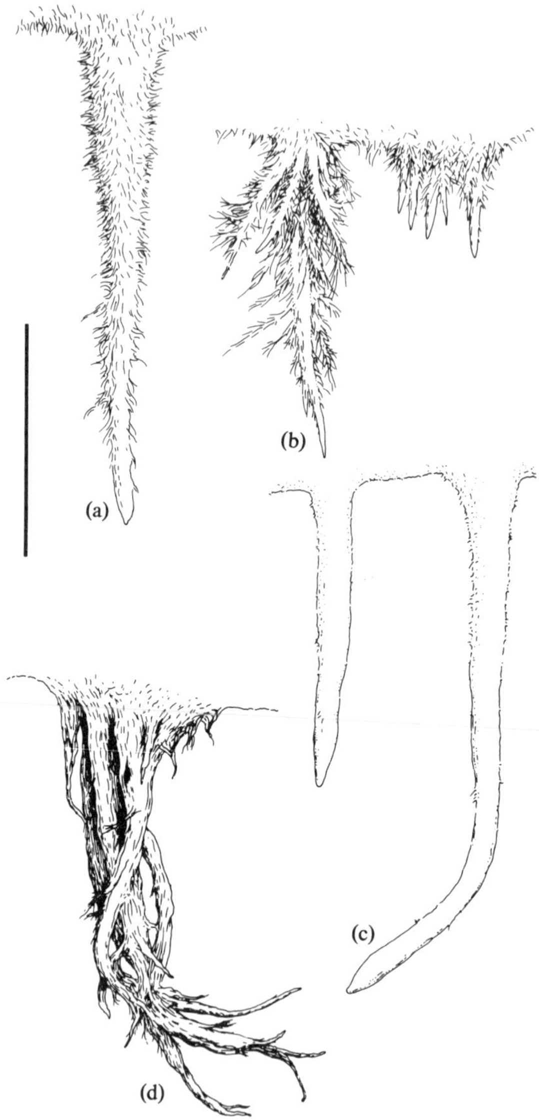

Fig. 20. *Peltigera* rhizines. (a) *P. membranacea*; (b) *P. canina*; (c) *P. praetextata*; (d) *P. horizontalis*. Scale = 1 mm.

P. rufescens is sometimes confused with fertile *P. didactyla* **(9),** which generally has smaller apothecia on entirely erect lobes, whereas in *P. rufescens* the thallus, at least in part, is usually prostrate. Sterile specimens of *P. didactyla* are sorediate. *P. membranacea* **(4)** and *P. canina* **(5)** have longer, more bullate-uneven thalli and down-turned, thin margins; those in *P. rufescens* are somewhat upturned and swollen; *P. rufescens* generally occurs in more basic, sunny habitats, often with *P. canina*.

8. P. ponojensis Gyelnik (1931)

Like 7, but upper surface tomentose or glabrous towards centre and the lower surface is whitish with pale, raised veins and pale simple rhizines.

On chalk grassland, calcareous soil an old lead mine spoil tips and on mica-schist boulders; very rare. S.W. England (Wiltshire, Fyfield Down), C. England (Derbyshire), C. Scotland (Breadalbane Mtns). Europe, Asia, North America.

The status of this species and its separation from *P. rufescens* needs further study.

9. P. didactyla (With.) Laundon (1984)

P. spuria (Ach.) DC. (1805)

Thallus rosette-shaped or consisting of a few irregular lobes, to 3 cm diam., rarely larger, but often in small, coalescing groups; lobes small, c. 1 cm wide, ear-shaped when young with ascending margins, becoming markedly or entirely ascending when older with apothecia held ± vertically; upper surface finely grey-white tomentose; when young, plants with laminal, rounded or irregular, pale, ulcerose, whitish soralia, to 2.5 mm diam., discrete, later becoming confluent, ± filled with pale blue-grey to brown coarsely granular soredia, disappearing with the formation of apothecia; lower surface with distinct, raised, pale cream or faintly flesh-ochraceous veins, anastomosing with well-developed interstices; rhizines simple, downy, occasionally ± bottle-brush-like; apices often somewhat penicillate. Apothecia ± oblong, saddle-shaped, red-brown or brown, delicately crenulate and denticulate at the margin. Ascospores 55–75 × 3.5–4.5 μm, 3- to 7-septate. No lichen products detected by t.l.c.

Characteristic of recently disturbed sites, e.g. cuttings and earth banks, roadsides, wall-tops, old soil in flower pots, old bonfire sites, grey dunes, lawns, pathsides, decaying logs, mine spoil heaps and quarries; common. Throughout British Is. Cosmopolitan.

Can be confused with the horizontally spreading *P. rufescens* **(7)**, also with *P. membranacea* **(4),** which is larger and has a bullate surface. *P. didactyla* is unusual in that a juvenile sorediate phase is superseded by a fertile, non-sorediate phase. As the thalli become fertile the soralia are reduced to non-sorediate scars and eventually disappear.

10. P. malacea (Ach.) Funck (1827)

Thallus to 10 cm diam., rather thick; lobes to 3 cm wide; margins raised, undulate; upper surface ± even, downy or greasy, bluish green-tinged, brownish when dry, deep olive-green or brownish when wet, smooth or often thinly scabrid-tomentose with erect hairs especially towards lobe margins; lower surface pale at margins, soon becoming brown-black, ± uniformly tomentose, veins and rhizines mostly absent, paler interstices few or absent. Apothecia orbicular, adpressed, brown-red, ± crenulate at margin. Ascospores 55–70 × 5–6 μm, 3- to 5-septate, elongate-fusiform. Tenuiorin, methyl gyrophorate, ± gyrophoric acid, unidentified terpenoids (as in *P. leucophlebia* **(1)**).

On oligotrophic soil amongst mosses; in acid dunes and conifer plantations on sand. Scotland (E. Sutherland; Fife). Widespread in the N. Hemisphere.

Distinguished by the continuous grey-brown tomentum of the underside.

11. P. lactucifolia (With.) Laundon (1984)

P. hymenina (Ach.) Delise ex Duby (1830)

Thallus to 10(-20) cm diam., wide-spreading; lobes to 2 cm wide, overlapping, ± ascending at apices and margins; margins not crisped or with folioles, small regenerative lobules occasionally present; upper surface ± matt, occasionally partly glossy, grey or slightly brownish, without or rarely with a few slash-like cracks; lower surface with ± well defined, pale to deep ochre veins extending towards margins of lobes, flattened, wide, somewhat ill-defined, with frequent pale interstices; rhizines pale, slender, simple, unbranched, occasionally splitting towards ends. Apothecia to 5 × 3 mm, pale brown or red-brown, ± on short extensions from the lobes. Asci (Fig. 34d). Ascospores 60–80 × 4–5 μm, 3- to 7-septate, oblong-elongate, saddle-shaped, acicular. Tenuiorin, methyl gyrophorate, ± gyrophoric acid, 7β-acetoxyhopan-22-ol (peltidactylin) (T1), 15α-acetoxyhopan-22-ol (dolichorrhizin) (T2), ± hopane-6α,22-ol (zeorin) (T3), ± hopane-7β,22-ol (T4).

On soil, mosses and rocks, sheltered banks, lawns and tree bases in damp situations; common. Throughout British Is. Europe, Africa, North America.

A very variable species with lobes ranging from small and very numerous and dissected to wide, entire and sparse; frequently sterile. The ochre-coloured veins with pale rhizines are diagnostic. For differences from *P. neckeri* **(12)** and *P. polydactyla* **(13)** see those species.

12. P. neckeri Hepp ex Müll. Arg. (1862)

Like **11**, but upper surface typically thinly white-grey-pruinose, often only towards the margins, shining, often with ± elongate, slash-like cracks with ± raised edges exposing medulla; lower surface pale at margins, black towards centre, with a few, wide, mostly diffuse, dark veins with white interstices, rhizines mostly few, diffuse, confluent. Apothecia mostly short-stalked, discs dark, brown-black. Tenuiorin, methyl gyrophorate, ± gyrophoric acid, 15α-acetoxyhopan-22-ol (dolichorrhizin) (T2), hopane-6α,22-diol (zeorin) (T3), hopane-7β,22-diol (T4) and two unidentified terpenoids.

On mosses on soil, including lead-contaminated soil, or rocks, grass tennis courts, ± basic dunes; local. Scattered in British Is. Widespread in N. Hemisphere.

The pruina is often only present on a few lobes and should be carefully searched for; the unusual chemistry is diagnostic and the apothecia are usually darker than in related species. The dark veins separate this species from *P. lactucifolia* **(11)** where they are persistently ochre-coloured.

13. P. polydactyla (Necker) Hoffm. (1790)

Like **11**, but ascending lobes are often markedly crisped-contorted, becoming squamulose; lower surface with very distinct, dull to dark brown anastomosing veins present to margins with numerous pale interstices; rhizines often confluent when young, pale to dark brown. Tenuiorin, methyl gyrophorate, 15α-acetoxyhopan-22-ol (dolichorrhizin) (T2), hopane-6α,22-diol (zeorin) (T3) and ± gyrophoric acid.

Amongst mosses, on tree bases and rocks; local. N. England, Scotland. Widespread in N. Hemisphere.

14. P. elisabethae Gyelnik (1927)

Thallus forming multilobate patches to 10 cm diam. or more; upper surface grey-brown in exposed situations, metallic-shiny, medulla thick, the upper part peeling near margins to form crisped and curled isidioid schizidia, the surface often ± cracked as in *P. neckeri* **(12)**; lobes 0.5–1 cm wide, margins ± crisped, forming, when well developed, flattened or nodular schizidia, often with isidia or lobules; lower surface dark brown-black, without veins, but with small whitish interstices centrally; rhizines dark, fasciculate, rather thick. Apothecia rare (not observed in British material), horizontal, with a flat, rounded or oval disc as in *P. horizontalis*. Ascospores 25–45 × 2–3 μm, 3-septate. Tenuiorin, hopane-6α,22-diol (T3) and ± hopane-7β,22-ol (T4), gyrophoric acid, methyl gyrophorate.

On mossy rocks. C. Scotland (Breadalbane Mtns; Angus, Caenlochan). Europe, Asia, North America.

Characterized by the thick thallus, crisped marginal schizidia, and the felted, dark underside with scattered, pale interstices. *P. horizontalis* **(16)** has distinct anastomosing veins and entire margins to the lobes.

15. P. degenii Gyelnik (1927)

Thallus wide-spreading, to 10(-20) cm diam.; upper surface bluish grey, occasionally brownish, glossy, tomentum absent, marginal isidia and lobules often present, smooth or ± bullate-wrinkled; lower surface white-cream, with narrow, ± conspicuous, elevated, whitish to pale yellow-brown veins and slender, simple or little-branched, pale cream or greyish brown rhizines with a ± smooth surface, occasionally slightly pubescent. Apothecia round, on erect, short, narrow lobes, often recurved. Lichen products not detected.

On mossy rocks, fallen trunks in ravines, rotting logs and tree bases. Scotland (Fife; S. & C. Highlands), N. Ireland (Londonderry). Widspread in the N. Hemisphere.

The glabrous upper surface, uniformly pale, raised veins and simple rhizines as well as the absence of lichen products are diagnostic.

16. P. horizontalis (Huds.) Baumg. (1790)

Thallus forming wide-spreading patches to 10 cm or more diam.; upper surface bluish-grey, often tinged brown, ± glossy; lobes to 2 cm wide, margins ± entire or indented, rarely sublobulate and crisped; lower surface whitish towards margins becoming darker towards centre, with conspicuous anastomosing dark brown-black veins with numerous starkly white, elongate interstices; rhizines fasciculate, often brown-black or black (Fig. 20d). Apothecia ± rounded, occasionally broader than long, horizontal, discs flattened. Ascospores 30–46 × 6–7 μm., 3-septate. Tenuiorin, methyl gyrophorate, ± gyrophoric acid, hopane-6α,22-diol (T3), hopane-7β,22-diol (T4) and unidentified terpenoids.

On mossy trunks of old trees, rotting logs and stumps, and mossy rocks, often in sheltered valleys in woodlands with a long history of ecological continuity. Throughout W. British Is., extending locally eastwards. Widespread in the N. Hemisphere.

A distinctive old woodland indicator species, characterized by the rounded apothecia held horizontally. The dark brown anastomosing veins bearing fasciculate dark rhizines are also diagnostic; *P. elisabethae* **(14)** differs in having schizidia and a more evenly felted underside.

17. P. collina (Ach.) Schrader (1801)

Thallus forming radiating patches to 10 cm diam., usually less; lobes 1–1.5 cm wide, elongate, radiating or ± randomly arranged (in old specimens), adpressed for most of length; margins ± raised, undulate, ± entirely sorediate; upper surface grey, blue-grey, rarely brownish, smooth, ± shiny, occasionally somewhat scabrose, rarely pruinose towards margins; soralia marginal, pale grey or slightly bluish, rather coarsely granular and partly pigmented, finger-like and partly corticate when young; lower surface with pale brown veins and ± simple rhizines. Apothecia small, to 2 mm diam., rare, rounded or ± elongate, eventually saddle-shaped, on short, ascending lobes, dark brown or blackish. Ascospores 65–85 × 4–5 μm, 3(-5)-septate. Tenuiorin, methyl gyrophorate, ± 7β-acetoxyhopan-22-ol (T1), hopane-6α,22-diol (T3), hopane-6α,7β,22-triol (T6), and unidentified terpenoids.

On mossy trunks and rocks. N. & W. British Is. Europe, Asia, Africa, North & South America.

The only *Peltigera* with marginal soralia. When sterile may be confused with *Nephroma parile*, which is almost naked and lacking veins on the lower surface and has a different chemistry.

18. P. lepidophora (Nyl.) Bitter (1904)

Thallus to 1.5 cm diam., level with or ± sunken into the substratum; lobes to 1 cm wide, concave to ear-shaped with ± ascending, entire or ± eroded, sometimes inrolled margins; upper surface smooth, naked in part or ± thinly tomentose; isidia to 1 mm diam., numerous, button-like, resembling cephalodia, crowded and contiguous, sometimes overlapping or ± widely dispersed; lower surface with often rather indistinct, whitish to pale grey, rarely pale brown anastomosing veins and discrete unbranched pale rhizines. Apothecia not known in British Is. No lichen products detected by t.l.c.

On flat, mossy rock ledges in a river gorge. C. Scotland (E. Perth). Europe, Asia, North America, Hawaii.

Distinguished by the small size and numerous cephalodia-like isidia on the upper surface. Rather similar to young stages of *P. didactyla* **(9)** which is sorediate.

19. P. scabrosa Th. Fr. (1861)

Thallus wide-spreading, to 2–5(-10) cm diam.; lobes 3–5 cm wide, subcoriaceous; margins ± raised, rounded; upper surface roughened, bluish green to brownish green when moist, not tomentose; lower surface with pale brown, often indistinct veins and short, penicillate rhizines. Apothecia not seen in British material.

On acid soil, montane; very rare. N. Scotland (Highlands, Ross, Orkney Is.). Widespread in the N. Hemisphere.

Characterized by the roughened upper surface.

PERTUSARIA DC. (1805)

Thallus crustose, superficial, moderately thick to thick or thin and ± immersed, continuous to rimose-cracked, fissured-areolate or warted. **Soralia** or **isidia** frequent. **Cortex** thin, cartilaginous, paraplectenchymatous, of thick-walled, septate, agglutinated hyphae. **Photobiont** chlorococcoid. **Ascomata** apothecia, either ± sessile with an expanded, open disc, or almost closed and perithecia-like and 1 or several immersed within a fertile wart. **Disc** sometimes occluded with granular soralia and appearing soralia-like or pruinose. **Hamathecium** of paraphyses, lax, branched and richly anastomosed. **Asci** 1- to 8-spored, broadly cylindrical, apex with a broad ocular chamber, outer sheath K/I+ blue, otherwise K/I−, with an inner extensible layer, *Pertusaria*-type (Fig. 34e). **Ascospores** to more than 250 μm, thick-walled; wall uniform or at times with radiating canals to the surface, or ornamented. **Conidiomata** pycnidia, rare. **Conidia** straight, ± acicular or bacilliform. **Chemistry**: xanthones, fatty acids, depsides or depsidones present in many species. **Ecology**: varied, if on rocks, not on limestones and other strongly calcareous substrata.

The related arctic-alpine *Varicellaria rhodocarpa* (Körber) Th. Fr. (1871), characterized by ascospores 200–400 × 70–130 μm, 1-septate and a C+ red (lecanoric acid) thallus was reported last century as amongst mosses from N.E. Scotland (Braemar, summit of Ben Avon); *Pertusaria carneopallida* (Nyl.) Anzi ex Nyl. (1868) occuring in Scandinavia and the Alps, characterized by a thin, immersed thallus with flat, *Lecanora*-like, warted, pinkish brown disc with a lacerate border was reported from N. Scotland (Argyll) last century. The records of both these species are based on material from outside the British Is.

LITERATURE: Dibben (1980), Hanko (1983).

1	Isidia or soralia present; apothecia often absent 2
	Isidia and soralia absent; apothecia often present 23
2(1)	Isidia present, fine granule-like to coarse and papilla-like 3
	Soredia present, delimited, punctiform to wide-spreading and effuse 11
3(2)	Thallus entirely composed of conspicuous, finger-like isidia; on mosses; montane 4
	Thallus clearly crustose, ± developing isidia; on rock or bark; lowland or montane 5
4(3)	Isidia 0.4–1 mm thick, uniformly white; asci 1-spored ..... **8. *dactylina***
	Isidia 0.3–0.5 mm thick, white, ± browned at the apices; asci 8-spored ..... **26. *oculata***
5(3)	Thallus white, grey or dark grey, C−; isidia distinct, not abrading into soredia 6
	Thallus green-yellow, C+ orange; isidia small, sometimes abrading into sorediate patches ..... **14. *flavida***
6(5)	Isidia to 2 × 0.1–0.5 mm, not dominant; thallus white to grey, taste mild; medulla KC− or yellow to red 7
	Isidia 2–3 × 0.5–2 mm, often predominant; thallus dark grey, taste bitter; medulla KC+ purple-pink ..... **23. *melanochlora***
7(6)	Thallus and medulla Pd+ orange, K+ red (crystals) 8
	Thallus and medulla Pd− or yellow, K− or + yellow-orange (no crystals) 9
8(7)	Isidia constricted at the base, rounded or top-shaped, ± browned at the apices, leaving non-sorediate pits when shed ..... **29. *pseudocorallina***
	Isidia not constricted at the base, elongate, not tinged brown at the apices, not leaving pits when shed ..... **5. *coccodes***
9(7)	Thallus K+ yellow to orange, on rock or bark 10
	Thallus K−; on mosses on the ground ..... **27. *panyrga***
10(9)	On rock; isidia hard, the surface ± shiny ..... **7. *corallina***
	On bark; isidia soft and brittle, the surface matt ..... **6. *coronata***
11(2)	Thallus ± yellow-grey to yellow-green, C+ orange 12
	Thallus white, grey to dark grey, C− or + red 14
12(11)	On rock; soralia present from the start 13
	On bark; soralia derived from eroded areas of isidia ..... **14. *flavida***
13(12)	Thallus thin, minutely cracked-areolate, grey, in patches ± yellow-grey; soralia 0.1–0.3 mm, fleck-like, pale grey-white, Pd + orange ..... **13. *amarescens***
	Thallus thick, smooth to coarsely warted-areolate, yellow- or green-grey; soralia 0.3–0.8 mm, ± rounded, pale yellow-green, Pd− ..... **12. *flavicans***
14(11)	Thallus and/or soralia C+ red 15
	Thallus and/or soralia C− 17

15(14) Thallus white to grey or pale brown; prothallus not distinctive; soralia 0.3–1.5 mm diam., white or pale brown 16
Thallus pale blue-grey; prothallus prominent, white; soralia 1–1.5 mm diam., concolorous, convex .... **18. *hemispherica***

16(15) On soil; thallus granular-warted, grey (becoming brown in the herbarium); soralia rounded, ± elevated, Pd+ orange .... **16. *geminipara***
On rock; thallus smooth, regularly areolate-cracked, grey-white; soralia rounded, white, ± flat, immersed; Pd− .... **20. *lactea***

17(14) Soralia KC− or yellow, orange or red; tastes mild 18
Soralia KC+ purple-violet; tastes bitter .... **2. *amara***

18(17) Soralia Pd+ yellow to rust-red, K− or yellow to red 19
Soralia Pd−, K−, rounded-disc-like to warted .... **1. *albescens***

19(18) Soralia Pd+ rust-red, K− or brownish (no crystals), rounded, ± punctiform on a thin, grey to dark grey thallus 20
Soralia Pd+ yellow to orange-red, K+ yellow to red (crystals) on a thick, warted, pale grey to grey thallus 21

20(19) On rocks .... **9. *dealbescens***
On bark or wood .... **30. *pupillaris***

21(19) Soralia present; isidia, if present, rare, inconspicuous, soon becoming sorediate 22
Soralia absent, with small, non-sorediate depressions; usually with peg-like brown-tipped isidia .... **29. *pseudocorallina***

22(21) Soralia conspicuous, white; thallus thick, rough coarsely cracked-areolate, pale grey to grey .... **10. *excludens***
Soralia inconspicuous, derived from the breakdown of coarse low papillae; thallus thin to medium thick, sparingly areolate-cracked, greenish grey .... **21. *lactescens***

23(1) Fertile warts soralia-like, each with a single apothecium; disc ± widely exposed, densely white-granular-pruinose 24
Fertile warts not resembling soralia, containing a single or several apothecia; disc pore-like or becoming ± exposed, not or lightly pruinose 27

24(23) Disc C− 25
Disc C+ carmine-red .... **32. *velata***

25(24) Thallus K−; on bark or associated mosses 26
Thallus K+ red (crystals); on rock .... **11. *monogona***

26(25) Fertile warts white, Pd+ rust-red, K+ yellow; disc wide; true exciple continuous .... **24. *multipuncta***
Fertile warts grey, concolorous with the thallus, Pd−, K−; disc narrow; true exciple irregular, crenate .... **25. *ophthalmiza***

27(23) On bark or overgrowing mosses and low vegetation 28
On rock 35

28(27) Overgrowing mosses and low vegetation; montane 29
On bark; montane or lowland 31

29(28) Disc pore-like; asci (2-)4-spored; epithecium K− 30
Disc expanded to 1 mm diam.; asci 1-spored; epithecium K+ violet .... **3. *bryontha***

30(29) Thallus Pd+ orange-red, K+ red (crystals); ascospores 70–125 × 25–50 μm **17. *glomerata***
Thallus Pd−, K− (no crystals); ascospores 55–70(-85) × 30–40 μm .... **33. *xanthostoma***

31(28) Asci 8-spored; epithecium K+ violet; disc ± pore-like often becoming expanded, one per wart .... **19. *hymenea***
Asci 2- to 4(-6)-spored; epithecium K−; disc remaining pore-like 32

32(31) Fertile warts usually conical, spreading at the base, rarely semiglobose and then pale pink, generally with a single apical pore-like disc; asci 2- or 4-spored; ascospores 55–125(-140) × 20–40(-45) μm 33
Fertile warts semi-globose, ± constricted at the base, with 1–7 pore-like discs; asci 2-spored; ascospores 140–230(-330) × 40–80(-90) μm ..... **28. *pertusa***

33(32) Asci 4-spored; ascospore wall smooth 34
Asci 2-spored; ascospore wall ± faintly ornamented ..... **31. *pustulata***

34(33) Thallus even, continuous, grey-white to yellow- or green-grey; fertile warts low, spreading at the base, widely dispersed on smooth bark ..... **22. *leioplaca***
Thallus irregular, in patches, grey-white; fertile warts semi-globose constricted at the base, often crowded; usually on old *Calluna* stems ..... **33. *xanthostoma***

35(2) Asci 8-spored; disc ± expanded or irregular-stellate; epithecium K+ violet 36
Asci <8-spored; disc remaining pore-like; epithecium K− or rarely violet 37

36(35) Apothecia *Aspicilia*-like, immersed, the warts inconspicuous; disc irregular-stellate; ascospores 25–42 × 11–24 μm; thallus dark grey ..... **4. *chiodectonoides***
Apothecia raised in distinct warts; disc pore-like to widened and rounded; ascospores 60–110(-120) × (20-)30–50 μm; thallus pale to dark green-grey or yellow-grey **19. *hymenea***

37(35) Ascospores colourless to dark brown; wall thickened, elaborately channelled; apothecia inconspicuous, little elevated 38
Ascospores colourless; wall undifferentiated; apothecia ± elevated 39

38(37) Ascospores colourless, K−; thallus yellow-green-grey, C+ orange ..... **15. *gallica***
Ascospores ± dark brown, K+ violet; thallus grey, C− ..... **21. *lactescens***

39(37) Thallus Pd+ yellow to orange, K+ yellow to red; fertile warts with several, 1–7, dark pore-like discs 40
Thallus Pd−, K−; fertile warts with a single, pale, pore-like disc ..... **33. *xanthostoma***

40(39) Thallus K+ red (crystals), grey-white ..... **29. *pseudocorallina***
Thallus K+ yellow to orange (no crystals), greenish grey ..... **28. *pertusa***

1. P. albescens (Huds.) M. Choisy & Werner (1932)

Thallus thin to moderately thick, pale grey to dark greenish grey, ± waxy, often zoned at the margin; upper surface smooth to coarsely warted, often rimose-cracked; taste mild; soralia rounded, scattered or ± contiguous, very variable, large, mostly more than 1(-4) mm diam., typically concave and marginate, disc-like, paler than the the thallus, or smaller, white to pale grey-white, coarsely granular-sorediate, more rarely convex becoming ± crowded, pseudocoralloid outgrowths. Apothecia to 4 mm diam., very rare, usually several combined in large sorediate-pruinose warts, concave, opening with deeply lacerate, white thalline margins; discs densely white-pruinose. Asci 1(-2)-spored. Ascospores 170–300 × 50–115 μm. Thallus Pd−, K−, KC−, C−, UV− (fatty acids).

On bark of parkland, woodland and wayside broad-leaved trees and, more rarely, conifers, rarely on siliceous rocks, porous sandstones, sometimes overgrowing mosses and plant detritus on the ground; very frequent and pollution tolerant. Throughout British Is. Europe, North America.

Characterized by the grey thallus, zoned margin and often large, concave or flat, KC− soralia. Very variable, especially in the form of the soralia which may be broad and discoid, smaller, convex and white, or irregular, contiguous and warted. *P. albescens* **var. *corallina*** (Zahlbr.) Laundon (1963) has coarsely nodulose-warted-granular, poorly defined soralia, often confluent in older parts of the thallus. *P. polythecia* (Taylor) Erichsen (1936), endemic to S.W. Ireland, morphologically closely resembles var. *corallina* and may be conspecific, but has a different series of fatty acids, similar to the range in *P. panyrga* **(27)**. Specimens with small, white soralia may be confused with *P. amara* **(2)**.

2. P. amara (Ach.) Nyl. (1873)

Like **1**, but thallus often smaller, less well-zoned at the margin, the soralia are typically smaller, sometimes punctiform, 0.5–1.5 mm diam., discrete or occasionally ± contiguous, rather regularly scattered over the thallus and with an extremely bitter taste due to picrolichenic acid. Apothecia very rare. Asci 1-spored. Ascospores 130–150 × 40–50 μm. Soralia Pd− or + red, K−, KC+ violet, C−, UV− (picrolichenic, ± protocetraric acids).

On a wide range of broad-leaved trees, rarely conifers, rarely on the ground and on low vegetation; common. Throughout British Is. Europe, North America.

A morph occasionally occurs on ± porous rocks which has a thin thallus and a markedly radiating prothallus. In the **f. *pulvinata*** (Erichsen) Almb. (1948) [syn. *P. pulvinata* Erichsen (1936)] the thallus forms ± compact cushions; upper surface warted-papillate, the apices and ridges breaking down into small, elevated, punctiform soralia, never forming discrete round soralia. On mossy rocks or rarely trees; infrequent. S. England (Devon; Hampshire, New Forest), Is. of Scilly, W. Ireland (Galway, Sheep's Head). C. Europe. The doubtful *P. polythecia* (see **1**), morphologically resembles this variety but is KC−.

3. P. bryontha (Ach.) Nyl. (1861)

Thallus thin to moderately thick, continuous, white-grey, smooth to wrinkled-nodulose, rarely tuberculate, margin entire, not zoned. Isidia absent, but small, coarse papillae occasionally present. Fertile warts abundant, crowded with a single expanded, roughened apothecial disc, (0.2-)1(-2.5) mm diam., yellow-brown to dark brown-black, at first sunken, later flat-convex, with a persistent, smooth or somewhat warted, irregular, flexuous thalline exciple; epithecium brown-black, K+ violet. Asci 1-spored. Ascospores (110-)150–210(-230) × (40-)60–90(-100) μm, wall (10-)15(-25) μm thick, smooth, with an outer, uniformly thickened wall and an unequal inner wall. Thallus Pd+ orange-red, K+ yellow, KC+ red, C+ red, UV± glaucous or orange (xanthone, stictic acid and accessory substances, gyrophoric acid).

Encrusting mosses and low alpine vegetation in ± calcareous habitats; very rare. N. Scotland (Highlands); only recently authentically recorded from Scotland (Coire Cheap, Ben Alder (1980)). European Alps, and arctic tundra.

Characterized by the papillate, grey thallus and abundant apothecia with an expanded disc. *P. oculata* **(26)** has a coralloid, isidiate thallus, apothecia with black discs and a Pd+ rust red medulla. A specimen of *P. bryontha* labelled as collected from Cairngorm last century contains only gyrophoric acid and a xanthone; this chemotype has otherwise only been reported from N. Scandinavia; this earlier Scottish record is probably based on material from that area.

4. P. chiodectonoides Bagl. ex Massal. (1856)

Thallus in small patches or sometimes wide-spreading, thin to moderately thick, margin ± entire, not zoned; upper surface dark grey, blue- or brown-grey, smooth or roughened, distinctly cracked-areolate; areoles (0.2-)0.6(-1.5) mm diam., ± flat. Fertile warts (0.2-)0.7(-1.5) mm diam., ± numerous and well dispersed, rarely confluent, flat. Apothecia 1–3(-8) per wart; discs (0.1-)0.3(-0.6) mm diam., black or dark grey, immersed, ± widened when mature, often roughened and grey-white-pruinose and appearing irregular or ± stellate with a ± paler, irregular, not or slightly raised, common margin; epithecium dark brown-black, K± deep violet. Ascospores (20-)25–42 × 11–24 μm, wall 3–4(-5) μm, even, smooth. Conidia 7–10 × 0.5–1 μm. Thallus Pd+ orange-red, K+ yellow, KC+ yellow, C−, UV− (stictic, ± norstictic and ± constictic acids).

On sunny, siliceous and ultrabasic rocks, especially serpentine and basalt, mainly in coastal sites; very local. S.W. England (Cornwall), W. Scotland north to Orkney. Europe, North America.

Characterized by the flat, dark grey, areolate-cracked, K+ yellow thallus, the fertile warts containing 1–3(-8) black discs and the K+ violet epithecium. Superficially resembling *Aspicilia cinerea* aggr. which, however, has more tuberculate, convex areoles and more elevated, rounded apothecia.

5. P. coccodes (Ach.) Nyl. (1857)

Thallus dull greenish or brownish grey, thin to moderately thick, continuous, often bordered by a pale brown or white prothallus, not zonate; upper surface smooth to ± warted, ± rimose-cracked; isidia often densely covering parts of the thallus, <0.5 mm wide, globose to ovoid-elongate, rather coarse, ± erect, smooth and firm, often with dark grey or brownish apices, mostly simple but occasionally branched-coralloid, when abraded leaving pits. Apothecia 1–1.5 mm diam., very rare, in smooth, convex-globular warts; disc punctiform. Asci 2-spored. Ascospores 110–145 × 40–58 μm. Thallus Pd+ orange, K+ yellow-red, KC−, C−, UV− (norstictic acid).

On ± well-lit, often somewhat nutrient-rich bark of deciduous trees in wayside and woodland sites, rarely on wood or smooth, siliceous rocks; frequent. Widespread in S. & E. British Is., rare in W. & N., scarce in Scotland. W. & C. Europe to C. Fennoscandia.

Characterized by the usually conspicuously isidiate thallus and K+ red reaction. Fertile thalli are often only sparingly isidiate. See also *P. coronata* **(6)**. *Ochrolechia subviridis* has softer, whitish isidia often dissolving into soredia and is K−, C+ orange-red. When on rocks, typically overhung by trees, *P. coccodes* is distinguished from *P. pseudocorallina* by the greener-grey thallus and the more regular, smooth, distinct isidia without ± brownish tips.

6. P. coronata (Ach.) Th. Fr. (1871)

Like **5**, but the isidia tending to be longer, more cylindrical, often thinner, decumbent and unorientated. Not known fertile in British Is. Thallus Pd+ orange, K+ yellow, KC+ yellow, C−, UV± pale to dark orange (coronaton, stictic, ± norstictic and ± constictic acids).

On wayside, more rarely woodland, broad-leaved trees; very local. S. & W. British Is., most frequent in C. Scotland. Europe.

7. P. corallina (L.) Arnold (1861)

P. areolata auct. brit., non (Ach.) Massal.

Thallus rather thick, pale to dark grey or whitish, continuous, rimose-cracked-areolate; prothallus often conspicuous, white; isidia *c.* 0.1–0.3 mm, abundant, often covering thallus, simple, cylindrical or becoming coralloid, erect or ± decumbent and unorientated. Fertile warts rare, 1.5–2 mm diam., convex, sharply constricted at base, often with an uneven to warted upper surface, enclosing 2–4(-11) apothecia; apothecial disc 0.3–0.5 mm, ± expanded, flat or ± convex, pale or dark brown, often faintly white-pruinose. Asci 2-spored. Ascospores 80–140(-230) × 40–80 μm; wall 2–3 μm, not zoned or striate. Thallus Pd+ yellow-red, K+ yellow, KC+ yellow, C−, UV− (thamnolic acid).

On dry, siliceous rocks, mainly in upland areas; very rarely on bark (*Betula*) and wood; common. Throughout British Is. Europe.

Characterized by the white-grey isidiate thallus, the isidia being cylindrical-coralloid and without brown apices, and by the presence of thamnolic acid. *P. pseudocorallina* **(29)** has a green-grey thallus with a brownish tinge, shorter, ± globose isidia with brown apices and is K+ red (crystals).

P. flavocorallina Coppins & Muhr (1992) is similar to *P. corallina*, but with longer, to 0.6 mm tall, yellowish isidia which are C+ orange, UV+ dark red (thiophaninic acid) and occurs on tops of ± calcareous, schistose boulders in E. Scottish Highlands (Angus, Caenlochan); known also from S. Sweden.

8. P. dactylina (Ach.) Nyl. (1863)

Thallus thin, white, smooth or uneven, mainly obscured by erect, crowded, isidia 1–2.5(-4) mm tall × 0.4–1 mm wide, robust, at first papillate, becoming columnar, mostly simple, at times 2- or 3-forked, occasionally becoming fused, ± constricted towards the base and blunt and rounded at apices, the apices I+ blue. Fertile warts very rare. Apothecia 1 per wart, immersed in the apices of the isidia; disc (0.2-)0.5(-1) mm diam., ± black, often ± white-pruinose and surrounded by a torn wart margin. Asci 1-spored. Ascospores 120–300 × 60–100 μm, wall 3–5(-10) μm thick, smooth. Thallus Pd+ red, K+ yellow, KC+ yellow, C−, UV− (fumarprotocetraric and ± protocetraric acids).

On soil, mosses and decaying vegetation, mostly above 900 m alt.; local. N. Scotland (Highlands). Arctic, Iceland, Greenland, Fennoscandia, Novaya Zemlya, North America.

Characterized by the robust, white isidia with pale tips, reacting Pd+ red and the montane, terricolous habitat. *P. oculata* **(26)** occurs in similar habitats but develops more numerous, slender, grey, coralloid isidia with characteristic black tips which are K+ violet.

9. P. dealbescens Erichsen (1936)

Thallus rather thin, even, at times ± fissured or cracked-areolate; prothallus indistinct, marginal, often paler; upper surface pale or dark grey; soralia ± evenly scattered, to 1 mm diam., white, punctiform, flat or somewhat, convex, rounded; isidia very rare, small, cylindrical. Apothecia unknown. Thallus Pd+ orange-rust red, K+ yellow-red, KC+ yellow-red, C−, UV− (fumarprotocetraric, succinprotocetraric and ± protocetraric acids).

On siliceous outcrops, scree and walls, mainly upland, also on bark and wood of *Pinus* and *Betula* in relict pine woodlands; fairly frequent. Widespread except in S.E. England. Europe. See p. 686.

Characterized by the small, pale, punctiform soralia and the presence of fumarprotocetraric acid. The soralia in *P. amara* **(2)** are KC+ violet and the thallus has a bitter taste; those in *P. lactea* **(20)** are C+ red; *P. excludens* **(10)** is K+ red and mild-tasting. Specimens on bark and wood in relict Scottish pine woodlands may be *P. borealis* Erichsen (1938).

10. P. excludens Nyl. (1885)

Thallus wide-spreading, rather thick, deeply rimose-cracked-areolate; prothallus ± distinct, concolorous or paler; areoles coarse, 0.5–2 mm, irregularly angular, flat or ± convex, sometimes somewhat papillate; upper surface white-grey, matt, becoming rugose-roughened; soralia 0.6–1(-1.5) mm diam., abundant, ± convex, arising from wide, low warts, coarsely granular, mostly paler than the thallus, occasionally tinged yellowish, discrete or becoming ± confluent, convex or somewhat irregularly excavate with irregular raised margins. Apothecia rare, arising within the soralia; disc pruinose, soon aborting. Ascospores unknown. Conidia 4–5 × 0.5 μm, straight. Soralia Pd+ yellow, K+ yellow-red, KC+ yellow-red, C−, UV± pale grey or pale grey-yellow (norstictic, ± salazinic, ± unknown fatty acids, ± UV+ yellow xanthone).

On sunny, siliceous rocks, especially granite, mainly coastal. W. British Is., to N. Scotland and W. Ireland. W. Europe, N. North America.

Characterized by the extensive, mainly thick, cracked, pale thallus with numerous, rounded, raised, soralia developing from the breakdown of coarse, low-convex tubercles. *P. dealbescens* **(9)** has a much thinner, not or indistinctly areolate, Pd+ orange-rust red, thallus. Possibly the sorediate counterpart of *P. monogona* **(11)**.

11. P. monogona Nyl. (1872)

Like **10**, but soralia and isidia absent, and fertile. Fertile warts numerous, arising from low tubercles, ± elevated, becoming 1.5–2.5 mm diam. Apothecia 1–4(-6) per wart, deeply immersed; disc flat, ± veiled, ± black, coarsely white-pruinose, surrounded by an irregular, often smooth and raised thalline exciple, the inner edge somewhat torn; epithecium with numerous crystals, K−. Asci 1-spored, often aborting. Ascospores 100–300 × 50–100 μm, wall rather thin, 4–5 μm, not striate or zoned. Thallus reactions and chemistry as in *P. excludens* **(10)**.

On sunny, coastal rocks. S.W. England, Is. of Scilly., Channel Is., Wales, W. Ireland. Atlantic Europe from Portugal to the Pyrénées, Macaronesia.

Characterized by the fertile warts in which apothecia are immersed with white-pruinose discs and by the K+ red thallus. The broad, thick-margined, rounded apothecia with a spreading, irregular, ± pruinose, veiled disc are also diagnostic; they are usually paler than the thallus.

12. P. flavicans Lamy (1878)

Thallus wide-spreading, rather thick, smooth to irregularly, coarsely warted, often ± deeply cracked-areolate with an indistinct margin, yellowish or greenish grey; areoles to 1 mm diam., irregularly round or angular, convex or flat; soralia ± numerous, 0.4–1 mm diam., discrete or ± confluent, paler than the thallus, greenish yellow, coarsely to finely granular. Apothecia rather rare, the small dark ostioles expanding with tumid margins; epithecium K+ violet. Asci 8-spored. Ascospores *c.* 65 × 27 μm. Thallus Pd−, K−, KC+ orange, C+ orange, UV+ orange, soralia UV+ bright orange (thiophaninic acid and ± *O*-methylmonochlornorlichexanthone).

On siliceous rocks, typically coastal but occasionally also inland; frequent. N. & W. British Is. Europe.

Characterized by the rather thick, yellowish green, UV+ orange, sorediate thallus. Other *Pertusaria* species which are C+ orange, UV+ orange include *P. flavida* **(14)** which is isidiate and grows on bark and the coastal species, *P. gallica* **(15)**, which lacks isidia and soralia; see also *P. amarescens* **(13)** and *Ochrolechia inversa*.

13. P. amarescens Nyl. (1874)

Like **12**, but thallus ± primrose-yellow to dull olive-yellow, thinner and more delicately rimose-areolate; areoles ± flat or roughened; soralia at first very small, pale and fleck-like, becoming ± diffuse and contiguous. Apothecia unknown. Thallus Pd+ orange, K+ yellow, KC+ orange, C+ orange, UV+

overall pale orange and ± spotted blue-white, (thiophaninic, stictic, menegazziaic, and ± norstictic acids and *O*-methylmonochlornorlichexanthone, UV+ orange and pale blue unknowns).

On ± basic, schistose rocks, upland; very local. C. Wales, Scotland.

The status of material named as this species in Britain is in need of further investigation. *P. amarescens* is the host of the lichenicolous fungus *Dactylospora saxatilis*.

14. P. flavida (DC.) Laundon (1963)

Thallus yellow-green, yellow-grey or yellow-brown, sometimes with a pale to dark grey margin, rather thick and uneven, ± coarsely rimose-cracked; upper surface coarsely warted, flat, smooth or somewhat roughened; isidia numerous, simple, ± globose or shortly cylindrical, occasionally dissolving into soredia, and obscuring the thallus. Apothecia very rare, 2–5 immersed in hemispherical, sorediate-isidiate warts; disc black-brown, punctiform. Asci (4-)8-spored. Ascospores 60–100 × 25–40 μm, wall 7–10 μm thick, up to 20 μm thick at the ends. Thallus Pd−, K−, KC+ orange, C+ orange, UV+ bright orange or yellow-orange (thiophaninic and an unidentified substance).

On smooth and rough bark of well-illuminated, mature, broad-leaved trees in open woodlands, parklands and wayside trees. Throughout E. British Is., rare to absent in the west. Europe, Macaronesia.

Distinguished by the thick, abundantly isidiate, yellow-tinged thallus and C+ and UV+ orange reactions. Three yellow-green, C+ orange, crustose species can cause confusion: *Lecanora expallens* has a less robust, thinner, distinctly sorediate thallus, contains usnic acid, zeorin and thiophanic acid, and is UV−; *Pyrrhospora quernea* has a brownish tinge, is ± continuously fine sorediate, and has isoarthothelin and thiophanic acid; *Ochrolechia inversa* has convex, bright, ± discrete, yellowish grey, efflorescent soralia and contains atranorin as well as arthothelin.

15. P. gallica B. de Lesd. (1925)

Thallus thin to thick, smooth, wax-like to ± wrinkled-uneven, sometimes rimose-cracked becoming coarsely warted at the centre; prothallus thin, pale; upper surface pale to dark yellow-grey to greenish grey. Fertile warts rare, inconspicuous, low, irregular, 0.5–1 mm wide, ± conical, containing a single apothecium; disc irregular, punctiform or somewhat radiating, blackish or brownish green; epithecium K−. Asci (1-)2(-3)-spored. Ascospores 100 × 70–110 μm, colourless, K−, wall striate in section, 12–15 μm thick, 3-zoned, outer surface patterned with crowded, minute oval pores. Thallus Pd+ orange-red, K+ yellowish, KC+ orange, C+ orange, UV+ deep orange-red (thiophaninic acid, stictic acid, ± norstictic, constictic and ± gyrophoric acids).

On sunny, sheltered or exposed granite, coastal rocks, rarely peaty soil or decayed *Armeria* tufts; locally abundant. Is. of Scilly, W. Cornwall. Portugal, Spain.

Very common on the Is. of Scilly, where it forms conspicuous patches in the xeric-supralittoral *Ramalina* zone on coastal rocks. The ascospore structure recalls *P. lactescens* **(21)** but the ascospores are persistently colourless.

16. P. geminipara (Th. Fr.) Knight (1883)

Ochrolechia geminipara (Th. Fr.) Vainio (1881)

Thallus greyish or yellowish white, of thickly crowded, ± rounded granules, 0.1–0.2 mm wide, which at times becomes ± papillate, apices bursting and sorediate; soralia contorted, coarsely granular becoming efflorescent, 0.5–1.5 mm diam. Apothecia unknown in British material, to 2.5 mm diam.; disc dark red to purplish black; thalline exciple thick, wavy or ± crenulate; epithecium red-brown, K± paler, dissolving. Asci 2-spored. Ascospores 22–40 × 15–20 μm, walls 2-zoned. Thallus Pd+ yellow-orange, K+ yellow, KC+ red, C+ red, UV− or deep orange (alectorialic, barbatolic acids, ± xanthones).

On exposed mountain summit heaths, mossy boulders; very rare. N. Scotland (W. Inverness, Cairngorm; Perth, Ben Lawers, Ben Alder). Arctic-alpine, Europe, North America.

Herbarium specimens turn red-brown in time and also stain packets pink-brown due to alectorialic acid. Morphs of *Trapeliopsis granulosa* are Pd− and K−; the colour and shape of the granules in the herbarium is reminiscent of *Trapelia involuta*.

17. P. glomerata (Ach.) Schaerer (1826)

Thallus thin to moderately thick; upper surface white-grey to yellow-white, occasionally tinged reddish, smooth, nodulose or tuberculate, often ± shining. Fertile warts 1(-1.5) mm wide, almost globose, often crowded, regular, ± constricted at base. Apothecia 1(-3) per wart; disc 0.1–0.2 mm diam., black ± protruding, punctiform ostiole-like; epithecium K+ violet. Asci (2-)4-spored. Ascospores 70–125 × 25–50 μm, with a thin outer wall, uniformly thickened, inner wall unequal, combined wall 5–8 μm thick laterally and 13–15 μm apically. Pycnidia occasional, immersed with a small, black ostiole. Conidia 8–11 × 0.5 μm, spindle-shaped. Thallus Pd+ yellow, K+ yellow-red, KC+ yellow-red, C−, UV+ glaucous (norstictic and ± stictic acid; no constictic acid).

Overgrowing mosses, decaying plant remains and soil in ± calcareous habitats above *c.* 900 m; very rare. N. Scotland (W. Inverness, Coire Cheap, Ben Alder); there are also records from last century Scotland (Perth, Ben Lawers and Craig Calliach). Arctic and subartic, Sudeten Mtns., Alps, Carpathians, Jura.

Characterized by the numerous, globose, regular, fertile warts with ± extended, punctiform ostiole-like discs, the K+ red thallus, and habitat.

18. P. hemisphaerica (Flörke) Erichsen (1932)

Thallus medium to thick, forming extended, conspicuous patches, pale bluish grey; prothallus distinct, zoned, white; upper surface smooth or uneven to warted; soralia 1–1.5 mm diam., markedly convex, often becoming confluent; soredia granular, pale to concolorous. Soralia Pd−, K−, KC+ red, C+ carmine-red, UV± glaucous (lecanoric acid, unidentified substance).

On rough, well-lit bark on mature, broad-leaved trees, often near the base, in open woodlands, parklands and waysides, very rarely on rocks, chiefly sandstone; frequent. S. England, Wales, Ireland, rarer in N. Scotland and W. areas. Europe.

The pale bluish-grey-tinged thallus with an often broad white margin and conspicuous, paler or concolorous, convex soralia, C+ carmine-red, UV± glaucous, are distinctive features. *Haematomma elatinum* has a softer, greenish grey, ± continuously granular, Pd+ yellow-orange, K+ yellow-orange, C−, UV− thallus. *Ochrolechia subviridis* has a darker green to grey-white thallus, Pd−, K−, C+ orange-red, and has less delimited or ± confluent, granular-pseudoisidate soralia. When growing on rocks *P. hemisphaerica* resembles *P. lactea* **(20)** in which the soralia are also C+ carmine-red but which has a white, regularly rimose-cracked thallus and only slightly convex, pure white, discrete soralia.

As apothecia are unknown in *P. hemisphaerica* the generic placement is speculative.

19. P. hymenea (Ach.) Schaerer (1836)

Thallus superficial, in small patches or wide-spreading, thin to moderately thick, continuous, somewhat waxy, margin ± entire, rarely zoned; upper surface grey to yellow-green-grey or dark grey, smooth, ± tuberculate, roughened, often with deep rimose cracks. Fertile warts to 2 mm diam., often numerous and crowded, obscuring the thallus, semi-globose, often distorted, ± constricted at base, containing 1(-4) apothecia; discs (0.3-)0.6(-1.8) mm diam., rounded or rather irregular, often widening, the surface brown- or grey-black, often ± grey-white-pruinose, surrounded by an uneven, often somewhat elevated thalline exciple; epithecium K+ violet. Asci 8-spored. Ascospores 60–110(-120) × (20-)30–50 μm, wall 6–9(-29) μm thick, zoned, not striate. Pycnidia immersed, ostiole pale; conidia 13–25 × 0.5 μm, straight or somewhat curved. Thallus Pd−, K−, KC+ orange, C+ yellow, UV ± pale orange (thiophaninic acid, ± *O*-methylmonochlornorlichexanthone and ± gyrophoric acid).

On bark of woodland, wayside and parkland broad-leaved trees, often in rather shaded sites, rarely on sheltered rocks in woodland; frequent. Throughout British Is. Europe, North America.

In shaded sites the apothecia of this rather variable species can develop widely exposed, pale discs which are *Lecanora*-like; in more exposed sites the disc is darker, less exposed and more punctiform. The thallus varies from yellow in warm, sheltered but sunny sites to greenish grey in shaded ones.

P. pertusa **(28)** has more rounded, smooth warts, each with several, dark, punctiform discs.

20. P. lactea (L.) Arnold (1872)

Thallus wide-spreading, rather thick, white to grey-white; prothallus, paler, often ± zoned; upper surface smooth or ± roughened, matt, mostly rather regularly rimose-areolate; areoles coarse, 0.3–0.8 mm diam., flat or slightly convex, irregularly rounded or angular; soralia 0.5–1.5 mm diam., white, ± convex, at times surrounded by a fragile thalline exciple, occasionally ± efflorescent, mostly discrete, rather widely but evenly scattered, rarely contiguous. Fertile warts 0.4–0.5(-0.7) mm wide, very rare, soralia-like, mostly with 1(-2) apothecia; disc pale brown, at first white-pruinose, often surrounded by an irregular, sorediate thalline exciple. Asci 1-spored. Ascospores 180–240 × 60–100 μm, walls rather thick, uniform, 8–10(-33) μm, not zoned or striate. Conidia 4.5 × 0.5 μm, cylindrical. Thallus Pd−, K−, KC+ red, C+ carmine-red, UV− or + faintly glaucous (lecanoric and variolaric acids).

On dry, mostly sunny, siliceous or, occasionally, slightly calcareous, coastal and upland rocks, boulders and scree; common. Throughout British Is. Europe.

Characterized by the white-grey, areolate thallus with neat, round, ± convex, white, C+ rose-red soralia; *P. hemisphaerica* **(18)**, which is also C+ red, has a pale blue-grey thallus with paler, but not white, markedly convex soralia and usually occurs on trees.

21. P. lactescens Mudd (1861)

P. urceolaria Nyl. (1873)

Thallus thin or rather thick, grey or yellowish grey, at first smooth and continuous, later becoming wrinkled and markedly rimose-areolate, the areoles ± roughened, occasionally ± finely granular with crowded, simple, ± globose, isidia-like papillae, *c.* 0.1 mm diam., which may leave eroded, granular-sorediate patches when dispersed. Apothecia common, 2- to 5-immersed within an areole, surrounded by a slightly raised, irregular, smooth or ± granular thalline exciple; disc ± exposed, brownish black, very irregular, much divided; epithecium K+ violet. Asci 2(3–4)-spored. Ascospores (70-)80–100(-140) × 55–85 μm, pale, soon becoming grey- or olive-black, wall 10–14 μm thick, striate in section, 3-layered, the thin outer layer, K−, the inner 2, K+ violet-purple, outer surface with numerous, minute pores. Thallus Pd+ yellow-orange, K+ yellow-red, KC+ yellow-red, C−, UV− (norstictic acid).

On siliceous, often granite, rocks and walls; ?extinct. Recorded last century from N. England (Cleveland and Ennerdale), Jersey (La Moye). ?S. Tyrol.

The apothecia are easily overlooked; they are best seen when moist and are only slightly elevated, the disc being much divided and occasionally incompletely star-shaped. The status and current distribution of this species remains unclear; sorediate specimens are referable to *P. lactescens* s. str., while non-sorediate material has been named *P. spilomanthoides* Nyl. (1881). A related species collected (1981) from N.E. England (Northumbria, Cockermouth) has somewhat larger but similar, dark-coloured, I+ purple ascospores, and a thallus which is Pd+ orange, K+ orange, C+ orange, UV+ bright orange (stictic, constictic acids, unidentified xanthone).

22. P. leioplaca DC. (1815)

P. leucostoma (Bernh.) Massal. (1852)

Thallus superficial or partly immersed, thin, margin ± entire and unzoned; upper surface grey-white to yellow- or green-grey, smooth, even to somewhat roughened, not or weakly rimose-cracked, densely packed with clusters of coarse crystals. Fertile warts (0.6-)1.5(-2.5) mm diam., scattered or occasionally 1- to 3-contiguous at the base, concolorous with thallus, semi-globose or conical with spreading base, often flat-topped; epithecium K−. Apothecia 1(-2) per warts, generally level to slightly immersed, often widely spaced; disc punctiform, ostiole-like. Asci (3-)4(-6)-spored. Ascospores (40-)50–80(-120) × (20-)25–40(-50) μm, wall rather thin, to 4–4.5 μm, uniform. Pycnidia rare. Conidia 0.5–1 × 7–10 μm. Thallus Pd+ orange-red or Pd−, K+ yellow, KC+ yellow, C−, UV± pale orange (coronaton, ± stictic, and ± norstictic and constictic acids).

Mainly on smooth bark of broad-leaved trees and shrubs in ± sheltered, shaded sites; often frequent. N. & W. British Is., rarer elsewhere. Europe, North America.

Characterized by the thin, continuous thallus, ± conical, ± flat-topped, fertile warts with a spreading base, enclosing a single apothecium opening by a minute punctiform disc, the 4-spored asci, and by the presence of coronaton and stictic acid. *P. pertusa* **(28)**, with 2-spored asci, has an identical chemistry but has a better developed thallus, and fertile warts which are constricted and not spreading at the base. The 8-spored *P. alpina* Hepp ex Ahles (1860) which has an identical chemistry and has been recorded from South Uist and Rhum, is doubtfully distinct .

23. P. melanochlora (DC.) Nyl. (1872)

Thallus rather thick, grey to dark grey, ± rimose-cracked, with abundant, crowded, coarse, ± cylindrical papillae, 2–3 mm tall × 0.5–2 mm thick, rarely ± branched, with blunt, flattened, pale apices, often abraded exposing a white medulla which may form a few, granular soredia. Fertile warts rare, arising from a papilla, later expanding to 2–3.5 mm wide. Apothecia often poorly developed, 1–5 per papilla; disc 1.5–2 mm wide, black, whitish pruinose, flat or convex, irregular, surrounded by a curved, wrinkled thalline exciple. Asci 1-spored. Ascospores 180–250(-300) × 70–200(-130) μm; wall *c.* 22 μm thick, smooth, unzoned. Conidia 4–5 × 0.5–1 μm, cylindrical. Thallus Pd+ orange-red, K+ yellow-red, KC+ violet, C−, UV− (picrolichenic and protocetraric acids).

On siliceous rocks; very rare. S.W. England (N. Devon), N. Wales. S.W. Europe, Mediterranean.

Like *P. amara* **(2)**, especially its form *pulvinata*, it has a characteristic bitter taste and KC+ violet reaction thallus; it is separated by the gross, distinctive, pale-tipped papillae.

24. P. multipuncta (Turner) Nyl. (1861)

Thallus wide-spreading, grey, thin to moderately thick; prothallus often present; upper surface ± smooth to coarsely warted, ± rimose-cracked, with ± elevated, flat, starkly white, sorediate warts, 0.5–1.5(-2) mm diam. Apothecia frequent, (1-)2–3(-5), ± concealed within each sorediate wart; disc *c.* 0.5 mm diam., pale to blackish brown, veiled in soredia and crystals. Asci 1-spored. Ascospores 90–170 × 30–70 μm, elongate-ellipsoid to cylindrical; wall *c.* 4.5 μm thick, uniform. Thallus and soralia Pd+ orange-red, K+ yellow, KC+ yellow, C−, UV− or faintly glaucous (physodalic and ± protocetraric acids).

On smooth, often leached, acid bark, particularly on twigs and branches, rarely on siliceous rocks. Throughout British Is., especially in the west. Europe.

Distinguished by the white, well-delimited, scattered, sorediate warts containing apothecia, 1-spored asci and chemistry; *P. albescens* **(1)** and *P. amara* **(2)** are nearly always sterile and have a different chemistry. See also *P. ophthalmiza* **(25)**.

The thallus is very variable; on sheltered, smooth bark it is often thin, continuous and even, but in more exposed habitats it is more robust and the surface coarsely warted.

25. P. ophthalmiza (Nyl.) Nyl. (1865)

Like **24**, but thallus smooth to very uneven-warted; soralia 0.5–1.2 mm diam., with a ± raised, irregular, tumid, segmented-crenulate margin, sparingly sorediate and torn on the inner side. Apothecia 1–2, sunken in soralia; disc partially obscured, ± greyish pruinose; epithecium K− or faintly purplish. Asci 1-spored. Ascospores (90-)130–150(-170) × 50–70 μm, ellipsoid or longly ellipsoid, wall 6–7 μm thick. Thallus and soredia Pd−, K−, KC−, C−, UV− (4 fatty acids).

On leached, acid bark and overgrowing mosses in humid sites, chiefly on *Betula* and conifers; local. N.W. Scotland. Scandinavia, Alps, Russia, North America, Japan.

26. P. oculata (Dickson) Th. Fr. (1871)

Thallus thin, inconspicuous, white-grey; isidia abundant, crowded, ± obscuring the thallus, 1–3mm long, 0.3–0.4 mm wide, at first wart-like, becoming elongate, simple or branched, cylindrical and often darkened, K+ purple, at apices, fragile, never producing soralia. Apothecia 1–2 mm diam., rare; disc wide, flat or slightly concave, black, non-pruinose; thalline exciple at first regular, smooth and well developed, later more irregular and ± excluded; epithecium K+ violet. Asci 8-spored. Ascospores (16-)23–28(-30) × 10–14 μm; wall uniform, 1.5–2 μm thick, smooth. Pycnidia occasional, at ends of isidia. Conidia 3.5 × 0.5 μm, short for the genus. Thallus Pd+ rust-red, K+ yellow-red, KC+ yellow-red, C ± pink, UV− (fumarprotocetraric and ± protocetraric and ± gyrophoric acids).

On gravelly and moorland soils, decaying vegetation or very rarely directly on rock, montane. N. Scotland (Highlands). Arctic, subarctic, Alps, Carpathian Mtns., Balkans.

The presence of fumarprotocetraric and gyrophoric acids distinguishes *P. oculata* from *P. dactylina* **(8)**, which also has thicker, white, and more robust isidia, 0.4–1 mm wide.

27. P. panyrga (Ach.) Massal. (1855)

Thallus grey-white or cream-coloured, thin or moderately thick, the margin ± entire and unzoned; upper surface smooth or wrinkled, matt and ± pruinose; isidia present, 0.2–1 mm wide × 0.6–2 mm tall, initially papillate, becoming cylindrical, often fusing above but never coralloid-branched. Apothecia 1(-3) immersed in each wart; disc 0.3–0.75 mm diam., brownish black and greyish pruinose; thalline exciple thick. Asci 1-spored. Ascospores 80–165 × 40–55 μm. Thallus Pd−, K−, KC−, C− (fatty acids, as in *P. polythecia* see p. 450).

Encrusting mosses on the ground; recorded last century [as *P. hutchinsiae* (Borrer) Leighton (1851)] from S.W. Ireland (Cork, hills near Bantry). Arctic, N. Fennoscandia, Greenland, North America.

The status of the Irish specimens is unclear but they are unlikely to be *P. panyrga* which has an arctic distribution.

28. P. pertusa (Weigel) Tuck. (1845)

?*P. colliculosa* Körber (1863)

Thallus thin to thick; prothallus sometimes distinct, pale to white, limiting; upper surface pale grey to greenish grey, ± shiny, smooth or sparingly rimose-cracked, uneven, wrinkled or warted. Fertile warts 0.8–2(-3) mm diam., often abundant, scattered to crowded, typically irregularly semiglobose, constricted at base and with a slightly flattened apex, each with several apothecia. Apothecia (1-)4–7(-15) per wart, immersed; disc black, punctiform. Asci (Fig. 34e) 2(-4)-spored. Ascospores (120-)145–230(-330) × (35-)40–80(-90) μm; wall 7–15 μm thick, ends up to 30 μm thick. Conidia 12–18 × 0.5 μm, acicular. Thallus Pd+ orange-red, K+ yellow, KC+ yellow, C−, UV± pale orange (coronaton, stictic acid, ± norstictic and ± constictic acids).

On smooth to rough bark, more rarely on rocks, especially walls; common. Throughout British Is. Europe.

The combination of well-developed, irregularly semi-globose, fertile warts with constricted bases each containing several apothecia, 2-spored asci, and presence of coronaton and stictic acid is distinctive. *P. hymenea* **(19)** differs in the 8-spored asci, a K+ violet epithecium, the fewer apothecia (often only one) per wart, wider discs, and different chemistry. *P. leioplaca* **(22)** with similar chemistry to *P. pertusa* has 4-spored asci and ± evenly spaced conical warts containing a single apothecium.

29. P. pseudocorallina (Liljeblad) Arnold (1887)

P. ceuthocarpoides Zahlbr. (1927)

Thallus rather thick, greyish white with a brownish tinge, ± smooth, wrinkled to coarsely warted; isidia usually present, scattered to crowded, rounded, 0.5–1 mm diam., constricted at base, apices brown-pigmented, rounded, often breaking off at the base leaving non-sorediate, crater-like depressions; when fertile, thallus warted, irregularly granular, often appearing eroded and without or with only a few isidia. Fertile warts scattered or ± crowded, 1-3 mm diam., rather rare, irregularly semi-globose, with 2–7 apothecia; discs punctiform, grey-black. Asci 2-spored. Ascospores 120–200 × 50–80 μm, wall 5–6 μm thick, uniform. Thallus Pd+ yellow, K+ yellow-red, KC+ yellow-red, C−, UV− (norstictic acid).

On dry, well-lit, siliceous rocks in both coastal and upland areas; often abundant. Throughout N. & W. British Is., rare in S.E. England. Scandinavia, C. & W. Europe.

A variable species, generally recognizable by the brown-tipped isidia and K+ red (crystals) thallus. *P. corallina* **(7)** has a uniformly pale grey, isidiate thallus, which is K+ yellow; the isidia are longer, cylindrical and are without brown apices. Fertile specimens of *P. pseudocorallina* ± lacking isidia were formerly referred to *P. ceuthocarpoides*. *P. monogona* **(11)**, which is also K+ red, has more open white-pruinose discs and a white or pale grey, deeply rimose-areolate thallus.

30. P. pupillaris (Nyl.) Th. Fr. (1871)

Thallus immersed to ± superficial and very thin; prothallus not distinct; upper surface grey, matt; soralia to 0.5 mm diam., punctiform, flat, pale greyish white or yellowish white, often evenly scattered, with coarse granular soredia. Fertile warts 0.5–1 mm diam., rather rare, prominent, scattered. Apothecia 1 per wart; disc red-black; thalline exciple at first surrounding the disc, thick, wavy and later becoming ± excluded; epithecium olive-green, K+ violet. Asci 8-spored. Ascospores 10–14 × 7–11 μm, ± globose to broadly ellipsoid, wall *c.* 1 μm thick, smooth. Thallus Pd+ orange-red to rust-red, K+ dirty reddish, KC+ yellow-red, C− (fumarprotocetraric acid).

On smooth-barked trees, especially *Corylus*, *Sorbus* etc., and on pine wood in sheltered, often ancient woodlands. N.W. & W. British Is., extending locally in the E. Fennoscandia, Pyrénées.

Characterized by the small, neat, punctiform, Pd+ orange-red soralia and the *Lecanora*-like apothecia. Formerly misidentified in the British Is. as *Lecidea cinnabarina*, which has K+ yellow soralia due to additional presence of atranorin. Has also been confused with morphs of *P. dealbescens* **(9)** growing on bark or wood which have a thicker thallus and larger soralia, and with *Schismatomma quercicola* which has *Trentepohlia* as a photobiont and pinkish grey (when fresh) soralia.

31. P. pustulata (Ach.) Duby (1830)

Thallus thin, ± superficial; prothallus sometimes present, ± paler, unzoned at the margin; upper surface grey or pale greenish grey, rather smooth, even, becoming fissured when old. Fertile warts 0.5–1 mm diam., flat, expanded at the base. Apothecia 1(-3) per wart; discs at first punctiform, later becoming expanded, black, surrounded by the torn margin of the warts. Asci 2-spored. Ascospores (39-)80–125(-140) × 20–40(-45) μm, wall 1.5–3 μm thick, smooth, uniform. Thallus Pd+ yellow-orange to orange-red, K+ yellow, KC+ yellow, C+ yellow (stictic acid, ± norstictic and ± constictic acids, and a xanthone).

On rough bark, especially in ancient woodlands; rare. S. England. Subatlantic-Mediterranean Europe, North America.

Characterized by the apothecia being surrounded by the torn warted margin, the 2-spored asci, and a distinct xanthone. *P. hymenea* **(19)** is 8-spored and contains thiophaninic acid.

32. P. velata (Turner) Nyl. (1861)

Thallus thin to rather thick, white to pale grey or glaucous, smooth to ± coarsely wrinkled and ± rimose-cracked, seldom completely areolate. Fertile warts 0.5–0.8(-1.2) mm diam., numerous, scattered or ± crowded, rounded, flat, ± constricted at base. Apothecia 1(-2) per wart; disc pale pink or reddish, but often ± densely white-pruinose, C+ red; thalline exciple well developed, 0.2–0.25 mm wide, smooth, or ± crenate, somewhat raised, persistent; epithecium with numerous crystals, K−. Asci 1(-2)-spored. Ascospores 150–280(-310) × (40-)57–90 μm, wall 9–15 μm thick, uniform. Thallus Pd−, K−, KC+ red, C+ carmine-red, UV− or tinged orange (lecanoric acid, ± lichexanthone).

On rough bark on sheltered but well-illuminated trunks of mature trees, particularly *Fagus*, *Quercus* and *Fraxinus* in ancient woodlands and parklands; very local. S. England (especially Hampshire, New Forest), N. Wales, S.W. Ireland. S. Sweden, Germany, Portugal, North & South America, Macaronesia, Africa, Asia, Australia.

Sometimes confused with *P. hemisphaerica* **(18)** which is sorediate and unknown fertile.

33. P. xanthostoma (Sommerf.) Fr. (1831)

Aspicilia poriniformis (Nyl.) Arnold (1870)

Thallus thin, grey or grey-white, areoles smooth or ± rimose-cracked, shiny; prothallus inconspicuous. Fertile warts 0.7–1.5 mm wide, ± irregularly semi-globose, constricted at the base, smooth, often crowded. Apothecia 1–3(-6) per wart; discs 0.05–0.1 mm wide, pore-like, colourless to pale yellow-brown or pink, often surrounded by a thick, shining pink to pale brown rim, below with a thin thalline cuff; epithecium ± red-brown, K± colourless. Asci 4-spored. Ascospores 55–75(-85) × 30–40 μm, colourless, becoming brownish when old, wall 1–1.5 μm thick, uniform, smooth. Thallus Pd−, K−, KC± pink, C−, UV± white (alectoronic and lichesterinic acids).

On *Calluna* stems and siliceous rock outcrops, old wall tops, usually above 700 m alt., but descending to sea level in the north. C. & N. Scotland (Highlands). Iceland, N. Fennoscandia, ?Alps.

Distinguished by the ± globular, fertile warts, flattened, pale brown with 1–6, pore-like discs. *Pertusaria xanthostoma* is the only species which contains alectoronic acid. The only British specimen of *P. trochiscea* from Scotland (Aberdeen, Peddie's Hill), on mosses overgrowing serpentine, is incorrectly identified; it is probably a race of *P. xanthostoma* without alectoronic acid (UV−, t.l.c. nil).

PETRACTIS Fr. (1846)

Thallus crustose, effuse, immersed. **Photobiont** *Scytonema* or *Trentepohlia*. **Ascomata** apothecia, perithecioid to urceolate, immersed, leaving pits in rock when decayed, globose or depressed-globose, developing below a circular thalline cover which splits radially from a central pore as the centrum expands; disc pore-like, exposed when wet, or if the thalline cover disintegrates with age. **True exciple** pale, thin, inconspicuous, detached from thalline cover when dry, composed of conglutinated, narrow hyphae. **Hymenium** colourless, I+ blue. **Hamathecium** of slender, simple, thinly septate paraphyses, scarcely swollen at the colourless apices; periphyses absent. **Asci** 8-spored, ± cylindrical; wall thin, with a single functional wall layer, I+ blue; without a distinct apical dome. **Ascospores** colourless, septate, perispore ± distinct. **Conidiomata** pycnidia, ± immersed. **Conidia** simple, colourless, bacilliform. **Chemistry**: lichen products not detected. **Ecology**: five species on calcareous rocks in temperate regions of the N. Hemisphere, West Indies.

The one species in British Is. has *Scytonema* as photobiont; the other four have *Trentepohlia*. Distinguished from *Belonia* and *Gyalecta* by the radially fissured, thalline covering layer of the apothecia, and lack of orange (carotenoid) globules in the paraphyses or excipular hyphae. *Hymenelia* and *Diploschistes* further differ in having simple or brown, muriform ascospores respectively. Other, at least superficially, similar genera, e.g. *Absconditella*, *Bryophagus*, *Ramonia* and *Thelotrema*, do not occur on calcareous rocks.

LITERATURE: Vězda (1965b).

1. P. clausa (Hoffm.) Krempelh. (1861)

Thallus whitish or yellowish white, sometimes with blue-grey flecks. Apothecia 0.4–0.7 mm diam., with usually whiter (than thallus), flat to convex, thalline covering layer which has (3-)4–6(-8) radiating fissures; disc pale yellow-orange when moist, sometimes drying reddish brown; thalline exciple 10–15 μm wide, ± expanding to 30 μm in uppermost part; true exciple very thin or inconspicuous, pale; hymenium 200–320 μm tall; paraphyses 1–1.5 μm wide. Asci 150–250 × 12–18 μm. Ascospores (15-)17–23(-35) × 5–7 μm (excluding perispore), 3-septate, fusiform, with a distinct perispore. Pycnidia not seen.

On highly calcareous rocks in humid situations, especially where partly shaded by trees or tall grasses etc., often on loose limestone or chalk stones on the ground; scattered but sometimes locally abundant. Throughout the British Is. S. Scandinavia to the Mediterranean, N. Africa, Japan.

Recognized by the immersed apothecia with a ± raised thalline covering layer with star-like fissures; the young apothecia can be misinterpreted as perithecia.

PHAEOCALICIUM A. Schmidt (1970)

Thallus immersed, inapparent. **Photobiont** absent, saprobic. **Ascomata** apothecia, stalked, black to brownish black. Head obovoid to lens-shaped or strongly compressed. **True exciple** well developed, persistent, consisting of dark brown, periclinally arranged hyphae. **Asci** cylindrical, formed singly from ascogenous hyphae with croziers, with a single wall-layer, apex strongly and uniformly thickened, persisting until the maturity of the ascospores when they deliquesce. **Ascospores** not aggregated into a dry spore mass, uniseriately arranged non- or 1-septate, ellipsoid with rounded apices, dark brown, smooth or warted. **Ecology**: mainly on twigs of vascular plants, more rarely on bryophytes, species are usually confined to one genus of host plants. **Chemistry**: lichen products not detected by t.l.c., some species with pigments that change colour upon addition of K or N.

In addition to the two species included below there is a single collection of *Phaeocalicium* sp. on *Alnus* (Outer Hebrides) and another on *Sorbus* (Argyll).

LITERATURE: Schmidt (1970), Tibell (1975, 1984).

1 Ascospores 1-septate; apothecia to 0.6 mm .... **1. *populneum***
Ascospores simple; apothecia to 1.3 mm .... **2. *praecedens***

1. P. populneum (Brond. ex Duby) A. Schmidt (1970)

Apothecia 0.3–0.6 mm, not flattened; stalk to 0.08 mm wide, black- or olivaceous brown, outer part consisting of colourless, swollen hyphae, inner part K+ red; head to 0.3 mm diam., globose, black. Ascospores 10–13 × 4–6 μm, 1-septate, smooth.

On *Populus* twigs and branches. Scotland, Ireland. Europe, North America.

P. praecedens **(2)** differs in the simple ascospores and taller apothecia with darker stalks.

2. P. praecedens (Nyl.) A. Schmidt (1970)

Apothecia 0.5–1.3 mm; stalk 0.05–0.1 mm wide, black, internally reddish brown, K+ green, not pruinose; head 0.15–0.35 mm wide, top- or lens-shaped. Ascospores 12–17 × 5–7 μm, simple, uniseriately arranged in ascus.

On *Populus tremula* twigs and branches; rare. N. and C. Scotland (E. Inverness, S. Aberdeen, Perth). Finland, North America.

P. populneum **(1)** differs in the 1-septate ascospores when mature and shorter apothecia with a paler stalk.

PHAEOGRAPHIS Müll. Arg. (1882)

Thallus crustose, thin, immersed or partially superficial, hyphae I+ blue in European species **Photobiont** *Trentepohlia.* **Ascomata** apothecia, simple, ± branched or stellate, elongated (lirellate), immersed or somewhat emergent. **Thalline exciple** usually inconspicuous. **True exciple** pale brown to opaque, black and brittle continuing below the hymenium or developed only laterally, lateral walls vertical to divergent. **Disc** expanded, black, matt, sometimes ± white-pruinose. **Hymenium** colourless, I−, with numerous crystal-like inclusions. **Hamathecium** of paraphyses, simple, apical cells clavate and brownish. **Asci** 8-spored, *Graphis*-type, K/I−. **Ascospores** colourless, soon becoming pale to deep red-brown within the ascus, I+ purple, transversely multiseptate with lens-shaped cells. **Conidiomata** pycnidia. **Conidia** ellipsoid, straight, or slightly curved. **Chemistry**: norstictic acid. **Ecology**: on smooth bark (or smooth plateaux of fissured bark) of deciduous trees, rarely conifer branches in woodlands; a primarily tropical genus of *c.* 200 species with four European representatives.

The generic concept is currently ascospore-based and will require revision as the tropical taxa become better understood.

Morphs of *Graphis scripta* with old, shrivelled, brown-walled ascospores can be distinguished by the thicker, more conspicuous true exciple and persistently slit-like disc.

LITERATURE: Poelt & Vězda (1981).

1 Thallus K+ yellow to red (crystals), whitish, pale grey or faintly yellowish, often powdery; ascospores 5- to 9-septate 2
Thallus K− or K+ brownish (no crystals), yellowish or olive greenish, usually glossy, rarely white-powdery; ascospores 3- to 5-septate 3

2(1) True exciple closed at the base; ascospores (5-)7- to 9-septate ..... **1. *dendritica***
True exciple open at the base; ascospores 5- to 7-septate ..... **2. *smithii***

3(1) Apothecia emergent; thalline exciple conspicuous, white; hymenium 95–120 μm tall; ascospores 24–35 μm long, 5-septate ..... **3. *lyellii***
Apothecia ± immersed; thalline exciple indistinct; hymenium 45-70 μm tall; ascospores 16–25 μm long, 3- to 5-septate ..... **4. *inusta***

1. P. dendritica (Ach.) Müll. Arg. (1882)

Thallus thin to rather thick, wide-spreading or in patches, pale grey sometimes tinged yellowish, smooth, even, to coarsely and deeply cracked-areolate, areoles less than 2.5 mm diam., irregular, surface whitish to pale grey, occasionally with a yellowish tinge, matt or slightly glossy, often white-powdery. Apothecia ± deeply immersed, very variable, without or ± with a distinct, ± raised, entire or notched thalline exciple, usually ± elongate, scattered, straight, curved or serpentine, commonly stellate, dendritically branched and radiating; disc to 0.16–0.30 mm wide, ends commonly acute, not or ± thinly white-pruinose; true exciple entire, continuous below the hymenium; opaque, lateral walls 10–15 μm wide (Fig. 21b); hymenium 90–120 μm tall. Ascospores 30–50 × 7–9.5 μm, (5-)7–9(-10)-septate. Thallus Pd+ yellow to orange, K+ red (crystals), KC+ red, C−, UV− (norstictic acid).

On wayside and woodland broad-leaved trees, especially in smooth bark plateaux of older trees,

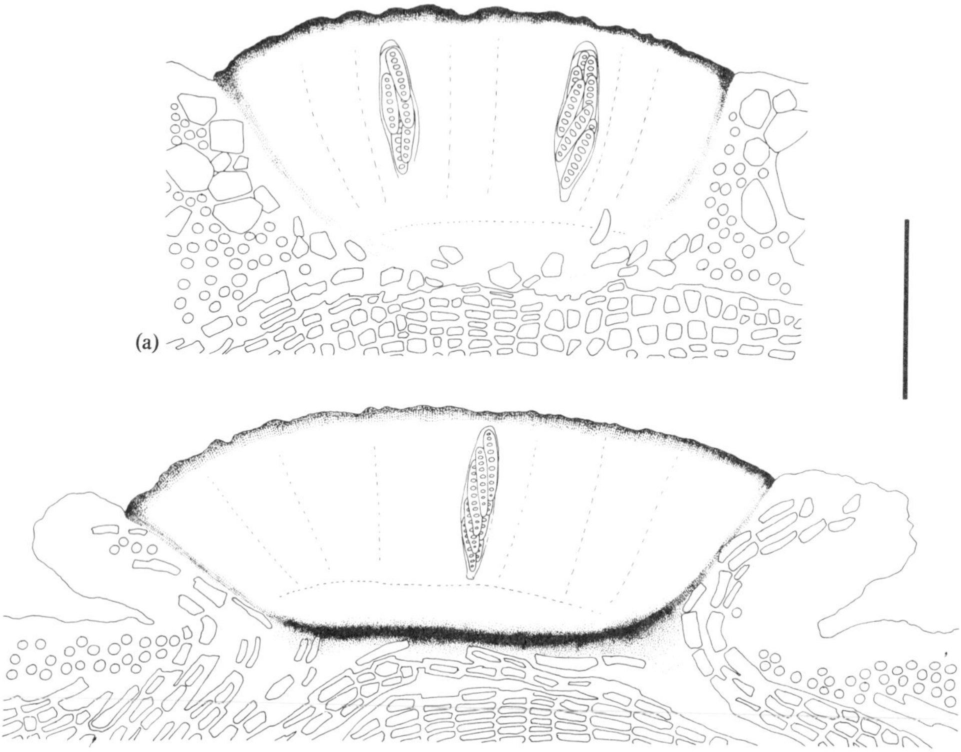

Fig. 21. *Phaeographis* apothecial sections. (a) *P. smithii*; (b) *P. dendritica*. Scale = 100 μm.

particularly *Quercus*, in moderate shade; locally frequent. W. England, rare in S.E. England, Wales, W. Ireland. Europe, North & South America, India, Japan.

Graphina anguina is similar in habit but has colourless, muriform spores, lirellae with a narrower disc, and is Pd –, K –. The apothecia in *P. dendritica* are very variable, ranging from short and almost square in outline or somewhat serpentine, to intricately dendroid-branched in large, radiating, stellate formations; the disc is usually ± widely exposed.

2. P. smithii (Leighton) B. de Lesd. (1910)

Like **1**, but with a smoother, often continuous or only sparingly rimose-cracked thallus. Apothecia often shorter, less branched, rarely radiating-dendroid; thalline exciple usually absent or mostly inconspicuous; true exciple colourless and open towards the base (Fig. 21a); hymenium 80–100 μm tall. Ascospores 26–40 × 7–9.5 μm, 5- to 7-septate. Thallus Pd+ orange, K+ red (crystals), KC+ red, C –, UV – (norstictic acid).

On mainly smooth bark of broad-leaved trees and shrubs, more rarely branches of conifers, in ancient and secondary woodland in moderate shade; locally frequent. Formerly widespread but now mainly in S.W. England, W. Wales, W. Scotland (N. to Is. of Skye), Ireland. Belgium.

A vertical cut across an epithecium in the field examined with a ×10 lens is sufficient to show the lack of the excipular tissue below the hymenium (Fig. 21a).

3. P. lyellii (Sm.) Zahlbr. (1905)

Thallus thin, wide-spreading or in patches or mosaics, yellowish, brown-yellow, olive-green, dark green, waxy, glossy, smooth or somewhat uneven, continuous or sparingly rimose-cracked, occasionally white powdery in part. Apothecia ± emergent, ± elongate, serpentine or flexuous, ± closely parallel, or in ± stellate clusters, rarely coarsely arthonioid, simple or sparingly branched; thalline exciple ±

conspicuous, white-powdery, mainly comprising large crystal clusters surrounding a thin, raised, black margin; true exciple entire, dark brown, lateral walls 25–35 μm wide, broadening to 60 μm above; disc 0.15–0.5 mm wide, broadly expanded, occasionally sparingly pruinose; hymenium 90–120 μm tall. Ascospores (20-)24–35 × 7–9 μm, (3-)5-septate. Conidia 8–10 × 4.5 μm, ellipsoid, straight or slightly curved. Thallus Pd−, K−, KC−, C−, UV− (no lichen products detected by t.l.c.).

On ± shaded, smooth bark of a wide range of broad-leaved trees, preferably young saplings or coppice in old woodlands, rarely on wayside trees; rare and very local. S.W. England (extinct E. of New Forest), W. Wales, S. Ireland. Portugal, France.

Recognized by the combination of thallus colour and white powdery margins to the apothecia.

4. P. inusta (Ach.) Müll. Arg. (1881)

Like **3**, but apothecia (0.1-)0.12–0.25 mm wide, ± immersed with an indistinct thalline exciple and often a narrower disc; true exciple entire, lateral walls 12–15 μm wide broadening to 25 μm above; hymenium 45–70 μm tall. Ascospores 16–25 × 7(-9) μm, 3- to 5-septate. Thallus Pd−, K−, KC−, C−, UV− (no lichen products detected by t.l.c.).

On broad-leaved trees, especially *Corylus* in old woodlands; rare and very local. S. England, W. Wales. Denmark, France, Germany, North America, India.

PHAEOPHYSCIA Moberg (1977)

Thallus foliose, sometimes ± shrubby, lobate. **Lobes** often radiating, short or usually elongate, mostly <1.5 mm wide, pale grey or greenish grey to dark brown, matt, not pruinose, not or only weakly maculate, without marginal cilia; underside whitish or more usually black with concolorous, simple rhizines. **Soredia** or **isidia** present in many species. **Upper** and **lower cortex** pseudoparenchymatous (lumina 3–7 μm diam.). **Photobiont** trebouxioid. **Ascomata** apothecia, laminal, ± sessile, usually with rhizines on lower surface; disc brown to black, epruinose. **Thalline exciple** present, smooth, rarely lobulate. **Epithecium** brown. **Hymenium** and **hypothecium** colourless. **Hamathecium** of paraphyses, slender, often forked above; apices clavate, pale brown with a thin dark brown cap. **Asci** ± cylindrical-clavate, 8-spored, *Lecanora*-type. **Ascospores** brown, 1-septate, thick-walled. **Conidiomata** pycnidia, immersed, with colourless walls except for brown zone around ostiole. **Conidiogenous cells** arranged in branched chains, short cylindrical, enteroblastic, acrogenous or pleurogenous. **Conidia** ellipsoid. **Chemistry:** cortex and medulla K− (atranorin absent), some species with yellow to orange-red K+ purple pigments (skyrin) or terpenoids. **Ecology:** nutrient-rich or -enriched substrata in well-lit situations.

LITERATURE: Maas Geesteranus (1952); Moberg (1977).

1 Thallus sorediate or isidiate; usually without apothecia 2
Thallus not sorediate or isidiate; usually with apothecia .... **3. *endococcina***

2(1) Lobes with marginal or laminal soralia 3
Lobes with marginal (or laminal at thallus centre) isidia 4

3(2) Medulla orange below, K+ purple; soralia marginal, rarely also laminal, ± lip-shaped .... **2. *endophoenicea***
Medulla white, or yellow to orange only in uppermost part of cortex; soralia mostly laminal, a few sometimes ± marginal, ± orbicular .... **1. *orbicularis***

4(2) Thallus < 1 cm diam., often ± shrubby; lobes 0.05–0.3 mm wide, ascending; underside pale; isidia often dissolving into granular soredia .... **5. *nigricans***
Thallus < 5 cm diam., not shrubby; lobes 0.2–0.5 mm wide, adpressed; underside black; isidia rarely sorediose .... **4. *sciastra***

1. P. orbicularis (Necker) Moberg (1977)

Thallus to 3 cm diam., orbicular, or irregular and confluent with others, ± closely adpressed; lobes 0.2–1.2 mm wide, usually radiating, separate to ± overlapping, pale grey or greenish grey, to grey- or dark brown, occasionally yellowish, sometimes weakly white-spotted in pale, wide-lobed morphs; soralia mostly orbicular, ± convex and laminal, sometimes marginal, grey to blackish or whitish (especially if abraded), occasionally yellowish; underside black with simple, black rhizines (some pale, or white-tipped at margin of lobes) that often project beyond the lobe margins; medulla white or yellow-orange in uppermost part. Apothecia occasional, to 1.5(-2.5) mm diam.; thalline exciple smooth or rarely lobulate. Ascospores 17–26 × 7–11 μm. Pycnidia frequent; conidia 2–4 × 1–1.5 μm.. Yellow or orange areas, K+ purple (skyrin), or no lichen products detectable.

On various nutrient-rich, or -enriched substrata; into urban areas on calcareous substrata (concrete, asbestos etc.); very common. Throughout British Is. Cosmopolitan.

Very variable. Morphs with conspicuous amounts of skyrin have been called var. *hueana* (Harm.) Clauz. & Roux (1985). Pale coloured, wide-lobed, corticolous morphs, sometimes erroneously called *Physcia labrata* Meresch. (1919), closely resemble *P. endophoenicea* **(2)** but that species has an orange-red lower medulla. A black underside distinguishes small, narrow-lobed morphs of *P. orbicularis* from *P. nigricans* **(5)** and *Hyperphyscia adglutinata*; the latter also has a prosoplechtenchymatous lower cortex.

2. P. endophoenicea (Harm.) Moberg (1977)

Like broad-lobed morphs of **1** (lobes 0.6–1.5 mm wide), but distinguished by the predominantly marginal, ± lip-shaped soralia (occasionally a few also laminal and orbicular) and confinement of orange-red pigment to the lower part of the medulla. Apothecia and pycnidia not present in British material. Medulla with orange-red pigment K+ purple (skyrin).

On bark, usually on spreading bases of trunks in parklands; rare or overlooked. England (Hereford, Oxford, Northumberland), Scotland. C. Europe to Fennoscandia.

3. P. endococcina (Körber) Moberg (1977)

Thallus to 3 cm diam., orbicular or confluent with others, closely adpressed, lobes 0.3–1 mm wide, ± radiating, ± separate, dark grey or grey-brown, not sorediate or isidiate, centre of thallus often with secondary lobules; underside black with black rhizines; medulla white, or usually (especially in older parts) with orange-red pigment in lower part. Apothecia to 1 mm diam., numerous; thalline exciple crenulate or lobulate. Ascospores 17–24 × 7–11 μm. Pycnidia frequent. Medulla with orange-red pigment, K+ purple (skyrin) and zeorin.

On ± nutrient-rich (bird-perch) boulders on a lake shore and in and by a stream. Scotland (Ben Lawers, Caenlochan). C. Europe to Fennoscandia, Nepal, North America.

Distinguished from *P. orbicularis* **(1)** and *P. sciastra* **(4)**, which may grow in similar situations, by the numerous apothecia, lack of soredia or isidia, and skyrin in lower medulla. *P. ciliata* (Hoffm.) Moberg (1977) is similar in being richly fertile without isidia or soredia but occurs exclusively on bark (usually of *Populus tremula*), and has an entirely white medulla (lichen products not detected by t.l.c.); a single 'British' specimen, supposedly from C. Scotland (Perth, Killin), is of doubtful origin as this is a boreal-alpine, continental species.

4. P. sciastra (Ach.) Moberg (1977)

Thallus to *c.* 5 cm across, ± orbicular or irregular, ± closely adpressed; lobes 0.2–0.5 mm wide, radiating, grey- to blackish brown, mostly separate; isidia marginal, but often also laminal and densely clustered in centre of thallus; underside black with blackish rhizines; medulla white. Apothecia not known in Britain. Pycnidia rare; conidia 2.5–4 × 1 μm. Lichen products not detected by t.l.c.

On ± nutrient-enriched (bird-perch) boulders protruding from, or at edge of, streams or lakes in upland districts; rather rare. N.W. England (Lake District), N. Scotland (Highlands), W. Ireland (Connemara). Arctic, N. Europe, North America, India.

Distinguished from *P. orbicularis* **(1)** and *P. endococcina* **(3)** by the dark isidia and absence of any orange pigmentation.

5. P. nigricans (Flörke) Moberg (1977)

Thallus to 1 cm diam., or confluent with other thalli, orbicular, loosely attached and often ± shrubby; lobes 0.05–0.3 mm wide, ascending, grey-brown to dark brown but grey-green and ± translucent when wet; isidia marginal, fragile, often dissolving into soredia; underside whitish, pinkish or pale brown-white, with sparse, generally pale rhizines; medulla very thin, white. Apothecia not known in Britain. Pycnidia rare. Lichen products not detected by t.l.c.

Mainly on calcareous stonework and asbestos-cement, more rarely on enriched, dust-impregnated tree bases in open parkland and wayside situations, lowland; common but easily overlooked. E. Britain, E. Ireland (Lambay Island). S. Europe to S. Fennoscandia, North America.

An inconspicuous but easily identified species, not to be confused with small specimens of *P. orbicularis* **(1)** which has a blackish undersurface.

PHLYCTIS Wallr. (1825)

Thallus crustose, smooth, or ± cracked-areolate or powdery, immersed or superficial, silver-grey to dark grey; prothallus often present, pale or whitish. **Photobiont** chlorococcoid. **Soralia** present or absent, discrete to confluent, leprose-farinose. **Ascomata** apothecia, immersed, scarcely emergent, the entire surface sometimes obscured by a white, coarse pruina. **Thalline exciple** irregularly crenate or indistinct. **True exciple** poorly developed. **Epithecium** colourless to ± pale brown, granular. **Hymenium** colourless, tall. **Hypothecium** colourless to pale brown, thin. **Hamathecium** of paraphyses, slender, mostly simple, sparingly branched near their apices, indistinctly septate with minute swollen tips. **Asci** broadly clavate, thin-walled, not strongly thickened at the apex, outer sheath I+ blue, tholus I−, 1- to 2(-4)-spored. **Ascospores** elongate, ellipsoid or sub-spherical, densely muriform, with a very thin perispore, sometimes apiculate, colourless to pale yellow-brown. **Conidiomata** pycnidia. **Conidiogenous cells** simple. **Conidia** short, colourless, bacilliform. **Chemistry**: various depsidones, only norstictic in the British species. **Ecology**: on wayside and woodland deciduous trees in often ± humid, well-lit situations, occasionally on rocks.

A temperate southern hemisphere genus of *c.* 20 species; probably congeneric with *Phlyctella* Kremp. (1876).

LITERATURE: Laundon (1970).

1 Thallus not sorediate; apothecia always present .... **1. *agelaea***
Thallus sorediate, discrete, becoming irregularly confluent; apothecia rare .... **2. *argena***

1. P. agelaea (Ach.) Flotow (1850)

Thallus whitish grey or grey, ± roughened, uneven, cracked, usually thin, determinate, forming small patches, occasionally coalescing to form more extensive areas; prothallus white, delimiting. Apothecia 0.2–0.5(-1.0) mm diam., irregular or ± rounded, scattered or ± contiguous, deeply embedded and veiled with coarse, powdery granules and white pruina of the thalline exciple, whitish, sometimes brownish with age. Asci 2(-4)-spored. Ascospores (35-)45–80(-91) × 11–32(-35) μm, ellipsoid, markedly muriform with ± apiculate ends, colourless, becoming tardily yellow-brown. Thallus Pd+ orange, KC+ red, K+ yellow-red (crystals), C− (norstictic acid).

On broad-leaved trees in moist coastal and inland sites, chiefly on the smooth trunks and branches of young trees, especially *Fraxinus*, *Populus* species, *Corylus*, *Salix* and *Acer pseudoplatanus*, in sheltered valley bottoms, copses, and bogs, very rarely on smooth, porous, acid to neutral rocks in sheltered sites; now very local and possibly declining. W. & S.W. British Is. Europe, from Scandinavia to Italy, Azores, North America.

The small, fleck-like, powdery, rounded apothecia may resemble soralia, the absence of which separates this species from *P. argena* **(2)**.

2. P. argena (Sprengel) Flotow (1850)

Thallus rather thin and smooth to thick and roughened, continuous or sparingly cracked, creamy white, silver-grey to dark grey, often paler at the margin, forming extensive patches sometimes surrounded by a ± conspicuous white prothallus. Soralia usually paler than thallus, occasionally tinged reddish or brownish, very irregular, shallow, ulcerose, often coalescing to form diffuse, irregular patches; soredia fine-powdery to coarsely granular. Apothecia rare, 0.2–0.4 mm diam., with a grey-black disc, irregular; thalline exciple ± elevated, ± granular-white-pruinose. Asci 1-spored. Ascospores (75-)100–140(-145) × 25–50 μm, markedly muriform, broadly ellipsoid, without apiculate apices. Pd+ orange-red, K+ red (crystals), KC+ red, C−, UV− (norstictic acid).

On broad-leaved trees, rarely conifers, in well-lit, wayside and woodland situations, occasionally on damp N.- and E.-facing, ± basic siliceous rocks and walls, memorials and window-sills of churches, rarely on limestone outcrops, pollution tolerant. Throughout British Is. Oceanic Scandinavia south to C. Europe.

Initially the irregular soralia may have ± raised, unevenly granular or subcrenate margins; in this state they resemble the apothecia of *Phlyctis agelaea* **(1)** and care is needed in identification of the two species. Older soralia are often very diffuse, irregular, and eroded and resemble browsed mollusc tracks across the surface of the thallus.

PHYLLOPSORA Müll. Arg. (1894)

Thallus crustose, of minute, pubescent squamules or granules, squamules ± corticate; prothallus usually distinct. **Photobiont** trebouxioid. **Ascomata** apothecia, convex and often tuberculate, disc often deeply fissured, pale to dark brown, pigment reddish brown, never greenish or purplish in K, not pruinose. **Thalline exciple** absent. **True exciple** often inconspicuous except occasionally when young, of radially orientated, thick-walled, gelatinized hyphae, some extending outwards into thallus. **Epithecium** poorly differentiated. **Hymenium** I− or faintly blue. **Hypothecium** massive, pale, of gelatinized, interwoven, narrow hyphae. **Hamathecium** of paraphyses, simple or sparingly branched, short-celled, each surrounded by a gel coat, apices not swollen, colourless. **Asci** 8-spored, clavate, *Bacidia*-type. **Ascospores** simple or thinly septate, ellipsoid to narrowly fusiform, colourless, without a thick perispore. **Conidiomata** pycnidia, resembling incipient apothecia. **Conidia** bacilliform, colourless, simple. **Chemistry**: including argopsin, atranorin, pannarin, zeorin, other terpenoids and fatty acids. **Ecology** : on bark, *c.* 40 species, mainly in the tropics, a few present in oceanic, temperate regions.

Similar to *Biatora*, but differing in the pubescent-squamulose thallus and in the slightly different ascus structure.

LITERATURE: Brako (1989), Coppins & James (1979b), Swinscow & Krog (1981).

1. P. rosei Coppins & P. James (1979)

Thallus of pubescent granules, or (especially at thallus edge) squamules, forming a ± thick, uneven pale grey- to blue-green crust, to 5 cm or more diam., granules 0.05–0.15 mm diam., squamules flattened, crenate to subdigitiform, 0.15–0.3 mm wide, weakly corticate; prothallus whitish, cottony, sometimes obscured in thick thalli. Photobiont cells 9–15 μm diam. Apothecia 0.3–1.5 mm diam., often absent, convex, often clustered or tuberculate, pale orange- to deep red-brown; true exciple inconspicuous except in some young apothecia, concolorous with or paler than the disc, colourless or reddish brown in section; hymenium 20–30 μm tall, colourless or tinged red to brownish above; hypothecium 200–230 μm tall, colourless or brownish red in parts. Paraphyses mostly simple, 2–3 μm wide; apical cell ± tapered. Asci 20–26 × 7–9 μm. Ascospores (7-)10–13(-15) × 2–3(-4) μm, simple or thinly 1(-3)-septate, narrowly ellipsoid to shortly fusiform. Pycnidia not known. Thallus Pd+ red, K− or faintly yellow, KC−, C− (argopsin).

On trunks of mature tree trunks, especially *Quercus*, often overgrowing mosses, characteristic of oceanic, rather 'acidic' *Lobarion* communities in old woodlands and sheltered stream valleys. S. England (Hampshire, New Forest), northwards in W. Britain to W.C. Scotland (Westerness), W. Ireland. France (Brittany), W. Pyrénéees.

When sterile, distinguished from *Catillaria pulverea* by the presence of minute squamules (at least at the edge of the thallus) and chemistry (t.l.c.). Other similar sterile crusts in such habitats also lack squamules, and are all Pd−, e.g. *Biatora epixanthoides*, *B. epixanthoidiza* and *Megalospora tuberculosa*.

PHYSCIA (Schreber) Michaux (1803)

Thallus foliose, lobate, often orbicular in outline with radiating lobes. **Lobes** short to elongate, mostly <3 mm wide; upper side whitish, bluish grey to dark grey, matt or slightly glossy, sometimes with white flecks owing to breaks in the algal zone, occasionally white-pruinose, with or without marginal cilia; underside whitish, pale tan, pale grey or pinkish, with few to many, simple or irregularly furcate rhizines. **Soredia** or **isidia** present in many species. **Upper cortex** pseudoparenchymatous. **Lower cortex** prosoplectenchymatous (lumina <2.5 μm wide) or pseudoparenchymatous (lumina 4–7 μm diam.). **Photobiont** trebouxioid. **Ascomata** apothecia, laminal, sessile or short-stalked, without rhizines below; disc brown to black, often white-pruinose. **Thalline exciple** present. **Epithecium** pale brown. **Hymenium** and **hypothecium** colourless. **Hamathecium** of paraphyses, slender, simple or forked above; apices clavate, pale brown with a thin, dark brown cap. **Asci** ± cylindrical-clavate, 8-spored; *Lecanora*-type. **Ascospores** brown, 1-septate, thick-walled. **Conidiomata** pycnidia, immersed with colourless walls except for dark ostiolar region (seen as black dots on lobe surface). **Conidiogenous cells** arranged in branched chains, short, cylindrical, enteroblastic, acrogenous or pleurogenous. **Conidia** simple, ± bacilliform, colourless. **Chemistry:** atranorin in the cortex, often also with zeorin or other terpenoids.

LITERATURE: Maas Geesteranus (1952), Moberg (1977).

1 Thallus sorediate or isidiate; apothecia usually absent 2
Thallus not sorediate or isidiate; apothecia usually present 8

2(1) Lobes with marginal cilia, ± ascending 3
Lobes without marginal cilia (but often with protruding rhizines), ± adpressed 4

3(2) Soralia lip-shaped .... **2. *tenella***
Soralia helmet-shaped .... **1. *adscendens***

4(2) With delimited soralia or marginal, granular soredia 5
Thallus surface with numerous isidia that may ± dissolve into granular soredia **10. *clementei***

5(4) Lobes whitish to pale grey, without distinct bluish tinge or white flecks; medulla K− or + yellow; lower cortex various 6
Lobes bluish grey with distinct white flecks; medulla K+ yellow; lower cortex prosoplectenchymatous .... **5. *caesia***

6(5) Soralia lip-shaped or lobes with marginal granular soredia; medulla K−; lower cortex various 7
Soralia mostly laminal, hemispherical; medulla K+ yellow; lower cortex pseudoparenchymatous .... **8. *tribacioides***

7(6) Soredia formed in lip-shaped soralia, mostly apical on lateral branches (occasionally laminal, crater-like soralia also present); lower cortex prospoplectenchymatous ... **7. *dubia***
Soredia usually developed on marginal protuberances, only rarely produced in lip-shaped soralia; lower cortex pseudoparenchymatous .... **9. *tribacia***

8(1) Lobes without marginal cilia, ± adpressed 9
Lobes with marginal cilia, ± ascending .... **3. *semipinnata***

9(8) Medulla K−; lobe surface without distinct white flecks .... **6. *stellaris***
Medulla K+ yellow (often weak); lobe surface with distinct white flecks .... **4. *aipolia***

1. P. adscendens (Fr.) H. Olivier (1882)

Thallus 2–4(-6) cm diam., orbicular, or confluent with others, loosely adpressed; lobes 0.3–1 mm wide, ascending, whitish to pale grey or rarely dark ash-grey, not pruinose, ± white-spotted in older parts, with 0.4–2 mm long pale marginal cilia that are often grey- or dark brown at their tips; underside whitish with sparse whitish, often brown-tipped, slender rhizines; soralia apical, helmet-shaped, margins of 'helmet' often upwardly recurved; old specimens sometimes with additional crater-like soralia bursting through upper surface of lobes; lower cortex prosoplectenchymatous. Apothecia to 2 mm diam., uncommon, ± stalked; disc sometimes thinly pruinose. Ascospores 16–23 × 7–10 μm. Pycnidia frequent, especially abundant on fertile thalli; conidia 4–6 × 1 μm. Cortex K+ yellow (atranorin); medulla K−.

On well-lit and nutrient-rich and -enriched substrata, including limestone, concrete, asbestos-cement, timberwork, tree trunks and branches; common. Throughout British Is. Europe, North & South America, E. Africa, India, Australia, New Zealand.

Immature or eroded specimens are easily confused with *P. tenella* **(2)** or *P. semipinnata* **(3)**, and may be impossible to determine with certainty. Mechanical hybrids with *P. tenella* **(2)** also occur rarely, having some lobes with helmet-shaped and some with lip-shaped soralia.

2. P. tenella (Scop.) DC. (1805)

Like **1**, but the soralia lip-shaped. Apothecia to 2.5 mm diam., more frequent than in *P. adscendens* **(1)**.

In similar habitats to *P. adscendens* **(1)**, but most usually on bark; common. Throughout British Is. Europe, North America, India.

A morph with a dark grey thallus and dark grey-brown marginal cilia on maritime rocks is sometimes recognized as subsp. *marina* (A. E. Nyl.) D. Hawksw. (1980). *Physcia dubia* **(7)** lacks marginal cilia.

3. P. semipinnata (Gmelin) Moberg (1977)

Like **1** and **2**, but lacking soralia and with abundant apothecia when mature. Apothecia often larger, to 3 mm diam., usually thinly pruinose. Thalli more frequently orbicular, and the lobes are more distinctly white-flecked, tend to be slightly narrower, and have abundant cilia on the ends.

Mostly on bark, especially on shrubs or tree branches, rarely on rocks; local and decreasing. W. British Is., rare elsewhere due to SO_2 pollution and excessive bark enrichment. Mediterranean - Atlantic Europe, North America, India.

The numerous marginal cilia and loosely attached lobes distinguish it from *P. aipolia* **(4)** and *P. stellaris* **(6)**. *Anaptychia ciliaris* has larger, minutely tomentose lobes that are K− and lack a lower cortex.

4. P. aipolia (Ehrh. ex Humb.) Fürnrohr (1839)

Thallus to 6(-10) cm diam., usually ± orbicular, ± closely adpressed; lobes 0.6–1.5(-2) mm, radiating, sometimes separate but usually partly overlapping, whitish to pale grey, often with a bluish tinge, usually distinctly white-flecked (best seen when moist), not pruinose or rarely weakly pruinose; bullate warts or small secondary lobes sometimes at centre of thallus and margins of all apothecia; underside whitish to pale tan or pale grey, with numerous, simple to irregularly forked, whitish to dark brown or grey rhizines sometimes protruding beyond the margins; lower cortex prosoplectenchymatous. Apothecia to 2.5(-3) mm diam., abundant, often crowded, disc often pruinose. Ascospores (15-)18–24(-26) × 7–10(-11) μm. Pycnidia frequent; conidia 4–6 × 1 μm. Cortex and medulla K+ yellow (atranorin, zeorin).

On ± nutrient-rich bark of tree trunks, branches and twigs, rarely on maritime rocks and old walls; common in unpolluted districts. Throughout British Is. Europe, North & South America, E. Africa, Australia, New Zealand.

P. stellaris **(6)** is similar, but ± lacks white flecks and has a K− medulla; *P. semipinnata* **(3)** has loosely attached lobes and marginal cilia (not to be confused with protruding rhizines). *Physconia distorta* can be similar in appearance but has a K− cortex, black, bottle-brush-like rhizines and larger ascospores. *P. phaea* (Tuck.) Thompson (1963) is similar but occurs always on nutrient-rich rocks (e.g. bird perches) and has markedly convex, rather narrow (*c.* 1 mm) lobes, ± smaller apothecia *c.* 1(-2) mm diam., and ascospores 15–21 × 7–10 μm; arctic-alpine and doubtfully British. Records (from Scotland) are misidentifications for saxicolous morphs of *P. aipolia* or are of dubious provenance.

5. P. caesia (Hoffm.) Fürnrohr (1839)

P. wainioi Räsänen (1921)

Like **4**, but with numerous soralia, the lobes slightly narrower (mostly 0.5–1 mm wide); apothecia rare; soralia to 2 mm diam., whitish or blue-grey (pigment in external soredia), variable in development, either laminal and markedly convex (sometimes eroded and crater-like), or apical on short lobe branches and then capitate or ± lip-shaped, sometimes marginal and lip-shaped. Structure and chemistry very similar to *P. aipolia* **(4)**.

On well-lit, nutrient-rich substrata, including tops of walls, memorials, concrete roofs, calcareous rocks, also on acid rocks, especially in coastal districts or boulders in or by upland streams and lake margins, less often on dust impregnated timberwork and tree bases; common and pollution-tolerant. Throughout British Is. Europe, N. & W. North America, E. Africa, W. Himalayas, Nepal, Australia, New Zealand, Antarctica.

Both laminal and apical/marginal soralia are often present on the same thallus, but the relative proportions vary. Extreme morphs with mostly apical/marginal lip-shaped soralia are sometimes distinguished as *P. subalbinea* Nyl. (1874) [syn. *P. wainioi*]; such morphs are prevalent in exposed situations and hence more common in N. Britain. Robust specimens of *P. dubia* **(7)** are distinguished by a K− medulla; and *P. tribacioides* **(8)** does not have white-flecked lobes and always has white, mostly laminal, soralia and a pseudoparenchymatous lower cortex.

6. P. stellaris (L.) Nyl. (1856)

Thallus to 3(-6) cm diam., usually ± orbicular, ± closely adpressed; lobes 0.5–1.5 mm wide, radiating, ± separate, white- to darkish grey, rarely with a bluish tinge, not or indistinctly white-flecked, not pruinose; bullate warts or secondary lobules sometimes at centre of thallus and on apothecia margins; underside whitish to pale brown-white or pale grey, with numerous simple or branched, whitish to dark brown or grey rhizines that often protrude beyond the lobe margins; lower cortex prosoplectenchymatous. Apothecia to 3(-4)mm diam., abundant, often crowded, disc sometimes pruinose. Ascospores 15–22 × 7–11 μm. Pycnidia frequent; conidia 4–6 × 1 μm. Cortex K+ yellow (atranorin), medulla K−.

On branches or twigs of deciduous trees in wayside or woodland edge situations, rarely on monuments or boulders; local, mainly in little-polluted areas. W. & N. Britain. Europe, North & South America, E. Africa, Saudi Arabia, Himalaya, Australia, New Zealand.

Easily confused with *P. aipolia* **(4)**, with which it often grows, but distinguished by its lack of white-flecked lobes, best seen when wet, K− medulla, and absence of zeorin. Differs from *P. semipinnata* **(3)** in the more adpressed lobes without cilia, although some rhizines often protrude beyond the lobe margins. Probably closely related to the sorediate *P. dubia* **(7)**.

7. P. dubia (Hoffm.) Lettau (1912)

Thallus to 5 cm diam., ± orbicular, or irregular in outline, ± closely adpressed but sorediate lobes often ± ascending; lobes 0.2–1 mm wide, usually radiating, very short and crowded to slender and distinctly separate, grey-white to darkish grey, not or only faintly white-flecked, occasionally weakly pruinose; soralia whitish to dark grey, usually lip-shaped and produced at apex of side lobes, rarely a few also laminal and crater-like (to 1 mm diam.) which may become confluent; underside white to pale tan, with rather sparse, whitish to brown, mostly simple rhizines; lower cortex prosoplectenchymatous. Apothecia to 2 mm diam., very rare (not seen in British material); thalline exciple often sorediate. Ascospores 16–24(-28) × 6–10 μm. Pycnidia rare; conidia 4–6 × 1 μm. Cortex K+ yellow (atranorin), medulla K−.

On various types of ± non-calcareous rocks and building materials in well-lit, nutrient-rich and -enriched situations, e.g. tops of memorials, roofs, bird-perch stones; occasional on timberwork and dust-impregnated bark. Throughout British Is. probably more common in the north and east. Europe, North America, E. Africa, W. Himalaya, Nepal.

Very variable, especially in the length, width and degree of separation of lobes; extreme morphs with slender, elongate lobes were previously called *P. teretiuscula* (Ach.) Lynge (1916). Like *P. tenella* **(2)** but that species has marginal cilia. Distinguished from weakly maculate morphs of *P. caesia* **(5)** with lip-shaped soralia, by the K− medulla. Rare morphs of *P. tribacia* **(9)** with lip-shaped soralia can be distinguished from broad-lobed morphs of *P. dubia* only by sections of the lower cortex.

8. P. tribacioides Nyl. (1874)

Thallus usually irregular in outline; lobes scarcely radiating, rather short, 0.5–1.5 mm wide, grey-white, never bluish, not pruinose, not distinctly white-flecked, closely adpressed, crowded and usually ± overlapping, marginal lobes usually with distinctly flared and crenately incised apices; soralia to 1.2 mm diam., mostly laminal, strongly convex, ± hemispherical, often confluent, white (never blue-grey), a few soralia sometimes apical and capitate on short side lobes; underside whitish to pale brown-white, with rather sparse, simple, whitish to brown rhizines; lower cortex pseudoparenchymatous. Apothecia to 1.6

mm diam., very rare. Ascospores 17–22 × 7–10 μm. Pycnidia not seen. Cortex and medulla K+ yellow (atranorin).

On nutrient-rich bark of wayside and parkland trees in well-lit situations, especially of *Acer pseudoplatanus*, *Malus*, *Sambucus* and *Ulmus*, declined with the demise of elms, exceptionally on bird perch stones on the coast; very rare and local. S.W. England, S. Wales (Pembroke), S.W. Ireland. S. Spain, W. France (Brittany), E. Africa, Nepal, Australia, New Zealand; North American reports refer to *P. americana* Merrill (1926).

Distinguished from morphs of *P. caesia* **(5)** with convex soralia by the ± non white-flecked lobes and pseudoparenchymatous lower cortex.

9. P. tribacia (Ach.) Nyl. (1874)

Thallus usually irregular in outline, lobes not distinctly radiating, 0.5–1.5 mm wide, whitish to bluish grey, often faintly pruinose, not white-flecked, usually appearing ± flattened, with marginal crenulations or horizontal lobules, the apices of which often dissolve into granular soredia; when soredia development is abundant ± lip-shaped soralia may be formed (usually when on rocks); underside whitish or with faint pinkish, creamy yellow or brownish tinge; rhizines rather sparse, mostly simple, white to pale brown. Lower cortex pseudoparenchymatous. Apothecia to 1.5 mm diam., very rare. Ascospores 15–24 × 9–11(-13) μm. Pycnidia rare. Cortex K+ yellow (atranorin), medulla K−.

On nutrient-rich bark of wayside and parkland trees and on ± vertical surfaces of ± calcareous, enriched or maritime rocks and walls. Frequent in S. & W. England, rarer northwards with scattered sites in E. & W. Scotland, especially on the coast. Europe, North America, E. Africa, India, Nepal, Australia.

Usually easily recognized, but rare morphs on rocks with lip-shaped soralia require sectioning to separate them from *P. dubia* **(7)**.

10. P. clementei (Sm.) Maas Geest. (1952)

Thallus to 3 cm diam., ± orbicular, often coalescing with others to form extensive swards; lobes 0.3–0.6(-1) mm wide, ± radiating, mostly contiguous and ± overlapping, closely adpressed, especially at apices, white to pale grey, not pruinose or white-flecked, centre with numerous, short, papilla-like isidia that often break down into granular soredia and form a ± continuous sorediate crust; underside whitish or with brownish tinge, with a few simple, whitish to pale brown rhizines; lower cortex pseudoparenchymatous. Apothecia to 1.6 mm diam., very rare; disc sometimes thinly pruinose. Ascospores 15–20 × 8–10 μm. Pycnidia rare. Cortex and medulla K+ yellow (atranorin, two unidentified terpenoids).

On nutrient-rich, well-lit trunks of wayside and parkland trees, and on steeply sloping, ± basic rock faces (often serpentines) near the sea, and inland in churchyards and on walls near the coast; very rare and local. S. England (especially the S.W.), with a few sites on rock northwards to Scotland (S. Argyll), S. Ireland. S. & W. Europe, Canary Islands, Azores, U.S.A. (California), Nepal, Australia.

Superficially resembles *Parmeliopsis aleurites* which has a Pd+ deep yellow to orange thallus with broader-lobes and occurs in different habitats.

PHYSCONIA Poelt (1965)

Thallus foliose, lobate, sometimes ± orbicular in outline, but often irregular and confluent with others, ± loosely adpressed (Fig. 36f). **Lobes** short to elongate, mostly <3 mm wide, without cilia on the margins, dull grey to dark brown, matt or slightly glossy, usually pruinose at least at tips; underside whitish to brown-black. **Soredia** present in some species. **Upper cortex** pseudoparenchymatous (lumina 3–7 μm diam.) or scleroplectenchymatous (lumina <2.5 μm wide). **Lower cortex** prosoplectenchymatous (lumina <2.5 μm wide), sometimes absent at lobe tips. **Photobiont** trebouxioid. **Ascomata** apothecia, laminal, without rhizines on lower surface; disc brown but usually white-pruinose. **Thalline exciple** present, persistent, often incurved, sometimes with lobules or soredia. **Epithecium** brown. **Hymenium** and **hypothecium** colourless. **Hamathecium** of paraphyses, simple, or branched above; apices clavate, light brown with a thin, dark brown cap. **Asci** cyclindrical-clavate, 8-spored,

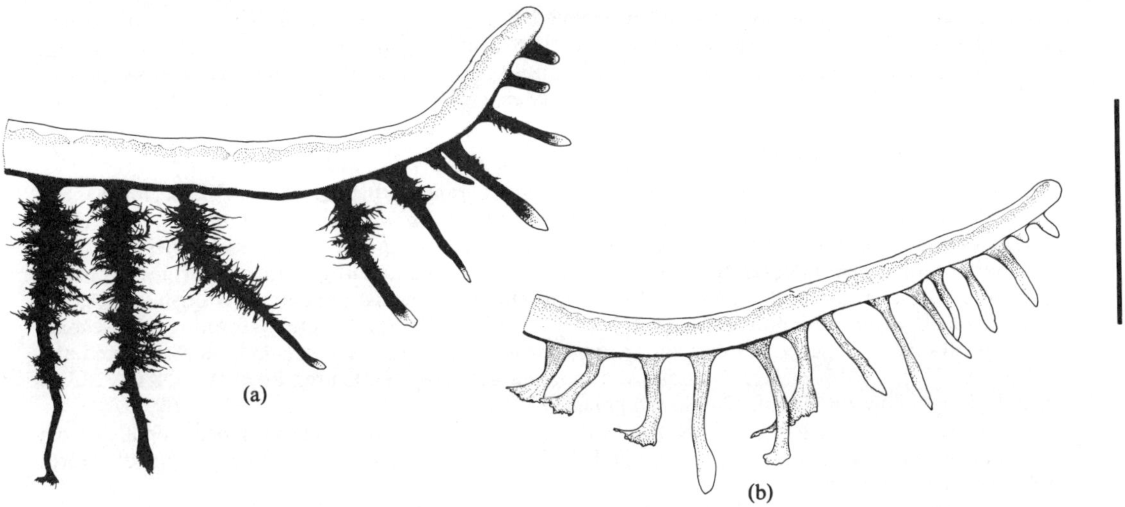

Fig. 22. *Physconia* rhizines. (a) *P. distorta*; (b) *P. grisea*. Scale = 1 mm.

Lecanora-type. **Ascospores** brown, finely warted, 1-septate, thick-walled, with a strong, thickened broad septum, not thickened at the apices. **Conidiomata** pycnidia, immersed in lobe surface or in small warts, walls colourless except for brownish ostiolar region. **Conidiogenous cells** arising in branched chains, short-cylindric, enteroblastic, acrogenous or pleurogenous. **Conidia** bacilliform, simple, colourless. **Chemistry:** lichen products not detected by t.l.c. or with one or more unidentified substances (atranorin absent).

The genus is primarily separated from *Physcia* by the absence of apical thickening in the ascospores and their thick and broad septum.

LITERATURE: Maas Geesteranus (1952), Moberg (1977).

1	Soredia present; apothecia rarely present	2
	Soredia absent; apothecia often present	**1. *distorta***
2(1)	Medulla white, K−	3
	Medulla yellowish, K+ yellow	**2. *enteroxantha***
3(2)	Rhizines simple, mostly pale	**4. *grisea***
	Rhizines bottle-brush-like, black	**3. *perisidiosa***

1. P. distorta (With.) Laundon (1984)

P. pulverulacea Moberg (1979)

Thallus (Fig. 36f) to 10(-15) cm diam., usually orbicular, ± loosely attached; lobes 0.6–2 mm wide, usually radiating, ± separate to overlapping, grey- to dark brown, usually white-pruinose at least at tips, sometimes with secondary lobules in centre of thallus; underside whitish at lobe tips but blackish towards the centre, with black bottle-brush-like rhizines often forming a dense mat (Fig. 22a); upper cortex scleroplectenchymatous. Apothecia to 5 mm diam., numerous, margin often with secondary lobules; disc often pruinose. Ascospores (25-)27–38 × 13–20 μm. Pycnidia frequent, immersed or in small warts 0.2–0.3 mm diam.; conidia 5–7 × 1–1.5 μm. Cortex and medulla Pd−, K−, KC−, C−.

On nutrient-rich or -enriched bark of tree trunks and branches, usually in well-lit situations, occasionally on enriched limestones or stonework, or birds' perching stones; widespread in areas not seriously affected by SO_2 pollution or inorganic fertilizers. Throughout British Is. N. and montane areas of S. Europe, E. Africa, Himalaya, Nepal, W. North America, Australia.

Sometimes confused with *Physcia aipolia* which has a K+ yellow cortex, simple or only forked rhizines, and smaller ascospores. *P. muscigena* (Ach.) Poelt (1965) also lacks soredia but has irregularly arranged, ± ascending lobes and a pseudoparenchymatous upper cortex; an arctic-alpine species growing over mosses, with one dubious British nineteenth century record from C. Scotland (Perth, Blair Atholl).

2. P. enteroxantha (Nyl.) Poelt (1966)

Thallus usually irregular in outline or confluent with others, rarely orbicular, to 5 cm diam.; lobes 0.6–2 mm wide, separate to partly overlapping, grey- or greenish brown, usually pruinose at lobe ends, sometimes totally pruinose, non-pruinose parts matt, marginal secondary lobules sometimes present in older parts; soralia marginal, rarely a few lip-shaped, sometimes extending on to lobe surface in central parts, often ± yellowish, sometimes soredia are isidiate and brownish; underside whitish at tips, but brown to black within, with numerous black, bottle-brush-like rhizines, sometimes forming a dense mat; medulla dull yellow; upper cortex pseudoparenchymatous. Apothecia very rare, margin often sorediate. Ascospores 25–37 × 16–21 μm. Pycnidia rare; conidia 4–6 × 1 μm. Cortex Pd−, K−, KC−, C−; medulla K+ yellow (unidentified yellow pigment).

On well-lit, nutrient-rich bark of tree trunks in parkland and wayside situations, occasionally on old walls and memorials; common. Throughout British Is, especially the north. Europe, Nepal, North America, New Zealand.

Usually recognized by the yellowish, K+ yellow medulla. *P. grisea* **(5)** additionally differs in the simple rhizines, and *P. perisidiosa* **(3)** in the mainly lip-shaped soralia, often striking mauve tinge, and scleroplectenchymatous upper cortex.

3. P. perisidiosa (Erichsen) Moberg (1977)

Thallus usually irregular in outline, ± loosely attached; lobes 0.5–1.2 mm wide, usually short, ± imbricate, dark brown, often pruinose (especially at tips), usually with a mauve or purplish tinge, non-pruinose parts ± glossy; soralia lip-shaped on central lobes, marginal lobes usually not sorediate, often confluent and densely isidiate in centre of thallus, usually with mauve tinge; upper cortex scleroplectenchymatous; medulla white; underside whitish and ecorticate at tips, becoming blackish and corticate within, with black bottle-brush-like rhizines. Apothecia very rare, thalline exciple often with sorediate lobules. Ascospores (24-)28–35 × 16–21 μm. Pycnidia rare; conidia 4–6 × 1 μm. Lichen products not detected by t.l.c.

On trunks of trees with basic bark, often overgrowing mosses, in parklands, avenues and woodland margins; occasionally on old walls or memorials; rather local. Throughout British Is. Generally scarcer than *P. grisea* **(4)** (in S. Britain) and *P. enteroxantha* **(2)** (in N. Britain), and less tolerant of polluted or excessively enriched conditions. Europe, E. Africa, W. Himalaya, Saudi Arabia, North America.

Usually distinguished from *P. grisea* **(4)** and *P. enteroxantha* **(2)** by the short, imbricate lobes with lip-shaped soralia, purple or mauve hue, and the structure of upper cortex. *P. grisea* **(4)** also has simple rhizines, and *P. enteroxantha* **(2)** a yellowish medulla. Superficially resembles *Pannaria mediterranea* which has *Nostoc* as photobiont.

4. P. grisea (Lam.) Poelt (1965)

Thallus irregular and confluent with others, sometimes orbicular to 8 cm diam., ± closely attached; lobes 0.6–2 mm wide, ± radiating and partly overlapping, grey to grey-brown, usually pruinose at least at tips, non-pruinose parts matt; fragile isidia or granular soredia present along lobe margins or on surface in centre of thallus, sometimes forming a dense, granular crust; upper cortex pseudoparenchymatous; medulla white; underside whitish, or pale brown in central parts, with whitish to brownish or grey, simple rhizines (Fig. 22b). Apothecia to 3 mm diam., uncommon; thalline exciple sorediate; disc usually pruinose. Ascospores 22–34 × 12–17 μm. Pycnidia frequent but inconspicuous; conidia 4–7 × 1–2 μm. Medulla K− (4 unidentified substances by t.l.c.).

On basic, usually dust-impregnated bark of tree trunks in parklands and waysides, also frequently on rocks, especially on calcareous walls and memorials; common in lowland areas, even where moderately polluted. Throughout British Is. but rare north of the C. Valley of Scotland and there apparently eastern. Europe, North America, Nepal, Australia, New Zealand.

Distinguished from other *Physconia* species by the simple rhizines, but often confused with the superficially similar *P. enteroxantha* **(2)** and *P. perisidiosa* **(3)**.

PILOPHORUS Th. Fr. (1857)

Primary thallus crustose, effuse, granular to granular-areolate. **Secondary thallus** of short pseudopodetia, with granular to granular-areolate surface. **Cephalodia** present. **Cortex** poorly developed. **Photobiont** chlorococcoid. **Ascomata** apothecia, terminal at apices of pseudopodetia, immarginate, black. **Thalline exciple** absent. **Hypothecium** dark brown. **Hamathecium** of slender, sparingly branched and anastomosed paraphyses, with scarcely swollen apices. **Asci** 8-spored, clavate, *Porpidia*-type. **Ascospores** simple, ellipsoid to fusiform, colourless, without a distinct perispore. **Conidiomata** pycnidia, apical on short pseudopodetia, often surrounded by a collar of granules. **Conidigenous cells** slender, acrogenous. **Conidia** sickle-shaped, simple, colourless. **Chemistry**: atranorin, ± zeorin. **Ecology**: on rocks, montane; *c*. 11 species mainly occurring on rocks in temperate regions, especially in the N. Hemisphere.

LITERATURE: Jahns (1970, 1981).

1. P. strumaticus Nyl. ex Crombie (1875)

Primary thallus of whitish or pale grey areoles; areoles granular to minutely squamulose, 0.1–0.3(-0.5) mm diam., scattered to confluent, or crowded into warty aggregations, *c*. 1 mm diam. Photobiont cells 7–14(-19) μm diam. Cephalodia 0.3–1.2 mm diam., dark brown, granular or wrinkled, containing *Stigonema*. Pseudopodetia to 2 mm tall, 0.3-0.5 mm diam., pin-like, unbranched, surface granular-areolate. Apothecia 0.5–1 mm diam., hemispherical to subglobose, epithecium dark olive-green, K−, N+ red; hymenium 70–80 μm tall, I− (except for I+ blue outer coats of asci); hypothecium dark brown, K−, N−, with underlying tissue of pseudopodetium reddish brown, K+ purple-reddish. Paraphyses 1.5–2.5 μm wide; apices ± slightly swollen to 3 μm wide. Ascospores (14-)16–21(-25) × 6.5–9.5 μm. Pycnidia 0.26–0.5 mm tall, 0.16–0.4 mm wide, ellipsoid to elongate-ampulliform or skittle-shaped, black, borne terminally on short pseudopodetia to 0.5 mm tall; conidiogenous cells 14–20 × 1.5 μm; conidia 7–11 × 0.8 μm. Thallus Pd− or faintly yellow, K+ yellow, KC−, C− (atranorin).

On the shaded sides or bases of damp, siliceous rock faces or boulders, less often on upper surfaces of low outcrops or small boulders; local. N. & W. British Is. W. Scandinavia.

Characterized by the pin-shaped pseudopodetia with black terminal apothecia giving a match-like appearance. Rare morphs of *Micarea lignaria* with stalked apothecia are similar but lack cephalodia and have a Pd+ red thallus and septate ascospores. Occasionally found sterile but can usually be identified by the characteristic, rather large, short-stalked pycnidia.

PLACIDIOPSIS Beltram. (1858)

Thallus squamulose or ± crustose. **Upper Cortex** pseudoparenchymatous, attached to the substratum by intricate colourless or brownish rhizoidal hyphae. **Photobiont** *Trebouxia*-like. **Ascomata** perithecia, immersed in the squamules. **Involucrellum** present or absent. **True exciple** of tangentially arranged, ± elongate cells, pigmented around the ostiole, the lower part pale or darkening. **Hamathecium** of periphyses; paraphyses absent. **Asci** 8-spored, thin-walled, clavate. **Ascospores** clearly 1-septate, biseriate, ellipsoid, colourless. **Conidiomata** unknown. **Chemistry**: lichen products not detected by t.l.c. **Ecology**: mostly on soil or mosses.

Closely related to *Catapyrenium*, but differing in having clearly 1-septate, not simple, ascospores.

LITERATURE: Breuss (1983), Breuss & Hansen (1988).

1 Squamules partly ascending, ± imbricate and divided, without pruina, somewhat tumid; rhizoidal hyphae colourless .... **1. *cartilaginea***
Squamules entirely firmly adpressed, rounded to slightly incised, often whitish pruinose; rhizoidal hyphae brown .... **2. *pseudocinerea***

1. P. cartilaginea (Nyl.) Vainio (1921)

P. custnani (Massal.) Körber (1863)

Squamules 1–2 mm diam., ascending, clustered or crowded, often ± imbricate, spreading, incised or divided, ± tumid (especially on the margins), sometimes ± nodulose-subcrustose, pale olive brownish to medium brown on upper side, green-olive when wet, often with ± eroded and snail-browsed whitish patches, underside dark brown, paler towards the margins; rhizoidal hyphae colourless. Perithecia frequent, 1–5 per squamule, immersed with a ± elevated, dark ostiole. True exciple pale to red-brown. Ascospores (12-)14–20 × 5.5–5.7 μm.

On ± consolidated, ± moist soil, associated with limestones, shell-sand dunes, mica-schist, serpentines, rarely on old mortar; widespread, but rare. Coastal and inland S.W. England, Wales, W. Scotland (up to 900 m in Highlands), W. Ireland (Mayo). Widespread but local throughout Europe, mainly in C. & S. Europe, northwards to S. Finland.

2. P. pseudocinerea Breuss (1983)

Squamules to 4 mm across, entirely closely adpressed, the margins not or obscurely lobulate, scattered or aggregated in small groups, rounded or slightly incised, light brownish and not or weakly pruinose and then with whitish grey tinge, faintly roughened, sometimes with thin intersecting lines forming a mosaic; underside blackish, with pigmented rhizoidal hyphae forming a dense dark hypothallus. Perithecia ± ¾-immersed; ostiole sometimes slightly elevated; true exciple at first colourless, becoming brownish and eventually blackish around the ostiole. Ascospores (14-)15–18(-20) × (7-)7.5–9.5(-10) μm, ellipsoid to slightly ovoid.

On calcareous soil or associated humus at high altitude, above 900 m. C. Scotland (Grampian, Ben Alder). Arctic-alpine.

Closely resembles *Catapyrenium cinereum* which has simple, more elongate, oblong-clavate ascospores; however, it should be noted that a thin pseudoseptum can occur in a proportion of the ascospores of this species which can lead to errors of identification with *P. pseudocinerea*.

PLACOPSIS (Nyl.) Lindsay (1866)

Thallus crustose, placodioid, with deeply incised, radiating, marginal lobes. **Cephalodia** often present. **Upper cortex** paraplectenchymatous, often overlain by a colourless, necrotic zone. **Photobiont** chlorococcoid. **Ascomata** apothecia, sessile. **Thalline exciple** present. **True exciple** thin, often visible in old apothecia, ± colourless, of elongated, thin-walled cells. **Epithecium** yellowish brown. **Hymenium** colourless, I+ blue. **Hypothecium** colourless. **Hamathecium** of paraphyses, branched, free, upper part ± moniliform, with minute, yellowish external granules. **Asci** 8-spored, mostly uniseriate, cylindrical to narrowly clavate, amyloid, with a K/I + pale blue tholus and a narrow, internal darker blue cap. **Ascospores** colourless, ellipsoid, simple, often somewhat pale pink with densely granular contents. **Conidiomata** pycnidia, immersed, causing slight swellings of the thallus, opening by a minute, dark ostiole to 0.1 mm diam. **Conidia** colourless, thread-like, simple. **Chemistry**: gyrophoric and ± lecanoric acids. **Ecology**: primarily on rocks and recently disturbed soils; *c.* 35 species mainly on siliceous rocks and alluvial soils in the S. Hemisphere.

LITERATURE: Coppins & James (1984), Hafellner (1984), Lamb (1947).

1. P. gelida (L.) Lindsay (1866)

Thallus dull whitish, pinkish or yellowish brown, closely adpressed to substratum, forming rosettes *c.* 1–5 cm diam., cracked to rimose-areolate in the central part, with deeply incised, ± radiating, marginal lobes, 0.5–1.5 mm wide; surface with reddish, pinkish or yellowish brown cephalodia, 0.5–3 mm diam., wart-like or placodioid. Soralia ulcerose (later excavate), greenish or greyish white, 0.3–0.8 mm diam., more numerous towards the centre of the thallus. Apothecia 0.8–1.6 mm diam., occasional, sessile; thalline exciple thick, entire; disc flat, rough, dark pink-, yellow- or red-brown, sometimes whitish pruinose. Ascospores 13–18 × 6–8.5 μm. Thallus C+ red (gyrophoric acid, ± trace lecanoric acid).

On damp, large, siliceous, often ± metal-rich boulders protruding from, and by, streams and on wall tops, scree, mine spoil-heaps, rarely on worked timber, uplands; common. N. & W. British Is. Arctic and temperate oceanic Europe, North America, Chile, E. Asia, Macaronesia, New Zealand.

Distinguished by the pinkish cream-coloured, closely adpressed, radiating thallus, often with pink-orange tuberculate cephalodia. The discs of the apothecia are often strikingly rose-pink when wet. *P. lambii* Hertel & V. Wirth (1987) is possibly an ecotype; the thalli lack cephalodia and are smaller, to only 2.5 cm diam., and often fragmented with only occasionally somewhat enlarged marginal lobes; apothecia rather rare. It is encountered on rocks rich in iron sulphide minerals, such as at old mining sites throughout the British Is.

Well-developed thalli of *Trapelia placodioides* may resemble *P. lambii* but lack any trace of the ± broadened, incised, ± radiating marginal lobes; *T. placodioides* also has smaller soralia, 0.2–0.3 mm diam.

PLACYNTHIELLA Elenkin (1909)

Thallus crustose, effuse, of minute, granular or isidiate-granular goniocysts or areoles, dark green-brown, at times tinged red or chestnut-brown. **Goniocysts** and **areoles** with an outer layer of brown-walled pseudoparenchymatous hyphae, colourless within. **Photobiont** ?*Chlorella*. **Ascomata** apothecia, adpressed to sessile, red-brown to black. **Thalline exciple** absent. **True exciple** brown, pseudoparenchymatous, often reflexed. **Hamathecium** of paraphyses, often branched, apices ± capitate, dark brown, gel matrix I+ green-blue. **Asci** 8-spored, cylindrical-clavate, *Trapelia*-type, apical dome not, or weakly I+ blue. **Ascospores** simple (to 1-septate), ellipsoid, often containing a single large oil drop and several smaller droplets. **Conidiomata** unknown. **Chemistry**: lichen products not detected by t.l.c. or with gyrophoric and ± lecanoric acids. **Ecology**: on acidic, lime-free substrata, particularly soils, bark and wood, less often on rocks.

Differs from *Trapelia* and *Trapeliopsis* in having a true exciple of brown-walled pseudoparenchymatous cells and paraphyses with irregularly capitate apices, each crowned with a dark brown apical cap.

LITERATURE: Coppins & James (1984), Coppins, James & Hawksworth (1987).

1 Thallus minutely granular-isidiate, coralloid or ± sub-gelatinous; granules <100 µm diam., C+ red or C− 2
 Thallus granular-warted; granules 100–300 µm diam., C− .... **2. *oligotropha***

2(1) Thallus of isidiate to minutely coralloid-elongated granules; not sub-gelatinous; C+ red (in squash preparations); true exciple ± persistent, disc flat or slightly convex **1. *icmalea***
 Thallus of rounded, never elongated, granules; ± sub-gelatinous, C− ; true exciple soon excluded, disc soon convex .... **3. *uliginosa***

1. P. icmalea (Ach.) Coppins & P. James (1984)

Thallus of isidiate to minutely coralloid granules, black-brown to red-brown, ± shining, 25–150(-200) µm tall, 25–50 µm diam., closely packed or sometimes scattered, effuse. Apothecia 0.2–0.6 mm diam., often absent, usually scattered when present, rarely crowded and confluent, usually with a distinct, often paler, true exciple. Ascospores 8–12 × 4–5 µm. Thallus (squashes) Pd−, K−, KC+ red, C+ red (gyrophoric, ± lecanoric, ± 5-*O*-methylhiasic acids).

On a wide range of acidic habitats, occurring on dead bark and wood of fallen trees and stumps, rotting fence-posts, plant debris, tree roots, ± horizontal branches, shaded siliceous rocks and humus-rich soils, also a primary colonizer, with *Trapeliopsis granulosa*, of heathland following burning; often abundant, sometimes dominant. Throughout British Is. Europe, North America.

Easily noticed when growing as a wide-spreading, brown, isidiate crust, but often occurs as less conspicuous, scattered, sterile, small clusters of coralloid granules amongst other lichens, including *P. uliginosa* **(3)**. The recently recognized *P. dasaea* (Stirton) Tønsberg (1992) is probably common; it differs in having a paler, often greenish thallus of soft, sorediose granules and absence of 5-*O*-methylhiasic acid.

2. P. oligotropha (Laundon) Coppins & P. James (1984)

Lecidea oligotropha Laundon (1960)

Thallus coarsely granular-verrucose; granules 100–300 μm diam., ± contiguous or dispersed, brown, often tinged yellowish green when wet, soredia absent. Apothecia 0.2–0.5 mm diam., brownish black, at first flat with a ± elevated true exciple, occasionally becoming convex and immarginate, surface rugose-roughened, matt; exciple deep red-brown, extending into a concolorous hypothecium; epithecium 10–15 μm, red-brown, granular; hymenium 55–60 μm, colourless or pale reddish brown. Ascospores 10–14 × 4.5–6 μm. Thallus Pd−, K−, KC−, C− (lichen products not detected by t.l.c.).

On peaty soil in open heathlands and woodland clearings, often around old stumps, but never on bark; local. Throughout British Is., but especially in E. and S.E. England. Europe, North America.

Distinguished from other species in the genus by the larger granules which become noticeably more prominent and paler when wet.

3. P. uliginosa (Schrader) Coppins & P. James (1984)

Lecidea uliginosa (Schrader) Ach. (1803), ?*Lecidea uliginascens* Stirton (1878)

Thallus of low, rounded granules, 25–100 μm diam., black-brown, granules never markedly vertically elongated or branched, often minutely wrinkled when dry (× 50), dark green to dark brown, ± gelatinous when wet. Apothecia 0.12–0.3(-0.4) mm diam., always abundant, often coalescing, pinkish to dark reddish brown (rarely blackish); true exciple present when young, usually soon disappearing as apothecium becomes convex. Ascospores 9–14(-16.5) × (4-)5–6(-7) μm. Thallus Pd−, K−, KC−, C− (lichen products not detected by t.l.c.).

On peaty, heathland soils, also on dead bark and wood of fallen trees and stumps, occasionally on bare sandy soil in conifer plantations and stable acid dunes; frequent. Throughout British Is. Europe, North America.

Characterized by the C−, subgelatinous, non-coralloid, granular thallus, and the numerous, concolorous, soon convex, apothecia. Less common than *P. icmalea* **(1)**.

P. hyporhoda (Th. Fr.) Coppins & P. James (1984) is similar but the apothecia are blackish, even when moist, with a prominent, ± incurved true exciple and a concave disc when young, the disc later becoming flat to slightly convex; hypothecium purple-violet in K; ascospores 9.5–12 × 4–6 μm. Recently found on soil in a lead mine in Wales (Cardigan).

PLACYNTHIUM (Ach.) Gray (1821)

Thallus closely attached, often rosette-shaped, grey, brown, olive or blackish, sometimes densely grey or bluish pruinose; margin variable, of distinct radiating, flattened to convex lobes, or stellate, spreading squamules, or without lobes, central area ± continuous to markedly areolate; prothallus absent, or ± well developed in some species, forming a ± extended, conspicuous, blue-black margin. **Lower surface** and associated hyphae dark blue-green. **Isidia** often ± densely present. **Photobiont** cyanobacterial (Rivulariaceae or Scytonemataceae), dispersed and not layered (homoiomerous). **Ascomata** apothecia. **True exciple** and disc dark brown to black, often shining, true exciple opaque, black, tinged brown or purple. **Hymenium** blue or green in upper part often tinged brown-violet. **Hypothecium** brownish or red-brown. **Hamathecium** of paraphyses, septate, simple or sparingly branched, apical cells pointed, thickened or not different in shape. **Asci** (4-)8-spored, cylindrical, *Peltigera*-type. **Ascospores** (0-)2- to 4(-7)-septate, ellipsoid to fusiform, often poorly developed, colourless. **Conidiomata** pycnidia, partly immersed, with dark ostioles. **Conidiophores** branched. **Conidia** bacilliform or dumb-bell-shaped. **Chemistry**: lichen products not detected by t.l.c., dark blue and purplish pigments in the apothecia. **Ecology**: on rocks frequently in damp situations but some species on dry, exposed limestone, very rarely on soil or bark; mainly North Temperate.

Immature thalli, poorly developed material and collections with a poorly developed thallus margin are frequently impossible to name. Most species show considerable variation in colour and morpology.

LITERATURE: Henssen (1963*c*).

1 Margin of thallus, at least in part, with distinct, minute to medium-sized, radially extended marginal lobes 2
Margin of thallus without distinct radially extended lobes 7

2(1) Thallus robust; marginal lobes 0.5–1 mm wide, conspicuous, ± flattened, fan-shaped or ± closely parallel; on moist acid rocks and seepage tracks 3
Thallus more delicate; marginal lobes 0.05–0.2 mm wide, ± inconspicuous, flattened, convex or terete to thread-like, often irregularly orientated; on ± basic rocks, often dry limestones 4

3(2) Marginal lobes ± flattened, *Phaeophyscia*-like, grey to brown, inner part of thallus with ± flattened, decumbent, squamule-like isidia or folioles; prothallus usually not well developed .... **2. *flabellosum***
Marginal lobes ± convex to terete *Vestergrenopsis*-like, dark olive to blackish, inner part with ± terete, erect, finger-like or coralloid isidia; prothallus usually distinct, blue-black or black .... **6. *pannariellum***

4(2) Thallus densely glaucous violet-grey-pruinose, or forming distinct rings or arcs 5
Thallus not pruinose, not forming rings or arcs 6

5(4) Thallus not or little grey-pruinose, forming distinctive, radiating rings or arcs, centre ± bare or with dark olive-brown, ± scattered, nodular-contorted areoles; ascospores 11–13 × 5–6 μm, 1-septate .... **8. *subradiatum***
Thallus densely violet glaucous-grey-pruinose, ± continuous, not forming arcs or rings, centre ± powdery-granular; ascospores 23–53 × 4–6 μm, 3- to 7-septate .... **3. *garovaglii***

6(4) Marginal lobes thread-like, convex, ± shining; thallus centre densely granular-isidiate; ascospores (11.5-)15–19(-21) × (4.5-)5.5 μm, (1-)3-septate .... **1. *asperellum***
Marginal lobes flattened to convex, finger-like, fimbriate, olive to blackish, matt; thallus centre uneven-nodulose; ascospores (10-)13–15 × 5–6.5 μm, 1-septate .... **4. *lismorense***

7(1) Ascospores (7-)9–18(-22) × 3.5–8(-9) μm, (0-)3-septate, ellipsoid or elongate-ellipsoid 8
Ascospores 30–45 × 4–5 μm, 3- to 5-septate, elongate-acicular .... **7. *pluriseptatum***

8(7) Ascospores (3.5-)4–5(-6) μm wide; thallus ± uniform in colour or ± glaucous-grey-pruinose, not mottled; prothallus often very distinct at the margin 9
Ascospores (5.5-)6–8(-9) μm wide; thallus minutely mottled grey-white and blue-black; prothallus indistinct at the margin .... **9. *tantaleum***

9(8) Ascospores (0-)1-septate .... **10. *tremniacum***
Ascospores (1-)3-septate .... **5. *nigrum***

1. P. asperellum (Ach.) Trevisan (1869)

Thallus to 2 cm diam., rosette-like, olive to black, shining; lobes very narrow, 1.5 × 0.05–0.2 mm wide, ± decumbent and radiating at the margin, the centre becoming densely and continuously granular or isidiate and forming coarse, thick areoles, 1–2 mm diam., which may themselves develop an independent border of narrow lobes; isidia nodular, granular-coralloid, or ± erect, extended-cylindrical and finger-like; prothallus seldom visible. Apothecia concolorous with the thallus, ± immersed to sessile; epithecium blue-green-black; hypothecium ± reddish brown, K± intensified. Asci (4-)8-spored. Ascospores (11.5-)15–19(-21) × (4.5-)5.5–7 μm, (1-)3-septate, ellipsoid or slightly clavate, ± indented at the septa.

On moist mica-schist; very local. C.N. Scotland (Perth, Ben Lawers Range). Circumpolar in the N. Hemisphere.

The very narrow, shining, marginal lobes forming a small, rosette-like thallus with a granular-isidiate, areolate centre are diagnostic. British collections are not typical in that they lack any extended marginal lobes. *Parmeliella triptophylla*, which sometimes grows on damp rocks, has flattened lobes and *Nostoc* as photobiont.

2. P. flabellosum (Tuck.) Zahlbr. (1925)

Thallus to 4 cm diam., of rosettes or parts of rosettes, bluish grey, dull grey, brown or olive, dark green when wet; conspicuously divided at the border into very narrow, flattened, *Phaeophyscia*-like lobes, to 3 × 0.3–0.5 mm wide, the apices to 1 mm wide, narrowly fan-shaped, flat, closely attached, surface smooth to lightly striate. Thallus centre ± densely covered by flat, grey-brown, small, scattered or ± contiguous, rarely overlapping, squamule-like lobules with notched or crenulate margins, sometimes forming coarse areoles 0.5–1 mm diam., which may become thickened (1–2 mm) due to repeated regeneration of lobes; prothallus inconspicuous. Apothecia unknown in British Is., small, concave, black. Ascospores (13-)15–18(-20) × 7–8 μm, (2-)3-septate.

On acid rocks in mountain streams, by lake margins, and associated with seepage tracks and frequently submerged; local. N. & W. British Is. Europe, North America.

The habitat, substantial dull grey-brown thallus with numerous, conspicuous, spreading, closely attached *Phaeophyscia*-like marginal lobes and flattened, squamule-like, inner lobules are distinctive. In habit, this species has a superficial resemblance to *Hyperphyscia adglutinata*.

3. P. garovaglii (Massal.) Malme (1918)

Thallus thick, seldom exceeding 2 cm diam., at least part of the margin obscurely lobed; prothallus not visible; marginal lobes, when present, very slender, 0.1–0.3 × 0.5–1 mm, centre with uneven, dense, flat to convex, nodular-scurfy, areoles 0.1–0.4 mm diam., the entire thallus powdery with a dense, violet or bluish white pruina, dull grey-brown when wet. Apothecia unknown in British Is., 0.3–0.6 mm diam., black, ± sessile. Ascospores 23–53 × 4–6 μm, 3- to 7-septate, elongate-fusiform.

On vertical to overhanging, dry, limestone rocks; rare. S.W. England (Somerset, Cheddar Gorge), N. England (Yorks), Wales, Scotland. S. & C. Europe, N. Sweden (to Gotland).

Distinguished by the thick, crumb-like, scurfy, uneven, pruinose, lead-grey thallus; the marginal lobes in British material are usually poorly developed.

Strongly pruinose specimens of *P. nigrum* commonly have a well-developed prothallus.

4. P. lismorense (Nyl. ex Crombie) Vainio (1909)

Thallus wide-spreading or in small patches, thin, ± with delicate, radiating, finger-like marginal lobes, best seen in young thalli, olive-black; prothallus ± indistinct; marginal lobes 0.1 mm wide, convex; inner lobes similar, distinct, elongate, decumbent, unorientated, overlapping, forming nodular areoles. Apothecia 0.2–0.7 mm, with thin, shining exciple; epithecium grey-green to blue-black; hypothecium pale violet-brown. Ascospores (10-)13–15 × 5–6.5 μm, 1-septate, broadly ellipsoid, often ± curved.

On hard limestone and other calcareous rocks; rare. W.C. Scotland (Is. of Mull & Lismore). Sweden, Alps, Carpathians.

This species is distinguished from *P. nigrum* chiefly by the more conspicuous, radiating marginal and distinct central lobes, the olivaceous tone and the absence of a well-developed, bluish black prothallus. The lobes in the inner part of the thallus are sometimes much reduced and indistinct, forming compacted, thin, smooth, ± scattered areoles; even so, no prothallus is visible.

5. P. nigrum (Huds.) Gray (1821)

Thallus to 12 cm diam., often smaller, wide-spreading, brown-black to jet black, sometimes ± grey-violet-pruinose; prothallus blue-black, conspicuous, sometimes fimbriate; thallus of small, crowded, flat, granular squamules, 0.4–1.5 mm across, with crenate or digitate margins, scattered over the prothallus or more usually crowded forming areoles 1–2.5 mm diam.; isidia often present, granular to coralloid. Apothecia 0.5(-1) mm across; true exciple black, shining, often becoming flexuose; disc brown to black, concave, later flat or weakly convex; epithecium blue-green. Ascospores (7-)9–18(-22) × 3.5–5.5(-6) μm, 1- to 3-septate, narrowly ellipsoid.

On a wide range of calcareous rocks, including memorials, stonework, also mortar, asbestos, cement, compacted soil and shell-sand and siliceous substrata indirectly influenced by lime, preferring somewhat damp niches or surfaces slow in drying out; frequent. Throughout the British Is. Circumpolar.

This abundant and very variable species is usually readily recognized by the conspicuous, blue-black prothallus and rather uniformly minutely squamulose thallus; it is the only British species of the genus regularly occurring in man-made habitats.

6. P. pannariellum (Nyl.) Magnusson (1936)

Thallus 1–5 cm diam., often forming wide-spreading regular patches as dark rosettes, often preceded by a conspicuous greenish-blue or bluish-black prothallus; margin strongly radiating, the lobes ± convex to terete, their ends fan-shaped, ± branched, ± radiating, to 0.7–1 mm wide; central lobes smaller fruticose, with numerous cylindrical, coralloid ± flattened isidia, sometimes forming a thickened densely isidiate, areolate crust, dull olive-brown, blackish-brown or black. Apothecia to 1 mm diam., dark brown. Ascospores 15–20 × 5–7 μm, (1-)3-septate.

On ± basic, semi-inundated rocks in streams and lakesides and in seepage tracks, also occasionally on mica-schist soils; rare. N. Wales, N. Scotland (Highlands). N. & C. Europe, Greenland.

The combination of distinct, radiating, marginal lobes, conspicous dark prothallus, and the central areas of mature thalli with a cover of finger-like to flattened isidia is distinctive. This species closely resembles *Vestergrenopsis isidiata*.

7. P. pluriseptatum (Arnold) Arnold (1889)

Thallus crustose without marginal lobes, brown to black-brown; surface strongly areolate, the surface of the areoles with ± conspicuous, low, rounded, granular papillae; prothallus poorly developed. Apothecia frequent, 0.3–0.6 mm diam.; disc concave, flat to convex, ± shiny. Ascospores 30–45 × 4–5 μm, 3- to 5-septate, fusiform-vermicular, mostly straight.

On montane limestones and mica schist often overgrowing mosses; very rare. N. Scotland (Perth, Ben Lawers; W. Inverness, Ben Alder). Europe, N. to Scandinavia.

The 3- to 5-septate, longer ascospores serve to distinguish this species from *P. asperellum* **(1)** and *P. nigrum* **(5)**.

8. P. subradiatum (Nyl.) Arnold (1884)

Thallus to 1.5 cm diam., forming ± complete rosettes or marginal lobes forming rings or narrow arcs, dying away from the centre, 2.5 cm × 3 mm broad, olive to dark-brown, matt, sometimes sparingly bluish-white pruinose. Margin of flattened to ± unevenly convex, radiating or parallel, tightly packed lobes, 2 × 0.05–0.2 mm wide, apices sometimes indented, dark olive-brown to blackish; prothallus inapparent; isidia sparse when present, to 0.4 mm long, globose or subcylindrical, erect, irregular, scattered or crowded into ± discrete, nodulose, contorted areoles at the centre of the thallus; older areoles may become secondarily lobed and generate a new thallus within the original one. Apothecia very rare, to 0.5 mm diam. Ascospores (9-)11–13(-15) × (5-)6–7(-8) μm, 1-septate, ellipsoid.

On dry to irrigated, well-illuminated, hard, often vertical limestone; local. N. & W. British Is. Europe from the Mediterranean coast to Scandinavia, W. Canada.

The dark, narrow, ring to arc-like development of the thallus on dry limestones is characteristic; specimens from moist habitats tend to lack pruina.

9. P. tantaleum (Hepp) Hue (1906)

Thallus thick, to 3 cm diam., of minute papilla-like granules, brownish-black; surface mottled, with minute grey-white spots on a blue-black surface (× 15); centre cracked-areolate, the areoles rather large, to 2 mm diam., on a black hypothallus; prothallus margin reduced or absent. Ascospores (9-)10–13(-20) × (5.5-)6–8(-9) μm, simple or 1-septate, ovoid to broadly ellipsoid.

On semi-irrigated, ± basic rocks, including mortar, rarely on exposed roots of *Taxus*; rare. England, Scotland. Alps, Scandinavia, North America.

Differs from *P. nigrum* **(5)** in the broader ascospores, reduced prothallus, 'maculate' surface and preference for more moist habitats. Sometimes regarded as a variety of *P. nigrum*.

10. P. tremniacum (Massal.) Jatta (1900)

Thallus less than 3 cm diam., formed entirely of small coralloid granules, generally ± cracked-areolate, ± white-pruinose or glaucous; margin not lobed. Ascospores 9–16 × 4–6 μm, simple to 1-septate.

On damp, shaded, hard limestone rocks, occasionally on mortar; local. N. & W. British Is. Europe.

Said to differ from *P. nigrum* **(1)** in the persistently 1-septate ascospores but the relationship between *P. tremniacum* and this species requires further investigation.

PLATISMATIA Culb. & C. Culb. (1968)

Thallus foliose, ± rosette-forming or wide-spreading, dorsiventral, lobe margins often ascending, wavy, sometimes crisped. **Upper** and **lower surfaces** with a well-developed prosoplectenchymatous cortex. **Lower surface** pale to ± black with a few, scattered rhizines. **Isidia** or **soredia** on margins and sometimes on upper surface. **Cortex** often I+ blue. **Photobiont** trebouxioid. **Ascomata** apothecia, very rare, marginal or submarginal. **Disc** brown, often perforate. **Thalline exciple** persistent. **Hymenium** I+ blue. **Hypothecium** thin, I+ blue, usually lacking algae below. **Asci** 8-spored, *Lecanora*-type. **Ascospores** less than 8 μm in length, subglobose-ellipsoid, simple, colourless. **Conidiomata** pycnidia, marginal, immersed. **Conidia** cylindrical, not swollen at apices. **Chemistry**: caperatic acid and atranorin.

Parmelia (*Parmotrema*) differs in often having marginal cilia and depsides and depsidones while *Cetraria*, in the absence of caperatic acid, thin I+ blue hymenium, smaller ascospores and conidia type. See also *Cetrelia*.

LITERATURE: Culberson & Culberson (1968).

1 Upper surface with a continuous coarse network of sharply angled ridges; summits of ridges and lobe margins isidiate, sometimes becoming granular-sorediate ...... **2. *norvegica***
Upper surface smooth, rugose or with a few or an incomplete network of low, rounded ridges; margins predominantly dissected or with simple or coralloid isidia, sometimes also granular-sorediate; isidia, if present on the upper surface, not confined to any ridges ...... **1. *glauca***

1. P. glauca (L.) Culb. & C. Culb. (1968)

Thallus 1–6(-15) cm diam, often forming extensive patches, rather thin; lobes to 1.5 cm wide, wavy, irregularly indented, margins ascending, ± entire or sub-lobulate, often with marginal simple to coralloid, much branched, clusters of isidia or, in part, granular soredia; upper surface pale to dull grey, often with a brown tinge or ± totally tinged brownish (in exposed sites), ± unchanged in colour when wet, smooth to wrinkled, or slightly ridged, without pseudocyphellae; lower surface entirely black, or ± brown or white (shade morphs), especially towards the margin, with few to many, scattered, simple or branched rhizines. Apothecia very rare, 5–9 mm diam., marginal; disc red-brown; thalline exciple thin, ± excluded at maturity. Ascospores 3.5–8.5 × 3–5 μm. Cortex K+ yellow (atranorin), medulla Pd−, K−, KC−, C−, UV− (caperatic acid).

On trees, rocks and soil, especially acidic, leached habitats; common. Throughout British Is. Europe, S. Greenland, North & South America, C. Asia, Macaronesia, Kenya, Tanzania.

A very variable species; the lobes may be elongate, ± entirely erect with much dissected, coralloid-lobulate-isidiate margins or, conversely, ± closely adpressed with only the margins upturned with granular-sorediate isidia. The isidia or soredia are mainly confined to the margins but, on occasion, may also occur in small patches or widely dispersed on the upper surface.

Cetraria chlorophylla has brown lobes turning olive-green when wet. *Cetrelia olivetorum* has scattered, dot-like pseudocyphellae on the upper surface (× 10), a K−, C± pink medulla and finely sorediate margins. *Parmelia perlata* and related species have marginal cilia and a different chemistry. For differences from *P. norvegica*, see **(2)**.

2. P. norvegica (Lynge) Culb. & C. Culb. (1968)

Like **1**, but with broader, more rounded lobes, to 2.5 cm, with rounded, sparingly indented margins; upper surface dull pale grey, not or sometimes ± tinged brownish, with a very pronounced network of low, sharp ridges with indistinct pseudocyphellae and simple or clustered isidia. Apothecia not seen.

Mainly in moss-lichen montane heaths of the *Arctoeto-Callunetum*, as well as on exposed, acid rocks and on well-lit, leached, low pH bark, especially of *Pinus*, *Betula* and *Quercus* in relict Caledonian forests; local and rare. N. and N.W. Scotland (Highlands). Scandinavia, Newfoundland, W. North America.

Distinguished from *P. glauca* **(1)** by the network of ± isidiate, sharp-angled ridges.

POELTINULA Hafellner (1984)

Thallus crustose. **Photobiont** chlorococcoid. **Ascomata** apothecia, angular to shortly lirelliform, occasionally forked, sometimes contorted, black. **Disc** slit-like to expanded. **True exciple** prominent and inflexed, dark reddish-brown, K ± reddish intensifying. **Hymenium** colourless, I+ blue. **Hypothecium** dark reddish brown. **Hamathecium** of thin, mostly simple, but sometimes anastomosed or branched near the apices, paraphyses, apices slightly swollen, some with a thin, blackish apical cap. **Asci** clavate, *Rhizocarpon*-type, 8-spored. **Ascospores** ellipsoid to oblong, 1-septate, at first colourless, soon becoming grey-green to violet-black, N+ red, with a distinct epispore and colourless ± gelatinous sheath. **Conidiomata** unknown. **Chemistry**: no lichen substances reported. **Ecology**: on rocks. **Distribution**: 2 or 3 species in Europe and North America.

Superficially like *Melaspilea*, but the ascus-type indicates a closer relationship to *Catolechia*, *Epilichen* and *Rhizocarpon.*

LITERATURE: Hafellner (1984), Renobales & Aguirre (1990).

1. P. cerebrina (DC.) Hafellner (1984)

Encephalographa cerebrina (DC.) Massal. (1856)

Thallus mostly rather thick, to 0.6 mm, chalky white, ± bluish tinged, continuous to areolate, sometimes eroded and undulating, effuse or weakly delimited. Photobiont cells 7–15 μm diam. Apothecia 0.5–2 mm diam. or elongate, scattered or in groups, sessile; epithecium olive-grey or olive-brown, K−, N+ red; hymenium 70–100 μm tall. Paraphyses 1.5–3 μm wide, apices with dark pigmented hood to 5 μm wide. Ascospores 15–23 × 8–12 μm, oblong-ellipsoid, slightly constricted at the septum, becoming violet-black but remaining ± colourless around the septum. Thallus Pd−, K−, KC−, C−.

On hard Carboniferous limestone, especially vertical surfaces; rare. N. England, N. Wales. European mountains (excluding Scandinavia), ?Canada (British Columbia).

Encephalographa interjecta (Leighton) Th. Fr. (1874) has been considered a synonym of *P. cerebrina*, but is referred to *Melaspilea* s. lat.

POLYBLASTIA Massal. (1852)

Thallus crustose, immersed or superficial, often inconspicuous, smooth to uneven, rarely subsquamulose. **Soredia** rarely present (see *P. efflorescens* **(15)**). **Photobiont** chlorococcoid. **Ascomata** perithecia, superficial or immersed in thalline warts, forming pits in limestone. **Involucrellum** present or absent. **Thalline exciple** in a few species. **True exciple** completely black and friable, or pale below. **Hamathecium** of periphyses; paraphyses absent at maturity. **Asci** (1-)2- or 8-spored, saccate, *Verrucaria*-type, thin-walled. **Ascospores** muriform or submuriform, ± ellipsoid or ovate, rarely reniform, colourless to deep brown, sometimes with a gelatinous perispore. **Conidiomata** pycnidia, rare, immersed. **Conidia** bacilliform, produced terminally from short conidiogenous cells. **Chemistry**: lichen products not detected. **Ecology**: on limestones, mica-schist, epidiorite, chalk pebbles, mortar, brick, on soil, overgrowing mosses, on siliceous rocks in and by streams and lakes, rarely on bark.

Closely related to *Staurothele*, in which the algal cells occur in the hymenium. *Polyblastia discrepans* Lahm (1867), recorded from N. Wales, has been transferred to *Merismatium discrepans* (Lahm) Triebel (1989) [syn. *M. lopadii* auct. brit. p.p.], a parasite of *Protoblastenia* species.

LITERATURE: Coppins & James (1978), Swinscow (1971).

1	On rocks, soil or mosses on the ground	2
	On bark	**3. *allobata***
2(1)	On rocks	3
	On soil or mosses on the ground	14

3(2) Ascospores colourless 4
Ascospores brown, a few colourless ones occasionally present 10

4(3) Ascospores >20 μm long; on ± calcareous substrata 5
Ascospores 12–20 μm long; on acid rocks in streams .... **13. *quartzina***

5(4) Ascospores submuriform, with few transverse septa and 1(-2) longitudinal septa 6
Ascospores muriform 7

6(5) Thallus immersed; perithecia mainly immersed; involucrellum absent; ascospores 25–50 × 12–25 μm .... **7. *dermatodes***
Thallus superficial; perithecia prominent; involucrellum present; ascospores 30–45(-68) × 15–20 μm .... **17. *verrucosa***

7(5) Ascospores <50 μm long; thallus immersed or superficial 8
Ascospores >50 μm long; thallus superficial 9

8(7) Thallus mainly immersed, at most a white-grey scurfy crust; perithecia immersed; involucrellum absent or hardly distinct .... **2. *albida***
Thallus mainly superficial, pale to dark grey; perithecia prominent; involucrellum present .... **5. *cupularis***

9(7) Thallus greenish or pale brownish grey, coarsely granular to smooth, widely cracked; perithecia immersed in thalline warts .... **15. *terrestris***
Thallus dark brown or dark grey-black, continuous or finely cracked; perithecia at times with a thalline exciple but not immersed in thalline warts .... **11. *inumbrata***

10(3) Perithecia prominent; involucrellum present; ascospores fully muriform, without thick perispore 11
Perithecia immersed; involucrellum absent; ascospores submuriform with thick, perispore .... **6. *deminuta***

11(10) On rocks in streams and lakes; thallus olivaceous .... **4. *cruenta***
On dry rocks; thallus pale to dark grey or brownish grey 12

12(11) Ascospores >50 μm long 13
Ascospores 25–40 μm long .... **12. *melaspora***

13(12) Ascospores 50–75 × 20–35 μm .... **11. *inumbrata***
Ascospores 60–100 × 30–60 μm .... **16. *theleodes***

14(2) Ascospores colourless 15
Ascospores brown 18

15(14) Asci (2-)8-spored 16
Asci 2-spored .... **1. *agraria***

16(15) Ascospores >30 μm long 17
Ascospores <30 μm long .... **14. *sendtneri***

17(16) Perithecia with thalline exciple; involucrellum present .... **15. *terrestris***
Perithecia lacking thalline exciple; involucrellum absent .... **8. *gelatinosa***

18(14) Asci (2-)8-spored; ascospores ellipsoid 19
Asci 1- or 2-spored; ascospores cylindrical-ellipsoid .... **10. *helvetica***

19(18) Ascospores 80–125 × 40–60 μm .... **18. *wheldonii***
Ascospores 20–30(-36) × 10–15 μm .... **9. *gothica***

1. P. agraria Th. Fr. (1864)

Thallus film-like, often inconspicuous, intermixed with minute granules, ± scurfy, green-grey, tinged brown in exposed sites, forming small patches, or ± granular and well delimited, or effuse. Perithecia minute, 0.1–0.2 mm diam., half immersed; involucrellum absent; true exciple black, slightly brownish below. Asci 2-spored. Ascospores 40–70 × 18–25 μm, colourless, becoming pale yellow-brown when old, muriform, elongate-ellipsoid.

Overgrowing mosses on ± calcareous dunes, also on bone; rare. N. England (Lancashire, Ainsdale), E. Scotland (E. Lothian, Bellhaven Bay), S.W. Scotland (Ayr, Ballantrae). Sweden.

Characterized by the 2-spored asci, colourless ascospores and habitat.

2. P. albida Arnold (1858)

Thallus immersed, occasionally ± superficial, scurfy or inconspicuous, sometimes minutely cracked-areolate, ± smooth, white to greyish white, ± bordered by a black prothallus, forming small colonies, rarely wide-spreading, with few or many empty pits. Perithecia 0.2–0.4(-0.5) mm diam., immersed in pits, scattered, numerous, often ± depressed; involucrellum absent; true exciple black or dark brown, base sometimes paler. Asci 8-spored. Ascospores 25–50 × 12–25 μm, colourless, muriform, ellipsoid.

On exposed, hard limestone rocks, particularly embedded in short turf, oolite detritus, occasionally on mortar of old walls. Throughout British Is. Europe.

Resembles several *Verrucaria*, *Staurothele* and *Polyblastia* species [especially *P. dermatodes* **(7)**], which cannot be reliably differentiated without microscopic examination.

3. P. allobata (Stizenb.) Zsch. (1918)

Thallus superficial, smooth and continuous to ± uneven, crowded and granular or scurfy, ± gelatinous, rarely minutely branched-squamulose or nodular, bright green when wet, brown- or grey-green when dry. Photobiont cells 4–7 μm. Perithecia 0.15–0.2 mm diam., often taller than wide, ± sessile to half immersed, sometimes tilted; involucrellum absent; true exciple blackened, often pale or pinkish at the ostiole when young. Asci 8-spored. Ascospores 30–35 (-44) × 10–15 μm, colourless, muriform, ellipsoid.

On smooth bark, amongst mosses, in crevices on sheltered trunks of old trees, particularly *Quercus, Ulmus* and *Fraxinus*, in long-established woodlands, sheltered gullies, rarely on wayside trees; local. S. & W. England, Wales, W. Scotland, Ireland. France, Switzerland.

Characterized by the habitat, and the muriform, colourless ascospores. *P. allobata* has recently been transferred to *Agonimia* as ***A. allobata*** (Stizenb.) P. James (1992); specimens with rudimentary, branched squamules may develop incipient papillae, a feature of that genus.

4. P. cruenta (Körber) P. James & Swinscow (1971)

Thallus superficial, film-like, ± gelatinous when wet, ± cracked when dry, dark brown, or grey- or green-brown, dark and oily when wet. Perithecia (Fig. 41f) prominent, 0.5–0.8 mm diam.; thalline exciple well developed; involucrellum black on outer side, paler on inside; true exciple colourless. Asci 8-spored. Ascospores 50–80 × 25–40 μm, muriform, ellipsoid, pale brown, some colourless ones often also present.

On inundated and ± submerged, siliceous rocks and stones in unpolluted streams, more rarely along sheltered lake margins, upland. N.W. England (Lake District), Wales, Scotland, Ireland. Scandinavia, North America.

Characterized by the aquatic habitat and the oily thallus when wet with prominent perithecia and a surrounding thalline exciple.

5. P. cupularis Massal. (1852)

Thallus superficial, pale to dark grey, sometimes brownish, ± minutely cracked-areolate, effuse or bordered by a thin, black prothallus, often appearing as a dark stain or evanescent. Perithecia prominent, 0.3–0.6 mm diam., $\frac{1}{4}$- to $\frac{1}{2}$-immersed; involucrellum generally well developed, ± pale below or brown tinged; ostiole sometimes depressed. Asci 8-spored. Ascospores 20–55 × 10–30 μm, colourless, muriform, broadly ellipsoid.

On hard limestones and basic rocks, rarely on consolidated chalk soil, especially on upland limestones, epidiorites and basic schists, favouring moist or sheltered sites. Throughout British Is. Europe.

P. albida **(2)** has perithecia immersed in pits and no involucrellum. A specimen from C. Scotland (E. Inverness, Ben Alder) overgrowing mosses, identified as *P. bryophila* Lönnr. (1858), is probably a morph of *P. cupularis* overgrowing mosses.

6. P. deminuta Arnold (1861)

Thallus mainly immersed, if evident finely granular, grey or brown-grey, ± bordered by a black prothallus or ± effuse, occasionally mosaic-forming. Perithecia 0.15–0.25 mm diam., often widely scattered, ± immersed in pits; involucrellum absent; true exciple black. Asci 8-spored. Ascospores 22–35 × 8–14(-16) μm, submuriform, 5- to 8- septate, ellipsoid, with a gelatinous sheath, grey-black or brownish, ± opaque.

On hard, exposed limestones, rarely on mortar of old walls; local. Throughout British Is. Europe, chiefly in the Mediterranean region.

Distinguished by the small perithecia and dark, submuriform ascospores with a gelatinous perispore.

7. P. dermatodes Massal. (1855)

Thallus immersed, rarely ± scurfy-cracked, thin, white or pale grey, effuse, occasionally with a thin, black prothallus. Perithecia 0.3–0.5 mm diam., immersed in pits, crowded but not confluent; involucrellum absent; true exciple black above, becoming paler below; ostiole often ± depressed, sometimes ± gaping. Asci 8-spored. Ascospores 25–50 × 12–25 μm, submuriform with 3–4 transverse and 1–2 longitudinal septa, ellipsoid or longly ellipsoid or ± clavate-extended, colourless.

On hard, sunny limestones, old mortar of walls and chalk nodules. Throughout British Is. N. Europe.

Resembles *Thelidium incavatum*.

8. P. gelatinosa (Ach.) Th. Fr. (1862)

Thallus ± gelatinous when wet, granular or ± smooth, ash grey, dark green-black or black-brown, effuse. Perithecia 0.2–0.4 mm diam., totally to ½-immersed; involucrellum absent; true exciple blackish, sometimes pale brownish inside; ostiole protruding; periphyses abundant, branched. Asci 8-spored. Ascospores 30–60(-75) × 15–25(-30) μm, muriform, ellipsoid, remaining colourless, rarely pale yellow-brown with age.

On calcareous, sandy or ± acid peaty soils, especially in dune systems, occasionally among mosses on tops of old mortar-filled walls, rarely on chalky soils. Throughout British Is. Europe, North America.

Resembles the brown-spored *P. wheldonii* **(18)**, with which it shares minute shiny structures discussed under that species.

9. P. gothica Th. Fr. (1865)

P. nigritella (Nyl.) Arnold (1870)

Thallus black, subgelatinous, granular, effuse. Perithecia ¼ to ½-immersed, 0.2–0.3 mm diam.; involucrellum absent. Asci 8-spored. Ascospores (15-)20–30(-36) × (8-)10–15 μm, submuriform, grey-black to red-brown.

On mica-schist soil between squamules of *Catapyrenium cinereum*. N. Scotland (Perth, Summit of Ben Lawers). Not collected since 1864. Sweden.

10. P. helvetica Th. Fr. (1865)

Thallus filmy, pale to blackish grey, subgelatinous when wet, effuse, often inconspicuous. Perithecia 0.3–0.5 mm diam., ± immersed, globose; involucrellum absent; true exciple black above, ± brownish below. Asci (1-)2-spored. Ascospores 70–135 × 30–60 μm, muriform, cylindrical-ellipsoid, soon becoming dark brown.

On base-rich, mossy soil, montane; very rare. N. Scotland (W. Inverness, Aonach Beag). Scandinavia, Alps.

P. gelatinosa **(8)** and *P. wheldonii* **(18)** differ in the (4-)8-spored asci.

11. P. inumbrata (Nyl.) Arnold (1870)

Thallus superficial, thick or thin, smooth to ± uneven-scabrose, greenish grey, brown-grey to dark brown, continuous or finely cracked, effuse. Perithecia 0.4–0.6 mm diam., elevated, ¼- to ½-immersed; thalline exciple often present; involucrellum 0.5–0.8 mm diam., well developed, black; true exciple ± opaque or transparent, colourless to pale yellow-brown. Asci 8-spored. Ascospores 50–75 × 20–35 μm, muriform, broadly ellipsoid, colourless or pale yellow-brown.

On dry or ± moist basic rocks, including schists and shales, on exposed mountain slopes; rare. N. Scotland (Highlands, Shetland Is., Cowal Peninsula). Scandinavia.

The ascospore colour is variable, they may remain colourless for a long time or soon become brown-tinged. Possibly a pale-spored variant of *P. theleodes* **(16)** but with shorter ascospores.

12. P. melaspora (Taylor) Zahlbr. (1921)

Thallus superficial, thick and uneven, warted-granular to ± smooth, sparingly cracked, pale to dark grey, rarely tinged brown, effuse or ± delimited, sometimes evanescent. Perithecia 0.4–0.7 mm diam., ¼- to ½-immersed; involucrellum well developed, 0.5–1.0 mm diam.; true exciple pale to ± dark brown (then semi-opaque). Asci 8-spored. Ascospores 25–40 × 15–25 μm, muriform, rounded-ellipsoid, soon becoming dark brown.

On base-rich (rarely siliceous) rocks in exposed situations, often on mica schist, also on slates associated with calcareous outcrops, usually above 700 m. N. Scotland (Highlands), Ireland; very rare elsewhere. Europe.

The perithecia are sometimes very conspicuous, contrasting with the pale thallus. *P. theleodes* **(16)** has larger ascospores.

13. P. quartzina Lynge (1928)

Thallus superficial, possibly lichenicolous and parasitic, smooth or finely granular, slightly rimose-cracked, pale to dark grey, green-grey, dark red-brown. Perithecia 0.15–0.25 mm diam., ½-immersed, often with a thin layer of thallus covering to ostiole; involucrellum well developed, spreading, opaque, black; true exciple colourless. Asci 8-spored. Ascospores 12–20 × 6–10 μm, submuriform, with 2–3 transverse and 0–2 longitudinal septa, ellipsoid, colourless.

On acid, semi-inundated rocks in sunny streams, on *Verrucaria aethiobola* and *Hymenelia lacustris*; rare. C. Scotland (Argyll, Kilchroan, Braemar, Grampian Mtns).

Characterized by the small perithecia, the well-developed involucrellum, small and submuriform ascospores, and its ecology.

14. P. sendtneri Krempelh. (1855)

Thallus irregularly delimited as scattered, ± convex granules, grey-white, occasionally tinged brown; granules coarse to subsquamulose, turgid-congested, waxy, cartilaginous. Perithecia 0.1–0.15(-0.2) mm diam., ½- to ¾-immersed, scattered and discrete, sometimes confluent, matt or ± shiny, often mottled brown-black; involucrellum 0.2–0.3 mm diam., opaque, black; true exciple pale to brownish, semi-opaque. Asci 8-spored. Ascospores 20–30 × 9–15 μm, colourless, muriform.

On exposed, micaceous soils and bryophytes, over 1000 m. N. Scotland (Highlands, Ben Lawers, Coire Cheap, Ben Alder). Scandinavia, Alps, North America.

Distinguished by the waxy, pale grey-brown, uneven thallus and the small, colourless ascospores. Superficially resembling some subcrustose species of *Catapyrenium* and *Placidiopsis*.

15. P. terrestris Th. Fr. (1860)

Thallus subsquamulose, warted, coarsely granular to ± smooth, ± widely cracked, pale grey-green or tinged brown, effuse. Perithecia 0.4–0.7 mm diam., ½ to almost fully enveloped by a thalline exciple; involucrellum 0.5–0.8 mm diam.; true exciple ± colourless or brownish, especially towards outer surface. Asci 8-spored. Ascospores 50–80 × 23–35 μm, colourless to pale yellow-brown, muriform, ± ellipsoid.

On ± basic, friable rocks and stones, derived soil, spreading to moribund bryophytes; rare. C. Scotland (Perth, Ben Lawers). Scandinavia, North America.

Similar to *P. theleodes* **(16)**, but characterized by the thick, subsquamulose thallus and colourless ascospores. *P. inumbrata* **(11)** has a darker thallus and perithecia are not so clearly immersed in thalline warts.

A similar species, ***P. efflorescens*** Coppins (1991), has less well-developed areoles, perithecia ¾- to ¼-immersed by thalline exciple, ascospores 36–50 × 22–33 μm and with soralia 0.2–1 mm diam., at times stalked and to 0.7 mm tall, pale greyish green (bright green when moist). It occurs on ± calcareous, schistose, overhanging rocks in E. Scottish Highlands and N. Ireland (Donegal).

16. P. theleodes (Sommerf.) Th. Fr. (1867)

Thallus superficial, usually thick, granular and warted, irregularly finely cracked or otherwise, continuous or dispersed in turgid granules, occasionally thin and inconspicuous, pale to dark grey. Perithecia 0.5–1.0 mm diam., ½-immersed, scattered, rarely confluent, often with a variably developed thalline exciple; involucrellum black, opaque; true exciple pale to dark brown or black, semi-opaque. Asci 8-spored. Ascospores 60–100 × 30–60 μm, muriform, ellipsoid, soon becoming dark brown.

On ± basic rocks in exposed situations, including old red sandstone, basalt, limestone, epidiorite and mica schists, sometimes on semi-inundated rocks by upland streams, rarely on soil; frequent, upland. Europe, North America.

Characterized by the large perithecia, large brown ascospores, and porridge-like, lumpy thallus.

17. P. verrucosa (Ach.) Lönnr. (1848)

Thallus superficial, thick and uneven-cracked, to ± inconspicuous, effuse, often patchy, pale grey. Perithecia 0.3–0.5 mm diam., ½-immersed, more emergent in thinner thalli, scattered, rarely contiguous; involucrellum 0.4–0.7 mm diam., well developed, arching over depressed ostiole, black, usually semi-opaque with a brownish tinge on inner side. Asci 8-spored. Ascospores 30–45(-68) × 15–20 μm, colourless, submuriform with 4–5 transverse and 1–4 longitudinal septa, elongate-ellipsoid, somewhat clavate.

On mica schist, 600 m. C. Scotland (Breadalbane, Killin, Creag na Calliach). Scandinavia, Alps.

Superficially similar to *Thelidium* species and *P. dermatodes* **(7)**. The latter has immersed perithecia and no involucrellum.

18. P. wheldonii Travis (1947)

Thallus granular, uneven-cracked, often invaded by gelatinous cyanobacteria, grey-black, often ± gelatinous when wet, effuse. Photobiont often appears to be a mixture. Perithecia 0.4–0.5 mm, ½- to ¾-immersed, ± shiny; involucrellum absent; true exciple blackened, ± brown on inner side. Asci (4-)8-spored. Ascospores 80–125 × 40–60 μm, strongly muriform, ellipsoid-ovoid, soon becoming dark red-brown or brown.

On base-rich soils and mosses associated with limestones, epidiorites and basic schists, also on calcareous sand dunes and mine spoil heaps; rare. Scattered throughout British Is. Endemic.

P. gelatinosa **(8)** has colourless, smaller ascospores, and *P. helvetica* **(10)** has (1-)2-spored asci and cylindrical, brown ascospores. As in *P. gelatinosa* **(8)**, minute, shiny, sterile perithecia-like structures are frequently present which may be the sclerotia of a non-lichenized fungus or, alternatively, are produced by these two *Polyblastia* species.

POLYCHIDIUM (Ach.) Gray (1821)

Thallus shrubby, of ± dichotomously branched ± terete filaments forming small interwoven, domed or straggling masses. **Cortex** well developed, cellular, 1-several cells thick, surrounding a central medullary strand of hyphae. **Photobiont** *Nostoc* or *Scytonema*. **Ascomata** apothecia, mostly lateral, brown, with a sunken disc. **Thalline exciple** absent. **True exciple** present. **Hamathecium** of paraphyses, unbranched, septate; apices capitate. **Asci** 8-spored, broadly cylindrical; apices thickened, I+ blue. **Ascospores** 1- to 2-celled, colourless, ellipsoid to spindle-shaped, thin or thick-walled. **Conidiomata** pycnidia, brown, lateral. **Conidiogenous cells** generally short. **Conidia** bacilliform. **Chemistry**: lichen products not detected by t.l.c..

LITERATURE: Henssen (1963a).

1 Filaments blue-grey; photobiont *Scytonema* .... **1. *dendriscum***
Filaments red-brown-black; photobiont *Nostoc* .... **2. *muscicola***

1. P. dendriscum (Nyl.) Henssen (1963a)

Thallus of short, slender, richly and intricately branched filaments, (30-)40–70(-75) μm wide, smooth or somewhat nodulose, forming shining blue-green, blue-grey or pale brown cushions in which grey colours predominate when dry; filaments with a single-layered cortex; cells 8–15(-18) × 6–11 μm, with ± thickened walls. Photobiont *Scytonema* in distinct bluish green chains in strands of 2–5 cells. Apothecia unknown in British Is. to 2 mm wide, orange- to red-brown; hymenium 80–150 μm tall; true exciple with pitted, thick-walled cells. Ascospores 11–14.5 × 5–7 μm, simple, ellipsoid. Pycnidia occasional; conidia 1.5–3 × 1 μm.

Associated with mosses and leafy liverworts, particularly *Frullania*, on shaded, sheltered trunks of *Corylus*, *Fraxinus*, *Quercus*, *Betula*, *Salix*, *Rhododendron* etc. in humid, undisturbed, boggy woodland; rare and local. W. Scotland, S.W. Ireland. Macaronesia, Brazil, Hawaii, New Caledonia..

This delicate species grows on stems of low shrubs in well-lit, boggy, undisturbed situations, often close to the ground; it superficially resembles some filamentous cyanobacteria.

2. P. muscicola (Swartz) Gray (1821)

Thallus of slender, sparsely to richly and divergently branched, brown-black, sometimes greyish in shade, filaments with shining, often nodulose apices; forming small, loosely inter-woven, decumbent to cushion-like masses and strands; filaments (50-)60–125(-135) μm wide, becoming progressively narrowed towards the apices. Cortex 1 to 3 cells thick; cells rounded or ± angular, 4–7 μm diam., the wall somewhat thickened, colourless or ± brown. Photobiont *Nostoc*, not in obvious chains. Apothecia occasional, lateral, to 2 mm wide; disc red-brown, brighter when wet; margin paler. Paraphyses simple, with a swollen apical cell, <4.5 μm wide. Ascospores 22–29 × 5–7 μm, 1-septate, spindle-shaped, sometimes becoming pale red-brown. Conidia 1.5–3.5 × 1 μm, bacilliform.

Among mosses in moist siliceous rocks and boulders, often in and near streams, occasionally on the ground, sometimes also at the base of mossy trees, especially *Fraxinus*; chiefly in uplands, rather local. N. & W. British Is. Europe, North America.

Resembles *Ephebe* in the loosely woven, dark filaments but is more red-brown, not olive-black, of a firmer texture, forming a coarser, more ascending, thallus.

POLYSPORINA Vězda (1978)

Thallus crustose, often inapparent, or on lichens. **Photobiont** chlorococcoid. **Ascomata** apothecia, black, perithecia-like, at first with a central pore, later with a thick, radially fissured margin surrounding a circular slit-like disc around a central umbo; old apothecia often gyrose-contorted. **Thalline exciple** absent. **True exciple** without algae; outer part dark brown; inner part colourless, of radiating, branched hyphae. **Epithecium** colourless to brown, but umbo region thick and dark brown. **Hymenium** colourless, zone below umbo without asci, I+ yellowish to reddish or bluish in parts, K/I+ blue. **Hypothecium** colourless. **Hamathecium** of richly branched and anastomosed paraphyses, 1–2 μm wide; apical cells shorter and ± broader, colourless or surrounded by brown pigment. **Asci** cylindrical-clavate, multispored (>100 spores); wall K/I− but outer coat K/I+ blue; apical dome distinct, K/I−. **Conidiomata** pycnidia, resembling small apothecia. **Conidia** minute, simple, colourless, ellipsoid. **Chemistry**: no lichen products reported. **Ecology**: mainly on hard rocks. **Distribution**: *c.* 6 species known, in temperate regions of N. & S. Hemispheres.

Distinguished from *Sarcogyne* by the umbonate to gyrose apothecia and richly branched and anastomosed paraphyses. The multispored asci separate the genus from *Mosigia*, *Rimularia* and *Lecidea umbonella* which have apothecia of a similar appearance but with 8-spored asci. The genus is well-defined but further study is required on the delimitation of its component species.

LITERATURE: Clauzade & Roux (1985), Poelt & Vězda (1981), Vězda (1978).

1 On siliceous, ± acid rock; ascospores 1–1.5 μm wide .... **1. *simplex***
On highly calcareous rock; ascospores 1.5–2 μm wide .... **2. *cyclocarpa***

1. P. simplex (Davies) Vězda (1978)

Thallus often inapparent, or thin and scurfy, pale grey or olivaceous, perhaps rarely thick and areolate. Apothecia partly immersed to sessile, scattered to clustered, often in lines following cracks in the rock surface, (0.2-)0.3–0.5(-1) mm diam., usually rounded or broadly elliptic, with a central umbo surrounded by a slit-like disc; true exciple divided by 4–8 fissures, sometimes contorted due to mutual pressure or regeneration; hymenium (60-)90–120(-200) μm tall. Asci (55-)65–100(-150) × (12-)15–17(-20) μm. Ascospores 3–5(-5.5) × 1–1.5 μm, narrowly ellipsoid to bacilliform.

On acid or weakly calcareous rocks, often in the xeric-supralittoral on rocky shores, cliffs and on pebbles in stable shingle, also inland on tops of boulders in and by streams, slate roofs and, especially, in churchards on granite memorials and granite chips, rarely on consolidated soil on sea-cliffs; especially common on the coast. W. Britain & Ireland, rarer elsewhere. Temperate (especially maritime) regions of N. & S. Hemispheres.

Specimens with ± immersed apothecia on a thick, grey to brown, ± areolate thallus from inland localities in N. England, N. Wales and Scotland, may be parasitic on unidentified (sterile) hosts, and referable to *P. dubia* (Magnusson) Vězda (1978).

2. P. cyclocarpa (Anzi) Vězda (1978)

Like **1**, but with broader ascospores, 4–5 × 1.5–2 μm, apothecia that more readily become gyrose-contorted and a different ecology.

On exposed, freely drained, montane limestone. C. & E. Scotland (C. & E. Highlands). Montane Europe, Pyrénées, E. Asia.

PORINA Müll. Arg. (1883)

Thallus crustose, immersed to superficial, often inconspicuous. **Photobiont** *Trentepohlia*. **Ascomata** perithecia, superficial or immersed, free or ± surrounded by a thalline exciple. **Involucrellum** present or absent. **True exciple** coloured or not. **Hamathecium** of paraphyses, *c.* 1 μm thick, simple, rarely forked; periphyses absent. **Asci** 8-spored, 2- to 3-seriate, cylindrical-clavate, thin-walled, only slightly thickened at the apex (Fig. 34a). **Ascospores** transversely septate, rarely also with 1(-3) longitudinal septa, colourless. **Conidiomata** pycnidia, ± sessile. **Conidia** short, cylindrical, macro- and micro- may be present. **Chemistry**: lichen products not detected by t.l.c.; exciple and pycnidial walls with variously coloured brown, reddish, purplish or greenish pigments. **Ecology**: mainly on rocks, less often on bark and soil.

The genus still includes disparate elements whose generic affinities are in need of critical evaluation; see also *Strigula* and *Zamenhofia*.

LITERATURE: Swinscow (1962, 1965b).

1 On soil, mosses and decaying plants 2
On rock or bark 4

2(1) Ascospores to 5-septate; thallus dark grey-purple-black 3
Ascospores 7- to 14-septate; thallus pale grey, pinkish grey or orange-pink .. **10. *heterospora***

3(2) Ascospores 5-septate ..... **16. *sudetica***
Ascospores 3-septate ..... **15. *mammillosa***

4(1) On rock 5
On bark 16

5(4) On calcareous rocks; thallus superficial to immersed 6
On siliceous rocks; thallus superficial; mostly in or near streams and lakesides 8

6(5) Ascospores 3(-7)-septate, if 7-septate then 18–22 μm in length 7
Ascospores (5-)7(-8)-septate, 28–40 μm in length ..... **9. *ginzbergeri***

7(6) Thallus grey-green to olive-brown, sometimes tinged rusty brown; ascospores 3(-7)-septate ..... **14. *byssophila***
Thallus pink-brown to reddish or rusty brown; ascospores 3-septate ..... **13. *linearis***

8(5)	Some ascospores with longitudinal septa	9
	All ascospores without longitudinal septa	10
9(8)	Ascospores with 5(-6) transverse and 1–3 longitudinal septa	**8. *interjungens***
	Ascospores with 7 transverse septa and 1(-2) longitudinal septa	**5. *guentheri***
10(8)	Perithecia pink or pink-grey, covered by a thalline layer	11
	Perithecia at least black in upper part	13
11(10)	Ascospores 3-septate, 16–30 × 3–8 μm	12
	Ascospores 6- to 7-septate, 30–50(-70) × 6–15 μm	**2. *ahlesiana***
12(11)	Ascospores 16–23 × 3–5 μm; perithecia 0.15–0.2 mm diam.	**12. *leptalea***
	Ascospores 20–30 × 5–8 μm; perithecia 0.4–0.5 mm diam.	**11. *lectissima***
13(10)	Ascospores 4- to 7-septate; perithecia often 2–7 contiguous	14
	Ascospores 3-septate; perithecia scattered	**4. *chlorotica***
14(13)	Perithecia *c.* 0.4 mm; ascospores less than 6 μm wide	15
	Perithecia 0.5–0.8 mm; ascospores more than 6 μm wide	**7. *grandis***
15(14)	Ascospores 35–45 × 5–6 μm; at the margin of lakes and in streams	**5. *guentheri***
	Ascospores 40–52 × 3–5 μm; on moist coastal rocks	**6. *curnowii***
16(4)	Ascospores 3-septate	17
	Ascospores 6- to 14-septate	19
17(16)	Perithecia black	18
	Perithecia orange to red-brown	**12. *leptalea***
18(17)	Thallus olive to red-brown or dark brown; perithecia 0.1–0.3 mm diam.	**1. *aenea***
	Thallus dark green-olive-brown; perithecia 0.2–0.3 mm diam.	**4. *chlorotica***
19(16)	Ascospores 7- to 14-septate, 40–80 × 10–12 μm	**10. *heterospora***
	Ascospores 6- to 7-septate, 25–40 × 3–5 μm	**3. *borreri***

1. P. aenea (Wallr.) Zahlbr. (1922)

Thallus mainly superficial, thin, forming small, ± delimited patches, sometimes almost disappearing, deep red-brown or dark brown, becoming ± olivaceous in shade; prothallus absent. Perithecia 0.1–0.3 mm diam., globose, emergent, mostly scattered, rarely two contiguous, neat; involucrellum present, black, exposed, shiny; true exciple semi-opaque, ± colourless with brownish inclusions; ostiole inconspicuous. Ascospores 13–17(-24) × 4.5–5 μm, 3-septate, ellipsoid, rounded or somewhat pointed at one or both ends. Macroconidia 12–16 × 4.5 μm, 1-septate, microconidia minute, 2 × 0.5 μm, straight or ± curved, ellipsoid with rounded ends.

On young, smooth bark, especially *Corylus* and *Fraxinus*, occasionally on smooth, older bark; widespread and locally frequent in non-polluted coppice and woodlands. Throughout British Is. Europe, North America.

P. chlorotica **(4)** on trees is more olivaceous, oily, continuous, and the perithecia are larger.

2. P. ahlesiana (Körber) Zahlbr. (1931)

Thallus superficial, ± thin, pale yellow-green, olivaceous or green-grey, smooth, even, occasionally wrinkled, flakey or ± rimose, continuous, forming small patches; ± a pale or whitish prothallus present; occasionally with loculi containing plaque-like crystals. Perithecia 0.3–0.5(-0.7) mm diam., single or occasionally 2 contiguous, usually frequent, 1–2 mm apart, low convex to ± hemispherical warts covered by a thalline exciple; involucrellum pink or grey-pink, translucent when wet; true exciple outer layers pinkish, pale or colourless internally; ostiole not obvious, pale pinkish or pale red-brown. Ascospores 30–50(-70) × 6–15 μm, (6-)7-septate, ± clavate-fusiform. Pycnidia pale grey-pink, partly emergent, inconspicuous; microconidia 6–7 × 0.5 μm.

On sheltered, shaded, moist, siliceous rocks, often in or near woodland streams or lakes, cave entrances, occasionally beneath several layers of boulders; rare. S.W. England, Wales, W. Scotland, S.W. & W. Ireland. S.W. Germany, Austria, ?W. Africa.

Characterized by the greenish grey thallus, pinkish perithecia which become translucent when moist, and the 7-septate spores. *P. lectissima* **(11)** has 3-septate spores and a darker thallus.

3. P. borreri (Trevisan) D. Hawksw. & P. James (1980)

Thallus mostly superficial, forming effuse, scattered patches or coalescing, thin, pale to blackish olive, occasionally brownish, smooth or sometimes ± granular, irregularly rimose. Perithecia (0.2-)0.3–0.5 mm diam., ¼ to ½-immersed, ½-1 mm apart, single or rarely contiguous; involucrellum present, shiny or matt, black, reaching half way down exciple; true exciple pale to dark purplish brown; ostiole inconspicuous. Ascospores 25–40 × 3–5 μm, (6-)7-septate, fusiform. Pycnidia *c.* 150 μm diam.; macroconidia 15–18 × 3–4 μm, 1-septate, cylindrical-fusiform, guttulate; microconidia 2–2.5 × 0.7 μm, ends ± rounded.

On rough bark of old trees, particularly *Quercus* and *Fraxinus*, in old woodlands and parklands, also on roughened areas of trunks of young trees (e.g. *Sorbus aucuparia*) and on old smooth bark, usually *Ilex* and *Fagus*, rarely spreading to bryophytes; widespread but local. S. & S.W. England, Wales, Scotland, Ireland. France, Switzerland, Germany, Austria, Italy, ?U.S.A.

P. aenea **(1)** has 3-septate ascospores and smaller perithecia. A narrow-spored variant, **var. *leptospora*** (Nyl.) D. Hawksw. (1992) [syn. *P. olivacea* var. *leptospora* (Nyl.) Keissler (1937)], has 7-septate ascospores 30–55 × 3–4 μm, and occurs on *Ilex* in S.W. Ireland (Killarney).

4. P. chlorotica (Ach.) Müll. Arg. (1884)

Thallus superficial, thin or disappearing, usually well-defined, continuous, ± smooth or cracked, olive, grey, dark green, brown or blackish. Perithecia 0.2–0.3 mm diam., half-immersed to sessile, ½-1 mm apart; involucrellum present, shiny, black, extending ± halfway down exciple; true exciple with outer layers brown-black, the inner layers ± colourless. Ascospores 16–25(-32) × 4–6 μm [25–35 × 3–4 μm in **f. *tenuifera*** (Nyl.) Swinscow (1962)], 3-septate, fusiform, rounded or pointed at one or both ends. Pycnidia *c.* 100 μm diam.; microconidia 2–4 × 1 μm.

On acidic to moderately basic rocks, loose pebbles and flints, either fully exposed or in deep shade and, at times, ± permanently inundated; rarely on the shaded, smooth trunks of trees, especially *Ilex*, in western areas. Throughout British Is., especially common in the north and west. Cosmopolitan.

Characterized by the thin, olivaceous-brownish or greenish black thallus with small, shiny, black perithecia and 3-septate ascospores. The involucrellum is pinkish in *P. lectissima* **(11)** and *P. leptalea* **(12)** which also differ in ascospore size.

5. P. guentheri (Flotow) Zahlbr. (1922)

Thallus superficial, thin, smooth, continuous or ± finely cracked, pale grey, often with a violet tinge, to dark purple-grey or blackish purple. Perithecia 0.4–0.7 mm diam., ¼- to ½-immersed, scattered or 2- to 5-confluent, occasionally fused and forming gnarled clusters; involucrellum smooth, ± shiny, occasionally ± nodular, black, ± purple or violet-tinged; true exciple colourless or straw-coloured; ostiole inconspicuous or slightly papillate. Asci 100–140 × 10–20 μm. Ascospores (30-)34–45 × 5–6 μm, (5-)7-septate, fusiform-clavate. Pycnidia usually present, sometimes predominating, black, opaque above, paler below; microconidia 4.5 × 0.7–1 μm, straight or slightly curved.

On moist, siliceous rocks by lakes and mountain streams, also in sheltered sites at low altitudes; rather local. Throughout British Is. Alps, S.W. Germany, Italy, Sardinia, U.S.S.R., North America.

P. guentheri **var. *lucens*** (Taylor) Swinscow (1962) has ascospores with 1(-2) longitudinal septa. It occurs on moist siliceous rocks by lakes and mountain streams, also in sheltered sites at low altitudes and is widespread and often common in W. Ireland, but rare in Wales and N.W. England (Lake District). Endemic.

For differences from *P. curnowii* (see **6**), *P. grandis* (see **7**) and *P. interjungens* (see **8**).

6. P. curnowii A. L. Sm. (1911)

Like **5**, but ascospores narrow-fusiform, 7-septate, 40–52 × 3–5 μm, lacking longitudinal septa.

Mainly on siliceous coastal rocks in sheltered, damp sites and seepage tracks; very local. S.W. England (Cornwall), Is. of Scilly, Channel Is., S.W. Ireland. France (Brittany).

7. P. grandis (Körber) Zahlbr. (1922)

Like **5**, but perithecia twice the size, and involucrellum ± incurved in the lower part. Asci 150–250 × 12–20 μm. Ascospores 35–50 × 6–10 μm, lacking longitudinal septa.

On hard siliceous rocks, especially by streams; rare. N. Scotland (Inverness). C. Europe.

8. P. interjungens (Nyl.) Zahlbr. (1922)

Like **5**, but ascospores 20–25(-30) × 6–10 μm, with more frequent, 0–3 longitudinal septa and fewer transverse septa.

On siliceous, damp, fully exposed or ± shaded rocks by rivers and lakes; rare. W. England (N. Devon, Cumbria), Wales, Scotland (Lanark, Ross). Sweden.

9. P. ginzbergeri Zahlbr. (1903)

Thallus mostly immersed, continuous or ± finely cracked, disappearing, inconspicuous, pale green-grey, rarely greyish black. Perithecia 0.25–0.4 mm diam., prominent, ± sessile or ¼ immersed; involucrellum shiny, black, purplish, K+ slate-black; true exciple ± blackened, semi-opaque. Ascospores (28-)35–40(-45) × 4–5.5 μm, (5-)7(-8)-septate, elongate-fusiform, sometimes tapered at one end.

On hard limestones in deep shade; rare. W. Ireland. France, Yugoslavia, Corfu.

P. linearis **(13)** has smaller, 3-septate ascospores.

10. P. heterospora (Fink) R. C. Harris (1980)

P. nucula auct. brit. non Ach. (1814)

Thallus in part immersed or ± superficial, thick, uneven, continuous, effuse, orange, orange-grey or pinkish, grey-white in the herbarium, coarsely granular to verrucose, sometimes ± corrugate-wrinkled. Perithecia 0.5–0.9 mm diam., subglobose, aggregated or scattered, largely covered by a thalline exciple containing locules with plaque-like crystals, immersed in thalline warts with only apex exposed; involucrellum grey-pink or pale yellow-brown where exposed; true exciple ± pale yellow. Ascospores 40–80 × 8–12 μm, 7- to 14-septate, fusiform-clavate.

On sheltered bases of mossy trees, especially *Quercus* in old woodland, also rarely overgrowing bryophytes in sheltered rock underhangs; locally frequent. S.W. Ireland. Europe, North America.

Characterized by the large, wart-like perithecia with pinkish apices, an orange-pink thallus fading grey-green in the herbarium, and the distinctive ascospores.

11. P. lectissima (Fr.) Zahlbr. (1903)

Thallus superficial, continuous, sometimes wrinkled and ± irregularly cracked, often forming extensive patches, pale ochre to deep brown or brownish green. Perithecia 0.4–0.5 mm diam., hemispherical, low convex to ± conical, wart-like, ± totally enveloped by a thalline exciple; involucrellum present, pink or dull reddish brown, brighter reddish when wet; true exciple pale pink-yellow. Ascus (Fig. 34a). Ascospores 20–30(-40) × 5–8 μm, 3-septate, fusiform. Pycnidia orange-pink, ± immersed; microconidia 3–4.5 × 0.7–1 μm, long ellipsoid or ± cylindrical.

On damp, siliceous rocks in hilly, ± well-wooded, shaded sites, often as extensive patches in seepage tracks or rocks receiving spray from streams; locally abundant. W. British Is. Germany, Alps, Italy, Finland, North America.

P. ahlesiana **(2)** has larger perithecia and 7-septate ascospores, and *P. leptalea* **(12)** has smaller perithecia and smaller, 3-septate ascospores.

12. P. leptalea (Durieu & Mont.) A. L. Sm. (1911)

Thallus mainly superficial, film-like or finely granular, often in small patches amongst other lichens, continuous, effuse, pale to dark olive-green, ± brownish, or reddish brown, often oily green when on rock. Perithecia 0.15–0.2 mm diam., hemispherical, low convex, scattered or 2 (rarely 3) contiguous, wart-like, mostly covered by a thin thalline exciple; involucrellum pale pink-orange or dull brown-red to brown-black; true exciple ± pale yellow or brown-yellow. Ascospores 16–23 × 3–5 μm, 3-septate, fusiform. Pycnidia *c.* 0.1 mm diam., dark brown-red; microconidia 1.7–2.5 × 1 μm, cylindrical, ± ellipsoid, straight to slightly curved or dumb-bell-shaped.

On smooth bark, especially of *Corylus* and *Ilex*, often near streams or associated with boggy ground in shaded sites, also on "buried" siliceous stones and low rock outcrops in scree or amongst vegetation; frequent but overlooked. W. Britain, Ireland. S.W. Germany, Alps, Finland, U.S.A. (New Hampshire).

P. aenea **(1)** and *P. chlorotica* **(4)** have a black involucrellum, and *P. lectissima* **(11)** has larger ascospores.

13. P. linearis (Leighton) Zahlbr. (1922)

Thallus immersed or partly superficial, thin, scurfy, effuse, pink-brown to reddish or rusty brown, sometimes mottled. Perithecia 0.2–0.4 mm diam., black, immersed in pits or ± superficial; involucrellum black-violet-purple, K+ blue-grey; true exciple ± pale brown or colourless. Ascospores 18–22 × 6–7 μm, 3-septate, elongate-fusiform, straight or slightly curved. Pycnidia ± immersed in pits, black; microconidia 4–4.5 × 1 μm, cylindrical or ± curved.

On ± sheltered and often shaded, hard limestones; locally common. Throughout British Is. Europe, North America.

14. P. byssophila (Körber ex Hepp) Zahlbr. (1903)

Like **13**, but thallus better developed, grey-green to olive-brown, sometimes tinged rusty brown. Ascospores 3(-7)-septate.

On hard limestone, usually in deep shade. W. Ireland (Clare). C. & S. Europe.

15. P. mammillosa (Th. Fr.) Vainio (1922)

Thallus smooth or ± granular, rather thick, nodular, warted, effuse, blackish grey, green-grey often tinged brown-purple. Perithecia 0.3–0.5 mm diam., prominent, globose, wart-like, contiguous or scattered, occasionally confluent, partly covered by a thalline exciple; involucrellum thick, outer layers black, matt, opaque, roughened, central layers bright purple-red, inner layers pinkish orange; true exciple yellow to pale orange. Ascospores 25–40 × 4–6 μm, 3-septate, fusiform. Pycnidia unknown.

On bryophytes, small woody plants, and soil in ± protected or sheltered soil crevices of mountain rocks, also at lower altitudes; rare. N. Scotland (Ben Lawers, Cairngorm, Caenlochan), W. Ireland (Connemara, Galway). Scandinavia, Switzerland, U.S.S.R., North America.

16. P. sudetica (Körber) Lettau (1912)

Like **15**, but ascospores (3-)5-septate, 25–32(-38) × 5–8 μm. Perithecial true exciple more intensely orange-coloured throughout.

On soil between heavy metal-rich scree. S.W. England (W. Cornwall, Botallack). C. Europe.

POROCYPHUS Körber (1855)

Thallus crustose, granular, areolate, subsquamulose to ± placodioid, or minutely shrubby, dark-coloured, gelatinous, not corticate, non-layered, hyphae short-celled and with a fan- or fountain-like arrangement spreading out from the base of the thallus, where they sometimes form a narrow, compact, pseudoparenchymatous layer. **Photobiont** *Calothrix*, with filaments often much disrupted and modified. **Ascomata** apothecia, developing as normal apothecia or from pycnidia (pycnoascocarps), immersed to ± sessile, ± concolorous with thallus. **Thalline exciple** persistent or becoming excluded to reveal the true margin; disc poriform, ± expanding with age. **True exciple** colourless, or brown in upper part, usually distinct above, but narrowing and often indistinct below, composed of ± parallel hyphae. **Hypothecium** colourless or pale brownish. **Hamathecium** of richly branched and anastomosed paraphyses, sometimes (in pycnoascocarps) intermixed with elongated conidiophores, gel I+ yellowish to blue. **Asci** ± cylindrical, thin-walled, I−, without thickened apex or amyloid structures, mostly 8-spored. **Ascospores** simple, ellipsoid, colourless, perispore absent. **Conidiomata** pycnidia, immersed in the thallus or thalline warts; wall colourless. **Conidiogenous cells** slender, cylindrical. **Conidia** acrogenous, globose to shortly ellipsoid, simple, colourless. **Chemistry**: lichen products not detected by t.l.c. **Ecology**: on usually inundated or moist rocks.

LITERATURE: Ellis (1981), Henssen (1963, 1974).

1 Thallus crustose, without a lobed margin, or minutely shrubby 2
Thallus placodioid, forming rounded patches with a distinctly lobed margin **2. *kenmorensis***

2(1) Thallus granular or minutely shrubby 3
Thallus areolate .... **1. *coccodes***

3(2) Ascospores 12.5–15 × 5–7.5 μm .... **3. *leptogiella***
Ascospores 12.5–24 × 7.5–10 μm .... **4. *rehmicus***

1. P. coccodes (Flotow) Körber (1855)

Thallus blackish, areolate; areoles to 2.5 mm wide and 1 mm thick, distinct and subsquamulose to flattened, confluent and indistinct, surface sometimes granular or minutely coralloid. Photobiont filaments mostly disrupted; cells *c.* 6–8 μm diam., arranged in short, vertically orientated chains or clusters. Apothecia to 0.3 mm diam., immersed to sessile, at first with a distinct thalline exciple which sometimes recedes to expose the paler, thin true exciple; disc poriform to expanded, then reddish brown; true exciple to 20 μm wide above, to 10 μm wide below; hymenium 100–120 μm tall, brownish in upper part; hypothecium 40–55 μm tall. Asci 55–80 × 9–11 μm, 8-spored. Ascospores 9–18 × 7–12 μm. Conidia 2–2.5 × 1–1.5 μm.

On damp, siliceous rock by lakes and streams, or on flushed rock faces, uplands. N. & W. British Is. Europe, Algeria, U.S.A.

Variable in appearance; *P. leptogiella* **(3)** and *P. rehmicus* **(4)** possibly represent only extreme modifications.

2. P. kenmorensis (Holl ex Nyl.) Henssen (1974)

Thallus ± placodioid, forming ± circular patches, to 6 cm diam., continuous to cracked-areolate but with irregular, ± knobbly marginal lobes up to 1.5 mm long and 0.5–0.7 mm wide. Apothecia arising from pycnidia, to 0.4 mm diam.; thalline exciple thin, soon receding; disc orange-brown; true exciple to 90 μm wide above, 10–15 μm below; hymenium 180–240 μm tall; hypothecium 45–160 μm tall. Asci 75–120 × 9.5–17.5 μm, 8-spored. Ascospores 11–19 × 8–9.5 μm. Conidia 2.5–3.5 × 1 μm.

On inundated siliceous rocks in or by lakes and streams; rare. S.W. England (S. Devon, Dartmoor), Scotland (Highlands), W. Ireland (Kerry, Galway, Connemara). Finland.

Distinguished by the marginal lobes of the thallus.

3. P. leptogiella (Nyl.) L. T. Ellis (1981)

Lemmopsis leptogiella (Nyl.) A. L. Sm. (1918)

Thallus black-brown, minutely shrubby with densely branching, cylindrical lobes. Photobiont cells 5–7.5 μm diam., arranged in clumps. Apothecia to 0.35 mm diam., developing on the lobes; thalline exciple irregularly lumpy, often thin. Paraphyses often swollen, ± globose apical cell(s) to 4 μm wide. Asci 70–80 μm tall, 8-spored. Ascospores 12.5–15 × 5–7.5 μm. Pycnidia unknown.

On shaded, ± basic rock. W. Ireland (Galway, Lough Inagh; Connemara).

Similar to *P. rehmicus* **(4)**, but said to differ by the swollen apices of the paraphyses, and smaller ascospores.

4. P. rehmicus (Massal.) Zahlbr. (1924)

Thallus dark olivaceous to blackish, granular to minutely shrubby. Photobiont filaments often remaining intact with cells, 13–14.5(-16.5) × 3.5–5.5(-9) μm; cells in disrupted (strongly lichenized) groups mostly smaller, 5.5–9 × 2.5–8 μm. Apothecia pycnoascocarps, to 0.5 mm diam., sessile; thalline exciple receding to expose the paler true exciple; disc reddish brown; true exciple 35–55 μm wide above, 10–20 μm wide below; hymenium 90–130 μm tall; hypothecium 20–90 μm tall. Paraphyses thin and branched, or replaced by modified, elongated conidiophores with swollen apices. Asci (55-)80–90 × 9–11 μm, 8-spored. Ascospores 11–17(-24) × (5.5)7–13 μm. Conidia 2.5 × 1 μm.

On calcareous mica-schist. C. Scotland (Perth, Ben Lawers). E. France, S. Germany, Switzerland..

This species and *P. leptogiella* **(3)** may be aberrant morphs of *P. coccodes* **(1)**.

PORPIDIA Körber (1855)

Huilia Zahlbr. (1930)

Thallus crustose, thick to inconspicuous, tartareous, continuous to areolate-cracked, grey, white or orange; prothallus present (especially between adjacent thalli) or absent, black or sometimes orange, thin; medulla I+ or I−; soredia present or absent, either forming irregularly and diffusely in cracks in thallus, or more commonly forming discrete, regular, round soralia. **Photobiont** green, chlorococcoid. **Ascomata** apothecia, scattered or clustered, sometimes in concentric rings, hemiangiocarpic, emerging from small pruinose dots on thallus, becoming sessile in some taxa, rather large, to 4 mm diam.; disc dark brown to black, pruinose or naked, with a distinct true margin (sometimes excluded with age). **Thalline exciple** absent. **True exciple** dark brown (outer edge dark green in some species) but sometimes a paler brown in innner part; comprised of conglutinated, radiating hyphae. **Epithecium** usually distinct, brown to greenish brown, rarely blue-green, N+ rose pink to purple. **Hymenium** 80–150 μm tall, colourless or pale greenish above, I+ blue. **Hypothecium** upper part (subhymenium) colourless, otherwise dark brown to brown-black, K− or K+ reddish in parts, hyphae irregularly arranged. **Hamathecium** of paraphyses, septate, branched-anastomosing, netted, slightly swollen at apices but without dark apical cups. **Asci** 8-spored, clavate to subcylindrical, *Porpidia*-type (Fig. 33g). **Ascospores** simple, rather large, ± ellipsoid, occasionally with one or both ends acuminate; perispore well developed, 2–7 μm thick. **Conidiomata** pycnidia, rare. Conidiogenous cells elongate-ampulliform, in a single layer, enteroblastic, acrogenous. **Conidia** bacilliform. **Chemistry**: β-orcinol depsidones or long-side chain orcinol depsides; e.g. stictic, norstictic, 2'-*O*-methylsuperphyllinic, 2'-*O*-methylmicrophyllinic, confluentic and 2'-*O*-methylperlatolic acids present or absent. **Ecology**: mainly on siliceous, rarely calcareous, rocks, pebbles and stonework, more rarely on bark, wood and artificial materials; predominantly in oroboreal and montane zones in temperate regions of both hemispheres.

Superficially similar to *Lecidea* s.str., but differs in the distinctive ascus type, and larger, thick-walled ascospores. Closely related to *Amygdalaria*, under which it should perhaps be subsumed. Also with a similar ascus type and crustose thallus are *Clauzadea* and *Farnoldia*, all species of which grow on highly calcareous substrata.

Identification to species within *Porpidia* is often difficult, requiring thin sectioning of ascomata and t.l.c. The delimitation of many species has been clarified by recent studies, but many collections remain difficult to place.

LITERATURE: Gowan (1988), Hertel (1985), Hertel & Knoph (1984), Rambold (1989), Schwab (1986).

1 Soredia present 2
 Soredia absent 5

2(1) Thallus mostly grey or whitish, not noticeably orange; medulla I+ violet or I− 3
 Thallus mostly orange, rarely with minor grey or olivaceous patches, cracks between areoles suffused orange-red; medulla I− .... **6. *flavocoerulescens***

3(2) Medulla I− 4
 Medulla I+ violet .... **15. *tuberculosa***

4(3) Thallus in small patches, often scurfy and indistinct, grey to grey-brown; soralia regularly rounded, punctiform to erose-excavate, Pd+ orange; soredia granular; stictic acid present .... **12. *soredizodes***
 Thallus spreading, thick, whitish to greenish grey; soralia irregular, crateriform, Pd−; soredia coarsely granular; 2'-*O*-methylsuperphyllinic and glaucophaeic acids **8. *glaucophaea***

5(1) Exciple and medulla K− 6
 Exciple and medulla K+ yellow to red .... **11. *platycarpoides***

6(5) Medulla I+ blue 7
Medulla I− 8

7(6) Thallus chalky white, finely cracked to irregularly areolate; ascospores 12–19 × 6–7 μm; on calcareous rocks .... **13. *speirea***
Thallus grey-brown, olivaceous or blackened, regularly areolate; ascospores (13-)16–24 × (6-)19–12(-14) μm; on acid rocks .... **1. *athroocarpa***

8(6) Thallus tinged orange or reddish 9
Thallus white, creamish, ashy grey, brownish, never orange or reddish 11

9(8) Exciple hyphae 3–7(-9) μm wide; thallus mostly 0.05–0.3 mm thick 10
Exciple hyphae 2–4 μm wide; thallus mostly 0.2–0.7 mm thick .... **7. *flavicunda***

10(9) Apothecia 0.3–1(-1.5) mm diam.; hymenium 60–80(-110) μm tall; ascospores 10–17 × 5–9 μm .... **5. *crustulata***
Apothecia 0.4–2(-4) mm diam.; hymenium (70-)80–120(-130) μm tall; ascospores (13-)16–23(-26) × (5-)6–11.5 μm .... **10. *macrocarpa***

11(8) Epithecium dull olive-green, olive-brown or brown; usually not on inundated rocks 12
Epithecium blue-green; on inundated rocks in or close to streams and lakes ... **9. *hydrophila***

12(11) Thallus inconspicuous to scurfy, *or* irregularly cracked, *or* if areoles ± discrete *then* apothecial disc black and epithecium with green or olive tinge 13
Thallus of discrete, white, hemispherical to bullate areoles; apothecia often sessile and constricted at base; disc brown, contrasting with black margin; epithecium brown .... **14. *superba***

13(12) Thallus white to light grey, usually thickish and cracked-areolate; confluentic acid or 2'-*O*-methylmicrophyllinate complexes by t.l.c. 14
Thallus inconspicuous to scurfy, *or* thickish and irregularly cracked, whitish to dark grey or olivaceous; stictic acid or no substances detected by t.l.c. .... **10. *macrocarpa***

14(13) Apothecia mostly remaining immersed; with confluentic acid 15
Apothecia soon sessile; with methyl 2'-*O*-methylmicrophyllinate .... **4. *contraponenda***

15(14) Ascospores 12–18 × 6–9 μm .... **2. *cinereoatra***
Ascospores 15–20(-23) × 8–10 μm .... **3. *musiva***

1. P. athroocarpa (Ach.) Hertel & Rambold (1985)

Lecidea athroocarpa (Ach.) Ach. (1803)

Thallus small to very large, irregularly spreading, distinctly cracked-areolate, yellowish brown, dark brown, greyish, olive brownish to ± blackened, ± glossy; areoles angular, subconcave, flat or subconvex, separated by deep cracks; medulla I+ blue-violet; prothallus black occasionally visible at margins. Apothecia (0.2-)0.5–1.2(-1.8) mm diam., rounded or deformed through mutual pressure, flat to subconvex at maturity, numerous, scattered or crowded in groups; immersed at first then flat, often aspicilioid, between areoles; disc black, roughened, not pruinose, the margins thin, raised, persistent, black; hymenium 70–110(-120) μm tall. Ascospores (12-)15–24(-26) × (6-)8–12(-14) μm. Conidia 5–12(-14) × 1–1.5(-2) μm. Medulla Pd−, K−, KC−, C− (confluentic and 2'-*O*-methylperlatolic acids).

On somewhat metal-rich or basaltic rocks and boulders and on schists, often on lichens (*Aspicilia* sp.), upland and montane; local. N. Wales, Scotland (Highlands). Europe, U.S.S.R., N. Africa, North America, China, New Zealand.

Old records from N. England (Yorkshire) and Ireland need confirmation as the species was confused with *Amygdalaria pelobotryon* in Britain during the last century. In the field resembles small morphs of *Lecidea fuscoatra*. Recently transferred to *Immersaria* as ***I. athroocarpa*** (Ach.) Rambold & Pietschm. (1989) (Rambold, 1989).

2. P. cinereoatra (Ach.) Hertel & Knoph (1984)

Huilia cinereoatra (Ach.) Hertel (1980), *Huilia albocaerulescens* auct. brit. p.p., non (Wulfen) Hertel (1975)

Thallus thick, unequal, cracked- to verrucose-areolate; areoles 0.5–1.5(-2) mm diam., flat to convex or occasionally hemispherical, obtuse-angular in outline to irregularly rounded, dirty creamish whitish or ashy grey; medulla I−; black prothallus sometimes visible between areoles. Apothecia 0.5–1.5 mm diam., numerous, irregularly scattered, solitary or 2–6 confluent forming irregular masses, often remaining ± immersed, sometimes some becoming adpressed, sessile, ± rounded to obtusely angular, ± flat; margin entire, matt or glossy black, sometimes eventually excluded; disc black, matt, finely grey- or white-pruinose at first, later glabrous and occasionally umbonate with plugs of sterile tissue; epithecium 6–12 μm thick, blue-green to dirty divaceous; exciple hyphae 3–5 μm thick; hymenium 90–105 μm tall, ± pale blue-green. Ascospores 12–18 × 6–9 μm. Medulla Pd−, K−, C− (confluentic acid, traces of 2'-*O*-methylperlatolic and 2'-*O*-methylmicrophyllinic acids).

On exposed, acid rocks; locally common. Upland British Is. Europe, North America.

Most British material of '*P. albocaerulescens*' belongs here. *P. speirea* **(13)** has an amyloid, I+ purple, medulla and grows on ± calcareous rocks. *P. contraponenda* **(4)** has mostly sessile apothecia and slightly larger ascospores, but t.l.c. is advised for their separation.

3. P. musiva (Körber) Hertel & Knoph (1984)

Like **2**, but differs in the more strongly convex, ± bullate areoles and larger ascospores, 15–20(-23) × 8–10 μm.

On upland, ± metal-rich rocks. N. Wales, C. England (Leicester, Charnwood), Scotland. Scandinavia, Europe, Japan.

Possibly an extreme variant of *P. cinereoatra* **(2)**; variation in convexity of areoles and ascospore size seems to be continuous in British material.

4. P. contraponenda (Arnold) Knoph & Hertel (1984)

Huilia albocaerulescens auct. brit. p.p., non (Wulfen) Hertel (1975)

Thallus continuous to ± regularly areolate-cracked, cracks narrow to gaping, chalky white or creamish; areoles polygonal, convex, surface occasionally shallowly papillate to verrucose, thick, tartareous; medulla I−; prothallus inconspicuous. Apothecia (0.3-)0.5–1.0(-1.3) mm diam., coal black, immersed at first, becoming sessile, solitary to 2- to 5-confluent, rounded to irregular through mutual pressure, margins swollen, slightly raised, shiny, disc shallowly concave at first then flat to subconvex, matt, shiny or ± granular, often white-pruinose; exciple hyphae 3–5 μm wide; epithecium greenish brown to olive-grey; hymenium 75–100(-120) μm tall; hypothecium brown-black. Ascospores 15–20(-25) × (5-)7–9(-11) μm. Medulla Pd−, K−, C−, UV+ ice blue (methyl 2'-*O*-methylmicrophyllinate and 2'-*O*-methylmicrophyllinic acid).

On sandstone and slate in upland regions. N. Wales, Scotland (Highlands). Europe, W. North America.

Often difficult to distinguish from *P. cinereoatra* **(2)** or *P. macrocarpa* **(10)**; t.l.c. is recommended for confirmation.

5. P. crustulata (Ach.) Hertel & Knoph (1984)

Huilia crustulata (Ach.) Hertel (1975), *Lecidea scutellata* W. Watson (1932)

Thallus effuse, thin, continuous to areolate or scurfy, whitish to creamish, or greenish to pale ashy-grey, rarely tinged orange, sometimes ± inconspicuous; medulla I−; prothallus occasionally developed, black or brownish. Soredia absent. Apothecia to 0.3–1(-1.5) mm diam., abundant, small, shiny, black, clustered in small groups or in ± concentric lines, slightly constricted at base; disc black, occasionally weakly pruinose, flat to convex; true exciple thin, black, ± glossy; exciple hyphae 4–8 μm wide; epithecium olivaceous to dark brown; hymenium 60–80(-110) μm tall. Ascospores 10–17 × 5–9 μm. Medulla Pd− or Pd+ orange, K− or K+ yellow, C− (lichen products not detected, or containing stictic acid, with traces of cryptostictic, constictic and norstictic acids).

On a wide range of siliceous rocks, stonework and pebbles, occasionally on worked wood and old sacking; common. Throughout British Is. Temperate to arctic-alpine Europe, North and South America, Japan, Kerguelen Is, Australia, New Zealand.

Close to *P. macrocarpa* **(10)** but distinguished by the smaller apothecia, small ascospores and shorter hymenium.

6. *P. flavocaerulescens* (Hornem.) Hertel & Schwab (1984)

Huilia flavocaerulescens (Hornem.) Hertel (1975), *Huilia melinodes* (Köerber) Hertel (1975)

Thallus moderately thick, distinctly areolate-cracked, orange or pale orange-grey or olivaceous; areoles angular, cracks deep, often tinged rusty orange, surface smooth, matt or slightly shiny; medulla I−, rarely + weakly bluish; prothallus black, in a distinct ± byssoid marginal zone. Soralia scattered, solitary or grouped 2–4 together at margins or in centre of areoles, 0.2–0.6(-1.0) mm diam., ± delimited by whitish, raised rim, whitish but often speckled dark grey; soredia coarsely granular; abundant on sterile specimens, fewer in fruiting material. Apothecia 0.2–2(-4) mm diam., solitary to 2- to 6(-10), clustered, immersed at first, then sessile, subconcave to flat, later convex, round to irregular through mutual pressure; true exciple persistent, distinctly raised, entire, sinuous, disc black, matt, shiny, roughened, sooty, ± thinly white-pruinose old apothecia with raised, gyrose plugs of sterile tissue; exciple hyphae 2–4 μm wide; epithecium brownish or greenish brown; hymenium 85–100(-150) μm tall, colourless or very pale brownish; hypothecium colourless to yellow-brown in upper parts, dark brown or dark red-brown below. Ascospores 9–14.5(-19.5) × 4–7.5 μm. Medulla Pd− or Pd+ orange, K− or K+ yellow, C− (confluentic, 2'-*O*-methylperlatolic and 2'-*O*-methylmicrophyllinic acids, sometimes with additional stictic, norstictic and cryptostictic acids).

On metal-rich siliceous rocks in upland areas. Throughout British Is. Europe, North America, Himalayas.

Similar, but fertile, non-sorediate material is referred to *P. flavicunda* **(7)**. Morphs of *P. tuberculosa* **(15)** with a partly rust-coloured thallus differ in having an I+ violet-blue medulla.

7. P. flavicunda (Ach.) Gowan (1988)

Like **6**, but without soredia. Densely fertile and with a prominent marginal black prothallus separating adjacent colonies. No detectable products by t.l.c.

On siliceous rocks; very rare. Scotland (E. Ross). Probably circumpolar.

In Europe, North America and Japan material of this species may also have substances of the confluentic and/or stictic acid complexes.

Differs from rusty orange morphs of *P. macrocarpa* **(10)** in the thicker, well-delimited thallus and narrower exciple hyphae.

8. P. glaucophaea (Körber) Hertel & Knoph (1984)

Huilia glaucophaea (Körber) Hertel (1875)

Thallus thick, tartareous, whitish to pale greenish grey, matt, continous to areolate-cracked, hummocky, verrucose or papillate; medulla I−. Soralia round to usually irregular, scattered, mostly forming along cracks in the thallus; soredia coarsely granular to minutely digitiform, whitish or greyish to ± pale greenish white. Apothecia 0.3–1.2(-2) mm diam., rare, scattered, solitary to 2- to 8-clustered, immersed at first, soon becoming sessile and ± constricted at base, round to irregular; true exciple thick, persistent, black, prominent, entire, to ± crenulate; disc black, often grey-pruinose, matt, papillate to ± gyrose and with ± prominent plugs of sterile tissue; exciple hyphae 2–4 μm wide; epithecium olive to brownish; hymenium 150–185 μm tall, colourless. Ascospores 15–23(-25) × (5-)7–10(-12) μm. Medulla and soralia Pd−, K−, C− (2'-*O*-methylsuperphyllinic and glaucophaeic acids).

On damp, siliceous rocks, on compacted soil amongst rocks in damp situations, sheltered sea cliffs to montane sites. Throughout British Is. Europe, E. North America.

Distinguished from *P. tuberculosa* **(15)** by the thicker, more verrucose thallus, I− medulla, more irregular soralia, which never have a bluish tinge, and chemistry (t.l.c.). Morphologically similar, unidentified material with medulla Pd+ orange, K+ yellow (stictic acid) is known from similar habitats in Scotland.

9. P. hydrophila (Fr.) Hertel & Schwab (1984)

Huilia hydrophila (Fr.) Hertel (1975)

Thallus continuous to irregularly areolate-cracked, cracking fine, rarely gaping between areoles, surface smooth, ± shiny, occasionally minutely wrinkled, creamish white to pale olivaceous white, pale greyish orange or brownish; medulla I− ; prothallus inconspicuous and pale, or dark between other lichens in mosaics. Apothecia (0.2-)0.5–1.2(-2.5) mm diam., round to irregular, scattered, immersed at first, soon becoming sessile; true exciple thick, prominent, crenulate, black, matt or shiny; disc flat to convex at maturity, subconcave to flat when young, black, matt or granular-roughened, occasionally white-pruinose, occasionally centrally papillate or with small irregular plugs of sterile tissue; exciple hyphae 3–5 μm wide; epithecium 12–25 μm thick, blue-green; hymenium (80-)100–120(-140) μm tall, bluish green in upper part. Ascospores (13.5-)18–23 × (5.5-)7–8(-9.5) μm. Pycnidia generally extremely abundant. Medulla Pd−, K−, C− (lichen products not detected by t.l.c.).

On inundated siliceous rocks by upland streams and lakes; locally abundant. N. & W. British Is., particularly W. Ireland. Sweden, the Faeroes, W. Germany.

Distinguished from the other members of the genus by the strikingly coloured blue-green epithecium.

10. P. macrocarpa (DC.) Hertel & Schwab (1984)

Huilia macrocarpa (DC.) Hertel (1975), *Huilia nigrocruenta* (Anzi) Hertel (1975)

Thallus very variable, ± obsolete (immersed) to continuous, rimose to ± areolate, thin to thick, surface roughed-subarachnoid to rugose, papillate, pale grey, greenish grey to olivaceous, frequently patchily or continuously "oxidized" orange to rust red; medulla I−; prothallus indistinct at margins of immersed thalli, but wavy, black, and ± distinct at margins of superficial thalli. Apothecia (Fig. 28a) 0.1–2(-3.5) mm diam., ± distinctly sessile, abundant, scattered or crowded, constricted at base; true exciple prominent, persistent, black, shiny, raised, thin to thick, entire to flexuose or crenulate; hyphae 3–9 mm wide; disc subconcave to flat or convex, black or brown-black, matt or shiny, occasionally grey-pruinose, often ± gyrose or with larger apothecia appearing to divide by formation of secondary margins within disc; epithecium pale brown to olive-brown; hymenium (70-)80–100(-120) μm tall; hypothecium and inner exciple sometimes with K+ reddish tinge. Ascospores (13-)16–20(-26) × (5-)6–11.5 μm. Medulla Pd− or + orange, K− or + yellowish (± stictic and cryptostictic acids).

On siliceous rocks and large boulders, less often on pebbles and loose stones in moorland and upland areas, rarely on wood and bark; common. Throughout British Is. Europe, North America, Asia, Australia, New Zealand.

A polymorphic species easily confused with *P. crustulata* **(5)**, *P. contraponenda* **(4)**, *P. flavicunda* **(7)**, *P. hydrophila* **(9)**, and *P. platycarpoides* **(11)**.

11. P. platycarpoides (Bagl.) Hertel (1987)

Huilia platycarpoides (Bagl.) Hertel (1985)

Thallus unequally granular-areolate, whitish or ashy grey; areoles ± convex, surface irregularly lumpy, minutely papillate or verrucose, cracks between areoles narrow to gaping; medulla I−; prothallus inconspicuous, or thinly developed at thallus edge. Apothecia large, 0.5–3 mm diam., scattered, sessile to slightly immersed; disc black, matt, or often thinly white-pruinose, occasionally ± umbonate at maturity, flat to convex; true exciple prominent, raised, swollen, often shiny; exciple hyphae 4–8 μm wide; epithecium olive-brown; hymenium 120–185 μm tall, colourless. Ascospores 15–23(-26) × 7–10(-12) μm. Medulla Pd+ yellow, K+ yellow to red (crystals), C−, UV+ orange (norstictic, connorstictic acids and 2 unidentified compounds).

On hard, siliceous rocks, maritime, but sometimes inland, especially on basaltic or ultrabasic rocks. W. & N. Britain and Ireland, extending to S. England (Sussex) on stonework. S. & W. Europe, Japan.

Close to *P. macrocarpa* **(10)**, but differs in the K+ red (crystals) reaction and more frequently pruinose apothecia. The less regularly areolate thallus, taller hymenium, and larger ascospores separate this species from *Lecidea lactea*.

12. P. soredizodes (Lamy ex Nyl.) Laundon (1989)

Huilia soredizodes (Lamy ex Nyl.) Hertel (1980)

Thallus small to irregularly spreading, dirty creamish grey to grey, often olivaceous or brownish, subcontinuous, rimose, subareolate to somewhat scurfy and indistinct; prothallus black, arachnoid, often visible between scattered areoles; medulla I−. Soralia 0.1–0.5 mm diam., scattered, round to irregular, punctiform or erose-excavate; soredia white, greenish white to somewhat greyish, farinose to granular. Apothecia rare, scattered, solitary, ± immersed to sessile, round to ± irregular; true exciple thick, black; hyphae 4–7 μm wide; disc black or brownish black, not pruinose, flat, matt or shiny; epithecium brown; hymenium 100–160 μm tall. Ascospores 15–19(-22) × 6–9 μm. Soralia Pd+ orange, K+ yellow (stictic acid).

On siliceous rocks, stonework, pebbles, and slate; probably common. Throughout British Is. Europe, Australia.

Differs from *P. tuberculosa* **(15)** in the generally smaller, thinner and darker thallus, and chemistry. *P. flavocaerulescens* **(6)** has a thicker, mostly rusty orange thallus. Unidentified material with similar chemistry, but a thick, warty thallus and irregular, coarse soralia is mentioned under *P. glaucophaea* **(8)**.

13. P. speirea (Ach.) Krempelh. (1861)

Lecidea speirea (Ach.) Ach. (1803); ?*L. subumbonella* Lamy (1883)

Thallus areolate-cracked, surface minutely granular-roughened, chalky, ± verrucose, white to creamish, to pale greenish grey; medulla I+ violet; prothallus indistinct or well developed, black. Apothecia (0.2-)0.3–1.2(-2.0) mm diam., common, often clustered centrally, or scattered, usually remaining ± immersed, black, often completely surrounded by a break in thallus surface or by a ± distinct thalline collar; disc concave to convex, often pruinose when young, and mature discs sometimes with small central plugs of sterile tissue; exciple hyphae 3–5 μm wide; epithecium greenish brown; hymenium 75–100 μm tall. Ascospores 12–19 × 6–7 μm. Medulla Pd−, K−, C− (confluentic with 2'-*O*-methylmicrophyllinic and 2'-*O*-methylperlatolic acids).

On hard limestones and other calcareous rocks (especially schists); local. Upland British Is. Europe, North America, Asia, Australia.

Often confused with *P. cinereoatra* **(2)** which grows on more acidic rocks, and in the field with *Rhizocarpon umbilicatum*. *P. superba* **(14)**, which can occur on weakly calcareous schists, has an I− medulla, a thallus with bullate areoles, and sessile apothecia.

14. P. superba (Körber) Hertel & Knoph (1984)

Thallus thick, verrucose to subverrucose-uneven, white, scabrid, with ± continuous or dispersed bullate areoles; medulla I−; prothallus black, between areoles or inconspicuous. Apothecia 0.5–2 mm diam., abundant, scattered among areoles, immersed at margins of areoles at first, later sessile and ± constricted at base, 0.5–2 mm diam., disc dark brown rarely thinly white-pruinose, flat to convex; true exciple thick, black; epithecium 20 μm thick, yellowish to orange-brown; hymenium 110–140 μm tall. Ascospores 17–25(-33) × 8–12(-14) μm. Medulla Pd+ orange or −, K+ yellow or − (stictic acid usually present but often in small amounts or patchily distributed).

Montane schistose, ± calcareous rocks; local. Scotland (Highlands). Europe, North America, Himalayas.

Distinguished by the white scabrid upper surface, often dispersed, ± bullate areoles, brown apothecial discs contrasting with the black margin, orange-brown epithecium, and chemistry.

15. P. tuberculosa (Sm.) Hertel & Knoph (1984)

Huilia tuberculosa (Sm.) P. James (1980)

Thallus irregularly areolate-cracked, surface smooth to roughened, occasionally minutely and irregularly papillate, white- to bluish grey or sordid creamish grey, sometimes rusty orange or red in parts; medulla I+ violet-blue; prothallus delimiting colonies black or brown, narrow, irregular or lacking. Soralia 0.1–1 mm diam., round to irregular, scattered to crowded, 1–4 or more per areole, shallowly erose or abraded, often with slightly raised rim; soredia farinose to granular, white, pale greyish or speckled bluish to ± blackened, I+ violet-blue. Apothecia (0.2-)0.5–1.5 mm diam., round to irregular through mutual pressure, solitary, scattered to 2–4 confluent, immersed at first, later sessile and ± constricted at base; true exciple ± prominent, persistent, slightly swollen, black, often shiny; hyphae 2.5–4 μm wide; disc flat to subconvex, black, matt, sometimes whitish to grey-pruinose, occasionally with small plugs of sterile tissue; epithecium 15–25 μm thick, brownish-olive; hymenium (70-)90–120(-135) μm, tall. Ascospores (10.5-)12–19(-22) × 7–9 μm. Medulla and soralia Pd−, K−, KC−, C− (confluentic and traces of 2'-*O*-methylmicrophyllinic and 2'-*O*-methylperlatolic acids (UV+ ice blue) very rarely with additional stictic and/or norstictic acids).

On siliceous rocks, walls, pebbles, rarely on worked timber; common. Throughout British Is. Europe, North America.

Very variable in appearance (e.g. colour, thallus thickness) reflecting its ability to colonize a wide range of substrata under widely differing environmental conditions. A few collections have a very weak or negative reaction of the medulla to iodine, but are otherwise indistinguishable. Other *Porpidia* species with sorediate thalli are *P. flavocaerulescens* **(6)**, *P. glaucophaea* **(8)** and *P. soredizodes* **(12)**.

PROTOBLASTENIA (Zahlbr.) Steiner (1911)

Thallus crustose, immersed to superficial and cracked, whitish to grey-green or pale brown. **Photobiont** chlorococcoid. **Ascomata** apothecia, ± flat to strongly convex, orange or orange-red to brown. **Thalline exciple** absent. **True exciple** poorly developed, of colourless branched hyphae. **Epithecium** with orange granular material (anthraquinone), K+ red. **Hymenium** colourless to orange, I+ blue. **Hypothecium** colourless to brown, occasionally ± violet. **Hamathecium** of paraphyses, 2–3.5 μm wide, rather conglutinate, sparingly branched and anastomosed, apices not or scarcely swollen. **Asci** 8-spored, cylindrical-clavate, *Porpidia*-type. **Ascospores** simple, colourless, ellipsoid or globose, without a thickened perispore. **Chemistry**: lichen products not detected in the thallus; apothecia with anthraquinones, principally parietin. **Ecology**: on calcareous rocky substrata.

LITERATURE: Poelt & Vězda (1977).

1 Ascospores ellipsoid, (8-)9–13(-17) μm diam. 2
Ascospores subglobose or globose, 5–8.5 μm diam. .... **2. *cyclospora***

2(1) Thallus immersed 3
Thallus superficial 4

3(2) Apothecia 0.2–0.5 mm diam., immersed in pits, flat or only slightly convex .... **3. *incrustans***
Apothecia 0.4–1.6 mm diam., not in pits, becoming convex .... **1. *calva***

4(2) Hypothecium colourless to yellowish; apothecia slightly to moderately convex, orange to orange-brown .... **4. *rupestris***
Hypothecium brown; apothecia becoming strongly convex, orange-brown to brown .... **5. *siebenhaariana***

1. P. calva (Dickson) Zahlbr. (1930)

Thallus immersed, inconspicuous, white to dirty grey. Apothecia 0.4–1.6 mm diam., not in pits at maturity, soon strongly convex, orange to dull orange or reddish-orange; hypothecium colourless or yellowish, rarely violet. Ascospores 8–15 × 5–8 μm, ellipsoid.

On hard limestones. Throughout British Is. Europe, North America, Asia.

Characterized by the immersed thallus and large, convex apothecia. *P. rupestris* **(4)**, with which it can grow and then remains distinct, generally has a less white thallus and less intensely pigmented and smaller, often flattened apothecia.

2. P. cyclospora (Hepp ex Körber) Poelt (1975)

Thallus immersed to ± superficial, sometimes finely cracked, inconspicuous, pale greyish to yellowish. Apothecia 0.4–0.85 mm diam., not in pits, becoming moderately convex, dark orange to orange-red; hypothecium colourless to pale orange-brown. Ascospores 5–8.5 × 5–7.5 μm, subglobose to globose.

On limestones; local. S.W. & N.W. England (N. Somerset, E. Gloucester, Westmorland). Austria, Scandinavia.

Distinguished by the globose spores; not easily separated from *P. rupestris* without microscope examination.

3. P. incrustans (DC.) Steiner (1911)

Thallus immersed, inconspicuous, whitish to grey or yellowish. Apothecia 0.2–0.5 mm diam., in pits, disc flat becoming slightly convex, yellow-orange; hypothecium colourless to yellowish. Ascospores 9–13 × 4.5–6 μm, ellipsoid.

On hard limestones. Throughout British Is. Europe, North America.

Distinguished from *P. calva* **(1)** and *P. rupestris* **(4)** by the small, often pale orange, apothecia immersed in pits in the rock.

4. P. rupestris (Scop.) Steiner (1911)

Thallus superficial, cracked and ± areolate, pale grey to dirty grey-brown or grey-green. Apothecia 0.3–0.9 mm diam., not in pits, slightly convex, or becoming moderately convex, ± dirty or dull orange; hypothecium colourless to yellowish. Ascospores 8–17 × 5–8 μm, ellipsoid.

On a wide range of calcareous substrata, including limestone, chalk pebbles, concrete, mortar and asbestos-cement; common. Throughout British Is. Europe, North America.

An important indicator species of calcareous substrata.

5. P. siebenhaariana (Körber) Steiner (1911)

Thallus superficial, ± smooth to uneven, cracked and ± areolate, whitish to pale dirty brown. Apothecia 0.45–1.5 mm diam., at first dull orange or orange-brown, often becoming dull brown to olive-brown and strongly convex; hypothecium brown (thick section), or light orange-brown (thin section). Ascospores 9.5–13 × 4.5–6 μm, ellipsoid.

On mica-schist, epidiorite and basalt. Scotland (Highlands). Subalpine and alpine Europe.

Differs from *P. rupestris* **(4)** in the darker hypothecium and the browner, more convex, apothecia; ascospore size does not seem to be as distinctive in British as in continental material.

PROTOPARMELIA M. Choisy (1929)

Thallus crustose, rimose, warted or areolate, corticate, rarely scurfy-isidiate and not corticate, often glossy, pale grey-brown to dark or chestnut brown. **Cortex** of branched, short-celled, anticlinal hyphae, often terminated by brown pigmented 'hoods', and usually overlain by a distinct, colourless epicortex; photobiont layer and medulla usually well defined. **Medulla** I−. **Photobiont** trebouxioid, cells sometimes dividing internally to form autospores. **Ascomata** apothecia, arising from within areoles or warts, immersed to sessile, usually glossy. Disc brown, often darker than margin, epruinose. **Thalline exciple** ± concolorous with thallus, 12–30 μm thick, with a medulla filled with algal cells and well-defined cortex (except *P. oleagina*) similar to that of the thallus. **True exciple** colourless. **Hymenium** I+ blue. **Hypothecium** colourless. **Hamathecium** of simple or occasionally forked, septate paraphyses; apices sometimes slightly wider, but usually each covered by a swollen gelatinous 'hood' which often contains a brown 'cap'. **Asci** 8-spored, clavate, *Lecanora*-type, with or without an ocular chamber, but always with a distinct, non-amyloid apical cushion. **Ascospores** simple (sometimes a few 1-septate when old), colourless, ellipsoid, or more usually fusiform- or oblong-ellipsoid to oblong; lacking a distinct perispore. **Conidiomata** pycnidia immersed, wall colourless except for brown pigmentation around the ostiole. **Conidiogenous cells** arising on branched conidiophores or in chains, ± cylindrical, enteroblastic, acrogenous or pleurogenous. **Conidia** bacilliform, short-acicular or curved and thread-like, aseptate, colourless. **Thalloconidia** present in one species (see **4**). **Chemistry**: gyrophoric, lobaric, stictic and norstictic acids, and zeorin, and several unidentified substances (incl. fatty acids) may be present.

Separated from *Lecanora* by the brown pigmentation, generally smaller (especially narrower) ascospores, straight conidia, and lack of atranorin. The apparent similarities to *Bryonora* warrant critical investigation, as does the taxonomy of the *P. montagnei/picea* complex.

1 On rocks 2
 On bark or wood 5

2(1) Ascospores ellipsoid or oblong with rounded ends 3
 Ascospores ellipsoid-fusiform .... **2. *badia***

3(2) Thallus mostly >4 cm across; cortex K−, medulla sometimes K+ 4
 Thallus forming patches mostly <4 cm across; cortex of thallus and apothecia (in sections) K+ yellow to red .... **1. *atriseda***

4(3) Medulla (at least upper part) C+ red, rarely C−; thallus with brown predominating, never blackish; on maritime rocks .... **3. *montagnei***
Medulla C−; thallus usually with grey or black predominating, ±with olivaceous tinge; mainly on inland rocks .... **4. *picea***

5(1) Thallus of subglobose areoles, not isidiate or sorediate .... **5. *ochrococca***
Thallus scurfy, granular-isidiate .... **6. *oleagina***

1. P. atriseda (Fr.) R. Sant. & V. Wirth (1986)

Lecanora atriseda (Fr.) Nyl. (1857)

Thallus forming small patches up to *c.* 3 cm across, closely associated with *Rhizocarpon geographicum* agg.; composed of discrete, hemispherical to ± globose, glossy, chestnut-brown areoles, 0.2–0.6 mm diam., seated on a black prothallus; small areoles sometimes arising from within the areoles of the *Rhizocarpon.* Apothecia 0.2–0.6(-0.8) mm diam., concolorous with the thallus, at first immersed and perithecia-like, eventually ± sessile; hymenium 45–50 µm tall. Ascospores 8–13 × 3.7–5 µm, oblong-ellipsoid to oblong. Pycnidia not seen. Sections (especially cortex) Pd+ yellow or orange, K+ yellow to red (crystals), KC−, UV− (norstictic acid, trace amounts of unidentified substances, including a fatty acid (? rangiformic acid); sometimes also with psoromic acid from the *Rhizocarpon*).

On hard acid rocks; rare. Wales, Scotland (Highlands). Scandinavia, C. European Mtns.

Recognized by the small, loose clusters of shiny, almost globose, red-brown areoles, ± oblong ascospores and K reaction.

2. P. badia (Hoffm.) Hafellner (1984)

Lecanora badia (Hoffm.) Ach. (1810), *Lecanora picea* (Dickson) Nyl. (1868)

Thallus very variable, thin and rimose-areolate to warted, or coarsely warted, to 2.5 mm thick, pale fawn-grey to dark brown, sometimes faintly olivaceous, usually ± glossy, effuse or rarely delimited by a dark prothalline line. Apothecia 0.4–2.2 mm diam., usually numerous, immersed to sessile, usually glossy; thalline exciple concolourous with thallus, mostly persistent; disc flat to convex, usually darker than the exciple; hymenium 50–60 µm tall. Ascospores 8–13 × 3–5 µm, ellipsoid-fusiform with distinctly pointed apices. Pycnidia often present, rarely abundant; conidia 8–11 × 0.7–1 µm. Sections Pd−, K−, KC+ pink (especially cortex), C−, usually UV+ white (lobaric acid, ± zeorin, and 3(-5) unidentified substances).

On hard (usually exposed) acid rocks in coastal and montane sites; frequent. Throughout upland British Is. and in a few lowland areas. Montane Europe, North America, temperate South America, Australia, New Zealand.

Very variable but usually easily recognized by the uniform or mottled dark brown colour and concolorous apothecia; distinguished from the rarer *P. montagnei* **(3)** and *P. picea* **(4)** complex by the sharply fusiform ascospores.

3. P. montagnei (Fr.) Poelt & Nimis (1987)

Lecanora psarophana Nyl. (1872), *L. nitens* auct., non Ach. (1814)

Like **2**, but thallus usually paler with numerous pycnidia, and often delimited (esp. when young). Best distinguished by its oblong-ellipsoid ascospores 9–13 × 2–3.5(-4.4) µm, and chemistry. There are three chemotype races (reactions are for upper medulla and cortex in sections): (1) gyrophoric and lobaric acids, C+ red, UV+ white; (2) lobaric acid, KC+ reddish, C−, UV+ white; (3) gyrophoric acid, C+ red, UV−; all chemotypes with some traces of unidentified substances (but different from those in *P. badia*). It should be noted that reactions of the lower part of the thallus can be unreliable as this tissue often belongs to another species overgrown by *P. montagnei.*

On maritime rocks, xeric-supralittoral, often overgrowing (? initially parasitic) on thalli of other crustose lichens, e.g. *Aspicilia cinerea* and *Diploschistes caesioplumbeus*; locally abundant. Channel Is. (chemotypes 1 & 2), S.W. England, S. Wales (Pembroke) (chemotype 3). S. Europe.

4. P. picea auct.

Lecanora picea auct. p.p., non (Dickson) Nyl. (1868)

Like **2**, but thallus usually darker, apothecia generally fewer and smaller, and ascospores 9–12 × 2.5–3 μm, oblong-ellipsoid to oblong. Sections K –, KC ± reddish, C –, UV + white (lobaric acid, various unidentified substances).

On hard acid rocks. Mainly in upland Britian.

Two entities (? species) are recognized in the British Is., based mainly on presence of unidentified substances: (1) with two of the substances found in *P. badia*; (2) with two of the substances found in *P. montagnei*, and a terpenoid (at A3, G2) - it also tends to have more swollen, often subsquamulose thalline warts. Entity (1) is known from N.E. Scotland (S. Aberdeen, near Braemar; Angus, Clova); also in the Harz Mtns., in Germany. Entity (2) is known only from S.W. & N.E. England (N. Devon, Cleveland). A third entity, also called '*P. picea*', that lacks lobaric acid (medulla UV –) is found in Scandinavia and C. Europe. A thorough chemotaxonomic survey of a geographically wide range of material is needed to resolve the taxonomy of this complex.

P. nephaea (Sommerf.) R. Sant. (1990) has chestnut-brown areoles on a black prothallus which at the thallus margin forms minute, radiating lobes; prothallus and margins of areoles sooty-black in places due to 2-celled to submuriform, coarsely warted, green-black thalloconidia, 9–14(-20) × 7–12 μm. Soralia sometimes present, 1 per areole, brown or abrading white, 0.2–0.3 mm diam. Apothecia rare, sessile, 0.3–0.8 mm diam., margin concolorous with areoles, disc black. Ascospores ellipsoid to oblong-ellipsoid, 7–10 × 3–3.5 μm. Pycnidia immersed; conidia curved, 12–14 × 0.8–1 μm. Medulla Pd + orange, K + yellow, KC –, C –, UV – (stictic acid). Recently discovered in siliceous rock underhangs in C. & E. Scotland (Mid- & E. Perth, Angus). Sweden, C. European mtns.

5. P. ochrococca (Nyl.) P. M. Jørg., Rambold & Hertel (1988)

Lecidea ochrococca Nyl. (1860), *Lecanora ochrococca* (Nyl.) Clauz. & Roux (1985)

Thallus of rounded, ± globose, shiny, chestnut-brown areoles 0.06–0.26 mm diam. Apothecia 0.3–1.1 mm diam., sessile and constricted below, flat and exciple raised or sometimes becoming convex with reflexed thalline exciple; thalline exciple concolorous with thallus, smooth or crenulate; disc usually more reddish; hymenium 35–50m μm tall. Ascospores (7-)8–12 × 2–3.5 μm, fusiform. Pycnidia rare; conidia 7–9 × *c.* 0.7 μm, shortly acicular, straight. Sections Pd –, K –, KC –, C – (lichen products not detected by t.l.c.).

On bark, or rarely wood of conifers (*Larix*, *Pinus*) and *Betula*, often in deep fissures; local. Scotland (Highlands). S. Norway, W. North America.

6. P. oleagina (Harm.) Coppins (1992)

Lecanora oleagina Harm. (1913)

Like **5**, but with a dull olivaceous brown, scurfy granular, ± isidiate thallus and apothecial margin; disc also olivaceous. Sections Pd –, K –, KC –, C –, UV – (traces of the two of the unidentified substances found in *P. montagnei* **(3)**, and one of the entities within the *P. picea* **(4)** complex).

Fertile on *Pinus* wood in native Scottish pinewoods; sterile on fence and gate rails in England (where it is host to *Sphinctrina anglica*), wood of ancient oak pollard. S and S.E. England, C. & N.E. Scotland. S. France, N. Spain.

PROTOTHELENELLA Räsänen (1943)

Thallus not or distinctly lichenized and then crustose, granular to cracked-areolate, margin effuse, usually bright green when fresh, whitish to grey-brown when dry, ± gelatinous when wet, anatomically hardly differentiated. **Photobiont** *Elliptochloris*. **Ascomata** perithecia, mostly single and scattered, immersed, globose to pear-shaped, dark brown to blackish. **Involucrellum** absent. **True exciple** mostly colourless towards the base, brown to greenish blue in upper part, of strongly anastomosing, to 1 μm wide, conglutinate hyphae with narrow lumina. **Hamathecium** of paraphysoids, persistent, strongly branched and anastomosed; periphysoids absent. **Asci** (6-)8-spored, ± cylindrical, thick-walled, (2–5 μm) with two functional wall layers, the outer wall layer I+ blue, the apex with an I+ blue, ± layered *Xylaria*-like apical apparatus (Fig. 9c). **Ascospores** colourless, transversely multiseptate, submuriform to strongly muriform, the outer wall distinctly thicker than the septa. **Ecology**: on acid substrata including rocks, soil, mosses and plant detritus or on rotten wood in moist situations.

Most of the British species formerly referred to *Microglaena* Körber (1855) *nom. illegit.* belong to this genus. It is distinguished from *Polyblastia* by the persistent paraphysoids and the structure of the asci. See also *Chromatochlamys*.

LITERATURE: Mayrhofer & Poelt (1985).

1 Ascospores elongate to elongate-ellipsoid; on wood, soil, decaying plants, mosses and other lichens 2
Ascospores broadly ellipsoid; on rocks .... **1. *corrosa***

2(1) Apothecia not in thalline warts but ± immersed in thallus; thallus greenish when fresh, C− 3
Apothecia in distinct whitish thalline warts; thallus greyish white, C+ red; ascospores 24–36 × 9–14 μm, muriform .... **2. *leucothelia***

3(2) Ascospores 22–33 × 7–10 μm, strongly muriform; 'perithecia' <0.3 mm diam. .... **3. *sphinctrinoidella***
Ascopsores 38–50 × 10–15 μm, submuriform; 'perithecia' 0.4–0.6 mm diam. **4. *sphinctrinoides***

1. P. corrosa (Körber) Mayrhofer & Poelt (1985)

Microglaena corrosa (Körber) Arnold (1877)

Thallus smooth to granular-uneven, warted, rarely ± continuously scurfy-rugose, often in ± dispersed areoles surrounding single or several apothecia, or disappearing, dirty yellowish, greenish or brownish white. Perithecia 0.2–0.4 mm diam., ± ½-immersed, scattered, rarely 2–3 contiguous, exposed part blackish, ± shiny; true exciple dark brown, opaque above, ± colourless or pale below. Ascospores 18–32 × 10–15 μm, muriform, broadly ellipsoid or ovoid. Thallus often C+ red.

On rather moist, shaded or sheltered mica-schists, epidiorite, in old quarry workings, semi-inundated rocks at streamsides, rarely on coarse granite; very local. Upland Britain, Ireland. C. & N. Europe, U.S.S.R., Canada.

The commonest British species of the genus.

2. P. leucothelia (Nyl.) Mayrhofer & Poelt (1985)

Microglaena leucothelia (Anzi) Arnold (1870)

Thallus finely granular to coarsely granular-warted, rarely partly and incompletely areolate, whitish; granules on a filmy background, some becoming fertile. Perithecia 0.3–0.5 mm, scattered to ± aggregated-contiguous, rounded, immersed in thalline warts with only black-brown ostiole apparent; true exciple brownish, semi-opaque above, colourless below. Ascospores 24–36 × 9–14 μm, muriform with up to 2 longitudinal walls per segment, ellipsoid-elongate, one or both ends with a small apiculus. Thallus C+ red (unidentified para-depside of orcinol type).

On decaying mosses and lichens or rotting wood, rarely on acid soil; very rare. N. Scotland (Ben Lawers). Finland, Sweden, Austria, Switzerland, Italy.

Apparently distinguished by the ± conspicuous, pale thallus, C+ red reaction and perithecia immersed in thalline warts.

3. P. sphinctrinoidella (Nyl.) Mayrhofer & Poelt (1985)

Microglaena sphinctrinoidella (Nyl.) Arnold (1870)

Thallus indistinct, thin, membrane-like or diaphanous, evanescent, pale grey or dirty white, greenish when wet. Perithecia 0.1–0.3 mm diam., sessile to ¼-immersed, black, ± shiny, rounded to ± pyriform, discrete or single, very rarely 2–4 contiguous; true exciple dark brown above, ± opaque, paler below, sometimes colourless, semi-opaque to translucent; ostiole sometimes extended as a short papilla. Ascospores 22–33 × 7–10 μm, elongate or elongate-ellipsoid, submuriform, with 7–9 transverse and 4–6 longitudinal septa.

Overgrowing hepatics on acid soils or rocks, mainly above 600 m. N. Scotland (Highlands, especially Breadalbane and Cairngorm Mtns.). Fennoscandia, Pyrénées, Alps, U.S.S.R.

Differs from *P. sphinctrinoides* **(4)** in the smaller perithecia and smaller, submuriform spores. The thallus may be almost absent and partially saprophytic with few associated photobiont cells.

4. P. sphinctrinoides (Nyl.) Mayrhofer & Poelt (1985)

Microglaena sphinctrinoides (Nyl.) Lönnr. (1858)

Thallus thin, film-like to finely warted, evanescent, dirty white when dry, greenish or pale brown, at times gelatinous when wet. Perithecia 0.4–0.6 mm diam., $\frac{1}{2}$ to ± entirely immersed, rounded or ± pyriform, scattered, blackish brown, paler when wet; true exciple red-brown, opaque above, ± colourless below; ostiole occasionally somewhat depressed. Ascospores 38–50 × 10–15 μm, ellipsoid to broadly spindle-shaped, strongly muriform, with to 4 longitudinal walls in middle part of spore.

Overgrowing bryophytes on acid soils or over schistose or granitic rocks, above 800 m. N. Scotland (Highlands, Breadalbane and Cairngorm Mtns.). Europe, from the Alps to N. Fennoscandia, U.S.S.R., Spitzbergen, Greenland, U.S.A.

See *P. sphinctrinoidella* **(3)** for differences; the large, strongly muriform, pale ascospores and totally immersed fruits are diagnostic for this species. *Chromatoclamys muscorum* has larger ascospores and an I− ascus apex.

PSEUDEPHEBE M. Choisy (1930)

Thallus shrubby, richly branched, prostrate, ± closely attached to substratum, attached by disc-like hapters, branches ± thread-like, terete. **Cortex** composed of longitudinally orientated hyphae which become prosoplectenchymatous to ± pseudoparenchymatous at surface. Medullary hyphae not ornamented. **Photobiont** chlorococcoid. **Ascomata** apothecia. **Thalline exciple** concolorous with thallus. **Asci** clavate, thick-walled, 8-spored. **Ascospores** ellipsoid, without distinct perispore, colourless at maturity. **Conidiomata** pycnidia, embedded in thallus and opening to surface through wide ostioles. **Conidia** simple, bacilliform, colourless. **Chemistry**: lichen products not detected by t.l.c. **Ecology**: on siliceous rocks, arctic-alpine.

Characterized by the low, prostrate habit and attachment by hapters. Distinguished from other alectorioid genera by the absence of lichen products, the distinctive cortical structure, and lack of isidia, soredia and pseudocyphellae.

LITERATURE: Brodo & Hawksworth (1977), Hawksworth (1972b).

1. P. pubescens (L.) M. Choisy (1930)

Thallus to 1 cm tall, prostrate, forming entangled mats, closely adpressed; branches to 0.2 mm diam., terete to slightly compressed; branching frequent, divergent isotomic - dichotomic, often interwoven; pale to dark brown-black, usually concolorous or paler at base, ± shiny. Apothecia frequent in Britain above *c.* 610 m; disc to 5.5 mm diam. Ascospores 7–12 × 6–8 μm. Pycnidia on tubercles, abundant, especially near the axils, to 0.4 mm diam., black; conidia 5–7 × 1 μm. Medulla Pd-, K-, KC-, C- (lichen products not detected by t.l.c.).

On exposed siliceous rocks in wind-swept, montane sites, often a component of the *Rhizocarpetum alpicolae*; locally abundant. N.W. England (Lake District), N. Wales (Snowdonia), Scotland (Highlands), W. Ireland (Connemara mtns); scarce elsewhere. Arctic-alpine with a wide circumboreal distribution in both the Southern and Northern Hemispheres.

Distinguished from all other British alectorioid species by its low, matted habit, wiry texture, negative reactions with Pd and K, and the absence of lateral spinules and soralia. *P. pubescens* **f. *subciliata*** (Nyl.) D. Hawksw. (1977) is distinguished from *f. pubescens* in the characteristic apical branching giving mats a net-like appearance, especially towards the margins.

PSEUDEVERNIA Zopf (1903)

Thallus foliose-shrubby, dorsiventral, with a single point of attachment; lobes strap-shaped, narrow, 1–4 mm wide, dichotomously branched. **Isidia** present. **Lower surface** without rhizines, at least partly black, channelled. **Photobiont** trebouxioid. **Medulla** compact. **Ascomata** apothecia, stalked. **Thalline exciple** persistent, concolorous with thallus. **Discs** usually ± concave, olive to red-dark brown, shiny. **Asci** 8-spored, clavate, *Lecanora*-type. **Ascospores** simple, colourless. **Pycnidia** immersed. **Conidia** 5.5–7 × 0.8 μm bifusiform. **Chemistry**: cortex contains atranorin, medulla a depside or depsidone. **Ecology**: on acidic substrata.

Characterized by the naked, channelled, often black lower surface. *Evernia* is only weakly dorsiventral, has a lax medulla, and contains usnic and divaricatic or evernic acids.

LITERATURE: Hawksworth & Chapman (1971), Maas Geesteranus (1947).

1. P. furfuracea (L.) Zopf (1903)

Thalli to 10 cm diam., composed of numerous strap-shaped, lobes 1–4 mm wide, ± dichotomously branched in one plane, branching widely divergent with short side branches; upper surface grey-white, often rough with isidia or small folioles; lower surface usually channelled, uniformly black or mottled black and brownish white or pinkish, with the incurved margins concolorous with upper surface. Apothecia to 1.5(-3) cm diam., lateral, on geniculate branches. Asci 30–40 × 14–16 μm. Ascospores 7.5–10 × 4–5.5 μm, ellipsoid. Cortex K+ yellow (atranorin); medulla Pd−, K−, KC−, C− (physodic acid) [var. *furfuracea*] or C+ red (olivetoric acid) [var. *ceratea* (Ach.) D. Hawksw. (1969)].

On exposed, well-lit bark and wood, on conifers, deciduous trees (particularly in slightly polluted sites) and fence posts, also on siliceous rocks; usually common. Throughout British Is. Europe, C. America, Bolivia, Ethiopia, E. Africa.

Resembles *Evernia prunastri*, but is characterized by the presence of isidia and dark lower surface. The olivetoric acid-containing race predominates in Britain and is particularly abundant in the north.

PSEUDOCYPHELLARIA Vainio (1890)

Thallus foliose, dorsiventral, spreading, ± rosette-forming or irregularly branched, loosely attached; lobes rounded or truncate, variously indented, often ascending at margins. **Upper surface** smooth or wrinkled, sometimes reticulately ridged, matt or dull. **Lower surface** tomentose, dotted with scattered pseudocyphellae (Fig. 31b). **Soredia** and **isidia** ± present or not. **Cortex** pseudoparenchymatous, of anticlinally arranged, ± isodiametric, thick-walled cells. **Photobiont** green (*Dictyochloropsis* or *Chlorella*-like) or blue-green (*Nostoc*). **Medulla** white or yellow. **Ascomata** apothecia, hemiangiocarpic, sessile or ± stalked, very rare or absent. **Thalline exciple** present. **Asci** *Peltigera*-type, 8-spored. **Ascospores** brown, 1- to 3-septate, fusiform-ellipsoid. **Conidiomata** pycnidia, very rare, *Lobaria*-type, globose or ovoid, 0.2–06. mm diam., walls dark brown at the ostiole, paler below. **Conidia** 3–5 × 0.7–1 μm, bacilliform. **Chemistry:** orcinol and β-orcinol derivatives, triterpenoids, terphenylquinones and pulvinic acid derivatives. **Ecology:** confined to humid, sheltered, oceanic woodlands.

Related to *Lobaria* and *Sticta*, with which *Pseudocyphellaria* species often grow, but characterized by the presence of dot-like pseudocyphellae on the lower surface, the emergent apothecia, and diverse chemistry. The genus is predominant in the Southern Hemisphere, particularly in temperate rain-forest areas of South America and New Zealand.

LITERATURE: Coppins & James (1979b), Galloway (1988).

1 Soralia and pseudocyphellae bright yellow; isidia absent 2
Soralia and pseudocyphellae white, grey or lilac-grey; isidia present or absent 3

2(1) Photobiont green; thalli bright green when wet; medulla bright yellow; soralia ± restricted to margins, bright yellow .... **1. *aurata***
Photobiont blue-green; thalli dull grey-brown when wet; medulla white or pale yellow in parts; soralia both laminal and marginal, bright yellow .... **2. *crocata***

3(1) Medulla C−, KC−; lobes elongate, surface ± smooth; soralia or isidia mostly marginal 4

Medulla C+ red, KC+ orange-pink; lobes rounded; surface ± reticulately-ridged; soralia both laminal and marginal .... **5. *norvegica***

4(3) Isidiate; isidia cylindrical, simple or branched .... **4. *lacerata***
Sorediate; soredia farinose to granular .... **3. *intricata***

1. P. aurata (Ach.) Vainio (1890)

Thallus 2–6 (-10) cm diam., forming rosettes or irregularly spreading; lobes 3–12 mm wide, rounded or indented at apices, often discrete at margins and overlapping centrally; upper surface pale yellow-green when wet, pale grey when dry, smooth, undulate, often scabrid-areolate in parts, ± pubescent at margins bearing conspicuous coarsely granular, bright yellow, ± linear, lip-like marginal soralia which occasionally also spread on to upper surface; medulla yellow; lower surface pale pink-brown, darker towards centre, thickly tomentose to the margins, ± wrinkled-uneven with numerous yellow pseudocyphellae, 0.1–0.6 mm diam. Photobiont green. Medulla and soralia Pd−, K−, KC−, C−, UV+ dull to bright orange or salmon-pink (pulvinic acid, pulvinic dilactone and calycin).

On trees, rocks and *Calluna* stems; formerly local, now decreasing and very rare. S. & S.W. England, Channel Is., Is. of Scilly, S.W. Ireland (S. Kerry, Blasket Is.). W. France, Portugal, Macaronesia, St. Helena, widespread in the tropics and temperate Southern Hemisphere.

Characterized by the bright yellow, predominantly marginal, linear soralia, yellow medulla, and green photobiont. *P. crocata* **(2)** commonly has both laminal and marginal soralia, a white to only pale yellow medulla, a blue-green photobiont, and Pd+ medulla.

2. P. crocata (L.) Vainio (1898)

Thallus 2–3(-10) cm diam., at first rosette-forming, soon irregularly spreading; lobes 5–15 mm wide, rounded or ± elongate, indented at apices, often discrete at margins and overlapping centrally; upper surface pale grey-brown (in shade) to dark red-brown, ± smooth or with a ± coarse, ridged network from which yellow soralia are derived, at first punctiform, later becoming confluent; medulla white; lower surface pale brown, darker brown towards centre, tomentose to margins, with frequent, conspicuous, yellow pseudocyphellae. Photobiont blue-green. Apothecia very rare, disc 1–2 mm diam., dark red-brown, margin thin, pale reddish brown, often wrinkled. Ascospores (18-)22–30 × 5–9(-12) μm, 1(-3)-septate, fusiform, brown. Medulla and soralia Pd+ orange, K+ yellow, KC−, C−, UV± dull orange (pulvinic acid, pulvinic lactone, calycin, tenuiorin, methylgyrophorate and stictic and constictic acids and hopane 6α, 7β, 22-triol).

On mossy trunks and branches of *Corylus*, *Fraxinus*, *Quercus*, *Sorbus aucuparia*, *Salix*, and old *Calluna* stems and mossy boulders in moist or boggy, sheltered, well-wooded or coastal sites; local. S.W. England (N.W. Cornwall, S. Somerset), W. Scotland, W. & S.W. Ireland. Cosmopolitan.

Characterized by the blue-green photobiont, white medulla, yellow pseudocyphellae and yellow soralia which are laminal and marginal and the presence of stictic acid (Pd+ orange). *P. aurata* **(1)** has a green photobiont, yellow medulla and soralia which are predominantly marginal and a very different chemistry (Pd−).

3. P. intricata (Del.) Vainio (1898)

Thallus 3–6(-10) cm, rosette-forming at first soon irregularly spreading; lobes to 1 cm wide, elongate, often deeply incised, with rounded apices; upper surface dark red-brown, ± greyish in shaded habitats, ± shiny, smooth, ± undulate or rarely weakly ridged; sorediate. Soralia farinose to coarsely granular-sorediate, whitish to grey-blue, predominantly marginal, subsequently becoming laminal, remaining ± punctiform on older lobes; medulla white; lower surface pale to dark brown, tomentose to margins with scattered, rather sparse, white pseudocyphellae (Fig. 31b). Photobiont blue-green. Medulla and soralia Pd−, K+ yellow, KC−, C−, UV± bluish white (± tenuiorin and ± methylgyrophorate and main terpenes, 7 β-acetoxyhopan 22-ol, hopane 15 α, 22-diol).

On mossy trunks and branches of *Corylus, Fraxinus, Quercus, Sorbus aucuparia* and *Salix,* on mossy boulders and, more rarely, old *Calluna* stems in moist, sheltered, well-wooded sites; locally frequent. W. Scotland, W. & S.W. Ireland; apparently extinct in S.W. England (S. Devon). Cosmopolitan.

Closely related to *P. norvegica* **(5)** but characterized by the KC− medulla, and the ± smooth upper surface and elongate lobes with predominantly marginal, pale grey soralia. Distinguished from *P. lacerata* **(4)** by the presence of soredia.

Immature specimens of *P. intricata* **(3)** and *P. norvegica* **(5)** may resemble *Nephroma parile*. In the latter, the underside is naked and smooth and lacks incipient pseudocyphellae.

4. P. lacerata Degel. (1941)

Like **3**, but differing in the distinct, simple or branched-coralloid, cylindrical isidia on the lobe-margins which in extreme cases may also spread on to the upper surface.

On mossy branches of broad-leaved trees in old woodland; very rare. N. Wales, S.W. Ireland. Azores, Madeira.

Distinctive when well developed, but intermediates occur and it is possible that a continuum may exist with the sorediate *P. intricata* **(3)**.

5. P. norvegica (Gyelnik) P. James (1979)

Like 3, but differing in the chemistry and particularly the distinctive KC+ orange, C+ red medulla reaction due to a range of depsides related to the barbatic acid complex the most important of which are 2'-*O*-methylnorbarbatic and 2'-*O*-methylnorobtusatic acids; tenuiorin and methylgyrophorate are absent and the two terpenes are the same as those in *P. intricata* **(3)**; medulla and soralia are more UV+ brighter ice-blue than in *P. intricata*. The thallus often has a coarse, incomplete ridged network on the upper surface and more rounded and wider lobes; soralia are usually laminal and marginal, and characteristically elongate and sigmoid on the upper surface and often have a pronounced lilac-mauve tinge when fresh.

On moist, sheltered tree trunks and mossy rocks, often near water, or in *Salix* carrs and *Corylus* stands. W. Scotland, extending locally to England, Wales and S.W. & N.W. Ireland. W. Norway, Madeira, Azores, S.C. Chile (commonly fertile in Valdivia).

Immature material, with ± exclusively marginal soralia, may be easily confused with *P. intricata*; in such cases the KC+ orange, C+ red medulla is diagnostic.

PSILOLECHIA Massal. (1860)

Thallus crustose or leprose, rarely granular-areolate, of ± distinct goniocysts interconnected by filamentous hyphae, bright yellow-green or dull green to whitish grey. **Photobiont** *Trebouxia*-like or *Stichococcus*. **Ascomata** apothecia, convex to ± globose, occasionally becoming tuberculate. **Thalline exciple** absent. **True exciple** poorly developed, a narrow zone at base of apothecia with protruding colourless hyphae. **Epithecium** often not clearly delimited. **Hymenium** laterally reflexed, variously pigmented, I+ blue. **Hypothecium** colourless or pale, I−. **Hamathecium** of paraphyses, simple or 1- or 2-forked, the apices not or slightly swollen. **Asci** 8-spored, cylindrical-clavate, apical dome K/I+ pale blue with dark blue apical tube diverging towards the apex, wall colourless with a K/I+ dark blue outer layer, *Porpidia*-like. **Ascospores** simple, oblong-ovoid or tear drop-like, colourless, without a perispore (Fig. 23). **Anamorphs**, where known, hyphomycetous with discrete enteroblastic conidiogenous cells arising directly on the thallus surface. **Conidia** ovoid to pyriform, colourless, adhering in chains, simple. **Chemistry**: with rhizocarpic or gyrophoric acids, or no lichen products detectable by t.l.c. **Ecology**: sheltered, ± humid situations.

LITERATURE: Coppins & Purvis (1987).

1	Thallus whitish to pale greenish or yellowish grey; without rhizocarpic acid	2
	Thallus bright yellow-green; with rhizocarpic acid	**1. *lucida***
2(1)	Apothecia dark brown to blue-black; thallus C−	**2. *clavulifera***
	Apothecia pale pink to pale brown, occasionally with a violet tinge; thallus C+ red	**3. *leprosa***

1. P. lucida (Ach.) M. Choisy (1949)

Thallus wide-spreading, bright yellow-green, leprose-granular and often rimose, rarely granular-areolate, effuse. Photobiont usually *Trebouxia*-like, very rarely with *Stichococcus*. Apothecia

Fig. 23. *Psilolechia leprosa* tear-drop-like (dacryoid) ascospores (from Coppins & Purvis, 1987). Scale = 10 μm.

mainly 0.1–0.3 mm diam. and convex, rarely tubeculate and to 0.5(-0.7) mm diam., yellow-green to yellow-orange, rather rare, often ± hidden in well-developed thalli; upper part of hymenium with yellow granules, the remaining tissues ± colourless. Ascospores 4–7 × 1–2 μm, oblong-ovoid. Anamorph unknown. Thallus Pd−, K−, KC−, C−, UV+ dull to bright orange (rhizocarpic acid).

Often conspicuous in dry, shady situations particularly on non-calcareous rocks and walls, rarely recesses in dry bark and wood, frequent in towns on brick and dressed stone; usually common. Throughout British Is. Europe, North America, Chile, Tasmania, New Zealand, Hawaii.

When sterile may be confused with *Chrysothrix chrysophthalma*, which differs in containing an additional unidentified substance with rhizocarpic acid and prefers more exposed situations; sterile specimens of *Chaenotheca furfuracea* have *Stichococcus*, contain vulpinic acid, and are UV± lemon-yellow or pale orange.

2. P. clavulifera (Nyl.) Coppins (1983)

Micarea clavulifera (Nyl.) Coppins & P. James (1980)

Thallus forming small patches, white to pale greenish grey, granular to granular-verrucose, effuse. Photobiont *Stichococcus*. Apothecia 0.1–0.3(-0.4) mm diam., frequent, dark brown to blue-black, ± globose to tuberculate, often surrounded by a basal white rim of protruding excipular hyphae; hymenium pale green to blue-green, especially in upper part, N+ purple-red; hypothecium ± colourless or pale green. Anamorph frequent, the thallus surface often with scattered conidiogenous cells 7–12 × 1–2 μm, ± cylindrical; conidia 7–15 × 2–2.3 μm, ± oblong. Thallus Pd−, K−, KC−, C−, UV− (lichen products not detected by t.l.c.).

On roots, stones and consolidated soil of dry underhangs on banks or the root systems of fallen trees; component of the *Micareetum sylvicolae*; rare. England (Wiltshire), S. Wales, Scotland, W. Ireland. N. & C. Europe, Tasmania.

Separated from *Micarea bauschiana* and *M. tuberculata* by the smaller, ± tear-like spores, the presence of *Stichococcus* and absence of pycnidia.

3. P. leprosa Coppins & Purvis (1987)

Thallus forming small patches, often merging and wide-spreading, to 30 cm, whitish to pale greenish or yellowish grey, leprose-granular to granular-verrucose, often rimose, effuse. Photobiont *Trebouxia*-like. Apothecia usually present, 0.1–0.5 mm diam., often becoming tuberculate, pale pink to pale brown, occasionally with a violet tinge; hymenium (22-)27–35 μm tall; hypothecium colourless. Ascospores 4.5–6.5(-7) × 1.5–2 μm, 0- to 1-septate (Fig. 23). Anamorph unknown. Thallus Pd −, K −, KC −, C+ red, UV ± glaucous (gyrophoric acid, ± porphyrilic acid).

On vertical, copper-rich rocks in sheltered walls of derelict mine builings, mineralized rock outcrops and on mortar and stonework near copper lightning conductors, window grills, etc. on church walls; frequent. Throughout Great Britain. Scandinavia, Greenland.

Locally abundant with few associated species and characterized by the C+ red reaction. Pale blue inclusions of copper oxalate are sometimes visible where abraded.

PSORA Hoffm. (1796)

Thallus squamulose; prothallus absent. **Upper cortex** thick, well developed, of anticlinally oriented, thick-walled hyphae with shortly cylindrical to round lumina; a well-developed epinecral layer is usually present and remnants of dead algae are present throughout cortex. **Lower cortex** of two types: (1) of ± periclinally arranged, long-celled, closely conglutinated, ± brown-pigmented hyphae and lacking crystals of calcium oxalate; or (2) of ± anticlinally arranged, short-celled, only partly conglutinated, colourless hyphae densely covered with crystals of calcium oxalate; lower cortex sometimes poorly developed to absent; often attached by a well-developed hyphal net (Fig. 39a). **Photobiont** chlorococcoid; algal layer continuous. **Medulla** well developed, I −. **Ascomata** apothecia, sessile, marginal or laminal, brown-black, usually flat or weakly convex when young, later becoming immarginate and strongly convex to semi-globose; pruina white, yellow or absent. **Thalline exciple** absent. **True exciple** annular, colourless to pale brown, of radiating hyphae, usually indistinct, becoming excluded. **Epithecium** red-brown or brown, containing anthraquinones (mainly parietin), K+ purple-red, N −. **Hymenium** I −. **Hypothecium** pale brown to colourless, filled with crystals of calcium oxalate. **Hamathecium** of paraphyses, simple or sparingly branched, strongly conglutinated, apical cell slightly swollen. **Asci** 8-spored, clavate, with a well-developed, I+ blue tholus containing a deeper I+ blue tube; ocular chamber absent. **Ascospores** colourless, simple, ellipsoid, smooth, without a thick perispore. **Conidiomata** pycnidia, frequent, immersed in squamules; ostiole colourless. **Conidia** elongate-bacilliform, simple, colourless. **Chemistry**: cortex, medulla and true exciple often contain calcium oxalate and lichen products.

Psora lurida differs from the above description in several important characters and will ultimately merit transfer to another genus.

LITERATURE: Schneider (1980), Timdal (1984b).

1 Squamules variously coloured, never pink or red; apothecia laminal or marginal, brown-black 2
Squamules bright pink-red-brown; apothecia marginal, black .... **1. *decipiens***

2(1) Squamules red-brown to dark brown; margin concolorous with upper side or sometimes slightly pruinose; lower surface brown; medulla C − 3
Squamules brownish to yellowish green, white-edged; lower surface pale brown to white; medulla C+ red or C − .... **4. *rubiformis***

3(2) Apothecia strongly convex to ± hemispherical; true exciple paler than disc; epithecium K+ red .... **2. *globifera***
Apothecia flat to weakly convex; true exciple darker than disc; epithecium K − .. **3. *lurida***

1. P. decipiens (Hedw.) Hoffm. (1794)

Squamules to 3–4 mm diam., round, scattered to contiguous, sometimes overlapping, adpressed, bright pink-red-brown, not pruinose or partly white-pruinose; margin usually pale, slightly up-turned, entire or somewhat serrated; underside white; lower cortex poorly developed or absent; attached by a hyphal net (Fig. 39a). Apothecia to 2 mm diam., sessile, marginal, black, not pruinose or ± white-yellow-pruinose. Ascospores 11–18 × 6–8 μm. Conidia 6–7 × 1 μm. Lichen products not detected by t.l.c.

On calcareous soils, chalk, shell-sand or mica-schist soils in open situations; local. Throughout British Is. Worldwide.

2. P. globifera (Ach.) Massal. (1852)

Squamules to 5 mm diam., elongated, adpressed to ascending, becoming imbricate, red-brown to dark brown, upper surface smooth, occasionally corrugated, not pruinose or more rarely faintly white-pruinose; margin concolorous with upper side or more rarely white-pruinose; underside pale brown to medium brown; lower cortex type (1). Apothecia to 2 mm diam., sessile, on the surface of the squamules, strongly globose, dark brown to black, not pruinose or more rarely faintly yellow-pruinose, often more distinct when wet; true exciple indistinct when young, paler than disc. Ascospores 10–13(-14) × (4-)5–6(-7) μm. Lichen products not detected from thallus by t.l.c.

On calcareous, and mica-schist soils, especially in rock crevices, montane. N. Scotland (Ben Lawers). Scandinavia, C. European Mtns, Greenland, North America.

P. lurida **(3)**, differs in having a flattened to weakly convex apothecial disc, with a more persistent exciple darker than the disc.

3. P. lurida (Ach.) DC. (1805)

P. petri (Tuck.) Fink (1935)

Squamules to 5 mm, rounded at apices, at times minutely lobed, ± concave, usually strongly imbricate, pale brown to dark brown, dull greenish when wet, matt, not pruinose; margin concolorous with upperside; the lower surface usually dark. Upper cortex with no algal remains; lower cortex type (1). Apothecia to 1(-1.5) mm diam., mostly single, marginal or on surface, dark brown, at first flat, becoming convex with an excluded exciple; epithecium brown, K−, N−, anthraquinones absent; hymenium I+ blue. Pycnidia immersed when young, soon sessile and marginal or on undersurface near margin of squamules; conidia 4–6 × *c.* 2 μm. Ascospores (8-)9–11(-15) × 6–8 μm, ellipsoid. Thallus Pd−, K−, C− (lichen products not detected by t.l.c.).

On calcareous soils, in crevices associated with limestone, sometimes directly on rock; rather local. Throughout British Is.

Differs from other *Psora* species in the type of upper cortex, the absence of calcium oxalate in the hypothecium, I+ blue hymenium, absence of anthraquinones in the hymenium, the type of pycnidia and the shape of the conidia. Specimens in shaded habitats are usually very pale in colour, and in exposed sites become blackish brown.

Resembles *Catapyrenium lachneum* which has perithecia, the tops of which are seen as small dark punctiform dots on the surface of the squamules. *Solenopsora holophaea* is similar but has one-septate spores, usually flatter, darker, more loosely organized squamules and more elevated apothecia which have a well-developed thalline exciple; it occurs on non-calcareous substrata.

4. P. rubiformis (Ach.) Hook. (1844)

Squamules to 4 mm diam., elongated, ascending and irregularly imbricate, thick, brownish to greenish yellow, upper surface smooth or cracked-areolate, not or white-pruinose towards the margin; margin yellow- to white-pruinose; underside ± white; lower cortex type (2). Apothecia to 2 mm diam., sessile, brownish black to black, not or more rarely yellow-pruinose. Ascospores 9–14 × 5–7 μm. Conidia 5–7 × 1 μm. Medulla C+ faintly red or C−, KC+ faintly red or KC− (usnic and ± gyrophoric acids).

On calcareous boulders and rock ledges. N. Scotland (near summit of Ben Lawers). Arctic, Iceland, Alps, N. Europe, North America (Rocky Mtns, Vermont).

Could be confused with the basal squamules of *Cladonia*, e.g. *C. symphycarpa*, which is K+ red (norstictic acid) and may occur in similar habitats; *Psora rubiformis* has thicker and coarser, more rounded squamules which are K−.

PSOROMA Ach. ex Michaux (1803)

Thallus small-squamulose; hypothallus indistinct. **Cephalodia** present, containing *Nostoc*. **Photobiont** ?*Myrmecia*. **Ascomata** apothecia, sessile, ± cup-shaped. **Thalline exciple** raised, persistent. **Hamathecium** of paraphyses, simple or branched near the apices, apices not thickened or capitate. **Asci** 8-spored, long-ellipsoid or cylindrical; apex with K/I+ blue central tube in a paler-staining tholus. **Ascospores** simple, colourless, often with 1–2 large oil droplets, ellipsoid, with warted-ridged epispore, ± apiculate. **Conidiomata** pycnidia, pale brown. **Conidia** simple, bacilliform, colourless. **Chemistry**: lichen products not detected by t.l.c.

A disparate genus with a predominantly southern temperate distribution, characterized by the presence of a green photobiont, external cephalodia, and warted-ridged ascospores. The affinities of this genus with *Pannaria* and *Parmeliella* are in need of critical investigation.

LITERATURE: Jørgensen (1978).

1. P. hypnorum (Vahl) Gray (1821)

Thallus small-squamulose, spreading irregularly; squamules 0.2–0.5 mm diam., thick, sometimes almost granular, green-grey to yellow-brown when dry, bright green when wet, on a thin, usually indistinct, pale hypothallus. Cephalodia frequent, sometimes few and inconspicuous, similar to thallus squamules, or smooth, rounded, pale red-brown, brown or bluish brown, scattered among the squamules or on the outer surface of the margin. Apothecia frequent, 1–3(-5) mm diam., disc red-brown to dark brown; thalline exciple often with numerous squamules, occasionally ± smooth, the lower part or underside ± densely covered by short, pale hairs; hymenium 100–120 μm tall. Paraphyses up to 4.5 μm wide, sub-moniliform; apices short-celled. Asci 70–90 × 11–15 μm. Ascospores 19–28 × 8–10 μm, including epispore 22–34 × 9–12 μm, ellipsoid or long-ellipsoid, with an apiculus at one or both ends; epispore distinctly warted-ridged.

Amongst mosses at bases of old trees, on acid rocks, acid dunes and on montane cliff-ledges, characteristic of leached substrata in humid situations, as by streams, lakesides, N-facing hill-sides, mountain summits or coastal heathlands; rather local and decreasing. Wales, Scotland, S.W. Ireland. Europe, North & South America, S. Georgia, Asia, S.E. Australia, New Zealand.

Resembles the more widespread *Pannaria pezizoides,* which has a blue-grey to red-brown thallus, is blackish when wet, contains *Nostoc,* lacks cephalodia, and has apothecia devoid of dense hairs on the lower side of the thalline exciple.

PSOROTICHIA Massal. (1855)

Thallus crustose, granular-areolate to subsquamulose, dark greenish to black, gelatinous, not corticate, pseudoparenchymatous ± throughout. **Photobiont** *Chroococcidiopsis* (‘*Xanthocapsa*’), cells single or in small groups surrounded by a gelatinous sheath which is brownish near the thallus surface. **Ascomata** apothecia, ± immersed, urceolate to flat or slightly convex. **Thalline exciple** present, concolorous with thallus. **True exciple**, when visible, usually paler than disc, in section colourless or pale brownish, well developed laterally especially in upper part, but narrowing below, open at base; disc red-brown to brown-black. **Hymenium** colourless, I+ blue. **Hypothecium** wedge-like. **Hamathecium** of simple to (especially above) sparingly branched paraphyses, with one to several swollen apical cells. **Asci** (4-)8-spored, cylindrical, thin-walled, I−, without apical thickening. **Ascospores** simple, ellipsoid, colourless, without a distinct perispore. **Conidiomata** pycnidia, immersed in small warts. **Conidia** bacilliform, simple, colourless. **Chemistry**: no lichen products reported. **Ecology**: on usually calcareous rocks. **Distribution**: *c.* 50 species, cosmopolitan.

Differs from *Lemmopsis* in the mature apothecia having the true exciple open at the base and not continuous with the hypothecium, though the significance of this feature requires further study. The genus is poorly understood and requires a modern revision.

LITERATURE: Ellis (1981).

1. P. schaereri (Massal.) Arnold (1869)

Thallus effuse, dark dull green (in shade) to dark brown or black, areolate, mostly *c.* 100 μm thick; areoles 0.3–0.7 mm across, often becoming detached at the edges and appearing subsquamulose, surface and margins of areoles often with granular isidia, 40–80 μm diam. Photobiont cells 5–7 μm diam. Apothecia 0.2–0.6 mm diam., at first immersed with a flat disc, later emergent with a $\pm$ convex disc; thalline exciple crenulate to granular, usually persistent but sometimes excluded, pseudoparenchymatous, to *c.* 75 μm wide; true exciple $\pm$ evident, paler than disc; disc reddish brown to black (then red-brown when wet), 25–50 μm wide at upper surface, but soon narrowing below, of $\pm$ parallel hyphae which in the upper part have ellipsoid lumina to 9 $\times$ 4.5 μm; hymenium 95–150 μm tall, colourless, or yellowish brown in uppermost part. Paraphyses rather few, simple to sparingly branched, 1.5–2 μm wide, upper 1- to several cells swollen to 5 μm wide. Asci 80–120 $\times$ 12–17 μm. Ascospores (12-)14–18(-20) $\times$ 7–9 μm, ellipsoid or $\pm$ ovoid. Conidia 3–4 $\times$ 1 μm.

On shaded limestone, often in seepage tracks, old mortared walls, or siliceous rocks subjected to calcareous flushing. Throughout British Is., but very rare in S.E. England, most common in Scotland. Europe, North America.

A variable species in respect of number of isidia produced and colour of thallus.

In addition there are two poorly understood species known only from single collections and the status of which is currently unclear: ***P. diffundens*** (Nyl.) Arnold (1869); thallus effuse, black, thin, composed of minute, granular-squamulose areoles; apothecia *c.* 0.3 mm diam., immersed, reddish; ascospores 15–21 $\times$ 7.5–11 μm; on greensand in S.E. England (Kent, near Maidstone) and with apparent affinities to *Poroscyphus*. ***P. pyrenopsoides*** (Nyl.) Forss. (1885): thallus, effuse, dark coloured, thin, granular-areolate; apothecia small with prominent exciples; asci 4- to 8-spored; ascospores 12–20 $\times$ 6–11 μm; on calcareous rocks in S.W. Ireland (Kerry, near Kenmare), also reported from France, but of uncertain affinities.

PTERYGIOPSIS Vainio (1890)

Thallus crustose, blackish, closely adpressed; rhizines absent; hyphae with $\pm$ round cells arranged in a fan-like fashion. **Photobiont** chroococcoid, cells present thoughout thallus. **Ascomata** apothecia, parietal, sessile; disc brown, punctiform. **Thalline exciple** $\pm$ paraplectenchymatous. **True exciple** inconspicuous. **Hymenium** gelatinous, I+ blue. **Hamathecium** of paraphyses, simple or branched-anastomosing, with thickened apices with external brown pigmentation. **Asci** 8-spored, thick-walled, I+ red-brown; apex thickened, I−. **Ascospores** simple, colourless. **Chemistry**: lichen products not detected. **Ecology**: on wet rocks.

LITERATURE: Jørgensen (1990).

1. P. coracodiza (Nyl.) Henssen (1990)

Thallus to 3–5 cm diam., areolate, $\pm$ scurfy, blackish, effuse. Photobiont cells 5–7 μm diam., with a brownish gelatinous sheath, K−, arranged in vertical rows towards the surface. Apothecia to 0.5 mm diam., blackish, at first $\pm$ immersed, finally distinctly superficial; thalline exciple conspicuous, shining, black, smooth, $\pm$ paraplectenchymatous, to 100 μm wide; true exciple hardly visible; hymenium 70–90 μm tall, colourless, upper parts olivaceous brown, N−, I+ blue; paraphyses simple. Asci 50–70 $\times$ 12–14 μm. Ascospores 8–14 $\times$ 6–9 μm, colourless, simple, broadly ellipsoid.

On irrigated, non-calcareous rocks, as in seepages or at the margins of rivers; rare. N. Wales (Merioneth, Barmouth), N.W. Ireland (W. Donegal). W. Europe.

A second species has recently been found: ***P. lacustris*** P. M. Jørg. & R. Sant. (1990) characterized by an olivaceous green, smooth, $\pm$ shiny, cracked, continuous thallus; photobiont cells 3–5 μm diam.; apothecia immersed; hymenium N+ blue-green, I−; paraphyses partly branched and anastomosing below. On boulder by lake-shore. N.W. Ireland (W. Donegal). Sweden.

PTYCHOGRAPHA Nyl. (1874)

Thallus crustose, indistinct, of minute, discrete, dark brown granules (goniocysts) scattered on or just below, the surface of the substratum; granules 20–50 μm diam. **Photobiont** chlorococcoid. **Ascomata** apothecia, narrowly elongated (lirellate) or sometimes becoming angular or rounded, black; disc slit-like, at least when young. **True exciple** brown-black, friable, of conglutinated hyphae. **Hypothecium** dark brown, merged with the exciple. **Hymenium** K/I+ blue. **Hamathecium** of scanty, simple or sparingly branched paraphyses, apices dark brown, conglutinated (forming an epithecium). **Asci** 8-spored, clavate, with a K/I+ blue outer fuzzy coat, the apical dome either *P. xylographoides* **(1)** thick and I−, or *P. flexella* **(2)** narrow and I+ blue. **Ascospores** ellipsoid, simple, colourless, without a thickened perispore. **Conidiomata** pycnidia. **Conidia** bacilliform, simple, colourless. **Chemistry**: Pd−, K−, KC−, C−, UV; no lichen products detected by t.l.c. **Ecology**: on hard or dry wood, but not recorded on worked timber.

Distinguished from *Xylographa* by the black and friable exciple, and from *Wadeana* by the 8-spored asci and ascus structure.

1 Apothecia persistently narrowly elongate with mostly (1-)2 parallel slits; ascospores 8.5–13 × 4.5–6.5 μm .... **1. *xylographoides***
Apothecia soon angular or rounded with expanded or ± gyrose disc; ascospores 6–8 × 3–4(-4.5) μm .... **2. *flexella***

1. P. xylographoides Nyl. (1874)

Thallus granules with cortex of uniformly brown, angular cells *c.* 4–7 μm wide. Photobiont cells 6–14 μm diam. Apothecia 0.3–1.4 × 0.1–0.3 mm, narrowly elongate, usually unbranched, running parallel with the grain of the wood; with (1-)2 (rarely more) longitudinal slits each indicating a parallel hymenium below, which are separated by dark brown tissue; hymenium 45–60 μm tall, colourless or greenish tinged along the edge of the exciple and epithecium, I+ yellowish, K/I+ blue. Paraphyses 1–2 μm wide. Ascospores 8.5–13 × 4.5–6.5 μm, ellipsoid. Pycnidia *c.* 40–50 μm diam., frequent, immersed to superficial, blackish, the wall dark brown; conidia 4–7 × 0.8 μm.

On wood of ± horizontal surfaces of fallen trees (*Betula*, *Quercus*, *Salix*, *Larix*, *Pinus*); local. S.W. England (S. Somerset), C. Wales (Radnor), N. Scotland (Highlands). ?Endemic.

Most likely to be confused with *P. flexella* **(2)**, and with *Xylographa abietina*, which has a pale true exciple (sections).

2. P. flexella (Ach.) Coppins (1980)

Thallus granules with cortex of rounded cells, each with a dark brown, cap-like outer surface. Photobiont cells 9–14(-17) μm diam. Apothecia at first narrowly elongate, to 0.4 × 0.1–0.2 mm, with slit-like disc, but soon angular or ± rounded, 0.16–0.6 mm diam., with an expanded or irregularly gyrose disc and a ± inrolled margin; hymenium 35–45 μm tall, colourless, I+ pale blue, K/I+ blue. Paraphyses 1–1.5 μm wide. Ascospores 6–8 × 3–4(-4.5) μm, broadly ellipsoid. Pycnidia not found.

On wood of stumps or standing trees (mainly *Pinus* and *Quercus*), usually on ± vertical surfaces, in old woodlands (especially native pinewoods); rare. C. & E. Scotland (Highlands). Europe.

Differs from *P. xylographoides* **(1)** in the more rounded apothecia and smaller ascospores; the further differences in ascus and goniocyst structure could justify its placement in a separate genus. Sometimes confused with simple-spored members of the non-lichenized genus *Durella* which, however, have an I− hymenium.

PYCNOTHELIA Dufour (1821)

Thallus of two types; **primary basal thallus** persistent, crustose, of non-corticate, rounded granules; **secondary ± vertical thallus** a pseudopodetium, fruticose, ± short-cylindrical, hollow, simple to richly branched, not forming cups, corticate, without squamules or soredia. **Cortex** not differentiated. **Photobiont** chlorococcoid. **Medulla** outer layer with algal cells forming a persistent pseudocortex; inner layer cartilginous, variable in thickness, not clearly defined. **Ascomata** apothecia, rare, arising on the tips of the pseudopodetia, ± peltate, often forming apical clusters. **Hamathecium** of paraphyses. **Asci** *Cladonia*-type. **Ascospores** simple or 1- to 3-septate, elongate-fusiform. **Conidiomata** pycnidia, frequent, apex red-brown or brown. **Conidia** thread-like, curved. **Chemistry**: various depsidones and(or) atranorin or fatty acids such as protolichesterinic acid. **Ecology**: on humic, sandy or clay soils, some species on rocks; predominantly in the Southern Hemisphere.

LITERATURE: Galloway & James (1989), Thomson (1968).

1. P. papillaria Dufour (1821)

Basal thallus often wide-spreading, of white, grey-white or cream-coloured, rounded, ± contiguous granules, 0.2–1 mm diam. Pseudopodetia to 0.5(-1.5) cm, scattered or ± obscuring the primary thallus, ± cylindrical or cone-shaped, apices ± rounded, simple to branched-coralloid, ± swollen, hollow and very fragile, surface smooth or with scattered, rounded granules, pycnidial ostioles frequent, dark brown, apical, not forming cups. Apothecia dark red-brown, rare. Ascospores 9–15 × 2–3.5 μm, 0- to 1-septate. Conidia 8–14 × 0.5 μm. Thallus Pd −, K + yellow, KC −, C −, medulla UV ± blue-white (atranorin, ± chloratranorin, protolichesterinic, lichesterinic and ± squamatic acids).

On acid peat and leached, sandy soil on heathlands; locally frequent. Throughout British Is. Europe, Greenland, North America.

Characterized by the tooth-like, whitish, swollen, hollow pseudopedetia on a closely adpressed, ± scattered, granular thallus.

PYRENOCOLLEMA Reinke (1885)

Thallus crustose, immersed or superficial, usually subgelatinous, the hyphae generally vertically orientated. **Photobiont** cyanobacterial (*Gloeocapsa*, *Hyella*, or *Nostoc*), cells orange or blue-green. **Ascomata** perithecioid, usually unilocular. **True exciple** dark brown; wall ± cellular, normally heavily melanized. **Involucrellum** usually absent. **Hamathecium** of sparingly to richly branched and anastomosed pseudoparaphyses, septate; I −. **Asci** with two functional wall layers and an internal apical beak, dehiscence fissitunicate, ovate to subcylindrical, usually long-stalked, I −, usually 8-spored. **Ascospores** colourless, oblong to ovoid-fusiform, 1-septate, the upper cell usually shorter and broader than the lower, a poorly defined gelatinous perispore sometimes present. **Conidiomata** pycnidia. **Conidiogenous cells** ± cylindrical, phialidic. **Conidia** bacilliform or ellipsoid. **Ecology**: on calcareous substrata in moist situations, on wet sand, or on acid rocks in freshwater or marine habitats.

Distinguished from *Arthopyrenia* in the ascomatal wall structure, general absence of an involucrellum, association with cyanobacteria, and the non-corticolous habitats, and from *Acrocordia* in ascus structure, smaller ascomata, and ascospores that are ± ovoid and lack an epispore. This genus is well circumscribed, but its component species are mostly poorly understood.

In studying littoral *Pyrenocollema* species, care should be taken to avoid confusion with the lichenicolous fungus ***Stigmidium marinum*** (Deakin) Swinscow (1965), which occurs on *Verrucaria* species, particularly *V. halizoa* and *V. mucosa*; it has perithecia 0.15–0.2 mm diam., lacks hamathecial filaments, and has 1-septate ascospores constricted at the septum, 10–15 × 4–6 μm.

LITERATURE: Fletcher (1975), Harris (1975), Kohlmeyer & Kohlmeyer (1979), Santesson (1939, 1992), Swinscow (1965a, b).

1	In terrestrial or freshwater habitats	2
	In marine habitats on rock, shells, barnacles, or seaweeds	9
2(1)	On turf, soil or bryophytes	3
	On rocks or mortar	6

3(2) Ascospores >25 μm long ... 4
Ascospores <25 μm long ... 5

4(3) Ascospores 26–37 μm long; in sand dune slacks ... **1. *arenisedum***
Ascospores 27–44 μm long; in montane habitats ... **3. *bryospilum***

5(3) Ascospores 21–22 × 8 μm; on sandy banks (inland) ... **2. *argilospilum***
Ascospores 15–19(-21) × 5–7 μm; in calcareous dune slacks or machair . **12. *subarenisedum***

6(2) Ascospores ± oblong, the lower cell not markedly narrowed ... **10. *saxicola***
Ascospores ovoid to ovoid-fusiform, the lower cell distinctly narrowed towards the apex . 7

7(6) Thallus immersed to superficial, pale coloured to reddish brown, never black; on calcareous rocks ... 8
Thallus superficial, black; on acid rocks in streams or lake margins ... **11. *strontianense***

8(7) Asci subcylindrical; pseudoparaphyses sparingly branched; on chalk stones or mortar, lowland ... **7. *monense***
Asci obclavate to obpyriform; pseudoparaphyses richly branched; on flushed limestones or calcareous schists, montane ... **4. *caesium***

9(1) Thallus immersed, in calcareous rocks, barnacles or shells, or superficial and smooth on siliceous rocks ... 10
Thallus superficial on *Pelvetia canaliculata*; perithecia single or 2–3 united by a common involucrellum ... **9. *pelvetiae***

10(9) Thallus immersed or smooth, lacking black ridges ... 11
Thallus superficial, with glossy, black, tar-like ridges; perithecia immersed ... **5. *elegans***

11(10) On calcareous rocks or shells; involucrellum absent or scarcely extended; perithecia not cracked around the ostiole ... 12
On siliceous rocks, involucrellum extended; perithecia cracked around the ostiole **8. *orustense***

12(11) Perithecia sessile and flattened, or immersed and globose, 0.2-0.25 mm diam; on calcareous rocks or shells ... **6. *halodytes***
Perithecia, immersed or protruding, 1–1.5 mm diam.; on chalk or limestone **13. *sublitorale***

1. P. arenisedum (A. L. Sm.) Coppins (1992)

Arthopyrenia areniseda A. L. Sm. (1911)

Thallus whitish or ashy grey, immersed. Photobiont cells yellowish orange (*Hyella*). Perithecia 0.16–0.2 mm diam., ± globose with a conical apex, mostly half to three-quarters immersed. Asci 100–105 × 19–20 μm, subcylindrical, 4- to 8-spored. Ascospores 26–37 × (8-)9–10 μm, ovoid-fusiform.

On damp turf in dune slacks. W. England (N. Devon, Cheshire, Lancashire). Endemic.

P. subarenisedum **(12)**, which occurs in similar situations, has smaller ascospores.

2. P. argilospilum (Nyl.) Coppins (1992)

Arthopyrenia argilospila (Nyl.) A. L. Sm. (1918)

Thallus dull olive-green, gelatinous when wet. Photobiont cells blue-green. Perithecia ± globose, semi-immersed. Ascospores 21–22 × 8 μm, oblong-ovoid.

On inland wet sand banks. W. England (Shropshire), not recorded this century. ?Endemic.

3. P. bryospilum (Nyl.) Coppins (1992)

Arthopyrenia bryospila (Nyl.) Arnold (1870)

Thallus indistinct, bluish. Perithecia ± conical. Asci 2- to 8-spored, oblong. Ascospores 27–44 × 8–12 μm.

On mosses and schistose soil in montane localities. N.E. Scotland (E. Inverness, Ben Alder; Perth, Ben Lawers).

This species may belong to the primarily lichenicolous genus *Cercidospora*, but requires a more critical study.

4. P. caesium (Nyl.) R. C. Harris (1987)

Arthopyrenia caesia (Nyl.) Zahlbr. (1921)

Thallus ashy or greenish grey or reddish brown, immersed or superficial, powdery. Photobiont cells yellowish or blue-green. Perithecia 0.25–0.4 mm diam., subconical to globose, mostly half-immersed. Pseudoparaphyses richly branched and anastomosed. Asci elongate-clavate to obpyriform. Ascospores 20–30 × 9–12 μm, ovoid to ovoid-fusiform, sometimes with additional pseudosepta.

On inundated or flushed limestone or calcareous schists; rare. Scotland (Highlands), W. Ireland. France, Switzerland, Norway, North America.

5. P. elegans R. Sant. (1992)

Arthopyrenia elegans R. Sant. (1985), *nom. inval.*

Thallus superficial, thin, brown, dull when dry, shiny and translucent when wet, with abundant black, convex, carbonaceous ridges and punctae, the ridges branched and irregularly stellate, typically 0.1 × 0.2–0.5 mm, flattening and broadening at the thallus margin to form a distinct edge. Photobiont cells orange (*Hyella*). Perithecia immersed in the black ridges, 0.1–0.3 mm diam., flattened and circular, with deeply depressed ostioles, base blackened. Ascospores clavate, the cells unequal, one end pointed, the other rounded, 11.5–18(-20) × 3.5–7 μm. Pycnidia also immersed in the black ridges.

On siliceous rocks in the littoral zone, as tiny patches amongst *Verrucaria striatula*, *Hildenbrandia* etc.; rare. S. & W. British Is. Spain, S. Scandinavia, N. & S. America.

Recalls a brown morph of *Verrucaria striatula* in the field, but is readily separated by the 1-septate ascospores and netted pseudoparaphyses.

6. P. halodytes (Nyl.) R. C. Harris (1987)

Arthopyrenia halodytes (Nyl.) Arnold (1891)

Thallus very thin, often glossy, yellow-brown, superficial on siliceous rocks, but immersed on limestone and shells (barnacles, limpets, etc.), dull brown, scabrid in sheltered sites, subgelatinous. Perithecia 0.2–0.25(-0.3) mm diam., sessile on siliceous rocks, flattened, with a black ring around the ostiole and a pale base (resembling lifebelts, especially when wet) globose and immersed on calcareous substrata, sometimes with a distinct involucrellum. Ascospores 12–20(-27) × 5–8(-10) μm, clavate, the cells unequal in size. Pycnidia 60–80 μm diam.; conidia 2.5–3.5 × 0.5–1 μm.

On limestones, chalk, barnacles, limpets and mussel shells in the littoral zone; common. Throughout British Is. Probably widespread on coasts in the Northern Hemisphere.

Sometimes treated in a broad sense to also include *P. elegans* **(5)**, *P. orustense* **(8)** and *P. sublitorale* **(13)**; since these taxa do, however, differ in clear morphological features and ecology they are recognized here pending a more critical evaluation of the variation in this complex. The softness of the substratum seems to some degree to determine the size of the perithecia. Specimens on chalk have generally larger perithecia than those on hard shells or limestone.

7. P. monense (Wheldon) Coppins (1992)

Arthopyrenia monensis (Wheldon) Zahlbr. (1922)

Thallus immersed to superficial, dull grey or green-grey, scurfy or filmy. Photobiont cells yellowish or blue-green. Perthecia 0.15–0.2 mm diam., $\frac{1}{2}$–$\frac{3}{4}$-immersed, ± globose. Pseudoparaphyses sparingly branched. Asci 80–115 × 16–19 μm, narrowly obpyriform to subcylindrical. Ascospores 17–30(-35) × 6–8 μm, ovoid to ovoid-fusiform.

On chalk stones in grassy banks and woodland clearings, or mortar of shaded walls, lowlands. England, Isle of Man.

P. caesium **(4)** differs in the more cylindrical asci and densely branching and anastomosing pseudoparaphyses.

8. P. orustense (Erichsen) A. Fletcher (1992)

Arthopyrenia orustensis Erichsen (1930)

Thallus similar to that of *P. halodytes* **(6)**. Perithecia 0.1–0.15 mm diam., subglobose, sessile, with a prominent, glossy, black exciple, cracking around the ostiole; involucrellum present, extending laterally. Ascospores 15–20 × 6–9 μm, ellipsoid.

On siliceous rocks on sheltered shores, especially shingle and salt-marshes. E. Britain. Baltic coasts.
Often included in *P. halodytes* **(6)** but differing in the well-developed involucrellum and occurrence on siliceous rocks.

9. P. pelvetiae (Sutherl.) D. Hawksw. (1988)

Arthopyrenia pelvetiae (Sutherl.) D. Hawksw. (1980)

Thallus superficial. Photobiont blue-green. Perithecia 1.2–3.4 mm diam., sometimes 2–3 united in a common stroma; involucrellum present, extending into the associated photobiont colonies. Asci 40–60 × 13–18 μm. Ascospores 11.5–16 × 5–6.5 μm. Pycnidia 40–70 μm diam.; conidia 2–2.5 × 1–1.5 μm.
On fronds of the seaweed *Pelvetia canaliculata*; no recent records. N. Scotland (Orkney Is.). France, Spain.
This species is of particular biological interest as forming a three-biont association in which the lichen grows epiphytically on a second photosynthetic partner, the seaweed *Pelvetia*. Not to be confused with the very common ***Mycosphaerella ascophylli*** Cotton (1907), which forms a systemic mycelium within *Pelvetia canaliculata* thalli and has immersed perithecia 55–130(-160) μm diam. and larger ascospores, 15–22.5(-25) × 4.5–5.5(-6.5) μm.

10. P. saxicola (Massal.) Coppins (1992)

Arthopyrenia saxicola Massal. (1865)

Thallus immersed to superficial, grey to reddish brown, often minutely mottled, sometimes clearly delimited. Photobiont cells yellowish (?*Gloeocapsa*), *Trentepohlia* often also present. Perithecia 0.1–0.2 mm diam., globose, half-immersed. Pseudoparaphyses richly branched. Asci 65–85 × 17–23 μm, clavate, obclavate to subcylindrical. Ascospores (19-)20–25 × 4.5–7 μm, ± oblong, the upper cell only slightly broader than the lower. Pycnidia frequent, 60–80 μm diam.; conidia 4–5 × 0.5 1 μm.
On shaded, hard limestones; local. Throughout England. Europe.
Distinguished by the relatively narrow and ± oblong ascospores, and the association with *Trentepohlia*; perhaps really a species of *Anisomeridium* and in need of further study.

11. P. strontianense (Swinscow) R. C. Harris (1987)

Arthopyrenia strontianensis Swinscow (1967)

Thallus superficial, filmy or cracked, dark grey to black. Photobiont cells yellowish. Perithecia 0.2–0.25 mm diam., globose, but covered by a thin layer of thallus. Asci obovate, 85–100 × 19–25 μm. Ascospores 18–25 × 8–12 μm, ovoid.
On hard acid rocks in streams, associated with, and possibly parasitic on, *Verrucaria* species. S.W. England, Scotland (Highlands). North America (Michigan).

12. P. subarenisedum (G. Salisb.) Coppins (1992)

Arthopyrenia subareniseda G. Salisb. (1953)

Thallus as in *P. arenisedum* **(1)**. Perithecia 0.08–0.2 mm diam, ± globose, semi-immersed. Asci 70–100 × 13–14 μm, subcylindrical. Ascospores 15–19(-21) × 5–7 μm, ovoid to ovoid-fusiform.
On soil in dune slacks and machair. N.W. England (Lancashire), W. Scotland (Colonsay). Endemic.
Distinguished from *P. arenisedum* **(1)** by the smaller ascospores.

13. P. sublitorale (Leighton) R. C. Harris (1987)

Arthopyrenia sublitoralis (Leighton) Arnold (1891)

Thallus entirely immersed, apparently white. Perithecia to 1.5 mm diam., on soft chalk, half-immersed by erosion of the substratum to 1 mm on limestone and then subsessile (even on hard rock); ostioles not prominent. Ascospores 12–20 × 5–10 μm.
On calcareous rocks amongst seaweeds in the sublittoral zone, especially ridges uncolonized by barnacles and on shells (barnacles and limpets); common. Throughout British Is. Scandinavia (outside the Baltic), New Zealand.
Distinguished from *P. halodytes* **(6)**, with which it is often united, by the white thallus and larger prominent perithecia.

PYRENOPSIS (Nyl.) Nyl. (1858)

Thallus crustose, effuse, granular-areolate or minutely squamulose, blackish but with reddish tinge (especially when wet), ± gelatinous when moist; cortex not differentiated, ± paraplectenchymous throughout. **Photobiont** *Chroococcidiopsis* or *Gloeocapsa*; cells or cell-clusters near the upper surface with reddish brown, K± purplish, gelatinous sheaths. **Ascomata** apothecia, perithecia-like; disc pore-like to urceolate, orange-brown to blackish. **Thalline exciple** prominent. **True exciple** very thin (mostly less than 15 μm wide), often inconspicuous. **Hymenium** colourless, or pale brownish above, I+ blue-green or reddish brown. **Hypothecium** colourless or pale brownish. **Hamathecium** of simple to branched paraphyses, which are often ± moniliform above. **Asci** clavate, 8- or multispored; apical dome I+ blue. **Ascospores** simple, ellipsoid to globose, colourless, without a distinct perispore. **Conidiomata** pycnidia, immersed; wall colourless. **Conidiogenous cells** in chains. **Conidia** shortly ellipsoid to bacilliform, simple, colourless. **Chemistry**: no lichen products reported. **Ecology**: on rocks, rarely on soil, in moist situations, *c.* 40 species, cool-temperae to arctic regions of both Hemispheres.

A poorly understood genus requiring a thorough revision. *Euopsis* has disc-like apothecia from an early stage, a different ascus structure, and usually more slender paraphyses (not moniliform), and *Psorotichia* has I− asci. Some collections of '*Pyrenopsis*' appear to represent members of the latter genus with perithecioid apothecia.

Most species formerly placed in *Pyrenopsidium* (Nyl.) Forss. (1885) are now referred here. The type species of that genus, *P. granuliforme* (Nyl.) Forss. (1885), which has been doubtfully reported from the British Is., has been shown to belong to *Cryptothele* Th. Fr. (1866).

LITERATURE: Henssen, Büdel & Titze (1987), Jørgensen & Henssen (1990).

1 Asci 8-spored 2
Asci mostly multispored .... **2. *grumulifera***

2(1) Ascospores ellipsoid 3
Ascospores subglobose to globose .... **5. *impolita***

3(2) Thallus areolate, or subsquamulose with areoles confluent or overlapping; ascospores mostly over 9 μm long and 4 μm wide 4
Thallus of discrete, squamulose areoles; ascospores 5–7 × 3 μm .... **3. *phylliscella***

4(3) Thallus deeply cracked-areolate to subsquamulose; apothecia becoming urceolate **1. *furfurea***
Thallus cracked-areolate; apothecia mostly perithecioid .... **4. *rhodosticta***

1. P. furfurea (Nyl.) Th. Fr. (1866)

Pyrenopsidium furfureum (Nyl.) Forss. (1885), *Pyrenopsidium homoeopsis* (Nyl.) Forss. (1885)

Thallus deeply cracked-areolate to subsquamulose, brown-black. Apothecia to 0.3 mm diam., perithecioid to urceolate; disc punctiform or slightly expanded; hymenium 75–100 μm tall. Paraphyses submoniliform, 2–2.5 μm wide. Ascospores 10–12(?-18) × 7–10 μm, broadly ellipsoid to subglobose. Pycnidia not seen.

On moist, mica-schist rocks, above 700 m alt.; very rare. N. Scotland (Perth, Ben Lawers). Fennoscandia.

Specimens with a well-developed thallus and expanded, urceolate apothecia resemble a small *Leptogium*.

2. P. grumulifera Nyl. (1861)

P. multispora Coppins (1988)

Thallus cracked-areolate, brown-black. Apothecia 0.1–0.2 mm diam., immersed, perithecioid with pore-like disc; hymenium 70–80 μm tall. Asci 40–55 × 15–20 μm, broadly clavate, mostly multispored (64 or more), a few 8-spored asci sometimes present. Ascospores 4.5–7 × 2–3 μm, oblong-ellipsoid. Pycnidia not seen.

On moist, mica-schist rocks, above 700 m; very rare. N. Scotland (Perth, Breadalbane Mtns). Finland, Greenland.

3. P. phylliscella Nyl. (1875)

Thallus of small, discrete, flattened, rounded squamules, dark brown. Apothecia minute, numerous and crowded, 2–15 per squamule, perithecioid. Asci 8-spored. Ascospores 5–7 × 3 μm, narrowly ellipsoid. Pycnidia not reported.

On quartzite boulders in a mountain stream, known only from the type gathering. N. Scotland (Perth). Endemic.

4. P. rhodosticta (Taylor) Mull. Arg. (1888)

P. fuscatula Nyl. (1858), ?*P. haematopsis* auct. brit., non (Sommerf.) Th. Fr. (1860), *P. subareolata* Nyl. (1861)

Thallus cracked-areolate, blackish. Apothecia to 0.2 mm diam., perithecioid, or disc becoming slightly expanded; hymenium 100–125 μm tall. Paraphyses to 2 μm wide; apical cells to 3 μm wide. Asci 70–100 × 12–15 μm, cylindrical-clavate, 8-spored. Ascospores (8-)10–15 × 4–6 μm, ellipsoid. Pycnidia to 100 μm diam.; conidia 2 × 1 μm.

On moist, acid rocks, in upland areas. Scotland, extending very locally to N. Wales and S.W. Ireland (Kerry). Fennoscandia, France.

A number of at least superficially similar entities are included here, all of which require critical study. The correct name of this species also requires further examination as the type of *P. rhodosticta* seems to have periphyses instead of paraphyses and is probably referable to *Cryptothele*.

5. P. impolita (Th. Fr.) Forss. (1885)

Like **4**, but ascospores (5-)8–10 × (5-)7–8 μm, subglobose to globose .

On flushed, siliceous rocks; very rare. C. Wales. Scandinavia.

PYRENULA Massal. (1852)

Thallus crustose, immersed or more rarely superficial, ± continuous, areolate in some immersed species, sometimes delimited by a ± black fringe; ± pseudocyphellate. **Photobiont** *Trentepohlia*. **Ascomata** perithecia, immersed to erumpent, ± globose to flattened, usually entire but dimidiate in some species, black. **Involucrellum** dark brown to black, spreading laterally to developed around and adhering closely to the exciple from which it is then not distinct, of fungal hyphae interspersed with bark cells and often crystals, K± reddish or violet. **True exciple** brown to pale brown, entire or not developed below the perithecial cavity, crystals often dense at the wall lining the locule; crystals of two types (1) colourless (calcium oxalate?) and K−, and (or) (2) orange (anthraquinones?) and K± orange or K± reddish purple; ostiole central or in some species lateral. **Hymenium** gelatin I− or + greenish blue in parts, ± oil drops or crystals. **Hamathecium** at first of branched and anastomosed, sparsely septate paraphysoids; ± unbranched paraphyses later developing from the base and replacing the paraphysoids; periphyses developed in the ostiolar canal. **Asci** cylindrical, long-stalked, multilayered in structure, thickened at the apex with an internal apical beak and also with a refractive subapical cap, I−, discharge fissitunicate, the different wall layers extending to differing extents. **Ascospores** ellipsoid to narrowly ellipsoid or broadly fusiform, the apices rounded to apiculate, 3-septate, very thick-walled, distoseptate, the cells appearing lenticular to angular, the apices sometimes attenuated, olivaceous to dark brown, smooth-walled, lacking a perispore. **Conidiomata** pycnidia, ± globose, black. **Conidia** ± thread-like to hooked, simple, colourless. **Chemistry**: ± lichexanthone or unidentified anthraquinones. **Ecology**: on ± smooth bark surfaces in humid situations.

A predominantly tropical genus of about 190 species, poorly represented in Europe.

The K+ reddish purple crystals inside the perithecial walls of *P. nitida* **(6)** and *P. nitidella* **(7)** are most easily demonstrated by placing hand sections in a drop of K on a slide when the colour diffuses into the medium if these are present.

In addition to the species keyed out below, there is a ninth rare and probably undescribed species in W. Scotland, N.W. England and S.W. Ireland with perithecia 0.7–1 mm diam., a generally laterally developed ostiole, and ascospores (18.5-)21–22(-25) × 8–10 µm.

LITERATURE: Harris (1973), Keissler (1938), Poelt & Vězda (1981).

1 Involucrellum not or slightly extended laterally, adhering to the true exciple; true exciple continuous below perithecial cavity ... 2
Involucrellum markedly extended laterally, separated from the true exciple; true exciple not developed below perithecial cavity ... **2. *coryli***

2(1) Thallus olive-green to brown or dark brown, yellowish brown or fawn; ± pseudocyphellae; perithecia ± globose ... 3
Thallus silvery to creamy white; pseudocyphellae absent; perithecia flattened ... **4. *laevigata***

3(2) Perithecia >0.4 mm diam. ... 4
Perithecia <0.4 mm diam. ... 6

4(3) Ascospores mainly <25 µm; K+ reddish purple crystals on the perithecial surface or lining perithecial cavity; ± with pseudocyphellae ... 5
Ascospores mainly >25 µm; K+ reddish purple crystals absent; pseudocyphellae abundant ... **5. *macrospora***

5(4) Perithecia 0.4–0.6 mm diam., only perithecial surface sometimes K+ purplish; pseudocyphellae absent ... **8. *occidentalis***
Perithecia 0.6–0.8(-1) mm diam.; K+ purplish crystals lining the perithecial cavity; ± with pseudocyphellae ... **6. *nitida***

6(3) Perithecial cavity lacking K+ purplish crystals inside the perithecial wall ... 7
Perithecial cavity with K+ purplish crystals inside the perithecial wall ... **7. *nitidella***

7(6) Thallus olive-green to pale brown or fawn, UV− or whitish; pseudocyphellae present; ascospores >25 µm ... **1. *chlorospila***
Thallus yellowish green to orange-rust coloured, UV+ yellow-orange; pseudocyphellae absent; ascospores <25 µm ... **3. *dermatodes***

1. P. chlorospila Arnold (1887)

Thallus olive-green to pale brown or fawn; pseudocyphellae 50–75 µm diam., white. Perithecia small, 0.2–0.3(-0.35) mm diam., often rather densely arranged, lacking K+ purplish crystals inside the wall, hamathecium including colourless K− crystals. Ascospores (25-)28–32(-35) × (9-)11–13(-14) µm, 3-septate. Thallus Pd+ faintly yellow, K+ yellow, KC−, C−, UV− or whitish (unidentified substance).

On ± smooth, shaded bark of deciduous trees, often with *P. macrospora* **(5)**; locally abundant. S. & W. British Is. France, Italy, Portugal.

Most pre-1980 records of *P. nitidella* **(7)** from Great Britain and Ireland refer to this species.

2. P. coryli Massal. (1852)

Thallus pale grey; pseudocyphellae ± absent; probably not lichenized. Perithecia 0.2–0.3 mm diam. and *c.* 0.1 mm tall, with a laterally spreading involucrellum; exciple not continuous below the perithecial cavity, K−. Ascospores 10–17 × 4–6 µm, 3-septate. Pycnidia scattered amongst the perithecia; conidia 20 × 0.5 µm, thread-like, colourless.

On *Corylus* bark; very rare. N. Wales (Merioneth), N. & W. Scotland (Westerness, Sutherland). C. & N. Europe, but rarely recorded this century.

Sometimes excluded from the genus and treated as *Mycopyrenula coryli* (Massal.) Vainio (1921) in view of the difference in the structure of the perithecia and the probably non-lichenized habit. As other fundamental microscopic characters are those of *Pyrenula*, this does not appear to be justified.

3. P. dermatodes (Borrer) Schaerer (1850)

Thallus yellowish green to ± orange-rust coloured, fading in the herbarium, superficial, smooth to ± areolate; pseudocyphellae absent. Perithecia 0.2–0.3 mm diam. (in section), ± completely immersed in the thallus, only the ostiole visible from above, lacking K+ purplish crystals inside the perithecial wall; hamathecium not richly interspersed with oil drops. Ascospores 14–19 × 6.5–7.5(-8.5) μm, 3-septate. Thallus Pd−, K+ orange-red in places, KC−, C−, UV+ vivid yellow-orange (lichexanthone).

On smooth bark of *Fagus*, *Ilex*, *Sorbus* and *Crataegus* in oceanic woodlands, rarely on shaded siliceous rocks; very rare. N. Scotland (W. Inverness, Knoydart), locally frequent in W. Ireland; not known elsewhere in Europe. South America, Macaronesia, China, India.

4. P. laevigata (Pers.) Arnold (1885)

Thallus silvery to creamy white; pseudocyphellae absent; prothallus generally absent. Perithecia 0.2–0.3(-0.35) mm diam., flattened; involucrellum often spreading somewhat laterally, lacking K+ purplish crystals inside the perithecial wall. Ascospores (14-)17–22(-26) × (7-)8–9(-11) μm, 3-septate. Pycnidia often interspersed with the perithecia; conidia 10 × 1 μm. Thallus Pd−, K+ yellow, KC−, C−, UV−.

On smooth bark, especially of *Corylus*, in oceanic woodlands; local. S.W. England (N. Devon), N. Wales (Merioneth), W. Scotland, S.W. Ireland (Clare, Kerry, Cork). Europe, E. North America.

The status of this taxon in the British Is. merits closer study, as the species is primarily continental.

5. P. macrospora (Degel.) Coppins & P. James (1980)

Thallus olivaceous or fawn to dark brown; pseudocyphellae 50–120 μm diam., white. Perithecia (0.4-)0.5–0.9(-1.2) mm diam., crystals inside the involucrellum K−. Ascospores (24-)27–33(-36) × (8-)10–13 μm, 3-septate. Thallus Pd+ faintly yellow, K+ yellow, KC−, C−, UV− or weak yellow, (unidentified substance).

On smooth, ± shaded bark of deciduous trees, often with *P. chlorospila* **(1)**; often common. Throughout S. and W. British Is. Strongly oceanic, from Portugal to S. Scandinavia.

The commonest British species of the genus. Most pre-1980 British records of *P. nitida* **(6)** from Great Britain and Ireland refer to this species.

6. P. nitida (Weigel) Ach. (1814)

Thallus olive-brown to dark brown or yellow-brown; pseudocyphellae usually absent, when present mainly 50–75 μm diam., white and often frequent. Perithecia 0.6–0.8(-1) mm diam., crystals inside the wall K+ reddish purple, oil drops sparse or absent in the hamathecium. Ascospores (17-)19–24(-26) × 6–8(-9) μm, 3-septate. Thallus Pd−, K+ orange-red, KC+ reddish, C−, UV− (unidentified anthraquinones).

On smooth, shaded bark of old deciduous trees, including *Carpinus* and *Fagus*; very rare. S. England (Buckingham, E. Kent, Hampshire, Sussex). Europe.

Most pre-1980 British records refer to *P. macrospora* **(5)**. *P. nitida* is primarily a continental species, as reflected by its distribution in Britain.

7. P. nitidella (Flörke ex Schaerer) Müll. Arg. (1885)

Thallus olive-brown to fawn; pseudocyphellae 50–75 μm diam., often sparse, white. Perithecia 0.2–0.3(-0.35) mm diam., crystals inside the wall K+ reddish purple, oil drops inconspicuous. Ascospores (20-)22–26(-28) × (8-)8.5–11(-12) μm, 3-septate. Thallus Pd−, K+ yellow to pale orange in parts, KC−, C−, UV− (unidentified anthraquinones).

On smooth, ± shaded bark of deciduous trees; ?extinct. N.E. England (N.E. Yorkshire), C. Scotland (E. Perth). Europe, widespread but much less frequent than *P. nitida*.

The present status of the species in Great Britain and Ireland requires re-evaluation; most pre-1980 records belong to *P. chlorospila* **(1)**.

8. P. occidentalis (R. C. Harris) R. C. Harris (1987)

P. neglecta auct. brit., non R. C. Harris (1973)

Thallus yellowish brown, with ± scattered, orange-brown flecks, which sometimes become ± continuous, rarely absent; pseudocyphellae absent. Perithecia 0.4–0.6(-0.7) mm diam., the surface K+

reddish purple in parts, lacking K+ purplish crystals inside the ascomatal wall. Ascospores (1-)3-septate, (17-)18–22 × (7-)8.5–10.5(-11) μm. Pycnidia often frequent; conidia ± hooked, 20 × 1 μm. Thallus Pd+ pale orange, K+ yellowish and + purplish in parts, KC− (unidentified anthraquinones), C−, UV+ golden yellow (lichexanthone).

On smooth bark, particularly *Corylus*, *Ilex* and *Sorbus*, in ± shaded, constantly humid woodland and sheltered stream valleys; local. N. Wales, W. Scotland, S.W. Ireland (Cork, Kerry). Norway, W. North America, South Africa.

This species has been much confused with *P. laevigata* **(4)** which is separated by the colour and UV− thallus as well as the lack of K+ reddish or purplish reactions of the perithecial surface and thallus flecks and the type of ascospore septation.

PYRRHOSPORA Körber (1855)

Thallus crustose, superficial, granular to subareolate, ± sorediate. **Photobiont** *Trebouxia*-like. **Ascomata** apothecia, reddish brown, convex. **Thalline exciple** absent. **True exciple** becoming excluded and scarcely apparent when mature, of conglutinated, thick-walled, radially orientated hyphae. **Epithecium** with brownish granules, K+ reddish purple (in section). **Hymenium** colourless, I+ blue. **Hypothecium** colourless. **Hamathecium** of paraphyses, simple or branched towards the apices, apices not conspicuously swollen, septate, constricted at the septa. **Asci** 8-spored, broadly clavate, *Lecanora*-type. **Ascospores** broadly ellipsoid, simple, colourless to slightly brownish, smooth, without a distinct perispore. **Conidiomata** unknown. **Chemistry**: anthraquinones in apothecia; thallus with ± arthothelin, thiophanic acid, or usnic acid. **Ecology**: on bark.

Pyrrhospora belongs to the Lecanoraceae on the basis of the ascus structure, even though no thalline exciple is present; it appears to be closely related to *Lecidella*.

LITERATURE: Poelt & Vezda (1981).

1. P. quernea (Dickson) Körber (1855)

Thallus thickly farinose to granular-sorediate, granules to *c.* 150 μm diam. arising over the surface of the thallus, even, often indistinctly areolate, yellowish or greenish fawn; prothallus generally present, forming a delimiting black line to 250 μm wide. Apothecia 0.4–1(-1.5) mm diam., strongly convex, often irregular in shape, dark reddish brown; true exciple excluded; epithecium interspersed with reddish brown granules, K+ dissolving, purplish. Ascospores (7-)8–12(-14) × (5-)6–7(-8) μm. Thallus Pd− or weakly yellowish, K−, KC+ orange, C+ orange, UV− or blackish orange (isoarthothelin, thiophanic acid and ± dichloronorlichexanthone); apothecia K+ reddish purple in section (1, 3, 8-trihydroxy-2-chloro-6-methylanthraquinone).

On moderately nutrient-rich, rough bark, occasionally wood, especially of *Quercus*, mostly in well-lit situations, very rarely on sandstone rocks. Throughout lowland Britain and Ireland, becoming scarcer in the N. and extreme Atlantic situations. Mediterranean and W. Europe, North America, Macaronesia.

Lecanora expallens has a thinner, less uniformly sorediate and much more distinctly yellowish green thallus, lacking a fawn tint, and is without a distinct black prothallus; its thallus contains usnic as well as thiophanic acid and zeorin. The only other species of *Pyrrhospora* in Europe is *P. rubiginans* (Nyl.) P. James & Poelt (1981), which has a more distinctly rimose, obviously sorediate thallus, which is C− and contains usnic acid and a similar anthraquinone; it occurs in upland areas of Norway and Sweden in dry siliceous rock recesses and could possibly occur in Scotland.

RACODIUM Fr. (1829)

Thallus filamentous, of elongated, straight hyphae longitudinally arranged closely surrounding the photobiont strands, not corticate, dark brown to black. **Photobiont** *Trentepohlia*. **Ascomata** and **conidiomata** not known. **Chemistry**: no lichen products recorded. **Ecology**: vertical, siliceous rock faces or underhangs.

LITERATURE: Hawksworth & Riedl (1977), Schade (1932).

1. R. rupestre Pers. (1794)

Thallus forming wefts or circular patches, the margin not delimited; hyphae 4–7 per photobiont filament, straight and parallel, unbranched, closely surrounding the photobiont, dark brown to black (Fig. 13b).

On vertical siliceous rock faces or underhangs (sometimes on more basic rocks), where the air is normally very humid, but where there is usually no direct wetting by rain, mostly upland; characteristic of the *Racodietum rupestris*, occasionally growing intermixed with *Cystocoleus ebeneus*; locally common. N. & W. British Is., extending into E. Sussex. Europe.

Distinguished from *Cystocoleus ebeneus* by the straight, not contorted or entwined, hyphae with more elongated cells which are slightly darker when seen under the microscope.

RAMALINA Ach. (1810)

Thallus shrubby, often tufted, erect to pendent, from a restricted or spreading holdfast (Fig. 39d), rarely unattached and straggling; branches single to numerous, generally markedly compressed and strap-shaped but not dorsiventral, rarely ± rounded, bilaterally symmetrical in cross section sometimes channelled, occasionally with fenestrations and pseudocyphellae, smooth or ridged, simple to much branched, the branching dichotomous to irregular, rarely palmate. **Cortex** plectenchymatous, thin, indistinct, overlying a ± well developed, ± cylindrical zone of prosoplectenchymatous tissue (absent in *R. lacera* **(9)**), invaded by the photobiont zone on its inner side. **Photobiont** trebouxioid. **Medulla** rarely dense and opaque, usually laxly arachnoid, sometimes completely absent when the branches are hollow. **Soralia** frequent. **Ascomata** apothecia, shortly stalked, apical or subapical, often on angled branches; disc pale green, brownish or pinkish yellow, sometimes whitish pruinose. **Thalline exciple** present, persistent to almost excluded. **Asci** elongate-clavate, *Bacidia*-type, 8-spored. **Ascospores** 1-septate, broadly ellipsoid or ± kidney-shaped, colourless. **Conidiomata** pycnidia; ostiole pale or blackened. **Conidiogenous cells** subcylindrical, enteroblastic, acrogenous. **Conidia** bacilliform, colourless. **Chemistry**: usnic acid, often with depsides, depsidones and aliphatic compounds.

Separated from *Evernia* and *Pseudevernia* by the thallus being generally uniformly green on all surfaces. *R. lacera* **(9)** may occasionally have a paler lower surface, but is never as white as in that of *Evernia*.

LITERATURE: Krog & James (1977), Krog & Østhagen (1980).

1 Thallus rounded to strap-shaped, (often >0.5 mm diam.); branches usually without hook-shaped ends; soredia present or absent 2
Thallus thread-like (often <0.5 mm diam.); lateral branches with hook-shaped ends; soredia absent .... **1. *chondrina***

2(1) Branches hollow 3
Branches ± solid 5

3(2) Soredia present; apothecia rare or absent 4
Soredia absent; apothecia usually present .... **7. *fastigiata***

4(3) Main branches ± rounded, less than 1.5 mm wide; soralia apical and lip-shaped, or as laminal ruptures in cortex; Pd+ orange, K+ orange-red .... **6. *portuensis***
Main branches compressed, mostly more than 1.5 mm wide; soralia within helmet-shaped vesicles or soredia produced through splits in the cortex; Pd−, K− .... **3. *canariensis***

5(2) Soredia or coarse corticate granules present; apothecia rare 6
Soredia absent; apothecia frequent 11

6(5) Soredia farinose or subgranular, <80 μm diam,, in ellipsoid soralia or irregularly spreading 7
Soredia with coarse, corticate granules, often >180 μm diam., singly or in clusters in numerous small laminal pseudocyphellae .... **11.** ***polymorpha***

7(6) Upper and lower cortex not splitting to reveal soredia (soredia superficial in discrete soralia) 8
Upper and lower cortex often splitting to reveal soredia .... **3.** ***canariensis***

8(7) Soralia ± oval or round, strictly delimited 9
Soralia spreading and ± irregular 10

9(8) Holdfast spreading; usually forming extensive swards; soredia ± granular in flat or excavate soralia; usually on rocks .... **5.** ***subfarinacea***
Holdfast simple; not forming extensive swards; soredia farinose in flat or slightly convex soralia; usually on bark .... **4.** ***farinacea***

10(8) Branches >1 cm wide, conspicuously compressed, or ± irregularly divided; surface ± smooth or with reticulate ridges, especially on the lower surface, matt .... **9.** ***lacera***
Branches <0.5 cm wide; sometimes ± terete; surface without reticulate ridges, cartilaginous, ± shiny .... **10.** ***pollinaria***

11(5) On rocks (rarely on wood or bark); chemical reactions various 12
On bark; Pd−, K− 13

12(11) Main branches often conspicuously compressed; surface often ± ridged; thallus not or rarely little blackened at the base, ostioles of pycnidia pale .... **12.** ***siliquosa***
Main branches mostly ± rounded; surface smooth, even; thallus frequently blackened at the base and often also above; ostioles of pycnidia blackened .... **13.** ***cuspidata***

13(11) Branches not channelled; surface ± wrinkled and ridged; branches >3 mm wide 14
Branches distinctly channelled, especially towards the base; surface ± smooth; branches <3 mm wide .... **2.** ***calicaris***

14(13) Thallus in ± erect tufts, to 5 cm, often hidden by numerous apical apothecia .. **7.** ***fastigiata***
Thallus pendent, to 20 cm; apothecia, if present, lateral .... **8.** ***fraxinea***

1. R. chondrina Steiner (1904)

Thallus to 10 × 5 cm, pendent or decumbent, arising from a narrow holdfast or often detached and straggling around other shrubby lichens; branches thread-like (<1 mm wide), mainly richly dichotomously branched, pale green-grey, often with discoloured blackish patches, smooth, terete, occasionally hook-shaped at apices; pseudocyphellae indistinct. Apothecia not observed in British material. Medulla Pd−, K−, KC−, C−, UV− (lichen products not detected by t.l.c.).

On coastal rocks; rare. S.W. Britain, S.W. Ireland. S. France, Canary Is., Madeira.

Not uncommon on exposed, steep N. to N.W.-facing, granite cliffs on the Is. of Scilly, where it grows entangled amongst *R. cuspidata* **(13)** resulting in it being easily overlooked. Previously confused with the continental European *R. thrausta* (Ach.) Nyl. (1860), a species characteristic of spruce twigs in moist, sheltered woodlands, which normally contains an undetermined medullary compound and has a less robust thallus with thinner branches which are less divergently branched. *R. thrausta* was reported from N. Scotland (Kincardine) in 1860 but those collections probably originated outside Great Britain.

2. R. calicaris (L.) Fr. (1824)

Thallus 5–8(-15) cm long, tufted, becoming pendent, grey-green, stiff and spiky when dry; branches 2–3(-10) mm wide, ± channelled, especially towards base surface shiny, smooth, texture horn-like; pseudocyphellae frequent, punctiform, indistinct. Apothecia frequent, ± marginal, subterminal, on angled branches. Ascospores 10–16 × 5–7 μm, broadly ellipsoid, straight. Medulla Pd−, K−, KC−, C−, UV− (± sekikaic acid complex).

On branches and twigs, especially of shrubs with a ± nutrient-rich bark, rarer on tree trunks, often near the coast in well-lit situations; locally abundant. W. British Is., extending locally to E. Scotland and S.E. England. Europe, S. Fennoscandia.

The number and width of branches vary; specimens with few, to 10 mm wide, branches can be confused with *R. fraxinea* **(8)** and straggling morphs may resemble *R. fastigiata* **(7)**; the overall tough texture of the thallus and channelling of the main branches, especially towards their bases are diagnostic.

3. R. canariensis Steiner (1904)

R. baltica auct. brit. p.p., ?non Lettau (1912)

Thallus forming pulvinate to ragged tufts, scattered or forming swards; branches pale grey-green or dull green, ± matt, broadening from base, sometimes tapered towards apices, compressed, internally solid or partly hollow with a lax, arachnoid medulla. Soralia mostly towards the apices, lateral or sub-terminal, in rupturing helmet-shaped vesicles, or released from irregular lacerations of cortex; soredia 20–40 μm diam., farinose. Apothecia very rare, marginal or on surface. Ascospores 15–21 × 6 μm, broadly ellipsoid or ± kidney-shaped. Medulla and soralia Pd−, K−, KC−, C−, UV± blue-white (divaricatic acid).

On bark with a relatively high pH in well-lit situations, also on rocks in dry underhangs and sheltered church walls, predominantly coastal. S., W. & S.E. Britain, C. & E. Ireland. Europe, N. to Norway, Mediterranean region, Azores, Canary Is.

A variable species, ranging from a single palmate lobe to a dense, ragged cluster of lacerate-contorted branches. Eroded specimens resemble *R. lacera* **(9)**, especially when on rocks; differences in anatomy, soredial development and chemistry then separate these species. The affinity of this species to *R. baltica* needs further investigation.

4. R. farinacea (L.) Ach. (1810)

Thallus 3–6 (-10) cm long, tufted, ± pendent, arising from a delimited holdfast, often subdivided into numerous, to 3 mm wide, flattened, occasionally slightly channelled, solid, branches, yellowish green to dark grey-green; surface matt, or ± shiny, smooth; texture firm. Soralia numerous, marginal or on surface, discrete, ± circular to elliptical, excavate to ± flat; soredia 20–30 μm diam., pale, farinose. Apothecia lateral, rare. Ascospores 8–15 × 5–7 μm, broadly ellipsoid. Chemistry: 4 chemotypes—medulla and soralia: (a) Pd+ orange-red, K− or orange, UV− (protocetraric acid); (b) Pd+ yellow-orange, K+ yellow-red, UV− (salazinic acid, ± norstictic acids); (c) Pd−, K−, UV+ blue-white (hypoprotocetraric acid); (d) Pd−, K−, UV− (lichen products not detected).

On a wide variety of substrata and in diverse habitats, ranging from trunks and twigs within shaded deciduous woodland to sunny, wind-exposed, isolated trees, hedgerows and scrub, more rarely on rocks and walls; common. Throughout British Is. W. Europe, becoming scarce in N. Fennoscandia.

Specimens in polluted habitats are often dark green, tufted, with short, decumbent, sometimes rather broad, contorted or recurved branches; conversely, in deep shade in woodland, material is often elongated, pale and rather sparingly branched with narrow branches. The least sensitive *Ramalina* to SO_2 pollution ($<60\ \mu g\ m^{-3}$) and to inorganic fertilizers. The chemotypes are sometimes given separate names as species or varieties but this is scarcely justifiable on the basis of current evidence.

5. R. subfarinacea (Nyl. ex Crombie) Nyl. (1872)

Like **4**, but differing mainly in the spreading holdfast and tendency to form swards, the predominantly saxicolous habit, the more erect, spiky appearance, the generally yellower colour of the thallus and the more granular, 25–60 μm diam., soredia in excavate soralia. Medulla and soralia generally K+ red (norstictic acid), but other chemotypes, as in *R. farinacea*, may occur.

On ± nutrient-rich boulders, rarely on old *Calluna* stems, predominantly coastal, but inland by freshwater lakes, on standing stones and church towers; frequent on rocky coasts. Throughout British Is. Fennoscandia southwards to Mediterranean region.

Extreme shade morphs may have very narrow branches with hook-shaped ends, resembling compact forms of *R. chondrina* **(1)**.

6. R. portuensis G. Samp. (1924)

Like **4**, with which it sometimes grows, but differing in the somewhat swollen, smooth, hollow branches, with often terminal, spreading, lip-shaped soralia and a constant chemistry of an unidentified medullary substance in addition to salazinic acid; medulla and soralia Pd+ yellow-orange, K+ yellow-red, UV−.

On sunny or ± sheltered, rather moist tree trunks, rarely old *Calluna* stems and rock outcrops, mainly near the coast; rare. S.W. England, Is. of Scilly (common), W. Wales, extending to Scotland (Is. of Skye), S.W. Ireland. Portugal, W. France (Brittany), Canary Is.

7. R. fastigiata (Pers.) Ach. (1810)

Thallus 2–5 cm, usually erect, densely tufted and richly branched, pulvinate, occasionally becoming lax and ± pendent, pale green to dull grey-green; branches terete or ± angular, ± flattened, ± hollow with a lax, arachnoid medulla; surface ± longitudinally striate, the fenestrations few to numerous. Apothecia often numerous, sometimes obscuring thallus; disc concave, becoming ± convex when mature. Ascospores 12–15 (-18) × 5–6 (-7) μm, mainly kidney-shaped, rarely broadly ellipsoid. Medulla Pd−, K−, KC−, C−, UV−, (evernic acid aggr.).

On trunks or twigs of well-lit, wind-exposed hedgerows or parkland trees with nutrient-rich bark, rarely on rocks, absent from hilly areas. Throughout British Is. Europe, usually oceanic.

Very sensitive to air pollution. This is also a polymorphic species; specimens are normally ± globose, with the top-shaped apothecial branches forming a rounded cluster all at ± the same level. In moist sites the thallus is paler, frequently fenestrate and rarely fertile. Very narrow straggling morphs occur with an almost transparent thallus when wet, and numerous angled branches with apothecia giving the thallus a prickly appearance. Coarse straggling morphs of *R. fastigiata* **(7)** could be confused with *R. calicaris* **(2)**, but the branches are not channelled and they contain evernic acid.

8. R. fraxinea (L.) Ach. (1810)

Thallus to 20(-30) cm, ± pendent, coarsely tufted, rarely monophyllous, grey-green or olive to green-black; branches to 3 cm wide, flattened, sparingly branched, ± twisted, tapering towards base and apices, flat or commonly ± channelled, often with an uneven, longitudinally and reticulately wrinkled surface, sometimes fenestrate or longitudinally lacerate; lateral branches usually arising near the distinct holdfast; pseudocyphellae frequent, inconspicuous, pale, rounded or oval. Apothecia usually frequent, marginal or laminal; disc cup-like, becoming flat or ± convex. Ascospores 10–17 × 4–7 μm, kidney-shaped. Medulla Pd−, K−, KC−, C−, UV− (lichen products not detected by t.l.c.).

On nutrient-rich bark in windy, well-lit sites; frequent but rare in woodlands. E. & S. England (much decreased), C. & E. Scotland, E. Ireland. Europe, mainly continental and rarer in the Mediterranean region.

Varies considerably in the size, width and number of branches; narrow-lobed morphs can be confused with *R. calicaris* **(2)**, but they are always finely and irregularly ridged, crumpled, matt, softer in texture, have reniform ascospores, and lichen products have not been detected in the medulla. Extremely sensitive to air pollution (not exceeding 35 μg m^{-3}).

9. R. lacera (With.) Laundon (1984)

R. duriaei (de Not.) Bagl. (1879)

Thallus ranging from a single, often ± convoluted lobe to ± densely tufted, forming ragged clusters, erect or ± pendent to partly decumbent; branches strap-like, flattened, concolorous or lower side paler, often with indented margins, with or without twisted and deformed marginal prolongations; surface matt, often ± peeling, smooth or with a coarse reticulum of anastomosing ridges. Medulla dense, without prosoplectenchymatous tissue. Soralia laminal and marginal; soredia 20–40 μm diam, farinose, derived from medulla, forming ± ulcerose patches. Apothecia not observed. Medulla and soralia Pd−, K−, KC−, C−, UV− (bourgeanic acid).

Mainly on wayside trees, hedgerows and scrub near the coast, also on rock in dry, sheltered situations; locally common. Throughout British Is. Europe, Mediterranean, Middle East.

A polymorphic species. *R. pollinaria* **(10)** may closely resemble narrow, branched morphs of *R. lacera*, but the former contains evernic acid aggr., has prosoplectenchymatous tissue in the medulla and often terminal, ± lip-shaped soralia containing granular soredia. See also under *R. canariensis* **(3)**.

10. R. pollinaria (Westr.) Ach (1810)

Thallus to 5 cm, usually much less, discrete or forming swards, pale green to pale yellow; branches usually numerous, ± flattened or partially terete, especially towards apices, much-branched and dissected with numerous small, nodulose proliferations, or branches more flattened, wavy with deeply incised or notched margins; surface smooth, shiny, appearing cartilaginous. Soralia mainly laminal, subterminal or terminal, more rarely on branch margins, irregularly spreading, often appearing lip-shaped at apices; soredia 50–70 μm diam., granular. Apothecia not seen. Medulla and soralia Pd−, K−, KC−, C−, UV± blue-white (± evernic acid aggr.).

On dry, sheltered underhangs on siliceous rock, or exposed tree roots, also on N. or E.-facing church walls; very rare. S. & N.W. England (W. Sussex, Cumberland), C. Wales (Brecon) to C. Scotland (Grampian mtns). C. Europe, Scandinavia.

This species varies in the degree of branching and in the disposition of soralia. Previously much confused with *R. lacera* **(9)** and *R. canariensis* **(3)**. Differing from *R. subfarinacea* **(5)** and *R. farinacea* **(4)** in the at least partly laminal and ulcerose soralia and chemistry.

11. R. polymorpha (Lilj.) Ach. (1810)

Thallus 3–6 cm, erect or ± decumbent, in dispersed tufts or often forming extensive swards, dark green; branches strap-shaped or variously longitudinally dissected, often ± tapered towards the ends, often lacerate, contorted-twisted; surface uneven, ± scabrid, ± longitudinally ridged, striate, commonly with scattered or confluent, concolorous or paler pseudocyphellae which may open into numerous fenestrations. Granules present, 60–180 μm diam., singly or mainly clustered around the pseudocyphellae, concolorous with thallus, ± smooth, corticate. Apothecia rare, terminal or subterminal in small thalli, lateral and marginal in larger morphs, ± persistently cup-like, rarely becoming convex; thalline exciple scabrid-verrucose. Ascospores 12–16 × 4–6 μm, broadly ellipsoid. Medulla Pd−, K−, KC−, C− (lichen products not detected by t.l.c.).

On nutrient-enriched rocks, particularly below bird-roosting or -nesting sites on cliffs; locally abundant. N.E. England, S. Wales (Pembroke, Skomer Is., rare on adjacent mainland), E. Scotland. Fennoscandia, France, C. & W. Pyrénées.

Characterized by the scabrid surface and numerous pseudocyphellae, each containing one or several, coarse, ± corticate granules. Mite-mutilated thalli of *R. siliquosa* **(12)** may bear a superficial resemblance but *R. polymorpha* is darker green and has characteristic pseudocyphellae; in most cases the absence of medullary lichen products in *R. polymorpha* is confirmatory.

R. capitata (Ach.) Nyl. (1879) differs in the smaller, 1–2 cm tall, erect thallus, narrow branches and soralia which are mainly concentrated towards or at the apices of the branches; reported last century on the basis of a single collection from C. Scotland (Grampian Mtns., summit of Beinn á Bhuird) which is doubtfully of British origin.

12. R. siliquosa (Hudson) A.L. Sm. (1918)

Thallus coarsely tufted, erect to pendent, stiff; branches usually numerous, simple to richly branched; main branches 2–9 mm wide, ± flattened, rarely partly rounded; secondary branching furcate; surface pale grey-green to yellow-green, rarely blackened at base or elsewhere, ± matt, unevenly wavy or frequently partly ridged, smooth or ± wrinkled, occasionally furrowed. Apothecia frequent, lateral, occasionally terminal. Ascospores 11–20 × 5–6 μm, broadly ellipsoid or slightly kidney-shaped. Pycnidia frequent, often clustered and forming nodulose irregularities, the ostioles ± flesh-coloured and transparent when wet. Chemistry: 4 chemotypes—medulla: (a) Pd+ orange-red, K−, UV− (protocetraric acid); (b) Pd+ orange, K+ red, UV− (salazinic acid); (c) Pd−, K−, UV+ blue-white (hypoprotocetraric acid); (d) Pd−, K−, UV− (acid deficient).

On siliceous rocky coasts in the xeric-supralittoral, often dominant and forming a distinct zone on sea-cliffs, more rarely inland on old churches or sarsen stones, occasionally on fence-posts, rarely on *Quercus* and *Fraxinus*. Throughout British Is. N. Norway to C. Portugal, extending to Baltic Sea coasts.

Very polymorphic; on vertical cliff faces often large, pendent and coarse, with a few, thick, flattened branches; on sloping rock faces and boulders the thalli are shorter and ± erect, the branches less flattened. In inland sites apothecia are rarely produced and the branches tend to become decumbent and recurved; chemotype (b) (salazinic acid) is the most frequently encountered at such sites. The chemotypes are sometimes given specific or varietal status but this does not seem to be justifiable on current evidence. See under *R cuspidata* **(13)**.

13. R. cuspidata (Ach.) Nyl. (1870)

Like **12**, but differing in the ± rounded, narrower (1–3 mm wide), main branches which are often markedly blackened at the base (Fig. 39d) and elsewhere; surface shiny, smooth, brownish yellow or pale green with a yellow tinge. Pycnidia blackened at ostioles. Chemistry: 3 chemotypes—medulla (a) Pd+ yellow-orange, K+ yellow-red, UV− (norstictic acid); (b) Pd+ yellow-orange, K+ yellow-orange, UV− (stictic acid); (c) Pd−, K−, UV− (acid deficient).

On siliceous coastal rocks, xeric-supralittoral but often in the lower part of the *Ramalina siliquosa* zone and below, rarely on wood of fence posts, coastal; local. W. British Is. extending locally to E. coasts of Scotland and Ireland. Portugal (Algarve), N. to Norway (Lofoten Is.), rare on Baltic Sea coasts.

Numerical taxonomic and protein profile studies have confirmed the distinctness of this species from *R. siliquosa* **(12)** despite the similar chemotypes. The two species also retain their identity when growing together.

RAMONIA Stizenb. (1862)

Thallus crustose, usually immersed, cortex not differentiated, non-layered, effuse. **Photobiont** *Trentepohlia*. **Ascomata** apothecia, urceolate, at first immersed with the exciple overtopping and hiding the hymenium except for a minute pore and perithecioid, later ± erumpent and the exciple splitting radially and exposing the sunken, pinkish or grey-white disc. **True exciple** well developed laterally, of small, thin-walled, angular cells, becoming rectangular along inner surface, ending in periphyses with ± pointed apices, ± colourless or dark brown in outer part. **Hymenium** colourless, I+ bluish at least in uppermost part. **Hamathecium** of paraphyses, slender, numerous, septate, unbranched, usually ± swollen above. **Asci** (2-)8- to 180-spored, clavate to ± cylindrical, apex rounded or with a short, blunt apiculus, with a single thin (<1 μm wall layer), I−, some species with a minute I+ blue, ring-structure in the apex. **Ascospores** simple, 1- to many-septate or muriform, variously shaped, colourless, smooth-walled, sometimes with a distinct perispore. **Conidiomata** unknown. **Chemistry**: lichen products not detected by t.l.c. **Ecology**: on bark or wood.

Gyalecta or *Pachyphiale* differ in having a true exciple composed of narrow, never distinctly angular, cells and lacking periphyses.

LITERATURE: Coppins (1987*a*).

1	Ascospores <6 μm wide, (3-)7- to 14- septate, elongate-fusiform or needle-shaped	2
	Ascospores >7 μm wide, muriform, ellipsoid or oblong	4
2(1)	Ascospores 27–43 μm, elongate-fusiform, (3-)7(-11)-septate	**2.** aff. ***luteola***
	Ascospores mostly >45 μm, needle-shaped, 8- to 14-septate	**1. *chrysophaea***
3(2)	Apothecia pale	**3. *dictyospora***
	Apothecia black	**4. *nigra***

1. R. chrysophaea (Pers.) Vězda (1966)

Apothecia 0.4–0.7 mm diam., with pore to 0.5 mm diam., pale; hymenium 90–120 μm tall. Ascospores (36-)45–74 × (3-)3.5–4(-5) μm, spindle-shaped, 8- to 14-septate Fig. 24d).

On rather 'spongy' bark of mature or old trees of *Quercus, Fraxinus* and *Ulmus* in shaded old woodland situations. Throughout British Is. France, Spain.

2. R. aff. **luteola** Vězda (1967)

Like **1**, but ascospores 24–43 × 4–6 μm, (3-)7(-11)-septate (Fig. 24c); hymenium 60–75 μm tall.

On bark of *Salix, Sambucus,* and *Ulmus*; widely distributed. England, Scotland. Czechoslovakia, Yugoslavia.

The British material appears to be distinct from *R. luteola* s.str. from Czechoslovakia in having slightly larger ascospores which do not have a distinct perispore.

Gyalecta derivata has 7- to 11-septate ascospores, sometimes 1–3 cells with a longitudinal septum, and *Pachyphiale fagicola* ascospores 15–21(-35) × 3.5–5 μm, 3- to 7-septate, and 16- to 32-spored asci.

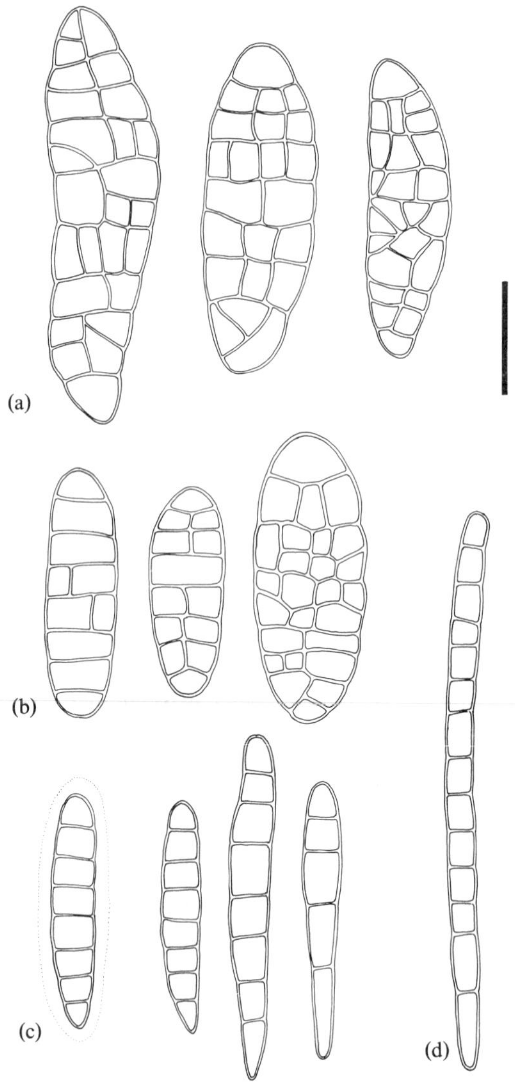

Fig. 24. *Ramonia* ascospores (after Coppins, 1987). (a) *R. nigra*; (b) *R. dictyospora*; (c) *R. luteola* (left, Czechoslovakia; right, Easterness); (d) *R. chrysophaea*. Scale = 10 μm.

3. R. dictyospora Coppins (1987)

Like **1**, but ascospores 21–38(-45) × 9–14(-15) μm, oblong or ellipsoid, at first (5-)7-septate, soon becoming muriform (Fig. 24b).

On *Acer pseudoplatanus*, *Fagus*, *Quercus* and *Ulmus* sometimes in wayside situations and secondary woodlands. S. England (Hampshire, Oxford), Scotland. ?Endemic.

The only *Ramonia* with pale apothecia and large muriform spores. For differences from *R. nigra* **(4)**.

4. R. nigra Coppins (1987)

Like **3**, but with black apothecia. Ascospores 28–45(-50) × 8–13 μm, oblong or oblong-ellipsoid, muriform (Fig. 24a).

On wood inside hollowed trunks of old *Fagus* and *Fraxinus* in ancient woodlands. N.W. England (Cumberland), S. England (Devon, Hampshire). ?Endemic.

Xerotrema megalaspora has 1-spored asci with very large, richly muriform ascospores, 90–140(-170) × 22–40 μm, and a different exciple structure.

RHAPHIDICYRTIS Vainio (1921)

Thallus crustose, superficial to immersed. **Photobiont** *Trentepohlia*-like. **Ascomata** perithecioid, black, arising singly, ± immersed, globose, ostiolate. **Involucrellum** including host tissue (clypeate). **True exciple** dark brown, entire, pseudoparenchymatous. **Hamathecium** of branched and anastomosed, remotely septate paraphysoids; periphyses also present in the ostiolar canal; centrum I+ blue (not deep blue). **Asci** narrowly cylindrical, bitunicate in structure, lacking an internal apical beak. **Ascospores** colourless, narrowly thread-like to sigmoid, multiseptate. **Conidiomata** unknown. **Chemistry**: characteristic lichen products absent. **Ecology**: on smooth bark.

This monotypic genus differs fundamentally from *Leptorhaphis* in the structure of the ascomata, the hamathecial tissues, and the exceptionally long and narrow ascospores. It differs from *Celothelium*, also with narrowly thread-like ascospores, in the solitary ascomata and I+ blue centrum.

LITERATURE: Aguirre-Hudson (1991).

1. R. trichosporella (Nyl.) Vainio (1921)

Thallus creamy-white to grey, smooth to somewhat powdery. Ascomata 150–200 μm diam.; involucrellum about 30 μm thick, expanded to about 60 μm thick near the ostiole; true exciple 15–30 μm thick. Paraphysoids 1.5–2 μm broad. Asci 75–95 × 9–15 μm, 8-spored. Ascospores (60-)75–95 × 2–3 μm, narrowly thread-like to ± sigmoid, 7- to 9-septate, the apices acute, arranged in a fascicle or spirally twisted in the asci.

On smooth bark of *Ilex* and *Betula* (also on *Carpinus* and *Quercus* in Europe), in ancient deciduous woodlands and associated with old-forest indicator lichens such as *Pachyphiale cornea* and *Thelotrema lepadinum*; rare. Wales, S.E. & N. Scotland, S.E. Ireland (Wicklow), N. Ireland (Fermanagh, Londonderry, Tyrone). Finland, Germany.

RHIZOCARPON Lam. ex DC. (1805)

Thallus crustose or absent in some parasitic species or sometimes parasitic on other lichens when young; when present usually truly areolate, more rarely partly immersed, continuous, rimose-cracked, warted, or somewhat effigurate at the margin, greenish yellow to yellow-green, white, grey, brown or rust red, rarely sorediate or isidiate. **Prothallus** usually present, black or whitish to brownish grey. **Upper cortex** usually well developed. **Lower cortex** absent. **Photobiont** chlorococcoid. **Medulla** K/I− or ± blue. **Ascomata** apothecia, black, ± concave to strongly convex, round to angular, attached to prothallus or on or at the edges of areoles. **Thalline exciple** absent. **True exciple** ± well developed, clearly evident when young, persistent or becoming excluded, of radiating hyphae, with a brownish or greenish black rim and usually a paler or rarely colourless inner part, frequently containing crystals. **Epithecium** brown or green, frequently containing crystals. **Hymenium** K/I+ blue. **Hypothecium** medium to dark brown, lacking crystals. **Hamathecium** of paraphysoids, strongly conglutinate, richly branched and anastomosed, apical cell slightly to distinctly thickened. **Asci** 1- to 8-spored, clavate to cylindrical, discharge fissitunicate, with a well-developed tholus which is I− in the lower part and I+ blue near the apex, lacking an ocular chamber, *Rhizocarpon*-type (Fig. 33h). **Ascospores** colourless to dark green or dark brown, 1-septate to muriform, ± ellipsoid, with a swollen perispore. **Conidiomata** pycnidia, known for a few species only, simple or chambered, associated with the prothallus or immersed in the areoles. **Conidia** cylindrical to acicular, terminal on elongate, flask-shaped conidiogenous cells. **Chemistry**: rhizocarpic acid in the cortex of all yellow-green species, usually with depsides, desidones or aliphatic compounds in the medulla. **Ecology**: predominantly on hard, acidic siliceous rocks; some species, especially those with whitish thalli, on limestone or other basic rocks; a few species parasitic or commensalistic on other crustose lichens on rocks.

Buellia species on rocks are distinguished by their simple, ± clavate paraphyses and generally smaller ascospores which lack a swollen perispore. *Poeltinula cerebrina*, which occurs on hard limestones, has 2-celled ascospores with a distinct perispore as in *Rhizocarpon*, but the apothecia are shortly lirelliform to angular with a thick, black, opaque exciple. *R. grande* (Flörke ex Flotow) Arnold (1871) and *R. transiens* Eitner (1911) have been incorrectly recorded from the British Is.

LITERATURE: Feuerer (1978), Poelt (1988), Poelt & Vězda (1977, 1981), Runemark (1956), Thomson (1967b), Timdal & Holtan-Hartwig (1988), Winchester (1986).

1	Apothecia present; isidia and soredia absent	2
	Apothecia absent; ± continuously finely isidiate-sorediate (blastidia)	**11. *furfurosum***
2(1)	Thallus conspicuous; not on *Pertusaria*	3
	Thallus inconspicuous; lichenicolous on *Pertusaria*	**1. *advenulum***
3(2)	Thallus ± bright yellow-green; ascospores soon becoming dark green-brown	4
	Thallus white, grey, brown or ochraceous; ascospores colourless to dark green-brown	11
4(3)	Ascospores 1-septate	5
	Ascospores 3-septate submuriform or muriform	8
5(4)	Ascospores 18–32 × 9–15 μm	6
	Ascospores 9–18 × 5–10 μm	**28. *superficiale***
6(5)	Epithecium greenish brown to green; medulla Pd+ yellow, K+ red (crystals), or Pd−, K−	7
	Epithecium brown without a greenish tinge; medulla Pd+ yellow, K−	**2. *alpicola***
7(6)	Medulla I+ distinctly blue	**9. *eupetraeoides***
	Medulla I− or faint blue	**15. *inarense***
8(4)	Medulla I+ markedly blue; epithecium brown or green, K+ purple-red or −; not on other lichens	9
	Medulla I− or faint blue; epithecium brownish black, K+ purple-red; parasitic on other lichens at least when young	**30. *viridiatrum***
9(8)	Ascospores distinctly muriform, with many longitudinal septa	10
	Ascospores transversely septate or, at most, with 1 or 2 longitudinal septa	**16. *intermediellum***
10(9)	Apothecium not surrounded by crescent-shaped areoles	**13. *geographicum***
	Apothecium partly surrounded by crescent-shaped areoles	**19. *lecanorinum***
11(3)	Ascospores 1-septate	12
	Ascospores 2- or more-septate or muriform	21
12(11)	Ascospores soon dark grey-green to dark brown	13
	Ascospores remaining ± colourless	15
13(12)	Medulla I−; ascospores 26–36 × 13–16 μm; hypothecium K−	14
	Medulla I+ blue; ascospores 12–16 × (5-)6–7 μm; hypothecium K− or + exuding red or magenta-red	**26. *simillimum***
14(13)	Epithecium olivaceous brown to greenish black, K− containing crystals refracting polarized light and dissolving in K; medulla K+ yellow	**17. *jemtlandicum***
	Epithecium reddish brown to dark brown, K+ red-purple, not containing crystals; medulla K− or ± yellow	**3. *badioatrum***

15(12) Medulla I+ blue 16
Medulla I− 17

16(15) Epithecium dark brown, K+ red or −, without crystals .... **23. *polycarpum***
Epithecium olive-green to grey-black, with crystals refracting polarized light, K− or turning intensely green .... **25. *richardii***

17(15) Thallus K− or ± yellow 18
Thallus K+ red .... **6. *cinereovirens***

18(17) Thallus usually >2 cm diam., white to dark brown; epithecium green to brown, K− ... 19
Thallus <1 cm diam. or becoming confluent, chalk-white; epithecium red-brown, K+ red-purple .... **5. *chioneum***

19(18) Thallus white to medium grey; epithecium olive-green to bright green; ascospores 14–20 × 6–10 μm 20
Thallus dark brown; epithecium dark brown; ascospores 25–36 × 12–15 μm **14. *hochstetteri***

20(19) Thallus pure white to greyish white; apothecial disc usually pruinose; exciple and hymenium containing crystals refracting polarized light; thallus K+ yellow **4. *caeruleoalbum***
Thallus greyish white to medium grey; apothecial disc epruinose; apothecium not containing crystals; thallus K− .... **10. *expallescens***

21(11) Asci 1 or 2(-4)-spored 22
Asci normally 8-spored 23

22(21) Ascospores remaining ± colourless .... **27. *subgeminatum***
Ascospores soon becoming dark brown .... **12. *geminatum***

23(21) Medulla K+ yellow or − 24
Medulla K+ red .... **22. *plicatile***

24(23) Medulla I− 25
Medulla I+ blue .... **8. *distinctum***

25(24) Ascospores 3-septate to submuriform 26
Ascospores truly muriform 27

26(25) Thallus rust-coloured; thallus K− .... **21. *oederi***
Thallus pale grey to dark brown, not rust-coloured; thallus K+ yellow .... **24. *postumum***

27(25) Epithecium with crystals dissolving in K; true exciple usually pruinose; thallus white to medium grey, K+ yellow 28
Epithecium usually not containing crystals; true exciple not pruinose; thallus medium grey, pale brown, or dark brown or ochraceous; thallus K− 29

28(27) Thallus chalk white, true exciple thick, pruinose; ascospores 18–28 × 10–16 μm .... **29. *umbilicatum***
Thallus pale greyish white, grey or brown; true exciple slightly pruinose or not; ascospores 20–50 × 13–24 μm .... **7. *concentricum***

29(27) True exciple thick, persistent, ± tumid; disc smooth or slightly roughened; thallus ± ochraceous; ascospores 30–40 × 14–18 μm .... **18. *lavatum***
True exciple thin; disc ± roughened; thallus grey, dark grey or brown-grey; ascospores 20–32 × 9–15 μm .... **20. *obscuratum***

1. R. advenulum (Leighton) Hafellner & Poelt (1976)

Thallus inconspicuous, immersed in the host lichen; vegetative hyphae I+ blue. Apothecia 0.4–0.7 mm diam., flat or slightly convex, scattered or in small groups on host thallus; true exciple well defined when young, evenutally becoming occluded, dark brown; epithecium brown; hymenium 80–90 μm tall, colourless; hypothecium dark brown. Ascospores 18–23 × 11–15 μm, 1-septate, guttulate.

On *Pertusaria pseudocorallina*, *P. flavicans* and other *Pertusaria* species; rare. N. Wales, N. Scotland (Outer Hebrides), Ireland. Endemic.

The only completely lichenicolous species of the genus in Britain.

2. R. alpicola (Anzi) Rabenh. (1861)

Thallus to 15 cm diam. often wide-spreading, areolate; prothallus well developed, black, sometimes faintly pruinose; areoles to 1.5(-3) mm diam., to 1.5 mm thick, bright yellow-green, matt, contiguous to widely scattered, usually secondarily cracked, rather thick, round to angular or irregular, flat to moderately convex, smooth to rarely minutely scabrid. Apothecia to 1.5 mm diam., black, not pruinose, round to angular, flat to weakly convex; true exciple indistinct, reddish brown, K+ purplish red; epithecium pale or partly dark brown, K+ faintly red, or locally olive-green to brown, K− or ± more intensely greenish; hymenium colourless; hypothecium dark brown K−; no crystals or granules in the apothecium. Ascospores 18–32 × 9–15 μm, 1-septate, infrequently with 1–2 additional, thin, transverse septa, dark brown. Medulla Pd+ yellow, K−, I− (rhizocarpic and psoromic acids).

On siliceous rocks in exposed situations, montane; local. N. Wales (Gwynedd, over 900 m), C. & N. Scotland. Scandinavia, C. European mtns., Alps to Balkans, N.W. North America, C. Asia.

The combination of large, dark, 1-septate ascospores, a brown epithecium and presence of psoromic acid separates *R. alpicola* from *R. eupetraeoides* **(9)** and *R. inarense* **(15)**, which have a greenish epithecium and are K+ red.

3. R. badioatrum (Flörke ex Sprengel) Th. Fr. (1874)

Thallus to 10 cm diam., areolate; prothallus usually well developed, black; areoles to 2 mm diam., medium to dark brown, often with a pinkish to red-brown tinge, sometimes partly greyish brown, matt, contiguous, thick, usually angular or ± scattered and crenulate, ± flat. Apothecia to 1 mm diam., black, not pruinose, round or angular, between the areoles, remaining ± flat; true exciple distinct, reddish brown in inner part, rim brownish black, K+ purple-red; epithecium reddish brown, K+ purple-red, sometimes with minute red granules dissolving in K; hymenium colourless; hypothecium medium brown, K−. Ascospores 26–36 × 13–16 μm, 1-septate, soon becoming dark brown. Medulla Pd− or Pd+ yellow, K− or K+ yellow (lichen products not detected by t.l.c., or ± stictic acid).

On siliceous rocks in open situations, montane. Scotland. C. & N. Europe, U.S.S.R., North America, Australia.

Some British material differs from the Scandinavian in containing stictic acid.

The soon dark brown, 1-septate ascospores, brown epithecium without crystals and K+ purplish, separate this species from *R. jemtlandicum* **(17)** which has an olive to green-black epithecium with crystals which is K−.

4. R. caeruleoalbum (Krempelh.) Zahlbr. (1927)

Thallus to 1 cm diam., areolate or rimose-cracked to partly continuous; prothallus poorly developed or absent, black, ± white-pruinose; areoles to 1.5 mm diam., white or whitish grey, dull, pruinose, contiguous, mainly angular, flat to weakly convex. Apothecia to 0.4 mm diam., black, mainly orbicular, flat to weakly convex; disc not or more rarely faintly pruinose; true exciple thick, persistent, and usually distinctly pruinose, dark greenish brown in the rim, inner part pale brown to colourless, containing crystals partly dissolving in K, K+ yellow; epithecium olive-green, containing crystals dissolving in K, K− (crystals K+ yellow, epithecium turning more brightly green); hymenium colourless; hypothecium dark brown, K−. Ascospores 16–20 × 8–10 μm, 1-septate, remaining colourless or only turning faintly green when old. Medulla Pd− or ± yellow, K+ yellow or K− (± stictic acid).

On epidiorite; very rare. C. Scotland (Angus, Caenlochan, Glen Isla). Scandinavia, C. Europe.

Superficially resembles some morphs of *R. concentricum* **(7)** with irregularly distributed apothecia, but has smaller, 1-septate, not muriform, ascospores. Scandinavian material is to 5 cm diam., and has apothecia to 1.5 mm diam.

5. R. chioneum (Norman) Th. Fr. (1874)

Thallus to 1 cm diam., or several thalli confluent, continuous to rimose-cracked, somewhat uneven, white, pruinose, matt; prothallus indistinct. Apothecia to 1 mm diam., black, ± immersed; disc usually not pruinose, orbicular, remaining ± flat; true exciple uneven, pruinose, persistent, dark brown in the rim, inner part paler brown, containing crystals dissolving in K, pigment K+ purple-red, crystals K+ yellow; epithecium dull grey to ± reddish brown, K+ red; hymenium colourless to pale brown, containing crystals at least partly dissolving in K; hypothecium brown, K−. Ascospores 12–16 × 6–10 μm, 1-septate, long remaining colourless, usually becoming unevenly green or brown with age. Thallus reactions negative (stictic acid in apothecium).

On epidiorite and other calcareous rocks, montane; very rare. C. & N. Scotland (Angus, Caenlochan, Glen Isla; Inverness, Ben Alder). Scandinavia, North America.

R. chioneum forms small, chalky-white, orbicular patches with ± immersed apothecia. The combination of small, 1-septate, colourless ascospores and a brown, K+ purple epithecium is diagnostic.

6. R. cinereovirens (Müll. Arg.) Vainio (1922)

Thallus to 6 cm diam., areolate; prothallus well developed, black; areoles to 0.5(-0.8) mm diam., pale grey to medium grey, often with a brown or mauve-grey tinge, matt, contiguous, or ± scattered, ± orbicular, moderately to strongly convex or sometimes partly hemispherical. Apothecia to 1 mm diam., often contiguous, black, not pruinose, round to somewhat angular, persistently flat to slightly convex; true exciple thin, persistent, brownish black in the rim, inner part paler brown, containing crystals dissolving in K, K+ red (forming acicular crystals); epithecium olive-green to greenish black, K− (turning more brightly green); hymenium colourless, containing crystals dissolving in K; hypothecium reddish brown, K−. Ascospores 15–24 × 6–9 μm, 1-septate, persistently colourless or becoming faintly brown with age. Thallus and exciple Pd+ yellow, K+ red (norstictic acid).

On pebbles, coastal shingle and around edges of late snow patches; very rare. N.E. Scotland (Sutherland, Culbin and Dornoch), N.W. Scotland (Westerness, Ben Nevis range). C. & N. Europe, U.S.S.R.

Distinguished by the rounded, convex areoles, green-black epithecium (K−), pale, 1-septate ascospores and norstictic acid.

7. R. concentricum (Davies) Beltr. (1858)

R. perlutum (Nyl.) Zahlbr. (1905)

Thallus to 5 cm diam., rimose to continuous, more rarely partly areolate; prothallus poorly developed, black; areoles to 0.5 mm diam., whitish grey to medium grey, matt, contiguous, mainly angular, flat. Apothecia to 1 mm diam., black; disc not pruinose, usually orbicular, flat; true exciple faintly to distinctly pruinose, persistent, brownish or greenish black in the rim, inner part pale brown to colourless, containing crystals partly dissolving in K, K+ yellow; epithecium olivaceous brown, containing crystals dissolving in K, K− (crystals K+ yellow); hymenium colourless; hypothecium dark brown, K−. Ascospores 20–50 × 13–24 μm, muriform, persistently colourless. Medulla Pd+ yellow, K+ yellow (stictic acid).

On ± basic rocks and walls and especially on hard acid rocks often in close proximity to basic rocks or mortar. Throughout British Is. Europe, North America, Australia.

The apothecia may be randomly dispersed on the thallus but are frequently arranged in ± concentric lines. *R. umbilicatum* **(29)** always has randomly scattered apothecia with a thicker, somewhat, chalky thallus and is restricted to pure limestone.

8. R. distinctum Th. Fr. (1874)

Thallus to 5 cm diam., areolate; prothallus well developed, black; areoles to 0.4 mm diam., dark brown, matt, contiguous or more rarely scattered, mainly angular, flat to weakly convex; medulla I+ blue. Apothecia to 0.6 mm diam., black, not pruinose, orbicular to angular, remaining ± flat; true exciple narrow, persistent, brownish black in the rim, inner part paler brown, K+ red; epithecium dark brown, K+ red; hymenium colourless; hypothecium dark brown, K−; no crystals or granules in the apothecium. Ascospores 16–27 × 8–13 μm, 3-septate to submuriform, persistently colourless or becoming greenish brown with age. Medulla Pd+ yellow, K+ yellow (stictic acid).

On siliceous rocks and walls, bricks, tiles, memorials; local. Throughout British Is. C. Europe, Fennoscandia, North America.

Distinguished by the I+ blue medulla, 8-spored asci, the presence of stictic acid and the pale, only tardily coloured, septate or submuriform ascospores.

9. R. eupetraeoides (Nyl.) Blomb. & Forss. (1880)

Thallus to 10 cm diam., areolate; prothallus well developed, black; areoles to 1.5 mm diam., bright yellow, matt, contiguous to somewhat scattered, usually cracked, rather thick, round to angular, flat to moderately convex; medulla I+ blue, sometimes I− in part. Apothecia to 1.5 mm diam., black, not pruinose, ± orbicular, flat to weakly convex; true exciple distinct to indistinct, olivaceous brown in the rim, pale brown to colourless in the inner part, containing crystals, pigment K− (intensifying green), crystals K+ red (acicular crystals); epithecium olivaceous green to brown, containing crystals, green pigment intensifying in K, crystals K+ red (acicular crystals); hymenium colourless to greenish, often subdivided by bands or columns of brownish hypothecial tissues; hypothecium dark brown, K−. Ascospores 18–32 × 10–15 μm, 1-septate, infrequently simple or with one or more additional, thin, transverse or oblique septa, dark brown. Medulla, true exciple and epithecium either Pd+ yellow, K+ yellow-red (norstictic and rhizocarpic acids), or Pd−, K− (bourgeanic and rhizocarpic acids).

On exposed sliceous rocks at high altitudes. N. Wales (Carneddau massif, 850–950 m), Scotland (Ben Nevis 1200–1340 m; Cairngorm mtns, Ben Macdhui 1300 m). N. Europe, circumpolar arctic-alpine.

The North Wales population is morphologically within the range of variation of the species but the medulla contains bourgeanic acid and reacts only faintly and patchily with Lugol's iodine; the I reaction led to it originally being designated *R.* aff. *inarense.*

Distinguished by the 1-septate ascospores, I+ blue medulla and green-brown epithecium. In *R. inarense* **(15)** the medulla is always I−.

10. R. expallescens Th. Fr. (1874)

Thallus to 5 cm diam., rimose-cracked to areolate; prothallus usually distinct, black; areoles to 0.5 mm diam., greyish white to medium grey, matt, contiguous or often scattered, ± angular, flat. Apothecia to 1 mm diam., black, not pruinose, orbicular, remaining ± flat or becoming moderately convex; true exciple narrow, brownish black in the rim, inner part paler brown, K−; epithecium bright green to blue-black, K−; paraphyses with a sharply delimited, green-pigmented cap; hymenium colourless; hypothecium brown, K−; no crystals or granules in the apothecium. Ascospores 14–17 × 6–9 μm, 1-septate, persistently colourless. Lichen products not detected by t.l.c.

On epidiorite. N. Scotland (Highlands; Sutherland, Ben Hope; W. Inverness, Ben Alder). Norway, Sweden, North America.

11. R. furfurosum Magnusson & Poelt (1955)

Thallus wide-spreading to 5 cm diam. or of several confluent thalli, ± rimose-cracked to continuous, sometimes partly obscurely areolate near the margin; prothallus poorly developed, black; areoles to 1 mm diam., contiguous, dull ochraceous, brownish grey to dark grey, often with a faint violet tinge, matt, mainly angular, densely and often completely covered with minute, at first, somewhat minutely warted, fragile blastidia. Apothecia to 0.8 mm diam., infrequent, black, not pruinose, orbicular; disc persistently flat or becoming weakly convex, sometimes umbonate; true exciple thick, persistent, dark brown in the rim, inner part paler brown, containing crystals partly dissolving in K, K+ yellow; epithecium medium brown, containing crystals dissolving in K, K+ yellow; hymenium colourless or pale brown; hypothecium dark brown, K−. Ascospores 14–24 × 7–12 μm, muriform, persistently colourless. Medulla and exciple Pd+ yellow, K+ yellow (stictic acid).

On often vertical, siliceous rock and old walls rich in iron sulphide minerals, especially old mine-workings where the species can be locally abundant. Throughout British Is. C. Europe.

Usually sterile and forming extensive, scurfy patches. The texture of thallus is somewhat reminiscent of *Lecidella prasinula* which is, however, C+ orange.

12. R. geminatum Körber (1855)

Thallus to 10 cm diam., areolate; prothallus well developed, black; areoles to 0.8 mm diam., pale to medium grey to red-brown or dark brown, matt, scattered to contiguous, orbicular to angular, flat to semi-globose. Apothecia to 1 mm diam., black, ± elevated, not pruinose, ± orbicular, flat to weakly convex; true exciple usually indistinct, brownish black in the rim, pale brown the inner part, usually containing crystals dissolving in K, brown pigment K+ red; epithecium reddish brown, usually containing crystals dissolving in K, pigment K+ red-purple; hymenium colourless; hypothecium dark brown, K−. Asci 2-spored. Ascospores 45–68 × 18–28 μm, strongly muriform, soon becoming dark greenish brown. Medulla Pd+ yellow or −, K+ yellow or K− with either stictic and ± norstictic acid, or no lichen products detected).

On damp siliceous rocks, especially near streams and on lake shores. N. & N.W. British Is. C. & N. Europe, North America.

Characterized by the 2-spored asci, the large dark ascospores and red-brown, K+ red-purple epithecium.

13. R. geographicum (L.) DC. (1805)

R. riparium Räsänen (1942) ?*R. frigidum* Räsänen (1944)

Thallus to 15 cm diam., areolate; hypothallus usually well developed, black; areoles 0.2–1.8(-2.5) mm wide, bright yellow-green or more rarely greenish, greyish, or orange-yellow, matt or shiny, usually contiguous, ± angular, flat to convex or rarely slightly concave, usually smooth; medulla I+ blue. Apothecia to 1.5 mm diam., black, not pruinose, round or angular, flat to slightly convex; true exciple thick to indistinct, dark brown in the rim, inner part paler brown, usually K+ red; epithecium red-brown (K+ purplish red) or brown to olive-green, (K− or ± green colour intensifying); hymenium colourless to faintly green or faintly brown; hypothecium dark brown, K−; no crystals or granules in apothecium. Ascospores (20-)22–40(-46) × 10–19(-22) μm, muriform with 6–20(-24) cells in optical section, dark greenish brown-black. Medulla Pd− or ± orange-yellow, K−, C− or ± red (rhizocarpic and either psoromic or barbatic acids, sometimes also ± gyrophoric acid).

On siliceous rocks, walls, acid roofing tiles and monuments in exposed sunny situations, from seashores to mountain summits; common and conspicuous in upland areas. Throughout British Is., rare in S.E. England. ?Cosmopolitan, Europe, North America, Macaronesia, N. Africa, India, Japan, Australia, New Zealand, Antarctic.

An extremely polymorphic species: *R. riparium* was separated by a brown to olive-green (K− or K+ more intensely green), not reddish-brown (K+ purplish red), epithecium, but this character is often difficult to assess unequivocally; the epithecial pigments may be faint or both reddish and greenish pigments in variable amounts can occur in some or all apothecia. The barbatic acid chemotype could prove to be worthy of formal taxonomic status. Various subspecies have been recognized but intergradations occur amongst these so that specimens cannot often be assigned with certainty to a particular one. Morphs referrable to subsp. *prospectans* (Räsänen) D. Hawksw. & Sowter (1969), with roundish apothecia and strongly fissured areoles, and subsp. *geographicum*, with angular apothecia and entire areoles, are predominant at middle elevations and on rocky seashores. At higher elevations, these grade into a highly variable complex which includes morphs resembling subsp. *diabasicum* (Räsänen) Poelt & Vězda (1980), with thick areoles and relatively many-septate ascospores (12–20 cells in optical section), and subsp. *frigidum* (Räsänen) Hertel (1976), with vivid yellow areoles and few-septate (5- to 11-celled) ascospores. A single specimen referrable to subsp. *arcticum* (Runemark) Hertel (1976) is known from Scotland (Ben Nevis); this has roundish apothecia and matt, pale yellow, convex areoles dispersed on a black to greyish prothallus. Population studies are required to assess the detailed nature and possible taxonomic significance of infra-specific variation in this complex.

14. R. hochstetteri (Körber) Vainio (1922)

Thallus to 5 cm diam., rimose to areolate; prothallus well developed, sometimes fimbriate, occasionally usually poorly developed and inconspicuous, black; areoles to 0.5(-0.7) mm diam., dark brown, usually with a greyish tinge, matt, mainly contiguous, angular, flat to weakly convex, often rather thin and even. Apothecia to 0.7(-1.5) mm diam., black, not pruinose, mainly round or sometimes slightly flexuose, remaining flat, true exciple narrow, persistent, brownish black throughout or paler brown in inner part, lacking crystals, K−; epithecium dark red-brown, olive brown to greenish black, rarely containing crystals dissolving in K, K− or K+ purplish red; paraphyses with a sharply delimited, dark brown pigment cap; hymenium colourless or green (especially in upper part); hypothecium dark brown, K−. Ascospores 25–36 × 12–15 μm, 1-septate, persistently colourless, only becoming faintly brown with age. Medulla Pd− and K−, or rarely Pd+ yellow and K+ yellow (no lichen products or rarely stictic acid).

On siliceous rocks and boulders, especially in moist or periodically irrigated sites, not uncommon, widespread in upland sites, sometimes on rock outcrops in woodland. Mainly W. & N. British Is. C. & N. Europe, Azores, North America.

The few specimens with stictic acid have thicker areoles, a taller hymenium, and somewhat larger ascospores than the acid-deficient chemotype. Distinguished by the dark thallus with black, flat apothecia, colourless, 1-septate ascospores and conspicuously dark-capitate paraphyses. Continental specimens are recorded as having ascospores 10–25 × 6–11 μm. The discrepancy in the reported ascospore dimensions of British and Continental populations of this species requires investigation.

15. R. inarense (Vainio) Vainio (1898)

Thallus to 5 cm diam., areolate; prothallus well developed, black, sometimes pruinose; areoles to 1 mm diam., bright yellow-green, matt or sometimes shiny, ± contiguous, rather thick, orbicular to angular, usually moderately convex. Apothecia to 1.2 mm diam., black, not pruinose, ± orbicular, weakly to strongly convex; true exciple persistent, dark brown in the rim, pale brown to colourless in the inner part, containing crystals, pigment K− (intensifying green), crystals K+ red (forming acicular crystals); epithecium olivaceous green, containing crystals, pigment K− (intensifying green), crystals K+ red (forming acicular crystals); hymenium colourless; hypothecium dark brown, K−. Ascospores 21–30 × 10–12 μm, 1-septate, dark brown. Medulla Pd+ yellow, K+ red (norstictic acid).

On acid rocks, montane; very rare. N. Scotland (Highlands; Sutherland, Ben Hope). Fennoscandia, Siberia, Greenland, Canada,

See also *R. alpicola* **(2)** and *R. eupetraeoides* **(9)**.

16. R. intermediellum Räsänen (1943)

Thallus to 2 cm diam., areolate, growing in the thallus of various crustose lichens when young, later becoming ± independent, often becoming ± moribund in central areas; prothallus ± indistinct, black; areoles 0.1–0.5(-0.7) mm diam., bright yellow, matt, contiguous or partly scattered, mainly angular, flat to moderately convex, smooth or minutely scabrid; medulla I+ blue. Apothecia to 0.5 mm diam., black, not pruinose, orbicular to angular, flat; true exciple rather thin, persistent, brownish black in the rim, paler in the inner part, K− or K+ faintly red; epithecium brown, K+ red; hymenium colourless to pale brown; hypothecium dark brown, K−; no crystals or granules in the apothecium. Ascospores 12–21 × 6–10 μm, (1-)3(-4)-septate to submuriform, dark greenish brown. Medulla Pd+ orange-yellow, K− (psoromic acid) or Pd−, K− (rhizocarpic acid only).

Parasitic, at least when young, on thalli of *Aspicilia*, *Buellia*, *Fuscidea*, *Lecidea*, *Polysporina*, *Rhizocarpon*, and *Tremolecia*, on exposed siliceous rocks; very rare. N. Wales (Gwynedd, Glyder Fach), N. Scotland (Glen Coe, near Loch Achtriochtan). N. Europe, Alps, circumpolar arctic-alpine.

The relationship to the central European *R. furax* Poelt & V. Wirth (1970) requires further study.

Distinguished by the I+ medulla, septate to sub-muriform and dark ascospores and the semi-parasitic habit when young.

17. R. jemtlandicum (Malme) Malme (1914)

Thallus to 10 cm diam., areolate; prothallus well developed, black; areoles to 1 mm diam., dark brown, usually with a faintly greyish tinge, matt, contiguous, irregularly rounded or angular, flat to weakly convex. Apothecia to 1.5 mm diam., black, not pruinose, orbicular, remaining ± flat; true exciple thick, persistent, brownish black in the rim, inner part paler brown, K−, containing crystals dissolving in K; epithecium olivaceous green, intensifying in K, containing crystals dissolving in K; hymenium colourless; hypothecium brownish black, K−. Ascospores 25–30 × 13–15 μm, 1-septate, soon becoming dark green or brown. Medulla Pd+ yellow, K+ yellow (stictic acid).

On siliceous rocks at *c.* 1000 m alt.; rare and very local. Scotland (Cairngorm; S. Aberdeen, Ben Avon). Fennoscandia, North America.

Identified by the dark, 1-septate ascospores, olive-brown epithecium, K− and containing crystals as well as the I− medulla. *R. badioatrum* **(3)**, with which it may occur, has a brown, K+ purplish, epithecium lacking crystal inclusions. Some records of *R. badioatrum* and *R. hochstetteri* **(14)** probably belong here. The circumpolar, arctic species, ***R. copelandii*** (Körber) Th. Fr. is similar with an olive-green, K− epithecium, soon darkening, 1-septate ascospores, and the medulla is I−. It differs in being Pd+ orange, K+ red (crystals), KC+ red, C− (norstictic acid) and has recently been recorded from Westerness (Ben Nevis, Knoydart).

18. R. lavatum (Fr.) Hazslin (1884)

Thallus to 10 cm diam., rimose; prothallus indistinct; areoles to 0.5(-1) mm diam., grey to pale ochraceous, matt, angular, flat. Apothecia to 1.5 mm diam., black, not pruinose, round, weakly concave to weakly convex; disc sometimes umbonate; true exciple thick, persistent, dark brown in the rim, inner part paler brown, K−; epithecium olivaceous green to olivaceous brown, K−; hymenium colourless; hypothecium reddish brown, K−; no crystals or granules in the apothecium. Ascospores 30–40 × 14–18 μm, muriform, persistently colourless or becoming faintly brown with age. Lichen products not detected by t.l.c.

On siliceous rocks subject to inundation by unpolluted streams and lakes, upland; local. N. & W. British Is. Norway, Sweden, North America.

R. lavatum is exclusively a lake- and stream-side species and is generally subject to temporary inundation; it has a rather obscurely thickened, tumid exciple and a ± pronounced ochraceous tinge to the thallus. *R. obscuratum* **(20)** has a pale to dark grey thallus and a narrower, more elevated exciple; the disc occasionally becoming rust-red.

19. R. lecanorinum Anders (1923)

Thallus to 4 cm diam., or thalli merging to form extensive cover, areolate; prothallus ± distinct, black; areoles to 1.2 mm diam., bright yellow to greenish yellow, matt, usually contiguous, mostly crescent-shaped, each segment partly to entirely surrounding an apothecium, or partly orbicular with or without ± central apothecium, rarely angular, flat to moderately convex, smooth; medulla I+ blue. Apothecia to 1 mm diam., black, not pruinose, ± round, flat to concave; true exciple persistent but usually rather indistinct, brownish black in the rim, the inner part colourless, K− or K+ faintly red; epithecium pale olivaceous brown, K− or intensifying green; hymenium colourless to faintly green; hypothecium thin, brownish black K−; no crystals or granules in the apothecium. Ascospores 27–45 × 11–19 μm, muriform, with 14–28 cells in optical section, dark brown. Medulla Pd+ reddish orange, K+ yellow (stictic, often in low concentration, and rhizocarpic acids).

On siliceous rocks, slate roofs and walls; local. N. England, Wales, Scotland. Fennoscandia, montane C. Europe, pan-temperate in N. Hemisphere.

The cresent-shaped areoles are diagnostic and differ from other similar species, e.g. *R. geographicum*, with angular areoles.

R. ferax Magnusson (1948) [syn. ?*R. drepanodes* Feuerer (1978)] recorded from Ireland, differs in the more abundantly septate ascospores and by the occurrence of psoromic, not stictic, acid as the main medullary substance. The material on which the Irish record is based has not been located.

20. R. obscuratum (Ach.) Massal. (1852)

Thallus to 10 cm diam., rimose-cracked to areolate; prothallus usually distinct, black, sometimes fimbriate; areoles to 0.5(-1) mm diam., grey, dark grey to dark brown, matt, angular to ± orbicular, flat to weakly convex. Apothecia to 1 mm diam., black, orbicular, weakly concave to weakly convex; disc often ± umbonate roughened, naked or rarely ± ochraceous pruinose; true exciple thick, persistent, brownish black in the rim, inner part sometimes pale brown or ± colourless, K−; epithecium olivaceous green or olivaceous brown, K−; hymenium colourless; hypothecium reddish brown, K−; no crystals or granules in apothecium. Ascospores 20–32 × 9–15 μm, muriform, persistently colourless or becoming faintly green-black or brown with age. Lichen products not detected by t.l.c.

On smooth siliceous rocks, stable shingle, flints and quartz, also monuments; very common locally. Throughout British Is. Europe, North America, Australia.

Some specimens may be difficult to separate from *R. lavatum* **(18)**. *R. roridulum* (Th. Fr.) H. Olivier (1881), characterized by a reddish brown epithecium and usually white-pruinose apothecial disc, appears to have been erroneously recorded from the British Is.; most material is referable to *R. obscuratum*. One specimen, from a seepage track on a rock outcrop in Scotland (Angus, Caenlochan) is possibly referable to *R. amphibium* (Fr.) Th. Fr. (1874), which has a reddish brown epithecium and non-pruinose apothecial disc. The polymorphic *R. obscuratum* group is in need of a critical revision.

21. R. oederi (Weber) Körber (1861)

Thallus to 5 cm diam., often forming extensive patches, rimose-cracked to areolate; prothallus indistinct; areoles to 0.2(-0.5) mm diam., ochraceous to orange-brown, matt, usually angular, flat to ± convex. Apothecia to 0.5 mm diam., not pruinose, angular to flexuous, persistently flat; disc usually umbonate or sub-gyrose, uneven; true exciple distinct, brownish black, K−; epithecium olive or greenish black, K−; hymenium colourless to faintly green in upper part; hypothecium dark brown, K−; no crystals or granules in the apothecium. Ascospores 12–18 × 3–7 μm, mainly 3-septate, persistently colourless. Lichen products not detected by t.l.c.

On siliceous rocks rich in iron minerals; locally frequent. Mainly N. & W. British Is. C. & N. Europe, North America.

Characterized by the rust-red thallus, gyrose-umbonate apothecia, and 3-septate, colourless ascospores. *Tremolecia atrata* has a similar, areolate thallus but the apothecia are smooth, shining with a concave disc and the ascospores are simple.

22. R. plicatile (Leighton) A. L. Sm. (1911)

R. rubescens Th. Fr. (1874)

Thallus to 5 cm diam., in small patches, areolate; prothallus rather indistinct, black; areoles to 0.5 mm diam., grey to pale brown, matt, ± contiguous, orbicular to angular, moderately to strongly convex or becoming warted. Apothecia to 0.8 mm diam., black, not pruinose, orbicular, persistently flat to slightly convex; true exciple persistent, brownish black in the rim, inner part paler brown, containing crystals dissolving in K, pigment K−; epithecium brown, K−; hymenium colourless, containing crystals dissolving in K; hypothecium dark brown, K−. Ascospores 24–32 × 10–14 μm, strongly muriform, remaining colourless. Medulla Pd+ yellow, K+ red (norstictic acid).

On hard acid rocks and pebbles, montane; rare. N.W. England (Cumbria), N. Wales (Merioneth, Dolgellau), N. Scotland (Perth). C. Europe, Norway, Sweden, North America.

23. R. polycarpum (Hepp) Th. Fr. (1874)

Thallus to 5 cm diam., areolate; prothallus well developed, black; areoles to 0.5 mm diam., dark brown, usually with a faint greyish or pinkish tinge, matt, contiguous or scattered, orbicular to angular, flat to weakly convex; medulla I+ blue. Apothecia to 0.7 mm diam., black, not pruinose, orbicular or angular, sometimes slightly flexuous, usually remaining flat; true exciple narrow, persistent, brownish black, K−; epithecium dark brown, K− or K± violet-red; hymenium colourless; hypothecium dark brown K−; no crystals or granules in the apothecium. Ascospores 17–30 × 8–13 μm, mainly 1-septate, persistently colourless or becoming pale brown with age. Medulla Pd+ yellow and K+ yellow, or Pd− and K− (stictic acid or, more frequently, lichen products absent).

On hard siliceous rocks, coastal and montane; local. N. & W. British Is. C. Europe, Fennoscandia, North America, Australia.

Differs from other species with 1-septate ascospores in the combination of I+ blue medulla, dark brown, K± reddish (never greenish) epithecium and colourless ascospores; in *R. badioatrum* **(3)** the medulla is I−.

24. R. postumum (Nyl.) Arnold (1870)

Thallus cracked to areolate, often very thin and sometimes partly immersed; prothallus indistinct; areoles to 0.2 mm diam., often scattered, minute, pale brown, sometimes with a grey tinge, matt, not pruinose, ± angular, smooth. Apothecia to 0.4 mm diam., scattered or in groups, brownish black to black, not pruinose, ± shining, orbicular; true exciple persistent, ± irregular, roughened or crenulate, sometimes gnarled, olivaceous brown or greenish black, K−; epithecium brown, containing crystals dissolving in K, K−; hymenium colourless; hypothecium red-brown to brown or partly olive-brown, K−. Ascospores 15–21 × 8–11(-13) μm, 3-septate, rarely submuriform, persistently colourless.

On ± basic rocks, montane; very rare. N. Scotland (Perth, Ben Lawers). Scandinavia, ?Canada.

Distinguished by the thin, often scattered areoles, the absence of any chemical reactions and the 3-septate or submuriform, colourless ascospores. The single British gathering has persistently 3-septate ascospores.

25. R. richardii (Nyl.) Zahlbr. (1926)

R. constrictum Malme (1919)

Thallus to 8 cm diam., areolate, often in small patches which sometimes coalesce; prothallus well developed, black; areoles to 0.5 mm diam., brown, usually with a faintly grey-violet tinge, matt, contiguous, angular, ± flat; medulla I+ blue. Apothecia to 1 mm diam., black, not pruinose, ± orbicular, remaining flat or becoming weakly convex; true exciple indistinct but usually persistent, brownish black in the rim, pale brown to colourless in inner part, containing crystals dissolving in K, K−; epithecium dark olivaceous green to greyish black, containing crystals dissolving in K, K− or intensifying green; hymenium colourless; hypothecium dark reddish brown, K−. Ascospores 22–27 × 11–14 μm, mainly 1-septate, but secondary pseudo-septa may occur in one or both cells of the primary division, persistently colourless, only becoming faintly grey-brown with age. Medulla Pd− or ± faintly yellow, K− or ± yellow, C− or ± red (gyrophoric, ± stictic and norstictic acids).

On maritime, siliceous rocks and pebbles, xeric-supralittoral or stable shingle; locally abundant. Throughout British Is. France, Norway, Sweden.

This species is distinguished by a combination of the thallus colour, an I+ blue medulla, K− epithecium, ± colourless, 1-septate ascospores (sometimes with secondary pseudosepta), and habitat.

26. R. simillimum (Anzi) Lettau (1912)

Thallus to 5 cm diam., areolate; prothallus well developed, black; areoles to 0.3 mm diam., dark brown, usually with a faintly greyish tinge, occasionally dark grey, matt, contiguous or scattered, orbicular to angular, flat; medulla I+ blue. Apothecia to 0.8 mm diam., black, not pruinose, orbicular or angular, sometimes slightly flexuose, remaining ± flat; true exciple narrow and persistent, brownish black, K± diffusing purplish red; epithecium brownish black, K− or K+ faintly red; hymenium colourless or faintly brown; hypothecium dark brown, K± diffusing purplish; without crystals or granules in the apothecium. Ascospores 12–16 × 6–8 μm, 1-septate, soon dark black-grey-brown. Lichen products not detected by t.l.c.

On sunny, acid, ± vertical outcrops; very rare. N.W. England (Westmorland, Langdale Pikes). Norway.

The single British record agrees well with the type material; in squash preparations in K the hypothecium ± exudes a reddish or magenta-purple colour; other diagnostic features are the small, dark ascospores and the I+ blue medulla. The anastomosing paraphysoids separate this species from *Buellia*.

27. R. subgeminatum Eitner (1911)

Thallus to 10 cm diam., areolate; prothallus well developed, black; areoles to 1 mm diam., dark brown, matt, scattered to contiguous, ± orbicular and flat. Apothecia to 1 mm diam., black, not pruinose, ± orbicular, flat to weakly convex; true exciple persistent, brownish black in the rim, pale brown to colourless in the inner part; epithecium olivaceous green, K− (intensifying green). Asci 1–2(-4)-spored. Ascospores 30–50 × 20–28 μm, strongly muriform, persistently colourless or becoming faintly green with age. Lichen products not detected by t.l.c.

On ± nutrient-rich, siliceous rocks and boulders ; rare. S.W. England (Devon, Dartmoor), N.W. England (Cumbria, Westmorland, Lancashire), C. Scotland (Perth).

R. geminatum **(12)** has ascospores which soon become dark brown and a K+ violet-purple epithecium.

28. R. superficiale (Schaerer) Malme (1914)

Thallus to 10 cm diam., prothallus well developed, black; areoles to 1.5(-2.5) mm diam., greenish yellow to bright yellow-green, matt, scattered to contiguous, angular to orbicular, entire to sparsely fissured, weakly concave to strongly convex, smooth or scabrid. Apothecia to 1(-1.4) mm diam., black, not pruinose, round to ± angular, flat to weakly convex; true exciple usually persistent, brownish black (K+ red) in the rim, inner part pale brown and containing crystals dissolving in K (K+ yellow or K+ red acicular crystals); epithecium brownish black (K+ faintly red) or partly to entirely greenish brown (K−, intensifying green); hymenium colourless; hypothecium dark brown K− ; no crystals or granules in the hymenium. Ascospores 11–18(-23) × (5-)6–8(-10) μm, 1-septate, dark brown. Medulla and exciple Pd+ orange, K+ yellow-red (norstictic and rhizocarpic acids).

On exposed siliceous rocks, montane; locally frequent. N. Wales (Gwynedd, Glyderrau, Yr Ole Wen), Scotland (Ben Nevis, Ben Hope). N. Europe, Alps, arctic-alpine, bipolar, extending to tropical high mountains.

The dark rim of the true exciple is often partly eroded exposing a paler inner part imparting a grey-white tone to the apothecial edge. This feature and the smaller ascospores distinguish the species from *R. eupetraeoides* **(9)** and *R. inarense* **(15)** which also are K+ red (crystals).

29. R. umbilicatum (Ramond) Flagey (1894)

Thallus to 5 cm diam., continuous or partly rimose, white, matt, pruinose; prothallus usually poorly developed, black, white-pruinose. Apothecia to 1.5 mm diam., black, mainly orbicular, flat or becoming moderately convex; disc not pruinose; true exciple thick, usually persistent, distinctly pruinose, brownish or greenish black in the rim, inner part pale brown to colourless, with crystals partly dissolving in K, K+ yellow; epithecium olive-brown, with crystals dissolving in K (crystals K+ yellow); hymenium colourless; hypothecium dark brown, K−. Ascospores 18–28 × 10–16 μm, muriform, persistently colourless. Medulla Pd+ yellow, K+ yellow (stictic acid).

On calcareous rocks, especially hard limestones; local. Mainly C. & N. British Is., widespread in Ireland. Europe, North America.

See also *R. concentricum* (**7**).

30. R. viridiatrum (Wulfen) Körber (1855)

Thallus to 2 cm diam., areolate, growing on the thallus of various crustose lichens, sometimes vanishing in the central part; prothallus indistinct or forming a narrow marginal band; areoles to 1(-1.2) mm diam., greenish yellow, matt, contiguous, flat to strongly convex, smooth, ± round to angular with irregularly crenulate margins. Apothecia to 1 mm diam., black, not pruinose, ± round, flat to strongly convex, often slightly elevated above areoles; true exciple usually becoming excluded, brownish black in the rim, paler in the inner part, K+ faintly red; epithecium brownish black, K+ purple-red; hymenium colourless to pale brown; hypothecium dark brown to grey-black, K−; no crystals or granules in apothecium. Ascospores 12–24 × 7–13 μm, muriform, with 5–9 cells visible in optical section, dark brown. Medulla Pd−, K− (rhizocarpic and unidentified fatty acids).

On acidic to moderately basic basalts, initially parasitic on *Aspicilia caesiocinerea* and perhaps other lichens (?*Lecidea* species); very local. S.W. England (Cornwall, Lizard Peninsula; Shropshire), W. & N. England, upland Wales, Scotland, Ireland. Pan-temperate montane, N. & S. Hemispheres, also at sub-alpine elevations in the tropics.

This species often occurs as scattered patches or areoles amongst other lichens.

RIMULARIA Nyl. (1868)

Thallus crustose, superficial, thin, continuous, subareolate to distinctly areolate, reddish brown to dark olivaceous. **Photobiont** chlorococcoid. **Ascomata** apothecia, immersed to sessile, flat to convex, black; disc umbonate or gryose. **Thalline exciple** absent. **True exciple** persistent, often contorted to gyrose, black. **Epithecium** brown. **Hymenium** I+ blue. **Hypothecium** dark brown to black, continuous with the true exciple, K−. **Hamathecium** of paraphysis-like hyphae, thin (< 1 μm wide), richly branched and anastomosed, septate; apices pigmented but not swollen. **Asci** 8-spored, cylindrical-clavate, *Rimularia*-type (Fig. 33i). **Ascospores** ellipsoid, non-septate, smooth- and thin-walled, initially colourless but becoming brown before dehiscence; perispore absent. **Conidiomata** pycnidial. **Conidia** bacilliform, simple, colourless. **Chemistry**: norstictic acid in one species. **Ecology**: on upland siliceous rocks.

Distinguished by the richly branched and anastomosed paraphyses, the thick-walled asci with a distinctive iodine reaction at the apex, and the large, finally brown ascospores, as well as the umbonate to gyrose apothecia which lack a thalline exciple. Species placed in *Mosigia* are possibly congeneric. *Lecidea insularis*, *L. furvella*, *L. gyrizans* and *L. mullensis* have been referred here, but none of these species have the thin paraphyses characteristic of *R. limborina*; they share bead-like ('monilioid') branched paraphyses. Distinguished from brown-spored *Fuscidea* species by the richly branched, and anastomosed paraphyses, and ascus structure. The iodine reaction of the asci also separates *Rimularia* from *Porpidia*.

LITERATURE: Hertel & Rambold (1990), Muhr & Tønsberg (1989).

1	Soredia present; on bark	**1. *fuscosora***
	Soredia absent; on rocks	**2. *limborina***

1. R. fuscosora Muhr & Tønsberg (1989)

Thallus effuse, thin, sorediate, immersed or superficial, when forming an uneven, continuous or areolate crust. Soralia dark brown, discrete, punctiform, to 0.6 mm diam., flat to convex. Apothecia to 0.4(-1.0) mm diam., sparse, sessile, black; true exciple persistent, sometimes flexuous, dark brown, occasionally developing a central column in the disc; disc flat to convex roughened or cracked; epithecium brown, rarely with a greenish tinge; hymenium 50–70 μm tall, colourless, subhymenium colourless or pale brown. Asci 40–65 × 20–30 μm, clavate. Ascospores (9.5-)11–16(-20) × (5-)7–11 μm, simple, ellipsoid, colourless. Pycnidia 0.04–0.08 mm wide, sessile, subglobose, black. Conidia 5–7 × 1 μm, bacilliform. Thallus and soralia Pd− or ± faintly yellow, K− or K+ yellow to red (norstictic acid).

On *Salix* bark (also *Alnus* and *Betula* in Scandinavia) in moist situations; very rare. W. Scotland (Argyll). Norway, Sweden.

2. R. limborina Nyl. (1868)

Lecidea limborina (Nyl.) Lamy (1878), *L. subgyratula* Nyl. (1873)

Thallus thin, contiguous or verrucose-areolate, sometimes nearly absent, ash to dark grey or chestnut to dark brown to black, matt; prothallus dark, sometimes visible between the areoles; medulla I−. Photobiont cells 5–12 μm diam., or some to 14 × 10 μm. Apothecia (0.3-)0.5–0.8 mm, dispersed or densely distributed, mostly sessile; hymenium 90–150 μm tall. Ascospores (14-)18–30(-40) × (5-)10–18(-21) μm, ellipsoid to almost subglobose. Pycnidia 80–100 μm diam., immersed to ½-emergent, black; wall dark brown, K−; conidia 3.5–5 × 0.5–0.1 μm.

On siliceous, especially basaltic, rocks, in uplands. N.E. England, N. Wales, N. Scotland (Aberdeen, Grampians mtns., Is. of Skye), S.W. Ireland. C. & N. Europe, North America.

RINODINA (Ach.) Gray (1821)

Thallus crustose, thick, thin or disappearing, pale to dark grey, ochraceous or brown, continuous, cracked-areolate, granular, rarely isidiate; prothallus absent, limiting, or entire and dark. **Photobiont** trebouxioid. **Ascomata** apothecia, immersed or sessile, usually frequent, contiguous or not. **Thalline exciple** concolorous with the disc or thallus, entire or crenulate, persistent or excluded at maturity, with or without a distinct cortex; cortex when present cellular, occasionally pigmented, sometimes columnar below or structure occluded with pigment, peripheral hyphae usually pigmented. **True exciple** colourless or rarely brown, prosoplectenchymatous. **Disc** brown to black, flat or convex at maturity, rarely pruinose. **Epithecium** brown or rarely blue-black (*R. colobina* **(6)**). **Hymenium** colourless, I+ blue. **Hypothecium** colourless, rarely brownish or dark brown, stipe not usually present, colourless or dark. **Hamathecium** of paraphyses, mostly unbranched except near apices, apices expanded, brown-capped. **Asci** 8-spored, clavate, *Lecanora*-type. **Ascospores** 1-(3-) septate, mainly double-walled, walls thickened, septum well developed at maturity, structure often complex, grey-green or pale to dark brown; surface smooth or ornamented (×1000). **Conidiomata** pycnidia, rarely present. **Conidia** bacilliform or in one species (*R. lecideina* **(16)**) thread-like. **Chemistry**: depsides, ± depsidones present in some species, but more chemical investigation needed. **Ecology**: on a wide variety of substrata.

R. arnoldii Mayrh. & Poelt (1979) has been incorrectly reported from the British Is., the material being referable to *R. interpolata* **(14)**. Special attention needs to be paid to the ascospore types (Fig. 25) in species identification; these are not always easy to distinguish and vary somewhat during maturation.

LITERATURE: Coppins (1989c), Coppins & James (1979b), Laundon (1986), Mayrhofer (1984), Mayrhofer & Poelt (1979), Sheard (1967, 1973).

1 On soil, mosses, or decaying plant debris 2
On rocks, wood, bark or on mosses growing on wood or trees 4

2(1) Ascospores with 2 locules; thallus K+ yellow or K− 3
Ascospores with >2 and typically 4 locules; thallus K− ..... **8. *conradii***

3(2) Ascospores >25 μm; thallus K− ..... **19. *mniaraea***
Ascospores <25 μm; thallus K+ yellow ..... **27. *subglaucescens***

4(1) On wood, bark or mosses growing on wood or trees 5
On rocks 16

5(4) Ascospores with predominantly two lumina 6
Ascospores with predominantly four lumina ..... **8. *conradii***

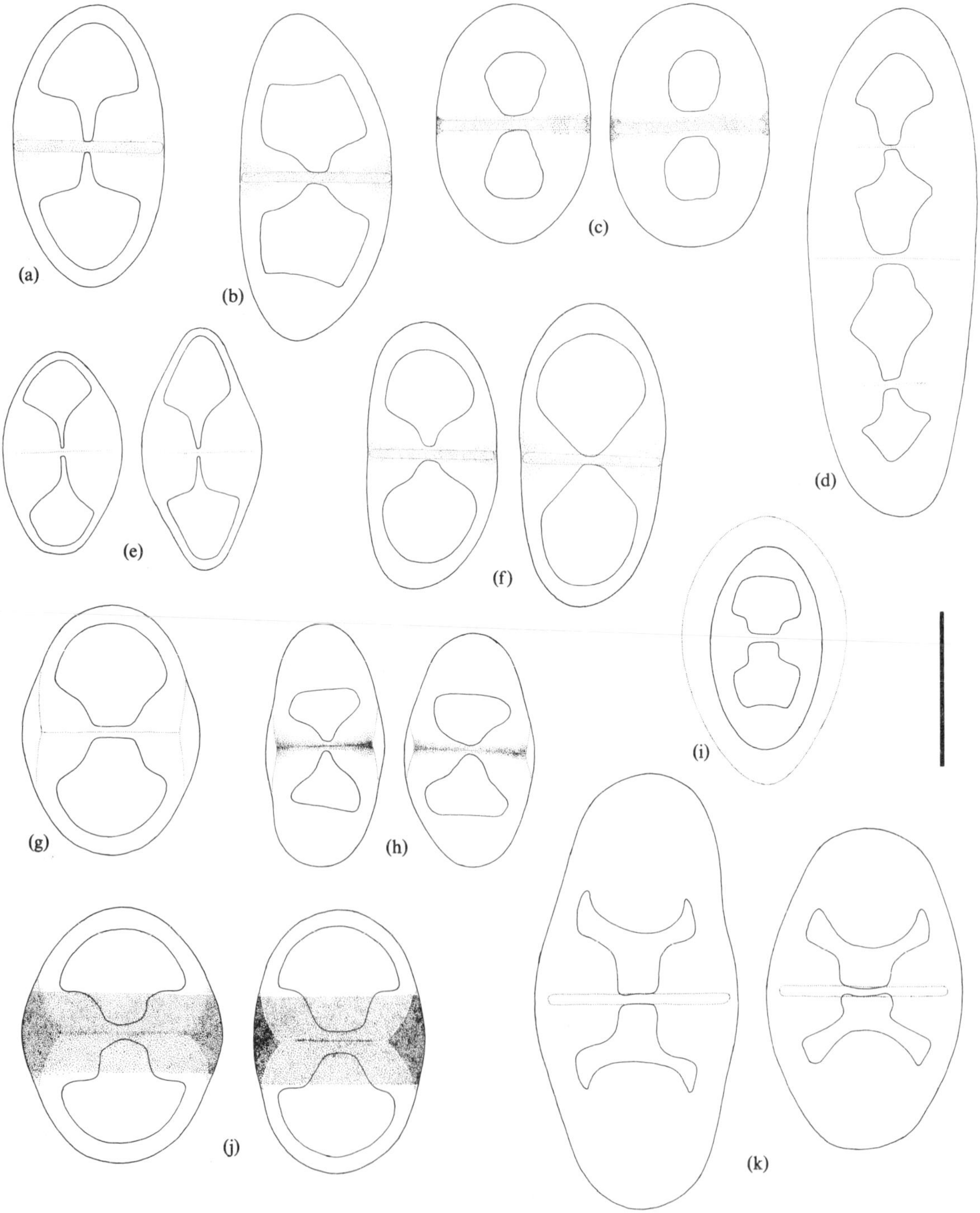

Fig. 25. *Rinodina* ascospores in 10% K. (a) *R. orculariopsis* (Physconia-type); (b) *R. luridescens* (Physcia-type); (c) *R. atrocinerea* (Pachysporaria-type); (d) *R. conradii* (Conradii-type); (e) *R. biloculata* (Orcularia-type); (f) *R. sophodes* (Milvina-type); (g) *R. dubyana* (Dubyana-type); (h) *R. gennarii* (Dirinaria-type) (i) *R. colobina* (Tunicata-type); (j) *R. bischoffii* (Bischoffii-type); (k) *R. oxydata* (Mischoblastia-type). Scale = 10 μm.

6(5) Exciple concolorous with the thallus; hypothecium colourless or pale brown 7
Exciple concolorous with the disc; hypothecium dark brown .... **3. *biloculata***

7(6) Epithecium red-brown or brown, K− and N− 8
Epithecium blue-black, K+ and N+ violet .... **6. *colobina***

8(7) Soredia present; isidia absent 9
Soredia absent; isidia present or absent 10

9(8) Soredia whitish or pale greenish yellow; Pd+ red .... **9. *efflorescens***
Soredia blue-grey; Pd− or faint yellow .... **12. *griseosoralifera***

10(8) Thallus not isidiate 11
Thallus composed entirely of coralloid or finger-like isidia .... **15. *isidioides***

11(10) Ascospores mostly straight, walls unequally thickened when mature 12
Ascospores slightly curved, walls remaining unthickened when mature .... **24. *pyrina***

12(11) Ascospores not swollen around septum in K 13
Ascospores swollen around septum in K .... **11. *gennarii***

13(12) Ascospores *Mischoblastia*- to *Pachysporaria*-type 14
Ascospores *Physcia*- or *Milvina*-type 15

14(13) Thallus very dark, K−; ascospores 22–32 × 11–19 μm .... **28. *teichophila***
Thallus pale to green-grey, K+ yellow; ascospores 14–22 × 8–12 μm .... **25. *roboris***

15(13) Thallus sordid whitish to pale grey, K+ yellow; apothecia sessile; thalline exciple without a distinct cortex, I−; ascospores *Physcia*-type .... **10. *exigua***
Thallus grey-green to olive brownish, K−; apothecia ± immersed; thalline exciple with a distinct cortex, I+ faint blue; ascospores *Milvina*-type .... **26. *sophodes***

16(4) Exciple concolorous with the disc 17
Exciple concolorous with the thallus 20

17(16) Ascospore not *Mischoblastia*-type, mainly less than 20 μm in length 18
Ascospores 19–25 × 9–14 μm, *Mischoblastia*-type .... **22. *oxydata***

18(17) Thallus or medulla C− 19
Thallus or medulla C+ orange-red .... **17. *luridescens***

19(18) Thallus K+ yellow (atranorin); ascospores with apical thickening; conidia bacilliform .... **20. *occulta***
Thallus K−; ascospores without apical thickenings; conidia thread-like .... **16. *lecideina***

20(16) On various crustose lichens, lichenicolous 21
Growing independently, not lichenicolous 22

21(20) Ascospores *Physcia*-type .... **23. *parasitica***
Ascospores *Milvina*-type .... **18. *milvina***

22(20) Thallus C+ orange-red 23
Thallus C− 25

23(22) Thallus thin, areolate, K+ yellow, pale to grey-brown; prothallus ± distinct, black 24
Thallus thick, rimose-cracked, K−, dark brown-grey-mauve; prothallus often indistinct .... **17. *luridescens***

24(23) Soredia present; apothecia rare .... **1. *aspersa***
Soredia absent; apothecia usually present .... **2. *atrocinerea***

25(22) Thallus K+ yellow 26
Thallus K− 29

26(25) Ascospores *Physcia*-type or close to *Physcia*-type 27
Ascospores *Pachysporaria*- or *Mischoblastia*-type 28

27(26) Ascospores 17–27 × 8–13 μm; thalline exciple corticate .... **7. *confragosa***
Ascospores 11–16 × 5–9 μm; thalline exciple indistinct .... **20. *occulta***

28(26) Ascospores of *Pachysporaria*-type .... **27. *subglaucescens***
Ascospores of *Mischoblastia*-type .... **22. *oxydata***

29(25) Ascospores of *Bischoffii*- or *Tunicata*-type; mainly on calcareous rocks 30
Ascospores of other types; mainly on siliceous rocks 32

30(29) Ascospores *Bischoffii*-type 31
Ascospores *Tunicata*-type .... **5. *calcarea***

31(30) Apothecia immersed in small depressions of the substratum; hymenium not interspersed with oil globules .... **13. *immersa***
Apothecia superficial; hymenium interspersed with oil globules .... **4. *bischoffii***

32(29) Ascospores not swollen around septum in K 33
Ascospores becoming swollen around septum in K .... **11. *gennarii***

33(32) Ascospores of other types, less than 20 μm in length 34
Ascospores intermediate between *Mischoblastia*- and *Pachysporaria*-type, 20–32 × 11–19 μm .... **28. *teichophila***

34(33) Ascospores *Physconia*-type 35
Ascospores *Milvina*-type .... **18. *milvina***

35(34) Thallus areolate, sometimes completely covered with apothecia; hymenium 90–110 μm tall; ascospores 17–23 × 8–12 μm, only strongly thickened near the septum, torus distinct, surface minutely roughened .... **21. *orculariopsis***
Thallus thin, rimose or rimose-areolate; hymenium 60–80 μm tall; ascospores 13–18 × 6–9 μm, not strongly thickened near the septum, torus diffuse, surface clearly roughened .... **14. *interpolata***

1. R. aspersa (Borrer) Laundon (1986)

Buellia aspersa (Borrer) P. James (1965)

Thallus poorly developed, small, scattered or ± contiguous, areoles greyish; prothallus usually conspicuous, sometimes fimbriate, black. Soralia *c.* 0.5 mm diam., greenish white (becoming yellowish white when dry), delimited, punctiform, flat, superficial somewhat elevated; soredia farinose. Apothecia unknown in British Is., to 1 mm diam.; thalline exciple well developed; hymenium 80–100 μm tall; hypothecium colourless; disc brown to black. Ascospores 15–20 × 13–20 μm, *Pachysporaria*-type (Fig. 25c). Soralia C+ orange-red (atranorin, gyrophoric acid).

On siliceous pebbles and flints, usually with *Porpidia* species, on stabilized coastal shingle beaches; local. S. England (Sussex, Hampshire, Kent, Suffolk), S. Scotland (Tayside). W. France, Germany, Sweden.

R. fatiscens (Th. Fr.) Vainio (1921), recorded from scattered localities in Europe, is based on fertile material of *R. aspersa*. Closely related to *R. atrocinerea* **(2)**.

2. R. atrocinerea (Hook.) Körber (1855)

Thallus wide-spreading, thin, rather smooth, whitish, pale grey to ochraceous, rarely brown-grey, distinctly areolate; areoles 0.45–1.25 mm wide, flat, discrete, scattered to ± contiguous; prothallus ± conspicuous, black. Apothecia 0.2–0.7(-1) mm diam., ± immersed, later becoming ± sessile; thalline exciple thin, even, scarcely prominent, concolorous with the thallus; hymenium 75–140 μm tall; hypothecium colourless; disc brown to brown-black, paler when moist, persistently flat or becoming weakly convex. Asci 50–90 × 19–23 μm. Ascospores 16–21 × 8.5–13.5 μm, with very thick, uniform walls, locules ± rounded, never angular, *Pachysporaria*-type (Fig. 25c). Pycnidia frequent; conidia 4–6 × 0.5–0.8 μm, bacilliform. Cortex Pd+ yellow, K+ yellow, C+ red (often faint and evanescent) (atranorin, chloroatranorin, gyrophoric acid).

On coarse, hard, siliceous rocks, often granite; locally abundant, especially near the coast. Predominantly in W. & S.W. British Is. Europe, Morocco.

The pale coloured, areolate thallus and contrasting black prothallus, together with the colour reactions, are diagnostic.

3. R. biloculata (Nyl.) Sheard (1967)

Thallus thin, pale grey, continuous or rarely rimose-cracked and flat, determinate or indeterminate; prothallus inconspicuous. Apothecia 0.25–0.3 mm diam., sessile, rarely contiguous; thalline exciple absent or poorly developed; true exciple brown or colourless, rarely continuous below hypothecium, *c.* 0.05 mm wide, concolorous with disc, or more rarely with thallus, entire and persistent; disc black, persistently flat; epithecium dark brown; hymenium 40–70 μm tall; hypothecium 15–60 μm high, dark brown, I+ blue. Asci 40–45 × 17–19 μm. Ascospores 11.5–19 × 6.5–9.5 μm, thin-walled, thickened only at the septum, *Orcularia*-type (Fig. 25e). Conidia 3.5–5 × 2 μm, ellipsoid. Lichen products not detected by t.l.c.

On bark, usually of *Sambucus nigra* or young *Ulmus*; very rare. Is. of Scilly, S. Devon, Ireland. Portugal.

Characterized by the pigmented true exciple, hypothecium, and *Orcularia*-type ascospores.

4. R. bischoffii (Hepp) Massal. (1855)

Thallus disappearing, pale grey or inconspicuous, of discrete, scurfy granules or rarely thin, grey to ochraceous, rimose-cracked and even. Apothecia to 0.45–0.6 mm diam., sessile, sometimes contiguous; thalline exciple 0.05–0.1 mm wide, entire but becoming excluded, at first concolorous with thallus, later becoming darker and concolorous with the disc; disc black, flat becoming convex; hymenium 85–115 μm tall, interspersed with oil droplets; hypothecium 75–100 μm tall, a colourless stipe sometimes developing below. Asci 50–70 × 18–22 μm. Ascospores 14.5–20 × 10–12.5 μm, thin-walled at apices with a heavily pigmented band around the central part, *Bischoffii*-type (Fig. 25j). Lichen products not detected by t.l.c.

On calcareous substrata, particularly hard limestones; local. S.E. England (Kent, Hythe), S.W. England (Severn Estuary), N. to C. England, N. Scotland (Highlands), Ireland. Europe, North America, N. Africa, Australia, New Zealand.

The only other species with this ascospore type is *R. immersa* **(13)**, distinguished by the immersed apothecia and lack of hymenial oil droplets.

5. R. calcarea (Arnold) Arnold (1879)

Thallus cracked-areolate, pale grey to grey-brown; areoles very thick and worm-like in places. Apothecia to 1 mm diam., at first ± immersed, becoming sessile; thalline exciple entire, slightly notched, concolorous with thallus; disc flat to slightly convex, dark brown to black, often irregular in outline; epithecium 10–20 μm tall, brown; hymenium 90–130 μm tall; hypothecium to 200 μm high, colourless. Ascospores 17–25 × 10–16 μm, wall two-layered, with a uniformly thickened outer wall and unevenly thickened inner wall, *Tunicata*-type (Fig. 25i). Lichen products not detected by t.l.c.

On limestone monuments; very rare; C. England (Northants). S. & C. Europe, N. Africa, Syria, Turkey, Uzbekistan.

Distinguished by the double-walled ascospores, only otherwise found in *R. colobina* (6).

6. R. colobina (Ach.) Th. Fr. (1871)

Thallus thin or disappearing, dark grey, continuous or more usually cracked, granular, indeterminate; prothallus inconspicuous. Apothecia 0.35–0.45 mm diam., sessile, infrequent, rarely contiguous; thalline exciple *c.* 0.05 mm wide, entire and persistent, concolorous with thallus or lighter; disc black becoming only slightly convex; epithecium blue-black, K+ violet, N+ violet; hymenium 60–95 μm tall; hypothecium 70–115 μm tall. Asci 50–70 × 13–19 μm. Ascospores 12.5–23.5 × 5–11.5 μm, thick-walled, lumina angular or irregular at first, becoming rounded, *Tunicata*-type (Fig. 25i). No lichen products detected by t.l.c.

On nutrient-rich bark, particularly *Fraxinus*; rare. S. Scotland (Perth, Blair Atholl), formerly in E. England (Cambridge, Suffolk). C. & S. Europe.

Characterized by the dark thallus, blue-black, K+ violet, epithecium and double-walled, *Tunicata*-type ascospores with a uniform thick outer wall.

7. R. confragosa (Ach) Körber (1855)

Thallus thick or thin, whitish, pale grey to ochraceous, rimose-cracked, and flat or of ± discrete warts, often with slightly indented edges, effuse; prothallus inconspicuous. Apothecia to 0.6–1.5 mm diam., sessile, sometimes contiguous; thalline exciple 0.05–0.1 mm wide, concolorous with thallus, entire and persistent; disc black, flat, becoming slightly convex; hymenium 80–100 μm tall; hypothecium 60–135 μm, tall, hypothecium and hymenium I+ blue. Asci 75–80 × 16–19 μm, inner walls uniform. Ascospores 17–27 × 8–13 μm, locules persistently angular with thick apical walls, *Physcia*-type (Fig. 25b). Thallus Pd+ yellow, K+ yellow (atranorin, chloratranorin, zeorin).

On coastal, siliceous, often serpentine, rocks, also to be expected on overhanging rocks at high altitudes; scattered. Throughout British Is. Europe, North America, Africa, Australia.

Easily overlooked for more delicate morphs of *Lecanora gangaleoides* or *Tephromela atra*. Morphotypes of *R. subglaucescens* **(27)** on rock can be distinguished by the *Pachysporaria*-type ascospores.

8. R. conradii Körber (1855)

Thallus thin and disappearing, pale grey to brown-grey, continuous, flat, cracked, or of ± discrete warts, effuse; prothallus inconspciuous. Apothecia 0.3–0.85 mm diam., sessile, often contiguous; thalline exciple *c.* 0.05 mm wide, concolorous with thallus, entire and persistent; disc often dark brown, soon becoming black, flat, sometimes becoming convex; epithecium brown or more usually red-brown; hymenium 90–130 μm tall; hypothecium 25–75 μm tall. Asci 65–95 × 23–28 μm. Ascospores 18–30 × 8.5–14 μm, with (2-)4(-6) lumina, lumina rounded, uniformly thick-walled, *Conradii*-type (Fig. 25d). No lichen products detected by t.l.c.

On soil, mosses, bark or wood, particularly on decaying turf, *Calluna* stems, dry moss or rabbit pellets in undisturbed localities, especially on or near the coast; local. W. & N. British Is., extending to Channel Is., E. Anglia, and Dunbarton. C. & S. Europe, North America, Australia, New Zealand.

The (1-)3(-5)-septate ascospores are diagnostic.

9. R. efflorescens Malme (1927)

Thallus effuse, often occurring as small patches among other lichens, composed of scattered to contiguous areoles; areoles 0.1–0.5 mm wide, brown-white or brown-grey, sometimes with a greenish tinge, matt, ± flat or slightly convex, usually remaining discrete and appearing as minute flattened squamules, *c.* 80–100 μm thick. Soralia scattered, sometimes becoming confluent but not forming a continuous leprose crust, erupting from the surface or margins of the areoles; soredia whitish, pale greenish or dull greenish yellow, often tinged brown. Apothecia not known in British Is., 0.3–0.5 mm diam., flat, narrowed at base; thalline exciple thick, concolorous with thallus; epithecium red-brown; hymenium 60–75 μm tall; hypothecium colourless. Asci *c.* 55–60 × 14–17 μm, 8-spored, clavate. Ascospores 15–22 × 7–10 μm, ?*Milvina*-type (Fig. 25f). Thallus Pd+ red, K−, C−, UV− (pannarin).

On acid bark, especially of horizontal branches, in moist woodlands and carrs, pollution-tolerant; locally frequent. Throughout British Is. Belgium, Scandinavia.

Often confused with *Fuscidea lightfootii*, which is usually fertile and Pd−, UV+ white-blue (divaricatic acid). Easily separated from *R. griseosoralifera* **(12)** which is sometimes Pd+ yellow but never Pd+ red.

10. R. exigua Gray (1821)

Thallus thin or rarely evanescent, pale to dark grey, continuous or cracked, flat or more rarely granular, effuse; prothallus inconspicuous. Apothecia 0.3–0.6 mm diam., sessile, frequent and often contiguous; thalline exciple *c.* 0.05 mm wide, concolorous with thallus, entire and persistent; disc black, rarely dark brown; epithecium dark brown; hymenium 70–110 μm tall; hypothecium 45–115 μm tall. Asci 45–65 × 13–20 μm. Ascospores 11.5–23 × 5.5–9.5 μm, *Physcia*-type (Fig. 25b). Atranorin may be present in low concentration.

On the rough bark of old tree trunks; fairly common. C. & S. England, C. Scotland, Ireland. Widespread in temperate regions of N. & S. Hemispheres.

R. sophodes **(26)** has a brownish olive thallus with ± immersed apothecia and is more characteristic of twigs, especially *Fraxinus*. Some records on enriched, dust-impregnated bark or wood may refer to *R. gennarii* **(11)**.

11. R. gennarii Bagl. (1861)

Thallus thin, pale to dark grey, often tinged olive or brown, continuous or more usually rimose-cracked, flat, sometimes inconspicuous, effuse or not; prothallus indistinct. Apothecia 0.35–0.6 mm diam., sessile, rarely ± immersed, frequent and often contiguous and crowded; thalline exciple *c.* 0.05 mm wide, concolorous with thallus, entire, persistent or occasionally ± occluded; disc dark brown-black, flat, becoming ± convex; hymenium 70–95 μm tall; hypothecium 25–95 μm tall, hymenium and hypothecium K+ blue. Asci 50–70 × 15–21 μm, inner walls uniform. Ascospores 11–20 × 7–11 μm, tending to swell around the middle in K, *Dirinaria*-type (Fig. 25h). Conidia 4–5 × 1 μm. No lichen products detected by t.l.c.

On ± calcareous, nutrient-rich and -enriched substrata, particularly rocks, mortar, concrete, asbestos-cement, more rarely on wood, particularly in coastal and urban sites, a pollution tolerant species; common. Throughout British Is. Widespread in temperate N. & S. Hemispheres.

The ascospores resemble the *Physcia*-type, found in *R. exigua* **(10)** and *R. confragosa* **(7)**, but swell medianly in K. *R. gennarii* is the commonest species of the genus on rock and man-made substrata.

12. R. griseosoralifera Coppins (1989)

Thallus effuse, wide-spreading or forming patches among other epiphytes, of scattered to contiguous areoles; areoles 0.08–0.2 mm wide, slightly convex to ± hemispherical, sometimes almost subsquamulose, usually inconspicuous and often obscured by soralia and superficial algae, etc., dull greenish white to pale grey. Soralia 0.14–0.3(-0.4) mm diam., numerous, discrete, sometimes a few confluent, but never forming a continuous leprose crust, efflorescent, each erupting from upper surface of an areole (never ulcerose), convex, blue-grey (fading in the herbarium). Apothecia 0.3–0.4 mm diam., very rare, sessile; thalline exciple *c.* 0.1 mm thick, whitish grey, sometimes sorediate. Ascospores 24–28.5 × 12–16.5 μm, ellipsoid, *Pachysporaria*-type (Fig. 25c). Thallus Pd− or ± faint yellow, K−, KC−, C−, UV−, I− (atranorin, zeorin).

On bark or amongst mosses overgrowing bark; local. S.W. England (Somerset), C. Wales, C. Scotland. Austria, Norway.

13. R. immersa (Körber) Zahlbr. (1906)

Thallus immersed, or rarely thin, finely granular, whitish to pale grey. Apothecia to 1 mm diam., immersed in deep depressions in the substratum, with isolated groups of algae in the thalline exciple; epithecium dark brown; hymenium *c.* 100 μm tall, without oil drops; hypothecium to 100 μm tall, colourless. Ascospores 15–20 × 9–13 μm, *Bischoffii*-type (Fig. 25j). No lichen products detected by t.l.c.

On limestone; local. S.W. England (S. Devon, Berry Head; Somerset, Dolebury Warren, Cheddar). S. Scandinavia, C. & S. Europe, Africa, Syria, Iran, Turkey.

Previously included within *R. bischoffii* **(4)** and separated from it only by the immersed apothecia and the hymenium lacking oil droplets.

14. R. interpolata (Stirton) Sheard (1973)

Thallus thin, rimose or rimose-areolate, pale grey to dark bown, sometimes with a mauve tinge, occasionally appearing as if frosted; areoles 0.25–0.8 mm diam.; prothallus ± well developed, black. Apothecia 0.3–0.4 mm diam., sessile, often widely scattered; thalline exciple persistent, entire, concolorous with thallus; disc flat to slightly convex, brown-black; epithecium 10–20 μm tall, dark brown; hymenium 60–80 μm tall; hypothecium 50–80 μm tall, colourless. Asci 45 × 13–17 μm. Ascospores 13–18 × 6–9 μm, narrowly ellipsoid, not so strongly thickened near septum but also slightly thickened apically, torus diffuse, outer surface, roughened, *Physconia*-type (Fig. 25). Thallus K− (no lichen products reported by t.l.c.).

On hard siliceous rocks, usually on sheltered, ± vertical surfaces of cliff-faces or underhangs. N. Scotland (Argyll, W. Sutherland, Ross, S. Aberdeen). N.W. Europe.

Previously confused with *R. occulta* **(20)**, which has a paler, K+ yellow thallus, and apothecia with only a distinct true exciple.

15. R. isidioides (Borrer) Olivier (1909)

Thallus effuse, composed of numerous, minute, coralloid, branched and thin, ± cylindric isidia. Isidia pale grey, sometimes adpressed or forming folded squamules. Apothecia 0.9–1.15 mm diam., sessile, frequent, sometimes contiguous; thalline exciple 0.1–0.15 mm thick, concolorous with thallus, entire or crenulate, persistent; disc black, persistently flat; epithecium red-brown; hymenium 130–160 μm tall; hypothecium 30–50 μm tall. Paraphyses 1–2.5 μm wide, apices to 4–5.5 μm wide. Asci 75–110 × 22–35 μm. Ascospores 21–31.5 × 11.5–17 μm, *Pachysporaria*-type (Fig. 25c). Thallus K+ yellow.

On bark or more rarely amongst mosses overgrowing bark of mature *Quercus* in ancient woodlands; locally frequent. England (Somerset; Hampshire, New Forest; Hereford, Cumberland, Westmorland), N. Wales, W. Scotland (Argyll), S.W. Ireland. W. France, Portugal, Tuscany.

When sterile distinguished from *Agonimia octospora* by the K+ yellow isidia, which are larger, more cylindrical and pointed and have a firm cortex without microscopic, peg-like extensions.

16. R. lecideina Mayrh. & Poelt (1979)

Thallus thin, superficial, dull pale grey-brown, cracked-areolate. Apothecia to 0.5 mm diam., sessile; true exciple conspicuous, with isolated algae; disc flat, black; epithecium 10–14 μm tall, dark brown; hymenium 70–90 μm tall; hypothecium *c.* 100 μm tall, brown. Ascospores 10–15 × 6.5–8.5 μm, thickened only in the region of the septum, otherwise *Dubyana*-type (Fig. 25g). Conidia thread-like. Lichen products not detected by t.l.c.

On siliceous rocks; rare. S.W. Ireland (Kerry). Corsica, Norway, Sweden, New Zealand.

Unique amongst *Rinodina* species in having thread-like conidia, but easily overlooked for *Buellia punctata* with which it is probably closely related.

17. R. luridescens (Anzi) Arnold (1872)

Thallus thick, deeply rimose-cracked-areolate, dark grey to black-brown, usually with a distinct purple tinge; areoles 1.25–2.7 mm wide, sometimes mosaic-forming; prothallus often limiting, dark. Apothecia 0.8–1.5 mm diam., ± immersed; thalline exciple 0.05–0.1 mm wide, darker than thallus, thin, eventually becoming excluded, frequent, often contiguous; disc black, flat becoming slightly convex; epithecium brown; hymenium 75–155 μm tall; hypothecium 70–155 μm tall, pale brown. Asci 50–65 × 18–23 μm. Ascospores 15–23 × 7.5–12 μm, ellipsoid, between *Physcia*- and *Milvina*-types (Figs. 25b & f). Thallus cortex C+ orange-red (gyrophoric acid, zeorin).

On maritime siliceous rocks in the xeric-supralittoral; locally common. W. British Is. Atlantic and Mediterranean sea-boards to the Black Sea.

A distinctive species with a thick grey-violet-black thallus and large immersed apothecia without a perceptible exciple.

18. R. milvina (Wahlenb.) Th. Fr. (1860)

Thallus usually matt, thick, areolate, grey, grey-brown to dark brown. Apothecia to 1 mm diam., crowded, slightly immersed to broadly superficial; thalline exciple entire, in part swollen, prominent; hymenium 90–120 μm tall; hypothecium to 200 μm tall, ± colourless; disc flat to slightly convex. Ascospores 16–22 × 8–12 μm, with a finely ornamented surface, *Milvina*-type. (Fig. 25f). Thallus Pd−, K−, KC−, C−.

On slaty, nutrient-rich rocks, particularly bird perches; rare. N.W. Scotland (Colonsay, Mingulay). Europe, North America, N. Africa, W. Asia, Mongolia.

R. interpolata **(14)** has smaller *Physcia*-type ascospores. The British material is poor.

19. R. mniaraea (Ach.) Körber (1855)

Thallus coarsely warted, granular, pale to dark grey or brownish; medulla partly orange. Apothecia 0.5–1.5 mm wide, flat, soon becoming convex; exciple excluded, dark brown-black; epithecium pale brown; hymenium 100–150 μm tall. Ascospores (22-)32–35(-38) × 11–16 μm, *Physcia*-type (Fig. 25b).

On mosses at *c.* 850 m alt., arctic-alpine; very rare. N. Scotland (W. Ross). Europe, North America.

The British specimen is referable to **var. *cinnamomea*** Th. Fr. (1860), characterized by a partly orange medulla. Easily recognized by the large, *Physcia*-type ascospores and convex apothecia. The arctic *R. turfacea* (Wahlenb.) Körber (1855) has similar ascospores and should be sought in similar habitats. It is distinguished by the persistently flat apothecia with a prominent thalline exciple.

20. R. occulta (Körber) Sheard (1967)

Thallus thin, whitish, grey, pale brown to ochraceous, areolate or rarely rimose; areoles to 0.2–0.7 mm wide, usually contiguous, flat, effuse; prothallus dark, entire, limiting. Apothecia 0.1–0.3 mm diam.; thalline exciple *c.* 0.05 mm wide, concolorous with the thallus, entire, persistent or sometimes excluded; disc black, flat or becoming convex; hymenium 55–85 μm tall; hypothecium 10–55 μm tall. Asci 45–50 ×

13–20 μm. Ascospores 11–16 × 5–9 μm, between the *Physcia-* and *Milvina*-types (Figs 25b & f). Thallus K+ yellow or − (atranorin).

On well-lit, hard siliceous rocks; rare. W. England (Shropshire), N. Wales, Scotland. Europe, Georgian S.S.R.

Superficially resembling *R. oxydata* **(22)** but with smaller ascospores of a different structure.

21. R. orculariopsis Mayrh. (1984)

Thallus thin, grey- to black-brown, membranous, partly cracked-areolate, at times completely covered with apothecia; prothallus black. Apothecia to 0.5 mm diam., sessile, not constricted at base, often crowded and clustered; thalline exciple thick, persistent; disc even, back-brown to black; true exciple ± distinct; epithecium dark brown; hymenium 90–110 μm tall; hypothecium colourless. Ascospores 17–23 × 8–12 μm, broadly ellipsoid, only strongly thickened near septum, torus distinct, surface microroughened, *Physconia*-type. Thallus Pd−, K−, C− (zeorin).

On quartzite cobble in churchyard wall, siliceous gravestones, standing stones and sarsens, in ± nutrient-rich crevices on coastal keratophyre; local. S. England (N. Wilts), N. England (Lancashire), S.W. Wales (Skomer), E. Scotland (Angus), N. Ireland (Donegal, Londonderry). S. Scandinavia, W. France.

Probably overlooked for *R. gennarii* **(11)** from which it differs in the ascospore type. *R. interpolata* **(14)** has more scattered apothecia, and smaller and narrower ascospores with a slightly thickened apical wall.

22. R. oxydata (Massal.) Massal. (1854)

Thallus thin, rarely thick and evanescent, pale grey to ochraceous, rimose or sometimes discrete or of contiguous areoles, flat, effuse; prothallus sometimes limiting, dark, entire. Apothecia to 0.3–0.45 mm diam., sessile, frequent and sometimes contiguous; thalline exciple 0.05–0.1 mm thick, frequently concolorous with disc, entire and persistent; disc dark brown, becoming black, persistently flat; epithecium, pale brown; hymenium 70–115 μm tall; hypothecium 20–90 μm tall, I−. Asci 60–90 × 18–26 μm. Ascospores 19–25 × 9–14 μm with thick walls, lumina not angular, *Mischoblastia*-type (Fig. 25k). Thallus K− or ± yellow (atranorin).

On somewhat shaded, hard siliceous rocks, especially metal-rich; scattered. W. & N. British Is. C. & S. Europe, Turkey, North America, Canary Is., E. Africa, Japan, Indonesia.

A rather variable species, which may be ochraceous when on iron-rich substrata. The only British species with a pigmented thalline exciple and *Mischoblastia*-type ascospores. The single collection of *R. trachytica* (Massal.) Bagl. & Car. (1880) was misidentified and is referable to this species.

23. R. parasitica Mayrh. & Poelt (1979)

Thallus darkish brown, forming small patches to *c.* 2 cm diam., on crustose lichens, areolate; areoles to 0.5 mm diam., rounded. Apothecia to 0.4 mm diam., sessile; thalline exciple concolorous with thallus; epithecium reddish brown; hymenium 70–80 μm tall; hypothecium to 75 μm tall. Paraphyses 1.5–2 μm wide, apices 4–6 μm wide. Ascospores 14.5–20 × 7.5–11 μm, *Physcia*-type (Fig. 25b). Thallus Pd−, K−, KC− (lichen products not detected by t.l.c.).

On thalli of *Aspicilia caesiocinerea* on tops of boulders, in mountain stream at *c.* 650 m alt.; very rare. C. Scotland (Angus, Caenlochan Glen). Europe, Norway, Finland, U.S.A. (Colorado), Mongolia.

In addition to its lichenicolous habit, this species is distinguished from *R. gennarii* **(11)** by the darker, brown thallus, and from *R. orculariopsis* **(21)** in ascospore type.

24. R. pyrina (Ach.) Arnold (1881)

Thallus minutely warted or cracked, effuse, whitish to grey, cortex I−. Apothecia 0.2–0.4 mm diam., usually numerous and crowded, flat, often becoming convex; exciple distinct; epithecium dark brown; hymenium *c.* 65 μm tall; hypothecium *c.* 40 μm tall. Ascospores 13–15 × 6–7 μm, thin-walled, *Physconia*-type and often slightly curved, mature ones lacking thickenings. Thallus Pd−, K−, C− (lichen products not detected by t.l.c.).

On wood; very rare. S. England (Sussex, Cuckmere). Lowland Europe.

Distinguished from *R. sophodes* **(26)** by the paler, more wide-spreading thallus and from *R. exigua* **(10)** by the smaller apothecia and thin-walled, somewhat curved ascospores.

25. R. roboris (Duf. ex Nyl.) Arnold (1881)

Thallus usually thick, continuous or more usually irregularly cracked, flat or more usually uneven-granular and often slightly squamulose and pinkish at margins, pale to green-grey, effuse; distinct prothallus absent. Apothecia to 1.0–1.6 mm, conspicuous, sessile, frequent, often contiguous; thalline exciple 0.1–0.5 mm wide, concolorous with thallus, entire, often becoming crenulate, persistent; disc dark brown, becoming black, persistently flat; hymenium 95–115 μm tall, I+ blue; hypothecium 95–175 μm tall, I−. Asci 50–85 × 18–22 μm, inner walls uniform. Ascospores 14–22 × 8–12 μm, with thick walls, locules rounded, *Pachysporaria*-type (Fig. 25c). Thallus K+ yellow (atranorin).

On well-lit bark, especially rough-barked trees, such as *Quercus* and *Fraxinus* also on *Fagus*; locally common. S. England, becoming rarer N. to Scotland (Argyll), W. Britain, Ireland. W. Europe.

Characterized by the usually pale grey, K+ yellow thallus, the flat, large apothecia with a relatively thick ± crenulate thalline exciple and by the *Pachysporaria*-type ascospores.

26. R. sophodes (Ach.) Massal. (1852)

Thallus in small patches, thick or thin, pale grey to olive-brown, irregularly cracked, flat, determinate; prothallus often present, thin, entire, grey-black. Apothecia to 0.45–1.15 mm diam., ± immersed, sometimes becoming ± sessile, frequent, usually contiguous; thalline exciple 0.05–0.15 mm thick, concolorous with thallus, entire, persistent; disc dark brown, becoming black, persistently flat; hymenium 85–130 μm tall, I+ blue; hypothecium 60–130 μm tall, I+ blue. Asci 50–70 × 15–20 μm, inner walls uniform. Ascospores 13–19 × 6.5–9 μm, with thin walls and rounded lumina, *Milvina*-type (Fig. 25f). Lichen products not detected by t.l.c.

Mainly on twigs of smooth-barked trees, especially *Fraxinus*, an early colonizer, usually associated with ageing leaf scars, most common in hilly and exposed districts; widespread. W. & N. British Is., rare or extinct in S. & E. English Midlands. Europe.

Distinguished from *R. exigua* **(10)** by the darker olive-brown, well-delimited thallus with crowded apothecia and the thin-walled, broadly ellipsoid ascospores.

27. R. subglaucescens (Nyl.) Sheard (1967)

Thallus wide-spreading, thick or thin, pale glaucous-grey to greenish, continuous, often uneven-granular or irregularly cracked, ± flat, effuse or not; prothallus inconspicuous or occasionally well developed, black. Apothecia to 0.7–1.25 mm diam., immersed or more usually sessile, sometimes contiguous; thalline exciple 0.05–0.2 mm wide, concolorous with thallus, entire, becoming distinctly crenulate, persistent; disc ± persistently flat, black; hymenium 90–100 μm tall; hypothecium 70–140 μm tall. Asci 60–95 × 15–25 μm, inner walls uniform. Ascospores 16.5–23 × 8.5–14 μm, with thick walls and rounded lumina, *Pachysporaria*-type (Fig. 25c). Thallus K− or ± yellow (atranorin, chloroatranorin, zeorin).

On ± sheltered sides of rocks, often near the ground, also soil and on dead *Armeria* tufts, coastal; locally common. S. & W. England, W. Ireland. W. & S. Europe.

Previously confused with *R. confragosa* **(7)** in Britain, but the thick, uniform ascospore walls distinguish it from that species.

28. R. teichophila (Nyl.) Arnold (1863)

Thallus thick to thin, irregularly rimose-cracked, flat or usually rather uneven, occasionally granular, effuse or not, dark grey to blackish, rarely pale grey to ochraceous; prothallus indistinct. Apothecia 0.7–1.25 mm diam., ± immersed, frequent, often contiguous; thalline exciple 0.05–0.10 mm wide, concolorous with thallus, entire, persistent; disc dark brown becoming black, persistently flat; epithecium brown; hymenium 80–145 μm tall, I+ blue; hypothecium 55–100 μm tall. Asci 75–90 × 20–30 μm. Ascospores 20–32 × 11–19 μm, thick-walled, lumina not angular, between *Pachysporaria* and *Mischoblastia*-types (Figs. 25c & k). Thallus K− or ± yellow.

On rocks, particularly stone walls, but also tree trunks; scattered. Throughout British Is. Temperate Europe, New Zealand.

Characterized by the dark thallus, concolorous, ± immersed apothecia, and large ascospores with a thick, internally uneven wall.

ROCCELLA DC. (1805)

Thallus shrubby, tufted, of strap-shaped or ± rounded branches, attached by a basal sheath. **Cortex** of hyphae perpendicularly oriented to upper surface, arising from outermost part of medulla and not always forming a distinct layer; in addition, irregular patches of hyphae oriented partly parallel to the surface may also occur on the cortex. **Photobiont** *Trentepohlia*. **Medulla** of densely interwoven hyphae. **Ascomata** apothecia, ± round or elongate. **Thalline exciple** present or absent. **Hypothecium** dark brown to black, dense, well developed. **Paraphysoids** branched, sparsely anastomosing, *c.* 2 μm thick. **Asci** 8-spored, thick-walled, fissitunicate, with an internal beak, and a small I+ blue ring in the thickened apex. **Ascospores** usually 3-septate, oblong-fusiform, colourless or pale brown. **Conidiomata** pycnidia. **Conidia** cylindrical, curved. **Chemistry**: various *para*-depsides and tridepsides, fatty acids, chromones. **Ecology**: coastal.

The two British species of this mainly tropical genus can be distinguished from *Ramalina siliquosa* by the pale mauve-grey (not green-grey) thallus, the large bluish white soralia and the C+ red reaction of the cortex or soralia. The S. European and Mediterranean *Roccella tinctoria* DC. (1805) was formerly used as source of the purple dye orchil, and the tropical *R. montagnei* Bél. (1846) for litmus. The genus is unexpectedly absent from Ireland.

LITERATURE: Darbishire (1898).

1 Thallus 5 cm long, ± rounded branches, ± tufted; cortex C+ red; medulla yellow-orange at point of attachment .... **1. *phycopsis***
Thallus to 20 cm long, flattened branches, ± pendent; cortex C−; medulla whitish at point of attachment .... **2. *fuciformis***

1. R. phycopsis Ach. (1810)

Thallus to 5 cm, usually erect and tufted, branches pale blue-grey, rounded, sometimes angular or slightly flattened, irregulary branched; medulla orange-yellow within the basal sheath. Soralia abundant, at first tubercle-like, later ± globose, efflorescent. Apothecia very rare, prominent, ± rounded to oblong, often deformed, without a thalline margin, disc black. Paraphysoids ± simple below, branched above. Ascospores (14-)18–21(-23) × 4–6 μm, becoming pale brown when old, straight or slightly curved. Conidia 12–17 × 1 μm, curved. Cortex C+ red, soralia C−, medulla C−, soralia and medulla UV+ glaucous (erythrin, roccellic acid and ± lecanoric acid).

On dry, vertical or overhanging rock faces of E.- or N.-facing aspect, occasionally on church towers and walls, memorials and dry sides of old, sheltered trees, on or near the coast; local. S. & S.W. England (Cornwall, Devon, Is. of Wight), Channel Is., Is. of Scilly, S. Wales (Pembroke). W. France, Spain, Portugal, N. Mediterranean region, Macaronesia.

A stunted-looking, shrubby lichen, usually abundantly sorediate and very rarely fertile. The orange-yellow medulla within the basal sheath is diagnostic and easily observed.

2. R. fuciformis (L.) DC. (1805)

Like **1**, but has a longer (to 20 cm), more flattened, wider (to 1.5 cm), strap-shaped, ± pendent thallus, often irregularly fringed with folioles; medulla not yellow within the basal sheath. Soralia blue-white, on the surface and margins, tubercle-like or in dense, ± confluent masses. Apothecia very rare, marginal or laminal, single or forming regular or gnarled clusters; disc to 1 mm diam., black, often white-pruinose, flat to ± convex, with thalline and true exciple. Paraphysoids ± anastomosing. Ascospores (18-)22–30 × 4–6 μm, elongate-fusiform. Cortex C−, soralia C+ red, medulla C−, soralia and medulla UV+ glaucous (erythrin, roccellic acid, lepraric acid).

On dry, or vertical or overhanging E.- or N.-facing rocks, not directly wetted by rain, in sunny, mostly frost-free coastal sites, rarely on the dry, sheltered side of old trees and old walls, church towers and walls near the coast; local. S.W. England (Devon, Cornwall), Channel Is., Is. of Scilly, S. Wales (Pembroke, Skomer Is.). W. France, Spain, Portugal, Macaronesia.

Often occurring with the previous species, attaining a much larger size, with characteristic, ribbon-like branches from *c.* 1–15 mm wide. Only the soralia react C+ red, the outer cortex and medulla remaining unchanged.

SAGIOLECHIA Massal (1854)

Thallus crustose, thin or endosubstratal, effuse. **Photobiont** *Trentepohlia*. **Ascomata** apothecia, ½-immersed to sessile. **Thalline exciple** absent. **True exciple** prominent, distinctly crenate, black, shining, when senescing may become pale to almost white or piebald; in section dark brown in outer part, colourless within. **Disc** black, concave to convex, somtimes umbonate or gyrose. **Epithecium** dark brown, K –. **Hymenium** colourless, I+ blue to reddish. **Hypothecium** colourless or dark brown. **Hamathecium** of paraphyses, septate, simple or sparingly branched, frequently anastomosed, apical 1–3 cells often swollen and bound by a dense matrix; periphyses present or not. **Asci** 8-spored (a few sometimes aborting), cylindrical, wall I+ blue, apical dome K/I –. **Ascospores** 3-septate at maturity, ellipsoid to fusiform, colourless, often with a thin perispore. **Conidiomata** pycnidia. **Conidiogenous cells** short-ampulliform, in a single layer, enteroblastic, acrogenous. **Conidia** bacilliform, colourless. **Chemistry**: no lichen products detected by t.l.c. **Ecology**: on calcareous rocks or on acidophilous bryophytes, plant debris or soil, montane.

Differs from *Gyalidea* in the mainly simple paraphyses, an I+ blue ascus wall, and photobiont, *Trentepohlia*.

LITERATURE: Vězda (1968a).

1 Asci mainly >100 μm in length; ascospores (10-)12–20(-25) × 6–9 μm; on limestone .... **1. *protuberans***
Asci mainly <100 μm in length; ascospores 16–24 × 5–7 μm; on bryophytes **2. *rhexoblephara***

1. S. protuberans (Ach.) Massal. (1854)

Thallus inconspicuous, immersed or partly superficial, a pale orange stain, greenish grey in herbarium, sometimes scurfy, effuse, not delimited. Apothecia (0.3-)0.4–0.7(-1) mm diam., frequently in lines or ± contiguous in groups of 2–3, partially immersed or ± emergent, appearing gyrose; true exciple thick, deeply divided or fluted into 3–8 segments, often star-like, well developed, <120 μm thick, in section black-brown, merging below with the lower part of the hypothecium; disc little exposed, with a central, convex, sterile plug; epithecium red- to black-brown; hymenium 100–115 μm tall; paraphyses *c.* 1 μm wide, apices *c.* 4 μm wide. Asci 100–140 × 10–20(-25) μm, elongate-clavate or cylindrical. Ascospores (10-)12–29(-25) × 6–9 μm, (1-)3-septate, one occasionally oblique, or septa quadrate.

On high altitude, hard limestone; very rare. N. Scotland (Highlands, Blair Atholl, Ben Alder). Europe, Greenland, North America, N. Asia.

Polysporina species differ in the multispored asci. The small, star-shaped apothecia of *S. protuberans* with a sterile central cone are diagnostic.

2. S. rhexoblephara (Nyl.) Zahlbr. (1913)

Thallus a thin film, rarely thicker, and thinly warted-uneven, grey, pale grey or grey-violet, sometimes with a brown tinge; forming haustoria within host cells. Apothecia (0.5-)0.7–2 mm diam., scattered or in clusters of 2–4, sessile; true exciple prominent, ± elevated, crenate, divided into 3–8 segments, in section black-brown, merging into the hypothecium; disc concave, becoming flat or irregularly convex, brown-black, shiny; epithecium red-brown or brown; hymenium 90–120 μm tall; hypothecium well developed, brown-black. Paraphyses *c.* 1 μm wide; apices 4.5 μm wide. Asci 70–110 × 10–12 μm, elongate-clavate. Ascospores 16–24 × 5–7 μm, 3-septate. Pycnidia *c.* 100 μm diam., half-immersed; conidia 3.5–4 × 1 μm.

On patches of mosses and hepatics overgrowing mica-schist and other weakly basic rocks and soil, semi-parasitic, rarely directly on soil, in sheltered, moist, cool sites in subalpine and alpine situations; rare. N. Scotland (Highlands). Arctic, Scandinavia, Tatra Mtns, North America.

S. rhexoblephara bears a superficial resemblance to *Lecidea hypnorum* but the latter species has a ± entire margin and different anatomical characters.

SARCOGYNE Flotow (1851)

Thallus crustose, superficial or more commonly immersed; **prothallus** absent. **Photobiont** chlorococcoid, including *Dictyochloropsis* and *Myrmecia*. **Ascomata** apothecia, sessile or immersed in pits in substratum, red-black to black, sometimes pruinose. **Thalline exciple** absent. **True exciple** black and friable, becoming excluded or persistent and then in some species crenulate, composed of compacted radiating hyphae, colourless internally, tips swollen and dark brown at the surface. **Epithecium** pale reddish brown to dark brown. **Hymenium** colourless, I+ blue. **Hypothecium** colourless to brown, often thick. **Hamathecium** of paraphyses, simple to sparsely branched, regularly septate, the end cells ± swollen and sometimes brown-capitate. **Asci** multispored, elongate-clavate to obclavate, often swollen in the lower third when mature, strongly thickened at apex, the apical dome K/I−; outer coat I+ blue. **Ascospores** narrowly ellipsoid, colourless, simple, smooth, lacking a distinct perispore. **Conidiomata**, when present, pycnidia, immersed in ± multilocular black warts. **Conidiogenous cells** subcylindrical, enteroblastic, acrogenous. **Conidia** subglobose to ellipsoid, simple, colourless. **Chemistry**: no lichen products detected by t.l.c., *Lecanora*-red in the apothecia. **Ecology**: on rocky substrata.

A cosmopolitan genus with about 10 species and numerous infraspecific taxa recognized in Europe alone but badly in need of a critical revision. Further taxa are expected to occur in the British Isles where the genus has never been studied in depth. *Polysporina* species, which also have multispored asci, can be easily distinguished from the non-pruinose species by the smaller and gyrose apothecia; they also differ in the branched and anastomosed paraphyses and ascus shape.

LITERATURE: Magnusson (1935).

1 Apothecial disc not pruinose; true exciple persistent; on hard siliceous rocks 2
Apothecial disc ± pruinose; true exciple becoming excluded; on calcareous substrata .... **3. *regularis***

2(1) Apothecia 1–3(-6) mm diam.; hypothecium brown; true exciple strongly crenulate **1. *clavus***
Apothecia (0.4-)0.5–1 mm diam.; hypothecium colourless to pale brown; true exciple not crenulate .... **2. *privigna***

1. S. clavus (DC.) Krempelh. (1861)

Thallus ± immersed and inconspicuous, sometimes appearing as a dark grey stain. Apothecia sessile, 1–3(-6) mm diam.; disc red-black to black, not pruinose; true exciple black, persistent, raised, distinctly crenulate, 0.3–0.7(-1) mm thick; hymenium 85–115 μm tall; hypothecium brown, 60–100 μm tall. Asci 45–80 × 12–15 μm, at least 200-spored. Ascospores 4–5(-6) × (1-)1.5–2 μm. Conidiomata in wart-like, multilocular stromata up to 1 mm diam.; conidia 2–3.5 × 1–1.5 μm.

On coastal siliceous rocks, especially granite, often following cracks, terrestrial-halophobic, rarely inland; rare. Channel Is., Is. of Scilly, S. Wales (Brecon), W. Scotland. Europe, North America, Africa.

The typification and nomenclature of this species are in need of a critical examination.

2. S. privigna (Ach.) Massal. (1854)

Thallus ± immersed and inconspicuous, sometimes grey or ± gelatinous. Apothecia sessile (0.4-)0.5–1 mm diam., disc pale red-brown to dark red, not pruinose; true exciple black, persistent, ± raised, not crenulate, sometimes angular by compression, to 0.3 mm thick; hymenium 60–85 μm tall; hypothecium colourless to pale brown. Asci 45–55 × 10–12 μm, *c.* 100-spored. Ascospores 3.5–5 × 1–1.5(-2) μm. Conidiomata unknown.

On hard siliceous rocks and walls, often in somewhat humid situations; now rare, many records dating from the last century. S.W. (S. Devon, Dartmoor) and N. England, N. Wales, Scotland, E. Ireland. Europe, North America, North Africa, Saudi Arabia.

S. clavus **(1)** has a taller hymenium and a crenulate true exciple.

3. S. regularis Körber (1855)

Thallus generally immersed and inconspicuous, white to grey and scurfy. Apothecia sessile or sunken into pits in the substratum (0.3-)0.4–1.5(-2) mm diam.; disc red-brown to black, generally densely blue-grey-pruinose but sometimes not; exciple black, often more densely pruinose than the disc and then white-grey, not or slightly raised, not crenulate, finally becoming excluded, to 50 μm thick; hymenium (65-)70–100(-110) μm tall; hypothecium ± colourless. Asci 60–85 × 14–18 μm, 100- to 200-spored. Ascospores 3–5(-6) × 1.5–2 μm. Conidiomata unknown.

On calcareous rocks, walls, chalk pebbles, sea-shells, asbestos-cement and old mortar, particularly in lowland areas. Throughout British Is. Europe, North America, Africa, Asia, ?New Zealand.

Morphs with red-brown, densely pruinose apothecia embedded in pits in the substratum, usually hard limestones, are sometimes distinguished as a separate species, "*S. pruinosa* auct.", by continental authors, the status and correct name of which remains unclear.

SARCOPYRENIA Nyl. (1858)

Thallus crustose, immersed or superficial, fawn-grey, often inapparent, effuse; **prothallus** white or whitish, rarely present. **Photobiont** chlorococcoid. **Ascomata** perithecia, simple, superficial. **Involucrellum** absent. **True exciple** of two walls, the outer thick, black, dimidiate, incurved at the base, the inner, thinner, dark brown, separated by a pale interstitial zone. **Hamathecium** of slender periphyses concentrated towards the ostiole, paraphyses soon gelatinising and disappearing. **Asci** 8-spored, *Verrucaria*-type, thin-walled, elongate-clavate to cylindrical, fusiform, soon disappearing. **Ascospores** simple, thread-like with swollen ends or dumb-bell-shaped, often twisted in the middle. **Conidiomata** unknown. **Chemistry**: lichen products not detected by t.l.c. **Ecology**: on rocks.

The genus is frequently overlooked, though colonies may be detected by the presence of prominent perithecia without any visible associated thallus; the vicinity is not colonized by algae or other crustose lichens and this is perhaps due to some allelopathic effect of the thallus below the surface of the substratum.

LITERATURE: Aguirre-Hudson (1991).

1. S. gibba (Nyl.) Nyl. (1858)

Thallus inconspicuous, reduced to a scurfy, effuse collar around the fruits or evanescent and immersed, rarely superficial, cracked-areolate and yellowish grey to reddish brown. Perithecia 0.5 (-1) mm diam., single or 2- to 3-grouped, black, hemispherical with a minute papillate ostiole, becoming sunken and depressed to appear ± laterally extended. Ascospores 30–40 × 3.0–3.5 μm, colourless, thin, elongated, with swollen ends.

In sunny exposed situations, particularly walls in churchyards and flat-topped chest tombs, also on oolite, black limestone, marble, concrete and basic sandstones, rarely on *Larix* and *Corylus* bark heavily impregnated with quarry-lime dust; locally frequent. Channel Is., S. & E. England, E. Scotland, E. Ireland. C. Europe, N. & S. France, Algeria.

This species is easily recognized by the double-walled exciple and the unique, twisted, dumb-bell-shaped ascospores.

SARCOSAGIUM Massal. (1856)

Thallus crustose, thin. **Photobiont** chlorococcoid. **Ascomata** apothecia, convex, ± translucent. **True exciple** initially swollen, robust, pale, of thickly interwoven hyphae within, outer parts paraplectenchymatous. **Hymenium** very tall. **Hamathecium** of paraphyses, simple or sparingly branched, lax in K, apices swollen. **Asci** multispored, cylindrical, thin-walled, K/I+ blue, without an apical dome. **Ascospores** ellipsoid to elongate-cylindrical, colourless, thin-walled, simple, occasionally 1-septate (visible in K/I). **Chemistry**: lichen products not detected by t.l.c. **Ecology**: on soil.

Characterized by the tall hymenium, slender and simple paraphyses, cylindrical K/I+ blue asci lacking an apical dome, and particularly, the multispored asci with ellipsoid to cylindrical-ellipsoid ascospores.

LITERATURE: Coppins & James (1979b), James (1971), Poelt & Vězda (1977).

1. S. campestre (Fr.) Poetsch & Schied. (1872)

Thallus ± gelatinous, especially when wet, greenish or blackish brown, occasionally with scattered green-grey granules. Apothecia 0.1–0.5 mm diam., sessile, ± barrel-shaped, dark red-brown, becoming ± translucent when wet, disc flat or slightly concave, at first closed, opening later; true exciple well developed, concolorous with disc but often thinly grey-pruinose; hymenium 120–170 μm tall; epithecium and hypothecium colourless or pale yellow. Ascospores 5–8 × 2–2.3 μm.

On ± base-rich or acid soils and decaying mosses, particularly on walls and mine-spoil heaps, rarely on the cut surface of tree stumps; very local. Throughout British Is.

An inconspicuous species in which the apothecia become paler and notably more apparent when wet. The **var. *macrosporum*** Coppins & P. James (1979) has consistently longer ascospores, 7–13 × 2–3 μm; it occurs amongst mosses on basic soils and dunes in England (Derbyshire, Leicestershire) and N. Wales (Anglesey).

SCHADONIA Körber (1859)

Thallus crustose, scurfy, verrucose to coralloid, effuse. **Photobiont** chlorococcoid. **Ascomata** apothecia, sessile, constricted towards base; disc black, concave to flat. **Thalline exciple** absent. **True exciple** prominent, even or somewhat flexuous, ± pseudoparenchymatous, cells ± isodiametric, radially orientated. **Hypothecium** dark. **Hamathecium** of paraphyses, thin, branched and anastomosed, not or slightly thickened at apex. **Asci** 2- to 8-spored, cylindrical-clavate with a strongly I+ blue tholus, *Bacidia*-type. **Ascospores** ellipsoid, richly muriform, colourless. **Conidiomata** unknown. **Chemistry**: lichen products not detected by t.l.c. **Ecology**: montane.

Closely similar to *Lopadium* but differing in the 2- to 8-spored asci, I+ dark blue tholus and branched-anastomosing paraphyses without swollen, pigmented apical cells.

1. S. fecunda (Th. Fr.) Vězda & Poelt (1980)

Thallus thin to moderately thick, ± continuous, unevenly minutely verrucose to granular, greyish brown. Apothecia 0.5–1.5 mm diam.; disc flat, matt, black; margin prominent, somewhat inrolled when young, shiny; epithecium overlain by a dark brown layer, K+ purplish-brown; hymenium 150–170 μm tall, colourless or in part pale yellow-brown; hypothecium deep brown, K+ deep red-brown; paraphyses thin, 1–1.5 μm thick, richly branched, anastomosed, apices not swollen. Asci (4-)8-spored, 90–140 × 30–50 μm. Ascospores (30-)40–50(-60) × 10–18(-20) μm, markedly muriform, ellipsoid to obovate, colourless.

Encrusting mosses and acid rocks, montane, at 1000 m; very rare. N. Scotland (Perth, Ben Lawers; W. Ross). Arctic, Scandinavia, Alps and the High Tatra Mtns.

SCHAERERIA Th. Fr. (1855)

Thallus crustose or squamulose, if crustose areolate, areoles often somewhat dispersed and convex, grey to reddish brown; **prothallus** often well developed, black. **Photobiont** *Trebouxia*. **Medulla** I-. **Ascomata** apothecia, immersed or sessile, black. **Thalline exciple** absent. **True exciple** persistent, black, darkish brown in section, K−, of ± globose cells, without algae. **Epithecium** bright green, at least in K, or partly violet and K+ bright green, occasionally discoloured brown (old material). **Hymenium** I+ weakly blue (ascus walls). **Hypothecium** brown, but upper part (subhymenium) usually colourless. **Hamathecium** of paraphyses that are lax in K, simple below but occasionally branched above; apical cells often swollen and sometimes ± moniliform. **Asci** ± cylindrical or rarely clavate, thin-walled, with a single wall layer, not thickened apically, only the outermost gelatinous layer I+ faint blue, discharge by splitting of the apex, *Schaereria-type* (Fig. 33j). **Ascospores** ± globose to ellipsoid, simple, uniseriately, ± biseriately or irregularly arranged in the asci, colourless, smooth, lacking a distinct perispore. **Conidiomata** pycnidia, immersed. **Conidiogenous cells** subcylindrical, enteroblastic, acrogenous. **Conidia** bacilliform, simple, colourless. **Chemistry**: gyrophoric acid. **Ecology**: on siliceous rocks, especially granite and schists, upland to montane, Northern Hemisphere.

The genus is distinguished from other squamulose lichens with *Lecidea*-like or *Aspicilia*-like apothecia by the thin- and single-walled cylindrical asci which lack any distinct apical thickening (tholus).

LITERATURE: Clauzade & Roux (1985), Hafellner (1984), Rambold (1989), Schneider (1980), Tønsberg (1992).

1 Thallus squamulose; squamules 1–2.5(-3) mm long, 1–1.5 mm thick; ascospores globose .... **1. *cinereorufa***
Thallus areolate; areoles <1 mm diam., usually <0.5 mm thick; ascospores ellipsoid .... **2. *fuscocinerea***

1. S. cinereorufa (Schaerer) Th. Fr. (1861)

Thallus squamulose, squamules compacted, 1–2.5(-3) × 0.5–1 mm, 1–1.5 mm thick, ± bullate, surface verrucose, not clearly lobate, grey-brown to deep reddish brown; prothallus black, well developed. Apothecia 0.5–1.2 mm wide, immersed to ± sessile, black, disc concave to flat; true exciple mostly persistent, dark brown, thick; hymenium 90–120 μm tall; hypothecium brown to brown-black. Asci 55–65 × 8–10 μm, cylindrical. Ascospores (6-)7–9(-10) μm diam., globose, uniseriately arranged. Pycnidia 60–100 μm diam., usually numerous, immersed; wall dark brown, K−; conidiogenous cells 6–11 × 1.5 μm; conidia 3.8–6 × 0.8 μm. Thallus and medulla Pd−, K−, KC−, C± pink (gyrophoric acid).

On granite, hard sandstones and schists, more rarely overgrowing mosses, in montane situations; local. N. England, Wales, Scotland (Highlands), ?S.W. Ireland (Kerry). C. & N. Europe, North America.

The squamules could be mistaken for those of an *Acarospora*, but are readily identified by the open discs of the apothecia and the larger and subglobose ascospores. The C+ pink reaction is variable, even within single specimens.

2. S. fuscocinerea (Nyl.) Clauz. & Roux (1985)

Lecidea fuscocinerea Nyl. (1852) non auct. brit., *S. tenebrosa* (Flotow) Hertel & Poelt (1980), *S. endocyanea* (Stirton) Hertel & G. Schneider (1980), *Aspicilia complanatoides* (A.L. Sm.) W. Watson (1935); incl. *S. fuscocinerea* var. *sorediata* (Houmeau & Roux) Coppins (1992)

Thallus crustose, grey-brown to brown, in particularly exposed situations sometimes black, areolate; areoles flat to ± convex or bullate, (0.2-)0.3–0.6(-0.8) mm diam. Soralia rare, blackish, convex, *c.* 0.1–0.4 mm diam. Apothecia black, immersed at first but becoming sessile; disc at first concave, then ± flat and finally convex; true exciple sometimes distinct at first but then often soon excluded, dark brown but thin in section; hymenium 90–110(-120) μm tall. Asci ± cylindrical, 55–75 × 10–12(-15) μm. Ascospores elongate-ellipsoid to broadly obpyriform, ± biseriately arranged, (10-)12–16(-18) × 5–6(-8) μm. Pycnidia not seen. Cortex and medulla Pd−, K−, KC−, C± pink (gyrophoric acid, ± unidentified compounds).

On hard siliceous rocks, especially schists, in sunny, exposed situations and often associated with yellow *Rhizocarpon* species. W. & N. Britain, from the Channel Is. to the Shetland Is.; extending locally to Ireland. Widespread in the Northern Hemisphere.

A morph from Arran with ascospores 14–17 × 6.5–8 μm and norstictic as well as gyrophoric acid has been reported and merits further study to clarify its status. *S. endocyanea*, described from W. Scotland (Mull), has sometimes been regarded as a separate species distinguished by reported differences in the colour of the epithecium, but falls within the range of *S. fuscocinerea*.

Often mistaken for *Lecidea gyrizans* and *L. mullensis*, with which it often grows, but those species have richly branched paraphyses, *Rimularia*-type asci, and Pd+ yellow or orange, C− thalli.

The recently described ***S. corticola*** Muhr & Tønsb. (1992) has inconspicuous areoles and mostly discrete, dark brown (abrading green) soralia, to 2 mm diam., and broadly ellipsoid to globose ascospores, (10-)13–16(-20) × (8-)9–12(-14.5) μm. Soralia C+ red (gyrophoric, 5-*O*-methylhiasic and ± lecanoric acids). Recorded once on trunk of roadside *Ulmus* in E. Sutherland. Norway and Sweden.

SCHISMATOMMA Flotow & Körber ex Massal. (1852)

Thallus crustose, mainly superficial; prothallus when present usually dark brown or black; with or without soralia, which predominate in some species. **Photobiont** *Trentepohlia*. **Ascomata** apothecia, ± immersed or sessile, rounded or somewhat irregular in shape. **Thalline exciple** present. **True exciple** poorly developed. **Epithecium** pale to dark brown. **Hymenium** I+ blue. **Hypothecium** dark olivaceous to black, unknown in some species. **Hamathecium** of branched and anastomosing paraphysoids. **Asci** 8-spored, clavate, with two functional wall layers, thickened at the apex, with an internal I+ blue ring, discharge fissitunicate. **Ascospores** fusiform to curved, multiseptate, cells sometimes unequal in size, colourless, sometimes brownish when old, thin-walled, smooth, sometimes coiled in the ascus. **Conidiomata** pycnidia, immersed. **Conidiogenous cells** ± elongate-ampulliform, lining the pycnidial cavity, enteroblastic. **Conidia** arising singly, bacilliform, straight or slightly curved to kidney-shaped, simple, colourless. **Chemistry**: ± orcinol *para*-depsides (e.g. lecanoric acid, erythrin), β-orcinol depsides (e.g. fumarprotocetraric acid, psoromic acid), chromones (e.g. lepraric acid), fatty acids, and unidentified substances; very variable. **Ecology**: on dry bark, and rocks.

A largely tropical genus with over 85 decribed species and in need of a modern revision. The genus is close to both *Enterographa* and *Lecanactis*, differing from both in the thalline exciple, and a poorly developed true exciple. Three species treated below have never been found with apothecia and their placement in the genus is based primarily on their chemistry. All species have *Trentepohlia* as the photobiont, demonstrated in the field or when fresh by scratching away the cortex or soralia when an orange mark remains. When thalli have been kept in the herbarium for a short time they become paler or brighter white than on collection; this is due to the rapid loss of orange pigmentation of the photobiont.

LITERATURE: Coppins & James (1979*a*), Coppins & James (1989), James & Coppins (1979).

1 Thallus sorediate, often abundantly so; apothecia absent 2
Thallus lacking soredia; apothecia present .... **3. *graphidioides***

2(1) Thallus and soredia Pd+ orange or yellow orange 3
Thallus and soredia Pd− 4

3(2) Thallus pale brownish grey to greyish white; soralia discrete, initially ± excavate, not or rarely confluent, pinkish grey; Pd+ orange .... **5. *quercicola***
Thallus pale pink or creamy white to glaucous grey; soralia initially ± raised, generally soon becoming confluent, pale pink or creamy; Pd+ yellow-orange .... **4. *niveum***

4(2) Thallus grey or brown, ± entirely sorediate, or with punctiform soralia 5
Thallus chalk-white, ± entirely sorediate, K+ yellow .... **1. *cretaceum***

5(4) Thallus thick, soft, ± entirely sorediate, brown or ochre-brown; UV+ glaucous; mostly on shaded rock .... **6. *umbrinum***
Thallus thin, pale mauve-grey; soralia punctiform, ± discrete; UV−; mostly on trees .... **2. *decolorans***

1. S. cretaceum (Hue) Laundon (1984)

S. virgineum D. Hawksw. & P. James (1974)

Thallus wide-spreading, thick and ± tartareous, whitish grey at first, becoming pure chalk-white due to the confluence of soralia; soralia at first ± delimited but becoming irregular and erose, then ± covering the whole thallus surface. Apothecia and pycnidia unknown. Thallus Pd−, K+ yellow, KC−, C−, UV± dull pale grey, tinged orange (unidentified substances related to lepraric acid, unidentified fatty acids).

On well-lit, dry sides of old wayside and woodland trees, especially with some slight nutrient enrichment, forming extensive patches. S. & E. Britain, extending to E. & N. Scotland and, very locally, Ireland. Denmark, France.

Distinguished from *S. niveum* **(4)** by the starkly white, Pd−, thallus.

2. S. decolorans (Turner & Borrer ex Sm.) Clauz. & Vězda (1965)

Thallus thin, lilac- to pinkish grey, becoming grey to brownish grey in the herbarium, ± rimose-cracked; a grey-black prothallus often ± present; soralia punctiform, 0.2–0.8(-1) mm diam., discrete or sparingly confluent, mauve-grey to pale lilac-grey, occasionally with an ochre-tinge. Apothecia and pycnidia unknown in the British Is. Thallus Pd −, K −, KC −, C −, UV − (unidentified fatty acids).

On the dry sides of young to old deciduous trees in wayside and woodland sites, tolerant of relatively high levels of agricultural chemicals and fertilizers, rarely on dry vertical rocks or walls; often abundant. S. & E. Britain, extending to the north and west, C. N. & E. Ireland. Central mediterranean Europe, S. Scandinavia, North America.

Apothecia are known only from a single site in France; the apothecia are 0.3–1 mm diam., rounded or often ± irregular, adpressed, convex, dark with a pale pink-grey pruina, the thalline exciple is poorly developed; paraphysoids densely anastomosed; ascospores 25–35 × 4–5 μm, 3-septate, curved or straight.

3. S. graphidioides (Leighton) Zahlbr. (1919)

Thallus white to pale grey, often cracked, with calcium oxalate crystals; prothallus when present black; soralia absent. Apothecia very irregular in shape, simple or branched, ± elongate, 0.3–1.5(-2) × 0.08–0.2 mm, immersed or slightly raised; thalline exciple ± white; epithecium reddish brown; hymenium 55–80 μm tall, I+ blue; hypothecium 40–115 μm tall, base root-like, dark red brown, K+ olivaceous green to black; paraphysoids sparsely branched, pale brown above. Ascospores 24–36(-40) × (2.5-)3–3.5 μm, 3-septate, narrowly fusiform to arcuate or sigmoid, coiled together in the ascus. Pycnidia 60–100 μm diam., immersed, black; conidia 4–5 × 1.3–1.5 μm, ± kidney-shaped. Thallus Pd −, K −, KC −, C −, I − (no lichen products detected by t.l.c.).

On trunks of deciduous trees (*Fagus*, *Fraxinus* and *Quercus*) in parklands or well-lit woodland with slightly nutrient-enriched bark; very rare. S.W. England (N. Somerset, Exmoor; Shropshire), ?N. Wales (Denbigh), Scotland (Peebles, Easterness), ?E Ireland (Dublin). S. Scandinavia, W. France, Germany, Italy, Portugal, Spain, ?Morocco.

S. pericleum (Ach.) Branth & Rostrup (1869) [*S. abietinum* (Humb.) Massal. (1852)] is doubtfully distinct, sharing an identical apothecial anatomy and spores, but is said to differ in having ± rounded to elongate-ellipsoid apothecia, becoming strongly convex, with a distinct, ± pruinose thalline exciple. A critical comparison of the range of variation in each is required to establish whether two or only one species is involved. See also *Opegrapha rufescens*.

4. S. niveum D. Hawksw. & P. James (1971)

Thallus wide-spreading, pale pink when fresh, soon becoming creamy white to glaucous grey in the harbarium; soralia at first punctiform, to 1 mm or more diam., rapidly becoming confluent and involving the entire surface, concolorous. Apothecia and pycnidia unknown. Thallus Pd+ bright orange-yellow, K −, KC −, C −, UV ± dull pale grey ± tinged orange (psoromic and 2-*O*-demethylpsoromic acids and ± an unidentified xanthone).

On bark on the dry sides of ancient trees, often forming extensive, pale pink patches, especially *Quercus*, and *Ilex* (in Ireland) in old woodlands, frequently associated with *Lecanactis premnea*, *L. abietina* and *S. decolorans*; common. S. & W. England, S. & C. Wales, extending very locally to N. Scotland (Inverness) and S.W. & N.W. Ireland (Kerry, Donegal). France (Brittany).

Haematomma ochroleucum var. *porphyrium* is separated by the distinct white and fibrous prothallus, greenish glaucous colour, the presence of *Trebouxia*, and the Pd − thallus. *S. quercicola* **(5)**, which has a similar or more intense Pd+ reaction, differs in the more pinkish or brownish thallus and the small, persistently discrete soralia.

5. S. quercicola Coppins & James (1989)

Thallus thin, pale brownish grey to greyish white, slightly verrucose, often overgrown with chlorococcoid algae; soralia scattered, 0.2–0.4(-0.5) mm diam., generally discrete and not confluent, pink grey when fresh but becoming whitish in the herbarium. Apothecia and pycnidia unknown. Thallus and soredia Pd+ orange, K −, KC −, C −, UV − (fumarprotocetraric acid, ± protocetraric acid and an unidentified substance).

On old trunks, particularly of *Quercus* and *Fagus*, often associated with *Haematomma elatinum* and *Thelotrema lepadinum*, widespread in ancient woodlands. N. & W. Great Britain, E. England (Suffolk), N. & S.W. Ireland. ?Endemic.

Has a superficial resemblance to *Pertusaria pupillaris*, which has a chlorococcoid photobiont and does not contain *Trentepohlia* (i.e. is not orange when scratched). Many British records of '*Lecidea cinnabarina*' refer to *S. quercicola*. The obligately lichenicolous *Arthonia invadens* occurs on this species.

6. S. umbrinum (Coppins & P. James) P. Jørg. & Tønsb. (1988)

Lecanactis umbrina Coppins & P. James (1979)

Thallus superficial, wide-spreading, only the outermost few mm at the edge corticate and there thin, dark brown, with a matt, slightly tomentose surface, surrounded by a conspicuous, black or purple-black, ± fibrose prothallus; central part of thallus entirely and continuously sorediate; soredia forming a thick, to 1 mm tall, uneven, soft, powdery crust, which, in part, becomes coarsely rimose-cracked, glaucous-grey-yellow-brown, sometimes with a distinct dark brown or yellow-ochre mottling, white or glaucous white when abraded. Soredia farinose to coarsely granular, at times partly tomentose. Apothecia and pycnidia not recorded in British material. Apothecia sessile, round, often somewhat irregular to lobate, to 1.5 mm diam.; disc black-brown, not pruinose, becoming ± convex; thalline exciple ± prominent, sometimes sorediate. Paraphysoids straight, only branched in the upper part. Ascospores 18–22 × 3–5 μm, 3-septate, fusiform, slightly curved. Conidia (10-)15–20(-30) × 1 μm, curved. Thallus Pd −, K −, KC −, C −, medulla and abraded soralia UV + ice-blue (schizopeltic acid, 2–3 unidentified UV + substances).

On sheltered, siliceous rocks, particularly on vertical faces and undersides of overhangs in old, moist, sheltered woodlands, rarely on bark (*Sorbus*) on dry undersides of overhanging boughs; very local. W. Britain, N.E. England (N. Northumberland). Norway.

Found fertile on a single occasion in Norway (Bergen, Milde).

SCLEROPHORA Chev. (1826)

Thallus crustose, immersed. **Photobiont** *Trentepohlia*. **Ascomata** apothecia, stalked; stalk short to long, pale or reddish bown. **True exciple** well developed, forming a distinct collar at base, lower part of collar ± extended and enclosing stalk, head globose to obconical. **Asci** cylindrical, with a single functional wall layer, I −, dissolving at early stages, formed singly from ascogenous hyphae with croziers. **Ascospores** globose, pale, wall colourless, minutely warted, accumulating in a well-developed, pale mass. **Conidomata** not seen. **Chemistry**: apothecia may have pruina containing ?anthraquinone (K + violet). **Ecology**: in dry bark crevices.

Separated from *Chaenotheca* by the different ontogeny and the absence of pigment in the ascospore walls.

LITERATURE: Middelborg & Mattsson (1987), Tibell (1984).

1 Stalk when wet with brown core, mazaedium K −; ascospores 2.5–4.0 μm .... **1. *peronella***
Stalk when wet translucent, mazaedium K + purple-violet (coloured crystals); ascospores 4–6 μm .... **2. *nivea***

1. S. peronella (Ach.) Tibell (1984)

Coniocybe peronella (Ach.) Tibell (1978)

Apothecia 0.5–0.8 mm tall, pale brown, darker towards centre; true exciple poorly developed, not raised, yellow-pruinose when young, becoming reddish brown on the lower surface of the head; stalk without pruina, pale, with brown core when wet. Ascospores 3–3.6 μm, globose; surface reticulately cracked.

In dry bark crevices and on wood of deciduous trees; rare. England (Yorkshire), C. & E. Scotland. C. Europe to S. Scandinavia.

2. S. nivea (Hoffm.) Tibell (1984)

Coniocybe pallida (Pers.) Fr. (1824)

Apothecia 0.4–0.7 mm tall; true exciple well developed, raised; lower surface of head and stalk pale yellow, translucent when wet, becoming grey when senescent, without pruina; young apothecia sessile, globose and yellow-pruinose. Ascospores 4–6 μm; surface with a cog-like ornamentation; spore mass pale brown when mature, K+ purple-violet (with formation of large crystals). Pruina K−.

In dry bark crevices and on wood of deciduous trees. N.E. England (Northumberland), C. and E. Scotland. C. Europe to S. Scandinavia.

Due to nomenclatural misunderstandings has often been confused with the non-lichenized *Roesleria subterranea* (Weinmann) Redhead.

SCLEROPHYTON Eschw. (1824)

Thallus crustose, thin, uniform. **Photobiont** *Trentepohlia*. **Ascomata** apothecia, immersed, often forming groups or lines. **Thalline exciple** absent. **True exciple** narrow, brown above, pale, below. **Hamathecium** of paraphysoids, thick, 2–2.5 μm wide, anastomosing. **Asci** 8-spored, fissitunicate. **Ascospores** *c.* 8-septate, elongate-fusiform or long clavate, black-grey to brown.

Resembles *Enterographa*, which has colourless ascospores.

LITERATURE: James (1974).

1. S. circumscriptum (Taylor) Zahlbr. (1905)

Thallus to 5 cm diam., often forming extensive, angular mosaics of smaller thalli, continuous, tartareous or finely scurfy, finely irregularly rimose-cracked, white, glaucous or mottled grey, often surrounded by a conspicuous, dark grey-black prothallus, surface with densely compacted crystals. Apothecia to 0.1 mm diam., dot-like, often very numerous, immersed, rounded or oval, sometimes contiguous, often in groups or dispersed in ± flexuose or radiating lines; disc flat, black, often ± densely white-grey-pruinose. Ascospores 17–26 × 5–6 μm, (4-)5(-7)-septate, ± constricted at each septum when mature, black-grey to dark brown, outer and cross-walls somewhat thickened. Thallus Pd+ yellow-orange, K−, KC−, C−, UV− (psoromic, 2′-*O*-demethylpsoromic acids).

In dry crevices and beneath sheltered underhangs of siliceous maritime rocks, intolerant of direct wetting by rain or sea spray; locally frequent. W. coasts of Britain from S.W. England (Dorset) to N. Scotland (Argyll, also Berwickshire) and Ireland. Coastal W. Europe from Brittany to Portugal, Mediterranean region, Madeira, Azores, ?S.W. Norway.

Easily recognizable in the field as the only species with a thin, white, even thallus growing in dry, sheltered crevices which, when scratched, is orange-yellow due to the photobiont; when fertile the thallus becomes pale grey, due to the numerous, tiny apothecia which become more apparent (black) when wet. It is the characteristic species of the *Sclerophyton circumscriptae* which includes *Dirina massiliensis* f. *sorediata*, *Roccella* species, *Lecanora tenera* and, in S. England, *Chiodecton myrticola*, *Arthonia endlicheri* and *Lecanora praepostera*.

The position of *S. circumscriptum* in an otherwise tropical genus confined to trees is in need of investigation.

SCOLICIOSPORUM Massal. (1852)

Thallus crustose, glaucous green, green, grey-white or brownish black, thin, not corticate, finely granular-scurfy to almost powdery, at times resembling growths of *Desmococcus*; prothallus absent, **Soralia** rare. **Photobiont** chlorococcoid, often forming goniocysts. **Ascomata** apothecia, at first flat, soon convex, sessile, constricted at base, whitish, yellowish, pale brown to black. **Thalline exciple** absent. **True exciple** evident in young apothecia, soon excluded, of radiating, branched and anastomosing hyphae, pale or colourless. **Epithecium** colourless, brownish, blue-green or olivaceous, sometimes interspersed with granules. **Hymenium** colourless, olivaceous, blue-green, brown or almost colourless above and then sometimes surrounded by numerous small granules, merging into exciple; I+ blue. **Hypothecium** colourless or pale, upper part of erect hyphae, randomly orientated below. **Hamathecium** of paraphyses, septate, branched and anastomosed in a gelatinous matrix, resembling excipular hyphae; apices not or only slightly thickened, sometimes a few with dark brown caps in young apothecia. **Asci** 8-spored, broadly clavate, or clavate-cylindrical, *Lecanora*-type with broad apical cushion. **Ascospores** acicular, sickle-shaped or bent spirally, rarely elongate-spindle-shaped and attenuated towards one end, straight or curved, colourless, 3- to multiseptate, the septa very thin, often indistinct. **Conidiomata** pycnidia, spherical, ± immersed to emergent, wall dark brown to colourless. **Conidia** of two types: macroconidia rod-shaped, straight or slightly curved; microconidia curved.

LITERATURE: James (1971), Vězda (1978).

1 Apothecia usually red-brown or blackish; epithecium not granular; ascospores 2–5 μm wide 2
Apothecia whitish; epithecium granular; ascospores 1–1.2 μm wide .... **2. *pruinosum***

2(1) Ascospores outside ascus spirally twisted, 2–3 μm wide .... **3. *umbrinum***
Ascospores outside ascus curved or straight, 4–5 μm wide .... **1. *chlorococcum***

1. S. chlorococcum (Graewe ex Stenhammar) Vězda (1978)

Thallus irregularly granular, granules scattered or contiguous, dirty grey-green, dark green or blackish green. Apothecia mostly 0.2–0.3 mm diam., convex, red-brown to dark brown-black, often shiny; epithecium pale brown to deep fuscous brown, occasionally with a greenish blue tinge, rarely entirely deep blue-green. Ascospores 20–40 × 4–5 μm, mostly 7-septate, elongate-spindle-shaped, tapered at one end, straight or curved.

On shaded, damp, ± nutrient-rich bark, often on branches or twigs, less often on wood and *Calluna* stems, pollution-tolerant, often with *Lecanora conizaeoides*; common. Throughout British Is.

Resembles, and often occurs with, the alga *Desmococcus olivaceus* and may form similarly large patches. Has wider ascospores than other *Scoliciosporum* species.

2. S. pruinosum (P. James) Vězda (1978)

Thallus thin, sub-leprose-granular or almost absent, glaucous to grey-white, often wide-spreading, effuse. Apothecia 0.1–0.3 mm diam., convex from initiation, becoming almost globose; true exciple obscured, sometimes becoming irregularly tuberculose; disc very pale pinkish, yellowish or brownish white, becoming browner when mature, ± white-pruinose; epithecium colourless or pale brownish yellow, with numerous colourless interspersed granules to 1 μm diam., dissolving in K. Ascospores (22-)25–32(-40) × 1–1.2 μm, spirally twisted in ascus, sigmoid-curved or more rarely straight on release; septa 3–5, very indistinct.

On dry bark in sheltered situations, especially *Quercus*. Mainly S. & S.W. Britain, extending to S.W. Scotland (Galloway). Pyrénées, Madeira.

Characterized by the tiny, white pruinose, whitish to pale pink, ± crowded apothecia, the densely granular epithecium and true exciple and the spiral twisting of the spores in the asci. Superficially recalls *Biatora sphaeroides* which has larger, not pruinose, more deeply coloured apothecia and wider, 1- to 3-septate ascospores. *Micarea* species such as *M. pycnidiophora* and *M. stipitata* have similar, but non-pruinose, apothecia which are always accompanied by numerous, white-pink, stalked, pycnidia; *M. alabastrites* is also similar but has a non-granular epithecium and differently shaped ascospores.

3. S. umbrinum (Ach.) Arnold (1871)

Thallus very variable, thin, scurfy-uneven, ± cracked, to rather thick, nodular-uneven, dark green-brown to black, occasionally rusty. Apothecia 0.3–0.8 mm diam., red-brown to dark brown-black; epithecium blue-green, green-brown or olive-brown, colours often intermixed, paler true margin sometimes persisting. Ascospores (15-)20–30(-40) × 2–3 μm, 3- to 7-septate, needle-shaped, spirally twisted.

On ± basic or siliceous rocks, memorials, metal-rich scree, rusty metal work, walls, branches, twigs, wood and man-made substrata in coastal and upland areas, inconspicuous but widespread and pollution-tolerant. Abundant. Throughout British Is. Europe, North America.

Very variable and with a wide ecological amplitude; often forming dark, poorly organized thalli amongst other lichens. The spirally twisted ascospores in the ascus are diagnostic and may give the illusion of a multispored ascus. Morphs on bark in E. England with pale apothecia and irregular, pale green soralia (KC+ red) may be referable to *S. sarothamni* (Vainio) Vězda (1978).

SIPHULA Fr. (1831)

Thallus shrubby, erect, sparingly branched, whitish. **Cortex** pseudoparenchymatous. **Photobiont** chlorococcoid. **Medulla** compact; hyphae ± longitudinally aligned. **Apothecia** and **conidiomata** unknown. **Chemistry**: *para-* and *meta*-depsides, dibenzofurans, and(or) chromones (e.g. siphulin). **Ecology**: a genus of *c.* 40 species characteristic of peaty soils and acid, leached bark in temperate rain-forests and wet alpine sites in the S. Hemisphere.

LITERATURE: Thomson (1984).

1. S. ceratites (Wahlenb.) Fr. (1831)

Thallus to 7 cm tall, forming compact tufts, scattered or ± continuous in extensive patches; branches to 2 mm diam., chalk-white, pale greyish or ivory, ± erect, not or sparingly branched, ± cylindrical or somewhat flattened, longitudinally plicate or furrowed, particularly towards a ± rooting base, smooth, ± rugose or scabrid-pruinose, solid; apices ± rounded, not markedly tapered. Cortex Pd−, K+ yellow to yellow-brown, KC+ yellow-orange to yellow-brown, C+ violet-yellow-brown, medulla UV± violet-glaucous or ± yellow (siphulin).

On shallow peat or gravelly soil, often in shallow, temporary rain hollows in bed-rock, montane; rare but locally frequent. N.W. Scotland. Fennoscandia, Siberia, North America, W. Greenland.

A distinctive species unlikely to be confused with any other British lichen; poorly developed material could be misidentified as *Cladonia uncialis,* which has a more yellow-green thallus composed of hollow podetia which are dichotomously branched with pointed apices and open axils. *Thamnolia vermicularis* differs in having pointed apices, hollow, extended branches, a ± prostrate habit and a different chemistry.

SOLENOPSORA Massal. (1855)

Thallus crustose, small squamulose or placodioid with lobed margins firmly attached, rhizines mostly absent but present in *S. holophaea* **(3)**; prothallus absent or blue-black. **Upper surface** corticate. **Photobiont** chlorococcoid; cells, mostly globose, a few ellipsoid. **Ascomata** apothecia, sessile. **Thalline exciple** present, sometimes becoming excluded. **Epithecium** yellowish brown to red-brown, sometimes interspersed with granules. **Hymenium** colourless to red-brown, I+ blue. **Hypothecium** colourless to yellowish brown or reddish brown or orange-brown. **Hamathecium** of paraphyses, simple, 1.5–2 μm wide, with swollen apices (4–5 μm wide) that often have an internally brown apical cap, simple, pigment within swollen tips. **Asci** 8-spored, clavate, ± *Catillaria*-type but sometimes with a small ocular chamber. **Ascospores** colourless, ellipsoid to fusiform, 1-septate, without a thick perispore. **Conidiomata** pycnidia, immersed in thallus. **Conidiogenous cells** Type VI. **Conidia** 3–4 × 0.8–1 μm, simple, colourless. **Chemistry**: various unidentified substances detected by t.l.c. **Ecology**: on rocks and soil in crevices in coastal or sunny areas, predominantly Mediterranean.

Distinguished from *Squamarina* and placodioid *Lecanora* species by the septate ascospores, and from *Psora* by the same character and also by the presence of a thalline exciple, at least when young. Related to *Lecania*.

1 Thallus placodioid, with ± distinct, radiating marginal lobes 2
Thallus small-squamulose; lobes not radiating 3

2(1) Upper surface whitish or pale grey, thickly white-pruinose; lobes flattened ..... **1. *candicans***
Upper surface olive-grey, margin pale bluish pruinose; lobes ± convex .... **2. *liparina***

3(1) Thallus red-brown or brown, distinctly squamulose; soredia absent; apothecia present .... **3. *holophaea***
Thallus pale, glaucous or grey-white, almost crustose; isidia-like lobules mixed with granular soredia; apothecia usually absent .... **4. *vulturiensis***

1. S. candicans (Dickson) Steiner (1915)

Thallus to 4(-5) cm diam., rosette-forming, placodioid, shortly radiating, orbicular or irregular, of ± indistinctly defined, flattened, ± contiguous lobes, overlapping secondary lobes sometimes present in older parts; upper surface uneven, pale whitish to bluish grey-white, matt, thickly white-pruinose, appearing roughened or scurfy, usually rimose-cracked towards the centre; prothallus blue-black. Apothecia to 1.2 mm diam., emergent, at first immersed, becoming ± sessile; disc flat, dark greyish blue to brownish black, faintly pruinose or roughened; thalline exciple ± persistent, concolorous with the thallus, eventually sometimes excluded, thickly crystalline (calcium oxalate); epithecium yellow-brown, ± granular but without crystals, granules not dissolving in K; apices of paraphyses mostly colourless, but a few swollen, brown-capitate apices present, especially in young apothecia; hymenium 50–60 μm tall, colourless; hypothecium colourless to pale straw. Ascospores (10-)14–18 × 3–4 μm, ellipsoid-elongate or somewhat clavate. Thallus Pd+ orange, K−, KC−, C−, UV+ grey (unidentified substances).

On sunny, exposed, hard, calcareous rocks, especially limestones, often near the ground, frequent on limestone monuments in churchyards. Almost throughout England, from Somerset and Wiltshire to Derby (Peak District) and N. Yorkshire, extending northwards to W.C. Scotland (Strathclyde, Lismore). W. Europe, Mediterranean, North America.

Characterized by the spreading, flattened, placodioid thallus, the emergent apothecia with a prominent thalline exciple and the complex chemistry. Recalling *Diploicia canescens*, which has narrower, low, convex lobes, is sorediate, and rarely with apothecia, the apothecia lacking a thalline exciple.

2. S. liparina (Nyl.) Zahlbr. (1919)

Like **1**, but lobes narrower, more distinctly strongly convex, often forming scattered clusters of partially loosely adpressed, ± indented lobes or small, compact rosettes; upper surface pale olivaceous grey, bluish pruinose at the periphery, sometimes nodular towards the centre. Apothecia sessile with a thick, smooth or ± scabrid thalline exciple, uneven, becoming ± excluded later; epithecium brown, granular, granules not dissolving in K. Ascospores (9-)13–16 × 4–4.5 μm, straight or slightly curved, not or ± constricted at septa. Thallus Pd+ orange, K−, KC−, C−, UV− (chemistry as in *S. candicans*).

On sunny, serpentine rocks near the coast. S.W. England (Cornwall, Lizard). Mediterranean.

S. candicans **(1)** has a more even, whitish pruinose thallus, more flattened lobes and occurs on hard, calcareous rocks.

3. S. holophaea (Mont.) G. Samp. (1921)

Thallus of numerous, overlapping, ± horizontal squamules, to 2.5 mm wide, often with rounded, indented, ± raised margins, irregular, not radiating; upper surface pale to deep red-brown, olive-greenish when wet, smooth, ± wavy, occasionally partly rugose; lower surface pale fawn, with scattered, pale rhizines. Apothecia soon emergent, becoming sessile; thalline exciple at first rather regular, later ± excluded; disc red-brown to blackish brown, concolorous or darker than margin; epithecium red-brown; hymenium *c.* 50 μm tall, ± red-brown; hypothecium colourless or tinged pale red-brown. Ascospores 12–16(-20) × 4(-5) μm, ellipsoid-elongate. Thallus Pd−, K−, KC−, C−, UV− (± atranorin, unidentified terpene).

On soil and soft rocks, chiefly on slightly sheltered clefts in vertical cliffs or old walls, chiefly on or near the coast; local. Throughout W. British Is., very rare in S.E. England. Mediterranean, Macaronesia.

Often confused with *Psora lurida*, which occurs on inland calcarous rocks, is without a thalline exciple, has ± globose, simple ascospores, and paraphyses which are very thick (3.5–5 μm) throughout and with a red-brown pigment external to the apices.

The placement of this species in *Solenopsora* is anomalous.

4. S. vulturiensis Massal. (1856)

Thallus very variable, usually of numerous, nodular, overlapping, rounded or slightly extended, ± radiating lobes, very crowded or discrete, ± scattered, often forming twisted, coralloid, ± isidiate clusters; upper surface whitish, glaucous, pale grey- to olive-green, ends ± whitish pruinose, opalescent grey-green when wet; lobes to 0.5 mm wide, apices often swelling and breaking down into coarse, granular soralia, paler or concolorous with the thallus. Apothecia rare, sessile, ± globose; thalline exciple ± persistent, thin and irregular, later becoming excluded; disc 0.3–0.6 mm diam., pinkish or red-brown, flat to convex; epithecium brown to reddish brown; hypothecium colourless to bright orange-brown. Ascospores 9–11.5(-14) × 4.5–5(-6) μm, simple (to 1-septate). Thallus Pd+ orange, K+ faint yellow, KC+ faint yellow, C−, UV− (± atranorin and unidentified substance).

In moist crevices, mud-covered rocks, decaying vegetation, and consolidated soil, mostly associated with sheltered, nutrient-enriched coastal rocks and damp banks, submesic to xeric-supralittoral; often locally frequent. S., W. & N. British Is. Mediterranean region northwards to S. Sweden & Norway, Macaronesia.

The development of soralia is very variable; the thallus may be ± entirely dissolved in granular soredia or (rarely) they may be completely absent. Resembles terricolous morphs, especially when fertile, of *Trapelia involuta* which have lobes which are less conspicuous, more adpressed, not nodular and C+ red. When densely sorediate *S. vulturiensis* may also superficially resemble *Lepraria* species with which it often grows; the small congested, sorediate lobes are usually diagnostic.

SOLORINA Ach. (1808)

Thallus foliose, wide-spreading, or in one species, reduced to a collar round the apothecium. **Upper surface** green-grey ± bright green when wet. **Lower surface** indistinctly veined, tomentose and with clusters of simple or ± branched rhizines. **Photobiont** *Coccomyxa,* internal or external, blue-grey cephalodia containing *Nostoc* also present. **Ascomata** apothecia, large, rounded, irregularly scattered, impressed to ± deeply immersed in the upper surface; disc dark red-brown, slightly to very deeply concave. **Thalline exciple** absent. **Hamathecium** of simple paraphyses, conglutinate, not or little swollen at the apex. **Asci** clavate (1-)2-, 4- or 8-spored, *Peltigera*-type. **Ascospores** red-brown or brown, ellipsoid to fusiform 1-(-5) septate with a median constriction, ± uniformly thickened, surface ornamented or ± warted. **Conidiomata** not known. **Chemistry:** lichen products mostly absent; one species with solorinic acid (orange pigment). **Ecology:** on soil, markedly calcicole, one species, *S. crocea* **(1),** mildly calcifuge.

This genus is easily recognizable by the presence of ± impressed apothecia which have given rise to the popular name "socket lichens".

LITERATURE: Gilbert (1975), Jahns (1988), Thomson & Thomson (1984).

1 Underside white or brownish white 2
 Underside of thallus vivid orange .... **1. *crocea***

2(1) Thallus ± wide-spreading, lobed; cephalodia internal 3
 Thallus confined to a narrow collar surrounding apothecia; cephalodia external, forming a dark mat amongst and under the apothecia .... **3. *spongiosa***

3(2) Thallus wide-spreading, 2–5 cm diam.; asci 4-spored .... **2. *saccata***
 Thallus restricted, <1 cm diam.; asci (1-)2-spored .... **4. *bispora***

1. Solorina crocea (L.) Ach. (1810)

Thallus forming large rosettes of rather thick lobes with ± rounded, scarcely indented apices; upper surface olive-green when moist, red-brown when dry, ± uneven or scabrid; lower surface bright orange, tomentose with a ± reticulate pattern of brown veins; external cephalodia absent, with internal cephalodia. Apothecia to 7 mm diam., rare, dark brown, ± round, not, or only little, depressed into the thallus; disc to 1 cm diam., flat or ± convex. Asci 6- to 8-spored. Ascospores 35–45 × 10–12 μm, brown, oblong, 1-septate. Medulla and lower surface K+ purple (solorinic acid).

A ± calcifuge soil lichen of late snow beds and leached windswept ridges at high altitudes, solifluxion areas, stonefields; locally abundant above 900 m, rarely descending to lower altitudes in river shingle sites. Scotland (Highlands), Ireland (Kerry). Arctic-alpine circumpolar, Alps, Pyrénées, Carpathians, Urals, Fennoscandia, North America, Greenland, Japan, New Zealand.

A highly distinctive species on account of the vivid orange lower surface. The structure of the thallus is unique among British lichens in having two discrete and ± continuous photobiont layers, an upper layer of *Coccomyxa* and a lower of *Nostoc*.

2. S. saccata (L.) Ach. (1810)

Thallus generally well developed, spreading, composed of conspicuous, rounded lobes with ± wavy margins; upper surface pale grey, ± tinged brown when dry, bright green when wet; naked to ± densely white-pruinose; lower surface white or pale, densely tomentose, rhizinose but not, or indistinctly, veined. Apothecia to 6 mm diam., frequent, brown-black, deeply sunk in depressions in upper surface. Asci normally 4-spored. Ascospores 32–50 × 18–27 μm, ellipsoid-oblong, 1-septate, with a thick, red-brown, furrowed and pitted epispore when mature.

On limestone outcrops, epidiorite, basalt, mica-schist, calcareous grassland, dunes, machair and walls; a well-marked calcicole locally abundant in uplands, rare in lowlands. British Is., except S.E. England and S.E. Ireland. Europe, North America, Japan.

The thallus incorporates a continuous layer of *Coccomyxa*, below which are scattered, internal cephalodia of *Nostoc* which can be seen on the underside as dark spots, occasionally spanning the entire thickness of the thallus so that their position is also denoted by blackish areas on the upper surface. At higher altitudes the thallus may be much reduced with a densely white-pruinose surface closely resembling that of *S. bispora* **(4)**.

3. S. spongiosa (Ach.) Anzi (1862)

Thallus with *Coccomyxa*, very reduced, restricted to a continuous or ± lacerate, narrow collar surrounding the scattered, small to moderate-sized, urceolate apothecia, to 5 mm diam., on a spreading cushion of dark grey-black, ± densely coralloid-warted or nodulose, gelatinous when wet, partially buried cephalodia. External cephalodia with *Nostoc*, becoming dark and sponge-like when wet. Asci (2-)4-spored. Ascospores 30–50 × 18–24 μm; epispore thickened and ornamented with pits.

In similar habitats to *S. saccata* **(2)** but less often in crevices, most records being from basic soil, old walls built with lime mortar, damp quarry floors and damp calcareous coastal dunes; local. Throughout Britain, mainly Pennines and Scotland (Highlands), also in N.E. Ireland. Arctic to temperate, circumpolar, Europe, North America, New Zealand.

Distinguished from *S. saccata* **(2)** by the dark, external cephalodia, and the mature ascospores which appear smoother in optical section. Occasional transitional morphs occur which are difficult to place and the status of the two species needs investigation. The apothecia are always accompanied by a narrow border or cuff of thallus containing *Coccomyxa*. The cephalodia, often partly buried, can exist independently but are never fertile.

4. S. bispora Nyl. (1860)

Thallus poorly developed, consisting of mostly scattered, ± rounded or irregular lobes, *c.* 5–10 mm diam. with a single urceolate apothecium, 2–5 mm diam. towards the centre of each lobe; upper surface pale grey to brown-grey, often pruinose. Cephalodia internal, rarely external. Asci normally 2-spored. Ascospores normally 1-septate, ± ellipsoid to broadly spindle-shaped. On the basis of ascospore size and number, three varieties are recognized: (a) var. *bispora* Nyl., asci 2-spored, ascospores 1-septate, (60-)90(-105) × (27-)40(-60) μm, rare; (b) **var. *macrospora*** (Harm.) H. Olivier (1921), as (a) but ascospores larger, (95-)110(-140) × 45–60 μm, rare; (c) **var. *monospora*** (Gyelnik) Frey (1952), asci 1-spored, ascospores 2-(3–5)-septate, 95–165 × 33–45 μm, very rare.

Base-rich cliffs, montane; locally abundant. Scotland (Highlands). Arctic to boreal, circumpolar, arctic-alpine in Europe, North America.

The smooth, collar-like thallus, which contains *Coccomyxa* and internal cephalodia, varies from pale grey to brown-green and tends to be strongly grey-white-pruinose remaining pale when moist. Morphs resembling *S. spongiosa* **(3)** with a supporting, cushion-like, coralloid thallus of dark external cephalodia are known. In the field *S. bispora* could be mistaken for small, high alpine morphs of *S. saccata* **(2)**, but is readily distinguished by its (1-)2-spored asci and much larger ascospores.

SPHAEROPHORUS Pers. (1794)

Thallus shrubby, erect or ± decumbent. **Branches** rounded or flattened, numerous, often rather brittle. **Cortex** chondroid, the cells with lumina 1–2 μm diam. **Photobiont** trebouxioid. **Medulla** of longitudinally orientated, thick-walled hyphae. **Ascomata** apothecia, apical or subterminal on branches, arising from ± globose thallus swellings, bursting irregularly exposing a purple-black, dry spore mass. **Thalline exciple** torn, thin. **Hypothecium** blackish brown. **Asci** 8-spored, cylindrical, *Calicium*-type; wall soon disintegrating. **Ascospores** simple, globose, with a dark, granular epispore, brown, dry. **Conidiomata** pycnidia. **Conidia** bacilliform. **Chemistry:** variable but sphaerophorin (UV+ blue) in all species. **Ecology**: mainly on acid rocks or acid, leached, mossy bark in humid situations.

1 Main branches persistently rounded; apothecia globose, terminal 2
Main branches conspicuously flattened; apothecia sub-terminal on lower surface near ends of branches .... **3. *melanocarpus***

2(1) Main branches notably wider than secondary branches; thallus grey, ± tinged or entirely red-brown; medulla I+ blue .... **1. *globosus***
Main and secondary branches similar in width; thallus grey or grey-brown, without a red-brown tinge; medulla I− .... **2. *fragilis***

LITERATURE: Ohlsson, K. E. (1973).

1. S. globosus (Huds.) Vainio (1903)

Thallus tufted, 1.5–5 cm, ± erect, forming spreading tufts or cushions, often richly, irregularly branched; main branches rather coarse, to 0.8–1.5 mm diam., rounded, branching anisotomic-dichotomous, with few to very numerous, thinner side branches, often appearing coralloid, with blunt apices, grey (shade), pale brown or brown-grey, often tinged red-brown, smooth. Apothecia infrequent, 1–3 mm diam., terminal, at first immersed in swollen, stalked, branch apices, soon bursting apically to ± expose a dry mass of dark brown-black ascospores, surrounded by a ± torn thalline exciple. Ascospores 8–10 μm diam., greenish to dark violet-blue-black. Medulla Pd− or ± yellow, K− or ± yellow, KC−, C−, I± blue (sphaerophorin, squamatic, ± thamnolic and ± hypothamnolic acids).

On rocks, often amongst other lichens and mosses, also on acid-barked, leached, mossy trees, more rarely on moorland soils; locally abundant. N. Scotland (Highlands), Ireland, extending locally to England and N. Wales. Europe, Arctic, Azores, North & South America, U.S.S.R., Japan.

Characterized by the distinct, coarse, rounded main branches and smaller secondary branches, a tendency to form lax to compacted, often ± pale or reddish brown tufts and the I± blue medulla. See also *S. fragilis* **(2)**.

2. S. fragilis (L.) Pers. (1794)

Like **1**, but forming dense, pale grey-brown tufts to 4 cm, usually without a red- brown tinge; branches ± erect or partially decumbent, branching sparingly isotomic-dichotomous, branches all of ± similar width, 0.6–0.8 mm wide, main branches not differentiated, surface dull grey-brown or grey, often somewhat roughened. Apothecia very rare, 1–2 mm wide, on ± elevated branches, subglobose, terminal, dry spore mass soon exposed. Ascospores 8–10 μm diam., deep violet-blue-black. Medulla Pd−, K−, KC−, C−, I− (sphaerophorin, ± hypothamnolic and ± squamatic acids).

On mountain outcrops, scree, boulders, generally at higher altitudes than *S. globosus*; locally common. N. Scotland (Highlands), extending locally to N. Wales, N. & S.W. England. Europe, North America, Arctic U.S.S.R., Japan.

S. globosus **(1)** has a more richly branched thallus with branches of unequal size and an I± blue medulla.

3. S. melanocarpus (Swartz) DC. (1815)

Thallus to 4 cm, of crowded, smaller, sterile, ± terete basal branches, and less frequent, coarser main branches, which are to 3 mm wide, decumbent, ± overlapping, markedly flattened, grey to grey-green above, paler or white below, sometimes red-purple towards base, divided digitately into flat branchlets with scattered or crowded, obtuse, ± rounded, lateral side branches, especially along the margins.

Apothecia infrequent, on the lower side near the apices of flattened main branches; margin often frilled or with short branchlets; dry spore mass soon exposed by irregular laceration of the thalline exciple. Ascospores 5.5–8 μm diam., grey-black. Medulla Pd+ orange, K+ yellow, KC−, C−, I− (sphaerophorin, stictic acid).

On mossy rocks, trees, and, occasionally, on soil, often in rather sheltered situations; local and rather rare. S.W., E. & N.W. England, N. Wales, W. Scotland. Europe, North & South America, E. Africa, Solomon Is., Asia, Australia, New Zealand.

The persistently flattened main branches with subterminal apothecia distinguish this species from *S. globosus* **(1)** and *S. fragilis* **(2)**. The thallus may sometimes be partly stained magenta-purple.

SPHINCTRINA Fr. (1828)

Thallus absent, or possibly rarely developing an autonomous thallus. **Ascomata** apothecia, sessile to stalked, black, without pruina; head globose to obovoid. **True exciple** well developed, constricted at the edge, consisting of dark brown to reddish brown or in part pale, ± periclinally arranged and interwoven hyphae; central part of stalk pale, outermost part dark brown or reddish brown. **Asci** cylindrical, formed singly from ascogenous hyphae with croziers, with a single functional wall layer, I−, disintegrating at a rather late stage. **Ascospores** globose to ellipsoid, simple, uniseriately arranged in the asci; wall thick, dark brown with a gelatinous coat (when semi-mature); mature spores smooth or ornamented with ridges or verrucae, accumulating in a black spore mass. **Conidiomata** unknown. **Ecology**: pathogenic or commensalistic on lichens, mainly on bark, sometimes also on rocks according to the host habitat.

With the exception of *S. turbinata* **(4)**, all species have become much rarer during the past 100 years in E. British Is.

LITERATURE: Hawksworth (1983), Löfgren & Tibell (1979).

1 Ascospores ellipsoid .... **3. *tubiformis***
Ascospores ± globose 2

2(1) Apothecia with a distinct stalk, stalk at least as tall as head; true exciple dull brown, K− 3
Apothecia sessile or with a very short stalk; true exciple in section reddish, K+ red **4. *turbinata***

3(2) Stalk approximately as tall as head, black; ascospores 7–10 μm, with a distinct ornamentation comprising minute warts .... **1. *anglica***
Stalk slightly taller than head, dark to pale brown; ascospores 4.5–6.5 μm, with an indistinct ornamentation .... **2. *leucopoda***

1. S. anglica Nyl. (1860)

Apothecia 0.19–0.36 mm tall, with a short, but distinct stalk; stalk 0.6–1.7 times as long as head, black or rarely dark brown; head 0.11–0.23 mm diam., ± globose, often slightly irregular, shiny, black; true exciple dull brown, in section K−. Ascospores 7–10 × 6.5–8.5 μm, ± globose, with surface ornamentation of minute warts.

On sterile thallus of *Protoparmelia oleagina* on wooden fencing and gates; very rare. S.E. England (Sussex). C. Europe, Scandinavia.

2. S. leucopoda Nyl. (1860)

Apothecia 0.2–0.3 mm tall, distinctly stalked; stalk 0.8–2.4 times as long as head, pale to dark brown, rarely black; head 0.12–0.19 mm diam., ± globose but often somewhat compressed vertically, regular and smooth, shiny dark brown, rarely black, stalk usually paler; true exciple dark brown, in section K−. Ascospores 4.5–6.5 × 4–6 μm, ± globose, with an indistinct or almost absent surface ornamentation.

Pathogenic or commensalistic on *Pertusaria pertusa*, or more rarely other *Pertusaria* species and *Diploschistes*, particularly on old deciduous trees. Channel Is., W. Ireland (Galway).

Characterized by the clearly-stalked, brown, shiny, top-like apothecia, the small, globose ascospores, and the K− apothecia.

3. S. tubiformis Massal. (1853)

Apothecia 0.15–0.36 mm tall, often ± immersed; stalk short to absent, 0–1.2 times as long as head, black or dark brown; head 0.14–0.36 mm diam., ± globose, often somewhat extended vertically and slightly irregular, shiny black or dark brown; true exciple dark brown in section, K−. Ascospores 10–15.5 × 6.5–9 μm, ellipsoid, the ends pointed, with surface ornamentation of longitudinally arranged, coarse ridges.

Commensalistic to pathogenic on *Pertusaria leioplaca*, sometimes on other *Pertusaria* species. S. England (S. Hampshire, New Forest). Europe, North, Central & South America, Africa, Asia, Fiji.

Recognized by the large and broadly ellipsoid ascospores with a distinctive surface ornamentation of longitudinally arranged and reticulately interconnected ridges.

4. S. turbinata (Pers.) de Not. (1846)

Apothecia 0.16–0.33 mm tall; stalk short to absent, 0–1 times as long as head, dark to pale brown, rarely black; head 0.16–0.36 mm diam., ± globose but often slightly irregular, shiny, black or dark brown; true exciple dark reddish brown in section, K+ red. Ascospores 5–7.5 × 4.5–6.5 μm, ± globose, surface ornamentation consisting of ± distinct minute warts, often with some irregular cracks.

Commensalistic to pathogenic on *Pertusaria pertusa*, more rarely on other *Pertusaria* species, mainly on old trees of *Quercus* and *Fagus*. Throughout British Is. Europe, North America.

The commonest *Sphinctrina* in the British Is., characterized by its virtually sessile apothecia and K+ red true exciple. A similar, undescribed species occurring on *Pertusaria leioplaca* in E. Perthshire (Glen Tilt; Milton Wood) has 4-spored asci with ascospores 12–16.5(-18) × 9.5–12 μm.

SPILONEMA Bornet (1856)

Thallus filamentous or minutely shrubby, blackish; filaments consisting of photobiont (*Stigonema* or *Hyphomorpha*) strands surrounded by a ± irregular network of periclinal hyphae; filaments attached to substratum by blue-green, N+ purple-red, rhizine-like hyphae. **Ascomata** apothecia, lateral, sessile, brown to blackish, convex; internally with green or violaceous, N+ reddish pigments. **Thalline exciple** absent. **True exciple** ± thin, soon excluded, of radiating pseudoparenchymatous cells. **Hymenium** blue-green to violet, I+ blue. **Hypothecium** pseudoparenchymatous. **Hamathecium** of rather stout, branched, septate paraphyses with pointed apices. **Asci** 8-spored, cylindrical apical dome, I+ blue. **Ascospores** simple, narrowly ellipsoid, thin-walled, colourless. **Conidiomata** pycnidia, lateral, sessile, convex to globose, black; wall green-black, N+ red. **Conidiogenous cells** short, catenate, acrogenous and pleurogenous. **Conidia** bacilliform, simple, colourless. **Chemistry**: no lichen products reported. **Ecology**: on moist siliceous rocks; non-European species also on bark, plant detritus, mosses etc. **Distribution**: cosmopolitan, with 4 species, 2 British.

Distinguished from *Ephebe* by the narrower filaments with blue-green rhizoidal hyphae, and the lack of a thalline exciple. May also be confused with *Thermutis*.

LITERATURE: Henssen (1963a).

1 Thallus of loosely interwoven filaments, mat-like; apothecia borne on surface of thallus mat; hypothecium pale .... **1. *paradoxum***
Thallus of erect filaments, cushion-forming; apothecia hidden within thallus cushion; hypothecium dark .... **2. *revertens***

1. S. paradoxum Bornet (1856)

Thallus filamentous, dark olivaceous to blackish, mat-forming, with loosely interwoven, branched filaments up to 50 μm wide and 4 mm long; hyphae 4–7 μm wide. Photobiont *Stigonema*; cells 9–16 × 9–12.5 μm. Apothecia to 1 mm diam., each borne laterally on a filament at the surface of the mat; true exciple often indistinct; hymenium 55–60 μm tall, blue-green or violaceous above; hypothecium 200–240 μm tall, colourless. Paraphyses 2–2.5 μm wide. Asci 40–50 × 7–9 μm. Ascospores 7–9 × 2.5(-3.5) μm. Pycnidia to 0.2 mm diam.; conidia 2.5–4 × 1 μm.

On moist, exposed acid rock; rare. N.W. England (Lake District), N. Wales, C. Scotland (Argyll, Breadalbane), Ireland. Europe, U.S.A., Brazil, Paraguay.

When sterile, easily mistaken for *Ephebe*, but distinguished by the narrower filaments and the blue-green rhizoidal hyphae.

2. S. revertens Nyl. (1865)

Thallus minutely shrubby, blackish, forming compact cushions 5–15 mm wide and 0.3–6 mm tall, with erect, branched filaments to 60 μm wide; hyphae 4–7 μm wide. Photobiont *Stigonema*; cells 10–20 × 9–14.5 μm. Apothecia to 0.5 mm diam., each borne laterally on a filament below the surface of the thallus cushion; true exciple violaceous; hymenium 45–55 μm tall, blue-green or partly violaceous above; hypothecium 100–180 μm tall, violaceous. Pycnidia to 0.2 mm diam.; conidia 2.5 × 1 μm.

On moist acid rocks. C. Scotland (Perth). N. & W. Europe, U.S.A., Japan.

Often sterile, and distinguished from *S. paradoxum* **(1)** in habit; the cushion-forming habit is reminiscent of *Thermutis velutina*, which has narrower photobiont filaments (*Scytonema*) with more scattered hyphae.

SPORASTATIA Massal. (1854)

Thallus crustose, areolate; prothallus distinct, black, sometimes forming a rim around each areole, delimiting. **Cortex** of apically pigmented, anticlinal hyphae, with a colourless epinecral layer. **Photobiont** chlorococcoid. **Medulla** I−. **Ascomata** apothecia, immersed, black, not pruinose, flat to slightly convex; disc often rough or wrinkled. **Thalline exciple** absent. **True exciple** narrow, thin, ± persistent, brown and K+ reddish at outer edge, colourless within, or brown throughout. **Epithecium** brownish (K+ reddish, N+ red) to dark greenish (K+ green intensifying, N+ red). **Hymenium** colourless. **Hypothecium** colourless, or pale brownish, K+ reddish. **Hamathecium** of mostly simple, septate, paraphyses, apical cell ± clavate; I+ blue. **Asci** clavate, 100 to 200-spored; apical dome uniformly K/I+ blue. **Ascospores** 2.5–4 × 2–3.5 μm, simple, subglobose to globose, colourless, without a distinct perispore. **Conidiomata** pycnidia. **Conidia** bacilliform, simple, colourless. **Chemistry**: cortex and medulla C+ red (gyrophoric acid). **Ecology**: on exposed hard siliceous rocks in montane and arctic regions of Northern Hemisphere.

Distinguished from *Polysporina* and *Sarcogyne* by the K/I+ blue apical dome of the ascus, and the generally better developed and delimited thallus.

1 Thallus bluish grey to yellow- or brownish grey, matt; marginal areoles scarcely elongate .... **1. *polyspora***
Thallus yellow-brown to brownish black, glossy; marginal areoles elongate and radiating .... **2. *testudinea***

1. S. polyspora (Nyl.) Grumm. (1963)

Thallus delimited, to 8 cm diam., pale bluish grey to yellow-grey or light brown-grey, matt; areoles to 1 mm diam., flat to convex, angular to rounded, scarcely elongate at thallus edge; epinecral layer thin; algal layer continuous; prothallus conspicuous only at thallus edge. Apothecia 0.2–1(-1.2) mm diam. Pycnidia not seen.

On exposed acid rocks above 800 m. N. Wales (Caernarvon, Glyder Fawr), N. Scotland (Angus, Caenlochan; Cairngorm mtns). Arctic-alpine, Europe, North America.

2. S. testudinea (Ach.) Massal. (1854)

Thallus delimited, to 4 cm diam., yellowish brown to more usually copper-brown to blackish brown, glossy; areoles to 0.6 mm diam., flat to convex, rounded in centre of thallus, but elongated and radiating at the margin; epinecral layer thick; algal layer discontinuous; prothallus usually forming a black rim around each areole, as well as around the whole thallus. Apothecia 0.2–0.6 mm diam. Pycnidia rare; conidia *c.* 3–6 × 1 μm.

On exposed acid rocks above 800 m. ?N. England (Yorkshire), N. Wales (Caernarvon, Snowdonia), N. Scotland (Cairngorm mtns). Arctic-alpine, Europe, North America.

SQUAMARINA Poelt (1958)

Thallus squamulose, lobes often pruinose. **Cortex** thick, sharply delimited. **Photobiont** chlorococcoid. **Medulla** thick, dense white. **Ascomata** apothecia, concave to flat or convex. **Thalline exciple** distinct at first, often becoming excluded. **Disc** yellowish brown to red-brown. **Epithecium** granular. **Hamathecium** of paraphyses. **Asci** 8-spored, *Bacidia*-type. **Ascospores** simple, colourless. **Conidiomata** pycnidia. **Conidia** filamentous, ± curved. **Chemistry**: usnic, ± psoromic acids, several depsidones and unknown substances. **Ecology**: on the ground, over mosses, or directly on rocks, almost exclusively calcicole.

LITERATURE: Poelt (1958), Poelt & Krüger (1970), Timdal (1983).

1 Thallus forming regular rosettes, central part densely pruinose; underside pale brown; medulla Pd− .... **2. *lentigera***
Thallus not forming regular rosettes, composed of ± irregular or imbricate lobes, pruinose or not; underside dark brown; medulla Pd+ yellow or Pd− .... **1. *cartilaginea***

1. S. cartilaginea (With.) P. James (1980)

Thallus variable, yellow-green to brownish green, of usually irregular to imbricate, concave to strongly convex, thickish squamules, strongly pruinose or not, loosely to firmly attached. Apothecia 3(-4) mm diam.; thalline exciple almost entire, becoming excluded; hymenium *c.* 70 μm tall; disc brownish to reddish brown, concave to flat, becoming irregularly convex. Paraphyses slightly widened towards apices. Ascospores (10-)12–14(-15) × (4-)4.5–6 μm, oblong-ellipsoid. Thallus Pd−, K−, KC+ yellowish, C−, medulla Pd+ yellow or Pd− (usnic, ± psoromic and ± 2-*O*-demethylpsoromic acids).

On soil, mosses and hard calcareous rocks, preferring shady conditions such as crevices of limestone cliffs and pavements, also on calcareous dunes; locally frequent but decreasing, particularly from coastal habitats. Mainly in N. & W. British Is. Europe, North America, N. Africa.

There are two chemotypes, possibly with different ecology and distribution patterns in the British Isles: **var. *cartilaginea*** with a Pd+ yellow medulla is apparently confined to coastal sites, while the rarer **var. *pseudocrassa*** (Mattick) D. Hawksw. (1980) with a Pd− medulla and which lacks psoromic acid also occurs inland.

2. S. lentigera (Weber) Poelt (1958)

Thallus to 6 cm diam., circular, whitish to pale brownish green, of coherent squamules, squamules slightly to densely pruinose, undulate, edges crenate, upwardly curved and producing distinctive white margins. Apothecia 1.5–2 mm diam.; thalline exciple entire, at first thick but becoming thinner and almost excluded; hymenium *c.* 120 μm tall; disc pale brown, concave then convex. Paraphyses slightly clavate. Ascospores 9–12 × 4–5 μm, ellipsoid. Thallus Pd−, K−, KC+ yellowish, C− (usnic acid).

On sandy, calcareous soils and dunes; rare and declining rapidly. E. England (E. Anglia, Breckland), formerly south coast and S.W. England (Brean Down). Europe, from Norway to the Mediterranean, S.E. U.S.S.R., North America, N. Africa, Asia.

STAUROTHELE Norman (1853)

Thallus crustose, immersed or superficial, often inconspicuous, smooth to uneven, rarely subsquamulose. **Photobiont** *Stichococcus* (? always), also present in hymenium as numerous globose-cuboid or cylindrical hymenial cells between the asci. **Ascomata** perithecia, superficial or immersed in thalline warts, forming pits in limestone. **Involucrellum** present or absent. **True exciple** completely black or pale below. **Hamathecium** of periphyses; paraphyses absent. **Asci**

2- to 8-spored, verrucarioid, without a distinct ocular chamber but sometimes with an apical 'plug' staining red with Congo Red. **Ascospores** muriform, colourless to deep brown. **Conidiomata** pycnidia, immersed. **Conidiogenous cells** short-ampulliform, acrogenous. **Conidia** bacilliform, simple, colourless. **Chemistry:** lichen products not detected. **Ecology**: on limestones, mica-schist, epidiorite, chalk pebbles, mortar, brick, and on siliceous rocks in and by streams and lakes.

The hymenial algal cells, though considerably smaller than those in the vegetative part of the thallus, are the same species. Hymenial algal cells occur elsewhere in British perithecioid lichens only in *Endocarpon* which has a squamulose thallus. *Polyblastia* differs from *Staurothele* in the absence of hymenial algal cells. Care is required in determining the presence and shape of hymenial algal cells and the number of ascospores per ascus.

Staurothele arctica Lynge (1937) has recently been found on Scottish snowbeds (West Inverness, Ben Nevis); it has a grey-brown, warty or immersed thallus; perithecia to 0.5 mm wide, sessile; asci 2-spored; ascospores 32–48 × 12–18 μm; hymenial algal cells 5–8(-12) × 4 μm, cylindrical. Elsewhere in Scandinavia and Greenland. For *S. geoica*, see p. 686.

LITERATURE: Swinscow (1963, 1965b).

1 Asci 8-spored; ascospores colourless or pale straw, rarely brownish when old 2
Asci 2- or 4-spored; ascospores pale to dark brown, rarely colourless 6

2(1) Hymenial algal cells cylindrical, 3–10 × 1–3 μm diam. 3
Hymenial algal cells globose-cuboid, 2–4.5 μm diam. 5

3(2) Involucrellum present; exciple brown-black throughout; perithecia in shallow pits, emergent to superficial 4
Involucrellum absent; lower part of exciple colourless to pale brown; perithecia immersed in deep pits .... **1. *bacilligera***

4(3) Thallus immersed, white or pale grey, often indistinct; perithecia often in shallow pits, half-immersed, surface ± white-pruinose; ascospores 25–32 μm long .... **6. *hymenogonia***
Thallus superficial, dark brown or black, often indistinct, uneven; perithecia prominent, superficial, not white-pruinose; ascospores (30-)35–45 μm long .... **10. *succedens***

5(2) Thallus immersed, white or pale grey, often indistinct; involucrellum absent; perithecia deeply immersed in pits .... **2. *caesia***
Thallus superficial, thick, areolate, yellowish to brownish grey; involucrellum present; perithecia mostly immersed in the areoles .... **8. *rugulosa***

6(1) Hymenial algal cells cylindrical, 3–10 × 1–3 μm diam. 7
Hymenial algal cells globose-cuboid, 2–4.5 μm diam. 8

7(6) Thallus immersed, pale grey, often indistinct; involucrellum absent; perithecia deeply immersed in pits .... **1. *bacilligera***
Thallus superficial, thick, on inundated siliceous boulders, dark brown; involucrellum ± present; perithecia immersed in thalline warts .... **3. *clopima***

8(6) Perithecia deeply immersed in pits 9
Perithecia ± prominent in thalline warts 10

9(8) Asci 4-spored; ascospores dark brown .... **9. *rupifraga***
Asci 2-spored; ascospores colourless .... **5. *guestphalica***

10(8) Perithecia emergent to prominent; involucrellum well developed .... **7. *rufa***
Perithecia wholly immersed in sub-globose thalline warts; involucrellum poorly developed .... **4. *fissa***

1. S. bacilligera (Arnold) Arnold (1885)

Thallus immersed, pale grey, scurfy or indistinct. Perithecia 200–350 μm diam., ± deeply immersed in pits; lower part of exciple colourless to pale brown, the upper part black; involucrellum absent. Asci

(2-)6- to 8-spored. Ascospores (25-)28–35 × 15–18(-22) μm, colourless or becoming brown. Hymenial algal cells 3–10 × 2–3 μm, cylindrical.

On hard upland limestones in exposed sites; very rare. N. England (Pennines), S. Scotland (Grampian, Braemar), Ireland. S. Sweden (Gotland), Jura, Alps to Balkans.

2. S. caesia (Arnold) Arnold (1885)

Thallus immersed, pale grey to bluish grey, scurfy or indistinct, sometimes with a black prothallus. Perithecia 300–700 μm diam., ± deeply immersed; involucrellum absent; outer part of exciple black with purple tints, the inner part pale. Asci 8-spored. Ascospores 30–40 × 18–25 μm, colourless to pale straw. Hymenial algal cells 2.5–4.5 μm diam., globose-cuboid.

On dry limestones in exposed situations, rarely on mortar; local. Throughout British Is. C. Europe to Balkans, Sweden (Öland, Gotland).

3. S. clopima (Wahlenb.) Th.Fr. (1881)

Thallus superficial, dark brown or dark reddish brown, thin to thick, deeply cracked-areolate to subsquamulose, effuse. Perithecia 200–400 μm diam., immersed in subglobose warts or areoles; involucrellum poorly developed, brown to black; exciple colourless to pale brown. Asci 2-spored. Ascospores 30–50 × 15–25 μm, dark brown. Hymenial algal cells, 6–10 × 3–4 μm, cylindrical.

On ± submerged granite boulders in a melt-water stream from a late snow patch. N. Scotland (Cairngorm mtns, Cairn Toul). Montane C. & N. Europe, North America.

Distinguished from *S. fissa* **(4)** by the cylindrical hymenial algal cells. The correct name for this species may be *S. areolata* (Ach.) Lettau (1912) as Wahlenberg's epithet appears to have been misapplied.

4. S. fissa (Taylor) Zwackh (1862)

S. catalepta sensu Malme (1919), *non* (Ach.) Blomb. & Forss. (1880)

Thallus superficial, dark greenish brown or dark reddish brown, thin, evenly rimose-cracked, smooth to uneven, deeply cracked-areolate, rarely subsquamulose. Perithecia 200–600 μm diam, immersed in thalline warts; involucrellum poorly developed, brown-black; exciple colourless to pale brown. Asci 2-spored. Ascospores (25-)35–50 × 14–22(-25) μm, dark brown. Hymenial algal cells 2–3 μm diam., globose-cuboid, the cells sometimes paired. Pycnidia immersed; conidia 3.5–4 × 1 μm.

On hard siliceous rocks by unpolluted streams and lakesides; locally abundant. Throughout British Is. N. Europe to C. European mtns, Alps to Balkans, North America, ? New Zealand.

Thalli which are thick, uneven to subsquamulose and have been incorrectly referred to *S. catalepta* by British authors, intergrade with *S. fissa*, which has a smooth, even, ± areolate thallus.

5. S. guestphalica (Lahm ex Körber) Arnold (1885)

Thallus immersed, pale grey or indistinct, wide-spreading, effuse. Perithecia 200–400 μm diam., deeply immersed in pits; involucrellum absent; exciple black throughout. Asci 2-spored. Ascospores 45–56 × 15–25 μm, colourless. Hymenial algal cells globose-cuboid.

On a calcareous deposit over an acidic maritime cliff, and on limestones; very rare. N.W. Scotland (W. Sutherland), W. Ireland, (Clare, Burren). C. Europe from N. Mediterranean to S. Sweden, ?Novaya Zemlya.

The only British species always with 2 colourless ascospores in each ascus.

6. S. hymenogonia (Nyl.) Th.Fr. (1865)

Thallus immersed, pale to medium grey, scurfy or indistinct, effuse. Perithecia 400–700 μm diam., half-immersed in shallow pits. Involucrellum dark brown-black, well developed, the surface matt or roughened, ± white-pruinose; exciple dark brown to black. Asci 8-spored. Ascospores 25–32 × 15–25 μm diam., colourless. Hymenial algal cells 3–10 × 2–3 μm, cylindrical, with many shorter cells.

On dry limestones, mortar, chalk pebbles; locally common. S.W. England, W. Britain from Devon to Scotland, Ireland. Scandinavia, C. Europe, North America.

7. S. rufa (Massal.) Zsch. (1913)

Thallus ± superficial, pale to dark grey, thin, cracked-areolate or sometimes immersed and indistinct, effuse. Perithecia 400–500 μm diam., prominent, emergent; involucrellum well developed; exciple pale to dark brown. Asci 2-spored. Ascospores (30-)35–45(-50) × 15–25 μm, dark brown. Hymenial algal cells 2–3 μm diam., globose-cuboid.

On dry liassic limestone. S. Wales (Glamorgan, Porthkerry). C. Europe, North America.

8. S. rugulosa (Massal.) Arnold (1897)

Thallus 400–600 μm thick, superficial, pale, yellowish ochre to greenish grey, deeply cracked-areolate, the areoles finely cracked, ± plicate, effuse or patchily determinate. Perithecia 400–600 μm diam., immersed in thalline areoles, one or two per areole; involucrellum well developed; exciple with dark brown outer part, the inner colourless. Asci 8-spored. Ascospores 25–34 × 12–19 μm, colourless. Hymenial algal cells 3–6 × 3–4 μm, globose or cuboid to slightly elongate.

On dry limestones in exposed situations and brick walls; very rare. S.W. England (Somerset), ?S. Wales (Pembroke), C. Scotland (Perth). S.C. Europe.

Superficially resembles *Verrucaria viridula*, characterized by the thick, cracked-areolate thallus. The semi-aquatic *S. fissa*, which may also have a well-developed thallus, has 2-spored asci with dark brown ascospores.

9. S. rupifraga (Massal.) Arnold (1880)

Thallus immersed, often indistinct, pale grey to pale blue-brown, scurfy, effuse. Perithecia 200–400 μm diam., immersed in pits; involucrellum absent; exciple with outer part dark brown to black. Asci (2-)4-spored. Ascospores 30–50(-52) × 18–25 μm, dark brown. Hymenial algal cells 2–3(-4.5) μm diam., globose or cuboid.

On hard limestones in exposed situations; very local. Throughout British Is. C. Europe to Balkans & N. Norway, North America.

The only British species with 4 brown ascospores per ascus.

10. S. succedens (Rehm ex Arnold) Arnold (1880)

Thallus superficial, finely wrinkled or minutely granular-isidiate, continuous, effuse or partly determinate, dark grey-black. Perithecia 300–500 μm wide × 400–900 μm tall, ± superficial; involucrellum incurved under base of perithecium; exciple outer part dark brown-black, inner part colourless. Asci 8-spored. Ascospores 35–45 × 15–25 μm, colourless. Hymenial algal cells 3–10 × 2–3(-4.5) μm, cylindrical but with some shorter cells.

On mica-schist, limestones, calcareous sandstones and epidiorite in damp situations; local. Scotland (C. Perth to Sutherland). N. Europe to Alps & C. European mtns.

The black superficial thallus and prominent, tall, perithecia distinguish this species from *S. hymenogonia* **(6)**. When wet the pale periphyses may project from the ostiole (×20 lens).

STEINIA Körber (1873)

Thallus crustose, often wide-spreading, membrane-like or slightly uneven-scurfy, effuse; prothallus absent. **Photobiont** chlorococcoid, cells ellipsoid. **Ascomata** apothecia, scattered, adpressed to sessile; disc dark brown-black, strongly convex. **Thalline exciple** absent. **True exciple** rudimentary, of ± undifferentiated vertically arranged hyphae. **Hymenium** I+ blue. **Hypothecium** pale to dark brown of ± vertically aligned hyphae. **Hamathecium** of paraphyses, sparse, simple or sparingly branched, in a ± uniform, gelatinous matrix, not swollen at the apices. **Asci** 16-spored, clavate-cylindrical, thin-walled, without a tholus, with a K/I+ blue, distinct apical pore. **Ascopores** globose or slightly ellipsoid, simple, colourless, thick-walled and distinct. **Conidiomata** pycnidia, black. **Conidiogenous cells** elongate-ampulliform, enteroblastic, acrogenous. **Conidia** simple, ellipsoid, colourless. **Chemistry**: lichen products not detected by t.l.c. **Ecology**: mainly on disturbed soils.

The generic affinities of this monotypic genus are unclear. It differs from *Lecidea* s.lat. in the poorly developed true exciple, its ascus-structure and multispored asci with ± globose spores.

LITERATURE: Faurel & Schotter (1954), Poelt & Vězda (1977).

1. S. geophana (Nyl.) B. Stein (1879)

Thallus thin, inconspicuous, rarely cracked, membrane-like or uneven-scurfy, pale grey to dull olive-green, ± subgelatinous when wet. Apothecia 0.2–0.6 mm diam., widely scattered, rarely contiguous; epithecium red-brown to brown, pigment in irregular granules, K− and not dissolving; hymenium 65–75 μm tall, colourless or patchily pale red-brown. Paraphyses *c.* 0.7 μm thick, often not extending to epithecium. Ascospores 5–7 μm diam.; wall 0.5 μm thick. Conidia 1–1.5 × 0.5 μm. Thallus Pd−, K−, KC−, C−, UV− (lichen products not detected by t.l.c.).

On ± moist, metal-rich, often recently disturbed soils, stable dunes, decaying wood, old leather and bones, flints on chalk, roofing tiles in lowland and upland sites; very local and ephemeral. Throughout British Is.

Superficially resembles *Aphanopsis coenosa* which has 8-spored asci and larger, ellipsoid ascospores.

STENOCYBE Nyl. ex Körber (1852)

Thallus absent. **Photobiont** absent. **Ascomata** apothecia, clearly stalked, black, to olivaceous brown; head obovoid. **True exciple** well developed, consisting of dark brown, periclinally arranged hyphae, in some species, a thickened margin surrounding a ± sunken pore-like disc. **Stalk** consisting of dark brown, periclinally arranged hyphae. **Asci** cylindrical, formed singly from ascogenous hyphae with croziers, uniformly thickened at the apex, with a single functional wall layer, ± persisting until the maturity of the ascospores. **Ascospores** simple to 3(-7)-septate, ellipsoid, rarely elongate-fusiform, uniseriately arranged in the asci, pale to dark brown-grey. **Chemistry**: lichen products not detected. **Ecology**: in humid and rather shaded situations on broad-leaved trees or hepatics, saprobic or parasitic; the species are mainly confined to a single host.

LITERATURE: Schmidt (1970).

1	Ascospores <20 μm, the cells equal; stalk <1 mm	2
	Ascospores >20 mm, the cells unequal; stalk >1 mm	3
2(1)	Stalk black, remaining opaque when moist; on *Alnus*	**2. *pullatula***
	Stalk brownish, becoming pale and transparent when moist; on *Populus*	**4. *tremulicola***
3(1)	On leafy liverworts; ascospores 30–40 × 12–15 μm	**1. *bryophila***
	On bark, nearly always *Ilex*; ascospores (35-)45–60(-90) × (11-)15–20 μm	**3. *septata***

1. S. bryophila W. Watson (1925)

Apothecia 1–1.5(-2) mm tall; stalk 0.08–0.12 mm wide, pale to dark olive- or grey-black; head 0.18–0.3 mm diam., narrowly obconical, dark grey to nearly black with a distinctly thick, cartilaginous, tumid, shining exciple margin surrounding a narrow, 0.1–0.2 mm diam., pore-like, black, roughened disc. Paraphyses *c.* 1 μm thick, ± numerous, branched or ± anastomosing, persistent. Asci 130–150 × 11–14 μm, cylindrical, wall persisting. Ascospores 30–40 × 12–15 μm, 3-septate, ellipsoid, wall thin or slightly thickened, comprising 2, large, dark red-brown, median cells, with large, rounded oil droplets and 2, much smaller, paler end cells.

Partially buried in old tufts of leafy liverworts (e.g. *Scapania punctata*, *Plagiochila tridenticulata* and *Microlejeunea ulicina*), especially on old trees and rocks in damp, humid situations; rather rare. W. British Is. N.W. Europe, Madeira.

Distinguished by the tough, shining, grey-black, golf-tee-like apothecia associated with tufts of leafy hepatics.

2. S. pullatula (Ach.) B. Stein (1879)

S. byssacea (Fr.) Körber (1855)

Apothecia very slender, 0.5–0.8 mm tall, black; stalk to 0.07 mm thick, simple, rarely branched; head 0.08–0.15 mm diam., ± matt, cup- to top-shaped, the asci somewhat extruded. Paraphyses absent. Asci soon disintegrating. Ascospores (10-)13–18(-20) × 4–5(-6) μm, at first simple, becoming 1- to 3-septate, when uniformly pale grey-brown.

On bark of decaying and living *Alnus glutinosa* twigs, particularly overhanging streams, lakes and bogs in unpolluted hilly districts; frequent. N. & W. British Is. France (Normandy, Brittany), montane C. & N. Europe, North America.

3. S. septata (Leighton) Massal. (1860)

Apothecia to 1.8 mm tall, black, single or sometimes 2–3 united at the base; stalk to 0.15 mm diam., sometimes branched; head narrowly clavate, shining with an elevated, ± incurving, inflexed exciple, black. Ascus walls soon disappearing. Ascospores (35-)45–60(-90) × (11-)15–20 μm, variable, (1-)3(-6)-septate, ellipsoid or elongate-ellipsoid, red-brown, the outer small cells and septa ± uniformly thickened, 2–2.5 μm, except for a small, paler, sometimes shortly extended nipple at one or both ends of these cells.

Mainly on bark of *Ilex* trunks, rarely also on *Betula, Quercus*, *Sorbus* and *Corylus*, especially in old woodlands; common. S. & W. British Is., extending to N. Scotland. France (Brittany).

Sometimes growing through the thalli of other lichens, e.g. *Graphis elegans* and *Thelotrema lepadinum*, and then appearing as a part of them.

4. S. tremulicola Norrlin ex Nyl. (1883)

Apothecia to 0.6 mm tall; stalk 0.07–0.08 mm diam., brownish black, semi-transparent, pale brown when moist; head 0.2–0.3 mm diam., ± globose or top-shaped, brown-black. Asci ± persistent, 65–70 × 4.5–5 μm, narrow-cylindrical. Ascospores 13–20 × 4–6 μm, ellipsoid, often slightly waisted, simple or 1-, sometimes thinly 3-septate, uniformly grey-brown or brown, not thickened.

On twigs of *Populus*. E.C. Scotland (E. Perth, Kindrogan). C. & N. Europe.

S. pullatula **(2)** differs in having a black stalk when wet and is on a different host (*Alnus*). Perhaps referable to *Phaeocalicium*.

STEREOCAULON Hoffm. (1796)

Primary thallus crustose, of basal granules ("phyllocladia") variously appearing warted, areolate or squamulose; in most species disappearing early. **Secondary thallus** shrubby, erect, solid-stemmed, often richly branched ("pseudopodetia"), corticate or not, covered in wart-like, coralloid or flattened, squamule-like, corticate phyllocladia which, depending on their distribition, result in the pseudopodetia appearing dorsiventral or not. **Cephalodia** frequent, containing *Nostoc* or *Stigonema*. **Photobiont** trebouxioid. **Medulla** lax. **Cartilaginous axis** of thick-walled, longitudinal hyphae present. **Ascomata** apothecia, terminal; disc usually becoming convex, pale to dark brown. **Thalline exciple** absent. **True exciple** raised to ± excluded. **Hypothecium** mostly colourless. **Hamathecium** of paraphyses, simple; apices mainly with a brown cap in K/I with a blue outer layer, and a blue apical dome ± with a central, darker blue tubular structure, *Porpidia*-like. **Asci** 8-spored, clavate to cylindrical. **Ascospores** colourless, thin-walled, ellipsoid, acicular or fusiform, 1- to 13-septate. **Conidiomata** pycnidia, mostly immersed at tips of phyllocladia, ovoid to spherical, darkened about ostiole. **Conidia** simple, thread-like to cylindrical, straight or curved. **Chemistry**: atranorin (K+ yellow) present in all species, also lobaric and stictic (with ± norstictic) acids, or fatty acids present in certain species. **Ecology**: mostly on rocks and montane, also occurring on metal-rich spoil heaps, on soil, shingle, amongst mosses on the ground and some near towns on brick.

Stereocaulon botryosum Ach. (1810), which contains porphyrilic acid and which has numerous pale blue-warted cephalodia, and *S. paschale* (L.) Hoffm. (1796) resembling *S.*

alpinum, but with the pseudopodetia not dorsiventrally arranged, and with wart-like phyllocladia, have been incorrectly recorded from the British Is.

LITERATURE: Lamb (1977, 1978).

1 Basal thallus persistent, of scattered or confluent phyllocladia or areole-like structures; pseudopodetia inconspicuous or absent 2
Basal thallus disappearing; pseudopodetia conspicuous 7

2(1) Basal thallus of ± granular or cylindrical phyllocladia or areole-like structures 3
Basal thallus of flattened, ± fan-shaped, erect phyllocladia, the lower surface sorediate .... **9. *nanodes***

3(2) Basal thallus of granules or areole-like structures with darkened centres 4
Thallus not as above 5

4(3) Thallus Pd− (lobaric acid and atranorin) .... **7. *leucophaeopsis***
Thallus Pd+ orange (stictic acid and atranorin) .... **8. *tornense***

5(3) Soredia absent; phyllocladia wart-like or cylindrical; cephalodia frequent 6
Soredia present, terminal, ± globose; phyllocladia wart-like; cephalodia infrequent .... **10. *pileatum***

6(5) Cephalodia dark reddish brown to black, containing *Stigonema* .... **2. *condensatum***
Cephalodia pale bluish brown, brownish violet or pinkish white, containing *Nostoc* **3. *glareosum***

7(1) Phyllocladia ± shield-like, with a well-defined darker grey-green zone in the centre 8
Phyllocladia not shield-like, without a darker centre 9

8(7) Pseudopodetia erect, not dorsiventral, Pd+ orange .... **15. *vesuvianum***
Pseudopodetia ± decumbent, dorsiventral, Pd− .... **13. *symphycheilum***

9(7) Soredia present 10
Soredia absent 11

10(9) Pseudopodetia dorsiventral, Pd+ orange; phyllocladia and terminal branches flattened, sorediate on lower surfaces and edges .... **12. *spathuliferum***
Pseudopodetia not dorsiventral, Pd−; phyllocladia and terminal branches not flattened; lateral branches often with terminal, granular, ± globose soralia .... **5. *delisei***

11(9) Apothecia mostly abundant; thallus Pd+ orange 12
Apothecia rare or absent; thallus Pd± pale yellow 13

12(11) Lower surface of pseudopodetia distinctly grey-tomentose; phyllocladia mostly flattened and squamule-like .... **14. *tomentosum***
Lower surface of pseudopodetia not tomentose; phyllocladia coralloid-elongate and finger-like .... **4. *dactylophyllum***

13(11) Pseudopodetia distinctly tomentose below 14
Pseudopodetia somewhat tomentose below .... **6. *evolutum***

14(13) Tomentum grey; cephalodia sparse, indistinct, containing *Stigonema* .... **11. *saxatile***
Tomentum pinkish white; cephalodia frequent, distinct, containing *Nostoc* .... **1. *alpinum***

1. S. alpinum Laurer (1827)

Primary thallus disappearing. Pseudopodetia to 1(-3) cm tall, dorsiventral, decumbent or erect, sparingly branched, forming low, flat cushions, lower part and undersurface with a distinct pink tomentum; phyllocladia to 0.5 mm wide, whitish grey, ± granular or dactyliform and attached together at base, flattened, often crowded, particularly towards the apices, but absent on lower surface. Cephalodia present on lower surface, hemispherical, purple, often grey-white tomentose, containing *Nostoc*. Apothecia terminal; disc dark brown, not observed in British material. Thallus Pd± yellow, KC+ violet [on filter paper] (atranorin, lobaric acid).

On stony ground or short mossy turf on exposed mountain summits and ridges; rare. N. Scotland (Highlands). Europe, Arctic, North & South America, U.S.S.R., India, ?China, Antarctica.

Characterized by the pink tomentum and pale cephalodia on the lower surface of the dorsiventral pseudopodetia, and the tendency to grow amongst mosses on the ground. Stunted morphs are often difficult to separate from *S. saxatile* **(11)**, which has grey tomentum and rarely has cephalodia.

2. S. condensatum Hoffm. (1796)

Primary thallus persistent, a basal crust of warty or ± cylindrical, grey, sometimes bluish, phyllocladia. Cephalodia 0.5–1 mm diam., hemispherical, with a rough surface, dark reddish brown to black, frequently interspersed amongst phyllocladia, also on pseudopodetia, containing *Stigonema*. Pseudopodetia to 1(-2) cm tall, infrequent or absent, erect, simple or sparingly branched, ± tomentose above with ± warted, occasionally incised, squamulose phyllocladia. Apothecia 1–2 mm diam., rather frequent, terminal on pseudopodetia or on basal phyllocladia. Thallus Pd± yellow, KC± violet (atranorin, ± lobaric acid).

On sandy and gravelly soil in montane situations, also at lower altitudes on acid coastal heaths and on metal-rich soils. S.E. & N. Scotland (Highlands) and N.E. coast, extending to Wales; formerly E. England (E. Anglia, N.E. Yorkshire, Durham). Europe, U.S.S.R., Greenland, North & South America, China, Japan.

See also *S. glareosum* **(3)**.

3. S. glareosum (Savicz) Magnusson (1928)

Like **2**, but usually taller, to 2.5 cm, with conspicuous, sparingly branched pseudopodetia which are grey tomentose on their lower surfaces and have smooth, pale bluish brown, brownish violet or pinkish white cephalodia (containing *Nostoc*) amongst basal phyllocladia and on pseudopodetia. Apothecia rare. Thallus Pd± yellow, KC+ violet [on filter paper] (atranorin, lobaric acid).

On sandy or gravelly soil subjected to periodic flooding; rare. Scotland (N. & C. Highlands). Europe, North & South America, U.S.S.R., China, Kashmir, Nepal.

Characterized by the pale cephalodia that contain *Nostoc*, but otherwise often difficult to distinguish from *S. condensatum* **(2)**.

4. S. dactylophyllum Flörke (1819)

Primary thallus disappearing. Pseudopodetia 1–4(-6) cm tall, erect or horizontally spreading, often forming tufts, irregularly branched, grey to greenish white, usually without tomentum on lower surface, very rarely slightly tomentose towards base of young branches, not dorsiventral. Phyllocladia to 1 mm, cylindrical, becoming coralloid branched, occasionally forming crust-like masses. Cephalodia to 0.3 mm diam., inconspicuous and rare, rough, brownish grey, containing *Stigonema*. Apothecia frequent, terminal on short, lateral branches; disc brown-black. Thallus Pd+ orange, KC− (atranorin, stictic acid, ± norstictic acid).

Firmly attached to siliceous, sometimes metal-rich, rocks; frequent. N. & S.W. England, N. Wales, Scotland, Ireland. Europe, U.S.S.R., Greenland, North America.

Characterized by the finger-like, much-branched phyllocladia, the tendency to form compact cushions covered with brown-black apothecia and the Pd+ orange reaction (stictic acid). The **var. *occidentale*** (Magnusson) Grumm. (1963) has darker grey, flattened phyllocladia with rounded margins, loosely packed into flattened masses with a ± level surface; it is particularly frequent in E. Scotland.

5. S. delisei Bory ex Duby (1830)

Primary thallus disappearing. Pseudopodetia 1–1.5(-3) cm tall, whitish grey, loosely tufted, often forming patches a few cm wide, sparingly branched above, sometimes slightly tomentose. Phyllocladia infrequent, flattened, ± squamulose, soon becoming sorediate; terminal, granular sorediate masses frequent. Cephalodia and apothecia rare. Thallus Pd± yellow, KC− (atranorin, rangiformic and norrangiformic acids).

On siliceous rocks in upland sites, often associated with heavy metals at old mining sites; local. N. England, N. Wales, N. Scotland (Highlands, Argyll, Jura). France, Scandinavia, Madeira, S. Africa.

Characterized by the conspicuous, terminal, granular sorediate masses, the flattened phyllocladia and Pd− reaction. Sorediate morphs of *S. vesuvianum* **(15)** differ in having peltate phyllocladia, with ± darkened centres, and are Pd+ orange.

6. S. evolutum Graewe (1865)

Primary thallus disappearing. Pseudopodetia 1–3 cm tall, grey, arched or prostrate, at least at the apices, forming compact cushions to 6 cm diam., irregularly branched from the base, dorsiventral. Phyllocladia flattened, deeply divided into ± cylindrical extensions, often overlapping and concealing pseudopodetia, the lower surface glabrous. Cephalodia indistinct, containing *Stigonema*. Apothecia terminal, rare. Thallus Pd± yellow, KC+ violet [on filter paper] (atranorin, lobaric acid).

Loosely attached to siliceous rocks, sometimes amongst mosses; locally frequent. N. & S.W. England, Wales, N. Scotland (Highlands). Europe, Madeira, Canary Is.

Characterized by the arching, dorsiventral stems, the flattened, deeply divided phyllocladia, and the glabrous lower surface. *S. saxatile* **(11)** has a grey-tomentose lower surface and is mainly restricted to high altitudes on Scottish mtns. When stunted, can be distinguished from stunted morphs of *S. dactylophyllum* **(4)** by the Pd−, KC+ violet reactions [best carried out using an acetone extract on filter paper].

7. S. leucophaeopsis (Nyl.) P. James & Purvis (1985)

Toninia leucophaeopsis (Nyl.) Th. Fr. (1874)

Primary thallus crustose, areolate; areoles to 1 mm diam., grey, spreading, with slightly tumid margins and a darker central area, remaining conspicuous, especially when wet; soralia, when present, usually confined to the margins of the areoles. Cephalodia absent, but cyanobacteria are often abundant between the areoles. Pseudopodetia absent. Apothecia rather frequent; disc to 1.5 mm diam., black. Ascospores (25-)30–38(-40) × 6–7 μm, 3- to 5-septate, ± sub-clavate with one end ± tapered, rarely fusiform. Thallus Pd± yellow, KC+ violet [on filter paper] (atranorin, lobaric acid).

On usually metal-rich rocks, with Cu, Mn, Pb and Zn, occasionally on calcareous substrata, in hills; very local. Mainly W. Britain. Norway, Sweden, C. Europe.

Characterized by the basal, shield-like areoles with dark centres and in this resembling the basal phyllocladia in young material of *S. vesuvianum* **(15)**, which, however, produces pseudopodetia. Distinguished from *S. tornense* **(8)** in the restriction of soredia to the margins of the areoles and the presence of lobaric acid and atranorin.

8. S. tornense (Magnusson) P. James & Purvis (1985)

Primary thallus crustose, areolate; areoles to 0.5 mm diam., grey, with tumid, raised margins and a small darker central area which is sometimes obliterated by the margins even when wet; soralia, when present, involving the whole surface of an areole. Cephalodia absent, although cyanobacteria are often abundant between the areoles. Pseudopodetia absent. Apothecia rather frequent; disc to 1.5 mm diam., black. Ascospores (18-)21–25(-30) × 5–8 μm, (1-)3-septate, regularly fusiform with equally rounded apices. Thallus Pd+ orange, KC− (atranorin, stictic acid).

On pebbles in turf-areas of late snow-lie; very rare. N. Scotland (Highlands). Norway, Sweden, Jan Meyen Is.

Characterized by the swollen areoles with dark centres. Distinguished from *S. leucophaeopsis* **(7)** by the tendency of individual areoles to dissolve into convex, finely granular soralia (somewhat resembling those in the apices of the pseudopodetia of *S. pileatum* **(10)**, and the Pd+ orange (not yellow) thallus.

9. S. nanodes Tuck. (1859)

Primary thallus persistent, of ± erect, to 0.5 cm tall, grey, flattened phyllocladia, scattered or forming swards, at first widened at apices and appearing fan-shaped, later becoming irregularly branched in one plane; lower surface sorediate. Pseudopodetia to 1 cm tall, rather rare, sparingly branched, the branches flattened, sorediate below; soredia occasionally delimited in ± globular soralia. Cephalodia infrequent, bluish grey, on pseudopodetia and amongst basal phyllocladia. Apothecia very rare. Thallus Pd± yellow, KC+ violet [on filter paper] (atranorin, lobaric acid).

On sheltered rocks, often in rather damp situations, particularly frequent on old mine spoil heaps associated with Pb/Zn deposits, also on roadside walls in urban areas; rather frequent. Throughout British Is. Europe, North America, U.S.S.R..

Characterized by the persistent, basal, ± erect and flattened phyllocladia, which have sorediate lower surfaces. *S. pileatum* **(10)** has wart-like basal phyllocladia and the pseudopodetia bear terminal, globose soralia.

10. S. pileatum Ach. (1810)

Primary thallus persistent, of grey, ± wart-like, granular basal phyllocladia to 3 mm, spreading irregularly on rock. Pseudopodetia 2–5 mm tall, mostly unbranched, with a terminal globose soralium; phyllocladia infrequent, wart-like or ± cylindrical, sometimes branched. Cephalodia frequent, grey- to violet-brown, amongst basal phyllocladia and on pseudopodetia, containing *Stigonema*. Apothecia rare; disc red-brown. Thallus Pd+ pale yellow, KC+ violet [on filter paper] (atranorin, lobaric acid).

On damp, siliceous rocks, particularly on loose stones, often on mine spoil heaps, also on brick and road-side walls in industrial and urban areas; rather frequent. Throughout British Is. Europe, Canada, U.S.S.R..

Characterized by the persistent, wart-like, basal phyllocladia and small pseudopodetia with terminal, globose soralia. *S. nanodes* **(9)** has flattened phyllocladia, sorediate below.

11. S. saxatile Magnusson (1926)

Primary thallus disappearing. Pseudopodetia 1–2 cm, ± firmly attached, forming dense tufts to 7 cm across, stout, dorsiventral, with a grey tomentum towards apices, richly branched in upper part. Phyllocladia abundant, dark grey, almost completely covering pseudopodetia, flat to ± convex, deeply incised at margins, appearing ± digitately squamulose. Cephalodia rare, inconspicuous, containing *Stigonema*. Apothecia rare, terminal. Thallus Pd± yellow, KC+ violet [on filter paper] (atranorin, lobaric acid).

On rocks, often on gravelly soil in high gullies or on mountain summits; locally frequent. N. England (Cumbria), N. Scotland (Highlands). Fennoscandia, Belgium, Netherlands, Germany, Czechoslovakia, Poland, U.S.S.R., North America.

Characterized by the dorsiventral pseudopodetia covered in a grey tomentum and the flattened, digitate, squamule-like phyllocladia. *S. tomentosum* **(14)** also has a grey tomentum, but is Pd+ orange. *S. alpinum* **(1)** has a pinkish tomentum, and more frequent cephalodia with *Nostoc*.

12. S. spathuliferum Vainio (1909)

Primary thallus disappearing. Pseudopodetia 1–3 cm, irregularly branched, terete and thickened basally with ± flattened, mainly decorticate, pale grey-white side branches. Phyllocladia wart-like or flattened, with spathulate apical extensions, the lower surface often coarsely granular-sorediate, occasionally terminal, ± globose soralia may also occur. Cephalodia mostly distinct, bluish grey, usually stalked and uneven, containing *Nostoc*. Apothecia terminal, not observed in British material. Thallus Pd+ orange, KC− (atranorin, stictic acid).

On rocks in the vicinity of waterfalls and in snowbeds in montane situations; very rare. N. Scotland (Highlands). Fennoscandia, Iceland, U.S.S.R., Greenland, North America.

13. S. symphycheilum Lamb (1961)

Primary thallus disappearing, of small peltate squamules with pale margins. Pseudopodetia to 1.5 cm, ± prostrate, dorsiventral, sparingly branched, firmly attached to rock, without tomentum. Phyllocladia on upper side of pseudopodetia, grey-green, often concave, with a grey-white, wavy and swollen margin. Soredia terminal, ± globose. Cephalodia dark brown-black, with a roughened surface, containing *Stigonema*, commonly on the lower surface of the pseudopodetia and on the substratum between basal phyllocladia. Apothecia terminal, rare. Thallus Pd± yellow, KC+ violet [on filter paper] (atranorin, lobaric acid).

On rocks containing Cu minerals, in montane situations; very rare. N.W. England (Lake District, Coniston). Fennoscandia, North America, U.S.S.R.

British specimens are poorly developed, lacking well-defined terminal soralia. Characterized by the flattened phyllocladia with darkened centres and the presence of lobaric acid, Pd− or pale yellowish. *S. vesuvianum* **(15)** tends to be more erect, is Pd+ orange, and does not have basal, ± flattened phyllocladia which may bear soredia and which are frequently interspersed by cephalodia. For differences from *S. vesuvianum* var. *symphycheileoides* see **15**.

14. S. tomentosum Fr. (1824)

Primary thallus disappearing. Pseudopodetia to 3 cm tall, prostrate or erect, loosely attached, dorsiventral, the lower surface thickly grey tomentose. Phyllocladia wart-like or flattened and ±

squamulose, grey, overlapping and thickly covering the pseudopodetia; soredia absent. Cephalodia indistinct, containing *Nostoc*, concealed in the tomentum of the lower surface. Apothecia 0.3–0.6 mm wide, rather frequent, terminal on short side branches. Thallus Pd+ orange (atranorin, stictic and ± norstictic acids).

On rocks in gravelly situations in heaths; ?extinct. E. Scotland. Europe, Iceland, Greenland, U.S.S.R., North & South America, China.

Characterized by the dorsiventral podetia and grey tomentose lower surface. The Pd+ orange reaction distinguishes it from *S. saxatile* **(11)**.

15. S. vesuvianum Pers. (1811)

Primary thallus disappearing. Pseudopodetia to 1.5(-4) cm tall, erect or decumbent, often forming compact tufts, richly branched or unbranched, ± woody at the base, grey-white, tomentum absent; phyllocladia wart- or shield-like, occasionally becoming flattened or concave, sparse or crowded, at least some with greenish grey central parts surrounded by a paler grey-white, often swollen and ± crenulate margin. Soredia occasionally present as terminal, ± globose soralia at apices of erect pseudopodetia which in their upper parts mostly lack phyllocladia. Cephalodia rare, verrucose, dark olive-brown to black, containing *Stigonema*. Apothecia 0.5–1 mm diam., rare, on lateral branches. Thallus Pd+ orange (atranorin, stictic and ± norstictic acids).

Mostly firmly attached to well-lit siliceous rocks, including those containing heavy metals, also colonizing walls and brick near towns; frequent, especially the north and west. Throughout British Is. Europe, Iceland, Spitsbergen, Greenland, U.S.S.R., North & South America, Macaronesia, Africa, Japan, New Zealand, Tasmania.

Very variable and best characterized by the wart-like to peltate phyllocladia with dark grey centres, though these may be sometimes difficult to find or are rarely absent. Tolerant of sulphur dioxide pollution and extending in lowland England on man-made substrata. Sorediate morphs [**var. *nodulosum*** (Wallr.) Lamb (1969)] may resemble *S. delisei* **(5)**, but the phyllocladia lack dark centres and are Pd−.

S. vesuvianum **var. *symphycheileoides*** Lamb in Vězda (1975) strongly resembles *S. symphycheilum* **(13)** in morphology but shares the characteristic chemistry of *S. vesuvianum*. On wall tops in urban areas and rusting metal of disused railway tracks. S.E. & E. England. Czechoslovakia, U.S.A., Japan.

STICTA (Schreber) Ach. (1803)

Thallus foliose, dorsiventral, loosely attached, single- to multi-lobed; lobes ± rounded, variously indented. **Upper surface** smooth or wrinkled, ± matt. **Lower surface** tomentose, pitted by scattered cyphellae (Fig. 31a). **Soredia** and **isidia** present or absent. **Cortex** pseudoparenchymatous. **Photobiont** chlorococcoid or blue-green (*Nostoc*). **Ascomata** apothecia, sessile, mostly very rare or absent. **Thalline exciple** present. **Hamathecium** of simple paraphyses. **Asci** *Peltigera*-type, 8-spored. **Ascospores** colourless to pale brownish, 1- to 3-septate, ± fusiform. **Conidiomata** pycnidia, *Lobaria*-type, globose to ovoid; wall red-brown at ostiole, paler below. **Conidia** bacilliform or ampulliform. **Chemistry**: lichen products not detected by t.l.c. **Ecology**: on bark, more rarely rocks, characteristic of humid, sheltered woodlands.

Similar in morphology and habitat to *Lobaria* and *Pseudocyphellaria*, but distinguished by the presence of cyphellae on the lower surface, a medullary chemistry deficient in depsides, depsidones or triterpenoids, but rich in methylamine and trimethylamine and polyhydric alcohols, the trimethylamine often imparting a fishy smell to damp thalli particularly when handled. Very sensitive to SO_2 air pollution ($<30\ \mu g\ m^{-3}$).

LITERATURE: James & Henssen (1976).

1 Upper surface pale green or glaucous-green, bright green when wet; photobiont green; without isidia or soredia .... **1. *canariensis***
Upper surface brown-grey, rarely green-brown, darker when wet; photobiont blue-green; thallus with isidia, soredia or folioles 2

2(1)	Isidia or folioles present	3
	Soredia present	**4. *limbata***
3(2)	Margin dissected, fringed with ± flattened isidia or folioles which also occur on upper surface	**1. *canariensis***
	Margin ± smooth; isidia ± cylindrical in coralloid clusters on surface and margin	4
4(3)	Thallus distinctly multi-lobed and branched into lobes	**3. *sylvatica***
	Thallus ± single-lobed or slightly indented	**2. *fuliginosa***

1. S. canariensis (Bory) Bory ex Delise (1823)

S. dufourii Delise (1823)

Green algal morph [*S. canariensis* s.str.] with thallus 1–5(-15) cm diam., often forming extensive patches; lobes to 1 cm diam., elongate and dichotomously branched, discrete or overlapping, apices ± truncate, axils ± rounded; upper surface pale green or glaucous-green, sometimes tinged brown (exposed sites), bright green when wet, ± shiny, smooth, often with a median channel; lower surface flat to somewhat ribbed, whitish to red-brown, ± darkening to black-brown towards centre, ± uniformly felted-tomentose with small, scattered, rounded, white cyphellae, 0.3–0.6 mm diam. Apothecia rare; disc to 7 mm, red-brown; thalline exciple paler, entire or ± crenulate, of small, ± rounded cells, ± uniformly thickened, colourless walls; epithecium red-brown, K−, conglutinated; hypothecium dense, pale grey-brown, not clearing in K. Paraphyses *c.* 3 μm wide, not swollen at the apex. Ascospores colourless, (20-)23–28(-32) × 6–7(-8) μm, 1(-3)-septate. Pycnidia level or slightly elevated, with a dark brown apex. Conidia 5–7 μm, straight, both ends slightly swollen.

On damp, shaded rocks and trees; very local. Wales, W. Scotland, S.W. & W. Ireland. Portugal, Macaronesia.

Cyanobacterial morph [*S. dufourii* s.str.] with thallus 1–5 cm diam., often forming extensive colonies, lobes to 2 cm diam., overlapping, margins rounded, irregularly incised, very thin and wrinkled, ± minutely dissected into small, terete or often flattened, branched isidia, to 0.5 mm diam., often aggregated into tight clusters, 0.2–1 mm diam., occasionally becoming densely imbricate and covering entire surface of older parts of the thallus; upper surface pale grey to dark grey-brown, matt, sometimes white-marbled; lower surface finely white-tomentose, pale brown-white, ± reticulately ridged, with white, scattered cyphellae. Apothecia unknown. Pycnidia unknown.

Amongst mosses on trees and rocks in perpetually moist, very sheltered, deeply shaded sites, often under a canopy of herbaceous vegetation, e.g. *Pteridium*; locally frequent, more widespread in British Is. than the green morphotype. S.W. & N.W. England, N.W. Wales, W. Scotland, W. Ireland. Norway, France, Spain, Portugal, Macaronesia.

Composite thalli of the blue-green and green morphs occur scattered throughout W. British Is. within the distribution of the blue-green morph. These are easily distinguished when wet as contrasting bright green leaflets arising from dark blue-black thalli. The two morphs are the response of a single mycobiont to different photobionts and thus for the reason that all lichen names are based on the mycobiont alone a single name has to be applied to two seemingly very disparate morphs sharing the same mycobiont. Such pairs of morphs are not uncommon in the *Peltigeraceae*.

The blue-green morph differs from *S. sylvatica* **(4)** in the dissected margin fringed with flattened isidia.

2. S. fuliginosa (Hoffm.) Ach. (1803)

Thallus 3–5(-10) cm diam., ± single-lobed; lobes to 3(-4) cm wide, rounded, often ascending, when young, ± down-turned when mature; upper surface dark brown-grey, ± even; isidia ± cylindrical, in coralloid clusters on surface and margins, often darker than the thallus, scattered; lower surface brown, tomentose, with conspicuous, rounded, oval or ± angular, scattered cyphellae. Apothecia very rare, to 4 mm diam., red-brown; thalline exciple paler, crenate or often ciliate, with ± numerous, short, pale tomentose hairs. Ascospores 30–33 × 5.5–8.5 μm, colourless, 1-septate, slightly curved, apices ± pointed.

On mossy trees and damp rocks in humid, sheltered situations, in ancient, oceanic woodlands. Abundant in Scotland, locally frequent elsewhere. W. British Is., S. to Devon and Cornwall. Norway, France, Africa, South America, Australia, New Zealand.

For differences from *S. sylvatica* **(3)** see below.

3. S. sylvatica (Huds.) Ach. (1803)

Like **2**, but with a marked tendency to become multi-lobed with ascending margins and narrower lobes, and is characteristically shallowly reticulately ridged or uneven with more clustered isidia along the ridges and elsewhere on the surface. Apothecia unknown.

On mossy trees and rocks in humid, sheltered situations, especially in old woodlands but also on mossy rocks outside or even by lanes; British distribution as *S. fuliginosa* **(2)**. W. Norway, W. France, Portugal, Madeira.

Immature specimens are often difficult to distinguished from *S. fuliginosa* **(2)**, but older specimens have a more lobed, uneven-margined thallus.

4. S. limbata (Sm.) Ach. (1803)

Thallus 1–3 cm wide, ± single-lobed, ± ascending; lobes irregularly incised or notched, indistinctly lobed, apices rounded, edged with pale brown to blue-grey farinose soredia, occasionally also spreading to upper surface; upper surface grey to pale, dull brown-grey; lower surface pale brown becoming darker towards centre, tomentose, with scattered, pale cyphellae (Fig. 31a). Apothecia very rare, to 3 mm diam., shortly stalked; disc red-brown; thalline exciple roughened or somewhat downy. Ascospores (25-)27–36(-40) × (6-)7–9(-10) μm.

On mossy bark and rocks in old woodlands, and parklands; locally frequent. W. British Is., extending locally to E. England (Sussex, Northumberland) and C. & E. Scotland. Europe, North America, Macaronesia, N. Africa, Australia, New Zealand.

The only sorediate *Sticta* in the British Is.; small specimens can be confused with *Nephroma parile* which is naked below and lacks cyphellae.

STRANGOSPORA Körber (1860)

Thallus crustose, usually thin and poorly developed, ± continuous, without a well-defined margin; prothallus absent. **Photobiont** chlorococcoid. **Ascomata** apothecia, biatorine, convex, colourless, pale yellow to brown or black, rarely pruinose. **True exciple** usually very thin to almost absent, persistent in some tropical species, of parallel, brownish, ochre-coloured to colourless hyphae. **Hymenium** variously coloured, I+ deep blue. **Hypothecium** colourless or pale grey-yellow, often cloudy-oily. **Hamathecium of** paraphyses, branched and ± anastomosed, in a gelatinous matrix, the apices free, usually unbranched, ± vertical and unthickened. **Asci** clavate, with a strongly thickened, I+ blue wall, particularly at the apex, without gelatinous outer layer, and a strongly I+ blue apical dome, multispored. **Ascospores** globose, simple, thin-walled, colourless. **Conidiomata** pycnidia, small, globose, colourless to brownish. **Conidia** ellipsoid. **Ecology**: on bark or wood, sometimes overgrowing bryophytes.

Differs from *Biatorella* in that the asci lack a gelatinous outer layer and have a strongly I+ blue apical dome.

LITERATURE: James (1971), Magnusson (1935), Poelt & Vězda (1977).

1 Apothecia brightly coloured or brownish or black ... 2
  Apothecia pale yellowish or pinkish brown or pallid, translucent when wet .... **1. *delitescens***

2(1) Apothecia pale orange to scarlet-red ... 3
  Apothecia brown or blackish ... 4

3(2) Apothecia pale orange to deep orange, densely pruinose, K+ red-purple .... **4. *ochrophora***
  Apothecia scarlet-red, not pruinose, K− or intensified ... **2. *microhaema***

4(2) Apothecia blackish or brown-black; epithecium pale blue-grey, olive-grey, pale brown or violet; paraphyses 0.5–1 μm wide ... 5
  Apothecia reddish brown, becoming darker and blackish when mature; epithecium yellow- to red-brown; paraphyses 1.5–2 μm thick... **5. *pinicola***

5(4) True exciple well developed; epithecial pigment N−; pycnidia sessile to ± stalked; conidia 1.5–2 μm diam. .... **6. *deplanata***
True exciple thin to almost absent; epithecial pigment N+ reddish; pycnidia ± immersed, globose; conidia 2.5–3.5(-4) × 1–1.5 .... **3. *moriformis***

1. S. delitescens (Arnold) Coppins (1992)

Thallus pale or grey, sparse to absent, usually of scattered granules, occasionally minutely rimose. Apothecia 0.2–0.4(-0.5) mm diam., convex, adpressed, ± dispersed, white to pale yellowish or pinkish brown, translucent when wet; true exciple poorly developed, inapparent; epithecium indistinct, colourless, not granular; hymenium (50-)65–70 μm tall, colourless; hypothecium 35–65 μm tall, yellowish grey, becoming clearly visible in K, densely packed, I+ blue. Paraphyses 1–1.5 μm wide, very distinct in a gel matrix, scarcely branched, ± flexuose. Asci 50–55 × 20–22 μm, 100(-150)-spored, elongate-clavate; wall 2–2.5 μm thick, rather uniform. Ascospores 4–4.5(-5) μm.

On *Corylus* bark; very rare. W. Scotland (Argyll, Is. of Skye). C. & N. Europe,

This very inconspicuous species is distinguished by the pale apothecia, which become completely translucent when wet.

2. S. microhaema (Norman) R. Anderson (1975)

Thallus wide-spreading, inconspicuous, often evanescent, of scattered or ± contiguous, minute, low granules, often concentrated around apothecia giving them a spurious thalline exciple, pale green-grey or grey. Apothecia 0.05–0.2(-0.3) mm, convex to globose, ± sessile, scattered or in small, ± contiguous groups (1–3), scarlet to coral red, becoming brighter and somewhat translucent when wet; true exciple poorly developed, inapparent, hyphae densely anastomosed; hymenium (60-)70–85(-95) μm tall, ± golden-yellow in parts, upper 10–20 μm together with epithecium bright golden-yellow, without granular inclusions, K−; hypothecium 25–50 μm, colourless to pale yellowish, of closely compacted, short-celled anastomosed hyphae in a gel matrix, I+ dark blue. Paraphyses 1(-1.5) μm, richly branched and anastomosing, particularly towards the apices (clearly visible in N), orange-yellow in part. Asci 60–85 × 18–30 μm, 50- to 70-spored, broadly clavate-ellipsoid, wall at apex 4–8 μm thick, sometimes with golden-yellow granular contents. Ascospores (3-)4–4.5 μm.

On *Quercus* and *Alnus* bark in boggy, ancient woodland; very rare. N.W. Scotland. N. & C. Europe. North America.

The small orange-red to scarlet apothecia and the internal orange-red pigmentation of the paraphyses and ascus contents are diagnostic.

3. S. moriformis (Ach.) B. Stein (1879)

Thallus greyish, irregularly granular, sometimes scurfy, often sparse to absent. Apothecia 0.2–0.5 mm diam., adpressed, black, rarely dull blue-black or black-brown, not noticeably paler or translucent when wet; true exciple thin to almost absent, inapparent; hymenium to 110 μm tall, upper layers and epithecium translucent aeruginose blue-grey (K+ pale greenish grey), or in part pale olive-grey (K−), rarely pale brown (K−) or violet, not granular and never K+ purple, usually N+ mauve-purple. Paraphyses 0.5–1 μm wide, densely anastomosed. Asci 40–50 × 7–23 μm, 100- to 200(-300)-spored. Ascospores 1.5–2.5 μm. Pycnidia 50–70 μm diam., ± globose, colourless to brownish; walls dark green; conidia 2.5–3.5(-4) × 1–1.7 μm, shortly oblong.

On decorticated conifers, especially worked timber, rarely on bark, local. Throughout Britain. N. & C. Europe. North America.

4. S. ochrophora (Nyl.) R. Anderson (1975)

Thallus pale grey, immersed or effuse, ± minutely granular. Apothecia 0.1–0.5 mm diam., adpressed to ± sessile, pale ochre to deep orange-red, surface roughened, ± densely orange-pruinose, convex, mainly scattered; true exciple 10–20 μm tall, red-orange-brown-yellow, I+ blue; hymenium 75–85(-125) μm, pale red-orange-yellow, the upper 15–30 μm, together with epithecium, densely granular, the granules minute, golden-yellow or ochre, K+ dissolving, scarlet-red or purple; hypothecium 50–150 μm, ± ochre-coloured, pale beneath, I+ blue. Paraphyses 1–1.5 μm wide, anastomosed, hardly visible in water but clearly visible in K; apices 3–3.5 μm wide. Asci 55–60 × (15-)20–25 μm, 100- to 200-spored, swollen-clavate or oval; upper wall 2.5–3.5 μm wide. Ascospores 4–4.5 μm. Apothecia K+ scarlet- or purplish red.

Overgrowing mosses or on bark of broad-leaved trees (including *Quercus*, *Fraxinus*, *Ulmus*, *Sambucus niger*, *Populus*, *Salix*), in well-wooded sites or occasionally wayside trees in sheltered sites; local. W. British Is., very rare in E. England (Suffolk) and E. Scotland (Fife). N. C. & S. Europe, Macaronesia, N. & S. America.

The apothecia are sometimes confused with small tufts of free-living *Trentepholia*, which are similar in colour and texture. The apothecia are very brittle and are easily broken when touched with a dissecting needle.

5. S. pinicola (Massal.) Körber (1860)

Thallus wide-spreading, thin, evanescent, minutely warted, or rather thick, ± delimited, granular-warted, pale to dark grey or grey-brown; thalline granules dispersed or contiguous, ± convex, irregular. Apothecia 0.2–0.5 mm diam., often crowded or sometimes in groups of two or three, dark red-brown, brighter when wet, immersed, becoming sessile; true exciple thin to almost absent, if well developed, the outer 12–15 μm brownish with radiating hyphae and inner 12–15 μm colourless with parallel hyphae, I– ; hymenium 55–65 μm tall, mainly colourless, the upper 12–20 μm and epithecium yellow-brown to brown, often with a reddish tinge, semi-transparent, not granular, K– or K+ red-brown, N+ pale brown or orange-brown, with a gelatinous coat 12 μm thick; hypothecium 50–100 μm, colourless or pale straw, of small, compacted indistinct isodiametric cells. Paraphyses 1.5–2 μm wide, densely anastomosed towards the base, often sparingly so above, indistinct in water, but visible in K. Asci 40–50 × 15–20 μm, *c.* 100-spored, broadly clavate; wall 3 μm thick at sides and 6–8 μm at apex. Ascospores 1.5–2.5 μm. Pycnidia *c.* 50 μm diam., very rare, globose, ± colourless; conidia 3–4 × 1 μm, ellipsoid.

On bark of broad-leaved trees, rarely on conifer wood; local in polluted areas. E. & C. England, E. Scotland. Europe.

See also *S. deplanata* **(6)**.

6. S. deplanata (Almq.) Clauz. & Roux (1985)

Thallus effuse, thin, partly immersed, ± minutely arachnoid, whitish. Apothecia 0.2–0.5 mm diam., not pruinose, sessile, adpressed, at first ± flat with indistinct margin, becoming convex, dark brown to blackish, paler when wet, surface ± roughened; true exciple well developed, 33–45 μm wide, distinct in sections, colourless or outer part brownish or olivaceous, K+ dull brownish, N–, hyphae branched, 0.7–1 μm wide but sometimes swollen to 3 μm at edge; hymenium 50–60 μm tall, colourless but ± pale brown, red-brown or usually olivaceous in upper part, concolorous with outer exciple, I– ; hypothecium well developed, pale straw-yellow, of densely interwoven, compacted hyphae. Paraphyses 0.5–1 μm wide, indistinct in water, apices sometimes to 3 μm wide, simple or occasionally branched; in K some paraphyses appear to have an individual gel coating, giving an overall width of 5–8 μm. Asci 45–50 × 15–20 μm, longly clavate, apex strongly thickened. Ascospores 2–3 μm. Pycnidia 60–120(-140) μm diam., to 200 μm tall, sessile to short-stalked, reddish brown to black but superficial conidia often give a white-pruinose appearance, wall brown, K– ; conidia 1.5–2 μm diam., broadly ovoid to globose. Thallus Pd–, K–, KC–, C–.

On dry fissured bark of large *Salix* and *Fraxinus* trunks; very rare. C. Scotland (Perth, Craig Tulloch). S.E. Norway, C. & S. Sweden, Poland, U.S.S.R., Germany, Czechoslovakia, Hungary.

Distinguished from bark morphs of *S. moriformis* **(3)** by the well-developed exciple of narrower hyphae the massive hypothecium, epithecial pigment, smaller ascospores, larger sessile to ± stalked pycnidia and ± globose conidia. *S. pinicola* **(5)** has a red- to yellow-brown epithecium.

STRIGULA Fr. (1823)

Discosiella H. Sydow & Sydow (1912) [anamorph]

Thallus crustose, thin, usually at least partly immersed, cortex not differentiated. **Photobiont** *Trentepohlia* (or *Cephaleuros* in tropical species). **Ascomata** perithecia, globose, or upper part conical. **Involucrellum** present in most species, blackish brown. **True exciple** dark brown or (especially below) pale brown or colourless. **Hamathecium** of simple or sparingly branched, slender, (*c.* 1 μm wide) long-celled paraphysoids; gel I+ blue. **Asci** cylindrical or cylindrical-clavate, thin-walled laterally but with a conspicuous apical dome penetrated by a narrow ocular chamber, discharge fissitunicate 8-spored, I – (Fig. 34b). **Ascospores** colourless, septate, ellipsoid, ovoid or fusiform, often constricted at mid-septum, smooth, without a conspicuous perispore. **Conidiomata** pycnidia, globose or conical, immersed to $\frac{1}{2}$-emergent, black at least in upper (exposed) part. **Conidiogenous cells** mostly sessile, often percurrently proliferating, acrogenous. **Conidia** colourless, of one of two types: (a) macroconidia [*Discosiella*] 1- or more septate, oblong to fusiform but often ± curved at proximal end with a subterminal point of attachment, one or both ends sometimes with an inconspicuous mucoid appendage; (b) microconidia, simple, ellipsoid or narrowly ellipsoid. **Chemistry**: lichen products not reported; UV – . **Ecology**: on bark or mosses, more rarely soil or rock, commonly on leaves in the tropics.

Separated from similar genera by the macroconidia which are obliquely attached to the conidiogenous cells. *Anisomeridium* and *Arthopyrenia* further differ in their usually richly branched paraphyses, and asci which have an apical dome without, or with, a broad, ocular chamber. In *Porina* the asci have a narrow apex without an apical dome and a thin, refractive (chitinoid) apical thickening and are not distinctly fissitunicate.

LITERATURE: Santesson (1952), Swinscow (1962, 1965b).

1 Ascospores at least 3-septate 3
Ascospores mainly 1-septate .... **3. *taylorii***

2(1) Ascospores 3-septate .... **1. *jamesii***
Ascospores (4-)6- to 7(-9)-septate .... **2. *stigmatella***

1. S. jamesii (Swinscow) R. C. Harris (1980)

Thallus thin and filmy, effuse, whitish to pale grey. Perithecia 0.1–0.2 mm diam., *c.* $\frac{1}{2}$-immersed, usually rather crowded; involucrellum usually distinct; true exciple colourless, or pale brown above. Asci (Fig. 34b) 45–60(-70) × 9–10 μm. Ascospores 12–16(-19) × 4–5 μm, 3-septate, the second cell from the top usually widest. Pycnidia 40–100 μm diam.; macroconidia 12–20(-22) × 2–3.5 μm, 3-septate, oblong-fusiform; microconidia 2.5–4 × 1–1.4 μm.

Usually on smoothish or rough bark of tree trunks (especially *Ulmus*) or exposed roots in sheltered woodland, rarely over mosses on crumbling mortar; local. S. & N.E. England, Wales, S. & E. Scotland, S. Ireland. Denmark.

Porina aenea has a darker, brown or olive-brown thallus and a narrowed ascus apex. *S. affinis* (Massal.) R. C. Harris (1980), which has larger perithecia and ascospores and broader macroconidia, has not been correctly reported from British Is.

A similar undescribed species with asci 47–65 × 6–7 μm, ascospores 10–13 × 2.3–3.5 μm, 1-septate, ovoid with a constriction at the septum, and macroconidia 8–11 × 1.8–2 μm, 1-septate, is known from shaded bark of mature *Ulmus* in stream valleys in Scotland (Midlothian, E. Sutherland). A second potential undescribed species with smaller ascospores, 9.3–10.5 × 2.3–2.5 μm, has been collected on rough bark of old *Betula* in S.W. Ireland.

2. S. stigmatella (Ach.) R. C. Harris (1980)

Thallus thin, mostly immersed, effuse, white to pale grey. Perithecia 0.3–0.4 mm diam., *c.* $\frac{1}{2}$-immersed, partly covered by a thin layer of thallus; involucrellum closely clasping the exciple; true exciple pale brown or colourless. Asci 70–80 × 12–15 μm. Ascospores 25–35(-50) × 5–8 μm, (4-)6–7(-9)-septate, fusiform, ± constricted at mid-septum. Pycnidia *c.* 100 μm diam.; macroconidia 20–30 × 4–6 μm, 5- to 7-septate, oblong-fusiform.

On bark of old broad-leaved trees, or over mosses on tree bases; ?extinct. S. England (Sussex). Europe, U.S.A. (Michigan), W. Canada.

S. stigmatella **var. *alpestris*** (Vězda) Coppins (1980) has a thicker and often ± verrucose thallus, perithecia which are $\frac{1}{2}$-to $\frac{3}{4}$-immersed, the involucrellum not delimited from the true exciple, which is pale brown but blackish brown and thickened around the ostiole, and asci 100–110 × 13–15 μm. It occurs on mosses over epidiorite or mica-schist above 600 m and is rare in N. Scotland (Angus, Caenlochan; Perth, Ben Chonzie, Ben Lawers). It is reported from Czechoslovakia (Tatra Mtns).

3. S. taylorii (Carroll ex Nyl.) R. C. Harris (1980)

Thallus 25–75 μm thick, immersed, smooth or finely cracked, in irregular, ± delimited patches, brown or olivaceous brown, sometimes bordered by a narrow, black prothallus. Perithecia 0.15–0.2(-0.3) mm diam., ½-immersed to ± sessile, semi-globose to distinctly conical; involucrellum distinct; true exciple colourless, or pale brown above. Asci 50–75 × 8–12 μm, cylindrical-clavate. Ascospores (18-)20–32 × (3.5-)4–5 μm, 1(-3)-septate, fusiform markedly constricted at septum and often breaking into two part-spores. Pycnidia 80–100 μm diam., usually numerous; macroconidia 11–16.5 × 2.5–3 μm, 1-septate; microconidia 3–4.5 × 1.2–1.8 μm.

On usually ± smooth bark of mature trees (especially *Acer* and *Fraxinus*) in sheltered situations; local. W. Britain, W. Ireland. Europe; distribution uncertain due to confusion with a similar (? undescribed) species with shorter macroconidia, 9–11 μm long.

Usually distinguished in the field from *Porina aenea* or *P. borreri* by the numerous pycnidia scattered among the larger, often ± conical perithecia. See also under *S. jamesii* **(1)**.

SYNALISSA Fr. (1825)

Thallus shrubby, with erect, coralloid branches forming small cushions, black when dry, red-brown and gelatinous when wet; not corticate, the hyphae forming ± angular patterns, enclosing the photobiont clusters; centrally with a hyphal core devoid of photobiont cells. **Photobiont** *Gloeocapsa*. **Ascomata** apothecia, terminal, at first ± globose with a prominent thalline exciple surrounding a poroid disc; disc later ± expanding. **Hymenium** I−. **Hamathecium** of thin, slender paraphyses. **Asci** cylindrical, 8- or multispored, thin-walled, I−. **Ascospores** broadly ellipsoid to globose, simple, colourless. **Conidiomata** pycnidia, terminal, immersed; wall colourless. **Conidiogenous cells** acrogenous. **Conidia** bacilliform, simple, colourless. **Chemistry**: no lichen products reported. **Ecology**: mainly on rocks. **Distribution**: *c*. 5 species, mainly in Europe and North America.

Distinguished from *Lempholemma* by the *Gloeocapsa* photobiont, and from other genera (e.g. *Pyrenopsis*) with similar photobionts by the shrubby habit.

1. S. symphorea (Ach.) Nyl. (1856)

S. ramulosa (Hoffm.) Körber (1855)

Thallus forming small, dense cushions or tufts to 3 mm tall, consisting of stout, erect or decumbent branches with blunt, swollen ends in which apothecia or pycnidia may develop. Photobiont cells 5–6 μm diam., often in small groups of 2–3; in outer part of thallus and thalline exciple, with reddish sheaths, I+ violet. Apothecia 0.2–0.5(-0.8) mm diam.; disc often remaining pore-like, sometimes expanding and then reddish brown; thalline exciple to 150 μm wide; true exciple indistinct; hymenium to 125 μm tall, colourless; hypothecium to 40 μm tall, colourless to pale brown, of interwoven hyphae. Paraphyses to 1 μm wide. Asci to 100 μm tall, 8- to 24-spored. Ascospores 7–10(-12) × 6–9 μm. Conidia 3–4 × 1–1.5 μm.

On limestone rocks, usually in small crevices growing among other lichens (e.g. *Catapyrenium* species, *Psora lurida*, *Toninia caeruleonigricans*); local. S.W. England, W. Scotland (Argyll). Europe, N. Africa, North America.

Often confused with some *Lempholemma* species (e.g. *L. botryosum*) which, however, have *Nostoc* as photobiont.

TELOSCHISTES Norman (1853)

Thallus shrubby, lobes flattened, dorsiventral and sparingly branched to ± rounded and ± richly branched, often with fibrils or cilia, attached by a basal sheath, rhizines absent. **Upper surface** ± entirely bright yellow-orange, sometimes pale below, grey-green to ash-grey in some non-British species. **Soralia** sometimes present. **Cortex** of periclinally arranged, conglutinated hyphae. **Photobiont** trebouxioid. **Medulla** of ± parallel, vertically aligned, loosely interwoven hyphae. **Ascomata** apothecia, laminal, terminal or marginal. **Thalline exciple** present, persistent, sometimes ciliate. **Disc** yellow to orange. **Epithecium** granular, yellow, K+ crimson. **Hymenium** and **hypothecium** colourless. **Hamathecium** of paraphyses septate, ± unbranched; apices ± capitate. **Asci** 8-spored, elongate-clavate, of several layers, apically ± thickened with a broad internal beak, inner part of apex and external cap I+ blue, non-fissitunicate, *Teloschistes*-type. **Ascospores** colourless, ellipsoid, polarilocular. **Conidiomata** pycnidia, round, deep orange-red, immersed, usually multilocular; **conidiogenous cells** parenchymatous. **Conidia** 2.5–3.5 × 0.5–0.8 μm, rod-shaped, simple, colourless. **Chemistry:** physcion (parietin).

1 Thallus forming small, compact, orbicular clumps to 2 cm diam., of flattened, radiating lobes; soralia absent, usually abundantly fertile .... **1. *chrysophthalmus***
Thallus forming lax, entangled clumps, 3–10 cm diam., of ± terete, slender branches; soralia present; apothecia not observed .... **2. *flavicans***

1. T. chrysophthalmus (L.) Th. Fr. (1860)

Thallus 1–2 cm diam., forming small, compact, ± orbicular tufts, 0.5–1 cm tall; lobes 0.5–2.5 mm wide, radiating from centre, ascending or ± erect, dorsiventral, irregularly branched with ± numerous marginal fibrils; upper surface orange-yellow, matt, smooth or ± slightly longitudinally ridged; lower surface white or greyish, often longitudinally veined and wrinkled. Apothecia ± stalked, marginal or terminal; disc 1–6 mm diam., orange-yellow, ± flat; thalline exciple thin, often with abundant, concolorous or grey fibrils, 0.5–1.5 mm long. Ascospores 10–15 × 5–8 μm. Pycnidia frequent, immersed below shallow, orange-red warts, *c.* 0.1 mm diam.

On well-lit, nutrient-rich twigs of shrubs and small trees, particularly Rosaceae, in sunny, sheltered situations, formerly in old orchards and hedgerows; very rare. S. England (Devon, Sussex), Channel Is. (Jersey), S.W. Ireland. Cosmopolitan in drier, sunny, warm, temperate areas of both Hemispheres.

This attractive species was formerly more widespread in S. England and it has not been seen since 1966.

2. T. flavicans (Swartz) Norman (1853)

Thallus richly and intricately branched, forming ± lax, entangled tufts, 3–10 cm diam., often rather loosely attached; branches 0.5–1 mm diam., rounded or somewhat flattened or angled, slender, matt, smooth or rarely slightly pubescent, ± dichotomously branched, uniformly golden-yellow to orange-red, sometimes paler below and becoming overall green-grey in shaded habitats, with few to numerous scattered fibrils; soralia yellow, tubercle-like, rounded or ± elongate, frequent. Pycnidia frequent, immersed in orange-red tubercles, 0.15–0.2 mm diam. Apothecia not observed in British material, 1–4 mm diam., lateral; thalline exciple sparingly ciliate.

On acid rocks, soil, wood, twigs and bark usually in well-lit and wind-exposed sites, occasionally in more sheltered, inland, boggy sites and *Salix* carr, especially on or near the coast; local and decreasing. S.W. England (Dorset westward), S.W. Wales, S.W. Ireland. Cosmopolitan in dry, sunny, warm temperate areas of both Hemispheres.

A very distinctive species sometimes resembling *Usnea* in habit but yellow-orange or green-yellow in colour. Formerly a more widespread species, extending into C. England and S. Scotland; its more recent decline has been attributed to the loss of habitat and, especially, the spread of air pollution.

TEPHROMELA M. Choisy (1929)

Thallus crustose, warted or cracked-areolate, pale coloured, ± glossy; medulla I−; **prothallus** occasionally visible between areoles, often forming a dark border to the thallus. **Photobiont** chlorococcoid. **Ascomata** apothecia, immersed, adpressed to sessile, disc black, cup-like to strongly convex. **Thalline exciple** present or absent. **True exciple** thin, ± inconspicuous. **Epithecium** or **hymenium** with purplish or greenish, N+ red pigments. **Hypothecium** ± ochraceous below. **Hamathecium** of paraphyses, simple or sparingly branched, each with a gelatinous coat, swelling strongly in water; apices not swollen, but often with a pigmented hood. **Asci** clavate, 8-spored, *Bacidia*-type. **Ascospores** simple, colourless, ellipsoid, without a distinct perispore, ± thick-walled. **Conidiomata** pycnidia, immersed; wall colourless except for green pigmentations around the ostiole. **Conidiogenous cells** in chains, pleurogenous. **Conidia** oblong-ellipsoid to shortly thread-like, ± straight, simple, colourless. **Chemistry**: various depsidones in the medulla and atranorin, and, or usnic acid in the cortex. **Ecology**: mostly on rocks, sometimes on bark or wood, *c.* 13 species, cosmopolitan.

Distinguished from *Lecanora* s.str. and *Lecidea* s.str. by the combination of a poorly developed true exciple, *Bacidia*-type asci, paraphysis structure, and conidiogenous cells in chains.

LITERATURE: Hertel (1977), Hertel & Rambold (1985), Rambold (1989).

1	Soredia present	2
	Soredia absent	3
2(1)	Thallus with discrete, scattered soralia; soralia UV+ white	**5. *pertusarioides***
	Thallus ± continuously sorediate; all parts UV−	**4. *grumosa***
3(1)	Thalline exciple absent; hymenium colourless; epithecium green	4
	Thalline exciple usually present; hymenium and epithecium purple-violet	**3. *atra***
4(3)	Medulla Pd+ yellow; cortex KC+ red	**2. *armeniaca***
	Medulla Pd−; cortex KC± yellow	**1. *aglaea***

1. T. aglaea (Sommerf.) Hertel & Rambold (1985)

Lecidea aglaea Sommerf. (1826)

Thallus rather thick, warted-areolate, wide-spreading, to *c.* 10 cm diam.; areoles to 2 mm diam., mostly contiguous, rarely scattered, convex or flat, at times wrinkled, greenish, brownish or yellowish white; prothallus black, indistinct. Photobiont cells 7–14 μm diam. Apothecia 0.5–2(-3) mm diam., black, immersed or ± adpressed; disc at first flat, later becoming irregularly convex; thalline exciple absent; true exciple excluded, much reduced, visible in upper part only as a narrow green-black zone continuous with epithecium; epithecium blue-green to green-black; hymenium 60–80 μm tall, ± colourless; hypothecium colourless above, pale yellowish brown below. Ascospores 8–14 × 5–8 μm. Conidia 5.5–9 × 1–2 μm. Cortex Pd−, K+ yellow, KC+ yellow, C− (atranorin, usnic, and bourgeanic acids).

On siliceous, exposed rocks in uplands. Throughout British Is. N. Europe, C. European mtns., Arctic, North America, N. Asia, Nepal.

2. T. armeniaca (DC.) Hertel & Rambold (1985)

Lecidea armeniaca (DC.) Fr. (1825)

Thallus thick or moderately thick, areolate, wide-spreading, to 10 cm diam.; areoles 0.4–2(-4) mm wide, ± angular, contiguous or rarely scattered, convex or flat, greyish or yellowish green when fresh becoming red-brown or brownish yellow in the herbarium, shiny, often minutely wrinkled or becoming secondarily cracked; prothallus black, inconspicuous, sometimes visible between the areoles. Photobiont cells 7–18 μm. Apothecia 0.8–2(-4.5) mm diam., immersed; disc black; thalline exciple absent; true exciple ± inapparent; epithecium bluish green-black, in part brownish; hymenium 45–55 μm tall; hypothecium colourless, but ± pale ochraceous in lower part. Ascospores 9–12 × 3.5–4 μm. Pycnidia not seen; conidia 6.5–10 × 1.5–1.8 μm. Cortex Pd−, K± yellow or orange-red, KC+ red, C−; medulla Pd+ yellow, K−, KC± red, C− (alectorialic acid and ± protocetraric acid).

On exposed siliceous, rocks, above *c.* 1000 m alt.; rare. Scotland (Highlands, Cairngorm mtns, Breadalbane mtns, Ben Nevis). C. European Mtns., Arctic, Fennoscandia, Iceland, Greenland, North America, Himalaya, Australia.

Specimens in the herbarium become deeper reddish brown and the paper may be discoloured pinkish due to the alectorialic acid. Distinguished from *T. aglaea* **(1)** by the narrower ascospores and chemistry.

3. T. atra (Huds.) Hafellner ex Kalb (1983)

Lecanora atra (Huds.) Ach. (1810)

Thallus rather thick, warted-areolate, often secondarily cracked into larger 'areoles', grey-white or greyish green, wide-spreading to 30 cm or more diam.; areoles 0.3–1.5 mm diam., mostly ± contiguous and fused, irregular, often warted-wrinkled; prothallus blackish. Photobiont cells 8–17(-19) μm diam. Apothecia 1–2.5 mm diam., round or irregular, immersed or sessile, black; disc flat or concave; thalline exciple conspicuous, persistent, swollen, ± entire to flexuose at maturity; epithecium dark red-brown; hymenium 50–60 μm tall, dark purplish brown or purple-violet, pale purple-violet in upper part, ochraceous (K+ yellow intensifying) in lower part; hypothecium dark. Ascospores 10–15 × 5–8 μm. Conidia 12–21(-24) × 1 μm, shortly thread-like. Cortex Pd−, K+ yellow, KC+ yellow, C−; medulla UV+ ice-white (atranorin, α-collatolic and ± alectoronic acids).

On siliceous and slightly calcareous, nutrient-rich rocks and walls, rarely timberwork and trees; common. Throughout British Is. Cosmopolitan.

4. T. grumosa (Pers.) Hafellner & Roux (1985)

Lecanora grumosa (Pers.) Du Rietz (1926)

Like 3, but surface ± totally covered to edge in dense dark bluish grey (pigment K−, N+ red), ± granular soredia, *c.* 50–100 μm diam; thallus usually delimited by a blue-black prothallus. Photobiont cells 6–15 μm diam. Apothecia rare; thalline exciple sorediate. Pycnidia not seen. Thallus and soredia Pd−, K± yellow, KC−, UV− (atranorin and unidentified substance).

On exposed, siliceous rocks, rarely on timberwork, mainly upland; local. British Is. N. & C. Europe.

5. T. pertusarioides (Degel.) Hafellner & Roux (1985)

Like 3, but with large (0.4–2 mm diam.), scattered, hemispherical, white or blue-grey spotted soralia. Photobiont cells 7–17 μm diam. Soralia Pd−, K+ yellow, KC± pinkish, UV+ ice-white (atranorin and α-collatolic acid).

On slate in a quarry and vertical mica-schist cliff; very rare. N. Wales (Caernarvon, Betws-y-Coed), C. Scotland (Perth, Beinn Heasgarnich). Sweden, C. European mtns.

Perhaps only a sorediate morph of *T. atra* **(3)**.

THAMNOLIA Ach. ex Schaerer (1850)

Thallus fruticose, prostrate or decumbent, not or sparingly branched, the branches cylindrical, the base usually dying. **Cortex** paraplectenchymatous, of longitudinally orientated hyphae. **Photobiont** trebouxioid. **Medulla** thin, of longitudinally orientated hyphae, the interior hollow. **Apothecia** and **pycnidia** unknown. **Chemistry**: with depsidones (baeomycesic, squamatic or thamnolic acids). **Ecology**: in montane exposed heaths in both Hemispheres; absent from Africa.

A monotypic genus with an extraordinarily wide world distribution for a species not known to produce propagules other than thallus-fragments.

1. T. vermicularis (Swartz) Schaerer (1850)

Thallus decumbent, straggling, or ± erect, occasionally ± densely tufted, simple or sparingly, irregularly branched, to 5 cm tall, 1–2 mm wide, chalk-white, cylindrical, hollow, the apices pointed, smooth, rarely with few short, lateral, pointed branches. Soredia and isidia absent. Thallus Pd+ yellow, K+ yellow, KC−, C−, UV+ white (baeomycesic and squamatic acids).

On short, exposed montane heaths, often amongst *Rhacomitrium lanuginosum* and *Salix herbacea* above 850 m, rarely occurring down to sea-level on dunes in N. Scotland; locally frequent. N.W. England (Lake District), N. Wales, N. Scotland (Highlands). Scandinavia, Europe, U.S.S.R., North America, New Guinea, Australia, New Zealand.

Only the **var. *subuliformis*** (Ehrh.) Schaerer (1850) is known in the British Is.; var. *vermicularis* differs in having thamnolic acid alone and neither baeomycesic nor squamatic acids. The chalk-white, worm-like, usually unbranched, hollow and prostrate thalli are unlikely to be confused with any other species; *Siphula ceratites* has a shorter, erect, compact thallus. Several lichenicolous fungi occur on *T. vermicularis*, which are sometimes mistaken for its own ascomata; these include the gall-forming *Thamnogalla crombiei* (Mudd) D. Hawksw. (1980) described from Scotland with narrowly ellipsoid, simple ascospores 8–10(-11) × 3–5 μm.

THELENELLA Nyl. (1855)

Thallus crustose, superficial to partially immersed. **Photobiont** chlorococcoid. **Ascomata** perithecia, ± immersed in warts, globose to ± top-shaped. **Involucrellum** absent. **True exciple** brownish above, colourless below. **Hamathecium** of paraphysoids, richly branched and anastomosed; periphysoids absent. **Asci** (4-)8-spored, ± cylindrical, fissitunicate, without an indentation in the apical dome, I− ; wall uniformly thick (Fig. 9b). **Ascospores** uniseriate in the ascus, muriform, colourless, rarely becoming pale yellow-brown, ellipsoid or longly ellipsoid, often somewhat broadened towards one end. **Chemistry**: lichen products not detected by t.l.c. **Ecology**: on bark.

Chromatochlamys has periphysoids and the apical dome of the asci has a distinct indentation. *Protothelenella* has asci with an I+ blue apical apparatus and *Elliptochloris* as photobiont.

LITERATURE: Eriksson (1981), Mayrhofer (1987b), Morgan-Jones & Swinscow (1965).

1. T. modesta (Nyl.) Nyl. (1855)

Microglaena modesta (Nyl.) A. L. Sm. (1911)

Thallus uneven-warted, ± smooth to deeply cracked-areolate, thick or disappearing, yellow-grey to pink- or red-brown, fading to pale grey or grey in the herbarium. Apothecia 0.3–0.5 mm diam., rounded, discrete or ± aggregated, $\frac{3}{4}$ to fully immersed but appearing as ± prominent warts; true exciple brownish above, pale or colourless below; ostiole dark brown. Paraphyses 1–1.5 μm wide. Ascospores (20-)25–42 × (9-)11–17 μm.

On ± rough bark of wayside or parkland *Fraxinus*: very rare. N.E. Scotland (Angus, Glamis Village); also last century in S.W. England (Gloucester, Barnsley Park). W. Germany, France, Greece, Italy, Poland, Portugal, Romania, Sweden, Africa.

In the herbarium superficially resembles *Pertusaria pertusa* as the original pink thallus soon fades to grey.

THELIDIUM Massal. (1855)

Thallus crustose, immersed or superficial, continuous or rimose-cracked, pale grey-green to dark brown. **Soralia** rarely present (see **7**). **Photobiont** chlorococcoid. **Ascomata** perithecia, superficial, immersed in thallus or forming pits in limestone. **Involucrellum** present or absent (Figs 41d-e). **True exciple** dark brown throughout, or colourless below, comprised of periclinally orientated cells. **Hymenium** I+ red. **Hamathecium** of periphyses; paraphyses absent. **Asci** wall thickened above when young, I−, with small ocular chamber, lacking apical apparatus, discharge fissitunicate (Fig. 34f); 8-spored. **Ascospores** ellipsoid to oblong-ellipsoid, colourless, smooth, without a thick perispore, 1- to 3(-5)-transseptate, occasionally a small proportion 1(-2) longiseptate. **Chemistry**: lichen products not detected by t.l.c. **Ecology**: on limestone, sandstone, mica-schist and other rocks, rarely on soil.

Polyblastia differs in the consistently submuriform to muriform ascospores and *Verrucaria* has persistently simple ascospores. All species treated, with the exception of *T. fumidum* **(2)** and *T. impressum* **(3)**, have a dark brown to dark greenish brown pigment in the exciple and involucrellum, which is ± K−.

T. decipiens **(1)** and *T. incavatum* **(4)** are the only British species to form well-defined pits in limestone; both lack an involucrellum.

The name *T. cataractarum* (Hepp.) Lönnr. (1859) is of uncertain application and the British records refer to various species. *T. minimum* (Massal. ex Körber) Arnold (1871) has not been correctly reported from the British Is.

1 Involucrellum present, not forming well-defined pits in limestone 2
Involucrellum absent, some species forming well-defined pits in limestone 6

2(1) Mature ascospores 1-septate 3
Mature ascospores 3-septate ..... **7. *papulare***

3(2) Ascospores >18 μm long 4
Ascospores (9-)10–15 × 6–9 μm ..... **3. *impressum***

4(3) Involucrellum variously developed; thallus without dark basal layer, pigments K− or almost so 5
Involucrellum well developed, continuous with dark basal layer of thallus, both with K+ greyish pigment; thallus superficial, dark brown ..... **2. *fumidum***

5(4) Perithecia not covered by a thalline layer; involucrellum well developed ... **9. *pyrenophorum***
Perithecia covered almost to apex by a thalline layer; involucrellum well developed, or thin and poorly developed ..... **8. *pluvium***

6(1) Perithecia immersed in pits in limestone; thallus immersed 7
Perithecia not in pits, prominent or partly immersed in thallus; thallus superficial 8

7(6) Ascospores 1-septate, 21–37 × 10–16(-17) μm ..... **1. *decipiens***
Ascospores 3(-4)-septate, (28-)31–55(-61) × 11.5–17 μm ..... **4. *incavatum***

8(6) Ascospores 1-septate, (15-)17–29 × 7–12 μm ..... **6. *minutulum***
Ascospores 3-septate, 25–37 × 9–15 μm ..... **5. *microcarpum***

1. T. decipiens (Nyl.) Krempelh. (1861)

Thallus immersed, a pale greyish or pale brownish stain (similar in colour to substratum), continuous or finely cracked. Perithecia 3.1–4 mm diam., almost completely immersed in pits; apex separated from side of pit by a fine crack. Involucrellum absent. Ascospores 21–37 × 10–16(-17) μm, 1-septate.

On limestone, marble, and calcareous schist; common. Throughout British Is. Europe, North America.

2. T. fumidum (Nyl.) Hazslin (1884)

T. auruntii auct. brit., non (Massal.) Krempelh. (1861)

Thallus superficial, dark brown to blackish brown, cracked, up to 200 μm thick, with a thick dark basal layer. Perithecia 2.6–6.5 mm diam., forming projections, often ± covered by a thalline layer below, free at the apex; involucrellum well developed, enclosing the perithecium and continuous with the dark basal layer of the thallus; pigment purplish or reddish brown, K+ greyish. Ascospores 24–34 × (8-)9.5–15 μm, 1-septate.

On damp mica-schist rocks above 550 m altitude; rare. N. Scotland (Highlands). Pyrénées, Romania.

Distinguished from other British species in the dark thallus with a dark basal layer, and from all species, except *T. impressum* **(3)**, in the K+ grey pigment of the involucrellum.

3. T. impressum (Stizenb.) Zsch. (1920)

Thallus immersed, pale grey to brownish grey, similar in colour to the substratum, continuous or with a few fine cracks. Perithecia 0.2–0.45 mm diam., forming projections; ostiole sometimes in a depression; involucrellum forming a thick cap to the perithecium, reaching at least half-way down perithecium, and often almost to base, pigment purplish or reddish brown, K+ greyish. Ascospores (9-)10–15 × 6–9 μm, 1-septate.

On limestone; uncommon. Throughout British Is. Spain, Portugal.

Characterized by the small ascospores and well-developed involucrellum with a K+ greyish pigment.

4. T. incavatum Mudd (1861)

Thallus immersed, forming a pale grey to dirty cream-coloured stain, continuous or finely cracked. Perithecia 0.15–0.45 mm diam., almost completely immersed in pits, the apex separated from side of pit by a fine crack; involucrellum absent. Ascospores 33–48(-50) × 11.5–17 μm, 3(-4)-septate, occasionally with 1(-2) longitudinal septa.

On hard limestones; common. Throughout British Is. W. & S. Europe, Norway, Sweden, Canada.

The relationship between this species and *Polyblastia dermatodes*, which has consistently submuriform spores, is in need of critical evaluation.

5. T. microcarpum (Davies ex Leighton) A. L. Sm. (1911)

T. subgelatinosum (Sm.) Zsch. (1920)

Thallus superficial, pale grey-green to dark brown, occurring as numerous small patches or ± continuous, thin, to 50(-95) μm thick, not cracked or rarely slightly cracked. Perithecia 0.1–0.3 mm diam., prominent to about half-immersed in thallus; ostiole inconspicuous; involucrellum absent. Ascospores (23-)25–35(-37) × (8.5-)9–15 μm, 3-transseptate, rarely with 1 longitudinal septum.

On limestone, chalk, mica-schist and various siliceous rocks, very rarely on calcareous soil, in moist or shaded situations and beside streams, often on pebbles; local. Throughout British Is. France, Austria, Spain, Sweden.

T. zwackhii (Hepp) Massal. (1855) is the correct name for this species.

T. fontigenum Massal. (1856) differs in having perithecia 0.19–0.22 mm diam., largely immersed in soft calcareous deposits. An involucrellum is absent or very weakly developed at apex; ascospores 32–39 × 12–15 μm., 3-septate. In dune-slacks in N. Devon, and on damp tufa in Glamorgan. France. The status of this species is in need of further study.

6. T. minutulum Körber (1863)

T. margaceum (Leighton) Zsch. (1920), *T. terrestre* W. Watson (1917), *T. mesotropum* (Nyl.) A. L. Sm. (1911), *T. hospitum* Arnold (1882)

Thallus superficial, pale grey-green to dark brown, occurring either as numerous small patches or ± continuous and then often cracked, thin, 30–160 μm thick. Perithecia (0.08-)0.1–0.25(-0.3) mm diam., very prominent or to half-immersed in thallus; ostiole inconspicuous; involucrellum absent (Fig. 41d). Ascus (Fig. 34f). Ascospores (15-)17–29 × 7–12 μm, 1-septate.

On limestone, chalk, sandstone, mica-schist and brick, often on pebbles and usually in shaded situations, rarely on soil; probably frequent. Throughout British Is. Scandinavia, W. France.

Variable in the degree of development of the thallus and in the size of the perithecia and ascospores.

7. T. papulare (Fr.) Arnold (1885)

Thallus immersed and greyish, to superficial and grey-brown to light brown or mid-brown, often cracked. Perithecia (Fig. 41e) (0.2-)0.4–0.65(-0.75) mm diam., forming conspicuous projections; ostiole often visible as a paler spot, often in a depression; involucrellum conspicuous, forming a thick cap, or reaching half-way down perithecium, rarely reaching almost to its base. Ascospores (28-)31–55(-61) × 13–21 μm, 3(-4)-septate, occasionally with 1(-2) longitudinal septa.

On limestone, mica-schist, calcareous sandstone and on acidic rocks, often in damp situations; occasional. Mainly in N. & W. Britain.

Variable in size of the ascospores and perithecia, and in the degree of development of the thallus. Distinguished from other British species by the combination of large perithecia, conspicuous involucrellum and mainly 3-septate ascospores. *T. papulare* **f. *sorediatum*** Coppins (1991) has scattered, irregularly linear soralia developing along cracks in the thallus, at times becoming rounded and confluent, dark brown, abrading pale green; on vertical schistose ± calcareous montane rocks, rare; Scotland (Perth, Breadalbane mtns and Angus, Caenlochan). A poorly understood species, ***T. microbolum*** (Tuck. in Drummond) Hasse (1913), with perithecia, 0.1–0.25 mm diam., and ascospores 23–36 × 9–14.5 μm, has been recorded from Scotland (Highlands) and Ireland (W. Galway, Howth), Canada, Norway. It differs from *T. microcarpum* **(5)** in possessing an involucrellum.

8. T. pluvium A. Orange (1991)

Thallus superficial, pale grey-green to darkish brown, continuous or somewhat cracked, to 75 μm thick. Perithecia 0.2–0.45(-0.5) mm diam., forming projections, covered by a thalline layer almost to apex; involucrellum relatively thin and weakly developed, present around upper half of exciple and scarcely spreading, or reaching nearly to level of base of exciple and spreading ventrolaterally. Ascospores (20-)23–36 × 9–16 μm, 1-septate.

On acidic or slightly basic rocks in or beside streams and on dripping rocks. N. England, Wales, N. Scotland (Highlands), W.C. Ireland (W. Galway). Norway.

Distinguished from *T. fumidum* **(2)** by the usually paler thallus and by the different pigment in the involucrellum; and from *T. pyrenophorum* **(9)** by the distinct thalline covering to the perithecium, when it resembles certain morphs of *Verrucaria hydrela*.

9. T. pyrenophorum (Ach.) Mudd (1861)

T. auruntii (Massal.) Krempelh. (1861)

Thallus superficial, thin, pale grey to pale brownish grey, continuous or somewhat cracked, occasionally ± immersed. Perithecia (0.25-)0.3–0.65(-0.7) mm diam., forming conspicuous projections, not covered by thallus; ostiole sometimes in a depression; involucrellum conspicuous, forming a well-developed apical cap, overarching up to half-way down perithecium, rarely almost to base. Ascospores 18–29(-31) × (8-)9–14(-15) μm, 1-septate.

On limestone and mica-schist. N. & W. British Is. Europe, Norway, Sweden, North America.

The similar *T. aeneovinosum* (Anzi) Arnold (1868) differs in having ascospores (26-)30–40(-44) × 12–17 μm, and the grey to brown superficial thallus; no British material of this widespread continental species has been traced.

THELOCARPON Nyl. (1853)

Thallus of raised warts, sometimes forming a thin crust, usually minute and evanescent, very inconspicuous, sessile or immersed in the substratum, globose to conical. **Photobiont** chlorococcoid, with algal cells only present around base of thalline warts, enclosed in a ± well-developed algal sheath around the apothecia, sometimes in the lower exciple, or absent in some lichenicolous species. **Ascomata** apothecia, ± perithecioid with a narrow opening, globose to conical; wall colourless or pale. **Hamathecium** of periphyses around the ostiole and on the side walls, occasionally resembling paraphyses, simple, branched or absent; paraphyses present or absent, simple, branched or anastomosed. **Asci** 50- to 300-spored, flask-shaped, often narrowed above, thin-walled, apex not or slightly thickened, I− or with an I+ blue internal tholus or meniscus, persisting long after ascospore release (Fig. 33p). **Ascospores** colourless, small or minute, simple, 1–2 guttulate or pseudoseptate, ellipsoid or ± elongate. **Conidiomata** pycnidia, infrequent, formed within algal sheath or in separate warts; wall entire, colourless. **Conidiogenous cells** simple, acrogenous. **Conidia** oblong, simple. **Chemistry**: pulvinic acid derivatives. **Ecology**: on limestones, mica-schist, flints and brick, on rotten wood and soil, leather, and on decaying thalli of other lichens (e.g. *Peltigera*, *Solorina*, *Baeomyces*).

LITERATURE: Poelt & Hafellner (1975), Salisbury (1966).

1 Paraphyses present 5
Paraphyses absent 2

2(1) Algal sheath present 3
Algal sheath absent .... **3. *intermediellum***

3(2) Asci I+ pale blue; hymenial jelly I−; ascospores not constricted 4
Asci I−; hymenial jelly I+ yellow-red; ascospores slightly constricted .... **6. *magnussonii***

4(3) Ascospores 2–4 × 1.5–2 μm; thalline warts usually yellow-pruinose .... **7. *olivaceum***
Ascospores 6–10 × 3–3.5 μm; thalline warts pale globules when moist, not yellow-pruinose .... **8. *pallidum***

5(1)	Paraphyses branched	6
	Paraphyses simple	7
6(5)	Apothecia not widened at apex; periphyses present	**4. *laureri***
	Apothecia abruptly widened at apex; periphyses absent	**5. *lichenicola***
7(5)	Asci I+ blue	8
	Asci I−	**1. *epibolum***
8(7)	Asci I+ dark blue; hymenial jelly I−; pseudosepta present	**9. *superellum***
	Asci I+ pale blue; hymenial jelly I+ red; pseudosepta absent	**2. *impressellum***

1. T. epibolum Nyl. (1866)

Thallus of inconspicuous, scattered warts, 0.1–0.16 mm diam., 0.15–0.2 mm high, conical to globose, yellow-pruinose with dark centres, slightly immersed. Algal sheath absent, generally not lichenized. Apothecia *c.* 100 μm diam., the wall 15–30 μm thick, colourless; hymenial jelly I+ yellow-red, rarely bluish, K/I+ blue. Periphyses and paraphyses equal in length to asci, thin, simple. Asci (Fig. 33p) 110–135 × 18–21 μm, I−. Ascospores 4–7(-12) × 1.7–2.5(-3) μm, simple, oblong, occasionally slightly constricted. Vulpinic acid, pulvinic dilactone, and (?)pulvinic acid reported.

On decaying lichens (e.g. *Baeomyces*, *Coriscium*, *Solorina*, *Peltigera*), rotten wood and soil, uplands; local. Throughout Britain. C. & N. Europe, North America.

Two varieties have been described: var. *epibolum*, ascospores 4–5.5(-6) × 1.7–2 μm.; var. *epithallinum* (Leighton) G. Salisbury (1953), ascospores (6-)8–10(-12) × 2.5–3 μm.

The combination of simple persistent paraphyses, yellow-pruinose apothecia without an algal sheath and asci which are I− is diagnostic.

2. T. impressellum Nyl. (1867)

Thallus of scattered warts, 0.2(-0.37) mm diam., 0.2 mm high, globose, truncated and depressed above, yellow pruina mainly on upper flattened surface. Algal sheath absent. Apothecia 130–190 μm tall, punctiform, the wall colourless; hymenial jelly I+ yellow-red. Paraphyses taller than asci, the apices crowded together forming the disc. Asci I+ pale blue. Ascospores 6–8.5(-11) × 4–4.5 μm, oblong-ellipsoid, simple or 1- to 2-guttulate.

On rock or humus, especially in calcareous dunes; rare. N. & W. British Is. C. & N. Europe.

3. T. intermediellum Nyl. (1865)

Thallus of scattered, slightly immersed warts, 0.15–0.25(-0.3) mm diam., globose, with yellow pruina, the wall 30–45 μm thick above, narrowing to 15–25 μm below. Algal sheath usually ± absent. Apothecia *c.* 150 μm diam. Periphyses 20–25 μm long, much divaricately branched, around ostiole and on the sides, easily detached. Paraphyses absent. Asci (80-)100–160(-220) μm long, I+ pale blue; bases (and scanty hymenial jelly) I+ yellow-red. Ascospores 3–4(-5) × 1–1.5 μm, oblong, slightly constricted, simple or biguttulate.

On rotten wood, old leather, moist rocks, generally associated with algae; rare. Throughout British Is. C. & N. Europe, North America.

The algal sheath is very rudimentary and thin, and may sometimes be absent. When present it is composed of loose hyphal filaments, and yellow pruina mixed with very few algal cells.

4. T. laureri (Flotow) Nyl. (1855)

Thallus of scattered or crowded warts, 0.15–0.3(-0.4) mm diam., 0.15–0.2(-0.25) mm high, globose or hemispherical, yellow-pruinose, sometimes darker in the centre. Algal sheath present, 25–60(-100) μm thick. Apothecia 90–150(-180) μm diam., globose, the wall 12 μm thick, colourless. Periphyses and paraphyses as long as asci, forked, branched and entangled. Asci 75–150(-170) μm long, I+ pale blue; bases (and scanty hymenial jelly) I+ yellow-red. Ascospores 1.5–4(-6) × 1.5–2 μm, subglobose to ellipsoid, simple or 1- to 2-guttulate.

On wood, burnt ground, litter, brick, leather, etc.; rare. Throughout British Is. C. & N. Europe, North America.

The only species with persistent, branched paraphyses; the ascospores, which can be almost globose, are also diagnostic.

5. T. lichenicola (Fuckel) Poelt & Hafellner (1975)

Thallus absent, not lichenized. Apothecia 0.1–0.35 mm wide and high, only slightly immersed in the host, when fully developed hymenium fully exposed to form small discs; disc concave; true exciple raised, 15–20(-25) μm wide, green-yellow-brown; hymenium 65–75 μm tall, I−. Periphyses absent. Paraphyses not or sparingly branched, 1μm thick. Asci 60–65(-75) × 12–14 μm, cylindrical-clavate, I+ pale blue (apex deep blue). Ascospores 4–6(-7.5) × 1.5–2 μm, ellipsoid, simple.

On sandstone pebbles and *Baeomyces rufus*, also on fallen pine trunks; rare. N. England (Co. Durham), S. Scotland (S. Aberdeen and Fife). C. Europe.

The absence of a lichenized thallus and disc-like fruits formerly led to this being placed in *Ahlesia* Fuckel (1870). Some British specimens have very small apothecia (*c.* 0.1 mm diam.) and may be referable to *T. strasseri* Zahlbr. (1902).

6. T. magnussonii G. Salisb. (1953)

Thallus very thin, with crowded, conical warts, 0.2–0.25 mm diam., almost black when dry, brown-green when moist, not yellow-pruinose. Algal sheath present, 20–35 μm thick. Apothecia 130–180 μm diam., globose, sessile; wall 15–18 μm thick, colourless; hymenial jelly I+ yellow-red, K/I+ blue. Periphyses 6–10 μm long, slightly branched, sited round ostiole and on the sides. Paraphyses absent. Asci (90-)120–150 × 30–40 μm; apices pointed, often projecting through the ostiole, I−. Ascospores 3.5–4.5(-5) × 1.2–1.8 μm, oblong, slightly constricted, simple or biguttulate.

On old leather or loose stones; rare. N. England (S. Lancashire), N. Wales (Flintshire), N.W. Scotland (Is. of Skye). Endemic.

Similar to *T. olivaceum* **(7)** and *T. intermediellum* **(3)** in squash preparations but differing in the absence of yellow pruina and I reactions.

7. T. olivaceum B. de Lesd. (1914)

Thallus of scattered or crowded, hemispherical warts, 0.15–0.27 mm diam., 0.12–0.17 mm high, totally yellow-pruinose or grey to discoloured brown and yellow pruina confined to around the ostiole. Algal sheath present, 50–70 μm thick. Apothecia 100–150 μm diam., globose, wall 6–12 μm thick, colourless. Periphyses 20–25 μm long, divaricately branched, sited around the ostiole and on the sides. Paraphyses absent. Asci 70–120 μm long, I+ pale blue, bases (and scanty hymenial jelly) I+ yellow-red. Ascospores (2-)2.5–3.5(-4) × 1.5–2 μm, oblong-ellipsoid, sometimes nearly spherical, not constricted, simple. Pycnidia rare, sited in algal sheath, *c.* 35 μm diam., globose, the wall colourless; conidia 4.5 × 1 μm.

On old leather and brick; rare. W. England (Shropshire), N.W. England (S. Lancashire). C. Europe.

Similar to *T. pallidum* **(8)**, which has larger, 6–10 μm long ascospores and is not, or very sparingly, pruinose.

8. T. pallidum G. Salisb. (1953)

Thallus when moist composed of globose warts 0.2–0.3(-0.4) mm, almost colourless to pale translucent yellow, sometimes sparingly yellow-pruinose at the apex; warts inapparent when dry. Algal sheath present, 35–40 μm thick. Apothecia 150–250 μm diam., globose, the wall ± 15 μm thick, colourless. Periphyses 6–10 μm long and slightly branched, or ± 20 μm long and divaricately branched, sited around the ostiole and on the sides. Paraphyses absent. Asci 110–180 × 25–30 μm, I+ pale blue, bases (or scanty hymenial jelly) I+ yellow-red. Ascospores (6-)7–9(-10) × 3.0–3.5 μm, oblong-ellipsoid, simple or 1- to 2-guttulate.

On brick, chalk, and damp mortar. England (S. Lancashire, Sussex), Wales (Cardigan). Endemic.

Distinguished from other species with an algal sheath by the large ascospores.

9. T. superellum Nyl. (1866)

Thallus of scattered warts, 0.2–0.3(-0.4) mm diam., 0.25–0.35(-0.5) mm high, slightly immersed, conical-globose, yellow-pruinose. Algal sheath absent. Apothecia *c.* 200 μm diam.; wall 35–60 μm thick, colourless. Periphyses and paraphyses as long as asci, to 1 μm thick, simple. Asci 150–250(-300) μm long, I+ dark blue. Ascospores (6-)8–13 × (2.5-)3.5–5(-6) μm, oblong, simple or usually with a pseudoseptum, two guttules I+ brick-red.

On decaying algae on stones and soil; very rare. N. Wales (Caernarvon), Scotland (Perth, Blair Atholl; N. Ebudes, Canna). C. & N. Europe, North America.

T. superellum most closely resembles *T. vicinellum* Nyl. (1885), which has similar reactions to iodine but differs in the much larger size of its parts; it occurs in Italy.

THELOMMA Massal. (1860)

Thallus crustose, warted, grey, corticate. **Cortex** formed by hyphae perpendicular to the surface, sometimes with swollen and brownish apices. **Photobiont** *Trebouxia*. **Medulla** I+ blue-black. **Ascomata** apothecia, immersed in warts. **True exciple** reduced laterally, forming a basal collar at the edge of the hypothecium. **Hypothecium** blackish brown, thick. **Asci** cylindrical, formed singly, with a single functional wall layer, dissolving at an early stage. **Ascospores** 1-septate, dark brown-black, ellipsoid, with an almost smooth surface, forming a well-developed dry, black, spore mass. **Conidiomata** not seen. **Chemistry**: vulpinic acid derivatives and usnic acid. **Ecology**: on wood.

Closely resembles *Cyphelium*, which has sessile or immersed apothecia with a dark brown-black, laterally well-developed exciple forming a distinct raised rim, and 1-septate or submuriform ascospores in which the secondary wall usually has concentrically arranged layers in the electron microscope.

LITERATURE: Tibell (1976).

1. T. ocellatum (Körber) Tibell (1976)

Thallus coarsely warted, well developed with elevated, flattened, ± crenulate warts or ± areolate, grey; isidia frequent, ± globose, forming irregular, but well-delimited, blue- or brown-black ± convex, soralia-like clusters. Apothecia immersed, unknown in British specimens. Ascospores 22–28 × 12–15 μm. Ascospore mass with rhizocarpic acid and epanorin, and the thallus usnic acid.

On fence posts and gate rails; local. England (East Anglia, W. Sussex, Shropshire, Northumberland), Scotland (Midlothian; E. Highlands). Europe, North America.

In Europe sterile material occurs predominantly in nutrient-enriched sites, whereas abundantly fertile material is more characteristic of high altitudes.

THELOPSIS Nyl. (1855)

Thallus crustose, ± superficial, thin, grey or orange-red, effuse; prothallus absent. **Photobiont** *Trentepohlia*. **Ascomata** perithecia, partly immersed or sessile, ± tough, pale reddish brown to black, not black and friable. **Hymenium** colourless, I+ blue, becoming red-brown. **Hamathecium** of paraphyses thread-like, unbranched, persistent. **Asci** cylindrical to oblong, multispored; walls uniformly thin, without an apical apparatus. **Ascospores** ellipsoid or shortly fusiform, (1-)3-septate. **Conidiomata** pycnidia, very rare, pale. **Conidia** thread-like, straight or slightly curved. **Chemistry**: lichen products not reported.

Recognized by the tough, semi-gelatinous perithecia, persistent, simple paraphyses, and asci with very numerous, septate ascospores.

LITERATURE: Vězda (1968b).

1 Perithecia pale brown to red-brown, surface smooth; thallus ± grey-brown .... **1. *rubella***
Perthecia black, surface roughened; thallus bright orange to red-brown, fading to grey-green in the herbarium .... **2. *melathelia***

1. T. rubella Nyl. (1955)

Thallus inconspicuous, waxy, grey to greyish green or pale brownish, often wide-spreading and diffuse. Perithecia scattered, discrete, sub-gelatinous when moist, at first ± immersed, then erumpent and sessile, often retaining part of the thallus in a peripheral basal ring recalling a false thalline exciple, pale pink-brown, red-brown to dark brown, often paler around the ostiole; exciple < 50 μm wide. Periphyses simple, 4- to 5-celled, to 35 μm long, 1.5 μm wide. Asci (100-)150–200 × (16-)18–25 μm, 100- to

150-spored. Ascospores (10-)12–16(-18) μm × 4–8 μm, ellipsoid-oblong, ends rounded, (1-)3-septate, without a distinct perispore.

On trunks of mature, broad-leaved trees, particularly *Quercus*, *Fagus* and *Fraxinus*, with bark of pH > 5.0; mostly in the *Lobarion* in ancient woodlands or parklands where it is an important indicator species; often in furrows or intermittent rain tracks in association with *Metzgeria furcata*; local. Throughout Britain, from S. England (E. Sussex to Cornwall) to Scotland (West Ross, Skye, and Nairn), absent from most of C. & E. England and E. Scotland. Europe from S.W. Norway to S. France and C. Italy, E. to Czechoslovakia and Yugoslavia, Azores.

The perithecia are unusual among British lichenized ascomycetes in being relatively short-lived and are soon replaced by new fruits when shed. Perithecia are usually most abundant in winter and early spring. *Porina leptalea* also has red perithecia, but is usually on smooth bark, has smaller, more orange fruits and 8-spored asci.

2. T. melathelia Nyl. (1864)

Like **1**, but thallus normally orange-red to red-brown when fresh, becoming grey-green in the herbarium; perithecia black, ± wrinkled, half-immersed, 0.5–0.8 mm diam.; asci 200–250 × 15–25 μm; ascospores 11–20 × 4–7 μm, with a thickened perispore.

Overgrowing bryophytes and plant detritus, often ± in sheltered crevices in basic schistose rocks and epidiorite on upland crags; very rare. N. Scotland (Highlands, Breadalbane mtns). Norway, C. Europe mtns.

The bright orange thallus, black, striate perithecia, and multispored asci are diagnostic. *T. flaveola* Arnold (1873), with yellowish perithecia and simple ascospores and *T. isiaca* Stizenb. (1895), with perithecia immersed in greyish white warts and 1-septate ascospores, occur on trees in continental Europe.

THELOTREMA Ach. (1803)

Thallus crustose, ± superficial, sometimes immersed, usually shades of cream or fawn; cortex continuous or exfoliating, sometimes with crystals internally; prothallus absent. **Photobiont** *Trentepohlia*. **Ascomata** apothecia, urceolate to ± perithecioid, immersed, often in raised warts; disc concave to flat, often gaping, black, sometimes pruinose. **Thalline exciple** distinct, entire or eroding, usually incurved. **True exciple** colourless or brown above, closely adhering to the thalline exciple or more commonly free from it, then distinct when viewed from above as an internal ring, sometimes striate, composed of ± densely conglutinated, intertwined short-celled hyphae. **Epithecium** colourless to brown or black. **Hymenium** colourless, I−. **Hypothecium** colourless. **Hamathecium** of filamentous, unbranched, rarely septate paraphyses; periphysoids lining the upper inner surface of the true exciple. **Asci** subcylindrical, with a single functional wall layer, abruptly thickened at the apex, sometimes with a minute internal apical bark, I−, 1- to 8-spored. **Ascospores** narrowly ellipsoid to broadly fusiform, septate or muriform, colourless or brown, the walls often thick and laminated with the cells ± lenticular, smooth, lacking a distinct perispore, I+ purplish. **Conidiomata**, when present, pycnidial. **Conidiogenous cells** unbranched. **Conidia** bacilliform, simple, colourless. **Chemistry**: very variable, a wide range of depsidones, anthraquinones, and unidentified substances reported; however, no lichen substances are detected by t.l.c. in many species. **Ecology**: on bark, wood or rarely rocks in humid situations, cosmopolitan.

As re-defined on the basis of exciple structure rather than ascospore colour and septation, *Thelotrema* s.str. comprises about 100 mainly tropical species. The non-British *Myriotrema* Fée (1824) and *Ocellularia* Meyer (1815) lack periphysoids on the inner face of the exciple; the exciple is colourless to pale reddish brown in *Myriotrema*, and black and friable in *Ocellularia*.

LITERATURE: Bailey & James (1977), Hale (1980, 1981), Poelt & Vězda (1981), Salisbury (1972).

1	Ascospores muriform	2
	Ascospores only transversely septate	**4. *subtile***

2(1) Ascospores smoky brown to brown when mature 3
Ascospores remaining colourless when mature .... **2. *lepadinum***

3(2) Asci 1(-2)-spored; true exciple free; ascospores (50-)90–120(-160) × (14-)30–40(-50) μm, with 16–22(-24) transverse septa .... **3. *monosporum***
Asci usually 8-spored; true exciple ± united with the thallus; ascospores 30–50 × 12–16(-18) μm, with 7–9(-11) transverse septa .... **1. *isidioides***

1. T. isidioides (Borrer) R. Sant. (1980)

Thallus greyish brown to yellowish brown, at first ± scurfy but finally cracked-areolate, areoles ± convex, effuse, prothallus absent. Apothecia 0.2–0.3 mm diam., ± perithecioid, half to almost completely immersed, one per areole; true exciple remaining united with the thalline exciple, brown above. Asci usually 8-spored. Ascospores 30–50 × 12–16(-18) μm, elongate-ellipsoid, brown when mature, muriform, with 7–9(-11) transverse and 1–3 longitudinal septa. Thallus Pd+ orange, K+ orange, KC+ orange, C− (stictic, hypostictic, hypoconstictic acids).

On hard, siliceous rocks; apparently extinct. S.W. Ireland (W. Cork, Glengariff). Known only from the original collection made last century.

2. T. lepadinum (Ach.) Ach. (1803)

Thallus creamy to fawn or pale ochraceous, ± continuous, in section interspersed with crystals. Apothecia (0.6-)1–2 mm diam., immersed in hemispherical warts, urceolate, ostiolar opening to 0.5 mm wide; thalline exciple entire; true exciple free, ± conspicuous when viewed through the ostiole, colourless, not striate; hymenium (120-)140–190(-200) μm tall. Asci (1-)2–4(-8)-spored. Ascospores (30-)60–135(-150) × (10-)15–25(-33) μm, broadly fusiform, remaining colourless when mature, with (8-)10–15(-19) transverse and 1–3(-5) longitudinal septa, I− or weakly purple. Thallus and medulla Pd−, K− (internal crystals ± yellow changing to red), C− (no lichen products detected by t.l.c.).

On bark of deciduous trees, in old woodland, rarely on siliceous rocks, declining due to woodland disturbance and air pollution. Throughout the British Is. Probably cosmopolitan.

The shape of the apothecia is very variable; in dry situations on smooth bark they are regularly volcano-like, smooth with a narrow opening. In moist, often shaded, sites the apothecia are larger and become more open and ulcerose with a thick, prominent, somewhat scurfy, rim.

3. T. monosporum Nyl. (1861)

Thallus ochraceous to brownish grey, ± continuous. Apothecia (0.7-)-0.8–1.1(-1.5) mm diam., immersed in hemispherical warts, somewhat flattened at their summits, urceolate, the ostiolar opening usually narrow, 0.2–0.4(-0.7) mm wide; true exciple colourless, inapparent when seen through the ostiole; thalline exciple entire; hymenium (100-)140–200 μm tall. Asci 1(-2)-spored. Ascospores (50-)90–120(-160) × (14-)30–40(-50) μm, cylindrical or broadly fusiform with rounded ends, grey-brown when mature, with 16–22(-24) transverse and numerous longitudinal septa, I−. Thallus and medulla Pd−, K−, KC−, C−, UV− (no lichen products detected by t.l.c.).

On bark, especially twigs of *Corylus*, occasionally *Fraxinus*, in sheltered, boggy, undisturbed ancient woodlands; very local. W. Scotland, W. Ireland. Unknown elsewhere in Europe, probably pantropical; S. North America, Indonesia, Sri Lanka, Pacific Islands, Australia, New Zealand.

Closely resembling *T. lepadinum* **(2)**, but with a grey-green thallus with apothecia with a narrower opening; microscopic examination is necessary for certain identification. The large, dark ascospores are visible with a ×20 lens and are often ± scattered over the surface of the thallus and around the ostioles.

4. T. subtile Tuck. (1858)

Thallus cream, white or pale ochraceous, somewhat immersed, ± continuous, in section interspersed with crystals. Apothecia 0.4–0.7(-0.8) mm diam., flattened, not immersed in warts, erumpent, urceolate; ostiolar opening to 0.3 mm wide; disc pruinose; true exciple free, colourless, ± striate and covering the disc; hymenium 90–110 μm tall. Asci (4-)6–8-spored. Ascospores (30-)35–50(-54) × (7-)8–9(-10) μm, broadly fusiform, remaining colourless when mature, with (7-)9–11(-12) transverse and no longitudinal septa, I+ purple. Thallus Pd−, K−, C− (no lichen products detected by t.l.c.).

On smooth bark, mainly of *Corylus*, but also *Salix* species, *Sorbus aucuparia*, *S. aria*, *Alnus*, *Betula*, *Fagus* and *Fraxinus* in sheltered woodlands and boggy areas; locally abundant. W. Scotland, W. & S.W. Ireland. Sweden; mainly subtropical, S. North America, Azores, Indonesia, Pacific Islands.

The relatively flattened ascomata with a well-developed, striate true exciple and pruinose disc, and the eroding thalline margin are useful field characters separating this species from *T. lepadinum* (**2**). *T. suecicum* (Magnusson) P. James (1977), has (2-)4–6(-8), notably ellipsoid ascospores per ascus, each 20–30(-35) × 8–10 μm with (4-)6–8 cells and no longitudinal septa; it is known only from S.W. Norway.

THERMUTIS Fr. (1825)

Thallus filamentous, blackish; filaments ± erect, cushion-forming, without any special attachments to the substratum, each comprising a filament of the photobiont with the fungal hyphae loosely interwoven in the gelatinous sheath of the photobiont, penetrating the photobiont cells by short, capitate haustoria. **Photobiont** *Scytonema*. **Ascomata** apothecia, lateral, reddish brown. **Thalline exciple** absent. **True exciple** distinct, of loosely interwoven hyphae. **Hymenium** I+ blue. **Hamathecium** of simple, septate paraphyses; apices swollen. **Asci** cylindrical, 8(-12)-spored, thin-walled; apical dome I+ blue. **Ascospores** simple, ellipsoid, colourless. **Conidiomata** pycnidia, lateral, reddish brown. **Conidiogenous cells** narrow, acrogenous. **Conidia** simple, globose, colourless. **Chemistry**: no lichen products reported. **Ecology**: on rocks.

A monotypic genus distinguished from *Spilonema* by the distinct true exciple, different photobiont, and lack of any greenish, N+ red pigment in the apothecia and pycnidia.

LITERATURE: Henssen (1963a).

1. T. velutina (Ach.) Flotow (1850)

Thallus forming brown-black to black cushions to 1 cm diam.; filaments to 15 μm thick, simple or sparingly branched. Apothecia 0.2–0.5 mm diam., sessile; disc flat; true exciple swollen, 50–70 mm wide, brown at outer edge, colourless within; hymenium 100–150 μm tall, colourless, or brownish in upper part; hypothecium 80–130 μm tall, colourless. Asci 60–100 × 5.5–7 μm. Ascospores 9–15 × 5–7 μm. Conidia 1–1.5 μm diam.

On acid and basic rocks in moist gullies and ravines, also in the spray zone of waterfalls, montane; rare. N.W. England (Lake District), N. Wales, Scotland. N. & C. Europe.

Easily mistaken for non-lichenized *Scytonema* cushions, but readily identified when sterile by the thin hyphae with capitate haustoria ramifying through the photobiont's sheath. The identity of the photobiont and narrowness of the filaments distinguish it from *Spilonema* and small specimens of *Ephebe* and *Polychidium*. *Cystocoleus* and *Racodium* have *Trentepohlia* as photobiont which completely lacks a gelatinous sheath.

THROMBIUM Wallr. (1831)

Thallus crustose, membranous, subgelatinous when wet, indeterminate. **Photobiont** *Leptosira*. **Ascomata** perithecia, entirely immersed, the ostiole dark. **Involucrellum** absent. **True exciple** brown-black, widened towards the ostiole. **Hamathecium** of paraphyses, unbranched, thin, persistent; periphyses absent. **Asci** (Fig. 35b) elongate-cylindrical or narrowly clavate, thin-walled, with a well-defined K/I+ blue apical cap; tholus K/I+ blue with narrow, cylindrical axial mass; 8-spored. **Ascospores** simple, colourless, thin-walled, without a distinct perispore. **Conidiomata** not known. **Chemistry**: lichen products not detected by t.l.c. **Ecology**: on soil.

Distinguished by the combination of the perithecia being simple, the colourless, simple ascospores and, particularly, the persistent paraphyses; *Verrucaria* lacks hamathecial tissues from an early stage. For *T. thelostomum* (Ach. ex Harriman) A.L. Sm. see p. 686.

LITERATURE: Ellis (1981), Swinscow (1964, 1982).

1. T. epigaeum (Pers.) Wallr. (1831)

Thallus wide-spreading, inconspicuous or a yellow- or grey-green film, smooth or slightly uneven, subgelatinous when wet, effuse, indistinct. Perithecia 250–400 μm diam., visible only as a grey to black ostiole; true exciple wall 20–30 μm wide, pale to dark brown, becoming 50–100 μm towards the ostiole. Paraphyses few to abundant, *c.* 1 μm thick, unbranched. Asci (Fig. 35b) 130–170 μm tall, cylindrical, ± clavate at apices, thin-walled. Ascospores (15-)18–25(-30) × 5–10(-12) μm, ellipsoid or longly ellipsoid, often somewhat clavate, with numerous oil droplets.

On recently disturbed, consolidated, neutral to acid soil of sheltered cuttings and earth banks by roadsides, especially in well-wooded sites, also cliff tops; very local. W. British Is. Europe, Azores, Chile.

The thallus is host to the lichenicolous fungus *Leightoniomyces phillipsii*.

TOMASELLIA Massal. (1856)

Thallus inconspicuous or slightly bleaching the bark, immersed. **Photobiont** probably absent. **Ascomata** circular, ellipsoid or irregular in outline, black, ± flattened, composed of few to many perithecia-like locules each with its own, often raised, ostiole (best seen when wet); overlying wall dark brown, K−, composed of compacted hyphae and bark cells; locules separated by colourless to darkish brown compacted hyphae. **Hamathecium** of branched pseudoparaphyses, I−. **Asci** with two functional wall layers, discharge fissitunicate, obpyriform to obclavate, 8-spored, I−. **Ascospores** clavate or sole-shaped, 1- or 3-septate, constricted at middle septum, upper half at least slightly wider than lower, colourless or brown and finely warted; perispore thin or indistinct. **Conidiomata** pycnidia, black. **Conidiogenous cells** ± cylindrical. **Conidia** bacilliform. **Chemistry**: lichen products not detected by t.l.c. **Ecology**: on smooth bark of trees and shrubs.

Differs from *Arthopyrenia* in the multilocular ascomata. *Mycoporum* differs in having submuriform ascospores.

1	Ascomata regularly circular or broadly ellipsoid; ascospores soon 3-septate	**2. *gelatinosa***
	Ascomata elliptical or irregular; ascospores 1-septate, sometimes 3-septate	2
2(1)	Ascospores soon dark brown; on *Alnus*	**1. *diffusa***
	Ascospores colourless, or some old spores brownish; on *Ilex*	**3. *lactea***

1. T. diffusa (Leighton) Lahm (1885)

Ascomata (0.2-)0.4–1 mm diam., circular to ellipsoid, with (1-)4–12 locules; tissue between locules often darkish brown. Pseudoparaphyses 1.5–2 μm wide; the cells 6–13 μm long. Asci 50–55 × 21–25 μm. Ascospores 19–24 × 6–7 μm, 1-septate, at first colourless but soon becoming dark brown, often with a distinct perispore, 1–1.5 μm thick in K.

On young trunks and thick branches of *Alnus*; local. N. Wales, N. Scotland (Highlands). Germany.

Inconspicuous owing to the dark colour of *Alnus* bark. Distinguished from forms of *Arthopyrenia punctiformis* with clustered ascomata by the dark brown ascospores.

2. T. gelatinosa (Chev.) Zahlbr. (1922)

Ascomata 0.3–1.1 mm diam., ± circular, occasionally broadly ellipsoid, with 10–55 locules; locules separated by colourless or pale brown tissue. Pseudoparaphyses 1.5–2.5(-3) μm wide, the cells 3–10 μm long. Asci 52–70 × 18–26 μm. Ascospores 19–24(-26) × 5–7 μm, 3-septate, colourless or a few old spores brownish; perispore indistinct.

On young trunks and branches of various trees and shrubs, especially *Corylus*; local. N. & W. British Is. W. Europe, N.E. North America.

Distinguished by the more regularly circular ascomata and ascospores that are soon 3-septate.

3. T. lactea (Ach.) R. C. Harris (1980)

Ascomata 0.25–0.6(-0.8) mm, ± circular or more usually irregular in outline, with (1-)2–15 locules, sometimes decaying at centre leaving an irregular circle of locules; tissue between locules colourless or pale brown. Pseudoparaphyses 1.2–1.7(-2) μm wide, the cells *c.* 7–12 μm long. Asci 55–75 × 17–24 μm. Ascospores 18–26 × 5–7(-8) μm, 1-septate, the upper cell broader and with ± pointed apex, each cell biguttulate, colourless, but a few old spores sometimes 3-septate and/or brownish. Pycnidia rare, 30–60 μm diam; conidia 3–4 × 0.8 μm.

On lower parts of trunks of mature *Ilex* in old woodlands; rare and local. S. England, W. Scotland, W. Ireland. Canary Is, North, Central & South America.

TONINIA Massal. (1852)

Kiliasia Hafellner (1984)

Thallus crustose, ± immersed, rimose-cracked, areolate or squamulose, sometimes absent or in other lichens. **Upper cortex** absent to well developed, often consisting of a layer of dead tissue and a lower, stainable layer; lower layer composed of anticlinally oriented, thin- to thick-walled hyphae with rounded to narrowly cylindrical lumina; both layers incorporating remnants of the photobiont (seen in chlor-zinc-iodine) and often crystals of calcium oxalate; sometimes with pseudocyphellae. **Photobiont** green, chlorococcoid when present forming a continuous layer or of ± scattered groups of cells; sometimes absent and not lichenized or loosely associated with cyanobacteria. **Medulla** usually well developed, of loosely interwoven, non-I+ blue hyphae, sometimes with calcium oxalate. **Lower cortex** absent to well developed, sometimes developing extensive root-like branched outgrowths penetrating the substratum. **Ascomata** apothecia, black, often white- or grey-pruinose, usually weakly concave to flat when young, later often becoming convex. **Thalline exciple** absent. **True exciple** raised and distinct at first but often finally excluded, of radiating, thick-walled, conglutinated hyphae with rounded to narrowly cylindrical lumina, colourless within with a green or brown outer margin to uniformly dark brown throughout. **Epithecium** green (K−, N+ violet), dull brown (K−, N−), reddish brown (K+ red, N−), or grey [K+ violet, N+ violet (green and dull brown pigments sometimes mixed)], often containing crystals of calcium oxalate. **Hymenium** K/I+ blue. **Hypothecium** colourless to dark brown, rarely containing crystals of calcium oxalate. **Hamathecium** of paraphyses, straight, simple or sparingly branched and anastomosed, not conglutinated, thin-walled, apical cell distinctly swollen and covered by a ± well-developed, gelatinous pigment cap. **Asci** 8-spored, clavate, surrounded by a gelatinous, I+ blue sheath, with a well-developed, I+ blue tholus containing a deeper-staining tube and a well-developed ocular chamber. **Ascospores** simple to 7-septate, colourless, ellipsoid, bacilliform to acicular, smooth, without a thickened perispore. **Conidiomata** pycnidia, black, immersed to partly protruding. **Conidiogenous** cells lining the pycnidial cavaity. **Conidia** bacilliform to thread-like. **Chemistry**: terpenoids occur in some species; all species (except **6** and **13**) Pd−, K−, KC−, C−, and UV−. **Ecology**: often lichenicolous, at least when young, more rarely on mosses, or free-living from the first, mainly on ± basic rock and soil in exposed habitats; most species, at least when young, are associated with cyanophilic lichens.

The above description applies to *Toninia* s. str., and the following species, with ± conglutinated paraphyses, with or without a swollen apical cell and pigment cap, asci with or without an ocular chamber and different chemistry, are retained here only temporarily: *T. cumulata* **(4)**, *T. lobulata* **(6)**, *T. pulvinata* **(11)**, *T. squalescens* **(13)** and *T. tumidula* **(15)**. *Toninia alutacea* (Anzi) H. Olivier (1911) and *T. candida* (Web. ex Wigg.) Th. Fr. (1867) are both known from single collections, made last century, reputed to be from near the summit of Ben Lawers, and their presence in the British Is. is in need of confirmation. *Toninia tristis* (Th. Fr.) Th. Fr. (1874) is erroneously reported from the British. Is. *T. episema* is treated under *Catillaria*.

LITERATURE: Timdal (1991).

1 Paraphyses not conglutinated, each with a distinct swollen apical cell covered by a pigmented cap; asci with a well-developed ocular chamber 2
Paraphyses ± conglutinated, with or without a swollen apical cell and pigment cap; asci with or without an ocular chamber 11

2(1) Epithecium grey, K+ violet 3
Epithecium green or brown, K− 7

3(2) Hypothecium pale brown to colourless, ± transparent 4
Hypothecium reddish brown, ± opaque 5

4(3) Squamules bullate to columnar; pseudocyphellae ± inconspicuous, white, not to moderately pruinose; pruina farinose .... **7. *lurida***
Squamules flattened to bullate, marginal squamules often elongated; pseudocyphellae absent; usually densely pruinose; pruina granular .... **12. *rosulata***

5(3) Squamules irregularly convex to bullate, not to entirely pruinose; pruina farinose 6
Squamules ± convex but not bullate, usually entirely white-pruinose; pruina granular .... **5. *diffracta***

6(5) Squamules strongly convex to extended bullate-nodulose and partly vertically flattened, often imbricate .... **9. *opuntioides***
Squamules weakly convex to shortly bullate but not vertically flattened or regularly imbricate .... **2. *caeruleonigricans***

7(2) Hypothecium and inner part of exciple dark reddish brown, opaque 8
Hypothecium and inner part of exciple pale brown to colourless, transparent ... **14. *squalida***

8(7) Ascospores > 3-septate, ellipsoid to fusiform 9
Ascospores < 7-septate, acicular .... **10. *oribata***

9(8) Epithecium green, N+ violet or reddish 10
Epithecium brown, N− .... **3. *cervina***

10(9) Thallus crustose, areolate to rimose-cracked; areoles without maculae .... **8. *mesoidea***
Thallus squamulose; squamules often with pale maculae .... **1. *aromatica***

11(1) Ascospores simple to 1(-3)-septate; thallus pale to dark grey, C− 12
Ascospores (5-)7-septate; thallus yellow-brown to buff, C+ orange .... **11. *pulvinata***

12(11) Ascospores (0-)1-septate; squamules ± even-coloured, not overlapping 13
Ascospores (0-)2 to 3-septate; squamules ± with darker centres and pale margins, ± overlapping .... **6. *lobulata***

13(12) Apothecia <1 mm diam., not clustered 14
Apothecia <0.3(-0.5) mm diam., clustered, blackberry-like .... **4. *cumulata***

14(13) Hypothecium pale brown to colourless; epithecium olive-green .... **13. *squalescens***
Hypothecium and epithecium dark brown .... **15. *tumidula***

1. T. aromatica (Sm.) Massal. (1852)

T. fusispora (Körber) Th. Fr. (1867)

Thallus squamulose; squamules to 4 mm diam., widely scattered or contiguous, rounded or ± irregular; upper surface flat or irregularly convex, pale grey to dark brown, often tinged green, pruinose or not, dull, usually with faint, paler spots along more convex areas and ridges, without regular fissures and pseudocyphellae, not pruinose; margin concolorous with upper side; upper cortex 60–90 μm thick, lacking a well-defined epinecral layer, without calcium oxalate crystals; lower cortex poorly developed or absent. Photobiont zone continuous. Medulla lacking crystals. Apothecia to 1.5 mm diam., ± persistently flat and marginate, not pruinose or sometimes faintly white-pruinose. True exciple confluent with hypothecium, uniformly dark reddish brown, K−, N−; epithecium olivaceous green to bright green, lacking crystals, K−, N+ violet; hymenium 70–80 μm tall, colourless to pale brown;

hypothecium dark reddish brown. Ascospores 12–23 × 3–6 μm, (1-)3-septate, bacilliform with abruptly rounded ends. Lichen products not detected by t.l.c.

On ± basic soft rock, old mortar, especially in fissures or seepage tracks, also on consolidated soil, shell-sand dunes, very rarely on bark, on coastal and inland rocks usually in close association with or on cyanobacterial lichens; common. Throughout British Is. Throughout N. Hemisphere, New Zealand.

Very variable, but distinguished by the characteristic pale, irregular-white flecks on the upper surface of the squamules. Differs from *T. cervina* **(3)** mainly in the pigment in the epithecium and shape and colour of the squamules. The exclusively coastal *T. mesoidea* **(8)** has a more crustose thallus and is not white-flecked.

T. plumbina (Anzi) Hafellner & Timdal (1991) [syn. *Bacidia plumbina* (Anzi) R. Sant. (1960)] is parasitic on *Degelia plumbea*, lacks a visible thallus and has longer, (1-)3(-5)-septate ascospores; Lake District and W. Ross. W. Europe and Canary Is.

2. T. caeruleonigricans (Lightf.) Th. Fr. (1874)

Thallus squamulose; squamules to 3 mm diam., scattered to contiguous, sometimes irregularly imbricate, rounded or irregularly lobed, weakly convex to swollen; upper surface dull olive-green to brown, usually ± densely white-, or bluish pruinose but sometimes not pruinose, dull to slightly shiny, smooth or sometimes with shallow fissures in the cortex, without pseudocyphellae; margin concolorous with upper side or often more densely white-pruinose; underside white to pale brown; upper cortex 20–60 μm thick, including and up to 20 μm thick epinecral layer, often containing (or overlain by) calcium oxalate crystals; lower cortex resembling upper, but thinner and lacking crystals. Photobiont zone continuous. Medulla without crystals. Apothecia to 3 mm diam., weakly concave to weakly convex, not to densely white-pruinose; true exciple raised when young, often later becoming excluded, medium to dark brown throughout, K− and N−, or K+ violet and N+ violet (additional grey pigment); epithecium grey, often containing crystals of calcium oxalate, K+ violet, N+ violet; hymenium 70–80 μm tall, colourless; hypothecium medium brown in lower part, dark brown in upper part. Ascospores 14–25 × 2.5–5 μm, 1-septate, fusiform. Lichen products not detected by t.l.c. or with traces of fatty acids or unidentified substances.

On calcareous soil, especially dunes or on soil and in fissures in calcareous rocks, occasionally on old mortar, usually associated with cyanobacterial lichens, at least when young; locally frequent. Throughout British Is. Cosmopolitan.

Very variable and easily confused with *T. diffracta* **(5)**, *T. lurida* **(7)** and *T. opuntioides* **(9)**. Specimens with fatty acids are less frequent but more widely distributed than the other race which is rare.

The correct name is ***T. sedifolia*** (Scop.) Timdal (1991).

3. T. cervina Lönnr. (1858)

T. kolax Poelt (1957)

Thallus squamulose; squamules to 2 mm diam., scattered to imbricate, flat to weakly convex, irregularly crenulate; upper surface dark grey-brown, partly faintly pruinose or not, dull or somewhat shiny, often with shallow fissures, lacking maculae and pseudocyphellae, edge concolorous with upper side; upper cortex 50–120 μm thick,. with a well-developed epinecral layer, often containing crystals of calcium oxalate; lower cortex poorly developed or absent. Photobiont layer continuous. Medulla without crystals. Apothecia to 1 mm diam., ± persistently flat, faintly white-pruinose or not; true exciple persistent, confluent with hypothecium below, uniformly dark reddish brown, K−, N−; epithecium brownish black, sometimes with a faint green tinge, without crystals, K−, N− or faintly violet (green pigment); hymenium 70–80 μm tall, colourless to pale brown; hypothecium dark reddish brown. Ascospores 10–18 × 4–6 μm, (1-)3-septate, bacilliform. Lichen products not detected by t.l.c.

Overgrowing cyanobacterial lichens, especially *Placynthium*, on calcareous rock; local. Throughout British Is. Europe, North America.

Differs from *T. aromatica* **(1)** in the brown epithecium, sometimes with a green tinge, and brown, often somewhat imbricate, squamules without white flecks.

The correct name is ***T. verrucarioides*** (Nyl.) Timdal (1991).

4. T. cumulata (Sommerf.) Th. Fr. (1874)

Thallus squamulose; squamules to 4 mm diam., scattered or forming a ± continuous crust, flat to weakly convex, crenulate to lobate; upper surface greyish white to medium grey, not pruinose, dull, lacking fissures and pseudocyphellae, edge concolorous with upper side; upper cortex 20–30 μm thick, poorly developed, packed with crystals which dissolve in K, K+ yellow; lower cortex poorly developed or absent. Photobiont zone continuous. Medulla containing crystals dissolving in K, K+ yellow. Apothecia to 0.3(-0.5) mm diam., marginal on the squamules when young, soon forming large, to 3 mm diam. agglomerations between the squamules, flat to weakly convex, dark brown, not pruinose; true exciple persistent but usually indistinct from above, brown in the inner part, dark brown at the rim, containing crystals dissolving in K, K+ yellow, N−; epithecium brown, lacking crystals, K−, N−; hymenium 40–60 μm tall, colourless to pale brown; hymenium I−. Ascospores 12–20 × 4–6 μm, 1-septate, ellipsoid to fusiform. Thallus K+ yellow (atranorin, lobaric acid).

On soils associated with ± basic schistose rocks, montane. N. Scotland (Mid-Perth, Ben Lawers; S. Aberdeen, Ben Avon). Europe, North America.

Distinguished by the blackberry-like fruits, and 0- to 1-septate ascospores.

5. T. diffracta (Massal.) Zahlbr. (1901)

Thallus squamulose; squamules to 3 mm diam., scattered when young, later contiguous, weakly convex to hemispherical; upper surface grey, densely white-pruinose or more rarely partly not pruinose; pruina granular, dull, smooth or with shallow fissures, lacking pseudocyphellae, the edges concolorous with upper side, the underside white to pale brown; upper cortex 40–70 μm thick, lacking an epinecral layer but overlain by an amorphous layer of calcium oxalate crystals; lower cortex resembling upper cortex, but thinner and without crystals. Photobiont zone continuous. Medulla without crystals. Apothecia to 2 mm diam., weakly concave to weakly convex, partly to entirely white-pruinose or rarely not; true exciple ± persistent, brown, K− and N−, or K+ violet and N+ violet (additional grey pigment); epithecium grey, containing crystals of calcium oxalate, K+ violet, N+ violet; hymenium 70–80 μm high, colourless to pale brown; hypothecium lower part brown, upper part dark brown. Ascospores 14–20 × 3–5 μm, 1-septate, fusiform. Unidentified terpenoids or lichen products not detected by t.l.c.

On calcareous rock, mainly in fissures on steep to overhanging rock faces, associated with cyanobacterial lichens when young. S. England, Wales, Scotland, Ireland. W., C. & S. Europe.

British material has often been incorrectly referred to *T. candida* (see p. 603), which forms rosette-shaped thalli with a densely pruinose upper surface, and is always without regular fissures in the cortex. Three chemical races of *T. diffracta* occur in the British Is.: one lichen product deficient, and two with different series of terpenoids.

6. T. lobulata (Sommerf.) Lynge (1926)

Thallus warted-granular to minutely-squamulose; squamules 0.2–1 mm wide, mostly contiguous, compacted, horizontal, ill-formed to crenate-lobed, ± overlapping, whitish grey to grey or fawn, not pruinose, edges ± pale, underside pale white; upper cortex thin, 5–10 μm thick, folded, with an epinecral layer, without crystals. Photobiont zone discontinuous, confined to zones under exposed areas of cortex. Medulla of ± isodiametric cells, with thin or ± thickened colourless or pale walls. Apothecia 0.3–1 mm diam., abundant, discrete or often aggregated, black to brown-black, sessile, rounded or occasionally ± sublobulate, convex, surface matt or ± shiny, not pruinose; true exciple excluded from the first, thin, red-brown, opaque, merging with a similarly coloured hypothecium, K+ intensifying N+ purple-red; epithecium ± pale greenish grey, K−, N+ reddish, without crystals; hymenium 70–90 μm tall, colourless or pale red-brown. Paraphyses conglutinated, indistinct, simple, occasionally branched, cells towards apices to 2.5 μm wide, submoniliform, shortly septate. Asci 60–70 × 14–15 μm, wall thick, to 7 μm at apex. Ascospores 14–20(-26) × 3–5(-6) μm, (0-)2- to 3-septate, ellipsoid-fusiform or clavate, colourless or very pale brown, with a finely warted epispore. Zeorin by t.l.c.

On soil associated with limestone, epidiorite, chalk and mortar, also on consolidated calcareous dunes, shell sand, mica-schist, upland and some lowland areas; local. Throughout British Is. C. & S. Europe, Scandinavia, North America.

The squamules often have a characteristic darker centre with a whitish edge. Anatomically close to *Bacidia sabuletorum*, differing in the ± squamulose thallus and shorter ascospores.

7. T. lurida (Arnold) H. Olivier (1911)

Thallus squamulose; squamules to 2 mm diam., scattered or contiguous, ± hemispherical when young, later bullate to columnar, sometimes branched; upper side dark olivaceous green to dark brown, usually sparingly pruinose, rarely densely pruinose or not, dull to shiny, smooth, with punctiform to irregular but often indistinct pseudocyphellae; edge concolorous with upper side; underside paler than upperside, not usually pruinose; upper cortex 20–50 μm thick, including an epinecral layer, to 20 μm thick, partly disintegrating into pseudocyphellae, sometimes containing (or overlain by) calcium oxalate crystals (especially in the pseudocyphellae); lower cortex resembling upper, thinner and lacking crystals. Photobiont layer continuous (sometimes broken under pseudocyphellae). Medulla without crystals. Apothecia to 3 mm diam., flat to weakly convex, not pruinose; true exciple raised when young, later often excluded, pale grey to colourless in inner part, dark grey in the rim, K+ violet, N+ violet; epithecium grey, not containing crystals, K+ violet, N+ violet; hymenium 70–80 μm tall, colourless; hypothecium pale brown to colourless. Ascospores 12–18 × 3.5–5 μm, 1-septate, fusiform. Lichen products not detected by t.l.c., or with traces of fatty acids.

On calcareous soil, usually associated with cyanobacterial lichens when young; very rare. E. England (Cambridge, Suffolk). Europe.

Distinguished from *T. caeruleonigricans* **(2)** by the more regularly bullate to columnar thallus, presence of pseudocyphellae (often inconspicuous), and ± colourless hypothecium and inner part of the exciple.

The correct name is ***T. physaroides*** (Opiz) Zahlbr. (1926).

8. T. mesoidea (Nyl.) Zahlbr. (1926)

Thallus crustose, rimose-cracked to areolate; areoles to 0.5 mm diam., flat, thin, rounded or irregular, sometimes crenulate, brown, olive-brown or somewhat greenish grey, not pruinose, not white flecked; upper cortex ± absent, to 40 μm thick, poorly developed, lacking crystals; lower cortex absent. Photobiont layer filling the entire inner part of the areoles or ± restricted to upper part. Medulla (when present) without crystals. Apothecia to 0.8 mm diam., flat, later becoming weakly to moderately convex, not pruinose; true exciple distinct when young, later becoming excluded, dark reddish brown towards the rim, somewhat paler in inner part; epithecium green, lacking crystals K −, N+ violet; hymenium 60–70 μm tall, colourless; hypothecium dark reddish brown throughout or faintly greenish in upper part, K −, N − or ± violet (green pigment). Ascospores 14.5–17 × 5.5–6 μm, 3-septate, narrowly ellipsoid to shortly bacilliform. Lichen products not detected by t.l.c.

On ± basic coastal rocks and serpentine (but not limestone); local. England, Channel Is., Wales, Ireland. France (Brittany), Portugal.

Resembles *T. aromatica* **(1)**, which may also occur on coastal rocks, but differs mainly in the rimose-cracked to scattered-areolate and never squamulose thallus, and the absence of white flecks. The ascospores tend to be shorter, but those of *T. aromatica* can be very variable in size.

9. T. opuntioides (Vill.) Buschardt (1979)

Thallus squamulose; squamules to 4 mm diam., strongly convex to bullate when young, later vertically flattened and often imbricate; upper surface olivaceous brown to dark brown, usually partly white-pruinose (especially in convex parts) or sometimes not, slightly shiny, with shallow fissures in the cortex (when well developed), lacking pseudocyphellae; edge usually more densely white-pruinose; under surface white to pale brown; upper cortex 40–90 μm thick, including an thick epinecral layer to 50 μm, sometimes containing, or overlain by, calcium oxalate crystals; lower cortex resembling the upper, thinner and lacking crystals. Photobiont layer continuous. Medulla lacking crystals. Apothecia to 4 mm diam., weakly concave to weakly convex, usually moderately white-pruinose; true exciple distinct when young but later often excluded, medium to dark brown throughout or with a dark grey rim, K − and N − or K+ violet and N+ violet (grey pigment); epithecium grey, usually containing crystals of calcium oxalate, K+ violet, N+ violet; hymenium 70–80 μm high, colourless; hypothecium medium brown in lower part, dark brown in upper part. Ascospores 15–25 × 3–5 μm, 1-septate, fusiform. Unidentified terpenoid in t.l.c. turning bright yellow on charring.

On calcareous soil and rock, mainly among mosses or in fissures on rock walls, apparently always associated with cyanobacterial lichens when young, also on shell sand; very rare. S.W. England (Somerset, Cleeve Hill), N.W. Scotland (Mull). C. & W. Europe.

T. caeruleonigricans **(2)** differs in the more regularly bullate, irregularly sausage-shaped, often imbricate, squamules with a more cracked upper surface and usually with a more distinct white margin. Special care is required when identifying this species on morphology alone; however, the unknown terpenoid is diagnostic.

10. T. oribata (Nyl.) P. James (1965)

Thallus crustose to subsquamulose, of small proliferating granules forming a thick, ± continuous crust; upper surface greyish brown, not pruinose, dull, lacking fissures and pseudocyphellae; upper cortex varying from almost absent to 20 μm thick, irregular, lacking crystals; lower cortex absent. Photobiont layer filling inner part of granules or restricted to upper part. Medulla, when present, lacking crystals. Apothecia to 1.5 mm diam., flat but later becoming weakly convex, not pruinose; true exciple distinct when young, later becoming excluded, dark brown in inner part, with a green edge in section, K−, N+ violet; epithecium green, lacking crystals, K−, N+ violet; hymenium 70–80 μm tall, colourless; hypothecium dark brown. Ascospores 25–38 × 3–4 μm, (3-)7-septate, acicular. Lichen products not detected by t.l.c.

On calcareous rock, apparently associated with cyanobacterial lichens; not seen this century. N. Scotland (Mid-Perth, near summit of Ben Lawers). Czechoslovakia, Italy, Norway.

Resembles *T. aromatica* **(1)** in apothecial anatomy but the ascospores are acicular and mainly 7-septate. The correct name is ***T. coelestina*** (Anzi) Vězda (1961).

11. T. pulvinata (Taylor) H. Olivier (1911)

Thallus small-squamulose; squamules to 4 mm diam., weakly to strongly convex, sometimes nodulose, composed of aggregations of proliferating granules; upper surface pale yellow to brownish yellow, not pruinose, dull, lacking fissures and pseudocyphellae; upper cortex 30–50 μm thick, poorly differentiated from the photobiont layer, lacking an epinecral layer, with crystals which dissolve in K; lower cortex absent. Photobiont layer continuous. Medulla with crystals dissolving in K. Apothecia to 1 mm diam., weakly to strongly convex, shiny, not pruinose; true exciple inapparent or indistinct when young, confluent with hypothecium, brownish black, K−, N−; epithecium colourless, lacking crystals, K−, N−; hymenium 60–80 μm tall, dark brown; hypothecium brownish black. Ascospores 20–38 × 3–5 μm, mainly 7-septate, acicular. Two unidentified lichens products detected by t.l.c., UV−.

On mosses on siliceous, ± vertical rock and boulder faces. W. British Is. Norway, Madeira.

The correct name is ***T. thiopsora*** (Nyl.) H. Olivier (1911).

12. T. rosulata (Anzi) H. Olivier (1911)

Thallus squamulose; squamules to 4 mm diam., flat to bullate, marginal squamules often elongated; upper surface grey, densely white-pruinose, dull, with shallow fissures, without pseudocyphellae; edge concolorous with upper side; underside white to pale brown; upper cortex 40–100 μm thick, lacking an epinecral layer, overlaid by an amorphous layer of calcium oxalate crystals; lower cortex resembling the upper, thinner and lacking crystals. Photobiont layer continuous. Medulla without crystals. Apothecia to 3 mm diam., weakly concave to weakly convex, often lobed, usually faintly white-pruinose; true exciple ± persistent, pale grey to colourless in inner part, medium grey in the rim, K+ violet, N+ violet; epithecium grey, containing crystals of calcium oxalate, K+ violet, N+ violet; hymenium 70–80 μm tall, colourless; hypothecium pale brown to colourless. Ascospores 14–22 × 3–5 μm, 1-septate, fusiform. Lichen products not detected by t.l.c.

On calcareous soil and rock, montane, apparently always associated with cyanobacterial lichens when young, montane; not recorded this century. N. Scotland (Mid-Perth, Ben Lawers, Craig Calliach). C. & N. Europe, Greenland.

Differs from *T. diffracta* **(5)** and *T. caeruleonigricans* **(2)** in the more rosette-shaped thallus and the colourless to pale brown hypothecium and inner part of the exciple.

13. T. squalescens (Nyl.) Th. Fr. (1874)

Thallus squamulose; squamules to 1 mm diam., forming a ± continuous crust, strongly convex to bullate; upper surface grey, often tinged brown, not pruinose, dull, without fissures and pseudocyphellae; edges concolorous with upper side; upper cortex poorly differentiated from the photobiont layer, 15–25 μm thick, lacking crystals; lower cortex poorly developed or absent. Photobiont layer continuous. Medulla lacking crystals. Apothecia to 1 mm diam., strongly convex, black, shiny, not pruinose; true exciple inapparent even when young, very thin, sometimes ± absent; epithecium olivaceous green, K−, N−, lacking crystals; hymenium 30–40 μm tall, colourless; hypothecium colourless to pale brown, I+ blue. Ascospores 7–12 × 4–6 μm, 1-septate, ellipsoid. Thallus K+ yellow (stictic acid).

On mosses on acid rocks, mainly on rock faces, montane; very local. N. Scotland (Highlands). Europe, North America.

14. T. squalida (Ach.) Massal. (1852)

T. havaasii Magnusson (1949)

Thallus squamulose; squamules to 2 mm diam., often proliferating and forming a ± continuous crust, weakly concave to weakly convex; upper surface brown to dark brown, sometimes with a grey tinge, not pruinose, dull, often with shallow fissures, lacking pseudocyphellae; edges concolorous with upper surface; upper cortex 25–100 μm thick, including a 20–30 μm thick epinecral layer, lacking calcium oxalate; lower cortex pale brown, poorly developed. Photobiont layer discontinuous. Medulla lacking crystals. Apothecia to 1.5 mm diam., flat to weakly convex, not pruinose; true exciple indistinct, dark (sometimes greenish) brown in the rim, pale brown to colourless in inner part, K−, N− or ± violet (green pigment); epithecium green, lacking crystals, K−, N+ violet; hymenium 70–80 μm tall, colourless; hypothecium colourless to pale brown. Ascospores 17–45 × 3–5 μm, 3- to 7-septate, acicular. Lichen products not detected by t.l.c.

On calcareous rock, mainly amongst mosses in seepage tracks, in alpine areas; rare. N. Scotland (Highlands). Europe, North America, Macaronesia.

A rather variable species characterized by the green epithecium, pale hypothecium and 3- to 7-septate acicular ascospores.

15. T. tumidula (Sm.) Zahlbr. (1926)

Thallus squamulose; squamules to 8 mm diam., scattered to contiguous, weakly to moderately convex, rounded to slightly lobate; upper surface pale greenish grey to medium brown, usually densely white-pruinose, dull, smooth or with a few shallow fissures, lacking pseudocyphellae; edges concolorous with upper surface or white; underside white to medium brown; upper cortex 50–100 μm thick, sometimes including a thin epinecral layer, filled with crystals of calcium oxalate; lower cortex poorly developed or absent. Photobiont layer continuous. Medulla filled with crystals of calcium oxalate. Apothecia to 1.5 mm diam., arising at the edges of squamules, flat when young, later often becoming convex, not pruinose; true exciple distinct when young but becoming excluded, confluent with hypothecium, dark brown throughout, K−, N−; epithecium brown, lacking crystals, K−, N−; hymenium 70–80 μm high, colourless to pale brown, hypothecium dark brown. Ascospores 10–16 × 3–5 μm, ellipsoid, simple or 1-septate. Thallus K± yellow (atranorin).

On calcareous rock, mainly in fissures of limestone; very rare. S.W. England. Europe, N. Africa.

T. caeruleonigricans **(2)** has ascospores 14–25 × 2.5–5 μm and the paraphyses are not conglutinate.

TORNABEA Østh. (1980)

Thallus shrubby, often entangled and spreading, ± dichotomously divided, divergent, rounded or often ± flattened and angular. **Cortex** of conglutinated, thick-walled hyphae mostly orientated longitudinally, sometimes covered with a colourless layer, present on both surfaces. **Photobiont** trebouxioid. **Ascomata** apothecia, sessile, lateral. **Thalline exciple** persistent. **Hymenium** colourless. **Hamathecium** of paraphyses, ± thickened, branched apices brown. **Asci** 8-spored, elongate-clavate, *Lecanora*-type. **Ascospores** brown, ellipsoid, polarilocular. **Chemistry**: no lichen products detected. **Ecology**: on rocks and twigs in well-lit situations.

Closely resembles *Anaptychia* in the polarilocular, brown ascospores and non-pseudoparenchymatous cortex, but the lobes are entirely corticate, are not dorsiventral, and lack the marginal rhizines present in the ± shrubby *Anaptychia* species.

LITERATURE: Kurokawa (1962b).

1. T. scutellifera (With.) Laundon (1984)

Tornabenia atlantica (Ach.) Kurok. (1962)

Thallus 2–3 cm tall, tufted, glaucous grey, pale grey-brown to reddish brown; branches to 1 mm diam., dichotomously or ± irregularly branched, locally entangled, tomentose or finely pubescent, rounded or

± flattened and angular, occasionally somewhat channelled, gradually narrowing towards apices. Apothecia rare, sessile, to 1 mm diam.; disc black-brown, convex; thalline exciple thin, entire, grey. Thallus Pd−, K−, KC−, C− (lichen products not detected by t.l.c.).

On rocks, sand, trunks of old trees and sides of old barns in coastal districts; not observed this century and probably extinct. S. England (Sussex). Atlantic Europe, Mediterranean region, Africa, Canary Is., N.W. Himalaya.

Distinguished from related genera by the radial structure. The cortex comprises hyphae which run parallel with the long axis of the branch which give rise to short lateral hyphae to the surface, giving the plant a soft, pubescent appearance. This species was formerly locally frequent in S. England (Hastings-Chichester area); landslip of cliffs, increasing pollution and over-zealous collecting may have hastened its extinction.

TRAPELIA M. Choisy (1929)

Thallus crustose to squamulose. **Upper cortex** of entangled hyphae, only present in species with distinctly squamulose thalli. **Photobiont** chlorococcoid. **Ascomata** apothecia, yellow-brown to almost black, constricted at base, adpressed, at first globose and appearing perithecia-like, then splitting at apex and margin appearing ± lobed. **Thalline exciple** pronounced or absent. **True exciple** of colourless hyphae, the gel-matrix occasionally weakly pigmented, but never dark brown, varying from relatively weak and of short-celled, thin-walled hyphae (*T. coarctata* **(1)**) to well-developed, and of long-celled, pachydermatous hyphae (*T. mooreana* **(5)**). **Hamathecium** of paraphyses, very delicate, branched and anastomosed, usually colourless throughout, the apices never abruptly swollen, the upper parts in some species (in the epithecium) may appear slightly swollen due to an external coating of pigment. **Asci** (Fig. 33m) clavate-cylindrical, thin-walled; apical dome I−, or weakly I+ blue, 8-spored. **Ascospores** simple, ellipsoid, colourless or faint pink, smooth, without a distinct perispore. **Conidiomata** pycnidia, immersed. **Conidia** cylindrical to thread-like. **Chemistry**: gyrophoric and traces of lecanoric acid often present. **Ecology**: on siliceous rocks, wood, and soil.

Trapelia is very close to *Trapeliopsis*, the latter apparently differing in its more robust apothecia with more massive excipular and 'hypothecial' tissues, and, usually, a proportionally shallower hymenium and smaller ascospores. The delimitations of these genera, which have been subsumed by some authors, requires clarification by detailed anatomical and ontogenetic studies. The species are especially difficult to separate when sterile and the key to *Trapeliopsis* should also be tried for such material. *Placynthiella* has a true exciple of brown-walled pseudoparenchymatous cells and paraphyses with irregularly capitate apices, each crowned with a dark brown apical cap.

Lecidea lichenicola, which occurs on chalk pebbles, may also belong to *Trapelia*; it differs primarily in its ecology.

LITERATURE: Coppins & James (1984), Hertel (1969b, 1977).

1	Soredia present	2
	Soredia absent	4
2(1)	Thallus inconspicuous, uneven, of ± scattered areoles or squamules, dark grey, olivaceous, pink-brown to brownish; margin diffuse	3
	Thallus conspicuous, flat, ± placodioid and continuous, white or pink- tinged; margin discrete, effigurate	**6. *placodioides***
3(2)	On bark or mosses over bark	**3. *corticola***
	On siliceous rocks, stones, or plant debris	**4. *obtegens***
4(1)	Thallus C+ red; apothecia, when present, with a thin, not raised true exciple	5
	Thallus C−; apothecia, when present, with a thick, raised true exciple	**5. *mooreana***
5(4)	Thallus effigurate or subsquamulose; areoles often overlapping	**2. *involuta***
	Thallus crustose, ± cracked; areoles not overlapping	**1. *coarctata***

1. T. coarctata (Sm.) M. Choisy (1932)

Thallus thin, whitish, pale grey or pinkish, rarely pale green, smooth or ± rugose, continuous or cracked, never distinctly effigurate at the edge, the areoles mostly contiguous or sometimes scattered; prothallus ± present, white. Apothecia 0.2–0.8 mm diam., rose-pink to red-brown; thalline exciple thin, white, smooth to crenate, ± forming a halo-like rim, at least when young. Asci (Fig. 33m). Ascospores 15–25 × 7–13 μm. Thallus Pd −, K −, KC+ red, C+ red (gyrophoric acid).

On siliceous rocks and stones, brickwork, occasionally on consolidated clay soil; abundant. Throughout British Is. Europe, North America, India, Japan, Indonesia, Australia, New Zealand.

A very variable species distinguished by the even, smooth to areolate, non-effigurate thallus. Immature apothecia may appear as scattered, starkly white, punctiform spots.

2. T. involuta (Taylor) Hertel (1973)

Like **1**, but thallus of whitish, sometimes brown-tinged, smooth or minutely rugose, effigurate or subsquamulose, ± convex, often ± overlapping areoles; marginal squamules 0.2–0.4 mm wide and 0.04–0.15(-0.2) mm thick, contorted, sometimes in irregular, knotted clusters, often sterile.

On siliceous rocks, rarely on worked timber, and occasionally on peaty turf amongst rocks, especially those on the coast. Throughout British Is. Europe, North America, Indonesia.

The apothecia, as in *T. coarctata* (**1**), have a thin, whitish thalline exciple.

3. T. corticola Coppins & P. James (1984)

Thallus effuse, immersed, of inconspicuous, dull green to greenish brown, ± scattered, non-effigurate areoles to 0.15 mm diam. Soralia (0.1-)0.15–0.25(-0.4) mm diam., punctiform, rarely confluent, markedly convex, pale buff to pale greenish buff, often erupting from a single areole; soredia farinose. Apothecia 0.07–0.1(-0.2) mm diam., very rare, sessile; disc flat when mature, pinkish brown; true exciple very thin, paler than the disc. Ascospores 11–14 × 4.8–6.5 μm. Soralia Pd −, K −, KC+ red, C+ red (gyrophoric acid).

On *Alnus* and *Quercus* in sheltered woodlands by streams, also on other acid-barked trees and rarely on wood; locally frequent. W. British Is. N.W. France, Tuscany, Canary Is., Azores, Madeira, Chile.

The inconspicuous, greenish or brownish areoles, numerous, small, punctiform soralia, and apothecia without a white thalline, halo-like exciple, suggest *T. corticola* is related to *T. obtegens* (**4**). Easily confused with *Micarea coppinsii* (see note under *M. peliocarpa*) which has brighter green soralia and a micareoid photobiont. *Rinodina efflorescens*, which occurs in similar habitats, has soralia which are Pd+ red, C −.

4. T. obtegens (Th. Fr.) Hertel (1970)

Like **3**, but with pale green ± suffused ochraceous brown, often convex areoles, excavate to slightly convex, slightly larger soralia, 0.2–0.4 mm diam., and larger apothecia, 0.2–0.5 mm diam. which though often present, are usually few and scattered.

On siliceous rocks or stones, more rarely on plant debris. Throughout British Is. Europe, Madeira, Azores, North America.

5. T. mooreana (Carroll) P. James (1975)

Thallus granular-verrucose, sometimes secondarily cracked, pale fawn, yellow-grey to ochraceous. Apothecia 0.4–1.5 mm diam., robust, sessile, with a thick, elevated exciple and concave disc, dull brown-black. Ascospores 15–21 × 6–10 μm. Pycnidia *c.* 50–80 μm diam., often numerous, immersed in thalline warts; conidia 5–6 × 0.7 μm, bacilliform. Apothecia in section and pycnidia C+ red (gyrophoric acid).

On sheltered, ± metal-rich, siliceous rocks, especially basalts and pebbles in moorland; very local. Upland Britain. Europe, North and South America, Indonesia.

The large, sessile apothecia with a thick exciple and chemical tests are diagnostic.

6. T. placodioides Coppins & P. James (1984)

Thallus forming small, ± delimited patches or an extensive crust to 15 cm across, white- or pink-tinged, matt; edge usually with discrete, ± effigurate, flat or ± convex areoles 0.2–0.4(-0.6) mm diam., giving the thallus a placodioid appearance, towards centre areoles coalesce, the thallus then becoming secondarily cracked. Soralia 0.2–0.3 mm diam., numerous, pale greenish or yellowish white, rarely ochraceous, usually developing from sides of areoles or cracks in thallus; soredia 20–30 μm diam., farinose. Apothecia very rare, as in *T. coarctata* **(1)**. Thallus and soralia C+ red (gyrophoric acid).

On siliceous rocks, mine-spoil heaps and walls; common. Throughout Britain, especially in N. and upland areas. Europe, North America.

Closely related to *T. coarctata* **(1)** and *T. involuta* **(2)** in the apothecial characteristics and whitish or pinkish thallus, but differing from both in possessing soralia. See also *Placopsis lambii*.

TRAPELIOPSIS Hertel & G. Schneider (1980)

Thallus granular, or squamulose to small-foliose. **Upper cortex** of entangled hyphae only present in representatives with distinctly squamulose thalli. **Photobiont** chlorococcoid (including *Chlorella* and *Pseudochlorella*). **Ascomata** apothecia, yellow-brown to almost black, constricted at base, adpressed, disc-like from start which expands without splitting. **Thalline exciple** pronounced or absent. **True exciple** of colourless hyphae, the gel-matrix occasionally weakly pigmented, never dark brown. **Hamathecium** of paraphyses, delicate, branched and anastomosed, usually colourless throughout and never with abruptly swollen apices, although in some species their upper parts (in the epithecium) may appear slightly swollen due to an external coating of pigment. **Asci** 8-spored, clavate-cylindrical, thin-walled; apical dome not, or weakly, I+ blue. **Ascospores** simple, colourless, smooth, ellipsoid, without a thickened perispore. **Conidiomata** pycnidia, immersed. **Conidia** cylindrical to thread-like. **Chemistry**: gyrophoric and traces of lecanoric acid often present.

Doubtfully distinct from *Trapelia*; non-fertile specimens cannot be definitely referred to either genus and the key under *Trapelia* should also be tried for such sterile specimens.

LITERATURE: Coppins & James (1984), Hertel (1969b).

1 Thallus, and soralia if present, C+ red 2
Thallus, and soralia if present, C− 6

2(1) Thallus coarse; if placodioid or subsquamulose marginal areoles mostly <0.5 mm wide; isidia absent 3
Thallus squamulose; marginal lobes ± distinct, 0.5–1.6 mm wide; isidia usually present .... **9. *wallrothii***

3(2) Thallus without orange-red patches; all parts K− 4
Thallus with irregular orange-red patches, K+ purple .... **7. *pseudogranulosa***

4(3) Thallus of granular-warted areoles, areoles mostly >0.1 mm diam., often effigurate at edge of thallus or when scattered; soralia when present ± discrete; true exciple distinct at least when young 5
Thallus minutely granular, granules <0.1 mm diam., never effigurate; soralia, when present, often confluent; true exciple not apparent .... **8. *viridescens***

5(4) Soralia, when present, grey-green to dark green, often tinged blue-green; ascospores 7–9.5 × 2.5–4 μm .... **1. *flexuosa***
Soralia, when present, whitish to brownish yellow, sometimes tinged grey-green; ascospores 9–14 × 4–6 μm .... **6. *granulosa***

6(1) Thallus squamulose; squamules with ascending, usually sorediate apices 7
Thallus crustose, minutely warted or granular, sorediate or not 8

7(6) Squamules 0.7–2.5 mm long .... **4. *glaucolepidea***
Squamules 0.3–0.8 mm long .... **5. *percrenata***

8(6) Apothecia dark green-grey to grey-black; epithecium green .... **2. *gelatinosa***
Apothecia pale to reddish brown; epithecium ± colourless to reddish brown . **3. *aeneofusca***

1. T. flexuosa (Fr.) Coppins & P. James (1984)

Lecidea aeruginosa Borrer (1831)

Thallus greenish grey to grey-green, white to dark blue-grey, of scattered to crowded, uneven, granular areoles; areoles 0.08–0.25 mm diam., but scattered or marginal areoles often ± flattened and effigurate and up to 0.4 mm diam. Soralia 0.2–0.4 mm diam., at first discrete, sometimes becoming confluent; soredia farinose to finely granular, grey-green to dark blue-green. Apothecia 0.2–0.7 mm diam., dark greenish grey to green-black, rarely pale, or pinkish (in extreme shade), flat to slightly convex; hymenium 40–50 μm tall; asci 30–50 μm tall. Ascospores 7–9.5 × 2.5–4 μm. Thallus Pd−, K−, KC+ red, C+ red (gyrophoric acid).

On hard wood, rarely on plant debris or sandstone rocks; often abundant. Throughout British Is. Europe, North America.

Resembles *T. granulosa* **(6)** which has larger ascospores and also grows on soil and rotting wood.

2. T. gelatinosa (Flörke) Coppins & P. James (1984)

Lecidea gelatinosa Flörke (1809)

Thallus thin and membranous, or minutely granular, dark green-brown to green-grey, without or often (especially when sterile or sparingly fertile) with pale green soralia, at first 0.2–0.7 mm diam., but often becoming very conspicuous, irregular and confluent. Apothecia 0.2–1(-1.6) mm diam., adpressed; exciple excluded or as a thin, pale rim not exceeding the level of the disc, dark green-grey to grey-black; epithecium green, K+ brown. Ascospores 8–14 × 4.5–6 μm. Thallus Pd−, K−, KC−, C−. The pale green colour of the soralia is in marked contrast with the darker thallus; occasionally no soralia are present.

On shaded, peaty or clay soil, mainly in banks or cuttings with overhanging herbs or small shrubs; rather local, especially in upland districts. Throughout British Is. Europe, North America.

3. T. aeneofusca (Flörke ex Flotow) Coppins & P. James (1984)

Like **2**, but apothecia pale to reddish brown, and epithecium is ± colourless to reddish brown, K−; very rare. N. Wales (Caernarvon). C. Europe.

4. T. glaucolepidea (Nyl.) G. Schneider (1980)

Thallus squamulose; squamules scattered or ± contiguous, 0.7–2.5 mm wide with ascending, usually sorediate apices, grey-white to glaucous-grey, mostly with pale grey or greenish grey, lip-shaped soralia. Apothecia 1.0–2.5 mm diam., sessile, at first flat and exciple ± distinct, later becoming ± convex and the exciple excluded; disc pale dull grey to dark grey, or dull reddish brown, margin often paler. Ascospores 8–11 × 4–5 μm. Thallus and soralia Pd−, K−, KC−, C−; medulla and soralia UV± glaucous (unidentified UV+ substance).

On bare, peaty turf and edge of peat hags, rarely old ant hills on moorlands; local in uplands. British Is. Scandinavia.

Resembles a diminutive *Hypogymnia physodes* on peat.

5. T. percrenata (Nyl.) G. Schneider (1980)

Like **3**, but squamules smaller, 0.3–0.8 mm wide, and soralia lip-shaped or ± capitate. Apothecia not seen in British material, similar but somewhat smaller (to 1.2 mm diam.) and darker.

On shaded, moist, rotting wood of fallen trunks and tops of old fence posts; local. S. England (N. Devon), C. Wales, Scotland. Scandinavia.

Easily mistaken for small, sorediate, *Cladonia* squamules.

6. T. granulosa (Hoffm.) Lumbsch (1983)

Lecidea granulosa (Hoffm.) Ach. (1803)

Thallus in small patches or coalescing to form more extensive patches, usually whitish to pale grey, sometimes in part pinkish, more rarely greenish grey; of ± crowded, granular to verrucose areoles, 0.12–0.5 mm diam., effigurate and less often flattened when scattered or at the edge of the thallus. Soralia 0.3–0.6 mm diam., irregular and granular, whitish to cream-yellow, sometimes tinged grey-green, sometimes very few or absent. Apothecia frequent, 0.3–1(-1.5) mm diam., at first flat but often becoming convex and the exciple ± excluded, occasionally tuberculate and to 1.7 mm diam., very variable in colour, pale pink to reddish brown, or pale to dark green-grey, very often a piebald mixture of these hues; hymenium 70–80 μm tall. Epithecium pale to greenish. Ascospores 9–14 × 4–6 μm, a few occasionally becoming 1-septate. Thallus Pd−, K−, KC+ red, C+ red (gyrophoric acid).

Usually on acid soils, moribund bryophytes or plant debris, and on decayed wood, more rarely on shaded, acid rocks; common. Throughout British Is. Cosmopolitan.

A primary colonizer of recently burnt heathland, often with *Placynthiella icmalea*. Thalli which lack soralia are often richly fertile; on sorediate plants, though often present, apothecia are normally more sparse.

7. T. pseudogranulosa Coppins & P. James (1984)

Thallus effuse, often wide-spreading and forming large patches to 20 cm diam., grey- or greenish white, in part patchily orange-pigmented, K+ purple, of granular areoles that soon coalesce forming a faintly cracked, granular crust; areoles 0.04–0.2 mm diam. Soralia 0.2–1.6 mm diam., greenish white, in part orange-pigmented, at first convex and discrete, later coalescing and forming coarse, irregular patches; soredia 18–25 μm diam., farinose. Apothecia 0.4–1(-1.6) mm diam., rare, ± adpressed, flat with a shallow, wavy true exciple; disc greenish grey to grey-black. Ascospores 10–12.5 × 3.5–6.0 μm. Thallus, soralia and apothecia Pd−, C+ red, UV−; orange-pigmented areas of thallus and soralia K+ purple, UV± deep orange-red (gyrophoric acid and an unidentified anthraquinone).

On various humid, sheltered, acidic substrata, especially peat or over decaying bryophytes, rotting wood and plant debris, often in old woodlands; frequent. Throughout British Is., especially in N. & W. N. Europe, Madeira, North America, Chile.

Characterized by the minutely granular thallus, rather large, farinose, C+ red soralia, and irregular, patchy orange coloration reacting K+ purple.

8. T. viridescens (Schrader) Coppins & P. James (1984)

Lecidea viridescens (Schrader) Ach. (1803)

Thallus of coherent, minutely granular, whitish, pale grey or pale greenish areoles, mostly to 0.1 mm diam., which are never effigurate. Soralia pale greenish, irregular and soon confluent (or soralia effuse from the beginning) often giving the thallus a leprose appearance. Apothecia 0.2–0.6 mm diam., or tuberculate and to *c.* 1 mm diam., black or greenish black, flat to convex; exciple inapparent. Ascospores 9–12 × 4–5 μm.

On decaying trunks and stumps; very rare. N.E. Britain. Europe, North America.

9. T. wallrothii (Flörke ex Sprengel) Hertel & G. Schneider (1980)

Thallus closely adpressed, distinctly but unevenly scattered or ± crowded-squamulose, ± effigurate, with ± distinct, short, flat to ± convex marginal lobes, 0.5–1.6 mm wide and 0.1–0.3 mm thick, whitish or pale grey; often with scattered clusters of coarse, irregular isidia-like protuberances which, following abrasion, leave small circular scars, 0.5–1 mm diam. Apothecia 1–2 mm diam., often absent, with pink-brown to dull green-grey, often faintly pruinose disc, and flexuose, persistent, paler exciple. Ascospores 8–14 × 4–5 μm. Thallus Pd−, K−, KC+ red, C+ red (gyrophoric acid).

On soil and turf, usually amongst low, protruding rocks, earth banks and walls; local. W. coastal districts of British Is., very rare inland. Norway, France, Madeira, Canary Is., Brazil.

Distinguished by the ± scattered, occasionally somewhat weakly radiate, pale areoles and the coarse clusters of irregular, elongate isidia. Sometimes the thallus forms a compact, uneven, ± continuous, warted crust which is abundantly fertile. Robust morphs of *Trapelia involuta* are sometimes similar but the lobes are narrower, usually more convex and the apothecia have a whitish thalline rim and longer spores, 15–26 × 7–12 μm.

TREMOLECIA M. Choisy (1953)

Thallus crustose, superficial, closely areolate; prothallus black. **Medulla** I−. **Photobiont** trebouxioid. **Ascomata** apothecia, immersed in the areoles; discs black, generally concave, saucer-shaped. **Thalline exciple** absent. **True exciple** well developed, somewhat raised, composed of radially arranged, thick or dark brown hyphae, ± black and friable. **Epithecium** greenish tinged. **Hymenium** I+ blue. **Hypothecium** conspicuous, dark brown. **Hamathecium** of branched and anastomosed, regularly septate paraphyses, not markedly swollen at the apex. **Asci** (Fig. 33k) *Tremolecia*-type, clavate, with a thin I+ blue outer gelatinous coat, the tholus well developed, I− or ± weakly blue, without a distinct ocular chamber or internal apical beak, 8-spored. **Ascospores** ellipsoid, simple, colourless, smooth, lacking a distinct perispore. **Conidiomata** pycnidial, immersed in the areoles. **Conidia** bacilliform. **Chemistry**: unidentified rust-coloured pigment in the cortex. **Ecology**: on hard siliceous rocks.

This genus appears to be monotypic, species formerly placed here having been transferred to either *Farnoldia* or the non-British *Melanolecia* Hertel (1981), both of which have ascospores with a distinct perispore. The former genus is also distinguished by the sessile ascomata and the distinctly I+ blue tholus with a densely stained apical tube (*Porpidia*-like), the latter by the asci having a weakly I+ blue tholus and much longer conidia.

LITERATURE: Hertel (1977), Vainio (1934).

1. T. atrata (Ach.) Hertel (1977)

Thallus rust-coloured, more rarely ± grey-black or ochre-coloured in exposed situations, I−. Apothecia 0.1–0.5(-0.7) mm diam., numerous, angular when contiguous, shining; hymenium 65–80 μm tall; true exciple and hypothecium K± diffusing purplish. Ascospores 10–15(-17) × 6–9(-10) μm. Conidia 3–6 × 1–1.5 μm. Thallus and medulla Pd−, K−, KC−, C−, UV− (unidentified rust-coloured pigment in the cortex).

On exposed, hard, siliceous rocks, mainly in upland areas, especially those which are ± iron-rich, also mine-spoil tips forming small but distinctive patches; locally common. Throughout British Is. Cosmopolitan, in arctic and montane situations in both Hemispheres, Europe, North & South America, E. Africa, Himalayas, Japan, Australia, New Zealand.

Rhizocarpon oederi, which occurs in similar habitats and also has rust-coloured, areolate thalli as well as apothecia with a more erumpent exciple which becomes folded giving the disc a roughened, gyrose appearance, an I+ blue medulla, and 3-septate to submuriform, often pale brown ascospores. Occasionally, in very exposed situations, *T. atrata* has much reduced, rust-coloured areoles and the thallus is ± shining black due to the predominating, well-developed prothallus.

TRIMMATOTHELE Norman ex Zahlbr. (1903)

Thallus crustose. **Photobiont** chlorococcoid. **Ascomata** perithecia, black. **Hamathecium** without paraphyses; periphyses present; gel I− or I+ yellowish, K/I+ blue. **Asci** ± clavate, multispored, *Verrucaria*-type. **Ascospores** ± ellipsoid, simple, colourless. **Conidiomata** not known. **Chemistry**: no lichen products reported. **Ecology**: on rocks.

Similar to *Verrucaria* except for the multispored asci. A little-known genus, with two European species, and much in need of revision.

1. T. perquisita (Norman) Norman ex Zahlbr. (1903)

Thallus to 120 μm thick, brownish (but often overgrown by a blackish cyanobacterium), brown, matt, cracked and secondarily areolate. Photobiont cells 5–10 × 5–8 μm, dividing by binary fission. Perithecia *c.* 0.3 mm diam., $\frac{1}{2}$–$\frac{3}{4}$ immersed in thallus, simple; wall dark brown throughout, 70–80 μm thick above, 50–60 μm wide at sides and base. Periphyses to 30 μm long, *c.* 1.5 μm wide. Asci 50–65 × 13–17 μm, *c.* 65-spored, cylindrical-clavate, K/I−. Ascospores 6–9 × 2–3.5 μm, ellipsoid to oblong-ellipsoid.

On crystalline limestone at *c.* 1000 m; very rare. N. Scotland (E. Inverness, Ben Alder). N. Scandinavia, French Alps.

The ascospores of the Scottish collection are slightly longer than those cited for other European material.

TYLOTHALLIA P. James & Kilias (1981)

Thallus thick, areolate, the areoles convex and warty, often mosaic-forming, white to green-grey, purple-grey or dull green-white; prothallus black, often well developed. **Photobiont** chlorococcoid, cells *c.* 6–12 μm diam. **Ascomata** apothecia, to 1.5 mm diam., at first flat, becoming convex and at times tuberculate, irregularly thickened with a warty surface, dark blue-grey to black; disc sometimes thinly pruinose, especially when young. **True exciple** distinct at first but becoming excluded, of richly branched coherent hyphae, 1–1.7 μm wide, each surounded by a gel coat that swells to 10 μm wide in K, K+ yellowish to orange, Pd+ light yellow. **Hamathecium** of paraphyses, 1.5–2 μm wide, numerous, slender, often branched and anastomosed, the apices mostly slightly irregularly swollen, not regularly capitate, I+ blue. **Asci** *Biatora*-type. **Ascospores** (0-)1-septate, oblong, without a distinct perispore, colourless. **Conidiomata** pycnidia, immersed often with a residual thalline rim. **Conidiogenous cells** ± cylindrical, enteroblastic, acrogenous. **Conidia** ellipsoid or shortly oblong, simple, colourless. **Chemistry**: atranorin, 2'-*O*-methylperlatolic, norstictic, and fumarprotocetraric acids. **Ecology**: on siliceous rocks.

A monotypic genus similar to *Cliostomum*, but with more coherent, more branched and anastomosed paraphyses and with immersed pycnidia with colourless walls, except for a greenish (not purplish) pigment around the ostioles.

LITERATURE: Kilias (1981).

1. T. biformigera (Leighton) P. James & Kilias (1981)

Catillaria biformigera (Leighton) Magnusson (1949)

Thallus thin to thick, irregularly warted, conspicuously rimose-cracked, ± dull green-grey. Apothecia to 1.5 mm diam., rounded or irregular to lobate, ± convex, blue-black to black, occasionally thinly greyish pruinose, sometimes with thickish, often flexuose proper margin, scattered or sometimes ± clustered; true exciple colourless to pale red-brown within, K± intensified or orange-brown, the outer edge pale to dark brown, K± olivaceous-black, N+ red, Pd+ yellow-orange; epithecium black-blue-green or in part dark olive- to fuscous-brown, minutely granular, K± greenish or blackish, intensifying, N+ red, this and outer edge of exciple ± with scattered, minute, interspersed crystals; hymenium 50–70 μm tall, usually tinged bluish; hypothecium massive, downwardly extended, pale. Ascospores 10–15 × (2.5-)3–5 μm, elongate-ellipsoid with abruptly rounded ends. Cortex and medulla Pd− to Pd+ orange, K− to K+ yellow to orange, UV− or faint bluish white (atranorin, 2'-*O*-methylperlatolic acid, 2 unidentified depsides, norstictic and fumarprotocetraric acids).

On shaded, dry sides of ± vertical, siliceous, coastal rocks and walls, more rarely on dry, sheltered, vertical rock faces or underhangs in inland montane areas (to 640 m); rather local. N. & W. British Is. W. Europe, North America.

A very variable species often forming extensive mosaics.

UMBILICARIA Hoffm. (1789)

Thallus foliose, single- or multi-lobed, dorsiventral, attached at a single point by a central or excentric strand (umbilicate; Fig. 36g), soft, pliable and somewhat leathery when wet, brittle when dry. **Upper surface** pale grey-brown to black, smooth to warted-areolate, sometimes folded or reticulate-ridged with a raised central area, margins sinuous, entire or incised, rhizinate or not. **Lower surface** smooth or warted-areolate, sometimes pitted, rarely with plate-like structures, black or pale brown-pink; rhizines present or absent, not attaching thallus. **Isidia** and **soredia** occasionally present. **Upper cortex** at times overlain by an amorphous zone. **Photobiont** chlorococcoid. **Medulla** loose or compact, not always clearly differentiated from lower cortex. **Ascomata** apothecia, ± irregular, immersed, sessile or stalked; disc black, flat or ± convex, mostly gyrose (Fig. 28d), occasionally smooth with a central, protruding button of sterile tissue. **Thalline exciple** absent. **Asci** 8-spored, elongate-clavate, thick-walled, apical dome I+ blue. **Ascospores** ellipsoid, simple and colourless or muriform, becoming brown. **Conidiomata** pycnidia, uni- or multi-loculate. **Conidiogenous** cells in chains, enteroblastic, acrogenous or pleurogenous. **Conidia** shortly cylindrical. **Thallospores** present in non-isidiate, non-sorediate, rarely fruiting species, originating from lower cortex or on rhizines, single-celled or of several cells. **Chemistry:** gyrophoric acid frequent, umbilicaric, stictic and norstictic acids occasional.

Lasallia differs in the pustulate thallus, 2-spored asci, and larger ascospores.

LITERATURE: Hestmark (1990), Llano (1950).

1 Upper surface without isidia 2
Upper surface with scattered or crowded, rounded to flattened, isidia ..... **3. *deusta***

2(1) Soredia or coarse granules present on margins 3
Soredia and coarse granules absent 4

3(2) Rhizines abundant, pale; lower surface ± smooth, pale brown-black ..... **5. *hirsuta***
Rhizines sparse, dark or absent; lower surface markedly warted-areolate, dark brown-black ..... **4. *grisea***

4(2) Lower surface with abundant rhizines, scattered or forming a dense black felt 10
Lower surface without rhizines 5

5(4) Upper surface wrinkled or coarsely warted-areolate 6
Upper surface ± smooth or minutely areolate, sometimes with a prominent, ± ridged central area 7

6(5) Lower surface black; thallospores single-celled ..... **7. *nylanderiana***
Lower surface brown-black; thallospores absent ..... **6. *hyperborea***

7(5) Upper surface grey-brown to black, sometimes rough with a raised, sharply ridged central area 8
Upper surface red-brown, smooth, without a central, raised, ridged area, multi-lobed ..... **8. *polyphylla***

8(7) Marginal cilia absent 9
With abundant marginal cilia ..... **2. *cylindrica***

9(8) Thallus perforate or lacerate towards margins; lower surface with scattered shred-like lamellae ..... **12. *torrefacta***
Thallus not perforate or lacerate towards margins; lamellae absent on lower surface ..... **10. *proboscidea***

10(4) Upper surface grey to grey-brown; lower surface not densely black-felted 11
Upper surface red-brown, shining; lower surface densely felted with short black rhizines ..... **9. *polyrrhiza***

11(10) Apothecia mostly abundant, smooth with protruding central sterile button; thallus monophyllous 12
Apothecia absent or gyrose; thallus single- or multi-lobed 13

12(11) Lower surface grey-brown to black, coarsely warted-areolate ..... **11. *spodochroa***
Lower surface pale-grey, often pinkish, smooth or slightly warted-areolate ..... **1. *crustulosa***

13(11) Thallus not perforate or lacerate towards margins; shredded lamellae absent on lower surface 14
Thallus perforate or lacerate towards margins; lower surface with scattered, flattened, shredded lamellae ..... **12. *torrefacta***

14(13) Marginal cilia present; central areas of thallus not ridged and pruinose ..... **2. *cylindrica***
Marginal cilia absent; central areas of thallus ± ridged and pruinose ..... **10. *proboscidea***

1. U. crustulosa (Ach.) Frey (1931)

Thallus 3–5(-8) cm diam., single-lobed; upper surface pale to dark grey, sometimes with a brown or pink tinge, mostly smooth, at times minutely cracked-areolate, ± pruinose and partly folded towards centre; lower surface usually pale grey-white, tinged pink, smooth, with abundant, mostly dark brown-black, densely packed rhizines. Apothecia to 4 mm diam., frequent, smooth with a central protruding 'button' of sterile tissue. Ascospores 18–24 × 11–21 μm. Thallus Pd−, K−, KC+ red, C+ red (gyrophoric acid).

On siliceous rocks; locally frequent. N.W. England (Lake District). W. Alps, France, Pyrénées, Scandinavia.

Characterized by the grey, ± smooth thallus which often has smooth apothecia with a central protruding 'button' of sterile tissue. *U. spodochroa* **(11)** has a rough, dark brown-black lower surface and the upper surface is usually whitish brown.

2. U. cylindrica (L.) Delise ex Duby (1830)

Thallus 2–5(-8) cm diam., single- or multi-lobed, lobes rounded, ± wavy and contorted with long, to 5 mm, branched, black marginal cilia; upper surface pale to dark grey, sometimes tinged brownish, smooth or wrinkled, the centre often slightly ridged, folded and ± white-pruinose; lower surface rather smooth, grey-white, often tinged pink-brown towards centre, with sparse or densely packed rhizines, especially towards margins. Apothecia frequent, with black, gyrose discs. Ascospores 9–15 × 3–9 μm. Medulla Pd+ orange or Pd−, K+ red or K−, KC−, C− (± norstictic acid).

On siliceous rocks in exposed, well-lit sites, predominantly montane; locally abundant. N.W. England (Lake District), N.E. England (Cheviots), N. Wales, Scotland, scattered in Ireland (particularly Connemara). Europe, Arctic, North America, Himalayas, Australia (Tasmania), New Zealand.

Very variable, occasionally almost devoid of marginal cilia and rhizines, or alternatively densely rhizinose with abundant marginal rhizines. Morphs with coarsely toothed or lacerate margins occur.

3. U. deusta (L.) Baumg. (1790)

Thallus 2–4(-8) cm diam., single- to multi-lobed, margins often downwardly recurved, ± rounded and entire or variously lacerate; upper surface brown-black with abundant, irregularly scattered or clustered, cylindrical, branched or flattened-foliose isidia; lower surface brown-black, with deep depressions; rhizines absent. Apothecia not observed in British material. Medulla Pd−, K−, KC+ red, C+ red (gyrophoric acid).

On siliceous rocks, often near water, upland, rarely lowland (Leicester). England, Wales, Scotland. Europe, Arctic, North America, U.S.S.R., Asia.

Characterized by the recurved margin, ± flattened isidia on the upper surface, and absence of apothecia.

4. U. grisea Hoffm. (1796)

Thallus 1–5(-8) cm diam., single-lobed, flat or wavy, lobe margins rounded or lacerate; upper surface pale grey often tinged brown and whitish towards centre, distinctly rimose-areolate, pruinose or scabrid, becoming sorediate at margins; lower surface dark brown-black, markedly warted-areolate; rhizines very sparse or absent. Apothecia not observed in British material. Medulla Pd−, K−, KC+ red, C+ red (gyrophoric acid).

In a seepage track on siliceous coastal rocks; very rare. Channel Is. (Jersey, St. Brelade's Bay). Sweden, France, Portugal.

Similar to *U. hirsuta* **(5)** but usually thicker and more rigid and the lower surface is markedly warted-areolate, dark brown-black and the rhizines sparse or absent.

5. U. hirsuta (Swartz ex Westr.) Hoffm. (1796)

Thallus 2–5 cm diam., single-lobed, lobe margins often downwardly recurved, ± rounded or lacerate; upper surface grey or brownish grey, finely cracked-areolate, becoming finely sorediate at the margins; lower surface pale brown to black, usually darker towards centre, smooth or slightly areolate-warted with numerous brown-black rhizines. Apothecia not observed in British material. Medulla Pd−, K−, KC+ red, C+ red (gyrophoric acid).

In crevices and seepage channels on siliceous rock-faces; very local. N. Wales, E. Scotland. Europe, Greenland, North & South America. Himalaya.

Similar to *U. grisea* **(4)** but the lower surface is typically paler, has numerous rhizines, is less strongly cracked-areolate below, and the thallus is typically thinner and less rigid.

6. U. hyperborea (Ach.) Hoffm. (1796)

Thallus 2–4(-6) cm diam., single-lobed, rather thin, margins rounded or lacerate; upper surface grey-brown to dark brown, at times paler towards centre, coarsely wrinkled and folded or ± warted-granular, except towards margins which are often smooth or areolate; lower surface uniformly brown-black, smooth or occasionally dimpled; rhizines mostly absent or very sparse. Apothecia (Fig. 28d) frequent; disc black, gyrose. Ascospores 7–20 × 3–8 μm. Medulla Pd−, K−, KC+ red, C+ red (gyrophoric acid).

On siliceous boulders on mountains, usually above 860 m; local. N. Scotland (Highlands). Fennoscandia, Switzerland, Iceland, Spitzbergen, Greenland, North & South America, New Zealand.

Characterized by the coarsely wrinkled lobes, the typically dark brown upper surface and uniformly brown-black lower surface. *U. arctica* (Ach.) Nyl. (1859) is similar but the thallus is characteristically thick and rigid and the lower surface is grey-brown or pinkish and only distinctly darker around the holdfast; although recorded last century from granitic rocks in Scotland (Grampian mtns) and in S.W. Ireland (Kerry, Brandon Mtn.) the material in question probably originated from outside the British Is. For differences see also *U. nylanderiana* **(7)**.

7. U. nylanderiana (Zahlbr.) Magnusson (1937)

Like **6**, but differs in the very dark or black lower surface, the smaller ascospores (7–14 × 5–7 μm) (apothecia not observed in British material) and the single-celled thallospores on the rhizines.

On a stone wall; very rare. S.E. Scotland (Selkirk). Spitzbergen, Scandinavia, Germany, Switzerland, Italy, North & South America.

8. U. polyphylla (L.) Baumg. (1790)

Thallus 2–4 cm diam., multi-lobed, rather thin; lobe margins ascending, wavy, irregularly lacerate; upper surface dark brown to brown-black, smooth, ± shining; lower surface black, smooth, without rhizines. Thallospores present, composed of several cells. Apothecia very rare; disc black, gyrose. Ascospores 12–19 × 4–7 μm. Medulla Pd−, K−, KC+ red, C+ red (gyrophoric acid).

On ± nutrient-enriched siliceous boulders in exposed, well-lit sites in montane situations; locally abundant. Upland British Is., extending to C. England (Leicester, Charnwood Forest). Widespread in Europe, Greenland, North & South America, Falkland Is., E. South Africa, Asia, Australia, New Zealand.

The most SO_2-tolerant species of the genus.

9. U. polyrrhiza (L.) Fr. (1825)

Thallus 2–5 cm diam., single- or multi-lobed, rather thick; lobes wavy or flat, occasionally folded, margins ± entire; upper surface dark brown, sometimes tinged reddish, smooth, with abundant, erect, marginal rhizines which may also arise from surface cracks; lower surface black, densely felted with black, often branched, rhizines. Thallospores present, composed of several cells. Apothecia very rare; disc black with radiating ridges from a central point; true exciple becoming excluded. Ascospores 8–9 × 3.5–6 μm. Medulla Pd−, K−, KC+ red, C+ red (gyrophoric acid).

On ± nutrient-enriched, siliceous rocks in exposed, well-lit sites in montane situations; local. S.W. (Dartmoor) & N. England, N. Wales, Scotland (throughout Southern Uplands and Highlands). Fennoscandia, France, Germany, Portugal, North & South America, Falkland Is.

10. U. proboscidea (L.) Schrader (1794)

Thallus (Fig. 36g) 2–4(-6) cm diam., single-lobed, rather thin and fragile; upper surface grey-black or dark brown-black, ± rugose, strongly wrinkled and reticulately ridged towards centre which is often elevated and white-pruinose; lower surface pale brown-grey to dark brown, smooth, occasionally with sparse, scattered rhizines. Apothecia rather frequent; disc black, gyrose. Ascospores 10–17 × 3.5–7 μm. Medulla Pd+ yellow or Pd−, C+ red (gyrophoric and ± norstictic acids).

On siliceous boulders in well-lit, exposed sites in montane situations; locally frequent. N. England, Wales, Scotland. Arctic, Scandinavia, North America, Russia, Siberia, China, Japan, Australia.

Characterized by the ridged, boss-like centre which is often pruinose.

11. U. spodochroa (Ehrh. ex Hoffm.) DC. (1805)

Thallus 2–5 cm diam., single-lobed; upper surface grey-brown, often whitish, smooth; lower surface pale brown-black, rough, warted-areolate, with abundant, branched, dark brown-black rhizines. Apothecia frequent, particularly towards margins of thallus; disc black, smooth with a thick rim and a central protruding 'button' of sterile tissue. Thallus Pd−, K−, KC+ red, C+ red (gyrophoric acid).

In seepage tracks on sheltered boulders close to a sea loch. N. Scotland (Sutherland), S.E. Ireland (Wicklow ?extinct). Europe, China.

Resembles *U. crustulosa* **(1)**, which has a smooth non-rhizinate, pale grey-white lower surface and a grey upper surface without a whitish tinge.

12. U. torrefacta (Lightf.) Schrader (1794)

Thallus 2–4 cm diam., single-lobed, often crumpled, the margin often finely dissected and perforated by small holes, which may also occur (but less frequently) towards the centre of thallus; upper surface brown, smooth or warted-areolate; lower surface pale brown to black, with irregular, thin, ± flattened, lamellae which are often lacerate or perforate. Apothecia frequent; disc black, gyrose. Ascospores 7–12 × 5–7 μm. Medulla Pd+ orange or Pd−, K + yellow or K− (stictic or gyrophoric acids).

On ± nutrient-enriched, siliceous boulders in exposed, well-lit sites in montane areas; often locally abundant. S.W. (Dartmoor) & N. England, N. Wales, Scotland. Europe, North America, China.

USNEA Hill (1753)

Thallus shrubby, erect to longly pendent or decumbent, in most species attached by a compact holdfast, branching irregular, branches rounded to angled, even and smooth or ± pitted with a network of ridges, shiny or matt, occasionally with annulations or jointed, then divided into segments which may be swollen and ± constricted at their point of attachment, with or without fibrils, tubercles, papillae, pseudocyphellae, isidia, pseudoisidia and soredia. **Outer cortex** present. **Photobiont** trebouxioid. **Medulla** lax or compact. **Central axis** a cartilaginous strand of solid, tough, compacted, longitudinally arranged hyphae, white, rarely pinkish or yellow. **Ascomata** apothecia, lateral or terminal; disc rounded, flat or ± concave. **Thalline exciple** persistent, often with short, ray-like projections. **Asci** 8-spored, elongate-clavate, *Lecanora*-type. **Ascospores** simple, ellipsoid, colourless. **Conidiomata** pycnidia, mainly terminal or subapical. **Conidia** simple, colourless, bacilliform, bifusiform, thickened at one or both ends, straight or curved. **Chemistry**: usnic acid in the cortex of all species, the medulla with range of depsides, depsidones or fatty acids. **Ecology**: on trees and rocks in well-lit situations.

The 'beard lichens' somewhat resemble species of *Evernia* and *Ramalina*, which also grow as grey-green or yellow-green tufts on trees, but these in most cases have a flattened, less branched thallus and lack an elastic, cartilaginous central axis. Some *Usnea* species are extremely variable and in many instances t.l.c., ecological and distributional data are essential for identification.

Specimens assigned to *U. foveata* Vainio (1928) are referable to *U. hirta* and *U. wasmuthii*. Reports of *U. cavernosa* Tuck. (1850) are based on specimens probably collected outside the British Is.

LITERATURE: Clerc (1984a,b, 1987a,b).

1 Thallus yellowish, green-grey, dark grey, only turning brown when moribund or in the herbarium 2
Thallus red-brown or grey-green mottled red-brown at least on the main branches **15. *rubicunda***

2(1) Older parts of thallus not or only slightly constricted at the site of attachment of the main and lateral branches, never in sausage-like segments; surface variously ornamented with tubercles, papillae, soralia and isidia; medulla Pd−, or yellow to rust-red 3
Older parts of thallus of conspicuous, inflated, sausage-like segments; surface smooth, mainly without ornamentation; medulla Pd+ rust-red .... **1. *articulata***

3(2) Axis or adjacent part of medulla pigmented, rarely white 4
Axis and medulla always white, never pigmented 5

4(3) Axis and inner medulla pale to primose-yellow, CK−; thallus soft, pliant, ± erect, shrubby, pale green; medulla lax, cortex thin, fragile; psoromic acid present ... **19. *wirthii***
Axis ± rose-pink, CK+ yellow-orange; thallus scrawny, harsh, ± pendulous, dark grey-green; medulla compact, cortex thick, tough; barbatic, diffractaic acids present .... **2. *ceratina***

5(3) Thallus without soralia 6
Thallus with soralia 7

6(5) Thallus with abundant papillae and fibrils; remaining erect when moist, frequently fertile; medulla K+ yellow .... **8. *florida***
Thallus without papillae but with ± numerous spine-like isidia; very flaccid when wet; rarely fertile; medulla K− (rarely ± red) .... **13. *hirta***

7(5) Medulla and/or soralia Pd+ rust-red, K− 8
Medulla and/or soralia Pd−, K−, or Pd+ yellow to orange-red and K+ yellow-red 9

8(7) Soralia conspicuous, excavate to irregularly efflorescent-excrescent, often exceeding width of branches, without isidia; cortex thin, medulla lax; lateral branches narrowed at point of attachment .... **11. *glabrata***
Soralia very small, punctiform, superficial, smaller than half width of branches, with fragile isidia when young; cortex thick, medulla compact; lateral branches not constricted at point of attachment .... **18. *subscabrosa***

9(7) Thallus pendulous; main branches ± parallel 10
Thallus ± erect, shrubby or only subpendulous; main branches ± diverging 12

10(9) Soralia absent, or when present, punctiform, inconspicuous; thallus pendulous throughout 11
Soralia conspicuous, pale or whitish, rounded, weakly excavate; isidia absent; thallus often ± bushy below with several, pendulous, flexuose, extended main branches [see also *U. subscabrosa* **(18)**] .... **12. *glabrescens***

11(10) Fibrils few or absent along main branches, irregularly distributed; soralia tuberculate; isidia sparse or absent towards the base of main branches; medulla ± lax; branches often with annular cracks .... **3. *chaetophora***
Fibrils usually numerous, ± regularly distributed along the main branches (fish bone-like); soralia rare, not tuberculate; isidia ± dense on main branches especially towards the base; branches without annular cracks .... **6. *filipendula***

12(9) Thallus ± erect and stiff when wet; often with ± discrete soralia; isidia if present, in clusters; medulla compact or lax; main branches round in section 13
Thallus limp and flaccid when wet; without soralia; usually with numerous, irregularly scattered or crowded, spine-like isidia; medulla lax; main branches ± indented-honeycombed, not round in transverse section .... **13. *hirta***

13(12) Branches constricted at their point of attachment or fusiform, and/or segments ± swollen, articulated 14
Branches never constricted, segments cylindrical, never swollen 17

14(13) Soralia conspicuous, larger than half of branch diameter, infrequent, often ± discrete, isidiate or not when young 15
Soralia minute, punctiform, smaller than half of branch's diameter, numerous, often coalescing to form larger soralia-like patches, isidiate when young .... **4. *cornuta***

15(14) Soralia isidiate, at least when young, sparse; branch apices straight or curved; medulla K+ yellow to yellow-orange (norstictic and/or stictic acids) 16
Soralia never isidiate even when young; branch apices often markedly flexuose, twisted, resembling skeletal fingers, crowded with large soralia; medulla K+ yellow to red (salazinic and bourgeanic acids) .... **5. *esperantiana***

16(15) Basal part pale, usually with numerous, pale annular cracks, especially near the holdfast; cortex matt (transverse section); medulla usually dense; soralia irregular in shape, usually richly isidiate; branching isotomic-dichotomous .... **7. *flammea***
Basal part jet-black, without conspicuous pale annular cracks; cortex shiny (transverse section); medulla lax; soralia regular, circular; isidiate only when young; branching distinctly sympodial .... **9. *fragilescens***

17(13) Basal part jet black, at least for 1–2 mm with a few inconspicuous annulations; fibrils distributed over entire thallus; papillae usually numerous; lobaric acid absent 18
Basal part concolorous with the branches or paler, never conspicuously blackened, with numerous, conspicuous annulations; fibrils absent except towards apices; papillae sparse or absent; lobaric acid mostly present .... **7. *flammea***

18(17) Isidia never present; soralia ± excavate 19
Isidia usually present (at least on juvenile soralia), sometimes sparse, very fragile; soralia flat to ± tuberculate; 20

19(18) Thallus erect, rarely sub-pendulous, laterals numerous, of ± equal length and appearing like a fish-bone; soralia irregular, often crowded and coalescing, deeply excavate, reaching central axis, stictic acid (in quantity), norstictic and ± diffractaic acids. .... **10. *fulvoreagens***
Thallus ± densely branched towards the base with a few, extended, unbranched and mainly lateral-free, pendulous, flexuose main branches; soralia very regular, rounded, scarcely excavate, usually not reaching central axis, never exceeding ½ branch diam. (or only on very thin branches at the extremities); either with norstictic acid (in quantity) and ± salazinic or stictic acids, or psoromic acid [see also *U. esperantiana* **(5)**]. .... **12. *glabrescens***

20(18) Thallus blackened at base only, or if above without annular cracks; soralia various, but rarely transversely ellipsoid; soralia Pd− to yellow or ± instant yellow-orange 21
Thallus ± extensively blackened towards base and often also on main branches, here and above with numerous, distinct annulations and cracks; soralia flat or slightly excavate, transversely ellipsoid or irregularly rounded; soralia Pd− to + yellow (after one minute); salazinic acid only .... **14. *madeirensis***

21(20) Soralia slightly tuberculate and raised, irregularly rounded; isidia abundant even when mature; holdfast without small, longitudinal cracks; soralia Pd+ yellow-orange (instantly), K+ yellow-orange (thamnolic acid), or Pd−, K− (squamatic acid, rare); occasionally fertile .... **16. *subfloridana***
Soralia flat to slightly excavate, when mature ± oval, longitudinally stretched; isidia present only on young soralia; holdfast with small (×50) longitudinal cracks; soralia Pd− to yellow (after one minute) (salazinic and barbatic acids); not known fertile .... **17. *wasmuthii***

1. U. articulata (L.) Hoffm. (1796)

Thallus to 1 m or more in favoured sites, pendent or scrambling, main branches to 3 mm diam., the older portions strongly inflated, ± cylindrical, forming sausage-like segments (Fig. 26a), young or terminal branches less inflated, adjacent segments interconnected by a cartilaginous axil or with annulations, sparingly branched at base, more richly irregularly branched towards apices, with few, ± flexuose lateral branches, often in a loose, flexuose tassel; cortex thin (5–7% of diam.), surface pale green-grey, uniformly smooth or rarely, in part, minutely roughened-papillate, or ± with scattered, slightly raised, white, comma-like pseudocyphellae; medulla wide, lax, cottony-arachnoid, especially in inflated parts. Apothecia unknown in Britain. Medulla Pd+ red, K−, C− (fumarprotocetraric, and ± protocetraric acid).

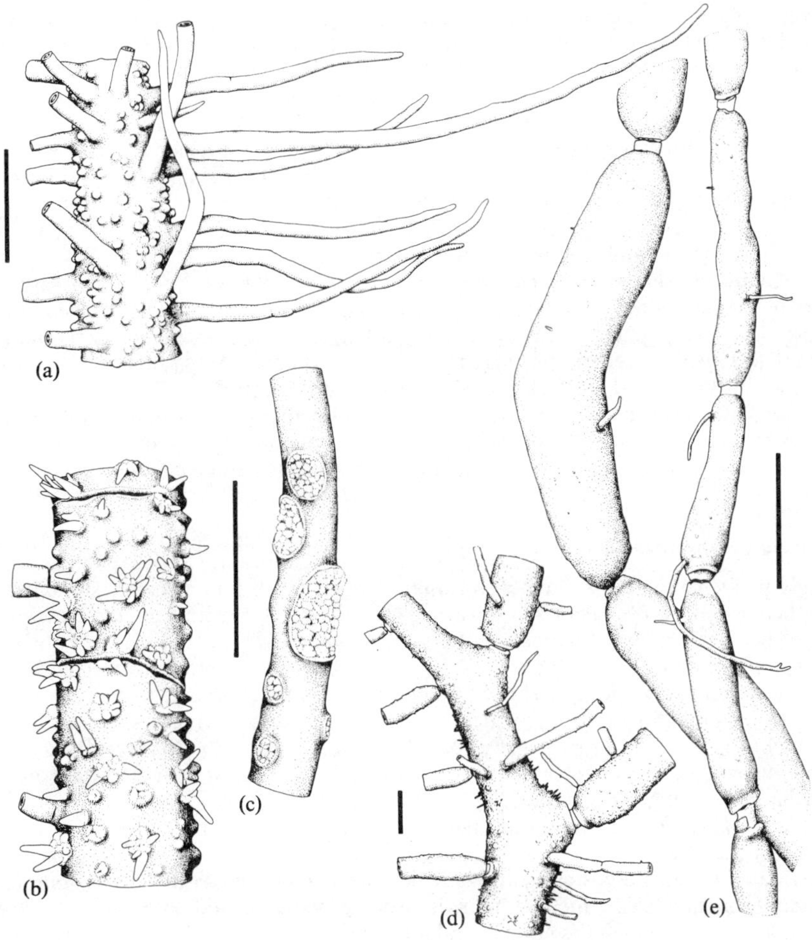

Fig. 26. Thallus features in *Usnea*. (a) Papillae and fibrils (*U. florida*); (b) tubercles giving rise to clusters of pseudoisidia (*U. rubicunda*); (c) soralia (*U. glabrescens*); (d) inflated branches constricted at base; (e) main branches constricted into segments (*U. articulata*). Scales a-d = 1 mm, e = 5 mm.

In well-lit, xeric situations, especially on inclined branches in crowns of trees, or rarely scrambling over low vegetation, also on the ground in a few coastal sites; locally abundant. S.W. England, Wales, N. to Is. of Man, Ireland, formerly extending to N. England. Europe, E. & N. Africa, Saudi Arabia, Atlantic & Mediterranean Islands.

Characterized by the conspicuous articulations of the older, main branches producing a "string of sausages" effect. One of the most sensitive species to sulphur dioxide air pollution known and absent at levels exceeding 30 $\mu g\ m^{-3}$. Young specimens which lack the characteristic articulations could be confused with *U. flammea* **(7)** which is more erect, has a bottle-brush appearance and a K+ yellow-red medulla.

2. U. ceratina Ach. (1810)

Thallus normally ± stiffly pendent, often long and straggling, to 30 cm long, juvenile specimens more erect and tufted, rigid and coarse, pale or ± blackened at the base, main branches coarse, to 1.5 mm diam., of uniform thickness, only tapering at apices, typically rather straight when well developed or somewhat twisted, often ± angular; branching usually sparse, occasionally with numerous irregularly arranged lateral branches, with scattered or sometimes abundant single or clusters of thin fibrils to 1 cm, arising at right-angles to branches, surface dark grey-green, conspicuously coarsely warted with numerous white, raised, ± hemispherical tubercles, these often bursting to form coarsely granular, ulcerose soralia, and, occasionally, secondarily corticate isidia, main branches also with smaller, low, translucent, paler papillae; medulla and axis very compact, pale to deep pink, rarely white. Apothecia rare, lateral on main branches, disc with few marginal rays. Medulla Pd −, K −, CK + yellow-orange, C + yellow-orange (diffractaic and barbatic acids and ± several accessory substances).

On acid-barked old trees, particularly *Quercus* and *Fagus* in relict woodlands and parklands where it is characteristic of well-lit sites on trunks of ancient trees at margins and in glades, also often on inclined or horizontal trunks and boughs in thin tree canopies; locally frequent. S. & S.W. England, Wales, Scotland (Dumfries, Lochwood), Ireland. Europe, North America.

Characterized by the scrawny, coarse, scabrid texture, ± coloured axis, and conspicuous white tubercles often bursting with coarse soredia. Young plants can be identified by their pink medulla (almost white when the pigment is in low concentration) and the presence of barbatic acid (CK + yellow-orange). Richly branched morphs may resemble *U. filipendula* **(6)** but that species is usually more regularly branched, fish-bone-like, lacks conspicuous, white, hemispherical tubercles which become eroded and sorediate, has a K + red medulla and is usually abundantly spinose-isidiate on the main branches, especially towards the holdfast. Small, erect, pollution-damaged specimens superficially resemble *U. cornuta* **(4)** which, however, has a lax medulla, persistently colourless axis and a different chemistry.

3. U. chaetophora Stirton (1883)

Thallus longly pendent, to 30 cm, branches rather slender, 0.3–0.8 mm diam., similar in thickness throughout their length, only narrowing towards the apices, often with annular cracks forming ± cylindrical segments which may become slightly swollen and narrowed towards their ends; branching frequent towards the base, less so towards the apices, fibrils rarely present, irregularly distributed, surface dark grey-green, becoming darker in older parts, smooth, shining, with ± evenly scattered, white, punctate or comma-like tubercles which may become ulcerose and produce isidia, numerous low papillae in patches occasionally present on the main branches occasionally elongating into isidia; soralia, when present, distinctly tuberculate, punctiform, smaller than half the branch diameter; medulla ± lax, white. Medulla Pd + orange, K + red, C −, (salazinic acid).

On *Betula* and *Larix*, often in rather open situations; very rare. Scotland (Highlands, particularly E. Grampian mtns). Europe.

Characterized by the numerous annulations which are not swollen as in *U. articulata* **(1)**, a ± lax medulla, punctiform tubercles which sometimes become nodulose and sorediate. *U. chaetophora* is similar to *U. filipendula* **(6)** which has more lateral fibrils and is often densely isidiate towards the base of the main branches near the holdfast; its status is in need of critical investigation.

4. U. cornuta Körber (1859)

U. inflata (Duby) Motyka (1937)

Thallus 2–6(-10) cm, ± erect, tufted, rarely subpendent, main branches to 1.5 mm diam., inflated, with few to numerous, short, lateral branches clearly constricted and often annulated at the point of attachment to the main stems, mostly curved, often widely divergent to 90° in terminal branches, branching anisotomic dichotomous, ± isotomic towards apices, surface dull grey-green becoming ± dark brown in the herbarium, base usually concolorous or paler, rarely blackened, cortex often smooth and ± shiny-glassy in lower parts of main branches, more rarely corrugate-areolate, sparsely to densely papillate above; scattered or clustered isidia-like structures present together with ± roughened, whitish grey, erose, ± erumpent soralia; soralia minute, 60–200 μm, numerous, often becoming confluent and spreading; soredia coarsely granular, often mixed with a few to many secondarily corticate isidia; papillae absent on the upper branches; medulla very lax. Two chemotypes occur: (a) Pd + yellow to orange-red, K + blood-red (salazinic, ± protocetraric and ± constictic acids); (b) Pd + yellow-orange, K + yellow to orange-red (stictic, norstictic, menegazziaic, constictic and ± salazinic acids); the two chemotypes are equally represented.

On mossy, woodland or wayside trees, more rarely siliceous rocks; widely distributed and locally frequent. W., S.W. and E. (rare) British Is. W. Europe.

This common and variable species is characterized by the shrubby habit, the often divergent and claw-like appearance of the main and terminal branches, the dark green colour (yellow-green in very exposed sites), the ± inflated, widely divergent branches which are markedly constricted and annulated at their point of attachment, the presence of small papillae, the minute soralia which coalesce to form extensive soralia-like areas with granular soredia, few to many, fragile, secondarily corticate isidia and spinules, and, occasionally, naked, smooth areas. *U. fragilescens* **(9)** differs in the size and density of soralia, blackened base and chemistry.

5. U. esperantiana Clerc (1992)

U. fulvoreagens auct. brit. p.p., non (Räsänen) Räsänen (1935)

Thallus 4–6 cm, erect, rarely subpendent, shrubby, main branches 1–2 mm diam., conspicuous, lateral branches disproportionately thinner, 0.3–0.5 mm diam., numerous, entangled, often irregularly branched, markedly flexuose and contorted at apices, surface grey-green, not blackened at the base, main branches thickly scabrid-papillate, fibrils numerous over the whole thallus, secondary branches smooth, becoming ± eroded towards the tips with numerous, ± rounded, white, flat to somewhat excavate, irregularly shaped soralia, often recurved and resembling skeletal fingers, especially when heavily sorediate; medulla somewhat lax to compact. Apothecia very rare. Medulla Pd+ red, K+ red, C− (salazinic, constictic and bourgeanic acids).

On twigs and small branches of broad-leaved trees, hedgerows, carrs, orchards and garden shrubs; locally frequent. S. & W. England, Ireland. Europe.

A small, distinctive, rather delicate species with twisted terminal branches which are often heavily sorediate and eroded.

6. U. filipendula Stirton (1881)

Thallus to 30 cm long, pendent, often rather narrow and composed of 4–6 main branches arising from the point of attachment, branches rather thin, 0.2–0.6 mm diam., mostly even in thickness, usually with many short, to 1 cm long fibrils arising at right-angles from the main branches, regularly but unevenly distributed along the entire length of the branches in a fish-bone-like arrangement, branching variously sympodial or dichotomous, upper surface grey-green, often darker in the older parts and blackened towards the base, main branches typically densely papillate and with tubercles, these may produce isidia or isidiate soredia and can also occur on side branches where they are usually less frequent, but are absent towards the apices which narrow to a point, often laterally fractured into ± cylindrical segments. Medulla Pd+ orange, K+ red, C− (salazinic acid).

On *Pinus*, *Larix* and *Betula*, chiefly in exposed mountain woods; locally abundant. S.W. & N. England, Wales, Scotland, N. Ireland. Europe.

A very variable, pendulous species differing in the abundance and shape of the tubercles and papillae, the formation of isidia and soredia, the frequency of associated fibrils, the degree of branching, the presence of cylindrical segments, and the colour which may range from pale yellow-grey to dark green-black. Specimens with few fibrils are often difficult to separate from *U. chaetophora* **(3)**. Attempts to define new entities from this polymorphic species have proved unsuccessful; in the British Isles the various morphs appear to represent a continuum.

7. U. flammea Stirton (1881)

Thallus to 3(-8) cm long, erect to subpendulous, sometimes scrambling, forming sparsely branched, ± lax to compact, dense tufts, main branches several, often arising 3–5 mm above the annulated stalk above the holdfast, rather stout, to 1 mm diam., sparingly to richly branched, the branches often originating at ± the same level forming a ± brush- or tassel-like tuft of finer secondary branches towards the apices, surface matt to slightly shiny, pale green- to yellow-grey, not blackened at the base, surface matt or somewhat smooth and shining, usually with ± numerous, pale annular cracks, especially towards the holdfast, forming non-constricted, but sometimes worm-like segments, scarcely narrowed at their point of attachment; papillae often scarce or absent in patches on main branches, when present warted, sometimes eroded to form soralia; soralia flat or slightly tuberculate, efflorescent, rounded, with clusters of very fragile, easily abraded isidia; medulla moderately compact to ± lax, yielding to the finger nail. Apothecia very rare. Medulla Pd+ yellow-orange, K+ yellow to red-orange, C− (stictic, menegazziaic, constictic and ± norstictic and ± lobaric acids).

On mossy trees, decorticated wood, rocks, soil, shrubs and *Calluna* stems, especially in wind-swept coastal areas; widespread and locally frequent. S. & W. Britain, Ireland. France, Spain, Portugal, Mediterranean.

Characterized by the primarily erect habit, pale annulations and pale holdfast area and the irregular branching and pollarded appearance arising from the coarse main branches and finer secondary branches terminating at about the same level. Sometimes the main branches may arise together some distance (to 1 cm) from the holdfast. Lobaric acid, which occurs in about 80% of the specimens, has not been recorded in any other European *Usnea* species. This is the most frequent species on rock faces, especially near the coast.

8. U. florida (L.) Wigg. (1780)

Thallus 2–5(-10) cm tall, ± erect, forming often dense, shrubby tufts, main branches to 1 mm diam., often curved, often with a few inconspicuous annulations, branching irregular, thinner branches often contorted; surface grey-green, blackened at the base, main stems covered with low, inconspicuous, dense papillae and numerous fibrils to 1 cm long (Fig. 26a), mostly arising at right angles from branches and becoming curved. Apothecia frequent, arising at apices of main and major lateral branches; disc 0.5–1 cm diam., at first ± concave, becoming flat, smooth or wrinkled, with abundant, often branched, marginal projections, to 5 mm long. Ascospores 8.5–11 × 5.5–7 μm, ellipsoid. Medulla Pd+ orange, K+ yellow, C− (thamnolic and alectorialic [in apothecia] acids).

On twigs and branches in the canopy of broad-leaved trees, rarely on trunks, also on shrubs, palings and fence-posts, mostly in exposed, well-lit sites; locally frequent. S. & W. British Is. extending locally to Scotland (Dumfries, Lochwood). Europe.

A distinctive and attractive species, typically richly fertile with abundant papillae and fibrils and always lacking isidia and soralia. The fertile counterpart of *U. subfloridana* **(15)**; however, occasionally richly fertile specimens of the latter with only a few soralia occur, suggesting a continuum exists between these two species. Juvenile forms of *U. florida* and *U. subfloridana* are often indistinguishable

9. U. fragilescens Havaas ex Lynge (1921)

Thallus 3–5(-8) cm tall, mostly subpendulous to ± erect, flaccid, main branches 1.5–2 mm diam., main branches originating over the whole of the length with relatively few, thinner, side branches which often arise at ± right angles from these main branches, primary branches much elongate, distinct, thick, often divided into long, somewhat inflated, segments, often constricted and annulate where they join the main stem, giving a characteristic overall combed appearance; fibrils few or absent, axils at an angle of 60–90°; surface very pale grey to green, often ± pellucid when wet, brown to jet-black 1–2 mm at ± attenuated base, shiny (resembling broken glass), smooth; papillae usually numerous, evenly and densely distributed, low, broader than high, often inapparent; soralia even, ± flat to slightly concave, large, conspicuous, distinctively rounded, discrete, widely spaced; isidia irregularly present, mostly absent on mature soralia; medulla lax. Medulla Pd+ red, K+ yellow-red, C− (stictic, menegazziaic and ± cryptostictic, ± norstictic and ± constictic acids).

On broad-leaved trees in moist woodland and *Salix* carr, rarely on rocks; rather rare and local, most frequent in W. Scotland. W. British Is. Oceanic Europe (Norway, France).

Var. *mollis* (Vain.) Clerc (1987) differs in being subdichotomously and more richly branched, especially towards the base, the stalk-like holdfast is not attenuated towards the base, the primary branches are shorter and not fusiform, the soralia are often densely isidiate; also the variety is more consistently on trees and is more frequent than var. *fragilescens* which is known from a single gathering (Cumberland). *U. fragilescens* is a distinctive species, usually with a pallid, subpendulous thallus with combed main branches; it is larger than *U. cornuta* **(4)**, has larger and more discrete soralia and a different branching pattern and a blackened holdfast area.

10. U. fulvoreagens (Räsänen) Räsänen (1935)

Thallus 2–10 cm tall, erect, rarely subpendulous; branching mostly isotomic-dichotomous; main branches to 1.5 mm diam., cylindrical, not inflated, often richly branched with numerous laterals (fibrils) mostly of equal length, giving a fish-bone appearance to the branches, often densely papillose; surface grey-green or yellow-grey, blackened at the basal part; soralia conspicuous, non-isidiate, deeply excavate, typically reaching the central axis, mostly bursting from low pustules (tubercle-like structures but much wider than high) and a tearing off of the adjacent cortex, broader than branch's diameter and irregular in shape. Medulla Pd+ yellow-orange, K+ yellow-red, C−, (norstictic, stictic and ± diffractaic acids).

On wayside trees; rare. Ireland. Europe.

Resembles *U. glabrescens* **(12)**, which usually does not have deeply excavate soralia (except the psoromic acid strain), the mostly circular soralia which are derived from pseudocyphellae which enlarge and not by sloughing of the adjacent cortex; this species also lacks fibrils or has only a few fibrils near the basal part.

11. U. glabrata (Ach.) Vainio (1915)

Thallus to 3 cm tall, ± erect, tufted, main branches 1–1.5 mm diam., inflated, markedly articulate and constricted at point of attachment, richly branched with frequent fibrils ± attached at right angles, surface pale grey-yellow, becoming deep brown in the herbarium, not blackened at the base, cortex smooth, shiny throughout without distinct papillae, tubercles and isidia; soralia conspicuous, mainly situated towards apices, usually exceeding the width of the branch, mostly erumpent, convex-globular, coarsely granular, whitish, sorediate, without isidia; medulla very lax, axis thin. Medulla Pd+ red, K−, C− (fumarprotocetraric, and ± protocetraric acids).

On deciduous trees and shrubs, often in swards of many individuals; very rare, and only recorded this century in E. Kent (Dungeness). E. England, E. Scotland. Europe, North America, Mexico, Asia.

A distinctive, but easily overlooked species characterized by the small size, smooth, shiny surface, lax medulla and coarse, granular soredia. Fumarprotocetraric acid is otherwise only present in the much larger *U. articulata* **(1)**. The branching habit, lax medulla, and articulated lateral branches recall *U. cornuta* **(4)**, but *U. glabrata* differs in the smaller size, absence of papillae, soralia never forming isidia, smoother cortex, and chemistry. The eroded terminal branches are similar to those in *U. esperantiana* **(5)**, but that species has papillose main branches, different soralia, a more compact medulla and a different chemistry.

12. U. glabrescens (Nyl. ex Vainio) Vainio (1925)

Thallus 3–10(-15) cm tall, ± erect at the base, usually becoming ± pendulous towards the apices, main branches to 1.5 mm diam., often very richly branched and crowded towards the base, with long side branches, fibrils few or absent; surface grey-green or yellow-grey, blackened at the base, main branches with evenly spaced, densely papillate, thinner branches, smooth; soralia (Fig. 26c) conspicuous, frequent, rounded, even to slightly tuberculate or regularly shaped and distinctively rounded, never larger than half branch diameter, discrete or occasionally forming paler eroded patches, exclusively sorediate (isidia absent). Three chemotypes occur: (a) medulla Pd+ orange, K+ red, C− (norstictic and stictic acids); or (b) (norstictic and salazinic acids); (c) medulla Pd+ yellow, K−, C− (psoromic acid).

On deciduous trees, in damp situations, especially *Salix* carr; rare and local. Throughout W. British Is. N. & E. Europe.

Differs from subpendulous morphs of *U. subfloridana* **(16)** in the small, rounded, ± excavate soralia always without isidia as distinct from the abrasions left by shed isidia in *U. subfloridana*, and in chemistry. *U. glabrescens* tends to be subpendulous and often bushy from the base with several extended, hanging, little-branched, flexuose, main branches with scattered, distinct, rounded, barely excavate soralia. The chemotype with psoromic acid may merit separate recognition. *U. fulvoreagens* **(10)** also resembles *U. glabrescens*, but the soralia are deeply excavate, irregularly shaped, the terminal branches often twisted and contorted, fibrils numerous over the whole thallus, and the medulla contains norstictic, stictic and ± diffractaic acids; most British reports of this species refer to *U. esperantiana* **(5)**.

13. U. hirta (L.) Wigg. (1780)

Thallus 1–4 cm tall, ± tufted, erect, rarely subpendulous, notably flaccid when wet, main branches to 1 mm diam., angular in cross-section, ± slightly swollen, surface often with incomplete, net-like ridges with ill-defined depressions, richly branched, papillae absent, fibrils few or absent, grey-green, yellow-green or blackish grey, not blackened at base; isidia abundant, scattered or densely clustered, short spinule-like, sometimes arising from the ridges; medulla rather lax. Two chemotypes occur: (a) [common] medulla Pd−, K−, C− (± fatty acids); (b) [rare] medulla Pd+ orange, K+ red, C− (norstictic and ± fatty acids with affinities to the murolic acid complex).

Most frequent on conifers, more rarely on wayside and woodland, acid-barked, deciduous trees, especially *Betula* and *Quercus*, decorticated wood and *Calluna* stems; locally frequent. Predominantly E. & N. British Is., rare in W. & S. England. Europe.

Distinguished by the ± angular main stems, the flaccid and limp habit when wet, absence of papillae or tubercles, abundant, long, spinule-like isidia and the absence of a blackened base. A very variable species consisting of few ± pendulous, contorted strands to small, erect when dry, densely tufted and compact thalli. *U. subfloridana* **(16)** and *U. wasmuthii* **(17)** are not flaccid when wet, have a different chemistry, isidia usually in clusters, have papillae, are without depressions in main branches and have a blackened basal part.

14. U. madeirensis Motyka (1964)

Thallus often ± robust, erect, shrubby to ± pendulous, 6 to 12 cm long, richly irregularly branched, branches rounded, tapered, never constricted at the point of attachment; papillae and fibrils few; surface pale green, main branches often ± coarsely papillate towards the holdfast, matt, basal part of thallus black and here and above with numerous annular cracks; medulla compact, narrow, cortex thick; soralia even, ± flat to slightly tuberculate, raised and then concave, large, conspicuous, punctate to transversely elliptical to irregularly rounded, remaining discrete, moderately dense to dense but rarely becoming confluent; isidia irregularly present, generally absent on mature soralia. Medulla Pd+ yellow-red, K+ yellow-red (salazinic acid).

On deciduous trees, shrubs and rock outcrops in windy situations; very rare. S. England (S. Devon), W. Scotland (Kintyre), recorded last century also from Cornwall and Kent. S.W. Europe, Macaronesia.

Differs from *U. wasmuthii* **(17)** in the soralia type, the more freqent, distinct annulations on the main branches and in the often extended, blackened base and the usually paler thallus colour; *U. wasmuthii* additionally contains barbatic acid. *U. flammea* **(7)** has a smoother cortex, a more lax medulla, is not blackened at the base, and contains stictic acid.

15. U. rubicunda Stirton (1881)

Thallus 3–10(-20) cm long, at first ± erect, sometimes becoming longly pendulous, main branches to 1.5 mm diam.; often rather straight, irregularly branched with few to abundant fibrils to 1 cm long arising at right angles from the branches, surface reddish brown, more rarely ± green-grey with ± numerous red-brown flecks, often irregularly articulate, but segments not conspicuously inflated, tubercles abundant and conspicuous on main branches, often eroded at apices forming coarsely granular soredia which are occasionally intermixed with a few secondarily corticate isidia (Fig. 26b), papillae rather frequent on main branches but absent on fibrils. Two chemotypes occur: (a) medulla Pd+ orange, K+ yellow-red, C− (stictic, constictic, and ± norstictic and psoromic acids); (b) medulla Pd+ orange, K+ red, C− (salazinic and ± norstictic acid).

On deciduous trees and shrubs; locally frequent. Mainly in S. & W. England, Wales, W. Scotland. Europe.

Characterized by the reddish brown colour; pale coloured, shade morphs may resemble *U. ceratina* **(2)** which is Pd−. The salazinic acid chemotype is rare.

16. U. subfloridana Stirton (1882)

Thallus 2–8(-20) cm tall, at first erect, ± tufted, sometimes ± becoming pendulous, main branches to 1.5 mm diam., cylindrical, not inflated, branching irregular, isotomic-dichotomous with numerous laterals of unequal length; surface pale to dark grey-green, often conspicuously blackened at the base, with minute transverse cracking without interconnecting longitudinal cracks (×50), main branches often densely papillose, especially towards the base, occasionally with white, rounded or oval abraded areas, terminal parts and fibrils smooth with scattered papillae and pseudocyphellae giving rise to ± tuberculate, irregularly rounded soralia with coarsely granular soredia, sometimes intermixed with secondarily corticate isidia which tend to persist on mature soralia; medulla dense. Apothecia occasional. Two chemotypes occur: (a) medulla Pd+ yellow-orange (instantly), K+ yellow, C−, UV− (thamnolic, and ± alectorialic acid, especially associated with apothecia); (b) medulla Pd−, K−, C−, UV+ whitish blue (squamatic acid).

On trees, more rarely on mossy rocks, also *Salix* carrs; common. Widespread in British Is.

The most abundant and widespread *Usnea* species in Britain, and also the most air pollution tolerant.

The thallus is very variable in the degree of branching, some individuals being reduced to a single main stem with a few lateral branches, and others being extremely richly branched and tufted. A few terminal branches may be ± pendulous, the rest of thallus remaining erect and shrubby; very rarely entire thalli may be shortly pendulous. The combination of a dense medulla, lack of articulation, blackened base, nature of the soralia and chemistry define this polymorphic species; the squamatic acid race is rare and mostly confined to Scotland. See also *U. wasmuthii* **(17)**.

17. U. wasmuthii Räsänen (1932)

Like **16**, but soralia slightly excavate, ulcer-like, longitudinally stretched (at least when mature), isidiate only when young, basal blackened part of thallus with faint, transverse and longitudinal cracking (× 50) forming a fine rimose areolation. Soralia Pd −, becoming yellowish after one minute (salazinic acid diffusing from inner medulla), medulla K + yellow-orange or red (salazinic and barbatic acids).

On deciduous trees; local. Throughout N. & W. British Is. N. Europe.

Easily confused with *U. subfloridana* **(16)** which has been regarded as a chemotype and with which it often grows; any difficulties in identification are best resolved by t.l.c.; fertile specimens are invariably *U. subfloridana*, *U. wasmuthii* has yet to be recorded with apothecia.

18. U. subscabrosa Nyl. ex Motyka (1937)

Thallus to 4(-6) cm tall, pale green, erect to subpendulous (4–10 cm) or scrambling (up to 25 cm), forming loose, rather irregular tufts, main branches rather stout, hard, to 1 mm diam., richly branched especially towards the ends, lateral branches finer, not constricted at the base; cortex thin, 8–15% of diam., surface pale green- to yellow-grey, not blackened at the base, smooth and shining, at times with inconspicuous annulations; soralia minute, even and flat with respect to cortex, numerous, mainly on terminal branches, isidiate, isidia rapidly becoming abraded; medulla compact. Medulla Pd + red, K −, C − (protocetraric acid).

Amongst low vegetation (*Calluna*) and on rocks; locally frequent. Is. of Scilly. Macaronesia.

A rather coarse, scrambling species with conspicuous main branches with a tendency to form loose, disorganized tufts. It has a dense, compact medulla, a relatively smooth, hard cortex ± mottled with minute, punctiform soralia, bearing short, fragile isidia which are often poorly developed; extensive areas of the cortex may remain smooth and ± shining except for the soralia. *U. glabrata* **(11)** is erect and shrubby, contains fumarprotocetraric acid, has a lax medulla, a smooth, shining cortex and large soralia. The habit recalls *U. flammea* **(7)** which has a semi-lax medulla and contains stictic acid.

19. U. wirthii Clerc (1984)

Thallus to 4 cm long, pale green, yellowish, becoming red-brown in the herbarium, paler towards the base, often with minute purple spots (× 10) on main branches; branches rounded, rather stout, tapered and often ± constricted at the base, distinctly segmented by conspicuous, annular cracks bordered by white rings of medullary tissue on the main branches indicating conspicuous gaps between the segments at the point of attachment and on main stems, papillae numerous, low, inconspicuous on main branches, fibrils sparse or absent; pseudocyphellae numerous on ± eroded papillae, becoming later sorediate; isidia absent or rare; soralia numerous, especially towards the apical parts of thallus, slightly excavate, soredia rather coarse; medulla lax, outer part white, inner part pale lemon to primose yellow adjoining axis, axis yellow-white to yellow. Apothecia unknown. Soralia Pd + yellow, K + brownish yellow, C −; medulla Pd + faint yellow, K + yellow-orange, C + yellow-orange (psoromic and ± 2-*O*-demethylpsoromic acids and ± 2 unidentified substances).

On *Salix* in carrs, rarely on mossy boulders; local. S.W. England (Cornwall), Wales (Pembroke), W. Ireland (Co. Leitrim, Lough Scur). France, Portugal, Spain, Corsica, Madeira.

U. cornuta **(4)** differs in having a white, not pale yellow, inner medulla and axis, containing salazinic or stictic acids and has a different type of soralia. In *U. wirthii* the main and secondary branches may be distinctly inflated and constricted at their point of attachment, with a very lax medulla; the presence of psoromic acid, which is concentrated in the soralia, is diagnostic. It may superficially resemble *U. glabrata* **(11)** which has a sward-forming, smaller thallus, with branches which have a smooth, shining cortex without papillae and large, convex, efflorescent soralia, often wider than the branches; it contains fumarprotocetraric acid and lacks the yellow pigment of the axis and inner medulla.

VERRUCARIA Schrader (1794)

Bagliettoa Massal. (1853), *Protobagliettoa* Servít (1955)

Thallus crustose, superficial or immersed, hyphae often vertical and palisade-like in species on rocks, black ridges (juga) or spots (punctae) present on thallus surface in some aquatic species. **Photobiont** *Coccobotrys*, *Dilabifilum* (*Pseudopleurococcus*), *Heterococcus*, *Myrmecia*, or other genera, unicellular, sometimes interspersed between hyphae in vertical rows. **Ascomata** perithecia, immersed in pits, the thallus or erumpent to superficial, ± globose. **True exciple** colourless to brown or black, often dense and brittle. **Involucrellum** present in most species. **Hamathecium** of periphyses, lining the ostiole, and pseudoparaphyses-like filaments gelatinized early leaving mucilage and oil globules, absent in mature ascomata. **Asci** clavate to elongate-ellipsoid, with two distinct wall layers, discharge often obscure, *Verrucaria*-type, or the asci deliquescing leaving spore packets enclosed in mucilage, 8-spored. **Ascospores** subglobose to ellipsoid or somewhat fusiform, colourless, thin-walled, smooth, non-septate (sometimes with "pseudosepta"). **Conidiomata** unknown in most species. **Chemistry**: no lichen products have been detected. **Ecology**: on calcareous and siliceous rocks, building materials, compacted sand and soil, including periodically inundated maritime and freshwater species, very rarely on bark, cosmopolitan but optimally developed in temperate regions.

The genus is easily recognized on rocks or compacted sand by the discrete, scattered perithecia, absence of paraphyses when mature, and particularly the simple ascospores. *Thelidium* is distinguished by the presence of 1-septate ascospores, and *Polyblastia* by the muriform ascospores. The little-known *Leucocarpopsis* (q.v.) has ascomata with a colourless exciple and involucrellum and may be synonymous, and *Thrombium* occurs on clay soils and has paraphyses and an ± or blue ascus apex.

The genus is probably heterogeneous as currently conceived. Some authors have taken up *Bagliettoa* for the *V. baldensis* group, distinguished by the involucrellum having radiating grooves, but this hardly seems a generic criterion alone and is not followed here pending a more thorough study.

In identification of the species it is important to pay particular attention to the texture of the thallus, immersion of the perithecia, whether any pits are formed in the rock, the nature of any involucrellum, and especially the size of the ascospores. Ascospore sizes can vary considerably within species and mature spores for several different perithecia should be measured whenever practicable. The species concepts for the maritime and freshwater species are now reasonably clear, but considerable work remains to be done on those of calcareous substrata.

V. phaeosperma Arnold (1874), a lichenicolous species recorded from limestone in W. Ireland (Burren), has brown, guttulate ascospores; it is omitted here as it is certainly not a member of this genus.

The first divisions in the key are based on the ecology of the species. While this generally presents little problem, some caution is required; for example, freshwater seepages running down rocks on the seashore may support freshwater and not maritime species of *Verrucaria*.

LITERATURE: *Maritime*: Fletcher (1975), Lamb (1948), Santesson (1939). *Aquatic*: Clauzade & Roux (1985), Hawksworth (1988), Swinscow (1968). *Calcicolous*: McCarthy (1988a, b), McCarthy & Mitchell (1988), Orange (1991), Poelt & Vězda (1981). *Lichenicolous*: Zehetleitner (1978).

1 Maritime or aquatic 2
Not maritime or aquatic 22

2(1) Maritime, on rocks or compacted sand on seashores 3
Aquatic, on rocks in or by the margins of lakes and rivers 14

3(2) Littoral; thallus translucent, subgelatinous, green, brown or black, never white or grey 4
Supralittoral; thallus never translucent or subgelatinous, white, grey or brown-black, never green 11

4(3) Thallus thin, brown; perithecia sessile, <0.2 mm diam. 5
Thallus thickish, green or shiny black; perithecia often immersed, or >0.3 mm diam. 7

5(4) Ascospores 7–10 × 4–7 μm 6
Ascospores 12–14 × 2–3 μm .... **41. *sandstedei***

6(5) Thallus brown, usually black-dotted; perithecia <0.2 mm diam., sessile .... **13. *ditmarsica***
Thallus green or olive brown; perithecia >0.3 mm, half-immersed .... **20. *halizoa***

7(4) Thallus with glossy black ridges 8
Thallus smooth, lacking black ridges 10

8(7) Thallus not areolate, the ridges prominent 9
Thallus areolate, the ridges hardly prominent, bordering the areoles .... **12. *degelii***

9(8) Ridges minute, fingerprint-like, <100 μm wide .... **2. *amphibia***
Ridges bead-like, often branched, >100 μm wide .... **43. *striatula***

10(7) Thallus black, regularly areolate, not subgelatinous; perithecia half-immersed; ascospores >12 μm in length .... **29. *maura***
Thallus green to green-black, smooth or weakly cracked, subgelatinous; perithecia immersed; ascospores <10 μm in length .... **30. *mucosa***

11(3) Thallus white, grey or brown; perithecia partly immersed to ± sessile 12
Thallus black, regularly areolate; perithecia immersed .... **29. *maura***

12(11) Thallus white or grey; prothallus black or absent 13
Thallus brown, regularly areolate; prothallus black .... **18. *fusconigrescens***

13(12) Thallus white-grey, areolate; prothallus black; spore ends ± pointed ...... **23. *internigrescens***
Thallus mauve-grey, effuse; prothallus inconspicuous; spore ends abruptly rounded; perithecia erumpent to ± sessile .... **39. *prominula***

14(2) Ascospores <15 μm long 15
Ascospores >15 μm long 16

15(14) Thallus lacking black sterile punctae; involucrellum extending laterally into the thallus; ascospores 6–10(-11) × (4-)5–7 μm .... **3. *aquatilis***
Thallus with black sterile punctae; involucrellum poorly developed and not extending laterally into the thallus; ascospores (7-)10–12(-15) × (5-)6–8(-10) μm ...... **40. *rheitrophila***

16(14) Thallus shades of deeper brown, green, or greenish black 17
Thallus pale brownish white, pale greyish white or pinkish; ascospores ± fusiform, (17-)20–25(-27) × (6-)7–9(-11) μm .... **38. *praetermissa***

17(16) Ascospores mainly >25 μm in length 18
Ascospores mainly <25 μm in length 20

18(17) Thallus subgelatinous, not or scarcely cracked, green to brownish green, brown, dark grey or almost black 19
Thallus not subgelatinous, repeatedly cracked, grey to greyish brown .... **25. *latebrosa***

19(18) Thallus green to brownish green or almost black; involucrellum not or scarcely extending laterally; ascospores (20-)25- 30(-33) × (10-)11–12(-16) μm; on calcareous rocks .... **16. *elaeomelaena***
Thallus dark brown to dark grey, sometimes greenish tinged; involucrellum spreading laterally; ascospores (20-)25–35(-40) × (0.9-)11–15(-20) μm; on siliceous rocks **28. *margacea***

20(17) Thallus subgelatinous, green to greenish black, not or irregularly cracked 21
Thallus not subgelatinous, brown to reddish brown, cracked and often almost areolate .... **1. *aethiobola***

21(20) Involucrellum well developed and extending laterally; perithecia >0.3 mm wide; ascospores 18–25 × (7-)8–10 μm .... ***22. hydrela***
Involucrellum poorly developed, not extending laterally; perithecia not <0.3 mm wide; ascospores 11–22(-28) × 6–8(-10) μm .... ***17. funckii***

22(1) On soil 23
On rocks or exceptionally bark 25

23(22) Perithecia without an involucrellum, simple, 0.1–0.15 mm diam.; goniocysts present 24
Perithecia with an involucrellum, compound, 0.25–0.5(-0.75) mm diam.; goniocysts absent .... ***31. muralis***

24(23) Goniocysts ± colourless; perithecia at first with a superficial colourless layer; ascospores often with a gelatinous appendage .... ***7. bryoctona***
Goniocysts brown; perithecia lacking a superficial colourless layer; ascospores without gelatinous appendage .... ***45. xyloxena***

25(22) Thallus ± immersed, usually continuous, occasionally faintly rimose, or absent 26
Thallus distinctly superficial 38

26(25) Perithecial exciple continuous below the asci, or with an entire involucrellum 27
Perithecial exciple not continuous below the asci, or with a spreading or deeply extended involucrellum 29

27(26) Ascospores 13–24 μm long 28
Ascospores 23–35(-40) μm long .... ***21. hochstetteri***

28(27) Thallus grey, grey-green or grey-brown; perithecia 0.15–0.2(-0.25) mm diam. .... ***32. murina***
Thallus white to pale grey; perithecia 0.25–0.4 mm diam. .... ***10. calciseda***

29(26) Ascospores mostly >15 μm long 30
Ascospores 7.5–12 μm long .... ***24. knowlesiae***

30(29) Ascospores (11-)15–28 × (4.5-)6–15 μm 31
Ascospores 23–40 × 11–20(-24) μm .... ***36. papillosa***

31(30) Ascospores 11–20(-23) × (4.5-)6(-10) μm 32
Ascospores 15–25(-28) × 8–15 μm .... ***31. muralis***

32(31) Perithecia 0.1–0.3(-0.4) mm diam. 33
Perithecia 0.3–0.6 mm diam. .... ***15. dufourii***

33(32) Thallus not intersected by prothalline lines 34
Thallus intersected by dark brown prothalline lines .... ***11. cyanea***

34(33) Involucrellum extending laterally but scarcely downwards, usually with fine fissures radiating from the ostiole 35
Involucrellum dimidiate or extending to level of true exciple-base, without fissures radiating from the ostiole 36

35(34) Thallus white to pale grey, matt .... ***6. baldensis***
Thallus dark grey, dark grey-green or blue-grey, glossy .... ***36. parmigerella***

36(34) True exciple colourless or brown 37
True exciple black .... ***32. murina***

37(36) Ascospores 11–21(-23) × (4.5-)6–9 μm; involucrellum 0.15–0.3(-0.4) mm diam. ***37. pinguicula***
Ascospores 11–17 × 4–7.5 μm; involucrellum 0.15–0.25(-0.3) mm diam. .... ***14. dolosa***

38(25) Ascospores 6.5–11 μm long 39
Ascospores mostly >12 μm long 41

39(38) Perithecia compound; involucrellum present 40
Perithecia simple; involucrellum absent .... ***42. simplex***

40(39) Perithecia (0.08-)0.1–0.15 mm diam.; thallus olive-brown to dark brown .... ***8. bulgarica***
Perithecia 0.2–0.3 mm diam.; thallus black .... ***4. aranensis***

41(38) Ascospores 22–40 μm long 42
Ascospores 10–28(-30) μm long 45

42(41) Perithecia compound; involucrellum present 43
Perithecia simple; involucrellum absent .... **44. *viridula***

43(42) Thallus thick 44
Thallus thin, continuous to areolate, pale grey-green or grey-brown .... **35. *papillosa***

44(43) Thallus areolate to subsquamulose, pale brown to olive-green .... **27. *macrostoma***
Thallus areolate; areoles delimited by black prothallus lines .... **19. *glaucina***

45(41) On lichens 46
Not on lichens 47

46(45) Ascospores 10–18 × 5–7 μm; mainly on *Caloplaca* .... **26. *latericola***
Ascospores 19–26 × 6–8 μm; mainly on *Aspicilia* .... **5. *aspiciliicola***

47(45) Ascospores 16–27(-30) × 8–14 μm 48
Ascospores 10–20(-23) × 6–9(-10) μm 50

48(47) Thallus grey, grey-green or grey-brown 49
Thallus dark green to black .... **33. *nigrescens***

49(48) Perithecia simple; involucrellum absent .... **34. *ochrostoma***
Perithecia compound; involucrellum present (if thallus areolate see **19**) .... **31. *muralis***

50(47) Perithecia ⅔ to entirely immersed 51
Perithecia semi-immersed to superficial 52

51(50) Thallus 0.2–1.0(-1.5) mm thick; areoles with a black edge .... **19. *glaucina***
Thallus 0.1–0.3 mm thick; areoles without a black edge .... **9. *caerulea***

52(50) Thallus predominantly brown; involucrellum 0.15–0.3(-0.4) mm diam.; ascospores 11–21(-23) × (4.5-)6–9 μm .... **37. *pinguicula***
Thallus predominantly green; involucrellum 0.15–0.25(-0.3) mm diam.; ascospores 11–17 × 4–7.5 μm .... **14. *dolosa***

1. V. aethiobola Wahlenb. (1803)

V. aquilella Nyl. (1876)

Thallus superficial, to 0.2 mm thick, brown to reddish brown or occasionally greyish, irregularly cracked to almost areolate, not gelatinous; prothallus absent or poorly developed. Perithecia 0.2–0.4 mm diam., semi- to almost entirely immersed, often several in each areole; involucrellum not spreading, thick, dark brown to black; true exciple continuous, colourless or dark brown. Ascospores (14-)18–27(-30) × 8–12(-15) μm, elongate-ellipsoid.

On semi-inundated, hard, siliceous rocks, exceptionally on limestone, by lakes and rivers, often associated with *Hymenelia lacustris*; scattered, locally frequent. Throughout British Is., mainly in the N. & W. Europe, North America.

Distinguished from other aquatic species by the distinctive, brown to reddish brown and cracked thallus. Greyish areas sometimes occur on the same thallus and range to almost entirely grey morphs; these appear to be merely environmental modifications.

2. V. amphibia R. Clem. (1807)

Thallus superficial, very variable, thin, subgelatinous, bright grass-green, translucent when wet, becoming obscured by fine, black, glossy ridges 20–50 μm wide, which coalesce to resemble fingerprints, the ridges developed extensively in strong light, the entire thallus then glossy black with fine ridges radiating at the edge; prothallus, when present, pale brown. Perithecia sessile, 0.4–0.5 mm, weakly fluted, cylindrical with a flat or concave top and crenellate rim; involucrellum thick and extending, usually coalescing with the thalline ridges. Ascospores 12–15 × 7–10 μm, ellipsoid.

On maritime rocks, forming an often extensive band below *V. maura* **(29)** in the littoral zone, often forming dark patches 3–5 cm diam. on vertical, sunny faces with *Pyrenocollema halodytes*; common. W. Britain, extending locally to E. coast and shaded shores. S. Norway to the Mediterranean.

3. V. aquatilis Mudd (1861)

Thallus superficial, to 50 μm thick, greenish to greenish black, smooth, not cracked, subgelatinous; prothallus absent. Perithecia 0.2–0.3 mm diam., immersed with only the ostiole evident; involucrellum extending laterally within the thallus; true exciple poorly developed, colourless to very pale brown. Ascospores 6–10(-11) × (4-)5–7 μm, broadly ellipsoid to subglobose.

On inundated or immersed, hard, siliceous rocks, exceptionally on limestone, by lakes and rivers; locally abundant. N. & W. Britain. C. Europe, Scandinavia.

Readily recognized amongst those occurring on rocks inundated by freshwater by the small and distinctly rounded ascospores and the absence of black ridges on the thallus.

4. V. aranensis McCarthy (1988)

Thallus superficial, black, green-black when wet, cracked, rarely areolate. compound, semi-immersed; involucrellum 0.25–0.3 mm diam., hemispherical, glossy black; the thallus grows over part of the involucrellum leaving the 100–140 μm wide apex bare; centrum 0.15–0.2 mm diam., globose; true exciple brown-black. Ascospores 7–11 × 5.5–7.5 μm.

On shaded, Carboniferous limestone; rare. W. Ireland (Co. Galway, Aran Is.). Endemic.

5. V. aspiciliicola R. Sant. (1984)

V. aspiciliae Zehetl. (1978), non (Lahm) Stizenb. (1882)

Thallus superficial, 5–10 mm wide, grey, areolate, retaining the habit of its host; areoles 0.2–1.0 mm wide, flat, angular, smooth, with a black rim. Perithecia simple, immersed; apex 0.1–0.14 mm diam., flat to convex, glossy black; centrum 0.16–0.21 mm diam., globose to pyriform; true exciple colourless to dark brown. Ascospores 19–26 × 6–8 μm.

On *Aspicilia calcarea*, *A. caesiocinerea* and *Hymenelia lacustris*; very rare. S.W. England (Somerset), Wales (Caernarvon), C. Scotland (Clackmannan), W. Scotland (Skye). S. Europe, Sweden, Hungary.

6. V. baldensis Massal. (1852)

Thallus immersed, white to pale grey (sometimes greenish in shade), continuous, smooth, usually matt, delicately pitted, with or without a dark brown, delimiting prothalline line or a groove. Perithecia (Fig. 41c) numerous, compound, almost entirely immersed; involucrellum 0.2–0.3 mm diam., lid-like, black, spreading laterally but scarcely downwards, flat to slightly convex, usually with 3–6 fine fissures radiating from the ostiole; centrum 200–400 μm diam., globose to pyriform; true exciple colourless to brown. Ascospores 14–21 × 6–10 μm.

Mainly on hard limestones, also on mortar; often abundant. Throughout British Is. Sweden, C. & S. Europe, U.S.S.R., Australia, New Zealand.

Distinguished by the closeness of the perithecia with a fissured involucrellum.

See also *V. parmigerella* **(36)**.

7. V. bryoctona (Th. Fr.) A. Orange (1991)

V. psammophila Erichsen (1939)

Thallus granular-warted, grey-green, of unpigmented goniocysts 15–40 μm diam. Perithecia 0.1–0.3 mm diam., ± semi-immersed; involucrellum absent, at first with a superficial unpigmented layer; centrum I+ red; true exciple dark brown to reddish, sometimes paler below. Ascospores 19–27(-29) × 5–7(-8) μm, frequently with a small gelatinous appendage, becoming 1(-3)-septate when over mature.

On basic or calcareous soils, usually associated with or over acrocarpous mosses; widespread in semi-natural or man-made habitats, dunes, waste ground and wall tops. England (N. Devon to Yorkshire), C. & S. Wales, Ireland (Dublin). Denmark, Finland, Sweden, Austria.

V. xyloxena **(45)** differs in having brown goniocysts, no distinct superficial colourless layer to the perithecia, and the ascospores always lacking appendages.

8. V. bulgarica Szat. (1930)

Thallus superficial, thin, olive-brown to dark brown, continuous to areolate; areoles 0.15–0.30(-0.40) mm wide, angular, sometimes peeling. Perithecia compound, semi-immersed to almost superficial; involucrellum (0.08-)0.1–0.15 mm diam., thin, extending to base-level of true exciple, occasionally incurved; centrum 0.07–0.13 mm diam., globose; true exciple colourless to brown, rarely brown-black. Ascospores 7–10(-11) × 4–5(-7) μm.

On damp, somewhat shaded *Ulmus* bark, England (W. Sussex, Sidlesham) and on shaded, calcareous pebbles, Wales (Glamorgan, near Barry). Bulgaria.

9. V. caerulea DC. (1805)

Thallus superficial, 100–300 μm thick, blue-grey (rarely grey, grey-green or grey-brown), rimose to finely areolate, occasionally with a dark brown prothallus. Perithecia compound, $\frac{2}{3}$ to entirely immersed; involucrellum 0.15–0.3 mm diam., dimidiate, black; apex flat to slightly convex; ostiole often depressed; centrum 0.15–0.25 mm diam., globose; true exciple colourless to black. Ascospores 14–20 × 4–7 μm.

On hard, rather shaded limestones; local. Throughout British Is. Norway, Sweden, C. & S. Europe, U.S.S.R., U.S.A. (California).

10. V. calciseda DC. (1805)

Thallus immersed, white to pale grey, occasionally darker, smooth, continuous, with fine cracks radiating from the perithecia, delicately pitted, occasionally with a dark brown prothallus. Perithecia 0.15–0.4 mm diam., almost entirely immersed, simple, black; apex 0.15–0.3 mm diam., convex; true exciple black. Ascospores 13–24 × 7–13 μm.

On hard limestones; local. Throughout British Is. Norway, Sweden, C. & S. Europe, U.S.S.R., Ethiopia, Cuba, U.S.A. (California), Australia.

11. V. cyanea Massal. (1853)

Thallus ± immersed, grey-green to grey-brown, smooth, continuous, sometimes farinose, often with shallow pits, delimited and divided by dark brown prothalline lines. Perithecia compound, $\frac{1}{2}$-$\frac{3}{4}$ immersed; involucrellum 0.12–0.3 mm diam., convex, spreading laterally and scarcely downwards to dimidiate, dark brown to black; centrum 0.1–0.2 mm diam., globose; true exciple colourless to pale brown. Ascospores 12–19(-20) × 5–8 μm.

On exposed, calcareous rocks; local. N. England, W. Ireland. Norway, Sweden, C. & S. Europe.

12. V. degelii R. Sant. (1939)

Thallus superficial, to 1 mm thick, the surface dull blackish green, white inside, scabrid, finely and regularly areolate; areoles with thickened, black, edges, with black ridges forming a network on the areoles; prothallus white. Perithecia immersed, *c.* 0.1 mm diam., globose; true exciple pale; involucrellum small, black, plate-like, not enveloping the true exciple. Ascospores 7–10 × 4–7 μm.

Forming patches 1–2 cm diam. amongst other maritime lichens, on sheltered shores. N. Scotland (Sutherland, Caithness). Scandinavia.

This species can be confused with *V. amphibia* **(2)**, but is closer to *V. mucosa* **(30)**.

13. V. ditmarsica Erichsen (1937)

Thallus superficial, very thin, subgelatinous, brown and translucent when wet, developing tiny, black, non-glossy, black and friable papillae, especially in the sun, which can coalesce, becoming scabrid and dirty brown-black, not distinctly ridged, areolate or otherwise patterned; prothallus not distinct. Perithecia sessile, to 0.2 mm diam.; true exciple colourless or pale brown, with age becoming brown or black at the base; involucrellum thin, completely enclosing the true exciple. Ascospores 7–10 × 4–7 μm, kidney- or "D"-shaped, relatively thick-walled.

On intertidal shingle on mudflats, estuaries etc. or in the littoral zone on sheltered rocky shores; very common in shade and shelter. Mainly in E. Britain. Baltic region, N.E. U.S.A.

Easily overlooked as *Pyrenocollema halodytes* in the field.

V. erichsenii, from the Baltic, has a dark brown true exciple, spreading involucrellum and vague, thin black ridges, but British specimens so-named are within the range of *V. ditmarsica*.

14. V. dolosa Hepp (1860)

V. mutabilis auct. p.p., non Borrer ex Leighton (1851)

Thallus superficial or almost absent, pale to dark green or olive-brown, diffuse and granular to definite and continuous, rimose or areolate. Perithecia compound, numerous, semi-immersed to almost superficial; involucrellum 0.15–0.25(-0.30) mm diam., hemispherical to conical, extending to base-level of true exciple, black; centrum 0.1–0.18(-0.2) mm, globose; true exciple colourless, rarely brown. Ascospores 11–17 × 4–7 μm.

On limestones, flint, chalk, calcareous and non-calcareous sandstones; rare. Throughout British Is., including on urban wasteland. Denmark, C. Europe, Canada (Newfoundland).

15. V. dufourii DC. (1805)

Thallus ± immersed, continuous, rimose, rarely areolate, smooth or uneven, occasionally with shallow pits, grey, grey-green, grey-brown or indistinct, frequently delimited by a dark brown prothalline line or a groove. Perithecia compound, $\frac{1}{3}$-$\frac{1}{2}$ immersed; apex flat or excavate; involucrellum 0.3–0.6 mm diam., dimidiate or extending to base-level of true exciple; centrum 0.2–0.4 mm diam.; true exciple colourless to pale brown, rarely dark brown. Ascospores 12–23 × 6–10 μm.

Mainly on hard limestones; local. Throughout British Is. Norway, Sweden, C. & S. Europe, U.S.S.R., Australia.

16. V. elaeomelaena (Massal.) Arnold (1868)

Thallus superficial, subgelatinous, green to brownish green or almost greenish black, smooth or occasionally cracked; prothallus not distinct. Perithecia (0.2-)0.3–0.4(-0.5) mm diam., immersed at first but later becoming erumpent in thalline warts; involucrellum generally poorly developed, not or scarcely spreading beyond the vicinity of the perithecia; true exciple often flattened, dark brown or black above but mostly colourless, remaining colourless below. Ascospores (20-)25–30(-33) × (10-)11–12(-16) μm, elongate-ellipsoid, rounded at both ends.

On inundated, calcareous rocks and pebbles in lowland streams. England (Dorset, Gloucestershire, Westmorland, W. Yorkshire), N.E. Ireland (Antrim). Europe.

This species has been confused with the commoner *V. funckii* **(17)**, the ascomata of which remain almost entirely immersed, the ascospores are narrower and often shorter and it occurs only on upland siliceous rocks.

17. V. funckii (Sprengel) Zahlbr. (1921)

Thallus superficial, subgelatinous, dark greenish black, smooth but generally with irregular cracks; prothallus indistinct, creamy or black. Perithecia remaining immersed, only the ostioles visible as small black dots from above, mainly 0.2–0.3 mm diam.; involucrellum dark brown, scarcely extending beyond the true exciple from which it is scarcely distinguishable; true exciple dark brown above and colourless below. Ascospores 11–22(-28) × 6–8(-10) μm, narrowly ellipsoid, rounded at the apices.

On inundated or submerged siliceous rocks in upland streams and rivers and on lake shores. N. & W. Britain, W. Ireland. Europe, North America.

This species has been confused with *V. elaeomelaena* **(16)**, but the habit most strongly recalls *V. hydrela* **(22)**, distinguished by the wide-spreading involucrellum and more narrowly ellipsoid ascospores.

18. V. fusconigrescens Nyl. (1872)

V. collematodes auct. brit., non Garov. (1865)

Thallus superficial, thin, medium to dark brown, paler (in shade), regularly areolate; areoles flat, sometimes widely separated, dull and somewhat scabrid; prothallus black and friable. Perithecia sessile, to 0.5 mm diam., often arising directly on the prothallus, or half-immersed in areoles; true exciple pale; involucrellum convex, extending beyond the true exciple. Ascospores 16–26 × 10–15 μm, ellipsoid.

On sunny, siliceous rocks in the supralittoral zone on seashores; common. Mainly in W. British Is. Europe, North America.

Superficially similar to *V. nigrescens* **(33)**, which occurs on calcareous substrata away from the shore, and which generally lacks the black prothallus, and has $\frac{1}{2}$-$\frac{3}{4}$ immersed, non-sessile perithecia. The status of *V. fusconigrescens* requires further study.

19. V. glaucina auct. brit., non Ach. (1810)

Thallus 0.2–1.0(-1.5) mm thick, superficial, pale to dark grey, grey-brown, olive-brown or dark brown, deeply areolate, delimited by a brown-black prothallus that is also visible beneath the 0.1–0.4 mm wide areoles (thin section) and usually forms a black rim to these. Perithecia compound, numerous, entirely immersed; apex flat; involucrellum 0.1–0.25 mm diam., apical, dimidiate or extending to and merging with prothallus, black; centrum 0.1–0.2 mm diam., globose; true exciple colourless to black. Ascospores 10–18(-23) × 5–8 μm.

On calcareous rocks, especially limestone, also on walls and mortar; common. Throughout British Is. C. & S. Europe, U.S.S.R., U.S.A. (Mid-west & W.), Australia.

Subsp. *canella* (Nyl.) A. L. Sm. (1911) is distinguished by the more elongated ascospores 20–30(-32) × (5-)7–8(-10) μm. On calcareous rocks and mortar; uncommon. England, Wales. Norway, Sweden, U.S.A. This may be identical to *V. aspicilicola* **(5)**.

20. V. halizoa Leighton (1871)

V. mackenzie-lambii Erichsen (1940)

Thallus superficial, thin, pale olive-green or pale brownish green in the sun, translucent when wet; often bordered by the black prothallus. Perithecia to 0.25 mm diam., sessile, hemispherical; true exciple brownish, pale at the base; involucrellum extending widely, with a clear gap between it and the true exciple. Ascospores 7–10 × 4–7 μm.

On rocky seashores, usually amongst *V. mucosa, V. striatula* and the red, crustose alga *Hildenbrandia*, forming patches 1–2 cm diam. in crevices and shade in the mid-littoral. Throughout W. British Is. N. Europe, North America.

The least variable of the maritime *Verrucaria* species. Occasionally infected with the lichenicolous fungus *Stigmidium marinum*, with colourless 1-septate ascospores.

21. V. hochstetteri Fr. (1831)

Thallus ± immersed, grey, grey-green, grey-brown or indistinct, occasionally with a red tone, smooth to uneven, sometimes farinose, continuous or faintly rimose especially about the perithecia, rarely areolate, occasionally with a dark brown prothallus. Perithecia (Fig. 41b) simple, $\frac{3}{4}$ to entirely immersed; apex 0.2–0.3(-0.5) mm diam., black, flat to slightly convex; centrum 0.3–0.6(-0.8) mm diam., globose, pyriform or widely ellipsoid; true exciple brown to black, the basal wall often remaining in pits. Ascospores 23–35(-40) × (10-)12–21 μm.

On calcareous rocks and mortar; common. Throughout British Is. Scandinavia, C. & S. Europe, U.S.S.R., N. & W. Africa, E. Asia, New Caledonia, Australia.

V. foveolata (Flörke) Massal. (1852) [syn. *V. dolomitica* (Massal.) Krempelh. (1861)] may be distinct from this species, separated primarily by the less immersed perithecia and a stouter true exciple, but is not recognized here pending further studies on the circumscription of *V. hochstetteri.*

22. V. hydrela Ach. (1814)

?*V. hibernica* Servít (1955), non Zsch. (1927)

Thallus superficial, to 50 μm thick, green to blackish green, subgelatinous, smooth, generally uncracked; prothallus dark brown to black, occasionally apparent at the margins. Perithecia immersed in the thallus, 0.3–0.5(-0.65) mm diam., in low hemispherical projections; involucrellum well developed, spreading laterally into the thallus; true exciple colourless or pale brown, continuous. Ascospores 18–25 × (7-)8–10 μm, ellipsoid.

On siliceous rocks in or by rivers and streams, which are regularly or permanently submerged, more rarely on flushed or damp rocks; usually upland. N. & W. British Is., extending to S. England (Hampshire, New Forest, Is. of Wight). Europe, North America.

Superficially most similar to *V. funckii* **(17)** but may also be confused with *V. margacea* **(28)**, which also has a wide-spreading involucrellum, but is distinguished by the longer ascospores.

23. V. internigrescens (Nyl.) Erichsen (1928)

Thallus superficial, dirty white to pale grey, not subgelatinous, uneven, with black streaks, scabrid, cracked to irregularly areolate; prothallus black. Perithecia to 0.2 mm diam., half-immersed; true exciple pale below; involucrellum black, clasping the upper half of the true exciple. Ascospores 17–23 × 10–15 μm.

On shaded, damp or seepage siliceous rock in the supralittoral zone on seashores; common. Throughout British Is. Europe, North America.

This species could be confused with the freshwater *V. aethiobola* **(1)**, with perithecia to 0.4 mm diam. and the involucrellum spreading, or the calcicolous *V. murina* **(32)** or *V. muralis* **(31)**.

24. V. knowlesiae McCarthy (1988)

Thallus immersed and indistinct to ± superficial, dark grey-green, continuous to rimose. Perithecia compound, semi-immersed to almost superficial; involucrellum 0.15–0.25(-0.3) mm diam., hemispherical to conical, glossy black, extending to base-level of true exciple, may spread laterally; centrum 0.1–0.15 mm diam., globose; true exciple brown to brown-black. Ascospores 7.5–12 × 5.5–7 μm.

On shaded, carboniferous limestone. N. England (N.E. Yorkshire, Loftus), W. Ireland (Galway, Aran Is.). Endemic.

25. V. latebrosa Körber (1855)

Thallus superficial, grey to greyish brown, not subgelatinous, repeatedly cracked, irregularly areolate in parts; prothallus black, evident at the margins; photobiont cells dispersed, not in vertical columns. Perithecia (0.2-)0.3–0.5(-0.55) mm diam., erumpent, about half-exposed, conspicuous, sometimes becoming confluent; involucrellum black, closely adpressed to the exciple, not spreading; true exciple integrated with the involucrellum, rounded, dark brown to black above, colourless to brownish below. Ascospores (25-)28–35(-40) × (12-)14–15(-18) μm, widely ellipsoid.

On semi-inundated, calcareous rocks by rivers and lakes; rare. N. Wales (Denbigh), W. Scotland (Lismore), W. Ireland (Galway, Leitrim).

Distinguished from *V. margacea* **(28)**, which occurs on inundate, siliceous rocks, in that the thallus is not subgelatinous, predominantly grey, and the involucrellum is not wide-spreading.

26. V. latericola Erichsen (1943)

V. granulosaria Clauz. & Zehetl. (1978), *V. insularis* auct. brit., non (Massal.) Jatta (1909)

Thallus crustose and rimose to areolate or subsquamulose or placodioid (the habit largely dependent on that of the host), grey-green to grey-brown, frequently white- to grey-pruinose; areoles 0.2–1.6 mm wide, rounded or angular, contiguous or scattered, flat to convex. Perithecia simple, numerous, almost entirely immersed; apex 70–140 μm diam., flat to convex, brown to black; centrum 0.1–0.18 mm diam., globose to pyriform; true exciple colourless. Ascospores 10–19 × 4.5–7.5 μm.

On lichens, mainly calcicolous *Caloplaca* species, also on *Lecanora albescens* and *Diplotomma alboatrum*; rare. S.W. England (Somerset), W. Ireland (Clare, Galway). S. France, Spain.

27. V. macrostoma Dufour ex DC. (1805)

?*V. griseorubens* Migula (1930)

Thallus superficial, areolate to subsquamulose, pale brown to olive-green; areoles 0.4–1.5 mm wide, flat to convex, smooth to verruculose; sterile thalli may have soralia or isidia on the margins of areoles [f. *furfuracea* B. de Lesd. (1949); syn. *V. viridula* f. *tectorum* auct. brit., non (Massal.) Laundon (1967)]. Perithecia compound, $\frac{1}{2}$-$\frac{3}{4}$ immersed; apex flat to convex; involucrellum 0.3–0.6 mm diam., dimidiate, black; centrum 0.2–0.4 mm diam., globose; true exciple dark brown to black. Ascospores 25–37 × 10–20 μm.

On calcareous rocks, walls and mortar; locally abundant. Throughout British Is. Norway, Sweden, C. & S. Europe, Algeria.

28. *V. margacea* (Wahlenb.) Wahlenb. (1812)

V. leightonii Hepp (1853)

Thallus superficial, dark brown to dark grey, sometimes greenish tinged, to about 50 μm thick, smooth, not or rarely cracked when dry, sometimes breaking up into irregular patches, subgelatinous, without black ridges; black prothallus occasionally present. Perithecia (0.3-)0.35–0.5(-0.7) mm wide, immersed except around the ostiole; involucrellum spreading laterally into the thallus; true exciple dark brown above but often colourless below. Ascospores (20-)25–35(-40) × (9-)11–15(-20) μm, ellipsoid, sometimes 1-pseudoseptate.

On permanently submerged to semi-inundated, hard, siliceous rocks by lakes and rivers, mainly in upland areas. N. & W. British Is., extending locally to C. & E. England (Leicester, N. Norfolk).

Distinguished from other aquatic species recorded in the British Is. except *V. latebrosa* **(25)** in the relatively massive ascospores.

29. V. maura Wahlenb. (1803)

Thallus superficial, to 1 mm thick, black, dull or rarely shiny, regularly cracked at the edge, becoming extensively areolate, the areoles flat with angular edges (like cracked mud), black-dotted (visible best at the brown edge), the dots creating a rough, almost scabrid texture, brown and only weakly rimose in shade, but the dots are then more obvious. Perithecia half-immersed, 0.3–0.5 mm diam.; true exciple pale to dark brown at the base; involucrellum extended and wide-spreading. Ascospores 12–15 × 7–10 μm, ovate.

On rocky seashores, including shingle, as a black band in the upper littoral zone, sometimes several hundred metres above sea level on wind-exposed seacliffs; very common. Throughout British Is., but absent from S.E. & E. England. Cosmopolitan.

30. V. mucosa Wahlenb. (1803)

Thallus superficial, to 1 mm thick, shiny, olive-green, becoming dark green to black in the sun, smooth and even, weakly translucent when wet, firm and subgelatinous; prothallus chalky white. Perithecia to 0.15 mm diam., immersed; exciple colourless; involucrellum a pale brown ring bordering the ostiole. Ascospores 7–10 × 4–7 μm, relatively thick-walled.

On rocky seashores in the mid-littoral zone, whether sunny or shaded, with barnacles and macroalgae, rarer on sheltered shores and intolerant of silt, forming ± extensive patches; very common. Throughout British Is., but absent from S.E. & E. England. Cosmopolitan.

Often infected with the lichenicolous fungus *Stigmidium marinum.*

31. V. muralis Ach. (1803)

Thallus immersed to superficial, very variable, pale grey, grey-green, grey-brown or reddish brown, indistinct, continuous, finely rimose to areolate, delicately pitted. Perithecia ¾ immersed to almost superficial, compound; involucrellum 0.25–0.4(-0.5) mm diam., dimidiate or, occasionally, extending to exciple-base level, black; centrum 0.2–0.35 mm diam., globose; true exciple colourless to pale brown. Ascospores 15–25(-28) × 8–15 μm.

On calcareous rocks, walls and mortar, rarely on hard limestones; common. Throughout British Is. Scandinavia, C. & S. Europe, U.S.A., U.S.S.R., N. Africa, Australia.

32. V. murina Leighton (1851)

Thallus ± immersed, dark grey, grey-green or grey-brown, rarely with a reddish tinge, continuous, but occasionally faintly rimose, sometimes delicately pitted. Perithecia ⅓ to almost entirely immersed; involucrellum 0.15–0.2(-0.25) mm diam., thin, black, extending to base-level of true exciple, incurved beneath it, or entire; centrum 0.1–0.18 mm diam., globose; true exciple black. Ascospores 14–20(-25) × 6–9(-10) μm.

On hard limestones; local. Throughout British Is. Sweden, C. & S. Europe, U.S.S.R.

33. V. nigrescens Pers. (1795)

Thallus superficial, dark brown, green-black or black (frequently bright green when wet), areolate; areoles 0.2–0.8 mm wide, usually smooth, flat to slightly convex, occasionally with sorediate or isidiate margins; prothallus black, also visible beneath the areoles (thin section), occasionally forming a black rim to the areoles. Perithecia compound, ½–¾ immersed; apex flat to hemispherical; involucrellum 0.2–0.4 mm diam., dimidiate or extending down to the prothallus, black; centrum 0.15–0.25 mm diam., globose; true exciple dark brown. Ascospores (17-)19–27(-30) × 8–14 μm.

On calcareous rocks, walls and mortar, more rarely on siliceous rocks; common. Throughout British Is. Scandinavia, C. & S. Europe, U.S.S.R., North America, N. Africa, Australia.

34. V. ochrostoma (Borrer ex Leighton) Trevisan (1860)

Thallus superficial, pale yellow-brown, grey, grey-brown or dark brown, areolate; areoles 0.3–1.2 mm wide, 0.1–0.2 mm thick, flat or, more commonly, convex, rounded or angular. Perithecia simple, entirely immersed; apex 180–220 μm diam., convex, brown to black; centrum 0.3–0.4 mm diam., globose; true exciple brown-black. Ascospores 18–23 × 8–12(-14) μm.

On mortar; rare. S. England (Sussex, near Henfield; Kent, Dungeness).

35. V. papillosa Ach. (1810)

Thallus immersed and indistinct to superficial and diffuse, rimose or areolate, pale grey-green or grey-brown, sometimes patchily reddish brown, with or without a dark brown, delimiting prothalline line. Perithecia compound, $\frac{1}{2}$-$\frac{2}{3}$ immersed; apex convex, flat or excavate; involucrellum (0.4-)0.6(-1.0) mm diam., dimidiate, black; centrum 0.3–0.6 mm diam., globose; true exciple dark brown to black. Ascospores 24–40 × 14–24 μm.

On sheltered calcareous sandstone and schist; rare. S. Wales (Dyfed, near Broadhaven), N.E. Scotland (Easterness, S. Clunes). Sweden, Switzerland, Pyrénées, North America.

36. V. parmigerella Zahlbr. (1919)

Thallus immersed, grey, dark grey-green or blue-green, glossy, smooth, continuous, delicately pitted. Perithecia compound, almost entirely immersed, numerous but usually sterile; involucrellum 0.15–0.22 mm diam., lid-like, flat to slightly convex, spreading laterally but scarcely downwards, black, usually with 3–6 fine fissures radiating from the ostiole; centrum 0.2–0.3 mm diam., globose to pyriform; exciple colourless to pale brown. Ascospores 14–21 × 6–8 μm.

On hard, exposed limestone; locally abundant. S.W. England (S. Devon), W. Ireland (Clare, Galway, Sligo, Leitrim). C. & S. Europe.

Differs from *V. baldensis* **(6)** in thallus colour and smaller involucrellum.

37. V. pinguicula Massal. (1856)

Thallus immersed, or more commonly superficial, continuous, finely rimose to areolate, pale to dark brown, grey-green or grey-brown, smooth to verruculose, frequently delimited by a dark brown prothallus. Perithecia compound, $\frac{1}{3}$-$\frac{1}{2}$ immersed; involucrellum 0.15–0.3(-0.4) mm diam., dimidiate or extending to base-level of true exciple, black; centrum 0.1–0.2(-0.25) mm diam., globose; true exciple colourless to brown. Ascospores 11–21(-23) × (4.5-)6–9 μm.

On hard limestones; local. N. England (Yorkshire). C. & S. Europe, U.S.A. (New York).

38. V. praetermissa (Trevisan) Anzi (1864)

V. amylacea auct. brit., non Hepp ex Arnold (1858)

Thallus superficial, pale brownish white, pale greyish white, or pinkish (then K ± reddish purple), to 0.15 mm thick, smooth or irregularly cracked, not gelatinous, prothallus where visible black. Perithecia (0.2-)0.25–0.3(-0.4) mm diam., immersed, the thallus only slightly raised in the vicinity of the ostiole; involucrellum not spreading, thick dark brown to black; true exciple continuous, usually ± colourless throughout. Ascospores (17-)20–25(-27) × (6-)7–9(-11) μm, elongate-ellipsoid to broadly fusiform.

On submerged or ± continuously inundated, hard, siliceous rocks, exceptionally on limestones, in lakes and rivers; local. N. Great Britain and Ireland, extending to S.W. England (S. Devon) and S.C. Wales (Brecknock). C. Europe.

Perhaps the most distinctive freshwater *Verrucaria* species, easily recognized by the pale colour of the thallus dotted with the black tops of the ascomata, and the elongate to almost fusiform ascospores.

39. V. prominula Nyl. (1861)

Thallus superficial, often poorly developed, white to pale grey, always with a mauve tinge, discontinuous, soft and crumbling, not subgelatinous, scabrid; prothallus indistinct. Perithecia to 0.8 mm diam., black, prominent, sometimes 2 or 3 contiguous; true exciple black, even at the base; involucrellum closely clasping the true exciple. Ascospores 12–15 × 7–10 μm, ellipsoid, the ends abruptly truncate.

On vertical, shaded faces, or in dry, crevices on soft, siliceous rocks, especially schists, most frequently in the mesic to xeric-supralittoral zones on seashores. W. British Is., rare elsewhere. Europe, North America.

This species may be confused with *V. internigrescens* **(23)**, *V. murina* **(32)** or *V. muralis* **(31)**, but is distinguished by the thallus colour, perithecial details, ascospore shape, and substratum preference.

40. V. rheitrophila Zsch. (1922)

V. kernstockii Zsch. (1927)

Thallus superficial, to 100 μm thick, green or greyish green, becoming brownish in the herbarium, smooth, subgelatinous, with black spots *c.* 30 μm diam.; prothallus, when present, white or black. Perithecia 0.1–0.2 mm diam., immersed, the ostiolar region level with the thallus surface; involucrellum poorly developed and represented only by a thickened upper exciple; true exciple ± continuous, black above but brown to colourless below. Ascospores (7-)10–12(-15) × (5-)6–8(-10) μm, widely ellipsoid.

On semi-inundated hard siliceous rocks in lakes and streams, particularly on slates at low altitudes; local. W. Britain, extending locally to S.E. England (W. Sussex). Europe, North America.

The small ascospores separate this species from others in this habitat, except for *V. aquatilis* **(2)** in which the thallus lacks sterile black spots, the perithecia are present in raised parts of the thallus surface, and these have a well-developed, spreading involucrellum; the ascospores also tend to be longer than those of *V. aquatilis* **(3)**.

41. V. sandstedei B. de Lesd. (1911)

V. lorraine-smithii Knowles (1913)

Thallus similar to that of *V. ditmarsica* **(13)**, dark brown to blackish, very thin, subgelatinous, effuse. Perithecia to 0.15 mm diam., sessile, hemispherical, often flattened; true exciple pale at the base; involucrellum rough, extending beyond the true exciple. Ascospores 12–14 × 2–3 μm, narrowly ellipsoid.

On sheltered, littoral, siliceous seashore rocks, amongst barnacles; very rare. S. England (along the Channel coast), S. Wales, E. Ireland (Dublin).

V. lorraine-smithii was originally cited as having ascospores 16–21 × 3–5 μm.

42. V. simplex McCarthy (1988)

Thallus superficial, dark green, sparingly rimose, smooth to uneven, pulpy-translucent when wet. Perithecia 0.1–0.13 mm diam., simple, ± superficial, black, the lower part often bounded by a thalline collar; centrum 0.09–0.12 mm diam., globose; true exciple brown to brown-black. Ascospores 6.5–10.5 × 4–6 μm.

On shaded mortar; very rare. N.E. England (Northumberland, Morpeth). Endemic.

43. V. striatula Wahlenb. (1803)

Thallus variable, grass-green, translucent when wet, subgelatinous, developing black spots which coalesce to form glossy, branched and swollen ridges 100–200 μm wide, sometimes obscuring the green tissue so that the thallus appears black; ridges develop in the sun and thalli from shaded crevices are often only weakly spotted or devoid of them. Perithecia to 0.3 mm diam., sessile, becoming excavate to one side when old, the ostioles large; true exciple usually pale at the base; involucrellum clasping the true exciple, then extending a little at the base. Ascospores 7–10 × 4–7 μm.

In the mid-littoral zone on rocky seashores, between *V. amphibia* **(2)** and *V. mucosa* **(30)**, often amongst the red alga *Hildenbrandia prototypus*, on littoral siliceous or calcareous rocks; very common. Throughout British Is., but absent from S.E. & E. England. Europe, North America. Cosmopolitan.

This species can be distinguished from *V. amphibia* **(2)** by the shorter ascospores, perithecial shape and coarse black ridges, and from shade morphotypes of *V. halizoa* **(20)**, by the perithecial shape and greener thallus.

44. V. viridula (Schrader) Ach. (1803)

Thallus superficial, pale to dark brown, grey-green or grey-brown, rimose to areolate, often forming extensive colonies, occasionally delimited by a dark brown prothallus; areoles (0.2-)0.4–1.0(-2.0) mm wide, angular, flat, smooth or uneven. Perithecia simple, almost entirely immersed; apex 0.3–0.5(-0.8) mm diam., flat to slightly convex, dark brown to black; centrum 0.3–0.5(-0.7) mm diam., globose to widely ellipsoid; true exciple brown-black. Ascospores 22–35(-39) × 10–17 µm.

On calcareous rocks, walls and mortar; very common. Throughout British Is. Scandinavia, C. Europe, Greece, U.S.S.R., U.S.A.

The sorediate morphs often referred to this species are now considered to belong to *V. macrostoma* **(27)**.

45. V. xyloxena Norm. (1867)

V. melaenella Vainio (1921)

Thallus usually thin, brown to blackish brown or almost black, indistinct to granular-areolate, of brown goniocysts 15–35 µm diam., a brown prothallus sometimes visible. Perithecia (0.1-)0.14(-0.26) mm diam., simple, semi-immersed to almost superficial, crowded; centrum 0.08–0.13 mm diam., globose, I+ red; true exciple black and friable. Ascospores (15-)16–23 × 6–8(-9) µm, rarely 1-septate.

On calcareous soils, especially stable dunes; rare. E. England (W. Suffolk, Thetford Heath). Denmark, Sweden, Finland, France, U.S.S.R

VESTERGRENOPSIS Gyelnik (1940)

Thallus placodioid, forming rosettes; lobes stellate-radiating, closely adpressed. **Isidia** present or absent. **Upper cortex** not distinctly differentiated; hyphae short-celled in upper part of thallus, periclinally orientated in parallel along the length of the lobes, looser and with more elongated cells in a lower medullary zone. **Lower cortex** absent, underside pale. **Photobiont** *Scytonema*; cells in chains. **Ascomata** apothecia, hemiangiocarpic, sessile. **Thalline exciple** present, concolorous with thallus, entirely paraplectenchymatous, with included photobiont cells. **True exciple** much reduced, inconspicuous. **Epithecium** red-brown. **Hymenium** colourless to ± pale brown, I+ blue. **Hypothecium** pale yellow-brown, of interwoven hyphae. **Hamathecium** of paraphyses, simple, rarely forked towards the apices, ± conglutinated, upper cells enlarged, ± moniliform. **Asci** (8-)12- to 16-spored, clavate, without an I+ blue tholus. **Ascospores** colourless, simple, sometimes with plasma bridges, ellipsoid to subglobose. **Conidiomata** unknown. **Chemistry**: lichen products not detected by t.l.c.

LITERATURE: Henssen (1967), Gyelnik (1940).

1. V. elaeina (Wahlenb. ex Ach.) Gyelnik (1940)

Thallus to 4 cm diam., olive-brown, lobes 0.2–0.4 mm wide, widening to 0.8 mm at apices, elongate, fan-like, flat, longitudinally grooved; upper surface striate, matt, nodulose-uneven towards the centre, nodules 0.2–0.3 mm wide, coarse, at first globose to cylindrical, becoming ± flattened, erect or ± decumbent, conspicuous, at times forming small, simple or branched, secondary, overlapping lobules. Apothecia to 0.8 mm diam., frequent, scattered, rounded; disc black-brown, matt. Ascospores 7–10 × 4–6 µm.

On basalt rocks in a seepage track; very rare. W. Scotland (Glencoe; Skye). Scandinavia, Greenland, arctic Canada.

Distinguished from *Placynthium* species e.g. *P. pannariellum*, by the pale underside of the thallus. The radiating, stellate thallus with grooved and striate lobes and coarse, ± flattened nodules are diagnostic. *V. isidiata* (Degel.) E. Dahl (1950), the isidiate counterpart of *V. elaeina*, is not yet known from the British Is. and should be sought in similar habitats; it has ± numerous, finger-like isidia up to 1 × 0.1 mm, ± thickly dispersed in central area of thallus.

VEZDAEA Tsch.-Woess & Poelt (1976)

Thallus crustose, of corticate granules (goniocysts), usually developing below cuticle of bryophytes, etc., or below cortex of lichens. **Photobiont** *Leptosira*. **Ascomata** apothecia, rounded or irregular. **True exciple** and **hypothecium** absent. **Hamathecium** of paraphyses, branched, at times characteristically entwined around asci. **Hymenium** lacking gelatin. **Asci** 8-spored, ± cylindrical, thick-walled, apex thickened, I+ blue, except for an apical pore. **Ascospores** simple to 11-septate, colourless, with smooth or roughened walls. **Conidiomata** unknown, but conidiogenous cells producing simple, colourless conidia may be present. **Chemistry**: lichen products not detected by t.l.c. **Ecology**: on plant debris, bryophytes, lichens, rocks or often metal-rich soil.

The apothecia are short-lived and material in best condition is collected in the winter and spring. Differs from *Micarea*, in which hymenial gelatin, a true exciple, distinct conidiomata, and lichen products occur; the asci also have a different structure.

LITERATURE: Coppins (1987b), James (1971).

1 Ascospores ellipsoid or oblong, simple or 1- to 3-septate, <30 μm long 2
Ascopores acicular, 7- to 11-septate, >60 μm .... **1. acicularis**

2(1) Apothecia broadly adpressed, never stalked; ascospores simple or 1- to 3-septate, smooth or warted 3
Apothecia usually stalked, or at least strongly constricted below, the stalk occasionally hidden by goniocysts; ascospores becoming 1-septate, smooth .... **3. leprosa**

3(2) Ascospores non-septate 4
Ascospores mostly 1(-3)-septate .... **2. aestivalis**

4(3) Ascospores warted; goniocysts with spines to 4–15 μm long; paraphyses short, not entwining individual asci .... **5. rheocarpa**
Ascospores smooth; goniocysts with spines to 2 μm long; paraphyses long, entwining individual asci .... **4. retigera**

1. V. acicularis Coppins (1987)

Thallus thin, effuse, of dull green, ± globose goniocysts. Goniocyst surface smooth or with shallow, blunt papillae to 2 μm tall. Apothecia 0.2–0.4 mm diam., scattered or 2- to 3-confluent, convex and adpressed but sometimes short top-shaped when young, grey-white to pink-brown, matt, ± tomentose. Paraphyses rarely reaching surface of hymenium. Ascospores 60–85 × 2–2.5 μm, 7- to 11-septate, acicular; upper apex rounded; lower end tapered to a narrow apex.

On lead-contaminated ground; rare. N.E. England (Northumberland), S. Wales (Brecon, Cardigan), Scotland (Dumfries, Stirling). N. France, Belgium.

Readily distinguished by the long, acicular ascospores. *Bacidia* species have a well-developed exciple, mostly simple, septate, apically swollen paraphyses, hymenial jelly and a different ascus structure.

2. V. aestivalis (Ohl.) Tsch.-Woess & Poelt (1976)

Goniocysts ± contiguous, with short conical processes. Apothecia 0.3–1 mm diam., convex, adpressed or immersed in thallus, grey to dull reddish brown, often ± tomentose. Paraphyses extremely contorted and anastomosed, entwining individual asci. Ascospores 15–19 × 5–7 μm, 1(-3)-septate, verrucose when mature.

On mosses and plant remains amongst or on ± calcareous rocks and mortar in shaded or moist sites, also amongst mosses on trees with a base-rich bark; common. Throughout British Is. Europe.

The largest and most conspicuous species of the genus. Superficially similar to *Bacidia sabuletorum*, which has a well-developed hypothecium and exciple, ± simple paraphyses which do not clasp the asci, larger, septate ascospores and different asci. An undescribed species from lead- and zinc-contaminated soil in Wales (Cardigan) has larger, 1-septate ascospores, 19–24 × 7–8 μm, and smaller apothecia (0.1–0.2 mm diam.).

3. V. leprosa (P. James) Vězda (1977)

Thallus conspicuously minutely granular, dull dark green, goniocysts slightly warted, rarely with spines. Apothecia 0.3–1 mm diam., ± sessile or crowded amongst thalline granules, convex to globose, pale pink- or red-brown. Paraphyses abundant, flexuose, not entwining individual asci. Ascospores 10–15.5 × 2.5–4 μm, ± cylindrical or long-ellipsoid with rounded apices, simple or 1-septate.

On disturbed soils and decaying vegetation of transient, open habitats under maritime influence or contaminated by metals such as iron, lead and zinc, particularly beneath old, galvanized fencing wires, also on wall tops; common. Throughout British Is.

4. V. retigera Poelt & Döbb. (1977)

Goniocysts with short spines, to 2 μm. Apothecia rather small, mostly below 0.3 mm diam., semi-globose. Paraphyses 0.7–1.5 μm wide, numerous, entwining individual asci. Ascospores 15–22 × 7–11 μm, persistently simple, the wall smooth.

On decaying bryophytes and *Peltigera* species, especially on lead-rich soils. N. & W. British Is.

V. rheocarpa **(5)** has somewhat larger and distinctly warted ascospores, sparse, short paraphyses that do not become entwined around individual asci, and usually longer goniocyst spines.

5. V. rheocarpa Poelt & Döbb. (1977)

Goniocysts with long, pointed spines to 4–15 μm, spines best developed on the exposed surfaces of superficial goniocysts. Apothecia indeterminate in size and form. Paraphyses rather sparse and short, not entwining individual asci, never reaching the height of mature asci. Ascospores (18-)20–24 × 9–11 μm, simple, distinctly warted.

On lead-contaminated soil, moribund *Marchantia* thalli and mossy trunks of old *Quercus* and *Ulmus*. S.W. (Devon, Somerset) and N. England (Derbyshire, Peak District), Wales, E. Scotland. Europe.

Rather similar to *V. retigera* **(4)** which has smaller, smooth ascospores, numerous, long paraphyses that become entwined around individual asci and usually shorter goniocyst spines.

WADEANA Coppins & P. James (1978)

Thallus immersed, ± inconspicuous; prothallus absent. **Photobiont** *Trentepohlia*. **Ascomata** apothecia, narrowly-elongate (lirelliform), erumpent, ± superficial, oval to elongate-ellipsoid, black, disc ± exposed, simple or branched. **True exciple** ± raised, persistent, black-brown, slightly extended laterally at the base, but not generally below the hypothecium, of compacted gelatinized hyphae with red-brown walls. **Epithecium** indistinct. **Hypothecium** pale to red-brown, this and all apothecial tissues tending to be red-brown, intensifying in N. **Hamathecium** of branched and occasionally anastomosed, delicate paraphysoids, somewhat moniliform, swollen and red-brown at the tips; periphysoid filaments developing on the inside margin of the exciple, more rarely at its upper margin. **Asci** clavate-ellipsoid, rather evenly thin-walled, slightly thickened at the apex with an I+ deep blue cap, 100- to 200-spored. **Ascospores** ellipsoid or ± globose, simple, colourless, smooth-walled. **Conidiomata** unknown. **Chemistry**: ? an unidentified xanthone in one species. **Ecology**: on rough bark of deciduous trees, Europe.

Separated from *Lithographa* by the presence of periphysoid filaments, the exciple not being well developed under the hypothecium, and the multispored asci; and from *Opegrapha* by the simple spores and non-fissitunicate asci. The red-brown colour of the apothecial tissues is best studied in water mounts; it is K+ dark brown.

LITERATURE: Coppins & James (1978).

1 Apothecia 0.5–2.5 × 0.3–0.5 mm, usually distinctly elongate with attenuated ends; hymenium 100–120 μm tall; ascospores ellipsoid, 5–8 × 3–4 μm .... **1. *dendrographa***
Apothecia 0.25–0.8 × 0.2–0.3 mm, shortly elongate or oval with ± rounded ends; hymenium 50–60 μm tall; ascospores ± globose, 2-3 μm diam. .... **2. *minuta***

1. W. dendrographa (Nyl.) Coppins & P. James (1978)

Apothecia 0.5–2.5 × 0.3–0.5 mm, distinctly elongate with attenuated or sometimes ± rounded ends, not or irregularly branched; true exciple 60–85(-100) μm wide, swollen and persistent, lip-like; hymenium 100–120 μm tall. Asci 78–85 × 12–19 μm, 100- to 200-spored. Ascospores 5–8 × 3–4 μm, ellipsoid. Thallus Pd−, K−, KC−, C−, UV± orange-yellow (? an unidentified xanthone detected by t.l.c.).

On rough, nutrient-rich bark of mature trees in little-disturbed, wayside and old woodland habitats, particularly on *Fraxinus*, *Quercus* and *Ulmus*, often with *Cryptolechia carneolutea*; local. Extreme S. & S.W. England (Sussex to Cornwall), N.W. England (Cumbria), W. Scotland (to Skye). Portugal, Italy (Tuscany).

2. W. minuta Coppins & P. James (1978)

Apothecia 0.25–0.8 × 0.2–0.3 mm, shortly elongate or oval, sometimes clustered, not or 1-branched; true exciple persistent, 30–40(-60) μm wide; hymenium 50–60 μm tall. Asci (35-)40–45 × 12–18 μm, *c.* 100-spored. Ascospores 2–3 μm diam., ± globose. Thallus Pd−, K−, KC−, C−, UV−; no lichen products detected by t.l.c.

On rough bark of mature trees with nutrient-rich bark, particularly *Fraxinus* and *Ulmus*; very rare. S. England (Surrey; S. Hampshire, New Forest), N. Scotland (Perth, Inverness). Norway (Bergen).

W. dendrographa (**1**) has bigger, more conspicuous apothecia with larger internal proportions throughout.

XANTHORIA (Fr.) Th. Fr. (1860)

Thallus yellow-orange, foliose, horizontal or partly ascending, lobate, not pruinose, the underside pale, smooth, with thick, simple, pale rhizines, but these often poorly developed, or small attachment discs present. **Isidia** and **soredia** present in some species. **Cortex** plectenchymatous, continuous on both surfaces. **Photobiont** trebouxioid. **Medulla** distinct, white. **Ascomata** apothecia, laminal, orbicular to contorted; disc smooth, orange, margin involute, entire or crenulate, often paler than the disc, usually persistent. **Thalline exciple** present. **Hypothecium** colourless. **Asci** 8-spored, clavate, *Teloschistes*-type (Fig. 33l). **Ascospores** colourless, polarilocular, ellipsoid. **Conidiomata** pycnidia, laminal, immersed in hemispherical warts, inconspicuous, concolorous with thallus, multilocular. **Conidiogenous cells** lining the locules, ± barrel-shaped, enteroblastic, acrogenous. **Conidia** mostly small, ellipsoid, but subcylindrical in *X. fallax* (see **1**), colourless. **Chemistry:** anthraquinones, especially physcion (K+ dark violet-red).

Xanthoria is one of the best-known lichen genera on account of its conspicuous, yellow, leafy thallus, growing especially in habitats associated with man, or else coastal sites. It is K+ dark violet-red, a colour test which distinguishes it from *Candelaria*, which is K−. Placodioid species of *Caloplaca* differ in not being readily separable from the substratum.

1 Soredia absent 2
  Soredia present, situated on edge or underside of lobes .... **1. *candelaria***

2(1) Thallus with narrow lobes, the apices up to 1 mm across 3
  Thallus with wide lobes, the apices (0.5)-1–7 mm across 4

3(2) Thallus with finger-like, convex lobes spreading outwards beyond apothecia; on rock **2. *elegans***
  Thallus with finely divided lobe-ends often obscured by crowded apothecia; on bark and wood, rare on rock .... **6. *polycarpa***

4(2) Lobes rounded at apices, scarcely indented; thallus of ± flat, regular rosettes 5
  Lobes strap-shaped, deeply divided; thallus often irregular, hummock-forming, of overlapping lobes .... **5. *ectaneoides***

5(4) Thallus isidiate; isidia irregularly globose-capitulate, continuous; apothecia few, peltate, urceolate to concave, with involute, roughened, uneven thalline exciple .... **4. *calcicola***
  Thallus not isidiate, but sometimes with flattened lobules (on damaged thalli); apothecia numerous, ± sessile, usually with raised, smooth thalline exciple .... **3. *parietina***

1. X. candelaria (L.) Th. Fr. (1861)

Thallus in small, compact clusters, often forming extensive swards, of numerous, ± crowded, compacted, small, yellow-orange lobes, *c.* 1 mm wide, broadening towards the apices, ± horizontal at first, soon curving upwards, the margins becoming truncated, flexuose and ± erect; underside greyish white, with scattered rhizines, often visible from above at the edges of the lobes. Soredia farinose to minutely granular, on the edges of the lobes, sometimes extending to the underside, occasionally appearing almost continuously sorediate. Apothecia occasional, to 4 mm diam., ± peltate, with conspicuous thalline exciple and sunken red- or brownish orange discs when young, becoming flat, the exciple excluded, and contorted when old. Ascospores (10-)11–15(-18) × 8–10(-11) μm.

On nutrient-rich and -enriched substrata, especially decaying tops of old wooden gate-posts used as bird perches and tops of vertical monuments, also on bark, less frequent on natural outcrops of acid and calcareous rocks and walls usually in association with bird colonies or farm animals; common. Throughout British Is. Europe, North & South America, N. Africa, Antarctica.

Easily identified by the crowded, flexuose, ascending, eroded, sorediate lobe-ends. In very shaded situations the thallus may be reduced to minute, narrow, almost nodular, terete, spiky lobes with scattered, granular soredia.

Xanthoria ulophyllodes Räsänen (1931) [*X. fallax* (Hepp) Arnold (1980)] has lip-like marginal soredia but differs in having non-eroded, slightly swollen lobes, with ± widened, down-turned margins, which lie flat; somewhat reminiscent of *Physcia tenella* in habit. On eutrophicated bark; local. S.E. England to E. Scotland; a single fragmentary collection from a boulder at the entrance to a cave (Is. of Scilly) may represent a distinct species.

2. X. elegans (Link) Th. Fr. (1860)

Thallus to 4 cm diam., forming ± regular pale orange to dark reddish orange rosettes, sometimes incomplete; lobes ± uniformly 0.5–1.0 mm wide throughout their length, ± nodulose, ± discrete for much of their length or contiguous or ± overlapping at the centre, markedly plicate, strongly convex. Apothecia usually abundant throughout the central area of the thallus, to 1.5 mm diam., often crowded but rarely contorted, sessile, orbicular, flat, with persistent thalline exciple. Ascospores (9-)11–16 × 5–7(-9) μm, ellipsoid or ovoid.

On well-lit, nutrient-rich acid and calcareous rocks used as bird perches, on mountain rocks, but also in man-made habitats, especially on roofs of slate and asbestos-cement; local. Throughout British Is. Bipolar, widespread in the N. Hemisphere (S. to the mountains of Africa) and parts of the S. Hemisphere, extending far into the arctic and antarctic.

Recognized by the bright reddish orange rosettes of narrow, ± discrete, finger-like, convex lobes. It frequently occurs as isolated individuals, not forming colonies. *Caloplaca saxicola*, which often grows in similar habitats, has shorter, more adpressed lobes, is often pruinose, and lacks a lower cortex.

3. X. parietina (L.) Th. Fr. (1860)

Thallus to 15 cm diam., often forming extensive patches, forming ± regular, yellow-orange (grey when in shade) adpressed, somewhat wrinkled rosettes; lobes ± overlapping, plicate, broadening towards the apices, ± indented, the apices 1–7 mm wide, rounded or somewhat notched and ± flat; coarse, flat, subsquamulose to ligulate lobules sometimes present towards the centre of the rosette, usually scattered but occasionally abundant, apparently the result of regeneration. Apothecia usually numerous, to 4 mm diam., scattered to clustered, sessile to peltate (sometimes ± stalked when on twigs), orbicular to contorted, concave when young, becoming ± flat when mature, with prominent, raised, smooth, concolorous thalline exciple; old apothecia on moribund thalli becoming convex and immarginate. Ascus (Fig. 33l). Ascospores (10-)12–16 × (6-)7–9 μm.

On a wide variety of nutrient-rich and -enriched substrata, including wayside trees and roofing tiles, the wood of farmyard fences, on both inland and coastal acid and calcareous rocks, old bones in drift on beaches and stable dunes, the principal component of the *Xanthorion*; very common. Throughout British Is. Cosmopolitan, but probably introduced into some areas (e.g. S. Australia).

This conspicuous lichen is usually easily identified by its plain yellow rosettes with plentiful apothecia of varied age. Thalli in sheltered, shaded habitats (e.g. in bushes of *Sambucus niger*), are green-grey, often with only a hint of the characteristic yellow-orange coloration. Rocks near the sea and farmhouse roofs are often coloured a brilliant yellow by this lichen. Morphs with lobules are sometimes difficult to distinguish from *X. calcicola* **(4)** and *X. ectaneoides* **(5)**.

4. X. calcicola Oxner (1937)

X. aureola auct. pro parte, non (Ach.) Erichsen (1928)

Like **3**, but upper surface becoming densely, coarsely isidiate throughout the central area of the thallus; isidia 0.1–0.7 mm diam., ± simple, typically erect, irregularly globose-capitate or peg-like, sometimes becoming flattened and lobule-like, often crowded and partially obscuring the thallus. Apothecia scattered and few, ± stalked, urceolate to concave, with involute, roughened, thalline exciple.

Lowlands, on especially ± calcareous nutrient-rich stonework, brickwork, tiles, monuments, rare on bark and wood. E. Britain, extending locally to the west. Europe, from S. Scandinavia to the Mediterranean, eastwards to the Ukraine, Middle East.

Characterized by the central area of the mature thallus being covered with typically convex, coarse granules, and with few, scattered, cup-shaped, urceolate apothecia with involute, roughened margins. The flattened regeneration lobules of damaged thalli of *X. parietina* **(3)** can resemble the isidia of *X. calcicola* and are sometimes a cause of confusion, the former are usually more randomly scattered, and the numerous apothecia usually sessile, ± flattened, smooth and with only slightly raised margins. When the two species grow together *X. calcicola* is often also deeper orange in colour.

5. X. ectaneoides (Nyl.) Zahlbr. (1931)

Like **3**, but thallus forming irregular to circular, rather loose hummocks of spreading, overlapping, crenulate to strap-shaped lobes. Lobe-ends ligulate and axils well developed, apices of lobes 0.5–2 mm wide, truncated to rounded and ± flat. Apothecia occasionally plentiful but often absent.

On exposed, nutrient-rich, usually ± vertical, acid and calcareous stone, chiefly on sea-cliffs, also on tall, ruined buildings in wind-swept, flat, open country near the coast and on calcareous rocks near mountain summits; locally common. Highland Britain, Ireland, rare elsewhere. W. Europe, Islas Baleares, Madeira.

Characterized by the often sterile thallus of crenulate-indented to strap-shaped lobes. *X. ectaneoides* can sometimes be confused with a morph of *X. parietina* **(3)**, which has flattened lobules, rarely becoming ± ligulate and densely overlapping; such lobules are invariably smaller than the lobes of *X. ectaneoides* and are borne on thalli of a ± regular rosette-shape characteristic of *X. parietina*. *X. ectaneoides*, reproducing mainly by fragmentation, and *X. calcicola* **(4)** chiefly reproducing by isidia, are interpreted as secondary species derived from the primary species, *X. parietina*. *X. ectaneoides* has been considered an ecotype of *X. parietina* and its status merits further investigation.

6. X. polycarpa (Hoffm.) Th. Fr. ex Rieber (1891)

Thallus to 2 cm diam., of small, ascending crenulate lobes, greyish yellow to yellow (grey when in shade), forming small, orbicular hummocks; lobes broadening towards the apices, which are finely divided into many tiny lobe-ends, under 1 mm across, occasionally almost terete, nodulose. Soredia and isidia absent. Upper surface with crowded apothecia, often completely covering and obscuring the underlying thallus. Apothecia to 4 mm diam., contorted, erect, peltate, often with short stalks, flat and orbicular when young with prominent, raised margins, becoming contorted when old, the thalline exciple sometimes becoming excluded; thalline exciple greyish yellow (grey when in shade); disc brownish orange. Ascospores 11–15 × 6–8 μm.

On living and, especially, dead bark, most commonly in the axils of dead twigs on shrubs with wide-spreading branches, especially *Populus* species, *Salix* species, *Larix* and *Sambucus nigra*, also on wooden fences, very rare on coastal flints and rock outcrops in farmland, chiefly in wind-swept, flat, open country bordering the sea, and on trees and shrubs in boggy places inland; local. Throughout British Is. Europe, North America.

The hummocks of crowded apothecia obscuring the finely divided thallus make this lichen unmistakable; the thallus is much paler or greyish in shade although the apothecial disc often remain bright orange.

XEROTREMA Sherw. & Coppins (1980)

Thallus inconspicuous or immersed. **Photobiont** absent, or loosely associated with unidentified chlorococcoid algae and *Trentepohlia*, not or facultatively lichenized. **Ascomata** apothecia, erumpent, dark brown to black, at first ± closed but becoming urceolate, opening entire to dentate. **Thalline exciple** absent. **True exciple** of irregularly septate hyphae in a gelatinous matrix, black. **Epithecium** brownish. **Hymenium** colourless, I+ blue. **Hypothecium** poorly developed, colourless. **Hamathecium** of thread-like, septate paraphyses, simple below and branched at the apices; periphyses lining the inner surface of the true exciple, simple. **Asci** at first cylindrical, becoming elongate-clavate, the wall ± even in thickness, with a single functional wall layer, not thickened at the apex, I−, 1-spored. **Ascospores** muriform, very large, colourless, I−, smooth, lacking a distinct perispore. **Conidiomata** unknown. **Chemistry**: lichen products unknown. **Ecology**: on wood.

A monotypic genus of the Odontotremataceae distinguished by the massive muriform ascospores, I− hymenium, the absence of any apical thickening in the ascus and brown epithecium. The true exciple structure separates it from *Ramonia* species, which also have clearly lichenized thalli.

LITERATURE: Sherwood & Coppins (1980).

1. X. megalospora Sherw. & Coppins (1980)

Apothecia 0.3–0.6 mm diam., the margins dentate, with coarse, radial striations; disc urceolate, dark brown. Asci 140–180 × 35–40 μm, 1-spored. Ascospores 90–140(-170) × 22–40 μm, composed of numerous cells individually 3–4 μm diam.

On hard, dry, bleached *Pinus* branch. E.C. Scotland (Perth, Black Wood of Rannoch). North America (Oregon).

XYLOGRAPHA (Fr.) Fr. (1838)

Thallus crustose, immersed or thinly superficial, effuse, sometimes with scattered, brown goniocysts. **Photobiont** chlorococcoid. **Ascomata** apothecia, semi-immersed to superficial, narrowly elongate, lirelliform, roundish to linear, usually aligned parallel to the grain of the wood, brown to dark brown; disc ± flat. **Thalline exciple** absent. **True exciple** thin, brown at least in outer part, not black. **Epithecium** brown, rarely with an olivaceous tinge. **Hymenium** colourless, I+ blue. **Hypothecium** colourless. **Hamathecium** of paraphyses, simple or sparingly branched and anastomosed, the apices brown, sometimes branched, gradually widening. **Asci** 8-spored, clavate-cylindrical, ± *Trapelia*-type; in K/I, wall blue, apical dome distinct, with convex base, light blue laterally surrounding a broad, colourless apical cushion; no ocular chamber. **Ascospores** simple, colourless or a few old ascospores becoming grey-brown, ellipsoid to oblong-ellipsoid, without a distinct thickened perispore. **Conidiomata** pycnidia, ± globose, ± immersed, brown-black. **Conidia** narrowly falcate, colourless. **Chemistry**: no detectable lichen products, or with stictic acid, norstictic acid, or unidentified substances. **Ecology**: on wood, rarely on bark; *c.* 8 species in temperate regions of both hemispheres.

Similar genera on wood are the apparently closely related, though non-lichenized, *Agyrium* with convex, dull orange to orange-brown apothecia without a raised exciple, and the lichenized *Ptychographa* which has more intensely pigmented, black apothecia with mainly slit-like discs.

LITERATURE: Laundon (1963).

1 Soralia absent 2
 Soralia present .... **3. *vitiligo***

2(1) Apothecia 0.3–1.7(-2.7) mm long, elliptic to linear; ascospores 11–16 × 5–7 μm **1. *abietina***
 Apothecia 0.2–1.1 mm long, roundish to narrowly elliptic; ascospores 9–12.5 × 4.5–6 μm .... **2. *trunciseda***

1. X. abietina (Pers.) Zahlbr. (1922)

?*X. laricicola* Nyl. (1875)

Thallus largely immersed, light greyish, usually minutely speckled with brown goniocysts 17–40 μm diam. Apothecia elliptic to linear, semi-immersed to superficial, 0.3–1.7(-2.7) × 0.1–0.28 mm; disc pale brown to dark brown; true exciple thin, often prominent when young, later slightly elevated or inconspicuous, concolorous with disc or usually paler, thin, brown in outer part, colourless within. Ascospores 11–16 × 5–7 μm, ellipsoid to oblong-ellipsoid. Pycnidia frequent but inconspicuous, resembling large goniocysts. Thallus Pd ± yellow, K ± yellow, C − (stictic acid often with traces of other members of complex, including norstictic acid, or no lichen products detected by t.l.c).

On slowly-decaying wood, especially of conifers. N. Wales (Merioneth), Scotland, S.W. Ireland (N. Kerry). Europe, North America.

The type of *X. laricicola* appears to be merely a morph of this species, or possibly of *X. trunciseda* **(2)** on old *Larix* bark; its thallus formed a thin greyish crust, but the algae are still organized in brown goniocysts. The apothecia are short, perhaps due to the lack of grain in the substratum.

2. X. trunciseda (Th. Fr.) Minks ex Redinger (1938)

Like **1**, but apothecia 0.2–1.1 × 0.1–0.3 mm, roundish to narrowly elliptic, pale brown to dark brown. Ascospores 9–12.5 × 4.5–6 μm. Thallus Pd −, K −, C − (unidentified substance, or no substances detected by t.l.c.).

In similar habitats to **1**, but often preferring slightly more moist conditions. Scotland, N. Ireland. Sweden.

3. X. vitiligo (Ach.) Laundon (1963)

Thallus immersed, pale greyish; without brown goniocysts. Soralia present, erumpent, mostly discrete, roundish to narrowly elliptic, 0.2–1 × 0.18–0.42 mm, ± flat, brown to dark grey or indigo, cream when abraded. Soredia 19–30 μm diam., with brown pigment which is unchanged in K. Apothecia present or absent, superficial, roundish to narrow-elliptic, 0.24–0.96 × 0.16–0.4 mm; disc ± flat, pale brown to dark brown; true exciple narrow, slightly elevated, usually paler than disc, brown. Ascospores 10–14(-16.5) × 4–7.5 μm, ellipsoid. Pycnidia usually present, inconspicuous. Soralia Pd+ orange, K+ yellow, C − (stictic acid, usually with traces of other members of complex, including norstictic acid).

On slowly decaying wood, especially of conifers. S.W. England (Somerset), Wales, Scotland, S.W. Ireland (Kerry), N. Ireland (Fermanagh). Europe, North America.

Distinguished from other species of the genus by the soralia. When sterile it may be confused with various sorediate species, from which it may be distinguished by the reactions of the soralia and the presence of stictic acid. T.l.c. is sometimes required to separate sterile specimens from *Buellia griseovirens*.

ZAMENHOFIA Clauz. & Roux (1985)

Thallus crustose, superficial, isidiate. **Photobiont** *Trentepohlia*. **Ascomata** perithecia, superficial or immersed, free or ± surrounded by an alga-containing thalline layer. **True exciple** coloured or not, in some species with a layer reacting K+ red. **Involucrellum** present or absent. **Hamathecium** of paraphyses, numerous, persistent, simple, rarely branched; periphyses present, well developed, numerous. **Asci** 8-spored, elongate-ellipsoid, 2- to 3-seriate, thin-walled, not thickened at the apex, I −, not functionally fissitunicate. **Ascospores** 5- to 17-transversely septate, ellipsoid to elongate-fusiform, colourless; wall distinct, swelling markedly in K. **Conidiomata** unknown. **Chemistry**: lichen products not detected by t.l.c. **Ecology**: on bark in ancient woodlands.

Formerly included within *Porina*, but characterized by the densely isidiate thallus and presence of periphyses. All species are often sterile, but are readily identified by differences in their isidia.

LITERATURE: Clauzade & Roux (1985), James (1971), Sérusiaux (1990), Swinscow (1962).

1 Thallus orange-brown to pale green-brown or yellow-green; ascospores 22–30(-33) × 4–6(-7) μm, 3-septate, or sterile 2
Thallus dull grey, often with a bluish or purplish tinge when fresh; ascospores (35-)40–57(-63) × (8-)9–13(-15) μm, (5-)9–11(-12)-septate .... **1. *coralloidea***

2(1) Isidia cylindrical, normally bright orange, parallel, densely packed, little-branched, without or with a very narrow and ill-defined cortex .... **2. *hibernica***
Isidia forming ± distinct, minutely branched, shrub-like, pale greenish yellow, ± loose, globose aggregations with a wide, well-defined, algal-free cortex .... **3. *rosei***

1. Z. coralloidea (P. James) Clauz. & Roux (1985)

Porina coralloidea P. James (1971)

Thallus smooth, soon becoming ± densely coralloid-isidiate, continuous, effuse; isidia to 1.5 mm tall, simple to sparingly coralloid, scattered or closely packed, blue-brown, black-grey-brown when wet, becoming dull purplish brown in herbarium, with pale apices when young; *Trentepohlia* cells 8–15 × 6–9 μm. Perithecia 0.3–0.4 × 0.4–0.6 mm, ½-¾ immersed, rare, widely scattered and sometimes not obviously associated with thallus; involucrellum present, closely adpressed, black; true exciple 30–45 μm thick, pale to dark brown-black, especially towards outer edge. Ascospores (35-)40–57(-63) × (8-)9–13(-15) μm, (5-)9 to 11(-12)-septate, ellipsoid, rarely clavate-fusiform.

On medium aged to very old trees, especially *Quercus*, rarely on *Fraxinus* and *Fagus*, in old forests or parklands; local. S. England (especially in Hampshire, New Forest), N.W. England (Lake District), Wales, W. Scotland. Brittany, W. Pyrénées, N. Spain, Canary Is. Azores.

When sterile the dark purplish brown thallus resembles that of *Catillaria atropurpurea* which occurs in similar habitats; the minute isidia with pale tips of *Z. coralloidea* are, however, diagnostic. The thallus may be very thin with only a few, very scattered, characteristic isidia. *Z. hibernica* **(2)** has a greenish or pale orange-brown thallus and larger ascospores with more transverse septa.

2. Z. hibernica (P. James & Swinscow) Clauz. & Roux (1985)

Porina hibernica P. James & Swinscow (1962)

Thallus partly immersed, of ± scattered to confluent granules forming a thin, continuous, effuse, scurfy, non-areolate crust giving rise to ± richly branched, coralloid isidia; isidia lacking cortex or, when present, very narrow and ill-defined, bright orange-brown, green-orange (in shade) becoming pale brown in the herbarium. *Trentepohlia* cells 7–12 μm diam. Perithecia to 0.7 mm diam., ½-¾ immersed, prominent, black, ± globose, discrete or occasionally 2- to 6- aggregated, forming contorted clusters; involucrellum present, often vestigial, rather dark, semi-opaque; true exciple black in outer part, paler within; ostiole a shallow apical depression, not conspicuous. Ascospores (55-)60–90(-95) × 5–7(-8) μm, (7-)12 to 16(-17)-septate, elongate with fusiform ends.

On sheltered, rather shaded trunks of old *Quercus* in ancient woodland and parkland sites; locally frequent. S. England (Hampshire, New Forest; Cornwall, Bocconoc Park), N. Wales, S.W. Scotland (Wigtown, Castle Kennedy), S.W. Ireland (Killarney). W. Pyrénées, France, Spain.

Z. coralloidea **(2)** has a darker blue-brown thallus and smaller ascospores with fewer transverse septa. See also *Z. rosei* **(3)**.

3. Z. rosei (Sérusiaux) P. James (1992)

Porina rosei Sérusiaux (1990)

Like **2**, in the densely isidiate thallus, but isidia very fragile, tending to form ± distinct, minutely branched, shrub-like clusters, pale greenish yellow, with a wide, well-developed, algal-free cortex of well-defined, globose to ± elongate mycobiont cells, 2.5(-8) × 2–3 μm. *Trentepohlia* cells 8–10(-17) μm diam. Perithecia unknown in Britain, 0.2–0.3 mm diam., 0.23–0.26 mm tall, orange-brown. Ascospores 22–30(-33) × 4–6(-7) μm, 3-septate.

On old trees, especially *Quercus*, in ancient forests; very rare. S. England (Hampshire, New Forest). France (Vercors & W. Pyrénées), Spain (Navarra), Madeira.

The perithecia resemble those of *Porina leptalea*, but that species lacks isidia, and has smaller perithecia and 3-septate ascospores not exceeding 23 μm in length.

AUTHORS OF GENERIC ACCOUNTS

Absconditella	B. J. Coppins
Acarospora	O. W. Purvis & P. W. James
Acrocordia	O. W. Purvis
Adelolecia	D. L. Hawksworth
Agonimia	O. W. Purvis
Agyrium	O. W. Purvis & P. W. James
Alectoria	O. W. Purvis
Allantoparmelia	O. W. Purvis
Amygdalaria	D. L. Hawksworth
Anaptychia	O. W. Purvis
Anisomeridium	B. J. Coppins & P. W. James
Aphanopsis	O. W. Purvis
Arctomia	O. W. Purvis
Arthonia	B. J. Coppins
Arthopyrenia	B. J. Coppins
Arthothelium	B. J. Coppins
Arthrorhaphis	O. W. Purvis
Aspicilia	O. W. Purvis & B. J. Coppins
Bacidia	B. J. Coppins
Bactrospora	P. W. James & O. W. Purvis
Baeomyces	P. W. James & D. H. Dalby
Bellemerea	O. W. Purvis
Belonia	O. W. Purvis
Biatora	B. J. Coppins
Biatorella	T. Duke & P. W. James
Blarneya	O. W. Purvis
Brigantiaea	O. W. Purvis
Bryonora	O. W. Purvis
Bryophagus	O. W. Purvis & P. W. James
Bryoria	O. W. Purvis
Buellia	A. Orange, B. J. Coppins & C. Scheidegger
Byssoloma	E. Sérusiaux
Calicium	O. W. Purvis
Caloplaca	J. R. Laundon
Candelaria	P. W. James & O. L. Gilbert
Candelariella	P. W. James & O. L. Gilbert
Carbonea	D. J. Galloway & P. W. James
Catapyrenium	O. Breuss & B. W. Fox
Catillaria	B. J. Coppins
Catolechia	D. L. Hawksworth
Cavernularia	O. W. Purvis
Celothelium	D. L. Hawksworth & B. Aguirre-Hudson
Cetraria	O. W. Purvis
Cetrelia	O. W. Purvis
Chaenotheca	O. W. Purvis & P. W. James
Chaenothecopsis	B. J. Coppins
Chiodecton	O. W. Purvis & P. W. James
Chromatochlamys	O. W. Purvis & P. W. James
Chrysothrix	O. W. Purvis
Cladonia	O. W. Purvis & P. W. James
Claurouxia	D. L. Hawksworth
Clauzadea	D. L. Hawksworth
Clauzadeana	D. L. Hawksworth
Cliostomum	B. W. Fox
Coccotrema	P. W. James

Coelocaulon	J. R. Laundon
Collema	O. W. Purvis & P.W. James
Cornicularia	O. W. Purvis
Cryptolechia	O. L. Gilbert
Cyphelium	O. W. Purvis
Cystocoleus	D. H. Dalby
Degelia	P. W. James & O. W. Purvis
Dermatocarpon	B. J. Coppins & B. W. Fox
Dictyonema	O. W. Purvis & B. J. Coppins
Dimerella	O. W. Purvis & B. J. Coppins
Diploicia	O. W. Purvis
Diploschistes	D. L. Hawksworth
Diplotomma	O. W. Purvis
Dirina	O. W. Purvis
Eiglera	B. J. Coppins
Endocarpon	B. W. Fox
Enterographa	P. W. James & F. S. Dobson
Eopyrenula	B. J. Coppins
Ephebe	O. L. Gilbert
Epigloea	J. C. David
Epilichen	D. L. Hawksworth
Euopsis	B. J. Coppins & P. M. Jørgensen (based on NCLF account)
Evernia	O. W. Purvis
Farnoldia	D. L. Hawksworth
Fellhanera	B. J. Coppins
Foraminella	O. W. Purvis
Fulgensia	O. L. Gilbert
Fuscidea	O. W. Purvis, L. H. Skjoldahl & T. Tønsberg (based on NCLF account)
Gomphillus	P. W. James & F. H. Brightman
Graphina	A. Orange
Graphis	A. Orange & P. W. James
Gyalecta	P. W. James & R. G. Woods
Gyalidea	P. W. James & R. G. Woods
Gyalideopsis	V. Giavarini & P. W. James
Haematomma	B. J. Coppins & F. H. Brightman
Halecania	B. J. Coppins
Herteliana	D. L. Hawksworth
Heterodermia	O. W. Purvis
Hymenelia	B. J. Coppins & O. W. Purvis
Hyperphyscia	B. J. Coppins
Hypocenomyce	O. W. Purvis & P.W. James
Hypogymnia	O. W. Purvis
Icmadophila	P. W. James & F. H. Brightman
Ionaspis	B. J. Coppins & F. S. Dobson
Julella	B. J. Coppins & P. W. James
Koerberiella	D. L. Hawksworth
Lasallia	O. W. Purvis
Lecanactis	O. W. Purvis & F. Rose
Lecania	P. W. James & O. W. Purvis
Lecanora	D. L. Hawksworth & D. H. Dalby
Lecidea	D. L. Hawksworth & B. J. Coppins
Lecidella	O. W. Purvis & P. W. James
Lecidoma	O. W. Purvis
Lemmopsis	O. L. Gilbert
Lempholemma	B. J. Coppins, O. L. Gilbert & P. M. Jørgensen (based on NCLF account)
Lepraria	J. R. Laundon
Leprocaulon	O. W. Purvis

Leproloma	J. R. Laundon
Leproplaca	J. R. Laundon
Leptogium	B. J. Coppins & O. W. Purvis
Leptorhaphis	D. L. Hawksworth & B. Aguirre-Hudson
Leucocarpia	B. J. Coppins
Leucocarpopsis	D. L. Hawksworth
Lichenothelia	D. L. Hawksworth
Lichina	O. W. Purvis & A. Fletcher
Lithographa	O. W. Purvis
Lobaria	O. W. Purvis
Lopadium	O. W. Purvis
Macentina	A. Orange
Massalongia	O. W. Purvis
Megalaria	B. J. Coppins
Megalospora	F. Rose & P. W. James
Megaspora	O. W. Purvis & P. W. James
Melaspilea	D. L. Hawksworth
Menegazzia	O. W. Purvis
Micarea	B. J. Coppins
Microcalicium	O. W. Purvis
Miriquidica	B. J. Coppins & O. W. Purvis
Mniacea	B. J. Coppins
Moelleropsis	O. W. Purvis & P.W. James
Mosigia	J. C. David
Mycoblastus	M. F. Watson & P. W. James
Mycocalicium	O. W. Purvis
Mycoglaena	B. J. Coppins
Mycomicrothelia	D. L. Hawksworth
Mycoporum	B. J. Coppins
Nephroma	F. J. White
Normandina	O. W. Purvis
Ochrolechia	O. W. Purvis
Omphalina	A. Orange & R. Watling
Opegrapha	A. Pentecost & P. W. James
Ophioparma	P. W. James & F. H. Brightman
Orphniospora	O. W. Purvis & B. J. Coppins
Pachyphiale	F. Rose & P. W. James
Pannaria	O. W. Purvis
Parmelia	O. W. Purvis & P.W. James
Parmeliella	O. W. Purvis
Parmeliopsis	O. W. Purvis
Parmentaria	O. W. Purvis
Peltigera	O. W. Purvis & P. W. James
Pertusaria	O. W. Purvis & P. W. James
Petractis	B. J. Coppins & O. L. Gilbert
Phaeocalicium	O. W. Purvis
Phaeographis	B. J. Coppins & P. W. James
Phaeophyscia	B. J. Coppins
Phlyctis	P. W. James & C. J. B. Hitch
Phyllopsora	O. W. Purvis & B. J. Coppins
Physcia	B. J. Coppins
Physconia	B. J. Coppins
Pilophorus	O. W. Purvis & B. J. Coppins
Placidiopsis	O. Breuss
Placopsis	O. W. Purvis
Placynthiella	O. W. Purvis
Placynthium	O. L. Gilbert & P. W. James

Platismatia	O. W. Purvis
Poeltinula	O. L. Gilbert & B. J. Coppins
Polyblastia	O. W. Purvis & P. W. James
Polychidium	B. J. Coppins & P. W. James
Polysporina	D. J. Galloway & B. J. Coppins
Porina	O. W. Purvis & P. W. James
Porocyphus	B. J. Coppins & O. L. Gilbert
Porpidia	D. J. Galloway & B. J. Coppins
Protoblastenia	A. Orange
Protoparmelia	B. J. Coppins
Protothelenella	O. W. Purvis & P. W. James
Pseudephebe	O. W. Purvis
Pseudevernia	O. W. Purvis
Pseudocyphellaria	O. W. Purvis
Psilolechia	B. J. Coppins & O. W. Purvis
Psora	O. W. Purvis & E. Timdal
Psoroma	O. W. Purvis
Psorotichia	B. J. Coppins, O. L. Gilbert & P. M. Jørgensen (based on NCLF account)
Pterygiopsis	O. W. Purvis
Ptychographa	B. J. Coppins
Pycnothelia	O. W. Purvis
Pyrenocollema	B. J. Coppins
Pyrenopsis	B. J. Coppins, O. L. Gilbert & P. M. Jørgensen (based on NCLF account)
Pyrenula	D. L. Hawksworth
Pyrrhospora	D. L. Hawksworth
Racodium	D. H. Dalby
Ramalina	O. W. Purvis & P. W. James
Ramonia	O. W. Purvis
Rhaphidicyrtis	D. L. Hawksworth & B. Aguirre-Hudson
Rhizocarpon	O. W. Purvis, P. W. James, J. Holtan-Hartwig, E. Timdal & S. C. Clayden (based on NCLF account)
Rimularia	J. C. David
Rinodina	B. W. Fox & O. W. Purvis
Roccella	O. W. Purvis
Sagiolechia	B. J. Coppins & P. W. James
Sarcogyne	D. L. Hawksworth
Sarcopyrenia	C. J. B. Hitch
Sarcosagium	O. W. Purvis
Schadonia	O. W. Purvis & P. W. James
Schaereria	D. L. Hawksworth
Schismatomma	D. L. Hawksworth
Sclerophora	O. W. Purvis
Sclerophyton	O. W. Purvis
Scoliciosporum	O. W. Purvis
Siphula	O. W. Purvis
Solenopsora	O. W. Purvis & P. W. James
Solorina	O. L. Gilbert
Sphaerophorus	O. W. Purvis
Sphinctrina	O. W. Purvis
Spilonema	B. J. Coppins, O. L. Gilbert & P. M. Jørgensen (based on NCLF account)
Sporastatia	B. J. Coppins & O. L. Gilbert
Squamarina	M. R. D. Seaward
Staurothele	O. W. Purvis & P. W. James
Steinia	O. W. Purvis
Stenocybe	O. W. Purvis
Stereocaulon	O. W. Purvis
Sticta	O. W. Purvis

Strangospora	T. Duke & B. J. Coppins
Strigula	B. J. Coppins
Synalissa	O. L. Gilbert & B. J. Coppins
Teloschistes	O. W. Purvis
Tephromela	O. W. Purvis
Thamnolia	O. W. Purvis
Thelenella	O. W. Purvis & P. W. James
Thelidium	A. Orange
Thelocarpon	M. F. Watson, P. W. James & D. M. Moore
Thelomma	O. W. Purvis
Thelopsis	F. Rose & P. W. James
Thelotrema	D. L. Hawksworth
Thermutis	V. Giavarini
Thrombium	O. W. Purvis
Tomasellia	B. J. Coppins
Toninia	E. Timdal & P. W. James
Tornabea	O. W. Purvis
Trapelia	O. W. Purvis
Trapeliopsis	O. W. Purvis
Tremolecia	D. L. Hawksworth
Trimmatothele	B. J. Coppins
Tylothallia	B. W. Fox & P. W. James
Umbilicaria	O. W. Purvis
Usnea	P. W. James, P. Clerc & O. W. Purvis
Verrucaria	D. L. Hawksworth, P. M. McCarthy & A. Fletcher
Vestergrenopsis	P. W. James & P. M. Jørgensen (based on NCLF account)
Vezdaea	O. W. Purvis
Wadeana	D. L. Hawksworth
Xanthoria	J. R. Laundon
Xerotrema	D. L. Hawksworth
Xylographa	A. Orange
Zamenhofia	P. W. James & O. W. Purvis

NCLF = *Norwegian Crustose Lichen Flora* (in preparation).

Glossary

acicular, needle-shaped (Fig. 27w).
acrogenous, (of conidia) formed at the apex of conidiophores or conidiogenous cells.
acuminate, gradually narrowing to a point.
allantoid, slightly curved with rounded ends (Fig. 27q).
ampulliform, flask-like with a narrow neck (Fig. 27s).
amyloid, stained blue by iodine.
anamorph, an asexual stage or morph characterized by the production of conidia.
anastomosing, a joining together to form a vein-like net (Fig. 37f).
annular, ring-like.
anisotomic, with unequal branching (Fig. 5a).
anticlinal, perpendicular to the surface.
apical cushion, colourless zone running through an amyloid apical dome (Fig. 2c).
apical dome, the thickened inner part of the ascus wall in the ascus apex (Fig. 2b); tholus.
apiculate, with a short projection at one end.
apiculus, a short projection at one end.
apothecium (pl. **apothecia**), a cup or saucer-like ascoma in which the hymenium is exposed at maturity (Figs. 28, 32).
applanate, flattened.
arachnoid, covered with, or formed of, delicate interwoven hairs or fibres.
arcuate, arc-like.
areole, an island-like portion of a crustose thallus separated from adjacent areoles (crazy-paving-like).
articulate, jointed.
ascoma (pl. **ascomata**), any ascus-containing structure.
ascospore, reproductive spore formed within an ascus.
ascus (pl. **asci**), a sac-like cell in which ascospores are produced (Figs. 33-35).
aseptate, simple, without cross walls.
aspicilioid, apothecia $\pm$ immersed in the thallus.
attenuate, narrowed.
bacilliform, cylindrical with rounded ends, *Bacillus*-like (Fig. 27h).
basidioma (pl. **basidiomata**), a basidium-containing structure.
biatorine, apothecia lacking a true exciple when mature, and which are pale or $\pm$ coloured, soft in consistency, and generally strongly convex.
biseriate, in two rows.
bitunicate, with two functional wall layers; see fissitunicate
blastidium (pl. **blastidia**), a rounded lichen propagule produced by the budding of a thallus (Fig. 29e).
botryose, grouped like grapes.
bryophilous, growing on mosses and liverworts.
bullate, having bubble or blister-like swellings.
byssoid, made up of delicate threads, cotton-like.
canaliculate, channelled.
capitate, having a well-formed head (Fig. 27u).
carbonaceous, black and friable.
catenulate, in chains.
cephalodium (pl. **cephalodia**), delimited region within, or a warty, squamulose, or shrubby structure on the surface of, a lichen thallus containing a photobiont (cyanobacterium) different to that characteristic of the rest of the thallus (Fig. 30).
chemodeme, see chemotype.
chemosyndrome, a biogenetically meaningful set of major and minor natural metabolic products produced by a species.
chemotype, a group of chemically differentiated individuals of a species of unknown or of no taxonomic significance.
chondroid, cartilaginous.
cilium (pl. **cilia**), a hair-like out-growth.
clavate, club-like (Fig. 27t).
clypeus, stromatic growth, comprising fungal hyphae and host tissue.

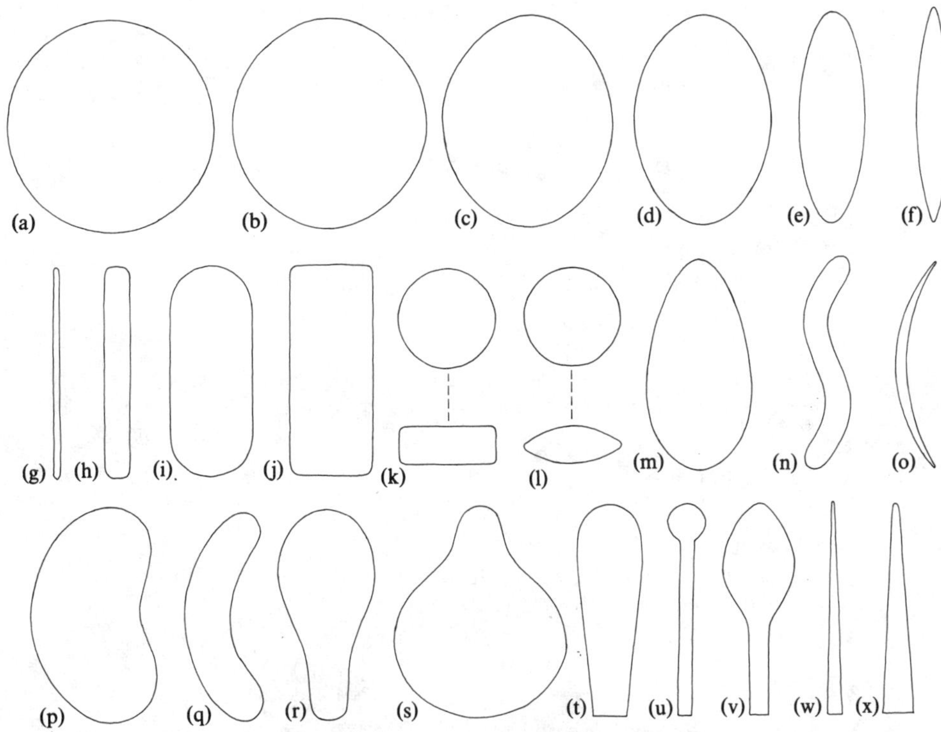

Fig. 27. Shape terminology (after Hawksworth *et al.*, 1983). (a) Globose; (b) subglobose; (c) broadly ellipsoid; (d) ellipsoid; (e) narrowly ellipsoid; (f) fusiform; (g) filiform; (h) bacilliform; (i) oblong, rounded ends; (j) oblong, truncate ends; (k) discoid; (l) lenticular; (m) ovoid; (n) sigmoid; (o) falcate; (p) reniform; (q) allantoid; (r) pyriform; (s) ampulliform; (t) clavate; (u) capitate; (v) spathulate; (w) acicular; (x) subulate.

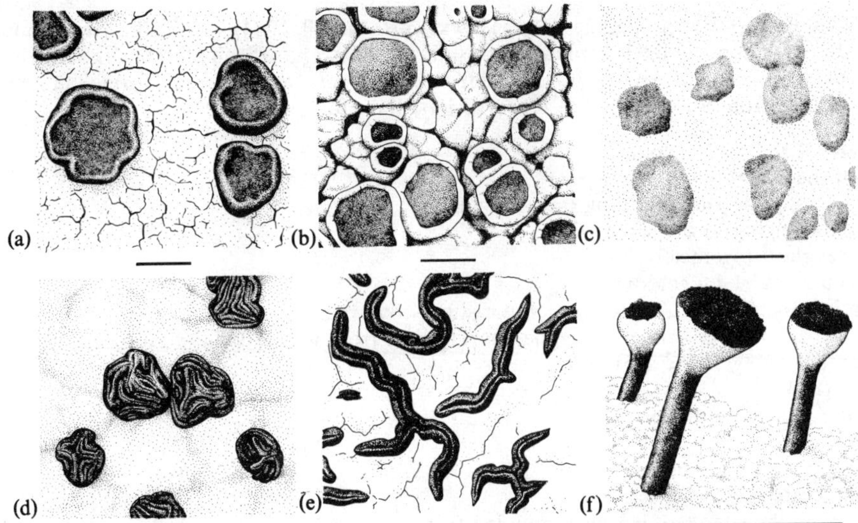

Fig. 28. Types of apothecia. (a) Lecideine, lacking a thalline exciple (*Porpidia macrocarpa*); (b) lecanorine, with a thalline exciple (*Lecanora gangaleoides*); (c) arthonioid (*Arthonia spadicea*); (d) gyrose (*Umbilicaria hyperborea*); (e) lirellate (*Opegrapha vulgata*); (f) stalked (*Calicium viride*). Scales = 1 mm.

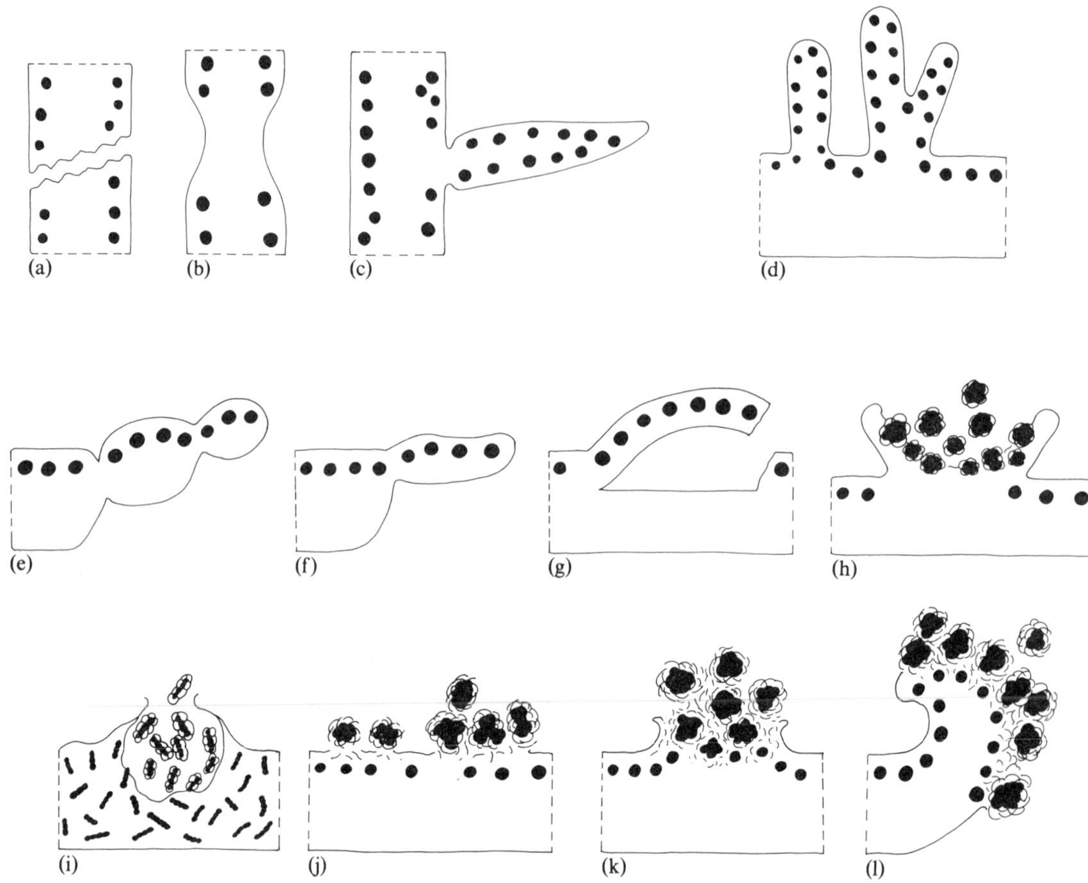

Fig. 29. Types of vegetative propagules (after Hawksworth & Hill, 1984). (a) Thallus fragmentation; (b) fragmentation region; (c) lateral spinule; (d) isidia; (e) blastidia; (f) phyllidium; (g) schizidium; (h) goniocysts; (i) hormocysts; (j) soredia formed from eroded surface; (k) soredia formed in soralium; (l) soredia formed from lower cortex of recurved lobe (labriform).

concolorous, of the same colour.
concrescent, becoming jointed.
confluent, coming together; running into one another.
congeneric, belonging to the same genus.
conglutinate, glued together.
conidiogenous, producing conidia.
conidioma (pl. **conidiomata**), multi-hyphal, conidium-containing structure.
conidiophore, a simple or branched hypha bearing cells from which conidia are produced.
conspecific, belonging to the same species.
corrugate, wrinkled.
cortex, outer layer of a thallus.
corticolous, on bark.
corymbose, arranged in clusters.
crateriform, crater-like.
crenate, having the edge toothed with rounded teeth.
crenulate, delicately crenate.
crozier, the hook of an ascogenous hypha before ascus-development.
crustose, crust-like (Fig. 36 a-d).
cupulate, cup-like.

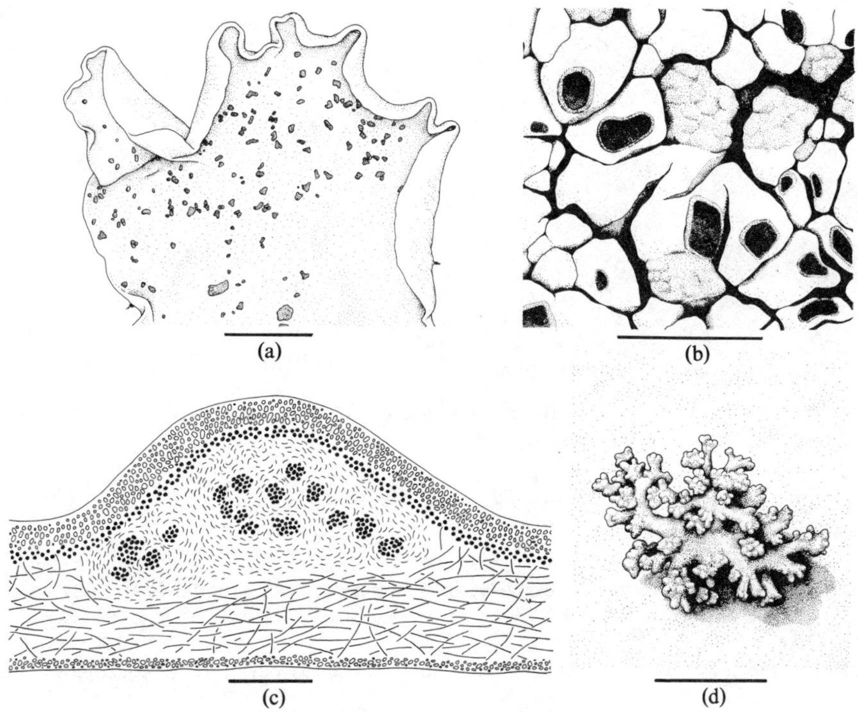

Fig. 30. Cephalodia. (a) *Peltigera britannica*; (b) *Amygdalaria pelobotryon*; (c) *Nephroma arcticum*, internal cephalodium; (d) *Lobaria amplissima* young shrubby cephalodium. Scale a = 5 mm, b = 1 mm, c = 100 μm, d = 500 μm.

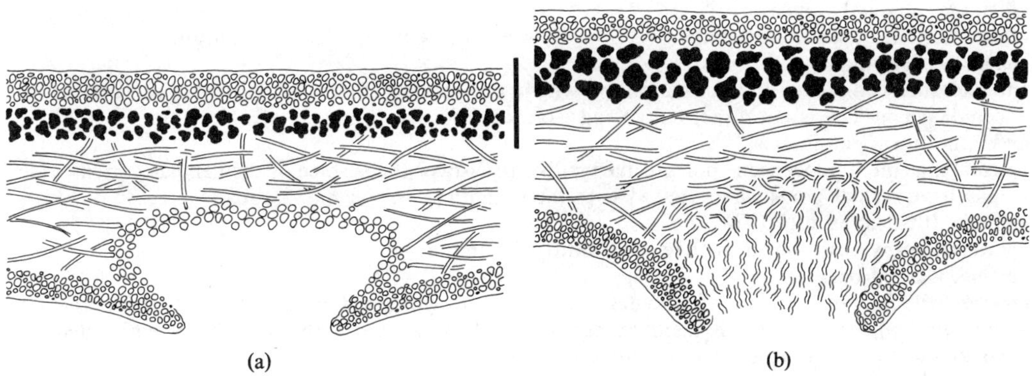

Fig. 31. Cyphella and pseudocyphella in section. (a) Cyphella (*Sticta sylvatica*); (b) pseudocyphella (*Pseudocyphellaria intricata*). Scale = 100 μm.

cyphella (pl. **cyphellae**), break in the lower (rarely upper) cortex of a lichen thallus which is roundish or ovate and in section appears as a cup-like structure lined with a layer of loosely connected, frequently globular, cells formed from the medulla (Fig. 31a).
dacryoid, tear-like.
dactyliform, finger-like.
decumbent, resting on the substratum with the ends turned up.
decorticate, having no cortex, but formerly with one (cf. ecorticate).
dentate, toothed.

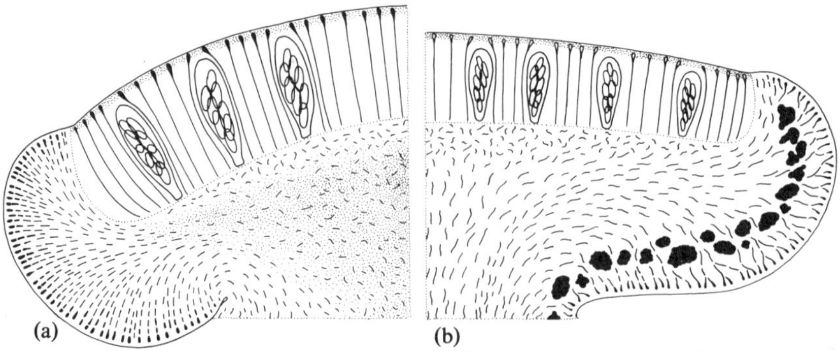

Fig. 32. Structure of apothecia. (a) Lecideine; (b) lecanorine. Showing exciple types, hypothecium, generative layers, paraphyses, asci, ascospores.

denticulate, having small teeth.
determinate, well-defined, definite.
dichotomous, branching, frequently successively, into two more or less equal arms; (cf. anisotomic, isotomic).
digitate, finger-like.
dimidiate (of a perithecial wall), the outer wall covering only the top part and not continuous below the asci.
dimorphic, having two forms, e.g. both a crustose and fruticose thallus as in *Cladonia*.
distal, situated away from the centre of a body, or from the point of origin; terminal.
distoseptate, having the individual cells delimited by a wall formed within the outer wall.
doliiform, barrel-like in form.
e-, lacking, without.
effigurate, obscurely lobed.
effuse, stretched out flat, especially as a spreading growth without a distinct margin.
endophloeodal, immersed in bark.
enteroblastic, conidia in the formation of which the existing inner or neither wall layer of the conidiogenous cell is not directly involved; formed from the inside.
entire, not torn; continuous.
epicortex, a thin homogeneous polysaccharide-like layer over the surface of the cellular cortex, which may have regular pores when viewed with the scanning electron microscopy.
epilithic, (thallus) on rocks, superficial.
epinecral, horny dead fungal hyphae with indistinct lumina.
epiphloeodal, on bark
epispore, the fundamental and often outer wall of a spore which determines its shape (cf. perispore).
epithecium, upper part of the hymenium, consisting of the apices of the paraphyses embedded in a gelatinous substance which is often coloured (Fig. 32).
erose, eroded.
erumpent, bursting through the surface (Fig. 43g).
evanescent, short-lived.
eutrophication, nutrient-enrichment.
excavate, hollowed out (Fig. 43h).
exciple, tissue forming the margins or walls of an ascoma (Figs. 32, 41).
fabiform, bean-shaped.
falcate, curved like the blade of a scythe or sickle (Fig. 27o).
farinose, like flour.
fascicle, a little group or bundle.
faveolate, honey-combed.
fenestrate, having windows or openings.

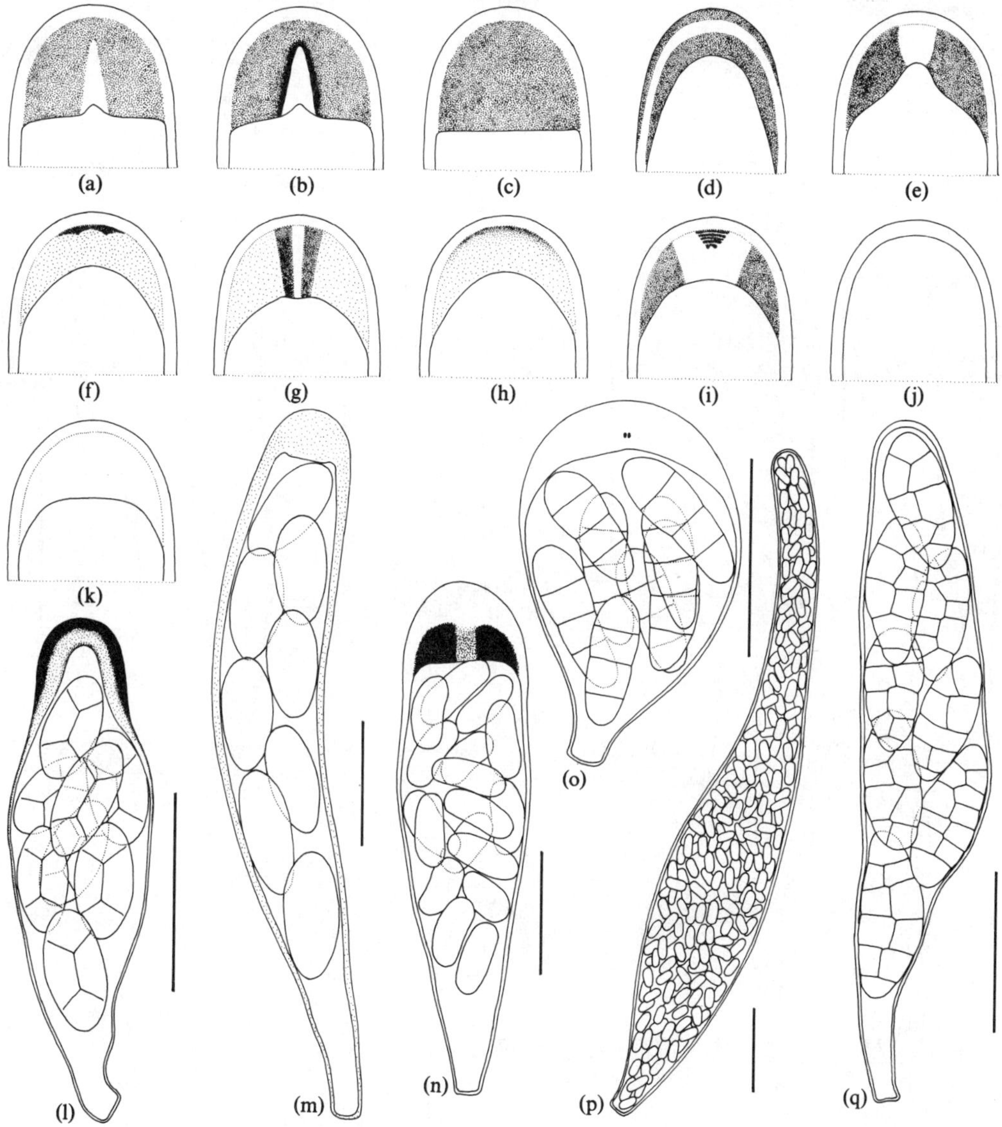

Fig. 33. Ascus structure (1). (a-k) ascus apex types and (l-q) examples. (a) *Bacidia*-type; (b) *Biatora*-type; (c) *Catillaria*-type; (d) *Fuscidea*-type; (e) *Lecanora*-type; (f) *Lecidea*-type; (g) *Porpidia*-type; (h) *Rhizocarpon*-type; (i) *Rimularia*-type; (j) *Schaereria*-type; (k) *Tremolecia*-type; (l) *Xanthoria parietina* (*Teloschistes*-type); (m) *Trapelia coarctata*; (n) *Candelariella vitellina*; (o) *Arthonia radiata*; (p) *Thelocarpon epibolum*; (q) *Gyalecta truncigena*. All in I (Lugol's) after pretreatment with K. Scale l-q = 20 mm.

fibril, short simple branches, perpendicular to the main branches.
filamentose (filamentous), lichen thalli in which the photobiont is thread-like and surrounded by hyphae or cells of the mycobiont (Figs. 13, 44a).
filiform, thread-like (Fig. 27g).
fimbriate, delicately toothed or fringed.
fissitunicate, ascus discharge involving the separation of wall layers (Fig. 34).
fistulose, hollow.

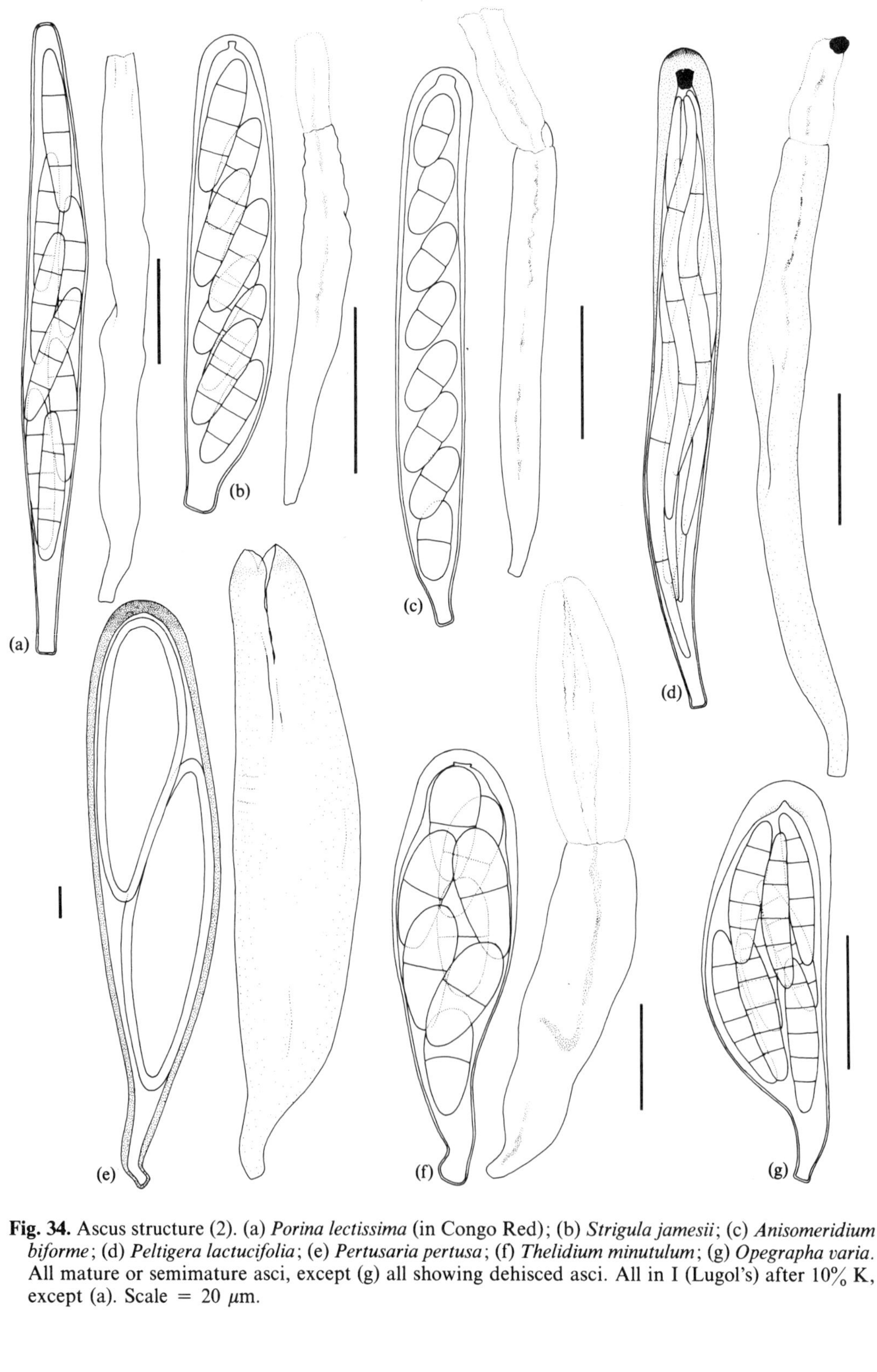

Fig. 34. Ascus structure (2). (a) *Porina lectissima* (in Congo Red); (b) *Strigula jamesii*; (c) *Anisomeridium biforme*; (d) *Peltigera lactucifolia*; (e) *Pertusaria pertusa*; (f) *Thelidium minutulum*; (g) *Opegrapha varia*. All mature or semimature asci, except (g) all showing dehisced asci. All in I (Lugol's) after 10% K, except (a). Scale = 20 μm.

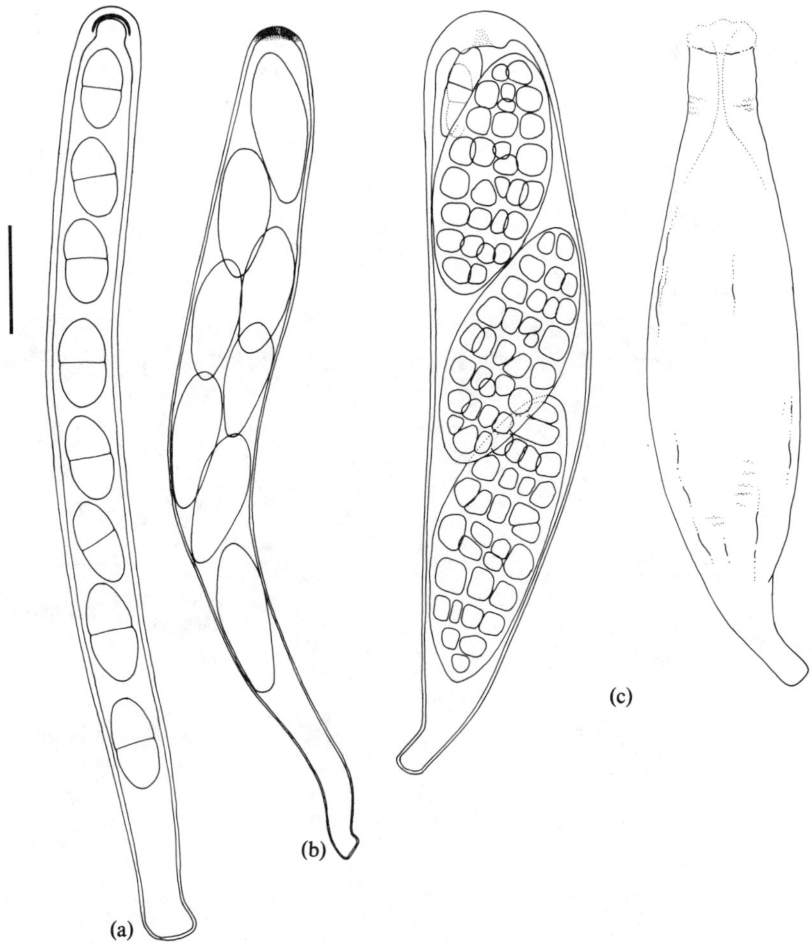

Fig. 35. Ascus structure (3). (a) *Acrocordia conoidea* (in Congo Red); (b) *Thrombium epigaeum* (in Lugol's); (c) *Graphina ruiziana* (in Lugol's, mature and dehisced asci). Scale = 20 μm.

flexuous (flexuose), wavy.
foliicolous, living on leaves.
foliole, a small leaf-like structure on the surface of a foliose lichen.
foliose, leaf-like, of lichen thalli with an upper and lower cortex separable from the substratum (Figs. 36g, h).
foveate, pitted.
foveolate, delicately pitted.
furcate, forked.
fusiform, spindle-like; narrowing towards the ends (Fig. 27f).
fruticose, shrub-, beard- or worm-like, of lichen thalli ± radially symmetrical in cross section (Figs. 36h-j).
fuzzy coat, outermost layer of ascus wall, often staining I+ blue (Fig. 2a).
glabrous, smooth, not hairy.
globose, spherical or almost so (Fig. 27a).
goniocyst, a group of algal cells surrounded by hyphae forming a rounded structure which is not a soredium (Fig. 29h).

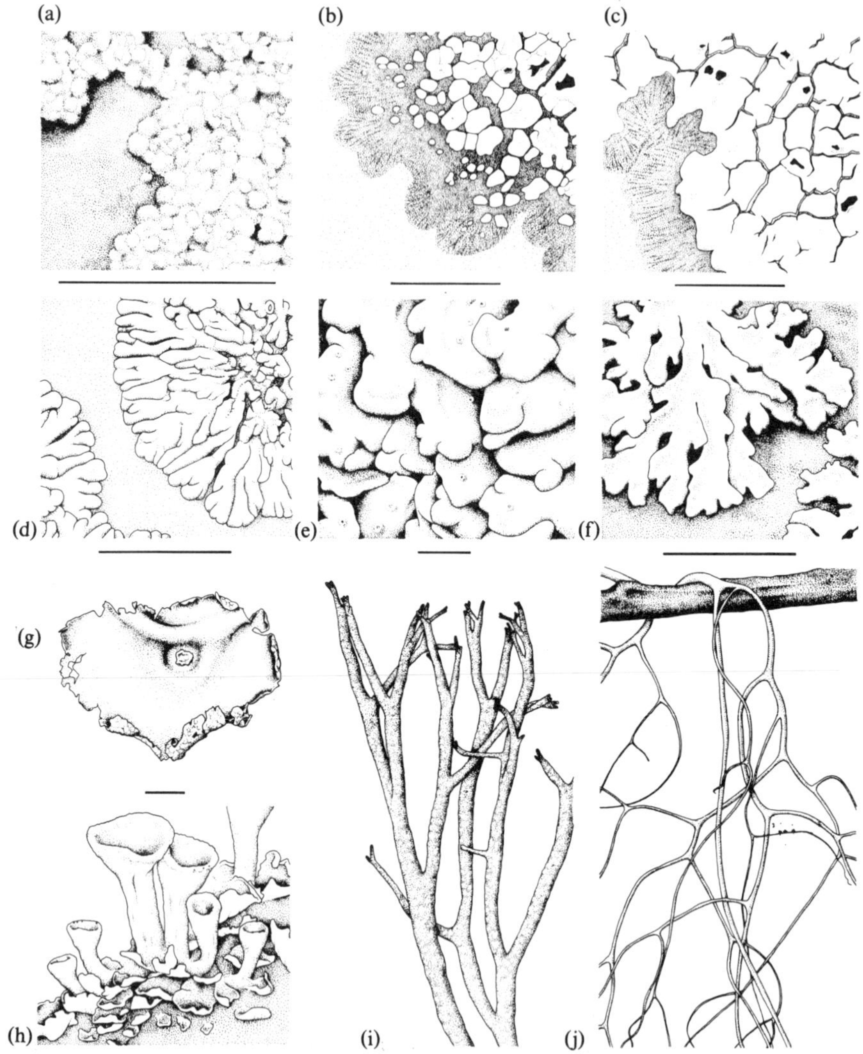

Fig. 36. Thallus types. (a) crustose: leprose (*Leproloma diffusum*); (b) crustose: areolate (*Buellia ocellata*); (c) crustose: rimose (*Buellia subdisciformis*); (d) crustose: placodioid (*Caloplaca thallincola*); (e) squamulose (*Endocarpon adscendens*); (f) foliose (*Physconia distorta*); (g) foliose: umbilicate, underside (*Umbilicaria proboscidea*); (h) fruticose: cup-like (*Cladonia humilis*); (i) fruticose: shrub-like (*Cladonia furcata*); (j) fruticose: beard-like (*Bryoria capillaris*). Scales: a-c, e = 1 mm; d, f-j = 5 mm.

guttule, small drop or drop-like spot in spores.

guttulate, (of spores), having one or more oil-like drops inside.

gyrose, circularly folded or brain-like (Fig. 28d).

halonate (of a spore), having a thick transparent outer coat (perispore).

hamathecium, a neutral term for all kinds of hyphae or other tissues between asci, or projecting into the locule or ostiole of an ascoma (Fig. 38).

hapter, aerial organ of attachment of some shrubby lichens formed by a secondary branch.

hemiangiocarpic (of an apothecium), opening before quite mature.

hepaticolous, on liverworts.

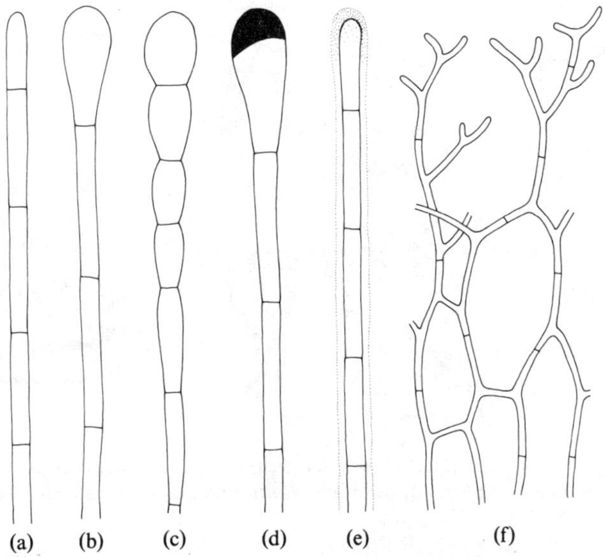

Fig. 37. Hamathecial elements. (a) septate; (b) apex swollen, capitate; (c) moniliform; (d) apex swollen with pigmented cap; (e) apex not swollen with pigment in gel-like matrix surrounding apex; (f) branched and anstomosing.

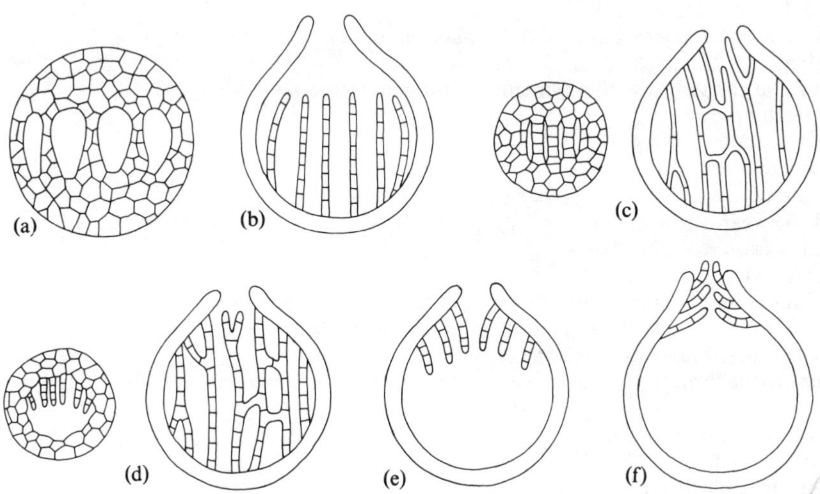

Fig. 38. Hamathecium types (adapted from Eriksson, 1981). (a) Interascal pseudoparenchyma; (b) paraphyses; (c) paraphysoids (trabeculate pseudoparaphyses); (d) pseudoparaphyses (cellular pseudoparaphyses); (e) periphysoids; (f) periphyses.

heteromerous, layered.

hispid, hairy.

holdfast, a part of a thallus adopted for attachment (Fig. 39d).

holoblastic, conidia in the formation of which the existing wall layer of the conidiogenous cell is directly involved; formed from the whole.

homoiomerous, not layered.

hormocyst, a vegetative propagule produced in a special hormocystangium comprising a few photobiont cells and fungal hyphae (Fig. 29i).

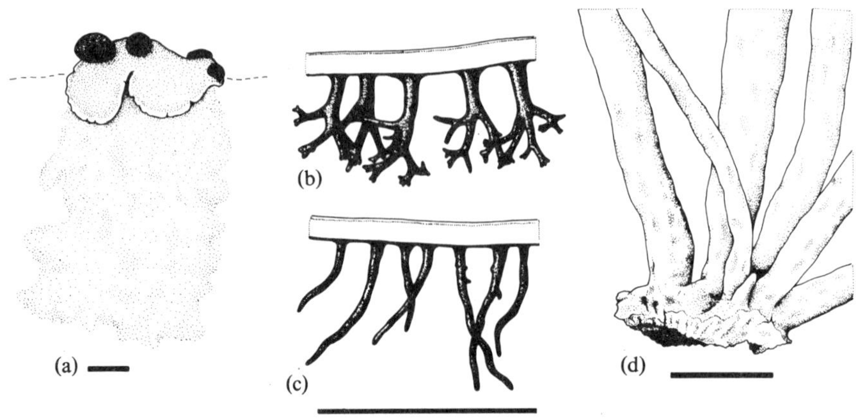

Fig. 39. Attachment organs. (a) Hyphal net (*Psora decipiens*); (b) branched rhizines (*Parmelia laevigata*); (c) simple rhizines (*Parmelia crinita*); (d) holdfast (*Ramalina cuspidata*). Scales: a-c = 1 mm; d = 5 mm.

hyaline, ± transparent, colourless.
hymenium, the layer of tissue in which the asci arise (Fig. 32).
hyphophore, erect stalked or peltate asexual sporophore.
hypophloeodal, immersed in bark.
hypothallus, a layer of spongy tissue without photobiont cells on the underside of the thallus of certain macrolichens.
hypothecium, the tissue below the hymenium and generative layer (Fig. 32).
imbricate, overlapping.
immarginate, without a margin.
indeterminate, effuse.
innate, immersed.
interthecial, between the asci.
intraparietal, within a wall or walls.
isidiose, isidia-like.
isidium, a photobiont-containing protuberance of the cortex in lichens which may be warty, cylindrical, clavate, scale-like, coralloid, simple, or branched (Fig. 29d).
isotomic, with equal branches (Fig. 5b).
involucrellum, tissue forming the upper part of some perithecia, distinct from and surrounding the exciple (Fig. 40).
involute, with margins rolled in.
labriform, lip-shaped (Fig. 43e).
lacerate, with irregularly cut or torn margins.
lacuna, a depression or pit.
lacunose, with depressions.
lamina, the main part of the thallus in foliose lichens.
lecanorine, with a thalline exciple (Figs. 28b, 32b).
lecideine, without a thalline exciple (Figs. 28a, 32a).
lenticular like a double convex lens in form (Fig. 27).
leprose, having the surface of the whole thallus entirely granular without any cortex even in young stages (Fig. 36a).
lichenicolous, growing on lichens.
lignicolous, growing on wood.
ligulate, strap-like.
lirella (pl. **lirellae**), long, narrow apothecium (Fig. 28e).
lobate, lobed.

lobulate, having small lobes.
locule, cavity, especially in a stroma or perthecium.
lumen, central cavity of a cell or other structure.
macroconidium, the larger conidium of a species that has more than one type of conidium.
macronematous (of conidiophores), morphologically different from the vegetative hyphae (cf. micronematous)
macula, a very small, light-coloured spot on the upper surface of the thallus, often caused by the grouping of the algae beneath the surface.
maculate, with macula.
marginate, with a well-marked edge.
mazaedium, a dry powdery ascospore mass in which the ascospores become free from the asci.
medulla, the loose layer of hyphae below the cortex and algal layer.
melanized, containing dark brown pigments.
mesoconidium, an intermediate-sized conidium in a species with three conidium types.
microconidium, the smallest conidium in a species with two or more conidium types.
mischoblastia, see (Fig. 25k).
mischoblastiomorph, see (Fig. 25k).
moniliform, in bead-like chains (Fig. 37c).
monoblastic, producing a blastic conidium at one point.
monophyllous, single-lobed.
morph, a particular shape or form.
morphotype, a group of morphologically differentiated individuals of unknown or of no taxonomic significance.
mucronate, ending in a short, sharp point.
muriform, divided by transverse and vertical or oblique cross walls.
muscicolous, on mosses.
necral, a layer of dead tissue, giving a white, granular appearance.
ob-, inversely.
ocular chamber, the finger-like protrusion of the inner part of a bitunicate ascus into the inner tunica; apical beak (Fig. 2d).
-oid, resembling, like.
ostiole, the opening of a perithecioid ascoma or pycnidial conidioma.
palisade, hyphae $\pm$ perpendicular to the surface.
palmate, hand-shaped.
papilla, small, wart-like proturberance.
paraphysis (pl. **paraphyses**), a slender jointed filament growing upward and basally attached in an ascoma (Figs. 37, 38b).
paraphysoid, interascal or pre-ascal tissue stretching and coming to resemble pseudoparaphyses; very often remotely septate, anastomosing and very narrow (Fig. 38c).
paraplectenchyma, tissue consisting of cell-like hyphae with $\pm$ isodiametric lumina.
parasymbiotic, commensalistic.
pectinate, like the teeth of a comb.
pellicle, a delicate outside membrane.
peltate, plate-like with a stalk from the centre of the lower surface.
penicillate, like a little brush.
periclinal, curved in the direction of, or parallel to, the surface or the circumference.
periphysis (pl. **periphyses**), hair-like projection inside, the ostiole of a perithecium or pycnidium (Figs. 38f, 40).
periphysoid, short interascal filaments growing down from the top of a peritheciod ascoma (Fig. 38e).
perispore, a colourless, often gelatinous layer enveloping a spore outside the main spore wall (cf. epispore).
perithecium, a subglobose or flask-like ascoma in which the hymenium is not exposed (Figs. 40–41).
photobiont, a photosynthetic symbiont which may be either a green alga, or a cyanobacterium (blue-green alga) (Fig. 42).
phyllidium, a dorsiventral leaf-like growth constricted at the point of attachment (Fig. 29f).
phyllocladia, squamule-like structures occurring in *Stereocaulon*.
placodioid, crustose thalli with radiating lobe margins (Fig. 36d).

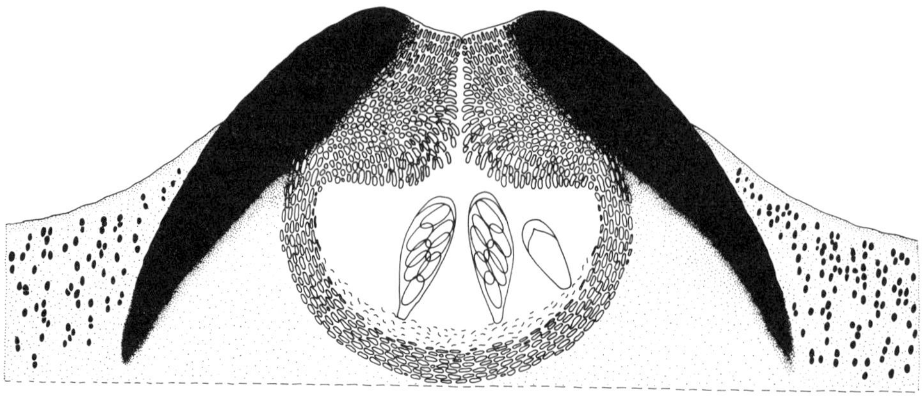

Fig. 40. Structure of perithecia. Showing involucrellum, exciple, subhymenium, asci and periphyses.

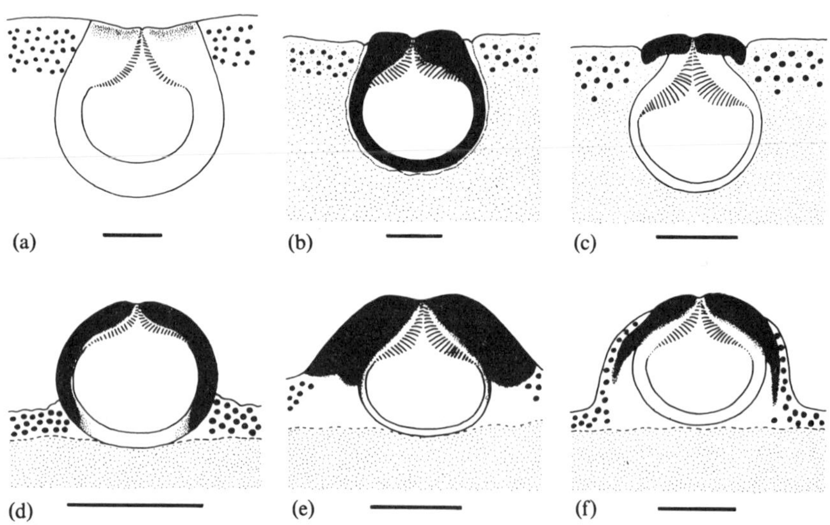

Fig. 41. Types of perithecia, omitting asci and subhymenial tissues. (a) immersed in foliose lichenized thallus, involucrellum absent, exciple pale (*Dermatocarpon miniatum*); (b) immersed in limestone, involucrellum absent, exciple black and friable (*Verrucaria hochstetteri*); (c) immersed in limestone, involucrellum lid-like ('dimidiate'), exciple pale except at the apex (*Verrucaria baldensis*); (d) prominent, involucrellum absent, exciple black and friable above, pale at base (*Thelidium minutulum*); (e) involucrellum well-developed, exciple pale below (*Thelidium papulare*); (f) immersed in thalline wart, involucrellum present, exciple pale below (*Polyblastia cruenta*). Scale = 200 μm.

plectenchyma, a thick tissue formed by hyphae becoming twisted and fixed together.
plicate, folded into pleats.
pleurogenous, (of conidia) formed on the side of conidiophores.
plica, a fold of skin, membrane, or lamella.
podetium, lichenized stem-like portion (stipe) of an apothecium bearing the hymenial discs and sometimes conidiomata in a fruticose lichen.
polarilocular (of ascospores), bicellular with the two cells separated by a thick central perforated septum (Fig. 7).

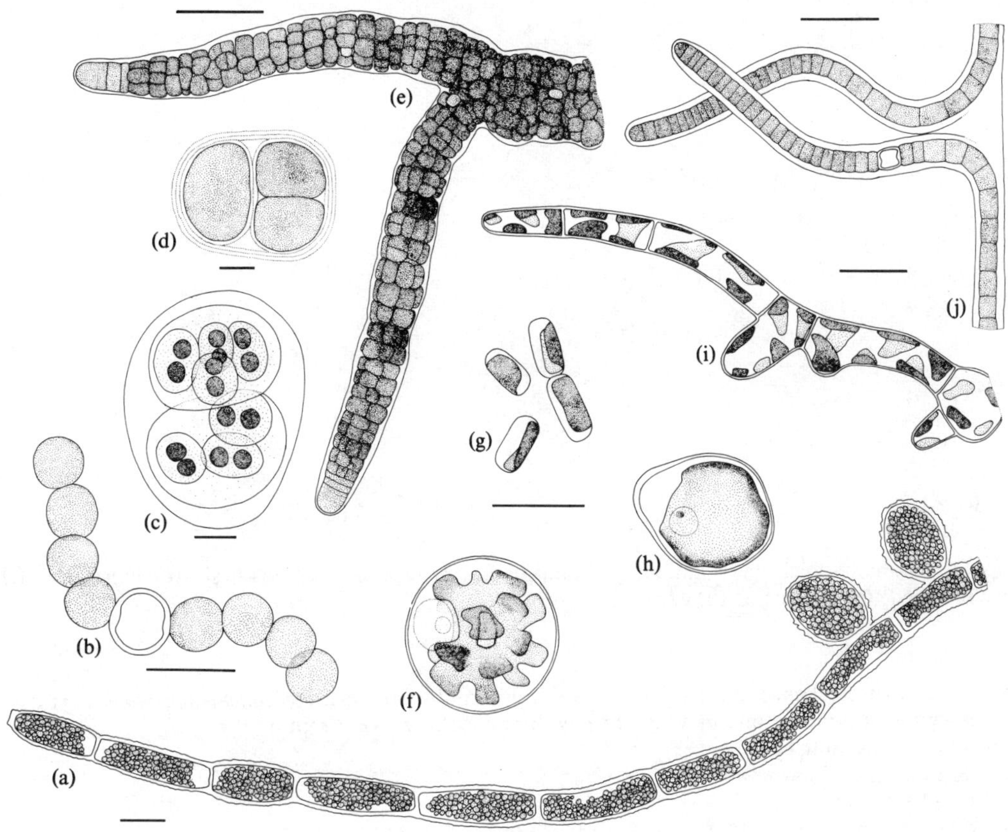

Fig. 42. Photobionts, examples of selected genera (redrawn from a variety of sources). (a) *Trentepohlia* (Chlorophyceae); (b) *Nostoc* (Cyanobacteria); (c) *Gloeocapsa* (Cyanobacteria); (d) *Chroococcus* (Cyanobacteria); (e) *Stigonema* (Cyanobacteria); (f) *Trebouxia* (Chlorophyceae); (g) *Stichococcus* (Chlorophyceae); (h) *Myrmecia* (Chlorophyceae); (i) *Heterococcus* (Xanthophyceae); (j) *Scytonema* (Cyanobacteria). Scale a-d, f-i = 10 μm; e, j = 100 μm.

polyblastic (of conidiogenous cell), producing blastic conidia at several points.
polyphyllous, multi-lobed.
polytomic, forking in groups of several equal branches.
poriform, pore-like.
prosoplectenchyma, a plectenchyma composed of cells with longish lumina.
prothallus, initial structure of hyphae without algae from which a lichenized thallus develops, often visible along the edge of the thallus or areoles.
pruina, a frost-like or flour-like surface covering.
pseudocyphella (pl. **pseudocyphellae**), dot-like to fusiform or irregular area pale areas of the thallus where the medulla comes to the surface (Fig. 31b).
pseudo-, false, spurious, looking like.
pseudoisidium, an outgrowth from the surface of a lichen thallus resembling an isidium but lacking algae.
pseudoparaphysis (pl. **pseudoparaphyses**), hyphae originating below the level of the asci and growing downwards between the developing asci, finally becoming attached to the base of the cavity and often also then free in the upper part; often regularly septate, branched and anastomosing and broader (Fig. 38d).
pseudoparenchymatous, a cell-like structure formed from hyphae.

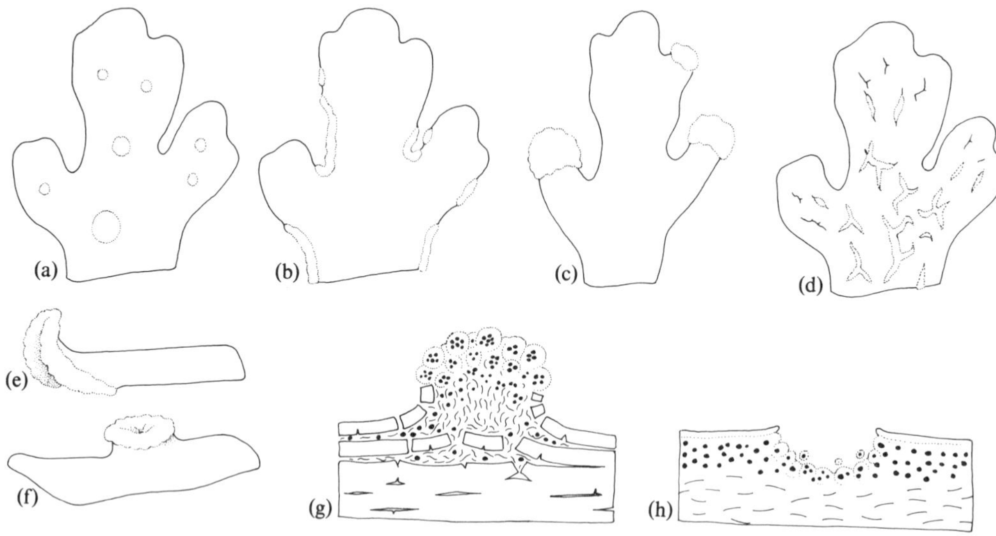

Fig. 43. Soralium types. (a) Laminal; (b) marginal; (c) terminal; (d) fissural; (e) labriform; (f) cuff-shaped; (g) erumpent; (h) excavate.

pseudopodetium, erect or ascending, not dorsiventral, shrubby parts in *Stereocaulon* and *Pilophorus* thalli.
pseudoseptum, a protoplasmic or vacuolar membrane looking like a septum.
pubescent, having soft hairs.
puncta, small spots.
pustule, blister-like.
pycnidium (pl. **pycnidia**), a frequently ± flask-shaped, ostiolate conidioma.
pyrenocarpous, with perithecioid ascoma.
pyriform, pear-like (Fig. 27r).
pycnoascocarp, ascoma arising from a pycnidium.
reniform, kidney-like (Fig. 27p).
reticulate, like a net, netted.
rhizine, a root-like hair or thread of hyphae acting as an attachment organ (Figs. 39b–c).
rimose, irregularly cracked in all directions.
rugose, roughened.
saccate, sac-like.
saxicolous, on rocks.
scabrid, rough with delicate and irregular projections.
schizidium, a propagule formed by upper layers of a lichen thallus splitting off as scale-like segments from the main lobes (Fig. 29g).
scleroplectenchymatous, a plectenchyma composed of very thick-walled conglutinate cells.
scyphus, cup.
septum, cell wall or partition.
sessile, having no stem, arising on the surface.
sigmoid, curved like the letter S (Fig. 27n).
simple, (of spores) without septa.
sinuate, having rounded angles.
soleiform, shaped like the sole of a shoe.
soralium (pl. **soralia**), a structure or region of a thallus bearing soredia (Fig. 43).
soredium (pl. **soredia**), a non-corticate combination of photobiont cells and fungal hyphae, the 'granules' variously appearing flour-like, castor sugar-like or granular with a × 10 lens (Fig. 29j–l).
spathulate, spoon-like (Fig. 27v).
spinule, a small spine constricted at the base (Fig. 29c).

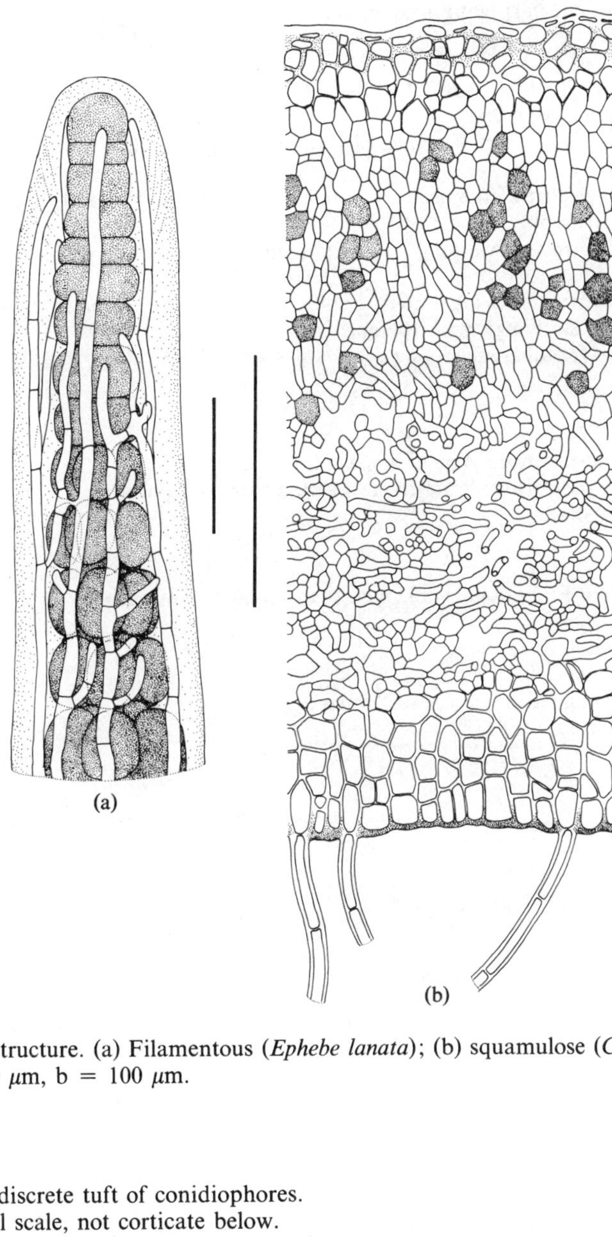

Fig. 44. Thallus structure. (a) Filamentous (*Ephebe lanata*); (b) squamulose (*Catapyrenium lachneum*). Scales: a = 20 μm, b = 100 μm.

sporodochium, a discrete tuft of conidiophores.
squamule, a small scale, not corticate below.
squamulose, having small scales (Figs. 36e, 44b).
squarrose brush-like, of rhizines.
stellate, star-like.
stipe, stalk of an ascoma or basidioma.
stipitate, stalked.
striate, marked with delicate lines, grooves, or ridges.
stroma, a thick compact wad of (usually blackish) hyphae, often covering or supporting a group of ascoma.
sub-, under, below; frequently in the sense of approximating to the condition qualified.
subulate, slender and tapering to a point; awl shaped (Fig. 27x).
tartareous, having a thick rough crumbling surface.

terete, circular in section but narrowing to one end.
terricolous, on the ground.
tetrachotomous, 4-branched.
thalline exciple an exciple which contains algal as well as fungal cells; usually of the same colour and consistency as the thallus (Fig. 28b).
thalloconidia, conidia arising from the surface of a thallus or prothallus.
thallus, the vegetative body.
tholus, the thickened inner part of the ascus wall in the ascus apex (Fig. 2b); apical dome.
tomentum, felt-like mat of hyphal projections.
trama, the layer of hyphae in the central part of a lamella of an agaric.
trichotomous, 3-branched.
true exciple, an exciple which lacks algal cells, usually of a different colour to the thallus (Fig. 28a).
truncate, ending abruptly, as though with the end cut off.
tubercule, small wart-like protuberance.
turbinate, top-like.
turgid, swollen.
ulcerose, ulcer-like.
umbilicus, the central hold-fast occurring in some foliose lichens.
umbo, a central raised tissue; often sterile tissue in the centre of apothecia.
umbonate, having an umbo(s).
undulate, wavy.
uniseriate, in one row.
unitunicate (of asci), with one functional wall layer, the wall layers not splitting apart at discharge; non-fissitunicate.
urceolate, deeply concave.
verruca, a wart-like swelling.
verrucose, having small rounded processes or "warts".
verruculose, delicately warted.

References

Aguirre-Hudson, B. (1991). A taxonomic study of the species referred to the ascomycete genus *Leptorhaphis* Körber. *Bull. Br. Mus. nat. Hist., Bot.*: **21**: 85–192.

Aguirre, B. & Hawksworth, D. L. (1987). The circumscription, biology and relationships of the genus *Leptorhaphis* Körber. *Bibliotheca lich.* **25**: 249–255.

Ahmadjian, V. & Hale, M. E., eds (1974) ["1973"]. *The Lichens*. 697 pp. New York: Academic Press.

Ahti, T. (1961). Taxonomic studies on reindeer lichens (*Cladonia*, subgenus *Cladina*). *Suomal. eläin-ja Kasvit. Seur. van. Kasvit. Julk.* **32**(1): 1–160.

Ahti, T. (1966). Correlation of the chemical and morphological characters in *Cladonia chlorophaea* and allied lichens. *Ann. Bot. fenn.* **3**: 380–390.

Ahti, T. (1978). Nomenclatural and taxonomic remarks on European species of *Cladonia*. *Ann. Bot. fenn.* **15**: 7–14.

Ahti, T. (1980). Nomenclatural notes on *Cladonia* species. *Lichenologist* **12**: 125–133.

Ahti, T. (1984). The status of *Cladina* as a genus segregated from *Cladonia*. *Beih. Nova Hedwigia* **79**: 25–62.

Ahti, T. & Henssen, A. (1965). New localities for *Cavernularia hultenii* in eastern and western North America. *Bryologist* **68**: 85–89.

Alstrup, V. & Hawksworth, D. L. (1990). The lichenicolous fungi of Greenland. *Meddr Grønl., Biosci.* **31**: 1–90.

Bailey, R. H. (1974). Distribution maps of lichens in Britain. Map 8. *Cyphelium inquinans*. *Lichenologist* **6**: 169–171.

Bailey, R. H. & James, P. W. (1977). Distribution maps of lichens in Britain. Map 23. *Thelotrema lepadinum* (Ach.) Ach. *Lichenologist* **9**: 175–179.

Barr, M.E. (1986) ["1985"]. On *Julella*, *Delacourea*, and *Decaisnella*, three dictyosporous genera described by J. H. Fabre. *Sydowia* **38**: 11–19.

Beltman, H. (1978). Vegetative Strukturen der Parmeliaceae und ihre Entwicklung. *Bibliotheca lich.* **11**: 1–193.

Brako, L. (1989). Re-evaluation of the genus *Phyllopsora* with taxonomic notes and introduction of *Squamacidia*, gen. nov. *Mycotaxon* **35**: 1–19.

Breuss, O. (1983). Eine neue *Placidiopsis*-Art (Lichenes, Verrucariaceae) aus Norwegen und den Alpen. *Plant Syst. Evol.* **142**: 247–250.

Breuss, O. (1990). Die Flechtengattung *Catapyrenium* (Verrucariaceae) in Europa. *Stapfia* **23**: 1–153.

Breuss, O. & Hansen, E. S. (1988). The lichen genera *Catapyrenium* and *Placidiopsis* in Greenland. *Plant Syst. Evol.* **159**: 95–105.

Brodo, I. M. (1973). The lichen genus *Coccotrema* in North America. *Bryologist* **76**: 260–270.

Brodo, I. M. (1984). The North American species of the *Lecanora subfusca* group. *Beih. Nova Hedwigia* **79**: 63–185.

Brodo, I. M. & Hawksworth, D. L. (1977). *Alectoria* and allied genera in North America. *Op. bot. Soc. bot. Lund* **42**: 1–164.

Brodo, I. M. & Hertel, H. (1987). The lichen genus *Amygdalaria* (Porpidiaceae) in North America. *Herzogia* **7**: 493–521.

Brown, D. H., Hawksworth, D. L. & Bailey, R. H., eds (1976). *Lichenology: Progress and Problems*. 551 pp. London: Academic Press.

Cannon, P. F., Hawksworth, D. L. & Sherwood-Pike, M. A. (1985). *The British Ascomycotina: An annotated checklist*. 302 pp. Kew: Commonwealth Mycological Institute.

Clauzade, G. & Roux, C. (1981). Les *Acarospora* de l'Europe occidentale et de la région Méditerranéenne. *Bull. Mus. Hist. nat. Marseille* **41**: 41–93.

Clauzade, G. &. Roux, C. (1984). Les genres *Aspicilia* Massal. *et Bellemerea* Hafellner et Roux. *Bull. Soc. bot. Centre-Ouest, n.s.* **15**: 127–141.

Clauzade G. & Roux C. (1985). Likenoj de okcidenta Eŭropo, illustrita determinlibro. *Bull. Soc. bot. Centre-Ouest, n. sér., num. spéc.* **7**: 1–893. [Supplements *loc. cit.* **18**:177–214 (1987); *Bull. Soc. linn. Provence* **40**:73–110 (1989).]

Clémençon, H. (1982). Kompendium der Blätterpilze. Europäische omphalinoide Tricholomataceae. *Z. Mykol.* **48**: 195–237.

Clerc, P. (1984a). Contribution à la révision de la systématique des Usnées (Ascomycotina, *Usnea*) d'Europe I. - *Usnea florida* (L.) Wigg. emend. Clerc. *Cryptogamie, bryol. lichén.* **5**: 333–360.

Clerc, P. (1984b). *Usnea wirthii* - a new species of lichen from Europe and North Africa. *Saussurea* **15**: 33–36.

Clerc, P. (1987a). Systematics of the *Usnea fragilescens* aggregate and its distribution in Scandinavia. *Nordic J. Bot.* **7**: 479–495.

Clerc, P. (1987b). On the morphology of soralia in the genus *Usnea*. *Bibliotheca lich.* 25: 99–107.

Coppins, B. J. (1976). Distribution patterns shown by epiphytic lichens in the British Isles. In *Lichenology: Progress and Problems* (D. H. Brown, D. L. Hawksworth & R. H. Bailey, eds.):249–278. London: Academic Press.

Coppins, B. J. (1983). A taxonomic study of the lichen genus *Micarea* in Europe. *Bull. Br. Mus. nat. Hist., Bot.* **11**: 17–214.

Coppins, B. J. (1984). Key to crustose pyrenocarpous lichens on limestone and associated substrata. *Bull. Br. lichen Soc.* **54**: 36–45.

Coppins, B. J. (1985). A new *Micarea* from the Scottish Highlands. *Lichenologist* **17**: 99–101.

Coppins, B. J. (1987a). The genus *Ramonia* in the British Isles. *Lichenologist* **19**: 409–417.

Coppins, B. J. (1987b). The genus *Vezdaea* in the British Isles. *Lichenologist* **19**: 167–181.

Coppins, B. J. (1988a). Notes on the genus *Arthopyrenia* in the British Isles. *Lichenologist* **20**: 305–325.

Coppins, B. J. (1988b). Two new *Micarea* species from Europe. *Notes R. Bot. Gdn. Edinb.* **45**: 161–169.

Coppins, B. J. (1989a). Notes on the Arthoniaceae in the British Isles. *Lichenologist* **21**: 195–216.

Coppins, B. J. (1989b). On some species of *Catillaria* s. lat. and *Halecania* in the British Isles. *Lichenologist* **21**: 217–227.

Coppins, B. J. (1989c). *Rinodina griseosoralifera*, a new corticolous sorediate lichen from Western Europe. *Lichenologist* **21:** 169–172.

Coppins, B. J. (1989d). *Micarea pseudomarginata*, a new saxicolous lichen from the British Isles. *Notes R. Bot. Gdn Edinb.* **45**: 397–400.

Coppins, B. J. & Bennell, A.P. (1979). Thallus surface features in *Agonimia tristicula*. *Lichenologist* **11:** 107–108.

Coppins, B. J. & James, P. W. (1974). Distribution maps of lichens in Britain. Maps 9–10. *Cyphelium notarisii* and *C. tigillare*. *Lichenologist* **6**: 172–177.

Coppins, B. J. & James, P. W. (1978). New or interesting British lichens II. *Lichenologist* **10**: 179–207.

Coppins, B. J. & James, P. W. (1979a). New or interesting British lichens III. *Lichenologist* **11**: 27–45.

Coppins, B. J. & James, P. W. (1979b). New or interesting British lichens IV. *Lichenologist* **11**: 139–179.

Coppins, B. J. & James, P. W. (1979c). A British species of *Dictyonema*. *Lichenologist* **11**: 103–105.

Coppins, B. J. & James, P. W. (1984). New or interesting British lichens V. *Lichenologist* **16**: 241–264.

Coppins, B. J. & James, P.W. (1989). The identity of '*Lecidea cinnabarina*' in the British Isles. *Lichenologist* **21**: 237–242.

Coppins, B. J., James, P. W. & Hawksworth, D. L. (1987). The generic names *Placynthiella* Elenkin, *Saccomorpha* Elenkin and *Placynthiella* Gyelnik. *Lichenologist* **19**: 93–95.

Coppins, B. J., James, P. W. & Hawksworth, D. L. (1992). New species and combinations in "The Lichen Flora of Great Britain and Ireland". *Lichenologist* **24**(4).

Coppins, B. J. & Purvis, O. W. (1987). A review of *Psilolechia*. *Lichenologist* **19**: 29–42.

Coppins, B. J. & Vězda, A. (1977). *Macentina*, a lichen genus new to Europe. *Lichenologist* **9:** 47–49.

Creveld, M. (1981). Epilithic lichen communities in the alpine zone of southern Norway. *Bibliotheca lich.* **17**: 1–287.

Culberson, W. L. (1967) ["1966"]. Chemistry and taxonomy of the lichen genera *Heterodermia* and *Anaptychia* in the Carolinas. *Bryologist* **69**: 472–487.

Culberson, W. L. & Culberson, C. F. (1968). The lichen genera *Cetrelia* and *Platismatia* (Parmeliaceae). *Contr. U.S. natn. Herb.* **34**: 449–558.

Cuberson, W. L. & Culberson, C. F. (1979) ["1978"]. *Cetrelia cetrarioides* and *C. monachorum* (Parmeliaceae) in the New World. *Bryologist* **81**: 517–523.

Dahl, E. & Krog, H. (1973). *Macrolichens of Denmark, Finland, Norway and Sweden*. 184 pp. Oslo, Bergen & Tromsø: Universitetsforlaget.

Darbishire, O. V. (1898). Monographia Roccelleorum. *Bibliotheca bot.* **45**: 1–103.

David, J. C. & Hawksworth, D. L. (1989). *Lauderlindsaya*, a new genus in the Verrucariales for *Sphaerulina chlorococca* (Leighton) R. Sant. *Sydowia* **41**: 108–121.

Degelius, G. (1934). Über *Dermatocarpon rivulorum* (Arn.) DT. et Sarnth. und *D. arnoldianum* Degel. n. sp. *Nyt Mag. Naturvid.* **75**: 151–161.

Degelius, G. (1954). The lichen genus *Collema* in Europe. *Symb. bot. upsal.* **13**(2): 1–499.

Degelius, G. (1974). The lichen genus *Collema* with special reference to the extra-European species. *Symb. bot. upsal.* **20**(2): 1–215.

Dennis, R. W. G. (1981). *British Ascomycetes. Addenda and Corrigenda.* Vaduz: J. Cramer. 40 pp.

Dibben, M. J. (1980). The chemosystematics of the lichen genus *Pertusaria* in North America north of Mexico. *Publs Biol. Geol. Milwaukee Publ. Mus.* **5**: 1–162.

Döbbeler, P. (1984). Symbiosen zwischen Gallertalgen und Gallertpilzen der Gattung *Epigloea* (ascomycetes). *Beih. Nova Hedwigia* **79**: 203–239.

Döbbeler, P. & Poelt, J. (1977). *Absconditella celata* spec. nov., eine Flechtenart aus Lappland. *Herzogia* **4**: 363–366.

Dobson, F. S. (1992). *Lichens. An illustrated Guide to the British and Irish Species.* Third edition. 16 + 376 pp. Slough: Richmond Publishing.

Eigler, G. (1969). Studien zur Gliederung der Flechtengattung *Lecanora. Dissnes. Bot., Lehre* **4:** 1–195.

Elix, J. A. (1979). A taxonomic revision of the lichen genus *Hypogymnia* in Australasia. *Brunonia* **2**: 175–245.

Elix, J. A. & Hale, M. E. (1987). *Canomaculina, Myelochora, Parmelinella, Parmelinopsis* and *Parmotremopsis*, five new genera in the Parmeliaceae (lichenized Ascomycotina). *Mycotaxon* **29**: 233–244.

Ellis, L. T. (1981). A revision and review of *Lemmopsis* and some related species. *Lichenologist* **13**: 123–139.

Eriksson, O. (1981). The families of bitunicate ascomycetes. *Opera bot.* **60**: 1–220.

Eriksson, O. E. & Hawksworth, D. L. (1988). Notes on ascomycete systematics - Nos. 552–732. *Syst. Ascom.* **7**: 59–101.

Eriksson, O. E. & Hawksworth, D. L. (1991). Outline of the ascomycetes - 1990. *Syst. Ascom.* **9**: 39–271.

Esnault, J. & Roux, C. (1987). *Amygdalaria tellensis* (lichens), nouvelle espèce du Tell Algérien. *An. Jard. bot. Madrid* **44**: 211–225.

Esslinger, T. L. (1977). A chemosystematic review of the brown *Parmeliae. J. Hattori bot. Lab.* **42**: 1–211.

Esslinger, T. L. (1978). A new status for the brown *Parmeliae. Mycotaxon* **7**: 45–54.

Faurel, L. & Schotter, G. (1954). Remarques à propos du *Pleolecis geophana* (Nyl.) Clements. *Bull. Soc. Hist. nat. Afr. N.* **25**: 126–133.

Feuerer, T. (1978). Zur Kenntnis der Flechtengattung *Rhizocarpon* in Bayern. *Ber. bayer. bot. Ges.* **49**: 59–135.

Fletcher, A. (1975). Key for the identification of British marine and maritime lichens I. Siliceous rocky shore species. *Lichenologist* **7**: 1–52.

Foucard, T. (1990). *Svensk Skorplavs Flora.* 306 pp. Stockholm: Interpublishing.

Frey, E. (1933). Cladoniaceae (unter Ausschluss der Gattung *Cladonia*), Umbilicariaceae. *Rabenh. Krypt.-Fl.* **9**, 4(1): 1–426.

Galløe, O. (1927–72). *Natural History of the Danish Lichens.* 10 vols. 1397 plates, 7010 figures. Copenhagen: H. Aschehoug.

Galloway, D. J. (1985). *Flora of New Zealand. Lichens.* 662 pp. Wellington: Government Printer.

Galloway, D. J. (1988). Studies in *Pseudocyphellaria* (lichens) I. The New Zealand species. *Bull. Br. Mus. nat. Hist., Bot.* **17**: 1–267.

Galloway, D.J. & James, P.W. (1987). *Metus*, a new austral lichen genus and notes on an Australasian species of *Pycnothelia. Notes R. bot. Gdn. Edinb.* **44**: 561–579.

Galun, M., ed. (1988). *Handbook of Lichenology.* 3 vols. 297 + 181 + 147 pp. Boca Raton, Florida: CRC Press.

Gams, H. (1967). *Kleine Kryptogamenflora.* Vol. 3. *Flechten* (*Lichenes*). viii + 244 pp. Jena: G. Fischer.

Gauslaa, Y. (1985). The ecology of *Lobarion pulmonariae* and *Parmelion caperatae* in *Quercus* dominated forests in south-west Norway. *Lichenologist* **17**: 117–140.

Gilbert, O. L. (1975). Distribution maps of lichens in Britain. Maps 19–22. *Lichenologist* **7**: 181–192.

Gilbert, O. L. (1978). *Fulgensia* in the British Isles. *Lichenologist* **10**: 33–45.

Gilbert, O. L., Henderson, A. & James, P. W. (1981). Citrine-green taxa in the genus *Candelariella. Lichenologist* **13**: 249–251.

Gilenstam, G. (1969). Studies in the genus *Conotrema. Arkiv för Bot.* II **7**: 149–179.

Gowan, S. P. (1989). The lichen genus *Porpidia* (Porpidiaceae) in North America. *Bryologist* **92**: 25–59.

Greuter, W., Burdet, H. M., Chaloner, W. G., Demoulin, V., Grolle, R., Hawksworth, D. L., Nicolson, D. H., Silva, P.C., Stafleu, F. A., Voss, E. G. & McNeill, J. (1988). International Code of Botanical Nomenclature adopted by the Fourteenth International Botanical Congress, Berlin, July-August 1987. *Regnum Vegetabile* **118**: i - xiv, 1–328.

Gyelnik, V.K. (1940). *Vestergrenopsis. Rabenh. Krypt.-Fl.* **9:** 265–268.

Hafellner, J. (1978). *Catolechia* Flotow ex Massalongo emend. Körber und *Epilichen* Clements ex Hafellner - zwei nahe verwandte Flechtengattung. *Nova Hedwigia* **30**: 673–692.

Hafellner, J. (1979). *Karschia*, Revision einer Sammelgattung an der Grenze von lichenisierten und nichtlichenisierten Ascomyceten. *Beih. Nova Hedwigia* **62**: 1–248.

Hafellner, J. (1984). Studien in Richtung einer natürlicheren Gliederung der Sammelfamilien Lecanoraceae und Lecideaceae. *Beih. Nova Hedwigia* **79**: 241–371.

Hafellner, J. (1989). Die europäischen Mycobilimbia-Arten-eine erste Übersicht (lichenisierte Ascomycetes, Lecanorales). *Herzogia* **8**: 53–59.

Hafellner, J., Mayrhofer H. & Poelt, J. (1979). Die Gattungen der Flechtenfamilie Physciaceae. *Herzogia* **5**: 39–79.

Hafellner, J. & Poelt, J. (1979). Die Arten der Gattung *Caloplaca* mit pluriloculären Sporen (*Meroplacis, Triophthalmidium, Xanthocarpia*). *J. Hattori bot. Lab.* **46**: 1–41.

Hakulinen, R. (1954). Die Flechtengattung *Candelariella* Müller Argoviensis, mit besonderer Beruksichtigung ihres Auftretens und ihrer Verbreitung in Fennoskandien. *Suomal. eläin-ja Kasvit. Seur. van. Kasvit. Julk* **27**(3): 1–127.

Hale, M. E. (1965a). A monograph of *Parmelia* subgenus *Amphigymnia*. *Contr. U.S. natn Herb.* **36**: 193–358.

Hale, M. E. (1965b). Studies on the *Parmelia borreri* group. *Svensk bot. Tidskr.* **59**: 37–48.

Hale, M. E. (1975). A revision of the lichen genus *Hypotrachyna* (Parmeliaceae) in tropical America. *Smithsonian Contr. Bot.* **25**: 1–73.

Hale, M. E. (1976). A monograph of the lichen genus *Parmelina* (Parmeliaceae). *Smithsonian Contr. Bot.* **33**: 1–60.

Hale, M. E. (1980). Generic delimitation in the lichen family Thelotremataceae. *Mycotaxon* **11**: 130–138.

Hale, M. E. (1981). A revision of the lichen family Thelotremataceae in Sri Lanka. *Bull. Br. Mus. nat. Hist., Bot.* **8**: 227–332.

Hale, M. E. (1983). *The Biology of Lichens.* Third edition. 190 pp. London: Edward Arnold.

Hale, M. E. (1986). *Flavoparmelia*, a new genus in the lichen family Parmeliaceae (Ascomycotina). *Mycotaxon* **25**: 603–605.

Hale, M. E. (1987). A monograph of the lichen genus *Parmelia* Acharius sensu stricto (Ascomycotina: Parmeliaceae). *Smithsonian Contr. Bot.* **66**: 1–55.

Hale, M. E. (1990). A synopsis of the lichen genus *Xanthoparmelia* (Vainio) Hale (Ascomycotina, Parmeliaceae). *Smithsonian Contr. Bot*: **74**: 1–250.

Hanko, B. (1983). Die Chemotypen der Flechtengattung *Pertusaria* in Europa. *Bibliotheca lich.* **19**: 1–297.

Hanko, B., Leuckert, C. & Ahti, T. (1986) ["1985"] Beiträge zur Chemotaxonomie der Gattung *Ochrolechia* (Lichenes) in Europa. *Nova Hedwigia* **42**: 165–199.

Harding, P. T. & Rose, F. (1986). *Pasture - Woodlands in Lowland Britain.* 89 pp. Huntingdon: Institute of Terrestial Ecology.

Harris, R. C. (1973). The corticolous pyrenolichens of the Great Lakes region. *Michigan Bot.* **12**: 3–68.

Harris, R. C. (1975). *A taxonomic revision of the genus* Arthopyrenia Massal. *s. lat.* (*Ascomycetes*) *in North America.* Ph.D. thesis, University of Michigan.

Harris, R. C. (1989). A sketch of the lichen family Pyrenulaceae (Melanommatales) in eastern North America. *Mem. N.Y. bot. Gdn* **49**: 74–107.

Harris, R. C. & Buck, W. R. (1978). Lichens of the Mackinac Region II. *Candelariella* Müll. Arg. *Michigan Bot.* **17**: 155–161.

Hawksworth, D. L. (1972a). *Lichens 1961–1969.* [*Index of Fungi Supplement.*] ix + 78 pp. Kew: Commonwealth Mycological Institute.

Hawksworth, D. L. (1972b). Regional studies in *Alectoria* (Lichenes) II. The British species. *Lichenologist* **5**: 181–261.

Hawksworth, D. L. (1978). Notes on British lichenicolous fungi, II. *Notes R. bot. Gdn Edinb.* **36**: 181–197.

Hawksworth, D. L. (1980). Two little-known members of the Mycocaliciaceae on polypores. *Trans. Br. mycol. Soc.* **74**: 650–651.

Hawksworth, D. L. (1981a). *Lichenothelia*, a new genus for the *Microthelia aterrima* group. *Lichenologist* **13**: 141–153.

Hawksworth, D. L. (1981b). The lichenicolous Coelomycetes. *Bull. Br. Mus. nat. Hist., Bot.* **9**: 1–98.

Hawksworth, D. L. (1983). A key to the lichen-forming, parasitic, parasymbiotic and saprophytic fungi occurring on lichens in the British Isles. *Lichenologist* **15**: 1–44.

Hawksworth, D. L. (1985). A redisposition of the species referred to the ascomycete genus *Microthelia*. *Bull. Br. Mus. nat. Hist., Bot.* **14**: 43–181.

Hawksworth, D. L. (1988). The variety of fungal-algal symbioses, their evolutionary significance, and the nature of lichens. *Bot. J. Linn. Soc.* **96**: 3–20.

Hawksworth, D. L. (1989). Notes on aquatic species of *Verrucaria* in the British Isles. *Lichenologist* **21**: 23–28.

Hawksworth, D. L., ed (1991). Improving the stability of names: needs and options. *Regnum Vegetabile* **123**: 1–358.

Hawksworth, D. L. & Ahti, T. (1990). A bibliographic guide to the lichen floras of the world (second edition). *Lichenologist* **22**: 1–78.

Hawksworth, D. L. & Chapman, D. S. (1971). *Pseudevernia furfuracea* (L.) Zopf and its chemical races in the British Isles. *Lichenologist* **5**: 51–58.

Hawksworth, D. L., Coppins, B. J. & James, P. W. (1980). *Blarneya*, a lichenized hyphomycete from southern Ireland. *Bot. J. Linn. Soc.* **79**: 357–367.

Hawksworth, D. L., Coppins, B. J. & Rose, F. (1974). Changes in the British lichen flora. In *The Changing Flora and Fauna of Britain* (D. L. Hawksworth, ed.): 47–78. London & New York: Academic Press.

Hawksworth, D. L. & David, J. C. (1989). *Family Names*. [*Index of Fungi Supplement*.] 75 pp. Kew: CAB International Mycological Institute.

Hawksworth, D. L. & Dibben, M. J. (1982). *Cryptolechia*, an earlier name for *Gyalectina*. *Lichenologist* **14**: 98–100.

Hawksworth, D. L. & Eriksson, O. E. (1986). The names of accepted orders of ascomycetes. *Syst. Ascom.* **5**: 175–184.

Hawksworth, D. L. & Hill, D. J. (1984). The *Lichen-Forming Fungi*. 158 pp. Glasgow & London: Blackie.

Hawksworth, D. L., James, P. W. & Coppins, B. J. (1980). Checklist of British lichen-forming, lichenicolous and allied fungi. *Lichenologist* **12**: 1–115.

Hawksworth, D. L. & Riedl, H. (1977). Proposal to conserve the generic name *Racodium* Fr. *Taxon* **26**: 208.

Hawksworth, D. L. & Seaward, M. R. D. (1977). *Lichenology in the British Isles 1568–1975*. ix + 231 pp. Richmond: Richmond Publishing.

Hawksworth, D. L. & Sherwood, M. A. (1981). A reassessment of three widespread resinicolous discomycetes. *Can. J. Bot.* **59**: 357–372.

Hawksworth, D. L., Sutton, B. C. & Ainsworth, G. C. (1983). *Ainsworth & Bisby's Dictionary of the Fungi*. Seventh edition. xii + 445 pp. Kew: Commonwealth Mycological Institute.

Hennipman, E. (1978). De nederlandse Cladonias (Lichenes). *Wet. Meded. K. ned. natuurh. Veren.* **124**: 1–80.

Henssen, A. (1963a). Eine Revision der Flechtenfamilien Lichinaceae und Ephebaceae. *Symb. bot. upsal.* **18**(1): 1–123.

Henssen, A. (1963b). The North American species of *Massalongia* and generic relationships. *Can. J. Bot.* **41**: 1331–1346.

Henssen, A. (1963c). The North American species of *Placynthium*. *Can. J. Bot.* **41**: 1687–1724.

Henssen, A. (1967). A study of the genus *Vestergrenopsis*. *Can. J. Bot.* **41**: 1259–1306.

Henssen, A. (1968). *Thyrea radiata*, eine *Lempholemma*-Art mit Hormocystangien. *Ber. dtsch. bot. Ges.* **81**: 176–182.

Henssen, A. (1969). Eine Studie über die Gattung *Arctomia*. *Svensk bot. Tidskr.* **63**: 126–138.

Henssen, A. (1974). New or interesting cyanophilic lichens II. *Lichenologist* **6**: 106–111.

Henssen, A. (1976). Studies in the developmental morphology of lichenized ascomycetes. In *Lichenology: Progress and problems* (D. H. Brown, D. L. Hawksworth & R. H. Bailey, eds): 107–138. London: Academic Press.

Henssen, A. (1987). *Lichenothelia*, a genus of microfungi on rocks. *Bibliotheca lich.* **25**: 257–293.

Henssen, A., Büdel, B. & Titze, A. (1987). *Euopsis* and *Harpidium*, genera of the Lichinaceae (Lichenes) with rostrate asci. *Botanica Acta* **101**: 49–55.

Henssen, A., & Jahns, H. M. (1973) ["1974"]. *Lichenes. Eine Einführung in die Flechtenkunde*. 467 pp. Stuttgart: G. Thieme.

Hermann, S., Leuckert, C. & Poelt, J. (1973). Zur Kenntnis der Flechtengruppe *Lecanora radiosa s. ampliss*. *Willdenowia* **7**: 9–30.

Hertel, H. (1967). Revision einiger calciphiler Formenkreis der Flechtengattung *Lecidea*. *Beih. Nova Hedwigia* **24**: 1–155.

Hertel, H. (1968). Beiträge zur Kenntnis der Flechtenfamilie Lecidiaceae I. *Herzogia* **1:** 25–39.
Hertel, H. (1969a). Beiträge zur Kenntnis der Flechtenfamilie Lecidiaceae II. *Herzogia* **1: 321–329.**
Hertel, H. (1969b). Die Flechtengattung *Trapelia* Choisy. *Herzogia* **1**: 111–130.
Hertel, H. (1970a). Beiträge zur Kenntnis der Flechtenfamilie Lecideaceae III. *Herzogia* **2:** 37–62.
Hertel, H. (1970b). Parasitische lichenisierte Arten der Sammelgattung *Lecidea* in Europa. *Herzogia* **1**: 405–438.
Hertel, H. (1971). Beiträge zur Kenntnis der Flechtenfamilie Lecideaceae IV. *Herzogia* **2:** 231–261.
Hertel, H. (1973). Beiträge zur Kenntnis der Flechtenfamilie Lecideaceae V. *Herzogia* **2:** 479–515.
Hertel, H. (1975a). Ein vorläufiger Bestimmungschlüssel für die kryptothallinen schwarzfrüchtigen, saxicolen Arten der Sammelgattung *Lecidea* (Lichenes) in der Holarktis. *Decheniana* **127**: 37–78.
Hertel, H. (1975b). Beiträge zur Kenntnis der Flechtenfamilie Lecideaceae VI. *Herzogia* **3:** 365–406.
Hertel, H. (1977). Gesteinsbewohnende Arten der Sammelgattung *Lecidea* (Lichenes) aus Zentral-, Ost-, und Südasien. *Ergebn. Forsch Unternehmens Nepal Himalaya* **6**: 145–378.
Hertel, H. (1981). Beiträge sur Kenntnis der Flechtenfamilie Lecideaceae VIII. *Herzogia* **5:** 449–463.
Hertel, H. (1983). Über einige aus *Lecidea* und *Melanolecia* (Ascomycetes lichenisati) auszuzchliessende Arten. *Mitt. Bot. StSamml. München* **19**: 441–447.
Hertel, H. (1984). Über saxicole lecideoide Flechten der Subantarktis. *Beih. Nova Hedwigia* **79**: 399–499.
Hertel, H. (1985). New, or little-known New Zealand lecideoid lichens. *Mitt. Bot. StSamml. München* **21**: 307–337.
Hertel, H. & Knoph J. G. (1984). *Porpidia albocaerulescens* eine weit verbreitete, doch in Europa seltene und vielfach verkannte Krustenflechte. *Mitt. Bot. StSamml. München* **20**: 476–488.
Hertel, H. & Leuckert, C. (1969). Über Flechtenstoffe und Systematik einiger Arten der Gattungen *Lecidea*, *Placopsis* und *Trapelia* mit C+ rot reagierendem Thallus. *Willdenowia* **5**: 369–383.
Hertel, H. & Rambold, G. (1985). *Lecidea* sect. *Armeniaca*: lecideoide Arten der Flechtengattungen *Lecanora* und *Tephromela* (Lecanorales). *Bot. Jahrb.* **107**: 469–501.
Hertel, H. & Rambold, G. (1987). *Miriquidica* genus novum Lecanoracearum (Ascomycetes lichenisati). *Mitt. bot. StSamml. München* **23**: 377–392.
Hertel, H. & Rambold, G. (1988). *Lecidea mosigii* (Koerb.) Anzi-eine Art der Gattung *Orphniospora* Koerb. (Fuscideaceae, Teloschistales). *Mitt. bot. StSamml. München* **27**: 111–123.
Hertel, H. & Rambold, G. (1990). Zur Kenntnis der Familie Rimulariaceae (Lecanorales). *Bibliotheca lich.* **38**: 145–189.
Hestmark, G. (1990). Thalloconidia in the genus *Umbilicaria*. *Nordic J. Bot.* 9: 547–574.
Hillmann, J. (1936). Parmeliaceae. *Rabenh. Krypt.-Fl.* **9**, 5(3): 1–309.
Holien, H. (1989). The genus *Bryoria* sect. *Implexae* in Norway. *Lichenologist* **21**: 243–258.
Howard, G. E. (1970). The lichen genus *Ochrolechia* in North America north of Mexico. *Bryologist* **73**: 93–130.
Imshaug, H. A. & Brodo, I. M. (1966) Biosystematic studies on *Lecanora pallida* and some related lichens in the Americas. *Nova Hedwigia* **12:** 1–59.
Imshaug, H. A. & Harris, R. C. (1969). *Parmentaria chilensis* Fée. *Lichenologist* **4**: 77–82.
Inoue, M. (1981a). A taxonomic study on the Japanese species of *Fuscidea* (lichens). *Hikobia, Suppl.* **1**: 161–176.
Inoue, M. (1981b). A preliminary revision of extra-Japanese species of *Fuscidea* (lichens). *Hikobia, Suppl.* **1**: 177–181.
Inoue, I. (1984). Japanese crustose lichen genera formerly reported under *Lecidea sensu lato* I. *Amygdalaria* Norman. *J. Hattori bot. Lab.* **56**: 321–330.
Jahns, H. M. (1970). Remarks on the taxonomy of the European and North American species of *Pilophorus* Th. Fr. *Lichenologist* **4**: 199–213.
Jahns, H. M. (1981). The genus *Pilophorus*. *Mycotaxon* **13**: 289–330.
Jahns, H. M. (1983). *Collins Guide to the Ferns, Mosses and Lichens of Britain and North and Central Europe*. 272 pp. London: Collins.
Jahns, H. M. (1988). The establishment, individuality and growth of lichen thalli. *Bot. J. Linn. Soc.* **96**: 21–29.
James, P. W. (1970). The lichen flora of shaded acid rock crevices and overhangs in Britain. *Lichenologist* **4**: 309–322.
James, P. W. (1971). New or interesting British lichens I. *Lichenologist* **5**: 114–148.
James, P.W. (1974). Distribution maps of lichens in Britain. Map 17. *Sclerophyton circumscriptum* (Tayl.)Zahlbr. *Lichenologist* **6:** 197–199.

James, P. W. (1975). The genus *Gyalideopsis* Vězda in Britain. *Lichenologist* **7**: 155–161.
James, P. W. (1982). Key to *Parmelia* in Great Britain. *Bull. Br. lichen Soc.* **51**: 27–36.
James, P. W. & Coppins, B. J. (1979). Key to British sterile crustose lichens with *Trentepohlia* as phycobiont. *Lichenologist* **11**: 253–262.
James, P. W., Hawksworth, D. L. & Rose, F. (1977). Lichen communities in the British Isles: a preliminary conspectus. In *Lichen Ecology* (M. R. D. Seaward, ed.): 295–413. London etc.: Academic Press.
James, P. W. & Henssen, A. (1976). The morphological and taxonomic significance of cephalodia. In *Lichenology: Progress and Problems* (D. H. Brown, D. L. Hawksworth & R. H. Bailey, eds): 27–77. London: Academic Press.
James, P. W. & White, F. J. (1987). Studies on the genus *Nephroma* I. The European and Macaronesian species. *Lichenologist* **19**: 215–268.
Jørgensen, P. M. (1978). The lichen family Pannariaceae in Europe. *Op. bot. Soc. bot. Lund* **45**: 1–123.
Jørgensen, P. M. (1989). Notes on the lichen genus *Ionaspis* in Scandinavia. *Graphis Scripta* **2**: 118–121.
Jørgensen, P. M. (1990). The lichen genus *Pterygiopsis* in northern Europe. *Lichenologist* **22**: 213–217.
Jørgensen, P. M. & Henssen, A. (1990). Proposal to conserve *Pyrenopsis* (Nyl.) Nyl., and nomenclatural comments on some related genera (lichenized fungi). *Taxon* 39: 343–348.
Jørgensen, P. M. & James, P. W. (1983). Studies on some *Leptogium* species of western Europe. *Lichenologist* **15:** 109–125.
Jørgensen, P. M. & James, P. W. (1990). The lichen genus *Degelia*. *Bibliotheca lich.* **38:** 253–276.
Jørgensen, P. M. & Ryvarden, L. (1970). Contribution to the lichen flora of Norway.*Årbok Univ. Bergen, mat.-naturv. ser.* **1969**(10): 1–24.
Jørgensen, P. M. & Tønsberg, T. (1988). On some crustose lichens with *Trentepohlia* from shaded overhangs in coastal Norway. *Nordic J. Bot.* **8**: 293–304.
Jørgensen, P. M., Vězda, A. & Botnen, A. (1983). *Clathroporina calcarea*, a misunderstood lichen species, and a note on the genus *Clathroporina* in Europe. *Lichenologist* **15**: 45–55.
Kärnefelt, I. (1979). The brown fruticose species of *Cetraria*. *Opera Bot.* **46**: 1–150.
Kärnefelt, I. (1986). The genera *Bryocaulon*, *Coelocaulon* and *Cornicularia* and formerly associated taxa. *Opera Bot.* **86**: 1–90.
Kärnefelt, I. (1990). Isidiate taxa in the Teloschistales and their ecological and evolutionary significance. *Lichenologist* **22**: 307–320.
Keissler, K. (1936–38). Pyrenulaceace bis Mycoproaceae, Coniocarpineae. *Rabenh. Krypt.-Fl.* **9**, 1(2): 1–846.
Keissler, K. (1958–60). Usneaceae. *Rabenh. Krypt.-Fl.* **9**, 5(4): 1–755.
Kershaw, K. A. (1985). *Physiological Ecology of Lichens*. 293 pp. Cambridge: Cambridge University Press.
Kilias, R. (1981). Revision gesteinwöhnender Sippen der Flechtengattungen *Catillaria* Massal. in Europa. *Herzogia* **5**: 209–448.
Knoph, J. G. (1990). Untersuchungen an gesteinsbewohnenden xanthonhaltigen Sippen der Flechtengattung *Lecidella* (Lecanoraceae, Lecanorales) unter Berücksichtigung von aussereuropäischen Proben exklusive Amerika. *Bibliotheca lich.* **36:** 1–183.
Kohlmeyer, J. & Kohlmeyer, E. (1979). *Marine Mycology. The Higher Fungi*. 690 pp. New York: Academic Press.
Krog, H. (1974). Taxonomic studies in the *Hypogymnia intestiniformis* complex. *Lichenologist* **6**: 135–140.
Krog, H. (1980). On *Bryoria chalybeiformis* and some related species. *Lichenologist* **12**: 243–245.
Krog, H. (1982). *Punctelia*, a new lichen genus in the Parmeliaceae. *Nordic J. Bot.* **2**: 287–292.
Krog, H. & James, P. W. (1977). The genus *Ramalina* in Fennoscandia and the British Isles. *Norw. J. Bot.* **24**: 15–43.
Krog, H. & Østhagen, H. (1980). The genus *Ramalina* in the Canary Islands. *Norw. J. Bot.* **27**: 255–296.
Krog, H., Østhagen, H. & Tønsberg, T. (1980). *Lavflora. Norske busk- og bladlav*. 312 pp. Oslo, Bergen & Tromsø: Universitetsforlaget.
Kurokawa, S. (1962a). A monograph of the genus *Anaptychia*. *Beih. Nova Hedwigia* **6**: 1–115.
Kurokawa, S. (1962b). A note on the lichen genus *Tornabenia* Trev. *J. Jap. Bot.* **37**: 289–294.
Kurokawa, S. (1973). Supplementary notes on the genus *Anaptychia*. *J. Hattori bot. Lab.* **37**: 563–607.
Lamb, I. M. (1947). A monograph of the lichen genus *Placopsis* Nyl. *Lilloa* **13**: 151–288.
Lamb, I. M. (1948). Antarctic pyrenocarp lichens. *Discovery Repts* **25**: 1–30.
Lamb, I. M. (1963). *Index Nominum Lichenum*. xiv + 809 pp. New York: Ronald Press.

Lamb, I. M. (1977). A conspectus of the lichen genus *Stereocaulon* (Schreb.) Hoffm. *J. Hattori bot. Lab.* **43**: 191–355.

Lamb, I. M. (1978). Keys to the species of the lichen genus *Stereocaulon* (Schreb.) Hoffm. *J. Hattori bot. Lab.* **44**: 209–250.

Lamb, I. M. & Ward, S. (1974). A preliminary conspectus of the species attributed to the imperfect lichen genus *Leprocaulon* Nyl. *J. Hattori Bot. Lab.* **38**: 499–553.

Laundon, J. R. (1963). The taxonomy of sterile crustaceous lichens in the British Isles 2. Corticolous and lignicolous species. *Lichenologist* **2**: 101–151.

Laundon, J. R. (1965). Lichens new to the British flora: 3. *Lichenologist* **3**: 65–71.

Laundon, J. R. (1970). Lichens new to the British flora: 4. *Lichenologist* **4**: 297–308.

Laundon, J. R. (1974). *Leproplaca* in the British Isles. *Lichenologist* **6**: 102–105.

Laundon, J. R. (1976). Lichens new to the British flora: 5. *Lichenologist* **8**: 139–150.

Laundon, J. R. (1979). Deceased lichenologists: their abbreviations and herbaria. *Lichenologist* **11**: 1–26.

Laundon, J. R. (1981). The species of *Chrysothrix*. *Lichenologist* **13**: 101–121.

Laundon, J. R. (1986). Studies in the nomenclature of British lichens II. *Lichenologist* **18**: 169–177.

Laundon, J. R. (1989). The species of *Leproloma* - the name for the *Lepraria membranacea* group. *Lichenologist* **21**: 1–22.

Laundon, J. R. (1992a). New British species of *Caloplaca*. *Lichenologist* **24**: 1–5.

Laundon, J. R. (1992b). *Lepraria* in the British Is. *Lichenologist* **24**.

Lawrey, J. D. (1984). *Biology of Lichenized Fungi.* 408 pp. New York: Praeger Publishers.

Lettau, G. (1932–7). Monographische Bearbeitung einiger Flechtenfamilien. *Beih. Repert. Spec. nov. Regni Veg.* **69**: 1–250.

Leuckert, C., Knoph, J. G., Ziegler, H. G. & Hertel, H. (1989). Chemotaxonomische Studien in der Gattung *Lecidella* (*Lecanorales, Lecanoraceae*) I. *Lecidella carpathica* und *Lecidella viridans* - Untersuchungen an mittel- und südeuropäischen Proben. *Herzogia* **8:** 265–272.

Leuckert, C. & Poelt, J. (1989). Studien über die *Lecanora rupicola*-Gruppe in Europa (Lecanoraceae). *Nova Hedwigia* **49**: 121–167.

Lindsay, D. C. (1971). Notes on Antarctic lichens: III. *Cystocoleus niger* (Huds.) Hariot. *Brit. Antarc. Surv. Bull.* **24**: 119–120.

Llano, G. A. (1950). *A Monograph of the Lichen family Umbilicariaceae in the Western Hemisphere*. 281 pp. Washington: Office of Naval Research.

Löfgren, O. & Tibell, L. (1979). *Sphinctrina* in Europe. *Lichenologist* **11**: 109–137.

Lumbsch, H. T. (1988). The identity of *Diploschistes gypsaceus*. *Lichenologist* **20**: 19–24.

Lumbsch, H. T. (1989). Die holarktischen Vertreter der Flechtengattung *Diploschistes* (Thelotremataceae). *J. Hattori bot. Lab.* **66**: 133–196.

Lumbsch, H. T., Kothe, H. W. & Elix, J. A. (1988). Resurrection of the lichen genus *Pleurosticta* Petrak (Parmeliaceae: Ascomycotina). *Mycotaxon* **33**: 447–455.

Maas Geesteranus, R. A. (1947). Revision of the lichens of the Netherlands I. Parmeliaceae. *Blumea* **6**: 1–199.

Maas Geesteranus, R. A. (1952). Revision of the lichens of the Netherlands. II Physciaceae. *Blumea* **7**: 206–287.

McCarthy, P. M. (1988a). New and interesting species of *Verrucaria* I. *Lichenologist* **20**: 1–10.

McCarthy, P. M. (1988b). New and interesting species of *Verrucaria* II. *Lichenologist* **20**: 245–251.

McCarthy, P. M. & Mitchell, M. E. (1988). *Lichens of the Burren Hills and the Aran Islands.* 123 pp. Galway: Officina Typographica.

Magnusson, A. H. (1929). A monograph of the genus *Acarospora*. *K. svenska VetenskAkad. Handl.*, III, **7**(4): 1–400.

Magnusson, A. H. (1930). New or otherwise interesting *Lecanora* species. *Acta Horti gothoburg.* **6**: 1–20.

Magnusson, A. H. (1933). A monograph of the lichen genus *Ionaspis*. *Acta Horti gothoburg.* **8**: 1–47.

Magnusson, A. H. (1935). Acarosporaceae und Thelocarpaceae. *Rabenh. Krypt.-Fl.* **9**, 5(1): 1–318.

Magnusson, A. H. (1939). Studies in species of *Lecanora*, mainly the *Aspicilia gibbosa* group. *K. svenska VetenskAkad. Handl.*, III, **17**(5): 1–182.

Magnusson, A. H. (1944). Studies in the *ferruginea*-group of the genus *Caloplaca*. *Göteborgs K. Vetensk. -o. VitterSamh. Handl.* 6B, **3**: 1–71.

Magnusson, A. H. (1952). Keys to the species of *Lecidea* in Scandinavia and Finland I-II. *Svensk bot. Tidskr.* **46**: 178–198, 313–323.

Magnusson, A. H. (1956). A second supplement to the monograph of *Acarospora* with keys. *Göteborgs K. Vetensk. -o. VitterhSamh. Handl.*, 6B, **6**(17): 1–34.

Mayrhofer, H. (1984). Die Saxicolen Arten der Flechtengattungen *Rinodina* und *Rinodinella* in der Alten Welt. *J. Hattori bot. Lab.* **55**: 327–493.
Mayrhofer, H. (1987a). Ergänzende Studien zur Taxonomie der Gattung *Protothelenella. Herzogia* **7**: 313–342.
Mayrhofer, H. (1987b). Monographie der Flechtengattung *Thelenella. Bibliotheca lich.* **26**: 1–106.
Mayrhofer, H. & Poelt, J. (1979). Die saxicolen Arten der Flechtengattung *Rinodina* in Europa. *Bibliotheca lich.* **12**: 1–186.
Mayrhofer, H. & Poelt, J. (1985). Die Flechtengattung *Microglaena* sensu Zahlbruckner in Europa. *Herzogia* **7**: 13–79.
Mayrhofer, M. (1987). Studien über die saxicolen Arten der Flechtengattung *Lecania* in Europa I. *Halecania* gen. nov. *Herzogia* **7**: 381–406.
Mayrhofer, M. (1988). Studien über die saxicolen Arten der Flechtengattung *Lecania* in Europa II. *Lecania* s.str. *Bibliotheca lich.* **28**: 1–133.
Meyer, S. L. F. (1982). Segregation of the new lichen genus *Foraminella* from *Parmeliopsis. Mycologia* **74**: 592–598.
Middelborg, J. & Mattsson, J. (1987). Crustaceous lichenized species of the Caliciales in Norway. *Sommerfeltia* **5**: 1–71.
Miles, C. J. (1980). The genus *Melaspilea* Nyl. in Britain. 98 pp. M.Sc. thesis, University of Reading.
Mitchell, M. E. (1971). *A Bibliography of Books, Pamphlets and Articles relating to Irish Lichenology, 1727–1970.* 76 pp. Galway: Privately printed.
Moberg, R. (1977). The lichen genus *Physcia* and allied genera in Fennoscandia. *Symb. bot. upsal.* **22**(1): 1–108.
Moberg, R. (1982). *Lavar. En falthandbok.* 237 pp. Stockholm: Interpublishing.
Morgan-Jones, G. & Swinscow, T. D. V. (1965). Pyrenocarpous lichens 7. On the genus *Microglaena* Körb. *Lichenologist* **3**: 42–54.
Muhr, L.-E. & Tønsberg, T. (1989). *Rimularia fuscosora*, a new corticolous sorediate lichen from north western Europe. *Nordic J. Bot.* **8**: 649–652.
Nordin, I. (1972). Caloplaca *sect.* Gasparrinia *i Nordeuropa.* 184 pp. Uppsala: Skriv Service.
Ohlsson, K. E. (1973). *A Revision of the Lichen Genus Sphaerophorus.* 270 pp. University Microfilms, Ann Arbor: Michigan.
Orange, A. (1989). *Macentina stigonemoides* (Verrucariaceae), a new lichenized species from Great Britain and Ireland. *Lichenologist* **21**: 229–236.
Orange, A. (1991). Notes on some terricolous species of *Verrucaria. Lichenologist* **22**: 3–10.
Oshio, M. (1968). Taxonomical studies on the family Pertusariaceae of Japan. *J. Sci. Hiroshima Univ.*, B(2), **12**: 81–163.
Ozenda, P. & Clauzade, G. (1970). *Les Lichens. Etude biologique et flore illustrée.* 801 pp. Paris: Masson & Cie.
Parmasto, E. (1978). The genus *Dictyonema* ('Thelephorolichenes'). *Nova Hedwigia* **29**: 99–144.
Pentecost, A. & Coppins, B. J. (1983). Key to *Opegrapha* in Great Britain. *Bull. Br. lichen. Soc.* **53**: 27–35.
Petersen. R. H. & Kantvilas, G. (1986). Three lichen-forming fungi from Tasmania. *Aust. J. Bot.* **34**: 217–222.
Poelt, J. (1958). Die lobaten Arten der Flechtengattung *Lecanora* Ach. sensu ampl. in der Holarktis. *Mitt. bot. StSamml. München* **2**: 411–589.
Poelt, J. (1969). *Bestimmungsschlüssel europäischer Flechten.* (71) + 757 pp. Lehre: J. Cramer.
Poelt, J. (1983). *Bryonora* eine neue Gattung der Lecanoraceae. *Nova Hedwigia* **38**: 73–111.
Poelt, J. (1988). *Rhizocarpon* Ram. em. Th. Fr. sugen. *Rhizocarpon* in Europe. *Arctic Alpine Res.* **20**: 292–298.
Poelt, J. & Hafellner, J. (1975). Schlauchpforten bei der Flechtengattung *Thelocarpon. Phyton, Horn* **17**: 67–77.
Poelt, J. & Krüger, U. (1970). Die Verbreitungsverhältnisse der Flechtengattung *Squamarina* in Europa. *Feddes Repert.* **81**: 187–201.
Poelt, J. & Vězda, A. (1977). Bestimmungsschlüssel europäischer Flechten. Ergänzungsheft I. *Bibliotheca lich.* **9**: 1–258.
Poelt, J. & Vězda, A. (1981). Bestimmungsschlüssel europäischer Flechten. Ergänzungsheft II. *Bibliotheca lich.* **16**: 1–390.
Rambold, G. (1989). A monograph of the saxicolous lecideoid lichens of Australia (excl. Tasmania). *Bibliotheca lich.* **34**: 1–345.
Rambold, G., Hertel, H. & Triebel, D. (1990). *Koerberiella wimmeriana* (Lecanorales, Porpidiaceae) and its lichenicolous fungi. *Lichenologist* **22**: 225–240.

Rambold, G. & Triebel, D. (1989). *Gelatinopsis, Geltingia* and *Phaeopyxis*—three helotialean genera with lichenicolous species. *Notes R. Bot. Gdn. Edinb.* **46**: 375–389.

Redhead, S. A. & Kuyper, T. W. (1987). Lichenized agarics: taxonomic and nomenclatural riddles. In *Arctic & Alpine Mycology* (G. A. Laursen, J. F. Ammirati & S. A. Redhead, eds) **2**: 319–348. New York & London: Plenum Press.

Redhead, S. A. & Kuyper, T. W. (1988). *Phytoconis*, the correct generic name for the basidiolichen *Botrydina*. *Mycotaxon* **31**: 221–223.

Redinger, K. (1936–38). Arthoniaceae, Graphidaceae, Chiodectonaceae, Dirinaceae, Roccellaceae, Lecanactidaceae, Thelotremaceae, Diploschistaceae, Gyalectaceae und Coenogoniaceae. *Rabenh. Krypt.-Fl.* **9**, 2(1): 1–404.

Renobales, G. & Aguirre, B. (1990). The nomenclature and systematic position of the genus *Encephalographa*. *Syst. Ascom.* **8**: 87–92.

Richardson, D. H. S. (1975). *The Vanishing Lichens*. 231 pp. Newton Abott: David & Charles.

Riedl, H. (1971). Zur Kenntnis von *Polyblastiopsis* A. Zahlbr. und *Mycoglaena* v. Höhnel (= *Winteria* Rehm). *Öst. bot. Z.* **119**: 41–67.

Rogers, R. W. & Hafellner, J. (1988). *Haematomma* and *Ophioparma*: two superficially similar genera of lichenized fungi. *Lichenologist* **20**: 167–174.

Rose, F. (1976). Lichenological indicators of age and environmental continuity in woodlands. In *Lichenology: Progress and Problems* (D. H. Brown, D. L. Hawksworth & R. H. Bailey, eds): 279–307. London: Academic Press.

Roux, C. (1984). *Clauzadeana* Roux, nova likengenro. *Bull. Soc. linn. Provence* **35**: 99–102.

Runemark, H. (1956). Studies in *Rhizocarpon* I-II. *Op. bot. Soc. bot. Lund* **2**(1): 1–152, (2): 1–150.

Ruoss, E. (1987). Chemotaxonomische und morphologische Untersuchungen an den Rentierflechten *Cladonia arbuscula* und *C. mitis*. *Bot. Helv.* **97**: 239–263.

Ruoss, E. (1989). Verzweigung als Unterscheidungsmerkmal bei Reintierflechten (*Cladonia* subg. *Cladina*). *Herzogia* **8**: 125–136.

Ruoss, E. & Ahti, T. (1989). Systematics of some reindeer lichens (*Cladonia* subg. *Cladina*) in the Southern Hemisphere. *Lichenologist* **21**: 29–44.

Salisbury, G. (1966). A monograph of the lichen genus *Thelocarpon* Nyl. *Lichenologist* **3**: 175–196.

Salisbury, G. (1972). *Thelotrema* Ach. sect. *Thelotrema* I. The *T. lepadinum* group. *Lichenologist* **5**: 262–274.

Salisbury, G. (1975). *Leucocarpopsis*, a new lichen genus. *Nova Hedwigia* **26**: 351–352.

Santesson, J. (1970) ["1969"]. Neuere Probleme der Flechtenchemie. *Vortr. GesGeb. Bot. [Dtsch. bot. Ges.]* **4**: 5–21.

Santesson, R. (1939). Amphibious pyrenolichens I. *Ark. Bot.* **29A**(10): 1–67.

Santesson, R. (1952). Foliicolous lichens I. A revision of the taxonomy of the obligately foliicolous, lichenized fungi. *Symb. bot. upsal.* **12**(1): 1–590.

Santesson, R. (1992). *Pyrenocollema elegans*, a new marine lichen. *Lichenologist* **24**: 7–11.

Schade, A. (1932). Die Verbreitung von *Racodium* rupestre Pers. und *Coenogonium nigrum* (Huds.) Zahlbr. in Sachsen. *Beih. bot. Zbl.* **49**: 421–437.

Schauer, T. (1965) Die holz- und rindenbewöhnenden Arten der Flechtengattung *Buellia* s. str. im Nordalpenraum. *Mitt. bot. StSamml. München* **5**: 609–626.

Scheidegger, C. (1987). *Buellia uberior* und *B. miriquidica* (Physciaceae, Lecanorales), zwei lichenicole Krustenflechten auf *Schaereria tenebrosa*. *Bot. Helv.* **97**: 99–116.

Schmidt, A. (1970). Anatomisch-taxonomische Untersuchungen an europäischen Arten der Flechtenfamilie Caliciaceae. *Mitt. StInst. allg. Bot. Hamburg* **13:** 111–166.

Schneider, G. (1979). Die Flechtengattung *Psora* sensu Zahlbruckner versuch einer Gliederung. *Bibliotheca lich.* **13**: 1–291.

Schwab, A. J. (1986). Rostfärbene Arten der Sammelgattung *Lecidea* (Lecanorales) Revision der Arten Mittel- und Nordeuropas. *Mitt. Bot. StSamml. München* **22**: 221–476.

Seaward, M. R. D., ed. (1977). *Lichen Ecology*. 550 pp. London: Academic Press.

Seaward, M. R. D. (1984). Census catalogue of Irish lichens. *Glasra* **8**: 1–32.

Seaward, M. R. D. (1988). Progress in the study of the lichen flora of the British Isles. *Bot. J. Linn. Soc.* **96**: 81–95.

Seaward. M. R. D. (1992). Distribution maps of lichens in Britain. Map 28. Progress map. *Lichenologist* **24**: 104.

Seaward, M. R. D. & Hitch, C. J. B., eds (1982). *Atlas of the Lichens of the British Isles*. Vol. 1. 196 pp. Cambridge: Institute of Terrestrial Ecology.

Sérusiaux, E. (1991). *Porina rosei* sp. nov., une espèce nouvelle de europe occidentale. *Cryptogamie, bryol. lichén.* **12**: 31–39.

Sheard, J. W. (1964). The genus *Buellia* de Notaris in the British Isles (excluding section *Diploicia* (Massal.) Stiz.). *Lichenologist* **2**: 225–262.

Sheard, J. W. (1967). A revision of the lichen genus *Rinodina* (Ach.) Gray in the British Isles. *Lichenologist* **3**: 328–367.

Sheard, J. W. (1973). *Rinodina interpolata* (Stiz.) Sheard, a new combination in the British and Scandinavian lichen floras. *Lichenologist* **5**: 461–463.

Sherwood, M. A. (1977). The ostropalean fungi II. *Schizoxylon*, with notes on *Stictis*, *Acarosporina*, *Coccopeziza* and *Carestiella*. *Mycotaxon* **6**: 215–260.

Sherwood, M. A. & Coppins, B. J. (1980). *Xerotrema*, a new genus of odontotremoid fungi from Scotland. *Notes R. bot. Gdn Edinb.* **38**: 367–371.

Sierk, H. A. (1964). The genus *Leptogium* in North America north of Mexico. *Bryologist* **67**: 245–317.

Sipman, H. J. M. (1983). A monograph of the lichen family Megalosporaceae. *Bibliotheca lich.* **18**: 1–241.

Smith, A. L. (1918, 1926). *A Monograph of the British Lichens.* 2 vols. Second edition. 520 + 447 pp. London: British Museum (Natural History).

Swinscow, T. D. V. (1960). *Cavernularia hultenii* Degelius in Scotland. *Lichenologist* **1**: 179–183.

Swinscow, T. D. V. (1961). Pyrenocarpous lichens: 2. *Gongylia* Körb. in the British Isles with first British record of *G. incarnata*. *Lichenologist* **1**: 242–250.

Swinscow, T. D. V. (1962). Pyrenocarpous lichens: 3. The genus *Porina* in the British Isles. *Lichenologist* **2**: 6–56.

Swinscow, T. D. V. (1963). Pyrenocarpous lichens: 4. Guide to the British species of *Staurothele*. *Lichenologist* **2**: 152–171.

Swinscow, T. D. V. (1964). Pyrenocarpous lichens: 6. The genus *Thrombium* in the British Isles. Species of *Belonia* Körb. in Britain. *Lichenologist* **2**: 276–283.

Swinscow, T. D. V. (1965a). Pyrenocarpous lichens: 8. The marine species of *Arthopyrenia* in the British Isles. *Lichenologist* **3**: 55–64.

Swinscow, T. D. V. (1965b). Pyrenocarpous lichens: 9. Notes on various species. *Lichenologist* **3**: 72–83.

Swinscow T. D. V. (1966). The ascocarp of *Anthracothecium pyrenuloides* (Mont.) Müll. Arg. *Lichenologist* **3**: 274–275.

Swinscow, T. D. V. (1968). Pyrenocarpous lichens: 13. Freshwater species of *Verrucaria* in the British Isles. *Lichenologist* **4**: 34–54.

Swinscow, T. D. V. (1970). Pyrenocarpous lichens: 14. *Arthopyrenia* Massal. sect. *Acrocordia* (Massal.) Müll. Arg. in the British Isles. *Lichenologist* **4**: 218–223.

Swinscow, T. D. V. (1971). Pyrenocarpous lichens: 15. Key to *Polyblastia* in the British Isles. *Lichenologist* **5**: 92–113.

Swinscow, T. D. V. (1982). *Thrombium thelostomum*: a further revision. *Lichenologist* **14**: 97.

Swinscow, T. D. V. & Krog, H. (1976). The genera *Anaptychia* and *Heterodermia* in East Africa. *Lichenologist* **8**: 103–138.

Swinscow, T. D. V. & Krog, H. (1981). The genus *Phyllopsora*, with a report on East African species. *Lichenologist* **13**: 203–247.

Tehler, A. (1983). The genera *Dirina* and *Roccellina* (Roccellaceae). *Opera bot.* **70**: 1–86.

Thomson, J. W. (1967a). The lichen genus *Baeomyces* in North America north of Mexico. *Bryologist* **70**: 285–298.

Thomson, J. W. (1967b). Notes on *Rhizocarpon* in the arctic. *Nova Hedwigia* **14**: 321–481.

Thomson, J. W. (1968) ["1967"]. *The Lichen Genus* Cladonia *in North America.* 172 pp. Toronto: University of Toronto Press.

Thomson, J. W. (1984). *American Arctic Lichens.* Vol. 1. *The Macrolichens.* xii + 504 pp. New York: Columbia University Press.

Thomson, N. F. & Thomson, J. W. (1984). Spore ornamentation in the lichen genus *Solorina*. *Bryologist* **87**: 151–153.

Tibell, L. (1971). The genus *Cyphelium* in Europe. *Svensk bot. Tidskr.* **65**: 138–164.

Tibell, L. (1975). The Caliciales of boreal North America. *Symb. bot. upsal.* **21**(2): 1–128.

Tibell, L. (1976). The genus *Thelomma*. *Bot. Notiser* **129**: 221–249.

Tibell, L. (1977). Lavordningen Caliciales i Sverige. Inledning och släktet *Calicium*. *Svensk bot. Tidskr.* **71**: 239–259.

Tibell, L. (1978). The genus *Microcalicium*. *Bot. Notiser* **131**: 229–246.

Tibell, L. (1980). The lichen genus *Chaenotheca* in the Northern Hemisphere. *Symb. bot. upsal.* **23**(1): 1–65.
Tibell, L. (1981). Comments on Caliciales Exsiccatae II. *Lichenologist* **13**: 51–64.
Tibell, L. (1982). Caliciales of Costa Rica. *Lichenologist* **14**: 219–254.
Tibell, L. (1984). A reappraisal of the taxonomy of Caliciales. *Beih. Nova Hedwigia* **79**: 597–713.
Tibell, L. (1987). Australian Caliciales. *Symb. bot. upsal.* **27**(1): 1–279.
Timdal, E. (1983). The genus *Squamarina* in Scandinavia. *Lichenologist* **15**: 169–179.
Timdal, E. (1984a). The genus *Hypocenomyce* (Lecanorales, Lecideaceae) with special emphasis on the Norwegian and Swedish species. *Nordic J. Bot.* **4**: 83–108.
Timdal, E. (1984b). The delimitation of *Psora* (Lecideaceae) and related genera, with notes on some species. *Nordic J. Bot.* **4**: 525–540.
Timdal, E. (1991). A monograph of the genus *Toninia* (Lecideaceae, Ascomycetres). *Opera bot.* **110**: 1–137.
Timdal, E. & Holtan-Hartwig, J. (1988). A preliminary key to *Rhizocarpon* in Scandinavia. *Graphis Scripta* **2**: 41–54.
Tønsberg, T. (1990). *Japewia subaurifera*, a new lichen genus and species from north- west Europe and western North America. *Lichenologist* **22**: 205–212.
Tønsberg, T. (1992). The sorediate and isidiate, corticolous, crustose lichens in Norway. *Sommerfeltia* **14**: 1–331.
Triebel, D. & Rambold, G. (1988). *Cecidonia* und *Phacopsis* (Lecanorales): zwei lichenicole Pilzgattungen mit cecidogenen Arten. *Nova Hedwigia* **47**: 279–309.
Vainio, E. A. (1921–34). Lichenographia fennica I-IV. *Acta Soc. Fauna Flora fenn.* **49**(2): 1–274, **53**(1): 1–340, **57**(1): 1–138, **57**(2): 1–531.
van den Boom, P. P. G. (1992). The saxicolous species of the lichen genus *Lecania* in The Netherlands, Belgium and Luxembourg. *Nova Hedwigia* **54**: 229–254.
Verseghy, K. (1962). Die Gattung *Ochrolechia*. *Beih. Nova Hedwigia* **1**: 1–146.+
Vězda, A. (1958). Československé druhy rodu *Gyalecta* a *Pachphiale* s Klíšem a prehledem evropskych druhu. *Sborn. Vys. Zemšd. Lesn. Brnŏ,* C 1958(1): 21–56.
Vězda, A. (1959). K taxonomii, rozšíreni a ekologii lišejníku *Belonia russula* Kbr. ve strední Evrope. *Príridov. Čas. slezsky* **20**: 241–253.
Vězda, A. (1961). Systematische Studien über die Flechtengattung *Bacidia* DNot. emend. Zahlbr. I. Neue oder wenig bekannte Art. *Čas. Slezsk. Mus. Opavé,* A, **10**: 103–111.
Vězda, A. (1965a). Flechtensystematische Studien II. *Absconditella*, eine neue Flechtengattungen. *Preslia* **37**: 237–245.
Vězda, A. (1965b). Flechtensystematische Studien I. Die Gattung *Petractis* Fr. *Preslia* **37**: 127–143.
Vězda, A. (1966). Flechtensystematische Studien IV. Die Gattung *Gyalidea* Lett. *Folia geobot. phytotax. bohemoslavica* **1**: 311–340.
Vězda, A. (1968a). Flechtensystematische Studien VI. Die Gattung *Sagiolechia* Massal. *Folia geobot. phytotax. bohemoslavica* **2**: 383–396.
Vězda, A. (1968b). Taxonomische Revision der Gattung *Thelopsis* Nyl. (lichenisierte Fungi). *Folia geobot. phytotax. bohemoslavica* **3**: 363–406.
Vězda, A. (1969a). Neue taxa und Kombinationen in der Familie Gyalectaceae (lichenisierte Fungi). *Folia geobot. phytotax bohemoslavica* **4**: 443–446.
Vězda, A. (1969b). *Leucocarpia* gen. nov., eine neue Gattung der Flectenfamilie Verrucariaceae. *Herzogia* **1**: 187–194.
Vězda, A. (1978). Neue oder wenig bekannte Flechten in der Tschechoslowakei II. *Folia geobot. phytotax. bohemoslavica* **13**: 397–420.
Vězda, A. (1979). Flechtensystematische Studien IX. Beiträge zur Kenntnis der Familie Asterothyriaceae (Discolichens). *Folia geobot. phytotax. bohemoslavica* **14**: 43–94.
Vězda, A. (1986). Neue Gattungen der Familie Lecideaceae s.lat. (Lichenes). *Folia geobot. phytotax. bohemoslavica* **21**: 199–219.
Vězda, A. (1987). Foliicole Flechten aus Zaire (III). Die Gattung *Byssoloma* Trevisan. *Folia geobot. phytotax. bohemoslavica* **22**: 71–83.
Vězda, A. & Pišút, I. (1984). Zwei neue Arten der Flechtengattung *Absconditella* (lichenisierte Stictidaceae, Ostropales) in der Tschechoslowakei. *Nova Hedwigia* **4**: 341–346.
Vězda, A. & Vivant, J. (1975). *Absconditella pauxilla* Vězda & Vivant sp. n., un nouveau lichen des Pyrénés-Atlantiques. *Folia geobot. phytotax. bohemoslavica* **10**: 205–208.

Vitakainen, O. (1982). A key to *Peltigera* in Great Britain. *Bull. Br. lichen Soc.* **50**: 28–36.
Wade, A.E. (1965). The genus *Caloplaca* Th. Fr. in the British Isles. *Lichenologist* **3**: 1–28.
Watling, R. (1981). Lichenicolous agarics. *Bull. Br. lichen Soc.* **49**: 28–31.
Wetmore, C. M. (1960). The lichen genus *Nephroma* in North and Middle America. *Publs Mus. Mich. St. Univ., Biol.* **1**: 369–452.
Wetmore, C. M. (1970). The lichen family Heppiaceae in North America. *Ann. Mo. bot. Gdn* **57**: 158–209.
White, F. J. & James, P. W. (1985). A new guide to microchemical techniques for the identification of lichen substances. *Bull. Br. Lichen Soc.* **57** (Suppl.): 1–41.
White, F. J. & James, P. W. (1987). A chemical checklist of British lichens: Part 2 (*Nephroma* and *Peltigera*). *Bull. Br. lichen Soc.* **60**: 42–47.
Winchester, V. (1986). A tabular key to yellow *Rhizocarpon* species in Britain. *Bull. Br. lichen Soc.* **58**: 23–25.
Wirth, V. (1980). *Flechtenflora.* 552 pp. Stuttgart: E. Ulmer.
Wirth, V. (1987). *Die Flechten Baden-Württembergs Verbreitungsatlas.* 528 pp. Stuttgart: E. Ulmer.
Wunder, H. (1974). Schwarzfrüchtige, saxicole Sippen der Gattung *Caloplaca* (Lichenes, Teloschistaceae) in Mitteleuropa, dem Mittelmeergebeit und Vorderasien. *Bibliotheca lich.* **3**: 1–186.
Yoshimura, I. (1971). The genus *Lobaria* of eastern Asia. *J. Hattori bot. Lab.* **34**: 231–364.
Zahlbruckner, A. (1921–40). *Catalogus Lichenum Universalis.* 10 vols. Leipzig: Borntraeger.
Zehetleitner, G. (1978). Über einige parasitische Arten der Flechtengattung *Verrucaria.* Nova Hedwigia **29**: 683–734.
Zschacke, H. (1933–34). Epigloeaceae, Verrucariaceae und Dermatocarpaceae. *Rabenh. Krypt.-Fl.* **9**, 1(1): 44–695.

Addendum

The occurrence of the following species in the British Is. is dubious and they are not treated elsewhere in the Flora:

Bacidia hemipolia (Nyl.) Malme (1865)
Bacidia herbidula (Nyl.) Zahlbr. (1926) [doubtful species]
Bacidia microcarpa (Th. Fr.) Lettau (1912)
Catillaria erysiboides (Nyl.) Th. Fr. (1874)
Catillaria melaenida (Nyl.) Arnold (1870) = *Micarea melaenida* (Nyl.) Coppins (1983)
Pertusaria rupicola (Fr.) Harm. (1898)
Placynthium majus Harm. (1913)
Usnea lapponica Vainio (1925)
Verrucaria pseudomemnonia Zsch. (1924)

Bacidia killiasii (Hepp) D. Hawksw. (1983), a lichenicolous fungus on *Peltigera* is also not treated.

Additional synonyms not listed elsewhere include:-

Arthonia stictarum (Tul.) Th. Fr. (1874) = *Plectocarpon lichenum* (Sommerf.) D. Hawksw. (1983) [lichenicolous fungus]
Aspicilia alpina (Sommerf.) Arnold (1871) = *Bellemerea alpina* (Sommerf.) Clauz. & Roux (1984)
Aspicilia cinereorufescens (Ach.) Massal. (1852) = *Bellemerea cinereorufescens* (Ach.) Clauz. & Roux (1984)
Parmeliopsis ambigua (Wulfen) Nyl. (1866) = correct name for *Foraminella ambigua* (Wulfen) Fricke Meyer (1982)
Parmeliopsis hyperopta (Ach.) Arnold (1884) = correct name for *Foraminella hyperopta* (Ach.) Fricke Meyer (1982)
Lecanora intercedens Magnusson (1930) = *Rimularia badioatra* (Krempelh.) Hertel & Rambold (1990)
Lecidea albohyalina auct brit., non (Nyl.) Th. Fr. (1875) = *Lecania cyrtellina* (Nyl.) Sandst. (1912)
Lecidea assimilata Nyl. (1861) = *Micarea assimilata* (Nyl.) Coppins (1983)
Lecidea humigena Taylor (1847) = *Aphanopsis coenosa* (Ach.) Coppins & P. James (1984)

Lecidea muddii Salwey ex Mudd (1853) = *Halecania ralfsii* (Salwey) M. Mayrh. (1987)
Lecidea nigroglomerata Leighton ex Crombie (1871) = *Lecidella bullata* Körber (1861)
Lecidea rhizobola Nyl. (1865) = *Acarospora rhizobola* (Nyl.) Alstrup (1986)
Lecidea supersparsa Nyl. (1865) = *Carbonea supersparsa* (Nyl.) Hertel (1983)
Lecidea vitellinaria Nyl. (1852) = *Carbonea vitellinaria* (Nyl.) Hertel (1983)
Lecidea vorticosa (Flörke) Körber (1855) = *Carbonea vorticosa* (Flörke) Hertel (1983)
Leptogium azureum auct brit., non (Swartz) Mont. (1840) = *L. cochleatum* (Dickson) P. Jørg. & P. James (1983)
Pertusaria aspergilla (Ach.) Laundon (1992) = correct name for *P. dealbescens* Erichsen (1936)
Rhizocarpon riparium subsp. *lindsayanum* (Räsänen) Thomson (1967) = *R. geographicum* (L.) DC. (1805)
Verrucaria conturmatula Nyl. (1879) = *Polycoccum microsticticum* (Leighton) Arnold (1891) [lichenicolous fungus]
Xanthoria parietina var. *ectanea* auct., non (Ach.) Kickx f. (1867) = *Xanthoria ectaneoides* (Nyl.) Zahlbr. (1931)

Multiclavula vernalis (Schw.) Petersen (1967)

Thallus a small mycelial pad on algae-covered soil. Basidiomata to 14 × 2.5 mm, of simple clubs, subclavate to cylindrical, bluntly rounded apically, at times minutely roughened, ivory, flesh-coloured to pale orange; stipe to 6 × 1.5 mm, pale yellow-orange, flesh solid, white. Basidiospores 6.1–8.3 × 2.9–3.6 μm.

Characterized by the presence of pale clubs, clamp connections and 4 basidia with sterigmata. This genus of basidiolichen has recently been found in N. Scotland, on peaty soil in damp gullies on acid rocks (Outer Hebrides, North Harris; Shetland Is.). Europe, E. North America, Tasmania (Petersen & Kantvilas, 1986).

Staurothele geoica Zsch. (1918)

Thallus thin, dark brown, granular. Perithecia 0.3–0.36 mm diam., *c.* 1/3–2/3-immersed; involucrellum absent. Exciple brown. Hymenial algal cells globose-shortly oblong. Asci 8-spored. Ascospores (28-)30–37 x (13-)16–18 μm, colourless. On compacted, calcareous soil; rare. Scotland (Mid Perth, Killin; W. Inverness, near Fort William). Switzerland.

Thrombium thelostomum (Ach. ex Harriman) A.L. Sm. (1911)

Thallus reddish brown, thin, minutely cracked-areolate, suborbicular. Photobiont blue-green. Apothecia perithecia-like, *c.* 300 μm diam., hemispherical, reddish brown, enveloped by a thick thalline exciple, becoming widely depressed around dark ostiole. Pseudoparaphyses branched, apices not swolllen. Ascospores 17–20 × 9–12.5 μm. On semi-inundated rocks by streams; local. England (Durham, Lancashire, Somerset) and Scotland (Argyll). Its systematic position is unclear.

Index

Names of species are arranged alphabetically by epithet. Names accepted in this account are in roman type, synonyms in *italic* type. The main page references for genera and species are in **bold** type. Illustrations are in *italics*. Species in keys are not indexed, apart from a few newly described which are not fully described in the main text.